Role of Herbal Medicines

Amal Kumar Dhara • Subhash C. Mandal
Editors

Role of Herbal Medicines

Management of Lifestyle Diseases

Springer

Editors
Amal Kumar Dhara
Department of Pharmacy
Contai Polytechnic, Government of West Bengal
Kolkatta, West Bengal, India

Subhash C. Mandal
Department of Pharmaceutical Technology
Jadavpur University
Kolkata, West Bengal, India

ISBN 978-981-99-7702-4 ISBN 978-981-99-7703-1 (eBook)
https://doi.org/10.1007/978-981-99-7703-1

This Springer imprint is published by the registered company Springer Nature Singapore Pte Ltd.
The registered company address is: 152 Beach Road, #21-01/04 Gateway East, Singapore 189721, Singapore

Paper in this product is recyclable

Preface

Modern drifts towards the utilization of herbal medicines in different healthcare aspects necessitate the replacement of synthetic molecules with phytoconstituents because of their biocompatibility, biodegradability, economic extraction process, and ready availability from natural sources. The emergence of lifestyle diseases, e.g. type II diabetes, cancers, obesity, cardiovascular disorders, and liver cirrhosis, is exponentially increasing day by day. The major causes of modern lifestyle diseases are a lack of physical activity, consumption of imbalanced diet and junk foods, unhealthy habits like alcoholic beverages consumption, tobacco smoking, late-night sleep, and so on.

Despite significant developments in the field of science and technology, the majority of the population relies on phytomedicines in both Third World and developed countries for their healthcare. Considering the valuable ancient approaches for the utilization of herbal constituents and the application of modern scientific knowledge for evaluating (both therapeutic and toxicological) and standardizing phytoconstituents can help to find a better way for the management of lifestyle diseases.

The book *Role of Herbal Medicines: Management of Lifestyle Diseases* provides a systematic insight into the inclusive herbal medicines' chemistry, extraction, purification, toxicity studies, characterization, standardization, bioactivity, and applications in the management of lifestyle diseases. This volume is a collection of four different sections (namely introduction; extraction and chemical characterization; bioactivity and applications; pharmacovigilance, clinical trial, and regulatory) containing 31 chapters contributed by academicians and researchers from across the globe. A concise overview of the contents of each chapter has been provided to give readers a glimpse of the book's contents.

Chapter 'Introduction to Lifestyle Diseases and Role of Herbal Medicines' describes different types of lifestyle diseases such as diabetes, obesity, COPD, and liver cirrhosis and the role of herbal medicines in the management of these diseases.

Chapter 'The Importance and Scope of Medicinal Plants Suggested in Traditional Medicine in the Holistic Care of Occupational Lifestyle Disorders with Special Mention to Insulin Resistance Associated Clinical Syndromes' focuses on lifestyle disorders (LsD), the pathology of insulin resistance in LsD, and a few medicinal plants extensively used in traditional medicine practices to treat lifestyle diseases and the associated IR states.

Chapter 'Extraction of Phytoconstituents for Lifestyle Diseases' provides an overview of traditional and modern extraction techniques that are used for the extraction of herbal drugs.

Chapter 'Chemical Characterization of Phytoconstituents for Lifestyle Diseases' provides an overview of different phytoconstituents frequently involved in the management of lifestyle diseases and the various chemical techniques used for their characterization, namely mass spectrometry, UV-visible spectroscopy, infrared spectroscopy, nuclear magnetic resonance spectroscopy, X-ray crystallography, and so forth.

Chapter 'Role of Medicinal Plants in the Management of Diabetes Mellitus' describes the significance of bioactive phytoconstituents from various plant sources in exhibiting antidiabetic activity and also explains how herbal medicines act through different pathways and their multifaceted mechanisms of action.

Chapter 'Herbal Medicines for the Management of Obesity' addresses how lifestyle causes obesity and its potential role in generating various diseases and disorders, such as atherosclerosis, diabetes mellitus, cardiovascular problems, metabolic syndrome, lung diseases, neurological disorders, arthritis, osteoarthritis, kidney disease, sleep disorders, obstructive sleep apnoea, asthma, blood lipid disorders, insulin resistance, depression, anxiety, infertility, urinary incontinence, stroke, and cancer. This chapter also covers some recent studies on plants, plant products, and plant-originated compounds having anti-obesity activity with a focus on their anti-obesity-related activities.

Chapter 'Herbal Medicines for the Management of Diseases in the Heart, Circulation, and Blood' gives an insight into the cardio-protective potential of various medicinal plants such as *Emblica officinalis, Rauwolfia serpentina, Terminalia arjuna,* and *Ziziphus oxyphylla*. It also encompasses the therapeutic potential of many phytomolecules such as taxifolin, caffeic acid, quercetin, rosmarinic acid, arjunolic acid, andrographolide, and others against cardiovascular disorders.

Chapter 'Role of Lifestyle Modification and Diet in the Prevention of Cancer' describes different natural mechanisms that can lower the risk of developing different types of cancer and how lifestyle modifications as well as changes in nutritional diet can modulate different regulatory players associated with cancer signalling axes. This chapter also highlights the mitigating roles of low fibre, omega-3 and omega-6 fatty acids, red meat, flax seeds, fruits, and vegetables in protecting different cancer types ranging from breast to lung to prostrate carcinoma.

Chapter 'Herbal Medicine for the Treatment of COPD' presents a brief review on molecular mechanisms involved in chronic obstructive pulmonary disease (COPD), the natural history of herbal medicine, novel and marketed herbal medicine available for the treatment of COPD, and secondary metabolites involved in the inhibition of COPD.

Chapter 'Herbal Medicines for the Treatment of Liver Cirrhosis' focuses on several natural compounds implicated in reversing the development of liver cirrhosis via numerous molecular pathways. In-depth descriptions of various phytochemicals in preclinical stages are provided. The successful therapeutic regimens of herbal compounds imparting to multiple cell lines, and well-established animal models are discussed in detail.

Chapter 'Plants Affecting Serotonergic Neurotransmission' provides a general overview of the serotonergic system, associated disorders, and natural compounds involved in the modulation of the serotonergic system with potential therapeutic applications.

Chapter 'Herbal Medicines for Management of Alzheimer's Disease' reviews some of the important and widely studied plants and their constituents against Alzheimer's disease.

Chapter 'Herbal Medicine for the Management of Anxiety, Depression, and Insomnia' intends to provide a comprehensive view of mechanisms of herbs used in the management of anxiety, depression, and insomnia.

Chapter 'Role of Herbal Medicines for the Treatment of Arthritis' gives an insight into the promising anti-arthritis effects of varieties of phytoconstituents, different animal models and clinical studies of rheumatoid arthritis and osteoarthritis, and the important mechanisms responsible for the anti-arthritis properties of these herbal medicines.

Chapter 'Role of Herbal Medicines for the Treatment of Chronic Kidney Disease' focuses on the promising anti-chronic kidney disease effects of different herbal medicines.

Chapter 'Herbal Medicines for the Management of Irritable Bowel Syndrome and Constipation Problem' addresses the aetiopathogenesis of IBS and constipation, gut microbiota dysbiosis in IBS and related disorders, diet-microbiota interaction in IBS and related disorders, and efficacy of herbal medicines in the management of irritable bowel syndrome, constipation, and related disorders.

Chapter 'Herbal Medicines for the Management of Diseases on Vitamin Deficiency' describes the dietary sources of vitamins, deficiency disorders of vitamins and their herbal remedies, the qualitative and quantitative procedures of vitamin analysis as well as different analytical approaches for the analysis.

Chapter 'Herbs for Autoimmune Diseases' focuses on providing an update on herbs reported in the management and treatment of autoimmune diseases, with detailed, in-depth understanding of their molecular level mechanism of action and modulation of various cell signalling pathways.

Chapter 'Role of Herbal Medicine in Boosting Immune System' reviews the use of herbal medicines in boosting the immune system. The chapter covers why people prefer herbal medicines, role of some herbal plants in enhancing the immune system, and role of some individual natural products in boosting the immune system.

Chapter 'Medicinal Plants, Antioxidant Potential, and Applications to Aging' highlights some important plant material and isolated compounds that are reported to have the ability to slow down the rate of ageing and are used worldwide. Majority of them exhibited anti-ageing effects through their antioxidant potential.

Chapter 'Role of Natural Polysaccharides in the Management of Lifestyle Diseases' outlines numerous attributes and cellular mechanisms of several plant polysaccharides (PPS) in metabolic disorders and also elucidates the interaction of PPS with various genes and proteins involved in molecular signalling pathways.

Chapter 'Role of Antioxidant Vitamins and Minerals from Herbal Source in the Management of Lifestyle Diseases' highlights the oxidative stress, its generation in different pathological condition of diseases, and their mechanism in promoting lifestyle diseases. It also describes remedies and lists antioxidant vitamins and minerals, along with their modes of antioxidant action, which are essential to alleviate lifestyle diseases and lead a disease-free life.

Chapter 'Role of Nutraceuticals in the Management of Lifestyle Diseases' discusses different types of nutraceuticals and their significance in the management of various lifestyle disorders.

Chapter 'Eating and Healing: Traditional Food as Medicine' highlights the traditional systems of medicines, food, spices, condiments, and herbs with medicinal value and probiotics and prebiotics.

Chapter 'Nanotechnology in Delivery and Targeting of Phytochemicals for Lifestyle Diseases' provides an overview of main problems connected with the delivery of phytopharmaceuticals, main properties of different types of plant drugs, and properties and limitations connected to the application of numerous lipid, polymer, and inorganic nanoparticles as phytodrug carriers. It also provides information about up-to-date investigated combinations of phytopharmaceuticals and nanocarriers.

Chapter 'Pharmacovigilance of Herbal Medicines for Lifestyle Diseases' reviews basic knowledge of pharmacovigilance and adverse drug reactions, along with the safe dosage range of phytochemicals in lifestyle diseases, viz. obesity, diabetes, and atherosclerosis.

Chapter 'Standardization of Herbal Medicines for Lifestyle Diseases' highlights the significance of herbal medicine standardization in ensuring consistent quality, safety, and efficacy of herbal products. Also it describes qualitative and quantitative methods, along with fingerprint profiling for the characterization of herbal samples and bioactive compounds.

Chapter 'Intellectual Property Rights, Naturally Derived Bioactive Compounds, and Resource Conservation' describes recent scientific advancements and comprehends the interrelationship between IPR, bioactivity of natural origin, and conservation of resources in order to establish a smooth avenue for traditional drugs and medicine thereof.

Chapter 'Clinical Trial on and Regulatory Considerations for Herbal Medicines for Lifestyle Diseases' highlights regulations by the FDA and its mechanisms and also describes several ongoing clinical trials aimed at generating precise evidence for their acceptance and greater acceptability.

Chapter 'Market Overview of Herbal Medicines for Lifestyle Diseases' focuses on the market analysis of herbal medicines used in lifestyle diseases, which demonstrates a rising trend in the adoption of herbal medicines, with significant potential for continued growth in the foreseeable future.

Chapter 'Future Prospective of Herbal Medicines for Lifestyle Diseases' addresses the role of some herbal medicines in the treatment of major non-communicable diseases (NCDs), current challenges, and some future perspectives for herbal medicines.

At the end, we the editors convey our sincere thanks to all the distinguished authors for contributing significant chapters in a timely manner, which facilitated to bring out this volume on time. We specially thank the entire team at *Springer Nature* for the invaluable support in organizing the intelligent editing of the book. We also would put on record our special thanks to Dr. Naren Aggarwal (Editorial Director—Books, Asia, Medicine and Life Sciences), Dr. Emmy Lee (Editor, Biomedicine & Life Science Books), Dr. Sudha Ramachandran (Project coordinator—Books), and Dr. Chaitra Jawalkoti (Editorial Advisor) for their priceless support right from the beginning to the completion of this book. We gratefully acknowledge all the permissions we received for reproducing the copyright materials from different sources. Finally, we cannot overlook the sacrifices and support from our family members during the preparation of the book. All our friends, colleagues, and students who have helped in the process of editing of this book deserve our great appreciations. Contributing authors, the publishers, and we the editors will be extremely pleased if our endeavour fulfils the needs of academicians, researchers, students, pharmaceutical experts, biomedicine experts, and herbal drug formulators.

Kolkata, West Bengal, India Amal Kumar Dhara
Kolkata, West Bengal, India Subhash C. Mandal

Contents

Editors and Contributors

About the Editors

Amal Kumar Dhara, PhD has over 25 years of teaching and research experience in the field of Pharmacology and Toxicology, especially in the field of neuropharmacology, anti-inflammatory, antiulcer, free radical scavenging activities of Indian indigenous plants. He has earned his PhD in Pharmaceutical Sciences from Jadavpur University, Kolkata, West Bengal, India. Currently, Dr. Dhara is working as a Lecturer in Pharmacy at Contai Polytechnic, Govt. of West Bengal, India. Dr. Dhara has also served as Production Executive at M/S Deys Medical Stores Mfg (UP) Ltd., U.P., India, for more than 3 years. To date, he has authored over 10 research and review publications in various high-impact peer-reviewed journals, 14 book chapters, and 4 books to his credit. Dr. Dhara has participated and presented his research work at several conferences in India and abroad. He is a life member of the Indian Pharmaceutical Association, Indian Pharmacological Society, and is a Registered Pharmacist.

Subhash C. Mandal, PhD is a Professor at the Division of Pharmacognosy, Department of Pharmaceutical Technology, Jadavpur University, Kolkata, India. Professor Mandal is the recipient of prestigious awards including Endeavour Research Award, Government of Australia; Distinguished Education & Research Award, American Association of Indian Pharmaceutical Scientists (AAiPS), United States; Talented Scientist Award, University of Colombo, Sri Lanka; SAARC Fellowship Award, University Grants Commission (UGC), Bangladesh; Fast Track Young Scientist Award, Department of Science and Technology, UGC Research Award, India; and Outstanding University Teachers' Award, Department of Higher Education, Government of West Bengal, India. He has supervised more than 30 doctoral and 25 M. Pharm scholars and has more than 300 research publications, several patents, books, and book chapters to his credit. Professor Mandal has delivered more than 50 research presentations around the globe, has chaired many international conferences, and successfully completed more than 10 government-funded research projects. He is a reviewer and board member of several high-impact journals and is a recognized host scientist for many international research programmes with developing countries sponsored by the Government of India. He has visited more than 30 countries for various scientific deliberations, collaborations, and exchange programmes.

Contributors

Md Abubakar Department of Pharmacology and Toxicology, National Institute of Pharmaceutical Education and Research (NIPER)-Guwahati, Guwahati, Assam, India

Department of Pharmacology and Toxicology, National Institute of Pharmaceutical Education and Research (NIPER)-Hajipur, Hajipur, Bihar, India

Niyati Acharya Department of Pharmacognosy, Institute of Pharmacy, Nirma University, Ahmedabad, Gujarat, India

Priyanka Adhikari Center for GMP Extraction Facility (Department of Biotechnology), National Institute of Pharmaceutical Education and Research (NIPER), Guwahati, Assam, India

School of Medicine, University of Maryland Marlene and Stewart Greenebaum Comprehensive Cancer Center, Baltimore, MD, USA

Bruno Dupon A. Ambamba Department of Biochemistry, Faculty of Science, University of Yaoundé I, Yaoundé, Cameroon

Vahid Reza Askari International UNESCO Center for Health-Related Basic Sciences and Human Nutrition, Mashhad University of Medical Sciences, Mashhad, Iran

Arun Baidya Department of Pharmaceutical Technology, Jadavpur University, Kolkata, West Bengal, India

Sparsha Bandekar Satara College of Pharmacy, Satara, Maharashtra, India

Sugato Banerjee Department of Pharmacology and Toxicology, National Institute of Pharmaceutical Education and Research, Kolkata, India

Tanmoy Banerjee Department of Pharmaceutical Technology, Jadavpur University, Kolkata, West Bengal, India

Jagdish S. Bankar Department of Pharmaceutical Sciences, Rashtrasant Tukadoji Maharaj Nagpur University, Nagpur, Maharashtra, India

Dhunusmita Barman Department of Pharmacology, NEF College of Pharmaceutical Education and Research, Nagaon, Assam, India

Souvik Basak Dr. B. C. Roy College of Pharmacy and AHS, Durgapur, India

Ashok Behera Faculty of Pharmacy, School of Pharmaceutical and Populations Health Informatics, DIT University, Dehradun, India

Samir Bhargava Faculty of Pharmacy, School of Pharmaceutical and Populations Health Informatics, DIT University, Dehradun, India

Bedanta Bhattacharjee Department of Pharmacology, Girijananda Chowdhury Institute of Pharmaceutical Science, Tezpur, Assam, India

Bhavna Faculty of Pharmacy, School of Pharmaceutical and Populations Health Informatics, DIT University, Dehradun, India

Surbhi S. Bhope Department of Pharmaceutical Sciences, Rashtrasant Tukadoji Maharaj Nagpur University, Nagpur, Maharashtra, India

Kajal N. Bondre Department of Pharmaceutical Sciences, Rashtrasant Tukadoji Maharaj Nagpur University, Nagpur, Maharashtra, India

Rituparna Chaki Dr. B.C. Roy College of Pharmacy and Allied Health Sciences, Durgapur, India

Moumita Chatterjee V. Sivaram Research Foundation, Bangalore, Karnataka, India

Sunayna Choudhary Department of Pharmacognosy, M.M. College of Pharmacy, M.M. (Deemed to be University), Mullana, Ambala, Haryana, India

Anchal Dabas Department of Applied Chemistry, Delhi Technological University, Delhi, India

Zora Dajić Faculty of Agriculture, University of Belgrade, Belgrade, Serbia

S. Dangat Department of Pharmacognosy and Phytochemistry, Bombay College of Pharmacy, Mumbai, Maharashtra, India

Anish Kumar Das Department of Natural Products, National Institute of Pharmaceutical Education and Research (NIPER) - Kolkata, Kolkata, West Bengal, India

Amal Kumar Dhara Department of Pharmacy, Contai Polytechnic, Government of West Bengal, Contai, West Bengal, India

Amit Kumar Dixit Central Council for Research in Ayurvedic Sciences (CCRAS)-Central Ayurveda Research Institute, Ministry of Ayush, Govt. of India, Kolkata, West Bengal, India

Amit Kr. Dubey Faculty of Pharmacy, School of Pharmaceutical and Populations Health Informatics, DIT University, Dehradun, India

Subhajit Dutta Functional Genomics and Metabolism Research Unit, Department of Biochemistry and Molecular Biology, University of Southern Denmark, Odense, Denmark

Ronin Institute for Independent Scholarship, Montclair, NJ, USA

Sharmistha Ganguly University Department of Botany, Ranchi University, Ranchi, Jharkhand, India

Geetanjali Department of Chemistry, Kirori Mal College, University of Delhi, Delhi, India

Shatabdi Ghose Department of Pharmacology, Guwahati, Assam, India

Madhurima Ghosh Department of Pathology and Cancer Screening, Chittaranjan National Cancer Institute, Kolkata, India

Nilanjan Ghosh Department of Pharmaceutical Technology, Jadavpur University, Kolkata, West Bengal, India

Shailendra S. Gurav Department of Pharmacognosy, Goa College of Pharmacy, Goa University, Panaji, Goa, India

Amit Kumar Halder Dr. B. C. Roy College of Pharmacy and Allied Health Sciences, Dr. Meghnad Saha Sarani, Bidhannagar, Durgapur, West Bengal, India

Md Saquib Hasnain Department of Pharmacy, Palamau Institute of Pharmacy, Chianki, Daltonganj, Jharkhand, India

Prakash R. Itankar Department of Pharmaceutical Sciences, Rashtrasant Tukadoji Maharaj Nagpur University, Nagpur, Maharashtra, India

Siddhi Jain Department of Pharmacology and Toxicology, National Institute of Pharmaceutical Education and Research (NIPER), Guwahati, Assam, India

Donald Ulrich K. Kagho Department of Chemistry, Higher Teacher Training College, University of Yaoundé I, Yaoundé, Cameroon

Sanmoy Karmakar Department of Pharmaceutical Technology, Jadavpur University, Kolkata, West Bengal, India

Ceyda Sibel Kılıç Faculty of Pharmacy, Department of Pharmaceutical Botany, Ankara University, Ankara, Turkey

Simeon F. Kouam Department of Chemistry, Higher Teacher Training College, University of Yaoundé I, Yaoundé, Cameroon

Ankit Kumar Department of Pharmaceutical Technology, Jadavpur University, Kolkata, West Bengal, India

J. Kumar University Department of Botany, Ranchi University, Ranchi, Jharkhand, India

K. Jayaram Kumar Department of Pharmaceutical Sciences and Technology, Birla Institute of Technology, Mesra, Ranchi, Jharkhand, India

Sonali Maheshwari Institute of Pharmacy, Nirma University, Ahmedabad, India

Jai Malik Department of Pharmacognosy, University Institute of Pharmaceutical Sciences—UGC Centre of Advanced Study, Panjab University, Chandigarh, India

Keya Mallick Department of Pharmacology and Toxicology, National Institute of Pharmaceutical Education and Research, Kolkata, India

Subhash C. Mandal Department of Pharmacognosy and Phytochemistry, Shri Vile ParleKelavani Mandal's Institute of Pharmacy, Dhule, Maharashtra, India

Pharmacognosy and Phytochemistry Research Laboratory, Division of Pharmacognosy, Department of Pharmaceutical Technology, Faculty of Engineering & Technology, Jadhavpur University, Kolkata, India

Department of Pharmacognosy, K.L.E.'s College of Pharmacy, Nehru Nagar, Belagavi, Karnataka, India

Vilma Matuleviciute Klaipėda University, Klaipeda, Lithuania

The Excelsior School of Herbal Medicine, Belfast, Northern Ireland

Milan Milivojević Faculty of Technology and Metallurgy, University of Belgrade, Belgrade, Serbia

Jayashree Mondal Department of Pharmaceutical Technology, Jadavpur University, Kolkata, West Bengal, India

Titus Alfred M. Msagati Institute for Nanotechnology and Water Sustainability, College of Science, Engineering and Technology, University of South Africa, Florida Science Campus, Johannesburg, South Africa

A. Mukne Department of Pharmacognosy and Phytochemistry, Bombay College of Pharmacy, Mumbai, Maharashtra, India

V. G. M. Naidu Department of Pharmacology and Toxicology, National Institute of Pharmaceutical Education and Research (NIPER), Guwahati, Assam, India

Center for GMP Extraction Facility (Department of Biotechnology), National Institute of Pharmaceutical Education and Research (NIPER), Guwahati, Assam, India

Parvathy G. Nair CCRAS-National Research Institute for Panchakarma, Ministry of Ayush, Govt. of India, Thrissur, Kerala, India

Pratibha P. Nair VPSV Ayurveda College, Kottakal, Kerala, India

Ajay G. Namdeo Department of Pharmaceutical Sciences, Hemvati Nandan Bahuguna Garhwal University (A Central University), Srinagar, Uttarakhand, India

Satheesh Kumar Nanjappan Department of Natural Products, National Institute of Pharmaceutical Education and Research (NIPER) - Kolkata, Kolkata, West Bengal, India

Vilas D. Nasare Department of Pathology and Cancer Screening, Chittaranjan National Cancer Institute, Kolkata, India

Anuj Nautiyal Department of Pharmacy, SGRR University, Dehradun, India

Amit Kumar Nayak Department of Pharmaceutics, School of Pharmaceutical Sciences, Siksha 'O' Anusandhan (Deemed to be University), Bhubaneswar, Odisha, India

Bathelemy Ngameni Department of Pharmacognosy and Pharmaceutical Chemistry, Faculty of Medicine and Biomedical Sciences, University of Yaoundé 1, Yaoundé, Cameroon

Yvan Anderson T. Ngandjui Department of Chemistry, Higher Teacher Training College, University of Yaoundé I, Yaoundé, Cameroon

Institute for Nanotechnology and Water Sustainability, College of Science, Engineering and Technology, University of South Africa, Florida Science Campus, Johannesburg, South Africa

Department of Organic Chemistry, Faculty of Science, University of Yaoundé I, Yaoundé, Cameroon

Sunday Oritsetimenyin Otimenyin Department of Pharmacology, Faculty of Pharmaceutical Sciences, University of Jos, Jos, Nigeria

Ivana Pajic-Lijaković Faculty of Technology and Metallurgy, University of Belgrade, Belgrade, Serbia

Samir Ranjan Panda Department of Pharmacology and Toxicology, National Institute of Pharmaceutical Education and Research (NIPER), Guwahati, Assam, India

Jayesh S. Pande Department of Pharmaceutical Sciences, Rashtrasant Tukadoji Maharaj Nagpur University, Nagpur, Maharashtra, India

Shweta Parihar Department of Pharmacognosy, University Institute of Pharmaceutical Sciences—UGC Centre of Advanced Study, Panjab University, Chandigarh, India

A. Parthiban Center for GMP Extraction Facility (Department of Biotechnology), National Institute of Pharmaceutical Education and Research (NIPER), Guwahati, Assam, India

Kalpana Patil Department of Pharmacognosy and Phytochemistry, Shri Vile Parle Kelavani Mandal's Institute of Pharmacy, Dhule, Maharashtra, India

Pharmacognosy and Phytochemistry Research Laboratory, Division of Pharmacognosy, Department of Pharmaceutical Technology, Faculty of Engineering & Technology, Jadhavpur University, Kolkata, India

Department of Pharmacognosy, K.L.E.'s College of Pharmacy, Nehru Nagar, Belagavi, Karnataka, India

Tiyasa Paul Department of Pharmaceutical Sciences and Technology, Birla Institute of Technology, Mesra, Ranchi, Jharkhand, India

Arun N. Prakash Department of Pharmacology and Toxicology, National Institute of Pharmaceutical and Research (NIPER)-Guwahati, Guwahati, Assam, India

Aabir Pramanik Institute of Pharmacy, Nirma University, Ahmedabad, India

Neethu Prasad Department of Pharmacology and Toxicology, National Institute of Pharmaceutical and Research (NIPER)-Guwahati, Guwahati, Assam, India

Satyendra K. Prasad Department of Pharmaceutical Sciences, Rashtrasant Tukadoji Maharaj Nagpur University, Nagpur, Maharashtra, India

Eswara Rao Puppala School of Medicine, University of Maryland Marlene and Stewart Greenebaum Comprehensive Cancer Center, Baltimore, MD, USA

Mohamed Rahamathulla Department of Pharmaceutics, College of Pharmacy, King Khalid University, Abha, Saudi Arabia

Vafa Baradaran Rahimi Department of Cardiovascular Diseases, Faculty of Medicine, Mashhad University of Medical Sciences, Mashhad, Iran

Vimalavathini Ramesh Department of Pharmacology, College of Pharmacy, Mother Theresa Post Graduate and Research Institute of Health Sciences, Puducherry, India

Vijay Singh Rana Faculty of Pharmacy, School of Pharmaceutical and Populations Health Informatics, DIT University, Dehradun, India

Alok Ranjan Center for GMP Extraction Facility (Department of Biotechnology), National Institute of Pharmaceutical Education and Research (NIPER), Guwahati, Assam, India

Lolita Rapoliene Klaipėda University, Klaipeda, Lithuania

Baltic Medics Port Clinic, Klaipeda, Lithuania

Saptapadi Saha Department of Pharmaceutical Technology, Jadavpur University, Kolkata, West Bengal, India

Ram Kumar Sahu Department of Pharmaceutical Sciences, Hemvati Nandan Bahuguna Garhwal University (A Central University), Chauras Campus, Tehri Garhwal, Uttarakhand, India

K. Sandhanam Department of Pharmacology, SRM College of Pharmacy, SRM Institute of Science and Technology, Chengalpattu, Tamil Nadu, India

Shuvam Sar Department of Pharmaceutical Technology, Jadavpur University, Kolkata, West Bengal, India

Arnab Sarkar Department of Pharmaceutical Technology, Jadavpur University, Kolkata, West Bengal, India

Sathi Sarkar Department of Natural Products, National Institute of Pharmaceutical Education and Research (NIPER) - Kolkata, Kolkata, West Bengal, India

K. Sawate Department of Pharmacognosy and Phytochemistry, Bombay College of Pharmacy, Mumbai, Maharashtra, India

Neeraj Kr. Sethiya Faculty of Pharmacy, School of Pharmaceutical and Populations Health Informatics, DIT University, Dehradun, India

Satish Shilpi Faculty of Pharmacy, School of Pharmaceutical and Populations Health Informatics, DIT University, Dehradun, India

P. Shirodkar Department of Pharmacognosy and Phytochemistry, Bombay College of Pharmacy, Mumbai, Maharashtra, India

Meenakshi Singh Centre for GMP Extraction Facility, Sponsored by Department of Biotechnology, National Institute of Pharmaceutical Education and Research (NIPER)-Guwahati, Guwahati, Assam, India

Ram Singh Department of Applied Chemistry, Delhi Technological University, Delhi, India

Manmohan Singhal Faculty of Pharmacy, School of Pharmaceutical and Populations Health Informatics, DIT University, Dehradun, India

Saroj Singhmura Dr. B. C. Roy College of Pharmacy and AHS, Durgapur, India

Kayalvizhie Subbarayan Department of Pharmacology, College of Pharmacy, Mother Theresa Post Graduate and Research Institute of Health Sciences, Puducherry, India

Kavimani Subramanian Department of Pharmacology, College of Pharmacy, Mother Theresa Post Graduate and Research Institute of Health Sciences, Puducherry, India

N. P. Syamprasad Department of Pharmacology and Toxicology, National Institute of Pharmaceutical Education and Research (NIPER), Guwahati, Assam, India

Rostan M. Talla Department of Organic Chemistry, Faculty of Science, University of Yaoundé I, Yaoundé, Cameroon

Joseph Tchamgoue Department of Organic Chemistry, Faculty of Science, University of Yaoundé I, Yaoundé, Cameroon

Department of Chemistry, Higher Teacher Training College, University of Yaoundé I, Yaoundé, Cameroon

Jean Claude Tchouankeu Department of Organic Chemistry, Faculty of Science, University of Yaoundé I, Yaoundé, Cameroon

Sittarthan Viswanathan Department of Pharmacology, College of Pharmacy, Mother Theresa Post Graduate and Research Institute of Health Sciences, Puducherry, India

Raju Wadekar Department of Pharmacognosy and Phytochemistry, Shri Vile ParleKelavani Mandal's Institute of Pharmacy, Dhule, Maharashtra, India

Pharmacognosy and Phytochemistry Research Laboratory, Division of Pharmacognosy, Department of Pharmaceutical Technology, Faculty of Engineering & Technology, Jadhavpur University, Kolkata, India

Department of Pharmacognosy, K.L.E.'s College of Pharmacy, Nehru Nagar, Belagavi, Karnataka, India

Pratiksha P. Wagh Department of Pharmaceutical Sciences, Rashtrasant Tukadoji Maharaj Nagpur University, Nagpur, Maharashtra, India

Parul Yadav Department of Applied Chemistry, Delhi Technological University, Delhi, India

Introduction to Lifestyle Diseases and Role of Herbal Medicines

Amal Kumar Dhara and Amit Kumar Nayak

Abstract

Lifestyle diseases are medical conditions and are primarily caused by unhealthy habits and practices. The different risk factors responsible for lifestyle diseases include poor diet, lack of physical activity, unrestricted alcohol consumption, smoking and chronic stress, and anxiety. Commonly occurring lifestyle diseases are type 2 diabetes mellitus, obesity, chronic obstructive pulmonary disease (COPD), cardiovascular disorders, chronic liver disease, mental illness, and lung, breast, ovarian, and colorectal cancers. Lifestyle modification and use of herbal medicines can help in the management of lifestyle diseases. The current chapter briefly describes the different types of lifestyle diseases and their management particularly the role of herbal medicines.

Keywords

Lifestyle diseases · Management · Herbal medicines

1 Introduction

Lifestyle diseases are medical conditions that predominantly arise from individuals' everyday habits and behaviors [1, 2]. Certain practices that discourage individuals from engaging in physical activity and instead encourage a sedentary lifestyle can contribute to various health problems, ultimately leading to chronic non-communicable diseases (NCDs) that may cause significant risks to one's life [3, 4]. There are two types of lifestyle factors that are associated with diseases: (1) Modifiable factors, such as dietary habits, physical activity levels, addictive behaviors (such as smoking and drinking), and stress levels, can be altered or adjusted, and (2) Non-modifiable factors, such as age and heredity, gender, ethnicity or race are inherently unalterable [5–8].

The progression of modern civilization has revealed a notable trend wherein an increasing number of young individuals engage in smoking and drinking, despite knowing the detrimental effects these behaviors have on their health [9, 10]. The prevalence of fast food and junk food is displacing our nutritious and traditional well-balanced meals, resulting in a decline in the consumption of fresh fruits and vegetables [11, 12]. Additionally, the preference for processed and packaged food is supplanting the inclusion of these healthier options. Furthermore, the consumption of soft drinks is increasingly substituting milk intake. Our inclination is to utilize transportation, such as buses or cars, rather than engaging in pedestrian travel, even for brief distances [13]. The development of machines to reduce physical labor in the workplace is an increasingly prevalent trend. The aforementioned practices have contributed to the prevalence of overweight and obesity within our population. Obesity refers to the condition of having a body weight that exceeds the normal or ideal weight in relation to an individual's height [14]. Obesity serves as a primary contributing factor to various lifestyle diseases, including but not limited to diabetes, hypertension, heart disease, and other related conditions [15, 16].

2 Prevention of Lifestyle Diseases

Individuals must modify their behaviors to promote a healthier lifestyle. Key practices encompass engaging in consistent physical activity such as [17]: (1) walking, practising yoga, participating in dance activities, performing aerobics, and cycling; (2) Utilization of staircases in lieu of lifts or escalators; (3) Adhering to a well-rounded dietary regimen during appropriate meal periods; (4) The consumption of processed and packaged foods that are high in sugar, fats, salt, and calo-

A. K. Dhara (✉)
Department of Pharmacy, Contai Polytechnic, Government of West Bengal, Contai, West Bengal, India

A. K. Nayak
Department of Pharmaceutics, School of Pharmaceutical Sciences, Siksha 'O' Anusandhan (Deemed to be University), Bhubaneswar, Odisha, India

A. K. Dhara, S. C. Mandal (eds.), *Role of Herbal Medicines*, https://doi.org/10.1007/978-981-99-7703-1_1

ries, while lacking in fiber, high-quality protein, minerals such as iron and calcium, and vitamins, should be minimized; (5) It is recommended to consume whole grains such as cereals (wheat, whole wheat flour) and millets (sorghum, pearl millet, finger millet, etc.), while minimizing the intake of refined foods; (6) Consuming a daily quantity of 400–500 g of fruits and vegetables in season is recommended; (7) It is recommended to consume an ample amount of water; (8) Implementing yoga or meditation techniques can be beneficial in mitigating stress levels in one's daily existence. (9) It is advisable to refrain from engaging in smoking and consuming alcohol. It is advisable to reduce the amount of time spent engaging in sedentary activities such as watching television, using mobile phone (WhatsApp, Facebook, Twitter, etc.) and instead prioritize participation in outdoor games and activities, such as gardening or playing sports like football, cricket, or badminton.

3 History of Herbal Medicine

Ancient Chinese and Egyptian papyrus writings mention the therapeutic applications of plants date back to about 3000 BC, demonstrating that using plants for medicinal purposes predates recorded history. Indigenous cultures, including African and Native American societies, incorporated the utilization of herbs within their healing rituals. In contrast, while other cultures adopted traditional medical systems, such as Ayurveda and Traditional Chinese Medicine, which prominently featured herbal therapies. According to a recent estimation by the World Health Organization, approximately 80% of individuals globally depend on herbal medicines as a component of their primary health care [18, 19]. There has been a significant rise in the utilization of herbal supplements within the last three decades. In the USA, herbal supplements are categorized as dietary supplements.

Scientists worldwide are diligently endeavoring to ascertain the mechanisms of action underlying herbal medications [18, 20]. In numerous instances, researchers encounter uncertainty regarding the precise active component within a given herb that contributes to its therapeutic efficacy in addressing a specific medical condition. Whole plants consist of numerous components that have the potential to synergistically generate a desirable outcome. Using herbal medicines in scientific contexts is very important due to their potential efficacy in treating diverse conditions [18, 20, 21]. Sometimes using of whole plant may produce lesser adverse effects than conventional pharmaceutical interventions. The assumption that herbs are inherently safe due to their natural origin is not universally valid. Certain medical conditions may render certain herbs inappropriate for individuals. Due to the absence of regulatory oversight, herbal products frequently exhibit mislabeling tendencies, potentially harboring undisclosed additives and contaminants [22–24]. Certain herbs have the potential to induce allergic reactions or interact with conventional pharmaceuticals, while others can be toxic when misused or administered in excessive quantities [25, 26]. Engaging in self-medication by consuming herbs independently poses an elevated risk, thus emphasizing the significance of seeking guidance from a specialist in phytomedicines or a Pharmacist before utilization of herbal remedies. An important example of the adverse effects associated with Evening primrose (*Oenothera biennis*) involves an elevated likelihood of seizures in individuals with seizure disorders, as well as an increased risk of bleeding in individuals with bleeding disorders or those who are concurrently taking blood-thinning medications, such as warfarin. Numerous herbal substances possess the potential to interact with prescription medications, thereby eliciting undesirable or hazardous reactions [27, 28]. For instance, an exceptional level of herb-drug interaction has been observed in individuals undergoing cancer treatment. Certain herbal supplements, particularly those originating from Asian nations, can potentially contain elevated concentrations of heavy metals, such as mercury, lead, and cadmium which should be carefully taken into consideration before administration for the treatment of certain disease conditions.

Various herbal preparations can be found in the commercial market, encompassing teas, syrups, oils, liquid extracts, tinctures, and dry extracts as pills or capsules [18]. Teas are prepared by steeping dried herbs in hot water for a few minutes or boiling herbs in water, separating the liquid from the solid components by straining. Syrups, derived from concentrated extracts and incorporated into sweetened formulations, are frequently employed to alleviate sore throats and coughs. Plant-derived oils are commonly extracted and employed as massage rubs, either in their pure form or as constituents of ointments or creams. Tinctures and liquid extracts are bioactive components derived from herbs solubilized in a liquid medium, typically water, alcohol, or glycerol. Tinctures are commonly prepared with a concentration ratio of 1:5 or 1:10, indicating that one herb unit is combined with five to ten units (by weight) of the liquid. Liquid extracts are characterized by a higher concentration compared to tinctures, typically exhibiting a 1:1 concentration ratio. The most concentrated form of an herbal product, known as a dry extract, is generally available in the market in the form of tablet, capsule, or lozenge, with a concentration ratio ranging from 2:1 to 8:1. The most commonly used herbs and their uses are mentioned in Table 1.

Table 1 Some common herbs and their uses

Name of herbs	Uses	References
Ginkgo (*Ginkgo biloba*)	Used in circulatory disorders, enhance memory, in treating dementia (including Alzheimer disease)	[29]
Kava kava (*Piper methysticum*)	Elevates mood, enhance well-being and contentment, and produce a feeling of relaxation, treat anxiety, insomnia, and related nervous disorders	[30]
Saw palmetto (*Serenoa repens*)	Used in treatment of benign prostatic hyperplasia (BPH), a noncancerous enlargement of the prostate gland	[31]
St. John's wort (*Hypericum perforatum*)	Antidepressant effect	[32, 33]
Valerian (*Valeriana officinalis*)	Sleep problems	[34]
Echinacea preparations (*Echinacea purpurea*)	Improve the body's natural immunity, treat colds	[35]

4 Lifestyle Diseases

4.1 Obesity

Obesity is a metabolic disorder characterized by excess fat accumulation in the body due to consuming more energy than one expends [14]. Obesity alone is considered a distinct risk factor in its own right for the occurrence of metabolic syndrome. This condition is closely linked to the emergence of various significant medical issues, including hypertension, non-insulin dependent diabetes mellitus or type 2 diabetes (T2DM), dyslipidemia, respiratory disorders, and sleep apnea [36]. Furthermore, obesity is also associated with the development of life-threatening cardiovascular disease (CVD), stroke, fatty liver disease, specific forms of cancer and osteoarthritis [37]. For more than 2000 years, herbal therapy has been used to cure ailments, and its effectiveness has been established. Numerous studies have demonstrated the efficacy of herbal medicine in managing obesity, although the underlying mechanisms remain unclear. Examples of antiobesity herbs are mentioned in Table 2.

In recent years, there have been significant advancements in herbal medicine research for weight reduction treatments [43]. The efficacy of herbal medicine in treating obesity has been demonstrated through clinical investigations, while animal experiments have begun elucidating the potential mechanisms underlying the various herbal remedies [53, 54]. Nevertheless, certain constraints must be acknowledged: (1) Certain herbal remedies also exhibit toxicity and, thus, should be employed with caution. (2) According to clinical reports, herbal medicines used to treat obesity are associated with minimal side effects, establishing their acceptable safety profile. Nevertheless, there have been documented instances of adverse reactions, including a reported incident of sudden death attributed to the consumption of green tea. Hence, it is imperative to establish regulations governing the utilization of traditional Chinese medicines. (3) The drug composition of herbal medicine exhibits a high degree of complexity, rendering difficulty in determining its mechanism(s) of action as compared to Western medicine.

4.2 Chronic Obstructive Pulmonary Disease (COPD)

COPD is characterized by a multitude of physiological processes like chronic bronchitis, emphysema, and airway remodeling, affecting all lung parts, including small and large airways and parenchyma, which contribute to chronic airway obstruction [55, 56]. Chronic bronchitis is a clinical manifestation resulting from a persistent elevation in bronchial secretions [58]. It is characterized by submucosal glandular hypertrophy and hyperplasia in addition to a productive cough. Emphysema, persistent inflammation, chronic bronchitis, and asthmatic bronchitis are frequently linked to COPD [57]. Many epidemiological studies have demonstrated that COPD can lead to the development of lung cancer [58–61]. Comorbidities are often associated with exposure to cigarette smoke, and individuals diagnosed with COPD exhibit a significantly elevated risk of mortality due to lung cancer, which is eightfold higher compared to the general population [62, 63].

Several herbs, such as clove, mint, lilies, roses, rosemary, ephedra, fennel, and sage, have been used for the treatment of acute and chronic inflammatory diseases [64, 65]. Herbal cough syrup, formulated by extracts from *Hedera helix* and *Thymus vulgaris,* is one of the most significant drugs on acceptance in several European countries [66, 67]. Oral administration of curcumin is also reported for significantly reducing COPD in rats' cell by reducing the levels of IL-6, IL-8, and TNF-α after treatment. This shows that curcumin plays significant role in treating and preventing COPD [68–72]. *Echinacea purpurea* conventionally been used to avert upper respiratory tract infections [73]. One of the most studied herbal formulations for the treatment of COPD is Bufei Yishen formula I, II, and III. The main ingredients in the formulation of different medicinal plants including *Panax ginseng*, *Astragalus tibetanus*, *Cornus officinalis*, *Lycium barbarum*, *Schisandra arisanensis*, *Fritillaria thunbergia*, *Perilla frutescens*, *Citrus sinensis*, *Epimedium acuminatum*, *Paeonia anomala*, *Pheretima aspergillum,* and *Ardisia japonica* [74–77].

Table 2 Antiobesity herbal medicines

Drugs	Sources	Uses	Mechanisms	References
Rhizoma coptidis (Huang Lian)	*Coptis chinensis*, *Coptis deltoidea*, *Coptis teeta* Wall	Reduce body weight gaining and blood glucose and plasma cholesterol levels	Inhibits adipogenesis	[38, 39]
Panax ginseng C. A. Mey (Ren Shen)	*Panax ginseng*	Reduce body weight	Attenuates fat accumulation, suppress lipid accumulation and reactive oxygen species (ROS) production, and improve insulin resistance	[40, 41]
Radix Lithospermi (Zicao)	*Arnebia euchroma*	Reduce abdominal fat thickness	Inhibits lipid accumulation, induce lipolysis, and regulate lipid metabolism	[42, 43]
Ephedra sinica Stapf. (Ma Huang)	*Ephedra sinica*	Weight loss	Modulates gut microbiota, reduce weight, and improve glucose intolerance. Increasing peroxisome proliferator-activated receptor alpha (PPAR-α) and adiponectin activity and reducing tumor necrosis factor-alpha (TNF-α) activity	[44, 45]
Rheum palmatum L. (Da-Huang)	*Rheum palmatum*, *Rheum tanguticum*, *Rheum officinale*	Reduce body weight	Suppresses lipid accumulation and down regulate adipogenic factors	[43, 46]
Green Tea (Lvcha)	*Camellia sinensis*	Loss of body weight	Activates the nuclear factor erythroid-2-related-factor-2 (Nrf2) pathway, up regulation of neprilysin, prevention of gut dysbiosis, regulating metabolic balance in the body, inhibiting fat accumulation and cholesterol synthesis, and reducing abdominal fat	[47–49]
Astragalus membranaceus (Fisch.)	*Astragalus membranaceus*	Reduced metabolic stress-induced increase of body weight	Alleviates glucose intolerance/insulin resistance	[50, 51]
Tripterygium wilfordii	*Tripterygium wilfordii*	Reduce body weight	Relief of endoplasmic reticulum (ER) stress and increased leptin sensitivity	[52]

4.3 Cancers

Cancer is a multifactorial disease and is still leading cause of death till date [78]. Most of the cancers are sporadic in nature, with only 10% being genetic and inherited in an autosomal dominant manner. Recent studies and evidence suggests a drop in mortality rates in recent years. Cancer is a collection of diseases characterized by the unregulated invasion and proliferation of cells [79]. Normal cells become cancer cells through a multistep process involving changes to both the metabolic phenotype and molecular signature. The incidence of female breast cancer has exceeded that of lung cancer, making it the most frequently diagnosed malignancy [80]. It is estimated that there are approximately 2.3 million new cases of female breast cancer, accounting for about 11.7% of all cancer diagnoses. The following types of cancers are included: 10% colorectal, 11.4% lung, 5.6% stomach, and 7.3% prostate. Lung cancer remained the predominant form, accounting for approximately 1.8 million fatalities, constituting 18% of all cancer-related deaths. Subsequently, colorectal cancer accounted for 8.3% of cases, while liver cancer constituted 9.4%, female breast cancer accounted for 6.9%, and stomach cancer represented 7.7% of the total cases. Studies revealed that, changes in lifestyle and diet can prevent 30–40% of all cancers [81]. Obesity, diet with deficient nutrition such as high sugar food preparations and refined flour products, which affect the metabolism of glucose, low fiber intake, red meat, and an imbalance of omega 6 and omega 3 fats are the different factors causing increased risk of developing cancer [81, 82]. Phytomedicines have been employed in numerous countries for the therapeutic management of cancer [83, 84]. Globally, the scientific community has documented over 3000 plant species that possess properties advantageous to anticancer activity. Consuming plenty of vegetables and fruits as well as flax seed (mainly its lignan fraction), will decrease the risk of developing cancer. Folic acid, vitamin B-12, selenium, vitamin D, chlorophyll, and antioxidants such carotenoids (α-carotene, lutein, lycopene, cryptoxanthin, β-carotene) are protective constituents in a cancer preventive diet [81]. According to the report, people who consume the most dietary fiber may experience a slight decrease in the risk of acquiring breast cancer and a reduction in the incidence of colon cancer [85]. The presence of phytochemicals like isothiocyanate and phenolic compounds in plant-based food helps to detoxify the enzymes, metabolizes the carcinogens, and improves the immune system [86–88].

4.4 Psychiatric Disorders

With the rapid development of science, technology, and economy the life style has been dramatically changed and cause psychiatric disorders appears to be common in recent times, this may be attributed to the changes in life style over the decades [89]. Depressive disorders, being one of the most pervasive manifestations of mental illness, exert a significant impact on both individuals and society at large [90, 91]. Psychiatric disorders have been identified as significant contributors to the prevalence of suicide and ischemic heart disease, earning the rank of the second most prominent cause of disability globally [92]. Depression, anxiety, and insomnia frequently co-occur as comorbid psychiatric illnesses within the intricate landscape of mental health disorders. Insomnia and anxiety disorders have been found to negatively impact the functioning of both the immune and cardiovascular systems. Psychiatric disorders significantly affect an individual's occupational functioning, daily activities, overall quality of life, and subjective well-being [93, 94].

Herbal therapy is often used to treat mental illnesses [95–97]. Because these herbs impact several target areas, they activate more than one receptor in most situations. Herbal medicines employed in treating psychiatric disorders exert their effects through the modulation of brain chemistry or restoring hormonal equilibrium [97]. Table 3 provides a comprehensive list of medicinal plants that manage psychiatric conditions.

4.5 Diabetes

Diabetes mellitus (DM) is a highly prevalent worldwide health problem [98]. According to a report by the International Diabetes Federation (IDF) in 2017, the global prevalence of diabetes exceeded 425 million individuals. Furthermore, projections indicate that by 2045, the number of individuals affected by diabetes is expected to reach approximately 629 million [99]. DM is a persistent metabolic disorder characterized by elevated blood glucose levels, commonly called hyperglycemia [98, 100]. This particular disease exhibits associations with various chronic metabolic conditions, including but not limited to obesity, dyslipidemia, hypertension, and cardiovascular complications [101]. Several complications associated with diabetes include neuropathy, retinopathy, heart disease, diabetic nephropathy, Alzheimer's disease, and diabetes foot problems [102]. This disease has two distinct classifications: diabetes mellitus type 1 (DMT1) and diabetes mellitus type 2 (DMT2). Nevertheless, in pregnant females lacking a prior diagnosis of this disorder, gestational diabetes mellitus (GDM) may manifest [103, 104]. In general, T1DM arises from the total insulin deficiency resulting from the destruction of pancreatic β cells, leading to hyperglycemia. The T cells of the individual's immune system are responsible for eliminating the pancreatic cells.

T2DM is more prevalent among individuals, accounting for approximately 90% of diabetes cases worldwide [105, 106]. This disease is attributed to an insufficiency in the synthesis and efficacy of insulin in regulating blood glucose levels and insulin resistance. Age, obesity, dyslipidemia, hypertension, cardiovascular complications, and genetic predisposition are just a few factors that distinguish this type of diabetes [107]. Additional factors, such as tobacco use, consumption of heavily processed foods, a lack of physical activity, and disrupted sleep patterns, have reduced insulin sensitivity, elevating the risk of developing type 2 diabetes mellitus (T2DM) [108].

Gestational diabetes mellitus (GDM) refers to short-term hyperglycemia during pregnancy, typically observed in the latter part of the second or early stages of the third trimester of pregnancy [103, 104]. Gestational diabetes, a form of diabetes, has been observed to impact approximately 15% of pregnant women, potentially leading to adverse outcomes for maternal and fetal health [109]. Several factors contribute to the development of GDM, including being overweight or obese, inheriting T2DM genetically, and being of advanced age during pregnancy [103].

The role of nutrition in preventing or delaying T2DM is significant. Several studies have indicated that a dietary pattern characterized by low consumption of simple carbohydrates and fat and regular physical activity may be more effective in managing T2DM than medication [110]. The recommended dietary regimen for individuals diagnosed with T2DM encompasses consuming a variety of fruits and vegetables, excluding those that are high in sugar content [111, 112]. Additionally, including whole grains, nuts, and legumes is advised. Also, it is advisable to eliminate highly processed foods, refined grains, and beverages with high sugar content. Nevertheless, the quantity and composition of food consumed will be contingent upon the specific requirements of an individual. Engaging in physical activity is a preventive and remedial strategy for individuals with DM [113]. Many plant species possess antidiabetic properties, which can be attributed to their phenolic compounds and antioxidant content [114]. Several compounds in this context include phenolic acids, flavonoids, anthocyanins, saponins, carotenoids, terpenes, and polysaccharides [115]. Table 4 shows the plants with antidiabetic properties.

4.6 Herbs as Immunomodulators

The modern lifestyle and consumption of high-calorie foods are key causes of many metabolic and inflammatory illnesses. Consuming nutraceuticals is another way to boost our health [126, 127]. Recently, there has been a concerning escalation in the consumption of high-calorie sweeteners. As a means of mitigating the consequences of this trend, the utilization of sugars in confectioneries and various food products has been substituted with artificial sweeteners [128]. Nevertheless, it has been observed that artificial sweeteners can lead to sig-

Table 3 A comprehensive list of medicinal plants that manage psychiatric conditions

Plant name/family	Constituents	Pharmacological effects
Adiantum capillus-veneris L./ Adiantaceae	Carvacrol, hexadecanoic acid, thymol, quercetin -3-O-glucoside, isoadiantol	Anxiety, anti-inflammatory
Agrimonia eupatoria L./Rosaceae	Apigenin, naringin, gallic acid	Antioxidant, anxiety, insomnia
Angelica archangelica L./ Apiaceae	Flavonoids (quercetin, alkaloids, coumarins (osthole, angelicin)	Anxiety, insomnia, anti-inflammatory, hypnotic
Angelica sylvestris L./Apiaceae	Flavonoids (apigenin-7-glucoside	Anticancer, antioxidant, anxiety, insomnia,, hypnotic
Artemisia arborescens L./ Asteraceae	Chrysoeriol, apigenin, artemetine	Cytotoxic, anxiety, insomnia, anti-inflammatory, hypnotic
Artemisia vulgaris L./Asteraceae	Eupatilin, apigenin, jaceosidin, luteolin, galangin	Antioxidant, anxiety, insomnia, anti-inflammatory, hypnotic
Ballota nigra L./Solanaceae	Flavonoids (apigenin, ladanein, luteolin), diterpenes, betaines, pinene, linalool, cadinol	Anxiolytic, insomnia, anti-inflammatory, sedative-hypnotic antidepressant
Borago officinalis L./ Boraginaceae	Flavonoids (quercetin), gallic acid, kaempferol, vitamin C	Anxiolytic, sedative-hypnotic antidepressant, carcinogenic
Centranthus ruber L./ DC. Valerianaceae	Valtrate (valepotriates), flavonoids	Antioxidant, anxiolytic, insomnia, sedative-hypnotic antidepressant
Citrus aurantium L./Rutaceae	Kaempferol, rutin, hesperidin, naringenin, quercetin, synephrine	Anxiolytic, insomnia, sedative-hypnotic antidepressant
Crataegus laevigata (Poir.) DC./ Rosaceae	Proanthocynidins, fisetin, quercetin, lupeol	Anxiolytic, sedative-hypnotic, anticancer
Cynodon dactylon L./Pers Poaceae	Quercetin, rutin, kaempferol, catechin, myricetin, carotene, violaxanthin, luteolin	Anxiolytic, sedative-hypnotic, antidepressant
Foeniculum vulgare Miller/ Apiaceae	Rutin, quercetin, apigenin, caffeic acid	Hypnotic, anxiolytic, antidepressant, antiamnesic
Hyoscyamus niger L./Solanaceae	Chlorogenic acid, rutin, quercetin, scopolamine	Sedative-hypnotic, calming agent (anxiolytic) antidepressant
Laurus nobilis L./Lauraceae	Pinene, terpineol, quercetin, kaempferol, luteolin, apigenin	Anxiolytic, analgesic, sedative-hypnotic
Malva neglecta Wallr./Malvaceae	Isoquercetin, quercetin, rutin, kaempferol, apigenin, luteolin, chrysin, coumarin, gallic acid, vanillin	Anxiolytic, sedative-hypnotic
Matricaria camomilla L. (*Chamomilla recutita* L. Rauschert)/Asteraceae	Quercetin, rutin, kaempferol, apigenin, luteolin	Anxiolytic, antidepressant, sedative-hypnotic
Ocimum basilicum L./Lamiaceae	Isoquercetin, quercetin, rutin, kaempferol, apigenin, catechin, naringin, genistein	Anxiolytic, anti-inflammatory, sedative-hypnotic
Olea europaea L./Oleaceae	Luteolin, diosmetin quercetin, rutin, kaempferol, apigenin, oleuropein	Anxiolytic, neuroprotective, sedative-hypnotic, anti-tumor
Opuntia ficus-indica L. Mill./ Cactaceae	Isoquercetin, quercetin, rutin, kaempferol, apigenin, isorhamnetin	Antioxidant, anxiolytic, antibacterial, anti-inflammatory, analgesic, sedative-hypnotic, antidepressant. Anticancer, neuroprotective, anti-viral
Origanum majorana L./ Lamiaceae	Quercetin, rutin, apigenin, naringenin, luteolin, scutellarein	Anxiolytic, sedative-hypnotic
Papaver rhoeas L./Papaveraceae	Isoquercetin, quercetin, rutin, kaempferol, apigenin	Antioxidant, anxiolytic, anti-inflammatory, sedative
Papaver setigerum DC./ Papaveraceae	Quercetin, rutin, apigenin	Anxiolytic, antifungal, sedative
Papaver somniferum L./ Papaveraceae	Isoquercetin, quercetin, rutin, kaempferol, apigenin, luteolin	Anxiolytic, sedative-hypnotic
Rosmarinus officinalis L./ Lamiaceae	Ferulic acid, rosmarinic acid, chorogenic acid, isoquercetin, quercetin, rutin, kaempferol, apigenin, chlorogenic acid, rosmaric acid	Anxiolytic, neuroprotective, sedative-hypnotic
Salix alba L./Salicaceae	Quercetin, rutin, salicin	Anxiolytic, anti-inflammatory,
Solanum nigrum L./Solanaceae	Quercetin	Anxiolytic, anti-inflammatory, antidiabetic
Sonchus oleraceus L./Asteraceae	Quercetin, rutin, apigenin, apigenin, luteolin	Anxiolytic, gastroprotective, anti-inflammatory, anticancer

Table 3 (continued)

Plant name/family	Constituents	Pharmacological effects
Thalictrum aquilegiifolium L./ Ranunculaceae	Thalifoline	Antioxidant, anxiolytic, antifungal
Valeriana officinalis L./ Valerianaceae	Valerenic acid	Insomnia, restlessness

Table 4 Plants with antidiabetic properties

Common names	Biological sources	Mechanism of actions	References
Turmeric	*Curcuma longa* L.	Curcumin is an antioxidant, antidiabetic, anticancer, and anti-inflammatory compound, in addition to reducing hypertension and scleroderma. This compound improves insulin sensitivity, antioxidant protection of pancreatic beta cells, and hyperlipidemia. It also has actions related to the suppression of hepatic gluconeogenesis due to the hypoglycemic effect of curcuminoids	[116]
Tea tree	*Camellia sinensis*	This plant has polysaccharides that lower blood glucose levels and showed α-glucosidase inhibition effects. In addition, it contains flavonoids such as catechins that improve glucose uptake in cells	[47, 48, 117]
Ginkgo Japanese walnut	*Ginkgo biloba*	The leaf extract of this plant can intervene in fasting plasma glucose levels. The compounds obtained are quercetin, genistein, kaempferol, and isorhamnetin. they are associated with antioxidant and antidiabetic effects since they can modify the absorption, distribution and metabolism of glucose	[29, 118]
Moringa	*Moringa oleífera*	Some of the compounds identified in moringa are flavonoids (kaempferol, quercetin, quercitrin, isoquercitrin, rutin, catechin, and epicatechin) and phenolic acids (gallic, ellagic, and chlorogenic acids). Both leaves and seeds are used for the treatment of type 2 diabetes, however, the leaves show significantly higher levels of phenolic compounds compared to the content of the seeds. Compounds such as chlorogenic acid, rutin, quercetin, and catechin have shown antidiabetic action	[119]
Ginseng	*Panax ginseng*	It is a plant that contains bioactive compounds such as ginsenosides, phenolic compounds, and polysaccharides. Compound K (CK) is a biotransformed secondary ginsenoside of complex and hepatoprotective ginsenosides. Regarding its antidiabetic characteristics, it has been found that CK regulates the metabolism of lipids and glucose, protects from the inflammatory response and oxidative stress, as well as modulates insulin resistance in adipose tissue. It can also eliminate gluconeogenesis by inhibiting the expression of glucose-6-phosphatase and phosphoenolpyruvate carboxykinase	[120, 121]
Stevia sweet herb	*Stevia rebaudiana* Bertoni	The extract of this plant contains diterpenes such as stevioside and its steviolaglycon. These terpenes have the ability to increase glycolysis and suppress gluconeogenesis, which gives it antihyperglycemic properties. In particular, stevioside is attributed insulinotropic properties and a glucagonostatic effect in type 2 diabetics. The possible point of action of stevioside has been determined to be the glucose transport system in skeletal muscle	[122, 123]
Jambolán Java plum	*Syzygium cumini* L.	The jambolan is a tropical tree, of which its fruits, cortex and seeds have reported antidiabetic effects. The extract from its leaves has the ability to normalize the levels of glucose and triacylglycerols in the blood, inhibit the effect of α-amylase, as well as control oxidative stress, thanks to its content of quercetin, kaempferol, luteolin, and (Epi) catechin	[124]
Arjun	*Terminalia arjuna*	The extracts of this plant have the ability to inhibit α-amylase through compounds such as shahidin, epicatechin, quercetin, isoolumbin, ellagic acid, and luteolin	[125]

nificant adverse effects over an extended period of time, prompting their substitution with natural non-carbohydrate sweeteners. Polyols such as sorbitol, mannitol, and xylitol are promising natural sweeteners in various fruits and vegetables [129]. Of the options considered, xylitol emerged as an up-and-coming candidate for providing parenteral nutrition. Additionally, it has been found to have advantageous effects in managing diabetes, pulmonary infections, otitis media, and osteoporosis. This molecule exhibits significant potential for preventing dental caries and excels in its exceptional capabilities for modulating the immune system.

5 Clinical Studies and Regulatory Considerations of Herbal Medicines

There is a high demand for these herbal medications worldwide. Still, there are quality, safety, and effectiveness difficulties [19]. Many fatalities were reported due to improper usage or severe adverse effects. Therefore, implementing regulatory policies is the only way this can be achieved. There is a significant demand for clinical research studies on herbal medicines; however, the available data on clinical quality, safety, and efficacy are currently inadequate. Regulatory perspectives such as Good Manufacturing Practice (GMP) have been established to tackle these challenges. GMP serves as the quality standard set by the European Medicines Agency (EMA) and the Food and Drug Administration (FDA) in the USA [22–24]. The herbal biomolecules are labeled and marketed under New Drug Approval (NDA) or Abbreviated New Drug Approval (ANDA) regulatory policies. These policies are in place to ensure that the herbal medicines meet the necessary standards in terms of quality, safety, and efficacy. The adherence to "Good Laboratory Practice" (GLP) guidelines in preclinical and clinical testing as well as the conformance to the "Good Clinical Practice" (GCP) standard during clinical investigations, hold significant importance in the process of applying for approval of a new drug.

6 Conclusion

The incidence of lifestyle diseases are gradually increasing and are primarily caused by unhealthy habits and practices. The different risk factors responsible for lifestyle diseases include poor diet, lack of physical activity, unrestricted alcohol consumption, smoking and chronic stress, and anxiety. Commonly occurring lifestyle diseases are type 2 diabetes mellitus, cardiovascular disorders, obesity, chronic obstructive pulmonary disease (COPD), chronic liver disease, mental illness and lung, breast, ovarian, and colorectal cancers. Management of lifestyle diseases involve good habits, i.e., nutritious diet, physical activity on regular basis, avoiding excessive alcohol consumption and smoking, stress management and happy life. Our earth is decorated with plethora of plants and most of the plants have some medicinal value. Herbal medicines play important role in the management of lifestyle diseases. In the traditional medicinal system varieties medicinal plants including turmeric, (Curcuma *longa*), ginger (*Zingiber officinale*), garlic (*Allium sativum*), green tea (Camellia *sinensis*), *Ginkgo biloba*, etc. have been mentioned to be used for the management of lifestyle diseases. It is also important to remember that all herbal medicines are not devoid of adverse effects. Thus, clinical trial, pharmacovigilance, etc. are the important factors to be considered before the use of herbal medicines clinically.

References

1. Mathur P, Mascarenhas L. Life style diseases: keeping fit for a better tomorrow. Indian J Med Res. 2019;149(Suppl):S129–35.
2. Pappachan MJ. Increasing prevalence of lifestyle diseases: high time for action. Indian J Med Res. 2011;134(2):143–5.
3. Jaikumar S, Dutta S, Sood N. Impact of lifestyle diseases on income and household consumption: evidence from an emerging economy. Health Mark Q. 2021;38(1):35–49.
4. Erdmann E. Worldwide spread of lifestyle diseases and its importance for cardiovascular events. Drugs Today (Barc). 2006;42 Suppl C:5–8.
5. Ng R, Sutradhar R, Yao Z, Wodchis WP, Rosella LC. Smoking, drinking, diet and physical activity-modifiable lifestyle risk factors and their associations with age to first chronic disease. Int J Epidemiol. 2020;49(1):113–30.
6. Salzberg L. Risk factors and lifestyle interventions. Prim Care. 2022;49(2):201–12.
7. Börnhorst C, Russo P, Veidebaum T, Tornaritis M, Molnár D, Lissner L, Mårild S, De Henauw S, Moreno LA, Floegel A, Ahrens W, Wolters M. The role of lifestyle and non-modifiable risk factors in the development of metabolic disturbances from childhood to adolescence. Int J Obes. 2020;44(11):2236–45.
8. Budreviciute A, Damiati S, Sabir DK, Onder K, Schuller-Goetzburg P, Plakys G, Katileviciute A, Khoja S, Kodzius R. Management and prevention strategies for non-communicable diseases (NCDs) and their risk factors. Front Public Health. 2020;8:574111.
9. Myers MG, Kelly JF. Cigarette smoking among adolescents with alcohol and other drug use problems. Alcohol Res Health. 2006;29(3):221–7.
10. Harvey J, Chadi N, Canadian Paediatric Society, Adolescent Health Committee. Preventing smoking in children and adolescents: recommendations for practice and policy. Paediatr Child Health. 2016;21(4):209–21.
11. Güven Y, Öncü E. The relationship between junk food consumption, healthy nutrition, and obesity among children aged 7 to 8 years in Mersin, Turkey. Nutr Res. 2022;103:1–10.
12. Fuhrman J. The hidden dangers of fast and processed food. Am J Lifestyle Med. 2018;12(5):375–81.
13. Vartanian LR, Schwartz MB, Brownell KD. Effects of soft drink consumption on nutrition and health: a systematic review and meta-analysis. Am J Public Health. 2007;97(4):667–75.
14. Lin X, Li H. Obesity: epidemiology, pathophysiology, and therapeutics. Front Endocrinol (Lausanne). 2021;12:706978.

15. Pi-Sunyer X. The medical risks of obesity. Postgrad Med. 2009;121(6):21–33.
16. Jiang SZ, Lu W, Zong XF, Ruan HY, Liu Y. Obesity and hypertension. Exp Ther Med. 2016;12(4):2395–9.
17. Neville RD. Exercise is medicine: some cautionary remarks in principle as well as in practice. Med Health Care Philos. 2013;16(3):615–22.
18. Ekor M. The growing use of herbal medicines: issues relating to adverse reactions and challenges in monitoring safety. Front Pharmacol. 2014;4:177.
19. Parveen A, Parveen B, Parveen R, Ahmad S. Challenges and guidelines for clinical trial of herbal drugs. J Pharm Bioallied Sci. 2015;7(4):329–33.
20. Dhara AK, Nayak AK. Introduction to herbal biomolecules. In: Mandal SC, Nayak AK, Dhara AK, editors. Herbal biomolecules in healthcare applications. Cambridge, MA: Academic Press, Elsevier Inc.; 2021. p. 1–19.
21. Karimi A, Majlesi M, Rafieian-Kopaei M. Herbal versus synthetic drugs; beliefs and facts. J Nephropharmacol. 2015;4(1):27–30.
22. Forte JS, Raman A. Regulatory issues relating to herbal products—part 2: safety and toxicity. J Med Food. 2000;3(1):41–57.
23. Forte JS, Raman A. Regulatory issues relating to herbal products—part 3: quality and its determination. J Med Food. 2000;3(1):59–69.
24. Barkat MA, Goyal A, Barkat HA, Salauddin M, Pottoo FH, Anwer ET. Herbal medicine: clinical perspective and regulatory status. Comb Chem High Throughput Screen. 2021;24(10):1573–82.
25. Başaran N, Paslı D, Başaran AA. Unpredictable adverse effects of herbal products. Food Chem Toxicol. 2022;159:112762.
26. Poppenga RH. Herbal medicine: potential for intoxication and interactions with conventional drugs. Clin Tech Small Anim Pract. 2002;17(1):6–18.
27. Williamson EM. Drug interactions between herbal and prescription medicines. Drug Saf. 2003;26(15):1075–92.
28. Chen XW, Serag ES, Sneed KB, Liang J, Chew H, Pan SY, Zhou SF. Clinical herbal interactions with conventional drugs: from molecules to maladies. Curr Med Chem. 2011;18(31):4836–50.
29. Biernacka P, Adamska I, Felisiak K. The potential of *Ginkgo biloba* as a source of biologically active compounds—a review of the recent literature and patents. Molecules. 2023;28(10):3993.
30. Kuchta K, Hladikova M, Thomsen M, Nahrstedt A, Schmidt M. Kava (*Piper methysticum*) extract for the treatment of nervous anxiety, tension and restlessness. Drug Res (Stuttg). 2021;71(2):83–93.
31. Franco JV, Trivisonno L, Sgarbossa NJ, Alvez GA, Fieiras C, Escobar Liquitay CM, Jung JH. Serenoa repens for the treatment of lower urinary tract symptoms due to benign prostatic enlargement. Cochrane Database Syst Rev. 2023;6(6):CD001423.
32. Sharpley AL, McGavin CL, Whale R, Cowen PJ. Antidepressant-like effect of Hypericum perforatum (St John's wort) on the sleep polysomnogram. Psychopharmacology. 1998;139(3):286–7.
33. Canenguez Benitez JS, Hernandez TE, Sundararajan R, Sarwar S, Arriaga AJ, Khan AT, Matayoshi A, Quintanilla HA, Kochhar H, Alam M, Mago A, Hans A, Benitez GA. Advantages and disadvantages of using St. John's wort as a treatment for depression. Cureus. 2022;14(9):e29468.
34. Shinjyo N, Waddell G, Green J. Valerian root in treating sleep problems and associated disorders—a systematic review and meta-analysis. J Evid Based Integr Med. 2020;25:2515690X20967323.
35. Manayi A, Vazirian M, Saeidnia S. Echinacea purpurea: pharmacology, phytochemistry and analysis methods. Pharmacogn Rev. 2015;9(17):63–72.
36. Jehan S, Myers AK, Zizi F, Pandi-Perumal SR, Jean-Louis G, McFarlane SI. Obesity, obstructive sleep apnea and type 2 diabetes mellitus: epidemiology and pathophysiologic insights. Sleep Med Disord. 2018;2(3):52–8.
37. Fruh SM. Obesity: risk factors, complications, and strategies for sustainable long-term weight management. J Am Assoc Nurse Pract. 2017;29(S1):S3–S14.
38. Tan HL, Chan KG, Pusparajah P, Duangjai A, Saokaew S, Mehmood Khan T, Lee LH, Goh BH. Rhizoma coptidis: a potential cardiovascular protective agent. Front Pharmacol. 2016;7:362.
39. Kwon OJ, Noh JW, Lee BC. Mechanisms and effect of Coptidis rhizoma on obesity-induced inflammation in silico and in vivo approaches. Int J Mol Sci. 2021;22(15):8075.
40. Li Z, Ji GE. Ginseng and obesity. J Ginseng Res. 2018;42(1):1–8.
41. Zhang L, Virgous C, Si H. Ginseng and obesity: observations and understanding in cultured cells, animals and humans. J Nutr Biochem. 2017;44:1–10.
42. Siavash M, Naseri M, Rahimi M. *Arnebia euchroma* ointment can reduce abdominal fat thickness and abdominal circumference of overweight women: a randomized controlled study. J Res Med Sci. 2016;21:63.
43. Liu Y, Sun M, Yao H, Liu Y, Gao R. Herbal medicine for the treatment of obesity: an overview of scientific evidence from 2007 to 2017. Evid Based Complement Alternat Med. 2017;2017:8943059.
44. Song MK, Um JY, Jang HJ, Lee BC. Beneficial effect of dietary Ephedra sinica on obesity and glucose intolerance in high-fat diet-fed mice. Exp Ther Med. 2012;3(4):707–12.
45. Kim BS, Song MY, Kim H. The anti-obesity effect of Ephedra sinica through modulation of gut microbiota in obese Korean women. J Ethnopharmacol. 2014;152(3):532–9.
46. Fang JY, Huang TH, Chen WJ, Aljuffali IA, Hsu CY. Rhubarb hydroxyanthraquinones act as antiobesity agents to inhibit adipogenesis and enhance lipolysis. Biomed Pharmacother. 2022;146:112497.
47. Brimson JM, Prasanth MI, Kumaree KK, Thitilertdecha P, Malar DS, Tencomnao T, Prasansuklab A. Tea plant (*Camellia sinensis*): a current update on use in diabetes, obesity, and cardiovascular disease. Nutrients. 2022;15(1):37.
48. Sánchez M, González-Burgos E, Iglesias I, Lozano R, Gómez-Serranillos MP. The pharmacological activity of *Camellia sinensis* (L.) Kuntze on metabolic and endocrine disorders: a systematic review. Biomol Ther. 2020;10(4):603.
49. Yang CS, Zhang J, Zhang L, Huang J, Wang Y. Mechanisms of body weight reduction and metabolic syndrome alleviation by tea. Mol Nutr Food Res. 2016;60(1):160–74.
50. Huang YC, Tsay HJ, Lu MK, Lin CH, Yeh CW, Liu HK, Shiao YJ. *Astragalus membranaceus*-polysaccharides ameliorates obesity, hepatic steatosis, neuroinflammation and cognition impairment without affecting amyloid deposition in metabolically stressed APPswe/PS1dE9 mice. Int J Mol Sci. 2017;18(12):2746.
51. Wang L, Zheng W, Yang J, Ali A, Qin H. Mechanism of *Astragalus membranaceus* alleviating acquired hyperlipidemia induced by high-fat diet through regulating lipid metabolism. Nutrients. 2022;14(5):954.
52. Zhang C, Sun PP, Guo HT, Liu Y, Li J, He XJ, Lu AP. Safety profiles of *Tripterygium wilfordii* Hook F: a systematic review and meta-analysis. Front Pharmacol. 2016;7:402.
53. Hasani-Ranjbar S, Nayebi N, Larijani B, Abdollahi M. A systematic review of the efficacy and safety of herbal medicines used in the treatment of obesity. World J Gastroenterol. 2009;15(25):3073–85.
54. Payab M, Hasani-Ranjbar S, Shahbal N, Qorbani M, Aletaha A, Haghi-Aminjan H, Soltani A, Khatami F, Nikfar S, Hassani S, Abdollahi M, Larijani B. Effect of the herbal medicines in obesity and metabolic syndrome: a systematic review and meta-analysis of clinical trials. Phytother Res. 2020;34(3):526–45.
55. Wang Y, Xu J, Meng Y, Adcock IM, Yao X. Role of inflammatory cells in airway remodeling in COPD. Int J Chron Obstruct Pulmon Dis. 2018;13:3341–8.

56. Petty TL. The history of COPD. Int J Chron Obstruct Pulmon Dis. 2006;1(1):3–14.
57. Kim V, Criner GJ. Chronic bronchitis and chronic obstructive pulmonary disease. Am J Respir Crit Care Med. 2013;187(3):228–37.
58. Shen Y, Huang S, Kang J, Lin J, Lai K, Sun Y, Xiao W, Yang L, Yao W, Cai S, Huang K, Wen F. Management of airway mucus hypersecretion in chronic airway inflammatory disease: Chinese expert consensus (English edition). Int J Chron Obstruct Pulmon Dis. 2018;13:399–407.
59. Parris BA, O'Farrell HE, Fong KM, Yang IA. Chronic obstructive pulmonary disease (COPD) and lung cancer: common pathways for pathogenesis. J Thorac Dis. 2019;11(Suppl 17):S2155–72.
60. Yang IA, Relan V, Wright CM, Davidson MR, Sriram KB, Savarimuthu Francis SM, Clarke BE, Duhig EE, Bowman RV, Fong KM. Common pathogenic mechanisms and pathways in the development of COPD and lung cancer. Expert Opin Ther Targets. 2011;15(4):439–56.
61. Wang ZL. Association between chronic obstructive pulmonary disease and lung cancer: the missing link. Chin Med J. 2013;126(1):154–65.
62. Chatila WM, Thomashow BM, Minai OA, Criner GJ, Make BJ. Comorbidities in chronic obstructive pulmonary disease. Proc Am Thorac Soc. 2008;5(4):549–55.
63. Forder A, Zhuang R, Souza VGP, Brockley LJ, Pewarchuk ME, Telkar N, Stewart GL, Benard K, Marshall EA, Reis PP, Lam WL. Mechanisms contributing to the comorbidity of COPD and lung cancer. Int J Mol Sci. 2023;24(3):2859.
64. Ghasemian M, Owlia S, Owlia MB. Review of anti-inflammatory herbal medicines. Adv Pharmacol Sci. 2016;2016:9130979.
65. Yatoo MI, Gopalakrishnan A, Saxena A, Parray OR, Tufani NA, Chakraborty S, Tiwari R, Dhama K, Iqbal HMN. Anti-inflammatory drugs and herbs with special emphasis on herbal medicines for countering inflammatory diseases and disorders—a review. Recent Patents Inflamm Allergy Drug Discov. 2018;12(1):39–58.
66. Sierocinski E, Holzinger F, Chenot JF. Ivy leaf (*Hedera helix*) for acute upper respiratory tract infections: an updated systematic review. Eur J Clin Pharmacol. 2021;77(8):1113–22.
67. Patil SM, Ramu R, Shirahatti PS, Shivamallu C, Amachawadi RG. A systematic review on ethnopharmacology, phytochemistry and pharmacological aspects of *Thymus vulgaris* Linn. Heliyon. 2021;7(5):e07054.
68. Aggarwal BB, Gupta SC, Sung B. Curcumin: an orally bioavailable blocker of TNF and other pro-inflammatory biomarkers. Br J Pharmacol. 2013;169(8):1672–92.
69. Jain SK, Rains J, Croad J, Larson B, Jones K. Curcumin supplementation lowers TNF-alpha, IL-6, IL-8, and MCP-1 secretion in high glucose-treated cultured monocytes and blood levels of TNF-alpha, IL-6, MCP-1, glucose, and glycosylated hemoglobin in diabetic rats. Antioxid Redox Signal. 2009;11(2):241–9.
70. Peng Y, Ao M, Dong B, Jiang Y, Yu L, Chen Z, Hu C, Xu R. Anti-inflammatory effects of curcumin in the inflammatory diseases: status, limitations and countermeasures. Drug Des Devel Ther. 2021;15:4503–25.
71. Moghaddam SJ, Barta P, Mirabolfathinejad SG, Ammar-Aouchiche Z, Garza NT, Vo TT, Newman RA, Aggarwal BB, Evans CM, Tuvim MJ, Lotan R, Dickey BF. Curcumin inhibits COPD-like airway inflammation and lung cancer progression in mice. Carcinogenesis. 2009;30(11):1949–56.
72. Safari S, Davoodi P, Soltani A, Fadavipour M, Rezaeian A, Heydari F, Khazeei Tabari MA, Akhlaghdoust M. Curcumin effects on chronic obstructive pulmonary disease: a systematic review. Health Sci Rep. 2023;6(3):e1145.
73. Weber W, Taylor JA, Stoep AV, Weiss NS, Standish LJ, Calabrese C. Echinacea purpurea for prevention of upper respiratory tract infections in children. J Altern Complement Med. 2005;11(6):1021–6.
74. Guo R, Pittler MH, Ernst E. Herbal medicines for the treatment of COPD: a systematic review. Eur Respir J. 2006;28(2):330–8.
75. Lyu YR, Lee SW, Kim SY, Han HB, Yang WK, Kim SH, Jung IC, Kwon OJ, Kim AR, Kim J, Lee MY, Park YC. Herbal medicines for the treatment of chronic obstructive airway diseases (asthma or chronic obstructive pulmonary disease): a prospective observational study. Evid Based Complement Alternat Med. 2022;2022:3485757.
76. Kwon CY, Lee B, Lee BJ, Kim KI, Jung HJ. Herbal medicine compared to placebo for chronic obstructive pulmonary disease: a systematic review and meta-analysis. Front Pharmacol. 2021;12:717570.
77. Liao YN, Hu WL, Chen HJ, Hung YC. The use of Chinese herbal medicine in the treatment of chronic obstructive pulmonary disease (COPD). Am J Chin Med. 2017;45(2):225–38.
78. Panigrahi BK, Nayak AK. Carbon nanotubes: an emerging drug delivery carrier in cancer therapeutics. Curr Drug Deliv. 2020;17(7):558–76.
79. Krieghoff-Henning E, Folkerts J, Penzkofer A, Weg-Remers S. Cancer—an overview. Med Monatsschr Pharm. 2017;40(2):48–54.
80. Sung H, Ferlay J, Siegel RL, Laversanne M, Soerjomataram I, Jemal A, Bray F. Global cancer statistics 2020: GLOBOCAN estimates of incidence and mortality worldwide for 36 cancers in 185 countries. CA Cancer J Clin. 2021;71(3):209–49.
81. Donaldson MS. Nutrition and cancer: a review of the evidence for an anti-cancer diet. Nutr J. 2004;3:19.
82. Clemente-Suárez VJ, Beltrán-Velasco AI, Redondo-Flórez L, Martín-Rodríguez A, Tornero-Aguilera JF. Global impacts of western diet and its effects on metabolism and health: a narrative review. Nutrients. 2023;15(12):2749.
83. Chaudhary T, Chahar A, Sharma JK, Kaur K, Dang A. Phytomedicine in the treatment of cancer: a health technology assessment. J Clin Diagn Res. 2015;9(12):XC04–9.
84. Zimmermann-Klemd AM, Reinhardt JK, Winker M, Gründemann C. Phytotherapy in integrative oncology-an update of promising treatment options. Molecules. 2022;27(10):3209.
85. McRae MP. The benefits of dietary fiber intake on reducing the risk of cancer: an umbrella review of meta-analyses. J Chiropr Med. 2018;17(2):90–6.
86. Porro C, La Torre ME, Tartaglia N, Benameur T, Santini M, Ambrosi A, Messina G, Cibelli G, Fiorelli A, Polito R, Messina G. The potential role of nutrition in lung cancer establishment and progression. Life (Basel). 2022;12(2):270.
87. Johnson IT. Phytochemicals and cancer. Proc Nutr Soc. 2007;66(2):207–15.
88. Nishino H, Murakoshi M, Mou XY, Wada S, Masuda M, Ohsaka Y, Satomi Y, Jinno K. Cancer prevention by phytochemicals. Oncology. 2005;69(Suppl 1):38–40.
89. Firth J, Solmi M, Wootton RE, Vancampfort D, Schuch FB, Hoare E, Gilbody S, Torous J, Teasdale SB, Jackson SE, Smith L, Eaton M, Jacka FN, Veronese N, Marx W, Ashdown-Franks G, Siskind D, Sarris J, Rosenbaum S, Carvalho AF, Stubbs B. A meta-review of "lifestyle psychiatry": the role of exercise, smoking, diet and sleep in the prevention and treatment of mental disorders. World Psychiatry. 2020;19(3):360–80.
90. Arvind BA, Gururaj G, Loganathan S, Amudhan S, Varghese M, Benegal V, Rao GN, Kokane AM, Chavan BS, Dalal PK, Ram D, Pathak K, Lenin Singh RK, Singh LK, Sharma P, Saha PK, Mehta RY, Shibukumar TM, NMHS Collaborators Group. Prevalence and socioeconomic impact of depressive disorders in India: multisite population-based cross-sectional study. BMJ Open. 2019;9(6):e027250.

91. Karrouri R, Hammani Z, Benjelloun R, Otheman Y. Major depressive disorder: validated treatments and future challenges. World J Clin Cases. 2021;9(31):9350–67.
92. Kessler RC, Bromet EJ. The epidemiology of depression across cultures. Annu Rev Public Health. 2013;34:119–38.
93. Zimmerman M, Morgan TA, Stanton K. The severity of psychiatric disorders. World Psychiatry. 2018;17(3):258–75.
94. Hui C, Morcillo C, Russo DA, Stochl J, Shelley GF, Painter M, Jones PB, Perez J. Psychiatric morbidity, functioning and quality of life in young people at clinical high risk for psychosis. Schizophr Res. 2013;148(1–3):175–80.
95. Liu L, Liu C, Wang Y, Wang P, Li Y, Li B. Herbal medicine for anxiety, depression and insomnia. Curr Neuropharmacol. 2015;13(4):481–93.
96. Mischoulon D. Popular herbal and natural remedies used in psychiatry. Focus (Am Psychiatr Publ). 2018;16(1):2–11.
97. Sarris J. Herbal medicines in the treatment of psychiatric disorders: 10-year updated review. Phytother Res. 2018;32(7):1147–62.
98. Padhi S, Nayak AK, Behera A. Type II diabetes mellitus: a review on recent drug based therapeutics. Biomed Pharmacother. 2020;131:110708.
99. Cho NH, Shaw JE, Karuranga S, Huang Y, da Rocha Fernandes JD, Ohlrogge AW, Malanda B. IDF Diabetes Atlas: global estimates of diabetes prevalence for 2017 and projections for 2045. Diabetes Res Clin Pract. 2018;138:271–81.
100. American Diabetes Association. Diagnosis and classification of diabetes mellitus. Diabetes Care. 2009;32(Suppl 1):S62–7.
101. Petrie JR, Guzik TJ, Touyz RM. Diabetes, hypertension, and cardiovascular disease: clinical insights and vascular mechanisms. Can J Cardiol. 2018;34(5):575–84.
102. Chandy A, Pawar B, John M, Isaac R. Association between diabetic nephropathy and other diabetic microvascular and macrovascular complications. Saudi J Kidney Dis Transpl. 2008;19(6):924–8.
103. Buchanan TA, Xiang AH, Page KA. Gestational diabetes mellitus: risks and management during and after pregnancy. Nat Rev Endocrinol. 2012;8(11):639–49.
104. Lende M, Rijhsinghani A. Gestational diabetes: overview with emphasis on medical management. Int J Environ Res Public Health. 2020;17(24):9573.
105. Galicia-Garcia U, Benito-Vicente A, Jebari S, Larrea-Sebal A, Siddiqi H, Uribe KB, Ostolaza H, Martín C. Pathophysiology of type 2 diabetes mellitus. Int J Mol Sci. 2020;21(17):6275.
106. Reed J, Bain S, Kanamarlapudi V. A review of current trends with type 2 diabetes epidemiology, aetiology, pathogenesis, treatments and future perspectives. Diabetes Metab Syndr Obes. 2021;14:3567–602.
107. Martín-Timón I, Sevillano-Collantes C, Segura-Galindo A, Del Cañizo-Gómez FJ. Type 2 diabetes and cardiovascular disease: have all risk factors the same strength? World J Diabetes. 2014;5(4):444–70.
108. Wu Y, Ding Y, Tanaka Y, Zhang W. Risk factors contributing to type 2 diabetes and recent advances in the treatment and prevention. Int J Med Sci. 2014;11(11):1185–200.
109. Plows JF, Stanley JL, Baker PN, Reynolds CM, Vickers MH. The pathophysiology of gestational diabetes mellitus. Int J Mol Sci. 2018;19(11):3342.
110. Bonsembiante L, Targher G, Maffeis C. Type 2 diabetes and dietary carbohydrate intake of adolescents and young adults: what is the impact of different choices? Nutrients. 2021;13(10):3344.
111. Evert AB, Dennison M, Gardner CD, Garvey WT, Lau KHK, MacLeod J, Mitri J, Pereira RF, Rawlings K, Robinson S, Saslow L, Uelmen S, Urbanski PB, Yancy WS Jr. Nutrition therapy for adults with diabetes or prediabetes: a consensus report. Diabetes Care. 2019;42(5):731–54.
112. Petroni ML, Brodosi L, Marchignoli F, Sasdelli AS, Caraceni P, Marchesini G, Ravaioli F. Nutrition in patients with type 2 diabetes: present knowledge and remaining challenges. Nutrients. 2021;13(8):2748.
113. Bhaskarabhatla KV, Birrer R. Physical activity and diabetes mellitus. Compr Ther. 2005;31(4):291–8.
114. Deka H, Choudhury A, Dey BK. An overview on plant derived phenolic compounds and their role in treatment and management of diabetes. J Pharmacopuncture. 2022;25(3):199–208.
115. Tran N, Pham B, Le L. Bioactive compounds in anti-diabetic plants: from herbal medicine to modern drug discovery. Biology (Basel). 2020;9(9):252.
116. Fu YS, Chen TH, Weng L, Huang L, Lai D, Weng CF. Pharmacological properties and underlying mechanisms of curcumin and prospects in medicinal potential. Biomed Pharmacother. 2021;141:111888.
117. Zhao T, Li C, Wang S, Song X. Green tea (*Camellia sinensis*): a review of its phytochemistry, pharmacology, and toxicology. Molecules. 2022;27(12):3909.
118. Boateng ID. Polyprenols in Ginkgo biloba; a review of their chemistry (synthesis of polyprenols and their derivatives), extraction, purification, and bioactivities. Food Chem. 2023;418:136006.
119. Pareek A, Pant M, Gupta MM, Kashania P, Ratan Y, Jain V, Pareek A, Chuturgoon AA. *Moringa oleifera*: an updated comprehensive review of its pharmacological activities, ethnomedicinal, phytopharmaceutical formulation, clinical, phytochemical, and toxicological aspects. Int J Mol Sci. 2023;24(3):2098.
120. Potenza MA, Montagnani M, Santacroce L, Charitos IA, Bottalico L. Ancient herbal therapy: a brief history of *Panax ginseng*. J Ginseng Res. 2023;47(3):359–65.
121. Zhou G, Wang CZ, Mohammadi S, Sawadogo WR, Ma Q, Yuan CS. Pharmacological effects of ginseng: multiple constituents and multiple actions on humans. Am J Chin Med. 2023;51(5):1085–104.
122. Borgo J, Laurella LC, Martini F, Catalán CAN, Sülsen VP. *Stevia* genus: phytochemistry and biological activities update. Molecules. 2021;26(9):2733.
123. Yadav SK, Guleria P. Steviol glycosides from Stevia: biosynthesis pathway review and their application in foods and medicine. Crit Rev Food Sci Nutr. 2012;52(11):988–98.
124. Ochieng MA, Ben Bakrim W, Bitchagno GTM, Mahmoud MF, Sobeh M. *Syzygium jambos* L. Alston: an insight into its phytochemistry, traditional uses, and pharmacological properties. Front Pharmacol. 2022;13:786712.
125. Kumar V, Sharma N, Saini R, Mall S, Zengin G, Sourirajan A, Khosla PK, Dev K, El-Shazly M. Therapeutic potential and industrial applications of *Terminalia arjuna* bark. J Ethnopharmacol. 2023;310:116352.
126. Cencic A, Chingwaru W. The role of functional foods, nutraceuticals, and food supplements in intestinal health. Nutrients. 2010;2(6):611–25.
127. Chen Y, Michalak M, Agellon LB. Importance of nutrients and nutrient metabolism on human health. Yale J Biol Med. 2018;91(2):95–103.
128. Sharma A, Amarnath S, Thulasimani M, Ramaswamy S. Artificial sweeteners as a sugar substitute: are they really safe? Indian J Pharmacol. 2016 48(3):237–40.
129. Lenhart A, Chey WD. A systematic review of the effects of polyols on gastrointestinal health and irritable bowel syndrome. Adv Nutr. 2017;8(4):587–96.

The Importance and Scope of Medicinal Plants Suggested in Traditional Medicine in the Holistic Care of Occupational Lifestyle Disorders with Special Mention to Insulin Resistance Associated Clinical Syndromes

Parvathy G. Nair, Pratibha P. Nair, and Amit Kumar Dixit

Abstract

Lifestyle disorders (LsD) involve chronic pathologies of slow and steady progress owing to the conglomeration of multiple interrelated dynamics, viz. behavioural, environmental, physiological, and genetic factors. Key metabolic changes that increase the risk of LsD while one's lifetime consists of hypertension, overweight, dyslipidaemia, and hyperglycaemia. The affordability-cost complex in screening and treatment monitoring of LsD dramatically impacts patients' life satisfaction and quality of life domains.

When LsD and their molecular level deviations are evaluated, some striking facts relating to the shared pathogenesis of autophagy, autoimmunity, and, much more importantly, their derivate the Insulin resistance (IR). IR is identified as a crucial pathological factor in the incidence and progression of various LsD, for instance, type 2 diabetes mellitus, non-alcoholic fatty liver disease, obesity, polycystic ovarian syndrome, dyslipidaemia, cardiovascular disease, sleep apnoea, and hormone-sensitive cancers. It is a matter of disquiet that there are no predefined clinical practice guidelines or consensus in treating patients with IR. However, there is sufficient evidence that glycaemic targets are not achieved in patients with severe IR despite higher insulin doses. Interlinked phenomena of autoimmunity, autophagy, auto-inflammation, and gut dysbiosis significantly add to the challenges in targeting IR in IR-associated syndromes.

In this era of rapidly emerging diseases and healthcare challenges, awareness and interest in traditional medicines (TMs) and their role and relevance worldwide are entirely renewed. The integrative approaches accommodating TMs and standard care protocols should explore lifestyle manipulations, and administration of immune-modulating, anti-inflammatory, and adaptogenic drugs in distinct IR endotypes. Integrative approaches accommodating herbs should effectively target gut microbiota revitalization, develop operative management strategies for a functionally regressed autophagy phase, counter auto-inflammation, and bring in immune modulation to enhance tissue compatibility and insulin sensitivity in tissues. TMs suggest various herbs and polyherbal preparations for managing metabolic syndrome, obesity, neoplastic presentations, neurodegenerative disorders, and endocrine pathologies like PCOS associated with IR. The current book chapter focuses on LsD, the pathology of insulin resistance in LsD, and a few medicinal plants extensively used in traditional medicine practices to treat lifestyle diseases and the associated IR states. An attempt has been made to review the mode of action specific medicinal plants and their monomers in improving IR parameters and associated LsD states in various experimental models and clinical trials.

Keywords

Herbal medicine · Occupational diseases · Lifestyle diseases · Traditional medicines · Ayurveda

Parvathy G. Nair and Pratibha P. Nair contributed equally with all other contributors.

P. G. Nair
CCRAS-National Research Institute for Panchakarma, Ministry of Ayush, Govt. of India, Thrissur, Kerala, India

P. P. Nair
VPSV Ayurveda College, Kottakal, Kerala, India

A. K. Dixit (✉)
Central Council for Research in Ayurvedic Sciences (CCRAS)-Central Ayurveda Research Institute, Ministry of Ayush, Govt. of India, Kolkata, West Bengal, India

A. K. Dhara, S. C. Mandal (eds.), *Role of Herbal Medicines*, https://doi.org/10.1007/978-981-99-7703-1_2

1 Background

World Health Organization has identified non-communicable diseases [NCDs] as the leading cause of morbidity and mortality (71% of deaths yearly). Among these, cardiovascular diseases (17.9 million deaths each year), cancer (9 million deaths each year), respiratory disease (3.9 million deaths each year), and diabetes (1.6 million deaths each year) rank high on the charts. Furthermore, diseases such as gastrointestinal, endocrine, haematological, dermatological, neurological, hepatic, and renal conditions, trauma, disabilities such as blindness and deafness, and mental ailments come under the umbrella term of NCDs [1]. If not all, most of them are pretty interlinked and share pathological deviations at molecular levels. This set of disorders can be termed lifestyle disorders [LsD]. Lifestyle diseases are categorized mainly as a direct result of people's habitual behaviour and, in particular, a maladaptive or inappropriate relationship with their immediate environment. Unhealthy diets, nutritional irregularities, inappropriate physical activities, and alcohol and tobacco consumption are emphasized as significant risk factors for LsD [2]. It is evident that despite living in relatively improved living conditions and consciously adopting the so-called protective lifestyles, the disease burden of [LsD], such as cancer, diabetes, cardiovascular disorders, depression, and degenerative disorders, is huge.

LsD involves chronic pathologies of slow and steady progress owing to the conglomeration of multiple interrelated dynamics, viz. behavioural, environmental, physiological, and genetic factors [1]. Key metabolic changes that increase the risk of LsD while one's lifetime consists of hypertension, overweight, dyslipidemia, and hyperglycaemia [3]. The affordability-cost complex in screening and treatment monitoring of LsD dramatically impacts patients' life satisfaction and quality of life domains.

2 Insulin Resistance

Certain concomitant or interlinked deviated processes happen at subtler levels which, with their cumulative effects, transpire into different symptoms or a disease or a syndrome. Curiously, the fundamental errors happening subtly in seemingly different disorders may often share characteristics. Later, due to many personalized complexities, the final manifested disorder may invariably vary. When LsD and their molecular level deviations are evaluated, some striking facts relating to the shared pathogenesis of autophagy, autoimmunity, and, much more importantly, their derivate the insulin resistance (IR). IR has been scrutinized due to its strong link with the clinic-metabolic association in otherwise healthy individuals [4]. IR is identified as a crucial pathological factor in the incidence and progression of various LsD, for instance, type 2 diabetes mellitus, non-alcoholic fatty liver disease, obesity, polycystic ovarian syndrome, dyslipidemia, cardiovascular disease, sleep apnoea, and hormone-sensitive cancers [5, 6]. The risk factors of LsD include raised plasma glucose, lipid deregulation (in the form of elevated triglycerides, reduced high-density lipoproteins, and elevated low-density lipoproteins), hypertension, and obesity, all of which share the common denominator of IR. This grouping is referred to as metabolic syndrome [MS] or syndrome X, or even IR syndrome [5].

This means that IR if screened relatively earlier is a candidate space for intervention to adopt early protective lifestyle behaviours, appropriate metabolic error-reversing medicaments, and advocating a balanced physical activity that would delay or halt further pathological mechanisms that might eventually lead to life-threatening LsD and associated distress.

2.1 Shared Pathology of Autoimmunity, Autoinflammation, Impaired Autophagy, and Gut Dysbiosis in IR

It should be understood that converse to the conventional thinking of demarcating type 1 and type 2 diabetes, for instance, type 1 as autoimmune and type 2 as peripheral insulin-resistant variants, recent research studies explain the grey area wherein autoimmunity and insulin resistance are associated with either form though in varying grades. Most non-communicable disease spectrum consists of intermediary phenotype expression owing to maladaptive epigenetic mechanisms linked with inflammatory processes, autoimmunity, and IR. Currently, endotypes in NCDs are viewed in an immunological continuum triggered and shaped by specific epigenetic deviations [7]. Heterogeneity is present even in diabetes, cancer, NAFLD, and obesity. Analysis of autoimmunity and IR in different permutations and combinations expressed as personalized characteristics help in the differential understanding of LsD, especially their similarities and dissimilarities. This is important in precision medicine; wherein protective and curative therapies are strategically tailored to meet clinical endpoints.

Abnormal feeding patterns in obese people with diabetes are associated with metabolic deregulation in the adipose tissue, which triggers low-grade chronic inflammation. This is expressed as increased amounts of inflammatory markers, for instance, free fatty acids (FFA), tumour necrosis factor (TNFα), interleukin 6 (IL-6), macrophage chemo-attractant protein (MCP-1) that furthermore recruits pro-inflammatory leucocytes and M1 macrophages and culminates in IR. Sufficient evidence reinstates the role of T&B cells and leukocytes in IR, viz., dendritic cells, neutrophils, and NK cells. CD 4+ and CD 8+ T cells permeate adipose tissues after a high-fat diet, perhaps quite earlier, and a parallel pro-inflammatory state sets in the adipose tissue (for instance, infiltration of macrophages), which reinforces the autoimmune inflammatory mechanisms in IR. The fact that T cell targeting immunotherapy improves insulin sensitivity in peripheral tissues is thus quite obvious. Thus, pro-inflammatory autoimmune phenotypes are candidate pathogenesis initiators in IR. Studies suggest that in diseases associated with IR, viz. type 2 diabetes, adipose tissue, and pancreas are plausible primary sites of auto-antigen dependent priming of T and B lymphocytes, especially under bodily metabolic stressful states. Similarly, in IR subjects with obesity, an enhanced major histocompatibility complex class II expression is noted.

Furthermore, interleukin-2 and interferon-γ production, too, sets in. Sequentially, adipocytes transform into antigen-presenting cells (APCs). Likewise, a hypothesis exists that there is a presence of IgG antibodies formed specifically for adipocyte-derived antigens. This pathology even exists in the pancreas, at islets. Therefore, authors believe there is an equal chance that (i) adipose-reactive T cells generated in the adipocyte migrate to other metabolically active tissues, such as the liver or pancreas, where they induce IR, or (ii) islet-reactive T cells generated in the pancreas, migrate to adipocyte and other metabolically active tissues where they induce IR either by augmenting chronic tissue inflammation or by targeting protective elements for the development of IR [7]. Also, evidence states that the anti-islet antibodies' presence is causally linked with IR. Curiously autophagy process via lysosomes as well as specific autophagy proteins, improve insulin sensitivity giving leads to the fact that functional regression in autophagy culminates in IR [8].

Gut microbiota is a candidate contributor to metabolic errors. Those individuals with IR present with definite disproportionate bacterial phyla that hamper intestinal permeability and increase lipopolysaccharide (LPS) absorption, which, on reaching circulation, activates inflammatory pathways. Thereby impairment in insulin signalling pathways and defective functioning of insulin receptors set in manifested as IR [9].

Thus, considering the interlinked and tangled occurrences and complex mechanisms at gross and cellular levels that underline the pathogenicity of IR and IR associated with other clinical conditions, a team of inter-professionals needs to integrate for designing appropriate protective and curative guidelines entirely dependent on the prescribed endotype in respective subjects. Also, healthcare research that goes parallel to clinical practices should be planned so that patient-preferred outcomes are judiciously incorporated as primary objectives in addition to marker-assisted methods (surrogates).

3 Understanding Insulin

Insulin is a peptide hormone produced by the beta cells of islets of Langerhans of the pancreas. Cells in the human body need glucose (sugar) for energy, and insulin helps to transport this glucose from the blood into the cells. Previously, insulin was viewed as an anabolic hormone related only to glucose metabolism, but later studies suggest it has much broader pleiotropic effects.

The synthesis and secretion of insulin can be affected by both nutrient and non-nutrient secretagogues. Glucose is the primary stimulus for insulin release in the body, but other than that, beta cells also respond to other nutrients like free fatty acids and amino acids. The entry of glucose into beta cells causes glucokinase to phosphorylate into glucose-6-phosphate. Glycolysis then metabolizes it to produce pyruvate, NADH, and ATP. Furthermore, the closure of ATP-sensitive potassium channels causes membrane polarization and activation of voltage-dependent calcium channels, the consequent intracellular influx of calcium triggers insulin secretion. Non-nutrient secretagogues include various other hormones, such as melatonin, oestrogen, leptin, growth hormone, incretin, and glucagon-like peptide-1 also regulate insulin secretion in the body. They stimulate insulin secretion via neural stimuli like adrenergic and cholinergic pathways. Approximately 0.25–1.5 units per hour of insulin are secreted by beta cells during the fasting state, which is sufficient enough for supplying glucose to cells. This basal insulin secretion accounts for about 50% of insulin secreted by the body in a day. The nutrient-stimulant insulin concentration contributes to the rest of the fraction [10, 11].

Insulin mediates energy availability to the body in both fasting and fed states. It works synchronized to facilitate energy storage during fasting and ensure energy expenditure and uptake during the fed state. In healthy individuals, the amount of insulin secreted is proportionate to the body's metabolic demand. During fasting, there will be low insulin concentration in the portal circulation. To maintain the optimum concentration, the insulin clearance rate of the liver also becomes considerably high during fasting. The glucose production in such conditions will be modulated to meet the requirement of essential glucose-dependent tissues like neurons and red blood cells. In response to the fall of insulin in the blood, another hormone, glucagon, gets activated, which facilitates glycogenolysis and gluconeogenesis. In the absence of insulin, cells become deprived of taking up glucose and searching for alternate fuels resulting in lipolysis with the release of non-esterified fatty acids. In the fed state, there will be an increased concentration of insulin and glucose in the circulation. This inhibits glycogenolysis and gluconeogenesis but promotes energy accumulation through glycogenesis and lipogenesis [12, 13].

Insulin reaches multiple organ tissues like the liver, muscles, and adipose tissues through the extracellular insulin receptor on their plasma membrane. The endocrine effect of insulin on these organs depends on the interstitial insulin concentration, which is further influenced by the exogenous insulin secretion and clearance from circulation. Insulin has a profound effect on carbohydrate and lipid metabolism and protein and mineral metabolism. It also stimulates the liver to store a significant fraction of glucose absorbed from the small intestines as glycogen Insulin, through its receptors, influences multiple physiological processes in the body, including reproduction, mitogenesis, cognition, and lifespan. It is a tyrosine kinase receptor with 2 alpha and 2 beta glycoprotein subunits bound together with a disulfide bond. Following insulin binding to the extracellular domain, the receptor activates a complex intracellular signalling network through insulin-responsive substrate (IRS) proteins. IRS binds to other signalling molecules, mediating various cellular functions through different pathways. Derangements in insulin signalling in the liver and beta cells are recently being highlighted as a "new biology of diabetes,“ causing insulin resistance and glucose intolerance [5]. Figure 1 corresponds to the functional range of insulin—the anabolic hormone [14].

3.1 Defining Insulin Resistance (IR)

Insulin has intricate effects on cell growth and differentiation, and metabolism. IR is recognized as a compromised biological response to insulin stimulation of target tissues, principally the liver, muscle, and adipose tissue. IR impairs glucose disposal, resulting in a compensatory rise in beta cell insulin production and hyperinsulinemia. The metabolic magnitudes of IR, as cited above, relate to the occurrence of but are not limited to, hypertension, hyperglycaemia visceral adiposity, dyslipidaemia, hyperuricemia, raised inflammatory indicators, endothelial dysfunction, and a prothrombic state, the preponderate consequence of IR being type 2 diabetes mellitus, and non-alcoholic fatty liver disease (NAFLD) [15].

IR syndrome is the term used to collectively describe abnormalities that may occur in insulin-resistant individuals. This includes glucose intolerance, dyslipidaemia, endothelial dysfunction, elevated procoagulant factors, hemodynamic changes, elevated inflammatory markers, abnormal uric acid metabolism, increased ovarian testosterone secretion, and sleep disorder. The clinical conditions individuals with insulin resistance are at high risk of developing include type 2 diabetes mellitus, obesity, metabolic syndrome, lipodystrophies,

Fig. 1 Multiaxial functional dimension of insulin

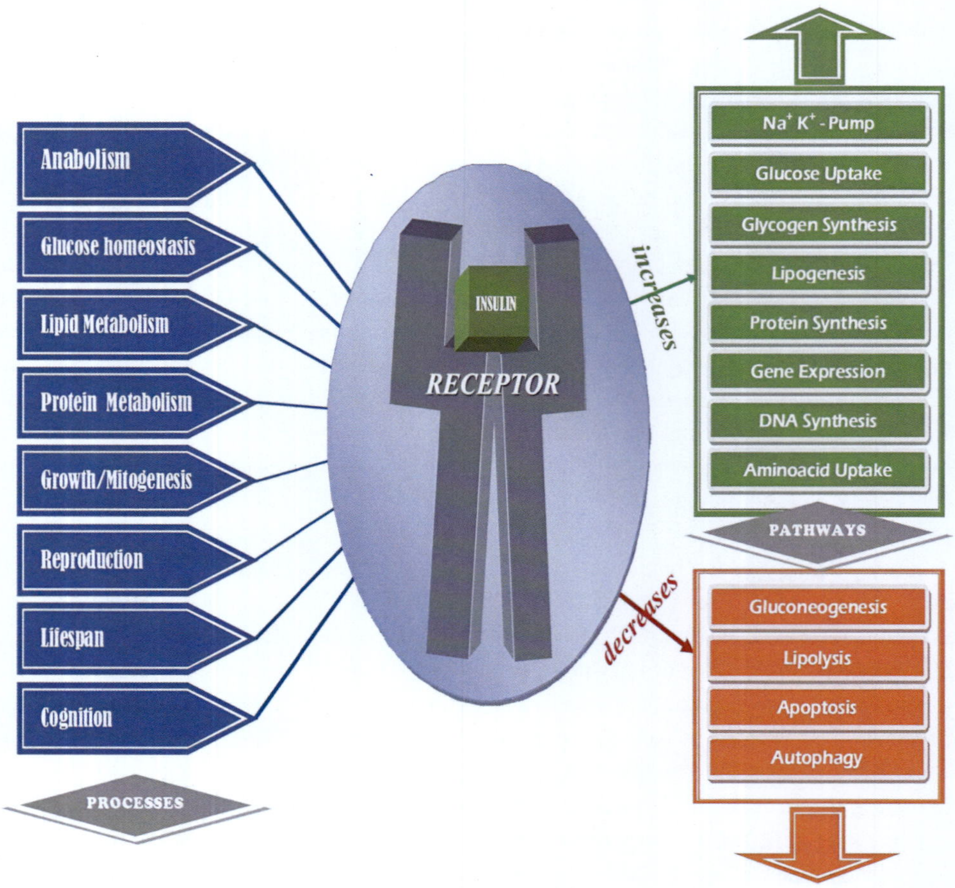

polycystic ovary syndrome, certain types of cancers, and chronic infections. Acanthosis nigricans, visceral obesity, acne, hirsutism, and hepatic steatosis are the major clinical markers of IR. Figure 2 details the pathogenic magnitude of IR that influences the occurrences of various pathological conditions.

3.2 IR Causative Factors

The aetiology of IR includes both genetic and environmental factors. A gene profiling study reports the association of more than 180 genes with insulin sensitivity in skeletal muscle cells [16]. Environmental factors include increased calorie intake, reduced physical activity, smoking, and regular intake of some drugs. Mostly combined genetic and environmental factors contribute to the development of IR. Hormonal disorders like Cushing's syndrome, acromegaly, and hypothyroidism can also cause IR. IR can be developed temporarily, i.e., gestational diabetes developed due to certain hormones secreted by the placenta. The condition is reversible, but the off springs of these women are at high risk of developing adolescent obesity and type 2 DM in the future. Studies also suggest that unrecognized glucose intolerance and insulin resistance begin in these women before conception, especially in those who are obese before conception. Lifestyle measures, including diet and physical activity, avoid obesity in women before conception and can significantly prevent gestational diabetes and related complications. Increased supplementation of micronutrients like zinc, magnesium, etc., to women in the reproductive age group is also highlighted to play a positive role [17, 18]. Table 1 comprehends the plausible risk and causative factors of IR.

3.3 Gross Pathological Process in IR

The primary sites affected by insulin resistance include muscle, adipose tissues, and hepatic tissues. Maximum glucose uptake happens in muscles with every calorie intake and its conversion to glucose. However, with an excessive calorie load, the muscles cannot up take all the glucose produced. This excess glucose returns to the liver triggering de novo lipogenesis (DNL). DNL increases the concentration of triglycerides and free fatty acids, causing ectopic fat deposition in the liver, muscle, and adipose tissue [19]. The glucolipotoxicity induces metabolic stress in the beta cells of the pancreas and other insulin-sensitive tissues. This metabolic stress causes the tissues to produce and release pro-inflammatory mediators like cytokines and chemokines. These abnormal pro-inflammatory mediators cause apoptosis, amyloidosis, and fibrosis in beta cells of the pancreas, impairing insulin resistance. In the peripheral insulin-sensitive tissues of the body, activation of various other inflammatory happens, impairs the phosphorylation of insulin signalling pathways resulting in systemic insulin resistance. The oxidative stress induced due to glucolipotoxicity is the primary culprit which impairs glucose secretion and induces insulin resistance in peripheral tissues [20, 21].

Once IR sets in, endogenous insulin production has a compensatory rise. Insulin, an anabolic hormone, gradually gains weight and further intensifies IR. As the condition progresses, a functional decline in pancreatic beta cell activity is noted against the increase of insulin put forth by an established IR. This is manifested as hyperglycaemia. If the situation consistently persists, the patient is diagnosed with type 2 diabetes mellitus [15]. If not found to respond to antiglycaemic agents, the patient is put on exogenous insulin. Physicians believe patients requiring more than 1 unit/kg/day of exogenous insulin to maintain glycaemic control are to be recognized as IR. Patients requiring more than 200 units of exogenous insulin daily are considered severely IR.

As mentioned earlier, significant sites of primary IR expression include muscular tissues, liver, and adipose tissues. Muscle accounts for a major share of glucose clearance. Thus, of all sites, muscles correspond to the first and foremost site of IR activity via the immune-mediated inflammatory pathways and release of excess free fatty acids, which eventually transpires into ectopic lipid deposition [22]. If not properly disposed of due to impaired muscular glucose uptake, the excess glucose returns to the liver, raising lipogenesis and subsequent increase of circulating free fatty acids that further contribute to the pathogenicity of IR. Likewise, visceral adipose tissue suffers the brunt due to the functional decline of insulin, which subsequently results in reduced lipolysis. These increases circulating free fatty acids (FFA). Higher levels of circulating FFAs, in turn, affect both hepatic and muscle metabolism, further aggravating insulin resistance.

Strikingly IR and reduced uptake of insulin by muscle end up in accelerated glucose delivery to the liver, which in turn triggers lipogenesis and ectopic lipid deposition. This whole phenomenon is linked to the inflammatory mechanism. A parallel rise in glucose production in the liver and a postprandial glucose rise. These mechanisms when crosses the threshold at cellular levels, and toxicity sets in that further hastens IR [15].

3.4 Indicators of IR

Although not clinically relevant, the marker for investigating IR is identified as the hyperinsulinemic-euglycemic glucose clamp technique. Clinically useful surrogates include HOMA-IR, HOMA2, QUICKI, serum triglyceride, and triglyceride/HDL ratio. Furthermore, IR is also quantified with

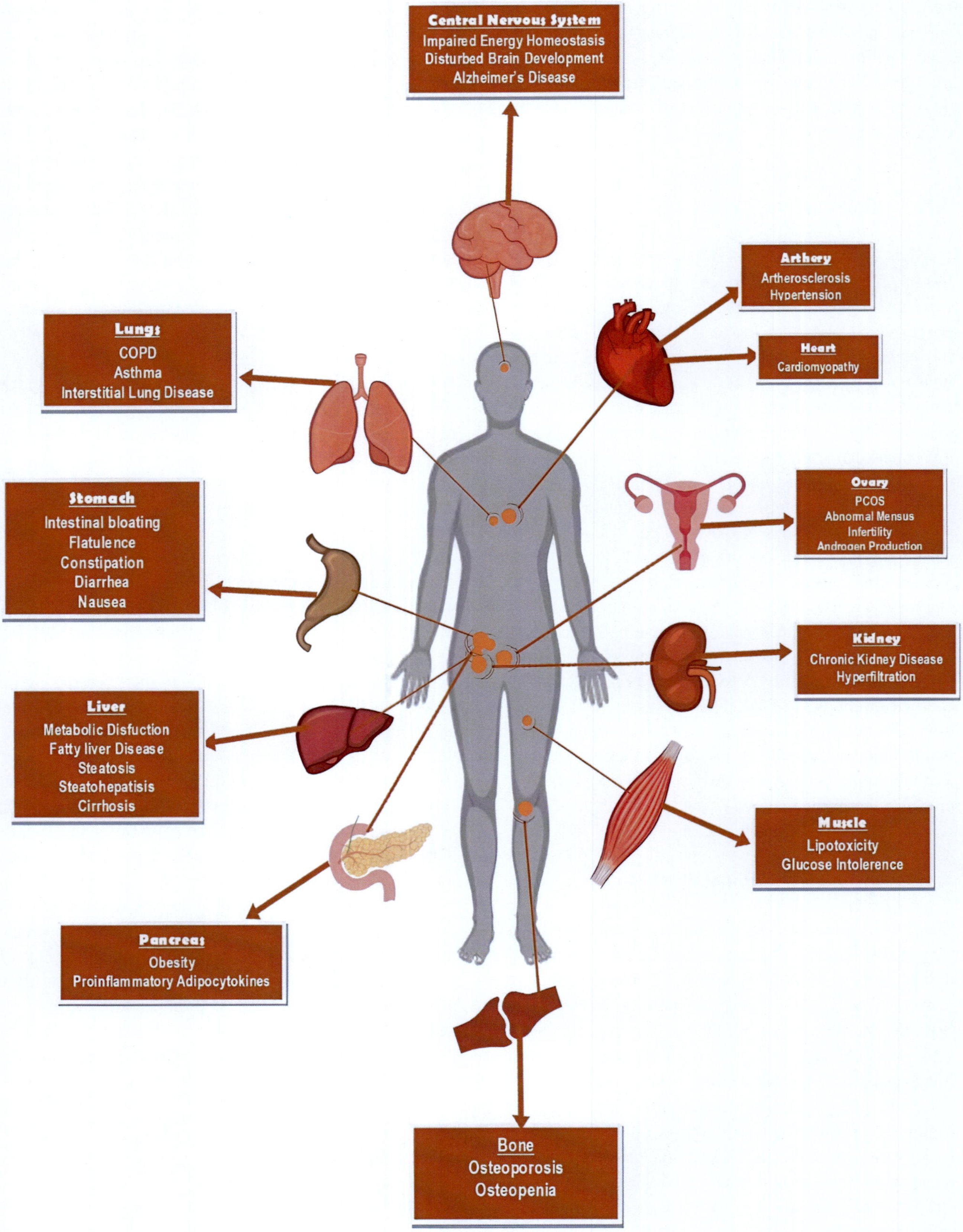

Fig. 2 The pathogenic magnitude of IR at varied biological systems and their derivate disorders

Table 1 The plausible risk-causative factors and common presentation of IR

Specific physiologic abnormalities which increase the risk of insulin resistance syndrome	Other factors	Common presentations
Impaired glucose tolerance or impaired fasting glucose	Body mass index (BMI) greater than or equal to 25 kg/m^2	The asymptomatic patients with obesity, hypertension, or hyperlipidemia
Abnormaluric acid metabolism	Diagnosis of CVD, PCOS, NAFLD, or Acanthosis nigricans	Metabolic syndrome prediabetes or type 2 diabetes mellitus Type A or type B insulin resistance PCOS (menstrual irregularities, hirsutism, acne, and alopecia)
Dyslipidemia (increased triglycerides, decreased HDL-C, or small, dense LDL)	A family history of T2DM, hypertension, or CVD	Xanthelasma or xanthomas Acanthosis nigricans
Hemodynamic changes such as elevated blood pressure	Sedentary lifestyle	Symptomatic microvascular disease (retinopathy, neuropathy, ornephropathy)
Prothrombic factors (PAI-1, fibrinogen)	Age greater than 40 years	Macrovascular disease (Stroke, PAD, and CAD)
Markers of inflammation (CRP, WBC, etc.)		
Endothelial dysfunction		

serum glucose and insulin response to a glucose challenge [15]. It is a fact that, more often than not, physicians resort to clinical endpoints such as 'how the patient feels, functions, or survives (without unusual/ distressing events)', rather than trying to understand the pathological process reversal that eventually results in clinical endpoints.

4 Shared Mechanism of Insulin Resistance in Diverse Clinical Syndromes

4.1 IR Targeted Basic Metabolic Errors, Shared Molecular and Histological Mechanisms

When it comes to human pathogenic mechanisms, mathematical predictive models of individual histocompatibility are more often than not absurd. Molecular mechanisms and signalling pathways integrate and influence each other. So specific pathognomy deviations may be shared in different functional systems. This is precisely the case with IR. Impairment of the signalling pathway, insulin receptor defects, and defective insulin secretion can all contribute to insulin resistance.

4.2 IR and Metabolic Syndrome

Metabolic syndrome (MetS) refers to a conglomeration of several clinical signs, viz. increased waist circumference, elevated triglycerides, reduced high-density lipoprotein (HDL), elevated fasting glucose levels, and elevated blood pressure. Mets have profound health implications as they predispose individuals to cardiovascular and cerebrovascular disorders and associated morbidity and mortality. The fundamental pathogenic process in MetS is adipose tissue dysfunction and resultant IR. As cited earlier, inflammatory materials released from an impaired adipocyte adversely alter peripheral insulin sensitivity. Sequentially the vicious cycle of impaired insulin signalling mechanisms, receptors deficits, and altered insulin secretion reinforces IR. Earlier signs of MetS and associated errors manifest as autonomic and vascular changes. Visceral adipocytes react critically when compared to subcutaneous fat cells, though both are involved in MetS and IR.

MetS unfavourably impacts several body tissues. IR in MetS causes microvascular insult, which inclines a patient towards endothelial irregularities, hypertension, and vessel wall inflammation, ending in vascular resistance. Endothelial damage can impact the body's homeostasis causing atherosclerotic disease, fatty liver, and hypertension. Additionally, hypertension undesirably affects bodily physiology and results in peripheral vascular disease, cardiomyopathies, and renal functional impairment [23].

4.3 IR and Endocrine Errors and Poly Cystic Ovarian Syndrome

Hormones maintain an antagonistic effect on insulin secretion and signalling pathways. As a result of these aggressive reactions, there is a remarkably decreased glucose utilization at the periphery compared to its synthesis in the liver. A cumulative effect of this progression hampers insulin secretion and establishes insulin resistance. Therefore, in endocrine disorders, such as growth hormone deficiencies, acromegaly, Cushing's syndrome, thyroid-parathyroid dysfunctions, adrenal disorders, hypogonadism, or neuroendocrine, tumours impair glucose metabolism and utilization due to progressive IR. Most importantly, polycystic ovarian syndrome (PCOS), a distressing condition causally associated with infertility, is quite a concern these days. PCOS-related IR is an essential cause of non-insulin-dependent diabetes mellitus in women. Excess serine phosphorylation of the insulin receptors is a candidate cause of IR in PCOS. Serine or threonine kinase, an extrinsic factor controlling insulin receptor signalling, plausibly causes this pathogenic process of excess phosphorylation. Through its receptors, insulin enhances ovarian and adrenal steroidogenesis and pituitary LH discharge. The fact that PCOS has a menarche age of onset, the ontogeny of insulin signalling pathways, and other metabolic faults are to be screened for genetic errors. PCOS women are predisposed to cardiovascular disease owing to lipid abnormalities, dysfibrinolysis, and IR [15]. PCOS is currently managed with metformin, incretins, and glitazones. Integrated practices are genuinely needed in these conditions that appropriately manage insulin sensitivity and glucose metabolism [24].

4.4 IR, Neuro-Inflammation, and Neurodegenerative Disorders

In addition to regulating glucose metabolism and energy utilization at the periphery, insulin can cross the blood–brain barrier and influence neuronal survival and growth, dopaminergic transmission, and maintenance of synapses and pathways involved in cognition. Brain insulin resistance and insufficiency, along with neuro-inflammation, results in definite degenerative pathologies in atrophied cortices, nuclei, and white matter and hastened astrocytosis, gliosis, and microvascular pathologies. Markers such as hyper-phosphorylated tau (pTau)-containing cytoskeletal lesions, increased amyloid-beta (Aβ42) deposits in plaques, vessels, and neurons, and increased ubiquitin immune-reactivity in degenerating neurons are candidate signatures of neurodegeneration, for instance, the Alzheimer's disease. It is a concern that brain glucose and oxygen metabolism issues are not incorporated into the cluster of neurodegenerative indicators. This results in a limited understanding of interconnected pathogenic mechanisms in degenerative conditions. Thus, it limits the opportunity to fully understand the disease's natural course and plan holistic therapeutic strategies efficacious in decelerating the progression. Increased expression of inflammatory markers near degenerative plaques or proteins indicates inflammation hastens degeneration. Chronic inflammation exacerbates IR in Alzheimer's (AD) and Parkinson's' disease (PD) [25]. Also, there is sufficient evidence that IR and degenerative disorders such as PD share similar dysregulated pathways. There is evidence that even without type 2 diabetes manifestation, peripheral IR is found in PD. All this information suggests that the insulin signalling pathway may be a candidate target for disease modification concerning neurodegenerative conditions [26].

4.5 IR and Cancer

IR and the subsequently enhanced bioavailability of insulin-like growth factor (IGF-I) are antagonists to the hepatic synthesis of sex hormone-binding globulin. At the same time, these hormones stimulate the ovarian synthesis of sex steroids. Sex steroids affect breast epithelium and endometrium. It can promote cell proliferation and inhibit apoptosis. Thus hormone-mediated tumour initiation and progression are causally associated with IR. Moreover, IR in adipocytes stimulates systemic inflammation, overproduction of inflammatory cytokines and other markers, and reactive oxygen species (ROS) that hastens mutagenesis and carcinogenesis [27].

4.6 Psycho-Neuro-Immuno-Endocrinal Axis, Stress, and IR

As science advances, newer interactive mechanisms and interconnected pathways come into light that has clinical significance. Psycho-neuro-immuno-endocrinal axis thus describes the influence of neural, psychological (stress-related), and endocrine mechanisms on immune responses and immune responses triggering neural, psychological, and endocrine conditions. The role of stress-related conditions in the manifestation of IR, metabolic syndrome, cardiovascular diseases, and psychiatric diseases is depicted in Fig. 3.

4.7 Prognosis and Complications Associated with IR

The prognosis of IR categorically depends on the particular variant of the disease (endotype), the severity of IR, histocompatibility of metabolic tissues, genes, the onset of complications, and personalized characteristic responses to therapeutic measures. The stages may vary from mild IR in asymptomatic individuals to subjects with definite cardiovascular or cerebrovascular deficits or degeneration and accompanying morbidity and mortality or with loss of reproductive capability associated with PCOS.

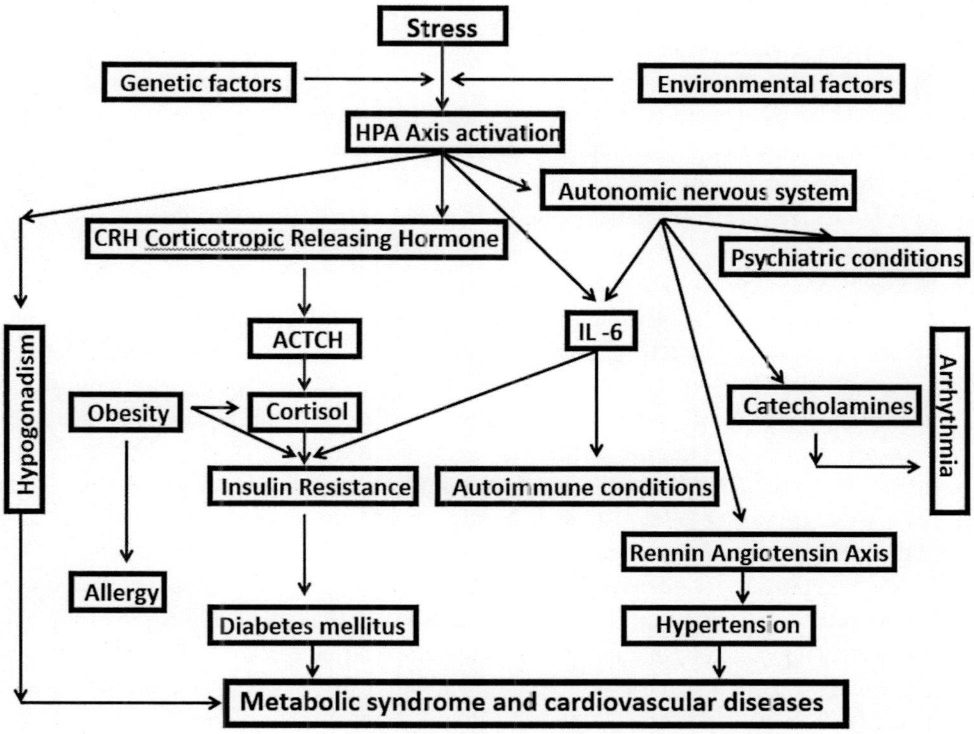

Fig. 3 The complex multi-axial interaction in psycho-neuro-endocrine-immune pathways

5 Current Treatment Strategies for Insulin Resistance and Scope for Integrative Medicine Accommodating Herbal Drugs

5.1 Current Management Strategies Adopted in IR-Associated Clinical Conditions

It is a matter of disquiet that there are no predefined clinical practice guidelines or consensus in treating patients with IR. Insulin therapy has been the first line of management. However, there is sufficient evidence that despite higher insulin doses, glycaemic targets are not achieved in patients with severe IR [28]. Moreover, they suffer brunt in the form of weight gain or sudden episodes of hypoglycaemia. Metformin extended release is now recommended, which is also advocated in IR and IR-linked disorders [29]. Glucagon-like peptide one inhibitor (GLP-1) receptor agonists and metformin have joined the race lately for better glycaemic control, especially in obese patients. The same is true with Sodium-glucose co-transporter two (SGLT2) inhibitors and pramlintide. Dipeptidyl peptidase-4 inhibitors (DPP-4) extend the activity of endogenous GLP-1 and gastric inhibitory polypeptide (GIP) by preventing their breakdown. Thiazolidinediones (TZDs) increase insulin sensitivity by enhancing insulin-dependent glucose disposal in muscle and adipose tissue and decelerating hepatic glucose yield. It is of prime importance that, more often than not patient specific characters dictate the drug of choice or elective procedures. For instance, gastric sleeve, banding, and bypass surgery are indicated in IR individuals with obesity. The excess fat loss associated with bariatric surgery decreases insulin resistance [28].

Management strategies adopted in pre-diabetic and diabetic states that mark IR generally intend to instil a balanced lifestyle through monitored physical activity and calorie restriction—reduced high glycaemic index dietary articles. Recent research advances give evidence in the form of proof of principle for methods, viz. enhancing phosphorylation and perpetuating the crucial kinases activity of the insulin receptor and its protein substrates following insulin stimulation. Also, enhancing the action of phosphatidylinositol 3-kinase and other essential providers of insulin signalling and metabolic pathways are being tested. The causal role of pro-inflammatory cytokines, adipocyte hormones, fatty acids, the vicious cycle of glucotoxicity, and the protective mechanisms offered by certain metabolic enzymes, vitamins-minerals, and co-factors in IR-associated states are under scrutiny for identifying feasible therapeutic targets [30]. IR seems inadequate despite these advances, as is evident from the rise of NCDs and other global IR-associated conditions. The aforecited interlinked phenomena of autoimmunity, autophagy, auto-inflammation, and gut dysbiosis significantly add to the challenges in targeting IR in IR-associated syndromes.

5.2 Adverse Effects of Commonly Prescribed Medicaments

Metformin, the most commonly prescribed medicine to manage IR, is also reported to cause gastrointestinal side effects. Many studies link it to cause mild to moderate renal failure, and it is contraindicated in those individuals with elevated serum urea, creatinine, and high glomerular filtration rate. Other categories of drugs, like glucagon-like peptide 1 receptor agonists and sodium-glucose co-transporter 2 inhibitors, also have serious side effects in long-term usage, including changes in renal function and an increased chance of genital infections, especially in women with a history of mycotic infections. Drugs like thiazolidinediones improve insulin sensitivity in type 2 DM and PCOS cases but have been linked with severe side effects, including heart failure and fractures. The long-term outcomes of many of these drugs, including metformin in pregnancy, are not studied well yet [28]. Another major challenge is that the responsiveness of various ethnic populations to many of these medicines varies considerably. Ethnic differences considerably affect insulin sensitivity as well as treatment/dosage choices. Research suggests that a precision medicine approach is required to get optimum results in managing IR [31, 32]. Episodic hypoglycaemia is another issue concerning oral antidiabetics and insulin usage. The risk is more with the usage of sulphonylurea usage and comparatively less for metformin. However, metformin, in some instances, has been associated with inducing lactic acidosis significantly when the clinical conditions worsen [33].

5.3 Scope of Traditional Systems of Medicines (TMs) as an Arm of Integrative Medicine in IR-Associated States

Tackling such non-linear and complex functional deviations in IR, in this era of rapidly emerging diseases and healthcare challenges, awareness and interest in traditional medicines (TMs) and their role and relevance worldwide are entirely renewed. Universal health coverage is achieved only by an inclusive attitude of incorporating all available healthcare practices in the mainstream. In addition, with more people from developing countries opting for traditional healthcare systems to treat common and sporadic diseases and prevent lifestyle disorders, there is a clear trend in health seeker behaviour.

The pattern and predictors of traditional, complementary, and alternative medicine use in non-communicable and chronic ailments are associated with various disease-specific demographic factors and endotypes [34]. The integrative approaches accommodating TMs and standard care protocols should explore lifestyle manipulations, and administration of immune-modulating anti-inflammatory, and adaptogenic drugs in distinct IR endotypes. Integrative approaches and herbs should effectively target gut microbiota revitalization, develop operative management strategies for a functionally regressed autophagy phase, counter autoinflammation, and bring in immune modulation to enhance tissue compatibility and insulin sensitivity in tissues. Most importantly, precision medicine in the form of personalized therapeutic approaches in varied endotype-driven presentations of IR especially targeting the clinical reversal of symptoms in conditions such as DM, NAFLD, PCOS, obesity, etc., seems to be the most novel and robust method.

Patient community turnover to TMs for managing LsD and related functional deviations wherein the quality of life, recovery and rehabilitation, correction of metabolic errors, and management of complications such as neuropathies, depression, infertility, etc., are the categorically expected outcomes. As the current approaches in IR that focus on therapies that address gene mutation, polymorphism, and individualistic markers seem indeterminate in many aspects, traditional medicine like Ayurvedic Sciences, which are advocated and categorically accepted by masses in LsD and related inflammatory, autoimmune, and degenerative pathologies should come up with an integrative research mind-set. Such holistic systems of medicine addressing patient distress should also rigorously validate the proof of mechanisms by accommodating surrogate markers to clinical endpoints such as gene signatures, metabolome, genotype-phenotype interactions, and epigenetics when it comes to conditions involving IR. Thus, targeted multifaceted precision-based personalized medicine in an integrative mode seems the correct solution.

6 Herbal Medicine for the Management of IR and Other Associated States of LsD

TMs suggest various herbs and polyherbal preparations for managing metabolic syndrome, obesity, neoplastic presentations, neurodegenerative disorders, and endocrine pathologies like PCOS associated with IR. In Ayurvedic herbal medicine, the drugs are used as a whole, where therapeutic efficacy is expected to come from the combined activity of various phytochemicals. Despite promising clinical efficacy, the complexity of formulations and the inability to explain the mode of action have been the primary reasons limiting the R&D of these traditional medicine practices. Much research has focused on developing new drugs from traditional herbal medicine practices by isolating various active ingredients. The discovery of guggulsterone from *Commiphora mukul*, boswellic acid from *Boswellia serrata*, and reserpine from *Raufolwia serpentina* show how herbal medicine-based empirical practices have guided innovative drug discovery [35]. The current book chapter here after focus on a few medicinal plants extensively used in traditional medicine practices to treat occupational lifestyle diseases and the associated IR. An attempt has been made to review the mode of action of a few medicinal plants and their monomers in improving IR and associated occupational lifestyle disorders in various experimental models and clinical trials.

6.1 *Curcuma longa* L.

C. longa, commonly known as turmeric, is an herb used in traditional Indian medicine for its broad therapeutic activity. Apart from medicinal uses, it is also used in households as a spice and food preservative. Matured rhizomes of *C. longa* contain a yellow-coloured substance, a combined form of resin and oil called curcumin. Curcumin is the principal bioactive constituent responsible for its pharmacological activity [36]. In the chronic high-fat diet-fed (HFD) mouse model, long-term dietary curcumin administration prevented weight gain and obesity. It also improved glucose disposal by stimulating insulin sensitivity. Curcumin supplementation could also prevent HFD-impaired insulin-stimulated protein kinase B (PKB) phosphorylation in adipocyte tissues and the liver. It was concluded that the antioxidant and anti-inflammatory activity of curcumin on adipocytes were the underlying mechanism that reduced IR, obesity, and prevented diabetes in the animal models [37]. Adiponectin, an adipocyte-derived hormone, has been reported to reduce IR by reducing triglyceride levels in the muscle and liver of obese animal models [38]. Studies have shown that in skeletal muscle cells, adiponectin-induced autophagy could alleviate IR and metabolic dysfunction [39]. Oral administration of curcumin has been reported to increase the expression of adipocytes and thereby improve insulin sensitivity in animal models [40]. Insulin receptors and insulin-linked glucose transporters in the brain are responsible for various cognitive functions, including learning and memory. Experiments have shown that oral administration of curcumin could improve glucose homeostasis and could increase the gene expression of cholinergic, Glut 3, and insulin receptors in the cerebellum of streptozotocin (STZ) induced diabetes mellitus (DM) rats [41]. Bio-enhanced turmeric extract supplementation has also been reported to decrease IR and increase beta cell function in experimental models.

Decreasing IR increased glucose uptake and fatty acid oxidation in the skeletal muscles of treated animals, causing weight loss. Compared to the regular extract, the bio-enhanced extracts treated group showed an increased concentration of curcumin and its derivatives in the pancreas [42]. In a randomized, double-blind, placebo-controlled clinical trial of 100 individuals with type 2 DM curcuminoids (500 mg/day) daily dose along with piperine (5 mg/day), the administration showed a beneficial effect on the glycaemic and hepatic parameters. The serum concentration of insulin, HbA1c, and homeostasis model assessment estimated insulin resistance (HOMA-IR) significantly reduced in both the study groups. Weight and BMI reduction was significant in the curcuminoids-treated group [43]. Systematic reviews and meta-analysis studies of RCTs recommend that curcumin be administered as an adjunct to patients with type 2 DM to improve insulin resistance and control glycaemic control. The study results were more significant to individuals habituated in Asian and Middle Eastern countries [44].

6.2 *Emblica officinalis* Gatertn

E. officinalis, also known as the Indian gooseberry or amla, is a drug extensively used in indigenous traditional practices like Ayurveda for its medicinal properties and nutritional values. Medicinal properties are attributed to different parts of *E. officinalis*, but fruits are the most preferred choice for therapeutic purposes. Fruits are rich in phytoconstituents such as polyphenols, tannins, flavonoids, and vitamins responsible for diverse pharmacological activity [45]. Hyperinsulinemia accompanied by IR leads to derangement of the lipid metabolism in the body. It leads to the overproduction of very low-density lipoprotein (VLDL), low-density lipoprotein (LDL), and triglycerides, and a decrease in high-density lipoproteins [46]. In the post-menopausal animal model with atherogenic dyslipidaemia, amla extract supplementation prevented IR, decreased LDL, and increased HDL levels. The study suggested that amla can be protective in preventing coronary heart disease in vulnerable post-menopausal groups by increasing insulin sensitivity [47]. Combinations of turmeric and amla are used in traditional medicine to manage DM. Standardized extract of both drugs containing curcuminoids (23.89%), gallic acid (5.27%), and tannins (25.44%) administered in high-fat diet and STZ-induced type 2 DM rat models significantly ameliorated the IR. The result was consistent with all three models of IR assessment, i.e., in HOMA-IR, quantitative insulin sensitivity check index (QICKI), and Matsuda index [48]. Research findings support that reactive oxygen species (ROS) play a crucial role in the pathogenesis of IR and other related chronic diseases [49]. Amla extracts have exhibited good antioxidant activity in DPPH radical scavenging and ROS inhibition assay suggesting its protective role against IR [50]. Many studies are reported on the crucial role amla fruit constituents like gallic acid and ellagic acid play in activating the insulin signalling pathways, reducing IR, and improving signal transduction

[51]. A randomized clinical trial (RCT) that administered *E. officinalis* extract to newly diagnosed DM patients with dyslipidaemia reported its anti-diabetic and lipid-lowering potentials. Fruit extract at a dose of 2 g/day exhibited superior activity to Metformin [52].

6.3 *Cinnamomum verum* J. Presl

C. verum, commonly called true cinnamon, is a small tree belonging to the Lauracea family. The tree's inner bark is the useful part traditionally used in treating many conditions like DM. It is also used as a spice in culinary practices to add colour and flavour to food [53]. Much research has been carried out on the biological activity of *C. verum* extract and its phytocompounds. A bioactive compound extracted from the aqueous extract has been reported to increase insulin signalling by decreasing the activity of insulin receptor kinase inhibitors in rat epididymal fat cells. Cinnamon compounds are supposed to regulate protein phosphorylation and dephosphorylation in intact adipocytes, thereby regulating IR [54]. Oral administration of cinnamon improved glucose infusion rate by upregulating the insulin signalling pathway in the skeletal muscles of Wistar rats. In the study, it was seen that the cinnamon extract did not affect insulin secretion. However, it significantly increased the tyrosine phosphorylation level of the insulin receptor β subunit (IR-β) and insulin receptor substrate-1 (IRS-1) [55]. In HFD-induced models of IR, cinnamon extract (CE) significantly reduced hyperinsulinemia, dyslipidaemia, and plasma adiponectin. IR has been strongly associated with the overexpression of lipid metabolism-related biomarkers. CE regulated lipid metabolism by reducing the chylomicron-apolipoprotein (apo) B48, VLDL-apoB100 particles, sCD36, plasma FABP4, and RBP4 levels. At the molecular level, CE increased the IRS 1, IRS 2, and adipose mRNA expression in adipose tissue. CE also suppressed the overexpression of Lpl, Fas, and Fabp4 genes, which are usually the early markers of metabolic syndrome [56]. CD36, a complex multifunctional protein, is another essential macrophage activation marker in IR conditions. The soluble form of CD36 (sCD36) is elevated in metabolic syndrome and is a surrogate marker in conditions like atherosclerosis [57]. CE reduced the increased levels of sCD36 in adipose tissue, highlighting the protective role of cinnamon [55]. Clinically cinnamon administration decreased serum glucose, total cholesterol, triglycerides, and LDL in individuals with type 2 DM [58]. In another pilot study, oral administration of cinnamon for 8 weeks to 15 non-diabetic women with PCOS improved insulin sensitivity and reduced IR significantly compared to the placebo group. Large sample prospective RCTs are required to validate the findings of such studies [59].

6.4 *Momordica charantia* L.

M. charantia (MC), also called bitter gourd, is a medicinal plant with a long history of usage in traditional medicine. The unripe fruit of the plant is bitter and is a commonly used vegetable. Cucurbitane-type triterpenoids isolated from MC are extensively studied for their anti-diabetic activities [60]. Many biologically effective novel proteins and enzymes have been isolated from MC fruit. These have been reported to be effective in various metabolic dysfunctions and exhibit cytotoxic activity against various cancer cells [61]. In diabetic animal models, MC extract reduced serum glucose levels by increasing glucose utilization in the liver. The study also reported that the extract showed no insulin secretagogue activity [62]. In another study, MC aqueous extract significantly lowered blood glucose levels compared to insulin in KK-Ay mice with IR. It increased the GLUT 4 protein content in mouse muscles and significantly reduced hyperinsulinemia by increasing the peripheral utilization of glucose. However, in normal mice, repeated administration of MC did not affect the blood glucose levels. This suggests that MC may directly activate insulin receptors in peripheral muscles to increase glucose uptake [63]. The adenosine monophosphate-activated protein kinase (AMPK) is a metabolic sensor that is involved in activities that regulate lipid, glucose, and energy imbalances. It controls various key molecules of metabolic pathways and is recognized as a potential drug target for treating metabolic dysfunctions [64]. Terpenoid fractions extracted from the stem of MC has shown to activate AMPK in cells. This further increased the tyrosine phosphorylation of IRS-1 and triggered the insulin signalling pathway [65]. Peroxisome proliferator-activated receptor (PPAR γ) in the adipose tissue plays a crucial role in glucose homeostasis and differentiation of fat cells. Synthetic PPAR γ ligands are usually employed to activate PPAR γ, increasing insulin sensitivity. Administration of MC extracts to HFD-fed C57BL/6J mice increased the PPARγ mRNA expressions in adipose tissues. It also acted as a ligand of PPAR γ and increased insulin sensitivity. MC extracts also regulated the PPAR **α**-mediated pathway to upregulate the genes responsible for fatty acid oxidation and cellular uptake of free fatty acids [66]. Leptins are a group of proteins secreted by the white adipose tissue that regulate metabolic activities, energy expenditure in the body. Serum leptin levels are usually elevated in conditions like obesity and act as a predictive marker of metabolic syndrome [67]. MC extracts could significantly reduce S. leptin levels and thus may be effectively used to manage metabolic syndrome triggered by visceral obesity [66]. In a preliminary open-label uncontrolled clinical study, MC supplementation for 3 months significantly reduced the incidence of metabolic syndrome. The results remained consistent after 1 month of drug cessation.

Waist circumference was also considerably reduced in the study population following MC supplementation [68]. A few clinical trials have reported MC to be effective in DM, dyslipidaemia, and obesity [69].

6.5 *Trigonellum foenum-graecum* L.

T. foenum-graecum (fenugreek) is a small plant belonging to the Fabaceae family. Fenugreek leaves are used as a vegetable in culinary practices, whereas seeds are used as a spice or condiment. Traditional medicinal practices have recorded much therapeutic activity of the plant, especially as a digestive aid. Many recent studies have found it helpful in treating DM, dyslipidaemia, cancer, and obesity. It is rich in dietary fibre and contains bio constituents such as saponins, alkaloids, steroidal compounds, flavonoids, and amino acids that exhibit diverse pharmacological activity [70]. In animal models of IR, fenugreek extract (FE) lowered serum insulin, glucose, and triglyceride levels [71]. Adiponectin, a biologically active protein produced by adipose tissues, plays a pivotal role in regulating energy metabolism, cell proliferation, and insulin sensitivity in the body. It also exhibits anti-inflammatory and anti-atherogenic activities. In conditions like obesity, excessive fat accumulation decreases adiponectin production, causing different metabolic complications [72]. FE enhanced adiponectin levels and caused considerable weight reduction in experimental animals. It also increased PPAR γ expression at molecular levels to increase fatty acid uptake and insulin sensitivity [71]. Three polyphenolic stilbenes rhaponticin (RHAc), desoxyrhaponticin (dRHAc), and rhapontigenin (RHAg) isolated from FE have been reported to improve insulin sensitivity and mitochondrial function in IR 3T3-L1 adipocytes. Lipid drop accumulation is a common phenomenon during the differentiation of preadipocyte 3T3-L1 cells into adipocytes. This is usually presented as increased triglyceride levels. FE stilbenes treatment reduced triglyceride levels during cell differentiation suggesting its role in reducing lipid accumulation. They also increased glucose uptake by activating protein kinase B (AKT) and AMPK signalling pathways. IR causes mitochondrial dysfunction and decreases ATP production. Stilbenes improved mitochondrial membrane potential and increased ATP in adipocytes. The free radical scavenging potential of the compounds was highlighted as one of the reasons which promoted the biological activities of these compounds. Among the three compounds, RHAc showed maximum activity [73]. In mice, intragastric administration of FE reduced hyperglyacemia in HFD and STZ-induced type 2 DM. Reactive oxygen species (ROS) play a significant role in developing IR and related complications. In the study, FE increased the superoxide dismutase (SOD) activity for clearing excessive ROS production and inhibiting lipid peroxidation [74]. Aqueous FE was administered orally to HFD-fed C57BL/6J, type 2 DM models to study its preventive and curative effects. HFD models with and without DM decreased serum insulin and glucose by reducing IR. Elevated lipid parameters were also significantly reduced [75]. In HFD and fat emulsion-administered models of hyperlipidaemia, fenugreek-controlled lipid profile by inhibiting cholesterol absorption and enhancing its excretion. Fenugreek contains polysaccharides like galactomannan which increases the viscosity of the small intestine and thereby inhibits the reabsorption of cholesterol and bile salt [76]. A double-blind placebo-controlled clinical study that administered FE to 25 newly diagnosed DM patients with fasting sugar less than 200 mg/dL showed considerable hypoglycaemic activity with improved insulin sensitivity. Serum triglyceride levels were decreased in the study population [77]. Another RCT that administered fenugreek as adjuvant therapy to Metformin in 12 patients with type 2 DM reduced blood glucose levels. It significantly increased HDL/LDL ratio compared to Glibenclamide treated group. Although it reduced IR, HbA1c, LDL, and triglycerides, the results were non-significant compared to the standard drug. Rich dietary fibre and polysaccharides in fenugreek delay gastric emptying and inhibit glucose absorption. It was also found helpful in reducing body weight. Nevertheless, the study's sample size seems significantly smaller, and a large prospective RCT is required to validate the findings [78].

6.6 *Eugenia jambolana* Lam.

E. jambolana, called black plum or Jamun, is an evergreen tree from the Myrtacea family. Indian traditional medicine practices use all plant parts to treat various diseases. Jamun fruits are advised to be ideal for inclusion in the diet of diabetic patients, and seeds are used explicitly by traditional practitioners in the treatment of DM [79]. Interestingly before the discovery of insulin, *E. jambolana* preparations were frequently used as an antidiabetic drug in European countries [80]. In HFD-induced obese mice models, jamun extract (JE) reduced IR, body weight, and white adipose mass. It also prevented liver steatosis by reducing the expression of genes related to lipogenesis. Cholesterol, triglyceride, and free fatty acid levels were also significantly reduced. Gut microbial dysbiosis is a major causative factor of metabolic disorders. Studies suggest that in obese individuals, dysbiosis is seen in the bacterial flora with an increased proportion of Firmicutes and a decrease in Bacteroidetes. This derangement alters intestinal permeability to cause increased absorption of lipopolysaccharides and activate inflammatory pathways. JE administration helped to restore the ratio of Firmicutes to Bacteroidetes. Prevotella, Bacteroides, and Alloprevotella, which are typically short-chain fatty acid producers, were also enhanced in experimental models. These help to improve glucose metabolism and reduce obesity and liver steatosis [81]. Administration of JE to HFD-fed IR albino rat models reversed hyperinsulinemia, hyperlipidaemia, and hyperglycaemia. It could significantly reduce the IR compared to the control group [82]. In the long-term and short-term models of DM, the ethyl acetate fraction of JE showed good antihyperglycemic activity. The gallic acid and other polyphenolic compounds present in the fraction contributed to its antidiabetic activity. It was also noticed that the apoptotic cells in pancreatic islets were significantly reduced in the study group due to the potent antioxidant activity of JE. Hexokinase (Hex-1), an enzyme crucial to carbohydrate metabolism, is diminished in diabetic patients. After 35 days of administration, JE improved the expression of Hex-1 in experimental animals and restored carbohydrate metabolism [83]. Patients usually take JE as an adjuvant, along with other oral antihyperglycemic medications. Drugs reducing hyperinsulinemia can significantly treat conditions such as DM, obesity, hypertension, and coronary heart diseases. In high fructose diet-fed animal models, an aqueous extract combination of *E. jambolana* and *M. charantia* prevented hyperinsulinemia and hyperglycaemia. However, alcoholic extracts were found to be ineffective [84]. An herb-drug interaction study reported that the administration of JE along with Sitagliptin reduced the systemic exposure of Sitagliptin. It helped reduce Sitagliptin's side effects without compromising its therapeutic efficacy [85]. An open-label RCT administered *E. jambolana* seed powder to 30 newly diagnosed cases of type 2 DM significantly reduced blood glucose levels and IR. It also reduced HDL cholesterol in the study population and thus may be considered a suitable choice for preventing atherosclerosis and other cardiovascular complications associated with DM [86].

6.7 Gymnema sylvestre (Retz.) R.Br.

G. sylvestre (GS) is an essential medicinal plant in the Asclepiadaceae family. It is an effective herb used in Ayurvedic medicine to treat DM. Leaves are the foremost useful part. Triterpene saponins isolated from leaves are extensively studied for their broad therapeutic activity. Leaves on intake numb the taste buds on the tongue and interfere with the perception of sweet taste. This helps to limit the consumption of sweet food later. Various studies have reported that *G. sylvestre* extracts and their isolated compounds possess antidiabetic, antihyperlipidemic, anticancerous, and anti-obesity properties. It is also consumed as herbal supplements, beverages, and teas to regulate DM and related complications [87, 88]. In animal models of metabolic syndrome, deacyl gymnemic acid isolated from GS decreased systolic blood pressure, hyperglycaemia, hyperinsulinemia, and IR [89]. In an affinity ultrafiltration method coupled with liquid chromatography-mass spectrometry, nine sapogenin compounds in GS extract inhibited α-glucosidase [90]. Apart from antihyperglycemic activity, GS has shown anticancerous activity in human skin melanoma cells. It was found that GS treatment increased the mortality of melanoma cells dose-dependent. It also increased the expression of cytochrome c, caspase 3, PARP, Bax genes responsible for apoptosis. On the other hand, the expression of anti-apoptotic mRNAs of the Bcl2 gene was considerably decreased. This study suggested that GS may be a drug of choice for inclusion in the palliative care of DM patients with cancer [91]. Another STZ-induced type 2 DM study showed that GS extract increased insulin secretion in rat pancreas by increasing insulin mRNA expression. It also promoted endogenous β-cell regeneration by upregulating pancreatic regenerative markers such as Pdx1, Neurogenin 3, MafA, and NeuroD1. E-cadherin (ECAD) is a critical adhesion molecule found in islets of Langerhans which can influence insulin production. GS administration upregulated ECAD, influencing pancreas structure and stimulating insulin secretion [92]. A double-blind placebo-controlled RCT that supplemented GS to 24 patients with metabolic syndrome reduced body weight, body mass index (BMI), and VLDL cholesterol. It showed no significant changes in insulin sensitivity and secretion in those patients [93]. Another RCT reported that GS supplementation to 30 patients with impaired glucose tolerance reduced HbA1c and increased insulin sensitivity. Improvements were noted in the anthropometric measures after treatment and reduced LDL cholesterol [94] (Table 2).

Table 2 The reported activity of medicinal plants and their monomers in IR and associated occupational lifestyle disorders

Study drug	Dose/duration	Type of study	Key findings	Reference
Curcumin	4 g/kg, 2 days a week for 28 weeks	In vivo	Reduced IR by potent antioxidant and anti-inflammatory activity	[37]
	3% by weight admixture with a standardized diet	In vivo and in vitro	Improved insulin sensitivity by increasing serum adiponectin levels	[40]
	60 mg/kg suspension for 14 days	In vivo	Neuroprotective activity by upregulation of cholinergic, Glut 3, insulin receptors in the cerebellum of STZ-induced DM rats	[41]
Turmeric extract	Bio-enhanced extracts (BTE-30 mg/kg and 60 mg/kg) and regular (RTE-30 mg/kg) for 30 days	In vivo	BTE showed significant anti-diabetic activity and weight reduction by decreasing IR and increasing beta cell function	[42]
Curcumin C3 complex	Curcuminoids (500 mg/day) with piperine (5 mg/day) for 3 months	Clinical	Reduction in the serum concentration of insulin, HOMA-IR and HbA1c	[43]
Indian gooseberry extract	100 mg/kg of body weight	In vivo	Decreased dyslipidemia, IR and prevented metabolic alterations in ovariectomized high fructose-fed rat models	[47]
EmbliQur (*E. officinalis* fruit and *C. longa* rhizome extract in proportion 2:1)	500/1000 mg/kg for 28 days	In vivo	Both doses significantly reduced IR, blood glucose levels, and insulin. A dose of 1000 mg/kg was more effective than 500 mg/kg in reducing IR	[48]
Sabbery	Amla fruit extract with 100/kg β-glucogallin	In vitro	Antidiabetic activity by inhibiting pancreatic α-amylase and salivary α-amylase enzymes. Exhibited significant antioxidant potential	[50]
E. officinalis extract with 10% β-glucogallin	1 g/2 g for 90 days	Clinical	Significant reduction in FBS, PPBS, HbA1c, and lipid profile	[52]
Cinnamon (purified fraction)	1:30/1:300 final dilution fraction	In vitro	Inhibit protein tyrosine phosphatases-1 activity and increase insulin receptor activity	[54]
Cinnamon extract	30/300 mg/kg BW for 3 weeks	In vivo	Increased glucose uptake in skeletal muscles by IR- β and IRS phosphorylation in a dose-dependent manner	[55]
Cinnamon extract	50 mg/kg for 8 weeks	In vivo and in vitro	Decreased the level of various biomarkers related to IR. Increased mRNA levels of components involved in the insulin signalling pathway	[56]
Cinnamon	1,3,6 gm/day for 40 days	Clinical	Reduced serum glucose, total cholesterol, LDL, and triglycerides in a dose-dependent manner	[58]
Cinnamon capsule	333 mg, three times a day for 8 weeks	Clinical	Improved insulin sensitivity in non-diabetic women with PCOS	[59]
M. charantia fruit alcoholic extract	500 mg/kg of BW	In vivo	Hypoglycaemic activity by increasing the glycogen synthesis in liver	[62]
M. charantia aqueous extract	100 mg/kg BW for 3 weeks	In vivo	Decreased IR by increasing GLUT 4 content in muscles	[63]
Triterpenoid compounds purified from the stem extract of *M. charantia*	Diluted fractions from 5 mg/mL compound dissolved in dimethylsulfoxide	In vitro	Activated AMPK to increase the IRS-1-linked tyrosine phosphorylation	[65]
M. charantia aqueous P extract and methanolic G extract	P extract—0.5 g, 1 gm/kg of BW G extract—0.2 g, 1 gm/kg of BW	In vivo	Regulation of PPAR γ/PPAR α pathways to improve insulin sensitivity and plasma lipid profile, respectively	[66]
Wild bitter gourd powder	4.8 gm daily for 3 months	Clinical	Decreased metabolic syndrome incidence rate. Reduction in waist circumference	[68]
Fenugreek extract	Intragastric injection of 1000 mg/kg/day for 2 weeks	In vivo	Significantly increased insulin sensitivity and adiponectin levels. Decreased weight gain and triglycerides	[71]
Stilbenes isolated from fenugreek seed extract	0–100 µmol/L	In vitro	Reduced lipid accumulation during differentiation of preadipocyte cell lines into adipocytes. Increased glucose uptake by AKT and AMPK phosphorylation. Improved mitochondrial function	[73]

(continued)

Table 2 (continued)

Study drug	Dose/duration	Type of study	Key findings	Reference
Fenugreek seed extracts (E1, flavonoid and E2, stilbene glycoside)	Intragastric administration of 80 mg/kg/day for 4 weeks	In vivo	Exhibited anti-diabetic and antioxidant activities	[74]
Fenugreek seed hydro-alcoholic extract	2 g/kg/day	In vivo	Decreased IR and lipid parameters	[75]
Fenugreek seed hydro-alcoholic extract	1gm/day	Clinical	Decreased IR and triglycerides	[77]
Fenugreek	2 g/day for 12 weeks	Clinical	Significant increase in HDL/LDL ratio	[78]
Jamun fruit extract	100 mg/kg for 8 weeks	In vivo	Reduced IR, cholesterol, free fatty acids, and triglycerides. Restoring gut microbial dysbiosis	[81]
E. jambolana aqueous and alcoholic extract	1000 mg/kg for 14 days	In vivo	Reversed hyperglycaemia, hyperinsulinemia, and hyperlipidemia	[82]
Ethyl acetate fraction of *E. jambolana* seeds	200 mg/2 mL distilled water/kg of body weight/day for 35 days	In vivo	Decreased degeneration of pancreatic beta cell DNA and increased glucose tolerance	[83]
Combined extract of *M. charantia* fruit and *E. jambolana* kernel	100, 200, and 400 mg/day for 15 days	In vivo	Aqueous extract prevented hyperglycaemia and hyperinsulinemia. Alcoholic extract was ineffective	[84]
E. jambolana powder	5gm twice daily for 6 months	Clinical	Reduced IR and hyperglycaemia. Elevated HDL levels	[86]
Deacyl gymnemic acid	50/100/200 mg/kg for 20 days	In vivo	Decreased systolic blood pressure, S. glucose, insulin, and IR	[89]
Ethanolic extract of *G. sylvestre*	0–300 µg/mL	In vitro	Anticancerous activity by increasing the expression of apoptotic genes cytochrome c, caspase 3, PARP, Bax, and downregulation of anti-apoptotic gene Bcl2	[91]
Gymnema extract	150 mg/kg for 30 days	In vivo	Promoted endogenous pancreatic β-cell regeneration by upregulating markers Pdx1, Neurogenin 3, MafA, and NeuroD1	[92]
G. sylvestre	600 mg/day for 12 weeks	Clinical	Decreased BMI, VLDL, and body weight. Insulin sensitivity and secretion remained unaffected	[93]
	300 mg two times a day for 12 weeks	Clinical	Reduced HbA1c, body weight, BMI, and LDL	[94]

7 Conclusion

The world has witnessed an irrefutable transition of disease burden from infectious diseases and infant-maternal mortality to lifestyle disorders. Considering universal health coverage, an inclusive approach in integrative medicine accommodating herbal drugs has much scope in lifestyle disorders. There is rising evidence regarding the possibilities of herbal medicines in different states of IR and associated clinical syndromes commonly found in lifestyle disorders. In addition to standard care medicines that focus on the patient's routine biochemical parameters, herbal drugs that target multiple pathways, as in IR and associated autophagy, autoimmunity–autoinflammation, gut dysbiosis, and glucose intolerance, shall be a breakthrough in lifestyle disorder management. However, this approach demands safety-efficacy profiling of herb-standard drug interactions. Hence pre-clinical and clinical trials that judiciously highlight the objective of integrative medicine in IR and related clinical states should be systematically planned and executed.

Acknowledgments The authors are thankful to Prof. Rabinarayan Acharya, Director General, Central Council for Research in Ayurvedic Sciences, Ministry of AYUSH, Government of India, G. Babu. Director, Central Ayurveda Research Institute, Kolkata and Dr. D. Sudhakar, Director, National Ayurveda Research Institute for Panchakarma, Kerala for their support, motivations and encouragement.

Conflict of Interest None declare.

Consent for Publication All authors have given consent for publication.

References

1. Budreviciute A, Damiati S, Sabir DK, Onder K, Schuller-Goetzburg P, Plakys G, Katileviciute A, Khoja SKR. Management and prevention strategies for non-communicable diseases (NCDs) and their risk factors. Front Public Health. 2020;26(8):574111. https://doi.org/10.3389/fpubh.2020.574111.

2. Sharma M, Majumdar PK. Occupational lifestyle diseases: an emerging issue. J Occup Environ Med. 2009;13(3):109–12. https://doi.org/10.4103/0019-5278.58912.
3. Noncommunicable diseases. https://www.who.int/news-room/fact--sheets/detail/noncommunicable-diseases. Accessed 27 Jul 2023.
4. Bhattacherjee S, Guha S, Banerjee I, Nath I, Banerjee R, Ray K. A study of insulin resistance and its clinico-metabolic associations among apparently healthy individuals attending a tertiary care hospital. Ann Med Health Sci Res. 2014;4(5):823. https://doi.org/10.4103/2141-9248.141572.
5. Kelly GS. Insulin resistance: lifestyle and nutritional interventions. Altern Med Rev. 2000;5(2):109–32.
6. Krentz A. Insulin resistance: a clinical handbook. Wiley; 2002. p. 202. https://www.wiley.com/en-us/Insulin+Resistance%3A+A+Clinical+Handbook-p-9780470698389. Accessed 27 Jul 2023.
7. Petrelli A, Giovenzana A, Insalaco V, Phillips BE, Pietropaolo M, Giannoukakis N. Autoimmune inflammation and insulin resistance: hallmarks so far and yet so close to explain diabetes endotypes. Curr Diab Rep. 2021;21(12):54. https://doi.org/10.1007/s11892-021-01430-3.
8. Kuramoto K, Kim YJ, Hong JH, He C. The autophagy protein Becn1 improves insulin sensitivity by promoting adiponectin secretion via exocyst binding. Cell Rep. 2021;35(8):109184. https://doi.org/10.1016/j.celrep.2021.109184.
9. Caricilli AM, Saad MJA. The role of gut microbiota on insulin resistance. Nutrients. 2013;5(3):829–51. https://doi.org/10.3390/nu5030829.
10. MacDonald PE, Joseph JW, Rorsman P. Glucose-sensing mechanisms in pancreatic β-cells. Philos Trans R Soc B Biol Sci. 2005;360(1464):2211. The Royal Society. https://doi.org/10.1098/RSTB.2005.1762. Accessed 22 Jul 2022.
11. Wilcox G. Insulin and insulin resistance. Clin Biochem Rev. 2005;26(2):19. Australasian Association for Clinical Biochemistry and Laboratory Medicine. /pmc/articles/PMC1204764/. Accessed 22 Jul 2022.
12. Fu Z, Gilbert ER, Liu D. Regulation of insulin synthesis and secretion and pancreatic beta-cell dysfunction in diabetes. Curr Diabetes Rev. 2013;9(1):25. NIH Public Access. https://doi.org/10.2174/15733998130104. Accessed 21 Jul 2022.
13. Mann E, Sunni M, Bellin MD. Secretion of insulin in response to diet and hormones. Pancreapedia: The Exocrine Pancreas Knowledge Base. 2020. https://doi.org/10.3998/panc.2020.16.
14. D'Alessandris C, Lauro R, Presta I, Sesti G. C-reactive protein induces phosphorylation of insulin receptor substrate-1 on Ser307 and Ser612 in L6 myocytes, thereby impairing the insulin signalling pathway that promotes glucose transport. Diabetologia. 2007;50(4):840–9. https://doi.org/10.1007/s00125-006-0522-y.
15. Freeman AM, Pennings N. Insulin resistance—StatPearls—NCBI Bookshelf. https://www.ncbi.nlm.nih.gov/books/NBK507839/. Accessed 1 Aug 2022.
16. Parikh HM, Elgzyri T, Alibegovic A, Hiscock N, Ekström O, Eriksson KF, et al. Relationship between insulin sensitivity and gene expression in human skeletal muscle. BMC Endocr Disord. 2021;21(1):1–12. BioMed Central Ltd. https://doi.org/10.1186/S12902-021-00687-9/TABLES/2. Accessed 31 Jul 2022.
17. Catalano PM, Kirwan JP, Haugel-De Mouzon S, King J. Gestational diabetes and insulin resistance: role in short- and long-term implications for mother and fetus. J Nutr. 2003;133(5 Suppl. 1):1674S–83. American Society for Nutrition. https://doi.org/10.1093/jn/133.5.1674s.
18. Mlinar B, Marc J, Janež A, Pfeifer M. Molecular mechanisms of insulin resistance and associated diseases. Clin Chim Acta. 2007;375(1–2):20–35. Elsevier. https://doi.org/10.1016/J.CCA.2006.07.005.
19. Courtney CH, Olefsky JM. Insulin resistance. In: Mechanisms of insulin action: medical intelligence unit. StatPearls Publishing; 2022. p. 185–209. https://doi.org/10.1007/978-0-387-72204-7_10. Accessed 31 Jul 2022.
20. Rehman K, Akash MSH. Mechanisms of inflammatory responses and development of insulin resistance: how are they interlinked? J Biomed Sci. 2016;23(1):1–18. BioMed Central. https://doi.org/10.1186/S12929-016-0303-Y. Accessed 31 Jul 2022.
21. Akash MSH, Rehman K, Chen S. Role of inflammatory mechanisms in pathogenesis of type 2 diabetes mellitus. J Cell Biochem. 2013;114(3):525–31. Wiley. https://doi.org/10.1002/JCB.24402. Accessed 31 Jul 2022.
22. Zhang X, Shao H, Zheng X. Amino acids at the intersection of nutrition and insulin sensitivity. Drug Discov Today. 2019;24(4):1038–43. Elsevier Current Trends. https://doi.org/10.1016/J.DRUDIS.2019.02.008.
23. Swarup S, Goyal A, Grigorova Y, Zeltser R. Metabolic syndrome. In: StatPearls. StatPearls Publishing; 2020. http://www.ncbi.nlm.nih.gov/pubmed/29083742. Accessed 18 Feb 2021.
24. Rogowicz-Frontczak A, Majchrzak A, Zozuliska-Ziolkiewicz D. Insulin resistance in endocrine disorders-treatment options. Endokrynol Pol. 2017;68(3):334–50. https://doi.org/10.5603/EP.2017.0026.
25. de la Monte SM. Insulin resistance and neurodegeneration: progress towards the development of new therapeutics for Alzheimer's disease. Drugs. 2017;77(1):47–65. https://doi.org/10.1007/s40265-016-0674-0.
26. Athauda D, Foltynie T. Insulin resistance and Parkinson's disease: a new target for disease modification? Prog Neurobiol. 2016;145–146:98–120. https://doi.org/10.1016/J.PNEUROBIO.2016.10.001. Accessed 27 Jul 2023.
27. Arcidiacono B, Iiritano S, Nocera A, Possidente K, Nevolo MT, Ventura V, et al. Insulin resistance and cancer risk: an overview of the pathogenetic mechanisms. Exp Diabetes Res. 2012;2012:789174. https://doi.org/10.1155/2012/789174.
28. Church TJ, Haines ST. Treatment approach to patients with severe insulin resistance. Clin Diabetes. 2016;34(2):97. https://doi.org/10.2337/DIACLIN.34.2.97. Accessed 1 Aug 2022.
29. Pala L, Barbaro V, Dicembrini I, Rotella CM. The therapy of insulin resistance in other diseases besides type 2 diabetes. Eat Weight Disord. 2014;19(3):275–83. https://doi.org/10.1007/s40519-014-0139-y.
30. Wolosowicz M, Prokopiuk S, Kaminski TW. Recent advances in the treatment of insulin resistance targeting molecular and metabolic pathways: fighting a losing battle? Medicina (Kaunas); 2022;58(4). https://doi.org/10.3390/MEDICINA58040472. Accessed 27 Jul 2023.
31. Franceschi R. Precision medicine in diabetes, current research and future perspectives. J Pers Med. 2022;12(8):1233. Multidisciplinary Digital Publishing Institute. https://doi.org/10.3390/JPM12081233. Accessed 1 Aug 2022.
32. do Vale Moreira NC, Ceriello A, Basit A, Balde N, Mohan V, Gupta R, et al. Race/ethnicity and challenges for optimal insulin therapy. Diabetes Res Clin Pract. 2021;175:108823. Elsevier. https://doi.org/10.1016/J.DIABRES.2021.108823.
33. Bodmer M, Meier C, Krähenbühl S, Jick SS, Meier CR. Metformin, sulfonylureas, or other antidiabetes drugs and the risk of lactic acidosis or hypoglycemia: a nested case-control analysis. Diabetes Care. 2008;31(11): 2086–91. American Diabetes Association. https://doi.org/10.2337/DC08-1171. Accessed 1 Aug 2022.
34. Peltzer K, Pengpid S, Puckpinyo A, Yi S, Anh LV. The utilization of traditional, complementary and alternative medicine for non-communicable diseases and mental disorders in health care patients in Cambodia, Thailand and Vietnam. BMC Complement Altern Med. 2016;16(1):1–11. https://doi.org/10.1186/s12906-016-1078-0.
35. Katiyar C, Gupta A, Kanjilal S, Katiyar S. Drug discovery from plant sources: an integrated approach. Ayu. 2012;33(1):10. Wolters

Kluwer—Medknow Publications. https://doi.org/10.4103/0974-8520.100295. Accessed 23 Jul 2023.
36. Fuloria S, Mehta J, Chandel A, Sekar M, Rani NNIM, Begum MY, et al. A comprehensive review on the therapeutic potential of Curcuma longa Linn. In relation to its major active constituent curcumin. Front Pharmacol. 2022;13:820806. Frontiers Media S.A. https://doi.org/10.3389/FPHAR.2022.820806/BIBTEX.
37. Shao W, Yu Z, Chiang Y, Yang Y, Chai T, Foltz W, et al. Curcumin prevents high fat diet induced insulin resistance and obesity via attenuating lipogenesis in liver and inflammatory pathway in adipocytes. PLoS One; 2012;7(1). https://doi.org/10.1371/JOURNAL.PONE.0028784. Accessed 23 Jul 2023.
38. Yamauchi T, Kamon J, Waki H, Terauchi Y, Kubota N, Hara K, et al. The fat-derived hormone adiponectin reverses insulin resistance associated with both lipoatrophy and obesity. Nat Med. 2001;7(8):941–6. https://doi.org/10.1038/90984. Accessed 23 Jul 2023.
39. Ahlstrom P, Rai E, Chakma S, Cho HH, Rengasamy P, Sweeney G. Adiponectin improves insulin sensitivity via activation of autophagic flux. J Mol Endocrinol. 2017;59(4):339–50. Bioscientifica Ltd. https://doi.org/10.1530/JME-17-0096. Accessed 23 Jul 2023.
40. Weisberg SP, Leibel R, Tortoriello D V. Dietary curcumin significantly improves obesity-associated inflammation and diabetes in mouse models of diabesity. Endocrinology. 2008;149(7):3549. The Endocrine Society. https://doi.org/10.1210/EN.2008-0262. Accessed 23 Jul 2023.
41. Peeyush KT, Gireesh G, Jobin M, Paulose CS. Neuroprotective role of curcumin in the cerebellum of streptozotocin-induced diabetic rats. Life Sci; 2009;85(19–20):704–10. https://doi.org/10.1016/J.LFS.2009.09.012. Accessed 24 Jul 2023.
42. Sayeli VK, Shenoy AK. Antidiabetic effect of bio-enhanced preparation of turmeric in streptozotocin-nicotinamide induced type 2 diabetic Wistar rats. J Ayurveda Integr Med. 2021;12(3):474–9. Elsevier. https://doi.org/10.1016/J.JAIM.2021.04.010.
43. Panahi Y, Khalili N, Sahebi E, Namazi S, Simental-Mendía LE, Majeed M, et al. Effects of curcuminoids plus piperine on glycemic, hepatic and inflammatory biomarkers in patients with type 2 diabetes mellitus: a randomized double-blind placebo-controlled trial. Drug Res (Stuttg). 2018;68(7):403–9. https://doi.org/10.1055/S-0044-101752. Accessed 24 Jul 2023.
44. Zhang T, He Q, Liu Y, Chen Z, Hu H. Efficacy and safety of curcumin supplement on improvement of insulin resistance in people with type 2 diabetes mellitus: a systematic review and meta-analysis of randomized controlled trials. Evid Based Complement Alternat Med. 2021;2021. https://doi.org/10.1155/2021/4471944. Accessed 24 Jul 2023.
45. Variya BC, Bakrania AK, Patel SS. Emblica officinalis (Amla): a review for its phytochemistry, ethnomedicinal uses and medicinal potentials with respect to molecular mechanisms. Pharmacol Res. 2016;111:180–200. Academic Press. https://doi.org/10.1016/J.PHRS.2016.06.013.
46. Howard B V. Insulin resistance and lipid metabolism. Am J Cardiol. 1999;84(1 A):28–32. Elsevier Inc. https://doi.org/10.1016/S0002--9149(99)00355-0. Accessed 24 Jul 2023.
47. Koshy SM, Bobby Z, Hariharan AP, Gopalakrishna SM. Amla (Emblica officinalis) extract is effective in preventing high fructose diet-induced insulin resistance and atherogenic dyslipidemic profile in ovariectomized female albino rats. Menopause; 2012;19(10): 1146–55. https://doi.org/10.1097/GME.0B013E31824E5BF7. Accessed 21 Aug 2022.
48. Panda V, Deshmukh A, Singh S, Shah T, Hingorani L. An Ayurvedic formulation of Emblica officinalis and Curcuma longa alleviates insulin resistance in diabetic rats: Involvement of curcuminoids and polyphenolics. J Ayurveda Integr Med. 2021;12(3):506. Elsevier. https://doi.org/10.1016/J.JAIM.2021.05.005. Accessed 21 Aug 2022.
49. McKeegan K, Mason SA, Trewin AJ, Keske MA, Wadley GD, Della Gatta PA, et al. Reactive oxygen species in exercise and insulin resistance: working towards personalized antioxidant treatment. Redox Biol. 2021;44:102005. Elsevier. https://doi.org/10.1016/J.REDOX.2021.102005.
50. Majeed M, Majeed S, Mundkur L, Nagabhushanam K, Arumugam S, Beede K, et al. Standardized Emblica officinalis fruit extract inhibited the activities of α-amylase, α-glucosidase, and dipeptidyl peptidase-4 and displayed antioxidant potential. J Sci Food Agric. 2020;100(2):509–16. Wiley. https://doi.org/10.1002/JSFA.10020. Accessed 24 Jul 2023.
51. Huang H-Z, Qiu M, Lin J-Z, Li M-Q, Ma X-T, Ran F, et al. Potential effect of tropical fruits Phyllanthus emblica L. for the prevention and management of type 2 diabetic complications: a systematic review of recent advances. Eur J Nutr; 2021;60(7): 3525–42. https://doi.org/10.1007/S00394-020-02471-2. Accessed 24 Jul 2023.
52. Majeed M, Mundkur L, Paulose S, Nagabhushanam K. Novel Emblica officinalis extract containing β-glucogallin vs. metformin: a randomized, open-label, comparative efficacy study in newly diagnosed type 2 diabetes mellitus patients with dyslipidemia. Food Funct. 2022;13(18):9523–31. Royal Society of Chemistry. https://doi.org/10.1039/D2FO01862D. Accessed 24 Jul 2023.
53. Singh N, Rao AS, Nandal A, Kumar S, Yadav SS, Ganaie SA, et al. Phytochemical and pharmacological review of Cinnamomum verum J. Presl-a versatile spice used in food and nutrition. Food Chem; 2021;338. https://doi.org/10.1016/J.FOODCHEM.2020.127773. Accessed 25 Jul 2023.
54. Imparl-Radosevich J, Deas S, Polansky MM, Baedke DA, Ingebritsen TS, Anderson RA, et al. Regulation of PTP-1 and insulin receptor kinase by fractions from cinnamon: implications for cinnamon regulation of insulin signalling. Horm Res. 1998;50(3):177–82. S. Karger AG. https://doi.org/10.1159/000023270. Accessed 25 Jul 2023
55. Qin B, Nagasaki M, Ren M, Bajotto G, Oshida Y, Sato Y. Cinnamon extract (traditional herb) potentiates in vivo insulin-regulated glucose utilization via enhancing insulin signaling in rats. Diabetes Res Clin Pract. 2003;62(3):139–48. Elsevier Ireland Ltd. https://doi.org/10.1016/S0168-8227(03)00173-6. Accessed 25 Jul 2023.
56. Qin B, Polansky MM, Anderson RA. Cinnamon extract regulates plasma levels of adipose-derived factors and expression of multiple genes related to carbohydrate metabolism and lipogenesis in adipose tissue of fructose-fed rats. Horm Metab Res. 2010;42(3):187–93. https://doi.org/10.1055/S-0029-1242746. Accessed 25 Jul 2023.
57. Handberg A, Levin K, Højlund K, Beck-Nielsen H. Identification of the oxidized low-density lipoprotein scavenger receptor CD36 in plasma. Circulation. 2006;114(11):1169–76. Lippincott Williams & Wilkins. https://doi.org/10.1161/CIRCULATIONAHA.106.626135. Accessed 25 Jul 2023.
58. Khan A, Safdar M, Ali Khan MM, Khattak KN, Anderson RA. Cinnamon improves glucose and lipids of people with type 2 diabetes. Diabetes Care. 2003;26(12):3215–8. American Diabetes Association. https://doi.org/10.2337/DIACARE.26.12.3215. Accessed 25 Jul 2023.
59. Wang JG, Anderson RA, Graham GM, Chu MC, Sauer M V., Guarnaccia MM, et al. The effect of cinnamon extract on insulin resistance parameters in polycystic ovary syndrome: a pilot study. Fertil Steril; 2007;88(1):240–3. https://doi.org/10.1016/J.FERTNSTERT.2006.11.082. Accessed 25 Jul 2023.
60. Çiçek SS. Momordica charantia L.—diabetes-related bioactivities, quality control, and safety considerations. Front Pharmacol. 2022;13:904643. Frontiers Media S.A. https://doi.org/10.3389/FPHAR.2022.904643/BIBTEX.
61. Raina K, Kumar D, Agarwal R. Promise of bitter melon (Momordica charantia) bioactives in cancer prevention and therapy. Semin Cancer Biol. 2016;40–41:116. NIH Public Access. https://doi.org/10.1016/J.SEMCANCER.2016.07.002. Accessed 25 Jul 2023.

62. Sarkar S, Pranava M, Marita RA. Demonstration of the hypoglycemic action of Momordica charantia in a validated animal model of diabetes. Pharmacol Res. ; 1996;33(1): 1–4. Academic Press doi:https://doi.org/10.1006/PHRS.1996.0001.
63. Miura T, Itoh C, Iwamoto N, Kato M, Kawai M, Park SR, et al. Hypoglycemic activity of the fruit of the Momordica charantia in type 2 diabetic mice. J Nutr Sci Vitaminol (Tokyo); 2001;47(5):340–4. https://doi.org/10.3177/JNSV.47.340. Accessed 25 Jul 2023.
64. Srivastava RAK, Pinkosky SL, Filippov S, Hanselman JC, Cramer CT, Newton RS. AMP-activated protein kinase: an emerging drug target to regulate imbalances in lipid and carbohydrate metabolism to treat cardio-metabolic diseases: thematic review series: new lipid and lipoprotein targets for the treatment of cardiometabolic diseases. J Lipid Res. 2012;53(12):2490. American Society for Biochemistry and Molecular Biology. https://doi.org/10.1194/JLR.R025882. Accessed 25 Jul 2023.
65. Cheng HL, Huang HK, Chang CI, Tsai CP, Chou CH. A cell-based screening identifies compounds from the stem of Momordica charantia that overcome insulin resistance and activate AMP-activated protein kinase. J Agric Food Chem. 2008;56(16):6835–43. https://doi.org/10.1021/JF800801K. Accessed 25 Jul 2023.
66. Shih CC, Lin CH, Lin WL. Effects of Momordica charantia on insulin resistance and visceral obesity in mice on high-fat diet. Diabetes Res Clin Pract; 2008;81(2):134–43. https://doi.org/10.1016/J.DIABRES.2008.04.023. Accessed 25 Jul 2023.
67. Ghadge AA, Khaire AA. Leptin as a predictive marker for metabolic syndrome. Cytokine. 2019;121. https://doi.org/10.1016/J.CYTO.2019.154735. Accessed 25 Jul 2023.
68. Tsai CH, Chen ECF, Tsay HS, Huang CJ. Wild bitter gourd improves metabolic syndrome: a preliminary dietary supplementation trial. Nutr J. 2012;11(1):4. BioMed Central. https://doi.org/10.1186/1475-2891-11-4. Accessed 25 Jul 2023.
69. Liu Z, Gong J, Huang W, Lu F, Dong H. The effect of Momordica charantia in the treatment of diabetes mellitus: a review. Evid Based Complement Altern Med. 2021;2021. Hindawi Limited. https://doi.org/10.1155/2021/3796265. Accessed 25 Jul 2023.
70. Nagulapalli Venkata KC, Swaroop A, Bagchi D, Bishayee A. A small plant with big benefits: fenugreek (Trigonella foenum-graecum Linn.) for disease prevention and health promotion. Mol Nutr Food Res. 2017;61(6). https://doi.org/10.1002/MNFR.201600950. Accessed 26 Jul 2023.
71. Mohammadi A, Gholamhosseinian A, Fallah H. Trigonella foenum-graecum water extract improves insulin sensitivity and stimulates PPAR and γ gene expression in high fructose-fed insulin-resistant rats. Adv Biomed Res. 2016;5(1):54. Wolters Kluwer—Medknow Publications. https://doi.org/10.4103/2277-9175.178799. Accessed 21 Aug 2022.
72. Nigro E, Scudiero O, Monaco ML, Palmieri A, Mazzarella G, Costagliola C, et al. New insight into adiponectin role in obesity and obesity-related diseases. Biomed Res Int. 2014;2014. Hindawi Limited. https://doi.org/10.1155/2014/658913. Accessed 26 Jul 2023.
73. Li G, Luan G, He Y, Tie F, Wang Z, Suo Y, et al. Polyphenol stilbenes from fenugreek (Trigonella foenumgraecum L.) seeds improve insulin sensitivity and mitochondrial function in 3T3-L1 adipocytes. Oxid Med Cell Longev. 2018;2018. Hindawi Limited. https://doi.org/10.1155/2018/7634362.
74. Li X-Y, Lu S-S, Wang H-L, Li G, He Y-F, Liu X-Y, et al. Effects of the fenugreek extracts on high-fat diet-fed and streptozotocin-induced type 2 diabetic mice. Animal Model Exp Med. 2018;1(1):68. Wiley-Blackwell. https://doi.org/10.1002/AME2.12004. Accessed 26 Jul 2023.
75. Hamza N, Berke B, Cheze C, Le Garrec R, Umar A, Agli AN, et al. Preventive and curative effect of Trigonella foenum-graecum L. seeds in C57BL/6J models of type 2 diabetes induced by high-fat diet. J Ethnopharmacol. 2012;142(2):516–22. https://doi.org/10.1016/J.JEP.2012.05.028. Accessed 26 Jul 2023.
76. Wang Y, Zheng Y, Liu Y, Shan G, Zhang B, Cai Q, et al. The lipid-lowering effects of fenugreek gum, hawthorn pectin, and burdock inulin. Front Nutr. 2023;10:1–16. https://doi.org/10.3389/fnut.2023.1149094.
77. Gupta A, Rajeev G, Bechoo L. Effect of Trigonella foenum-graecum (fenugreek) seeds on glycaemic control and insulin resistance in type 2 diabetes mellitus: a double blind placebo controlled study. J Assoc Physicians India. 2001;49:1057–61.
78. Najdi RA, Hagras MM, Kamel FO, Magadmi RM. A randomized controlled clinical trial evaluating the effect of Trigonella foenum-graecum (fenugreek) versus glibenclamide in patients with diabetes. Afr Health Sci. 2019;19(1):1594. Makerere University Medical School. https://doi.org/10.4314/AHS.V19I1.34. Accessed 26 Jul 2023.
79. Baliga MS, Bhat HP, Baliga BRV, Wilson R, Palatty PL. Phytochemistry, traditional uses and pharmacology of Eugenia jambolana Lam. (black plum): a review. Food Res Int. 2011;44(7):1776–89. Elsevier. https://doi.org/10.1016/J.FOODRES.2011.02.007.
80. Helmstädter A. Antidiabetic drugs used in Europe prior to the discovery of insulin. Pharmazie. 2007;62(9):717–20. https://doi.org/10.1691/ph.2007.9.7094.
81. Xu J, Liu T, Li Y, Liu W, Ding Z, Ma H, et al. Jamun (Eugenia jambolana Lam.) fruit extract prevents obesity by modulating the gut microbiome in high-fat-diet-fed mice. Mol Nutr Food Res;2019;63(9). https://doi.org/10.1002/MNFR.201801307. Accessed 26 Jul 2023.
82. Bmhrk G, Sachidananda Adiga MN, Nagendra Nayak IM, Roy AD, Sampath Kumar AK. Evaluation of reversal effects of Eugenia jambolana seed extracts against high-fructose diet-induced insulin resistance in albino rats. Asian J Pharm Clin Res. 2018;11(12):572–6. Innovare Academics Sciences Pvt. Ltd. https://doi.org/10.22159/AJPCR.2018.V11I12.27548.
83. Jana K, Bera TK, Ghosh D. Antidiabetic effects of Eugenia jambolana in the streptozotocin-induced diabetic male albino rat. Biomark Genom Med. 2015;7(3):116–24. Elsevier. https://doi.org/10.1016/J.BGM.2015.08.001.
84. Vikrant V, Grover JK, Tandon N, Rathi SS, Gupta N. Treatment with extracts of Momordica charantia and Eugenia jambolana prevents hyperglycemia and hyperinsulinemia in fructose fed rats. J Ethnopharmacol. 2001;76(2):139–43. Elsevier. https://doi.org/10.1016/S0378-8741(01)00218-5.
85. Vora A, Varghese A, Kachwala Y, Bhaskar M, Laddha A, Jamal A, et al. Eugenia jambolana extract reduces the systemic exposure of Sitagliptin and improves conditions associated with diabetes: a pharmacokinetic and a pharmacodynamic herb-drug interaction study. J Tradit Complement Med. 2019;9(4):364. Elsevier. https://doi.org/10.1016/J.JTCME.2018.10.001. Accessed 26 Jul 2023.
86. Sahana DA, Shivaprakash G, Baliga R, Adhikari Prabha MR, Ganesh J, Pai MR. Effect of Eugenia jambolana on plasma glucose, insulin sensitivity and HDL-C levels: preliminary results of a randomized clinical trial. J Pharm Res. 2010;3:1268–70.
87. Kanetkar P, Singhal R, Kamat M. Gymnema sylvestre: a memoir. J Clin Biochem Nutr;2007;41(2):77. The Society for Free Radical Research Japan. https://doi.org/10.3164/JCBN.2007010. Accessed 27 Jul 2023.
88. Tiwari P, Mishra BN, Sangwan NS. Phytochemical and pharmacological properties of Gymnema sylvestre: an important medicinal plant. Biomed Res Int. 2014;2014:830285. https://doi.org/10.1155/2014/830285.
89. Bhansali S, Shafiq N, Pandhi P, Singh AP, Singh I, Singh PK, Sharma SMS. Effect of a deacyl gymnemic acid on glucose homeostasis & metabolic parameters in a rat model of metabolic syndrome. Indian J Med Res. 2013;137(6):1174–9.

90. Chen G, Guo M. Rapid screening for α-glucosidase inhibitors from Gymnema sylvestre by affinity ultrafiltration-HPLC-MS. Front Pharmacol. 2017;8:263626. Frontiers Research Foundation. https://doi.org/10.3389/FPHAR.2017.00228/BIBTEX.
91. Chakraborty D, Ghosh S, Bishayee K, Mukherjee A, Sikdar S, Khuda-Bukhsh AR. Antihyperglycemic drug Gymnema sylvestre also shows anticancer potentials in human melanoma A375 cells via reactive oxygen species generation and mitochondria-dependent caspase pathway. Integr Cancer Ther. 2013;12(5):433–41. https://doi.org/10.1177/1534735413485419.
92. Kannan P, Raghunathan M, Mohan T, Palanivelu SPK. Gymnemic acid ameliorates pancreatic β-cell dysfunction by modulating Pdx1 expression: a possible strategy for β-cell regeneration. Tissue Eng Regen Med. 2022;19(3):603–16. https://doi.org/10.1007/s13770-022-00435-7.
93. Zuñiga LY, González-Ortiz M, Martínez-Abundis E. Effect of Gymnema sylvestre administration on metabolic syndrome, insulin sensitivity, and insulin secretion. J Med Food. 2017;20(8):750–4. https://doi.org/10.1089/jmf.2017.0001.
94. Gaytán Martínez LA, Sánchez-Ruiz LA, Zuñiga LY, González-Ortiz M, Martínez-Abundis E. Effect of Gymnema sylvestre administration on glycemic control, insulin secretion, and insulin sensitivity in patients with impaired glucose tolerance. J Med Food. 2021;24(1):28–32. https://doi.org/10.1089/jmf.2020.0024.

Extraction of Phytoconstituents for Lifestyle Diseases

Joseph Tchamgoue, Yvan Anderson T. Ngandjui, Rostan M. Talla, Bruno Dupon A. Ambamba, Jean Claude Tchouankeu, and Simeon F. Kouam

Abstract

Diseases whose occurrence is mainly related to an individual's daily living habits are referred to as lifestyle diseases. Some of these diseases require treatment for several years, and others require lifelong treatment. These diseases attack men, women, and children in developed, middle-income, and poor countries and are ranked as one of the primary causes of morbidity and mortality worldwide. For centuries, several plants have been used for their curative and preventive abilities against human ailments including lifestyle diseases. In the past years, the therapeutic benefits of a good number of medicinal and edible plants have been studied underlining their role in preventing and managing lifestyle diseases, thanks to their bioactive compounds. In the present chapter, an overview of thirty-five frequently used plants for the management of different lifestyle diseases, their mechanism of action and their bioactive constituents are provided. Moreover, the suitable techniques for the extraction of phytoconstituents from these plants are detailed.

Keywords

Lifestyle diseases · Medicinal and edible plants · Mechanism of action · Phytoconstituents · Extraction techniques

1 Introduction

Lifestyle diseases (LSD) are diseases associated with daily habits and signifies adopting detrimental habits that detract from a healthy lifestyle [1]. Atherosclerosis, heart disease, stroke, obesity, hypertension, diabetes mellitus, hyperglycemia, asthma, and cancer are just a few LSD which can have long duration, slow progression and are likely to cause death. They result from the combination of physiological, behavioral, environmental, and genetic factors [2, 3]. The symptoms associated with these illnesses are not seen in the early stages due to the fact that they take a considerable number of times to develop. Some of these diseases have to be treated for several years, while others need everlasting management. These diseases attack men, women, and children in developed, middle-income, and poor countries. Consumption of tobacco and excessive use of alcohol are key risk factors, but there are many others such poor diets (consumption of foods high in fat, salt, and sugar), physical inactivity, poverty, and stress [4]. Worldwide, approximately 41 million people are killed by LSD each year, including 14 million early deaths between the ages of 30 and 70 [1]. Of these, cardiovascular diseases were responsible for 17.9 million deaths, and the number of deaths from cancers, chronic respiratory diseases, and diabetes mellitus were 9 million, 3.9 million, and 1.6 million, respectively. These four diseases account for 80% of deaths due to LSD. According to the World Health Organization (WHO), Europe has the highest LSD-related morbidity and mortality rate among the six regions of the world, with approximately 90% of all deaths. In the European Region, obesity and overweight are the most important risk factor for LSD, affecting more than 59% of adults and 29% of children.

J. Tchamgoue (✉)
Department of Organic Chemistry, Faculty of Science, University of Yaoundé I, Yaoundé, Cameroon

Department of Chemistry, Higher Teacher Training College, University of Yaoundé I, Yaoundé, Cameroon

Y. A. T. Ngandjui
Department of Chemistry, Higher Teacher Training College, University of Yaoundé I, Yaoundé, Cameroon

Institute for Nanotechnology and Water Sustainability, College of Science, Engineering and Technology, University of South Africa, Florida Science Campus, Johannesburg, South Africa

R. M. Talla · J. C. Tchouankeu
Department of Organic Chemistry, Faculty of Science, University of Yaoundé I, Yaoundé, Cameroon

B. D. A. Ambamba
Department of Biochemistry, Faculty of Science, University of Yaoundé I, Yaoundé, Cameroon

S. F. Kouam
Department of Chemistry, Higher Teacher Training College, University of Yaoundé I, Yaoundé, Cameroon

A. K. Dhara, S. C. Mandal (eds.), *Role of Herbal Medicines*, https://doi.org/10.1007/978-981-99-7703-1_3

Globally, 20% of adult deaths is related to unhealthy diet [5] and the WHO estimates that the number of deaths due to NCDs (non-communicable diseases) will rise to 55 million per year by 2030 [1]. For centuries, many plants have been used for their curative and preventive capacities against animal and human diseases owing to their content of bioactive compounds. The nutritional and therapeutic benefits and biological properties of a good number of medicinal and edible plants have been intensively investigated in the recent years. The role of these plants in the prevention and treatment of LSD has been broadly confirmed by several scientific studies. Numerous plant derived foods and remedies used in different cultural traditions around the world have countless beneficial effects on these diseases [6]. Scientific interest in studying plants for drug discovery has increased substantially due to the fact that the bioactive natural products contain in herbal plants have synergistic therapeutic ability that is important in the prevention and treatment of LSD [7]. In this chapter, a résumé of plants often used in the management of LSD and the techniques employed for the extraction of their phytoconstituents are presented.

2 Plants Frequently Used in the Treatment of LSD and Their Phytoconstituents

Since time immemorial medicinal plants have been used for the treatment of NCDs [8]. Interest in medicinal plants has been shown worldwide due to the safety and efficiency of plant derived products [9]. They are a reservoir of several classes of bioactive compounds that can interact with functional proteins and produce various pharmacological effects. Moreover, these compounds are scaffolds of the signaling pathways involved in an organism [10]. Several studies based on traditional assets have highlighted the beneficial effect of plants in the management of LSD [4, 5]. In the present section, the role of plants in the management of LSD will be discussed and the compounds identified in some of these plants will be highlighted.

2.1 Plants Used in the Treatment of Cardiovascular Diseases and Their Phytoconstituents

Cardiovascular diseases (CVD) represent the largest share of chronic non-communicable diseases and have as risk factors (RF), non-modifiable RF (age, sex, family history of CVD, ethnicity) and modifiable RF (physical inactivity, smoking, lifestyle, metabolic syndrome) [11].

Prior to the discovery of modern medicines, many plants were intensively used by humans for handling CVD and their complications [12]. Since the pathophysiology of CVD is heterogeneous, the cardioprotective properties of different plants may be imputed to their capacity to target CVD risk factors including their antioxidant, anti-atherosclerosis, anti-ischemic, and platelet aggregation inhibition activities [13]. It is worth mentioning that atherosclerosis is the most preponderant cardiometabolic risk [14].

2.1.1 *Allium sativum* (Amaryllidaceae)

Allium sativum commonly known as garlic is an indigenous plant of South Asia, Central Asia, and Northeastern Iran. It has been used for many years as a spice and has many properties that help treat conditions related to CVD. These properties include lowering cholesterol and low-density lipoprotein (LDL) levels, reducing the lipid content of blood vessel wall cells, and blood vessel growth. Previous studies have shown that garlic can help improve conditions related to atherosclerosis and hypertension and also has anti-atherosclerotic effects [15]. Garlic extract has been shown to impair sialidase activity in plasma and cause atherogenic LDL formation [16]. Furthermore, garlic downregulates lipid metabolism-related genes fatty acid synthase (FAS), acyl-CoA cholesterol acyltransferase (ACAT), acetyl-CoA carboxylase (ACC), HMGR, and sterol regulatory element binding protein-1c (SREBP-1c) [17]. Chemical analyzes of *A. sativum* bulbs reveal the presence of ajoene (E-ajoene, Z-ajoene), thiosulfinate (allicin), vinyldithiine (2-vinyl-(4H)-1,3-dithiine, 3-vinyl), etc. It also indicates the presence of sulfur compounds like sulfides (diallyl disulfide and diallyl trisulfide) [18]. In general, garlic was reported to contain a wide variety of organic sulfur compounds, selenium-containing natural products, saponins, amino acids, phenylpropanoids, steroids, flavonoids, fatty acids, and alkaloids.

2.1.2 *Glycyrrhiza* glabra (Fabaceae)

Licorice (*Glycyrrhiza glabra*) is an herb that has been used since ancient times in food and as medicine. Due to its very sweet taste it was named "sweet root." The ethanolic extract of licorice was reported to effectively prevent the progression of arteriosclerosis [19]. Its action is due to its ability to lower TC, LDL, and TG levels, increase HDL, and reduce aortic atherosclerotic lesions [19]. Recently, glabridin, a flavonoid isolated from this plant, was shown to inhibit LDL oxidation and reduce LDL [20]. Chemical studies have shown that licorice extract contains several organic acids, liquirtin, rhamnoliquirilin, liquiritigenin, prenyllicoflavone A, glucoliquiritin apioside, 1-methoxyphaseolin, shinpterocarpin, shinflavanone, licopyranocoumarin, glisoflavone, licoarylcoumarin, glycyrrhizin, isoangustone A, semilicoisoflavone B, licoriphenone, 1-methoxyficifolinol, kanzonol R, and some volatile components [21].

2.1.3 *Morusalba* (Moraceae)

Morus alba, commonly called "white mulberry" is common in Asia, Africa, America, and parts of Europe. Mulberries are generally taken as fruit, juices, or jams [22]. They have large amounts of active components that may be related with potentially health-promoting medicinal properties. Studies have shown that mulberries have diverse health benefits, inclusive of hepatoprotective effects, anti-cholesterol, and anti-obesity properties [22]. Additionally, the leaves of this plant are applied in folk medicine to treat hypertension and atherosclerosis. Mulberry leaf extract has cardioprotective [23], anticoagulant, and antithrombotic properties by stopping aggregation and platelet activation, thrombus formation, serotonin secretion, and thromboxane B2 formation [24]. Chemical investigation has revealed that *M. alba* leaves contain rutin, quercetin-3-β-D-glucose, isoquercetin, quercetin, and quercetin-3-*O*-glucose-6″-acetate [25]. In addition, mulberries were also reported to contain polyphenols and alkaloids [22].

2.1.4 *Curcuma longa* (Zingiberaceae)

Curcuma longa, well-known as turmeric or "Indian saffron," is a condiment also used in medicinal concoctions [26]. It is also applied as food coloring additive and dye in the textile industry because of its pronounced yellow color [27]. Turmeric has been demonstrated to possess a wide range of biological properties including against CVD and atherosclerosis [26]. Recent evidence has shown that the administration of 2.1 g of turmeric rhizome powder per day for 8 weeks meaningfully reduces BMI and plasma LDL, TC, and TG [28]. Chemical analysis showed that ar-turmerone, β-sesquiphellandrene, curcumenol, and curcumin are the main compounds of *Curcuma longa* [13]. Curcumin is the major active component of turmeric and was reported to possess anti-atherosclerotic, antioxidant, anti-inflammatory, and hepatoprotective properties [29]. Numerous studies have described the beneficial effect of curcumin in the mitigation of CVD via its anti-inflammatory capacities and its ability to repair chemokine secretion and chemokine receptors [30].

2.1.5 *Detarium microcarpum* (Fabaceae)

Detarium microcarpum (commonly known as sweet detar) is an indigenous African plant widespread in savannah regions. The fruits and leaves of *D. microcarpum* are mainly used for their nutritive and medicinal properties [31]. In folk medicine, all parts of *D. microcarpum* (leaves, fruits, stems, and roots) are used for the management of many diseases including CVD [31]. Results of a recent study highlighted the ability of the aqueous extract of the stem bark of this plant to prevent weight gain and regulate plasma lipid/lipoprotein levels in rats which was imputed to the presence of polyphenols [32]. Myricetin-3-*O*-rhamnoside, quercetin-3-*O*-glucoside, methyl gallate, lup-20(29)-ene-2α,3β-diol, lupeol, campesterol, protocatechuic acid, stigmasterol, vanillic, hexanedioic acid, sucrose, and myo-inositol are compounds reported to be responsible for the observed biological activities of *D. microcarpum* [33].

2.1.6 *Ginkgo biloba* (Ginkgoaceae)

Ginkgo biloba also called "maidenhair tree" is listed among the most ancient plant species. Its leaves extract is one of the most frequently used plants and is vulgarized for its alleged stimulant effect and likely healing and restorative abilities [34]. There is growing evidence of the potential role of the leaves extract of *G. biloba* in the treatment of CVD [35]. It was reported to exert vascular protective functions and preliminary results showed that it may be useful in the management (prevention and treatment) of CVD, particularly ischemic heart syndrome [36]. Recently, Wang and coauthors demonstrated that *G. biloba* extract improves atherosclerosis by rebalancing gut flora and microbial metabolism [37]. *G. biloba* contains Kaempferol, a natural flavonoid use to prevent and treat atherosclerosis [37].

2.1.7 *Quercus infectoria* (Fagaceae)

The different species of *Quercus* are native from Middle East (Turkey, Iran, and Iraq) but are now widespread in Europe, Asia, and North Africa. The Mazu medicine prepared from *Q. infectoria* is well described in the Perso-Arabic traditional medicine and have been reported to possess various actions such as hypoglycemic and anti-inflammatory [38]. The plant also has strong antioxidant potential through the inhibition of the production of Ox-LDL, thus avoiding the development of atherosclerosis. In addition, *Quercus infectoria* (known as Aleppo Oak) extract reduces total cholesterol, LDL, and triglyceride levels in plasma, highlighting its role in preventing and treating atherosclerosis [39]. *Q. infectoria* was reported to contain alkaloids, steroids, saponins, triterpenes, tannins, flavonoids, and phenols [40].

2.1.8 *Panax notoginseng* (Araliaceae)

Panax notoginseng commonly referred to as Chinese ginseng is abundantly found in China, Yunnan, and it is broadly used in cosmetics, healthcare, and other industries [41]. *P. notoginseng* possesses a large range of biological activities, including anti-atherosclerotic activity and cardiovascular protection [41]. Ginseng saponins have been used as a natural cure for CVD in Asian countries including Japan, South Korea, and China [42]. They upregulate the expression of hepatic X receptor alpha (LXRα) and cause subsequent overexpression of ABCA1 and ABCG1. In addition, they inhibit the DNA binding activity of NF-κB. As a receptor, LXRα controls macrophage functions involved in inflammation and lipid metabolism [43]. The ginsenoside derivative, ginsenoside Rd has the ability to block voltage-gated Ca^{2+} channels and significantly reduce the dimension of atherosclerotic plaques [44]. Fifty six types of saponins have been isolated and elucidated from Noto ginseng, 35 of which belong to the protopanaxadiol group and 21 to the protopanaxatriol group [45]. Phytochemical studies on *P. notoginseng* have shown that the olean-type saponins found in ginseng from Asia (*P. ginseng*) and America (*P. quinquefolius*) are missing in this species.

2.1.9 *Artemisia judaica* (Asteraceae)

Artemisia judaica known as "Beithran" in Arabic is a therapeutic and sweet-smelling plant that particularly grows in the southern Jordanian desert [46]. In Jordan, *A. Judaica* is widely used for the treatment of heart and inflammatory diseases, and atherosclerosis [46]. A 2016 report stated that controlled concentrations of its essential oil effectively inhibited lipopolysaccharide-induced NO production in macrophages, highlighting the potential anti-inflammatory assets of *A. judaica* [46]. Artemisinin, a compound in this plant, reduces atherosclerotic lesions by reducing macrophage inflammation through modulation of the AMPK/NF-κB/NLRP3 inflammasome signaling pathway [47]. Chemical analysis of this plant revealed eight flavones, namely apigenin, diosmetin, cirsimaritin, isovitexin, luteolin, luteolin-4′-methyl ether, 7-*O*-β-D-$^{4}C_{1}$-glucopyranoside, 8-methoxyapigenin, 7-*O*-β-D-$^{4}C_{1}$-galactopyranoside, and 8-methoxyluteolin 7-*O*-β-D-$^{4}C_{1}$-glucopyranoside [48] (Fig. 1 and Table 1).

2.2 Plants Frequently Used in the Treatment of Cancer and Their Phytoconstituents

Cancer is a major cause of increasing numbers of deaths in the world. It is the second most common lifestyle-related disease. Standard cancer treatment is usually based on the use

Fig. 1 Pictures of selected plants used in the treatment of CVD

Table 1 Cardioprotective potential, chemical constituents, and mechanisms of action of selected medicinal plants

Plant Species	Chemical constituents	Biological activities	Mechanism of actions	References
Allium sativum (Amaryllidaceae)	Saponines, flavonoids, phenylpropanoids, alkaloids, sulfur compounds, and sterols	Anti-atherosclerosis	Inhibition of sialidase activity in plasma causing LDL formation and downregulates lipid metabolism-related genes acyl-CoA cholesterol acyltransferase (ACAT), Acetyl-CoA carboxylase (ACC), fatty acid synthase (FAS),HMGR, and sterol regulatory element binding protein-1c (SREBP-1c)	[15–18]
Glycyrrhiza glabra (Fabaceae)	Flavonoids, isoflavonoids, coumarins, glycyrrhizic acid	Anti-atherosclerosis	Inhibition of LDL oxidation and LDL reduction	[19–21]
Morus alba (Moraceae)	Flavonoids, alkaloids and polyphenols	Cardioprotective	Blocking the activation and aggregation of platelet, serotonin secretion, thromboxane B2 formation, and thrombus formation	[22–25]
Curcuma longa (Zingiberaceae)	Polyphenols, flavonoids, triterpenoids	Anti-atherosclerosis and anti-inflammatory	Repair of chemokine secretion and chemokine receptors	[13, 26–30]
Detarium microcarpum (Fabaceae)	Polyphenols, phenolic acids, terpenoids, steroids, carbocyclic sugar, flavonoids	Antiatherogenic	Regulate plasma lipid/lipoprotein levels (LDL-c, VLDL-c, TC, TG, HDL-c, and non-HDL-c)	[31–33]
Ginkgo biloba (Ginkgoaceae)	Flavonoids	Anti-atherosclerosis	Rebalancing gut flora and microbial metabolism	[34–37]
Quercus infectoria (Fagaceae)	Polyphenols, steroids, triterpenes, saponins and alkaloids	Anti-atherosclerosis	Inhibition of oxidized LDL production and reduces total cholesterol, triglyceride and LDL levels in plasma	[38–40]
Panax notoginseng (Araliaceae).	Saponins, polyacetylenes, phytosterols, flavonoids	Anti-atherosclerosis and anti-inflammatory	Up-regulates expression of hepatic alpha X receptor (LXRα) and causes subsequent overexpression of ABCG1 and ABCA1. Inhibits the DNA binding property of NF-κB.	[41–45]
Artemisia judaica (Asteraceae)	Flavonoids, sesquiterpene lactone	Anti-atherosclerosis, anti-inflammatory	Inhibition of lipopolysaccharide-induced NO production in macrophages and modulation of the AMPK/NF-κB/NLRP3 inflammasome signaling pathway	[46–48]

of cytotoxic drugs, radiation therapy, chemotherapy, and surgery. However, plants play a role as an alternative treatment in the management of cancer [49]. They contain terpenoids, saponins, volatile oils, and flavonoid phytoconstituents, which give them anticancer activities. Several investigations have already shown the protecting effect of plants on cancers that occur due to lifestyle [49].

2.2.1 *Camellia sinensis* (Theaceae)

Camellia sinensis (tea plant) is native to Asia, but is now grown in subtropical and tropical regions in the world. It is one of the most consumed drinks in the world. White tea is made by steaming and drying the buds and young leaves of tea plants to minimize oxidation. The antioxidant and antiproliferative properties of this drink on cancer cells have already been described [50]. The extract improved the activity levels of intracellular caspase-3, 8, and 9 and protected 3T3-L1 cells from DNA damage by H_2O_2 [50]. Phytochemical studies identified several phenolic compounds from *C. sinensis* including flavonoids such as quamoreokchaside I-II, kamoreokchaside I, (-)-epigallocatechin, (-)-epicatechin, and (-)-epigallocatechin-3-*O*-gallate [178, 179].

2.2.2 *Annona muricata* (Annonaceae)

Annona muricata is a Central and Tropical South American native plant popularly known as "soursop," "graviola," or "guanabana" which has been globally used in folk medicine against many diseases. Its biological properties include antiviral, antihypertensive, antidiabetic, and anticancer activities highlighting its potential in the management of LSD [51]. The main active components of *A. muricate* are flavonoids, acetogenin, and alkaloids. The chemical screening of its leaf extract revealed the presence of coumarins, anthraquinones, flavonoids, terpenoids, saponins, lactones, glycosides, tannins, and phytosterols [51]. The reported anticancer property of *A. muricata* is due to its cytotoxicity against cancer cells including colorectal, prostate, and lung cancers [52]. A recent study demonstrated that the seeds, fruit, twigs and stems' extracts of *A. muricata* inhibited matrix metalloproteinases (MMPs) such as MMP-9andMMP-2, which are having primordial roles in the progression of cancer [53].

2.2.3 *Eleutherococcus senticosus* (Araliaceae)

Eleutherococcus senticosus also known as Siberian ginseng is native to Northeast Asia. It is primarily used as an adaptogen and has immune stimulating properties. Carbohydrates found in Siberian ginseng appear to have not only immunostimulatory effects, but also antitumor and anticancer properties [54]. The treatment of the lung cancer cells A-549 with the methanolic extract of *E. senticosus* displayed a concentration-dependent inhibition tendency, slightly suppressing the growth of QBC-939 cells. In addition, XWLC-05 cells (Yunnan lung cancer cell line), NEC cells (human nasopharyngeal cancer cell line) exhibited dose-dependent proliferative responses, and Beas-2 cells (human lung cancer cell line) exhibited dose-dependent responses. Weak inhibitory activity in colon cancer cell line (HCT-116) was also reported [54]. Phytochemical studies indicate that leaves of *E. senticosus* contain flavonoids, organic acid derivatives, and triterpene glycosides [55].

2.2.4 *Mangifera indica* (Anacardiacea)

Mangifera indica (Mango) is native to tropical Asia. Its active components are mainly present in the trunk bark, roots, heartwood, leaves, and fruits and have anti-allergic, antitumor, antioxidant, anti-inflammatory, radioprotective, immunomodulatory, antidiabetic, monoamine oxidase inhibition, antiviral, anti-resorptive, and antifungal properties [56]. The anti-breast cancer properties of extracts of different parts of *M. indica* have been previously reported [57]. Aqueous extract from the bark of *M. indica* has shown various anticancer and anti-inflammatory activities. *M. indica* was also reported to dose-dependently and significantly inhibit the spread of MDA-MB-231 cells. A related research report pointed out the capacity of *M. indica* nuclear extracts to decrease the MCF-7 (Luminal A; PR-positive, ER-positive, HER2-negative cells), MDA-MB-231, and MDA-MB-viability. *M. Indica* pulp and bark extracts were more often reported for their anti-breast cancer property as compared to its other parts [58]. Previous phytochemical profiles of pulp and bark extracts of *M. indica* displayed the presence of gallic acid and galloyl derivatives, and high levels of methyl gallate [59].

2.2.5 *Raphia vinifera* (Arecaceae)

Raffia vinifera popularly called "West African piassava palm," "bamboo palm," or "West African fibrous palm" belongs to the Raffia genus [60]. It is native to Cameroon, Togo, Ghana, Benin, Gambia, Nigeria, Central African Republic, and Democratic Republic of the Congo. A recent investigation of Nguenang and collaborators highlighted that botanical from *R. inifera* fruits were cytotoxic on 18 cancer cell lines and was found nontoxic on the result of its toxicological assessments [61]. The chemical components of this plant include sitosterol, diosgenin-3-*O*-β-D--glucopyranoside, and 26-*O*-β-D-glucopyranosyl-(22R, 25R)-3β,26-trihydroxyfrost-5-ene-3-*O*-β-D-glucopyranoside [62]. The pulp and rind of *R. vinifera* fruits contained high concentrations of alkaloids, saponins, and oxalates, medium concentrations of flavonoids, steroids, and tannins, and low concentrations of phytic acid, phenols, and glycosides which are source of its medicinal properties [63]. *R. inifera* provides steroidal saponins that are useful for tumor prevention and treatment of many types of cancer, with low toxic effects and high potency [64] (Fig. 2 and Table 2).

Fig. 2 Pictures of selected plants used in the treatment of cancer

Table 2 Anticancer potential, chemical constituents, and mechanisms of action of selected medicinal plants

Plant species	Chemical constituents	Biological activities	Mechanism of actions	References
Camellia sinensis (Theaceae)	Flavonoids and polyphenolic compounds	Antioxidant and antiproliferative	Improved the activity levels of intracellular caspase-3, -8, and -9. The extract protected 3T3-L1 cells from DNA damage by H_2O_2	[50, 178, 179]
Annona muricata (Annonaceae)	Acetogenin, alkaloids, and flavonoids	Antiproliferative property	Inhibition of matrix metalloproteinases (MMPs) such as MMP-2 and MMP-9 Disruption of MMPs, reactive oxygen species (ROS) generation, and the G0/G1	[51–53]
Eleutherococcus senticosus (Araliaceae)	Triterpene glycosides, organic acid derivatives and flavonoids	Antitumor and anticancer properties	Inhibition of colon cancer cell line (HCT-116)	[54, 55]
Mangifera indica (Anacardiacea)	Phenolic acid derivatives (high levels of methyl gallate)	Anticancer, antioxidant, and anti-inflammatory	Inhibition of the spread of MDA-MB-231 cells, decrease the viability of MCF-7 (Luminal A; ER-positive, PR-positive, HER2-negative cells), MDA-MB-231, and MDA-MB	[56–59]
Raphia vinifera (Arecaceae)	Saponins, alkaloids, flavonoids, and steroids	Antitumor	Cytotoxic effect on cancer cell lines	[60–64]

2.3 Plants Frequently Used in the Treatment of Chronic Respiratory Diseases and Their Phytoconstituents

Chronic respiratory diseases including obstructive pulmonary disease, interstitial lung disease, and asthma are listed by the WHO as one of the major chronic diseases posing great challenge to both public health and socio-economic growth. Reliance on conventional treatments has proven unsuccessful as many patients are poorly controlled with poor quality of life. This has prompted researchers to investigate natural products in order to search for new drugs which could improve treatment outcomes [65]. In this session, the roles some plants could play in the management of chronic respiratory diseases are presented as well as their phytoconstituents.

2.3.1 *Vitex rotundifolia* (Lamiaceae)

Vitex rotundifolia commonly called "beach vitex" or "round-leaf vitex" is a coastal herb that has traditionally been applied for the treatment of different ailments such as respiratory arrest, colds, headaches, and sore eyes [66]. *V. rotundifolia* aqueous extract was reported to have the ability to significantly inhibit goblet cell hyperplasia, OVA-induced eosinophilia, and smooth muscle mass production [67]. Casticin, isolated from *V. rotundifolia*, inhibits the expression of pro-inflammatory cytokines, chemokines, and ICAM-1 through suppression of PI3K/Akt, NF-κB, and MAPK signaling pathways in inflamed lungs. Reports have provided evidence that it has anti-inflammatory effects in blocking epithelial cells stimulated by IL-1β. Co-culturing MAPK, NF-κB, and PI3K inhibitors with casticin also resulted in more significant suppression of ICAM-1 expression in A549 inflammatory cells. The latter inhibits the phosphorylation of Akt, phosphatidylinositol-3-kinase (PI3K), and mitogen-activated protein kinase (MAPK) and translocates the p65 protein, a subunit of the nuclear transcription factor kappa-B (NF-κB) blocked nucleus [68]. Chemical analysis indicates the presence of flavonoids, phenolic acids, and terpenes in *V. rotundifolia* [66].

2.3.2 *Pistacia weinmannifolia* (Anacardiaceae)

Pistacia weinmannifolia is a rare and endangered plant endemic to Southwestern China and Northern Myanmar that has many economic uses. An in vitro investigation confirmed the anti-inflammatory properties of the root extract of *P. weinmannifolia* on NCI-H292 epithelial cells stimulated by phorbol myristate acetate or tumor necrosis factor α (TNF-α) by decreasing the expression of mucin A5 (MUC5AC) and interleukin (IL)-8, IL-6 which are closely linked to the pulmonary inflammatory response in the pathogenesis of COPD [69]. In addition, it downregulated nuclear factor κB activation and phosphodiesterase 4 expression in lung tissue [70]. Phytochemical analysis showed that *P. weinmannifolia* mainly contains phenolic compounds, including gallotanins, pistafolin A, and pistafolin B [70].

2.3.3 *Camphora chartophylla* (Lauraceae)

Camphora chartophylla previously called *Cinnamomum chartophyllum* is found mainly in subtropical and tropical regions, with certain species distributed in the Neotropics, Australia, and Africa [71]. Species of this genus are often used in community and folk medicine to treat several conditions such as indigestion, colds, coughs, and microbial infections [72]. *C. chartophylla* (known as bushy mat grass) protects human bronchial epithelial cells from oxidative stress by activating Nrf2 [73]. Systemic chemical investigations of the aerial parts of *C. chartophylla* revealed *α,β*-unsaturated keto groups, 3S-(+)-9-oxonerrolidol (NLD), and diphenyl, 3,3, which shares a phenolic group.

Sesquiterpenoids were also reported [73]. The latter activates Nrf2 and its downstream genes, NAD(P)H-quinone oxidoreductase 1 (NQO-1), and γ-glutamylcysteine synthase (γ-GCS), and induces nuclear translocation of Nrf2 in human lung epithelial cells and increased stabilization [73].

2.3.4 *Allium sativum* (Amaryllidaceae)

Crude garlic extracts were reported to possess anti-asthmatic properties and are utilized as anti-inflammatory agent [15]. Garlic extracts decreased the total number of inflammatory cells and eosinophil infiltration and decreased the production of IgE dermatophagoides pteronyssinus in serum and Th1/Th2/Th3 cytokine in bronchoalveolar fluid. Enzyme-linked immunosorbent assay demonstrated that garlic extracts downregulated the levels of cytokines and chemokines, namely IL-4, IL-5, and IL-13 related to Th2; but they simultaneously upregulated the expression of Th1-related IFN-γ, IL-12, and TGF-β in BALF [74].

2.3.5 *Moringa oleifera* (Moringaceae)

Moringa oleifera (drumstick tree) is a versatile plant eaten as food and known for its medicinal uses. It is a plant native to India, growing in the foothills of the Himalayas and parts of Africa [75]. The alcoholic extract of seed kernels of this plant was reported to significantly increase preconvulsion time in guinea pigs exposed to either acetylcholine or histamine aerosol. The authors suggested that the anti-asthmatic property of the seed kernels of *M. oleifera* might be imputed to its anti-inflammatory, antimicrobial, bronchodilator, and mast cell stabilization activities [76]. Seed powder administered to asthmatic patients was reported to reduce asthma attacks [77]. Chemical analysis of *M. oleifera* identified 2,6-dihydroxybenzoic acid, threonine, and fatty acids [78].

2.3.6 *Magnolia officinalis* (Magnoliaceae)

Magnolia officinalis commonly known as magnolia bark is used in Chinese folk medicine for the treatment of cough, diarrhea, and allergic rhinitis [180]. It is also used to relieve asthma and cough, to prevent cerebrovascular and cardiovascular diseases, and to treat depression and anxiety [79]. In response to formyl-L-methionyl-L-leucyl-L-phenylalanine/cytochalasin B, the methanolic extract of the stem bark of *M. officinalis* exhibited significant inhibitory effects on superoxide anion generation and elastase release by human neutrophils [80]. Chemical investigation has revealed lignans, alkaloids, and volatile oils to be the main of *M. officinalis* [79] (Fig. 3 and Table 3).

2.4 Plants Frequently Used in the Treatment of Diabetes and Hyperglycemia and Their Phytoconstituents

Type 2 diabetes is the most common form of diabetes and accounts for about 90% of all diabetes cases with insulin resistance and abnormalities in insulin secretion. The primary objective of antidiabetic treatment is to stimulate the

Fig. 3 Pictures of selected plants used in the treatment of chronic respiratory diseases

Table 3 Selected medicinal plants commonly used against chronic respiratory diseases, their chemical constituents and mechanisms of action

Plant species	Chemical constituents	Biological activities	Mechanism of actions	References
Vitex rotundifolia (Lamiaceae)	Flavonoids, phenolic acids, and terpenes	Anti-asthmatic	Inhibits the expression of pro-inflammatory cytokines, chemokines, and ICAM-1 through suppression of PI3K/Akt, NF-κB, and MAPK signaling pathways in inflamed lungs	[66–68]
Pistacia weinmannifolia (Anacardiaceae)	Phenolic compounds	Anti-inflammatory	Downregulated nuclear factor κB activation and phosphodiesterase 4 expression in lung tissue	[69, 70]
Cinnamomum chartophyllum (Lauraceae)	Phenolic compounds and sesquiterpenoids	Bronchoprotective	Activating Nrf2	[71–73]
Allium sativum (Amaryllidaceae)	Saponins, phenylpropanoids, flavonoids, alkaloids, sulfur compounds, and sterols	Anti-asthmatic and anti-inflammatory	Up-regulated the expression of Th1-related IFN-γ, IL-12, and TGF-β in BALF	[15, 74]
Moringa oleifera (Moringaceae)	Fatty acids, alkenes, phenolic, threonine	Anti-asthmatic	Reduction in the elevated release of histamine from the lungs of sensitized guinea pigs	[75–78]
Magnolia officinalis (Magnoliaceae)	Lignans, alkaloids and volatile oils	Anti-asthmatic	Inhibitory effects on superoxide anion generation and elastase release by human neutrophils	[180, 79, 80]

secretion of insulin, reinforce immunity, and/or reduce glucose in the blood. Although many synthetic antidiabetic compounds were discovered, their usefulness has been hampered due to their side effects and limited efficacy [81]. Plants occupy a prominent place in diabetes' management by providing important elements (secondary metabolites, fibers, vitamins, and minerals) that are able to modulate various metabolic pathways to relieve diabetic patients [82]. Thus, several investigations have demonstrated the antidiabetic efficacy of plants.

2.4.1 *Teucrium polium* (Lamiaceae)

Teucrium polium (aka *Teucrium capitatum* or felt germander) is rampant in stony and dry areas of nearly all Mediterranean countries, Europe, Northern Africa, and Southwest of Asia [83]. *T. polium* is utilized in Iranian herbal remedies for the treatment of many ailments including type 2 diabetes [83]. A recent study demonstrated the insulin-secreting ability of an "aqueous" extract of *T. polium* on BRIN-BD11 rat pancreatic cells [84]. At the same time, treatment with *T. polium* extract increased GLUT2 expression and glucokinase activity [84]. Chemical analysis indicates that quercetin is a flavonoid that naturally occurs as a glycoside like rutin (quercetin-3-rutinose) in *T. polyum* [85]. Another study showed that *T. polyum* essential oil was dominated by the presence of terpenes including spathulenol, caryophyllene (E),sabinene, limonene, myrcene, germacren-D, *β*-pinene, and *α*-pinene [86].

2.4.2 *Rauvolfia vomitoria* (Apocynaceae)

Rauvolfia vomitoria known as the "poison devil's-pepper" is found in Africa and Asia tropical regions [87]. *R. vomitoria* has been associated with various curative properties, including antidiabetic assets [87]. The hydromethanolic extract of *R. vomitoria* and its main component, phytol, were reported to lower blood sugar levels [87]. GC-MS analysis of hydromethanol extracts from leaves of *R. vomitoria* showed the occurrence of compounds such as propanamide, 2-methyl, 1H-indole-2-ethanol, *β*-(3-ethylidene-1-methyl-4--piperidinyl)-3-methyl, hexadecanoic acid, methyl ester, *n*-hexadecanoic acid, 9,12,15-octadecatrien-1-ol, (Z)9-octadecenoic acid, methyl ester, phytol, methyl stearate, and 9,12-octadecadienoic acid (Z,Z)-methyl ester [87].

2.4.3 *Baillonella toxisperma* (Sapotaceae)

Baillonella toxisperma popularly known as "African pearwood," "djave nut," or "moabi" is a large tree that grows in the rainforests of Africa. Its woods and bark are used for their medicinal properties, while its fruits are employed for culinary and cosmetic objectives, and the pulp is directly consumed [88]. The hydroethanolic extract of the fruit of *B. toxisperma* was reported to have antihyperglycemic potential by reducing glucose levels after sucrose and starch administration in rats with normal blood sugar level [89]. It was also reported to increase glucose absorption by muscle and yeast cells [90]. The hydroethanolic extract of the pulp of *B. toxisperma* pulp was reported to exhibit metal-chelating, reducing, antiradical, glucose-binding, and insulin-sensitizing activities [90]. The phytochemical analysis of this extract indicates a high content of flavonoids and polyphenols [90].

2.4.4 *Tetrapleura tetraptera* (Fabaceae)

Tetrapleura tetraptera (Aidan Tree) fruit is the most available and frequently used spices in Cameroon [91]. They are non-wood forest derived products and are known to possess numerous ethnobotanical uses (antidiabetic, antimalarial, anti-inflammatory, anticonvulsant, anticancer, etc.) in Central African countries [92]. Aqueous extracts of *T. tetraptera* fruit have hypoglycemic properties [93]. Phytochemical analyzes have shown the presence of terpenoids, steroids, saponins, alkaloids, flavonoids, tannins, and phenols in ethanolic and aqueous extracts of different plant parts of *T. tetraptera* [94].

2.4.5 *Alstonia boonei* (Apocynaceae)

Alstonia boonei, popularly known as God's tree, is a plant mostly found in many African countries [95]. Its bark aqueous extract was reported to possess antidiabetic effects and increase DNA methylation in the liver. The chemical screening of its stems and leaves highlighted the presence of alkaloids, flavonoids, tannins, phenols, saponosides, and terpenoids [96]. GC-MS exploration of the dichloromethane leaf extract showed that eugenol was the major component while the dichloromethane bark extract mainly contain 1,2-benzenedicarboxylic acid [97].

2.4.6 *Pseudarthria hookeri* (Fabaceae)

Pseudarthria hookeri commonly known as "*Zem lekouet*" in Western Cameroon is a tropical African plant widespread in South Africa, Ethiopia, Angola, and Cameroon. The leaves of this plant are used for the treatment of several ailments such as diarrhea, coughs, tuberculosis and malaria [98]. The ethnopharmacological investigation of this plant revealed its antidiabetic, estrogenic, cytotoxic, and antimicrobial properties [99]. Isolated flavonoids from this plant were assessed for their antidiabetic activity particularly for their ability to stimulate the secretion of insulin and were found to display significant insulin secretory activity higher than that of tolbutamide [98] (Fig. 4 and Table 4).

2.5 Plants Frequently Used in the Treatment of Hypertension and Their Phytoconstituents

Hypertension is a chronic condition in which arterial blood pressure increases. Constant hypertension is a risk factor for myocardial infarction, aneurysms, stroke, and heart failure. It is one of the major sources of chronic kidney disease [100]. A moderate increase in blood pressure reduces life expectancy. Medicinal plants have gained interest as a potential means of preventing and treating hypertension due to their multiple beneficial effects [101].

2.5.1 *Allium cepa* (Amaryllidaceae)

Allium cepa originated from Central Asia and many of its relatives still grow wild. It is widely cultivated and consumed throughout the world. Its bulb, which has a characteristic flavor, is the third most essential spice with a significant commercial value. Besides its culinary assets, it is also largely used as herbal medicine in several indigenous cultures. Several scientific investigations were done in other to ascertain its traditional claims [102]. The main

Fig. 4 Pictures of selected plants used in the treatment of diabetes and hyperglycemia

Table 4 Antidiabetic and anti-hyperglycemic potential, chemical constituents, and mechanisms of action of selected medicinal plants

Plant species	Chemical constituents	Biological activities	Mechanism of actions	References
Teucrium polium (Lamiaceae)	Flavonoids and monoterpenes	Antidiabetic	Insulin-secreting, increased GLUT2 expression, and glucokinase activity	[83–86]
Rauvolfia vomitoria (Apocynaceae)	Fatty acids	Antidiabetic	–	[87]
Baillonella toxisperma (Sapotaceae)	Polyphenols and flavonoids	Antihyperglycemic and hypoglycemic	Glucose-binding, insulin-sensitizing, increased glucose uptake by yeast and muscle cells	[88–90]
Tetrapleura tetraptera (Fabaceae)	Flavonoids, terpenoids, tannins, alkaloids, steroids, saponins, and phenols	Hypoglycemic	–	[91–94]
Alstonia boonei (Apocynaceae)	Phenols, flavonoids, alkaloids, tannins, saponosides, and terpenoids	Antidiabetic	Increase DNA methylation in the liver	[95–97]
Pseudarthria hookeri (Fabaceae)	Flavonoids	Insulin secretory	–	[98, 99]

pharmacological application is for preventing hypertension and arteriosclerosis through its hypotensive ability [103]. Flavonoids, saponins, sterols, and sulfur compounds were reported as its main components [103]. The sulfur compounds are well-known for their antiplatelet aggregation and hypotensive properties.

2.5.2 *Apium graveolens* (Apiaceae)

Apium graveolens, commonly called celery, is a biennial plant of the Apiaceae family [104]. Various parts of this plant are used in traditional medicine. *A. graveolens* was reported to possess hypolipidemic, antifungal, hepatoprotective, anticancer, diuretic, and hypotensive properties [105]. In mainland China, celery was prescribed for high blood pressure associated with pregnancy and menopause. It also lowers diastolic and systolic blood pressure [106]. Phytochemical profiling of the essential oils of *A. graveolens* stem (S) and leaves (L) revealed the presence of limonene, α-phellandrene, β-pinene, and α-pinene with the latter being the major component [107, 108].

2.5.3 *Olea europaea* (Oleaceae)

Olea europaea (olive plant) is a plant widely distributed in Africa, Mascarenes, Arabia, India, and China. *Olea europaea* (leaf, bark, and root) extract is widely used in Africa to treat various ailments, but little pharmacological research has been done. A randomized and actively controlled clinical study demonstrated the hypotensive effect and tolerance of the leaf extract of this plant in stage 1 hypertensive patients [109]. Studies have also revealed the antihypertensive, anti-atherosclerotic, diuretic, antidiarrheal, antioxidant, and hypoglycemic effects of olive leaf extracts. Phytochemical studies have shown isolated compounds such as oleuropein, oleanolic acid, ursolic acid, aesculin, and scopolina [109].

2.5.4 *Ocimum gratissimum* (Lamiaceae)

Ocimum gratissimum (African basil) is native to Africa, South Asia, and the Bismarck Islands. A recent study showed the antihypertensive properties of the plant, possibly due to its flavonoid content [110]. *O. gratissimum* was reported to reduce blood pressure in hypertensive rats in a dose-dependent way, its effect was ameliorated as collated to that of controls without nephrectomy. The improvement was primarily related to increased smooth vasorelaxation by *O. gratissimum* [111]. *O. gratissimum* leaves are rich in polyphenolic acids, flavonoids, and ellagic acid [112].

2.5.5 *Terminalia superba* (Combretaceae)

Terminalia superba is a 30–50 m tall tree distributed in the forests of West and Central Africa. In common medical practice in Central Cameroon, *T. superba* is considered to be a useful antihypertensive agent [113]. The intravenous injection of *T. superba* aqueous extract induced an antihypertensive response. When taken orally, the extract prevented the increase in blood pressure in rats with glucose hypertension [113]. Phytochemical examination of the hydroethanolic and ethanolic extracts of the bark of *T. superba* showed the occurrence of flavonoids, tannins (catechin and gallic acids), saponins, and anthracene derivatives [114] (Fig. 5 and Table 5).

2.6 Plants Used in the Management of Obesity and Their Phytoconstituents

Obesity prevalence keeps on rising, leading to a worldwide outbreak that displays no sign of abating [115]. It is a critical lifestyle-related disorder. There are many synthetic remedies for obesity, but with serious side effects. Taken into consideration the current situation, researchers have been focusing their efforts on the development of herbal based

Fig. 5 Pictures of selected plants used in the treatment of hypertension

Table 5 Antihypertension, chemical constituents, and mechanisms of action of selected medicinal plants

Plant species	Chemical constituents	Biological activities	Mechanism of actions	References
Allium cepa (Amaryllidaceae)	Sulfur compounds, flavonoids, and saponins	Antihypertension and hypotensive	Inhibitory action on platelet aggregation	[102, 103]
Apium graveolens (Apiaceae)	Monoterpenes	Hypotensive	Lower systolic and diastolic blood pressure	[104–107]
Olea europaea (Oleaceae)	Polyphenols and flavonoids	Phenolic compounds, coumarins, triterpenes	Antihypertensive and anti-atherosclerotic	[109]
Ocimum gratissimum (Lamiaceae)	Polyphenolic acids and flavonoids	Antihypertensive	Reduces blood pressure in DOCA salt-conscious hypertensive rats	[110–112]
Terminalia superba (Combretaceae)	Tannins, flavonoids, saponins, and free anthracene derivatives	Antihypertensive	Prevents the increase in blood pressure	[113, 114]

products for the treatment of obesity [116]. Growing research efforts demonstrate the ability of plants and their chemical constituents in preventing and managing obesity, in synergy or individually [117]. Formulations for obesity management have been developed using secondary metabolites such as flavonoids, polyphenols, saponins, and alkaloids derived from different plants around the world [117]. In this section, plants frequently used in treating and managing obesity will be presented.

2.6.1 *Salvia miltiorrhiza* (Lamiaceae)

Salvia miltiorrhiza (popularly called Sal) extract, a common herbal medicine has been shown to be active against obesity, hyperlipidemia, atherosclerosis, and dyslipidemia associated diseases [118]. A report highlighted a decrease in total cholesterol, serum triglycerides, glucose, hepatic lipid accumulation, low-density lipoprotein, adipocyte vacuolization, free fatty acids, and body weight as a result of treatment with *S. miltiorrhiza*. The observed results were related with elevated levels of lipid factors like hormone-sensitive lipase, cyclic adenosine monophosphate, and protein kinase A in both adipose tissue and liver, and the improvement of lipid metabolism and intestinal integrity [118]. The chemical exploration of *S. miltiorrhiza* revealed phenolic acids and their derivatives to be its main constituents. More specifically, the following compounds caffeic acid, protocatechuic acid, salvianolic acid A&B, protocatechuic aldehyde, ferulic acid, rosmarinic acid, isoferulic acid, przewalskin, cryptotanshinone, dihydrotanshinone I, and tanshinone I and IIA were identified in *S. miltiorrhiza* extracts [119].

2.6.2 *Ficus carica* (Moraceae)

Ficus carica originates from Mediterranean and western Asia where it was cultivated since antiquity. Nowadays, it is extensively grown worldwide as a fruit and ornamental plant. *F. carica* has already shown antioxidant, antidiabetic, and anti-obesity effects in vitro (due to its anti-lipase capacity) [120]. Another study highlighted a reduction in animal weight, locomotion, triglycerides, fat body weight (uterine fat, kidney fat, and mesenteric fat), LDL and VLDL levels, cholesterol, and greater HDL levels [121]. Phytochemical studies of the plant have revealed the presence of phenolic compounds and flavonoids, steroids, ceramides, pentacyclic triterpenes, and cerebrosides. In addition, the investigation of the most active fraction of *F. carica* by means of advanced analytical methods including LC-MS and NMR demonstrated that the fraction was rich in oligosaccharides and contained large amounts of oligomeric α-glucopyranosides [122].

2.6.3 *Hoodia gordonii* (Apocynaceae)

Hoodia gordonii is eaten worldwide for appetite suppression and its subsequent weight loss effects have already been documented. A study found that rats supplemented with *H. Gordonii* lost weight, which could be attributed to reduced adipocyte size and skeletal muscle fiber size [123]. Steroid glycosides are abundant in *H. gordonii* and hoodigoside L was found to be the major constituent of its organic extract [124]. However, 12-tigloyloxy-14-hydroxypregn5-en-20-one was identified as the constituent responsible for its appetite suppressant property [125].

2.6.4 *Moringa stenopetala* (Moringaceae)

Popularly called "African moringa" or "cabbage tree," *Moringa stenopetala* is a plant native to Kenya and Ethiopia. It is among the most domesticated edible plant with numerous nutritional and medicinal assets [126]. The extract of its seed oil was reported to considerably ameliorate resistin and leptin reduction, as well as metabolic disturbances and HFD-induced hematologic. These effects were exerted through antioxidant enzymes promotion, inflammatory cytokines, and lipid peroxidation reduction and iNOS protein expression [127]. Chemical composition revealed the occurrence of carbohydrates, phenols and flavonoids, alkaloids, saponins in *M. stenopetala* leaf powder [128].

2.6.5 *Origanum vulgare* (Lamiaceae)

Origanum vulgare, also known as wild marjoram, is a typical North American edible and medicinal plant [129]. A metabolite of an 80% ethanolic extract of *O. vulgare* prevents lipid droplet accretion in differentiating his 3T3-L1 preadipocytes in a concentration-dependent manner [130]. Moreover, the relative mRNA and protein expression levels of c/ebpα, pparγ,srebp-1c, and fas, associated with adipogenesis were lower in the *O. vulgare* group. Over hundred non-volatile and volatile compounds were identified in the oil and extracts of this plant with rosmarinic acid being the major phenolic acid [130] (Fig. 6 and Table 6).

3 Extraction Techniques of Phytoconstituents Involved in the Management of Lifestyle Diseases

3.1 Overview of Extraction Techniques

Medicinal plants produce various categories of bioactive natural products with different functional groups and polarities. The most important classes of bioactive compounds obtained after extraction from plant are terpenoids, alkaloids, and phenolics [131]. Extraction is a technique used pharmaceutically to separate bioactive parts of plant materials from the bioinactive or inert constituents by means of suitable solvents through usual procedures [132]. Several extraction processes are employed for the extraction of bioactive constituents in plants, namely distillation method (for volatile components), sublimation, pressing, and solvent extraction. The latter being the most utilized method for the separation of plant soluble natural products from the insoluble cellular marc [133, 134]. Several studies have reported that the extraction efficiency was affected by the plant materials size, its properties, the solvent's ratio in the raw constituents, temperature, and time of extraction. Increasing time lead to rise in the percentage and efficacy of extraction. In this method, the solubility, selectivity, safety, and cost of the solvent play an important role. Therefore, natural products with the same polarity as the solvents will lead to a greater extraction percentage. The solubility and stability of bioactive compounds are affected by a high temperature, which can result to the degradation of thermo-labile bioactive components in the solvents [135]. The extraction processes of plants natural products using solvents comprise infusion, maceration, soxhlet extraction, decoction, percolation, microwave and ultrasound-assisted extractions, supercritical fluid extraction, enzyme-assisted extraction, and pressurized liquid extraction [133, 134].

3.1.1 Maceration

In this method, the powdered plant raw material is soaked with a solvent at room temperature for 72 h and is shaken frequently to facilitate dissolution of the soluble matter. The mixture is then filtered and the solvent evaporated to afford the crude plant extract [135, 136].

3.1.2 Infusion

Infusion is an adequate method for the extraction of plants rich in volatile constituents or that easily release their bioactive

Fig. 6 Pictures of selected plants used in the treatment of obesity

Table 6 Anti-obesity, chemical constituents, and mechanisms of action of selected medicinal plants

Plant species	Chemical constituents	Biological activities	Mechanism of actions	References
Salvia miltiorrhiza (Lamiaceae)	Phenolic acids and derivatives	Anti-obesity	Elevate levels of lipid factors including HSL, cAMP, and PKAin adipose tissue Ameliorate lipid metabolism and gut integrity	[118, 119]
Ficus carica (Moraceae)	Phenolic compounds and flavonoids, steroids, ceramides, pentacyclic triterpenes, and cerebrosides	Anti-obesity	Anti-lipase, decrease in triglycerides, LDL, cholesterol, and VLDL levels. Increase in HDL levels	[120–122]
Hoodia gordonii (Apocynaceae)	Steroidal glycosides	Anti-obesity	Reduce adipocyte size and skeletal muscle fiber size was shown	[124, 125]
Moringa stenopetala (Moringaceae)	Carbohydrates, flavonoids, phenols, alkaloids, saponins	Anti-obesity	Promotes antioxidant enzymes Reduces lipid peroxidation, inflammatory cytokines, and iNOS protein expression	[126–128]
Origanum vulgare (Lamiaceae)	Phenolic acid derivatives	Anti-obesity	Inhibits lipid droplet accumulation in differentiating 3T3-L1 preadipocytes and reduces the relative expression levels of pparγ, c/ebpα, fas, and srebp-1c mRNAs and proteins associated with adipogenesis	[129, 130]

compounds in solvents (berries, dried flowers or herbs). The liquid is usually brought to another suitable temperature and then poured onto the plant material, which is then left to soak in the solvent for some time. Clarified solvents are obtained by filtration or the marc is simply retrieved from the liquid. The length of time plant materials are left in the solvent depends on the reason for which the infusion is prepared. The choice of the solvent depends on the nature of the compound to be extracted. Usually, infusion should be done for less than 30 min or until the liquid cools. The volume of the solvent depends on the quantity of the raw plant materials and required concentration of the resulting infusion [135, 137].

3.1.3 Percolation

This extraction method involves soaking for about 4 h the raw material in a closed container with the required amount of solvent. Due to an additional solvent which is poured on the raw materials, the maceration is left for 24 h in the closed container. The dripwise extract is poured out using an opening percolator. Additional solvent is added until the percolate is approximately three quarters of the required amount of the final product. The marc is squeezed and the resulting solvent is appended to the percolate. The required volume is produced using additional solvent and the filtration or decantation affords a mixed clarified liquid [134, 137].

3.1.4 Decoction

The bioactive no thermo-labile (no heat stable) natural products can be extracted using decoction method. This method is frequently used for the preparation of herbal drugs in traditional medicine. The process involves boiling for about 15 min to up to an hour the plant material depending on the plant's part and its bioactive constituents. The desired volume is obtained after cooling, straining, and addition of enough cold solvent through plant materials. Therefore, the leaves, roots, stems, and flowers are boiled for approx. 15 min, while other hard parts such as branches require a longer boiling time during which the evaporated solvent must be replaced [134].

3.1.5 Soxhlet Extraction or Hot Continuous Extraction

In this process, a powdered sample to be extracted is placed in a porous cartridge made of solid filter paper or cellulose and located in a chamber of the Soxhlet apparatus. The bottom flask is used to heat organic solvents to reflux and are evaporated in the sample cartridge. After condensation, the runoff results in the extraction of biological active constituents from the plant [135, 136, 138].

3.1.6 Microwave-Assisted Extraction (MAE)

The partition of analytes from the sample matrix into the solvent are facilitated in this method by the use of microwave energy. Microwave radiation with a frequency of 300 MHz to 300 GHz and a wavelength of 1 mm to 1 m acts directly on polarizable and polar materials such as solvent. A resulting plant sample is heated near the surface of the materials and conduction is used to transfer the heat. This method is used for the extraction different bioactive natural products from plant materials. It is a very rapid and cost-effective method as compared to other common methods [137].

3.1.7 Ultrasound-Assisted Extraction (UAE) or Sonication Extraction

UAE is known as a green extraction technique for plants secondary metabolites. Its waves frequency range from 20 to 2000 kHz [135, 139]. A green extraction technique of bioactive secondary metabolites is known as UAE. A complete extraction of bioactive natural products using UAE can be completed in minutes with higher extraction rates, streamlined processing, high capacity for the final product, while reduced input of solvent and post-extraction wastewater treatment disposal. It uses of fossil energy is reduced as compared to commonly used methods such as soxhlet, maceration, and distillation[139, 140]. The advantage of this method is that it significantly increases the solvent extraction of natural products from plants. In addition, the duration of the process is reduced and higher extraction yield and quality were obtained using UAE [139]. The extraction of bioactive components depends on the nature of the solvent, agitation, and heat. This method is suitable for extracting bioactive secondary metabolites from spices and herbs, proteins, and oil from soy and seeds, respectively [139, 141].

3.1.8 Supercritical Fluid Extraction (SFE)

This technique allows the extraction of targeted compounds using supercritical carbon dioxide as a solvent at high pressure and temperature above its critical values (7380 kPa and 31.1 °C) [134, 139]. The undesired materials can be removed from the product using this method [137]. Using supercritical fluid provides this method with numerous advantages compared to conventional techniques because, these solvents possess diverse physico-chemical assets including viscosity, dielectric constant, density, and diffusivity. Supercritical CO_2 diffuses easily through solid matrices and has low viscosity, high diffusivity, and enhanced transport abilities that greatly increase its extraction rates [142]. Supercritical fluid possesses low solubility for polar natural products. This problem of solubility of polar compounds can be solved by adding methanol or ethanol. In addition, CO_2 has low toxicity, is cheap and available [141].

3.1.9 Enzyme-Assisted Extraction (EAE)

It is an enzymatic process where a specific hydrolyzing enzyme, namely α-amylase, pectinase, or cellulase, is added during the extraction step [143]. In some plants, hydrogen or hydrophobic bonding is used to retain natural products in the polysaccharide-lignin network. Thus, a solvent extraction process is not accessible to secondary metabolites. In order to effectively free the bounded natural products and improve their total yield, an enzyme pretreatment is necessary [131]. It is a novel and effective method for the extraction of antioxidant natural products from natural resources [144]. The temperature, time of extraction, system's pH, concentration of the enzyme, and the size of substrate particles are numerous parameters for the effectiveness of the extraction process. Nevertheless, the bioactive compounds were reported to be extracted from plant materials by means of a mixture of cellulolytic enzymes and pectinolytic (1:2 ratio), at 40 °C, 4.0 pH, and for 2 h treatment [136, 145].

3.1.10 Pressurized Liquid Extraction (PLE)

The effective extraction of targeted compounds in this method requires using high pressure (2–20 MPa) and temperature (up to 200 °C) liquid solvent. The high solvent penetration into the matrix is due to the high diffusion and solubility rates of hydrophobic compounds into the solvent, which reduces solvent requirements and extraction time. Several research teams have described PLE as accelerated solvent or fluid extraction, "high-pressure solvent extraction (HSPE)," and "enhanced solvent extraction (ESE)." By comparison to the conventional soxhlet extraction, this method considerably decreases solvent and time consumption. It is a potential substitute to SFE [134, 136, 139, 146] (Table 7 and Scheme 1).

3.2 Specific Procedures for Extracting Some Classes of Compounds Involved in the Management of LSD

Several medicinal plants were reported to possess beneficial effects on lifestyle diseases. Their chemical and biological investigation have provided numerous compounds which have ascertain the claim pharmacological properties of these plants. As highlighted in the previous sections, these compounds belong to various classes of compounds including phenolics, alkaloids, and terpenoids. This section will focus on specific techniques for extracting these classes of compounds.

3.2.1 Extraction of Phenolic Compounds

This class of secondary metabolites represents one of the largest group of compounds found in plants. They have in common benzene ring substituted by one or more hydroxyl groups. Based on their basic structures, they can be divided in two main classes, namely flavonoids and non-flavonoids. The specific substituent in the basic structure gave rise to phenolic subclasses. Non-flavonoid phenolic compounds are divided into the following subgroups based on their carbon skeleton: coumarins, tannins, etc. [145].

3.2.1.1 Flavonoids

The general basic skeleton of flavonoids consists of 15 carbons grouped into two benzene rings (A and B) linked by three linear carbons (C_6-C_3-C_6) which could bridge with an oxygenated heterocycle to give a third ring (C) [148, 149]. Their classification is based on changes observed in C-ring heterocyclic structure and over 7000 flavonoids have been identified from natural sources [150]. Due to their interesting biological properties, the number of identified flavonoids keeps on rising. Based on their structure (which differs from the central structure of the flavans by the central motif of the heterocyclic pyran ring), they are allocated in seven subclasses including anthocyanins, flavones, isoflavones, flavonols, flavanonols, flavan-3-ols, and flavanones [148, 149]. Flavonoids could exist in free state but are often prenylated, methylated, acetylated, or glycosylated. In addition, several polyflavonoids have been identified [151]. The extraction and recovery of flavonoids has increased in recent years due to demographic trends toward better lifestyles and the inclusion of antioxidants in LSD management. Thus numerous flavonoid extraction techniques have been implemented to improve yield of extractions and reduce processing costs of these important bioactive natural products [148, 149]. Various extraction methods have been implemented, including infusion, decoction, maceration, percolation, steeping,

Table 7 Extraction methods of natural products [135]

Methods	Solvents	Pressure	Temperature	Time	Required volume	Compounds polarity
Maceration	Water, aqueous and non-aqueous solvents	Atmospheric	Ambient temperature	Long	Large	Depend on extracting solvent
Decoction	Water	Atmospheric	Under heat	Moderate	None	Polar compounds
Soxhlet extraction	Organic solvents	Atmospheric	Under heat	Long	Moderate	Dependent on extracting solvent
Percolation	Water, aqueous and non-aqueous solvents	Atmospheric	Ambient temperature, occasional heat	Long	Large	Dependent on extracting solvent
Pressurized liquid extraction	Water, aqueous and non-aqueous solvents	High	Under heat	Short	Small	Dependent on extracting solvent
Microwave-assisted extraction	Water, aqueous and non-aqueous solvents	Atmospheric	Ambient temperature	Short	Moderate	Dependent on extracting solvent
Ultrasound-assisted extraction	Water, aqueous and non-aqueous solvents	Atmospheric	Ambient temperature or under heat	Short	Moderate	Dependent on extracting solvent
Supercritical fluid extraction	CO_2	High	Near ambient temperatures	Short	None or small	Non-polar to moderate compounds
Enzyme-assisted extraction	Water, aqueous and non-aqueous solvents	Atmospheric	Ambient temperature or heated after enzyme treatment	Moderate	Moderate	Dependent on extracting solvent

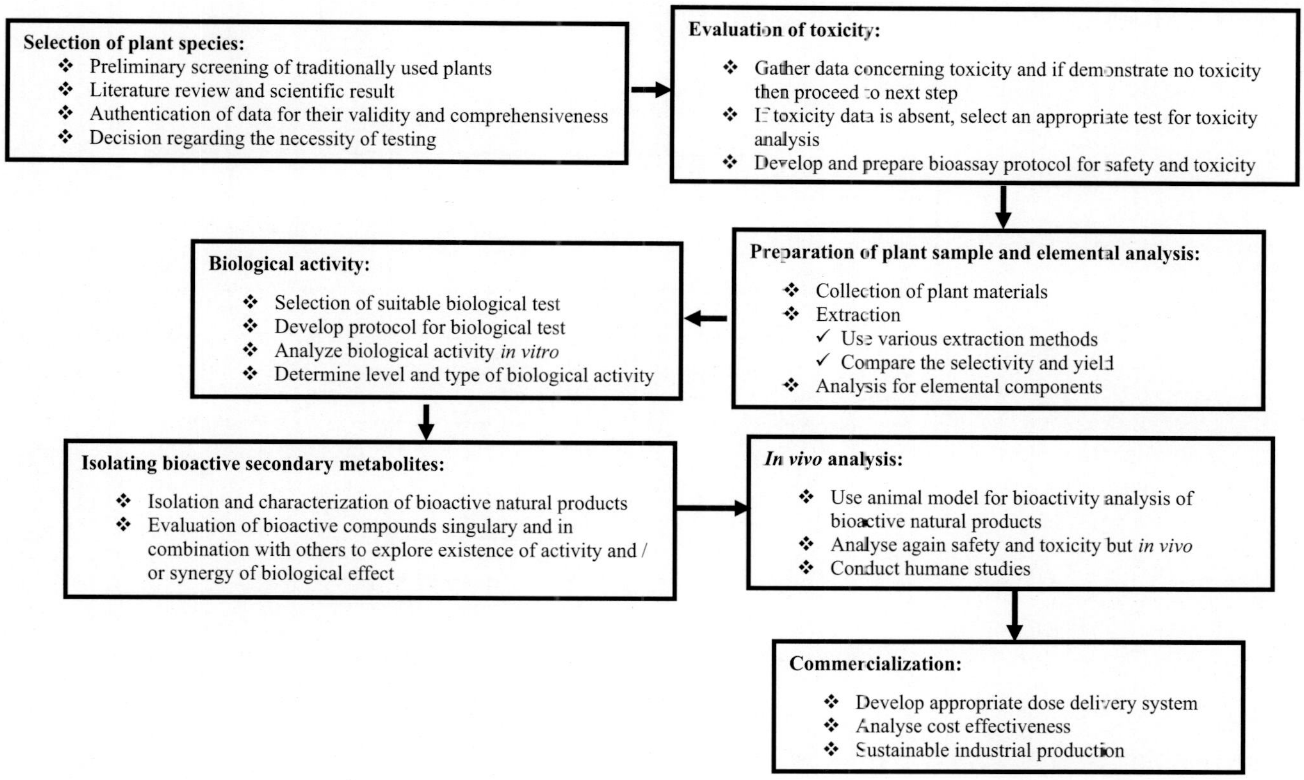

Scheme 1 The flow chart of medicinal plant study and position of extraction techniques [147]

and soxhlet. The latter is most often used and suitable technique for the extraction of flavonoids owing to its low solvent content, easiness, low cost, and ease of maintenance. The type of flavonoids extracted is influenced by the nature of the extraction solvent which affects the bioactivity of the obtained compounds. Thus, different solvents including ethanol, methanol, ethyl acetate, chloroform, petroleum ether, etc. were assessed for their flavonoid extraction potential and it resulted that ethanol and methanol are the most suitable solvents for the extraction of this class of compounds. Moreover, liquid–liquid and solid–liquid extractions are the most used techniques for extracting flavonoids [150]. Extraction methods involving heat might result in the denaturation of flavonoids and thus to an observed decrease in their bioactivity [148–150] (Schemes 2 and 3).

3.2.1.2 Coumarins

Coumarins are natural phenolic products found in specific plants families and genus. It comprises a benzene ringbonded to a lactone ring. They are derivatives of 1,2-benzopyrone and known for their pleasant vanilla odor [153–156]. There are four principal subclasses of coumarins, namely simple coumarins, pyranocoumarins, furanocoumarins, and pyrone-substituted coumarins. They can be alkoxylated, glycosylated, hydroxylated, and alkylated. They attract great attention owing to their pharmaceutical assets, physiological, antitumor, and bacteriostatic activity which makes these secondary metabolites attractive for further skeletal derivatization and screening as new therapeutics [153, 156]. Scientists have reported several methods for extracting coumarins and their glycosides. Extraction with high polarity solvents such as methanol, ethanol, and water have been proven to be more efficient. Greater amounts of free coumarins are obtained when water extraction is done at ambient temperature rather than at 100 °C. Extractions of courmarins and their glycosides with medium and low polarity solvents such as diethyl ether, EtOAc, and $CHCl_3$ are very poor. Regarding furanocoumarins, the best extraction is achieved with solvent reflux or soxhlet extraction techniques using boiling methanol as solvent. Similar results were obtained with H_2O, $CHCl_3$ at room temperature, EtOAc, and MeOH. The lower extraction yield of furanocoumarins was obtained with diethyl ether while the extraction with water gave similar results with methanol, making the latter a very interesting solvent for the extraction of furanocoumarins [154, 156].

3.2.1.3 Tannins

Tannins are phenolic compounds with high molecular weight and numerous hydroxyl groups. More than 8000 different tannins have been reported. In plant cells with molecular weights of up to 500 Da, they exist in free and bound forms. They have very diverse and different chemical structures. Phenolic acids such as gallic and ellagic acids can be obtained after hydrolysis of some tannins. Tannins are normally

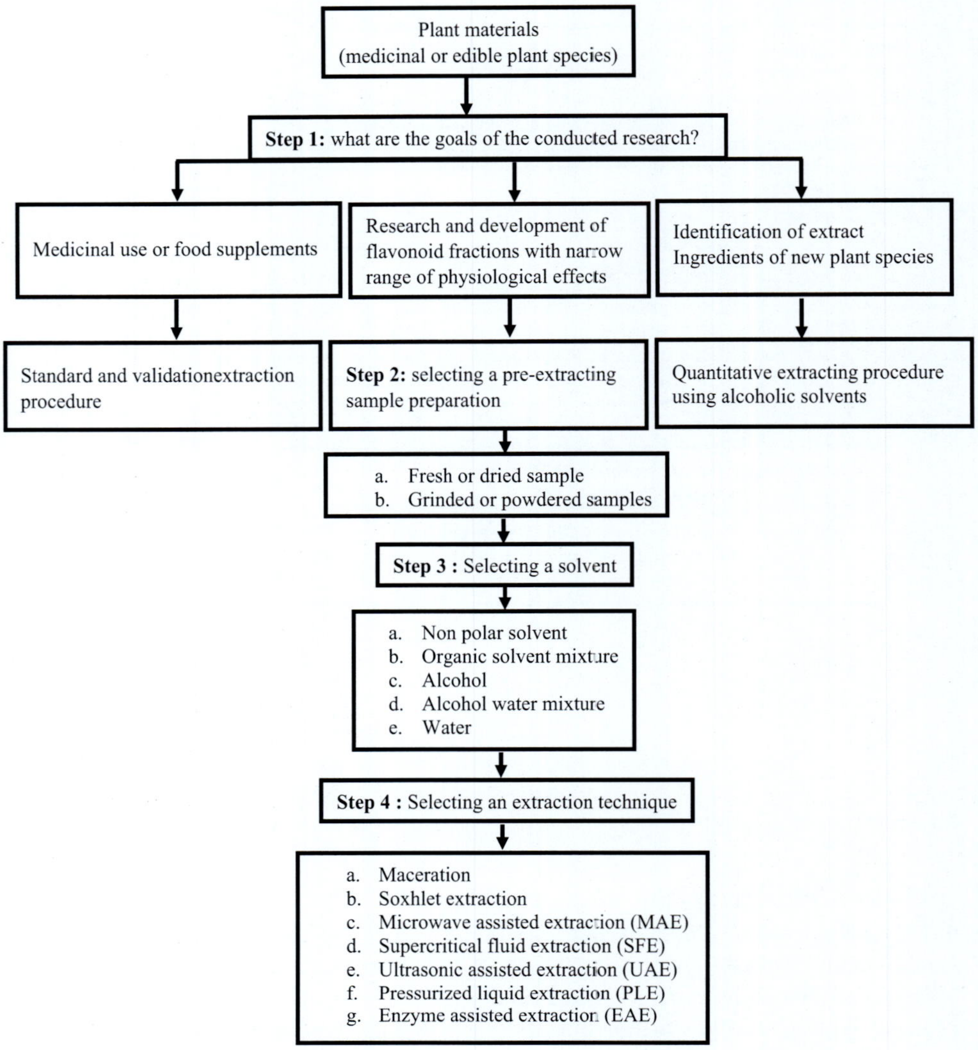

Scheme 2 Extraction flow chart of Flavonoids [136]

divided into two classes of chemical compounds, namely hydrolyzable and condensed tannins. Chemically, hydrolyzable tannins are heteropolymers composed of polyphenolic acids and their derivatives, esterified to a polyol while condensed tannins are polymeric flavonoids [157–159]. Tannin content generally ranges from 0.2 to 25 dry weight and depends different factors including extraction methods. Tannins were reported to possess several bioactivities such as antioxidant, antitumor, etc. which are important in the management of LSD. The number of aromatic rings and hydroxyl radicals greatly influences the biological potentials of tannins [157, 158].

The valorization of tannin extraction method remains the greatest challenge due to the fact that tannin extraction is not performed in a single process. The extracted tannins generally contain various kinds of impurities such as stilbenes, sugars, and minerals which are due to the extraction process. Processing factors such as size of particles, pressure, temperature, time, the amount of solid and extraction solvent, and the nature of the solvent impact the amount of impurities. These operational parameters need to be carefully controlled in order to extract tannins. Hot water or water with other solvents are generally utilized for the extraction of tannins from natural sources. Ethanol, methanol, acetone, ethyl acetate, ethyl ether, diisopropyl ether, sodium sulfite, and sodium hydroxide have been used as solvents with or without water by various scientists. Based on these studies, methanol extraction gave a better yield compared to water extraction. Yet, due to its simplicity and lower cost, extraction with hot water remains very common and widely employed in industries and laboratories for tannin extraction. The yield of extracted hydrolyzable and condensed tannins was reported to be higher using hot water. The temperature of water depends on the plant material, the extraction technique and the size of particles to be extracted [155].

3.2.2 Extraction of Alkaloids

Alkaloids are nitrogen-containing alkaline secondary metabolites and are well known for their physiological properties. They mostly originate from plants, particu-

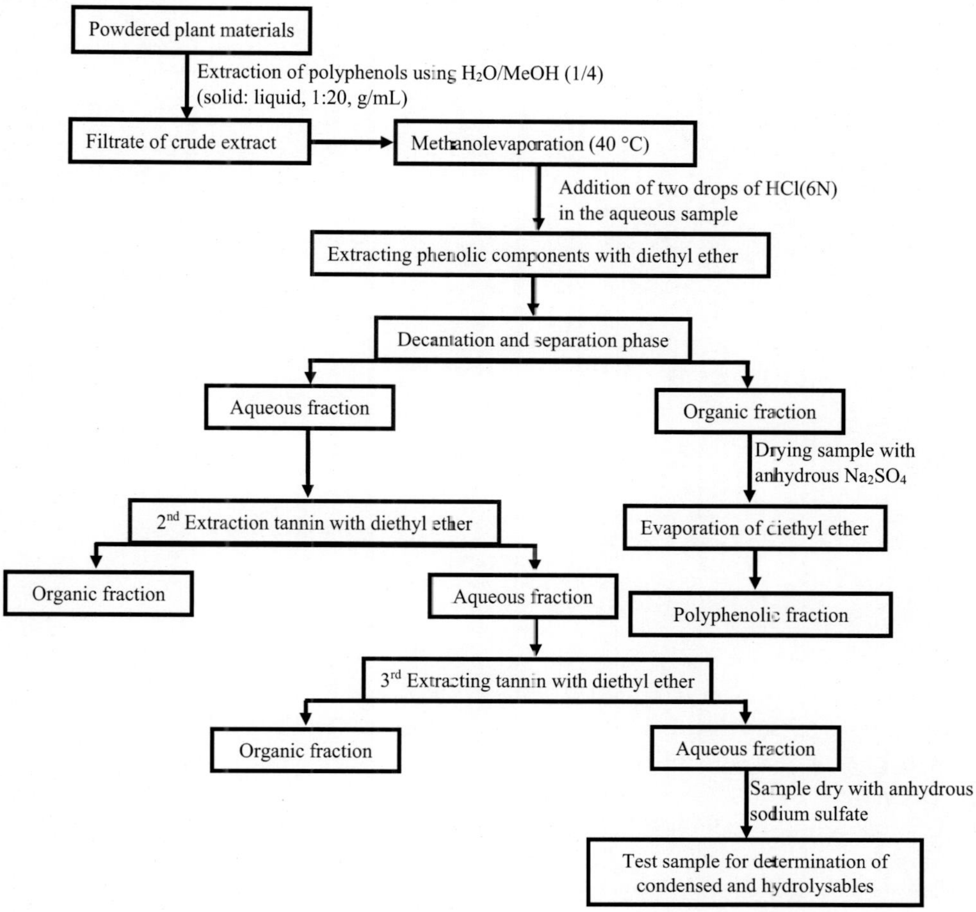

Scheme 3 Extraction flow chart of polyphenols and tannins [152]

larly higher plants [160, 161]. Alkaloids rich herbal drugs contain both active and toxic components. The weakly alkaline alkaloids exist in the free form, such as amide alkaloids. Alkaline alkaloids occur mainly as organic salts such as succinate, citrate, oxalate, tartrate, oxalate, etc. Few exist as inorganic salts such as morphine sulfate or berberine. There are other forms of N-oxide alkaloid glycosides [158]. Alkaloidal extracts have extensively been used in pharmaceutical formulations [159]. To improve their therapeutic effects, natural products should be efficiently extracted while eliminating the inactive and toxic constituents. In this light, alkaloids extraction has drawn more consideration due to their huge market demand, resulting in an upswing of research work on their extraction procedure [160, 162]. The choice of an extraction process principally depends on the type and occurrence of the alkaloids in plants. Generally, the solvent extraction process is mostly used due to their solubility, which is a significant basis for extraction. The frequently used technique include alcohol-solvent and acid-water extractions, percolation, soxhlet, lipophilic organic solvent extraction, ultrasound-assisted extraction, microwave extraction, supercritical fluid extraction, and enzymatic method, etc. [160, 162] (Scheme 4).

3.2.3 Extraction of Terpenoids

They are an important group of natural products by some plants and animals [163–165]. More than 80,000 types of terpenoids have been discovered to date. Thus, it is known that the structural diversity of terpenoids is responsible of the pharmacological applications of chemicals extracted from plants. Depending on the number of isoprene units, terpenoids are classified as saponins, cardiac glycosides, steroids, etc. [164, 165].

Saponins are natural bioorganic secondary metabolites containing one or more glycosidic bond on carbon C-3 between the aglycone and a sugar chain [166, 167]. The aglycone moiety, also called sapogenin or simply genin, is either a steroid, triterpenoid or an alkaloid [164]. They can be called mono-, di-, or tridesmosidic if one, two, or three sugar chains are linked to the sapogenin [163]. The extraction procedure of saponins from natural sources have been widely studied and this may be due to increased public health awareness. In general, traditional and green technologies are the two categories of extraction techniques used for extracting saponins from medicinal plants [166, 168].

Cardiac glycosides are a group of closely related natural substances with a strong specific and powerful effect on the heart muscle. Their action on the heart is through direct and

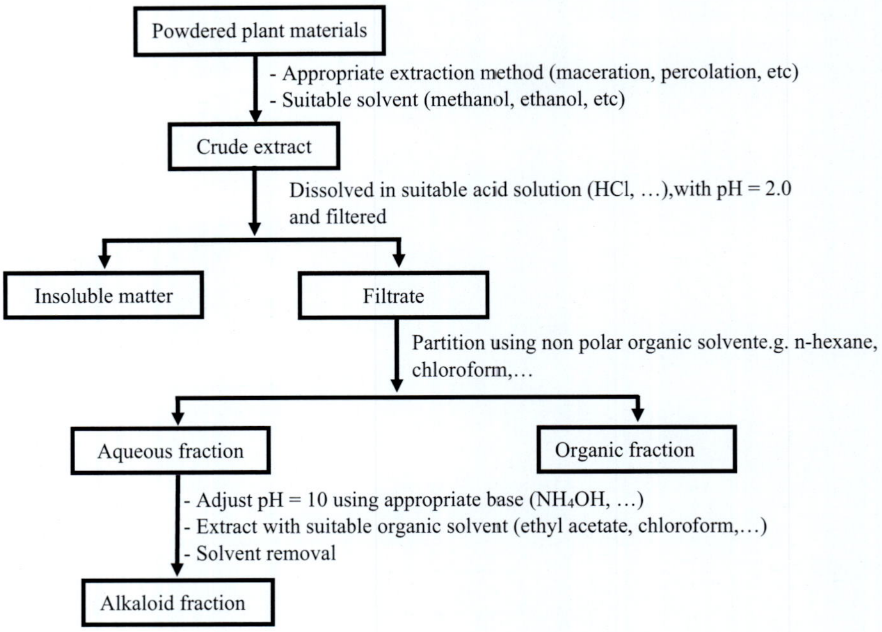

Scheme 4 Extraction flow chart of total alkaloids [160]

indirect mechanisms to increase the strength and speed of contraction. In high doses they can be toxic while smaller amounts could have useful stimulation effect on the heart. They have shown effects on myocardial tone, excitability and contractility, and diuretic activity [166]. Medicinal plants extracts rich in cardiac glycosides and their derived formulations were utilized since antiquity as medicines and poisons. The frequently employed method for extracting this class of compound is by protecting the raw plant material beforehand through maceration at 25–37 °C in toluene for several days in order to avoid enzymatic hydrolysis. An extensive extraction with a hydro-alcoholic mixture is further done and the resulting extract is evaporated under vacuum at 50 °C to a small volume. Fats are further removed by extracting with petroleum ether and aqueous glycoside syrup diluted in an equal volume of water. Acidic products including tannic acid and polyphenols are precipitated with prepared lead hydroxide and the resulting mixture filtered through hyflosuper gel. The pH of the filtrate is adjusted to 6 which is further concentrated in *vacuo* and successively extracted with petroleum ether, CHI_3, CHI_3/alcohol (2:1) and (3:2) [169, 170].

Steroids are primarily composed of lipophilic compounds with a variety of biological and physical properties. They contain natural polycyclic isopentenoids whose basic 1,2-cyclopentanophenthrene structure is modified on the side chain by the addition of lipophobic or hydrophilic or polar groups. Steroids are essential group of natural products found in most cells and rampant in medicinal plants. Several steroids have also already been synthesized. Regarding their extraction process from natural sources, their polarity and interaction with binding proteins must be considered [171–173]. The extraction solvent must completely break the bond between the protein and the steroid and facilitate quantitative extraction of the targeted steroids, leaving other steroids and non-desirable substances in the aqueous medium. Practically this is impossible as the solvent will also extract other compounds with analogue polarity which indeed affect the final yield. In the case of steroids linked to lipoproteins (e.g., vitamin D and cholesterol), the addition of chemical which will break down the lipoproteins prior to extraction might be necessary. Better extraction yields of bile acids and their conjugates linked to proteins are often obtained from alkaline media instead of acidic aqueous media. The extraction of more polar steroids such as sulfate conjugates, glucuronide, and polyhydroxy steroids is sometimes not efficient with very polar solvents [170] (Scheme 5).

3.2.4 Extraction of Others Classes of Compounds

Ceramides are specific bioactive sphingolipid metabolites formed from a fatty acid linked through an amide function to a long-chain amine called a sphingoid. The predominant long sphingoid chain bases in plants are the C_{18} amino alcohols. Long-chain dihydroxy-sphinganine and long-chain trihydroxy-4-hydroxysphinganine are found not only as generally minor components of complex sphingolipids, but also as long-chain free bases in plant tissues. Plant ceramides exist primarily in acylated, glycosylated, or polyhydric fatty acid forms and exhibit good biological activity. They have been identified as lipid compounds [174–176]. These amides reside in the cytoplasmic membrane where their role in cells development and growth are vital. Therefore, cancer, some skin diseases, and other dysfunctional cellular diseases can be caused by decreasing their levels. Ceramides and their glycosides are sphingolipids that constitute an important part (about 10%) of higher plant lipids. Thus far about 200 plant lipids have been described and several others are still on their

Scheme 5 Extraction flow chart of total saponins

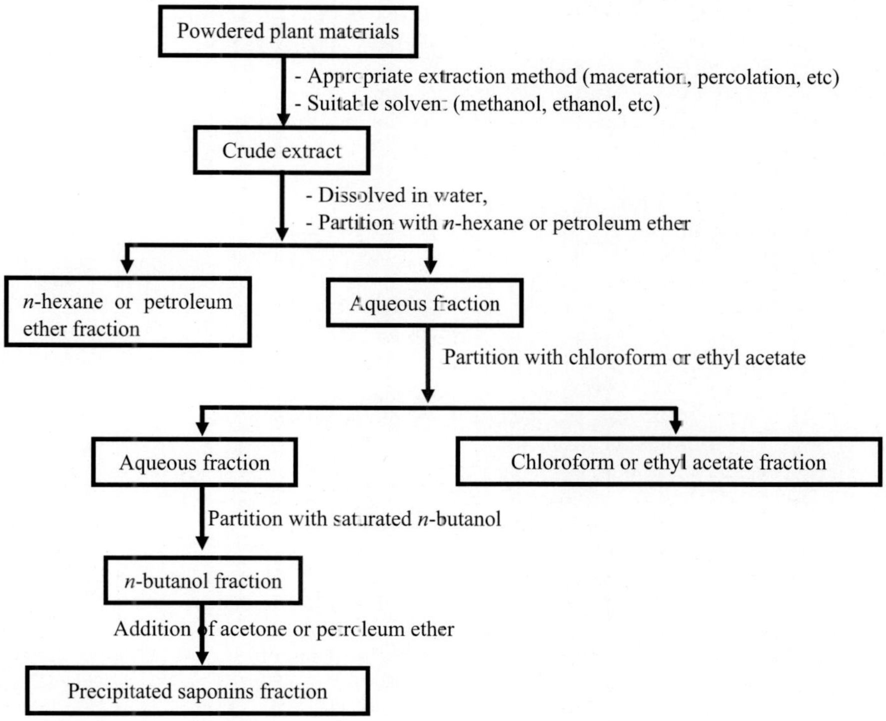

way. The diversity of their structures eases their role in different cellular processes resulting in their involvement in numerous activities such as the formation of membrane domains, structural integrity of membranes, and plant response to hypoxia and pathogen attack [173]. Besides commonly applied extraction procedures, supercritical fluid extraction (SFE) could also be used for extracting ceramides, but this method is very expensive. Thus, techniques such as ultrasonic-assisted extraction and enzyme-assisted extraction have become prominent [172].

Sulfur is an essential plant macronutrient that is a prominent component of amino acids and several metabolites. It also plays a great role in different cellular metabolic processes and is frequently responsible for the structure and biological activity of proteins [174, 175, 177]. Sulfur derive compounds play a vital role in plants life cycle and in their protection against pathogens and environmental stresses [175, 176]. Organo-sulfur compounds and particularly those from *Allium* species are low molecular weight molecules. These bioactive natural substances can be obtained using conventional or modern extraction methods, however, extraction by hydro-distillation or maceration with ethanol or water at ambient temperature have been proven to be very efficient [177].

4 Conclusion and Outlooks

This chapter has given insights into medicinal and edible plants with beneficial properties against lifestyle diseases, their mechanisms of action as well as their phytoconstituents. The most represented classes of compounds of these phytoconstituents were alkaloids, sulfur compounds, phenolic compounds (phenolic acids, flavonoids, and tannins), terpenoids, and steroids. The suitable extraction techniques of these phytoconstituents were also discussed. However, more investigations are required to quantify the bioactive components of these plants depending on the extraction method used.

Acknowledgment The authors are grateful to the Alexander von Humboldt Foundation (3.4-CMR-Hub) for the financial support to the Centre of Competence for the Study of Antimicrobial Natural Products from Fungi (CECANAPROF).

References

1. Kshirsagar MV, Ashturkar MD. Prevalence of lifestyle diseases in Maharashtra: a comparison between NFHS-5 and NFHS-4 surveys. J Family Med Prim Care. 2022;11(6):2474–8.
2. Umereweneza S, Kirkgulhorn T, Seetangkham S. Measures to prevent disease from lifestyle. Int J Health Sci Res. 2018;12(Suppl):20–8.
3. Tabish SA. Lifestyle diseases: consequences, characteristics, causes and control. J Cardiol Curr Res. 2017;9.
4. Module for multi-purpose workers—prevention, screening and control of common NCDS_2.pdf. [cited 2023 Jan 30]. https://main.mohfw.gov.in/sites/default/files/Module%20for%20Multi-Purpose%20Workers%20-%20Prevention%2C%20Screening%20and%20Control%20of%20Common%20NCDS_2.pdf.
5. Plant-based diets and their impact on health, sustainability and the environment: a review of the evidence. [cited 2023 Jan 30]. https://www.who.int/europe/publications/i/item/WHO-EURO-2021-4007-43766-61591.

6. Acquaviva R, Malfa GA, Di Giacomo C. Plant-based bioactive molecules in improving health and preventing lifestyle diseases. Int J Mol Sci. 2021;22(6):2991.
7. Singh S, Jain V, Jain S, Chandra K. Medicinal plants and phytochemicals in prevention and management of life style disorders: pharmacological studies and challenges. Asian J Pharm Clin Res. 2021 Dec;7:1–6.
8. Marrelli M. Medicinal plants. Plants. 2021;10:1355.
9. Salmerón-Manzano E, Garrido-Cardenas JA, Manzano-Agugliaro F. Worldwide research trends on medicinal plants. Int J Environ Res Public Health. 2020;17(10):3376.
10. Sülsen VP, Athanassopoulos CM, Padrón JM, Tamura RE. Editorial: natural compounds as scaffolds for the discovery of new anti-cancer drugs: focus on terpenoids and flavonoids. Front Pharmacol. 2022;13:984849.
11. Al-Mawali A, Jayapal SK, Morsi M, Al-Shekaili W, Pinto AD, Al-Kharusi H, et al. Prevalence of risk factors of non-communicable diseases in the Sultanate of Oman: STEPS survey 2017. PLoS One. 2021;16(10):e0259239.
12. Miller JS. The discovery of medicines from plants: a current biological perspective. Econ Bot. 2011;65(4):396–407.
13. Naveed M, Majeed F, Taleb A, Zubair HM, Shumzaid M, Farooq MA, et al. A review of medicinal plants in cardiovascular disorders: benefits and risks. Am J Chin Med. 2020;48(02):259–86.
14. Xu J, Kitada M, Ogura Y, Koya D. Relationship between autophagy and metabolic syndrome characteristics in the pathogenesis of atherosclerosis. Front Cell Dev Biol. 2021;9:641852.
15. Hosseini A, Hosseinzadeh H. A review on the effects of *Allium sativum* (Garlic) in metabolic syndrome. J Endocrinol Invest. 2015;38(11):1147–57.
16. Aksenov DV, Kaplun VV, Tertov VV, Sobenin IA, Orekhov AN. Effect of plant extracts on trans-sialidase activity in human blood plasma. Bull Exp Biol Med. 2007;143(1):46–50.
17. Kim MJ, Kim HK. Effect of garlic on high fat induced obesity. Acta Biol Hung. 2011;62(3):244–54.
18. Batiha G, Beshbishy A, Wasef L, Elewa Y, Al-Sagan A, El-Hack M, et al. Chemical constituents and pharmacological activities of garlic (*Allium sativum* L.): a review. Nutrients. 2020;12(3):872.
19. Asgary S, Jafari N, Madani H, Mahzoni P. Effect of Glycyrrhiza glabra extract on aorta wall atherosclerotic lesion in hypercholesterolemic rabbits. Pak J Nutr. 2007;6:313–7.
20. Belinky PA, Aviram M, Mahmood S, Vaya J. Structural aspects of the inhibitory effect of glabridin on LDL oxidation. Free Radic Biol Med. 1998;24(9):1419–29.
21. El-Saber Batiha G, Magdy Beshbishy A, El-Mleeh A, Abdel-Daim MM, Prasad Devkota H. Traditional uses, bioactive chemical constituents, and pharmacological and toxicological activities of *Glycyrrhiza glabra* L. (Fabaceae). Biomolecules. 2020;10(3):352.
22. Zhang H, Ma ZF, Luo X, Li X. Effects of mulberry fruit (*Morus alba* L.) consumption on health outcomes: a mini-review. Antioxidants (Basel). 2018;7(5):69.
23. Chan EWC, Lye PY, Wong SK. Phytochemistry, pharmacology, and clinical trials of *Morus alba*. Chin J Nat Med. 2016;14(1):17–30.
24. Kim DS, Ji HD, Rhee MH, Sung YY, Yang WK, Kim SH, et al. Antiplatelet activity of *Morus alba* leaves extract, mediated via inhibiting granule secretion and blocking the phosphorylation of extracellular-signal-regulated kinase and Akt. Evid Based Complement Alternat Med. 2014;2014:639548.
25. Kim GN, Jang HD. Flavonol content in the water extract of the Mulberry (*Morus alba* L.) leaf and their antioxidant capacities. J Food Sci. 2011;76(6):C869–73.
26. Kumar SSD, Houreld NN, Abrahamse H. Therapeutic potential and recent advances of curcumin in the treatment of aging-associated diseases. Molecules. 2018;23(4):835.
27. Leong XF. The spice for hypertension: protective role of *Curcuma longa*. Biomed Pharmacol J. 2018;11(4):1829–40.
28. Adab Z, Eghtesadi S, Vafa MR, Heydari I, Shojaii A, Haqqani H, et al. Effect of turmeric on glycemic status, lipid profile, hs-CRP, and total antioxidant capacity in hyperlipidemic type 2 diabetes mellitus patients. Phytother Res. 2019;33(4):1173–81.
29. Jabczyk M, Nowak J, Hudzik B, Zubelewicz-Szkodzińska B. Curcumin in metabolic health and disease. Nutrients. 2021;13(12):4440.
30. Hedayati-Moghadam M, Hosseinian S, Paseban M, Shabgah AG, Gholizadeh J, Jamialahmadi T, et al. The role of chemokines in cardiovascular diseases and the therapeutic effect of *Curcumin* on CXCL8 and CCL2 as pathological chemokines in atherosclerosis. Adv Exp Med Biol. 2021;1328:155–70.
31. Hassanin HAM, Koko M, Abdalla M, Mu W, Jiang B. *Detarium microcarpum*: a novel source of nutrition and medicine: a review. Food Chem. 2019;274:900–6.
32. Youovop AJ, Dieudonné M, Ngondi J, Oben J. The potential effect of aqueous extract of *Detarium microcarpum* bark on certain metabolic disorders associated with an atherogenic diet in rats. J Food Res. 2020;9(5):102–12.
33. Dogara AM. Biological activity and chemical composition of *Detarium microcarpum* Guill. and Perr—a systematic review. Adv Pharmacol Pharm Sci. 2022;2022:7219401.
34. Zhou W, Chai H, Lin PH, Lumsden AB, Yao Q, Chen C. Clinical use and molecular mechanisms of action of extract of *Ginkgo biloba* leaves in cardiovascular diseases. Cardiovasc Drug Rev. 2004;22(4):309–19.
35. Tian J, Popal MS, Liu Y, Gao R, Lyu S, Chen K, et al. *Ginkgo biloba* leaf extract attenuates atherosclerosis in streptozotocin-induced diabetic ApoE-/-mice by inhibiting endoplasmic reticulum stress via restoration of autophagy through the mTOR signaling pathway. Oxid Med Cell Longev. 2019;2019:8134678.
36. Mahady GB. *Ginkgo biloba* for the prevention and treatment of cardiovascular disease: a review of the literature. J Cardiovasc Nurs. 2002;16(4):21–32.
37. Wang Y, Xu Y, Xu X, Wang H, Wang D, Yan W, et al. *Ginkgo biloba* extract ameliorates atherosclerosis via rebalancing gut flora and microbial metabolism. Phytother Res. 2022;36(6):2463–80.
38. Ahmad W, Zeenat F, Hasan A, Abdullah A, Nargis A, Tarannum T. Mazu (*Quercus infectoria*): an overview. Indian J Unani Med. 2011;4:17–22.
39. Gholamhoseinian A, Shahouzehi B, Joukar S, Iranpoor M. Effect of *Quercus infectoria* and *Rosa damascena* on lipid profile and atherosclerotic plaque formation in rabbit model of hyperlipidemia. Pak J Biol Sci. 2012;15(1):27–33.
40. Shrestha S, Kaushik VS, Eshwarappa RSB, Subaramaihha SR, Ramanna LM, Lakkappa DB. Pharmacognostic studies of insect gall of *Quercus infectoria* Olivier (Fagaceae). Asian Pac J Trop Biomed. 2014;4(1):35–9.
41. Ji C, Liu J, Zhang Q, Li J, Wu Z, Wang X, et al. Multi-element analysis and origin discrimination of *Panax notoginseng* based on inductively coupled plasma tandem mass spectrometry (ICP-MS/MS). Molecules. 2022;27(9):2982.
42. Xie W, Meng X, Zhai Y, Zhou P, Ye T, Wang Z, et al. *Panax Notoginseng* saponins: a review of its mechanisms of antidepressant or anxiolytic effects and network analysis on phytochemistry and pharmacology. Molecules. 2018;23(4):940.
43. Fan JS, Liu DN, Huang G, Xu ZZ, Jia Y, Zhang HG, et al. *Panax notoginseng* saponins attenuate atherosclerosis via reciprocal regulation of lipid metabolism and inflammation by inducing liver X receptor alpha expression. J Ethnopharmacol. 2012;142(3):732–8.
44. Cai BX, Li XY, Chen JH, Tang YB, Wang GL, Zhou JG, et al. Ginsenoside-Rd, a new voltage-independent Ca^{2+} entry blocker, reverses basilar hypertrophic remodeling in stroke-prone renovascular hypertensive rats. Eur J Pharmacol. 2009;606(1–3):142–9.

45. Wang CZ, McEntee E, Wicks S, Wu JA, Yuan CS. Phytochemical and analytical studies of *Panax notoginseng* (Burk.) F.H. Chen. J Nat Med. 2006;60:97–106.
46. Abu-Darwish MS, Cabral C, Gonçalves MJ, Cavaleiro C, Cruz MT, Zulfiqar A, et al. Chemical composition and biological activities of *Artemisia judaica* essential oil from southern desert of Jordan. J Ethnopharmacol. 2016;191:161–8.
47. Jiang Y, Du H, Liu X, Fu X, Li X, Cao Q. Artemisinin alleviates atherosclerotic lesion by reducing macrophage inflammation via regulation of AMPK/NF-κB/NLRP3 inflammasomes pathway. J Drug Target. 2020;28(1):70–9.
48. Moharram FA, Nagy MM, El Dib RA, El-Tantawy MM, El Hossary GG, El-Hosari DG. Pharmacological activity and flavonoids constituents of *Artemisia judaica* L aerial parts. J Ethnopharmacol. 2021;270:113777.
49. Sagbo IJ, Otang-Mbeng W. Plants used for the traditional management of cancer in the eastern cape province of South Africa: a review of ethnobotanical surveys, ethnopharmacological studies and active phytochemicals. Molecules. 2021;26(15):4639.
50. Hajiaghaalipour F, Kanthimathi MS, Sanusi J, Rajarajeswaran J. White tea (*Camellia sinensis*) inhibits proliferation of the colon cancer cell line, HT-29, activates caspases and protects DNA of normal cells against oxidative damage. Food Chem. 2015;169:401–10.
51. Mutakin M, Fauziati R, Fadhilah F, Zuhrotun A, Amalia R, Hadisaputri Y. Pharmacological activities of soursop (*Annona muricata* Lin.). Molecules. 2022;27:1201.
52. Park SY, Kim EJ, Choi HJ, Seon MR, Lim SS, Kang YH, et al. Anti-carcinogenic effects of non-polar components containing licochalcone A in roasted licorice root. Nutr Res Pract. 2014;8(3):257–66.
53. Drishya G, Nambiar J, Shaji SK, Vanuopadath M, Achuthan A, Kumar A, et al. RECK and TIMP-2 mediate inhibition of MMP-2 and MMP-9 by *Annona muricata*. J Biosci. 2020;45(1):89.
54. Cichello SA, Yao Q, Dowell A, Leury B, He XQ. Proliferative and inhibitory activity of Siberian ginseng (*Eleutherococcus senticosus*) extract on cancer cell lines; A-549, XWLC-05, HCT-116, CNE and Beas-2b. Asian Pac J Cancer Prev. 2015;16(11):4781–6.
55. Wang YH, Meng Y, Zhai C, Wang M, Avula B, Yuk J, et al. The chemical characterization of *Eleutherococcus senticosus* and Ci-wu-jia tea using UHPLC-UV-QTOF/MS. Int J Mol Sci. 2019;20(3):E475.
56. Batool N, Ilyas N, Shabir S, Saeed M, Mazhar R. Mini-review—a mini-review of therapeutic potential of *Mangifera indica* L. Pak J Pharm Sci. 2018;31(4):1441–8.
57. Yap KM, Sekar M, Seow LJ, Gan SH, Bonam SR, Mat Rani NNI, et al. *Mangifera indica* (Mango): a promising medicinal plant for breast cancer therapy and understanding its potential mechanisms of action. Breast Cancer (Dove Med Press). 2021;13:471–503.
58. Lauricella M, Lo Galbo V, Cernigliaro C, Maggio A, Palumbo Piccionello A, Calvaruso G, et al. The anti-cancer effect of *Mangifera indica* L. peel extract is associated to γH2AX-mediated apoptosis in colon cancer cells. Antioxidants (Basel). 2019;8(10):422.
59. Yap KM, Sekar M, Seow LJ, Gan SH, Bonam SR, Mat Rani NNI, et al. *Mangifera indica* (Mango): a promising medicinal plant for breast cancer therapy and understanding its potential mechanisms of action. Breast Cancer Targets Ther. 2021;13:471–503.
60. Mogue Kamga S, Sonké B, Couvreur TLP. *Raphia vinifera* (Arecaceae; Calamoideae): misidentified for far too long. Biodivers Data J. 2019;7:e37757.
61. Nguenang G, Mbaveng A, Bonsou I, Fru C, Kuete V. Botanical from the fruits mesocarp of *Raphia vinifera* displays antiproliferative activity and is harmless as evidenced by toxicological assessments. Evid Based Complement Altern Med. 2022;2022:1–13.
62. Chi GF, Sop RVT, Mbaveng AT, Omollo Ombito J, Fotso GW, Nguenang GS, et al. Steroidal saponins from *Raphia vinifera* and their cytotoxic activity. Steroids. 2020;163:108724.
63. Youkparigha FO, Uzoekwe NM, Ogumo PM. Comparative assessment of phytochemical constituents of *Raphia Vinifera* P. Beauv and *Raphia Hookeri* G. Mann & H. Wendl. Int J Adv Res Bot. 2019 [cited 2023 Jan 15];5(4). https://www.arcjournals.org/pdfs/ijarb/v5-i4/2.pdf.
64. Nguenang GS, Mbaveng AT, Bonsou IN, Chi GF, Kuete V. Botanical from the fruits mesocarp of *Raphia vinifera* displays antiproliferative activity and is harmless as evidenced by toxicological assessments. Evid Based Complement Altern Med. 2022 [cited 2023 Jan 15];2022. https://www.hindawi.com/journals/ecam/2022/4831261/.
65. Chan Y, Ng SW, Dua K, Chellappan DK. Plant-based chemical moieties for targeting chronic respiratory diseases. In: Dua K, Löbenberg R, Malheiros Luzo ÂC, Shukla S, Satija S, editors. Targeting cellular signalling pathways in lung diseases. Singapore: Springer; 2021. p. 741–81. . [cited 2023 Jan 12]. https://doi.org/10.1007/978-981-33-6827-9_34.
66. LucyPi1. The coastal medicinal plant *Vitex rotundifolia*: a mini-review on its bioactive compounds and pharmacological activity [Internet]. 01:14:43 UTC [cited 2023 Jan 15]. https://www.slideshare net/LucyPi1/the-coastal-medicinal-plant-vitex-rotundifolia-a-minireview-on-its-bioactive-compounds-and-pharmacological-activity.
67. Bae H, Kim Y, Lee E, Park S, Jung KH, Gu MJ, et al. *Vitex rotundifolia* L. prevented airway eosinophilic inflammation and airway remodeling in an ovalbumin-induced asthma mouse model. Int Immunol. 2013;25(3):197–205.
68. Liou CJ, Huang WC. Casticin inhibits interleukin-1β-induced ICAM-1 and MUC5AC expression by blocking NF-κB, PI3K-Akt, and MAPK signaling in human lung epithelial cells. Oncotarget. 2017;8(60):101175–88.
69. Lee JW, Ryu HW, Lee SU, Kim MG, Kwon OK, Kim MO, et al. *Pistacia weinmannifolia* ameliorates cigarette smoke and lipopolysaccharide-induced pulmonary inflammation by inhibiting interleukin-8 production and NF-κB activation. Int J Mol Med. 2019;44(3):949–59.
70. Zhao X, Sun H, Hou A, Zhao Q, Wei T, Xin W. Antioxidant properties of two gallotannins isolated from the leaves of *Pistacia weinmannifolia*. Biochim Biophys Acta. 2005;1725(1):103–10.
71. Huang JF, Li L, van der Werff H, Li HW, Rohwer JG, Crayn DM, et al. Origins and evolution of cinnamon and camphor: a phylogenetic and historical biogeographical analysis of the Cinnamomum group (Lauraceae). Mol Phylogenet Evol. 2016;96:33–44.
72. Wang J, Su B, Jiang H, Cui N, Yu Z, Yang Y, et al. Traditional uses, phytochemistry and pharmacological activities of the genus *Cinnamomum* (Lauraceae): a review. Fitoterapia. 2020;146:104675.
73. Zhou MX, Li GH, Sun B, Xu YW, Li AL, Li YR, et al. Identification of novel Nrf2 activators from *Cinnamomum chartophyllum* H.W. Li and their potential application of preventing oxidative insults in human lung epithelial cells. Redox Biol. 2018;14:154–63.
74. Hsieh CC, Peng WH, Tseng HH, Liang SY, Chen LJ, Tsai JC. The protective role of garlic on allergen-induced airway inflammation in mice. Am J Chin Med. 2019;47(5):1099–112.
75. Ghimire S, Subedi L, Acharya N, Gaire BP. *Moringa oleifera*: a tree of life as a promising medicinal plant for neurodegenerative diseases. J Agric Food Chem. 2021;69(48):14358–71.
76. Goyal BR, Goyal RK, Mehta AA. Investigation into the mechanism of anti-asthmatic action of *Moringa oleifera*. J Diet Suppl. 2009;6(4):313–27.

77. Sangilimuthu A, Ramalingam S, Darsini D, Anitha J, Subban R. A review on phytoconstituents against asthma. Int J Pharm Sci Rev Res. 2015;30:7–16.
78. Shunmugapriya K, Vennila P, Thirukkumar S, Ilamaran M. Identification of bioactive components in *Moringa oleifera* fruit by GC-MS. J Pharmacogn Phytochem. 2017;6(3):748–51.
79. Niu L, Hou Y, Jiang M, Bai G. The rich pharmacological activities of *Magnolia officinalis* and secondary effects based on significant intestinal contributions. J Ethnopharmacol. 2021;281:114524.
80. Kuo WL, Chung CY, Hwang TL, Chen JJ. Biphenyl-type neolignans from *Magnolia officinalis* and their anti-inflammatory activities. Phytochemistry. 2013;85:153–60.
81. Egbuna C, Awuchi CG, Kushwaha G, Rudrapal M, Patrick-Iwuanyanwu KC, Singh O, et al. Bioactive compounds effective against type 2 diabetes mellitus: a systematic review. Curr Top Med Chem. 2021;21(12):1067–95.
82. Jacob B, Narendhirakannan RT. Role of medicinal plants in the management of diabetes mellitus: a review. 3 Biotech. 2019;9(1):4.
83. Bahramikia S, Hemmati Hassan Gavyar P, Yazdanparast R. *Teucrium polium* L: an updated review of phytochemicals and biological activities. Avicenna J Phytomed. 2022;12(3):224–40.
84. Mannan A. Molecular signaling pathways involved in the glucose-lowering effect of Teucrium polium [Thesis]. Curtin University; 2017 [cited 2023 Jan 14]. https://espace.curtin.edu.au/handle/20.500.11937/56434.
85. Dias MC, Pinto DCGA, Silva AMS. Plant flavonoids: chemical characteristics and biological activity. Molecules. 2021;26(17):5377.
86. Lograda T, Messaoud R, Chalard P, Figueredo G, Deghar A. Chemical analysis and antimicrobial activity of *Teucrium polium* L. essential oil from eastern Algeria. Am J Adv Drug Deliv. 2014;2(6):697–710.
87. Akpojotor P, Ebomoyi M. Investigating the anti-diabetic phytoconstituent(s) of *Rauvolfia vomitoria* leaves by gas chromatography-mass spectrometry (GC-MS). 2021;8:1–8.
88. Doucet JL. Le moabi, une espèce "phare" de l'exploitation forestière en Afrique Centrale. Parcs et Réserves. 2007;62(2):25–31.
89. Guy Roussel TN, Martin F, Janvier Aimé YF, Ferdinand Lanvin EE, Ruth Edwige DK, Boris AK, et al. Antihyperglycemic and antihyperlipidemic activities of hydroethanolic extract of the fruit of *Baillonella toxisperma* in streptozotocin-induced diabetic rats. Metabol Open. 2022;15:100199.
90. Roussel T, Laure NJ, Enyong OJ. antioxidant and glucose lowering effects of hydroethanolic extract of *Baillonella toxisperma* Pulp. J Food Res. 2020;9
91. Manga E, Takuissu NGR, Ziyyat A, Ngondi JL, Sindic M. Hypoglycaemic activity of preheated (roasting) *Aframomum citratum* (C. Pereira) K. Schum and *Tetrapleura tetraptera* (Schumach Thonn.) fruits beverage on Streptozotocin-induced rats. JPP. 2020;12(2):44–61.
92. Kemigisha E, Owusu EO, Elusiyan CA, Omujal F, Tweheyo M, Bosu PP. *Tetrapleura tetraptera* in Ghana, Nigeria and Uganda: households uses and local market. For Trees Livelihoods. 2018;27(4):243–56.
93. Ojewole JAO, Adewunmi CO. Anti-inflammatory and hypoglycaemic effects of *Tetrapleura tetraptera* (Taub) [Fabaceae] fruit aqueous extract in rats. J Ethnopharmacol. 2004;95(2–3):177–82.
94. Adusei S, Otchere JK, Oteng P, Mensah RQ, Tei-Mensah E. Phytochemical analysis, antioxidant and metal chelating capacity of *Tetrapleura tetraptera*. Heliyon. 2019;5(11):e02762.
95. Enyong J, Judith LN, Nguemto GT, Kingue BA, Ntentié RF, Arnold WT et al. Aqueous extract of Alstonia boonei bark reduces chronic hyperglycemia and prevents its complications through increase of hepatic global DNA methylation in diabetic wistar rats. Eur J of Med Plants. 2021;32(12):1–15.
96. Arogbodo JO. Phytochemical screening and antimicrobial effect of ethanolic leaf extract of Alstonia boonei de wild (Apocynaceae) on some selected pathogenic micro-organisms, (2019). http://ir.ucc.edu.gh/jspui/handle/123456789/5804 (accessed November 10, 2023).
97. Babatunde O. GC-MS analysis of leaf, stem-bark and root extracts of *Alstonia boonei*. AJPP. 2017;11(46):577–81.
98. Tchamgoue J, Hafizur R, Tchouankeu J, Kouam S, Adhikari A, Hameed A, et al. Flavonoids and other constituents with insulin secretion activity from *Pseudarthria hookeri*. Phytochem Lett. 2016;17:181–6.
99. Dzoyem JP, Tchamgoue J, Tchouankeu JC, Kouam SF, Choudhary MI, Bakowsky U. Antibacterial activity and cytotoxicity of flavonoids compounds isolated from *Pseudarthria hookeri* Wight & Arn. (Fabaceae). S Afr J Bot. 2018;114:100–3.
100. Kjeldsen SE. Hypertension and cardiovascular risk: general aspects. Pharmacol Res. 2018;129:95–9.
101. Othman NS, Che Roos NA, Aminuddin A, Murthy JK, Hamid A, A, Ugusman A. Effects of *Piper sarmentosum* Roxb. on hypertension and diabetes mellitus: a systematic review and meta-analysis. Front Pharmacol. 2022;13:976247.
102. Teshika JD, Zakariyyah AM, Zaynab T, Zengin G, Rengasamy KR, Pandian SK, et al. Traditional and modern uses of onion bulb (*Allium cepa* L.): a systematic review. Crit Rev Food Sci Nutr. 2019;59(sup1):S39–70.
103. Singh A, Singh B, Navneet. Role of medicinal plants and their bioactive compounds in obesity, hypertension, and cardiovascular diseases. In: Role of nutrigenomics in modern-day healthcare and drug discovery. Elsevier; 2023. p. 469–515.
104. Bhat M, Singh B, Chowdhary M, Singh S, Surmal O, Bhanwaria R, et al., Bhat MN, et al. Medicinal plants of district Kupwara used in the treatment of human diseases and their associated biological functions. In: Singh B, editor. Botanical leads for drug discovery. Singapore: Springer; 2020. p. 325–47. https://doi.org/10.1007/978-981-15-5917-4_15.
105. Al-Asmari AK, Athar MT, Kadasah SG. An updated phytopharmacological review on medicinal plant of Arab region: *Apium graveolens* Linn. Pharmacogn Rev. 2017;11(21):13–8.
106. Singh A, Singh B. Role of medicinal plants and their bioactive compounds in obesity, hypertension, and cardiovascular diseases. In: Role of nutrigenomics in modern-day healthcare and drug discovery. Elsevier; 2023. p. 469–515.
107. Thiam A, Gueye MT, Sanghare CH, Ndiaye EHB, Diop SM, Cissokho PS, et al. chemical composition and anti-inflammatory activity of *Apium graveolens* var. dulce essential oils from Senegal. Am J Food Technol. 2020;8(6):226–32.
108. Sbai H, Zribi I, DellaGreca M, Haouala R. Bioguided fractionation and isolation of phytotoxic compounds from *Apium graveolens* L. aerial parts (Apiaceae) (2017). Bioguided fractionation and isolation of phytotoxic compounds from Apium graveolens L. aerial parts (Apiaceae). S Afr J Bot. 2017;108:423–30.
109. Msomi N, Simelane M. *Olea europaea* subsp. africana (Oleaceae). In: Msomi NZ, Simelane MB, editors. Olea europaea subsp. africana (Oleaceae). Active ingredients from aromatic and medicinal plants; 2017. p. 159–74.
110. Shaw HM, Wu JL, Wang MS. Antihypertensive effects of *Ocimum gratissimum* extract: Angiotensin-converting enzyme inhibitor *in vitro* and *in vivo* investigation. J Funct Foods. 2017;35:68–73.
111. Interaminense LFL, Leal-Cardoso JH, Magalhães PJC, Duarte GP, Lahlou S. Enhanced hypotensive effects of the essential oil of *Ocimum gratissimum* leaves and its main constituent, eugenol, in DOCA-salt hypertensive conscious rats. Planta Med. 2005;71(4):376–8.
112. Ajayi AM, Martins DTO, Balogun SO, de Oliveira RG, Ascêncio SD, Soares IM, et al. *Ocimum gratissimum* L. leaf flavonoid-rich fraction suppress LPS-induced inflammatory response in RAW

264.7 macrophages and peritonitis in mice. J Ethnopharmacol. 2017;204:169–78.

113. Tom ENL, Demougeot C, Mtopi OB, Dimo T, Djomeni PDD, Bilanda DC, et al. The aqueous extract of *Terminalia superba* (Combretaceae) prevents glucose-induced hypertension in rats. J Ethnopharmacol. 2011;133(2):828–33.
114. Kougnimon E, Akpovi CD, Dah-Nouvlessounon D, Boya B, Baba-Moussa L, Frédéric L. Antioxidant and antibacterial activities of *Terminalia superba* Engl. and Diels (Combretaceae) bark extracts. Int J Curr Microbiol Appl Sci. 2018;7:2836–46.
115. Ahmad SR. Plant-based diet for obesity treatment. Front Nutr. 2022 [cited 2023;9. https://www.frontiersin.org/articles/10.3389/fnut.2022.952553.
116. Kapoor K, Madaan R, Kumar S, Bala R, Walia RP. Role of natural products in the treatment of obesity: nanotechnological perspectives. Curr Drug Metab. 2021;22(6):451–80.
117. Saad B, Ghareeb B, Kmail A. Metabolic and epigenetics action mechanisms of antiobesity medicinal plants and phytochemicals. Evid Based Complement Alternat Med. 2021;2021:9995903.
118. Ai ZL, Zhang X, Ge W, Zhong YB, Wang HY, Zuo ZY, et al. *Salvia miltiorrhiza* extract may exert an anti-obesity effect in rats with high-fat diet-induced obesity by modulating gut microbiome and lipid metabolism. World J Gastroenterol. 2022;28(43):6131–56.
119. Zhong GX, Li P, Zeng LJ, Guan J, Li DQ, Li SP. Chemical characteristics of *Salvia miltiorrhiza* (Danshen) collected from different locations in China. J Agric Food Chem. 2009;57(15):6879–87.
120. Mopuri DR, Islam M. Antidiabetic and anti-obesity activity of *Ficus carica*: *in vitro* experimental studies. Diabetes Metab. 2016;42:300.
121. Surendran S, Asdaq S, Putta P, Nerella M, Boggula N. Anti-obesity screening of figs (*Ficus Carica*) in animals fed on atherogenic and cafeteria diet. J Innov Develop Pharma Tech Sci. 2020; 3(06).
122. Raafat K, Wurglics M. Phytochemical analysis of *Ficus carica* L. active compounds possessing anticonvulsant activity. J Tradit Complement Med. 2019;9(4):263–70.
123. Smith C, Krygsman A. *Hoodia gordonii* extract targets both adipose and muscle tissue to achieve weight loss in rats. J Ethnopharmacol. 2014;155(2):1284–90.
124. Russell PJ, Swindells C. Chemical characterisation of *Hoodia gordonii* extract. Food Chem Toxicol. 2012;50(Suppl 1):S6–13.
125. van Heerden FR. Hoodia gordonii: a natural appetite suppressant. J Ethnopharmacol. 2008;119(3):434–7.
126. Kilany OE, Abdelrazek HMA, Aldayel TS, Abdo S, Mahmoud MMA. Anti-obesity potential of *Moringa olifera* seed extract and lycopene on high fat diet induced obesity in male Sprauge Dawely rats. Saudi J Biol Sci. 2020;27(10):2733–46.
127. Udofia NE, Misonge OJ, Mworia M, William N, Apiri MG. Chemical composition of *Moringaoleifera* Lam.and *Moringastenopetala* Bac. leaves from Kenya. Int J Plant Res. 2020;10(1):1–10.
128. Kumar V, Marković T, Emerald M, Dey A. Herbs: composition and dietary importance. In: Caballero B, Finglas PM, Toldrá F, editors. Encyclopedia of food and health. Oxford: Academic Press; 2016. p. 332–7. [cited 2023 Jan 16]. https://www.sciencedirect.com/science/article/pii/B9780123849472003767.
129. Lee HJ, Kim MJ. Metabolites of oregano (*Origanum vulgare*) seed and their anti-obesity effects on 3T3-L1 adipocytes through down-regulated adipogenesis. Food Sci Biotechnol. 2022;31(7):867–78.
130. Soltani S, Shakeri A, Iranshahi M, Boozari M. A review of the phytochemistry and antimicrobial properties of *Origanum vulgare* L. and subspecies. Iran J Pharm Res. 2021;20(2):268.
131. Handa SS, Khanuja S, Longo G, Rakesh DD. An overview of extraction techniques for medicinal and aromatic plants. In: Extraction technologies for medicinal and aromatic plants; 2008. p. 21–52.
132. Patra A, Abdullah S, Pradhan R. Review on the extraction of bioactive compounds and characterization of fruit industry by-products. Bioresour Bioprocess. 2022;9:14
133. Varma N. Phytoconstituents and their mode of extractions: an overview. Res J Chem Environ Sci. 2016;4(2):8–15.
134. Patel K, Panchal N, Ingle DP. Review of extraction techniques. Int J Adv Res Chem Sci. 2019;6(3):6–21.
135. Tzanova M, Atanasov V, Yaneva Z, Ivanova D, Dinev T. Selectivity of current extraction techniques for flavonoids from plant materials. Processes. 2020;8(10):1222.
136. Zia S, Khan MR, Shabbir MA, Aslam Maan A, Khan MKI, Nadeem M, et al. An inclusive overview of advanced thermal and nonthermal extraction techniques for bioactive compounds in food and food-related matrices. Food Rev Int. 2022;38(6):1166–96.
137. Chemat F, Rombaut N, Sicaire AG, Meullemiestre A, Fabiano-Tixier AS, Abert-Vian M. Ultrasound assisted extraction of food and natural products. Mechanisms, techniques, combinations, protocols and applications. A review. Ultrason Sonochem. 2017;34:540–60.
138. Nipornram S, Tochampa W, Rattanatraiwong P, Singanusong R. Optimization of low power ultrasound-assisted extraction of phenolic compounds from mandarin (*Citrus reticulata* Blanco cv. Sainampueng) peel. Food Chem. 2018;241:338–45.
139. Soria AC, Villamiel M. Effect of ultrasound on the technological properties and bioactivity of food: a review. Trends Food Sci Technol. 2010;21(7):323–31.
140. de Andrade LM, Kestekoglou I, Charalampopoulos D, Chatzifragkou A. Supercritical fluid extraction of carotenoids from vegetable waste matrices. Molecules. 2019;24(3):466.
141. Vongsak B, Sithisarn P, Mangmool S, Thongpraditchote S, Wongkrajang Y, Gritsanapan W. Maximizing total phenolics, total flavonoids contents and antioxidant activity of *Moringa oleifera* leaf extract by the appropriate extraction method. Ind Crops Prod. 2013;44:566–71.
142. Azwanida NN. A review on the extraction methods use in medicinal plants, principle, strength and limitation. Med Aromat Plants. 2015;4(196):2167-0412.
143. Selvamuthukumaran M, Shi J. Recent advances in extraction of antioxidants from plant by-products processing industries. Food Qual Saf. 2017;1(1):61–81.
144. Maier T, Göppert A, Kammerer DR, Schieber A, Carle R. Optimization of a process for enzyme-assisted pigment extraction from grape (*Vitis vinifera* L.) pomace. Eur Food Res Technol. 2008;227(1):267–75.
145. Chávez-González ML, Sepúlveda L, Verma DK, Luna-García HA, Rodríguez-Durán LV, Ilina A, et al. Conventional and emerging extraction processes of flavonoids. Processes. 2020;8(4):434.
146. Nortjie E, Basitere M, Moyo D, Nyamukamba P. Extraction methods, quantitative and qualitative phytochemical screening of medicinal plants for antimicrobial textiles: a review. Plants (Basel). 2022;11(15):2011.
147. Altemimi A, Lakhssassi N, Baharlouei A, Watson DG, Lightfoot DA. Phytochemicals: extraction, isolation, and identification of bioactive compounds from plant extracts. Plants (Basel). 2017;6(4):42.
148. Jain P, Joshi H. Coumarin: chemical and pharmacological profile. J Appl Pharm Sci. 2012;2:236–40.
149. Bourgaud F, Poutaraud A, Guckert A. Extraction of coumarins from plant material (Leguminosae). Phytochem Anal. 1994;5(3):127–32.
150. Tušek A, Samec D, Šalić A. Modern techniques for flavonoid extraction-to optimize or not to optimize? Appl Sci. 2022;12:11865.
151. Naima R, Hannache H, Oumam MM, Sesbou A, Charrier B, Pizzi AP, et al. Green extraction process of tannins obtained from Moroccan *Acacia mollissima* barks by microwave: modeling and

optimization of the process using the response surface methodology RSM. Arab J Chem. 2015;73.
152. Thada R, Chockalingam S, Dhandapani RK, Panchamoorthy R. Extraction and quantitation of coumarin from cinnamon and its effect on enzymatic browning in fresh apple juice: a bioinformatics approach to illuminate its antibrowning activity. J Agric Food Chem. 2013;61(22):5385–90.
153. Li Z, Li Q. Ultrasonic-assisted efficient extraction of coumarins from *Peucedanum decursivum* (Miq.) Maxim using deep eutectic solvents combined with an enzyme pretreatment. Molecules. 2022;27(17):5715.
154. Xuan Cuong D, Xuan Hoan N, Huu Dong D, Thi Minh Thuy L, Van Thanh N, Thai Ha H, et al. Tannins: extraction from Plants. Tannins—structural properties, biological properties and current knowledge. 2020. https://doi.org/10.5772/intechopen.86040.
155. Das AK, Islam MN, Faruk MO, Ashaduzzaman M, Dungani R. Review on tannins: extraction processes, applications and possibilities. S Afr J Bot. 2020;135:58–70.
156. Elgailani IE, Ishak C. Methods for extraction and characterization of tannins from some *Acacia Species* of Sudan. Pak J Anal Environ. 2016;17(1):43–9.
157. Zhang R, Wang R, Zhao S, Chen D, Hao F, Wang B, et al. Extraction, separation, antitumor effect, and mechanism of alkaloids in *Sophora alopecuroides*: a review. Separations. 2022;9(11):380.
158. Yubin JI, Miao Y, Bing W, Yao Z. The extraction, separation and purification of alkaloids in the natural medicine. J Chem Pharm Res. 2014;6(1):338–45.
159. Zhang W, Zhu D, Fan H, Liu X, Wan Q, Wu X, et al. Simultaneous extraction and purification of alkaloids from *Sophora flavescens* Ait. by microwave-assisted aqueous two-phase extraction with ethanol/ammonia sulfate system. Sep Purif Technol. 2015;141:113–23.
160. Li WX, Wang H, Dong AW. Systematic separation and purification of alkaloids from *Euchresta tubulosa* Dunn. by various chromatographic methods. Processes. 2019;7(12):924.
161. Bowman JM, Braxton MS, Churchill MA, Hellie JD, Starrett SJ, Causby GY, et al. Extraction method for the isolation of terpenes from plant tissue and subsequent determination by gas chromatography. Microchem J. 1997;56(1):10–8.
162. Chen L, Pang Y, Luo Y, Cheng X, Lv B, Li C. Separation and purification of plant terpenoids from biotransformation. Eng Life Sci. 2021;21(11):724–38.
163. Majinda RRT. Extraction and isolation of saponins. Methods Mol Biol. 2012;864:415–26.
164. El Aziz MMA, Ashour AS, Melad AG. A review on saponins from medicinal plants: chemistry, isolation, and determination. J Nanomed Res. 2019;8(1):282–8.
165. Cheok CY, Salman HAK, Sulaiman R. Extraction and quantification of saponins: a review. Int Food Res J. 2014;59:16–40.
166. Morsy N. Cardiac glycosides in medicinal plants. Aromatic and medicinal plants–back to nature. London: Intechopen; 2017. p. 29–45.
167. Jonas H, De Planas GM. Extraction and chromatographic purification of *Digitalis Cardiac* glycosides and their binding to plant pigments. Prep Biochem. 1974;4(5):411–34.
168. Makin H, Honour J, Shackleton C, Griffiths W. General methods for the extraction, purification, and measurement of steroids by chromatography and mass spectrometry. In: Steroid analysis; 2010. p. 163–282.
169. Patel SS, Savjani JK. Systematic review of plant steroids as potential antiinflammatory agents: current status and future perspectives. J Phytopharmacol. 2015;4(2):121–5.
170. Manley A, Collins A, Joynes A, Mellander PE, Jordan P. Comparing extraction methods for biomarker steroid characterisation from soil and slurry. Water Air Soil Pollut. 2020;231:1–5.
171. Wang N, Shi X, Zhang C, Zhou W, Zhu Z. Extraction and quantification of sphingolipids from hemiptera insects by ultra-performance liquid chromatography coupled to tandem mass spectrometry. Bio Protoc. 2021;11(4):e3923.
172. Wang G, Jia XJ, Song BB, Li R, Liu XF, Chen JP, et al. Extraction optimization, UHPLC-Triple-TOF-MS/MS analysis and antioxidant activity of ceramides from sea red rice bran. Foods. 2022;11(10):1399.
173. Nocedo-Mena D, Rivas-Galindo VM, Navarro P, Garza-González E, González-Maya L, Ríos MY, et al. Antibacterial and cytotoxic activities of new sphingolipids and other constituents isolated from *Cissus incisa* leaves. Heliyon. 2020;6(8):e04671.
174. Seeda A, Abou El-Nour EZ, Mervat G, Zaghloul S. Importance of sulfur and its roles in plants physiology: a review. Curr Sci Int. 2020;9:198–231.
175. Abdalla M, Mühling K. Plant-derived sulfur containing natural products produced as a response to biotic and abiotic stresses: a review of their structural diversity and medicinal importance. J Appl Bot Food Qual. 2019;92:204–15.
176. Mondal S, Pramanik K, Panda D, Dutta D, Karmakar S, Bose B. Sulfur in seeds: an overview. Plants. 2022;11:450.
177. Marcinkowska MA, Jeleń HH. Role of sulfur compounds in vegetable and mushroom aroma. Molecules. 2022;27(18):6116.
178. Rho T, Choi MS, Jung M, Kil HW, Hong YD, Yoon KD. Identification of fermented tea (*Camellia sinensis*) polyphenols and their inhibitory activities against amyloid-beta aggregation. Phytochemistry. 2019;160:11–8.
179. Calani L, Del Rio D, Luisa Callegari M, Morelli L, Brighenti F. Updated bioavailability and 48 h excretion profile of flavan-3-ols from green tea in humans. Int J Food Sci Nutr. 2012;63(5):513–21.
180. Luo H, Wu H, Yu X, Zhang X, Lu Y, Fan J, et al. A review of the phytochemistry and pharmacological activities of *Magnoliae officinalis* cortex. J Ethnopharmacol. 2019;236:412–42.

Chemical Characterization of Phytoconstituents for Lifestyle Diseases

Yvan Anderson T. Ngandjui, Joseph Tchamgoue, Donald Ulrich K. Kagho, Titus Alfred M. Msagati, Bathelemy Ngameni, and Simeon F. Kouam

Abstract

Medicinal plants have been largely used for the treatment of several illnesses including lifestyle diseases. The observed pharmacological properties of these plants are mainly attributed to the presence of secondary metabolites. Several of these secondary metabolites have been isolated, characterized, and assessed for their biological potential against lifestyle diseases. Reports on the chemical investigation of some medicinal plants involved in the management of lifestyle diseases such as diabetes, respiratory chronic diseases, cancer, and cardiovascular diseases have revealed that the main classes of compounds involved are phenolic compounds, terpenoids, steroids, alkaloids, etc. This chapter provides an overview of these phytoconstituents frequently involved in the management of lifestyle diseases and the various chemical techniques used for their characterization.

Y. A. T. Ngandjui (✉)
Department of Chemistry, Higher Teacher Training College, University of Yaoundé I, Yaoundé, Cameroon

Institute for Nanotechnology and Water Sustainability, College of Science, Engineering and Technology, University of South Africa, Florida Science Campus, Johannesburg, South Africa

Department of Organic Chemistry, Faculty of Science, University of Yaoundé I, Yaoundé, Cameroon

J. Tchamgoue
Department of Chemistry, Higher Teacher Training College, University of Yaoundé I, Yaoundé, Cameroon

Department of Organic Chemistry, Faculty of Science, University of Yaoundé I, Yaoundé, Cameroon

D. U. K. Kagho · S. F. Kouam
Department of Chemistry, Higher Teacher Training College, University of Yaoundé I, Yaoundé, Cameroon

T. A. M. Msagati
Institute for Nanotechnology and Water Sustainability, College of Science, Engineering and Technology, University of South Africa, Florida Science Campus, Johannesburg, South Africa

B. Ngameni
Department of Pharmacognosy and Pharmaceutical Chemistry, Faculty of Medicine and Biomedical Sciences, University of Yaoundé 1, Yaoundé, Cameroon

Keywords

Lifestyle diseases · Medicinal plants · Phytoconstituents · Characterization techniques

1 Introduction

Lifestyle diseases are diseases that are not spread directly from one person to another. There are well-known as non-communicable diseases (NCDs) and are caused by the way people live. Lifestyle diseases regroup mainly and increasingly diabetes, chronic respiratory diseases, cancers, and cardiovascular illness. Worldwide, NCDs are one of the important sources of death [1]. As said by WHO, NCD is responsible to 41 million deaths every year, which is corresponding to 74% of mortality globally. Among all NCD deaths, countries with low- and middle-income are most concerned by about 77% deaths. Diabetes accounts for less NCD deaths (2.0 million of deaths annually), followed by chronic respiratory diseases (4.1 million of deaths), cancers (9.3 million of deaths), and cardiovascular diseases (about 17.9 million of deaths) [1]. Factors such as tobacco, physical inactivity, unhealthy diets, and abusive use of alcohol increase the risk of either developing or dying from NCD which is linked to lifestyle choices.

Unfortunately, available medications usually used to fight against NCD are very expensive and have side effects [2, 3]. Consequently, the search for less toxic and cost-effective drugs is required [4]. In this regard, phytoconstituents from plants have been considered as an appreciated reservoir of lead compounds for drug development. The medicinal plants' potential is known since centuries and possess advantage to have minor secondary effects and low toxicity. Medicinal plants have acquired wide medical acceptance due to a more prominent information of how they make strides well-being and quality of life [5]. Many plants were reported to exhibit good therapeutics effects in the management of lifestyle diseases [6, 7]. In this light, phytochemicals from

A. K. Dhara, S. C. Mandal (eds.), *Role of Herbal Medicines*, https://doi.org/10.1007/978-981-99-7703-1_4

plants could be an alternative and hopeful approach to prevent and treat lifestyle diseases [8]. This chapter delivers an overview of phytoconstituents frequently involved in the management of lifestyle diseases and the various chemical techniques used for their characterization.

2 Phytoconstituents for the Management of Lifestyle Diseases

Medicinal plants have been largely used for the treatment of numerous illnesses including lifestyle diseases. Research into the chemical composition of these plants has shown that secondary metabolites are the main sources of the observed medicinal properties of plants. Reports on the chemical investigation of some medicinal plants involved in the management of lifestyle diseases such as diabetes, respiratory chronic diseases, cancer, and cardiovascular diseases have revealed that the main groups of phytoconstituents involved are phenolic compounds, terpenoids, steroids, alkaloids, etc.

2.1 Diabetes

Diabetes also called blood sugar is a chronic illness that arises when your blood glucose is too high. Diabetes is a multi-cause metabolic disease induced by hyperglycemia resulting from defects in insulin action, insulin secretion, or both. Insulin is a hormone formed in the pancreas that lets glucose from food to go in the body's cells where it is transformed into energy required by muscles and tissues to function. The patient suffering from diabetes does not absorb glucose correctly, and glucose remains flowing in the blood which is destructive for body tissues over time. This damage can lead to disabling and life-threatening health complications [9]. Medicinal plants and phytoconstituents have always had a crucial part in the management of diabetes. Among phytochemical groups, alkaloids, phenolic compounds, terpenoids, and steroids were reported to display antidiabetic properties [10–12].

2.1.1 Alkaloids

Alkaloids are frequently found in the plant kingdom; they are a group of naturally occurring organic nitrogen-containing compounds. Many alkaloids were obtained from numerous medicinal plants and are sorted into three main bunches counting true, pseudo-, and proto-alkaloids. For instance, berberine (**1**) [13], tecomine (**2**) [14], boschniakine (**3**), 5β-hydroxyskitanthine (**4**), casuarine 6-*O*-α-glucoside (**5**) [15], catharantine (**6**), vindoline (**7**), vindolinine (**8**) [16], cryptolepine (**9**) [17], harmane (**10**), jatrorrhizine (**11**), magnoflorine (**12**), palmatine (**13**) [18, 19], javaberine A (**14**), javaberine B (**15**) [20, 21], lepidine (**16**), semilepidine (**17**) [22], mahanimbine (**18**) [23], piperumbellactam A (**19**), piperumbellactam B (**20**), piperumbellactam C (**21**) [24], radicamines A (**22**) and B (**23**) [25], tecomine (**24**), and tecostanine (**25**) [26, 27] obtained from different plants were described to exhibit antidiabetic activities. The structures of some alkaloids which displayed significant antidiabetic properties are shown in Fig. 1.

2.1.2 Phenolic Compounds

They are molecules that are characterized by their structures having at least one phenol unit. Phenols are also a huge family of 10,000 plant compounds classified into several subclasses including flavonoids, catechins, anthocyanins, isoflavones, chalcones, curcuminoids, and phenolic acids [28–31]. Some phenolic compounds (Fig. 2) were reported to exhibit antidiabetic properties, viz. cyanidin-3-galactoside (**26**) [32], (-) epicatechin (**27**) [33], epigallocatechin-3-gallate (**28**) [34], (-)-3-*O*-galloylepicatechin (**29**), (-)-3-*O*-galloylcatechin (**30**) [35], hesperidin (**31**), naringin (**32**) [36], dorsamin F (**33**), 6-prenyl-3′-methoxyeriodictyol (**34**) [37], naringenin 7-O-β-D-glucoside (**35**) [38], kaempferitrin (**36**), kaempferol 3-O-β-D-(2-O-β-D-glucopyranosyl) galactopyranoside (**37**), kaempferol 3-O-β-D-glucopyranosyl(1 → 2)-O-[α-L-rhamno-pyranosyl(1 → 6)]-β-D-glucopyranoside (**38**), kaempferol 3-O-β-D-glucopyranosyl(1 → 2)-O-[α-L-rhamnopyranosyl(1 → 6)]-β-D-galactopyranoside (**39**), kaempferol 3-O-β-D-(2,6-di-O-α-L-rhamnopyranosyl) galactopyranoside (**40**) [39], marsupsin (**41**), and pterostilbene (**42**) [40], vitexin (**43**), isovitexin (**44**), isorhamnetin 3-O-β-D-rutinoside (**45**) [41], ferulic acid (**46**) [42], and bellidifolin (**47**) [43].

2.1.3 Terpenoids and Steroids

Terpenoids are bioactive compounds present naturally in several plants. They are precursors of steroids in both animals and plants, and are known to display potent antidiabetic activities [44]. Indeed, α-amyrin acetate (**48**) [45, 46], andrographolide (**49**) [47, 48], forskolin (**50**) [49], 3β-acetoxy-16β-hydroxybetulinic acid (**51**) [50], bassic acid (**52**) [51], corosolic acid (**53**) [52–54], elatosides E (**54**), G (**55**), H (**56**), and I (**57**) [55], escins-Ia, Ib, IIa, and IIb (**58–61**) [56], lactucain A (**62**), lactucaside (**63**) [57], pycnanthuquinone A (**64**), pycnanthuquinone B (**65**) [58], betavulgarosides II (**66**), III (**67**), and IV (**68**) [59] were reported to display antidiabetic activity (Fig. 3).

2.1.4 Miscellaneous

There are also secondary metabolites that do not belong to any of the aforementioned classes that have shown promising antidiabetic effects and that are worth to be listed, namely allicin (**69**) [60], chebulagic acid (**70**) [61], 4-hydroxyisoleucine (**71**) [62] (Fig. 4).

Fig. 1 Some alkaloids with potent antidiabetic property

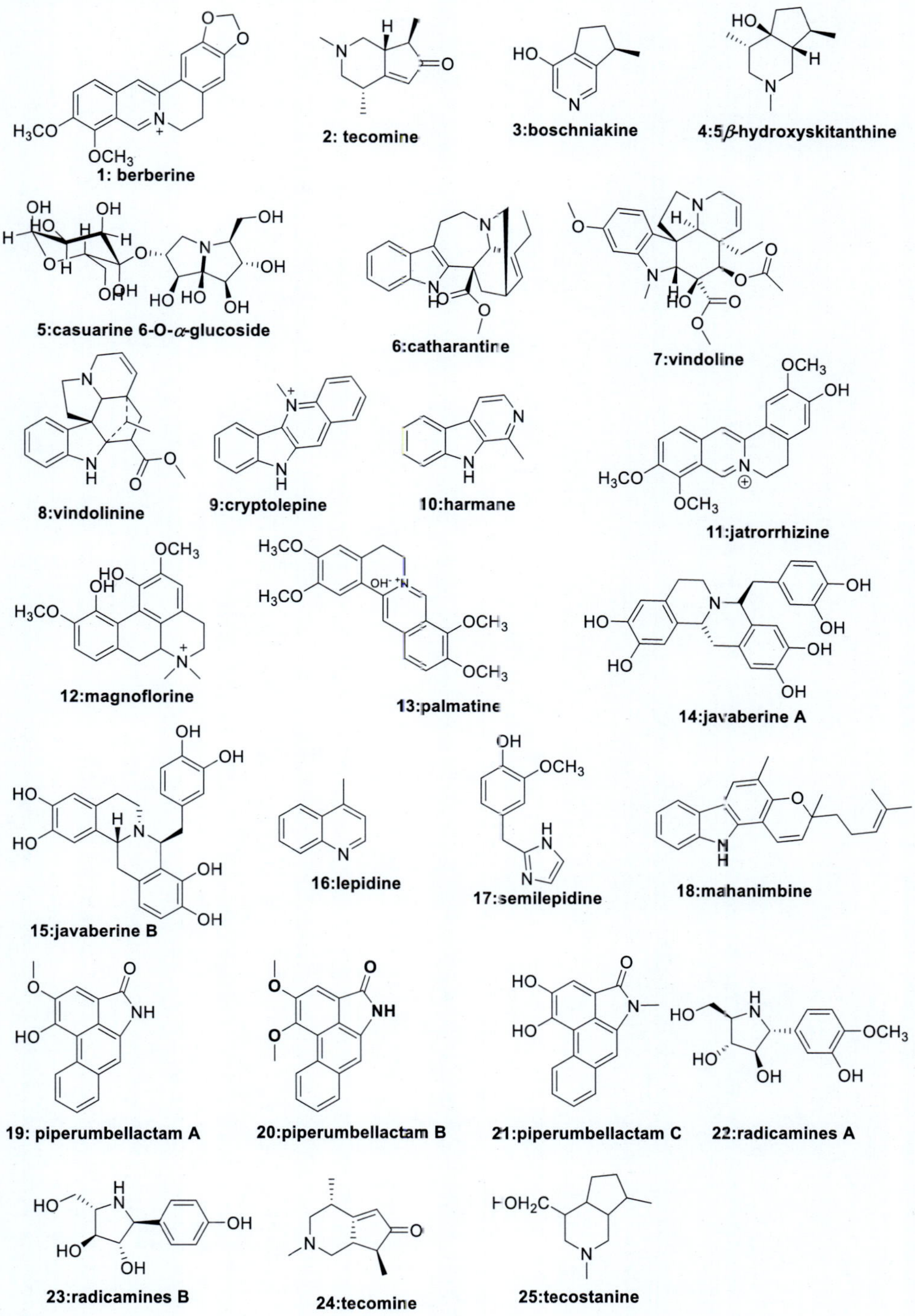

2.2 Cardiovascular Diseases

According to World Health Organization, "Cardiovascular diseases" is an expression generally used to describe disorders that affect the heart and blood vessels. CVDs comprise coronary artery diseases (CADs) such as myocardial infarction and angina. Other CVDs include aortic aneurysms, abnormal heart rhythms, congenital heart disease, carditis, cardiomyopathy, hypertensive heart disease, heart failure, peripheral artery disease, rheumatic heart disease, stroke, thromboembolic disease, venous thrombosis, and valvular heart disease. These diseases can touch one or many parts of your blood vessels and/or heart. A person may be asymptomatic (not feeling anything at all) or symptomatic (physically experiencing the disease). You may make lifestyle changes to manage cardiovascular diseases generally with the advices of your healthcare supplier because the sooner you detect cardiovascular diseases, the easier it is to treat [63]. Regarding the treatment of cardiovascular diseases, phytochemicals have long shown

Fig. 2 Some phenolic compounds with antidiabetic property

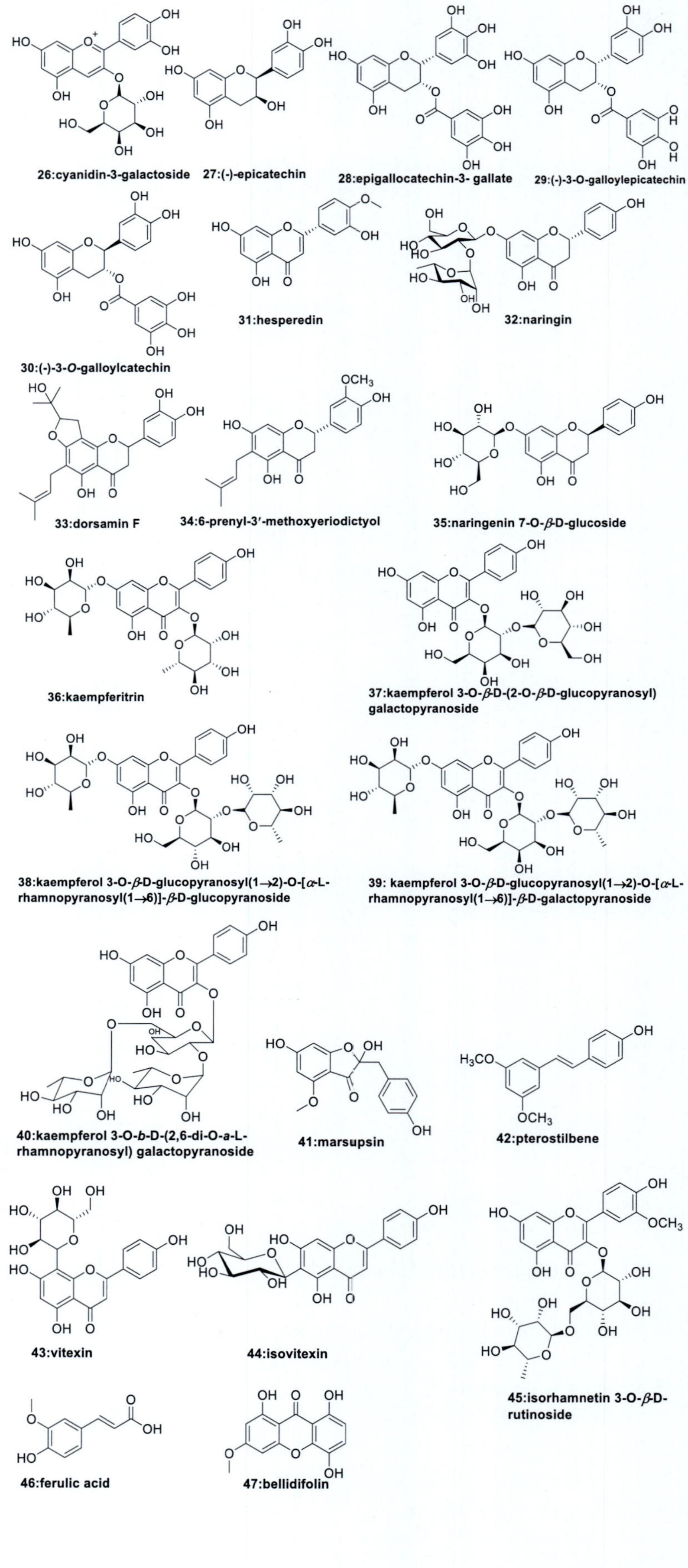

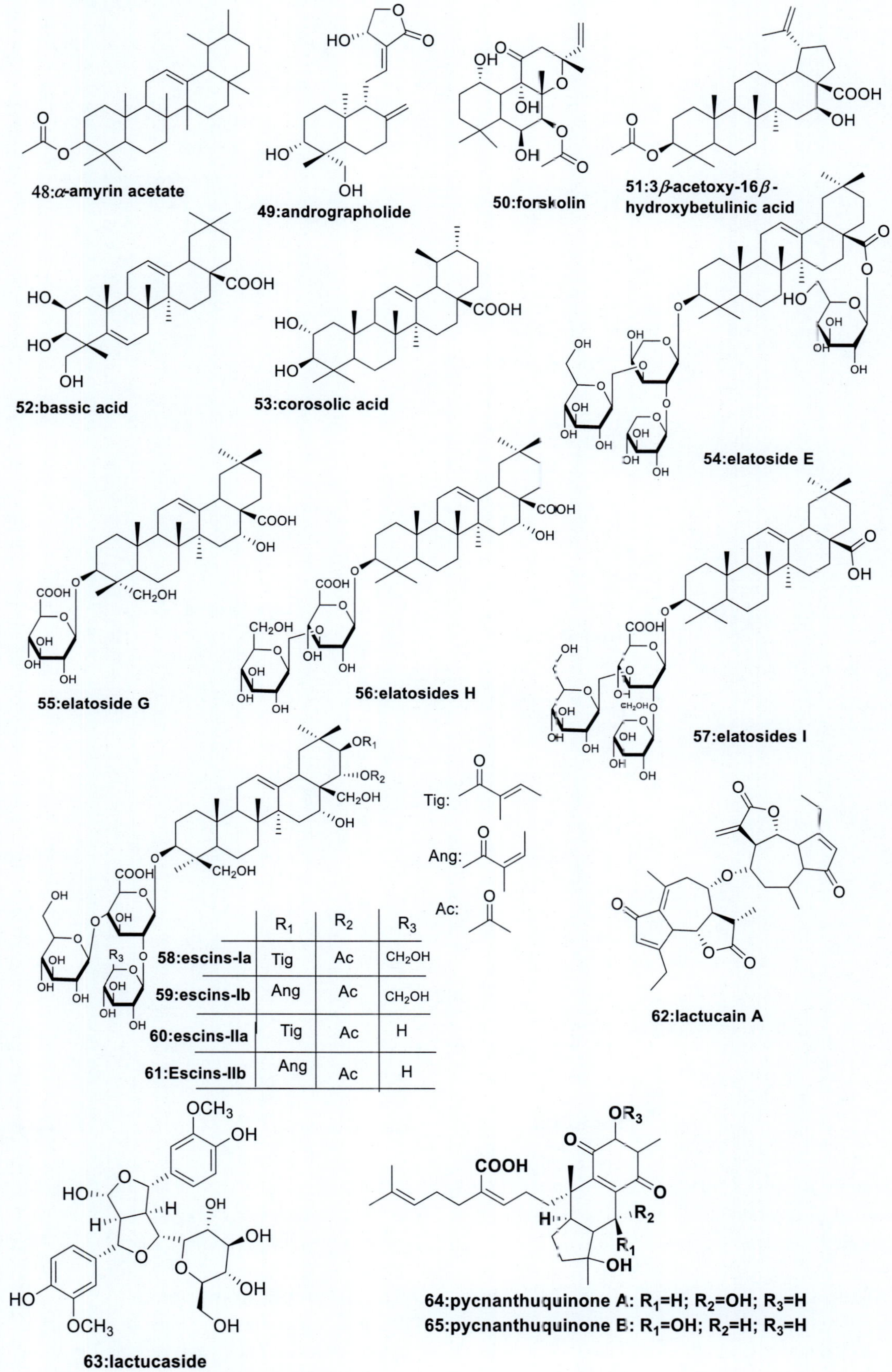

	R_1	R_2	R_3
58:escins-Ia	Tig	Ac	CH_2OH
59:escins-Ib	Ang	Ac	CH_2OH
60:escins-IIa	Tig	Ac	H
61:Escins-IIb	Ang	Ac	H

Fig. 3 Some terpenoids with antidiabetic property

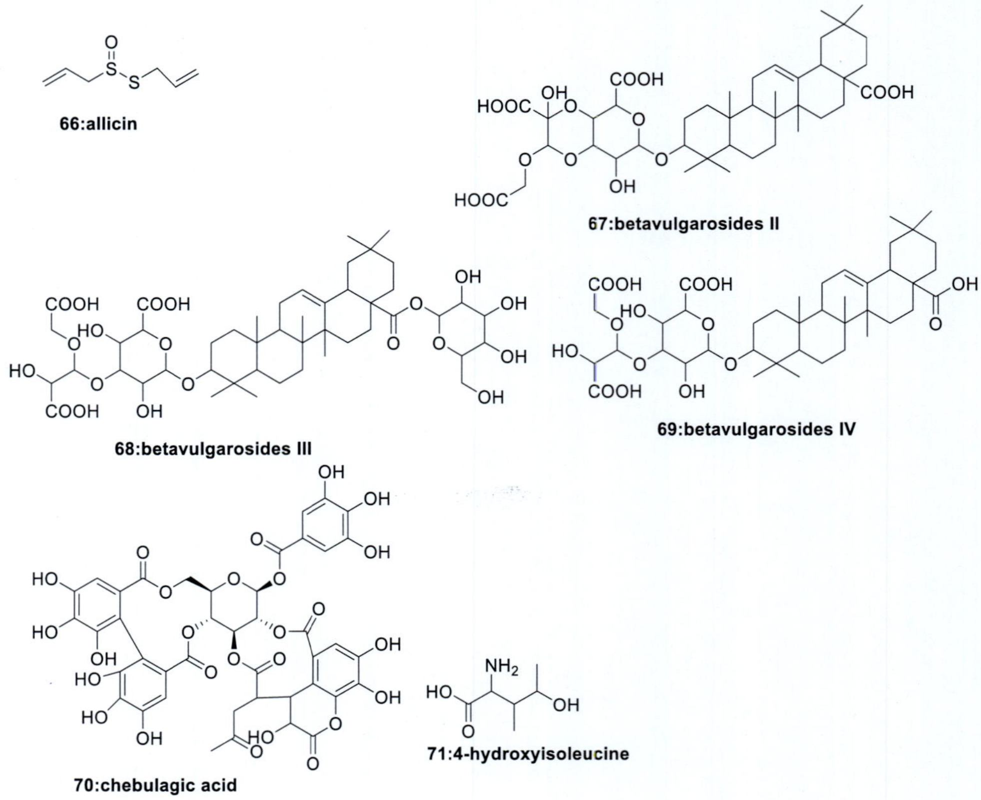

Fig. 4 Other compounds with potent antidiabetic property

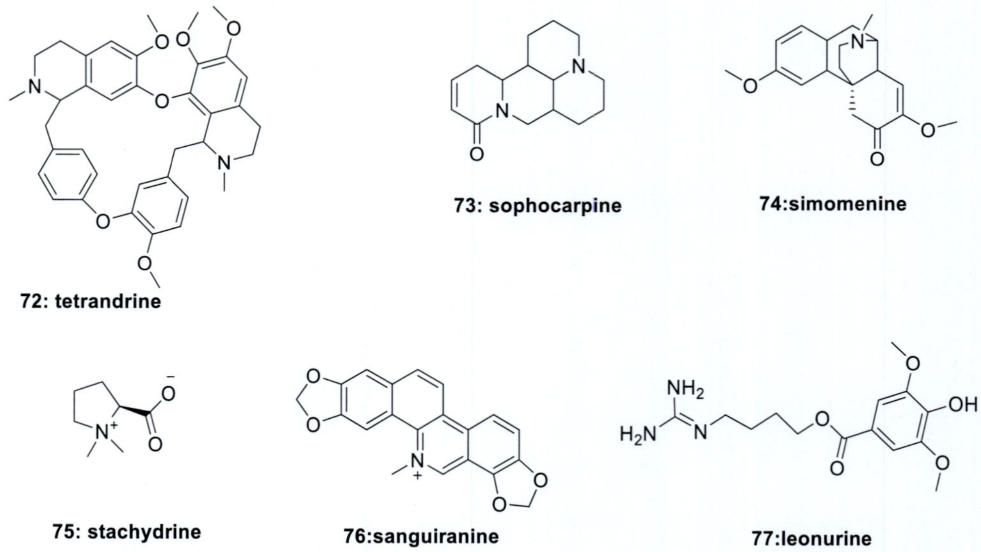

Fig. 5 Some alkaloids with anticardiovascular property

their efficiency against CVDs. These compounds have been classified in several classes including alkaloids, phenolic compounds, terpenoids, and steroids.

2.2.1 Alkaloids

Among the groups of secondary metabolites, we have alkaloids, found in nature and which have in their structure basic nitrogen atoms. Alkaloids can be found in the Acanthaceae, Apocynaceae, Papaveraceae, and Solanaceae families. Some alkaloids were shown to be effective against cardiovascular diseases or to have cardioprotective effect [64] such as tetrandrine (**72**) [65], berberine (**1**) [66, 67], sophocarpine (**73**) [68, 69], sinomenine (**74**) [70], stachydrine (**75**) [71–73], sanguiranine (**76**) [74], leonurine (**77**) [75–77] (Fig. 5).

2.2.2 Phenolic Compounds

Phenolic compounds are one of the largest clusters of phytochemicals. They were reported to show good activities against various diseases including cardiovascular diseases. Some phenolic compounds belonging to phenolic acids, stilbenes, flavonoids, and guaiacols were reported to exhibit good anticardiovascular activity including cur-

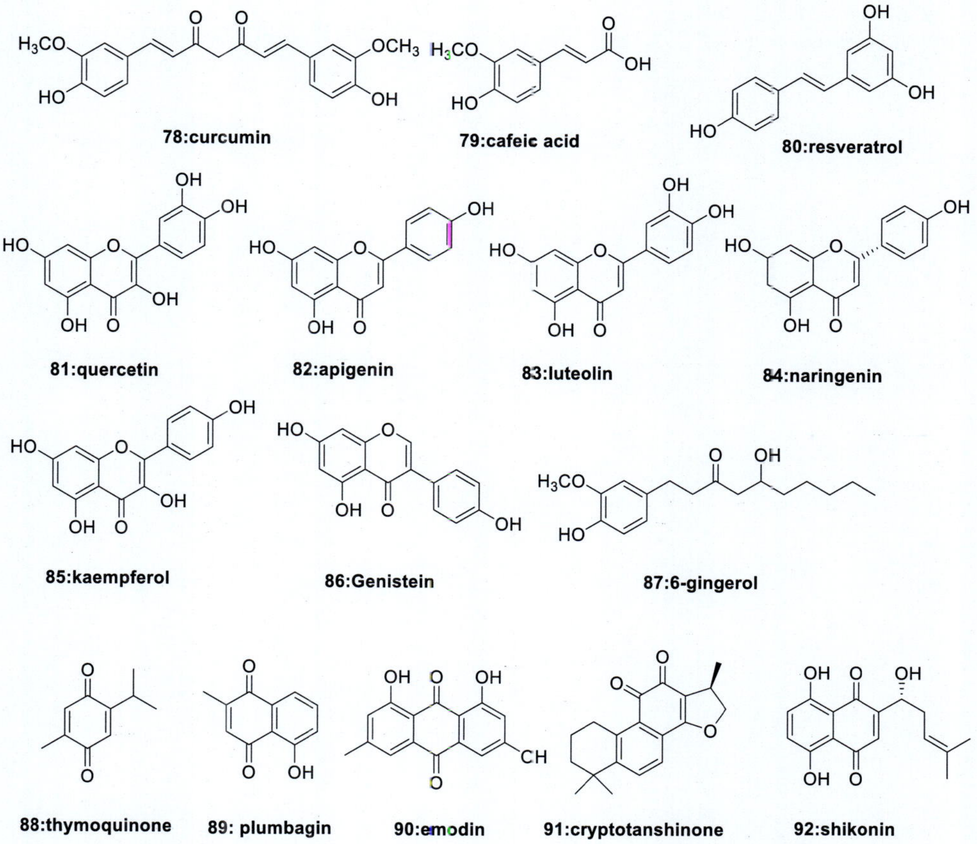

Fig. 6 Some phenolic compounds with anticardiovascular property

cumin (**78**) [78, 79], caffeic acid (**79**) [80], resveratrol (**80**) [81, 82], quercetin (**81**) [83–87], apigenin (**82**) [88–90], luteolin (**83**) [91–93], naringenin (**84**) [89, 94, 95], epigallocatechin-3-gallate (**28**) [96], kaempferol (**85**) [97], genistein (**86**) [98], 6-gingerol (**87**) [99], thymoquinone (**88**) [100], plumbagin (**89**) [101, 102], emodin (**90**) [103, 104], cryptotanshinone (**91**) [105, 106], and shinkonin (**92**) [107] (Fig. 6).

2.2.3 Terpenoids and Steroids

Terpenoids are sorted into structural types which are monoterpenoids, sesquiterpenoids, diterpenoids, triterpenoids, and carotenoid tetraterpenes [108]. Some phytoterpenoids were reported in the literature to possess cardiovascular properties including daucuside (**93**), daucusol (**94**) [109], triptolide (**95**) [110, 111], artemisinin (**96**) [112, 113], paeoniflorin (**97**) [114, 115], picroside II (**98**) [116], ginsenoside Re (**99**) [117, 118], diosgenin (**100**) [119], ginkgolide C (**101**) [120], sesamin (**102**) [121], germacranolide (**103**), heliangolide (**104**), eudesmane (**105**), α-spinasterol (**106**), β-amyrin (**107**) [122], alisol A (**108**), alisol B 23-acetate (**109**), cucurbitacin E (**110**) [123], rubiarbonol C (**111**) [124], rubiarbonone C (**112**) [125], β-sitosterol glycoside (**113**), asiatic acid (**114**) [126], stigmasterol (**115**), campesterol (**116**) [127, 128], betulinic acid (**117**) [129], boswellic acid (**118**) [130], celastrol (**119**) [131] (Fig. 7).

2.2.4 Miscellaneous

Other natural products not belonging to the above-mentioned classes were also reported compounds were reported to be used for the prevention of the CVDs like glucoraphanin (**120**), allicin, and diallyl disulfide (**121**) [132, 133] (Fig. 8).

2.3 Cancer

It is a large class of illnesses which is characterized by the unregulated and uncontrolled development of the body's cells that generally attack and as a rule destroy normal cells. Cancer can start in almost any tissue or organ of the body when abnormal cells grow uncontrollably, go beyond their habitual limits to occupy adjoining parts of the body and/or spread to other organs. Metastasizing is the latest process of cancer progression and is generally a major cause of death. Compounds isolated from natural sources have regularly had an important part in treating and preventing human illnesses since olden times including cancer [134]. Among natural products, alkaloids, phenolic compounds, and terpenoids were reported to possess antiproliferative activity.

2.3.1 Alkaloids

Alkaloids compounds isolated from plants were established to play a significant role in the suppression of cancer cells

Fig. 7 Some terpenoids with anticardiovascular property

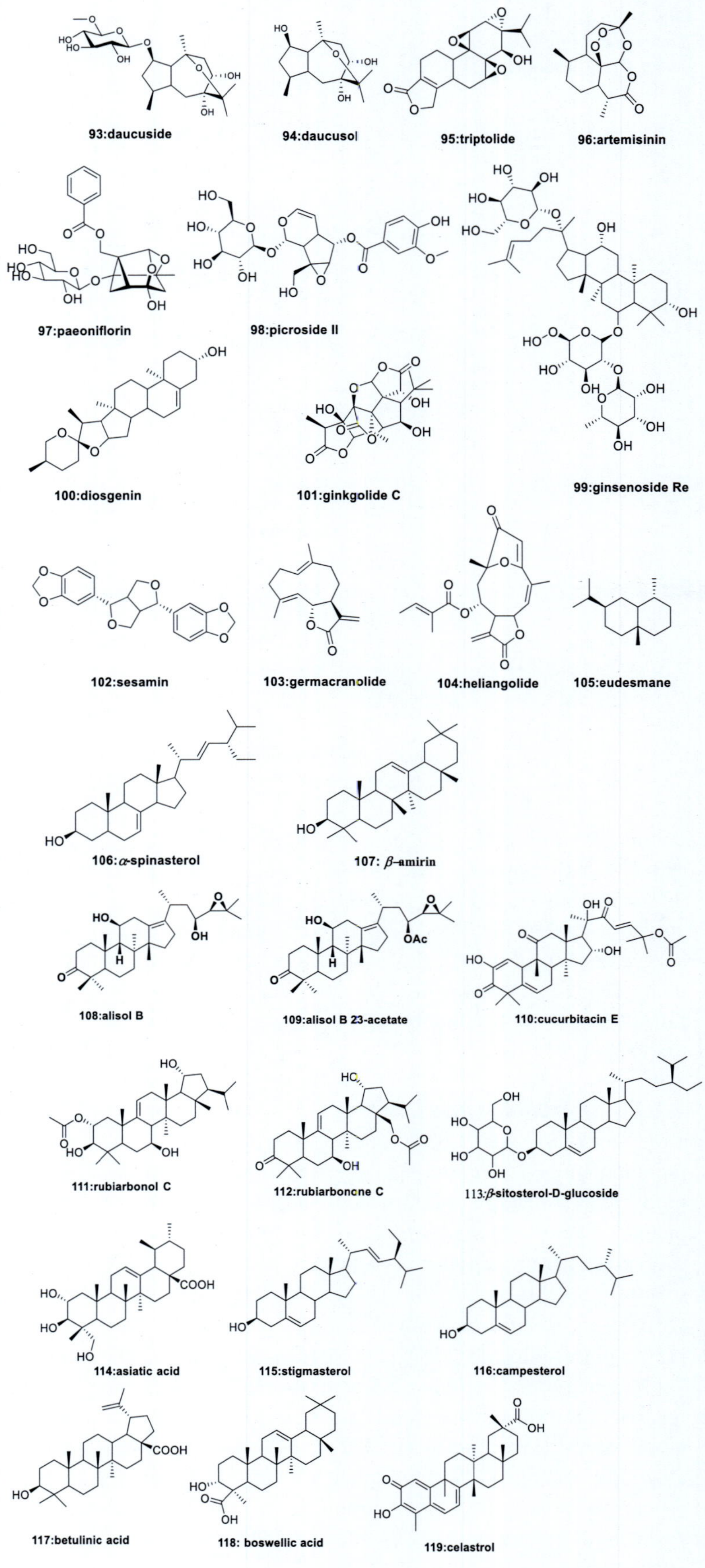

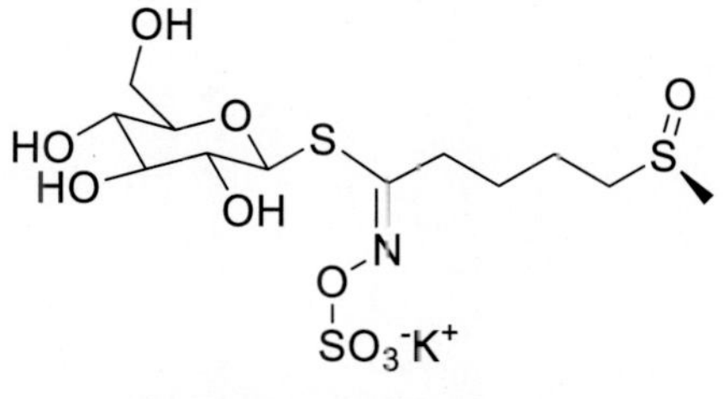

120:glucoraphanine

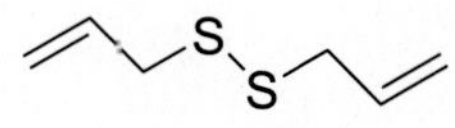

121:diallyl disulfide

Fig. 8 Other compounds with anti-cardiovascular property

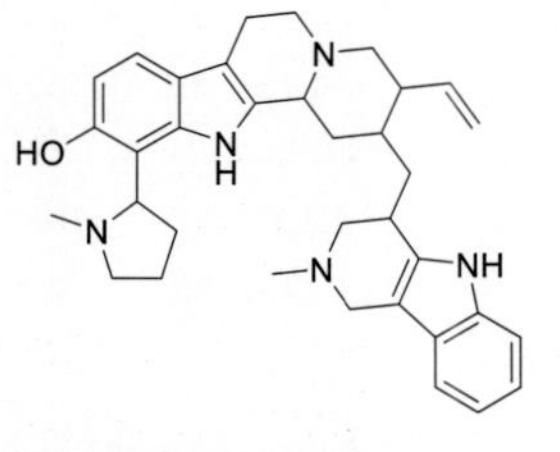

122:isostrychnopentamine

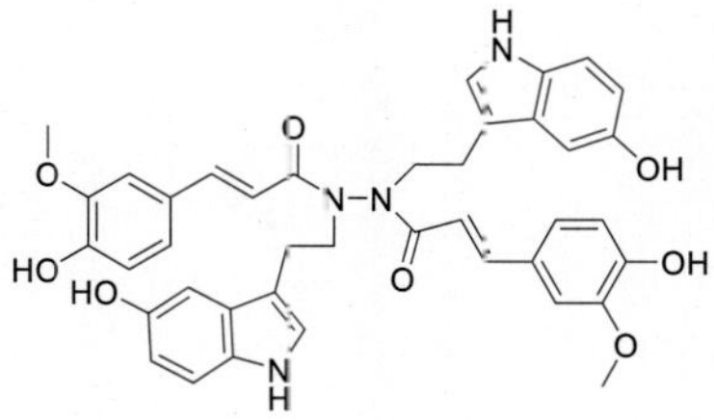

123:montamine

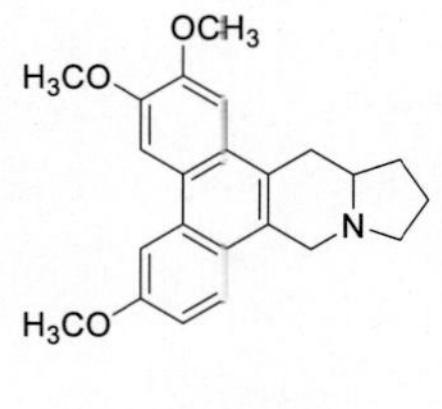

124:antofine

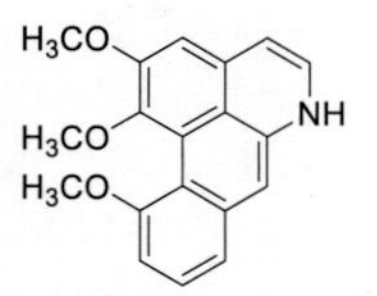

125: gramichunosin

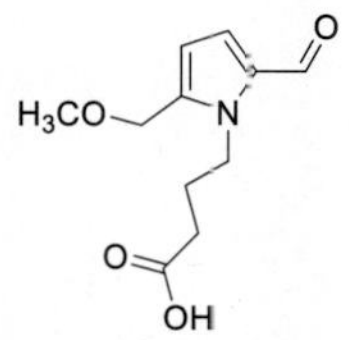

126:4-(2-formyl-5-(methoxymethyl)-1H-pyrrole-1-yl)butanoic acid

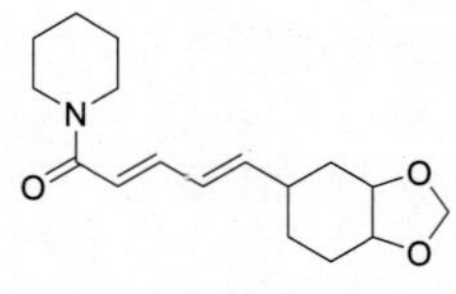

127:piperine

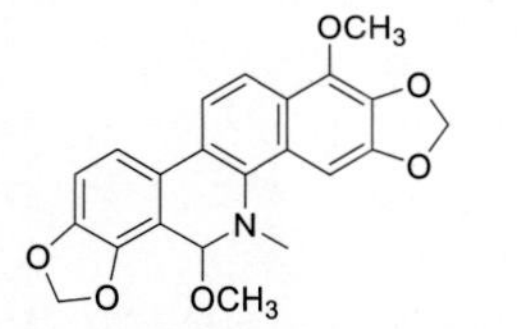

128: 8-methoxydihydrosanguinarine

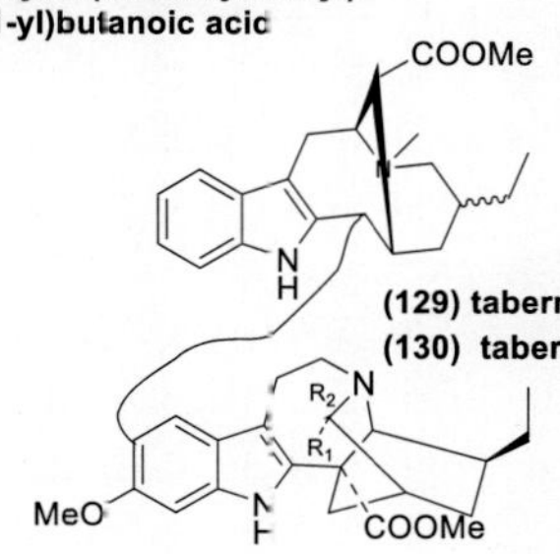

(129) tabernaelegantine B R_1=H, R_2=H, β ethyl
(130) tabernaelegantine D R_1=H, R_2=H, α ethyl

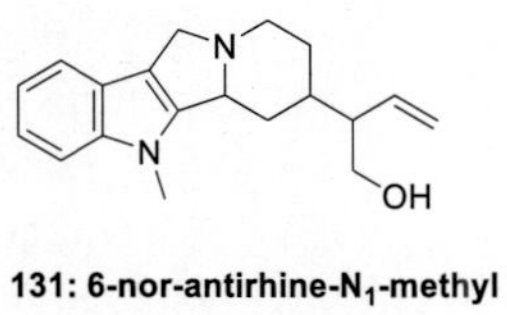

131: 6-nor-antirhine-N_1-methyl

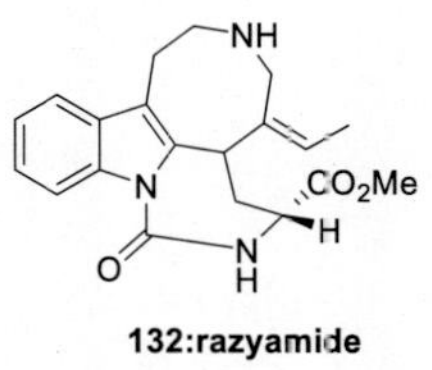

132:razyamide

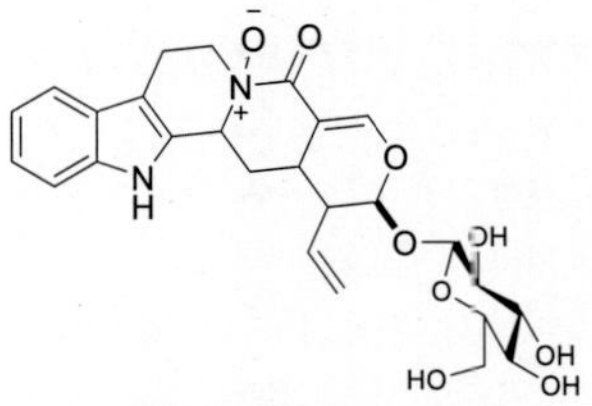

133:vincosamide-N-oxide

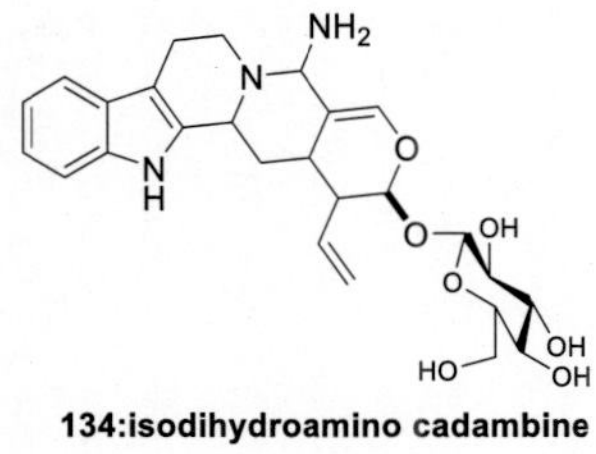

134:isodihydroamino cadambine

Fig. 9 Some alkaloids with anticancer activity

[135, 136]. Some secondary metabolites obtained from plants as isostrychnopentamine (**122**) [137], montamine (**123**) [138], antofine (**124**) [139], gramichunosin (**125**), 4-(2-formyl-5-(methoxymethyl)-1H-pyrrole-1-yl)butanoic acid (**126**) [6], piperine (**127**) [140], jatrorrhizine (**11**), 8-methoxydihydrosanguinarine (**128**) [141], tabernaelegantine B and D (**129–130**) [142], 6-*nor*-antirhine-N_1-methyl (**131**), razyamide (**132**) [143], vincosamide-*N*-oxide (**133**), and isodihydroamino cadambine (**134**) [144] were reported to exhibit anticancer activity (Fig. 9).

Fig. 10 Some phenolic compounds with anticancer activity

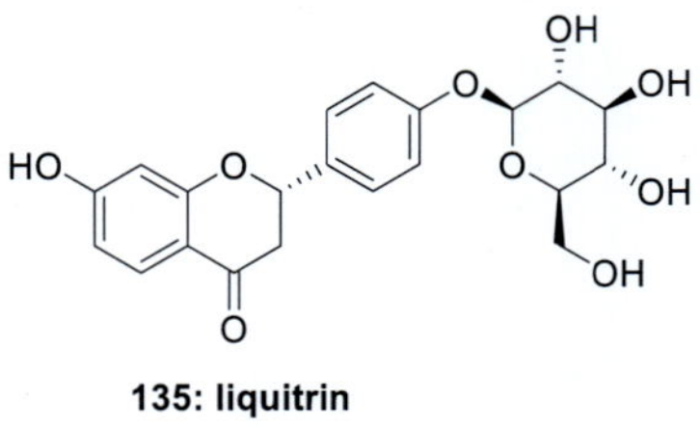

135: liquitrin

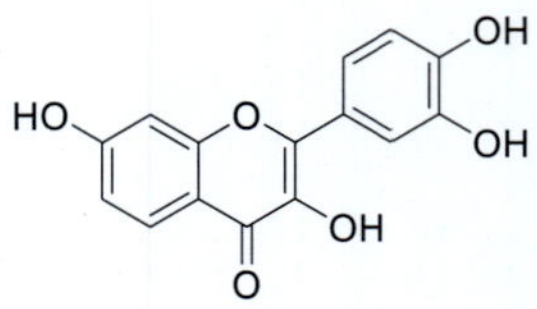

136:fisetin

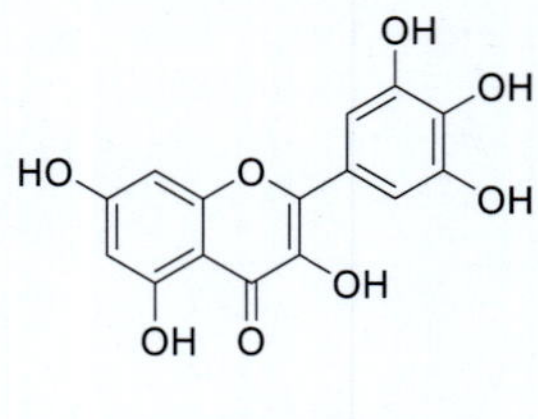

137:myricetin

2.3.2 Phenolic Compounds

Phenolic compounds are mostly found in fruits, red wine, coffee, cocoa, and seeds [30]. Some phenolic compounds were reported to display anticancer effects counting resveratrol (**80**) [145], curcumin (**78**) [146], quercetin (**81**) [52], apigenin (**82**), epigallocathechin-3-gallate (**28**) [147], liquiritin (**135**) [148], luteolin (**112**), fisetin (**136**), and myricetin (**137**) [149–151] (Fig. 10).

2.3.3 Terpenoids and Steroids

These classes of compounds account for the main class of secondary metabolites made by plants and were broadly considered as active agents counting anticancer agent [152]. Terpenoids which displayed anti-cancer activities are alisol A (**108**), alisol B 23-acetate (**109**) [153], euscaphic acids G (**138**), hederagenin (**139**), triregeloic acid (**140**) [154], hydroxylindestrenolide (**141**), lindestrenolide (**142**) [155], 3-epicyclomusalenol (**143**), cyclosadol (**144**) [156, 157], cauloside A (**145**), α-hederin (**146**), sapindoside B (**147**) [158], isoartemisolide (**148**) [159], α-cadinol (**149**) [160], linderolide G (**150**) [155], syreiteate A (**151**), syreiteate B (**152**) [161], 10-*O*-acetylmacrophyllide (**153**) [162], β-pinene (**154**) [163], withaferin A (**155**), and celastrol (**119**) [164] (Fig. 11).

2.4 Chronic Respiratory Diseases

Chronic respiratory diseases (CRDs) are chronic illnesses of the airways and other parts of the lung such as asthma, chronic and obstructive pulmonary diseases, cystic fibrosis, occupational lung diseases, pneumoconiosis, pulmonary sarcoidosis, and sleep apnea. CRDs affect seniors, adults, teens, and all ages-children, and pose major health care and economic strain across the world [165]. There are two most important risk factors for chronic respiratory illnesses which are outdoor and indoor air quality and tobacco smoke. Plants can be used against CRDs and they contain compounds which are able to show beneficial activity against CRDs. It is reported that secondary metabolites such as alkaloids, phenolic compounds, and triterpenes can produce desirable effects against chronic respiratory ailments.

2.4.1 Alkaloids

Alkaloids can be categorized into several subclasses depending on their structure, viz. indole, pyridines, pyrrolidines, pyrrolizidines, isoquinolines, quinolines, steroids, terpenoids and tropanes. Some alkaloids were reported to exhibit activities against respiratory chronic diseases such as picrine (**156**), 19-epischolaricine (**157**) [166], berberine (**1**) [167], evodiamine (**158**) [168], and antidesmone (**159**) [160] (Fig. 12).

2.4.2 Phenolic Compounds

Phenols are plant secondary metabolites commonly present in human diet [169]. Phenols represent ideal therapeutic agents for the management of respiratory diseases due to their potent anti-inflammatory and anti-oxidant activities [170, 171]. Resveratrol [172, 173], curcumin [174, 175], quercetin [176, 177], epigallocatechin-3-gallate (EGCG) [178, 179], apocynin (**160**), paeonol (**161**), zingerone (**162**) [180], ellargic acid (**163**) [181], protocatechuic acid (**164**) [182], tricetin (**165**), luteolin, apigenin-7-glucoside (**166**), baicalein (**167**), baicalin (**168**), hyperoside (**169**), quercitrin (**170**), morin (**171**), tectogenin (**172**), and eriodictyol (**173**) [183–185] were reported to display notable properties against CRDs (Fig. 13).

Fig. 11 Some terpenoids and steroids with anticancer activity

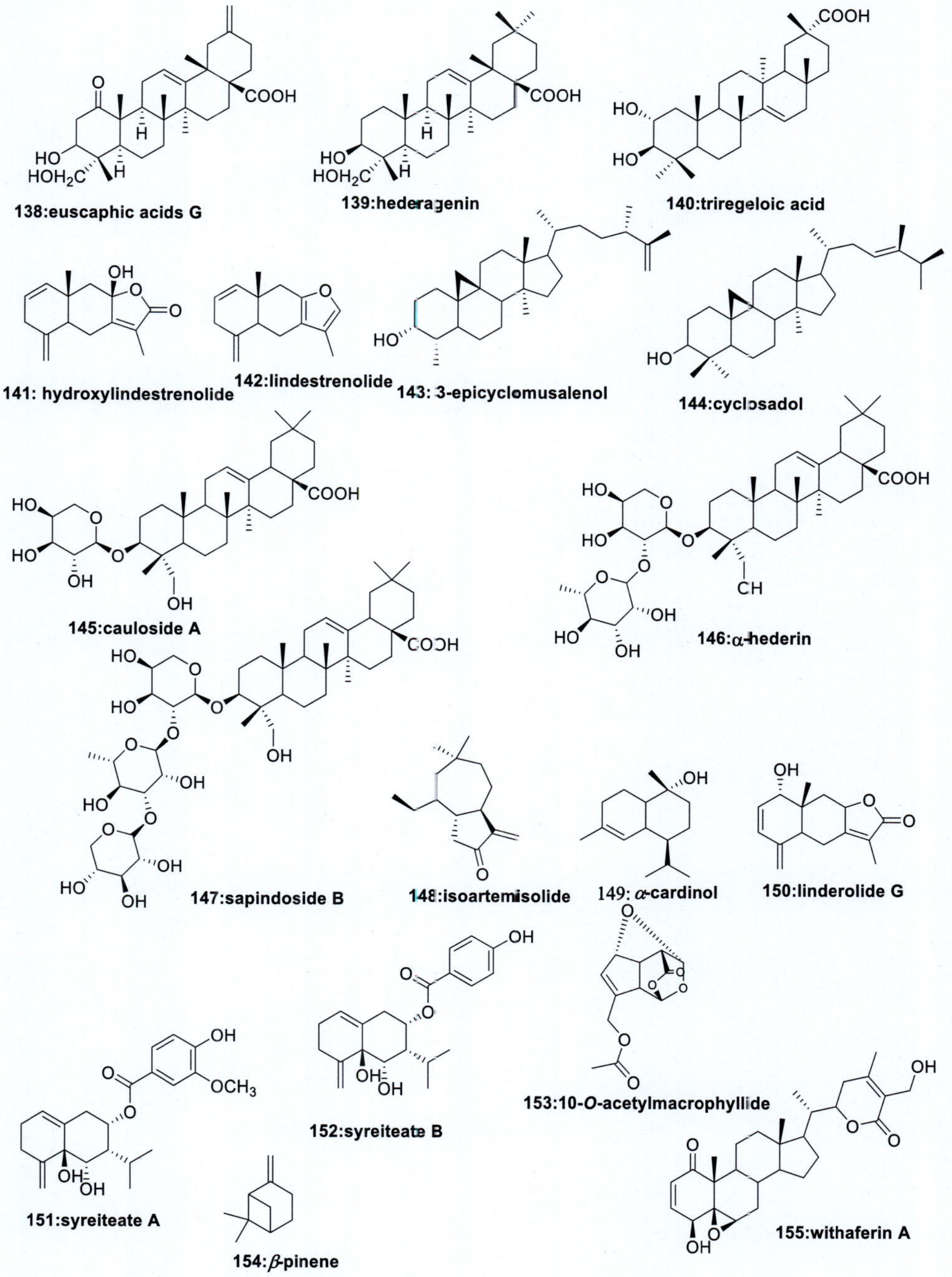

Fig. 12 Some alkaloids with anti-CRD property

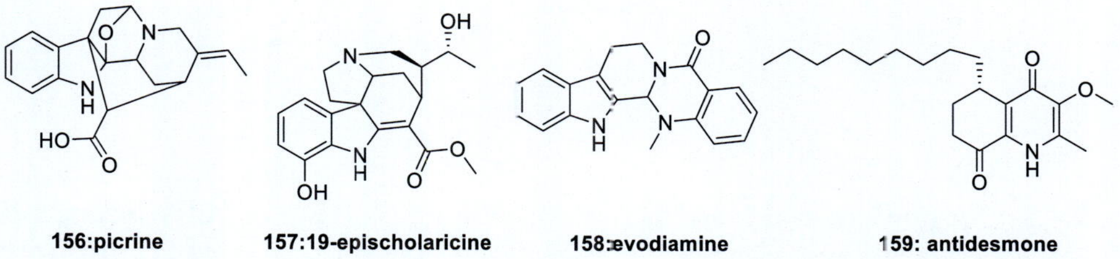

Fig. 13 Some phenolic compounds with anti-CRD property

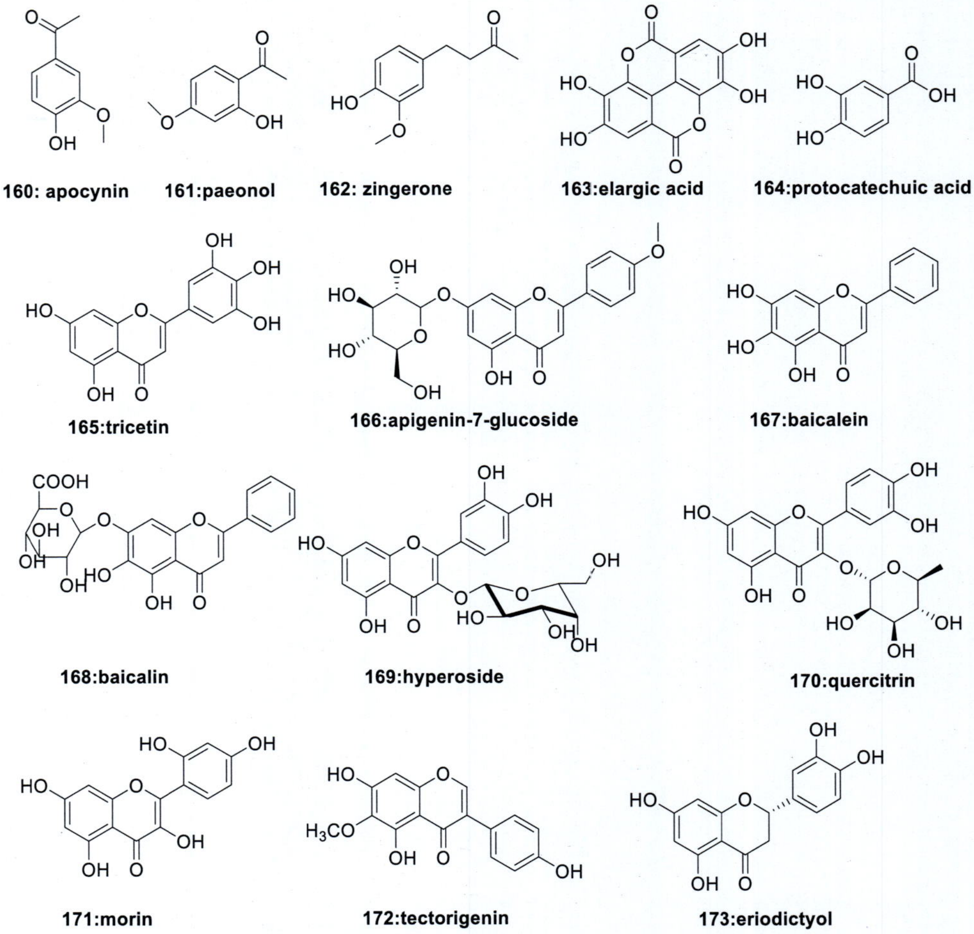

Fig. 14 Some terpenoids and steroids anti-CRD property

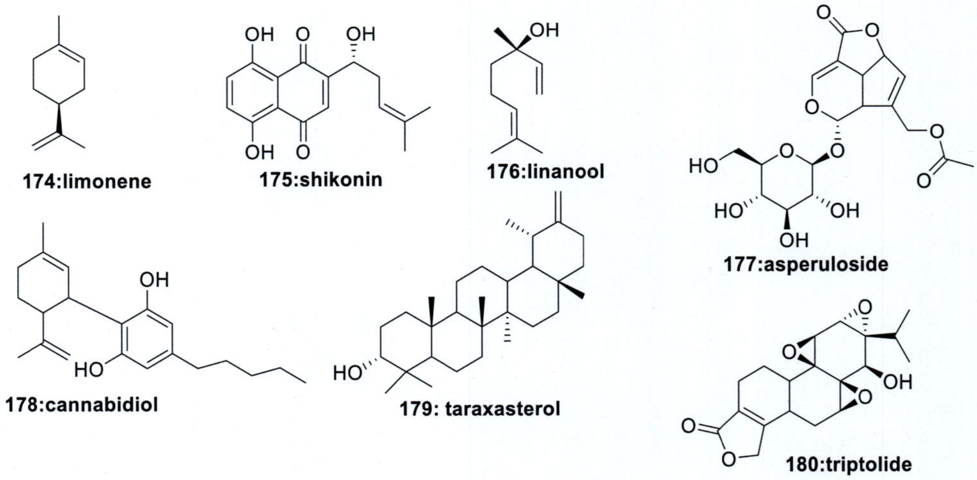

2.4.3 Terpenoids and Steroids

Several terpenoids were reported in the literature to play a significant role against some respiratory chronic diseases including limonene (**174**) [186], shikonin (**175**) [187, 188], linalool (**176**), asperuloside (**177**), cannabidiol (**178**), eriodictyol [189, 190], taraxasterol (**179**) [191], and triptolide (**180**) [192] (Fig. 14).

3 Characterization of the Main Groups of Compounds Involve in the Management of Lifestyle Diseases

The complete characterization of phytoconstituents obtained from natural products requires the determination of sample purity and physical properties, elementary composition, empirical formula, functional groups, and elucidation of structural formula and spatial relationships. It is obvious that the known molecule identification will include a lesser amount of advanced investigations. The criteria and procedural steps which will be chosen frequently depends on the availability of special equipment. Chemists and biologists make use of both qualitative and quantitative techniques for the characterization of phytoconstituents from plants including those involved in the management of lifestyle diseases. In this section, we will describe the main techniques used in the characterization of classes of compounds mostly involve in the management of these diseases.

3.1 Qualitative Estimation of Secondary Metabolites

Many interesting methods are used on plants' extracts to assess the existence of many groups of secondary metabolites. In order to show the presence of these compounds, different qualitative tests can be realized.

3.1.1 Some Qualitative Tests for Alkaloids

3.1.1.1 Test of Mayer

The reagent of Mayer is prepared from mercury and potassium iodide to form a solution of dipotassium tetraiodomercura (II) (K_2HgI_4). To perform the alkaloids identification, 1 mL of plant extract is mixed with 1 mL of potassium mercuric iodide solution (Mayer's reagent) in the test tube. After shaking to ensure complete mixing, the formation of a precipitate with cream color in the test tube indicates a positive result. The absence of this precipitate indicates a negative result that is the absence of alkaloids [193].

3.1.1.2 Test of Dragendorff

After adding 1 mL of the reagent of Dragendorff to 2 mL of extract, the presence of alkaloids can be indicated by the formation of an orange red precipitate observed [193].

3.1.1.3 Test of Hager

To perform the Hager's test, few drops of Hager's reagent are added to two 2 mL of extract. We observe the formation of a yellow precipitate which reveals the presence of alkaloids [193].

3.1.2 Some Qualitative Tests for Phenolic Compounds and Tannins

3.1.2.1 Test of Schinoda

To an alcoholic solution of the sample, few drops of concentrated hydrogen chloride and a pinch of magnesium shavings are added. The presence of flavonoids is indicated by effervescence and a deep pink (purplish) color [193].

3.1.2.2 Test of Ferric Chloride

This test is used to assess the existence of phenols and polyphenols in a plant extract. To an alcoholic solution of the sample, few drops of $FeCl_3$ are added and the change of color from yellow to purple designates the existence of phenols [194].

3.1.2.3 Test of Gelatin Hydrolysis

This test assesses the existence of phenolic compounds in a sample. It consists of the formation of a white precipitate due to the addition of 10% NaCl and 1% gelatin on dried extract dissolved in water [195].

3.1.3 Qualitative Identification of Ketone: 2,4-Dinitrophenyl Hydrazine Test

About 0.5 g of 2,4-dinitrophenyl hydrazine is dissolved in 2.5 mL aqueous solution of concentrated H_2SO_4 to give solution A. The sample is dissolved in a lowest quantity of 75% EtOH to give solution B. The appearance of a yellow/orange/red precipitate when solution A is added drop wise to solution B indicates the presence of aldehyde or ketone functional groups.

3.1.4 Qualitative Identification of Anthraquinones: Bornträger Test

A small amount of the sample (~0.5 mg) is dissolved in 5 mL of a solvent that is immiscible with water (chloroform, ether, methylene chloride, carbon tetrachloride). The formation of a red coloration in the aqueous phase when 10% of an aqueous alkaline solution is added (NH_4OH, NaOH or KOH) indicates the presence of anthraquinones [196].

3.1.5 Some Qualitative Tests for Terpenoids and Steroids

3.1.5.1 Test of Liebermann–Burchard

To a $CHCl_3$ solution of the sample to be analyzed, we add a few drops of acetic anhydride and concentrated H_2SO_4. The presence of triterpenes and steroids is indicated by a change of color to brick red for terpenoids and purple (or blue that rapidly changes to green) for steroids.

3.1.5.2 Test of Horizon

When 2 mL of trichloroacetic acid is added to 1 mL of an extract or organic sample, the existence of terpenoids is observed by the appearance of a red precipitate [193].

3.1.5.3 Test of Salkowski

The organic sample is shaken alongside with chloroform and concentrated solution of sulfuric acid (H_2SO_4), and added along the walls of the test tube. After some minutes, the red color which is observed indicates the presence of steroids [193].

3.1.6 Qualitative Test for the Presence of Saponins

In this qualitative test, few drops of a solution of sodium carbonate (Na_2CO_3) is added to 5 mL of extract in the test tube. The formation of foam after vigorous shaking indicates the presence of saponins [193].

3.1.7 Some Qualitative Tests for Carbohydrates

3.1.7.1 Test of Benedict

Eight to ten drops of the extract is added to 5 mL of Benedict's reagent, followed by heating for 5 min. The formation of dark red precipitate indicates the existence of carbohydrates [193].

3.1.7.2 Test of Molish

The sample to be analyzed is introduced into a test tube and dissolved in a solution of 1% ethanol in α-naphtol. A few drops of concentrated H_2SO_4 are added by letting it flow down the side of the tube. The appearance of a purple-red ring at the interface between the liquids indicates the presence of a sugar or sugars.

3.2 General Characterization Techniques of Phytoconstituents

After determining the nature of the organic sample by means of qualitative analyses, the complete characterization is generally done using spectroscopic methods such as UV, IR, mass, and NMR spectroscopy, and sometimes X-ray analysis. When necessary, synthesis can also be used [197].

3.2.1 Mass Spectrometry

This is a powerful analytical technique for the identification of known and unknown phytochemicals. Mass spectrometry consists of an ionization chamber in which the molecules being examined enter the gas phase and acquire electrical charge. In mass spectrometry, the symbol z characterizes the number of charges on the molecule and the symbol m means the molar mass of a molecule of analyte in grams per mole. The spectrum is a histogram of the relative amount of each type of analyte that has a particular mass-to-charge ratio which the symbol is m/z. Some ionization methods commonly used are chemical ionization (CI), fast atom bombardment (FAB), and field desorption (FD). Chemical ionization spectra can be recorded from both positive and negative ions [198]. In mass spectrometry, the relative molecular mass can be determined with high accuracy and an exact molecular formula can be determined with a knowledge of places where the molecule have been fragmented [199]. Moreover, the HPLC and MS combination facilitates rapid and accurate identification of chemical compounds in medicinal plants [200–203].

3.2.2 UV-Visible Spectroscopy

Spectrophotometry or UV-Vis Spectroscopy is a tool used to measure how much a chemical substance absorbs light. The experience consists of determining the intensity of light that passes through a sample with regard to the intensity of light through a reference sample or blank. This method can be utilized for numerous sample types counting glass, thin-films, solids, and liquids. The absorption of UV light or visible light by organic compounds is the principle of UV-Visible Spectroscopy which results in the production of distinct spectra. Spectroscopy is based on the interaction between matter and light. When the matter retains the light, it experiences excitation and de-excitation, coming about within the generation of a spectrum. This technique can be used for qualitative analysis and for the identification of certain groups of compounds which have chromophores in their structures [204]. For instance, phenolic compounds such as anthocyanins, phenols, polymer dyes, and tannins form complexes with iron that are detected by UV-visible spectroscopy [205, 206]. This technique does not take much time and is less expensive than other techniques [207].

3.2.3 Infrared Spectroscopy

Infrared (IR) spectroscopy through the absorption of infrared light is a very important technique for detecting the characteristic bonds of several functional groups. If infrared light is shone on a molecule, it is possible that the molecule will absorb the energy of the light. The energy absorbed can cause a bend or bond to stretch. That is called a bending or stretching vibration. These vibrations which correspond to the frequency of infrared light only occur at specific frequencies. The infrared light is absorbed when the frequency of IR light corresponds to the frequency of a particular vibrational mode. You can then know which frequencies are being absorbed by looking at your infrared spectrum. The different types of bonds vibrate at different frequencies and therefore absorb different frequencies of infrared light, which makes it possible to determine the functional groups present. The characteristic vibrational frequencies are known for the majority of commonly occurring chemicals [208]. Infrared spectra can be sorted into two principal regions:

- Diagnostic region—generally provides the clearest information and has fewer peaks. This region includes all signals from all bonds in a molecule. This region has the wavenumbers with value higher than 1500 cm^{-1};
- Fingerprint region—contains signals resulting from the vibrational excitation of most single bonds (bending and stretching). Values of wavenumbers in this region are less than 1500 cm^{-1}.

Infrared spectroscopy is used for the identification of different vibrational frequencies of single, double, and triple bonds present in an organic molecule [205, 207].

3.2.4 Nuclear Magnetic Resonance Spectroscopy

This technique is commonly known as NMR which gives chemical, physical, and biological properties of compounds. Chemists generally use one-dimensional and two-dimensional techniques to determine the structure of molecules [205]. The technique is linked to the magnetic properties of certain atomic nuclei of proton and carbon. For instance, ^{13}C-NMR and ^{1}H-NMR are used for the identification of different types of carbon and hydrogen present in natural products [204, 205].

3.2.5 X-Ray Crystallography

X-ray crystallography is a commonly used technique to determine the structure of an organometallic or organic molecule where the spatial arrangement of atoms around a chiral center is more readily determined. The X-ray crystallography technique can give a structure at a resolution of 0.083 Angstrom. A single crystal of a pure substance is an orderly range of molecules [209]. The underlying principle is that the crystalline atoms cause a beam of X-rays to diffract into many specific directions. A crystallographer can produce a three-dimensional picture of the density of electrons within the crystal by measuring the angles and intensities of these diffracted beams.

3.3 Characterization Techniques of Alkaloids

The identification of alkaloids depends on their classification. Known alkaloids can be identified by high performance liquid chromatography coupled and gas chromatography coupled to mass spectrometry (LC-MS and GC-MS). Known alkaloids can also be detected in complex mixtures by ^{1}H-NMR and ^{13}C-NMR which are more suitable because of the wide range of shifts [210]. In addition, 2D-NMR especially 2D-COSY seems very suitable because of their good sensitivity. For unknown alkaloids, the identification can be easily achieved if the alkaloid is a simple derivative of a known alkaloid that can be recognized by comparing MS and UV data.

3.3.1 UV and IR Characteristics of Alkaloids

UV is the oldest and still an important tool for the identification of alkaloids, especially indole and isoquinoline alkaloids which have a large number of different characteristic chromophores, providing information on the aromatic part of the molecule [211]. Identification of alkaloids by UV spectrophotometer was achieved by dissolving the isolates with solvent (chloroform for example) and measuring its absorbance at a wavelength of 200–400 nm.

IR is particularly useful for confirming the structure of known alkaloids. It can be useful for the identification of certain functional groups such as carbonyl groups but also for the determination of stereochemistry. For instance, in heteroyohimbine alkaloids, the appearance of Bohlman bands at about 2900 cm^{-1} is indicative of the stereochemistry of the C and D rings [212].

Identification of alkaloids using FTIR can be achieved by mixing isolates with potassium bromide using a mixing vibrator, then printing pellets and inserting then into the FTIR. The absorbance is then measured at a frequency of 4000–5000 cm^{-1}.

3.3.2 Mass Spectrometry (MS)

The mass spectroscopy of alkaloids has been widely reviewed in regard of the importance of the technique and its utility in the determination of the mass of the compounds [213, 214]. The direct coupling of mass spectroscopy with gas chromatography (GC) or high-pressure liquid chromatography (HPLC) turns these techniques into very powerful tools in the identification of alkaloids. GC-MS has the advantage that mass spectra also showing fragmentation, either by electron impact (EI) or chemical ionization (CI), are more easily obtained. In most of the interfaces available for LC-MS, no fragmentation spectra can be obtained; only the protonated molecule (M+ 1) can be observed in the positive ion mode. However, because of their great variety of structure, we have not given the general mass of alkaloids.

3.3.3 Nuclear Magnetic Resonance (NMR)

Many reviews on the NMR data of alkaloids have been published [215]. The ^{1}H-NMR spectrum gives an indication by showing features already known for other alkaloids, including the pattern of aromatic signals in indole alkaloids which can be used to obtain information on the type of indole alkaloid involved [210].

The fact that the signals of different protons overlap can complicate the interpretation of the ^{1}H-NMR spectra. The use of shifting reagents allows overlapping signals to be resolved. Trifluoroacetic acid has been shown to be a useful shifting reagent for alkaloids [216]. In $CDCl_3$ solution it forms a soluble ion pair with the alkaloids. The shifts obtained are particularly large for protons close to the protonated nitrogen, giving additional information on the structure and the stereochemistry of the alkaloid. Another way to obtain resolution of overlapping signals could be the application of the HOHAHA technique, which allows spectra to be obtained from separate spin systems, or the measurement of 2D *J*-resolved spectra.

The next step in the structure elucidation is the measurement of a ^{13}C-NMR spectrum. Each carbon emits a signal at a shift characteristic of its molecular environment, e.g., aromatic, vinylic, with hetero-atomic substituents. Alkaloids groups have great variety of structure and also many characteristic peaks in ^{1}H-NMR and ^{13}C-NMR spectroscopic. The value of these peak depending of each alkaloids class. For this reason, we will choose and present the general characteristic peak of one class of alkaloids involved in the management of lifestyle diseases.

3.3.4 Characterization of Monoterpene Indole Alkaloids

Among alkaloids, monoterpene indole alkaloids were reported to exhibited good activity against lifestyle diseases [143, 144].

3.3.4.1 Qualitative Identification of Monoterpene Indole Alkaloids

All alkaloids including monoterpene indole alkaloids react with reagents of Dragendorff, Mayer, and Wagner to give colored compounds.

3.3.4.2 Ultra Violet Spectral Properties

The UV spectrum of monoterpene indole alkaloids is carried out in MeOH and the characteristic absorption maxima are around 206–229 nm, 276–299 nm, 284–324 nm, and 367–392 nm [144, 217, 218].

3.3.4.3 Infrared Spectral Properties

The characteristics IR absorptions bands are in the region of 1660–1645 cm^{-1} for the six-member lactam ring, between 1597 and 1731 cm^{-1} for the carbonyl function and at 3145 and 3433 cm^{-1} for hydroxyl groups. In addition, absorptions bands at 3145–3364 cm^{-1} indicate the existence of N-H groups [144, 217, 219] (Table 1).

3.4 Characterization Techniques of Flavonoids

In general, phenolic compounds were reported to exhibit good activities against lifestyle diseases. Among phenolic compounds, flavonoids were the most reported to possess such activities [220, 221]. Flavonoids can be identified through both qualitative analysis and spectroscopic methods.

3.4.1 Qualitative Methods

Most flavonoids are positive to ferric chloride and Shinoda tests. They can also be positive to test of Molish if they possess sugar moiety in their structures [222].

3.4.2 Spectroscopic Methods

3.4.2.1 UV and IR Characteristic of Flavonoids

The UV spectrum of flavonoids is generally established in polar solvents such as MeOH or EtOH and two absorption maxima are observed, the range from 240 to 290 nm (band II) due to ring A and the range from 300 to 400 nm (band I) due to ring B, a cynnamoyle system. But these values highly depend on oxygenation pattern of a flavonoid (the position and the number of hydroxyl groups) [223, 224] (Scheme 1 and Table 2).

IR spectroscopy, compared to the above techniques shows little information, but important information on the elucidation of the structure of flavonoids. In fact, most isoflavones, hydroxylated flavones, chalcones or dihydrochalcones show strong maximum absorption bands between 3200 and 3500 cm^{-1} due to hydroxyl groups [223]. In addition, an intense absorption band characteristic of the carbonyl group (C=O) of these compounds is observed around 1680 cm^{-1} and is shifted to ca. 1620 cm^{-1} when the C=O is chelated with a hydroxyl. In the IR spectrum of flavonoids, an absorption band is also observed around 1600 cm^{-1} due to aromatic double bonds [223, 225].

3.4.2.2 Electronic Impact Mass Spectrometry (EI-MS)

The mass spectrometry data are really necessary in flavonoid structure elucidation. This technique is used for the determination of the molecular weight, for the establishment of the distribution of substituents between the A- and B-rings and also for the determination of the site of attachment and the nature of the sugar(s) in C-flavonoid and O-glucosides of flavonoids. The molecular weight of the basic flavonoid nucleus is 222 a.m.u. for the flavones, isoflavones, and aurones; 224 a.m.u. for the flavanones and chalcones; 238 a.m.u. for the flavonols, and 240 a.m.u. for the dihydroflavonols. From these basic molecular weights, the molecular weights of the elucidated compounds are deduced by adding the 16 mass units (a.m.u) for each -OH and 30 a.m.u. for each -OCH_3.

The loss of some molecular ion fragments (M^+) is very characteristic of flavonoid mass spectra (Scheme 2). The signals displayed accurately indicate the molecular mass per unit

Table 1 ^{1}H and ^{13}C-NMR of some monoterpene indole alkaloids derivatives

	6-nor-antirhine-N$_1$-methyl [143]		Razyamide [143]		Vincosamide-N-oxide [144]		Isodihydroaminocadambine [144]	
	^{1}H-NMR	^{13}C-NMR	^{1}H-NMR	^{13}C-NMR	^{1}H-NMR	^{13}C-NMR	^{1}H-NMR	^{13}C-NMR
1-CH$_3$ 1-NH	3.29 –	49.6 –	– 7.92 s	– –	–	–	–	–
2	–	137.1	–	133.0	–	137.6	–	134.9
3	4.08 brs	55.6	4.45 d (7.6)	27.7	4.96 m	73.9	4.56 brs	55.3
4	–	–	–	–	–	–	–	–
5	3.16 dd (9.0, 4.8) 2.92 brs	52.7	3.17 dd (11.0, 4.5) 2.65 brs	50.3	3.67 m 3.61 m	64.6	3.09	44.9
6	–	–	3.03 m 2.78 d (17.0)	20.3	2.97 m 2.62 m	22.1	2.97 m 2.68 m	22.3
7	–	107.5	–	107.8	–	110.2	–	110.4
8	–	128.3	–	126.3		128.6	–	128.8
9	7.38 d (7.8)	118.3	7.45 d (7.6)	118.1	7.38 d (7.6)	118.7	7.28 m	118.8
10	6.96 ddd (7.8, 7.8, 1.2)	119.4	7.08 dd (7.6, 7.6)	119.5	7.03 ddd (8.1, 7.3, 1.1)	120.3	6.98 t (7.3)	120.3
11	7.01 ddd (7.8, 7.8, 1.2)	121.4–	7.12 dd (7.6,7.6)	121.9	7.11 ddd (8.1, 7.3, 1.1)	122.5	7.07 t (7.3)	122.7
12	7.30 d (7.8)	111.8	7.28 d (7.6)	111.1	7.30 d (8.4)	112.4	7.27 m	112.5
13	–	137.3	–	136.6	–	134.7	–	137.9
14	2.14 m 2.10 m	32.1	2.64 brs 2.04 ddd (16.1, 10.3, 9.3)	33.6	2.45 m 1.98 m	27.3	2.05 m 2.45 m	27.5
15	1.66 m	32.0	3.82 brs	53.7	2.80 m	24.6	2.69 m	25.1
16	1.70 ddd (16.8, 9.0, 4.2) 1.55 dd (9.0, 4.2)	28.6	–	170.7	–	109.2	–	109.4
17	2.77 m 2.68 m	48.5	–	168.4	7.43 s	149.2	7.31 s	149.4
18	5.05 d (1.8) 5.04 d (1.8)	116.8	1.79 brs	13.1	5.43 dd (10.2, 2.0) 5.33 d (11.1)	120.9	5.33 m	120.8
19	5.69 ddd (18.6, 12.0, 9.6)	140.5	5.38 q (6.8)	121.9	5.63 m	134.1	5.51 ddd (9.6, 7.7, 2.9)	134.5
20	2.28 brs	49.6	–	133.0		45.1	4.96 dd (9.5, 2.9)	44.9
21	3.66 dd (10.8, 6.0) 3.62 dd (10.8, 6.0)	63.9	3.91 d (13.6) 3.14 brs	59.0	5.38 d (1.9)	98.0	5.30 br d (9.5)	98.2
22	–	–	–	–	–	169.0	3.31 brs	49.5
1′	–	–	–	–	4.62 d (7.8)	100.4	4.57 d (7.8)	100.6
2′	–	–	–	–	3.04 t (8.9)	74.2	2.97 dd (13.1, 8.3)	74.4
3′	–	–	–	–	3.27 m	77.8	3.33 m	78.4
4′	–	–	–	–	3.23 m	71.1	3.25 m	71.5
5′	–	–	–	–	3.29 m	77.9	3.24 m	78.1
6′	–	–	–	–	3.89 dd (12.1, 5.4) 3.69 dd (12.1, 2.1)	62.4	3.86 dd (12.1, 5.7) 3.65 dd (12.1, 2.2)	62.7
OCH$_3$	–	–	3.69 s	51.2	–	–	–	–
NH	–	–	7.92 s	–	–		–	–

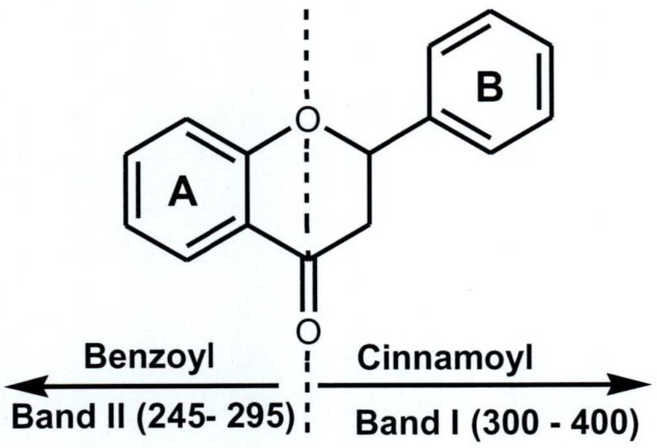

Scheme 1 Absorption bands of the benzoyl and cinnamoyl systems

Table 2 Ultraviolet-visible absorption ranges for flavonoids [223]

Band II (nm)	Band I (nm)	Flavonoid type
230–270	340–390	Chalcones
230–270	380–430	Aurones
245–275	310–330 shoulder	Isoflavone
	Ca. 320 peak	Isoflavones (5-deoxy-6,7-dioxygenated)
250–280	310–350	Flavones
250–280	330–360	Flavonols (3-OH substituted) oxydation
250–280	350–385	Flavonols (3-OH free)
270–280	465–560	Anthocyanidins and anthocyanins
275–295	300–330 shoulder	Flavonones and dihydroflavonols

charge (m/z) of the fragments they represent. The exact molecular mass for each fragment can be measured to the nearest 0.0001 mass unit if the mass spectrometer is capable of producing a high-resolution mass spectrum. This information allows the precise molecular formula to be calculated from the peak of the molecular ion and the fragment ions [223].

The scheme below highlights the main mechanism of fragmentation [226].

3.4.2.3 ^{1}H and ^{13}C-NMR

The ^{1}H-NMR spectrum appears important in some cases to distinguish the different subclasses of flavonoids such as flavanones, flavones, and isoflavones. From the following table some characteristic proton chemical shifts of flavonoids are listed (Table 3) [227, 228].

The ^{1}H-NMR spectroscopy also has typical applications:

- In defining the oxygenation pattern of the three rings of flavonoid and in determining the position and number of methyl groups;
- In distinguishing dihydroflavonols, flavonones, and isoflavones;
- In determining the number of sugars present and bound to either α- or β- (by ^{1}H-^{1}H coupling constant);
- In detecting of hydrocarbon side chains such as C-linked CH_3 and C- or O-linked prenyl.

The ^{13}C-NMR help to differentiate the *ortho* and *meta* dihydroxyl-substituted B-ring. In fact, when this ring carry

Scheme 2 Fragmentation pathways of flavonoids

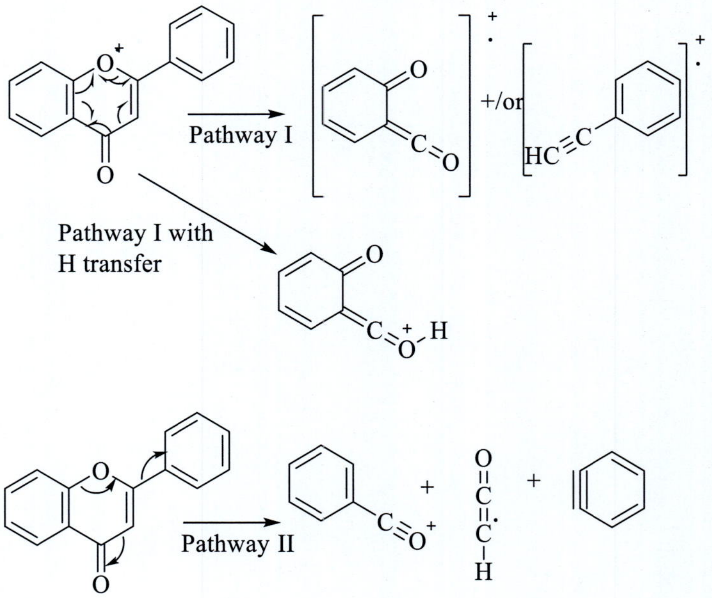

Table 3 Approximate proton chemical shifts of flavonoids [227]

Chemical shift (ppm)	Proton type
ca. 1.7	Prenyl [-CH_2-CH=C(CH_3)$_2$] methyl groups (other protons: 3.5 and 5.2 ppm)
ca. 2.0	Acetate (-OCO<u>CH_3</u>) and aromatic C-<u>CH_3</u>
2–3	H-3 of flavanones (two proton-multiplet)
4.2–6.0	H-2 dihydroflavonols: 5.0 ppm and H-2 of flavanones: 5–5.5 ppm
6.0–8.0	A- and B-ring protons
7.5–8.0	H-2 of isoflavones (singlet)
12–14	5-OH (usually observed when solvent = DMSO-d_6)

Table 4 ^{13}C-NMR chemical shift ranges of various flavonoid carbon types [230]

Usual chemical shift range (ppm)	**Carbon type**
210–170	Carbonyl (4-keto, acyl)
Aromatic and olefinic	
150–130 (with *ortho*/*para* oxygenation) 165–155 (no *ortho*/*para* oxygenation)	Oxygenated
125–90 (with *ortho*/*para* oxygenation) 135–125 (no *ortho*/*para* oxygenation)	Non-oxygenated
Aliphatic	
21 (CH_2), 122 (CH), 131 (C), 18 (CH_3)	Isopropenyl [-CH_2CH=C(CH_3)$_2$]
55–63 (60–63 for *O*-disubstituted)	O-CH_3
ca. 17–20	C-<u>CH_3</u>, CO<u>CH_3</u>
80–40 (epicatechin C-4, 28 ppm)	Non oxygenyted (C-2, 3 flavanones)
ca. 100	Methylenedioxy
83–69 (*C*-1 of *O*-glycoside, ca. 100 ppm)	Oxygenated (sugars)

only hydroxyl, signal is exhibited around 141–149 ppm, and in the case of *meta* substitution, the signal is shifted to low field around 151–159 ppm [223, 229] (Table 4).

The ^{13}C-NMR of flavonoids has also typical application in the structure elucidation mainly in:

- Establishing the number of oxygenated carbons on the flavonoid nucleus, the total number of carbon atoms per molecule and the number of carbons in the sugar moiety;
- Identifying of C- and *O*-linked sugars and;
- Determining inter-glycosidic linkage points, acyl substituents, and the site of acylation.

The sugar carbon resonances in the ^{13}C-NMR spectra of some flavonoid C- or *O*-glycosides are presented in the following table (Table 5) [231].

Various flavonoid aglycones have also characteristic signals in their ^{13}C-NMR spectra and some of them are given in Table 6.

3.5 Characterization Techniques of Terpenoids and Steroids

3.5.1 Mass Spectrometry

Several mass techniques can be used for the characterization of terpenoids including IE, FAB-MS, GC-MS, and HPLC-MS. For example, fast atom bombardment spectroscopy (FAB-MS) affords the exact molecular ion peak along with diagnostic fragmentation pattern of the terpenoid molecule. During the recent years, the structure determination of various terpenoids have been done using GC-MS, HPLC-MS spectrometry both coupled with computerized analysis and tandem MS spectral analysis and using FAB-MS in both positive and negative modes [232].

3.5.2 UV and IR Characteristics of Terpenoids

Functional groups, as olefinic bonds (double and triple bonds), carbonyl groups (aldehydes, ketones), carboxylic groups (carboxylic acid and its derivatives) present in terpenoids appear in the UV range between 200 and 350 nm. In terpenoid structures, isolated double and triple bonds do not absorb at wavelengths above 200 nm, while isolated ketones, aldehydes, carboxylic acids and their derivatives absorb at higher wavelengths with low intensities. Hence, both of the above-mentioned isolated chromophores are not of much value for the structure elucidation. However, UV data become valuable when the conjugated double bonds and α, β-unsaturated carbonyl group are present in the molecule [232].

The IR spectroscopy is commonly used for the identification and structural elucidation of new terpenoids. It gives information about functional groups [232].

3.5.3 Nuclear Magnetic Resonance Spectroscopy (NMR)

The ^{1}H-NMR and ^{13}C-NMR spectroscopy are the most important tools required for the structure elucidation of terpenoids. The combination of 1D and 2D NMR techniques including COSY, TOCSY, ROESY, 2D-IN ADEQUATE, HMQC, HMBC, HETCOR, and selective INEPT are a great value for the structure elucidation of various terpenoids such as saponins and glycosides consisting of sugar moieties [232]. Terpenoids were greatly implicated in the management of lifestyle diseases. Among these terpenoids, pentacyclic triterpenoids is among the subclass of triterpene which exhibited good activity in the management of lifestyle diseases.

Table 5 Sugar carbon resonances in the ^{13}C-NMR spectra of some flavonoid *C*- and *O*-glycosides [231]

Glycoside type	C-1	C-2	C-3	C-4	C-5	C-6
O-glucoside (7-linked)	100.2	73.3	76.6	69.8	77.4	60.9
C-glucoside (8-linked)	73.9	71.4	78.8	70.8	81.4	61.5
O-galactoside (3-linked)	102.3	71.3	73.4	68.0	75.8	60.8
C-galactoside (8-linked)	73.9	68.5	75.4	69.1	80.5	61.3
O-α-rhamnoside (3-linked)	101.9	70.4	70.6	71.5	70.1	17.3
C-α-rhamnoside (8-linked)	77.3	75.0	75.5	72.2	72.2	18.1
O-xyloside (2″-*O*-linked)	102.4	73.7	75.9	69.4	65.5	–
O-xyloside (6-linked)	74.6	70.3	78.5	70.0	70.0	–
O-α-arabinoside (3-linked)	101.8	71.7	70.8	65.9	64.1	–
O-α-arabinofuranoside (3-linked)	108.1	82.1	77.2	86.2	61.0	–
C-α-arabinoside (8-linked)	74.4	68.2	74.5	68.9	71.0	–
O-glucuronide(3-linked)	101.1	73.7	75.9	71.3	75.9	169.7
O-alloside (3-linked)	99.9	71.6	71.6	67.2	75.1	61.3
O-apioside (2″-*O*-linked)	109.0	76.5	79.1	74.0	64.4	–

3.5.4 Characterization of Pentacyclic Triterpenoids

3.5.4.1 Qualitative Analysis

Most triterpenoids including pentacyclic triterpenoids respond positively to sulfuric vanillin and Liebermann–Burchard tests.

3.5.4.2 IR Spectroscopy

The Infrared spectrum makes it possible to identify the chemical functions present such as ketones, alcohol, carboxylic acid acetate groups, etc. which characterize most triterpenes [233].

The vibration of the hydroxyl group (OH) is observed between 3200 and 3300 cm^{-1}; the carbonyl group (C=O) band is very intense around 1773 cm^{-1}.The symmetrical deformation vibration of the ester methyl appears between 1370 and 1390 cm^{-1} and the vibration of the carboxylate group is located between 1100 and 1300 cm^{-1} [234].

3.5.4.3 Mass Spectrometry

In mass spectroscopy, the fragments of the triterpenes are characteristics such as the peaks of fragmentation characteristic of the "Retro-Diels-Alder" mechanism by cleavage of the D/E and C/D cycles of the pentacyclic triterpenes [233]. Indeed, this method is advantageous in the elucidation of structures because it allows the identification of double bonds.

The retro Diels-Alder makes it possible to obtain, for pentacyclic triterpenes possessing a C-12 unsaturation such as urs-12-ene and olean-12-ene, a base peak at $m/z = 218$. When substituents are observed on rings D and E, the value of the base peak may change. There will be a base peak at $m/z = 203$ (218-Me) if there is loss of a methyl group during the electronic impact or values such as: $m/z = 248$ (203 + COOH), $m/z = 262$ (203 + COOMe), $m/z = 265$ (203 + COOH + OH) and $m/z = 276$ (203 + OAc) for substituents of the carboxyl, carbomethoxy, and hydroxyl type remaining on the ion of base peak. These data can, therefore, be used to predict the nature of substitutions on a skeleton [235].

Scheme 3 gives the different probable ions resulting from the Retro Diels-Alder type fragmentation of molecules of the urs-12-ene and olean-12-ene type [236].

Regarding pentacyclic triterpenoids of lupane and hopane types, there is be a base peak that will appear rather towards $m/z = 189$ or $m/z = 191$ when the substitution on the E ring is respectively isopropenyl or isopropyl [235].

3.5.4.4 ^{1}H and ^{13}C-NMR

The ^{1}H-NMR spectrum of triterpenes is not very explicit. However, it exhibits a series of characteristic peaks in the range 0.5–2 ppm, corresponding to the methyl groups. The profile (number, multiplicity, chemical shift) allows orientation regarding the type and series of the triterpene nucleus. The chemical shifts between 2 and 6 ppm indicate the presence of functional groups like the hydroxyl, double bond, and carbonyl [233].

The ^{1}H NMR is very characteristic and makes it possible to distinguish certain classes of pentacyclic triterpenes such as ursanes, oleananes, and lupanes. Indeed, the ursane and oleanene derivatives generally exhibit in the ^{1}H-NMR spectrum a large triplet between 5.1 and 5.5 ppm of coupling constant J = 3 Hz, characteristic of the vinyl proton H-12. Lup-20(29)-ene derivatives are generally characterized by the presence of two singlets of one proton each between 4.2 and 5.0 ppm, characteristic of the 02 vinyl protons on carbon C-29 [237].

Table 6 ^{13}C-NMR data of various flavonoid aglycones [223]

Flavonoid	C-2	C-3	C-4	C-5	C-6	C-7	C-8	C-9	C-10	C-1′	C-2′	C-3′	C-4′	C-5′	C-6′	CH_3
Flavones																
Apigenin	163.8	102.8	181.8	161.9	98.8	164.1	94.0	157.3	103.7	121.3	128.4	116.0	161.5	116.0	128.4	
Luteolin	164.5	103.3	182.2	162.1	99.2	164.7	94.2	157.9	104.2	122.1	113.8	146.2	150.1	116.4	119.3	
Tricetin	164.2	103.2	181.6	161.6	99.0	164.2	93.9	157.5	104.0	120.9	106.0	146.5	137.9	146.5	106.0	
Flavonols																
Kampferol	146.8	135.6	175.9	160.7	98.2	163.9	93.5	156.2	103.1	121.7	129.5	115.4	159.2	115.4	129.5	
Quercetin	146.9	135.5	175.8	160.7	98.2	163.9	93.3	156.2	103.1	122.1	115.3	145.0	147.6	115.6	120.0	
Patuletin	147.1	135.5	176.1	151.8	130.9	157.2	93.7	151.4	103.5	122.1	115.2	145.1	147.8	115.7	120.1	60.3
Isoflavones																
Genistein	153.6	122.4	180.2	162.1	98.6	164.3	93.7	157.5	104.6	121.4	130.0	115.2	157.6	115.2	130.0	
Chalcone	*C*-β	*C*-α	C=O	C-1′	C-2′	C-3′	C-4′	C-5′	C-6′	C-1	C-2	C-3	C-4	C-5	C-6	
Isoliquiritigenin	143.8	117.8	191.4	113.2	164.6	102.6	165.4	107.9	132.3	125.8	130.6	115.8	159.9	115.8	130.6	

Scheme 3 RDA fragmentation of urs-12-ene and olean-12-ene derivatives

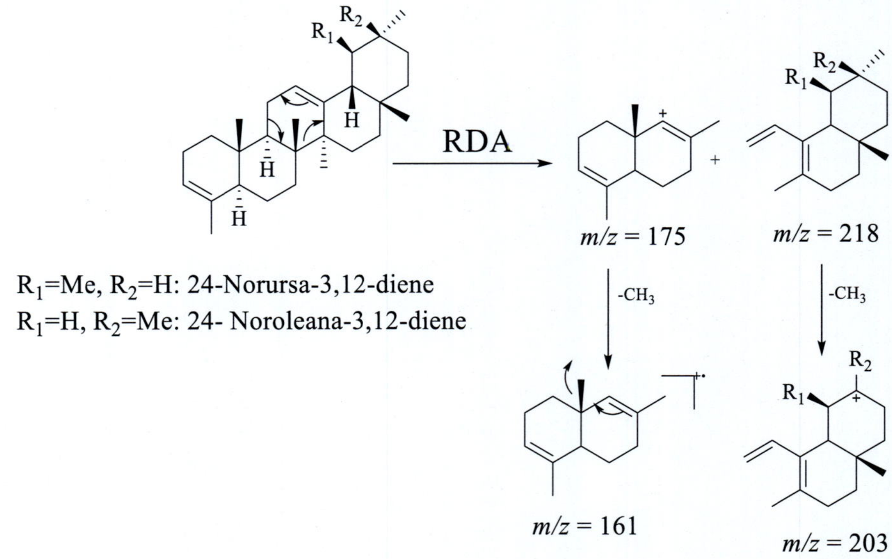

For C-3 substituted hydroxylated triterpenes, the H-3 proton generally appears between 3.1 and 3.8 ppm while that of acetoxylated triterpenes resonate between 4.2 and 4.6 ppm [238].

The methyl signals appearing between 0.5 and 2 ppm make it possible to differentiate the types of pentacyclic triterpenes. For example, the unoxidized triterpenes of the olean-12-enes series have 08 singlet signals of 03 protons attributable to 08 angular methyls while those of the urs-12-enes series possess 06 singlets of 03 protons and 02 doublets [239].

Regarding the ^{13}C-NMR spectrum, it is more valuable in the structural determination of triterpene nuclei. It allows us to clearly demonstrate the 30 carbon atoms which constitute the backbone of triterpenes, and in particular the carbonyl, carboxylic acid, and alcohol groups, by their specific chemical shifts that is around 210 ppm for a carbonyl (C=O), 170 ppm for a carboxylic acid (COOH) and 70 ppm for an alcohol (OH) [233].The evolution of this elucidation technique has made it possible to distinguish between the derivatives of the urs-12-ene and olean-12-eneseries. This is made possible by the number of quaternary carbon atoms (which are 05 in number for the urs-12-ene series and 06 for the olean-12-ene series) and the chemical shifts of the vinyl carbons C-12 and C-13 resonating respectively at 124.5 ppm (d) and 138 ppm (s) for urs-12-ene while for olean-12-ene the chemical shift of these carbons are at 122.5 ppm (d) and 143.5 ppm (s) [239].

3.5.5 Characterization Techniques of Saponins

The structural determination of the aglycones is similar to those of simple terpenoids. However, the ^{1}H-NMR spectrum in the pyridine solvent show the existence of anomeric proton between 4.3 and 6.9 ppm, as well as the methyl doublets of deoxy-6-hexoses in the range of 1.4–1.8 ppm. The high values of the coupling constants (J > 7.0 Hz) observed for the anomerous protons of D-glucose, D-galactose, and D-xylose is characteristic for the β-configuration while the small values (J = 1–1.5 Hz) observed for the L-rhamnose suggest α-configuration. ^{13}C-NMR spectrum analysis makes things even easier. The anomeric carbons resonate in a distinct region. This makes it possible to have an estimate of the amount of sugar present in the molecule. The configuration of the L-rhamnose anomere could also be confirmed from the chemical shifts of the C-3 and C-5 carbons. For β-L-rhamnopyranoside these carbons resonate at 75.4 and 73.5 ppm, while for α-L-rhamnopyranoside they are observed at 72.5 and 69.4 ppm, respectively [240, 241].

4 Conclusion and Outlooks

Phenolic compounds, alkaloids, terpenoids, and steroids were frequently reported in the literature to exhibit promising activities against lifestyle diseases such as diabetes, cardiovascular diseases, cancers, and chronic respiratory diseases. In this chapter, the main characterization techniques such as infrared (IR), UV-visible, mass spectrometry, and nuclear magnetic resonance (NMR) spectroscopy of these classes of secondary metabolites were discussed. In order to optimize the identification and quantification of these classes of compounds in plants extracts as well as their subsequent isolation, standardized methods using more advanced techniques (metabolomics and molecular networking) needs to be established.

Acknowledgments The authors are grateful to the financial support of the International Science Programme (ISP), the Natural Products

Research Network for Eastern and Central Africa (NAPRECA) and the University of South Africa (UNISA). The authors are also thankful to the Alexander von Humboldt Foundation (3.4-CMR-Hub) for the financial support to the Centre of Competence for the Study of Antimicrobial Natural Products from Fungi (CECANAPROF).

References

1. WHO. Noncommunicable diseases. 2022. [cited 2023 Jan 29]. https://Who.int/news-roon/fact-sheets/detail/noncommunicable-diseases.
2. Shoaib S. Cardioprotective effect of gemmotherapeutically treated *Withania somnifera* against chemically induced myocardial. Pak J Bot. 2010;42:1487–99.
3. Jahan N, Khalil-ur-Rahman, Ali S. Cardioprotective and antilipidemic potential of *Cyperus rotundus* in chemically induced cardiotoxicity. Int J Agric Biol. 2012;14:989–92.
4. Shaito A, Thuan DTB, Phu HT, Nguyen THD, Hassan H, Halabi S, Abdelhaby S, Nasrallah GK, Eid AH, Pintus G. Herbal medicine for cardiovascular diseases: efficacy, mechanisms, and safety. Front Pharmacol. 2020;11:422.
5. Panda VS, Suresh RN. Evaluation of cardioprotective activity of *Ginkgo biloba* and *Ocimum sanctum* in rodents. Altern Med Rev. 2009;14:161–71.
6. Kim KH, Moon E, Kang KS, Kim SY, Choi SU, Lee KR. Alkaloids from *Acorus gramineus* rhizomes and their biological activity. J Braz Chem Soc. 2015;26:3–8.
7. Donnelly LE, Newton R, Kennedy GE, Fenwick PS, Leung RH, Ito K, Russell RE, Barnes PJ. Anti-inflammatory effects of resveratrol in lung epithelial cells: molecular mechanisms. Am J Physiol Lung Cell Mol Physiol. 2004;287:L774–83.
8. Islam SU, Ahmed MB, Ahsan H, Lee YS. Recent molecular mechanisms and beneficial effects of phytochemicals and plant-based whole foods in reducing LDL-C and preventing cardiovascular disease. Antioxidants. 2021;10(5):784.
9. American Diabetes Association (ADA). Diagnosis and classification of diabetes mellitus. Diabetes Care. 2009;32:S62–7.
10. Malviya N. Antidiabetic potential of medicinal plants. Acta Pol Pharm. 2010;67:113–8.
11. Aggarwal N, Aggarwal S. A review of recent investigations on medicinal herbs possessing antidiabetic properties. J Nutr Disorder Ther. 2011;1:102.
12. Kayarohanam S, Kavimani S. Current trends of plants having antidiabetic activity: a review. J Bioanal Biomed. 2015;7:55–65.
13. Chopra RN, Nayar SL, Chopra IC. Glossary of Indian medicinal plants. New Delhi: Council of Scientific and Industrial Research; 1996.
14. Costantino L, Laura R, Renato P, Tiziana B, Pompeo P, Fabio G. Isolation and pharmacological activities of the *Tecoma stans* alkaloids. Il Farmaco. 2003;9:781–5.
15. Kiyoteru T, Shinichi T, Junichi K, Shunivhi Y, Kazuo L, Kinzo W, Samisoni J, Taraiasi V, Bill A. Hyperglycemia inhibitor and method for producing the same. Japanese Patent JP 2005132837; 2005.
16. Chattopadhyay RR. A comparative evaluation of some blood sugar lowering agents of plant origin. J Ethnopharmacol. 1999;67:367–72.
17. Luo J, Fort DM, Carlson TJ, Noamesi BK, nii-Amon-Kotei D, King SR, Tsai J, Quan J, Hobensack C, Lapresca P, Waldeck N, Mendez CD, Jolad SD, Bierer DE, Reaven GM. *Cryptolepsis sanguinolenta*: an ethnobotanical approach to drug discovery and the isolation of a potentially useful new antihyperglycaemic agent. Diabet Med. 1998;15:367–74.
18. Sharma R, Amin H, Galib, Prajapati PK. Antidiabetic claims of *Tinospora cordifolia* (Willd.) Miers: critical appraisal and role in therapy. Asian Pac J Trop Biomed. 2015;5:68–78.
19. Kharal P, Patil BM, Mandar BK, Dey YN, Duyu T. Network pharmacology-based assessment to elucidate the molecular mechanism of anti-diabetic action of *Tinospora cordifolia*. Clin Phytosci. 2019;5:1–9.
20. Shimoda H, Nishida N, Ninomiya K, Matsuda H, Yoshikawa M, Javaberine A. New TNF-alpha and nitric oxide production inhibitor, from the roots of *Talinum paniculatum*. Heterocycles. 2001;55(11):2043–50.
21. Catthareeya T, Papirom P, Chanlun S, Kupittayanant S. *Talinum paniculatum* (jacq.) gertn: a medicinal plant with potential estrogenic activity in ovariectomized rats. Int J Pharm Pharm Sci. 2013;5(2):478–85.
22. Attia ES, Amer AH, Hasanein MA. The hypoglycemic and antioxidant activities of garden Cress (*Lepidium sativum*, L.) seed on alloxan-induced diabetic male rats. Nat Prod Res. 2019;33:901–5.
23. Dineshkumar B, Mitra A, Mahadevappa M. Antidiabetic and hypolipidemic effects of mahanimbine (carbazole alkaloid) from Murraya koenigii (Rutaceae) leaves. Int J Phytomed. 2010;2:22–30.
24. Tabopda TK, Ngoupayo J, Liu J, Mitaine-Offer AC, Tanoli SA, Khan SN, Ali MS, Ngadjui BT, Tsamo E, LacailleDubois MA, Luu B. Bioactive aristolactams from *Piper umbellatum*. Phytochemistry. 2008;69:1726–31.
25. Shibano M, Tsukamoto D, Masuda A, Tanaka Y, Kusano G. Two new pyrrolidine alkaloids, radicamines A and B, as inhibitors of alpha-glucosidase from *Lobelia chinensis* Lour. Chem Pharm Bull. 2001;49(10):1362.
26. Hammouda Y, Amer MS. Antidiabetic effect of tecomine and tecostanine. J Pharm Sci. 2006;55:1452–4.
27. Jung M, Park M, Lee HC, Kang YH, Kang ES, Kim SK. Antidiabetic agents from medicinal plants. Curr Med Chem. 2006;13:1203–8.
28. Perron NR, Brumaghim JL. A review of the antioxidant mechanisms of polyphenol compounds related to iron binding. Cell Biochem Biophys. 2009;53(2):75–100.
29. Li AN, Li S, Zhang YJ, Xu XR, Chen YM, Li HB. Resources and biological activities of natural polyphenols. Nutrients. 2014;6(12):6020–47.
30. Mojzer BE, Knez Hrnčič M, Škerget M, Knez Ž, Bren U. Polyphenols: extraction methods, antioxidative action, bioavailability and anticarcinogenic effects. Molecules (Basel, Switzerland). 2016;21(7):901.
31. Arranz S, Chiva-Blanch G, Valderas-Martínez P, Medina-Remón A, Lamuela-Raventós RM, Estruch R. Wine, beer, alcohol and polyphenols on cardiovascular disease and cancer. Nutrients. 2012;4(7):759–81.
32. Akkarachiyasit S, Piyawan C, Sirintorn Y, Sirichai A. Inhibitory activities of cyanidin and its glycosides and synergistic effect with acarbose against intestinal -glucosidase and pancreatic—amylase. Int J Mol Sci. 2010;11(9):3387–96.
33. Chakravarthy BK, Gupta S, Gambhir SS, Gode KD. The prophylactic action of (-)-epicatechin against alloxan induced diabetes in rats. Life Sci. 1981;29:2043–7.
34. Li T, Liu J, Zhang X, Ji G. Antidiabetic activity of lipophilic (-)-epigallocatechin-3-gallate derivative under its role of alpha-glucosidase inhibition. Biomed Pharmacother. 2007;61:91–6.
35. Bhandari MR, Anurakkun NJ, Hong G, Kawabata J. α-Glucosidase and -amylase inhibitory activities of Nepalese medicinal herb Pakhanbhed (*Bergenia ciliata*, Haw.). Food Chem. 2008;106:247–52.
36. Jung UJ, Lee MK, Jeong KS, Choi MS. The hypoglycaemic effects of hesperidin and naringin are partly mediated by hepatic

glucose regulating enzymes in C57BL/KsJ-db/db mice. J Nutr. 2004;134:2499–503.
37. Tchamgoue J, Hafizur RM, Tchouankeu JC, Kouam SF, Adhikari A, Hameed A, Green IR, Choudhary MI. Flavonoids and other constituents with insulin secretion activity from *Pseudarthria hookeri*. Phytochem Lett. 2016;17:181–6.
38. Choi JS, Suh SS, Young HS, Park HJ. Hypolipemic and hypoglycaemic activities of *Prunus davidiana* in high fat-fed rats. Arch Pharm Res. 1991;1:44–7.
39. Zang Y, Sato H, Igarashi K. Anti-diabetic effects of a Kaempferol glycoside-rich fraction from unripe soybean (Edamame, Glycine max L. Merrill. "Jindai") leaves on KK-A mice. Biosci Biotechnol Biochem. 2011;75:1–8.
40. Manickam M, Ramanathan M, Jahromi MA, Chansouria JP, Ray AB. Antihyperglycaemic activity of phenolics from *Pterocarpus marsupium*. J Nat Prod. 1997;60:609–10.
41. Chen YG, Li P, Li P, Yan R, Zhang XQ, Wang Y, Zhang XT, Ye WC, Zhang QW. α-Glucosidase inhibitory effect and simultaneous quantification of three major flavonoid glycosides in *Microctis folium* yan-gan chen. Molecules. 2013;18:4221–32.
42. Ohnishi M, Matuo T, Tsuno T, Hosoda A, Nomura E, Taniguchi H, Sasaki H, Morishita H. Antioxidant activity and hypoglycaemic effect of ferulic acid in STZ-induced diabetic mice and KK-Ay mice. Biofactors. 2004;21:315–9.
43. Basnet P, Kadota S, Shimizu M, Takata Y, Kobayashi M, Namba T. Bellidifolin stimulates glucose uptake in rat 1 fibroblasts and ameliorates hyperglycaemia in streptozotocin (STZ)-induced diabetic rats. Planta Med. 1995;61:402–5.
44. Rao AV, Gurfinkel DM. The bioactivity of saponins: triterpenoid and steroidal glycosides. Drug Metabol Drug Interact. 2000;17:211–35.
45. Ragasa CY, Tsai P, Shen CC. Terpenoids and sterols from the endemic and endangered Philippine trees, *Ficus pseudopalma* and *Ficus ulmifolia*. Philippine J Sci. 2009;138(2):205–9.
46. Narender T, Khaliq T, Singh AB, Joshi MD, Mishra P, Chaturvedi JP, Srivastava AK, Maurya R, Agarwal SC. Synthesis of alpha-amyrin derivatives and their in vivo antihyperglycemic activity. Eur J Med Chem. 2009;44(3):1215–22.
47. Zhang Z, Jiang J, Yu P, Zeng X, Larrick JW, Wang Y. Hypoglycemic and beta cell protective effects of andrographolide analogue for diabetes treatment. J Transl Med. 2009;7:62.
48. Jin L, Shi G, Ning G, Li X, Zhang Z. Andrographolide attenuates tumor necros factor alpha-induced insulin resistance in 3T3-L1 adipocytes. Mol Cell Endocrinol. 2011;332:134–9.
49. Wiedenkeller DE, Sharp GWG. Effects of forskolin on insulin release and cyclic AMP content in rat pancreatic islets. Endocrinology. 1983;113:2311–3.
50. Mbaze LM, Poumale HM, Wansi JD, Lado JA, Khan SN, Iqbal MC, Ngadjui BT, Laatsch H. α-Glucosidase inhibitory pentacyclic triterpenes from the stem bark of *Fagara tessmannii* (Rutaceae). Phytochemistry. 2007;68(5):591–5.
51. Naik SR, Barbosa FJM, Dhuley JN, Deshmukh V. Probable mechanism of hypoglycemic activity of bassic acid, a natural product isolated from *Bumelia sartorum*. J Ethnopharmacol. 1991;33(1–2):37–44.
52. Murakami A. Modulation of protein quality control systems by food phytochemicals. J Clin Biochem Nutr. 2013;52(3):215–227.
53. Judy WV, Hari SP, Stogsdill WW, Judy JS, Naguib YM, Passwater R. Antidiabetic activity of a standardized extract (Glucosol) from *Lagerstroemia speciosa* leaves in Type II diabetics. A dose-dependence study. J Ethnopharmacol. 2003;87(1):115–7.
54. Sundaram R, Naresh R, Shanthi P, Sachdanandam P. Antihyperglycemic effect of iridoid glucoside, isolated from the leaves of *Vitex negundo* in streptozotocin-induced diabetic rats with special reference to glycoprotein components. Phytomedicine. 2012;19(3–4):211–6.
55. Yoshikawa M, Yoshizumi S, Ueno T, Matsuda H, Murakami T, Yamahara J, Murakami N. Medicinal foodstuffs. I. Hypoglycaemic constituents from a garnish foodstuff "taranome," the young shoot of *Aralia elata* Seem: elatosides G, H, I, J, and K. Chem Pharm Bull. 1995;43:1878–82.
56. Yoshikawa M, Harada E, Murakami T, Matsuda H, Wariishi N, Yamahara J, Murakami N, Kitagawa I. Escins-Ia, Ib, IIa, IIb and IIIa bioactive triterpene oligoglycosides from the seeds of *Aesculus hippocastanum* L: their inhibitory effects on ethanol absorption, and hypoglycaemic activity on glucose tolerance test. Chem Pharm Bull. 1994;42:1357–9.
57. Hou CC, Lin SJ, Cheng JT, Hsu FL. Antidiabetic dimeric guianolides and a lignan glycoside from *Lactuca indica*. J Nat Prod. 2003;66:625–9.
58. Fort DM, Ubillas RP, Mendez CD, Jolad SD, Inman WD, Carney JR, Chen JL, Ianiro TT, Hasbun C, Bruening RC, Luo J, Reed MJ, Iwu M, Carlson TJ, King SR, Bierer DE, Cooper R. Novel antihyperglycemic terpenoidquinones from *Pycnanthus angolensis*. J Org Chem. 2000;65(20):6534–9.
59. Yoshikawa M, Murakami T, Kadoya M, Matsuda H, Muraoka O, Yamahara J, Murakami N, Medicinal foodstuffs. III. Sugar beet. (1): hypoglycaemic oleanolic acid oligoglycosides, beta-vulgarosides I, II, III, and IV, from the root of *Beta vulgaris* L. (Chenopodiaceae). Chem Pharm Bull. 1996;44:1212–7.
60. Dubey H, Singh A, Patole AM, Patole AM, Tenpe CR, Ghule BV. Allicin? A SUR2 opener: possible mechanism for the treatment of diabetic hypertension in rats. Rev Bras Farmacogn. 2012;22:1053–9.
61. Gao H, Huang YN, Xu PY, Kawabata J. Inhibitory effect on α-glucosidase by the fruits of *Terminalia chebula* Retz. Food Chem. 2007;105:628–34.
62. Narender T, Puri A, Shweta, Khaliq T, Saxena R, Bhatia G, Chandra R. 4-hydroxyisoleucine an unusual amino acid as antidyslipidemic and antihyperglycemic agent. Bioorg Med Chem. 2006;16(2):293–6.
63. Alves-Silva JM, Zuzarte M, Marques C, Salgueiro L, Girao H. Protective effects of terpenes on the cardiovascular system: current advances and future perspectives. Curr Med Chem. 2016;23(40):4559–600.
64. Vijayalakshmi A, Ravichandiran V, Velraj M, Hemalatha S, Sudharani G, Jayakumari S. Anti-anaphylactic and anti-inflammatory activities of a bioactive alkaloid from the root bark of *Plumeria acutifolia* Poir. Asian Pac J Trop Biomed. 2011;1:401–5.
65. Gao LN, Feng QS, Zhang XF, Wang QS, Cui YL. Tetrandrine suppresses articular inflammatory response by inhibiting pro-inflammatory factors via NF-κB inactivation. J Orthop Res. 2016;34:1557–68.
66. Yu Y, Zhang M, Hu Y, Zhao Y, Teng F, Lv X, Li J, Zhang Y, Hatch GM, Chen L. Increased bioavailable berberine protects against myocardial ischemia reperfusion injury through attenuation of NFκB and JNK signaling pathways. Int Heart J. 2018;59:1378–88.
67. Zhao Y, Tian X, Liu G, Wang K, Xie Y, Qiu Y. Berberine protects myocardial cells against anoxia-reoxygenation injury via p38 MAPK-mediated NF-κB signaling pathways. Exp Ther Med. 2019;17:230–6.
68. Li C, Gao Y, Tian J, Shen J, Xing Y, Liu Z. Sophocarpine administration preserves myocardial function from ischemiareperfusion in rats via NF-κB inactivation. J Ethnopharmacol. 2011;135:620–5.
69. Li J, Li L, Chu H, Sun X, Ge Z. Oral sophocarpine protects rat heart against pressure overload-induced cardiac fibrosis. Pharm Biol. 2014;52:1045–51.
70. Jiang C, Tong YL, Zhang D, Liu LZ, Wang JF. Sinomenine prevents the development of cardiomyopathy in diabetic rats by inhibiting infammatory responses and blocking activation of NF-κB. Gen Physiol Biophys. 2017;36:65–74.

71. Kuchta K, Volk RB, Rauwald HW. Stachydrine in *Leonurus cardiaca, Leonurus japonicus, Leonotis leonurus*: detection and quantifcation by instrumental HPTLC and ^{1}H-qNMR analyses. Pharmazie. 2013;68:534.
72. Chen HH, Zhao P, Zhao WX, Tian J, Guo W, Xu M, Zhang C, Lu R. Stachydrine ameliorates pressure overload-induced diastolic heart failure by suppressing myocardial fibrosis. Am J Transl Res. 2017;9:4250–60.
73. Zhao L, Wu D, Sang M, Xu Y, Liu Z, Wu Q. Stachydrine ameliorates isoproterenol-induced cardiac hypertrophy and fibrosis by suppressing inflammation and oxidative stress through inhibiting NF-κB and JAK/STAT signaling pathways in rats. Int Immunopharmacol. 2017;48:102–9.
74. Meng YY, Liu Y, Hu ZF, Zhang Y, Ni J, Ma ZG, Liao H-H, Wu Q-Q, Tang Q-Z. Sanguinarine attenuates lipopolysaccharide-induced inflammation and apoptosis by inhibiting the TLR4/NF-κB pathway in H9c2 cardiomyocytes. Curr Med Sci. 2018;38:204–11.
75. Yang D, Jia W, Zhu YZ. Leonurine, a potential agent of traditional Chinese medicine: recent updates and future perspectives. Nat Prod Commun. 2016;11:1757–61.
76. Liu XH, Xin H, Hou AJ, Zhu YZ. Protective effects of leonurine in neonatal rat hypoxic cardiomyocytes and rat infarcted heart. Clin Exp Pharmacol Physiol. 2009;36:696–703.
77. Wang R, Li D, Ouyang J, Tian X, Zhao Y, Peng X, Li S, Yu G, Yang J. Leonurine alleviates LPS-induced myocarditis through suppressing the NF-κB signaling pathway. Toxicology. 2019;422:1–13.
78. Hamer M, O'Donovan G, Stamatakis E. High-density lipoprotein cholesterol and mortality: too much of a good thing? Arterioscler Thromb Vasc Biol. 2018;38:669–72.
79. Lin K, Chen H, Chen X, Qian J, Huang S, Huang W. Efficacy of curcumin on aortic atherosclerosis: a systematic review and meta-analysis in mouse studies and insights into possible mechanisms. Oxid Med Cell Longev. 2020;1(520):747.
80. Shahbazi S, Sahrawat TR, Ray M, Dash S, Kar D, Singh S. Drug targets for cardiovascular-safe anti-inflammatory: in silico rational drug studies. PLoS One. 2016;11:e0156156.
81. Chen J, Cao X, Cui Y, Zeng G, Chen J, Zhang G. Resveratrol alleviates lysophosphatidylcholine-induced damage and inflammation in vascular endothelial cells. Mol Med Rep. 2018;17:4011–8.
82. Cong X, Li Y, Lu N, Dai Y, Zhang H, Zhao X, Liu Y. Resveratrol attenuates the inflammatory reaction induced by ischemia/reperfusion in the rat heart. Mol Med Rep. 2014;9:2528–32.
83. Nie J, Zhang L, Zhao G, Du X. Quercetin reduces atherosclerotic lesions by altering the gut microbiota and reducing atherogenic lipid metabolites. J Appl Microbiol. 2019;127:1824–34.
84. Sanchez M, Galisteo M, Vera R, Villar IC, Zarzuelo A, Tamargo J, Perez-Vizcaino F, Duarte J. Quercetin downregulates NADPH oxidase, increases eNOS activity and prevents endothelial dysfunction in spontaneously hypertensive rats. J Hypertens. 2006;24:75–84.
85. Calabriso N, Scoditti E, Massaro M, Pellegrino M, Storelli C, Ingrosso I, Giovinazzo G, Carluccio MA. Multiple anti-inflammatory and anti-atherosclerotic properties of red wine polyphenolic extracts: differential role of hydroxycinnamic acids, flavonols and stilbenes on endothelial inflammatory gene expression. Eur J Nutr. 2016;55:477–89.
86. Bhaskar S, Helen A. Quercetin modulates toll-like receptor-mediated protein kinase signaling pathways in oxLDL-challenged human PBMCs and regulates TLR-activated atherosclerotic inflammation in hypercholesterolemic rats. Mol Cell Biochem. 2016;423:53–65.
87. Vera M, Torramade-Moix S, Martin-Rodriguez S, Cases A, Cruzado JM, Rivera J, Escolar G, Palomo M, Diaz-Ricart M. Antioxidant and anti-inflammatory strategies based on the potentiation of glutathione peroxidase activity prevent endothelial dysfunction in chronic kidney disease. Cell Phys Biochem Int J Exp Cell Physiol Biochem Pharmacol. 2018;51:1287–300.
88. Pothineni NVK, Karathanasis SK, Ding Z, Arulandu A, Varughese KI, Mehta JL. LOX-1 in atherosclerosis and myocardial ischemia: biology, genetics, and modulation. J Am Coll Cardiol. 2017;69:2759–68.
89. Ciumarnean L, Milaciu MV, Runcan O, Vesa SC, Rachisan AL, Negrean V, Perné M-G, Donca VI, Alexescu T-G, Para I, Dogaru G. The effects of flavonoids in cardiovascular diseases. Molecules. 2020;25(18):4320.
90. Wang P, Sun J, Lv S, Xie T, Wang X. Apigenin alleviates myocardial reperfusion injury in rats by downregulating miR-15b. Med Sci Monit Int Med J Exp Clin Res. 2019;25:2764–76.
91. Xiao C, Xia ML, Wang J, Zhou XR, Lou YY, Tang LH, Zhang FJ, Yang JT, Qian LB. Luteolin attenuates cardiac ischemia/reperfusion injury in diabetic rats by modulating Nrf2 antioxidative function. Oxid Med Cell Longev. 2019;2(719):252.
92. Jia Z, Nallasamy P, Liu D, Shah H, Li JZ, Chitrakar R, Si H, McCormick J, Zhu H, Zhen W, Li Y. Luteolin protects against vascular inflammation in mice and TNF-alpha-induced monocyte adhesion to endothelial cells via suppressing IKppaBalpha/NF-kappaB signaling pathway. J Nutr Biochem. 2015;26:293–302.
93. Ding X, Zheng L, Yang B, Wang X, Ying Y. Luteolin attenuates atherosclerosis via modulating signal transducer and activator of transcription 3-mediated inflammatory response. Drug Des Dev Ther. 2019;13:3899–911.
94. Testai L, Piragine E, Piano I, Flori L, Da Pozzo E, Miragliotta V, Pirone A, Citi V, Di Cesare Mannelli L, Brogi S, Carpi S, Martelli A. The citrus flavonoid naringenin protects the myocardium from ageing-dependent dysfunction: potential role of SIRT1. Oxid Med Cell Longev. 2020;2019:4650207.
95. Liang JN, Zou X, Fang XH, Xu JD, Xiao Z, Zhu JN, Li H, Yang J, Zeng N, Yuan SJ, Pan R, Fu Y-J, Zhang M, Luo J-F, Wang S, Shan Z-X. The Smad3-miR-29b/miR-29c axis mediates the protective effect of macrophage migration inhibitory factor against cardiac fibrosis. Biochim Biophys Acta Mol Basis Dis. 2019;1865:2441–50.
96. Devika PT, Prince PSM. (-)-epigallocatechin gallate (EGCG) prevents isoprenaline-induced cardiac toxicity by stabilizing cardiac marker enzymes and membrane-bound ATpases. J Pharm Pharmacol. 2010;60:125–33.
97. Che J, Liang B, Zhang Y, Wang Y, Tang J, Shi G. Kaempferol alleviates ox-LDL-induced apoptosis by up-regulation of autophagy via inhibiting PI3K/Akt/mTOR pathway in human endothelial cells. Cardiovasc Pathol. 2017;31:57–62.
98. Gilbert ER, Liu D. Anti-diabetic functions of soy isoflavone genistein: mechanisms underlying effects on pancreatic β-cell function. Food Funct. 2013;4:200–12.
99. Son MJ, Miura Y, Yagasaki K. Mechanisms for antidiabetic effect of gingerol in cultured cells and obese diabetic model mice. Cytotechnology. 2015;67:641–52.
100. Diego A, Maria F, Yolanda E, Eduardo F, Marcelo A, Ivan P. Natural bioactive compounds as protectors of mitochondrial dysfunction in cardiovascular diseases and aging. Molecules. 2019;24:4259.
101. Panichayupakaranant P, Ahmad MI. Plumbagin and its role in chronic diseases. Adv Exp Med Biol. 2016;929:229–46.
102. Wang SX, Wang J, Shao JB, Tang WN, Zhong JQ. Plumbagin mediates cardioprotection against myocardial ischemia/reperfusion injury through Nrf-2 signaling. Med Sci Monit. 2016;22:1250–7.
103. Chen C, Liang Z, Chen Q, Li ZG. Irbesartan and emodin on myocardial remodeling in Goldblatt hypertensive rats. J Cardiovasc Pharmacol. 2012;60:375–80.

104. Ye B, Chen X, Dai S, Han J, Liang X, Lin S, Cai X, Huang Z, Huang W. Emodin alleviates myocardial ischemia/reperfusion injury by inhibiting gasdermin D-mediated pyroptosis in cardiomyocytes. Drug Des Dev Ther. 2019;13:975–90.
105. Pang H, Han B, Yu T, Peng Z. The complex regulation of tanshinone IIA in rats with hypertension-induced left ventricular hypertrophy. PLoS One. 2014;9:e92216.
106. Han JY, Fan JY, Horie Y, Miura S, Cui DH, Ishii H, Hibi T, Tsuneki H, Kimura I. Ameliorating effects of compounds derived from *Salvia miltiorrhiza* root extract on microcirculatory disturbance and target organ injury by ischemia and reperfusion. Pharmacol Ther. 2008;117:280–95.
107. Chen J, Xie J, Jiang Z, Wang B, Wang Y, Hu X. Shikonin and its analogs inhibit cancer cell glycolysis by targeting tumor pyruvate kinase -M2. Oncogene 2011;30:4297–4306.
108. Lyu X, Lee J, Chen WN. Potential natural food preservatives and their sustainable production in yeast: terpenoids and polyphenols. J Agric Food Chem. 2019;67:4397–417.
109. Zaini R, Clench MR, Le Maitre CL. Bioactive chemicals from carrot (*Daucus carota*) juice extracts for the treatment of leukemia. J Med Food. 2011;(14):1303–12.
110. Li R, Lu K, Wang Y, Chen M, Zhang F, Shen H, Yao D, Gong K, Zhang Z. Triptolide attenuates pressure overload-induced myocardial remodeling in mice via the inhibition of NLRP3 inflammasome expression. Biochem Biophys Res Commun. 2017;485:69–75.
111. Guo X, Xue M, Li CJ, Yang W, Wang S-S, Ma Z-J, Zhang X-n, Wang X-y, Zhao R, Chang B-c, Chen L-m. Protective effects of triptolide on TLR4 mediated autoimmune and inflammatory response induced myocardial fibrosis in diabetic cardiomyopathy. J Ethnopharmacol. 2016;193:333–44.
112. Wang KS, Li J, Wang Z, Mi C, Ma J, Piao LX, Xu GH, Li X, Jin X. Artemisinin inhibits inflammatory response via regulating NF-κB and MAPK signaling pathways. Immunopharmacol Immunotoxicol. 2017;39:28–36.
113. Xiong Z, Sun G, Zhu C, Cheng B, Zhang C, Ma Y, Dong Y. Artemisinin, an anti-malarial agent, inhibits rat cardiac hypertrophy via inhibition of NF-κB signaling. Eur J Pharmacol. 2010;649:277–84.
114. Wang QS, Gao T, Cui YL, Gao LN, Jiang HL. Comparative studies of paeoniforin and albiforin from *Paeonia lactifora* on anti-inflammatory activities. Pharm Biol. 2014;52:1189–95.
115. Zhou P, Yang X, Jia X, Yu J, Asenso J, Xiao F, Wang C, Wei W. Effect of 6′-acetylpaeoniforin on dinitrochlorobenzene-induced allergic contact dermatitis in BALB/c mice. Immunol Res. 2016;64:857–68.
116. Li JZ, Xie M, Mo D, Zhao XF, Yu S-Y, Liu L-J, Wu C, Yang Y. Picroside II protects myocardium from ischemia/reperfusion-induced injury through inhibition of the inflammatory response. Exp Ther Med. 2016;12:3507–14.
117. Chen RC, Wang J, Yang L, Sun GB, Sun XB. Protective effects of ginsenoside Re on lipopolysaccharide-induced cardiac dysfunction in mice. Food Funct. 2016;7:2278–87.
118. Wang QW, Yu XF, Xu HL, Jiang YC, Zhao XZ, Sui DY. Ginsenoside Re attenuates isoproterenol-induced myocardial injury in rats. Evid Based Complement Alternat Med. 2018;2018:8637134.
119. Chen CT, Wang ZH, Hsu CC, Lin HH, Chen JH. In vivo protective effects of diosgenin against doxorubicin-induced cardiotoxicity. Nutrients. 2015;7:4938–54.
120. Zhang R, Han D, Li Z, Shen C, Zhang Y, Li J, Yan G, Li S, Hu B, Li J, Liu P. Ginkgolide C alleviates myocardial ischemia/reperfusion-induced inflammatory injury via inhibition of CD40-NF-κB pathway. Front Pharmacol. 2018;9:109.
121. Su S, Li Q, Liu Y, Xiong C, Li J, Zhang R, Niu Y, Zhao L, Wang Y, Guo H. Sesamin ameliorates doxorubicin-induced cardiotoxicity: involvement of Sirt1 and Mn-SOD pathway. Toxicol Lett. 2014;224:257–63.
122. Chung SY. Method for preparing extract for the prevention and treatment of hyperlipidemia and obesity from the extract of *Aster spathulifolius* aerial part and composition containing the same. (US20090232912A1). 2009. [cited 2023 Jan 29]. https://patents.google.com/patent/US20090232912A1/en?oq=US2009232912+.
123. Xiuhua Z. Medicine composition for treating coronary heart disease and preparation method thereof. 2015. [cited 2023 Jan 29]. https://worldwide.espacenet.com/publicationDetails/biblio?FT=D&date=20151223&DB=&locale=en_EP&CC=CN&NR=105168396A&KC=A&ND=4.
124. Gao Y, Su Y, Huo Y, Mi J, Wang X, Wang Z, Liu Y, Zhang H. Identification of antihyperlipidemic constituents from the roots of *Rubia yunnanensis* Diels. J Ethnopharmacol. 2014;155:1315–21.
125. Wang M, Huang W, Hu Y, Zhang L, Shao Y, Wang M, Zhang F, Zhao Z, Mei X, Li T, Wang D, Liang Y, Li J, Huang Y, Zhang L, Xu T, Song H, Zhong Y, Lu B. Phytosterol profiles of common foods and estimated natural intake of different structures and forms in China. J Agric Food Chem. 2018;66:2669–76.
126. Ramachandran V, Saravanan R, Senthilraja P. Antidiabetic and antihyperlipidemic activity of asiatic acid in diabetic rats, role of HMG CoA: in vivo and in silico approaches. Phytomedicine. 2014;21:225–32.
127. Valsta L, Lemström A, Ovaskainen M-L, Lampi A-M, Toivo J, Korhonen T, Piironen V. Estimation of plant sterol and cholesterol intake in Finland: quality of new values and their effect on intake. Br J Nutr. 2004;92:671–8.
128. Klingberg S, Andersson H, Mulligan A, Bhaniani A, Welch A, Bingham S, Khaw K-T, Andersson S, Ellegard L. Food sources of plant sterols in the EPIC Norfolk population. Eur J Clin Nutr. 2008;62:695–703.
129. Rios MY, Lopez-Martinez S, Lopez-Vallejo F, et al. Vasorelaxant activity of some structurally related triterpenic acids from *Phoradendron reichenbachianum* (Viscaceae) mainly by NO production: ex vivo and in silico studies. Fitoterapia. 2012;83(6):1023–9.
130. Chan C, Mong M, Liu W, Huang C, Yin M. Three pentacyclic triterpenes protect H9c2 cardiomyoblast cells against high-glucose-induced injury. Free Radic Res. 2014;48(4):402–11.
131. Der Sarkissian S, Cailhier JF, Borie M, Stevens L-M, Gaboury L, Mansour S, Hamet P, Noiseux N. Celastrol protects ischaemic myocardium through a heat shock response with up-regulation of haeme oxygenase-1. Br J Pharmacol. 2014;171(23):5265–79.
132. Bai Y, Wang X, Zhao S, Ma C, Cui J, Zheng Y. Sulforaphane protects against cardiovascular disease via Nrf2 activation. Oxid Med Cell Longev. 2015;2015:407580.
133. Barbara VH, David K. Phytochemicals and cardiovascular disease. Circulation. 2021;95:2591–3.
134. Chin YW, Balunas MJ, Chai HB, Kinghorn AD. Drug discovery from natural sources. AAPS J. 2006;8:E239–53.
135. Burger T, Mokoka T, Fouché G, Steenkamp P, Steenkamp V, Cordier W. Solamargine, a bioactive steroidal alkaloid isolated from *Solanum aculeastrum* induces non-selective cytotoxicity and P-glycoprotein inhibition. BMC Complement Altern Med. 2018;18:137.
136. Tauchen J, Huml L, Bortl L, Doskocil I, Jarosova V, Marsik P, Frankova A, Peralta CZM, Zans C, Havlik M-EJ. Screening of medicinal plants traditionally used in Peruvian Amazon for in vitro antioxidant and anticancer potential. Nat Prod Res. 2019;33:2718–21.
137. Frédérich M, Tits M, Angenot L, Greimers R, Gielen J, Bours V, Merville M-P. Isostrychnopentamine, an indolomonoterpenic alkaloid from *Strychnos usambarensis*, induces cell

cycle arrest and apoptosis in human colon cancer cells. JPET. 2003;304(3):1103–10.

138. Shoeb M, MacManus SM, Jaspars M, Trevidu J, Nahar L, Kong-Thoo-Lin P, Sarker SD. Montamine, a unique dimeric indole alkaloid, from the seeds of *Centaurea montana* (Asteraceae), and its in vitro cytotoxic activity against the CaCo2 colon cancer cells. Tetrahedron. 2006;62:11172–7.
139. Oh J, Kim GD, Kim S, Lee SK. Antofine, a natural phenanthroindoloizine alkaloid, suppresses angiogenesis via regulation of AKT/mTOR and AMPK pathway in endothelial cells and endothelial progenitor cells derived from mouse embryonic stem cells. 2017;107:201–7.
140. Yaffe PB, Coombs PMR, Doucette CD, Walsh M, Hoskin DW. Piperine, an alkaloid from black pepper, inhibits growth of human colon cancer cells via G1 arrest and apoptosis triggered by endoplasmic reticulum stress. Mol Carcinog. 2015;54:1070–85.
141. Singh S, Verma M, Malhotra M, Prakash S, Singh TD. Cytotoxicity of alkaloids isolated from *Argemone mexicana* on SW480 human colon cancer cell line. Pharm Biol. 2016;54:740–5.
142. Liby KT, Yore MM, Sporn MB. Triterpenoids and rexinoids as multifunctional agents for the prevention and treatment of cancer. Nat Rev Cancer. 2007;7:357–69.
143. Abdul-Hameed ZH, Bawakid NO, Alorfi HS, Sobahi TR, Alburae NA, Abdel-Lateff A, Elbehairi SEI, Alfaifi MY, Alhakamy NA, Alarif WM. Monoterpene indole alkaloids from the aerial parts of *Rhazya stricta* induce cytotoxicity and apoptosis in human adenocarcinoma cells. Molecules. 2022;27:1422.
144. Mishra DP, Khan MA, Yadav DK, Rawat AK, Singh RK, Ahamad T, Hussain MK, Saquib M, Khan MF. Monoterpene indole alkaloids from *Anthocephalus cadamba* fruits exhibiting anticancer activity in human lung cancer cell line H1299. ChemistrySelect. 2018;3:8468–72.
145. Zhang W, Jiang H, Chen Y, Ren F. Resveratrol chemosensitizes adriamycin-resistant breast cancer cells by modulating miR-122-5p. J Cell Biochem. 2019;120(9):1628.
146. Deng L, Wu X, Zhu X, Yu Z, Liu Z, Wang J, Zheng Y. Combination effect of curcumin with docetaxel on the PI3K/AKT/mTOR pathway to induce autophagy and apoptosis in esophageal squamous cell carcinoma. Am J Transl Res. 2021;13(1):57–72.
147. Yuan CH, Horng CT, Lee CF, Chiang NN, Tsai FJ, Lu CC, Chiang J-H, Hsu Y-M, Yang J-S, Chen F-A. Epigallocatechin gallate sensitizes cisplatin-resistant oral cancer CAR cell apoptosis and autophagy through stimulating AKT/STAT3 pathway and suppressing multidrug resistance 1 signaling. Environ Toxicol. 2017;32(3):845–55.
148. Wei F, Jiang X, Gao HY, Gao SH. Liquiritin induces apoptosis and autophagy in cisplatin (DDP)-resistant gastric cancer cells in vitro and xenograft nude mice in vivo. Int J Oncol. 2017;51(5):1383–94.
149. Kubina R, Iriti M, Kabala-Dzik A. Anticancer potential of selected flavonols: fisetin, kaempferol, and quercetin on head and neck cancers. Nutrients. 2021;13:845.
150. Ha TK, Jung I, Kim ME, Bae SK, Lee JS. Anti-cancer activity of myricetin against human papillary thyroid cancer cells involves mitochondrial dysfunction-mediated apoptosis. 2017;91:378–84.
151. Byun S, Lee KW, Jung SK, Lee EJ, Hwang MK, Lim SH, Bode AM, Lee HJ, Dong Z. Luteolin inhibits protein kinase C (epsilon) and c-Src activities and UVB-induced skin cancer. Cancer Res. 2010;70:2415–23.
152. Prakash V. Terpenoids as cytotoxic compounds: a perspective. Phcog Rev. 2018;12:166–76.
153. Xu W, Li T, Qiu JF, Wu SS, Huang MQ, Lin L-G, Zhang Q-W, Chen X-P, Lu J-J. Anti-proliferative activities of terpenoids isolated from *Alisma orientalis* and their structure-activity relationships. Anticancer Agents Med Chem. 2015;15:228–35.
154. Fan D, Parhira S, Zhu GY, Jiang ZH, Bai LP. Triterpenoids from the stems of *Tripterygium regelii*. Fitoterapia. 2016;113:69–73.
155. Liu Q, Ahn JH, Kim SB, Lee C, Hwang BY, Lee MK. Sesquiterpene lactones from the roots of *Lindera strychnifolia*. Phytochemistry. 2013;87:112–8.
156. Wu J, Yang G, Zhu W, Wen W, Zhang F, Yuan J, An L. Anti-atherosclerotic activity of platycodin D derived from roots of *Platycodon grandiflorum* in human endothelial cells. Biol Pharm Bull. 2012;35:1216–21.
157. Kikuchi T, Akihisa T, Tokuda H, Ukiya M, Watanabe K, Nishino H. Cancer chemopreventive effects of cycloartane-type and related triterpenoids in in vitro and in vivo models. J Nat Prod. 2007;70:918–22.
158. Pasi S, Aligiannis N, Pratsinis H, Skaltsounis AL, Chinou IB. Biologically active triterpenoids from *Cephalaria ambrosioides*. Planta Med. 2009;75:163–7.
159. Wang S, Li J, Sun J, Zeng KW, Cui J-R, Jiang Y, Tu P-F. NO inhibitory guaianolide-derived terpenoids from *Artemisia argyi*. Fitoterapia. 2013;85:169–75.
160. Ou-Yang DW, Wu L, Li YL, Yang PM, Kong DY, Yang XW, et al. Miscellaneous terpenoid constituents of Abies nephrolepis and their moderate cytotoxic activities. Phytochemistry. 2011;72:2197–204.
161. Huang J, Han HY, Li GY, Wang HY, Zhang C, Zhang K, Tan Y, Li P-Y. Wang J-H. Two new terpenoid benzoates with antitumor activity from the roots of *Ferula dissecta*. J Asian Nat Prod Res. 2013;15:1100–6.
162. Chaipukdee N, Kanokmedhakul K, Kanokmedhakul S, Lekphrom R, Pyne SG. Two new bioactive iridoids from *Rothmannia wittii*. Fitoterapia. 2016;113:97–101.
163. Bakarnga-Via I, Hzounda JB, Fokou PV, Tchokouaha LR, Gary-Bobo M, Gallud A, Garcia M, Walbadet L, Secka Y, Dongmo PMJ, Boyom FF, Menut C. Composition and cytotoxic activity of essential oils from *Xylopia aethiopica* (Dunal) A. Rich, *Xylopia parviflora* (A. Rich) benth.) and *Monodora myristica* (Gaertn) growing in chad and Cameroon. BMC Complement Altern Med. 2014;14:125.
164. Calvaruso M, Pucci G, Musso R, Bravatà V, Cammarata FP, Russo G, Forte GI, Minafra L. Nutraceutical compounds as sensitizers for cancer treatment in radiation therapy. Int J Mol Sci. 2019;20:5267.
165. WHO. Chronic respiratory disease. 2009.
166. Zhao YL, Yang ZF, Wu BF, Shang JH, Liu YP, Wang XH, Luo XD. Indole alkaloids from leaves of *Alstonia scholaris* (L.) R. Br. protect against emphysema in mice. J Ethnopharmacol. 2020;259:112949.
167. Tewa XN, Laua NJX, Chellappanb DK, Madheswaranc T, Zeeshanc F, Tambuwalad MM, Aljabalie AAA, Balusamyf SR, Perumalsamyg H, Guptah G, Oliveri BG, Hsui A, Warki P, Reddyi K, Wadhwak R, Hansbrok PM, Dua K. Immunological axis of berberine in managing inflammation underlying chronic respiratory inflammatory diseases. Chem Biol Interact 2020;317:108947.
168. Wang Q, Cui Y, Wu X, Wang J. Evodiamine protects against airway remodelling and inflammation in asthmatic rats by modulating the HMGB1/NF-κB/TLR-4 signalling pathway. Pharm Biol. 2021;59:192–9.
169. Silberberg M, Morand C, Mathevon T, Bessong C, Manach C, Scalbert A, Remesy C. The bioavailability of polyphenols is highly governed by the capacity of the intestine and of the liver to secrete conjugated metabolites. Eur J Nutr. 2006;45:88–96.
170. Dragonieri S, Tongoussouva O, Zanini A, Imperatori A, Spanevello A. Markers of airway inflammation in pulmonary diseases assessed by induced sputum. Monaldi Arch Chest Dis. 2016;71:119–26.

171. Bargagli E, Olivieri C, Bennett D, Prasse A, Muller-Quernheim J, Rottoli P. Oxidative stress in the pathogenesis of diffuse lung diseases: a review. Respir Med. 2009;103:1245–56.
172. Lee M, Kim S, Kwon O-K, Oh S-R, Lee H-K, Ahn K. Anti-inflammatory and anti-asthmatic effects of resveratrol, a polyphenolic stilbene, in a mouse model of allergic asthma. Int Immunopharmacol. 2009;9:418–24.
173. Liu H, Ren J, Chen H, Huang Y, Li H, Zhang Z, Wang J. Resveratrol protects against cigarette smoke-induced oxidative damage and pulmonary inflammation. J Biochem Mol Toxicol. 2014;28:465–71.
174. Venkatesan N, Punithavathi D, Babu M. Protection from acute and chronic lung diseases by curcumin. In: The molecular targets and therapeutic uses of curcumin in health and disease. Springer; 2007. p. 379–405.
175. Biswas SK, McClure D, Jimenez LA, Megson IL, Rahman I. Curcumin induces glutathione biosynthesis and inhibits NF-κB activation and interleukin-8 release in alveolar epithelial cells: mechanism of free radical scavenging activity. Antioxid Redox Signal. 2005;7:32–41.
176. Nanua S, Zick SM, Andrade JE, Sajjan US, Burgess JR, Lukacs NW, Hershenson MB. Quercetin blocks airway epithelial cell chemokine expression. Am J Respir Cell Mol Biol. 2006;35:602–10.
177. Ying B, Yang T, Song X, Hu X, Fan H, Lu X, Chen L, Cheng D, Wang T, Liu D, Xu D, Wie Y, Wen F. Quercetin inhibits IL-1 beta-induced ICAM-1 expression in pulmonary epithelial cell line A549 through the MAPK pathways. Mol Bio Rep. 2009;36:1825–32.
178. Li G-X, Chen Y-K, Hou Z, Xiao H. Pro-oxidative activities and dose–response relationship of (-)-epigallocatechin-3-gallate in the inhibition of lung cancer cell growth: a comparative study in vivo and in vitro. Carcinogenesis. 2010;31:902–10.
179. Sadava D, Whitlock E, Kane SE. The green tea polyphenol, epigallocatechin-3- gallate inhibits telomerase and induces apoptosis in drug-resistant lung cancer cells. Biochem Biophys Res Commun. 2007;360:233–7.
180. Liu MH, Lin AH, Lee HF, Ko HK, Lee TS, Kou YR. Paeonol attenuates cigarette smoke-induced lung inflammation by inhibiting ROS-sensitive inflammatory signaling. Mediators Inflamm. 2014;2014:651890.
181. Favarin CD, Teixeira MM, de Andrade LE, de Freitas Alves C, Lazo Chica JE, Arterio Sorgi C, Faccioli LH, Rogerio PA. Anti-inflammatory effects of ellagic acid on acute lung injury induced by acid in mice. Mediators Inflamm. 2013;2013:164202.
182. Wei M, Chu X, Jiang L, Yang X, Cai Q, Zheng C, Ci X, Guan M, Liu J, Deng X. Protocatechuic acid attenuates lipopolysaccharide-induced acute lung injury. Inflammation. 2012;35:1169–78.
183. Ganesan S, Faris AN, Comstock AT, Chattoraj S, Chattoraj A, Burgess JR, Curtis JL, Martinez FJ, Zick S, Hershenson MB, Sajjan U. Quercetin prevents progression of disease in elastase/LPS-exposed mice by negatively regulating MMP expression. Respir Res. 2010;11:131.
184. Li L, Bao H, Wu J, Duan X, Liu B, Sun J, Gong W, Lv Y, Zhang H, Luo Q, Wu X, Dong J. Baicalin is antiinflammatory in cigarette smoke-induced inflammatory models in vivo and in vitro: a possible role for HDAC2 activity. Int Immunopharmacol. 2012;13:15–22.
185. Nie YC, Wu H, Li PB, Luo YL, Long K, Xie LM, Shen JG, Su WW. Anti-inflammatory effects of naringin in chronic pulmonary neutrophilic inflammation in cigarette smoke exposed rats. J Med Food. 2012;15:894–900.
186. Chi G, Wei M, Xie X, Soromou LW, Liu F, Zhao S. Suppression of MAPK and NF-κB pathways by limonene contributes to attenuation of lipopolysaccharide-induced inflammatory responses in acute lung injury. Inflammation. 2013;36:501–11.
187. Bai GZ, Yu HT, Ni YF, Li XF, Zhang ZP, Su K, Lei J, Liu BY, Ke CK, Zhong DX, Wang YJ, Zhao JB. Shikonin attenuates lipopolysaccharide-induced acute lung injury in mice. J Surg Res. 2013;182:303–11.
188. Yingkun N, Zhenyu W, Jing L, Xinyun L, Huimin Y. Stevioside protects LPS-induced acute lung injury in mice. Inflammation. 2013;36:242–50.
189. Zhu Y, He W, Gao X, Li B, Mei C, Xu R, Chen H. Resveratrol overcomes gefitinib resistance by increasing the intracellular gefitinib concentration and triggering apoptosis, autophagy and senescence in PC9/G NSCLC cells. Sci Rep. 2015;5:17730.
190. Ribeiro A, Ferraz-Paula V, Pinheiro ML, Vitoretti LB, Souza MDP, Quinteiro-Filho WM, Akamine AT, Almeida VI, Quevedo J, Dal-Pizzol F, Hallak JE, Zuardi AW, Crippa JA, Palermo-Neto J. Cannabidiol, a non-psychotropic plant-derived cannabinoid, decreases inflammation in a murine model of acute lung injury: role for the adenosine A(2A) receptor. Eur J Pharmacol. 2012;678:78–85.
191. San Z, Fu Y, Li W, Zhou E, Li Y, Song X, Wang T, Tian Y, Wei Z, Yao M, Cao Y, Zhang N. Protective effect of taraxasterol on acute lung injury induced by lipopolysaccharide in mice. Int Immunopharmacol. 2014;19:342–50.
192. Wei D, Huang Z. Anti-inflammatory effects of triptolide in LPS-induced acute lung injury in mice. Inflammation. 2014;37:1307–16.
193. Kancherla N, Dhakshinamoothi A, Chitra K, Komaram RB. Prelimanary analysis of phytoconstituents and evaluation of anthelminthic property of *Cayratia auriculata* (in vitro). Maedica (Bucur). 2019;14:350–6.
194. Wocheslander S, Groß F, Scholz B, Engel K. Quantitation of acyl chain oxidation products formed upon thermo-oxidation of phytosteryl/-stanylOleates and linoleates. J Agric Food Chem. 2017;65(11):2435–42.
195. Szydłowska-Czerniak A, Tułodziecka A. Optimization of ultrasound-assisted extraction procedure to determine antioxidant capacity of rapeseed cultivars. Food Anal Methods. 2014;8(3):778–89.
196. Fansworth NR, Fongs HNS, Tin-Wa N, Dobberstein RH. Phytochemical screening methods. Chicago: University of Ilinois Press; 1977.
197. Popova IE, Hall C, Kubátová A. Determination of lignans in flaxseed using liquid chromatography with time-of-flight mass spectrometry. J Chromatogr A. 2009;1216:217–29.
198. Verpoorte R, Schripsema J. Isolation, identification, and structure elucidation of alkaloids a general overview. In: Linskens, Jackson, editors. Modern methods of plant analysis, vol. 15. Alkaloids. Berlin: Springer; 1994.
199. Christophoridou S, Dais P, Tseng L-H, Spraul M. Separation and identification of phenolic compounds in olive oil by coupling high-performance liquid chromatography with postcolumn solid-phase extraction to nuclear magnetic resonance spectroscopy (LC-SPE-NMR). J Agric Food Chem. 2005;53:4667–79.
200. Beckey HD. 1-Theory of field ionization (FI) and field emission (FE). Bergama: Pergamon; 1971.
201. Beckey HD. 2–Field ionization sources. Bergama: Pergamon; 1971.
202. Beckey HD. 3-Application of the FI mass spectrometer to physicochemical problems. Bergama: Pergamon; 1971.
203. Beckey HD. 4-Qualitative analyses with the FI mass spectrometer. Bergama: Pergamon; 1971.
204. Kemp W. Energy and electromagnetic spectrum. In: Kemp W, editor. Organic spectroscopy. London: Macmillan Press; 1991. p. 1–7.
205. Banu SK, Cathrine L. General techniques involved in phytochemical Analysis. IJARCS. 2015;2:25–32.
206. Kemp W. Infrared spectroscopy. In: Organic spectroscopy. London: Macmillan Press Ltd.; 1991. p. 19–56.

207. Urbano M, Luque de Castro MD, Pérez PM, García-Olmo J, Gómez-Nieto MA. Ultraviolet-visible spectroscopy and pattern recognition methods for differentiation and classification of wines. Food Chem. 2006;97:166–75.
208. Griffiths PR, de Haseth JA. Fourier-transform infrared spectrometry. Science. 2007;222:297–302.
209. Massa W. Crystal structure determination (2nd completely updated ed.). New York: Springer; 2004.
210. Schripsema J, Verpoorte R. Rapid identification of trace amounts of indole alkaloids: analysis of the aromatic pattern from the ^{1}H-NMR spectrum. In: Atta-ur-Rahman, editor. Studies in natural products chemistry, vol. 9. Amsterdam: Elsevier; 1991. p. 163–99.
211. Sangster AW, Stuart KL. Ultra-violet spectra of alkaloids. Chem Rev. 1965;65:69–130.
212. Crab TA, Newton RF, Jackson D. Stereochemical studies of *N*-bridgehead compounds by spectral means. Chem Rev. 1971;71:109–26.
213. Hesse M. Progress in mass spectrometry, vol 1, parts 1 and 2. Mass spectrometry of indole alkaloids. Weinheim: Verlag Chemie; 1974.
214. Hesse M, Bernhard HO. Progress in mass spectrometry, vol 3. Mass spectrometry of alkaloids. Weinheim: Verlag Chemie; 1975.
215. Shamma M, Hindenlang DM. Carbon-13 NMR shift assignments of amines and alkaloids. New York: Plenum; 1979.
216. Schripsema J, Verpoorte R, Baerheim Svendsen A. Trifluoroacetic acid, a ^{1}H-NMR shift reagent for alkaloids. Tetrahedron Lett. 1986;27:2523–6.
217. Bankeu JJK, Kagho DUK, Fongang YSF, Toghueo RMK, Mba'ning BM, Feuya GRT, Fekam FB, Tchouankeu JC, Ngouela SA, Sewald N, Lenta BN, Ali MS. Constituents from *Nauclea latifolia* with anti-*Haemophilus influenzae* Type b inhibitory activities. J Nat Prod. 2019;82:2580–5.
218. Shigemori H, Kagata T, Ishiyama H, Morah F, Ohsaki A, Kobayash J. Nucleamides A-E, new monoterpene indole alkaloids from *Nauclea latifolia*. Chem Pharm Bull. 2003;51:58–61.
219. Bankeu JJK, Madjouka S, Feuya GRT, Fongang YSF, Siddiqui S, Iftikhar A, Mehreen L, Lenta BN, Sammer Y, Noungoué DT, Ngouela A, Shaiq MA. Pobeguinine: a monoterpene indole alkaloid and other bioactive constituents from the stem bark of *Nauclea pobeguinii*. Zeitschrift für Naturforschung C. 2018;73:335–44.
220. Bondonno CP, Croft KD, Ward N, Considine MJ, Hodgson JM. Dietary flavonoids and nitrate: effects on nitric oxide and vascular function. Nutr Rev. 2015;73:216–35.
221. Carocho M, Ferreira CFI. The role of phenolic compounds in the fight against cancer-a review. Anticancer Agents Med Chem. 2013;13:1236–58.
222. Bruneton J. Pharmacognosie, Phytochimie, Plantes médicinales. 4th ed. Paris: Technique et Documentation, Lavoisier; 2009. p. 1288.
223. Markham KR. Techniques of flavonoids identification. London: Academic Press; 1982.
224. Markham KR. Flavones, flavonols and their glycosides. In: Dey PP, Harborne JB, editors. Methods in plant biochemistry. Academic Press; 1989. p. 197–232.
225. Harborne JB, Williams CA. Advances in flavonoid research since 1992. Phytochemistry. 2000;55:481–504.
226. Agrawal K, Markham R. Carbon-13 NMR of flavonoids: introduction. In: Agrawal PK, editor. Studies in organic chemistry; 1989. p. 1–40.
227. Markhan KR, Mabry TJ. Ultraviolet-visible and proton magnetic resonance spectroscopy of flavonoids. In: Harborne JB, Mabry TJ, Mabry H, editors. The flavonoids. London: Chapman and Hall; 1975. p. 45–77.
228. Markham R, Geiger H. ^{1}H-NMR spectroscopy of flavonoids and their glycosides, in hexadeuterodimethylsulfoxide. In: The flavonoids. Edition Harborne. London: Chapman and Hall; 1993. p. 33–99.
229. Agrawal PK. Flavonoid structure and carbon-13 NMR spectroscopy. In: Agrawal PK, editor. Carbon 13-NMR of flavonoids. New York: Elsevier; 1989.
230. Levy GC, Nelson GL. Carbon-13 nuclear magnetic resonance for organic chemists. New York: Wiley; 1972. p. 80–1.
231. Markham KR, Chari VM. Carbon-13 NMR spectroscopy of flavonoids (Chap 2). In: Harborne JB, et al., editors. The flavonoids. Dordrecht: Springer; 1982. p. 19–134.
232. Manasa G. Structural elucidation of terpenoids by spectroscopic techniques. JCHPS. 2014;5:36–7.
233. Hamimed S. Caractérisation chimique des principes à effet antidermatophyte des racines d'*Anacyclus pyrethrum* L. Mémoire Présenté pour L'obtention du diplôme de Magister en chimie organique, option: Substances thérapeutique d'origine naturelle, faculté des sciences exactes département de chimie. Algérie: Universite Mentouri Constantine; 2009. p. 24–31.
234. Bory S, Fetizon M. Bull Soc Chim Fr. 1964;11:3458.
235. Ogunkoya L. Application of mass spectrometry in structural problems in triterpenes. Phytochemistry. 1980;20:121–6.
236. Simla B. Phytochemical investigations on Boswellia species: comparative studies on the essential oils, pyrolysates and boswellic acids of *Boswellia carterii* Birdw., *Boswellia serrata* Roxb., *Boswellia frereana* Birdw., *Boswellia neglecta* S. Moore and *Boswellia rivae* Engl. dissertation for the fulfillment of the requirements for the degree of Dr. rer. nat. Allemagne: University of Hamburg; 2005. p. 160–169.
237. Bruneton J. Pharmacognosie, phytochimie, plantes médicinales. 2nd ed. Paris: France Édition Tech. et Doc; 1999. p. 266–80.
238. Connolly J, Hill A. Methods in plants biochemistry, vol. 7. New York: Academy Press; 1991. p. 331–59.
239. Mahato S, Kundu A. 13C-MMR spectra of pentacyclic triterpenoids. A compilation and some salient features. Phytochemistry. 1994;37:1517–75.
240. Agrawal PK, Jain DC, Gupta RK, Thakur RS. Carbon-13 NMR spectroscopy of steroidal sapogenins and steroidal saponins. Phytochemistry. 1985;24:2479–96.
241. Sang S, Lao A, Wang H, Chen Z. Two new spirostanol saponins from *Allium tuberosum*. J Nat Prod. 1999;62:1028–9.

Role of Medicinal Plants in the Management of Diabetes Mellitus

Saroj Singhmura, Souvik Basak, and Nilanjan Ghosh

Abstract

Since ancient times, cure or prevention of diseases by medicinal plants was being practiced by indigenous people of different communities around the world through their traditional approaches. The development of civilization and humankind has shown the path of defining properties of medicinal plants and improvement in dealing with different diseases by the identification of responsible molecules. Diabetes mellitus (DM) represents a significant metabolic disorder of considerable concern. The disruption of insulin action in maintaining glucose homeostasis arises from elevated blood glucose levels, which can result from defects in insulin synthesis, secretion, binding to receptors, or increased insulin resistance. Other risk factors, such as obesity, urbanization, and genetic mutations, can further contribute to the development of DM. To counter this pathophysiological condition, several marketed remedies are available, but they limit their uses because of their expensive and several complications. On the other hand, the management of diabetes mellitus by herbal medicines is less cost-effective and the complications are also rare.

Keywords

Bioactive phytoconstituents · Glucose metabolism · Medicinal plants · Insulin resistance · Diabetes mellitus

1 Introduction

Diabetes mellitus is a persistent metabolic disorder marked by increased levels of blood glucose resulting from inadequate insulin production or ineffective utilization of insulin by the body and has become a global health concern. Globally, over 400 million people are currently affected by diabetes, and projections suggest this number may rise to 640 million by the year 2040. The impact of this disease is staggering, encompassing both public health expenses for diabetic patient care and the personal costs associated with severe complications. Approximately 90% of all diabetes cases are type 2 diabetes mellitus (T2DM) and are characterized by obesity, insulin resistance, and a gradual decline in insulin secretion. Genetic predisposition, age, obesity, improper dietary habits, and sedentary lifestyles are among the most significant factors contributing to the development of this condition. Thus, managing diabetes effectively is crucial to prevent complications and improve the quality of life for those living with the condition. The growing interest in natural and alternative therapies for diabetes management is a necessity of the time because of the side effects and economic burden of its management and hence medicinal plants have emerged as promising candidates in this regard. For centuries, traditional medicine systems have recognized the therapeutic potential of various plant species in controlling blood sugar levels and alleviating diabetes-related symptoms. The role of medicinal plants in managing diabetes mellitus is multifaceted. Within these medicinal plants, a rich assortment of bioactive compounds, including alkaloids, flavonoids, anthraquinone, saponins, glycosides, and carbohydrates, has been identified, exhibiting remarkable antidiabetic properties. Scientific research has focused on investigating the mechanisms of action of these compounds and their potential benefits in regulating glucose metabolism, enhancing insulin sensitivity, and reducing insulin resistance.

S. Singhmura · S. Basak
Dr. B. C. Roy College of Pharmacy and AHS, Durgapur, India

N. Ghosh (✉)
Department of Pharmaceutical Technology, Jadavpur University, Kolkata, India
e-mail: nilanjanghosh.phamacy@jadavpuruniversity.in

A. K. Dhara, S. C. Mandal (eds.), *Role of Herbal Medicines*, https://doi.org/10.1007/978-981-99-7703-1_5

2 Background: Blood Glucose Homeostasis and Its Faith in Diabetes Mellitus

Insulin maintains blood glucose homeostasis by taking into the cells, thus its level is maintained. If this regulation is disturbed, it results in a hyperglycemic state, and if not regulated, prolonged exposure may lead to chronic complications affecting the renal, neurological, gastrointestinal, genitourinary, cardiovascular system, and lower extremity [1]. In individuals without diabetes, the blood glucose concentration is meticulously controlled to provide a constant energy source for various tissues and organs. Insulin and glucagon, the primary hormones responsible for this equilibrium, are both secreted by the pancreas' β cells. Their release is triggered by a complex interplay of metabolic, hormonal, neuronal, and pharmacological factors, working through interconnected intracellular signaling pathways.

2.1 Pathways of Insulin Secretion

The secretion of insulin is a highly controlled process aimed at maintaining stable blood glucose levels during fasting and feeding periods. This delicate regulation relies on the orchestrated interaction between glucose and non-nutrient secretagogues, such as neurotransmitters and hormones.

2.1.1 Pathway of Insulin Secretion Induced by Glucose Stimulation

The homeostatic blood glucose level is upheld by insulin released by the beta (β) cells within the pancreatic islets of Langerhans. Insulin plays a pivotal role in governing the body's metabolism of carbohydrates, proteins, and fats. Its main function involves facilitating the uptake of carbohydrates, predominantly glucose, from the bloodstream into skeletal muscle, adipose tissue, and the liver. When foods containing carbohydrates are ingested, enzymatic processes trigger a rise in blood glucose levels, which is detected by beta (β) cells (Fig. 1). Subsequently, these cells take up glucose from the bloodstream via glucose transporters (GLUTs) and engage in glycolysis, leading to an increase in ATP production [2, 3]. Elevated levels of this leads to the suppression of the cell's K_{ATP} channels. [4, 5]. K_{ATP} channels are made up of two different proteins: sulfonylurea receptor (SUR) subunits and pore-forming (Kir6.x) subunits (Figs. 2 and 3). Channel activity is stimulated by Mg-nucleotide binding/hydrolysis at the nucleotide-binding domains (NBD1 and NBD2) of SUR subunit [6]. The closure of the pore in Kir6.2 is induced by the binding of ATP or ADP, and this effect does not depend on Mg^{2+}. Consequently, the closure of KATP channels results in depolarization and subsequent activation of L-type Ca^{2+} channels. This activation permits an influx of Ca^{2+} into the cell [5, 7–9], followed by exocytotic release of insulin [10]. This exocytosis of insulin from β-cells is rapid, taking place within the first 5–10 min, commonly known as the first phase of secretion. Following this, a sustained release, known as the second phase secretion, occurs. The significance of the second phase lies in its potential to become quantitatively significant, as it can be maintained for several hours in cases where elevated blood glucose levels persist [11]. In addition to the insulin secretion mechanism mentioned above, it is important to acknowledge that glucose can stimulate insulin release through other means. Notably, any alterations in the channels present in β-cells that impact membrane potential could also influence the movement of Ca^{2+} into the cytosol, thereby influencing insulin secretion [12–14].

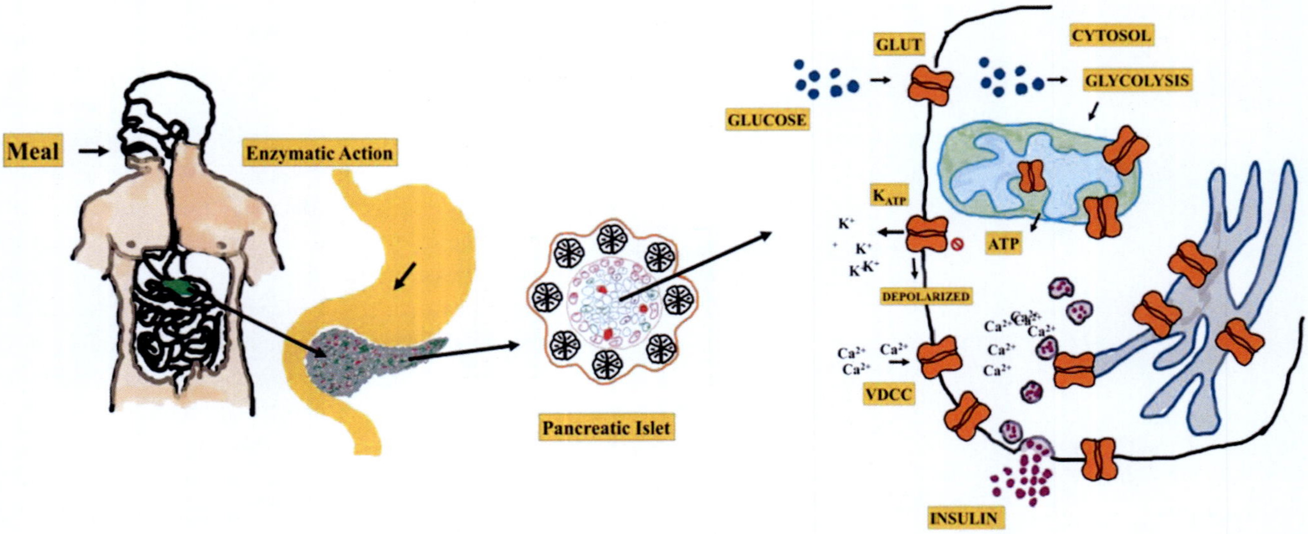

Fig. 1 Insulin secretion

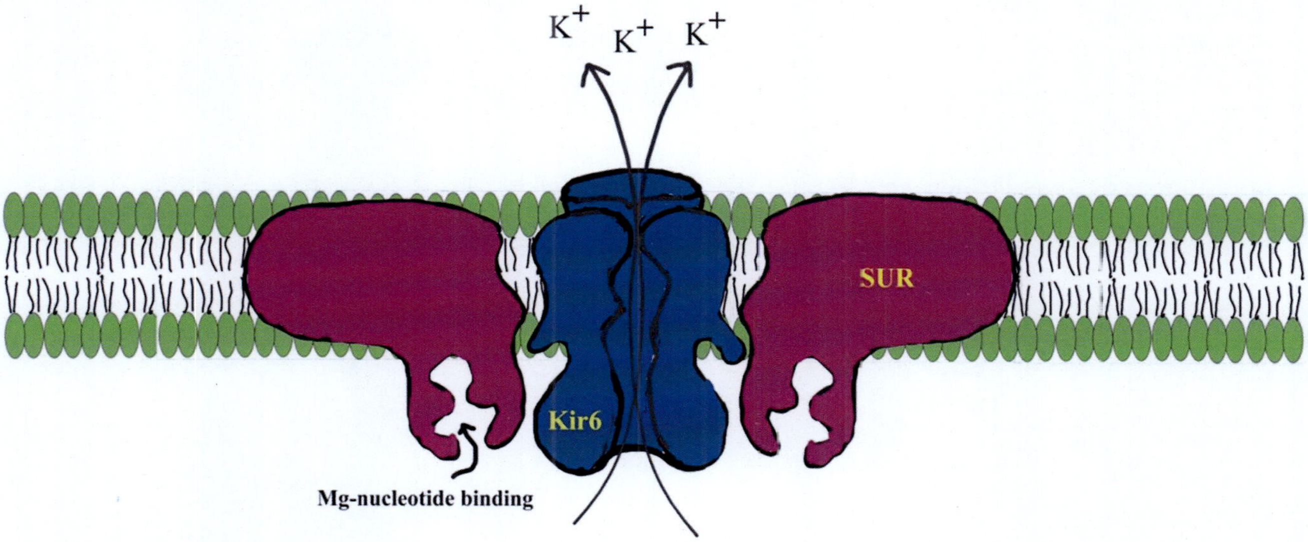

Fig. 2 K_{ATP} channel (stimulate)

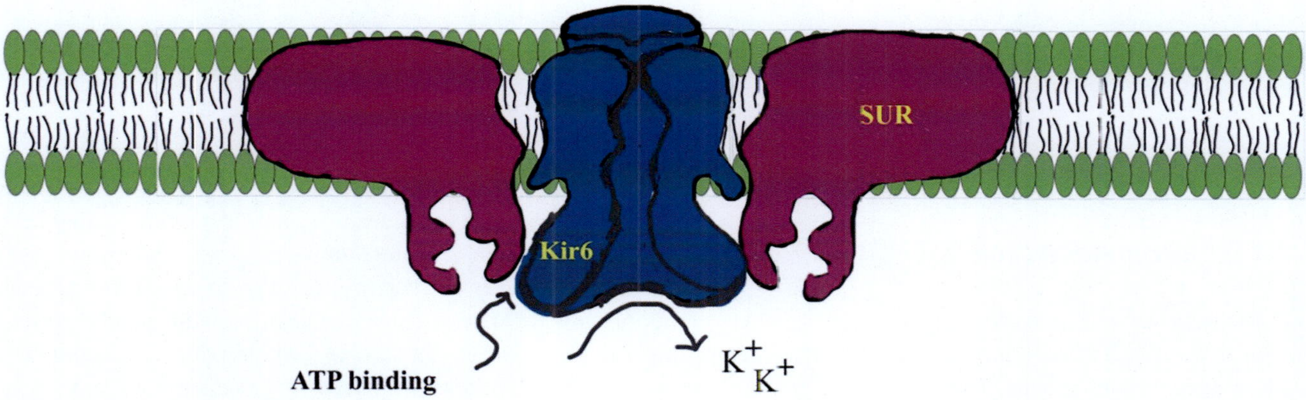

Fig. 3 K_{ATP} channel (closed)

2.2 Gut Hormone Pathway

Incretins are gastrointestinal hormones released in response to food intake. Glucagon-like peptide-1 (GLP-1) is secreted from enteroendocrine L cells, which are distributed throughout the small and large intestines. Similarly, glucose-dependent insulinotropic polypeptide (GIP) is secreted from enteroendocrine K cells in the duodenum and jejunum. Once released, these incretins bind to their specific G-protein-coupled receptors, leading to the activation of adenylate cyclase and subsequent elevation of intracellular cAMP levels [15], followed by activation of PKA (protein kinase A) and cAMP-binding proteins referred to as cAMP-regulated guanine nucleotide exchange factors (cAMP GEFs), also known as Epac2 (Exchange protein directly activated by cAMP 2) or cAMP-GEFII [16]. PKA activation is believed to facilitate exocytotic release of insulin [17–19].

Dipeptidyl peptidase-4 (DPP4) (Fig. 4) is an extensively distributed enzyme that exerts its actions through both a membrane-anchored transmembrane molecule and a soluble circulating protein [15]. DPP4 plays a significant role in glucose homeostasis. By breaking down these incretin hormones, DPP4 reduces their ability to promote insulin secretion, leading to decreased insulin release and impaired glucose clearance from the bloodstream. Overall, the regulatory role of DPP4 in incretin hormone metabolism and glucose homeostasis highlights its significance in maintaining proper blood glucose levels in the body.

Fig. 4 Dipeptidyl peptidase-4 (DPP-4) enzyme

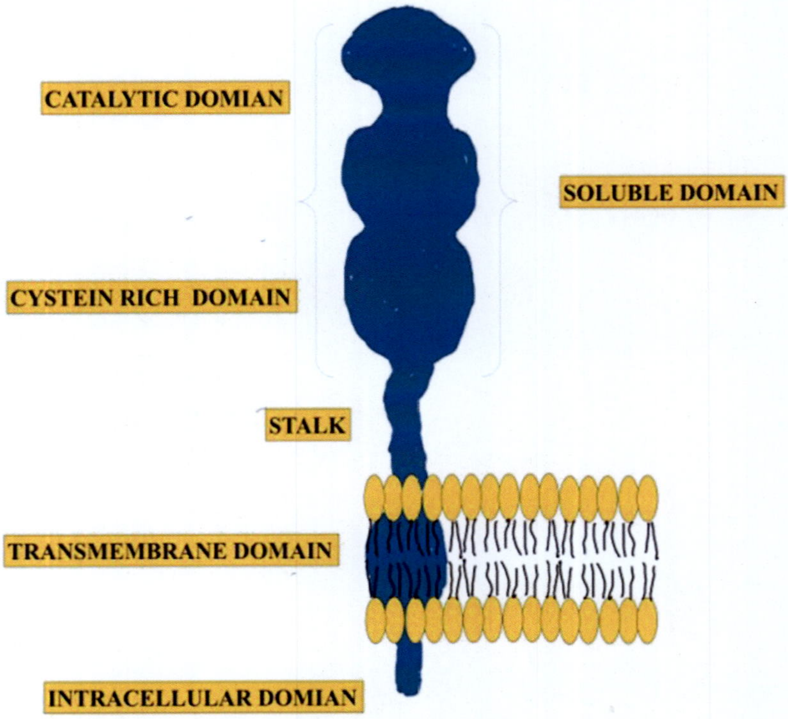

3 Regulation Glucose Level

3.1 Enzymatic Action

Carbohydrates constitute the primary components of the typical human diet. Among its major forms, starch and sucrose contribute around 70–80% of the body's energy requirements. The digestion of carbohydrates begins in the mouth and continues in the small intestine, resulting in the production of glucose. This glucose is then absorbed into the bloodstream through the enzymatic action within the intestinal walls and subsequently transported to various body parts, facilitated by the liver. Polysaccharides (starch) is converted into maltose, sucrose, and lactose (Disaccharides) and then finally into glucose, the main digested product, along with small amount of fructose and galactose (monosaccharides). The breakdown of starch initially occurs through α-amylase, an enzyme presents in saliva and pancreatic juices, which converts it into oligosaccharides. Subsequently, a membrane-bound enzyme called α-glucosidase, found in the small intestine's epithelium, catalyzes the cleavage of glucose from disaccharides and oligosaccharides, leading to postprandial hyperglycemia. Due to this process, inhibiting α-amylase and glucosidase can be an effective treatment for diabetes. By doing so, the absorption of glucose is delayed, which can be beneficial for individuals with impaired insulin response or production. In hepatocytes, glycolysis regulates hepatic glucose levels. Pancreatic β cells couple glycolysis with glucose-stimulated insulin secretion. Adipocytes utilize glycolysis to produce metabolites for lipogenesis and divert excessive fatty acids away from oxidation, reducing oxidative stress. In a proinflammatory state, adipocytes release prohyperglycemic factors, leading to hyperglycemia and insulin resistance. In hypothalamic neurons, glycolysis is involved in nutrient sensing, which influences feeding control. Dysregulation of glycolysis can contribute to the development of diabetes.

An imbalance in glucose-6-phosphatase activity can lead to hyperglycemia in diabetes. Aldose reductase (AR), a cytosolic enzyme, normally performs several essential functions, including detoxifying harmful aldehydes in extrahepatic tissues, producing fructose for sperm, maintaining osmoregulatory balance in the kidney, and reducing steroids and catecholamines. However, under certain conditions, AR catalyzes the NADPH-dependent reduction of glucose to sorbitol in the polyol pathway, resulting in an overproduction of reactive oxygen species (ROS). This increased ROS production can contribute to oxidative stress and other detrimental effects, potentially exacerbating the complications of diabetes [20]. Converted sorbitol poorly penetrates membranes, and thus if blood glucose rises, it causes intracellular accumulation along with its metabolites fructose, which creates osmotic swelling and cell dysfunction [21]. Overuse of polyol metabolism may cause hyperglycemia, sorbitol and osmotic buildup, which can all contribute to cataract development. [22–24]. Protein Tyrosine Phosphatase 1B (PTP1B) is a phosphatase situated on the cytoplasmic side of the endoplasmic reticulum (ER) [25]. Expressed in various tissues, including skeletal muscle, liver, adipose tissue, and brain, Protein Tyrosine Phosphatase 1B (PTP1B) plays a role in negatively regulating multiple cellular pathways, such as the response to insulin or endothelial nitric oxide (NO) production [17]. It functions as an essential enzyme in the dephosphorylation of tyrosine (Tyr)-phosphorylated proteins, making it a key player in the dephosphorylation of the insulin receptor (IR) and its downstream signaling components [25]. PTP1B has been identified as a key enzyme responsible for the dephosphorylation of both IR and its downstream signaling component, IRS-1. Disruptions in this pathway can lead to diverse metabolic disorders, including insulin resistance and type 2 diabetes.

3.2 Advance Glycation End Products

The accumulation of advanced glycation end products (AGEs) arises from the reaction between carbonyl groups of reducing sugars and amino groups of proteins, lipids, and nucleic acids. This cellular dysregulation occurs in diabetes, leading to various complications such as cardiomyopathy, neuropathy, retinopathy, and nephropathy, as well as contributing to atherosclerosis and aging. The presence of AGEs influences the expression of genes, intracellular signals, and the release of proinflammatory molecules and reactive oxygen species (ROS), further exacerbating the impact on cellular function and contributing to the development of diabetic complications and other age-related diseases.

3.3 Oxidative Stress

Hyperglycemia can initiate tissue damage through five significant mechanisms, all stemming from a single upstream event: mitochondrial overproduction of reactive oxygen species (ROS) [26].

1. Increased formation of AGEs: Elevated ROS levels in the mitochondria can lead to increased formation of AGEs. These harmful compounds can accumulate in tissues, causing oxidative stress, inflammation, and damage to proteins, lipids, and DNA [27].
2. Activation of Protein Kinase C (PKC): Hyperglycemia-induced ROS production which can activate certain enzymes, such as PKC. Overactivated PKC can further promote inflammation, disrupt cell signaling pathways, and contribute to tissue damage.
3. Increased polyol pathway activity: Mitochondrial-derived ROS can enhance the activity of the polyol pathway, leading to the accumulation of sorbitol in tissues. This can cause osmotic stress and damage to nerve cells, blood vessels, and other structures [28].
4. Altered cellular metabolism and mitochondrial dysfunction: High levels of ROS can disrupt cellular metabolism and lead to mitochondrial dysfunction. Impaired mitochondria affect the production of ATP, causing energy depletion and contributing to cellular damage [29].
5. Increased expression of the receptor for advanced glycation end products (RAGE). RAGE is a cell surface receptor that binds to AGEs formed as a result of prolonged exposure to high glucose levels. When blood glucose levels are elevated, the formation of AGEs is increased. These AGEs can bind to RAGE on the cell surface, leading to several detrimental effects [30]. Antioxidant therapy in diabetes involves the modulation of various cellular processes affected by oxidative stress and it can be achieved by neutralization of Reactive Oxygen Species (ROS). Antioxidants directly scavenge and neutralize reactive oxygen species (ROS) generated during oxidative stress. Restoration of antioxidant enzymes by natural antioxidants can help restore the activity of endogenous antioxidant enzymes, such as superoxide dismutase (SOD), catalase, and glutathione peroxidase. These enzymes play a crucial role in neutralizing ROS and maintaining cellular redox balance. Antioxidant therapy can enhance the activity of these enzymes, further bolstering the cellular defense against oxidative stress. Moreover, it can be done by inhibition of advanced glycation end products (AGEs) formation. Some antioxidants, such as alpha-lipoic acid, have been shown to inhibit the formation of AGEs. AGEs are harmful compounds that form when glucose reacts with proteins, and they contribute to tissue damage and diabetic complications. By reducing AGEs formation, antioxidants can help mitigate their detrimental effects on tissues. Antioxidants can further modulate inflammatory pathways by reducing the expression of pro-inflammatory molecules, such as cytokines and adhesion molecules. By suppressing inflammation, antioxidants can help protect tissues from inflammatory damage seen in diabetes-related complications. Antioxidant can improve mitochondrial function to reduce oxidative damage within these vital cellular organelles. Improving mitochondrial function contributes to overall cellular health and energy production and improves insulin sensitivity and glucose uptake in cells. By enhancing insulin signaling pathways, antioxidants may contribute to better glucose regulation in diabetes [31].

4 Antidiabetic Phytoconstituents and Their Mode of Action (Fig. 5)

Phytoconstituents, or phytochemicals, are natural chemical compounds that naturally occur in plants. These compounds play a vital role in the biological activities of plants and are responsible for their diverse medicinal properties. Researchers are actively isolating various phytoconstituents and exploring their potential to develop strategies for managing diabetes mellitus. Isolated phytoconstituents can be classified into several classes based on their chemical structures and properties, such as alkaloids, flavonoids, anthraquinones, glycosides, saponins, coumarins, and more. This categorization helps in better understanding and utilizing the diverse therapeutic potentials of these plant-derived compounds in the management of diabetes and related conditions (Tables 1 and 2).

4.1 Alkaloids

Among the prominent natural product categories, alkaloids are extensively isolated from medicinal plants and have demonstrated the ability to regulate glucose metabolism in animal models. This research is instrumental in comprehending and developing strategies for managing diabetes mellitus and its complications. Isolated alkaloids and their derivatives have exhibited various modes of antidiabetic activity. These include enhanced insulin secretion, improved digestion and absorption of carbohydrates, increased liver glycogen, regeneration of β-cells, inhibition of α-glucosidase, elevated glucose transporter concentration, and inhibition of digestive enzymes. These diverse mechanisms underscore the potential of alkaloids as valuable resources for the development of novel treatments in the realm of diabetes management.

4.1.1 Inhibition of Enzymes

Secondary metabolites isolated from medicinal plants can inhibit digestive enzymes (Sect. 3.1) and thereby help lower postprandial blood glucose levels [90]. These compounds prevent the formation of enzyme-substrate complexes, leading to reduced enzyme activity. For example, carbazole alkaloids from *Murraya koenigii* L. Spreng (Table 2) demonstrate α-glucosidase inhibition, and mahanimbine exhibits α-amylase and α-glucosidase inhibitory effects, potentially beneficial in managing diabetes-associated cardiovascular complications [90]. Several other alkaloids have been found to effectively stimulate glucose uptake in skeletal muscles, both insulin-dependent and insulin-independent, such as O-methylmurrayamine A [48], koenidine [75], mahanimbine [91], and murrayazoline are found to be effectively stimulating glucose uptake in skeletal muscles (measured in L6-GLUT4myc myotubes). These carbazole alkaloids increase in the surface GLUT4myc level [92]. Allylpropyl disulfide isolated from *Allium sativum*is involved in glycogen synthesis and insulin secretion. *Adhatoda vasica (Nees.)* isolated alkaloids like quinazoline, vasicine, and vasicinol have shown to inhibit rat intestinal α-glucosidase competitively [32]. Palmatine is a bioactive protoberberine alkaloid, which was isolated from *Coscinium fenestratum* stem extract that has effective alpha-amylase, alpha-glucosidase, and DPP-IV inhibitor activity, which may be helpful to reduce the postprandial glucose level [77]. *Oriciopsis glaberrima* showed an α-glucosidase inhibitory activity due to Oriciacridone C, 1,3,5-trihydroxy-4-(c,c-dimethylallyl)-acridone and oriciacridone F [93]. *Piper umbellatum* shows the α-glucosidase inhibitory activity due to the presence of three alkaloid piperumbellactam A, B, and C found in the fraction of the extract from the branches of the plant [94]. Indole alkaloids vindogentianine obtained from leaf extract in *Catharanthus roseus (L.) G*is are able to inhibit both the

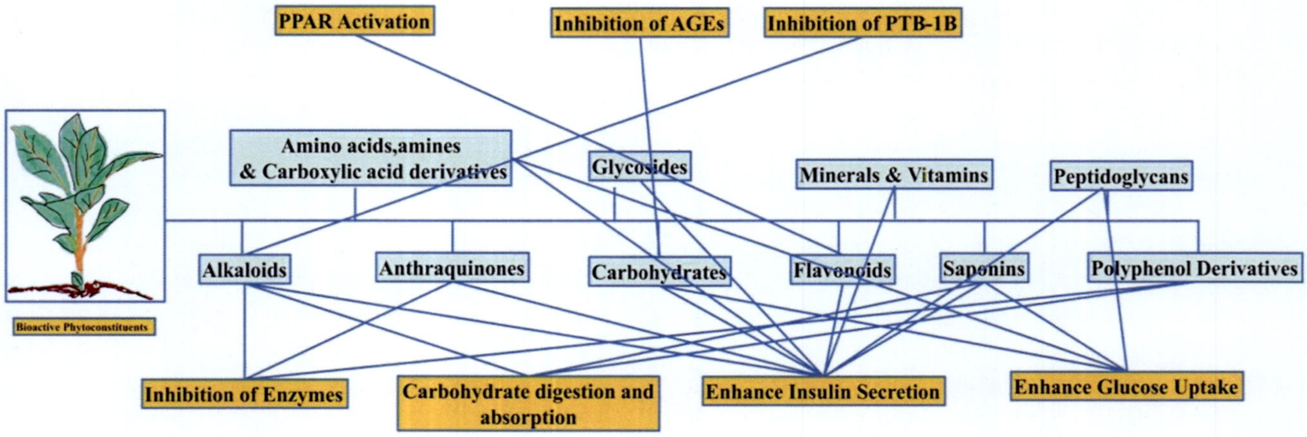

Fig. 5 Mode of action of phytoconstituents in the management of diabetes mellitus

Table 1 List of Antidiabetic medicinal plants and their mode of action

	Plant source	Parts used	Mode of action	References
1.	*Abies pindrow* (pinaceae)	Whole plant	Increased secretion of insulin	[32]
2.	*Acacia arabica* (leguminosae)	Seed	Increased secretion of insulin	[32]
3.	*Agrimony eupatoria* (leaves)	Seed	Increased secretion of insulin	[32]
4.	*Aloe barbadensis* (liliaceae)	Leaves	Increased secretion of insulin	[33]
5.	*Allium sativum*	Bulb	DPP-4 inhibition	[34]
6.	*Achyranthus aspera (amaranthaeeae)*	Whole plant	Carbohydrate digestion and absorption	[35]
7.	*Aegle marmelos*		Enhanced insulin release	[36, 37]
8.	*Annona squamosa* (annonaceae)	Leaves	Insulin secretagogue	[32]
9.	*Averrhoa bilimbi* (oxalidaceae)	Leaves	Insulin secretagogue	[38]
10.	*Azadirachta indica*	Leaves	Hypoglycemic effect	[36, 39]
11.	*Artocarpus heterophyllus Lam.*	Seed	α-glucosidase inhibitory activity, increased GLUT-2 activity	[34]
12.	*Bauhinia forficate link (leguminosae)*	Leaves	Hypoglycemic effect	[40]
13.	*Castanospermum australe*	Seed	DPP-IV inhibition	[41]
14.	*Cathanthrus roseus (apocynaeceae)*	Leaf	Free radical scavenging action	[42]
15.	*Bixa Orellana* (bixaceae)	Aerial parts	Enhanced insulin release	[32]
16.	*Boerhaavia diffusa* (nyctaginaceae)	Leaves	Enhanced insulin release	[43]
17.	*Camellia sinensis* (theaceae)	Leaves	Insulinotropic effect	[43]
18.	*Capsicum frutescens* (solanaceae)	Leaves	Insulinotropic effect	[43]
19.	*Cinnamomum zeylanicum* (lauraceae)	Dalchini	Insulinotropic effect	[43]
20.	*Coccinia indica*	Leaves	Insulinotropic effect	[36]
21.	*Cinnamomum tamala*	–	Blood glucose lowering effect	[36]
22.	*Dendrocalamus hamiltonii*	–	Blood glucose lowering effect	[36]
23.	*Eucalyptus globulus* (myrtaceae)	Leaves	Insulinotropic effect	[43]
24.	*Eugenia jambolana*	–	Blood glucose lowering effect	[36]
25.	*Ficus bengalensis* (moraceae)	Bark	Insulin secretagogues	[36, 44]
26.	*Ficus glomerata*		Blood glucose lowering effect	[36]
27.	*Ginkgo biloba*	Leaves	Insulin secretion	[45]
28.	*Moringa oleifera*	–	Blood glucose lowering effect	[36]
29.	*Momordica charantia*	–	Glucose utilization in the liver	[36, 46, 47]
30.	*Murraya koenigii L. Spreng (rutaceae)*	Leaves	α-glucosidase and α-amylaseinhibition	[48, 49]
31.	*Ocimum sanctum*	Inflorescence	Blood glucose lowering effect	[39]
32.	*Pterocarpus marsupium*	Hard wood	Blood glucose lowering effect	[36]
33.	*Premna integrifolia*	Leaves	Blood glucose lowering effect	[36]
34.	*Swertia chirayata* (gentianaceae)	Whole plant	Blood glucose lowering effect	[36]
35.	*Scoparia dulcis* (scrophulariaceae)	–	Hypoglycemic effect	
36.	*Sesbenia aegyptiaca*	–	Hypoglycemic effect	[36]
37.	*Tinospora crispa* (menispermaceae)	–	Blood glucose lowering effect	[36]
38.	*Tragia involucrata*	Whole plant	Blood glucose lowering effect	[36]
39.	*T. foenum-graecum*	Seed	Decreased insulin resistance	[36]
40.	*Terminalia bellirica*	Fruit pulp	Blood glucose lowering effect	[36]
41.	*Tinospora cordifollia*	Stem	Decreased insulin resistance	[36]
42.	*Trichosanthes cucumerina*	Seeds	Blood glucose lowering effect	[36]
43.	*Tribulus terrestris*	Aerial part	Increased insulin secretion and regeneration of β-cells in islets	[50]
44.	*Vinca rosea* (apocynaceae)	Leaves	Free radical scavenging action	[36]
45.	*Gymnema sylvestre*	Leaves	Blood glucose lowering effect	[36, 39]
46.	*Zingiber officinale* (zingiberaceae)	Rhizome	Blood glucose lowering effect	[36]

intestinal enzymes (α-amylase and glucosidase). Steroidal alkaloids, holaphylline and sarcovagine-D, isolated from chloroform extract of *Sarcococa saligna* possess hypoglycemic effect [95]. *Combretum dolichopetalum* shows promising antidiabetic activity in alloxan-induced diabetic mice when treated with isolated compound Echinulin and arestrictin B from its roots extract [96]. *Berberis aristata* shows dipeptidyl peptidase IV inhibiting activity due to Berberine [86]. Castanospermine, 7-deoxy-6-epi-castanospermine, and australine obtained from the seed extract of *Castanospermum australe* are responsible for showing DPP-IV inhibition activity in wistar rats [97].

Table 2 List of antidiabetic medicinal plants phytoconstituents

Active constituents	Plant source	References
Alkaloids		
Allylpropyl disulfide, allicin, alliin, diallyl trisulfide	*Allium sativum*	[34]
Carbazole derivatives: bisgerayafoline D, bismahanimbinol, bispyrayafoline, O-methyl mahanine, mahanine	*Murraya koenigii L. Spreng*	[51]
Betaine, achyranthine, β-ecdysone	*Achyranthus aspera*	–
Aegelin, marmesin, marmelosin	*Aegle marmelos*	[34, 37]
Gallic acid, catechin, caffeic acid, rutin, and quercetin	*Artocarpus heterophyllus*	[34]
Kaempferitrin	*Bauhinia forficate link*	[34]
Castanospermine, australine	*Castanospermum austral*	
Catharanthine, vindoline, vindolinene vinblastine, vincristine	*Cathanthrus roseus, Vinca rosea*	[36, 39]
Ginkgolides	*Ginkgo biloba*	
Barberin	*Tinospora cordifollia, Barberis aristata*	[36, 52]
Sotolon, trigonelline, gentianine	*Trigonella foenum*	[53, 54]
Harmine, pinoline	*Tribulus terrestris*	[55]
Leucocyanidin (3-O-ß-Dgalactosylcellobioside), glucoside of leucopelargonidin, 3-O-alfa-L-rhamnoside, leucodelphinidine	*Ficus bengalensis* (moraceae)	[44]
Anthraquinones		
Aloin, barbaloin, aloe-emodin, emodin, crysophanic acid	*Aloe vera, Cassia tora*	[56]
Vicine	*Momordica charantia*	[57]
Rhein, torachrysone, toralactone, alaternin	*Cassia tora*	[58]
Camphor, eugenol, trans-β-ocimene, geraniol, α-pinene, limonene, p-cymene, 1,8-cineole, thujone	*Ocimum canum, Coriandrum sativum, Artemisia roxburghiana, Syzygium aromaticum*	[58]
Gurmarin, betaine, choline, trimethylamine	*Gymnema sylvestre*	[58]
Alaternin, aurantio-obtusin, obtusifolin	*Senna obtusifolia*	[58]
Aurantioobtusin, catenarin	*S. obtusifolia*	[58]
Chrysophanol	*Rheum undulatum*	[59]
Aloe-emodin	*Cassia fistula*	[60]
2-hydroxy-3-methyl-anthraquinone	*Juncus setchuensis*	[61]
Lucidin primeveroside	*Morinda citrifolia*	[57]
Physcion, rhein	*Rheum emodi*	[62]
S-allyl cysteine, allicin, apigenin, alliin	*Allium sativum*	[34, 63]
Gurmarin, betaine, choline, trimethylamine	*Gymnema sylvestre*	[64]
(−) Hydroxycitric acid	*Garcinia cambogia, Gymnema sylvestr*	[65]
Ferulic acid	*Curcuma longa*	[66]
Isoleucine, alanine	*Aloe vera*	[15]
Polypeptide-P	*Momordica charantia*	[36]
S-methyl and S-allyl cysteine sulfoxide	*Alium sepa*	[38]
Nitrosamines	*Areca catechu*	[38]
Brevifolin	*Phyllanthus amarus*	[38]
Lectins, mistletoe lectin I, II, III, viscotoxin B, cycliton	*Viscum album*	[38]
Furfural, caprylic acid	*Agaricus campestris*	[38]
Procyanidins	*Grape seed*	[38]
Carbohydrates		
Glucomannan, cellulose, mannose	*Aloe vera*	[62]
Protein-bound polysaccharide	*Alpinia galangal, Aloe vera, Ocimum sanctum*	[67]
Guar gum, pectin	*Trigonella foenum, Citrus sinensis, Coccinia indica*	[68]
D-threitol, D-arabinitol, palmitic acid	*Hericium erinaceus*	[69]
Mucopolysaccharide	*Opuntia ficus indica*	[70]
Inulin, laevulin	*Taraxacum officinale*	[71]
Fructo-oligosaccharide	*Aureobasidium pullulans*	–
Glycosides		
Gymnemic acid and gymnemosides	*Gymnema sylvestre*	[52]

Table 2 (continued)

Active constituents	Plant source	References
Vin α-ginsenoside R3	*Panax quinquefolium*	[57]
Astragalin, scopolin	*Morus alba*	[72]
C-glycosides	*Trigonella foenum*	[73]
Momordin, momordicine, charantin	*Momordica charantia*	[36]
Tinosporine, cordifolide, tinosporide, cordifole, columbin	*Tinospora cordifollia, Tinospora crispa*	[52]
Cucurbitacin B, isocucurbitacin B	*Helicteres isora*	[2]
Momordin-a, luffin-a	*Luffa cylindrica*	[74]
Kotalanol, salacinol	*Salacia reticulate, Salacia oblonga*	[72]
Arbutin, eriolin	*Arctostaphylos uvaursi*	[72]
Citrullol, colocynthin, elaterin, elatericin B, colosynthetin	*Cifrullus colocynthis*	[43]
Leucocyanidin, pelarogonidin	*Ficus bengalensis*	[44]
Taraxacin	*Taraxacum officinale*	[71]
Flavonoids		
Chrysin, isoquercitrin	*Morus alba*	[75]
Epigallocatechin-gallate	*Camellia sinensis, Punica granatum*	[76]
Myrciaphenones A and B, myrciacitrins I and II	*Myrcia multiflora*	[77]
Myricetin-3-glucoside	*Abelmoschus moschatus*	[78]
Hesperidin, naringin	*Camellia sinensis*	[79]
Quercetin, quercetrin, apigenin, rutin, apigenin-7-O-glucoside	Multiple plant sources	[80]
Naringenin	Multiple plant sources	[79]
Soy isoflavones (genistein, diadzein)	*Glycin max*	[81]
Proanthocyanidins	*Vitis vinifera*	[25]
α-Terpineol, hexanol	*Agaricus campestris*	[43]
Kaempferitrin	*Bauhinia candicans, Bauhinia forficata*	[82]
Silymarin, silybin, silychristin, silidianin	*Silybum marianum*	[43]
Kaempferol, isorhamnetin	*Ginkgo biloba*	[83]
Amarogentin	*Swertia chirayita*	[84]
Tribulusamides A and B	*Tribulus terrestris*	[50]
Shamimin	*Biophytum sensitivum*	[17]
Matteuorien	*Matteuccia orientalis*	[85]
Polyphenol and derivatives		
Curcumin	*Curcuma longa*	[66]
Ellagic acid	Blackberries, raspberries, strawberries, etc.	[86]
Gallotannic acid	*Syzygium aromaticum*	[87]
Wedelolactone, dimethyl wedelolactone	*Eclipta alba*	[88]
Carvacrol, linalool	*Ocimum sanctum*	[43]
Saponins		
Asparosides A, B, C, and D	*Asparagus adscendens*	[84]
Lactucain C	*Lactuca indica*	[52]
Salacinol, kotalanol, ponkoranol	*Salacia reticulate*	[89]
Allo-aromadendrene	*Artemisia pallens*	[71]
Diosgenin	*Trigonella foenum graecum*	[73]
Trigonellin	*Trigonella foenum graecum*	[73]
Mulberrofuran-U	*Morus insignis*	[71]
Kotalagenin-16-acetate, diterpene, triterpens	*Salacia oblongaq, Croton cajucara*	[73]
Azorellanol	*Azorella compacta*	[36]
Sulfur containing compound		
Ajoene	*Allium sativum*	[34]

4.1.2 Inhibition of Aldose Reductase (AR)

The rhizome of *Coptis chinensis* was evaluated for AR inhibitory effects via RLAR and human recombinant AR (HRAR) inhibitory assays. Isolated compound protoberberine-type alkaloids majorly from the n-BuOH fraction which are epiberberine, coptisine, and groenlandicine are found to exhibit AR inhibitory effects [41]. The dioxymethylene group and its oxidized form in the D and A ring of protoberberinetype alkaloids are responsible for the AR inhibitory [98]. Isoquinoline alkaloids (berberine chloride and palmatine iodide) from *Coptis japonica* root-derived compound possess AR inhibitory effects [51]. Similarly, isoquinoline alkaloids (jatrorrhizine, palmatine, and magnoflorine) were isolated *Tinospora cordifolia* stem—was evaluated against lens AR isolated from male Wistar rats. Thus, alkaloids derivatives may be used as lead compounds in management of diabetic complications [80].

4.1.3 Inhibition of Protein Tyrosine Phosphatase-1B

As PTP-1B is negative regulator of insulin signalling pathway (Sect. 3), its inhibition can be the promising strategy in the development of the management of insulin resistance diabetic complications. One of the natural compounds of canthinone alkaloids from *Picrasma quassioides*, *Picrasma javanica*, *Ailanthus altissima*, *Simarouba amara*, *Eurycoma longifolia*, *Simabacuspidata*, and *Quassia amara*ispicrasidine has been shown to be a competitive inhibitor of PTP-1B [38]. Glucose uptake was assayed in β-TC6 pancreatic and C2C12 muscle cells, which has shown significant PTP-1B inhibitory effects [99], suggesting that hypoglycemic activity of vindogentianine obtained from *Catharanthus roseus (L.)* may be due to the enhancement of glucose uptake and PTP-1B inhibition, thus this could be used for management of obesity and type 2 diabetic conditions [100]. Rhizome of *Coptis chinensis Franch* has shown promising inhibitory activity towards PTP-1B by isolated protoberberine alkaloids coptisine, berberine, and epiberberine and an aporphine alkaloid magnoflorine [101]. Norditerpenoid alkaloids (nigelladine Anigelladine B, nigelladine C) from the seeds of *Nigella glandulifera Freyn* reduce the overexpression of PTP1B in L6 myotubes, showing effectiveness of the compound in improving the glucose metabolism by induction of phosphorylation of IRS-1, Akt, GSK-3b, and AMPK in the stimulation of insulin. Eventually compounds' effect was found to be blocked by wortmannin (PI3-kinase/Akt inhibitors), thus showing that the effect of compounds is dependent on PI3K/Akt signaling pathway [102].

4.1.4 Enhance Insulin Secretion

Quinolizidine alkaloids (lupanine, 13-a-OH lupanine, and 17-oxo-lupanine) from *Lupinus species* stimulate insulin secretion in a glucose-dependent manner, by blockage of β-cell KATP-sensitive channels [103]. Isoquinoline alkaloid (Berberine), the main active component of an ancient Chinese herb *Coptis chinensis French,* shows DPP IV and PTP-1B inhibitory effects [86]. Trigonelline from *Trigonella foenum graecum* and *Mirabilis jalapa L.* exhibits antidiabetic activity via enhancing insulin sensitivity [65, 73]. *Mangifera indica*-isolated compound mangiferin, a xanthone glucoside found in the leaves, has antidiabetic and antihyperlipidaemic properties by pancreatic and extrapancreatic mechanisms [104]. Aegelin, marmesin, and marmelosin are the major alkaloids *from Aegle marmelos* that have been found to cause regeneration of pancreatic β cells and insulin secretion [104]. *Tribulus terrestris*-isolated β carbolines (harmine, nor-harmine, pinoline) enhance insulin secretion by β-cell regeneration [89]. Castanospermine, epifagomine, and fagomine derived from *Xanthocercis zambesiaca are* involved in improved insulin secretion [55].

4.1.5 Inhibition of Advanced Glycation End Products

Berberine from *Rhizomacoptidis* and *Cortex phellodendri* has also shown to have the ability of inhibiting renal AGE levels in diabetic male Sprague-Dawley rats [59, 105]. Leonurine, from *Herbaleonuri,* was shown to have significant AGEs inhibition [106].

4.1.6 Antioxidant Activity

Oriciacridone C, 1,3,5-trihydroxy-4-(c,c-dimethylallyl)-acridone, and oriciacridone F from Stem Bark of *Oriciopsis glaberrima* showed moderate free radical scavenging activity against 1,1-diphenyl-2-picrylhydrazyl (DPPH) activity [93]. Piperumbellactam A, B, and C from *Piper umbellatum* show promising antioxidant activity [94]. Bisgerayafoline D, bismahanimbinol, bispyrayafoline, O-methyl mahanine, O-methyl mukonal, and mahanine from *Murraya koenigii* possess antioxidant activity [107]. Catharanthine, vindoline, and vindolinine, obtained from *Catharanthus roseus,* lower the blood sugar level and show free radical scavenging action [39, 108]. *Vinca rosea*-isolated compounds vinblastine and vincristine also activate free radical scavenging enzymes [42].

4.1.7 Enhancement of Glucose Uptake

Extracted from *Trigonella foenum graecum*, trigonelline, gentianine, and carpaine compounds exhibit the ability to downregulate the activity of fructose-1,6-bisphosphatase and inhibit the dephosphorylation of fructose-1,6-bisphosphate [36]. The alkaloid isolated from plant *Tinospora cordifolia* has been found to have a positive impact on phosphofructokinase and hexokinase activity, leading to enhanced glucose transport as well as improved absorption and carbohydrate digestion. In the case of gluconeogenesis, the phytoconstituent berberine plays a crucial role by decreasing the activity of glucose-6-phosphatase enzyme. Furthermore, compounds, like β ecdysone and betaine, achyranthine isolated from *Achyranthes aspera*, have been identified to participate in absorption and carbohydrate digestion processes. These bioactive constituents contribute to the efficient breakdown and utilization of carbohydrates in the body [53]. Castanospermine, epifagomine, and fagomine from *Xanthocercis zambesiaca are* involved in carbohydrate digestion and absorption [55].

5 Anthraquinones (AQs)

The AQs and their derivatives are the class of commonly available tricyclic aromatic quinones containing anthracenedione or 9,10-dioxoanthracene core and can exist in several possible isomeric. Effectiveness of AQs against DM and its complications has been studied and it has been found that it can act on different antidiabetic targets, suggesting their potential antidiabetic action for the diabetes management [35, 109]. Possible mechanisms of antidiabetic action involve upregulation of IRS-1, PI3K, and expression of GLP-1 level, PTP-1B, and DPP-4 inhibitions, activation of PPAR-γ, and inhibition of α-glucosidase activity.

5.1 α Amylase and Glucosidase Inhibitory Activity

Emodin, aloe emodin, alaternin, and questin isolated from *S. obtusifolia L.* and *R. emodi Wall* have been shown to inhibit α glucosidase [58, 110]. Rhein from *R. emodi* and *R. rhizoma* was shown to inhibit both α glucosidease and αamylase [62]. Physcion from *R. emodi* and *S. obtusifolia* was reported to inhibit α glucosidase activity.

5.2 Inhibition of DPP 4

Emodin isolated from *R. palmatum L.* inhibited the activity of DPP 4 [111]. Damnacanthol-3O-β-D-primeveroside and lucidin-3-O-β-D primeveroside, isolated from *Morinda citrifolia* L. roots, have been found to improve glycemic control. Physcion isolated from *Juncus setchuensis Buchen* was shown to reduce glucose production in H4IIE cells. Rhein isolated from *Rheum rhizoma* enhanced insulin-stimulated glucose uptake in Min6 cells and 3T3-L1 adipocytes [112].

5.3 Inhibition of PTP 1B

Emodin, aloe emodin, 2-hydroxyemodin-1 methyl ether, alaternin, and questin isolated from *S. obtusifolia*, Chrysophanol isolated from *S. obtusifolia* and rhubarb rhizome, Chrysophanol glucosides, chrysophanol-8-O-β-D glucopyranoside, chrysophanol triglucoside, and chrysophanol tetraglucoside, obtusin, chryso-obtusin, and chryso obtusin-2-O-β-D glucoside, physcion from *S. obtusifolia*, chrysophanol from *Cannabissativa* Linne, *Paeonia lactiflora Pallas*, *Citrus aurantium* Linne, *Magnolia officinalis* Rehd., *R. officinale,* and *Prunus armeniaca* Linne are found to inhibit PTP-1B activity.

5.4 Inhibition of Aldose Reductase Activity

Emodin from *S. tora L.* inhibits aldose reductase activity and advanced glycation end products formation [113].

5.5 Insulin Secretion

Aloin, barbaloin, isobarbaloine, aloetic acid, aloe-emodin, emodin, cinnamic acid, and crysophanic acid from *Aloe vera* and *Cassia tora* initiate insulin secretion/synthesis. *Momordica charantia* is a rich source of vicine which acts on insulin secretion and glycogen synthesis. Camphor, eugenol, trans-β-ocimene, geraniol, α-pinene, limonene, p-cymene, 1,8-cineole, and thujone, which help in pancreatic β-cell restoration and insulin secretion, are reported to be found in *Ocimum sanctum*, *Coriandrum sativum*, *Artemisia roxburghiana*, and *Syzygium aromaticum* [114].

6 Flavonoids

Flavonoids can be broadly categorized into three groups: the bioflavonoids, the iso-flavonoids (phytoestrogens), and the neo-flavonoids [71]. Numerous studies have indicated that flavonoids exhibit antidiabetic activity by influencing various cellular pathways, including those related to glucose transporters, hepatic enzymes, tyrosine kinase inhibition, AMPK, PPAR, and NF-κB.

6.1 Enhance Insulin Secretion

Epicatechin, derived from the bark of *Pterocarpus marsupium Roxb*, has been observed to inhibit cAMP phosphodiesterase, thereby modulating insulin secretion [82, 112]. Epigallocatechin gallate (EGCG) obtained from *Camellia sinensis* has been found to have the following effects; glucose-lowering effects, insulin-like effects—increases tyrosine phosphorylation of the insulin receptor and insulin receptor substrate-1 (IRS-1), and it reduces phosphoenolpyruvate carboxykinase gene expression in a phosphoinositide 3-kinase-dependent manner [115]. Citrus bioflavonoids (hesperidin and naringin) are obtained from *Camellia sinensis* which target glycogen synthesis, glycolysis, and gluconeogenesis [78]. Pancreatic β-cell restoration and insulin secretion were being observed due to quercetrin, apigenin, rutin, apigenin-7-O-glucoside, and naringenin which are extracted from *Panax notoginseng* [79], *Urtica dioica*, and *Bauhinia varigtla.* Soy isoflavones (genistein and diadzein) are obtained from *Glycin max* and are found to be involved in lipid and glucose metabolism by activation of PPARs [116]. Kaempferitrin obtained from *Bauhinia candicans* and *Bauhinia forficata* affects glycolysis [81]. Proanthocyanidins, α-terpineol, and hexanol obtained from *Vitis vinifera* and *Agaricus campestris* have insulinomimetic activity [76, 117, 118].

7 Saponins

Glycosidic compound chemically consists of two parts, aglycone and glycone; the aglycone part is also known as sapogenin which is classified to either triterpenoid, neutral, or alkaloid steroids. This natural phytoconstituent has been found to have antidiabetic action in the following manner to regulate the blood glucose level; activation of glycogen synthesis, suppression of the activity of disaccharides, modulating insulin signalling, regeneration of insulin action, and suppression of gluconeogenesis [119]. Gymnemic acids obtained from *Gymnema sylvestre* [120] might have following possible mechanisms: (1) it increases secretion of insulin; (2) it promotes regeneration of islet cells; (3) it increases utilization of glucose: it is shown to increase the activities of enzymes responsible for utilization of glucose by insulin-dependent pathways, increase in phosphorylase activity, decrease in gluconeogenic enzymes and sorbitol dehydrogenase; and (4) it causes inhibition of glucose absorption from intestine [72]. Ginsenosides isolated from *Panax ginseng* show antidiabetic action by modulating blood glucose levels by improving β-cell function and enhancing insulin sensitivity, glucose uptake by upregulating the expression of glucose transporters (GLUT), and suppression of oxidative stress [64].

8 Conclusion

The significance of bioactive phytoconstituents from various plant sources in exhibiting antidiabetic activity can be highlighted. Through extensive research and experimentation, numerous phytochemicals belonging to different categories have been identified for their potential in managing diabetes mellitus. Diverse parts of plants, including leaves, bark, seeds, and roots, have been utilized to isolate these beneficial phytoconstituents. The extraction of these compounds has provided valuable insights into their therapeutic effects on blood glucose regulation and diabetes control. Furthermore, it is evident that these phytoconstituents act through different pathways, indicating their multifaceted mechanisms of action. Some have shown a remarkable ability to increase insulin secretion from pancreatic beta cells, aiding in glucose uptake and utilization by various tissues. Others have been found to inhibit specific enzymes involved in glucose metabolism, thus contributing to lower blood glucose levels. However, despite the encouraging findings, further research at the molecular level is imperative to fully comprehend the intricate interactions between these phytochemicals and their cellular targets. This deeper understanding will pave the way for their effective translation into therapeutic agents for diabetes management.

In conclusion, the exploration of bioactive phytoconstituents as potential antidiabetic agents holds great promise in the quest for more accessible and sustainable solutions for diabetic individuals. By harnessing the power of nature's pharmacy, we can continue to uncover new avenues for supporting diabetes care and improving the overall well-being of those affected by this metabolic disorder. As research progresses, the possibility of integrating these natural remedies into conventional diabetes treatment becomes increasingly promising, offering hope for a healthier and brighter future for diabetic management.

References

1. Manual of Medicine. Am J Med Sci. 1904;127:720–1.
2. Kennedy HJ, Pouli AE, Ainscow EK, Jouaville LS, Rizzuto R, Rutter GA. Glucose generates sub-plasma membrane ATP microdomains in single islet β-cells. Potential role for strategically located mitochondria. J Biol Chem. 1999;274(19):13281–91. https://doi.org/10.1074/jbc.274.19.13281.
3. Malaisse WJ, Sener A. Glucose-induced changes in cytosolic ATP content in pancreatic islets. BBA-Mol Cell Res. 1987;927(2):190–5.
4. Cook DL, Hales N. Intracellular ATP directly blocks K+ channels in pancreatic B-cells. Nature. 1984;311(5983):271–3.
5. Meglasson MD, Matschinsky FM. Pancreatic islet glucose metabolism and regulation of insulin secretion. Diabetes Metab Rev. 1986;2(3–4):163–214.
6. Remedi MS, Nichols CG. Hyperinsulinism and diabetes. In: Ion channels in health and disease. Amsterdam: Elsevier; 2016. p. 199–221. https://doi.org/10.1016/B978-0-12-802002-9.00008-X.
7. Pan JY, Yuan S, Yu T, Su CL, Liu XL, He J, et al. Regulation of L-type Ca2+ channel activity and insulin secretion by huntingtin-associated protein 1. J Biol Chem. 2016;291(51):26352–63.
8. Way SC. Voltage-gated Ca2+ current in pancreatic B-cells. Pflugers Arch. 1985;404:385–7.
9. Ashcroft FM. Review series ATP-sensitive potassium channelopathies: focus on insulin secretion. J Clin Invest. 2005;115(8):2047–58.
10. Aung TTT, Xia MY, Hein PP, Tang R, Zhang DD, Yang J, et al. Chemical constituents from the whole plant of cuscuta reflexa. Nat

Prod Bioprospect. 2020;10(5):337–44. https://doi.org/10.1007/s13659-020-00265-x.
11. Curry DL, Bennett LL, Grodsky GM. Dynamics of insulin secretion by the perfused rat pancreas. Endocrinology. 1968;83(3):572–84.
12. Role of Ca2+ signaling in pancreatic β-cells.pdf.
13. Wang Z, Thurmond DC. Mechanisms of biphasic insulin-granule exocytosis - roles of the cytoskeleton, small GTPases and SNARE proteins. J Cell Sci. 2009;122(7):893–903.
14. Papadopoulos T, Abrahim A, Sergelidis D, Bitchava K. Ερευνητική. 2011;2(January):119–23.
15. Mulvihill EE, Drucker DJ. Pharmacology, physiology, and mechanisms of action of dipeptidyl peptidase-4 inhibitors. Endocr Rev. 2014;35(6):992–1019.
16. Seino S. Cell signalling in insulin secretion: the molecular targets of ATP, cAMP and sulfonylurea. Diabetologia. 2012;55(8):2096–108.
17. Yabe D, Seino Y. Two incretin hormones GLP-1 and GIP: comparison of their actions in insulin secretion and β cell preservation. Prog Biophys Mol Biol. 2011;107(2):248–56. https://doi.org/10.1016/j.pbiomolbio.2011.07.010.
18. Wheeler MB, Light PE, Manning Fox JE, Riedel MJ. Glucagon-like peptide-1 inhibits pancreatic ATP-sensitive potassium channels via a protein kinase A- and ADP-dependent mechanism. Mol Endocrinol. 2002;16(9):2135–44.
19. Shibasaki T, Takahashi H, Miki T, Sunaga Y, Matsumura K, Yamanaka M, et al. Essential role of Epac2/Rap1 signaling in regulation of insulin granule dynamics by cAMP. Proc Natl Acad Sci U S A. 2007;104(49):19333.
20. Oates P. Aldose reductase, still a compelling target for diabetic neuropathy. Curr Drug Targets. 2008;9(1):14–36.
21. Kinoshita JH, Nishimura C. The involvement of aldose reductase in diabetic complications. Diabetes Metab Rev. 1988;4(4):323–37.
22. Kim J, Kim CS, Sohn E, Lee YM, Jo K, Kim JS. Litsea japonica extract inhibits aldose reductase activity and hyperglycemia-induced lenticular sorbitol accumulation in db/db mice. Evid Based Complement Altern Med. 2015;2015:747830.
23. Ahmed AJ, Majeed SR, Obaid HM. Biochemistry and molecular cell biology of diabetic complications. Syst Rev Pharm. 2020;11(11):850–60.
24. Ramana KV. Aldose reductase: new insights for an old enzyme. Biomol Concepts. 2011;2(1–2):103–14.
25. Thiebaut PA, Besnier M, Gomez E, Richard V. Role of protein tyrosine phosphatase 1B in cardiovascular diseases. J Mol Cell Cardiol. 2016;101:50–7. https://doi.org/10.1016/j.yjmcc.2016.09.002.
26. Yaribeygi H, Sathyapalan T, Atkin SL, Sahebkar A. Molecular mechanisms linking oxidative stress and diabetes mellitus. Oxidative Med Cell Longev. 2020;2020:8609213.
27. Nowotny K, Jung T, Höhn A, Weber D, Grune T. Advanced glycation end products and oxidative stress in type 2 diabetes mellitus. Biomol Ther. 2015;5(1):194–222.
28. Giacco F, Brownlee M. Oxidative stress and diabetic complications. Circ Res. 2010;107(9):1058–70.
29. Bhatti JS, Bhatti GK, Reddy PH. Mitochondrial dysfunction and oxidative stress in metabolic disorders—a step towards mitochondria based therapeutic strategies. Biochim Biophys Acta Mol Basis Dis. 2017;1863(5):1066–77. https://doi.org/10.1016/j.bbadis.2016.11.010.
30. Mengstie MA, Chekol Abebe E, Behaile Teklemariam A, Tilahun Mulu A, Agidew MM, Teshome Azezew M, et al. Endogenous advanced glycation end products in the pathogenesis of chronic diabetic complications. Front Mol Biosci. 2022;9(September):1–11.
31. Matough FA, Budin SB, Hamid ZA, Alwahaibi N, Mohamed J. The role of oxidative stress and antioxidants in diabetic complications. Sultan Qaboos Univ Med J. 2012;12(1):556–69.
32. Patel OPS, Mishra A, Maurya R, Saini D, Pandey J, Taneja I, et al. Naturally occurring carbazole alkaloids from Murraya koenigii as potential antidiabetic agents. J Nat Prod. 2016;79(5):1276–84.
33. Sciences A, Abdulaziz K. Effect of aloes on blood glucose. J Ethnopharmacol. 1990;28:215–20.
34. Leone A. Antidiabetic phytochemicals from medicinal plants: prospective candidates for new drug discovery and development. Front Endocrinol. 2022;13(February):800714.
35. Akhtar MS, Iqbal J. Evaluation of the hypoglycaemic effect of Achyranthes aspera in normal and alloxan-diabetic rabbits. J Ethnopharmacol. 1991;31(1):49–57.
36. Kar A, Choudhary BK, Bandyopadhyay NG. Comparative evaluation of hypoglycaemic activity of some Indian medicinal plants in alloxan diabetic rats. J Ethnopharmacol. 2003;84(1):105–8.
37. Narender T, Shweta S, Tiwari P, Reddy KP, Khaliq T, Prathipati P, et al. Antihyperglycemic and antidyslipidemic agent from Aegle marmelos. Bioorg Med Chem Lett. 2007;17:1808–11.
38. Patel MB, Mishra S. Isoquinoline alkaloids from tinospora cordifolia inhibit rat lens aldose reductase. Phyther Res. 2012;26(9):1342–7.
39. Chattopadhyay RR. A comparative evaluation of some blood sugar lowering agents of plant origin. J Ethnopharmacol. 1999;67(3):367–72.
40. De Sousa E, Zanatta L, Seifriz I, Creczynski-Pasa TB, Pizzolatti MG, Szpoganicz B, et al. Hypoglycemic effect and antioxidant potential of kaempferol-3,7-O-(α)-dirhamnoside from Bauhinia forficata leaves. J Nat Prod. 2004;67(5):829–32.
41. Kumar S, Krishnan S, Kumar A, Kishore K, Murari K, Kumar B, et al. Phytomedicine antihyperglycemic activity with DPP-IV inhibition of alkaloids from seed extract of Castanospermum australe: investigation by experimental validation and molecular docking. Eur J Integr Med. 2012;20(1):24–31. https://doi.org/10.1016/j.phymed.2012.09.009.
42. Edwin Jarald E, Sheeja E, Motwani S, Dutt KR, Goel RK. Comparative evaluation of antihyperglycaemic and hypoglycaemic activity of various parts of Catharanthus roseus Linn. Res J Med Plant. 2008;2(1):10–5.
43. Patel DK, Prasad SK, Kumar R, Hemalatha S. An overview on antidiabetic medicinal plants having insulin mimetic property. Asian Pac J Trop Biomed. 2012;2(4):320–30. https://doi.org/10.1016/S2221-1691(12)60032-X.
44. Geetha BS, Mathew BC, Augusti KT. Hypoglycemic effects of leucodelphinidin derivative isolated from Ficus bengalensis (Linn.). Indian J Physiol Pharmacol. 1994;38(3):220–2.
45. Da M, Pinto S, Kwon Y, Apostolidis E, Maria F, Inés M, et al. Bioresource technology potential of Ginkgo biloba L. leaves in the management of hyperglycemia and hypertension using in vitro models. Bioresour Technol. 2009;100(24):6599–609. https://doi.org/10.1016/j.biortech.2009.07.021.
46. Paper O. Bitter gourd (Momordica charantia) extract activates peroxisome proliferator-activated receptors and upregulates the expression of the acyl CoA oxidase gene in H4IIEC3. J Biomed Sci. 2003;6:782–91.
47. Sarkar S, Pranava M, Marita R. Demonstration of the hypoglycemic action of momordica charantia in a validated animal model of diabetes. Pharmacol Res. 1996;33(1):1–4.
48. Dineshkumar B, Mitra A, Mahadevappa M. Antidiabetic and hypolipidemic effects of mahanimbine (carbazole alkaloid) from murraya koenigii (rutaceae) leaves. Int J Phytomed. 2011;2:22–30.
49. Pandey J. Naturally occurring carbazole alkaloids from Murraya koenigii as potential antidiabetic agents. J Nat Prod. 2016;79(5):1276–84.
50. El-shaibany A, Al-habori M, Al-tahami B. Anti-hyperglycaemic activity of Tribulus terrestris L aerial part extract in glucose-loaded normal rabbits. Trop J Pharm Res. 2015;14(December):2263–8.

51. Adhikari B. Roles of alkaloids from medicinal plants in the management of diabetes mellitus. J Chem. 2021;2021:2691525.
52. Singh SS, Pandey SC, Srivastava S, Gupta VS, Patro B, Ghosh AC. Chemistry and medicinal properties of Tinospora cordifolia (Guduchi). Indian J Pharm. 2003;35(2):83–91.
53. Khosla P, Gupta DD, Nagpal RK. Effect of trigonella Foenum graecum (Fenugreek) on serum lipids in-normal and diabetic rats. Indian J Pharm. 1995;27(2):89–93.
54. Kumar A, Bharti SK, Kumar A. Type 2 diabetes mellitus: the concerned complications and target organs. Apollo Med. 2014;11(3):161–6. https://doi.org/10.1016/j.apme.2014.01.009.
55. Cooper EJ, Hudson AL, Parker CA, Morgan NG. Effects of the β-carbolines, harmane and pinoline, on insulin secretion from isolated human islets of Langerhans. Eur J Pharmacol. 2003;482(1–3):189–96.
56. Yadav M, Khan KK, Beg MZ. Medicinal plants used for the treatment of diabetes by the baiga tribe living in Rewa District M. Pac Afr J. 2012;2(1):99–102. http://www.ajol.info/index.php/ajtcam/article/view/81432.
57. Kamiya K, Hamabe W, Harada S, Murakami R, Tokuyama S, Satake T. Chemical constituents of Morinda citrifolia roots exhibit hypoglycemic effects in streptozotocin-induced diabetic mice. Biol Pharm Bull. 2008;31(5):935–8.
58. Jung HA, Ali MY, Choi JS. Promising inhibitory effects of anthraquinones, naphthopyrone, and naphthalene glycosides, from Cassia obtusifolia on α-glucosidase and human protein tyrosine phosphatases 1B. Molecules. 2017;22(1):28.
59. Tang L, Wei W, Chen L, Liu S. Effects of berberine on diabetes induced by alloxan and a high-fat/high-cholesterol diet in rats. J Ethnopharmacol. 2006;108:109–15.
60. Anandjiwala S, Bagul MS, Parabia M, Rajani M. Evaluation of free radical scavenging activity of an ayurvedic formulation. Indian J Exp Biol. 2008;43:31–5.
61. Zhang QW, Lin LG, Ye WC. Techniques for extraction and isolation of natural products: a comprehensive review. Chin Med. 2018;13(1):1–26. https://doi.org/10.1186/s13020-018-0177-x.
62. Arvindekar A, More T, Payghan PV, Laddha K, Ghoshal N, Arvindekar A. Evaluation of anti-diabetic and alpha glucosidase inhibitory action of anthraquinones from Rheum emodi. Food Funct. 2015;6(8):2693–700.
63. Augusti KT, Sheela CG. Antiperoxide effect of S-allyl cysteine sulfoxide, an insulin secretagogue, in diabetic rats. Experientia. 1996;52(2):115–20.
64. Kanetkar P, Singhal R, Kamat M. Gymnema sylvestre: a memoir. J Clin Biochem Nutr. 2007;41(2):77–81.
65. García López PM, De La Mora PG, Wysocka W, Maiztegui B, Alzugaray ME, Del Zotto H, et al. Quinolizidine alkaloids isolated from Lupinus species enhance insulin secretion. Eur J Pharmacol. 2004;504(1–2):139–42.
66. Mehrabi M, Esmaeili S, Ezati M, Abassi M, Rasouli H, Nazari D, et al. Antioxidant and glycohydrolase inhibitory behavior of curcumin-based compounds: Synthesis and evaluation of anti-diabetic properties in vitro. Bioorg Chem. 2021;110(November):104720. https://doi.org/10.1016/j.bioorg.2021.104720.
67. van de Venter M, Roux S, Bungu LC, Louw J, Crouch NR, Grace OM, et al. Antidiabetic screening and scoring of 11 plants traditionally used in South Africa. J Ethnopharmacol. 2008;119(1):81–6.
68. Rath D, Kar DM, Panigrahi SK, Maharana L. Antidiabetic effects of Cuscuta reflexa Roxb. in streptozotocin induced diabetic rats. J Ethnopharmacol. 2016;192:442–9. https://doi.org/10.1016/j.jep.2016.09.026.
69. Liang B, Guo Z, Xie F, Zhao A. Antihyperglycemic and antihyperlipidemic activities of aqueous extract of Hericium erinaceus in experimental diabetic rats. BMC Complement Altern Med. 2013;13:253.
70. O'Neill AR, Rana SK. An ethnobotanical analysis of parasitic plants (Parijibi) in the Nepal Himalaya. J Ethnobiol Ethnomed. 2016;12(1):1–15. https://doi.org/10.1186/s13002-016-0086-y.
71. Hussain Z, Waheed A, Qureshi RA, Burdi DK, Verspohl EJ, Khan N, et al. The effect of medicinal plants of Islamabad and Murree region of Pakistan on insulin secretion from INS-1 cells. Phyther Res. 2004;18(1):73–7.
72. Sugihara Y, Nojima H, Matsuda H, Murakami T, Yoshikawa M, Kimura I. Antihyperglycemic effects of gymnemic acid IV, a compound derived from Gymnema sylvestre leaves in streptozotocin-diabetic mice. J Asian Nat Prod Res. 2000;2(4):321–7.
73. Subramanian SP, Prasath GS. Antidiabetic and antidyslipidemic nature of trigonelline, a major alkaloid of fenugreek seeds studied in high-fat-fed and low-dose streptozotocin-induced experimental diabetic rats. Biomed Prev Nutr. 2014;4(4):475–80. https://doi.org/10.1016/j.bionut.2014.07.001.
74. Current Medicinal Chemistry. http://books.google.com/books?hl=fr&lr=&id=W9LvB7-kG8sC&pgis=1.
75. Tachibana Y, Kikuzaki H, Lajis NH, Nakatani N. Comparison of antioxidative properties of carbazole alkaloids from Murraya koenigii leaves. J Agric Food Chem. 2003;51(22):6461–7.
76. Pinent M, Blay M, Bladé MC, Salvadó MJ, Arola L, Ardévol A. Grape seed-derived procyanidins have an antihyperglycemic effect in streptozotocin-induced diabetic rats and insulinomimetic activity in insulin-sensitive cell lines. Endocrinology. 2004;145(11):4985–90.
77. Assefa ST, Yang EY, Chae SY, Song M, Lee J, Cho MC, et al. Alpha glucosidase inhibitory activities of plants with focus on common vegetables. Plants. 2020;9(1):2.
78. Waltner-Law ME, Wang XL, Law BK, Hall RK, Nawano M, Granner DK. Epigallocatechin gallate, a constituent of green tea, represses hepatic glucose production. J Biol Chem. 2002;277(38):34933–40.
79. Jung UJ, Lee M-K, Jeong K-S, Choi M-S. Biochemical and molecular actions of nutritients the hypoglycemic effects of hesperidin and naringin are partly mediated by hepatic glucose-regulating enzymes in C57BL/KsJ-db/db mice 1. J Nutr. 2004;134:2499–503.
80. Lee HS. Rat lens aldose reductase inhibitory activities of Coptis japonica root-derived isoquinoline alkaloids. J Agric Food Chem. 2002;50(24):7013–6.
81. Mezei O, Banz WJ, Steger RW, Peluso MR, Winters TA, Shay N. Biochemical and molecular actions of nutrients soy isoflavones exert antidiabetic and hypolipidemic effects through the PPAR pathways in obese zucker rats and murine RAW 264.7 cells 1. J Nutr. 2003;133(March):1238–43. https://academic.oup.com/jn/article-abstract/133/5/1238/4558583.
82. Welch AA, Hardcastle AC. The effects of flavonoids on bone. Curr Osteoporos Rep. 2014;12(2):205–10.
83. Nooreen Z, Tandon S, Yadav NP, Ahmad A. New chemical constituent from the stem of Cuscuta reflexa Roxb. and its biological activities. Nat Prod Res. 2021;35(14):2429–32. https://doi.org/10.1080/14786419.2019.1669033.
84. Kanwal A, Kanwar N, Bharati S, Srivastava P, Singh SP, Amar S. Exploring new drug targets for type 2 diabetes: success, challenges and opportunities. Biomedicine. 2022;10(2):1–18.
85. Matthews DR. Diabetes mellitus: a fundamental and clinical text. Clin Endocrinol. 2000;53(4):543–4.
86. Uzor PF, Osadebe PO. Antidiabetic activity of the chemical constituents of Combretum dolichopetalum root in mice. EXCLI J. 2016;15:290–6.
87. Abima Shazhni JR, Renu A, Vijayaraghavan P. Insights of antidiabetic, anti-inflammatory and hepatoprotective properties of antimicrobial secondary metabolites of corm extract from Caladium x hortulanum. Saudi J Biol Sci. 2018;25(8):1755–61. https://doi.org/10.1016/j.sjbs.2018.03.013.

88. Kuate D, Kengne APN, Biapa CPN, Azantsa BGK, Wan Muda WAM, Bin. Tetrapleura tetraptera spice attenuates high-carbohydrate, high-fat diet-induced obese and type 2 diabetic rats with metabolic syndrome features. Lipids Health Dis. 2015;14(1):1–13.
89. Muruganandan S, Srinivasan K, Gupta S, Gupta PK, Lal J. Effect of mangiferin on hyperglycemia and atherogenicity in streptozotocin diabetic rats. J Ethnopharmacol. 2005;97(3):497–501.
90. Attjioui M, Ryan S, Ristic AK, Higgins T, Goñi O, Gibney ER, et al. Carbohydrate digestive enzymes involved in glucose release from the diet. J Nutr Sci. 2021;10:e5.
91. Fukui K, Nakayama M, Matsui M. Murraya koenigii. 1969;1353:790–1.
92. Joshi BS, Kamat VN, Gawad DH. On the structures of girinimbine, mahanimbine, isomahanimbine, koenimbidine and murrayacine. Tetrahedron. 1970;26(5):1475–82.
93. Okechukwu P, Sharma M, Tan WH, Chan HK, Chirara K, Gaurav A, et al. In-vitro anti-diabetic activity and in-silico studies of binding energies of palmatine with alpha-amylase, alpha-glucosidase and DPP-IV enzymes. Pharmacia. 2020;67(4):363–71.
94. Wansi JD, Wandji J, Mbaze Meva'a L, Kamdem Waffo AF, Ranjit R, Khan SN, et al. α-glucosidase inhibitory and antioxidant acridone alkaloids from the stem bark of Oriciopsis glaberrima ENGL. (Rutaceae). Chem Pharm Bull. 2006;54(3):292–6.
95. Tabopda TK, Ngoupayo J, Liu J, Mitaine-Offer AC, Tanoli SAK, Khan SN, et al. Bioactive aristolactams from Piper umbellatum. Phytochemistry. 2008;69(8):1726–31.
96. Ullah N, Ali A, Ahmad B, Iqbal N, Adhikari A. Biomedicine & pharmacotherapy evaluation of antidiabetic potential of steroidal alkaloid of Sarcococca saligna. Biomed Pharmacother. 2018;100:461–6. https://doi.org/10.1016/j.biopha.2018.01.008.
97. Al-Masri IM, Mohammad MK, Tahaa MO. Inhibition of dipeptidyl peptidase IV (DPP IV) is one of the mechanisms explaining the hypoglycemic effect of berberine. J Enzyme Inhib Med Chem. 2009;24(5):1061–6.
98. Jung HA, Yoon NY, Bae HJ, Min BS, Choi JS. Inhibitory activities of the alkaloids from coptidis rhizoma against aldose reductase. Arch Pharm Res. 2008;31(11):1405–12.
99. Sasaki T, Li W, Higai K, Koike K. Canthinone alkaloids are novel protein tyrosine phosphatase 1B inhibitors. Bioorg Med Chem Lett. 2015;25(9):1979–81. https://doi.org/10.1016/j.bmcl.2015.03.014.
100. Klaman LD, Boss O, Peroni OD, Kim JK, Martino JL, Zabolotny JM, et al. Increased energy expenditure, decreased adiposity, and tissue-specific insulin sensitivity in protein-tyrosine phosphatase 1B-deficient mice increased energy expenditure, decreased adiposity, and tissue-specific insulin sensitivity in protein-tyrosine. Mol Cell Biol. 2000;20(15):5479–89.
101. Tiong SH, Looi CY, Arya A, Wong WF, Hazni H, Mustafa MR, et al. Vindogentianine, a hypoglycemic alkaloid from Catharanthus roseus (L.) G. Don (Apocynaceae). Fitoterapia. 2015;102:182–8. https://doi.org/10.1016/j.fitote.2015.01.019.
102. Choi JS, Ali MY, Jung HA, Oh SH, Choi RJ, Kim EJ. Protein tyrosine phosphatase 1B inhibitory activity of alkaloids from Rhizoma Coptidis and their molecular docking studies. J Ethnopharmacol. 2015;171(1):28–36. https://doi.org/10.1016/j.jep.2015.05.020.
103. Tang D, Chen Q, Xin X, Aisa H. Anti-diabetic effect of three new norditerpenoid alkaloids in vitro and potential mechanism via PI3K/Akt signaling pathway. Biomed Pharmacother. 2017;87:145–52. https://doi.org/10.1016/j.biopha.2016.12.058.
104. Zhou J, Zhou S, Zeng S. Experimental diabetes treated with trigonelline: effect on β cell and pancreatic oxidative parameters. Fundam Clin Pharmacol. 2013;27(3):279–87.
105. Nojima H, Kimura I, Chen FJ, Sugihara Y, Haruno M, Kato A, et al. Antihyperglycemic effects of N-containing sugars from Xanthocercis zambesiaca, Morus bombycis, Aglaonema treubii, and Castanospermum australe in streptozotocin-diabetic mice. J Nat Prod. 1998;61(3):397–400.
106. Wu D, Wen W, Qi C, Zhao R, Lü J, Zhong C, et al. Phytomedicine Ameliorative effect of berberine on renal damage in rats with diabetes induced by high-fat diet and streptozotocin. Eur J Integr Med. 2012;19(8–9):712–8. https://doi.org/10.1016/j.phymed.2012.03.003.
107. Manuscript A. Function.
108. Uvarani C, Jaivel N, Sankaran M, Chandraprakash K, Ata A, Mohan PS. Axially chiral biscarbazoles and biological evaluation of the constituents from Murraya koenigii. Fitoterapia. 2014;94:10–20. https://doi.org/10.1016/j.fitote.2014.01.004.
109. Du H, Shao J, Gu P, Lu B, Ye X, Liu Z. Improvement of glucose tolerance by rhein with restored early-phase insulin secretion in db/db mice. J Endocrinol Investig. 2012;35(6):607–12.
110. Yen GC, Der Duh P, Chuang DY. Antioxidant activity of anthraquinones and anthrone. Food Chem. 2000;70(4):437–41.
111. Babu KS, Tiwari AK, Srinivas PV, Ali AZ, Raju BC, Rao JM. Yeast and mammalian α-glucosidase inhibitory constituents from Himalayan rhubarb Rheum emodi Wall.ex Meisson. Bioorg Med Chem Lett. 2004;14(14):3841–5.
112. Wang Z, Yang L, Fan H, Wu P, Zhang F, Zhang C, et al. Screening of a natural compound library identifies emodin, a natural compound from Rheum palmatum linn that inhibits DPP4. PeerJ. 2017;2017(5):1–14.
113. Choi SB, Ko BS, Park SK, Jang JS, Park S. Insulin sensitizing and α-glucoamylase inhibitory action of sennosides, rheins and rhaponticin in Rhei Rhizoma. Life Sci. 2006;78(9):934–42.
114. Mohammed A, Ibrahim MA. Antidiabetic potential of anthraquinones: a review. Phytother Res. 2019;34:436–504.
115. Hii CST, Howell SL. Effects of epicatechin on rat islets of Langerhans. Diabetes. 1984;33(3):291–6.
116. Small E. American ginseng. N Am Cornucopia. 2020;160:71–8.
117. Lemus I, García R, Delvillar E, Knop G. Hypoglycaemic activity of four plants used in Chilean popular medicine. Phyther Res. 1999;13(2):91–4.
118. Gray AM, Flatt PR. Insulin-releasing and insulin-like activity of agaricus campestris (mushroom). J Endocrinol. 1998;157(2):259–66.
119. Rankin JW, Andreae MC, Chen CYO, O'Keefe SF. Effect of raisin consumption on oxidative stress and inflammation in obesity. Diabetes Obes Metab. 2008;10(11):1086–96.
120. Barky A, El Hussein SA. Saponins-and-their-potential-role-in-diabetes-mellitus. Diabetes Manage. 2017;7:148–58.

Herbal Medicines for the Management of Obesity

Ceyda Sibel Kılıç

Abstract

Obesity is an important and ever-increasing health problem that is experienced by nearly every country in the world. Diets, increased physical activity, medications, bariatric surgery, and gastric botox are among the treatment and prevention options related to obesity. Some of these options are invasive and have important side effects, and thus, people seek effective options with less or no side effects. Natural products are considered to be among these efficient and safe options, and literally hundreds, even thousands of plants are being utilized for their weight loss activities. By all means, it is not possible to cover all of these plants, but we still would like to include some recent studies on plants, plant products, and plant-originated compounds having anti-obesity activity with a focus on their anti-obesity-related activities.

Keywords

Herbal medicines · Medicinal plants · Obesity

1 Introduction

Obesity is an important health problem that has been recognized by the World Health Organization (WHO), which used the word "epidemic" related to obesity nearly three decades ago [1, 2]. WHO states that 13% of the world population consists of obese adults [3]. Though Western diets with high energy have been recognized as the major contributing factor, obesity might develop due to certain diseases, certain medications, genetic factors, environmental factors, sedentary lifestyle, and less and/or untimely sleep with increasing prevalence and comorbidities [2, 4–7]. In short, obesity can be defined as the longtime imbalance of energy intake and consumption, in which excess energy resulting from this imbalance is deposited as triglycerides (TG) in white adipocytes [7].

Body mass index (BMI) is an important indicator of obesity severity. Obesity is illustrated by increasing ranges of BMI as shown in Fig. 1 [8]:

Obesity has the potential to result in various important diseases and disorders such as atherosclerosis, diabetes mellitus, hypertension, dyslipidemia, cardiac problems, metabolic syndrome, lung diseases, neurological disorders, arthritis, osteoarthritis, kidney disease, sleep disorders, obstructive sleep apnea, asthma, blood lipid disorders, insulin resistance, depression, anxiety, infertility, urinary incontinence, stroke, and cancer [2, 5, 9–15]. Furthermore, obesity was reported to contribute to the progression of COVID-19 following infection with SARS-CoV-2 [12]. In addition to being an important reason for worsening of life quality and increase rate in mortality, it has also become an important economic burden for the healthcare system [16].

Synthetic pharmaceutical medications are used in the treatment of obesity; however, they are associated with various complications such as gastrointestinal and cardiovascular side effects, and fat-soluble vitamin deficiency, and furthermore, some of these medications have been withdrawn from the market due to their serious adverse effects [7, 12–14].

Therefore, researchers started to focus on natural sources to find effective anti-obesity agents that have less or no side effects. For this purpose, natural sources such as plants, marine algae, fungi, and animal-originated products/compounds were investigated, and among them, medicinal plants, plant extracts, primary metabolites, and secondary metabolites isolated from them were examined intensively due to their long-term safety, metabolic activities, and mode of actions [17–20].

Among secondary metabolites, polyphenols consisting of phenolic acids, flavonoids and stilbenes, phytosterols, terpenoids, organosulfur compounds, and alkaloids have important anti-obesity effects; flavonoids have various subclasses

C. S. Kılıç (✉)
Faculty of Pharmacy, Department of Pharmaceutical Botany, Ankara University, Ankara, Türkiye
e-mail: erdurak@pharmacy.ankara.edu.tr

A. K. Dhara, S. C. Mandal (eds.), *Role of Herbal Medicines*, https://doi.org/10.1007/978-981-99-7703-1_6

Fig. 1 BMI indices depicting obesity (in kg/m^2)

in the nature such as flavonols, isoflavanones, isoflavonoids, flavones, flavans-3-ol, and anthocyanins [2, 21]. General structures of some of these groups are given in Fig. 2.

These secondary metabolites have different mode of actions, which contribute to their anti-obesity activities. Possible mode of actions found in the literature are summarized collectively in Fig. 3 [9, 13, 21–23].

These mode of actions can be summarized as follows:

- **Energy expenditure boosters/regulators:** Energy expenditure can be classified into physical activity, obligatory energy expenditure, and adaptive thermogenesis. Energy expenditure stimulants such as capsaicin, caffeine, and green tea work to increase energy expenditure with the establishment of non-shivering thermogenesis and thus dissipate excessive energy in the form of heat [2].
- **Bulk producers and laxatives:** Bulk producers such as dietary fibers result in a sense of fullness and also reduce appetite. Laxatives on the other hand make way for rapid excretion of foods and water that can contribute to weight loss [2].
- **Lipase inhibitors (pancreatic lipase inhibition):** Pancreatic lipase is a key enzyme of the lipid metabolism that is associated with the absorption of dietary triglycerides, and thus, when inhibited, fat hydrolysis will also be inhibited and fat absorption will be reduced [13, 24].
- **Appetite suppressants and controllers:** Appetite suppressants have the ability to control hunger centers that are found in the brain and end up in a sense of fullness/satiety [2].
- **Lipid metabolism regulators—thermogenesis:** Suppression of white adipose tissue (WAT) expansion and brown adipose tissue (BAT) activation is also an important approach in the fight against obesity [13]. WAT is responsible for energy storage since it converts excess energy and stores it in its cells in the form of TG. However, BAT has a vast number of mitochondria, consumes bioenergy, and emits it in the form of non-tremor thermogenesis as it maintains body temperature and energy consumption. Recently, a third type of adipocyte was identified in WAT. These newly found adipocytes are called beige adipocytes and resemble brown adipocytes with higher UCP1 levels. It was understood that the conversion of WAT into beige adipocytes could be achieved with exposure to cold or stimulation with medications and/or hormones and it could provide protection against obesity and insulin resistance [7].
- **Preadipocyte/adipocyte apoptosis:** Apoptosis of preadipocytes and adipocytes results in decrease in lipid mass, and this effect was found to be longer lasting compared to lipid mobilization and lipolysis [23].
- **Lipolysis:** Stimulation of hydrolysis of triglycerides (TG) diminishes fat depots [2]. Hypolipidemic activity acts on plasma triglycerides, total cholesterol (TC), and low-density lipoprotein cholesterol (LDL-C) and lowers their concentrations [13].
- **α-Amylase and α-glucosidase inhibitors:** α-Amylase and α-glucosidase enzymes function in the digestion of carbohydrates; thus, inhibition of these enzymes would result in inhibition of intestinal glucose absorption [13].
- **Adipocyte differentiation regulators/preventers:** These agents work to suppress late-phase adipocyte differentiation and thus suppress lipogenesis [2].

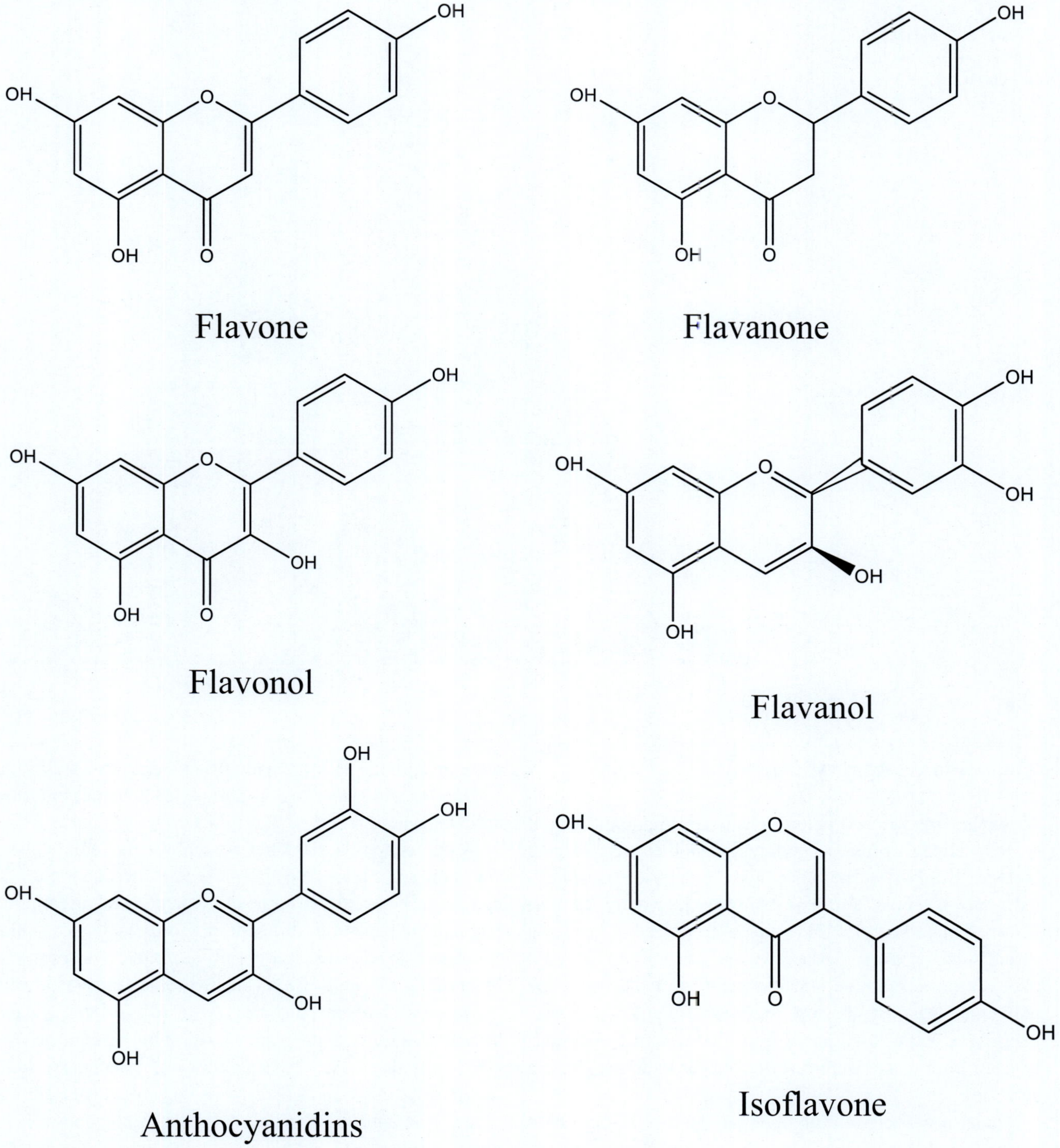

Fig. 2 Structures of some flavonoid subgroups (structures are drawn by Chem & Bio 3D 12.0)

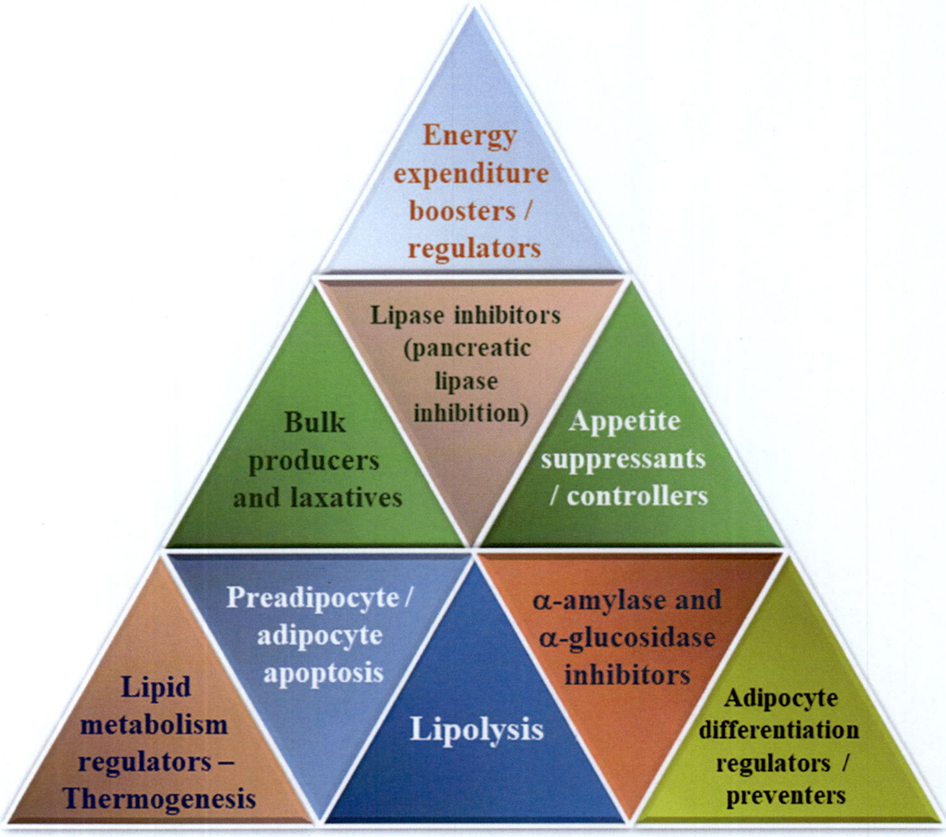

Fig. 3 Possible mode of actions of secondary metabolites contributing to anti-obesity activity

2 Plants Against Obesity

As we have mentioned before, plants are important natural sources having various secondary metabolites with varying biological activities; therefore, it is no surprise that plants have been examined for their anti-obesity activities as first-line treatment. With less side effects compared to synthetic medications or invasive processes such as bariatric surgery, plants have become the first treatment of choice for many individuals who are obese or overweight. For this purpose, plant parts, extracts prepared from plants, and compounds isolated from plants were investigated for their anti-obesity activities.

There are literally hundreds of reviews along with research studies performed on anti-obesity activities of different plant species. In this chapter, reviews were excluded and individual research studies performed on plants were included, but as you may acknowledge, it is not possible to include every single study, so relatively newer research studies were tried to be included. In Table 1, plants with effects that contribute to anti-obesity activity are provided with their Latin names, family names, and used parts. While Latin names are provided, recent changes with respect to plant and family names were checked from "The World Flora Online" website (formerly known as "The Plant List") and plants were given both with the names that were provided by the authors of the respective papers as synonyms and their accepted names.

When we look at the table, we can see that edible plants and some spice plants are also present in addition to medicinal plants. Furthermore, some of them are being used in traditional medicine for different purposes, some of which contribute to overall anti-obesity activity. For example, *Momordica charantia* from Cucurbitaceae family is a well-known remedy against diabetes [209, 210]. In Table 1, it can be seen that fruits of the plant also have hypophagic activity [137] and the leaves are inhibitors of pancreatic lipase enzyme that also contribute to its traditional and clinical usage as an anti-obesity remedy [138]. Photographs related to this plant species and the prepared traditional remedy are given in Fig. 4.

In addition to these aforementioned plant species, some plants are used in combination with different plants as mixtures, and some of them are even being sold with different brand names. As you can see in Table 2 below, plant species that are involved in these mixtures/combinations have a kind of individual anti-obesity activity. Therefore, combinations of them would yield more effects due to synergistic and/or additive effects. Yun [20] also mentioned this issue and stated that more effective treatments could

Table 1 Plant species having various effects contributing to anti-obesity activity

Plant name	Family name	Used part	Contribution to anti-obesity activity/mechanism of action	References
Acer truncatum Bunge	Sapindaceae	Leaves	Decreases BW, AW, and AW-BW ratio	Gao et al. [25]
Achyranthes aspera L.	Amaranthaceae	Seeds	Antihyperlipidemic	Khan et al. [26]
Aconitum heterophyllum Wall.	Ranunculaceae	Roots	Decreases BW along with BG, TG, TC; increases HDL levels (roots are first detoxified with an ayurvedic method)	Kumar et al. [27]
Acorus calamus L.	Acoraceae	Rhizomes	Inhibits pancreatic lipase, thus delays intestinal absorption of dietary fat	Athesh and Jothi [28]
Adenophora triphylla (Thunb.) A.DC.	Campanulaceae	Roots	Hypocholesterolemic, reduces cholesterol and increases cholesterol degradation	Choi et al. [29]
Adiantum capillus-veneris L.	Pteridaceae	Aerial parts	Reduces TAG and modulates pancreatic digestive enzymes	Kasabri et al. [30]
Adinandra nitida Merr ex. H.L.L.	Pentaphylaceae	Leaves	Anti-adipogenic	Yuan et al. [31]
Aegle marmelos (L.= Correa)	Rutaceae	Leaves	Anti-adipogenic	Karmase et al. [32]
Agave sisalana Perrine	Asparagaceae	Young leaves	Antihyperglycemic, antidyslipidemic, modulates insulin secretion	Chege et al. [33]
Agrimonia eupatoria L.	Rosaceae	Aerial parts	Anti-lipase	Şener et al. [34]
Akebia quinata (Thunb. Ex Houtt.) Decne.	Lardizabalaceae	Fruits	Improves serum lipid profiles, reduces differentiation of 3T3-L1 adipocytes, ameliorates excess BW and AW gain	Sung et al. [35]
Allium cepa L.	Amaryllidaceae	Leaves	Decrease in body weight, fat mass, glucose, and lipids	[36]
Allium fistulosum L.	Amaryllidaceae	Bulbs and roots	Downregulates expression of genes functioning in lipogenesis, thus inhibits fat accumulation and serum lipid levels, decreases adipocyte size, alters adipocytokine secretion	Sung et al. [37, 38]
			Improves TAG, TC, LDL-C, free fatty acids, HDL-c serum levels, and adiponectin levels	Sung et al. [39]
Allium hookeri Thwaites	Amaryllidaceae	Roots	Prevents BW gain, improves insulin resistance, decreases body fat mass	Yang et al. [40]
		Leaves	Decreases body weight and adipose tissue gains, reduces blood lipid profiles	Park et al. [41]
Allium sacculiferum Maxim	Amaryllidaceae	Aerial parts	Inhibition of adipogenesis in 3T3-L1 cells by decreasing lipid accumulation	Lee et al. [42]
Allium sativum L.	Amaryllidaceae	Bulbs (aged and black)	Reduces body fat, ameliorates weight control, lowers hyperlipidemia	Kim et al. [43]
Allium tuberosum Rottler ex Spreng	Amaryllidaceae	Aerial parts	Inhibits adipogenesis in 3T3-L1 cells by decreasing lipid accumulation	Lee et al. [42]
Alnus incana (L.) Moench	Betulaceae	Branch barks	Anti-adipogenic as a partial inhibitor for PPARγ (peroxisome proliferator activated receptor gamma), slows down adipogenesis	Martineau et al. [44]
Alnus nepalensis D. Don	Betulaceae	Leaves	Anti-adipogenic	Lamichhane et al. [45]
Aloe vera L.	Asphodelaceae	Whole plant	Decreases plasma levels of glucose and lipids	Pothuraju et al. [46]
		Leaves	Lowers BW (in diabetic rats)	Javaid and Waheed [47]
Amaranthus albus L.	Amaranthaceae	Aerial parts	Anti-lipase	Şener et al. [34]
Amaranthus dubius Mart. Ex Thell.	Amaranthaceae	Leaves	Decrease in BMI, food intake, and serum lipid levels	Nderitu et al. [48]

(continued)

Table 1 (continued)

Plant name	Family name	Used part	Contribution to anti-obesity activity/mechanism of action	References
Ammannia baccifera L. subsp. *aegyptiaca* (Willd.) Koehne (syn. for	Lythraceae	Aerial parts	Improves hyperglycemia and lipid profile	Swilam et al. [49]
Ananas comosus (L.) Merr.)	Bromeliaceae	Juice	Decreases WG, body serum lipid, hepatic lipid accumulation, reduces size and number of adipocytes, downregulates lipogenesis, and upregulates fatty acid oxidation at transcriptional level of genes related to lipid metabolism	El-Shazly et al. [50]
Annona muricata L.	Annonaceae	Leaves	Downregulates FTO (fat mass and obesity-associated protein) and upregulates STATA-3 genes that result in the reduction of food intake	Elekofehinti et al. [51]
Annona squamosal L.	Annonaceae	Leaves	Lipase inhibitory activity	Gorla and Shankar [52]
Anredera cordifolia (Ten.) Steenis	Basellaceae	Leaves	Inhibition of BW gain	Sukandar et al. [53]
Arachis hypogaea L.	Fabaceae	Nutshell	Anti-lipase, inhibits fat absorption in the gastrointestinal tract, reduces adipocyte lipolysis	Moreno et al. [54]
Artemisia campestris L.	Asteraceae	Leaves	Decrease in BW, hepatic TG, and TC levels	Belgacem et al. [55]
Artemisia princeps Pamp.	Asteraceae	Aerial parts	Suppresses hyperleptinemia, prevents hepatic lipid accumulation, antidiabetic	Yamamoto et al. [56]
Artocarpus lacucha Roxb. Ex Buch.-Ham.	Moraceae	Stem, bark	Anti-adipogenic	Lamichhane et al. [45]
Atractylodes lancea DC.	Asteraceae	Rhizomes	Inhibits lipase	Jiao et al. [57]
Bauhinia purpurea L.	Fabaceae	Bark	Decreases TC, TG, LDL, increases HDL	Ramgopal et al. [58]
Bauhinia variegate L.	Fabaceae	Bark	Decreases cholesterol, TG, VLDL levels, and body weight	Balamurgan and Muralidharan [59]
Blumea balsamifera DC.	Asteraceae	Leaves	Suppresses adipogenesis in 3T3-L1 preadipocytes	Kubota et al. [60]
Boesenbergia pandurata (Roxb.) Schltr. (syn. for *Boesenbergia rotunda* (L.) Mansf.)	Zingiberaceae	Rhizomes	Inhibits high-fat diet-induced weight gain, normalizes serum lipid parameters, attenuates excessive lipid accumulation in white adipocyte tissue and liver	Kim et al. [61]
Boerhavia diffusa L.	Nyctaginaceae	Roots	Reduces BW and lipid profiles	Singh et al. [62]
			Reduces BW and adiposity	Khalid et al. [63]
Bougainvillea spectabilis Willd.	Nyctaginaceae	Bracts (modified leaves)	Inhibits lipase and amylase, suppresses appetite, reduces the level of adipocytes	Kumar et al. [64]
Boussingaulti gracilis var. pseudobaselloides (Hauman) L.H. Bailey (syn. for *Anredera cordifolia* (Ten.) Steenis)	Basellaccac	Leaves	Anti-adipogenic via regulating AMPK activity and expression of genes involved in lipogenesis	Kim and Choung [65]
Brassica napus L.	Brassicaceae	Aerial parts	Anti-lipase	Jaradat et al. [66]
Brassica rapa subsp. *rapa* (syn. for *Brassica rapa* L.)	Brassicaceae	Seeds	Reduces BW, improves lipid levels	Eldamaty and Abd El-Maoula [67]
Caesalpinia pulcherrima (L.) Sw.	Fabaceae	Leaves	Reduces BW and decreases parametrial fat	Chichioco-Hernandez and Leonido [68]
Camellia nitidissima C.W. Chi (syn. for *Camellia petelotii* (Merr.) Sealy)	Theaceae	Flowers	Reduces food intake, increases feces total triglycerides and total cholesterol excretion	Zhang et al. [69]

Camellia sinensis (L.) Kuntze	Theaceae	Leaves	Anti-lipase	Sharma et al. [70]
			Decreases BW, TG, LDL-C, TC, atherosclerosis index, increases HDL-c	Kongchian et al. [71]
Caralluma acutangula (Decne) N.E. Br (syn. for *Desmidorchis retrospiciens* Ehrenb.)	Apocynaceae	Whole plant	Reduces BW, plasma TG, and glycemia	Dramane et al. [72]
			Inhibits pancreatic lipase and trypsin, reduces weight and food intake	Dramane et al. [73]
Carum carvi L.	Apiaceae	Seeds	Controls hunger, suppresses appetite, decreases carbohydrate intake	Kazemipoor et al. [74]
Cassia auriculata L. (syn. for *Senna auriculata* (L.) Roxb.)	Fabaceae	Aerial parts	Anti-lipase	Habtemariam [75]
		Leaves	Reduces weight gain and food intake, decreases serum levels of glucose, TG, TC, LDL, VLDL and increases HDL, decreases body fat stores	Garg and Singh [62]
Cassia fistula L.	Fabaceae	Leaves	Reduces BW and decreases parametrial fat	Chichioco-Hernandez and Leonido [68]
Caulerpa lentillifera J. Agardh (Green algae species)	Caulerpaceae	Thallus	Inhibits lipase	Nurkolis et al. [76]
Ceiba pentandra (L.) Gaertn.	Malvaceae	Leaves	Acts on the thermoregulation center found in the hypothalamus, resulting in thermogenesis and lipolysis	Patil et al. [77]
Ceiba speciosa (A. St.-Hil., A. Juss & Cambess.) Ravenna		Fixed oil of the seeds	Inhibits lipase	Rosselli et al. [78]
Chenopodium album L.	Amaranthaceae	Aerial parts	Anti-lipase	Şener et al. [34]
Cinnamomum zeylanicum Blume (syn. for *Cinnamomum verum* J.Presl)	Lauraceae	Bark	Reduces BW, body fat mass, improves lipid profile, decreases BG and leptin, and increases serum insulin levels	Shalaby and Saifan [79]
Citrullus colocynthis (L.) Schrad.	Cucurbitaceae	Fruits	Reduces BW and food intake	Sanadgol et al. [80]
Clerodendrum phlomidis L.f	Lamiaceae	Roots	Decreases BW and food intake, decreases absorption of dietary fats, inhibits pancreatic lipase activity	Chidrawar et al. [81]
		Leaves	Delays absorption of dietary sugar and fat by the intestines via inhibition of pancreatic amylase and lipase activity	Gopinathan and Naveenraj [82, 83]
Codonopsis lanceolata (Siebold & Zucc.) Benth. & Hook.f. ex Trautv.	Campanulaceae	Roots	Improves the restraint of excessive adipose formation and accumulation	Choi et al. [83]
Convolvulus pluricaulis Choisy (*convolvulus prostratus* Forssk.)	Convolvulaceae	Whole plant	Reduces BW and TC, LDL-C, and TG levels, increases HDL-c levels, inhibits intestinal lipid absorption	Sharma et al. [84]
Corchorus olitorius L.	Malvaceae	Leaves	Anti-lipase, inhibition of leptin resistance, increases adiponectin	Gomaa et al. [85]
Corydalis bungeana Turcz.	Papaveraceae	Whole plant	Reduces BW and lowers lipids	Fu et al. [86]
Cosmos caudatus Kunth	Asteraceae	Leaves	Reduces BW, visceral fat mass, plasma TC, LDL-C, and TG levels, increases HDL-c levels	Rahman et al. [87]
Crataegus aronia Bosc. (syn. for *Crataegus azarolus* var. *aronia* L.)	Rosaceae	Flowers, leaves, fruits	Inhibits enzymes related to carbohydrate and lipid digestion and absorption	Al-Hallaq et al. [88]
Crotalaria juncea L.	Fabaceae	Leaves	Prevents BW gain, decreases cholesterol, TG, LDL, VLDL, BG levels	Rajech et al. [89]
Cucurbita pepo L.	Cucurbitaceae	Leaves	Decrease in BMI, food intake, and serum lipid levels	Nderitu et al. [48]
Cuminum cyminum L.	Apiaceae	Seeds	Raises HDL levels, decreases LDL, cholesterol, TG levels	Akmal and Itrat [90]

(continued)

Table 1 (continued)

Plant name	Family name	Used part	Contribution to anti-obesity activity/mechanism of action	References
Cyperus rotundus L.	Cyperaceae	Rhizomes	Anti-adipogenic, reduces leptin and serum lipids	Majeed et al. [91]
		Tubers and rhizomes	Lipase inhibitor, decreases TG levels	Wongchum et al. [92]
Dalbergia latifolia Roxb.	Fabaceae	Barks	Decrease in BW, food intake, TC, protects the liver and the kidneys	Khalid et al. [93]
Desmostachya bipinnata L.	Poaceae	Aerial parts	Anti-hyperlipidemic	Deepika et al. [94]
Docynia indica (Colebr.) Decne	Rosaceae	Fruit	Hypolipidemic, hypoglycemic, decreases BW	Loan et al. [95]
Embelia ribes Burm.f	Primulaceae	Fruits	BW gain suppression, reducing lipids, improving insulin and leptin sensitivity	Bhandari et al. [96]
Elettaria cardamomum (L.) Maton	Zingiberaceae	Fruits	Inhibits pancreatic lipase and α-amylase enzymes	Al-Yousef et al. [97]
Elateriospermum tapos Blume	Euphorbiaceae	Fruits	Decreases calorie intake, reduces fat accumulation via suppression of AMPK activation in hypothalamus, reduces body weight	Bahari et al. [98]
Ephedra alata Decne.	Ephedraceae	Fruits	Lipase inhibition	Jaradat et al. [99]
Erigeron annuus (L.) Pers	Asteraceae	Leaves and stems	Inhibition of lipid accumulation, ROS formation, adipocyte differentiation	Choi et al. [100]
Eurycoma longifolia Jack	Simaroubaceae	Roots	Increases lipolysis	Lahrita et al. [101]
Ficus racemosa L.	Moraceae	Leaves	Lipase inhibitory activity	Gorla and Shankar [52]
Fagonia cretica L. (syn. for *Zygophyllum creticum* (L.) Christenh. & Byng)	Zygophyllaceae	Aerial parts	Inhibits pancreatic lipase and α-amylase	Patel and Kumar [102]
Ferula asafoetida H.Karst (syn. for *Narthex asafetida* Falner ex. Lindl.)	Apiaceae	Gum	Decreases BW, abdominal fat, and epididymal adipocyte size	Azizian et al. [103]
Ficus glumosa Delile	Moraceae	Leaves	Prevents elevation of TC, LDL-C, VLDL-c, also has atherogenic activity	Ntchapda et al. [104]
Ficus racemosa L.		Bark	Reduces BW	Nasution et al. [105]
Fraxinus mandshurica Rupr.	Oleaceae	Seeds	Reduces BW gain, attenuates lipid accumulation in serum and in liver	Guo et al. [106]
Garcinia atroviridis Griff. Ex T. Anderson	Clusiaceae	Fruits	Decreases BW, triglyceride, LDL cholesterol, TC, atherosclerosis index, increases HDL-c	Kongchian et al. [71]
Garcinia indica (Thouars) Choisy		Fruit juice	Anti-dyslipidemic	Nampoothiri et al. [107]
Garcinia pedunculata Roxb. ex Buch.-Ham.		Fruits	Attenuates hyperlipidemia	Sarma et al. [108]
Geranium thunbergii Siebold ex. Lindl. & Paxton	Geraniaceae	Leaves	Reduces weight gain and adipose tissue, improves serum TC, triglyceride, LDL-C levels	Sung et al. [37, 38]
Glycyrrhiza uralensis Fisch	Fabaceae	Roots	Induces subcutaneous adipocyte browning, augments thermogenesis, lipolysis, and fat oxidation, reduces lipogenesis	Lee et al. [109]
Gnidia glauca (Fresen.) Gilg. (syn. for *Lasiosiphon glaucus* Fresen.)	Thymelaeaceae	Leaves	Inhibits lipid peroxidation	Arika et al. [110]
Gymnema sylvestre (Retz.) R. Br. ex Sm.	Apocynaceae	Whole plant	Decreases plasma levels of glucose and lipids	Pothuraju et al. [46]
		Leaves	Decreases BW, leptin hormone, and lipid profile	Abdel-Rahman and Khateib [111]
Hancornia speciosa Gomes	Apocynaceae	Leaves	Controls postprandial hyperglycemia	Dos Santos et al. [112]
Helichrysum compactum Boiss.	Asteraceae	Aerial parts	Anti-lipase	Şener et al. [34]

Hibiscus sabdariffa L.	Malvaceae	Calyx	Anti-pancreatic lipase, anti-alpha-glucosidase	Krishnamurthy et al. [113]
Hoodia gordonii (Masson) Sweet ex Decne.	Apocynaceae	Aerial parts	Decreases food consumption and body mass	van Heerden et al. [114]
Hoodia pilifera (L.f.) Plowes				
Hordeum vulgare L.	Poaceae	Leaves	Reduces lipid absorption, inhibition of pancreatic lipase	Kato et al. [115]
Hydrangea serrata (Thunb.) Ser.	Hydrangeaceae	Leaves	Decreases BW and fat mass, reduces BMI, total abdominal fat area	Han et al. [116]
Hylocereus lemairei (Hook.) Britton & Rose (syn. for *Selenicereus monacanthus* (Lem.) D.R. Hunt)	Cactaceae	Pulp and peel	Inhibits pancreatic lipase and α-glucosidase enzymes	Lodi et al. [117]
Hypericum perforatum L.	Hypericaceae	Whole plant	Decreases BW, TC, LDL-C, TG, glucose, insulin	Husain et al. [118]
		Aerial parts	Reduces expression of Dgat1 gene responsible for triglyceride biosynthesis	Tokgöz and Altan [119]
Illicium verum Hook.f	Schisandraceae	Fruits	Decreases BW increase, serum TC, TG, LDL, and VLDL levels	Iftikhar et al. [120]
Irvingia gabonensis (Aubry-Lecomte ex O'Rorke) Baill.	Irvingiaceae	Bark	Reduces BW and fasting blood sugar	Omonkhua and Onoagbe [121]
Jacaranda decurrens Cham.	Bignoniaceae	Leaves	Reduces body mass and white adipose tissue, increases fluid intake and excretion of feces	Antunes et al. [122]
Jatropha tanjorensis J.L. Ellis & Saroja	Euphorbiaceae	Leaves	Lowers cholesterol	Amaechi et al. [123]
Kaempferia parviflora Wall. Ex Baker	Zingiberaceae	Rhizomes	Improves leptin resistance, prevents BW gain, decreases fat size and hepatic fat accumulation	Miyazaki et al. [124]
Kedrostis africana Cogn.	Cucurbitaceae	Tubers	Inhibits α-amylase, α-glucosidase, lipase	Unuofin et al. [125]
Kingiodendron pinnatum (Roxb. Ex DC.) Harms (syn. for *Prioria pinnata* (Roxb. Ex DC.) Breteler)	Fabaceae	Leaves	Inhibits α-amylase, lipoprotein lipase	Kumar et al. [126]
Lagerstroemia parviflora Roxb.	Lythraceae	Leaves	Antihyperglycemic	Tanquilut et al. [127]
			Alleviates hyperlipidemia	Banke and Tyagi [128]
Laminaria japonica J.E.Areschoug (syn. for *Saccharina japonica* (J.E. Areschoug) C.E. Lane, C. Mayes, Druehl & G.W. Saunders)	Laminariaceae	Thallus	Decreases BW gain, food intake, serum and liver lipid layers, adipocyte size and number	Jang and Choung [129]
Livistona carinensis (Chiou.) J. Dransf. & N.W. Uhl	Arecaceae	Fruit pericarps and seeds	Inhibits α-glucosidase and lipase	Hifnawy et al. [130]
Litsea monopetala Pers.	Lauraceae	Stem, bark	Anti-adipogenic	Lamichhane et al.[45]
Lycium barbarum L.	Solanaceae	Fruits	Decreases lipid peroxidation, LDL cholesterol, waist circumference	Zanchet et al. [131]
Lycium chinense Mill.		Leaves	Lowers BW and serum TG and LDL-C levels	Kang et al. [132]
Lysimachia foenum-graecum Hance	Primulaceae	Whole plant	Anti-adipogenesis activity, inducing lipid breakdown by acting on lipid metabolism	Seo et al. [133]
Macrotyloma uniflorum (Lam.) Verdc.	Fabaceae	Leaves, seeds	Reduces BW, decreases fat absorption via inhibiting pancreatic lipase	Vadivelu et al. [134]

(continued)

Table 1 (continued)

Plant name	Family name	Used part	Contribution to anti-obesity activity/mechanism of action	References
Malus domestica Baumg.	Rosaceae	Fruits	Anti-α-amylase activity, reduces carbohydrate absorption and dietary lipid	Utami et al. [135]
Manilkara zapota (L.) P. Royen	Sapotaceae	Leaves and fruit pulp	Decreases glycemia, insulin, leptin, cholesterol, TG levels and increases HDL-c	Barbalho et al. [136]
Momordica charantia L.	Cucurbitaceae	Fruits	Hypophagic	Bano and Akhter [137]
		Leaves	Inhibits pancreatic lipase	Karale et al. [138]
Morinda citrifolia L.	Rubiaceae	Leaves	Inhibits lipase, reduces weight gain, improves plasma lipid profiles (plasma TG and plasma LDL levels), improves plasma levels of insulin and leptin	Jambocus et al. [139]
Moringa oleifera Lam.	Moringaceae	Leaves	Inhibits α-lipase and α-amylase	Ogundipe et al. [140]
Moringa peregrina Fiori		Seed oil	Decreases BW and BMI, TC, TG, and LDL-C, increases HDL-c	Elarabany et al. [141]
Morus alba L.	Moraceae	Root bark	Controls appetite, manages BW, improves metabolic syndrome	Yimam et al. [142]
Murraya koenigii (L.) Spreng.	Rutaceae	Leaves	Anti-lipase activity	Birari et al. [143]
Myristica fragrans Houtt.	Myristicaceae	Fruits	Activates AMPK enzyme in differentiated C2C12 cells	Nguyen et al. [144]
Nasturtium officinale R.Br.	Brassicaceae	Aerial parts	Inhibits α-amylase and pancreatic lipase	Zor et al. [145]
Nelumbo nucifera Gaertn.	Nelumbonaceae	Leaves	Hypoglycemic, hypolipidemic	Zhang et al. [146]
Nigella sativa L.	Ranunculaceae	Seeds	Decreases BW and food intake, BG, serum cholesterol, TG, LDL, increases HDL	Bano et al. [147]
			Improves serum lipids and reduces BW (in patients with Hashimoto's thyroiditis	Farhangi et al. [148]
			induces weight loss via exerting effect of UCP-1 (uncoupling protein-1) at gene and protein levels	Mahmoudi et al. [149]
			Decreases BW, TC, TG, BG, increases HDL	Saiju et al. [150]
Nitraria retusa (Forssk.) Asch.	Nitrariaceae	Shoots	Lowers LDL, glucose, and TG levels, increases HDL, enhances lipid metabolism in the liver	Kalai et al. [151]
Ocimum basilicum L.	Lamiaceae	Leaves and flowers	Inhibits α-amylase enzyme, also has anti-lipase activity	Noor et al. [152]
Ocimum gratissimum L.		Leaves	Reduces BW gain and adipocytes	Chao et al. [153]
Oroxylum indicum (L.) Benth. ex Kurz	Bignoniaceae	Bark	Inhibits adipogenesis and pancreatic lipase	Mangal et al. [154]
Orthosiphon stamineus Benth. (syn. for *Orthosiphon aristatus* var. aristatus)	Lamiaceae	Leaves	Reduces BW gain, hypolipidemic	Seyedan et al. [155]
Peucedanum japonicum Thunb.	Apiaceae	Leaves	Inhibits adipocyte differentiation, stimulates glucose uptake via AMPK activation, downregulates adipogenic transcription factors	Choi et al. [156, 157]
Phytolacca americana L.	Phytolaccaceae	Berries	Suppresses appetite, increases metabolic rate, reduces excess BW, cholesterol, and TG concentrations	Kiran and Raju [158]
Picralima nitida T. Durand & H. Durand	Apocynaceae	Leaves	Antidiabetic	Teugwa et al. [159]
		Seeds	Alleviates dyslipidemia and hyperglycemia	De Campos et al. [160]

Piper betle L.	Piperaceae	Leaves	Inhibits pancreatic lipase	Daud et al. [161]
Piper sarmentosum Roxb.			Inhibits pancreatic lipase	Daud et al. [161]
Pistacia atlantica Desf.	Anacardiaceae	Roots	Lipase inhibitory activity	Ben Hmed et al. [162]
		Leaves	Inhibits α-amylase and α-glucosidase enzyme	Pekacar and Deliorman Orhan [163]
Pistacia vera L.		Leaves	Inhibits α-amylase and α-glucosidase enzyme	[164]
Plantago asiatica L.	Plantaginaceae	Seeds	Decreases abdominal white/brown adipocyte size, serum TC, TG, LDL-C, hepatic triglyceride, and hepatic TC	Ping et al. [165]
Plantago depressa Willd.				
Platonia insignis Mart	Clusiaceae	Seed butter	Shows positive effects on lipid profile, especially on HDL-c and LDL-C	De Morais Lima et al. [166]
Polygonum aviculare L.	Polygonaceae	Aerial parts	Reduces BW gain, adipose tissue weight, serum triglyceride leptin levels, adipocyte size, and lipogenic gene expression	Sung et al. [167]
Populus balsamifera L.	Salicaceae	Branch barks	Anti-adipogenic as a complete agonist for PPARγ, limiting clonal expansion of pre-adipocytes	Martineau et al. [44]
Portulaca oleracea L.	Portulacaceae	Aerial parts	Anti-lipase	Jaradat et al. [66]
			Reduces BW	Azizah et al. [168]
Prangos gaubae (Bornm.) Herrnst. & Heyn	Apiaceae	Aerial parts	Inhibits α-amylase, α-glucosidase, and lipase	Bahadori et al. [169]
Prunus mume Koehne (syn. for *Prunus mume* var. *tonsa* Rehder)	Rosaceae	Stem bark	Lowers serum TC, TAG, LDL-C, increases HDL-c	Xia et al. [170]
Psidium guajava L.	Myrtaceae	Leaves	Inhibits α-amylase, α-glucosidase	Simao et al. [171]
Psoralea corylifolia L. (syn. for *Cullen corylifolium* (L.) Medik.)	Fabaceae	Leaves	Inhibits pancreatic lipase	Karale et al. [138]
Pueraria thomsonii Benth. (syn. for *Pueraria montana* Subsp. *Thomsonii* (Benth.) M.R. Almeida)	Fabaceae	Flowers	Suppression of lipogenesis in the liver, increasing lipolysis in white adipose tissue, and thermogenesis in brown adipose tissue	Kamiya et al. [172]
Salvia plebeia R. Br.	Lamiaceae	Aerial parts	Attenuation of fat accumulation, prevention of BW increase, suppression of adipogenesis and lipogenesis	Choi et al. [156, 157]
Senna alata (L.) Roxb.	Fabaceae	Leaves	Reduces BW and decreases parametrial fat	Chichioco-Hernandez and Leonido [68]
Senna petersiana (Bolle) Lock		Leaves	Hypoglycemic, decreases body glucose and HbG levels, improves insulin sensitivity	Olofinsan et al. [173]
Sesbania grandiflora (L.) Poir	Fabaceae	Flowers	Antihyperlipidemic, reduces BW, BMI, obesity index, adiposity index	Nikam et al. [174]
Setaria italica (L.) P. Beauv.	Poaceae	Seeds	Hypoglycemic, reduces TC, TG, LDL, increases HDL	Rani et al. [175]
Smallanthus sonchifolius (Poepp.) H. Rob	Asteraceae	Roots	Suppresses BW, food intake, adipose tissue weight and size, serum TG levels, improves adipokine profiles	Honore et al. [176]
Smilax chinensis (F.T. Wang) P.Li & C.X.Fu	Smilacaceae	Roots	Reduces BW gain, fat depots, serum lipids, glucose, insulin, leptin levels, improves insulin resistance	Balamurgan [177]
Solanum nigrum Acerbi ex Dunal	Solanaceae	Leaves	Reduces BW, inhibits pancreatic lipase activity	Aabideen et al. [178]
Solenostemma argel (Delile) Hayne (syn. for *Solenostemma oleifolium* (Nectoux) Bullock & E.A. Bruce ex Maire)	Apocynaceae	Aerial parts	Controls weight gain, improves lipid profile, attenuates liver steatosis, decreases lipogenesis, modulates adipokine activities	El-shiekh et al. [179]

(continued)

Table 1 (continued)

Plant name	Family name	Used part	Contribution to anti-obesity activity/mechanism of action	References
Sonchus oleraceus L.	Asteraceae	Whole plant	Antidiabetic	Teugwa et al. [159]
Sonneratia caseolaris Engl.	Lythraceae	Fruit juice	Prevents increase in BW	Thuoc et al. [180]
Sphenostylis stenocarpa (Hochst. ex A. Rich) Harms	Fabaceae	Seeds	Reduction in BW, specific weight gain, and BMI	Ejere et al. [181]
Stellaria media (L.) Vill.	Caryophyllaceae	Whole plant	Reduced circulating lipid levels, adipocyte diameter, reduces BW, decreases food intake, decreases absorption of dietary fats, inhibits pancreatic lipase	Chidrawar et al. [81]
Stereospermum suaveolens DC. (syn. for *Stereospermum chelonoides* DC.)	Bignoniaceae	Roots	Reduces BW gain, food intake, serum TG, TC, LDL, VLDL	Kaveripakam et al. [182]
Stevia rebaudiana Bertoni	Asteraceae	Leaves	Antihyperlipidemic	Shaheen et al. [183]
Syzygium aromaticum (L.) Merr. & L.M. Perry)	Myrtaceae	Flower buds	Decreases BW, white adipocyte tissue	Jung et al. [184]
Tabebuia avellanedae Lorentz ex Griseb	Bignoniaceae	Inner bark of the tree	Prevention of adipocyte accumulation	Iwamoto et al. [185]
Tamarindus indica L.	Fabaceae	Pulp	Decreases plasma levels of TC, LDL, TG, reduces BW, increases HDL	Azman et al. [186]
		Fruits	Decrease in BMI, waist circumference,	Asgary et al. [187]
		Leaves	Antihyperlipidemic	Valaparla et al. [188]
			Anti-lipase	Wiyono et al. [189]
Taxillus chinensis (DC.) Danser	Loranthaceae	Stems	Reduces BW and appetite	Wang et al. [190]
Tetrapleura tetraptera (Schumach. & Thonn.) Taub.	Fabaceae	Stem bark	Decreases BW, fat weight, food intake, serum TG, TC, LDL-C levels	Dongmo et al. [191]
Thespesia populnea Sol. Ex Correa	Malvaceae	Stem bark	Hypophagic and hypolipidemic activity	Retnasamy and Adikey [192]
Thrinax parviflora Sw.	Arecaceae	Fruit pericarps and seeds	Inhibits α-glucosidase and lipase	Hifnawy et al. [130]
Toddalia aculeata Pers. (syn. for *Toddalia asiatica* (L.) Lam.)	Rutaceae	Leaves	Reduces BW, BMI, body fat content, and resistive index	Putro et al. [193]
Tournefortia paniculata Vent. (syn. for *Myriopus paniculatus* (Cham.) Feuillet)	Boraginaceae	Leaves	Reduces liver fat, glucose, and serum TG	Simao et al. [194]
Tribulus terrestris L.	Zygophyllaceae	Fruits	Increases ghrelin and adiponectin levels, attenuates leptin levels, improves glycemic and lipid profile	Goel et al. [195]
Tricholepis glaberrima DC.	Asteraceae	Aerial parts	Reduces BW, TC, TG, LDL, VLDL, increases HDL	Rani and Kiranmai [196]
Ulmus parvifolia Jacq.	Ulmaceae	Bark	Increases browning in white adipose tissue, increases energy expenditure and thermogenesis	Lee et al. [197]
Urtica urens L.	Urticaceae	Aerial parts	Anti-lipase	Jaradat et al. [66]
Vaccinium ashei J.M. Reade	Ericaceae	Pulp	Reduces BW, serum glucose, improves lipid profiles, attenuates epididymal adipocytes	Wu et al. [198]
Vaccinium myrtillus L.		Leaves	Reduces BW gain, visceral fat mass, serum TAG, free fatty acids, TC, LDL-C	Zagayko et al. [199]
Valeriana dageletiana Nakai ex. F. Maek (syn. for *Valeriana excelsa* subsp. *sambucifolia* (J.C. Mikan ex. Pohl) Holub	Caprifoliaceae	Aerial parts	Suppresses lipid accumulation in 3T3-L1 adipocytes, BW gain, food intake, serum TG, and cholesterol levels	Wang et al. [200]
Vateria indica C.F. Gaertn.	Dipterocarpaceae	Stem bark	Reduces BW, cholesterol, TG, LDL, VLDL	Smitha et al. [201]

Vernonia amygdalina Del. (syn. for *Gymnanthemum amygdalinum* (Delile) Sch. Bip)	Asteraceae	Leaves	Regulates appetite	Egedigwe et al. [202]
Vigna unguiculata (L.) Walp.	Fabaceae	Leaves	Decrease in BMI, food intake, and serum lipid levels	Nderitu et al. [48]
Wasabia japonica (Miq.) Matsum (syn. for *Eutrema japonicum* Koidz.)	Brassicaceae	Leaves	Decreases TG synthesis and accumulation in the liver, decreases TG accumulation in adipose tissue, and increases fatty acid oxidization	Yamasaki et al. [203]
Zingiber officinale Roscoe	Zingiberaceae	Rhizomes	Reduces BW, body fat mass, improves lipid profile, decreases BG and leptin, and increases serum insulin levels	Shalaby and Saifan [79]
		Fruits	Reduces BW gain, serum lipids, insulin, leptin levels	Nazish et al. [204]
		Roots	Inhibits BW gain and adipose tissue, increases lipid excretion via feces, downregulates mRNA expression of genes associated with adipogenesis	Kim et al. [205, 206]
		Fruits	Decrease in BW, increase in HDL levels	Patil and Kothavale [207]
Ziziphus mauritiana Lam.	Rhamnaceae	Bark	Decreases insulin resistance and BW gain and increases fecal fat mass, inhibition of pancreatic lipase	Deshpande et al. [208]

AW adipose weight, *BG* blood glucose, *BMI* body mass index, *BW* body weight, *HDL-c* HDL-cholesterol, *LDL-C* LDL-cholesterol, *TAG* triacylglycerol, *TC* total cholesterol, *TG* triglycerides

Fig. 4 (**a**) General appearance of *M. charantia*; (**b**) ripening fruit; (**c**) ripened and simultaneously opened fruit with seeds covered with red aril; and (**d**) squeezed fruit pulp maceration in olive oil

be obtained by combining multiple products or products having multiple activities in the reduction and prevention of diet-induced obesity.

Some of these mixtures are provided in Table 2 along with the Latin names of the plants and families involved, and the parts that are used and their anti-obesity activities.

In addition to studies performed on plant extracts, there are also studies performed on compounds with anti-obesity activity isolated from various plant species. Some of these studies are given in Table 3. The compound/agent, Latin names of the plants and families that they are isolated from (unless purchased from chemical companies as pure compounds), and their contribution to anti-obesity activity are provided.

Table 2 Plant species that are involved in some plant mixtures/combinations

Component of the mixture/formulation	Family name	Used part	Contribution to anti-obesity activity/mechanism of action	References
Moringa oleifera Lam.	Moringaceae	Leaves	Inhibits differentiation of preadipocyte, increases lipid breakdown in mature adipocytes	Krishnaraju et al. [211]
Murraya koenigii (L.) Spreng.	Rutaceae	Leaves		
Curcuma longa L.	Zingiberaceae	Rhizomes		
Alchemilla vulgaris Willd.	Rosaceae	Leaves	"Weighlevel tablets" Reduces BW, increases thermogenesis	Said et al. [212]
Olea europaea	Oleaceae	Leaves		
Mentha longifolia (L.) L.	Lamiaceae	Leaves		
Cuminum cyminum L.	Apiaceae	Seeds		
Anogeissus latifolia (Roxb. Ex DC.) Wall. Ex Guill. & Perr. (syn. for *Terminalia anogeissiana* Gere & Boatwr.)	Combretaceae	Barks	Decreases BW, alters lipid profile positively	Avanapu and Reddy [213]
Holostemma annularis (Roxb.) liede & Khanum (syn. for *Cynanchum annularium* (Roxb.) Liede & Khanum)	Apocynaceae	Whole plant		
Trichodesma amplexicaule Roth (syn. for *Trichodesma indicum* var. *amplexicaule* (Roth) T.Cooke	Boraginaceae	Roots		
Clerodendrum phlomidis L.f	Lamiaceae	Leaves	Inhibits pancreatic amylase and lipase activity and thus delays gastrointestinal absorption of dietary sugar and fat	Gopinathan and Naveenraj [82, 83]
Garcinia cambogia Roxb (syn. for *Garcinia cowa* Roxb.)	Clusiaceae	Fruits		
Rhamnus frangula L. (syn. for *Frangula alnus* Mill.)	Rhamnaceae	Bark	"Vitalplant" Decreases BW gain, plasma lipid concentrations as TC and LDL)	Brenesel et al. [214]
Mentha x piperita L.	Lamiaceae	Leaves		
Carum carvi L.	Apiaceae	Fruits		
Petroselinum crispum (Mill.) Fuss	Apiaceae	Fruits		
Hypericum perforatum L.	Hypericaceae	Tea	Lowers serum cholesterol, TAG, VLDL, LDL, increases HDL	Akinseye [215]
Matricaria chamomilla Blanco	Asteraceae			
Morus alba L.	Moraceae	Root bark	Suppresses appetite, decreases BW gain, calorie intake, cholesterol, and LDL	Yimam et al. [216]
Magnolia officinalis Rehder & E.H.Wilson	Magnoliaceae	Stem bark		
Ilex paraguariensis A. St.-Hil.	Aquafoliaceae	Leaves		
Citrus unshiu (Swingle) Marcow (syn. for *Citrus deliciosa* Ten)	Rutaceae	Peel	Inhibits pancreatic lipase, suppresses visceral fat accumulation and serum lipid parameters	Kim et al. [217, 218]
Diospyros kaki L.f.	Ebenaceae	Fruit		
Phyllostachys Morus alba pubescens Mazel ex. J. Houz. (Syn. for *Phyllostachys edulis* J. Houz.)	Poaceae	Leaves	Reduces serum lipid profile, glucose, leptin, and adiponectin levels and fat accumulation in the liver	Kim et al. [217, 218]
Scutellaria baicalensis Georgi	Lamiaceae	Roots		
Phyllanthus emblica L.	Phyllanthaceae	Fruits	Reduces BW, BMI, abdominal circumference, lipid profile	Nimmi and George [219]
Curcuma longa L.	Zingiberaceae	Rhizome		
Plumbago zeylanica L.	Plumbaginaceae	Roots		
Macrotyloma uniflorum (Lam.) Verdc.	Fabaceae	Seeds		
Cissus quadrangularis L.	Vitaceae	Stem	Reduces weight and serum cholesterol	Talreja et al. [220]
Achyranthes aspera L.	Amaranthaceae	Seeds		
Cinnamomum zeylanicum Blume (syn. for *Cinnamomum verum* J. Presl)	Lauraceae	Stem bark	Reduces BW, has pyretic effect resulting in hypolipidemia and thermogenesis	Kumar et al. [221]
Murraya koenigii (L.) Spreng.	Rutaceae	Leaves		

BW body weight, *BMI* body mass index, *TAG* triacylglycerol, *TC* total cholesterol

Table 3 Compounds confirmed to have anti-obesity activity

Name of the compound	Plant name	Family name	Contribution to anti-obesity activity/mechanism of action	References
Aloe vera polysaccharides	*Aloe vera* (L.) Burm.f.	Asphodelaceae	Alternative insulin sensitizer, lowers fasting BG levels	Kim et al. [205, 206]
Anthocyanins	*Glycine max* (L.) Merr.	Fabaceae	Reduces BW and food intake	Badshah et al. [222]
Asperuloside	–	–	Alters nutrient-sensing receptors	Ishaq et al. [223]
Berberine	–	–	Reduces BW, food intake, fat contents, serum leptin and glucose levels	Park et al. [224]
Cassiamin A	*Cassia siamea* Lam. (syn. for *Senna siamea* (Lam.) H.S. Irwin & Barneby	Fabaceae	Inhibits pancreatic lipase	Kumar et al. [225]
Chikusetsusaponins	*Panax japonicus* (T. Nees) C.A. Mey	Araliaceae	Inhibits pancreatic lipase activity and thus delays intestinal absorption of dietary fat	Han et al. [226]
Crocin, crocetin, safranal	*Crocus sativus* L.	Iridaceae	Anti-adipogenic	Jafari et al. [227]
Dioscin	–	–	Anti-adipogenic via AMPK/MAPK pathway, modulates BW and fat accumulation	Poudel et al. [228]
Euphorbiasteroid	*Euphorbia Panax japonicus lathyrus* L.	Euphorbiaceae	Inhibits adipogenesis of 3T3-L1 cells via AMPK pathway activation	Park et al. [229]
Eurycomanone	*Eurycoma longifolia* Jack	Simaroubaceae	Increases lipolysis	Lahrita et al. [101]
Gallic acid	–	–	Reduces serum phospholipid, TC, LDL-C, insulin, and leptin levels	Hsu and Yen [230]
Hispidin	*Alpinia zerumbet* (Pers.) B.L. Burtt & R.M. Sm.	Zingiberaceae	Inhibits pancreatic lipase and lipid accumulation, reduces TG content, and increases intracellular cAMP and glycerol release	Tu and Tawata [231]
Hydroxy-α-sanshool	*Zanthoxylum bungeanum* maxim	Rutaceae	Decreases serum and liver levels of TC, TG, LDL-C, increases HDL-c	Wang et al. [232]
Myricetin	–	–	Suppresses 3T3-L1 preadipocyte differentiation via reducing expression of C/EBP α and PPAR γ	Wang et al. [233]
			Decreases serum glucose levels, regulates lipid metabolism	Su et al. [234]
Myricitrin-5-methyl ether	*Illicium verum* Hook.f	Schisandraceae	Inhibits gastric lipase	Kamoun et al. [235]
Quassinoids	*Brucea javanica* Merr.	Simaroubaceae	Stimulates lipolysis	Lahrita et al. [236]
Polygonatum sibiricum polysaccharides	*Polygonatum sibiricum* Redoute	Asparagaceae	Decreases body mass, promotes hepatic lipid metabolism, inhibits progression of nonalcoholic fatty liver disease	Zheng et al. [237]
Resveratrol derivative (4[2-(3,5-dimethoxyphenyl)vinyl]pyridine	–	–	Inhibits 3T3-L1 adipocyte differentiation	Hwang et al. [238]
Scutellarein	*Erigeron breviscapus* (Vaniot) Hand.-Mazz.	Asteraceae	Improves obesity-associated hyperlipidemia	Lin et al. [239]
Shikonin	*Lithospermum erythrorhizon* Siebold. & Zucc.	Boraginaceae	Suppresses adipogenesis and lipogenesis, increases fatty acid oxidation	Gwon et al. [240]
Teasaponin (theasaponin E_1 and E_2)	*Thea sinensis* L. (syn. for *Camellia sinensis* (L.) Kuntze)	Theaceae	Inhibits pancreatic lipase and thus inhibits intestinal absorption of dietary fat	Han et al. [241]

BW body weight, *BG* blood glucose, *HDL-c* HDL-cholesterol, *LDL-C* LDL-cholesterol, *TC* total cholesterol, *TG* triglycerides

3 Challenges and Future Recommendation

Obesity is a significant health problem that is not specific to some countries; it threatens the world as a whole [242, 243]. Due to this fact, March 4 is known as the World Obesity Day as an awareness precaution. According to 2022 data of WHO, 1 billion people in the world are obese; 650 million of them are adults, 340 million are adolescents, and 39 million are children. WHO also predicts that 167 million people will be overweight or obese by the year 2025 (www.who.int). According to the World Obesity Atlas 2022, the countries with highest percentages of obese adults by 2030 are the Cook Islands, Nauru, and American Samoa, all having the percentage of 67.4% followed by Palau with 66.4% and Tuvalu with 64.1%. Countries with the lowest percentage are Vietnam with 3.4% followed by Bangladesh with 5.8% and Cambodia with 6.0% (www.worldobesity.org) [244]. Obesity rates continue to increase both in developed and underdeveloped countries, and efficient measures have to be taken. According to the WHO, approximately 80% of the world's population use traditional medicine for the treatment and prevention of diseases/disorders including both developed and underdeveloped countries (www.who.int). From these data, we can predict that medicinal plants, herbal medicines, and plant products will continue to be used in the near and far future for the treatment and prevention of various diseases and disorders and obesity will not be an exception.

References

1. Diab F, Zbeeb H, Baldini F, Portincasa P, Khalil M, Vergani L. The potential of Lamiaceae herbs for mitigation of overweight, obesity and fatty liver: studies and perspectives. Molecules. 2022;27:5043.
2. Mohamed GA, Ibrahim SRM, Elkhayat ES, El Dine RS. Natural anti-obesity agents. Bull Fac Pharm Cairo Univ. 2014;52:269–84.
3. Çınar EN, İlhan A, Tengilimoğlu Metin MM. Effects of dietary polyphenols on browning and brown adipose tissue activity. Manisa Celal Bayar Univ J Inst Health Sci. 2022;9(2):334–40.
4. Rahman MM, Islam MR, Shohag S, Hossain ME, Rahaman MS, Islam F, Ahmed M, Mitra S, Khandaker MU, Idris AM, Chidambaram K, Bin Emran T, Cavalu S. The multifunctional role of herbal products in the management of diabetes and obesity: a comprehensive review. Molecules. 2022;27:1713.
5. Semalty M, Adhikari L, Chauhan A, Mishra A, Semwal D, Kotiyal R, Semalty A. Obesity and herbal drug research: exploring the safer alternative and lead molecule. Curr Trad Med. 2017;3:74–92.
6. Wang H-N, Xiang J-Z, Qi Z, Du M. Plant extracts in the prevention of obesity. Crit Rev Food Sci Nutr. 2022;62(8):2221–34.
7. Zhang X, Zhang B, Zhang C, Sun G, Sun X. Effect of *Panax notoginseng* saponins and major anti-obesity components on weight loss. Front Pharmacol. 2021;11:601751.
8. Hosny H, Omran N, Handousa H. Edible seeds with potential anti-obesity impact: a review. Int J Plant Based Pharm. 2022;2(1):64–81.
9. Balaji M, Ganjayi MS, Kumar GENH, Parim BN, Mopuri R, Dasari S. A review on possible therapeutic targets to contain obesity: the role of phytochemicals. Obes Res Clin Pract. 2016;10:363–80.
10. Birari RB, Bhutani KK. Pancreatic lipase inhibitors from natural sources: unexplored potential. Drug Discov Today. 2007;12(19/20):879–89.
11. Fu C, Jiang Y, Guo J, Su Z. Natural products with anti-obesity effects and different mechanisms of action. J Agric Food Chem. 2016;64:9571–85.
12. Law SK, Wang Y, Lu X, Au DCT, Chow WYL, Leung AWN, Xu C. Chinese medicinal herbs as potential prodrugs for obesity. Front Pharmacol. 2022;13:1016004.
13. Marrelli M, Statti G, Conforti F. A review of biologically active natural products from Mediterranean wild edible plants: benefits in the treatment of obesity and its related disorders. Molecules. 2020;25:648.
14. Raoof GFA, Abd El Kareem AMZ. The golden role of natural products in obesity. Int J Pharm Res Health Sci. 2020;8(6):3248–55.
15. Vasudeva N, Yadav N, Sharma SK. Natural products: a safest approach for obesity. Chin J Integr Med. 2012;18(6):473–80.
16. Hu X, Tao N, Wang X, Xiao J, Wang M. Marine-derived bioactive compounds with anti-obesity effect: a review. J Funct Foods. 2016;21:372–87.
17. Castro M, Preto M, Vasconcelos V, Urbatzka R. Obesity: the metabolic disease, advances on drug discovery and natural product research. Curr Top Med Chem. 2016;16:2577–604.
18. Pavlos S, Nikiforou C. Medicinal plants against obesity: a met-analysis of literature. J Complementary Med Res. 2021;12(3):244–3.
19. Sapkota BK, Khadayat K, Aryal B, Bashyal J, Jaisi S, Parajuli N. LC-HRMS-based profiling: antibacterial and lipase inhibitory activities of some medicinal plants for the remedy of obesity. Sci Pharm. 2022;90:55.
20. Yun JW. Possible anti-obesity therapeutics from nature – a review. Phytochemistry. 2010;71:1625–41.
21. Sun N-N, Wu T-Y, Chau C-F. Natural dietary and herbal products in anti-obesity treatment. Molecules. 2016;21:1351.
22. Kazemipoor M, Radzi CWJWM, Cordell GA, Yaze I. Potential of traditional medicinal plants for treating obesity: a review. IPCBEE. 2012;39:1923. https://doi.org/10.48550/arXiv.1208/1923.
23. Saad B, Ghareeb B, Kmail A. Metabolic and epigenetics action mechanisms of antiobesity medicinal plants and phytochemicals. Evid Based Complement Alternat Med. 2021;2021:9995903. https://doi.org/10.1155/2021/9995903.
24. Bajes HR, Almasri I, Bustanji Y. Plant products and their inhibitory activity against pancreatic lipase. Rev Bras. 2020;30:321–30.
25. Gao L, Cao L, Tian M, Chen Z. Study on the weight-reducing effect of *Acer truncatum* leave extract in alimentary obesity rat. J Hyg Res. 2012;41(4):609–11.
26. Khan N, Akhtar MS, Khan BA, Braga VA, Reich A. Antiobesity, hypolipidemic, antioxidant and hepatoprotective effects of *Achyranthes aspera* seed saponins in high cholesterol fed albino rats. Arch Med Sci. 2015;11(6):1261–71.
27. Kumar JS, Ruchi J, Nilesh J. Antiobesity potential of *Aconitum heterophyllum* roots. Int J Pharm Bio Sci. 2019;10(3):65–74.
28. Athesh K, Jothi G. Pharmacological screening of anti-obesity potential of *Acorus calamus* Linn. In high fat cafeteria diet fed obese rats. Asian J Pharm Clin Res. 2017;10(4):384–90.
29. Choi H-J, Chung MJ, Ham S-S. Antiobese and hypocholesterolaemic effects of an *Adenophora triphylla* extract in HepG2 cells and high fat diet-induced obese mice. Food Chem. 2010;119:437–44.
30. Kasabri V, Al-Hallaq EK, Bustanji YK, Abdul-Razzak JJ, Abaza IF, Afifi FU. Antiobesity and antihyperglycaemic effects of *Adiantum capillus-veneris* extracts: *in vitro* and *in vivo* evaluations. Pharm Biol. 2017;55(1):164–72.

31. Yuan C, Huang L, Suh JH, Wang Y. Bioactivity-guided isolation and identification of antiadipogenic compounds in Shiya tea (leaves of *Adinandra nitida*). J Agric Food Chem. 2019;67:6785–91.
32. Karmase A, Jagtap S, Bhutani KK. Anti adipogenic activity of *Aegle marmelos* Correa. Phytomedicine. 2013;20:1267–71.
33. Chege BM, Nyaga NM, Kaur PS, Misigo WO, Khan N, Wanyonyi WC, Mwangi PW. The significant antidyslipidemic, hypoglycemic, antihyperglycemic, and antiobesity activities of the aqueous extracts of *Agave sisalana* juice are partly mediated via modulation of calcium signaling pathways. Heliyon. 2023;9:e12400.
34. Şener SÖ, Cılız E, Öztekin BN, Badem M, Özgen U. Investigation of selected medicinal plants for their anti-obesity properties. Turk J Pharm Sci. 2022;19(5):498–504.
35. Sung Y-Y, Kim D-S, Kim HK. *Akebia quinata* extract exerts anti-obesity and hypolipidemic effects in high-fat diet-fed mice and 3T3–L1 adipocytes. J Ethnopharmacol. 2015;168:17–24.
36. Momoh BJ, Okere SO, Anyanwu GO. The anti-obesity effect of *Allium cepa* L. on high fat diet induced obesity in male Wistar rats. Clin Complementary Med Pharmacol. 2022;2:100035.
37. Sung YY, Yoon T, Kim SJ, Yang W-K, Kim HK. Anti-obesity activity of *Allium fistulosum* L. extract by down-regulation of the expression of lipogenic genes in high-fat diet-induced obese mice. Mol Med Rep. 2011;4:431–5.
38. Sung YY, Yoon T, Kim SJ, Yang W-K, Kim HK. Anti-obesity effects of *Geranium thunbergii* extract via improvement of lipid metabolism in high-fat diet-induced obese mice. Mol Med Rep. 2011;4:1107–13.
39. Sung Y-Y, Kim D-S, Kim S-H, Kim HK. Aqueous and ethanolic extracts of welsh onion, *Allium fistulosum*, attenuate high-fat diet-induced obesity. BMC Complement Altern Med. 2018;18:105.
40. Yang MH, Kim N-H, Heo J-D, Rho J-R, Ock KJ, Shin E-C, Jeong EJ. Comparative evaluation of sulfur compounds contents and antiobesity properties of *Allium hookeri* prepared by different drying methods. Evid Based Complement Alternat Med. 2017;2017(2017):2436927. https://doi.org/10.1155/2017/2436927.
41. Park S, No K, Lee J. Anti-obesity effect of *Allium hookeri* leaf extract in high-fat diet-fed mice. J Med Food. 2018;21(3):254–60.
42. Lee JY, Seo KH, Lee EY, Ji Y-J, Lee YJ, Kang MH, Seong H-A, Kim HD. Antioxidant and anti-obesity potentials of Korean-native wild vegetables (*Allium* species). Horticulturae. 2021;7:541.
43. Kim I, Kim J-Y, Hwang Y-J, Hwang K-A, Om A-S, Kim J-H, Cho K-J. The beneficial effects of aged black garlic extract on obesity and hyperlipidemia in rats fed a high-fat diet. J Med Plant Res. 2011;5(14):3159–68.
44. Martineau LC, Herve J, Muhammad A, Saleem A, Harris CS, Arnason JT, Haddad PS. Anti-adipogenic activities of *Alnus incana* and *Populus balsamifera* bark extracts, Part 1: sited and mechanisms of action. Planta Med. 2010;76:1439–46.
45. Lamichhane G, Sharma G, Sapkota B, Adhikari M, Ghimire S, Poudel P, Jung H-J. Screening of antioxidant, antibacterial, anti-adipogenic, and anti-inflammatory activities of five selected medicinal plants of Nepal. J Exp Pharmacol. 2023;15:93–106.
46. Pothuraju R, Sharma RK, Rather SA, Singh S. Comparative evaluation of anti-obesity effect of *Aloe vera* and *Gymnema sylvestre* supplementation in high-fat diet fed C57BL6J mice. J Intercult Ethnopharmacol. 2016;5(4):403–7.
47. Javaid S, Waheed A. Anti-obesity effects of *Aloe vera* whole leaf and sitagliptin in diabetic rats. Ann Punjab Med Coll. 2020;14(4):295–8.
48. Nderitu KW, Mwenda NS, Macharia NJ, Barasa SS, Ngugi MP. Antiobesity activities of methanolic extracts of *Amaranthus dubius*, *Cucurbita pepo*, and *Vigna unguiculata* in progesterone-induced obese mice. Evid Based Complement Alternat Med. 2017;2017:4317321. https://doi.org/10.1155/2017/4317321.
49. Swilam N, Nawwar M, Radwan R, Mostafa E. Antidiabetic and anti-obesity acylated flavonol diglucoside from *Ammania baccifera* L. subsp. *aegyptiaca* (Willd.) Koehne waste. Res Square. 2020. https://doi.org/10.21203/rs.3.rs-116589/v1.
50. El-Shazly SA, Ahmed MM, Al-Harbi MS, Alkafafy ME, El-Sawy HB, Amer SAM. Physiological and molecular study on the anti-obesity effects pf pineapple (*Ananas comosus*) juice in male Wistar rat. Food Sci Biotechnol. 2018;27(5):1429–38.
51. Elekofehinti OO, Lawal AO, Ejelonu OC, Molehin OR, Famusiwa CD. Involvement of fat mass and obesity gene (FTO) in the anti-obesity action of *Annona muricata* Annonaceae: *in silico* and *in vivo* studies. J Diabetes Metab Disord. 2020;19:197–204.
52. Gorla US, Shankar KR. *In vitro* anti-obesity and anti-cancer activities of different extracts of *Annona squamosa* L. and *Ficus racemosa* L. leaves. World J Pharm Res. 2016;5(11):1184–91.
53. Sukandar EY, Kurniati NF, Nurdianti AN. Antiobesity effect of methanol extract of *Anredera cordifolia* (Ten) Steenis leaves on obese male Wistar rats induced by high-carbohydrate diet. Int J Pharm Pharm Sci. 2016;8(4):171–3.
54. Moreno DA, Illic N, Poulev A, Raskin I. Effects of *Arachis hypogaea* nutshell extract on lipid metabolic enzymes and obesity parameters. Life Sci. 2006;78:2797–803.
55. Belgacem A, Senejoux F, Felgines C, Fraisse D, Bitri L, Khemiri I. Anti-obesity effects of the n-butanol fraction of the methanolic leaf extract of *Artemisia campestris* from Tunisian pharmacopoeia in male Wistar rats. J Complementary Integr Med. 2022;19(2):365–73.
56. Yamamoto N, Kanemoto Y, Ueda M, Kawasaki K, Fukuda I, Ashida H. Anti-obesity and anti-diabetic effects of ethanol extract of *Artemisia princeps* inC57BL/6 mice fed a high-fat diet. Food Funct. 2011;2:45.
57. Jiao P, Tseng-Crank J, Cornellusen B, Yimam M, Hodges M, Hong M, Maurseth C, Oh M, Kim H, Chu M, Jia Q. Lipase inhibition and antiobesity effect of *Atractylodes lancea*. Planta Med. 2014;80:577–82.
58. Ramgopal M, Attitalla IH, Avinash P, Balaji M. Evaluation of antilipidemic and anti obesity efficacy of *Bauhinia purpurea* bark extract on rats fed with high fat diet. Acad J Plant Sci. 2010;3(3):104–7.
59. Balamurgan G, Muralidharan P. Antiobesity effect of *Bauhinia variegata* bark extract on female rats fed on hypercaloric diet. Bangladesh J Pharmacol. 2010;5:8–12.
60. Kubota H, Kojima-Yuasa A, Morii R, Huang X, Norikura T, Rho S-N, Matsui-Yuasa I. Anti-obesity effect of *Blumea balsamifera* extract in 3T3-L1 preadipocytes and adipocytes. Am J Chin Med. 2009;37(5):843–54.
61. Kim D-Y, Kim M-S, Sa B-K, Kim M-B, Hwang J-K. *Boesenbergia pandurate* attenuates diet-induced obesity by activating AMP-activated protein kinase and regulating lipid metabolism. Int J Mol Sci. 2012;13:994–1005.
62. Singh C, Virmani T, Gupta J, Virmani R, Gahlawat D, Geeta. Antiobesity potential of *Boerhavia diffusa* on animal model of obesity. World J Pharm Res. 2015;4(11):1196–206.
63. Khalid M, Alqarni MH, Shoaib A, Wahab S, Foudah AI, Aljarba TM, Akhtar J, Alamri MA, Ahmad S. Anti-obesity action of *Boerhavia diffusa* in rats against high-fat diet-induced obesity by blocking the cannabinoid receptors. Plan Theory. 2022;11:1158.
64. Kumar M, Kaushik D, Kaur J, Proestos C, Oz F, Kumar A, Anjali A, Elobeid T, Terzioğlu ME, Xiao J. Assessment of anti-obesity potential and techno-functional properties of *Bougainvillea spectabilis* Willd. Bracts. Separations. 2022;9:399.
65. Kim H, Choung S-Y. Anti-obesity effects of *Boussingaulti gracilis* Miers var. *pseudobaselloides* Bailey via activation of AMP-activated protein kinase in 3T3-L1 cells. J Med Food. 2012;15(9):811–7.

66. Jaradat N, Zaid AN, Zaghal EZ. Anti-lipase activity for *Portulaca oleracea*, *Urtica urens*, *Brassica napus* and *Lathyrus hierosolymitanus* wild plants from Palestine. Marmara Pharm J. 2017;21(4):828–36.
67. Eldamaty HSE, Abd El-Maoula LM. Effects of sprouted turnip seeds as a hypolipidemic and anti-obesity agents on adult human females. J Res Fields Specific Educ. 2021;7(34):1565–94.
68. Chichioco-Hernandez CL, Leonido FMG. Weight-lowering effects of *Caesalpinia pulcherrima*, *Cassia fistula* and *Senna alata* leaf extracts. J Med Plant Res. 2011;5(3):452–5.
69. Zhang H-L, Wu Q-X, Wei X, Qin X-M. Pancreatic lipase and cholesterol esterase inhibitory effect of *Camellia nitidissima* Chi flower extracts *in vitro* and *in vivo*. Food Biosci. 2020;37:100682.
70. Sharma NK, Ahirwar D, Jhade D, Jain VK. *In vitro* anti-obesity assay of alcoholic and aqueous extracts of *Camellia sinensis* leaves. Int J Pharm Sci Res. 2012;3(6):1863–6.
71. Kongchian A, Keawboonlert N, Boonrak T, Lookyee S, Buasri K, Surongkul N, Tangpong J. Anti-hyperlipidemia and anti-obesity properties of *Garcinia atroviridis* and *Camellia sinensis* extracts in high-fat diet mice. Walailak J Sci Tech. 2020;17(10):1126–38.
72. Dramane P, Adama H, Potarniche A, Kpemissi M, Sarpataki O, Noufou O, Blidaru AI, Taulescu M, Samson G, Marcus I, Andrei S, Sevastre B. *Caralluma acutangula* prevents body weight gain in rats feed on hyperlipidemic diet. Bull UASVM Vet Med. 2018;75(2):168–77.
73. Dramane P, Jotham Y-PN, Adama H, Nogma ES, Mabozou K, Sevavstre B. Anti obesogenic potential of *Caralluma acutangula* bioactive fractions. Sch Int J Biochem. 2020;3(8):183–92.
74. Kazemipoor M, Hamzah S, Hajifaraji M, Radzi CWBWM, Cordell GA. Slimming and appetite suppressing effects of caraway aqueous extract as a natural therapy in physically active women. Phytother Res. 2016;30:981–7.
75. Habtemariam S. Antihyperlipidemic components of *Cassia auriculata* aerial parts: identification through *in vitro* studies. Phytother Res. 2013;27:152–5.
76. Nurkolis F, Taslim NA, Ahabibi FR, Kang S, Moon M, Choi J, Choi M, Park MN, Mayulu N, Kim B. Ulvophyte green algae *Caulerpa lentillifera*: metabolites profile and antioxidant, anticancer, anti-obesity, and in vitro cytotoxicity properties. Molecules. 2023;28:1365.
77. Patil A, Thakurdesai PA, Pawar S, Soni K. Evaluation of ethanolic leaf extract of *Ceiba pentandra* for anti-obesity and hypolipidemic activity in cafeteria diet (CD) treated Wistar albino rats. Int J Pharm Sci Res. 2012;3(8):2664–8.
78. Rosselli S, Tundis R, Bruno M, Leporini M, Falco T, Candela RG, Badalamenti N, Loizzo MR. *Ceiba speciosa* (A. St.-Hil.) seeds oil: fatty acids profiling by GC-MS and NMR and bioactivity. Molecules. 2020;25:1007.
79. Shalaby MA, Saifan HY. Some pharmacological effects of cinnamon and ginger herbs in obese diabetic rats. J Ethnopharmacol. 2014;3(4):144–9.
80. Sanadgol N, Najafi S, Ghasemi LV, Sanadgol E, Motalleb G, Afsharimoghadam A. A study of the inhibitory effects of *Cirtullus colocynthus* (CCT) using hydro-alcoholic extract on the expression of cytokines: TNF-α and IL-6 in high fat diet-fed mice towards a cure for diabetes mellitus. J Pharmacogn Phytother. 2011;3(6):81–8.
81. Chidrawar VR, Patel KN, Shiromwar SS, Kshirsagar AD. Exploiting anti-obesity mechanism of *Clerodendrum phlomidis* against two different models of rodents. Int J Green Pharm. 2011;5(3):216–29.
82. Gopinathan S, Naveenraj D. Antiobesity potential of *Clerodendrum phlomidis* Linn and *Garcia cambogia* Linn – a comparative animal model study. World J Pharm Res. 2014;3(9):1083–111.
83. Choi H-K, Won E-K, Jang YP, Choung S-Y. Antiobesity effect of *Codonopsis lanceolate* in high-calories/high-fat-diet-induced obese rats. Evid Based Complement Alternat Med. 2013;2013:210297. https://doi.org/10.1155/2013/210297.
84. Sharma A, Verma S, Prasad SB. Evaluation of anti-obesity activity of *Convolvulus pluricaulis* extract. Int J Toxicol Pharmacol Res. 2014;6(4):148–52.
85. Gomaa AA, El-Sers DA, Al-Zokeim NI, Gomaa MA. Amelioration of experimental metabolic syndrome induced in rats by orlistat and *Corchorus olitorius* leaf extract; role of adipo/cytokines. J Pharm Pharmacol. 2019;71:281–91.
86. Fu M, Bao T, Yu H, Lisha A, Li H, Ba G, Cho S. Metabolomics investigation on anti-obesity effects of *Corydalis bungeana* on high-fat high-sugar diet-induced obese rats. Chin Herb Med. 2022;14:414–21.
87. Rahman HA, Sahib NG, Saari N, Abas F, Ismail A, Mumtaz MW, Hamid AA. Anti-obesity effect of ethanolic extract from *Cosmos caudatus* Kunth leaf in lean rats fed a high fat diet. BMC Complement Altern Med. 2017;17:122.
88. Al-Hallaq EK, Kasabri V, Abdalla SS, Bustanji YK, Afifi FU. Anti-obesity and antihyperglycemic effects of *Crataegus aronia* extracts: *in vitro* and *in vivo* evaluations. Food Nutr Sci. 2013;4:972–83.
89. Rajesh O, Kumar VR, Shankaraiah P. Anti-obesity and hypoglycemic effect of ethanolic extract of *Crotalaria juncea* in high fat diet induced hyperlipidemic and hyperglycemic rats. Int J Pharm Pharm Sci. 2014;6(2):739–42.
90. Akmal A, Itrat N. Hypolipidemic and antiobesity impact of *Cuminum cyminum* powder in overweight and obese females. Int J Innov Sci Res Technol. 2019;4(4):396–402.
91. Majeed M, Nagabhushanam K, Bhat B, Ansari M, Pansey A, Bani S, Mundkur L. The antiobesity potential of *Cyperus rotundus* extract containing piceatannol, scirpusin A and scirpusin B from rhizomes: preclinical and clinical evaluations. Diab Metab Syndr Obes Targets Ther. 2022;15:369–82.
92. Wongchum N, Dechakhamhu A, Panya P, Pinlaor S, Pinmongkhonkul S, Tanomtong A. Hydroethanolic *Cyperus rotundus* L. extract inhibits anti-obesity property and increases lifespan expectancy in *Drosophila melanogaster* fed a high-fat diet. J Herbmed Pharmacol. 2022;11(2):296–304.
93. Khalid M, Shoaib A, Akhtar J, Alqami MH, Badruddeen, Siddiqui HH, Foudah AI. Anti obesity prospective of *Dalbergia latifolia* (Roxb.) hydroalcoholic bark extract in high fat diet induced obese rats. Biotech. 2020;10:493.
94. Deepika MY, Devi MARL, Nagaraju B, Deepthi PR, Bakshi BV. Antiobesity activity and beneficial effects of methanolic extract of *Desmostachya bipinnata* in HFD and progesterone induced obesity in rats and mice. Int J Pharm Sci Res. 2016;7(11):4644–55.
95. Loan NTT, Tan HTMT, Tam VTHT, Luan CL, Huong LM, Lien DN. Anti-obesity and body weight reducing effect of *Docynia indica* (Wall.) Decne fruit extract fractions in experimentally obese mice. VNU J Sci Nat Sci Technol. 2011;27:125–33.
96. Bhandari U, Chaudhari HS, Bisnoi AN, Kumar V, Khanna G, Javed K. Anti-obesity effect of standardized ethanol extract of *Embelia ribes* in urine model of high fat diet-induced obesity. PharmaNutrition. 2013;1:50–7.
97. Al-Yousef HM, Alqahtani AS, Hassan WHB, Alzoubi A, Abdelaziz S. Chemical profile, *in vitro* antioxidant, pancreatic lipase, and alpha-amylase inhibition assays of the aqueous extract of *Elettaria cardamomum* L. fruits. J Chem. 2021;2021:5583001. https://doi.org/10.1155/2021/5583001.
98. Bahari H, Abidin AZ, Balan SS, Perumal KV, Roshi NS, Lotafi AHA, Danabala S, Manimaran M, Shafie NH, Abdullah MA, Jasni AS. The effects of *Elateriospermum tapos* against obese maternal rat in mitigating obesity development among their adult female offspring. Pharmacogn Mag. 2021;26(72):706–12.

99. Jaradat N, Dacca H, Hawash M, Abualhasan MN. *Ephedra alata* fruit extracts: phytochemical screening, anti-proliferative activity and inhibition of DPPH, α-amylase, α-glucosidase, and lipase enzymes. BMC Chem. 2021;15:41.
100. Choi Y-H, Lee O-H, Zheng Y, Kang I-J. *Erigeron annuus* (L.) Pers. Extract inhibits reactive oxygen species (ROS) production and fat accumulation in 3T3-L1 cells by activating an AMP-dependent kinase signaling pathway. Antioxidants. 2019;8(5):139.
101. Lahrita L, Hirosawa R, Kato E, Kawabata J. Isolation and lipolytic activity of eurycomanone and its epoxy derivative from *Eurycoma longifolia*. Bioorg Med Chem. 2017;25:4829–34.
102. Patel D, Kumar V. Phytochemical analysis and *in vitro* anti obesity activity of different fractions of methanolic extract of *Fagonia cretica* L. Int J Pharm Sci Drug Res. 2020;12(3):282–6.
103. Azizian H, Rezvani ME, Esmaeilidehaj M, Bagheri SM. Anti-obesity, fat lowering and liver steatosis protective effects of *Ferula asafetida* gum in type 2 diabetic rats: possible involvement of leptin. Iran J Diab Obes. 2012;4(3):120–6.
104. Ntchapda F, Djedouboum A, Talla E, Dongmo SS, Nana P, Adjia H, Nguimbou RM, Bonabe C, Gaimatakon S, Yanou NN, Dimo T. Hypolipidemic and anti-atherogenic effect of aqueous extract leaves of *Ficus glumosa* (Moraceae) in rats. Exp Gerontol. 2015;62:53–62.
105. Nasution R, Mustanir M, Marzuki I. Isolation compound anti-obesity from the bark ara (*Ficus racemosa*) of Aceh. Orient J Chem. 2016;32(5):2693–9.
106. Guo S, Zhao H, Ma Z, Zhang S, Li M, Zheng Z, Ren X, Ho C-T, Bai N. Anti-obesity and gut microbiota modulation effect of secoiridoid-enriched extract from *Fraxinus mandshurica* seeds on high-fat diet-fed mice. Molecules. 2020;25:4001.
107. Nampoothiri L, Sudra P, Dey A, Dhadhai S, Kureshi AA, Kumar S, Dhanani T, Singh R, Kumari P. Fruit juice of *Garcinia indica* Choisy modulates dyslipidemia and lipid metabolism in cafeteria diet based rat model. Ann Phytomed. 2021;10(1):78–85.
108. Sarma R, Kumari S, Elancheran R, Deon M, Devi R. Polyphenol rich extract of *Garcinia pedunculata* fruit attenuates the hyperlipidemia induced by high fat diet. Front Pharmacol. 2016;7:294.
109. Lee HE, Yang G, Han S-H, Lee J-H, An T-J, Jang J-K, Lee JY. Anti-obesity potential of *Glycyrrhiza uralensis* and licochalcone A through induction of adipocyte browning. Biochem Biophys Res Commun. 2018;503(3):2117–23.
110. Arika W, Kibiti CM, Njagi JM, Ngugi MP. In vitro antioxidant properties of dichloromethanolic leaf extract of *Gnidia glauca* (Fresen) as a promising antiobesity drug. J Evid Based Integr Med. 2019;24:1–17.
111. Abdel-Rahman AN, Khateib BRM. Anti-obesity and hypolipidemic effects of *Gymnema sylvestre* leaves on rats fed on high fat diet. JEDU. 2020;6(28):613–30.
112. Dos Santos UP, Tolentino GS, Morais JS, Souza KP, Estevinho LM, dos Santos EL. Physiochemical characterization, microbiological quality and safety, and pharmacological potential of *Hancornia speciosa* Gomes. Oxidative Med Cell Longev. 2018;2018(2018):2976985. https://doi.org/10.1155/2018/2976985.
113. Krishnamurthy ZMR, Ali IM, Dayoob M, Hussein SS, Khan NAK. *Hibiscus sabdariffa* extract as anti-aging supplement through its antioxidant and anti-obesity activities. Biomed Res Ther. 2020;7(1):3572–8.
114. Van Heerden FR, Horak RM, Maharaj VJ, Vleggaar R, Senabe JV, Gunning PJ. An appetite suppressant from *Hoodia* species. Phytochemistry. 2007;68:2545–53.
115. Kato E, Tsuruma A, Amishima A, Satoh H. Proteinous pancreatic lipase inhibitor is responsible for the antiobesity effect of young barley (*Hordeum vulgare* L.) leaf extract. Biosci Biotechnol Biochem. 2021;85(8):1885–9.
116. Han H-S, Chung K-S, Shin Y-K, Yu J-S, Kang S-H, Lee S-H, Lee K-T. Effect of standardized *Hydrangea serrata* (Thunb.) Ser. Leaves extract on body weight and body fat reduction in overweight or obese humans: a randomized double-blind placebo-controlled study. Nutrients. 2022;14:505247.
117. Lodi KZ, Cappelari MB, Pilatti GC, Fontana RC, Camassola M, Salvador M, Branco CS. Pre-clinical evidence for the therapeutic effect of Pitaya (*Hylocereus lemairei*) on diabetic intestinal microenvironment. Nat Prod Res. 2022;37(10):1735–41. https://doi.org/10.1080/14786419.2022.2110091.
118. Husain GM, Chatterjee SS, Singh PN, Kumar V. Hypolipidemic and antiobesity-like activity of standardized extract of *Hypericum perforatum* L. in rats. ISRN Pharmacol. 2011;2011:505247. https://doi.org/10.5402/2011/505247.
119. Tokgöz HB, Altan F. *Hypericum perforatum* L.: a medicinal plant with potential as a curative agent against obesity-associated complications. Mol Biol Rep. 2020;47:8679–86.
120. Iftikhar N, Hussain AI, Kamal GM, Manzoor S, Fatima T, Alswailmi FK, Ahmad A, Alsuwayt B, Alnasser SMA. Antioxidant, anti-obesity, and hypolipidemic effects of polyphenol rich star anise (*Illicium verum*) tea in high-fat-sugar diet-induced obesity rat model. Antioxidants. 2022;11:2240.
121. Omonkhua AA, Onoagbe O. Effects of long-term oral administration of aqueous extracts of *Irvingia gabonensis* bark on blood glucose and liver profile of normal rabbits. J Med Plant Res. 2012;6(13):2581–9.
122. Antunes KA, da Silva BD, dos Santos Rocha P, Casagrande JC, Argandona EJ, do Carmo Vieira M, Cardoso CAL, dos Santos EL, de Picoli Souza K. Antiobesity effects of hydroethanolic extract of *jacaranda decurrens* leaves. Evid Based Complement Alternat Med. 2016;2016(2016):4353604. https://doi.org/10.1155/2016/4353604.
123. Amaechi D, Yisa BN, Ekpe IP, Nwawuba PI, Rabbi A. Phytochemical screening, anti-obesity and hepatoprotective activities of ethanol leaf extract of *Jatropha tanjorensis* in Wistar rats. Asian J Appl Chem Res. 2022;12(4):20–6.
124. Miyazaki M, Izumo N, Yoshikawa K, Matsugami T, Miyadate Y, Hayazimu K, Watanabe Y. The anti-obesity effect of *Kaempferia parviflora* (KP) is attributed to leptin in adipose tissue. J Nutr Health Food Sci. 2019;7(2):1–9.
125. Unuofin JO, Otunola GA, Afolayan AJ. *In vitro* α-amylase, α-glucosidase, lipase inhibitory and cytotoxic activities of tuber extracts of *Kedrostis Africana* (L.) Cogn. Heliyon. 2018;4:e00810.
126. Kumar JK, Prasad AGD, Richard SA. Biochemical activity of endangered medicinal plant *Kingiodendron pinnatum*. Asian J Plant Sci Res. 2011;1(4):70–5.
127. Tanquilut NC, Tanquilut MRC, Estacio MAC, Torres EB, Rosario JC, Reyes BAS. Hypoglycemic effect of *Lagerstroemia speciosa* (L.) Pers. on alloxan-induced diabetic mice. J Med Plant Res. 2009;3(12):1066–71.
128. Banke A, Tyagi CK. Antiobesity activity of methanolic extract of *Lagerstroemia parviflora* Roxb. (leaves) on Wistar albino rat model. Int J Green Pharm. 2021;15(1):77–89.
129. Jang WS, Choung SY. Antiobesity effects of the ethanol extract of *Laminaria japonica* Areshoung in high-fat-diet-induced obese rat. Evid Based Complement Alternat Med. 2013;2013:492807. https://doi.org/10.1155/2013/492807.
130. Hifnawy MS, Issa MY, El-Seedi H, Mahrous AMK, Ashour RMS. Phytochemical study, nutritional evaluation and *in vitro* anti-obesity potential of fruits pericarp and seeds of *Livistona carinensis* and *Thrinax parviflora*. Jordan J Biol Sci. 2021;14(1):163–71.
131. Zanchet MZS, Nardi GM, Bratti LOS, Filippin-Monteiro FB, Locatelli C. *Lycium barbarum* reduces abdominal fat and improves lipid profile and antioxidant status in patients with metabolic syndrome. Oxidative Med Cell Longev. 2017;2017(2017):9763210. https://doi.org/10.1155/2017/9763210.

132. Kang N-H, Park W-J, Choi M-K. Anti-obesity and hypolipidemic effects of *Lycium chinense* leaf powder in obese rats. J Med Food. 2010;13(4):801–7.
133. Seo JB, Choe SS, Jeong HW, Choi SM, Park JY, Choi EW, Kim JB, Seen DS, Jeong J-Y, Lee TG. Anti-obesity effects of *Lysimachia foenum-graecum* characterized by decreased adipogenesis and regulated lipid metabolism. Exp Mol Med. 2011;43(4):205–15.
134. Vadivelu B, Arumugam VA, Subbarayan S, Alshatwi AA, Krishnamoorthy R. Effect of *Macrotyloma uniflorum* on antiobesity in rats fed with a high fat diet. Saudi J Biol Sci. 2019;26:1772–8.
135. Utami S, Endrini S, Nafik S, Lestari IML, Anindya D, Abu Bakar E, Rozy F, Said FF, Afifah E, Arumwardana S, Nufus H, Rihibiha DD, Kusuma HSW, Wibowo SHB, Widowati W. *in vitro* antioxidant and anti-obesity activities of freeze-dried *Canarium* sp., *Averrhoa bilimbi* L. and *Malus domestica*. Indones Biomed J. 2019;11(3):320–6.
136. Barbalho SM, dos Santos Bueno PCS, Delazari DS, Guiguer EL, Coqueiro DP, Araujo AC, de Souza MSS, Farinazzi-Machado FMV, Mendes CG, Groppo M. Antidiabetic and antilipidemic effects of *Manilkara zapota*. J Med Food. 2015;18(3):385–91.
137. Bano F, Akhter N. Hypophagic and weight reducing effect of *Momordica charantia* fruit by enhancing 5-HT neurotransmitter in rats brain. Pak J Pharm Sci. 2020;33(5):2439–43.
138. Karale P, Dhawale SC, Karale A. Quantitative phytochemical profile, antioxidant and lipase inhibitory potential of leaves of *Momordica charantia* L. and *Psoralea corylifolia* L. Indian J Pharm Sci. 2022;84(1):189–96.
139. Jambocus NGS, Ismail A, Khatib A, Mahomoodally F, Saari N, Mumtaz MW, Hamid AA. *Morinda citrifolia* L. leaf extract prevent weight gain in Sprague-Dawley rats fed a high fat diet. Food Nutr Res. 2017;61:1338919.
140. Ogundipe A, Adetuyi B, Iheagwam F, Adefoyeke K, Olugbuyiro J, Ogunlana O, Ogunlana O. *In vitro* experimental assessment of ethanolic extract of *Moringa oleifera* leaves as an α-amylase and α-lipase inhibitor. Evid Based Complement Alternat Med. 2022;2022:4613109. https://doi.org/10.1155/2022/4613109.
141. Elarabany N, Hamad A, AlSobeai SM. (2022) Evaluating anti-obesity potential, active components, and antioxidant mechanisms of *Moringa peregrine* seeds extract on high-fat diet-induced obesity. J Food Biochem. 2022;56:e14265.
142. Yimam M, Jioa P, Hong M, Brownell L, Lee Y-C, Kim H-J, Nam J-B, Kim M-R, Jia Q. *Morus alba*, a medicinal plant for appetite suppression and weight loss. J Med Food. 2019;22(7):741–51.
143. Birari R, Roy SK, Singh A, Bhutani KK. Pancreatic lipase inhibitory alkaloids of *Murraya koenigii* leaves. Nat Prod Commun. 2009;4(8):1089–92.
144. Nguyen PH, Le TVT, Kang HW, Chae J, Kim SK, Kwon K-I, Seo DB, Lee SJ, Oh WK. AMP-activated protein kinase (AMPK) activators from *Myristica fragrans* (nutmeg) and their anti-obesity effect. Bioorg Med Chem Lett. 2010;20:4128–31.
145. Zor M, Pekacar S, Deliorman OD. Phytochemical content, antioxidant and enzyme inhibitory effects of *Nasturtium officinale*. J Fac Pharm Ankara. 2022;46(1):114–28.
146. Zhang H, Chen G, Zhang Y, Yang M, Chen J, Guo M. Potential hypoglycemic, hypolipidemic, and anti-inflammatory bioactive components in *Nelumbo nucifera* leaves explored by bioaffinity ultrafiltration with multiple targets. Food Chem. 2022;375:131856.
147. Bano F, Wajeeh M, Baig N, Naz H, Akhtar N. Antiobesity, antihyperlipidemic and hypoglycemic effects of the aqueous extract of *Nigella sativa* seeds (Kalongi). Pak J Biochem Mol Biol. 2009;42(4):136–40.
148. Farhangi MA, Dehghan P, Tajmiri S. Powdered black cumin seeds strongly improves serum lipids, atherogenic index of plasma and modulates anthropometric features in patients with Hashimoto's thyroiditis. Lipids Health Dis. 2018;17:59.
149. Mahmoudi A, Samani KG, Farrokhi E, Heidarian E. Effects of *Nigella sativa* extracts on the lipid profile and uncoupling protein-1 gene expression in brown adipose tissue of mice. Adv Biomed Res. 2018;7:121.
150. Saiju P, Sharma A, Jain R, Jain N. Evaluation of promising antiobesity activity of ethanolic extract of *Nigella sativa* seed on high fat diet-induced obese rats. Int J Pharm Sci Res. 2020;11(12):6279–88.
151. Kalai FZ, Han J, Ksouri R, El Omri A, Abdelly C, Isoda H. Antiobesity effects of an edible halophyte *Nitraria retusa* Forssk in 3T3-L1 preadipocyte differentiation and in C57B6J/L mice fed a high fat diet-induced obesity. Evid Based Complement Alternat Med. 2013;2013:368658. https://doi.org/10.1155/2013/368658.
152. Noor ZI, Ahmed D, Rehman HM, Qamar MT, Froeyen M, Ahmad S, Mirza MU. *In vitro* anti-diabetic, anti-obesity and antioxidant analysis of *Ocimum basilicum* aerial biomass and *in silico* molecular docking simulations with alpha-amylase and lipase enzymes. Biology. 2019;8:92.
153. Chao P-Y, Chiang T-I, Chang I-C, Tsai F-L, Lee H-H, Hsieh K, Chiu Y-W, Lai T-J, Liu J-Y, Hsu L-S, Shih Y-C. Amelioration of estrogen-deficiency-induced obesity by *Ocimum gratissimum*. Int J Med Sci. 2017;14(9):896–901.
154. Mangal P, Khare P, Jagtap S, Bishnoi M, Kondepudi KK, Bhutani KK. Screening of six Ayurvedic medicinal plants for anti-obesity potential: an investigation on bioactive constituents from *Oroxylum indicum* (L.) Kurz bark. J Ethnopharmacol. 2017;197:138–46.
155. Seyedan A, Alshawash MA, Alshagga MA, Mohamed Z. Antiobesity and lipid lowering effects of *Orthosiphon stamineus* in high-fat diet-induced obese mice. Planta Med. 2017;83:684–92.
156. Choi R-Y, Nam S-J, Ham JR, Lee H-I, Yee S-T, Kang K-Y, Seo K-I, Lee J-H, Kim M-J, Lee M-K. Anti-adipogenic and antidiabetic effects of cis-3′,4′-diisovalerylkhellactone isolated from *Peucedanum japonicum* Thunb leaves *in vitro*. Bioorg Med Chem Lett. 2016;26:4655–60.
157. Choi S-I, Cho I-H, Han SH, Jeon Y-J, Choi J-G, Kim J-S, Lee J-H. Antiobesity effects of *Salvia plebeian* R.Br. extract in high-fat diet-induced obese mice. J Med Food. 2016;19(11):1048–56.
158. Kiran GR, Raju AB. Antiobesity effect of *Phytolacca* berry in rats. Environ Exp Biol. 2014;12:95–9.
159. Teugwa CM, Mejiato PC, Zofou D, Tchinda BT, Boyom FF. Antioxidant and antidiabetic profiles of two African medicinal plants: *Picralima nitida* (Apocynaceae) and *Sonchus oleraceus* (Asteraceae). BMC Complement Altern Med. 2013;13:175.
160. De Campos OC, Osaigbovo DI, Bisi-Adeniyi TI, Iheagwam FN, Rotimi SO, Chinedu SN. Protective role of *Picralima nitida* seed extract in high-fat high-fructose-fed rats. Adv Pharmacol Pharma Sci. 2020;2020:5206204. https://doi.org/10.1155/20205206204.
161. Daud D, Jalil UJ, Sahrol NA, Zulkefli SZ, Adam NA, Ishak NK, Nm M, Tawang A. Comparative anti-obesity and anti-diabetic properties of *Piper sarmentosum* and *Piper betle* aqueous extracts via *I-vitro* system. J Pharm Adv Res. 2021;4(11):1421–7.
162. Ben Hmed M, Rigane G, Ben Salem R, Zouari N, Cherif S. Antiobesity and inhibitory pancreatic lipase effects of bioactive compounds of *Pistacia atlantica* roots extract. Austin Pancreat Disorders. 2019;3(1):1013.
163. Pekacar S, Deliorman Orhan D. Investigation of antidiabetic effect of *Pistacia atlantica* leaves by activity-guided fractionation and phytochemical content analysis by LC-QTOF-MS. Front Pharmacol. 2022;13:826261.
164. Gök HN, Pekacar S, Deliorman Orhan D. Investigation of enzyme inhibitory activities, antioxidant and chemical properties of *Pistacia vera* leaves using LC-QTOF-MS and RP-HPLC. Iran J Pharm Res. 2022;21(1):e127033.
165. Ping L-J, Ren-Chao T, Xia-Meng S, Hao-Yue Z, Shuai S, Ai-Zhen X, Zheng-Tao W, Li Y. Comparison of main chemical composition of *Plantago asiatica* L. and *P. depressa* Willd. Seed extracts

and their anti-obesity effects in high-fat diet-induced obese mice. Phytomedicine. 2021;81:153362.
166. De Morais Lima G, da Silva Brito AK, de Farias LM, Rodrigues LARL, de Carvalho Pereira CF, Lima SKR, de Macedo Gonçalves Frota K, dos Santos Rizzo M, Nunes PHM, Lucarini M, Durazzo A, Arcanjo DDR, do Carmo de Carvalho e Martins M. Effects of "Bacuri" seed butter (*Platonia insignis* Mart.) on metabolic parameters in hamsters with diet-induced hypercholesterolemia. Evid Based Complement Alternat Med. 2021;2021:5584965. https://doi.org/10.1155/2021/5584965.
167. Sung YY, Yoon T, Yang W-K, Kim SJ, Kim D-S, Kim HK. The antiobesity effect of *Polygonum aviculare* L. ethanol extract in high-fat diet-induced obese mice. Evid Based Complement Alternat Med. 2013;2013:626397. https://doi.org/10.1155/2013/626397.
168. Azizah RN, Emelda A, Asmaliani I, Ahmad I, Fawwaz M. Total phenolic, flavonoid and carotenoids content and anti-obesity activity of purslane herb (*Portulaca oleracea* L.) ethanol extract. Pharm J. 2022;14(1):8–13.
169. Bahadori MB, Zengin G, Bahadori S, Maggi F, Dinparast L. Chemical composition of essential oil, antioxidant, antidiabetic anti-obesity and neuroprotective properties of *Prangos gaubae*. Nat Prod Commun. 2017;12(12):1945–8.
170. Xia D, Wu X, Yang Q, Gong J, Zhang Y. Anti-obesity and hypolipidemic effects of a functional formula containing *Prunus mume* in mice fed high-fat diet. Afr J Biotechnol. 2010;9(16):2463–7.
171. Simao AA, Marques TR, Marcussi S, Correa AD. Aqueous extract of *Psidium guajava* leaves: phenolic compounds and inhibitory potential on digestive enzymes. Ann Braz Acad Sci. 2017;89(3):2155–65.
172. Kamiya T, Sameshima-Kamiya M, Nagamine R, Tsubata M, Ikegucji M, Takagaki K, Shimada T, Aburada M. The crude extract from *Puerariae* flower exerts antiobesity and antifatty liver effects in high-fat diet-induced obese mice. Evid Based Complement Alternat Med. 2012;2012:272710. https://doi.org/10.1155/2012/272710.
173. Olofinsan KA, Salau VF, Erukainure OL, Islam MS. *Senna petersiana* (Bolle) leaf extract modulates glycemic hemostasis and improves dysregulated enzyme activities in fructose-fed streptozotocin-induced diabetic rats. J Ethnopharmacol. 2023;303:115998.
174. Nikam SA, Kolhe SU, Tembhurne SV. Preclinical evaluation of *Sesbania grandiflora* flower extract for antihyperlipidemic and antiobesity activity on experimental rats. J Drug Deliv Ther. 2019;9(3):537–43.
175. Rani KN, Swapna M, Pavani K, Jyothsna, Divya K, Rao V. Effect of anti-obesity activity of *Setaria italica* on rats by using food induced obesity method. Pharma Innov. 2020;9(7):280–4.
176. Honore SM, Grande MV, Rojas JG, Sanchez SS. *Smallanthus sonchifolius* (Yacon) flour improves visceral adiposity and metabolic parameters in high-fat-diet-fed rats. J Obes. 2018;2018:5341384. https://doi.org/10.1155/2018/5341384.
177. Balamurgan R. *Smilax chinensis* Linn. (Liliaceae) root attenuates insulin resistance and ameliorate obesity in high diet induced obese rat. South Indian J Biol Sci. 2015;1(1):47–51.
178. Aabideen ZU, Mumtaz MW, Akhtar MT, Raza MA, Mukhtar H, Irfan A, Raza SA, Nadeem M, Ling YS. Anti-obesity effect and UHPLC-QTOF-MS/MS based metabolite profiling of *Solanum nigrum* leaf extract. Asian Pac J Trop Biomed. 2022;12(4):164–74.
179. El-shiekh RA, Al-Mahdy DA, Mouneir SM, Hifnawy MS, Abdel-Sattar EA. Anti-obesity effect of argel (*Solenostemma argel*) on obese rats fed a high fat diet. J Ethnopharmacol. 2019;238:111893.
180. Thuoc DV, Mai NTM, Ha LTV, Hung LD, Tra DH, Hung NK, Hung NP. Evaluation of antibacterial, antioxidant and antiobese activities of the fruit juice of crabapple mangrove *Sonneratia caseolaris* (Linn.). Int J Agric Sci Nat Res. 2018;5(2):25–9.
181. Ejere VC, Ogbuke EF, Nnamonu EI, Ikele BC, Nweze BC. Evaluation of anti-obesity potentials of *Sphenostylis stenocarpa* ethanolic seed extract. Annu Res Rev Biol. 2018;26(3):1–9.
182. Kaveripakam SS, Adikay S, Retnasamy G. Anti-obesity efficacy of roots of *Stereospermum suaveolens* in high fat-induced obese rats. J Young Pharm. 2017;9(2):234–8.
183. Shaheen SM, Azad AKM, Ferdous W, Jashim MU. Antiobesity report of a sweet super leaves, *Stevia rebaudiana* (Bert.) in mice model. Pharmacol Online. 2019;2:1–8.
184. Jung CH, Ahn J, Jeon T-I, Kim TW, Ha TY. *Syzygium aromaticum* ethanol extract reduces high-fat diet-induced obesity in mice through downregulation of adipogenic and lipogenic gene expression. Exp Ther Med. 2012;4:409–14.
185. Iwamoto K, Fukuda Y, Tokikura C, Noda M, Yamamoto A, Yamamoto M, Yamashita M, Zaima N, Iida A, Moriyama T. The anti-obesity effect of Taheebo (*Tabebuia avellanedae* Lorentz ex Griseb) extract in ovariectomized mice and the identification of a potential anti-obesity compound. Biochem Biophys Res Commun. 2016;478:1136–40.
186. Azman KF, Amom Z, Azlan A, Esa NM, Ali RM, Shah ZM, Abdul Kadir KK. Antiobesity effect of *Tamarindus indica* L. pulp aqueous extract in high-fat diet-induced obese rats. J Nat Med. 2012;66:333–42.
187. Asgary S, Soltani R, Barzegar N, Sarrafzadegan N. Evaluation on the effects of *Tamarindus indica* L. fruit on body weight and several cardiometabolic risk factors in obese and overweight adult patients: a randomized controlled clinical trial. Int J Prev Med. 2020;11:24.
188. Valaparla GR, Sai GV, Kumar D. Assessment of anti-obesity activity of *Tamarindus* leaves on butter induced hyperlipidemia in mice. Asian J Pharm Pharmacol. 2022;8(3):100–3.
189. Wiyono T, Frediansyah A, Sholikhah EN, Pratiwi WR. UHPLC-ESI-MS analysis of Javanese *Tamarindus indica* leaves from various tropical zones and their beneficial properties in relation to antiobesity. J Appl Pharm Sci. 2022;12(8):137–47.
190. Wang Y, Deng M, Zhang S-Y, Zhou Z-K, Tian W-X. Parasitic loranthus from Loranthaceae rather than Viscaceae potently inhibits fatty acid synthase and reduces body weight in mice. J Ethnopharmacol. 2008;118:473–8.
191. Dongmo OLM, Ashu ELA, Tadjoua HT, Epoh NJ, Njina SN, Tapondjou LA, Telefo PB. Evaluation of anti-obesity and diuretic effects of aqueous extract of *Tetrapleura tetraptera* Taub. Stem bark on high fat diet-induced obese rats. Eur J Pharm Med Res. 2019;6(1):27–37.
192. Retnasamy G, Adikey S. Evaluation of anti-obesity activity of *Thespesia populnea* (L.) and flavonoid isolated quercetin on high fat diet-induced obese rats. J Exp Appl Anim Sci. 2016;2(1):46–58.
193. Putro PS, Indarto D, Purwanto B, Koeswondo W, Kusumaningrum S, Nguzum KY. The effects of leaves extract of *Toddalia aculeata* on body weight, body mass index, body fat content and resistive index in male rats with high-fat diet. Pharm J. 2022;14(6):771–7.
194. Simao AA, Ramos VO, Correa AD, Sousa RV, Marcussi S. Anti-obesity effects of the administration of *Tournefortia paniculata* Cham extract on Wistar rats subjected to a hypercaloric diet. Braz Arch Biol Technol. 2015;58(4):494–503.
195. Goel K, Gupta S, Singh R, Saini V, Chhabra P, Bansal S. Estimation of anti-obesity potential of *Tribulus terrestris* in Wistar rats (high fat diet inducted obesity). HIV Nurs. 2023;23(2):1014–22.
196. Rani VI, Kiranmai G. Anti-obesity activity of methanolic extract of *Tricholepis glaberrima*. Indo Am J Pharm Sci. 2017;4(9):2728–34.
197. Lee YY, Kim M, Irfan M, Yuk HJ, Kim D-S, Lee SE, Kim S-H, Kim S, Kim S-D, Rhee MH. *Ulmus parvifolia* Jacq. Exhibits antiobesity properties and potentially induces browning of white adipose tissue. Evid Based Complement Alternat Med. 2020;2020:9358563. https://doi.org/10.1155/2020/9358563.

198. Wu T, Jiang Z, Yin J, Long H, Zheng X. Anti-obesity effects of artificial planting blueberry (*Vaccinium ashei*) anthocyanin in high-fat diet-treated mice. Int J Food Sci Nutr. 2016;67(3):257–64.
199. Zagayko AL, Kolisnyk TY, Churnak OI, Ruban OA, Koshovyi OM. Evaluation of anti-obesity and lipid lowering properties of *Vaccinium myrtillus* leaves powder extract in a hamster model. J Basic Clin Physiol Pharmacol. 2018;29(6):697–703.
200. Wang Z, Hwang SH, Kim JH, Lim SS. Anti-obesity effect of the above-ground part of *Valeriana dagetiana* Nakai ex F. Maek extract in high-fat diet-induced obese C57BL/6N mice. Nutrients. 2017;9:689.
201. Smitha AS, Himanshu J, Gururaja MP, Prasanna SK, Ullas DP. Anti-obesity activity of *Vateria indica* Linn. Stem barks in rats. Res J Pharm Technol. 2018;11(12):5238–42.
202. Egedigwe CA, Ijeh II, Okafor PN, Ejike CECC. Aqueous and methanol extracts of *Vernonia amygdalina* leaves exert their anti-obesity effects through the modulation of appetite-regulatory hormones. Pharm Biol. 2016;54(12):3232–6.
203. Yamasaki M, Ogawa T, Wang L, Katsube T, Yamasaki Y, Sun X, Shiwaku K. Anti-obesity effects of hot water extract from wasabi (*Wasabia japonica*) leaves in mice fed high-fat diets. Nutr Res Pract. 2013;7(44):267–72.
204. Nazish I, Ansari SH, Arora P, Ahmad A. Antiobesity activity of *Zingiber officinale*. Pharm J. 2016;8(5):440–6.
205. Kim S, Lee M-S, Jung S, Son H-Y, Park S, Kang B, Kim S-Y, Kim I-H, Kim C-T, Kim Y. Ginger extract ameliorates obesity and inflammation via regulating microRNA-21/132 expression and AMPK activation in white adipose tissue. Nutrients. 2018;10:1567.
206. Kim K, Chung MH, Park S, Cha J, Baek JH, Lee S-Y, Choi S-Y. ER Stress attenuation by *Aloe*-derived polysaccharides in the protection of pancreatic β-cells from free fatty acid-induced lipotoxicity. Biochem Biophys Res Commun. 2018;500:797–803.
207. Patil SB, Kothavale SD. Antiobesity activity of *Zingiber officinale*. Res J Pharmacogn Phytochem. 2022;14(4):252–6.
208. Deshpande MS, Shengule S, Apte KG, Wani M, Piprode V, Parab PB. Anti-obesity activity of *Ziziphus mauritiana*: a potent pancreatic lipase inhibitor. Asian J Pharm Clin Res. 2013;6(2):168–73.
209. Çiçek SS. *Momordica charantia* L. – diabetes-related bioactivities, quality control, and safety considerations. Front Pharmacol. 2022;13:904643.
210. İnce İ, Çığırgil N, Gümüştaş B, Kozguş Güldü Ö, Karaman D, Medine Eİ, Güler G, Karasulu E. *Momordica charantia* L. meyveleri kullanılarak krem hazırlanması, kalite kontrolü ve *in vitro* yara iyi edici etkisinin araştırılması. Erzincan Univ J Sci Technol. 2019;12(1):38–48.
211. Krishnaraju AV, Sundararaju D, Srinivas P, Rao CV, Sengupta K, Trimurtulu G. Safety and toxicological evaluation of a novel anti-obesity formulation LI85008F in animals. Toxicol Mech Methods. 2010;20(2):59–68.
212. Said O, Saad B, Fulder S, Khalil K, Kassis E. Weight loss in animals and humans treated with "weighlevel", a combination of four medicinal plants used in traditional Arabic and Islamic medicine. Evid Based Complement Alternat Med. 2011;2011:874538. https://doi.org/10.1093/ecam/nen067.
213. Avanapu SR, Reddy DS. Anti-obesity effect of a polyherbal formulation in cafeteria and atherogenic diet induced obesity in rats. Indo Am J Pharm Res. 2013;3(8):6084–93.
214. Brenesel MD, Pilija V, Popovic T, Arsic A, Milic M, Kojic D, Milic N, Misan A. Antihyperlipidemic, antioxidant and weight-lowering effects of "Vitalplant". Open Life Sci. 2015;10:291–8.
215. Akinseye OR. The hypolipidemic activities of the tea extracts of St. John's wort tea, chamomile tea and their blend at different concentrations, orally induced on adult male Wistar rats. J Nat Sci Res. 2016;6(11):142–6.
216. Yimam M, Jiao P, Hong M, Brownell L, Lee Y-C, Hyun E-J, Kim H-J, Kim T-W, Nam J-B, Kim M-R, Jia Q. Appetite suppression and antiobesity effect of a botanical composition composed of *Morus alba*, *Yerba mate*, and *Magnolia officinalis*. J Obes. 2016;2016:4670818. https://doi.org/10.1155/2016/4670818.
217. Kim G-N, Shin M-R, Shin SH, Lee AR, Lee JY, Seo B-I, Kim MY, Kim TH, Noh JS, Rhee MH, Roh S-S. Study of antiobesity effect through inhibition of pancreatic lipase activity of *Diospyros kaki* fruit and *Citrus unshiu* peel. Biomed Res Int. 2016;2016:1723042. https://doi.org/10.1155/2016/1723042.
218. Kim D-S, Kim S-H, Cha J. Antiobesity effects of the combined plant extracts varying the combination ratio of *Phyllostachys pubescens* leaf extract and *Scutellaria baicalensis* root extract. Evid Based Complement Alternat Med 2016;2016:9735276. https://doi.org/10.1155/2016/9735276.
219. Nimmi OS, George P. Antiobesity and antioxidant effects of a new polyherbal formulation (PHF) in obesity induced Wistar rats. Indian J Tradit Knowl. 2017;16(2):297–302.
220. Talreja T, Kumar M, Sirohi P, Sharma T. Preparation and use of *Cissus quadrangularis* and *Achyranthes aspera* formulation in the management of weight loss. Pharma Innov 2017;6(3):143–51.
221. Kumar P, Singh S, Ahmad MI. Synergistic effect of *Cinnamomum zeylanicum* and *Murraya koenigii* formulation for antiobesity and hypolipidemic activity on wistar albino rats. Adv Trad Med. 2021;21:553–63.
222. Badshah H, Ullah I, Kim SE, Lee HY, Kim MO. Anthocyanins attenuate body weight gain via modulating neuropeptide Y and $GABA_{B1}$ receptors in rats hypothalamus. Neuropeptides. 2013;47:347–53.
223. Ishaq M, Tran M, Wu Y, Nowak K, Deans B, Xin JTZ, Loh HL, Ng WY, Yee CW, Southam B, Vicenzi S, Randall C, Yang C, Tan E, Pasupuleti M, Grewal AK, Ahmad T, Shastri M, Vicario C, Ronci M, Zuccarini M, Bleasel M, Scowen P, Raffaeli W, D'Andrea G, Chellappan DK, Jacobson G, Bissember AC, Smith JA, Eri R, Canales J, Iglesias M, Guben N, Caruso V. Asperuloside enhances taste perception and prevents weight gain in high-fat fed mice. Front Endocrinol. 2021;12:615446.
224. Park H-J, Jung EY, Shim I. Berberine for appetite suppressant and prevention of obesity. Biomed Res Int. 2020;2020:3891806. https://doi.org/10.1155/2020/3891806.
225. Kumar D, Karmase A, Hagtap S, Shekbar R, Bhutani KK. Pancreatic lipase inhibitory activity of cassiamin A, a bianthraquinone from *Cassia siamea*. Nat Prod Commun. 2013;8:1–4.
226. Han L-K, Zheng Y-N, Yoshikawa M, Okuda H, Kimura Y. Anti-obesity effects of chikusetsusaponins isolated from *Panax japonicus* rhizomes. BMC Complement Altern Med. 2005;5:9.
227. Jafari F, Emami SA, Javadi B, Salmasi Z, Tajarani-Najjaran M, Tajarani-Najjaran Z. Inhibitory effect of saffron, crocin, crocetin, and safranal against adipocyte differentiation in human adipose-derived stem cells. J Ethnopharmacol. 2022;294:115349.
228. Poudel B, Lim S-W, Ki H-H, Nepali S, Lee Y-M, Kim D-K. Dioscin inhibits adipogenesis through the AMPK/MAPK pathway in 3T3-L1 cells and modulates fat accumulation in obese mice. Int J Mol Med. 2014;34:1401–8.
229. Park S-J, Park JH, Davaatseren M, Kim HJ, Kim M-S, Hur HJ, Sung M-J, Hwang J-T, Yang HJ, Kwon DY. Euphorbiasteroid, a component of *Euphorbia lathyris* L., inhibits adipogenesis of 3T3-L1 cells via activation of AMP-activated protein kinase. Cell Biochem Funct 2015;33:220–5.
230. Hsu C-L, Yen G-C. Effect of gallic acid on high fat diet-induced dyslipidemia, hepatosteatosis and oxidative stress in rats. Br J Nutr. 2007;98:727–35.
231. Tu PTB, Tawata S. Anti-obesity effects of hispidin and *Alpinia zerumbet* bioactives in 3T3-L1 adipocytes. Molecules. 2014;19(10):16656–71.

232. Wang L, Zhang M, Zhang Q, Li L, Wang J, Zhu L, Wei D, Wei P, Wu C. Antiobesity, regulation of lipid metabolism, and attenuation of liver oxidative stress effects of hydroxyl-α-sanshool isolated from *Zanthoxylum bungeanum* on high-fat diet-induced hyperlipidemic rats. Oxidative Med Cell Longev. 2019;2019:5852494. https://doi.org/10.1155/2019/5852494.
233. Wang Q, Wang S-t, Yang X, You P, Zhang W. Myricetin suppress differentiation of 3 T3-L1 preadipocytes and enhances lipolysis in adipocytes. Nutr Res. 2015;35:317–27.
234. Su H-M, Feng L-N, Zheng X-D, Chen W. Myricetin protects against diet-induced obesity and ameliorates oxidative stress in C57BL/6 mice. J Zhejiang Univ Sci B. 2016;17(6):437–46.
235. Kamoun J, Rahier R, Sellami M, Koubaaa I, Mansuelle P, Lebrun R, Berlioz-Barbier A, Fioere M, Alvarez K, Abousaiham A, Carriere F, Aloulou A. Identification of a new natural gastric lipase inhibitor from star anise. Food Funct. 2019;10:469–78.
236. Lahrita L, Moriai K, Iwata R, Itoh K, Kato E. Quassinoids in *brucea javanica* are potent stimulators of lipolysis in adipocytes. Fitoterapia, 137: 104250 and licochalcone A through induction of adipocyte browning. Biochem Biophys Res Commun. 2019;503:2117–23.
237. Zheng L, Zhong F, Chen Z, Li G, Zhu Q. *Polygonatum sibiricum* polysaccharides protect against obesity and non-alcoholic fatty liver disease in rats fed a high-fat diet. Food Sci Human Wellness. 2022;11:1043–52.
238. Hwang JT, Kim S, Choi I, Choi SY. Resveratrol analog 4-[2-(3,5-dimethoxyphenyl)vinyl]pyridine reduces differentiation of the 3T3-L1 adipocyte. Pharm Biol. 2013;51(1):96–9.
239. Lin Y, Ren N, Li S, Chen M, Pu P. Novel anti-obesity effect of scutellarein and potential underlying mechanism of actions. Biomed Pharmacother. 2019;117:109042.
240. Gwon SY, Choi WH, Lee DH, Ahn JY, Jung CH, Moon BK, Ha TY. Shikonin protects against obesity through the modulation of adipogenesis, lipogenesis, and β-oxidation *in vivo*. J Funct Foods. 2015;16:484–93.
241. Han L-K, Kimura Y, Kawashima M, Takaku T, Taniyama T, Hayashi T, Zheng Y-N, Okuda H. Anti-obesity effects in rodents of dietary teasaponin, a lipase inhibitor. Int J Obes. 2001;25:1459–64.
242. Chidrawar VR, Patel KN, Sjeth NR, Shiiromwar SS, Trivedi P. Antiobesity effect of *Stellaria media* against drug induced obesity in Swiss albino mice. AYU. 2015;32(4):576–84.
243. Garg A, Singh R. Antiobesity activity of ethanolic extract of *Cassia auriculata* in high fat diet induced obese rats. Int J Pharm Pharm Sci. 2015;7(4):237–43.
244. Lobstein T, Brinsden H, Neveux M. World obesity atlas 2022. London: World Obesity Federation; 2022.
245. WFO. World Flora online. Published on the Internet. 2023. http://www.worldfloraonline.org. Accessed September 2022-April 2023.

Herbal Medicines for the Management of Diseases in the Heart, Circulation, and Blood

Jagdish S. Bankar, Kajal N. Bondre, Pratiksha P. Wagh, Surbhi S. Bhope, Jayesh S. Pande, Prakash R. Itankar, Satyendra K. Prasad, and Shailendra S. Gurav

Abstract

Cardiovascular diseases (CVDs) are the foremost reason for early death and associated disease complications; their prevalence is rising worldwide. As a result, CVDs are a measure of economic and health burden on the country. As per European statistics, the death rate is 45% of the population. Many allopathic drugs used to treat CVDs have side effects or adverse effects, and they are responsible for approximately 8% of hospital admissions in the USA. Thus, numerous therapeutic plants have typically been investigated for their potential impact on CVDs. It is commonly observed that herbal extracts are reached with many secondary plant metabolites and provide significant advantages for treating cardiovascular complications due to their safety profiles. The present chapter was undertaken to give an insight into the cardioprotective potential of various medicinal plants such as *Embilica officinalis*, *Rauwolfia serpentina*, *Terminalia arjuna*, *Ziziphus oxyphylla*, and so on. Many phytomolecules like taxifolin, caffeic acid, quercetin, rosmarinic acid, arjunolic acid, andrographolide, and others have therapeutic potential against CVDs. This chapter provides information about cardiovascular diseases and gives importance to medicinal plants in their management and specific mode of action.

Keywords

Cardiovascular diseases · *Embilica officinalis* · Hypertension · Medicinal plants · Phytomolecules · Taxifolin

J. S. Bankar · K. N. Bondre · P. P. Wagh · S. S. Bhope · J. S. Pande · P. R. Itankar · S. K. Prasad (✉)
Department of Pharmaceutical Sciences, Rashtrasant Tukadoji Maharaj Nagpur University, Nagpur, Maharashtra, India

S. S. Gurav (✉)
Department of Pharmacognosy, Goa College of Pharmacy, Goa University, Panaji, Goa, India
e-mail: shailendra.gurav@nic.in

Abbreviations

AT1R	Angiotensin II type 1 receptor
BDNF	Brain-derived neurotrophic factor
CAT	Catalase
cGMP	Cyclic guanosine monophosphate
ESI-MS/MS	Electrospray ionization mass spectrometry
GSH	Reduced glutathione
H9c2	Cardiomyoblast
HPLC	High performances liquid chromatography
I/R	Ischemic–reperfusion
IL-6	Interleukin 1
Keap1	Kelch-like ECH-associated protein 1
LDL	Low-density lipoprotein
NF-κB	Nuclear factor kappa B
Nrf2	Nuclear factor erythroid 2–related factor 2
ox-LDL	Oxidized low-density lipoprotein
SHR	Spontaneously hypertensive rats
SOD	Superoxide dismutase
TNF alpha	Tumour necrosis factor alpha

1 Background

The annual enumeration of cardiovascular diseases (CVDs) is approximately 17 million, around 31% of mortality worldwide. As a result, CVDs are a measure of economic and health burden. The present statistics of the American Heart Association revealed that, nearly 50% of the population nearby agonized from CVDs. As per European statistics, the death rate is 45% of the population. In India, heart disease begins 10–15 years earlier than in the West. Heart illness accounts for one-fifth of fatalities. The Government of India decided to reduce hypertension prevalence (high blood pressure) by 25% relative to 2025. Almost 220 million people in India suffer from hypertension. Thus, the Indian Government established the Indian Hypertension Control Initiative (IHCI) to expedite access to treatment options [1]. Heart attacks and

A. K. Dhara, S. C. Mandal (eds.), *Role of Herbal Medicines*, https://doi.org/10.1007/978-981-99-7703-1_7

strokes are typically due to several risk factors, including cigarette use, an unhealthy diet and obesity, indolence and alcohol abuse, hypertension, diabetes, and hyperlipidemia [2]. Synthetic drugs used to treat CVDs have side effects or adverse effects, and they are responsible for approximately 8% of hospital admissions in the USA. Associated toxicity issues with these drugs cause 100,000 deaths of people per year.

Consequently, the trend has increased; people move towards phytomedicine yearly because they believe it has minimal side effects [3]. Numerous therapeutic plants have typically been investigated for their potential impact on CVDs. Medicinal plants provide significant advantages for treating cardiovascular complications due to their safety profiles. In Asian medicine, it has been used for at least 3000 years [4]. This chapter overviews medicinal plant-based remedies, particularly for preventing and treating CVDs. Also, evidence on the anti-CVD effects and ethnopharmacological therapeutic potentials of several plants, including *Embilica officinalis*, *Rauwolfia serpentina*, *Terminalia arjuna*, *Ziziphus oxyphylla*, and other plants, are picked and reviewed. It has been investigated how these plants are used mainly concerning CVDs like myocardial infarction, hypertension, peripheral vascular diseases, coronary heart disease, cardiomyopathies, and dyslipidaemias [5].

2 Cardiovascular Diseases and Their Pathophysiology

2.1 Hypertension

It is a condition of health in which blood pressure is consistently exerted on the wall of the arteries and is characterized by systolic and diastolic blood pressure (BP) of more than 140 and 90 mmHg, respectively [6]. As per its type, primary hypertension has an unknown cause of BP elevation, but secondary hypertension is reported with some exact reasons like congenital narrowing of the aorta, cirrhosis of the liver, renal diseases, endocrine disorders, alcohol intake, obesity, medication (contraceptive pills, NSAIDs, cocaine, etc.), neurological disorders (head injury and brain tumours), and thyrotoxicosis. Normal physiological parameters such as cardiac output, degree of sodium, water retention in the kidney, and contraction of arteries and arterioles are significant determinants of total peripheral resistance [7] (Fig. 1).

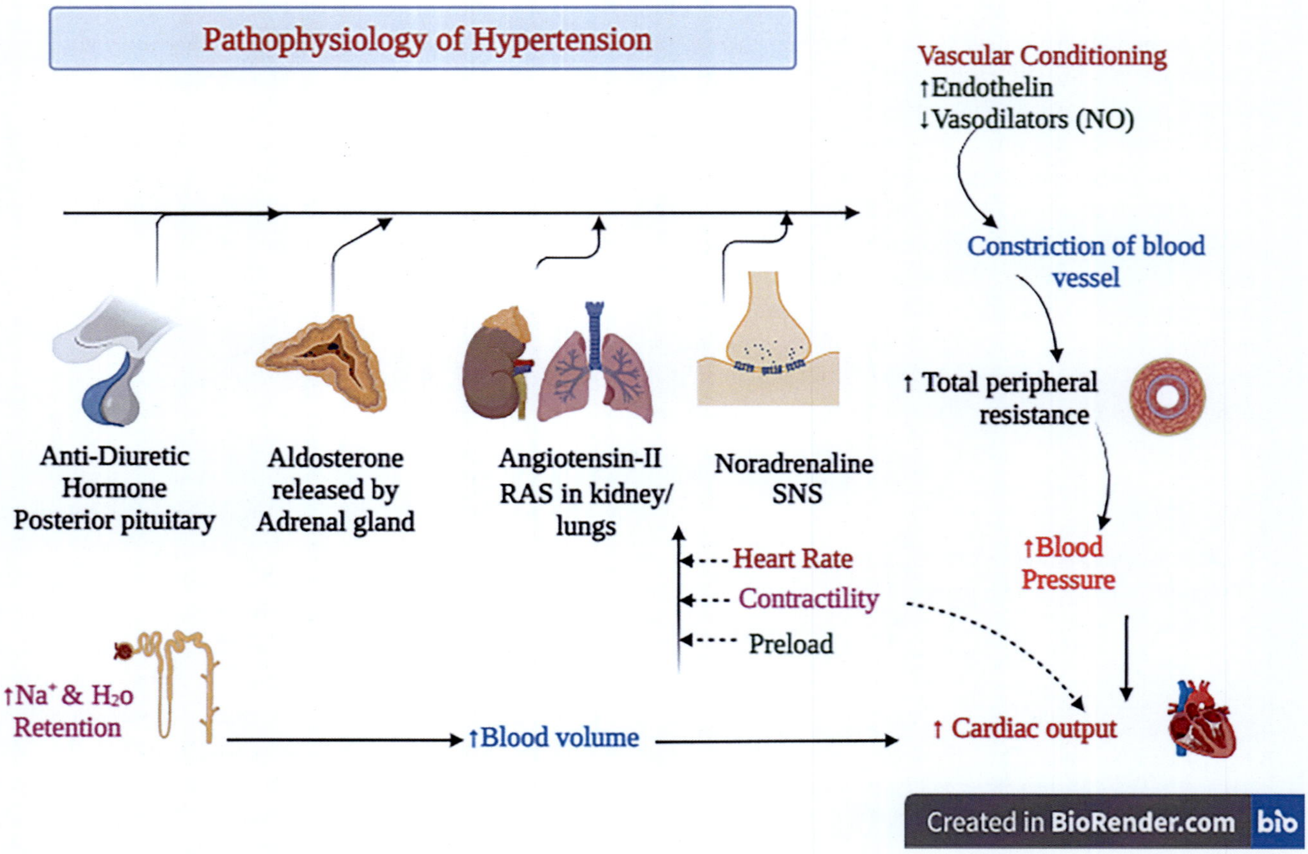

Fig. 1 Pathophysiology of hypertension

2.2 Ischaemic Heart Disease

Ischaemic heart disease (IHD) is a type of cardiac disability coming from the imbalance between the myocardial fulfilment and the requirement of O_2 carrying blood in the body. The narrowing or blockage of the coronary arteries is the initial cause of myocardial anoxia. IHD caused due to some events such as coronary artery disease (CAD) and vasospasm, which resist the blood flow in the myocardium. It is also termed 'atherosclerosis'. Later, it produces inflammation, which initiates proteolytic enzyme, making atherosclerotic cap rupture and tissue infarct. These all combined trigger/affect coronary microcirculation in the coronary artery [6].

2.3 Atherosclerosis

Low-density lipoprotein (LDL) accumulates in arteries leading to atherosclerosis, a lipid-driven disease that affects peripheral arteries (arterial elasticity) and causes coronary heart disease and strokes. Lipids start building up in the subendothelial area (intima) during the early stages of atherosclerosis because of the interplay between apolipoprotein, which is positive in charge, proteoglycan side chain with a negative charge, and extracellular matrix molecule made by smooth muscle cells (SMC) [8]. The macrophages, often known as 'foam cells', are lipid-filled arterial macrophages that eventually die and collect in the endothelium lining (intima). Along with other substances, the lining also contains smooth muscle cells, calcium and other mineral salts, and debris from different types of cells. Consequently, atheromas or atherosclerotic plaques are formed from the small initial lesions that become larger and thicker. These plaques make the vessel channel smaller and obstruct blood flow. In the aorta and major aortic branches, atherosclerotic lesions are frequently observed. The disorder is also known as coronary heart disease, coronary artery disease, or ischemic heart disease. When one or more coronary arteries are blocked, a section of the heart muscle will undergo apoptosis, known as a myocardial infarction or heart attack. Blood clots and stroke formation could be mediated by atherosclerotic lesions of the cerebral arteries [9] (Fig. 2).

2.4 Angina Pectoris

Angina pectoris is a cardiac-origin disease that results in chest pain. It is a common observation in medical practices; IHD predominates among 3–4% in the United Kingdom adults. According to the report, a total coronary angiogram was performed each year, and the result was 10% of new cases of angina found [10]. It may be categorized as stable angina because workload or demand and ischemia are predictable. Compared to unstable angina it is a magnified one with increasing frequency or intensity of episodes. The frequency and severity of plaque-related arterial constriction are not entirely consistent; it changes with a typical person's variances in arterial tone. The arterial tone is high in the morning; many people suffer from angina at the start of the day. During angina, the left ventricular (LV) diastolic pressure usually goes up; sometimes, pulmonary obstruction and breathing problems can also cause it. Although the precise process of ischemia induces discomfort is unknown, hypoxia metabolic products may stimulate the nerve cells [11].

2.5 Arrhythmia

A cardiac arrhythmia is a physiologically unjustified variation in the average heart and/or cardiac rhythm rate. Rhythm disturbances are mainly due to abnormality in generating impulses, signal conduction, or both. Bradyarrhythmias are caused by impaired intrinsic pacemaker function or conduction blockages, primarily in the node that produces impulse (AV node) or the His–Purkinje system. Abnormal mechanisms of automaticity and re-entry are reasons for most tachyarrhythmias [12].

2.6 Myocardial Infarction

Myocardial infarction (MI) frequently occurs when coronary atherosclerosis is built with a luminal thrombus [13]. It gives several combinations of inflammation and endothelial defects. In about 10% of cases, myocardial infarction may occur without detecting coronary artery obstruction. The prognosis and management in such cases differ from coronary artery occlusion [14]. In rare cases, MI is shown without any coronary artery blockage. Compared to coronary artery occlusion, the management and prospects in such conditions differ. MI is due to several factors like increasing cholesterol level (LDL), inflammation, excessive intake of alcohol, smoking during the hypertensive condition, and the presence of diseases like diabetes mellitus and family history as well [15].

2.7 Anaemia

In Greek, 'an' refers to lack/deficiency, and 'emia' means blood. Anaemia implies a lack of blood (RBCs count). The formation of red blood cells is mainly regulated by a glycoprotein hormone named erythropoietin (EPO), which is released by the peritubular cells of the kidney. The hypoxic condition of the tissue and level of haemoglobin is smoothly

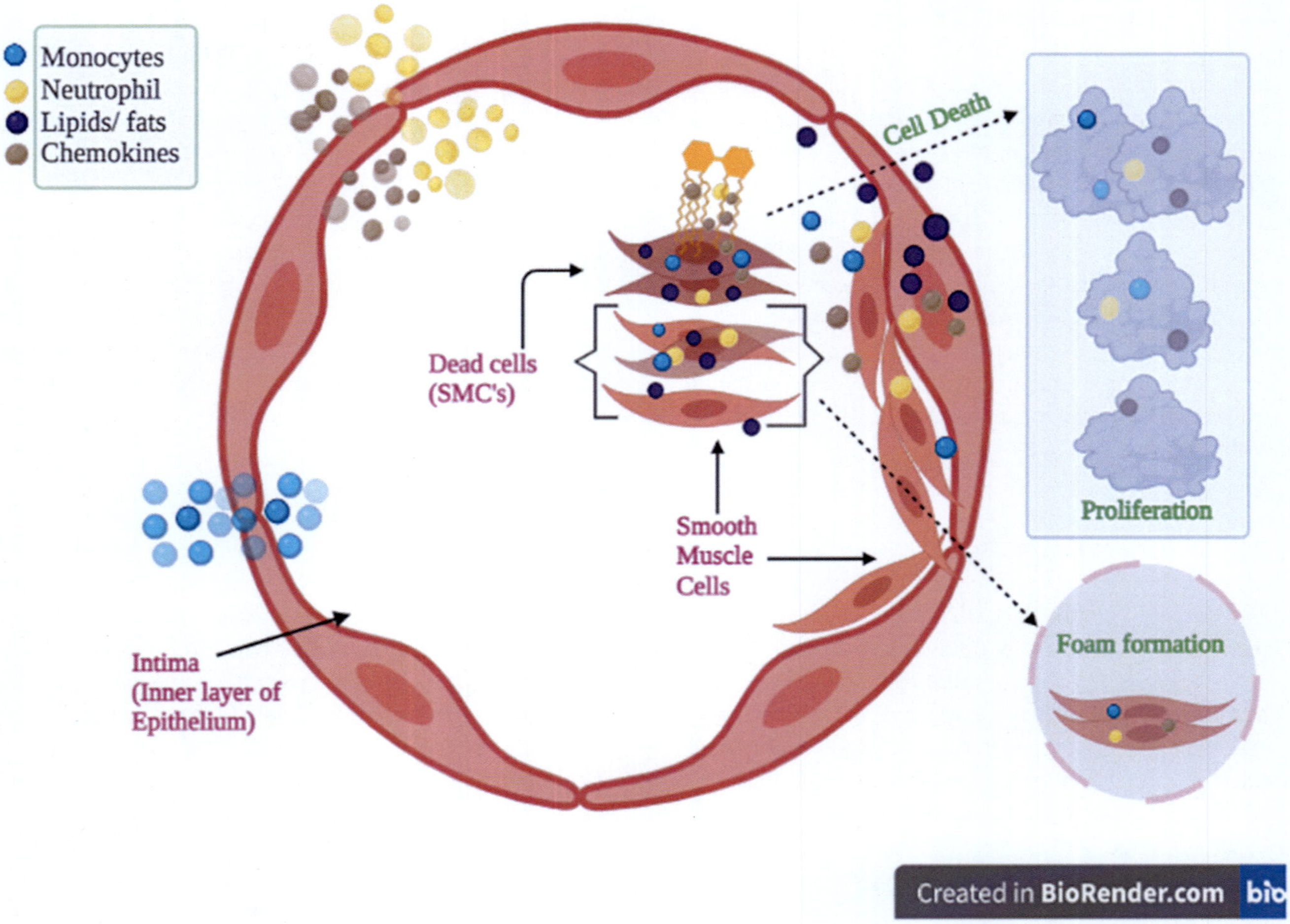

Fig. 2 Pathophysiology of atherosclerosis

associated with erythropoietin production. As the haemoglobin level falls below 7.0 g/dL, patients start experiencing several anaemia-related symptoms, while EPO levels in anaemic patients with renal failure are lower than expected/normal [16]. Some dietary and metabolite deficiencies like iron and vitamin B12 can lead to anaemia. Dietary iron deficiency, impaired iron absorption, and high requirement (in children as a growth requirement and in women to fulfil the minimal range that is affected due to pregnancy or periods) are general risk factors. Gastrointestinal bleeding and menstruation in women are the most frequent causes of iron deficiency anaemia. Iron deficiencies in various chronic inflammatory diseases include inflammatory bowel disease, chronic kidney disease, and congestive cardiac failure [17]. Cobalamin (Cbl) (vitamin B12) participates in two enzymatic activities in mammalian cells. Homocysteine (HCys) is converted into methionine, which is responsible for DNA formation in the methionine synthase process. DNA and RNA synthesis decreases, and cells grow more extensively than typical during early lysis due to vitamin B12/folic acid deficiency [18].

3 Therapeutic Significance of Medicinal Plants and their Role in CVD

Traditional medicine and ethnomedicine, the study of conventional drugs used by different ethnic groups, are prevalent today. In the past, natural resources were the primary supply of medication for traditional medicine [5]. People use modern medications more frequently than conventional medications because they have a faster effect. However, it is costly and has serious side effects. The anti-arrhythmic drug Amiodarone has the potential to have an impact on almost every organ system, including the lungs (pulmonary fibrosis, acute respiratory distress syndrome), thyroid gland (hypo- and hyperthyroidism), liver, gastrointestinal tract, eyes, skin, and nervous system (tremors) [19] and statins, which belong to the anti-hyperlipidemic medication class, also produce the most typical side effects, including headache, myalgia, dizziness, cardiomyopathy, and an increase in blood transaminase. These toxins frequently damage the kidneys and are hazardous [20]. As a result, individuals are more likely to use medicinal plants [21]. Due to a greater understanding of how herbs enhance health and quality of life, herbal medicine has acquired universal medical acceptance. According to growing evidence, using phytochemicals and complete nutrition derived from plants is a novel and potential CVD prevention method [22].

The traditional (herbal/alternative) systems of medicine mention more than 2000 plants, some of which give relief to people with cardiovascular disorders, particularly hyperlipidemia and ischemic heart disease. There can be no disputing the undeniable aspect that the usage of herbal medicines for preventive and therapeutic purposes is expanding globally due to the widespread assumption that natural goods are risk-free and have few adverse effects. A recent WHO research revealed that 80% of people in developing countries still rely on natural remedies from conventional medicine for health [23]. The prevention of low-density lipoprotein oxidation, which encourages vasodilatation, is a standard flavonoid action. Plant sterols reduce blood cholesterol uptake, thereby preventing CVD. By stimulating nuclear factor-erythroid factor 2-related factor 2 (Nrf2) and avoiding the production of cholesterol, plant sulphur compounds also prevent CVD [22]. The details of different medicinal plants playing critical roles in the treatment of CVD are represented below:

3.1 Acorus (*Acorus calamus* Linn., Acoraceae)

Shah et al. [24] investigated the activity of the *Acorus calamus* in ischemic cardiac conditions. Rhizomes of the *Acorus calamus* were used to perform the study, and as a model, male rabbits (1–1.5 kg) and bovine hearts were used. Isolated bovine coronary arterial rings were used to investigate the coronary vasodilator effect, and rabbit heart was used to study the effect on heart parameters. They looked at the effect of *Acorus calamus* extract at 0.01–10 mg/mL, ethyl acetate fraction and n-hexane fraction on coronary flow, pulse rate, and force of contraction. These findings suggested that the *Acorus calamus* has a coronary vasodilator action, which may be mediated by an endothelial-derived hyperpolarizing factor (EDHF). The partial cardiac suppression effect may be advantageous as total suppression can cause cardiac arrest.

Another study by Shah et al. [25] discovered that crude extract and ethyl acetate fraction inhibited the formation of phenylephrine peak in the calcium (Ca^{+2}) free medium. They used rabbits (1–1.5 kg; $n = 20$) and male Sprague–Dawley rats (200–250 g; $n = 50$) of either sex. The findings suggested that crude extract of *Acorus calamus* contains a combination of constituents which shows blood pressure lowering effect mediated by Ca^{+2} antagonism and nitric oxide pathways and vasoconstrictor effects to counteract the excessive vasodilatation. This study gives a base for the medicinal use of the *Acorus calamus* in hypertension.

3.2 Amla (*Emblica officinalis* Gaertn., Euphorbiaceae)

Amla is from the Indian subcontinent; however, countries like Pakistan, China, Malaysia, Southeast Asia, and Uzbekistan also cultivated and used it as a source of nutrition and therapeutic implications [26]. The hydroalcoholic extract of the herb *Emblica officinalis (E. officinalis)* has 9.4% polyphenols (notably 30% emblicanin A, 28% emblicanin B, 13% punigluconin, and 15% pedunculagin), 0.58% alkaloids, 0.11% flavonoids, and 0.85% vitamin C [27]. Deoxycorticosterone acetate 1% NaCl high salt (DOCA/HS-induced hypertension) induces an oxidative process that leads to cardiac dysfunction, renal impairment, and renovascular dysfunction; nevertheless, it may be advantageous due to its anti-oxidant effects. Hypertension induced by (20 mg/kg, s.c.) DOCA-salt was used twice a week for 5 weeks with the replacement of drinking water and 1% sodium chloride solution. Rats were cotreated for the same period with *E. officinalis* at effective therapeutic doses of 75, 150, and 300 mg/kg/day. Nitric oxide production in the blood increased along with decreased sodium (Na^+) and potassium (K^+) levels, all related to *E. officinalis* anti-hypertensive effects. Furthermore, endothelial nitric oxide synthase (eNOS), an alcoholic herb extract, significantly increased heart levels. As a sustained dose compared to DOCA control rats, *E. officinalis* considerably reduced oxidative stress, raised blood pressure, heart rate, and cardiovascular and nephrotic hypertrophy through regulating serum nitric oxide, electrolyte, an endogenous anti-oxidant, and enforced eNOS level [26].

3.3 Arjuna (*Terminalia arjuna* Roxb., Combretaceae)

Meghwani et al. [28] investigated *Terminalia arjuna* for its potential to prevent pulmonary hypertension (PH) in Wistar rats instigated by monocrotaline (MCT). Rats (150–200 g) were used in the study and arbitrarily separated into five groups; Sildenafil (175 mg/kg/day 3 days after MCT for 25 days), once MCT 50 mg/kg given by s.c., and aqueous extract of Arjuna (125,250 mg/kg/day through oral route) were all received. Arjuna and Sildenafil both prevented right ventricle hypertrophy and RVSP caused by MCT. Although Bcl2/Bax gene expression was substantially revealed to be suppressed in the right ventricle during MCT-induced PH, Nicotinamide Adenine Dinucleotide Oxidase expression was considerably boosted in the lung. In conclusion, the aqueous extract of *Terminalia arjuna* reduced PH due to its anti-oxidant capacity and effects on the thickness of the pulmonary artery.

3.4 Ashwagandha (*Withania somnifera.*, Solanaceae)

The given study was completed to determine the mechanism of *Withania somnifera* (WS) in myocardial damage. In the current exploration experiment, Wistar rats were divided into three groups, i.e. sham (LAD was not ligated), control, and WS-50 mg/kg. Each group received normal saline for 1 month. On the 31 days, rats from the control IR and WS-IR groups had their left anterior descending (LAD) coronary arteries obstructed for 45 min, followed by an hour of reperfusion. Cardiac necrosis was caused by post-ischemic reperfusion injury when the IR control group was compared to the sham group, which showed increased lipid peroxidation and decreased anti-oxidant capacity. The given study revealed that the anti-oxidant and anti-apoptotic properties of WS may support its heart-protective effect [29].

3.5 Betel Leaf (*Piper betel.*, Piperaceae)

Arya et al. [30] investigated the cardioprotective potential of *Piper betel* against isoproterenol (ISP)-induced myocardial infarction in rats. On days 28, 29, and 30, ISP (85 mg/kg, s.c.) was administered at intervals of 24 h, and functional and biochemical parameters were measured. In addition to increasing lipid peroxidation, ISP significantly decreased the myocardial anti-oxidants superoxide dismutase, catalase, peroxidases, reduced glutathione, and myocyte damage marker enzymes creatine phosphokinase-MB isoenzyme and lactate dehydrogenase; the results revealed that the hydroalcoholic extract of *P. betel* (150 and 300 mg/kg) showed the significant cardioprotective effect against ISP-induced myocardial infarction. This study suggested the cardioprotective potential of *P. betel*.

3.6 Bera or Mamyanu (*Ziziphus oxyphylla.*, Rhamnaceae)

Ziziphus oxyphylla (*Z. oxyphylla*) has long been used to treat hypertension. It contains more derivatives of quercetin, kaempferol, catechin, and cyclopeptide alkaloids. The hydroalcoholic extract of *Z. oxyphylla* (HAEZO) was prepared, and their anti-hypertensive effect was studied. NG-nitro-L-arginine methyl ester (L-NAME) was used to induce hypertension in rats. The study of the anti-hypertensive effects of HAEZO (200 and 400 mg/kg) and Kaempferol (100 mg/kg) was administered through intraperitoneal injections of L-NAME (185 umol/kg) in hypertensive rats. HPLC and ESI-MS/MS determined kaempferol, quercetin, catechin, ceanothic acid, zizybernalic acid, and oxyphylline F. After chronic HAEZO and kaempferol administration, systolic, diastolic, and mean blood pressure in L-NAME-induced hypertensive rats were effectively ($p < 0.001$) decreased. Serum nitric oxide and cGMP levels were significantly ($p < 0.001$) elevated. Oxidative stress-related inflammatory markers (MDA, CAT, SOD, GSH) were also enhanced by HAEZO and kaempferol administration. In hypertensive animals, the increased IL-6 and TNF- α concentrations were significantly reduced by HAEZO treatment. The anti-hypertensive effects of HAEZO also caused endothelial nitric oxide synthase to be upregulated and angiotensin-converting enzyme to be downregulated [31].

3.7 Candlenut Tree (*Aleurites moluccana.*, Euphorbiaceae)

The importance of an alcoholic extract from *Aleurites moluccana* leaves to lower lipid profile status was investigated using a high-fat diet supplemented with rats and Triton W-1339. These dried leaves (600 g) were cut into small pieces before being macerated at room temperature for 10 days to produce the methanol extract. The administration of leaf extract showed a significant reduction in body weight in Triton-induced hypercholesterolemic rats and serum cholesterol without any drop in the regular diet. The results provided the potential of this natural product to lower cholesterol by its capacity to bring down intestinal lipid utilization and inhibit the synthesis of hepatic cholesterol [32].

3.8 Cinnamon (*Cinnamomum verum.*, Lauraceae)

In one study carried out in 30 male Wistar rats (8-weeks-olds) weighing 250–300 g, long-term aerobic training and administration of cinnamon extract affecedt cardiac function, biochemical alterations, and lipid profile following exhaustive exercise. The study included a progressive increase in training speed and duration incorporated into an 8-week endurance training regimen. A catheter with a balloon tip was inserted into the left ventricles to measure the myocardial hemodynamics. Blood samples were taken to examine lipid profiles, biochemical indicators, and the lipid-peroxidation marker malondialdehyde (MDA). Simultaneously, physical exercise and cinnamon administration positively affected cardiac hemodynamics. Regular training and cinnamon supplementation reflect increased high-density lipoprotein (HDL) and HDL/LDL ratio and decreased serum total cholesterol level, low-density lipoprotein (LDL). Reduced serum MDA levels, improved blood lipid profile, where cinnamon and regular exercise also positively affected cardiac function [33].

3.9 Drumstick (*Moringa oleifera* Lam., Moringaceae)

Indians utilize the *Moringa oleifera* Lam. leaves as a hypocholesterolemic medication in people with obesity. Hence, the medical evidence for their usage in hypercholesterolemia was explored. In the experiment, 15 male Wistar rats were used, and it was determined that a high-fat diet increased the serum, liver, and kidney cholesterol by (115-103.2 mg/100 mL of serum), (9.4-8.8 mg/g wet weight), and (1.09-0.97 mg/g wet weight), i.e. 14.35%, 6.4%, and 11%, respectively. However, when *M. oleifera* aqueous crude leaf extract was administered, these increased levels were found to be reduced. The serum cholesterol effect was significant, and there was no significant effect on serum total protein. The aqueous crude extract showed an increase in serum albumin by 15.22% (46–53 g/L). The statistical significance of this value was also determined. The study concluded that *M. oleifera* leaves exhibited marked hypocholesterolemic action and have a sound pharmacological background in India [34].

3.10 Garlic (*Allium sativum*., Liliaceae)

Garlic (*Allium sativum*) is widely available and utilized worldwide as a spice, food ingredient, and home remedy. Garlic has been a component of folk medicine for thousands of years and has been used to treat many illnesses. Many studies have been published about preventing atherosclerosis, coronary thrombosis, and stroke. The garlic effect has been evaluated for its ability to prevent thromboxane production and platelet aggregation. The study determined the impact of garlic aqueous bulb extract on thromboxane and prostacyclin generation in whole rabbit blood and aorta in vitro and ex vivo. During blood clotting, a dose-dependent reduction of thromboxane synthesis was observed. Any concentration of garlic extract employed in the experiment did not affect prostacyclin production. The maximum concentration of garlic utilized in in vitro tests resulted in a small but negligible decrease in prostacyclin vascular synthesis. The aorta's ability to produce thromboxane was inhibited at all the measured garlic concentrations. After receiving an intraperitoneal injection of garlic (1 mL/kg) for a week, it was observed that the ex vivo thromboxane and prostacyclin enzymatic synthesis of these tissues showed a similar pattern. Garlic treatment significantly boosted the aortic production of prostacyclin as compared to control rabbits. The study findings suggested that garlic administration may selectively reduce platelet aggregation and production of thromboxane while protecting vascular prostacyclin synthesis [35].

3.11 Gentiana (*Gentiana floribunda*., Gentianaceae)

The investigation was done on *Gentiana floribunda* (*G. floribunda*) for its possible effect against hypertension and as a vasodilator. Hydroalcoholic extract of the whole plant was used in the study, and as a model, male Sprague–Dawley rats (240–260 g) were used. *G. floribunda* contains flavonoids, saponins, sterols, tannins, and terpenes, which cause falls in BP in anaesthetized rats in a dose-dependent (3.0–100 mg/kg) manner, and its vasodilator effect was endothelial-independent. Compared to verapamil, the drug showed a reduction in contraction induced by high K^+ (80 mM) and phenylephrine (1M) at concentrations between 1.0 and 10 mg/mL. Also, it shifted the Ca^{+2} dose–response curves to the right. These findings suggested that *G. floribunda* gives blood pressure-lowering action mediated by Ca^{+2} antagonism and is helpful in hypertension [36].

3.12 Ginko (*Gingko biloba*., Ginkgoaceae)

In the present research attempt, Tiana et al. [37] discovered the action of *Ginkgo biloba* (*G. biloba*) on endoplasmic reticulum stress (ERS) and autophagy. Using network pharmacology, the targets, metabolic routes of the *G. biloba* leaves and phytochemical constituents responsible for attenuating atherosclerosis (AS) are forecast. In this study, male $ApoE^{-/-}$ mice aged 6–7 weeks animal used. Streptozotocin (50 mg/kg/day) was administered to $ApoE^{-/-}$ mice intraperitoneally along with a high-fatty diet which caused diabetes. As indicated by network pharmacology, the active chemical constituents of *G. biloba* extract (GBE) were targeted at the NF-B and mTOR signalling pathways to lower AS. *G. biloba* also decreased blood glucose levels and inflammatory cytokines.

To assess clinical availability, Kubota et al. [38] examined the effects of *G. biloba* treatment on blood pressure, vascular tone, and calcium mobilization. Wistar Kyoto (WKY) and spontaneously hypertensive rats (SHR) were given either a control food or a diet containing 0.05–0.5% ginkgo for 30 days. While ginkgo powder was administered to WKY, it showed no effect but a significant drop in systolic blood pressure in SHR. As per network pharmacology prediction, chemical compounds of GBE targeted the mammalian target of rapamycin (mTOR) and NF-κB signalling pathways to lowered AS. GBE decreased plaque/lumen area and plaque lipid deposition area/intimal area while inhibiting CD68, MMP2, and MMP9 expression.

3.13 Ginseng (*Panax ginseng.*, Araliaceae)

Myocardial ischemia–reperfusion (I/R) injury has been prevented and treated with *Panax ginseng* (*P. ginseng*). The study set out to define the preventive effects of *P. ginseng* and its active components against myocardial I/R injury to give clinical evidence and novel perspectives for *P. ginseng* investigation and utilization in myocardial I/R injury. The study by Chen et al. [39] outlined the pathophysiological pathways underlying myocardial I/R injury and active components of *P. ginseng*, such as ginsenosides, ginseng polysaccharides, and phytosterols. Prevention and treatment of cardiac diseases using these active components of *P. ginseng* were broadly outlined, along with preclinical trials of myocardial I/R injury and probable mechanisms of action. The study concluded that active ingredients of *P. ginseng* might prevent myocardial I/R injury and control oxidative stress-related proteins, inflammatory cytokines, and apoptotic factors.

3.14 Guggul (*Commiphora mukul.*, Burseraceae)

Guggul is one of the most widely used herbs to boost cardiac and vascular health. The cardioprotective potential of *C. mukul* in rats with myocardial necrosis induced by isoprenaline is already established. *C. mukul* extract was given for 30 days orally at 100, 200, and 400 mg/kg doses. Isoprenaline (85 mg/kg; s.c.) was given to the animals, control and *C. mukul* pre-treatment groups consecutively on the 29th and 30th day, at intervals of 24 h. Administration of isoprenaline resulted in a significant ($p < 0.05$) rise in superoxide dismutase, catalase, glutathione peroxidase, glutathione, and a fall in malondialdehyde. *C. mukul* increased the myocardial anti-oxidant status and stopped the leakage of lactate dehydrogenase and creatine phosphokinase-MB from the heart. *C. mukul* pre-treatment was shown to reduce necrosis, edema, and inflammation through histological analysis. According to the data, *C. mukul* may be a preventative and therapeutic agent against oxidative stress linked to ischemic heart disease due to its anti-oxidant and anti-peroxidative activities. Moreover, a histopathological study revealed that the necrosis, edema, and inflammation were reduced after pre-treatment with *C. mukul*. The current research suggested that *C. mukul* may be a promising preventive and therapeutic agent against oxidative stress-related ischemic heart disease [40].

3.15 Haritaki (*Terminalia chebula.*, Combretaceae)

The present investigation evaluates the hypolipidemic effect of methanolic bark extract of *Terminalia chebula* (*T. chebula*) against high-fat diet-induced hypercholesterolemia in rats. Atherosclerosis, coronary heart disease, and stroke are cardiovascular diseases for which hypercholesterolemia is a significant risk factor. The methanolic extract of *T. chebula* at doses of 200, 400, and 600 mg/kg was utilized to evaluate its anti-hyperlipidemic effect in animals and fasting blood glucose levels after treatment with the same doses were assessed using commercially available kits. In the groups of animals treated with *T. chebula* extract and atorvastatin, the positive control showed a significant increase in serum levels of high-density lipoprotein (HDL) cholesterol in rats with diet-induced hypercholesterolemia. Also, acute toxicity studies showed no morbidity and mortality at different doses. Thus, the study confirmed the potential of *T. chebula* to treat hyperlipidemia [41].

3.16 Henbane (*Hyoscyamus niger.*, Solanaceae)

The research describes the anti-hypertensive and blood vessel dilatory properties of the crude extract of *Hyoscyamus niger* (*H. niger*) on rats under anaesthetic conditions. It showed dose-dependent (10–100 mg/kg) reduction of BP, rate, and force of atrial contractions in guinea pig atria. In the isolated rabbit aorta *H. niger* (0.01–1.0 mg/ml) relaxed contraction induced by phenylephrine (PE, 1 μM) and K^+ (80 mM) and it also suppressed the PE (1 μM) control peaks which are usually found in Ca^{+2} free medium. The vasodilator activity of *H. niger* in endothelium-intact rat aortic preparations was endothelium-independent as it was not opposed by N (omega)-nitro-L-arginine methyl ester. These findings suggested that *Hyoscyamus niger* reduced blood pressure by acting as a Ca^{+2} antagonist [42].

3.17 Kalmegh (*Andrographis paniculata.*, Acanthaceae)

Cardiac hypertrophy typically manifests as an expansion of the cardiac muscle and a rise in the volume of muscle cells that occur in response to hemodynamic stress, acute myocardial damage, infection, or abnormalities in the genes producing sarcomeric proteins [43]. An herbaceous plant known as *Andrographis paniculata (A. paniculata)* is often used as a medicinal food. It is used to treat endocrine disorders and hypertension in several regions of Asia as a decoction, powder, or tablet. The aqueous *A. paniculata* extract and its active constituent, andrographolide, have anti-inflammatory and anti-apoptotic properties. Three groups of test mice were divided and given different diets standard chow, an obese high-fat diet, or a high-fat diet treated with *A. paniculata* extract (2 g/kg/day, administered via gavage for a week). The results showed that the myocardial inflammation pathway-related proteins were elevated in the obese mice, potentially contributing to cardiac hypertrophy and myocardial apoptosis. The research strongly implies that obese persons may benefit from taking supplements containing *A. paniculata* extract to prevent and treat cardiovascular disease [44].

3.18 Orchid (*Orchis mascula.*, Orchidaceae)

Aziz et al. [45] explored the possible mechanisms of action for the medicinal use of *Orchis mascula* (OM) as a medication for hypertension and dyslipidaemia. The hydroalcoholic roots extract of OM is used as animal model hypertensive rats (strain: SHR/NCrlBR; hypertensive, non-stroke) and their normotensive controls, Wistar Koyoto (WKY) rats, Sprague–Dawley rats (170–200 g), and rabbits (1.5–2 kg) of either sex used. OM increased acetylcholine-induced relaxation and improved endothelial dysfunction. It considerably ($P < 0.05$) in hypertensive rats reduced systolic blood pressure. In a dose-dependent manner, the extract of OM at 10 and 30 mg/kg in normotensive anaesthetized rats lowered mean arterial pressure. OM also decreased serum triglycerides and high-fat diet and tyloxapol-induced hyperlipidaemia. In a second model, OM prevented weight gain. OM causes concentration-dependent relaxation of phenylephrine and high K^+ (80 mM)-induced contractions in the isolated rabbit aorta. This research demonstrated that OM showd action against hypertension and endothelial-modulating.

3.19 Pushkarmool (*Inula racemosa* Hook. f., Compositae)

For precordial chest discomfort, Ayurveda recommends *Inula racemosa* Hook. f. Mangathayaru et al. [46] aimed to investigate how *Inula racemosa* extract in alcohol (IrA) and hexane (IrH) affects high-fat diet-induced atherosclerosis in guinea pigs. In the aforementioned research, 5 groups were formed with 6 animals in each group and to them, pellets were given as a diet for 90 days along with 0.15% w/w cholesterol (positive control), 100 mg/kg of atorvastatin, or 10 mg/kg of atorvastatin acetate and 100 mg/kg of atorvastatin calcium also administered. To the control group, only a pellet diet was given. Compared to the positive control, IrA lowered total cholesterol, triglycerides, low-density lipoprotein cholesterol, and the atherogenic index while increasing high-density lipoprotein cholesterol. Sesquiterpene lactones in IrH and phenolics in IrA are probably responsible for the reported effects. The outcomes confirmed that *I. racemosa* has cardioprotective and anti-obesity activity by demonstrating its anti-atherogenic effect.

3.20 Radish (*Raphanus sativus.*, Brassicaceae)

Ghayur et al. [47] described the anti-hypertensive, endothelium-dependent vascular relaxant, and cardiac-modulatory activity of *Raphanus sativus* seed extract. In the study, Balb/c mice (20–25 g), Sprague–Dawley rats (170–200 g), and guinea pigs (500–600 g) of either sex were used. The aqueous extract of *R. sativus* was prepared, and its phytochemical testing was performed. Blood pressure was assessed in anaesthetized normotensive rats. It demonstrated that an atropine-sensitive pathway mediated a dose-dependent (0.1–3 mg/kg) decline in rats' blood pressure and heart rate. The extract reduced the force of contractions and rate in separated guinea pig atria in a dose-dependent manner (0.03–3.0 mg/mL). The aqueous extract reduced the contractions produced by phenylephrine in endothelium-intact rat aortas. In mice, the extract did not show any side effects up to a dose of 10 g/kg. The cardiovascular inhibitory effect of *R. sativus* was mediated by muscarinic receptor stimulation, which may justify its anti-hypertensive activity [47].

3.21 Rhubarb (*Rheum palmatum* Linn., Polygonaceae)

Evans et al. [48] evaluated the heart-protective actions of emodin via inhibiting histone deacetylase (HDAC)-dependent gene expression; for this reason, they primarily believed that emodin would reduce a pathological increase in cardiac size via prohibition of histone deacetylases enzyme and rhubarb which is rich in emodin would exhibit these effects. The data demonstrated that emodin altered gene expression globally and raised histone acetylation in cardiomyocytes. Additionally, they evaluated that emodin prevented hypertrophy induced by phorbol myristate acetate (PMA) and phenylephrine (PE) in newborn rat ventricular myocytes (NRVMs). Lastly, they proved the emodin's cardioprotective action by translating it to 10 weeks of age, C57BL/6 male and female mice models who received angiotensin II (Ang II) they proved its cardioprotective effect. The study lasted 14 days and showed the in vitro histone deacetylase enzyme activity inhibited by emodin and Turkish rhubarb containing emodin with fast-on, slow-off rates. Emodin partially prevented pathological cardiac hypertrophy by inhibiting histone deacetylase-dependent gene expression modifications.

In another study, Gehlken et al. [49] demonstrated the activity of rhubarb containing pectin. Pectin inhibits galectin-3-related endomyocardial fibrosis. They performed an in vitro study, evaluated the inhibitory capacity, and finally identified potent pectin. Compared to the well-known modified citrus pectin, findings showed that modified rhubarb pectin is the highest effective inhibitor of galectin-3. Cardiac fibrosis induced in mice by angiotensin II (AngII) infusion. In comparison to the control, Ang II infusion showd 4–5 times more fibrosis signal in the left ventricle tissue. Compared to Ang II alone, fibrosis decreased by 57% after receiving rhubarb pectin therapy but by 30% with modified citrus pectin. Natural rhubarb pectin, a galectin-3 inhibitor, significantly reduced myocardial fibrosis and maintained heart function in vivo. It also had a higher inhibitory ability than other pectins.

3.22 Sarpagandha (*Rauwolfia serpentina*., Apocynaceae)

By monitoring the structure of the liver and kidney tissues histologically, biochemical markers, and blood pressure in albino rats, the given study was done to examine the therapeutic effects of *Rauwolfia serpentina* on hypertension and hyperlipidaemia. Thirty albino rats were used for the analysis, which further separated into 5 groups (6 in each). Group 1 regular diet was given; in group 2 high-salt diet with only 8% NaCl and in group 3, atenolol (standard drug control) 50 mg/kg was given. It contains major secondary metabolites such as flavonoids, tannin, and alkaloids, which may have therapeutic potential against cardiovascular ailments. Due to the binding action of catecholamine in nerve cells, reserpine shows anti-hypertensive effects. Combined with a high-sodium diet, the methanolic plant extract was administered to groups 4 and 5 at 100 mg/kg body weight and 200 mg/kg body weight daily for four weeks. Compared to G2 and G3, *Rauwolfia serpentina* significantly lowered blood pressure in groups 4 and 5. In conclusion, *Rauwolfia serpentina* has therapeutic benefits that certainly lowered cholesterol and blood pressure by safeguarding the renal and liver tissues [50].

3.23 Tulsi (*Ocimum sanctum* Linn., Lamiaceae)

Ocimum sanctum has therapeutic potential with anti-oxidant characteristics, yet, nothing is known about its cardioprotective efficacy against myocardial infarction. Isoproterenol was administered at a dose of 85, 200, and 300 mg/kg twice at intervals of 24 h to induce myocardial infarction changes in anti-oxidant markers, lactate dehydrogenase (LDH), and morphological and histological alterations were analysed. Aqueous alcoholic extract of *O. sanctum* significantly decreased glutathione levels, superoxide dismutase, thiobarbituric acid reactive elements, and LDH. It also inhibited lipid peroxidation at 25, 50, 75, and 100 mg/kg concentrations. The optimum heart-protective effect of *O. sanctum* was reported in the current investigation at a 50 mg/kg dose. Histopathological findings supported the preceding outcomes even more. In this way, we presume that *O. sanctum* may be therapeutically and preventatively useful in managing myocardial infarction [51].

3.24 Viola (*Viola odorata* Linn., Violaceae)

Siddiqi et al. [52] studied the activity of hydroalcoholic leaf extract of *Viola odorata* (*V. odorata*) against hypertension and dyslipidaemia. Sprague–Dawley rats (180–200 g), guinea pigs (450–500 g), and Balb-C mice (20–25 g) of either sex were used in the research. Under anaesthesia, rats exhibit a dose-dependent reduction in mean systemic arterial blood pressure (0.1–1.0 mg/kg). When vasoconstriction was induced by phenylephrine (PE, 1M) and potassium (K^+, 80 mM), *V. odorata* showed a relaxation based on concentration along with a rightward shift of Ca^{+2} concentration–response curves and suppression of PE (1M) control peaks in Ca^{++}-free medium.

In tyloxapol-induced dyslipidaemia, *V. odorata* reduced total cholesterol and triglyceride levels. A high-fat diet-induced dyslipidaemia model reflected increased high-density lipoprotein cholesterol and significant reduction of total cholesterol, atherogenic index, low-density lipoprotein cholesterol, and preventing weight gain. As per the result, drug extract showed a decrease in blood pressure results probably due to pathways like inhibition of Ca^{+2} influx via membranous Ca^{+2} channels, the release of Ca^{+2} from intracellular stores and nitric oxide-mediated pathways. Inhibition of synthesis, absorption of lipids, and anti-oxidant activities are responsible for decreased body weight and anti-dyslipidaemic activity of the extract. The study provided pharmacological justification for using *V. odorata* as an anti-hypertensive and anti-dyslipidaemic medication.

4 Active Phytoconstituents for CVD

In addition to medicinal plants, several phytoconstituents are present in the medicinal plants showing very high cardioprotective activity. Table 1 represents detailed information on different phytoconstituents isolated from the specific parts of the medicinal plant, their dose, specific model, and use with the mechanism of action.

Table 1 Phytoconstituents reported for cardiovascular diseases

Name of the isolated compound	Name of medicinal plant and parts used for extraction	Pharmacological activity	Mechanism of action	Dose	Animal model	Study outcome	References
Allicin	Synthesized from diallyl disulfide	Anti-hypertensive	Act by reducing the vasoconstrictor effects of angiotensin II, oxidative stress, and increasing the expression of Nrf2, as well as by upregulating the AT1R and downregulating the Keap1 protein	40 mg/kg	Rats were used as a model for kidney disease	A study supported the utilization of allicin as an effective therapeutic agent for pathophysiological abnormalities that may be related to or linked with cardiac and renal algorithms	[53]
Trans-cinnamaldehyde	–	Anti-hypertensive	By promoting nitric oxide release via the nitric oxide synthase pathway	40 mg/kg	Spontaneously hypertensive rat model	The blood pressure in SHR-CP rats was lowered, and endothelial dysfunction was mitigated by trans-cinnamaldehyde	[54]
Caffeic acid and chlorogenic acid	–	Anti-hypertensive	By downregulating activities of an angiotensin-I converting enzyme, acetylcholinesterase, butyrylcholinesterase, and arginase	10 and 15 mg/kg, respectively	Cyclosporine-induced hypertension in rat model	Angiotensin-converting enzyme activity, acetylcholinesterase, butyrylcholinesterase, and arginase activity were all considerably decreased by caffeine and chlorogenic acid	[55]
Andrographolide	–	Myocardial infarction	Mostly through L-type Ca^{2+} channel inactivation and an elevation in cardiac transient outward K^+ currents	20 mg/kg/day	Isoproterenol-induced myocardial infarction in rats	There was the potential for andrographolide to serve as a medicine that can help to protect heart tissue in the case of myocardial infarction	[56]
Arjunolic acid	*Terminalia arjuna*/bark	Myocardial necrosis	Lowering the levels of myeloperoxidase, lipid peroxidation, ascorbic acid, glutathione, ceruloplasmin, and glutathione peroxidase	15 mg/kg	Myocardial necrosis has been driven by isoproterenol in rats	Arjunolic acid prevents myocardial necrosis-related damage	[57]
Rutin and digoxin		Myocardial infarction	Due to its potential to scavenge free radicals and inhibit Na^+-K^+-ATPase activity	100 mg/kg and 500 µg/kg, respectively	Ischemic reperfusion-induced myocardial infarction	According to the study's findings, rutin and digoxin can potentially treat cardiovascular problems since it can counteract I/R-induced myocardial functional alterations	[58]
Rosmarinic acid	–	Myocardial infarction	Through increasing the activity of glutathione, superoxide dismutase, interleukin-10, brain-derived neurotrophic factor, and other cardioprotective molecules	50 mg/kg	Myocardial infarction induced by isoproterenol	A potential treatment approach for cardiac protection over myocardial infarction is provided by the anti-inflammatory, protective effect against oxidation, anti-stress, and BDNF modulatory action by rosmarinic acid in comorbidly depressed rats	[59]

(continued)

Table 1 (continued)

Name of the isolated compound	Name of medicinal plant and parts used for extraction	Pharmacological activity	Mechanism of action	Dose	Animal model	Study outcome	References
Vasicine	–	Myocardial infarction	Cardioprotective potential via lowering oxidative stress, excessive autophagy, and regulating apoptosis by triggering the PI3K/Akt/mTOR signalling pathway	10 mg/kg	Isoproterenol-induced myocardial infarction in rats	The study revealed a concentration-dependent effect of the vaccine against myocardial infarction through the proposed mechanism	[60]
Vitexin	–	Chronic myocardial ischemia	Act via improving the anti-oxidant defence system, reducing Ca^{2+} excess, and suppressing cardiac apoptosis	6 mg/kg	Chronic myocardial ischemia/reperfusion injury model	The research outcomes indicated that vitexin is a cardioprotective compound that could serve as a valuable alternative to traditional treatment for individuals suffering from coronary heart disease	[61]
Taxifolin	–	Myocardial injury	Taxifolin may act by mediating the Nrf2/HO-1 pathway, decreasing reactive oxygen species, inflammation, and death of cells	50 mg/kg	Isoproterenol-induced cardiac injury	Taxifolin is an active component against ISO-induced myocardial injury and probably other heart-related problems	[62]
Arjunic acid	–	Cardioprotective	via improving the anti-oxidant capacity and reducing the activity of proteins involved in apoptosis in H9c2 cardiomyocytes	8 μg/mL	Cobalt chloride-induced hypoxia in rats	Arjunic acid has a cardioprotective activity by reducing mitochondrial damage, apoptosis, hypoxia injury, and damage from oxidative stress	[63]
Hesperidin	–	Anti-atherosclerotic	Hesperidin reduces atherosclerosis by improving insulin resistance, and lipid profiles, inhibiting the production of macrophage foam cells and having an anti-oxidative impact	200 mg/kg	Using mice lacking the LDL receptor, high-fat diet-induced atherosclerosis	Hesperidin may be used therapeutically to alleviate atherosclerosis and atherosclerosis associated with metabolic disruption, according to the established data results	[64]
Quercetin	–	Anti-atherosclerotic	Through minimizing oxidative stress, macrophage pyroptosis and associated mechanism promote NRF2 activation by competitively binding to the Arg483 site of KEAP1	0.1% (w/w)	Atherosclerosis-prone apolipoprotein E-deficient ($ApoE^{-/-}$) mice model	This work provided a novel understanding of the mechanisms by which quercetin protects from atherosclerosis	[65]
Zingerone	–	Anti-atherosclerotic	Anti-atherosclerotic effect by modulating lipid gene expression and promoting anti-oxidant potential	20 mg/kg	High cholesterol diet-induced atherosclerosis in Wistar rat	It was anticipated that using zingerone as an active ingredient could help prevent and treat diet-induced atherosclerosis and its health consequences	[66]
Salvianolic acid B	–	Anti-atherosclerotic	The yes-associated protein/TAZ/JNK signalling pathway controls both the production of oxidized low-density lipoprotein and inflammation, which are both inhibited as part of the mechanism	30 ug/mL	Done by in vitro analysis	Salvianolic acid B may slow the development of atherosclerosis, according to findings. Its potential to modulate the YAP/TAZ/JNK signalling pathway is associated with anti-atherosclerosis capabilities	[67]

Sophocarpine	–	Anti-atherosclerotic	By the modulation of NF-B signalling, sophocarpine lowered vascular endothelium apoptosis and inflammatory responses induced by oxidized low-density lipoprotein	100 μg/mL	Completed by in vitro analysis	The findings concluded that pre-incubation with sophocarpine mitigated the cytotoxic effects, DNA fragmentation, and apoptosis triggered by ox-LDL in endothelial cells in a dose-dependent approach	[68]
Crocin	–	Anti-arrhythmic	By decreasing Connexin-43 expression	50 mg/kg	Rat model for myocardial infarction caused by arterial ligation	A study indicates that crocin enhanced the expression of connexin-43 in an experimentally induced ischemic injury, which had a preventative effect on the arrhythmia	[69]
Clerofibrase from anti-coagulant fraction	*Clerodendrum colebrookianum*/ leaves	Anti-coagulant	The anti-coagulant impact is influenced by fibrinogenolytic action as well as anti-platelet activity	100 mg/kg	Formation of thrombus in mice by employing carrageen using supportive in vitro and in vivo analysis	The study revealed the pharmacological potential of potent anti-coagulant fraction/clerofibrase as a naturally obtained active agent for the prevention and/or management of cardiovascular disorders linked to blood clotting	[70]

5 Conclusion

In conclusion, we have attempted to represent cumulative data of all the medicinal plants' investigation strategies for various CVDs and their specific mode of action against a particular disease. Thus, the chapter will act as a referential source for scientific and general people to utilize natural resources to prevent conventional drugs' side effects in combating different cardiovascular diseases.

References

1. Hypertension in India. https://www.who.int/india/health-topics/hypertension.
2. Rastogi S, Pandey MM, Rawat AKS. Traditional herbs: a remedy for cardiovascular disorders. Phytomedicine. 2016;23(11):1082–9.
3. Karimi A, Majlesi M, Rafieian-Kopaei M. Herbal versus synthetic drugs; beliefs and facts. J Nephropharmacol. 2015;4(1):27–30.
4. Kamble JS, Raunakkumar C, Verma U, Pardhe HA. A review of ancient medicinal plants utilized in therapy of cardiovascular diseases. Indian J Pharm Pharmacol. 2021;8(1):1–4.
5. Shaito A, Thuan DTB, Phu HT, Nguyen THD, Hasan H, Halabi S, et al. Herbal medicine for cardiovascular diseases: efficacy, mechanisms, and safety. Front Pharmacol. 2020;11:422.
6. Ambrose JA, Singh M. Pathophysiology of coronary artery disease leading to acute coronary syndromes. F1000Prime Rep. 2015;7:8.
7. Hall JE, Granger JP, do Carmo JM, da Silva AA, Dubinion J, George E, et al. Hypertension: physiology and pathophysiology. Compr Physiol. 2012;2(4):2393–442.
8. Soehnlein O, Libby P. Targeting inflammation in atherosclerosis—from experimental insights to the clinic. Nat Rev Drug Discov. 2021;20(8):589–610.
9. WRC. Table of contents. Eur Univ Inst. 2012;2:2–5. https://eur-lex.europa.eu/legal-content/PT/TXT/PDF/?uri=CELEX:32016R0679&from=PT%0Ahttp://eur-lex.europa.eu/LexUriServ/LexUriServ.do?uri=CELEX:52012PC0011:pt:NOT.
10. Ford TJ, Berry C. Angina: contemporary diagnosis and management. Heart. 2020;106(5):387–98.
11. Fuchs RM, Becker LC. Pathogenesis of angina pectoris. Arch Intern Med. 1982;142(9):1685–92.
12. Antzelevitch C, Burashnikov A. Overview of basic mechanisms of cardiac arrhythmia. Cardiac Electrophysiol Clin. 2011;3(1):23–45.
13. Christia P, Frangogiannis NG. Targeting inflammatory pathways in myocardial infarction. Eur J Clin Investig. 2013;43(9):986–95.
14. Pelliccia F, Pasceri V, Niccoli G, Tanzilli G, Speciale G, Gaudio C, et al. Predictors of mortality in myocardial infarction and nonobstructed coronary arteries: a systematic review and meta-regression. Am J Med. 2020;133(1):73–83.
15. Mortensen MB, Kjolby M, Gunnersen S, Larsen JV, Palmfeldt J, Falk E, et al. Targeting sortilin in immune cells reduces proinflammatory cytokines and atherosclerosis. J Clin Invest. 2014;124(12):5317–22.
16. Jelkmann W. Regulation of erythropoietin production. J Physiol. 2011;589(6):1251–8.
17. Kumar A, Sharma E, Marley A, Samaan MA, Brookes MJ. Iron deficiency anaemia: pathophysiology, assessment, practical management. BMJ Open Gastroenterol. 2022;9(1):e000759.
18. Solomon LR. Disorders of cobalamin (vitamin B12) metabolism: emerging concepts in pathophysiology, diagnosis and treatment. Blood Rev. 2007;21(3):113–30.
19. Mankad P, Kalahasty G. Antiarrhythmic drugs: risks and benefits. Med Clin North Am. 2019;103(5):821–34. https://doi.org/10.1016/j.mcna.2019.05.004.
20. Shattat GF. A review article on hyperlipidemia: types, treatments and new drug targets. Biomed Pharmacol J. 2014;7(2):399–409.
21. Shabab S, Gholamnezhad Z, Mahmoudabady M. Protective effects of medicinal plant against diabetes induced cardiac disorder: a review. J Ethnopharmacol. 2021;265:113328.
22. Bachheti RK, Worku LA, Gonfa YH, Zebeaman M, Deepti PDP, et al. Prevention and treatment of cardiovascular diseases with plant phytochemicals: a review. Evid Based Complement Altern Med. 2022;2022:5741198.
23. Suroowan S, Mahomoodally F. Common phyto-remedies used against cardiovascular diseases and their potential to induce adverse events in cardiovascular patients. Clin Phytosci. 2015;1:1.
24. Shah AJ, Gilani AH. Aqueous-methanolic extract of sweet flag (Acorus calamus) possesses cardiac depressant and endothelial-derived hyperpolarizing factor-mediated coronary vasodilator effects. J Nat Med. 2012;66(1):119–26.
25. Shah AJ, Gilani AH. Blood pressure-lowering and vascular modulator effects of Acorus calamus extract are mediated through multiple pathways. J Cardiovasc Pharmacol. 2009;54(1):38–46.
26. Bhatia J, Tabassum F, Sharma AK, Bharti S, Golechha M, Joshi S, et al. Emblica officinalis exerts antihypertensive effect in a rat model of DOCA-salt-induced hypertension: role of (p) eNOS, NO and oxidative stress. Cardiovasc Toxicol. 2011;11(3):272–9.
27. Saldanha E, Pai RJ, George T, D'Souza S, Adnan M, Pais M, et al. Health effects of various dietary agents and phytochemicals (therapy of acute pancreatitis). In: Therapeutic, probiotic, and unconventional foods. Amsterdam: Elsevier; 2018.
28. Meghwani H, Prabhakar P, Mohammed SA, Seth S, Hote MP, Banerjee SK, et al. Beneficial effects of aqueous extract of stem bark of Terminalia arjuna (Roxb.), An ayurvedic drug in experimental pulmonary hypertension. J Ethnopharmacol. 2017;197:184–94.
29. Mohanty IR, Arya DS, Gupta SK. Withania somnifera provides cardioprotection and attenuates ischemia-reperfusion induced apoptosis. Clin Nutr. 2008;27(4):635–42.
30. Arya DS, Arora S, Malik S, Nepal S, Kumari S, Ojha S. Effect of *Piper betel* on cardiac function, marker enzymes, and oxidative stress in isoproterenol-induced cardiotoxicity in rats. Toxicol Mech Methods. 2010;20(9):564–71.
31. Shaukat B, Mehmood MH, Shah S, Anwar H. Ziziphus Oxyphylla hydro-methanolic extract ameliorates hypertension in L-NAME induced hypertensive rats through NO/cGMP pathway and suppression of oxidative stress related inflammatory biomarkers. J Ethnopharmacol. 2022;285:114825.
32. Pedrosa RC, Meyre-Silva C, Cechinel-Filho V, Benassi JC, Oliveira LFS, Zancanaro V, et al. Hypolipidaemic activity of methanol extract of Aleurites moluccana. Phyther Res. 2002;16(8):765–8.
33. Badalzadeh R, Shaghaghi M, Mohammadi M, Dehghan G, Mohammadi Z. The effect of cinnamon extract and long-term aerobic training on heart function, biochemical alterations and lipid profile following exhaustive exercise in male rats. Adv Pharm Bull. 2014;4(Suppl 2):515–20.
34. Ghasi S, Nwobodo E, Ofili JO. Hypocholesterolemic effects of crude extract of leaf of Moringa oleifera Lam in high-fat diet fed Wistar rats. J Ethnopharmacol. 2000;69(1):21–5.
35. Ali M, Mohammed SY. Selective suppression of platelet thromboxane formation with sparing of vascular prostacyclin synthesis by aqueous extract of garlic in rabbits. Prostaglandins Leukot Med. 1986;25(2-3):139–46.

36. Khan A, Mustafa MR, Khan AU, Murugan DD. Hypotensive effect of Gentiana floribunda is mediated through Ca++ antagonism pathway. BMC Complement Altern Med. 2012;12:121.
37. Tian J, Popal MS, Liu Y, Gao R, Lyu S, Chen K, et al. Ginkgo biloba leaf extract attenuates atherosclerosis in streptozotocin-induced diabetic ApoE-/-mice by inhibiting endoplasmic reticulum stress via restoration of autophagy through the mTOR signaling pathway. Oxidative Med Cell Longev. 2019;2019:8134678.
38. Kubota Y, Tanaka N, Kagota S, Nakamura K, Kunitomo M, Shinozuka K, et al. Effects of Ginkgo biloba extract on blood pressure and vascular endothelial response by acetylcholine in spontaneously hypertensive rats. J Pharm Pharmacol. 2010;58(2):243–9.
39. Chen J, Huang Q, Li J, Yao Y, Sun W, Zhang Z, et al. Panax ginseng against myocardial ischemia/reperfusion injury: a review of preclinical evidence and potential mechanisms. J Ethnopharmacol. 2023;300(September):115715. https://doi.org/10.1016/j.jep.2022.115715.
40. Ojha S, Bhatia J, Arora S, Golechha M, Kumari S, Singh AD. Cardioprotective effects of Commiphora mukul against isoprenaline-induced cardiotoxicity: a biochemical and histopathological evaluation. J Environ Biol. 2011;32(6):731–8.
41. Reddy MM, Dhas Devavaram J, Dhas J, Adeghate E, Starling EB. Anti-hyperlipidemic effect of methanol bark extract of Terminalia chebula in male albino Wistar rats. Pharm Biol. 2015;53(8):1133–40.
42. Khan AU, Gilani AH. Cardiovascular inhibitory effects of Hyoscyamus niger. Methods Find Exp Clin Pharmacol. 2008;30(4):295–300.
43. Liu Q, Chen Y, Auger-Messier M, Molkentin JD. Interaction between NFκB and NFAT coordinates cardiac hypertrophy and pathological remodeling. Circ Res. 2012;110(8):1077–86.
44. Hsieh YL, Shibu MA, Lii CK, Viswanadha VP, Lin YL, Lai CH, et al. Andrographis paniculata extract attenuates pathological cardiac hypertrophy and apoptosis in high-fat diet fed mice. J Ethnopharmacol. 2016;192:170–7.
45. Aziz N, Mehmood MH, Siddiqi HS, Mandukhail SUR, Sadiq F, Maan W, et al. Antihypertensive, antidyslipidemic and endothelial modulating activities of Orchis mascula. Hypertens Res. 2009;32(11):997–1003.
46. Mangathayaru K, Kuruvilla S, Balakrishna K, Venkhatesh J. Modulatory effect of *Inula racemosa* hook. f. (Asteraceae) on experimental atherosclerosis in guinea-pigs. J Pharm Pharmacol. 2009;61(8):1111–8.
47. Ghayur MN, Gilani AH. Radish seed extract mediates its cardiovascular inhibitory effects via muscarinic receptor activation. Fundam Clin Pharmacol. 2006;20(1):57–63.
48. Evans LW, Bender A, Burnett L, Godoy L, Shen Y, Staten D, et al. Emodin and emodin-rich rhubarb inhibits histone deacetylase (HDAC) activity and cardiac myocyte hypertrophy. J Nutr Biochem. 2020;79:108339.
49. Pozder Geb Gehlken C, van der Velde R, Meijers WC, Sillјé HHW, Muntendam P, Dokter MM, et al. Pectins from various sources inhibit galectin-3-related cardiac fibrosis. Curr Res Transl Med. 2022;70(1):103321.
50. Shah SMA, Naqvi SAR, Munir N, Zafar S, Akram M, Nisar J. Antihypertensive and antihyperlipidemic activity of aqueous methanolic extract of rauwolfia serpentina in albino rats. Dose-Response. 2020;18(3):1559325820942077.
51. Sharma M, Kishore K, Gupta SK, Joshi S, Arya DS. Cardioprotective potential of Ocimum sanctum in isoproterenol induced myocardial infarction in rats. Mol Cell Biochem. 2001;225(1):75–83.
52. Siddiqi HS, Mehmood MH, Rehman NU, Gilani AH. Studies on the antihypertensive and antidyslipidemic activities of Viola odorata leaves extract. Lipids Health Dis. 2012;11:6.
53. García-Trejo EMA, Arellano-Buendía AS, Argüello-García R, Loredo-Mendoza ML, García-Arroyo FE, Arellano-Mendoza MG, et al. Effects of allicin on hypertension and cardiac function in chronic kidney disease. Oxidative Med Cell Longev. 2016;2016:3850402.
54. Zhang L, Wu L, Liu X, Hisae Y, Katsumi I, Hiroko N, et al. Trans-cinnamaldehyde promotes nitric oxide release via the protein kinase-B/v-Akt murine thymoma viral oncogene -endothelial nitric oxide synthase pathway to alleviate hypertension in SHR. Cg-Leprcp/NDmcr rats. J Tradit Chin Med. 2018;38(4):548–55.
55. Agunloye OM, Oboh G, Ademiluyi AO, Ademosun AO, Akindahunsi AA, Oyagbemi AA, et al. Cardio-protective and antioxidant properties of caffeic acid and chlorogenic acid: mechanistic role of angiotensin converting enzyme, cholinesterase and arginase activities in cyclosporine induced hypertensive rats. Biomed Pharmacother. 2019;109:450–8.
56. Elasoru SE, Rhana P, de Oliveira BT, Naves de Souza DL, Menezes-Filho JER, Souza DS, et al. Andrographolide protects against isoproterenol-induced myocardial infarction in rats through inhibition of L-type Ca2+ and increase of cardiac transient outward K+ currents. Eur J Pharmacol. 2021;906:174194.
57. Sumitra M, Manikandan P, Kumar DA, Arutselvan N, Balakrishna K, Manohar BM, et al. Experimental myocardial necrosis in rats: Role of arjunolic acid on platelet aggregation, coagulation and antioxidant status. Mol Cell Biochem. 2001;224(1-2):135–42.
58. Singh H, Kaur P, Kaur P, Muthuraman A, Singh G, Kaur M. Investigation of therapeutic potential and molecular mechanism of vitamin P and digoxin in I/R-induced myocardial infarction in rat. Naunyn Schmiedeberg's Arch Pharmacol. 2015;388(5):565–74.
59. Verma H, Bhattacharjee A, Shivavedi N, Nayak PK. Evaluation of rosmarinic acid against myocardial infarction in maternally separated rats. Naunyn Schmiedeberg's Arch Pharmacol. 2022;395(10):1189–207. https://doi.org/10.1007/s00210-022-02273-9.
60. Jiang T, Zhang L, Ding M, Li M. Protective effect of vasicine against myocardial infarction in rats via modulation of oxidative stress, inflammation, and the PI3K/AKT pathway. Drug Des Devel Ther. 2019;13:3773–84.
61. Che X, Wang X, Zhang J, Peng C, Zhen Y, Shao X, et al. Vitexin exerts cardioprotective effect on chronic myocardial ischemia/reperfusion injury in rats via inhibiting myocardial apoptosis and lipid peroxidation. Am J Transl Res. 2016;8(8):3319–28.
62. Obeidat HM, Althunibat OY, Alfwuaires MA, Aladaileh SH, Algefare AI, Almuqati AF, et al. Cardioprotective effect of taxifolin against isoproterenol-induced cardiac injury through decreasing oxidative stress, inflammation, and cell death, and activating Nrf2/HO-1 in mice. Biomolecules. 2022;12(11):1–16.
63. Manu TM, Anand T, Pandareesh MD, Kumar PB, Khanum F. Terminalia arjuna extract and arjunic acid mitigate cobalt chloride–induced hypoxia stress–mediated apoptosis in H9c2 cells. Naunyn Schmiedeberg's Arch Pharmacol. 2019;392(9):1107–19.
64. Sun YZ, Chen JF, Shen LM, Zhou J, Wang CF. Anti-atherosclerotic effect of hesperidin in $LDLr^{-/-}$ mice and its possible mechanism. Eur J Pharmacol. 2017;815:109–17.
65. Luo X, Weng X, Bao X, Bai X, Lv Y, Zhang S, et al. A novel anti-atherosclerotic mechanism of quercetin: competitive binding to KEAP1 via Arg483 to inhibit macrophage pyroptosis. Redox Biol. 2022;57:102511.
66. Geng X, Liu H, Yuwen Q, Wang J, Zhang S, Zhang X, et al. Protective effects of zingerone on high cholesterol diet-induced

atherosclerosis through lipid regulatory signaling pathway. Hum Exp Toxicol. 2021;40(10):1732–45.

67. Yang Y, Pei K, Zhang Q, Wang D, Feng H, Du Z, et al. Salvianolic acid B ameliorates atherosclerosis via inhibiting YAP/TAZ/JNK signaling pathway in endothelial cells and pericytes. Biochim Biophys Acta Mol Cell Biol Lipids. 2020;1865(10):158779.
68. Fang H, Bo T, Zi X, Tan D, Liu X, Jiang C, et al. Sophocarpine exert protective effect against ox-LDL-induced endothelial damage via regulating NF-κB signaling pathway. Biosci Biotechnol Biochem. 2020;84(10):2104–12.
69. Li H, Li J, Wang J, Afzal O, Altamimi ASA, Nasar Mir Najib Ullah S, et al. Analysis of anti-arrhythmic impacts of crocin through estimation of expression of Cx43 in myocardial infarction using a rat animal model. ACS Omega. 2022;7(42):37164–9.
70. Gogoi D, Ramani S, Bhartari S, Chattopadhyay P, Mukherjee AK. Characterization of active anticoagulant fraction and a fibrin(ogen)olytic serine protease from leaves of Clerodendrum colebrookianum, a traditional ethno-medicinal plant used to reduce hypertension. J Ethnopharmacol. 2019;243:112099.

Role of Lifestyle Modification and Diet in the Prevention of Cancer

Moumita Chatterjee, Sharmistha Ganguly, and Subhajit Dutta

Abstract

Cancer is a multifactorial disease that is still leading the chart in terms of mortality rate. Almost 90% of the cancers are sporadic in nature, with only 10% being genetic and inherited in an autosomal dominant manner. Although accumulated evidence suggests a drop in mortality rates in recent years, we are still far away from fully understanding the pathophysiological mechanisms associated with such complex phenotypes. Moreover, anti-cancer medications possess a plethora of mild aftereffects without much relief. Hence, there is a need to explore the natural protective mechanisms that can limit the mortality rate with minimal or no adverse consequences. This book chapter entails a story citing different natural mechanisms that can lower the risk of developing different types of cancer. We focussed on lifestyle modifications and changes in nutritional diet that can modulate different regulatory players associated with cancer signalling axes. We have presented an avalanche of evidence ranging from molecular mechanisms to implications of different natural diets in cancer, thereby providing a strong correlation between lifestyle modifications and their mitigation. Through this chapter, readers will understand the mitigating roles of low fibre, omega-3- and omega-6 fatty acids, red meat, flax seeds, fruits, and vegetables in protecting different cancer types ranging from breast to lung to prostate carcinoma. Finally, the clinical evidence of lifestyle modifications will establish a positive association between reduced cancer progression with increased chances of survival.

Keywords
Lifestyle modification · Diet · Cancer

1 Introduction

Cancer is a group of diseases characterized by the unchecked invasion and growth of aberrant cells. Normal cells become cancer cells through a multistep process involving changes to both the metabolic phenotype and molecular signature. The "metastasis" or the concluding phase of the invasion process can be lethal [1]. The International Agency for Research on Cancer's GLOBOCAN 2020 estimated the status of cancer incidencence as well as mortality of cancer patients (Fig. 1a, b). They predicted around new 19.3 million cases worldwide (18.1 million excluding non-melanoma skin cancer) and reported more than 10 million deaths from cancer (9.9 million excluding non-melanoma skin cancer). Female breast cancer has surpassed lung cancer as the most commonly diagnosed malignancy, with an estimated 2.3 million new cases (11.7%). Colorectal (10.0%), lung (11.4%), stomach (5.6%), and prostate (7.3%) cancers are next in line. Lung cancer remained the most common type, with an estimated 1.8 million deaths (18%). It was then followed by colorectal (8.3%), liver (9.4%), female breast (6.9%) cancers, and stomach (7.7%) [2]. In the last few years, immuno-oncology and targeting actionable changes in oncogene-driven tumours have revolutionized cancer treatment paradigms. Both cancer remedy sectors carry on with significant obstacles. Evaluating the curative impact of druggable genomic alteration in biomarker-driven clinical trials is hampered because they are not only heterogeneous but also represent small subsets of patients in some tumour types. Even though the use of NGS (next-generation sequencing) technologies for molecular pre-screening in clinical research is growing, difficulties

M. Chatterjee
V. Sivaram Research Foundation, Bangalore, Karnataka, India

S. Ganguly
University Department of Botany, Ranchi University, Ranchi, Jharkhand, India

S. Dutta (✉)
Functional Genomics and Metabolism Research Unit, Department of Biochemistry and Molecular Biology, University of Southern Denmark, Odense, Denmark

Ronin Institute for Independent Scholarship, Montclair, NJ, USA
e-mail: subhajit@bmb.sdu.dk

A. K. Dhara, S. C. Mandal (eds.), *Role of Herbal Medicines*, https://doi.org/10.1007/978-981-99-7703-1_8

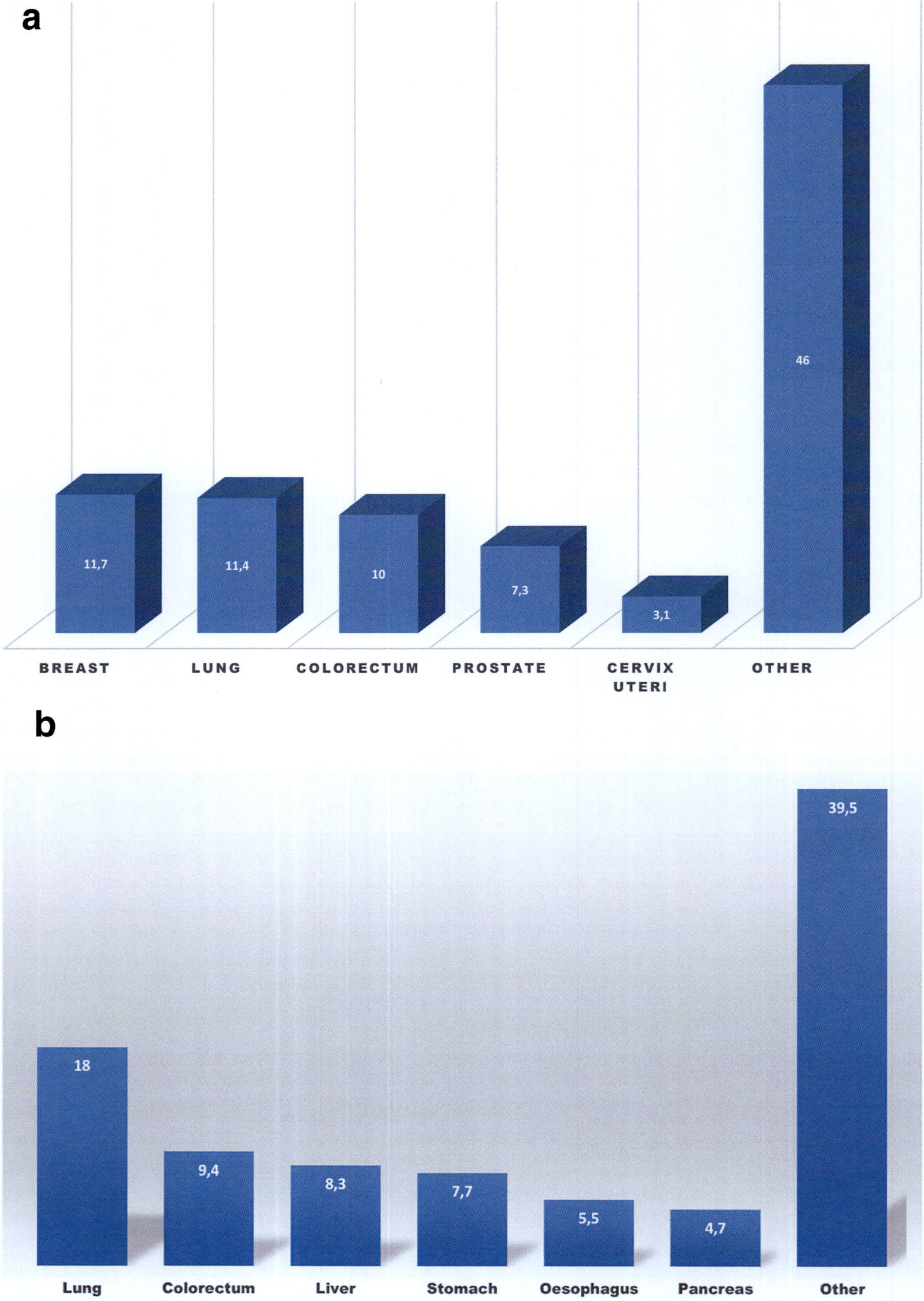

Fig. 1 (**a**) Occurrence percentage of different types of cancer worldwide (Source: Global Cancer Observatory, https://gco.iarc.fr/). (**b**) Mortality percentage of different types of cancer worldwide (Source: Global Cancer Observatory, https://gco.iarc.fr/)

with the clinical interpretation of vast genomic data prevent their widespread clinical application. The major obstacle to precision oncology's success is likely dealing with acquired resistance and tumour heterogeneity. Contrarily, only a small subset of patients benefit from immune checkpoint inhibitors (anti-cytotoxic T lymphocyte antigen-4 monoclonal antibodies and anti-programmed death cell protein-1/programmed death cell ligand-1 [PD-1/L1]) in terms of long-term survival. There are currently no robustly validated predictive markers that could aid in the identification of these subsets and the improvement of treatment delivery and patient selection. A combination of drugs targeting various molecular changes of cancer may be required to improve long-term survivance [3]. It is undeniably true that changes in lifestyle have contributed to an elevation of metabolic diseases including, obesity, diabetes, metabolic disorders, and even cancer [4, 5].

Only 5–10% of cancer occurrences can be traced to genetic flaws, whereas environmental and lifestyle factors cause the other 90–95%. Infectious agents, including hepatitis viruses, human papillomavirus (HPV), and *Helicobacter pylori*, as well as dietary variables, insufficient exercise, and excessive body mass, are a few important causes and risk factors for the cancers that are common in low- and middle-income countries (LMCs) [6]. Numerous factors have smaller but still noticeable effects. The variables differ according to how easily they can be changed [7, 8]. While calcium, milk, fibre, as well as whole grains decrease cancer risks, foods like processed meat and red meat can raise colorectal cancer threats. This can be elucidated by augmenting immunological receptivity, inflammation, and excess food intake, all of which are risk agents for colon cancer development [9]. A diet high in vegetables, whole grains, and fruits especially if it comprises plenty of soy products would elevate the isoflavones intake, thought to decrease cancer risk [10]. The six basic properties of cancer are acquired by human malignancies over the course of their multi-stage evolution. The distinctive traits act as a framework for organizing information in order to describe the complexity of the neoplastic disease. One of them is maintaining proliferative signalling, along with avoiding cell death inhibitors, replicative immortality, inducing angiogenesis, and promoting invasion and metastasis. These traits are characterized by genome instability, which generates the genetic variation that expedites their acquisition and inflammation, which supports many signature functions. In the last ten years, conceptual progress has added two more markers of potential universality to this list: reprogramming energy metabolism and preventing immunological destruction. Tumours reveal a second level of heterogeneity in addition to cancer cells: a variety of recruited, apparently normal cells that facilitate cancer cells acquire distinguishing characteristics [11]. Cancer cells do, in fact, vary from their healthy counterparts in that they exhibit a variety of metabolic reprogramming that is influenced by both the tumour microenvironment and intrinsic genetic abnormalities. As a result of this metabolic modification, ATP and building blocks for maintaining biosynthetic capacity and redox status balance are finally provided to cancer cells [12]. To understand cancer pathophysiology and the significance of lifestyle factors in reprogramming nutrient metabolism, this review will talk about changes in metabolic phenotype and molecular markers. It will also make a strong case for lifestyle modification to prevent and treat cancer.

2 Molecular Basis of Cancer

Cancer is a state that implies dynamic alterations in the structure of DNA, in accordance with centuries of research. The functions of numerous cancer-related genes have been made known through research on growing embryos and transgenic mice. Tumour suppressor genes and oncogenes, which comprise the majority of the altered genes discovered in cancer, code for parts of the pathways that control how social and proliferatively active cells behave in the body, particularly the mechanisms by which signals from a neighbouring cell can cause it to differentiate, divide, or die. Further cancer-critical genes protect the genome's integrity and preserve it. However, the genetic alterations that enable malignancies to metastasize, escape the primary tumour, and grow in other tissues remain largely unexplored.

By sequestering the retinoblastoma (Rb) protein [13] which controls the p53 protein and cell division [14], DNA viruses such as papillomaviruses are able to promote the development of cancer by acting as an emergency halt on cell division in cells that have experienced genetic damage and by calling a halt on cell division in senescent cells with shortened telomeres. The Rb protein (pRb) has been shown to be in charge of a critical G1 checkpoint that prevents cell development and entry into the S phase. Rb2/p130, Rb/p105, and p107 are three members of the retinoblastoma family that are collectively referred to as "pocket proteins". The pRb protein inhibits gene transcription by directly binding to the transactivation domain of E2F and binding to the promoter of these genes as a complex with E2F. This is required for the transition from G1 to S phases. pRb suppresses transcription by altering the structure of chromatin by interacting with proteins such as hBRM, HDAC1, SUV39H1, and BRG1, which are involved in nucleosome remodelling, histone acetylation/deacetylation, and methylation, respectively. Cell cycle dysregulation brought on by pRb action deficiency may result in malignant phenotypes. Numerous neoplasia, including mesothelioma, cervical cancer, and Burkitt's lymphoma associated with AIDS, exhibit functional pRb inactivation by viral oncoprotein binding [15, 16].

A nuclear transcription factor called p53 promotes apoptosis. Since the p53 gene possesses loss of function mutations in human malignancies (more than 50%), p53 has been considered as one of the traditional tumour suppressors. The dominant-negative inhibitor of wild-type p53 is mutant p53. Indeed, the oncogenic potential of mutant p53 exists. Malignant cells with p53 mutations occasionally exhibit a chemo-resistant phenotype. A variety of cellular stressors, such as DNA damage, cause p53 to build up in cell nuclei where pro-apoptotic activity can be seen. Activated p53 induced cell cycle arrest to allow DNA repair or apoptosis to limit the cell propagation with significant DNA damage via transactivation of its target genes involved in the induction of cell cycle arrest and/or apoptosis. Therefore, p53's DNA-binding ability and tumour-suppressing ability are closely related [17–19].

As the proliferation as well as metastatic spread of cancer cells depend on a suitable supply of nutrients, oxygen, and the elimination of waste products, new growth in the vascular network is vital. Angiogenesis and lymph angiogenesis are processes that create new blood and lymphatic vessels, respectively. Both activator and inhibitor chemicals control angiogenesis. Angiogenic activators and inhibitors have been found to exist in more than a dozen different protein species. The level of expression of angiogenic factors reflects the aggressiveness of tumour cells. The identification of angiogenic inhibitors is required to contribute to a decrease in the morbidity and death caused by carcinomas. Antiangiogenic therapy has so far been administered to thousands of patients. Antiangiogenic therapies have not shown long-term survival benefits, despite their theoretical efficacy [20–22].

The final phases of cancer development include tissue invasion and metastasis. It describes the way in which primary tumours spread out and infiltrate surrounding tissues before eventually making their way to a distant location to form new colonies of cancer. The metastasis process, which is also what causes cancer to spread, is to blame. Alteration in integrin expression can facilitate cancer cell invasion and is one typical example. In order to facilitate metastasis, cancer cells have the ability to switch the expression of integrins used by normal cells to other integrins such as α3β1 and αvβ3. These integrins can destroy stromal components made by extracellular proteases. Integrin antagonists, such as the αvβ3 and αvβ5 inhibitor cilengitide, have demonstrated promising effects in Phase II clinical studies, and it is currently being investigated in a Phase III trial in people with glioblastoma [23]. The significance of understanding how integrin antagonists affect the tumour and its surroundings is highlighted by these fascinating clinical advances [24, 25].

Generally, mutations that inactivate particular tumour suppressor genes as well as activate particular oncogenes are associated with the stages of tumour progression. However, various cancer types and even patients who ostensibly have the same form of the disease might have distinct combinations of mutations, which highlights the random nature of mutations. Although many of the same genetic defects are found repeatedly, this suggests that there are only a finite number of ways to circumvent our defences against cancer [26].

3 Nutrient Metabolism Reprogramming in Cancer

3.1 Implications of Different Nutrients on Cancer

According to estimates, changes in lifestyle and diet can prevent 30–40% of all cancers. Obesity, nutrient-deficient diets, such as concentrated sweets and refined flour products, which affect glucose metabolism, eating red meat, low fibre intake, and an imbalance of omega 6 and omega 3 fats are all factors in an increased risk of developing cancer. Consuming plenty of vegetables and fruits as well as flax seed (mainly its lignan fraction) will decrease the risk of developing cancer. The health benefits of cruciferous vegetables and *Allium* are particularly notable, with sprouts of broccoli having high sulphorophane concentration. Selenium, chlorophyll, vitamin D, vitamin B-12, folic acid, and antioxidants such as carotenoids (α-carotene, lutein, lycopene, cryptoxanthin, β-carotene) are protective constituents in a cancer-preventive diet. Ascorbic acid has few advantages when it is taken orally, but intravenously may be more helpful. Probiotics and oral digestive enzyme supplementation are effective anticancer dietary strategies [27]. The energy needs and macromolecular building blocks for maintaining cell development and survival are met by nutrients, which are essential resources. As they move through different developmental phases, cancer cells need a number of essential nutrients to meet their shifting metabolic requirements. Furthermore, variables that are impacted by both cell-intrinsic and microenvironment throughout the evolution of cancer determine nutrient requirements, for which a thorough characterization is yet lacking [28, 29]. For instance, although in other tissues or cancerous conditions, glutamine works as a signalling molecule for tumour propagation, glutamine is used as a primary carbon source to feed into metabolic pathways necessary for in vivo growth of certain tumour cells [30, 31].

3.1.1 Low Fibre

Unprocessed plant meals often contain a lot of fibre. There is one thing that eggs, dairy products, and meat have in common none of them contain fibre. Additionally, refined grain products lose nutritional fibre during processing. According to the report, people who consume the most dietary fibre may experience a slight decrease in the risk of acquiring breast cancer and a reduction in the incidence of colon cancer [32]. Little indication of publication bias was discovered, and dose–response analysis revealed that every increase in dietary fibre consumption of 10 g/day was related to a 4% [33] and 7% [34] decrease in the risk of breast cancer. Therefore, the ingestion of dietary fibre is strongly linked to a lower breast cancer risk, especially in postmenopausal women. High dietary fibre intake is linked to a lower risk of cardiovascular disease and all cancer-related mortality. These findings back up the existing advice that a healthy diet should include a high consumption of dietary fibre [35].

3.1.2 Glucose Metabolism

Gluconeogenesis, glycolysis, glycogenesis, and glycogenolysis are only a few of the mechanisms that have glucose metabolism. The theory that a high glycaemic diet is linked to cancer has been put to the test through case–control studies and prospective population studies. According to case–control studies, there is a consistently higher risk of developing gastric [36], upper aerodigestive tract [37], ovarian [38], endometrial [39], and colon cancers [40, 41] when a person has a high glycaemic load. Elevated colorectal cancer risk was linked to higher levels of fasting glucose, fasting insulin, glucose, and insulin levels after two hours following an oral glucose challenge, and a bigger waist circumference [42]. Diabetes has been related in numerous studies to an increased risk of pancreatic cancer [43, 44], colorectal cancer [45–47], and endometrial cancer [48]. It is obvious that severe glucose metabolism dysregulation is a risk factor for cancer. Refined sugar, meals containing refined sugar, and items made from refined flour should all be avoided and removed from a diet designed to prevent cancer. In premenopausal women or women under the age of 50, a high glycaemic load diet and high carbohydrate intake may increase breast cancer risk [49–51]. Through hyperinsulinemia, dietary carbohydrate, glycaemic load, and glycaemic index are hypothesized to affect the risk of colon cancer [52].

3.1.3 Omega-3-and Omega-6 Fatty Acids

Accumulated evidence in animal studies suggests the protective roles of omega-3 fatty acids (DHA, EPA, alpha-linolenic acid) to be a safeguard in cancer, while omega-6 fats such as arachidonic acid and linoleic acid have been found to be cancer-promoting fats [27, 53]. The report suggested a link between a higher N-3 to N-6 fat ratio and a lower breast cancer risk. The breast tumour suppressor genes BRCA2 and BRCA1 respond differently to long-chain N-3 and N-6 lipids. When N-3 lipids (EPA or DHA) are applied to breast cell cultures, the expression of these genes is raised but arachidonic acid has no impact [54]. Due to their numerous biological effects, omega-3 polyunsaturated fatty acids, i.e. PUFAs are frequently employed in the nutritional therapy of cancer patients and are regarded as immunonutrients. Omega-3 PUFAs are vital for maintaining membrane fluidity, cell shape, and cell signalling. They have anti-inflammatory and antinociceptive properties and take part in the healing process of inflammation. They can also stimulate the G protein-coupled receptors GPR120/FFA4 and GPR40/FFA1, respectively. Anorexia-cachexia syndrome, paraneoplastic syndromes, depression, and pain are among the problems that cancer patients experience [55]. Several studies also focus on omega-3 supplements in maintaining muscle mass and function in cancer patient even if they are undergoing active treatment, which supports their potential significance in cancer prevention. Omega-3 fatty acids may contribute to a decreased inflammatory response during chemotherapy; however, it is still unknown whether cancer treatment damage may be avoided. Finally, tiny trials revealed that omega-3 fatty acids boost chemotherapy response rates [56–58]. Total polyunsaturated and monounsaturated fats were all strongly correlated with postmenopausal breast cancer. High intakes of omega-6 fatty acids, in particular, were linked to an elevated risk after accounting for other forms of fat [59].

3.1.4 Red Meat

High content of processed meat intake was strongly correlated with risk of breast [60], colorectal, rectal, colon, and lung cancers. High content red meat intake was positively correlated with the risk of colon, endometrial, rectal, colorectal, hepatocellular carcinoma, and lung cancer [61]. Processed meat consumption and fresh red meat consumption both increased the colorectal cancer risk [62]. Similar findings were found for colon cancer, but insignificant relationships were seen for rectal cancer [63]. Eating high content of white meat may reduce the stomach cancer risk, whereas red or processed meat may raise that risk [64–66]. Low consumption of chicken and a high intake of red meat are linked to an elevated risk of oesophageal squamous cell cancer. High meat consumption, particularly processed meat, is likely to raise the risk of developing oesophageal adenocarcinoma [64, 67–69]. Additionally, there may be no connection between eating fish and the risk of oesophageal cancer [70].

3.1.5 Flax Seed

Alpha-linolenic acid [71], lignans, and phytoestrogens are present in flax seeds. They have both hormonal and anticancer properties. Flax has antiproliferative properties in the breast tissue of breast cancer-prone women and may offer protection from initial breast cancer. The risk of mortality may be lower for people with breast cancer [72]. The mammalian lignan precursor secoisolariciresinol-diglycoside and alpha-linolenic acid (ALA) are abundant in flaxseed, which has been demonstrated to have anti-carcinogenic properties during the early promotion stage of carcinogenesis. While the oil in flaxseed is more potent at the point when tumours have already formed, the secoisolariciresinol-diglycosidein flaxseed seems to be advantageous throughout the carcinogenesis' promoting phase [73]. Flaxseed downregulates metastasis of B16BL6 murine melanoma cells in C57BL/6 mice [74, 75]. In the TRAMP (transgenic adenocarcinoma mouse prostate) model, adding 5% flaxseed to the diet prevents prostate cancer from spreading and developing [76].

3.1.6 Fruits and Vegetables

A diet high in vegetables and fruits protects against cancer, which is one of the most important messages of contemporary nutrition science. High consumption of raw and cooked garlic may have a preventative impact on stomach and colorectal cancers [77]. Greater consumption of *Brassica* vegetables was linked to a considerably lower risk of breast cancer in Chinese women, as determined by the urine ITC (isothiocyanates) biomarker [78]. To lower the chance of developing lung cancer [79], ITC from cruciferous vegetables may trigger apoptosis or alter steroid metabolism [80]. There is a negative correlation between eating legumes and the risk of prostate cancer [81].

3.2 Implications of Nutritional Diet and Lifestyle Modifications in Different Cancer Types

3.2.1 Colorectal Cancer

Colorectal cancer is the second leading cause of cancer deaths in the United States, and the third in both genders. The involvement of lifestyle management in the cases of colorectal cancer has been a theme of discussion for the past few years where the diet consumed by people plays a vital role among which foods with high fibre content and refined sugars are mentionable. Studies suggested that the intake of refined carbohydrates instead of fibre-rich diet has drastically damaged the intestinal flora of the human body, finally affecting the stool bulk and time of transit of the faecal matters [82, 83]. The precursor lesions of colorectal cancers are called adenomas which act as assessment points for the risk of the disease. Few studies also mention an inverse relationship between fibre intake and coloreactal cancer [84]. The intake of refined sugars and high fat causes insulin resistance and hyperinsulinemia which becomes an aetiology for colorectal cancer [85, 86] besides IGF-1 being another one [87]. The mutation of p53 in colon polyps is related to the lifestyle aspects [88]. The subordinate bile acids are found to be carcinogenic which shows a drastic reduction and recurrence of adenomatous colorectal polyps by the intake of dietary fats, soluble fibres, and green vegetables along with vigorous physical exercise [82]. The dietary fibre enhances the fermentation process by the natural intestinal flora and helps in producing short fatty acid chains which have a regulatory action on the cell cycle, apoptosis, and cell proliferation in colorectal cancer [89]. They also show anti-inflammatory properties and diminishes the synthesis of IL-6, TNFα, and cyclooxygenase 2 (COX-2) gene expressions [90, 91]. Phytochemicals such as terpenes, carotenoids, thioethers as well as polyphenols are involved in the signalling pathways of cancer and hence intakes of food containing these like green leafy vegetables of Cruciferae family, garlic, onion, legumes, and citrus fruits are proved to have positive results in colorectal malignancy [92]. A European cohort study stated that doubling the total dietary fibres decreased the incidence of colorectal carcinoma by 40% [93]. Vitamin B_9 (folic acid) has a protective role in the incidence of adenomas and the threat of colorectal carcinoma [94]. It compensates for the loss of tumour suppressor gene DCC and attenuates the EGFR gene [95]. Vitamin B_6, which is found in many green vegetables, grains, and fruits, has a protective effect against colorectal cancer as it is found to reduce the risk by 49% for each 100 pmol/mL rise of serum pyridoxal phosphate [96]. Vitamin E has α-tocopherol which reduces the DNA damage in colorectal epithelium [97]. In studies, it is proved that quercetin reduces the mutation of genes in colonic mucosa [98]. Consumption of processed red meat increases the threat of colorectal malignancy as it contains a higher amount of haeme which is considered carcinogenic because it degrades down to release iron [99] which promotes the making of reactive oxygen species mainly H_2O_2 which acts as a main part in genetic mutation and dismantling cytokines [100]. Red meats also produce N-nitroso compounds which form DNA complexes with telomere-stabilizing proteins (TRF2) [101]. It also creates oxysterols and aldehydes which promote the proliferation of cells [102]. Catering red meat at elevated temperatures promotes the synthesis of heterocyclic amines which are carcinogenic in nature [103]. Grape seed extract has properties to reduce oxidative stress and also protects the cell cycle process [104]. Silibinin extracted from the plant *Sylibum marianum* and the major component of silymarin is proven to show chemoprotective roles in several cancers including colorectal cancer by targeting cell cycle, DNA mutation mechanism, processes of inflammation, metastatic signalling, and cell proliferation [105]. Curcumin extracted from the medicinal plant *Curcuma longa* promotes apoptosis by the inhibition of DNA mutation, cell proliferation, and cancer signalling pathways. It also upregulates the level of glutathione S-transferase and influences the concentration of reactive oxygen species resulting in upregulation of the p21 protein and thereby restricting the development of malignant cells [106, 107]. Epigallocatechin 3-gallate extracted from tea *Camellia sinensis* decreases the growth factors like IGF-1 and blocks cell proliferation [108]. It also downregulates the MAPK/kinase signalling pathways [109] and upregulates the production of p53 and p21 thereby promoting apoptosis [110]. The intake of probiotics containing *Saccharomyces boulardi*, *Bifidobacterium animalis*, and *Bacillus polyfermenticus* has been verified to have anti-cancerous and anti-inflammatory potencies. which decrease the threat of colorectal carcinoma [111–113]. Alcohol consumption not only has carcinogenic effect as it causes DNA strand breaks in colonic mucosa cells and lymphocytes [114] but also reduces the vitamin absorption into the body thereby increasing oxidative stress in the cells and promoting the procedure of metastatic and cell proliferation [82]. The intake of tobacco increases the risk of adenomas and the threat of colorectal malignancy as the aromatic and heterocyclic amines are its key ingredients which cause DNA aberration and mutation [115]. Obesity is another factor adding to the list as the adipose tissues produce several hormones which stimulate the T cells and promote insulin resistance [116] which again contributes to cell proliferation [117]. So, weight management becomes another way to prevent the threat of colorectal carcinoma. Some studies reveal that an increase in physical exercise affects the stages of adipohormone leptin which plays an important part in cancer signalling pathways by MAPK and PI3K/AKT cascade activation [118]. The process of DNA mutation, cell proliferation, and apoptosis pathways are affected by circadian rhythm and therefore the deprivation of sleep in night shift workers makes them 50% prone to colorectal cancer [119].

3.2.2 Breast Cancer

Today, most of the females are affected by breast cancer not only in the United Statesbut also in the European countries and considered to be the second most dreadful disease among females [120]. Just like prostate cancer, it is influenced and categorized under lifestyle disorder [121]. Breast cancer is an oestrogen-dependant process and is seen in post-menopause women. During this time, the oestrogen levels decrease without complete disappearance and continue to be synthesized by aromatase activity in adipose cells, making obesity the key factor to the problem [122]. Physical exercise, proper nourishment with regular consumption of vitamins and minerals, plus ceasing the habit of smoking and drinking act as a major part in the threat of breast melanoma [120]. The levels of insulin in serum are another factor responsible for influencing breast melanoma [123]. The communication between IGF and oestrogen receptor paths also affects breast melanoma risk [124]. The treatment of breast cancer is very pocket pinching and a simple lifestyle management can be easier, more effective, and economical in preventing the disease [120]. Alteration in lifestyle management is not only for the women who are at threat of this disease but likewise for the women who are diagnosed with it [125]. There are several major factors related to lifestyle management in preventing breast cancer, the high socioeconomic status being the vital one. The high socioeconomic status makes the women exposed to a Westernized lifestyle with the habit of smoking, drinking, fewer kids, and unhealthy diet [126]. It is been reported that women from Asian and African countries are less affected by breast cancer than women from American, European, and Australian countries [127, 128]. Taller women are at greater threat of breast melanoma than shorter women [129, 130]. Obesity is linked with insulin resistance which when gained in the middle age directly increases the threat of breast malignancy [131]. Some cohort studies showed that a significant weight loss and proper diet with physical activity have shown positive results for patients of breast malignancy [132, 133]. The increase in physical exercise in postmenopausal women decreases the threat of breast melanoma as it affects the levels of oestrogen, insulin, and IGF-1 [134–136]. Physical activity in post-diagnosed patients of breast cancer shows an increase in survival rates [137, 138]. Another lifestyle disorder which exposes urban women to the risk of breast cancer is smoking. Both active and passive smokers fall victim to this [139–143] especially teenagers. Studies revealed that the rate of recurrence is 6.7 times higher in females who smoke than the non-smokers [144]. Females who consume alcohol are highly prone to risk of breast melanoma [145–147]. Intake of folic acid attenuates the threat of this ailment [148]. Fat intake plus higher BMI directly influence the disease [49, 50, 149]. On one hand, the intake of red meat and caffeine induces the disease [150–152], whereas on the other hand adding vitamins and minerals like vitamin D and calcium to the daily diet attenuates it [153]. Breastfeeding has been decreased by urban women due to their professional life and other recreational involvement. Studies reported that breastfeeding for consecutive 12 months after childbirth decreases the threat of breast malignancy by 4.3% [154, 155]. Upsurge in melatonin levels due to night shifts at working place triggers the process of breast cancer in women [156, 157] and also the associated stress related to it [158]. In an eventual group study, it has been noted that the use of NSAIDs declines the danger of breast cancer [159–161]. Several studies also revealed the use of digoxin in the induction of breast cancer though the recurrence is not reported but the women prescribed with this drug are suggested for regular check-ups [162–164].

3.2.3 Melanoma

One of the human organs that is most vulnerable to environmental stimuli is the skin [165] which is the outermost layer organ. This fact leads us to hypothesize that a person's everyday habits may affect their chance of developing skin illnesses. The skin is also made up of many types of cells, including keratinocytes, melanocytes, immune cells, and adipocytes. Because of this, a person's everyday habits may have a significant impact on their chance of developing cutaneous cancers [166]. The risk of developing cutaneous malignancies, such as squamous cell carcinoma, malignant melanoma, Merkel cell carcinoma, and basal cell carcinoma, is known to be influenced by a number of daily lifestyle factors, including smoking, circadian rhythm disruption, fatty acids, obesity, dietary fibre, alcohol, and ultraviolet light [167, 168]. The epidemiology study found that working nights increases your risk of developing a number of cancers. The cutaneous malignancy risk factor is also related to shift work with circadian disruption [169, 170]. Some investigations revealed a link between current smoking and a higher occurrence of squamous cell cancer [171–173]. Ginsenoside Panoxatriol, rutin, vitamins C and E, and these substances all show promise in protecting skin cells from UVB-induced cellular damage [174–177]. Vitamins C and E demonstrated promise for photoprotection in human fibroblasts exposed to UVA radiation [178]. Higher levels of circulating vitamin D have been linked to thinner tumours and longer life in melanoma patients [179–181]. According to the findings, melanoma cell migration and invasion—a crucial stage in metastasis—can be prevented by EGCG (epigallocatechin-3-gallate), a significant green tea catechin, by focusing on the endogenous expression of COX-2 (cyclooxygenase 2), PGE(2) (prostaglandin E) receptors, and epithelial-to-mesenchymal transition utilizing melanoma cell lines Hs294t (non-BRAF-mutated) and A375 (BRAF-mutated) as an in vitro model [182]. GSPs (proanthocyanidins) extracted from grape seed inhibits endogenous expression of COX-2 in melanoma cell lines (A375 and Hs294t) [183]. By blocking COX-2, PGE2, and PGE2 receptors, berberine prevents melanoma cell lines (Hs294 and A375) from migrating, a crucial stage in invasion and metastasis [184]. Timosaponin AIII can prevent the migration of melanoma cells (WM-115 and B16-F10 cells lines), a crucial stage in the metastatic process, by suppressing the production of COX-2, NF-B, PGE2, and PGE2 receptors [185].

3.2.4 Prostate Cancer

The utmost predominant malignancy in males is prostate cancer which is mostly seen in the United States compared to Asia. However, the migration of Asian men to the United States for professional ventures and adaptation to the lifestyle in the United States makes them fall prey to the ailment which henceforth proves that lifestyle reasons are responsible for the initiation of the disease [186]. With an increase in adaptation to Western regime, the occurrence of prostate malignancy increases in underdeveloped countries [187]. According to the international data, a constructive association has been found between prostate cancer mortality rate and dietary fats which are found to be comparatively lower in men from Asia rather than in the United States and Europe [188]. Some studies have also failed to correlate dietary fats and the threat of prostate malignancy [189, 190]. The reasons of these results may be determined by the type of fats used up as omega-3 polyunsaturated fatty acids plus omega-9 monounsaturated fatty acids have a preventive nature in the progress of prostate cancer, whereas omega-6 fatty acids promote it [191]. Epidemiological studies reveal the positive aspects of physical exercise in the reduction of prostate cancer risk by 10–70% [192].

In a study, a bioassay was developed by using the serum to trigger the cultures of prostate cancer cells. A number of subjects were selected from the Pritikin Longevity Center Residential Program where the men were given low-fat diets including mainly fruits, veggies, and grains and were subjected to physical exercise. The serums were collected and studied by in vivo and in vitro methods. The food had different per cents of fats (10–15%), proteins (15–20%), and carbohydrates (65–75%). Plant-based proteins were given with non-fat dairy products where fishes were provided once a week. The cholesterol content of the diet was less than 100 mg, whereas tobacco, coffee, tea, and alcohols were prohibited. A treadmill test was done before every physical training, the heart rate training and walking (45–60 min) were practiced [193]. Eleven consecutive days of this routine reduced the serum-roused LNCaP prostate malignant cell growth in culture by 30%, and 14 years of nourishment and workout training in subjected males showed a reduction in the growth of LNCaP cells by 15%. LNCaP is an androgen-sensitive cell line derived from males suffering from prostrate adenocarcinoma [193]. Another androgen-dependent, LAPC-4, was seen to be reduced by nourishment and physical workout [194].

A significant drop in serum insulin and a rise in sex hormone-binding globulin were seen in males undergoing low-fat diet and physical training. The sex hormone-binding globulin lowers the free testosterone plus oestradiol stages in serum and acts as a factor to reduce the growth of LNCaP cells because this is androgen dependent and both testosterone and oestradiol bind to its receptors [195]. Another study reported that testosterone, insulin, and oestradiol can promote the LNCaP cell development individually. However, the nourishment and workout subjected individuals when treated with these hormones, a 50% reduction in LNCaP cell development was observed [193, 196].

One of the most potent mitogens for prostate is IGF-I which is a peptide development aspect formed from the liver and regulates the growth of cells, cell differentiation, and apoptosis process [197, 198]. There are six different proteins related to the IGF axes among which IGFBP-3 is abundant in nature and binds with 90% of the circulating IGF-I where no change was noted in males subjected to a special low-fat diet and exercise, but high after abstaining [199].

Accumulating evidence suggests that a low fat diet and physical training diminished the serum IGF-1 levels and increased IGFBP-1 levels relative to baseline. These changes are predicted to be the reason for modulating serum insulin levels and its effect on liver [199]. IGF-I has been stated to suppress apoptosis besides regulating the growth of LNCaP cells [200]. In the cell culture in vitro study, no sign of necrosis was established in foetal bovine serum of the control group and nourishment and workout subjected group [194]. Very little apoptosis was noticed in serum-stimulated LNCaP cells because of higher IGF-I and lower IGFBP-1 in the control group. This study also clarifies the reason why the higher threat of prostate malignancy is ubiquitous in the United States [194].

The exercise including calisthenics, swimming, and strenuous training of 40–50 minutes at a continuous stretch showed lower levels of serum insulin and IGF-I, whereas IGFBP-1 was advanced in equally exercising and exercise plus diet groups. The rate of apoptosis was higher in nourishment plus workout trials than in the other groups [201].

Smooth functioning of the p53 gene is a characteristic feature of every prostate cancer at its early stage which is found to become defective in the later stages [200]. IGF-1 supressess the normal function of p53 thereby activating the downstream signalling axes associated with cell cycle arrest, DNA repair and apoptosis [202].

In a study, patients with prostate malignancy were prescribed with very low-fat diet including 10% fats from soy, fish oil containing omega-3 fatty acid (3 g/day) and vitamin E (400 IU/day) followed by aerobic exercises like walking and jogging for 30–60 min (6 days/week) and yoga and breathing exercises (1 h/day) [203, 204]. The serum collected from these patients displayed a reduction in LNCaP cell growth by 60% in the nourishment plus workout set [204].

Intake of lycophene-rich food and tomatoes can diminish the threat of prostate malignancy and its growth [205–207]. Foods like guava, apricots, watermelon, pink grapefruit, brussels sprouts, cabbage, broccoli, cauliflower, watercress, kale, onions, leeks, garlic, scallions, and chives may also reduce the risk of this ailment [208–210]. Genistein and biochanin A are the precursors derived from soy that prohibit the growth of hormone-insensitive and sensitive cancer cell lines by decreasing the level of PSA in LNCaP prostate-cancer cell lines [211]. Genistein interferes with the tyrosine kinase growth receptors and receptors disturbing topoisomerase II, hindering angiogenesis, and stimulating apoptosis [211]. The levels of testosterone were noted to be decreasing in males from Japan who were taking genistein [212]. A slight change in the concentration of these hormones in the serum for months, years, and decades slows down the progress of prostate cancer [189, 190]. Several studies including men with Gleason scores below 7 who were prescribed soy in the diet and subjected to physical training gave positive results after one year of treatment for prostate cancer [211]. Flaxseed, canola oil, soybean oils, and flaxseed oil comprise alphalinolenic acid (ALA), which is an omega-3 fatty acid, and lignans which is a plant oestrogen. The lignans showed anti-cancerous activity in vitro in prostate cancer cell lines [76, 213]. A cohort study in the Netherlands showed that an increase in the intake of ALA reduced the threat of prostate malignancy [214]. Another research reported that the intake of flaxseed by males suffering from prostate cancer with a Gleason score below 6 showed reduced cancer cell proliferation and preferment of apoptosis in the prostate gland besides reducing the cholesterol levels [215]. Fish contains eicosapentaenoic acid (EPA), docosahexaenoic acid (DHA), and vitamin D. The consumption of fish oil lowers the threat of prostate malignancy [216–218]. Fish consumption includes preserved tuna, dark fish meat like mackerel, salmon, swordfish, bluefish, sardines, and others which reduced prostate cancer by 44% [219]. Nuts have elevated levels of omega-3-fatty acids, vitamin E, flavonoids, minerals like magnesium, potassium and selenium [210, 220] and are proved in lowering the threat of prostate malignancy [210]. Some epidemiological studies reported that physical workouts dipped the threat of prostate malignancy as obesity is linked with the disease [221]. Physical exercise boosts the mental health of the patient distressed from prostate malignancy and makes the body stronger to withstand the treatment the patient has to undergo during the therapy [222]. Reduction of the smoking habit has been reported to have positive results in curing prostate malignancy [223–225]. Some dietary supplements like the addition of zinc to diet show beneficial results in prostate malignancy risk [226].

3.2.5 Lung Carcinoma

This type of cancer has a high proliferation rate and low survival chances but can be restricted by nutritional diets and lifestyle modifications. Various nutritional diets can prevent and treat these diseases. Different lifestyle activities, like smoking cigarettes and intake of tobacco in various form, increases the oxidative stress in the internal organs thereby triggering the process of lung carcinoma [227]. There are different bioactive factors derived from adipose tissues which have a significant biological role in implicating lung cancer. Some adipocytokines induce the process which results in the development and proliferation of lung carcinoma. Besides, it should also be noted that all the adipokines are not tumour stimulants. Adiponectin regulates the cell growth and concentrations of inflammatory cytokines [228]. Increase in the levels of adiponectin by nutritional supplements has been used as a newly introduced therapeutic strategy for fighting this disease. Intake of foods rich in antioxidants and anti-inflammatory potential is showing fruitful results in the prevention of the deadly ailment [228]. Oxidative stress is the major factor in the carcinogenic process. It initiates, promotes, and progresses the process of cancer [229]. It is mainly the reactive nitrogen species (RNS) and reactive oxygen species (ROS) which are not compensated by the antioxidant system of the cells [230]. The enzymatic antioxidant defence system, which gets dismantled by ageing, malnutrition, and lack of physical exercise [231], is the root cause of higher ROS levels. The reactive oxygen species promote the expression of genes that control the differentiation and growth of the cell. The hyperactivity of the reactive oxygen species is generally caused due to mutation and chromosomal aberrations which leads to the degeneration of cells and stimulates carcinoma [232]. Smoking cigarettes creates inflammation in the lung epithelium which produces inflammatory cytokines and ROS resulting in oxidative stress which becomes the main aspect for the progression of tumours [233–237]. Several air pollutants activate the MAP kinase signalling pathway which produces inflammation in the lungs thereby causing carcinogenic effects in non-smokers [238]. Obesity on the other hand promotes carcinoma as the adipocytes trigger cell proliferation. Several studies support the hypothesis that adiponectin (Acrp 30), an oligomer, helps in preventing lung cancer by inhibiting cell growth and concentration of inflammatory cytokines [239]. In a study, it was revealed that the consumption of cereals, vegetables, fruits, and dietary fibres in high quantities lowers the threat of lung carcinoma [240]. The utmost effective is the ketogenic diet that increases the levels of adiponectin [241]. The addition of medicinal plants and bioactive compounds in the diet like anthocyanins, walnut, curcumin, soy, resveratrol, and dihydromyricetin increases the levels of adiponectin in blood which prevents lung cancer [242]. In earlier studies, it is established that the Mediterranean diet lowers the threat of cancer and reduces the death rate [243]. These include mainly vegetables, fruits, seeds and nuts, olive oil, legumes, dairy products, eggs, poultry, and wine which are rich antioxidants and decrease the concentration of reactive oxygen species and detoxify the system [244]. Foods with flavonoid contents like whole grains, olive oil, vegetables, fruit, nuts, coffee, tea, and red wine have high free radical scavenging property which protects the DNA from damage [245–248]. Cinnamal aldehyde from cinnamon oils overpowers the initiation of NF-$_{\kappa}$B in inflammation by TNFα [249]. Foods rich in carotenoids are natural antioxidants which neutralize the reactive oxygen species [250]. The presence of phytochemicals like isothiocyanate and phenolic compounds in plant-based food helps to detoxify the enzymes, metabolizes the carcinogens, and improves the immune system [251]. The intake of non-starchy food and food comprising selenium and quercetin lowers the threat of lung carcinoma [228]. A Danish study shows the higher the intake of nutritional food, the lower the threat of lung cancer which was supported by a similar study in Finland, where the intake of food rich in carotenoids lowered the threat of lung carcinoma [252, 253]. Olive oils and sunflower oils contain α-tocopherol which helps in the reduction of the risk of lung cancer [254]. Phytoestrogens and glucosinolate hydrolysed products exist in many fruits and vegetables and help to reduce the process of lung carcinogenesis [255, 256]. There are also roles of vitamin B, vitamin C, and phytochemicals in modulating epigenetics which provides promising interventions in the chemoprevention of lung carcinoma [257–260]. The intake of folate-rich foods shows positive results in smokers as it methylates the promoter of the tumour suppressor gene [261, 262]. Vitamins B_{12} and D, magnesium, manganese, niacin, and omega-3 fatty acids protect the methylation of gene and also prove to be potent anti-inflammatory agents that show an essential part in the protection of carcinogenesis of lungs [263, 264]. Figure 2 shows a schematic diagram related to the prevention of lung carcinoma. A decrease in the intake of red meat shows a positive effect in preventing lung cancer [265] as it produces polycyclic aromatic hydrocarbons (PAHs) and heterocyclic amines (HCAs) during the process of catering which have carcinogenic properties at higher temperatures [266, 267]. Bacteria in cheese and dairy products produce lactic acid and can influence the initiation of lung cancer by hampering the immune system [268].

4 Modifying Lifestyle Reduces the Risk of Developing Cancer

The underlying common denominator among most malignancies appears to be lifestyle, including smoking, a diet high in fat and processed sugar, and lack of exercise. The majority of cancers may be prevented by altering these lifestyle variables, which are simple to do. One could significantly alter growth factors and hormones known to be linked to cancer by consuming a diet high in fruits, vegetables, and whole grains with little to no meat, predominantly chicken or cold-water fish, and engaging in 45–60 min of daily exercise [82]. After adjusting for smoking behaviours and cigarette use, a large Norwegian study with 81,516 men and women over the course of 19 years reported that males who walked or cycled for at least 4 h per week had a 25% lower chance of developing lung cancer [269]. Beta-carotene has the capacity to act as an antioxidant; other micronutrients with this capability, in particular vitamins E and C and selenium, have also been reported to lower the chance of developing lung cancer [270]. About 50,620 incident malignancies accumulated throughout the 10.1 years of follow-up for a total of 755,459 individuals (median age, 62 years [range, 32–91 years]; 53% female). Seven out of 15 cancer types examined had a statistically significant lower risk in people engaged in the recommended levels of physical activity (7.5–15 MET h/week), including breast cancer (6–10% lower risk), colon cancer (8–14% lower risk in men), kidney cancer (11–17% lower risk), endometrial cancer (10–17% lower risk), liver cancer (18–27% lower risk), myeloma (14–23% lower risk), and

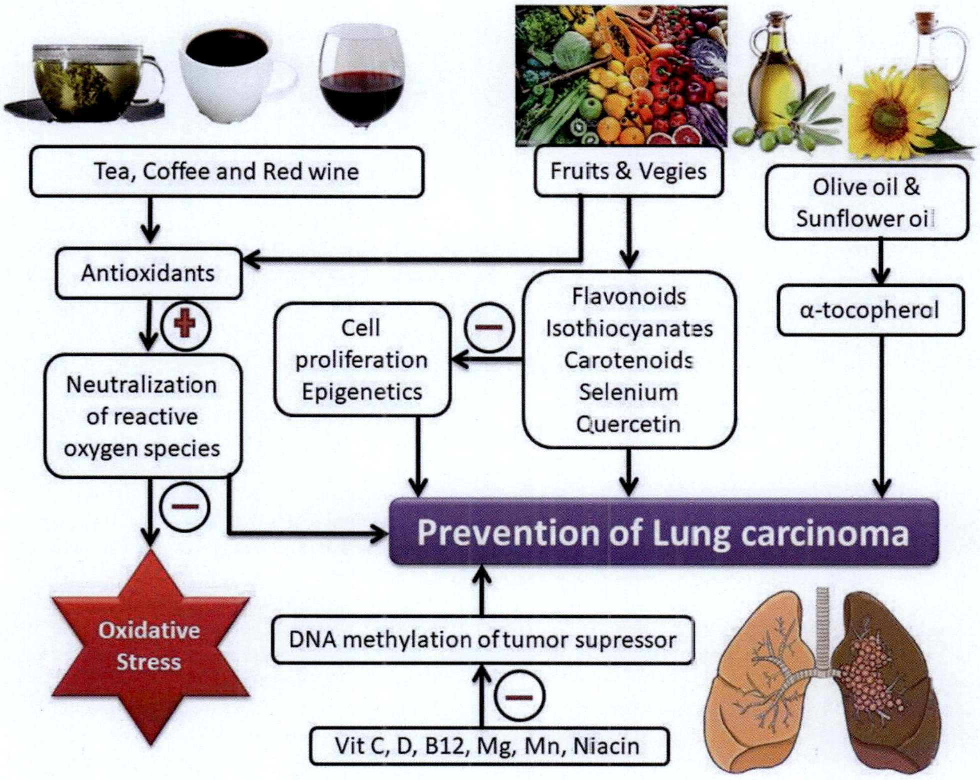

Fig. 2 Diagrammatic representation showing different sources of food preventing lung carcinoma ('+' = accelerates; '–' = retards)

non-Hodgkin lymph (11–18% lower risk in women) [271]. Recreational physical activity is linked to a lower risk of rectal cancer in older men and women as well as a significantly lower risk of colon cancer as time spent participating in it increases [272, 273]. However, there is evidence that suggests men who engage in high levels of vigorous activity may have a decreased chance of developing metastatic prostate cancer. Physical activity is unlikely to have a significant impact on the incidence of all cases of prostate cancer [274, 275]. Changes in lifestyle are crucial for preventing breast cancer. Such adjustments include losing weight, moving to a healthy diet, cutting out needless medications, etc. Additionally, lifestyle modifications can lower the recurrence rate and improve survival rates in women who have been diagnosed with and are receiving treatment for breast cancer [120, 276–278].

5 Summary

This chapter provides a vivid description of the protective mechanisms of lifestyle modifications and natural nutritional diets in mitigating cancer progression. This is evident that a diet high in carbohydrates and low fibre, red meat, smoking, and lack of physical activity plays a significant role in developing cancer. Intake diets possessing high fibres which decrease the chances of breast cancer and cardiovascular disease is recommended. Furthermore, a diet high in carbohydrates increases the incidence of breast, colon, ovarian, and endometrial cancers, while an increased consumption of red meat increases the prevalence of colorectal, endometrial, and lung cancers. Additionally, lignan and phytoestrogen from flaxseeds also play an important role in reducing the progression of breast and prostate cancer. Overall, adopting a diet containing whole grains, fruits and vegetables, high fibres, low carbohydrates, and daily exercise will potentially alter the hormones and growth factors associated with cancer signalling axes, finally lowering the progression of cancer.

References

1. National Institutes of Health (US); Biological Sciences Curriculum Study. In: NIH curriculum supplement series. Bethesda: National Institutes of Health (US); 2007. Available from https://www.ncbi.nlm.nih.gov/books/NBK20362/.
2. Sung H, Ferlay J, Siegel RL, Laversanne M, Soerjomataram I, Jemal A, Bray F. Global cancer statistics 2020: GLOBOCAN estimates of incidence and mortality worldwide for 36 cancers in 185 countries. CA Cancer J Clin. 2021;71(3):209–49. https://doi.org/10.3322/caac.21660.
3. Zugazagoitia J, Guedes C, Ponce S, Ferrer I, Molina-Pinelo S, Paz-Ares L. Current challenges in cancer treatment. Clin Ther. 2016;7:1551–66. https://doi.org/10.1016/j.clinthera.2016.03.026.
4. Shigeta H, Shigeta M, Nakazawa A, Nakamura N, Yoshikawa T. Lifestyle, obesity, and insulin resistance. Diabetes Care. 2001;24(3):608. https://doi.org/10.2337/diacare.24.3.608.

5. Weiderpass E. Lifestyle and cancer risk. J Prev Med Public Health. 2010;43(6):459–71. https://doi.org/10.3961/jpmph.2010.43.6.459.
6. Parkin DM. The global health burden of infection-associated cancers in the year 2002. Int J Cancer. 2006;118(12):3030–44. https://doi.org/10.1002/ijc.21731.
7. Anand P, Kunnumakkara AB, Sundaram C, Harikumar KB, Tharakan ST, Lai OS, Sung B, Aggarwal BB. Cancer is a preventable disease that requires major lifestyle changes. Pharm Res. 2008;25(9):2097–116. https://doi.org/10.1007/s11095-008-9661-9.
8. Sloan FA, Gelband H, Institute of Medicine (US) Committee on Cancer Control in Low- and Middle-Income Countries. Cancer control opportunities in low- and middle-income countries. Washington: National Academies Press; 2007. https://www.ncbi.nlm.nih.gov/books/NBK54030/.
9. Song M, Garrett WS, Chan AT. Nutrients, foods, and colorectal cancer prevention. Gastroenterology. 2015;148(6):1244–60. https://doi.org/10.1053/j.gastro.2014.12.035.
10. Park OJ, Surh YJ. Chemopreventive potential of epigallocatechin gallate and genistein: evidence from epidemiological and laboratory studies. Toxicol Lett. 2004;150(1):43–56. https://doi.org/10.1016/j.toxlet.2003.06.001.
11. Hanahan D, Weinberg RA. Hallmarks of cancer: the next generation. Cell. 2011;144(5):646–74. https://doi.org/10.1016/j.cell.2011.02.013.
12. Cairns RA, Harris IS, Mak TW. Regulation of cancer cell metabolism. Nat Rev Cancer. 2011;11:85–95. https://doi.org/10.1038/nrc2981.
13. Henley SA, Dick FA. The retinoblastoma family of proteins and their regulatory functions in the mammalian cell division cycle. Cell Div. 2012;7:10. https://doi.org/10.1186/1747-1028-7-10.
14. Bieging KT, Mello SS, Attardi LD. Unravelling mechanisms of p53-mediated tumor suppression. Nat Rev Cancer. 2014;14(5):359–70. https://doi.org/10.1038/nrc3711.
15. Giacinti C, Giordano A. RB and cell cycle progression. Oncogene. 2006;25(38):5220–7. https://doi.org/10.1038/sj.onc.1209615.
16. Umen JG, Goodenough UW. Control of cell division by a retinoblastoma protein homolog in *Chlamydomonas*. Genes Dev. 2001;15(13):1652–61. https://doi.org/10.1101/gad.892101.
17. Nakamura Y, Ozaki T, Niizuma H, Ohira M, Kamijo T, Nakagawara A. Functional characterization of a new p53 mutant generated by homozygous deletion in a neuroblastoma cell line. Biochem Biophys Res Commun. 2007;354(4):892–8. https://doi.org/10.1016/j.bbrc.2007.01.057.
18. Ozaki T, Nakagawara A. Role of p53 in cell death and human cancers. Cancer. 2011;3(1):994–1013. https://doi.org/10.3390/cancers3010994.
19. Rivlin N, Brosh R, Oren M, Rotter V. Mutations in the p53 tumor suppressor gene: important milestones at the various steps of tumorigenesis. Genes Cancer. 2011;2(4):466–74. https://doi.org/10.1177/1947601911408889.
20. Bugyik E, Renyi-Vamos F, Szabo V, Dezso K, Ecker N, Rokusz A, Nagy P, Dome B, Paku S. Mechanisms of vascularization in murine models of primary and metastatic tumor growth. Chin J Cancer. 2016;35:19. https://doi.org/10.1186/s40880-016-0083-5.
21. Nishida N, Yano H, Nishida T, Kamura T, Kojiro M. Angiogenesis in cancer. Vasc Health Risk Manag. 2006;2(3):213–9. https://doi.org/10.2147/vhrm.2006.2.3.213.
22. Veikkola T, Alitalo K. VEGFs, receptors and angiogenesis. Semin Cancer Biol. 1999;9(3):211–20. https://doi.org/10.1006/scbi.1998.0091.
23. Bello L, Francolini M, Marthyn P, Zhang J, Carroll RS, Nikas DC, Strasser JF, Villani R, Cheresh DA, Black PM. Alpha(v)beta3 and alpha(v)beta5 integrin expression in glioma periphery. Neurosurgery. 2001;49(2):380–9. https://doi.org/10.1097/00006123-200108000-00022.
24. Desgrosellier JS, Cheresh DA. Integrins in cancer: biological implications and therapeutic opportunities. Nat Rev Cancer. 2010;10(1):9–22. https://doi.org/10.1038/nrc2748.
25. Hamidi H, Ivaska J. Every step of the way: integrins in cancer progression and metastasis. Nat Rev Cancer. 2018;18:533–48. https://doi.org/10.1038/s41568-018-0038-z.
26. Alberts B, Johnson A, Lewis J, Raff M, Roberts K, Walter P. Molecular biology of the cell. 4th ed. New York: Garland Science; 2002. https://doi.org/10.1002/bmb.2003.494031049999.
27. Donaldson MS. Nutrition and cancer: a review of the evidence for an anti-cancer diet. Nutr J. 2004;3:19. https://doi.org/10.1186/1475-2891-3-19.
28. Da Ponte KF, Berro DH, Collet S, Constans JM, Emery E, Valable S, Guillamo JS. In vivo relationship between hypoxia and angiogenesis in human glioblastoma: a multimodal imaging study. J Nucl Med. 2017;58:1574–9. https://doi.org/10.2967/jnumed.116.188557.
29. Doglioni G, Parik S, Fendt SM. Interactions in the (pre)metastatic niche support metastasis formation. Front Oncol. 2019;9:219. https://doi.org/10.3389/fonc.2019.00219.
30. Cacace A, Sboarina M, Vazeille T, Sonveaux P. Glutamine activates STAT3 to control cancer cell proliferation independently of glutamine metabolism. Oncogene. 2017;36:2074–84. https://doi.org/10.1038/onc.2016.364.
31. Pan M, Reid MA, Lowman XH, Kulkarni RP, Tran TQ, Liu X, Yang Y, Hernandez-Davies JE, Rosales KK, Li H, et al. Regional glutamine deficiency in tumors promotes dedifferentiation through inhibition of histone demethylation. Nat Cell Biol. 2016;18:1090–101. https://doi.org/10.1038/ncb3410.
32. McRae MP. The benefits of dietary fiber intake on reducing the risk of cancer: an umbrella review of meta-analyses. J Chiropr Med. 2018;17(2):90–6. https://doi.org/10.1016/j.jcm.2017.12.001.
33. Chen S, Chen Y, Ma S, Zheng R, Zhao P, Zhang L, Liu Y, Yu Q, Deng Q, Zhang K. Dietary fibre intake and risk of breast cancer: a systematic review and meta-analysis of epidemiological studies. Oncotarget. 2016;7(49):80980–9. https://doi.org/10.18632/oncotarget.13140.
34. Dong JY, He K, Wang P, Qin LQ. Dietary fiber intake and risk of breast cancer: a meta-analysis of prospective cohort studies. Am J Clin Nutr. 2011;94(3):900–5. https://doi.org/10.3945/ajcn.111.015578.
35. Kim Y, Je Y. Dietary fibre intake and mortality from cardiovascular disease and all cancers: a meta-analysis of prospective cohort studies. Arch Cardiovasc Dis. 2016;109(1):39–54. https://doi.org/10.1016/j.acvd.2015.09.005.
36. Augustin LS, Gallus S, Negri E, La Vecchia C. Glycemic index, glycemic load and risk of gastric cancer. Ann Oncol. 2004;15:581–4. https://doi.org/10.1093/annonc/mdh130.
37. Augustin LS, Gallus S, Franceschi S, Negri E, Jenkins DJ, Kendall CW, Dal Maso L, Talamini R, La Vecchia C. Glycemic index and load and risk of upper aero-digestive tract neoplasms (Italy). Cancer Causes Control. 2003;14:657–62. https://doi.org/10.1023/a:1025676907942.
38. Augustin LS, Polesel J, Bosetti C, Kendall CW, La Vecchia C, Parpinel M, Conti E, Montella M, Franceschi S, Jenkins DJ, Dal Maso L. Dietary glycemic index, glycemic load and ovarian cancer risk: a case-control study in Italy. Ann Oncol. 2003;14:78–84. https://doi.org/10.1093/annonc/dkg022.
39. Augustin LS, Gallus S, Bosetti C, Levi F, Negri E, Franceschi S, Dal Maso L, Jenkins DJ, Kendall CW, La Vecchia C. Glycemic index and glycemic load in endometrial cancer. Int J Cancer. 2003;105:404–7. https://doi.org/10.1002/ijc.11089.
40. Franceschi S, Dal Maso L, Augustin L, Negri E, Parpinel M, Boyle P, Jenkins DJ, La Vecchia C. Dietary glycemic load and

colorectal cancer risk. Ann Oncol. 2001;12:173–8. https://doi.org/10.1023/a:1008304128577.

41. Slattery ML, Boucher KM, Caan BJ, Potter JD, Ma KN. Eating patterns and risk of colon cancer. Am J Epidemiol. 1998;148:4–16. https://doi.org/10.1093/aje/148.1.4-a.
42. Schoen RE, Tangen CM, Kuller LH, Burke GL, Cushman M, Tracy RP, Dobs A, Savage PJ. Increased blood glucose and insulin, body size, and incident colorectal cancer. J Natl Cancer Inst. 1999;91:1147–54. https://doi.org/10.1093/jnci/91.13.1147.
43. Calle EE, Murphy TK, Rodriguez C, Thun MJ, Heath CW Jr. Diabetes mellitus and pancreatic cancer mortality in a prospective cohort of United States adults. Cancer Causes Control. 1998;9:403–10. https://doi.org/10.1023/a:1008819701485.
44. Coughlin SS, Calle EE, Teras LR, Petrelli J, Thun MJ. Diabetes mellitus as a predictor of cancer mortality in a large cohort of US adults. Am J Epidemiol. 2004;159:1160–7. https://doi.org/10.1093/aje/kwh161.
45. Colangelo LA, Gapstur SM, Gann PH, Dyer AR, Liu K. Colorectal cancer mortality and factors related to the insulin resistance syndrome. Cancer Epidemiol Biomark Prev. 2002;11:385–91. https://pubmed.ncbi.nlm.nih.gov/11927499/.
46. La Vecchia C, Negri E, Decarli A, Franceschi S. Diabetes mellitus and colorectal cancer risk. Cancer Epidemiol Biomark Prev. 1997;6:1007–10. https://pubmed.ncbi.nlm.nih.gov/9419395/.
47. Sandhu MS, Luben R, Khaw KT. Self-reported non-insulin dependent diabetes, family history, and risk of prevalent colorectal cancer: population based, cross sectional study. J Epidemiol Community Health. 2001;55:804–5. https://doi.org/10.1136/jech.55.11.804.
48. Anderson KE, Anderson E, Mink PJ, Hong CP, Kushi LH, Sellers TA, Lazovich D, Folsom AR. Diabetes and endometrial cancer in the Iowa women's health study. Cancer Epidemiol Biomark Prev. 2001;10:611–6. https://pubmed.ncbi.nlm.nih.gov/11401910/.
49. Cho E, Spiegelman D, Hunter DJ, Chen WY, Colditz GA, Willett WC. Premenopausal dietary carbohydrate, glycemic index, glycemic load, and fiber in relation to risk of breast cancer. Cancer Epidemiol Biomark Prev. 2003;12(11):1153–8. https://pubmed.ncbi.nlm.nih.gov/14652274/.
50. Cho E, Spiegelman D, Hunter DJ, Chen WY, Stampfer MJ, Colditz GA, Willett WC. Premenopausal fat intake and risk of breast cancer. J Natl Cancer Inst. 2003;95(14):1079–85. https://doi.org/10.1093/jnci/95.14.1079.
51. Wen W, Shu XO, Li H, Yang G, Ji BT, Cai H, Gao YT, Zheng W. Dietary carbohydrates, fiber, and breast cancer risk in Chinese women. Am J Clin Nutr. 2009;89(1):283–9. https://doi.org/10.3945/ajcn.2008.26356.
52. Aune D, Chan DS, Lau R, Vieira R, Greenwood DC, Kampman E, Norat T. Carbohydrates, glycemic index, glycemic load, and colorectal cancer risk: a systematic review and meta-analysis of cohort studies. Cancer Causes Control. 2012;23(4):521–35. https://doi.org/10.1007/s10552-012-9918-9.
53. Hanson S, Thorpe G, Winstanley L. Omega-3, omega-6 and total dietary polyunsaturated fat on cancer incidence: systematic review and meta-analysis of randomised trials. Br J Cancer. 2020;122:1260–70. https://doi.org/10.1038/s41416-020-0761-6.
54. Bernard-Gallon DJ, Vissac-Sabatier C, Antoine-Vincent D, Rio PG, Maurizis JC, Fustier P, Bignon YJ. Differential effects of n-3 and n-6 polyunsaturated fatty acids on BRCA1 and BRCA2 gene expression in breast cell lines. Br J Nutr. 2002;87(4):281–9. https://doi.org/10.1079/bjnbjn2002522.
55. Freitas RDS, Campos MM. Protective effects of omega-3 fatty acids in cancer-related complications. Nutrients. 2019;11(5):945. https://doi.org/10.3390/nu11050945.
56. Laviano A, Rianda S, Molfino A, Rossi Fanelli F. Omega-3 fatty acids in cancer. Curr Opin Clin Nutr Metab Care. 2013;16(2):156–61. https://doi.org/10.1097/mco.0b013e32835d2d99.
57. Murphy RA, Mourtzakis M, Mazurak VC. n-3 polyunsaturated fatty acids: the potential role for supplementation in cancer. Curr Opin Clin Nutr Metab Care. 2012;15(3):246–51. https://doi.org/10.1097/mco.0b013e328351c32f.
58. Pardini RS. Nutritional intervention with omega-3 fatty acids enhances tumor response to anti-neoplastic agents. Chem Biol Interact. 2006;162(2):89–105. https://doi.org/10.1016/j.cbi.2006.05.012.
59. Wirfält E, Mattisson I, Gullberg B, Johansson U, Olsson H, Berglund G. Postmenopausal breast cancer is associated with high intakes of omega6 fatty acids (Sweden). Cancer Causes Control. 2002;13(10):883–93. https://doi.org/10.1023/a:1021922917489.
60. Farvid MS, Stern MC, Norat T, Sasazuki S, Vineis P, Weijenberg MP, Wolk A, Wu K, Stewart BW, Cho E. Consumption of red and processed meat and breast cancer incidence: a systematic review and meta-analysis of prospective studies. Int J Cancer. 2018;143(11):2787–99. https://doi.org/10.1002/ijc.31848.
61. Farvid MS, Sidahmed E, Spence ND, ManteAngua K, Rosner BA, Barnett JB. Consumption of red meat and processed meat and cancer incidence: a systematic review and meta-analysis of prospective studies. Eur J Epidemiol. 2021;36(9):937–51. https://doi.org/10.1007/s10654-021-00741-9.
62. Chiavarini M, Bertarelli G, Minelli L, Fabiani R. Dietary intake of meat cooking-related mutagens (HCAs) and risk of colorectal adenoma and cancer: a systematic review and meta-analysis. Nutrients. 2017;9(5):514. https://doi.org/10.3390/nu9050514.
63. Chan DS, Lau R, Aune D, Vieira R, Greenwood DC, Kampman E, Norat T. Red and processed meat and colorectal cancer incidence: meta-analysis of prospective studies. PLoS One. 2011;6(6):e20456. https://doi.org/10.1371/journal.pone.0020456.
64. Keszei AP, Schouten LJ, Goldbohm RA, van den Brandt PA. Red and processed meat consumption and the risk of esophageal and gastric cancer subtypes in The Netherlands Cohort Study. Ann Oncol. 2012;23(9):2319–26. https://doi.org/10.1093/annonc/mdr615.
65. Kim SR, Kim K, Lee SA, Kwon SO, Lee JK, Keum N, Park SM. Effect of red, processed, and white meat consumption on the risk of gastric cancer: an overall and dose-response meta-analysis. Nutrients. 2019;11(4):826. https://doi.org/10.3390/nu11040826.
66. Zhao Z, Yin Z, Zhao Q. Red and processed meat consumption and gastric cancer risk: a systematic review and meta-analysis. Oncotarget. 2017;8(18):30563–75. https://doi.org/10.18632/oncotarget.15699.
67. Cross AJ, Freedman ND, Ren J, Ward MH, Hollenbeck AR, Schatzkin A, Sinha R, Abnet CC. Meat consumption and risk of esophageal and gastric cancer in a large prospective study. Am J Gastroenterol. 2011;106(3):432–42. https://doi.org/10.1038/ajg.2010.415.
68. Qu X, Ben Q, Jiang Y. Consumption of red and processed meat and risk for esophageal squamous cell carcinoma based on a meta-analysis. Ann Epidemiol. 2013;23(12):762–70. https://doi.org/10.1016/j.annepidem.2013.09.003.
69. Salehi M, Moradi-Lakeh M, Salehi MH, Nojomi M, Kolahdooz F. Meat, fish, and esophageal cancer risk: a systematic review and dose-response meta-analysis. Nutr Rev. 2013;71(5):257–67. https://doi.org/10.1111/nure.12028.
70. Zhu HC, Yang X, Xu LP, Zhao LJ, Tao GZ, Zhang C, Qin Q, Cai J, Ma JX, Mao WD, Zhang XZ, Cheng HY, Sun XC. Meat consumption is associated with esophageal cancer risk in a meat- and cancer-histological-type dependent manner. Dig Dis Sci. 2014;59(3):664–73. https://doi.org/10.1007/s10620-013-2928-y.
71. Cunnane SC, Ganguli S, Menard C, Liede AC, Hamadeh MJ, Chen ZY, Wolever TM, Jenkins DJ. High alpha-linolenic acid flaxseed (Linumusitatissimum): some nutritional properties in humans. Br J Nutr. 1993;69(2):443–53. https://doi.org/10.1079/bjn19930046.

72. Flower G, Fritz H, Balneaves LG, Verma S, Skidmore B, Fernandes R, Kennedy D, Cooley K, Wong R, Sagar S, Fergusson D, Seely D. Flax and breast cancer: a systematic review. Integr Cancer Ther. 2014;13(3):181–92. https://doi.org/10.1177/1534735413502076.
73. Thompson LU, Rickard SE, Orcheson LJ, Seidl MM. Flaxseed and its lignan and oil components reduce mammary tumor growth at a late stage of carcinogenesis. Carcinogenesis. 1996;17(6):1373–6. https://doi.org/10.1093/carcin/17.6.1373.
74. Li D, Yee JA, Thompson LU, Yan L. Dietary supplementation with secoisolariciresinoldiglycoside (SDG) reduces experimental metastasis of melanoma cells in mice. Cancer Lett. 1999;142(1):91–6. https://doi.org/10.1016/s0304-3835(99)00158-5.
75. Yan L, Yee JA, Li D, McGuire MH, Thompson LU. Dietary flaxseed supplementation and experimental metastasis of melanoma cells in mice. Cancer Lett. 1998;124(2):181–6. https://doi.org/10.1016/s0304-3835(97)00470-9.
76. Lin X, Gingrich JR, Bao W, Li J, Haroon ZA, Demark-Wahnefried W. Effect of flaxseed supplementation on prostatic carcinoma in transgenic mice. Urology. 2002;60(5):919–24. https://doi.org/10.1016/s0090-4295(02)01863-0.
77. Fleischauer AT, Poole C, Arab L. Garlic consumption and cancer prevention: meta-analyses of colorectal and stomach cancers. Am J Clin Nutr. 2000;72(4):1047–52. https://doi.org/10.1093/ajcn/72.4.1047.
78. Fowke JH, Chung FL, Jin F, Qi D, Cai Q, Conaway C, Cheng JR, Shu XO, Gao YT, Zheng W. Urinary isothiocyanate levels, brassica, and human breast cancer. Cancer Res. 2003;63(14):3980–6. https://pubmed.ncbi.nlm.nih.gov/12873994/.
79. London SJ, Yuan JM, Chung FL, Gao YT, Coetzee GA, Ross RK, Yu MC. Isothiocyanates, glutathione S-transferase M1 and T1 polymorphisms, and lung-cancer risk: a prospective study of men in Shanghai, China. Lancet. 2000;356(9231):724–9. https://doi.org/10.1016/s0140-6736(00)02631-3.
80. Fowke JH, Gao YT, Chow WH, Cai Q, Shu XO, Li HL, Ji BT, Rothman N, Yang G, Chung FL, Zheng W. Urinary isothiocyanate levels and lung cancer risk among non-smoking women: a prospective investigation. Lung Cancer. 2011;73(1):18–24. https://doi.org/10.1016/j.lungcan.2010.10.024.
81. Diallo A, Deschasaux M, Galan P, Hercberg S, Zelek L, Latino-Martel P, Touvier M. Associations between fruit, vegetable and legume intakes and prostate cancer risk: results from the prospective SupplémentationenVitamines et MinérauxAntioxydants (SU.VI.MAX) cohort. Br J Nutr. 2016;115(9):1579–85. https://doi.org/10.1017/s0007114516000520.
82. Barnard RJ. Prevention of cancer through lifestyle changes. Evid Based Complement Alternat Med. 2004;1(3):233–9.
83. Martínez ME, Jacobs ET, Ashbeck EL, Sinha R, Lance P, Alberts DS, Thompson PA. Meat intake, preparation methods, mutagens and colorectal adenoma recurrence. Carcinogenesis. 2007;28(9):2019–27. https://doi.org/10.1093/carcin/bgm179.
84. Peters U, Sinha R, Chatterjee N, Subar AF, Ziegler RG, Kulldorff M, Bresalier R, Weissfeld JL, Flood A, Schatzkin A, Hayes RB, Prostate, Lung, Colorectal, and Ovarian Cancer Screening Trial Project Team. Dietary fibre and colorectal adenoma in a colorectal cancer early detection programme. Lancet. 2003;361(9368):1491–5. https://doi.org/10.1016/S0140-6736(03)13173-X.
85. Giovannucci E. Diet, body weight, and colorectal cancer: a summary of the epidemiologic evidence. J Women's Health. 2003;12:173–82. https://doi.org/10.1089/154099903321576574.
86. Komninou D, Ayonote A, Richie JP Jr, Rigas B. Insulin resistance and its contribution to colon carcinogenesis. Exp Biol Med. 2003;228:396–405. https://doi.org/10.1177/153537020322800410.
87. Yu H, Rohan T. Role of the insulin-like growth factor family in cancer development and progression. J Natl Cancer Inst. 2000;92(18):1472–89. https://doi.org/10.1093/jnci/92.18.1472.
88. Slattery ML, Curtin K, Ma K, Edwards S, Schaffer D, Anderson K, Samowitz W. Diet, activity, and lifestyle associations with p53 mutations in colon tumors. Cancer Epidemiol Biomark Prev. 2002;11(6):541–8.
89. Feregrino-Pérez AA, Berumen LC, García-Alcocer G, Guevara-Gonzalez RG, Ramos-Gomez M, Reynoso-Camacho R, Acosta-Gallegos JA, Loarca-Piña G. Composition and chemopreventive effect of polysaccharides from common beans (*Phaseolus vulgaris* L.) on azoxymethane-induced colon cancer. J Agric Food Chem. 2008;56:8737–44. https://doi.org/10.1021/jf8007162.
90. Kaczmarczyk MM, Miller MJ, Freund GG. The health benefits of dietary fiber: beyond the usual suspects of type 2 diabetes mellitus, cardiovascular disease and colon cancer. Metab Clin Exp. 2012;61:1058–66. https://doi.org/10.1016/j.metabol.2012.01.017.
91. Reddy BS, Hirose Y, Cohen LA, Simi B, Cooma I, Rao CV. Preventive potential of wheat bran fractions against experimental colon carcinogenesis: implications for human colon cancer prevention. Cancer Res. 2000;60:4792–7.
92. Surh YJ. Cancer chemoprevention with dietary phytochemicals. Nat Rev Cancer. 2003;3:768–80. https://doi.org/10.1038/nrc1189.
93. Bingham SA, Day NE, Luben R, Ferrari P, Slimani N, Norat T, Clavel-Chapelon F, Kesse E, Nieters A, Boeing H, Tjønneland A, Overvad K, Martinez C, Dorronsoro M, Gonzalez CA, Key TJ, Trichopoulou A, Naska A, Vineis P, Tumino R, Krogh V, Bueno-de-Mesquita HB, Peeters PH, Berglund G, Hallmans G, Lund E, Skeie G, Kaaks R, Riboli E, European Prospective Investigation into Cancer and Nutrition. Dietary fibre in food and protection against colorectal cancer in the European Prospective Investigation into Cancer and Nutrition (EPIC): an observational study. Lancet. 2003;361:1496–501. https://doi.org/10.1016/s0140-6736(03)13174-1.
94. Giovannucci E. Epidemiologic studies of folate and colorectal neoplasia: a review. J Nutr. 2002;132(8 Suppl):2350S–5S. https://doi.org/10.1093/jn/132.8.2350S.
95. Majumdar AP, Kodali U, Jaszewski R. Chemopreventive role of folic acid in colorectal cancer. Front Biosci. 2004;9:2725–32. https://doi.org/10.2741/1431.
96. Larsson SC, Orsini N, Wolk A. Vitamin B6 and risk of colorectal cancer: a meta-analysis of prospective studies. JAMA. 2010;303(11):1077–83. https://doi.org/10.1001/jama.2010.263.
97. Wozniak K, Arabski M, Malecka-Panas E, Drzewoski J, Blasiak J. DNA damage in human colonic mucosa cells induced by bleomycin and the protective action of vitamin E. Cell Mol Biol Lett. 2003;9:31–45.
98. Błasiak J, Arabski M, Pertyński T, Małecka-Panas E, Woźniak K, Drzewoski J. DNA damage in human colonic mucosa cells evoked by nickel and protective action of quercetin - involvement of free radicals? Cell Biol Toxicol. 2002;18(4):279–88. https://doi.org/10.1023/a:1016059112829.
99. Ishikawa S, Tamaki S, Ohata M, Arihara K, Itoh M. Heme induces DNA damage and hyperproliferation of colonic epithelial cells via hydrogen peroxide produced by heme oxygenase: a possible mechanism of heme-induced colon cancer. Mol Nutr Food Res. 2010;54:1182–91. https://doi.org/10.1002/mnfr.200900348.
100. Knobel Y, Weise A, Glei M, Sendt W, Claussen U, PoolZobel BL. Ferric iron is genotoxic in non-transformed and pre-neoplastic human colon cells. Food Chem Toxicol. 2007;45:804–11. https://doi.org/10.1016/j.fct.2006.10.028.
101. O'Callaghan NJ, Toden S, Bird AR, Topping DL, Fenech M, Conlon MA. Colonocyte telomere shortening is greater with dietary red meat than white meat and is attenuated by resistant starch. Clin Nutr. 2012;31:60–4. https://doi.org/10.1016/j.clnu.2011.09.003.
102. Biasi F, Mascia C, Poli G. The contribution of animal fat oxidation products to colon carcinogenesis, through modulation of

TGF-β1 signaling. Carcinogenesis. 2008;29(5):890–4. https://doi.org/10.1093/carcin/bgn106.

103. Sugimura T, Wakabayashi K, Nakagama H, Nagao M. Heterocyclic amines: mutagens/carcinogens produced during cooking of meat and fish. Cancer Sci. 2004;95:290–9. https://doi.org/10.1111/j.1349-7006.2004.tb03205.x.
104. Velmurugan B, Gangar SC, Kaur M, Tyagi A, Deep G, Agarwal R. Silibinin exerts sustained growth suppressive effect against human colon carcinoma SW480 xenograft by targeting multiple signaling molecules. Pharm Res. 2010;27(10):2085–97. https://doi.org/10.1007/s11095-010-0207-6.
105. Rajamanickam S, Agarwal R. Natural products and colon cancer: current status and future prospects. Drug Dev Res. 2008;69:460–71. https://doi.org/10.1002/ddr.20276.
106. Ye SF, Hou ZQ, Zhong LM, Zhang QQ. Effect of curcumin on the induction of glutathione S-transferases and NADP(H):quinone oxidoreductase and its possible mechanism of action. Yao XueXue Bao. 2007;42:376–80.
107. Yogosawa S, Yamada Y, Yasuda S, Sun Q, Takizawa K, Sakai T. Dehydrozingerone, a structural analogue of curcumin, induces cell-cycle arrest at the G2/M phase and accumulates intracellular ROS in HT-29 human colon cancer cells. J Nat Prod. 2012;75(12):2088–93. https://doi.org/10.1021/np300465f.
108. Khan N, Mukhtar H. Multitargeted therapy of cancer by green tea polyphenols. Cancer Lett. 2008;269(2):269–80. https://doi.org/10.1016/j.canlet.2008.04.014.
109. Larsen CA, Dashwood RH. (-)-Epigallocatechin-3-gallate inhibits Met signaling, proliferation, and invasiveness in human colon cancer cells. Arch Biochem Biophys. 2010;501(1):52–7. https://doi.org/10.1016/j.abb.2010.03.017.
110. Thakur VS, Ruhul Amin AR, Paul RK, Gupta K, Hastak K, Agarwal MK, Jackson MW, Wald DN, Mukhtar H, Agarwal ML. p53-Dependent p21-mediated growth arrest pre-empts and protects HCT116 cells from PUMA-mediated apoptosis induced by EGCG. Cancer Lett. 2010;296(2):225–32. https://doi.org/10.1016/j.canlet.2010.04.012.
111. Chen X, Kokkotou EG, Mustafa N, Bhaskar KR, Sougioultzis S, O'Brien M, Pothoulakis C, Kelly CP. *Saccharomyces boulardii* inhibits ERK1/2 mitogen-activated protein kinase activation both in vitro and in vivo and protects against Clostridium difficile toxin A-induced enteritis. J Biol Chem. 2006;281(34):24449–54. https://doi.org/10.1074/jbc.M605200200.
112. Matsumoto M, Kurihara S, Kibe R, Ashida H, Benno Y. Longevity in mice is promoted by probiotic-induced suppression of colonic senescence dependent on upregulation of gut bacterial polyamine production. PLoS One. 2011;6(8):e23652. https://doi.org/10.1371/journal.pone.0023652.
113. Park E, Jeon GI, Park JS, Paik HD. A probiotic strain of *Bacillus polyfermenticus* reduces DMH induced precancerous lesions in F344 male rat. Biol Pharm Bull. 2007;30(3):569–74. https://doi.org/10.1248/bpb.30.569.
114. Blasiak J, Trzeciak A, Malecka-Panas E, Drzewoski J, Wojewódzka M. In vitro genotoxicity of ethanol and acetaldehyde in human lymphocytes and the gastrointestinal tract mucosa cells. Toxicol In Vitro. 2000;14(4):287–95. https://doi.org/10.1016/s0887-2333(00)00022-9.
115. Leufkens AM, Van Duijnhoven FJ, Siersema PD, Boshuizen HC, Vrieling A, Agudo A, Gram IT, Weiderpass E, Dahm C, Overvad K, Tjønneland A, Olsen A, Boutron-Ruault MC, Clavel-Chapelon F, Morois S, Palli D, Grioni S, Tumino R, Sacerdote C, Mattiello A, Herman S, Kaaks R, Steffen A, Boeing H, Trichopoulou A, Lagiou P, Trichopoulos D, Peeters PH, van Gils CH, van Kranen H, Lund E, Dumeaux V, Engeset D, Rodríguez L, Sánchez MJ, Chirlaque MD, Barricarte A, Manjer J, Almquist M, van Guelpen B, Hallmans G, Khaw KT, Wareham N, Tsilidis KK, Straif K, Leon-Roux M, Vineis P, Norat T, Riboli E, Bueno-de-Mesquita HB. Cigarette smoking and colorectal cancer risk in the European Prospective Investigation into Cancer and Nutrition study. Clin Gastroenterol Hepatol. 2011;9(2):137–44. https://doi.org/10.1016/j.cgh.2010.10.012.
116. Gutierez DA, Puglisi MJ, Hasty AH. Impact of increased adipose tissue mass on inflammation, insulin resistance, and dyslipidemia. Curr Diab Rep. 2009;9(1):26–32. https://doi.org/10.1007/s11892-009-0006-9.
117. Calle EE, Kaaks R. Overweight, obesity and cancer: epidemiological evidence and proposed mechanisms. Nat Rev Cancer. 2004;4(8):579–91. https://doi.org/10.1038/nrc1408.
118. Cirillo D, Rachiglio AM, la Montagna R, Giordano A, Normanno N. Leptin signaling in breast cancer: an overview. J Cell Biochem. 2008;105:956–64. https://doi.org/10.1002/jcb.21911.
119. Hrushesky WJ, Lannin D, Haus E. Evidence for an ontogenetic basis for circadian coordination of cancer cell proliferation. J Natl Cancer Inst. 1998;90:1480–4. https://doi.org/10.1093/jnci/90.19.1480.
120. Hashemi SH, Karimi S, Mahboobi H. Lifestyle changes for prevention of breast cancer. Electron Physician. 2014;6(3):894–905. https://doi.org/10.14661/2014.894-905.
121. Ziegler RG, Hoover RN, Pike MC, Hildesheim A, Nomura AM, West DW, Wu-Williams AH, Kolonel LN, Horn-Ross PL, Rosenthal JF, Hyer MB. Migration patterns and breast cancer risk in Asian-American women. J Natl Cancer Inst. 1993;85(22):1819–27. https://doi.org/10.1093/jnci/85.22.1819.
122. Hankinson SE, Willett WC, Manson JE, Colditz GA, Hunter DJ, Spiegelman D, Barbieri RL, Speizer FE. Plasma sex steroid hormone levels and risk of breast cancer in postmenopausal women. J Natl Cancer Inst. 1998;90(17):1292–9. https://doi.org/10.1093/jnci/90.17.1292.
123. Hirose K, Toyama T, Iwata H, Takezaki T, Hamajima N, Tajima K. Insulin, insulin-like growth factor and breast cancer risk in Japanese women. Asian Pac J Cancer Prev. 2003;4(3):239–46.
124. Hamelers IH, Steenbergh PH. Interactions between estrogen and insulin-like growth factor signaling pathways in human breast tumor cells. Endocr Relat Cancer. 2003;10(2):331–45. https://doi.org/10.1677/erc.0.0100331.
125. Demark-Wahnefried W, Jones LW, Snyder DC, Sloane RJ, Kimmick GG, Hughes DC, Badr HJ, Miller PE, Burke LE, Lipkus IM. Daughters and mothers against breast cancer (DAMES): main outcomes of a randomized controlled trial of weight loss in overweight mothers with breast cancer and their overweight daughters. Cancer. 2014;120(16):2522–34. https://doi.org/10.1002/cncr.28761.
126. Pudrovska T, Anishkin A, Shen Y. Early-life socioeconomic status and the prevalence of breast cancer in later life. Res Aging. 2012;34(3):302–20. https://doi.org/10.1177/0164027511415632.
127. Jemal A, Bray F, Center MM, Ferlay J, Ward E, Forman D. Global cancer statistics. CA Cancer J Clin. 2011;61(2):69–90. https://doi.org/10.3322/caac.20107.
128. Parkin DM, Bray F, Ferlay J, Pisani P. Global cancer statistics, 2002. CA Cancer J Clin. 2005;55(2):74–108. https://doi.org/10.3322/canjclin.55.2.74.
129. Lahmann PH, Hoffmann K, Allen N, van Gils CH, Khaw KT, Tehard B, Berrino F, Tjønneland A, Bigaard J, Olsen A, Overvad K, Clavel-Chapelon F, Nagel G, Boeing H, Trichopoulos D, Economou G, Bellos G, Palli D, Tumino R, Panico S, Sacerdote C, Krogh V, Peeters PH, Bueno-de-Mesquita HB, Lund E, Ardanaz E, Amiano P, Pera G, Quirós JR, Martínez C, Tormo MJ, Wirfält E, Berglund G, Hallmans G, Key TJ, Reeves G, Bingham S, Norat T, Biessy C, Kaaks R, Riboli E. Body size and breast cancer risk: findings from the European Prospective Investigation into Cancer And Nutrition (EPIC). Int J Cancer. 2004;111(5):762–71. https://doi.org/10.1002/ijc.20315.

130. Van den Brandt PA, Spiegelman D, Yaun SS, Adami HO, Beeson L, Folsom AR, Fraser G, Goldbohm RA, Graham S, Kushi L, Marshall JR, Miller AB, Rohan T, Smith-Warner SA, Speizer FE, Willett WC, Wolk A, Hunter DJ. Pooled analysis of prospective cohort studies on height, weight, and breast cancer risk. Am J Epidemiol. 2000;152(6):514–27. https://doi.org/10.1093/aje/152.6.514.
131. Emaus MJ, van Gils CH, Bakker MF, Bisschop CN, Monninkhof EM, Bueno-de-Mesquita HB, Travier N, Berentzen TL, Overvad K, Tjønneland A, Romieu I, Rinaldi S, Chajes V, Gunter MJ, Clavel-Chapelon F, Fagherazzi G, Mesrine S, Chang-Claude J, Kaaks R, Boeing H, Aleksandrova K, Trichopoulou A, Naska A, Orfanos P, Palli D, Agnoli C, Tumino R, Vineis P, Mattiello A, Braaten T, Borch KB, Lund E, Menéndez V, Sánchez MJ, Navarro C, Barricarte A, Amiano P, Sund M, Andersson A, Borgquist S, Olsson A, Khaw KT, Wareham N, Travis RC, Riboli E, Peeters PH, May AM. Weight change in middle adulthood and breast cancer risk in the EPIC-PANACEA study. Int J Cancer. 2014;135(12):2887–99. https://doi.org/10.1002/ijc.28926.
132. Travier N, Fonseca-Nunes A, Javierre C, Guillamo E, Arribas L, Peiró I, Buckland G, Moreno F, Urruticoechea A, Oviedo GR, Roca A, Hurtós L, Ortega V, Muñoz M, Garrigós L, Cirauqui B, Del Barco S, Arcusa A, Seguí MA, Borràs JM, Gonzalez CA, Agudo A. Effect of a diet and physical activity intervention on body weight and nutritional patterns in overweight and obese breast cancer survivors. Med Oncol. 2014;31(1):783. https://doi.org/10.1007/s12032-013-0783-5.
133. Young A, Weltzien E, Kwan M, Castillo A, Caan B, Kroenke CH. Pre- to post-diagnosis weight change and associations with physical functional limitations in breast cancer survivors. J Cancer Surviv. 2014;8:539–47. https://doi.org/10.1007/s11764-014-0356-4.
134. Fairey AS, Courneya KS, Field CJ, Bell GJ, Jones LW, Mackey JR. Effects of exercise training on fasting insulin, insulin resistance, insulin-like growth factors, and insulin-like growth factor binding proteins in postmenopausal breast cancer survivors: a randomized controlled trial. Cancer Epidemiol Biomark Prev. 2003;12(8):721–7.
135. Irwin ML, Varma K, Alvarez-Reeves M, Cadmus L, Wiley A, Chung GG, Dipietro L, Mayne ST, Yu H. Randomized controlled trial of aerobic exercise on insulin and insulin-like growth factors in breast cancer survivors: the Yale Exercise and Survivorship study. Cancer Epidemiol Biomark Prev. 2009;18(1):306–13. https://doi.org/10.1158/1055-9965.EPI-08-0531.
136. Ligibel JA, Campbell N, Partridge A, Chen WY, Salinardi T, Chen H, Adloff K, Keshaviah A, Winer EP. Impact of a mixed strength and endurance exercise intervention on insulin levels in breast cancer survivors. J Clin Oncol. 2008;26(6):907–12. https://doi.org/10.1200/JCO.2007.12.7357.
137. Bradshaw PT, Ibrahim JG, Khankari N, Cleveland RJ, Abrahamson PE, Stevens J, Satia JA, Teitelbaum SL, Neugut AI, Gammon MD. Post-diagnosis physical activity and survival after breast cancer diagnosis: the Long Island Breast Cancer Study. Breast Cancer Res Treat. 2014;145(3):735–42. https://doi.org/10.1007/s10549-014-2966-y.
138. Schmid D, Leitzmann MF. Association between physical activity and mortality among breast cancer and colorectal cancer survivors: a systematic review and meta-analysis. Ann Oncol. 2014;25(7):1293–311. https://doi.org/10.1093/annonc/mdu012.
139. Avraham Z, Baron-Epel O, Boker LK. The relationship between passive smoking, breast cancer risk and n-acetyltransferase 2 (NAT2). Harefuah. 2014;153(3-4):171–5, 238.
140. Dossus L, Boutron-Ruault MC, Kaaks R, Gram IT, Vilier A, Fervers B, Manjer J, Tjonneland A, Olsen A, Overvad K, Chang-Claude J, Boeing H, Steffen A, Trichopoulou A, Lagiou P, Sarantopoulou M, Palli D, Berrino F, Tumino R, Vineis P, Mattiello A, Bueno-de-Mesquita HB, van Duijnhoven FJ, Bakker MF, Peeters PH, Weiderpass E, Bjerkaas E, Braaten T, Menéndez V, Agudo A, Sanchez MJ, Amiano P, Tormo MJ, Barricarte A, Butt S, Khaw KT, Wareham N, Key TJ, Travis RC, Rinaldi S, McCormack V, Romieu I, Cox DG, Norat T, Riboli E, Clavel-Chapelon F. Active and passive cigarette smoking and breast cancer risk: results from the EPIC cohort. Int J Cancer. 2014;134(8):1871–88.
141. Ilic M, Vlajinac H, Marinkovic J. Cigarette smoking and breast cancer: a case-control study in Serbia. Asian Pac J Cancer Prev. 2013;14(11):6643–7. https://doi.org/10.7314/apjcp.2013.14.11.6643.
142. Kawai M, Malone KE, Tang MT, Li CI. Active smoking and the risk of estrogen receptor-positive and triple-negative breast cancer among women ages 20 to 44 years. Cancer. 2014;120(7):1026–34. https://doi.org/10.1002/cncr.28402.
143. Rosenberg L, Boggs DA, Bethea TN, Wise LA, Adams-Campbell LL, Palmer JR. A prospective study of smoking and breast cancer risk among African-American women. Cancer Causes Control. 2013;24(12):2207–15. https://doi.org/10.1007/s10552-013-0298-6.
144. Bishop JD, Killelea BK, Chagpar AB, Horowitz NR, Lannin DR. Smoking and breast cancer recurrence after breast conservation therapy. Int J Breast Cancer. 2014;2014:327081. https://doi.org/10.1155/2014/327081.
145. Liu Y, Colditz GA, Rosner B, Berkey CS, Collins LC, Schnitt SJ, Connolly JL, Chen WY, Willett WC, Tamimi RM. Alcohol intake between menarche and first pregnancy: a prospective study of breast cancer risk. J Natl Cancer Inst. 2013;105(20):1571–8. https://doi.org/10.1093/jnci/djt213.
146. McDonald JA, Goyal A, Terry MB. Alcohol intake and breast cancer risk: weighing the overall evidence. Curr Breast Cancer Rep. 2013;5(3):114. https://doi.org/10.1007/s12609-013-0114-z.
147. Scoccianti C, Lauby-Secretan B, Bello PY, Chajes V, Romieu I. Female breast cancer and alcohol consumption: a review of the literature. Am J Prev Med. 2014;46(3):S16–25. https://doi.org/10.1016/j.amepre.2013.10.031.
148. Islam T, Ito H, Sueta A, Hosono S, Hirose K, Watanabe M, Iwata H, Tajima K, Tanaka H, Matsuo K. Alcohol and dietary folate intake and the risk of breast cancer: a case-control study in Japan. Eur J Cancer Prev. 2013;22(4):358–66. https://doi.org/10.1097/CEJ.0b013e32835b6a60.
149. Boyd NF, Stone J, Vogt KN, Connelly BS, Martin LJ, Minkin S. Dietary fat and breast cancer risk revisited: a meta-analysis of the published literature. Br J Cancer. 2003;89(9):1672–85. https://doi.org/10.1038/sj.bjc.6601314.
150. Ganmaa D, Willett WC, Li TY, Feskanich D, van Dam RM, Lopez-Garcia E, Hunter DJ, Holmes MD. Coffee, tea, caffeine and risk of breast cancer: a 22-year follow-up. Int J Cancer. 2008;122(9):2071–6. https://doi.org/10.1002/ijc.23336.
151. Ishitani K, Lin J, Manson JE, Buring JE, Zhang SM. Caffeine consumption and the risk of breast cancer in a large prospective cohort of women. Arch Intern Med. 2008;168(18):2022–31. https://doi.org/10.1001/archinte.168.18.2022.
152. Taylor EF, Burley VJ, Greenwood DC, Cade JE. Meat consumption and risk of breast cancer in the UK Women's Cohort Study. Br J Cancer. 2007;96(7):1139–46. https://doi.org/10.1038/sj.bjc.6603689.
153. Shin MH, Holmes MD, Hankinson SE, Wu K, Colditz GA, Willett WC. Intake of dairy products, calcium, and vitamin d and risk of breast cancer. J Natl Cancer Inst. 2002;94(17):1301–11. https://doi.org/10.1093/jnci/94.17.1301.
154. Stuebe AM, Willett WC, Xue F, Michels KB. Lactation and incidence of premenopausal breast cancer: a longitudinal study. Arch Intern Med. 2009;169(15):1364–71. https://doi.org/10.1001/archinternmed.2009.231.

155. Zheng T, Holford TR, Mayne ST, Owens PH, Zhang Y, Zhang B, Boyle P, Zahm SH. Lactation and breast cancer risk: a case-control study in Connecticut. Br J Cancer. 2001;84(11):1472–6. https://doi.org/10.1054/bjoc.2001.1793.
156. Schernhammer ES, Hankinson SE. Urinary melatonin levels and breast cancer risk. J Natl Cancer Inst. 2005;97(14):1084–7. https://doi.org/10.1093/jnci/dji190.
157. Schernhammer ES, Hankinson SE. Urinary melatonin levels and postmenopausal breast cancer risk in the Nurses' Health Study cohort. Cancer Epidemiol Biomark Prev. 2009;18(1):74–9. https://doi.org/10.1158/1055-9965.EPI-08-0637.
158. Wang L, Liao WC, Tsai CJ, Wang LR, Mao IF, Chen CC, Kao PF, Yao CC. The effects of perceived stress and life style leading to breast cancer. Women Health. 2013;53(1):20–40. https://doi.org/10.1080/03630242.2012.732680.
159. Agrawal A, Fentiman IS. NSAIDs and breast cancer: a possible prevention and treatment strategy. Int J Clin Pract. 2008;62(3):444–9. https://doi.org/10.1111/j.1742-1241.2007.01668.x.
160. Kwan ML, Habel LA, Slattery ML, Caan B. NSAIDs and breast cancer recurrence in a prospective cohort study. Cancer Causes Control. 2007;18(6):613–20. https://doi.org/10.1007/s10552-007-9003-y.
161. Zhao YS, Zhu S, Li XW, Wang F, Hu FL, Li DD, Zhang WC, Li X. Association between NSAIDs use and breast cancer risk: a systematic review and meta-analysis. Breast Cancer Res Treat. 2009;117(1):141–50. https://doi.org/10.1007/s10549-008-0228-6.
162. Ahern TP, Lash TL, Sorensen HT, Pedersen L. Digoxin treatment is associated with an increased incidence of breast cancer: a population-based case-control study. Breast Cancer Res. 2008;10(6):R102. https://doi.org/10.1186/bcr2205.
163. Ahern TP, Tamimi RM, Rosner BA, Hankinson SE. Digoxin use and risk of invasive breast cancer: evidence from the Nurses' Health Study and meta-analysis. Breast Cancer Res Treat. 2014;144(2):427–35. https://doi.org/10.1007/s10549-014-2886-x.
164. Biggar RJ, Wohlfahrt J, Oudin A, Hjuler T, Melbye M. Digoxin use and the risk of breast cancer in women. J Clin Oncol. 2011;29(16):2165–70. https://doi.org/10.1200/JCO.2010.32.8146.
165. Egawa G, Kabashima K. Skin as a peripheral lymphoid organ: revisiting the concept of skin-associated lymphoid tissues. J Investig Dermatol. 2011;131:2178–85. https://doi.org/10.1038/jid.2011.198.
166. Kabashima K, Honda T, Ginhoux F, Egawa G. The immunological anatomy of the skin. Nat Rev Immunol. 2019;19:19–30. https://doi.org/10.1038/s41577-018-0084-5.
167. Rigel DS. Cutaneous ultraviolet exposure and its relationship to the development of skin cancer. J Am Acad Dermatol. 2008;58(5 Suppl 2):S129–32. https://doi.org/10.1016/j.jaad.2007.04.034.
168. Sawada Y, Nakamura M. Daily lifestyle and cutaneous malignancies. Int J Mol Sci. 2021;22(10):5227. https://doi.org/10.3390/ijms22105227.
169. Schernhammer ES, Razavi P, Li TY, Qureshi AA, Han J. Rotating night shifts and risk of skin cancer in the nurses' health study. J Natl Cancer Inst. 2011;103(7):602–6. https://doi.org/10.1093/jnci/djr044.
170. Yousef E, Mitwally N, Noufal N, Tahir MR. Shift work and risk of skin cancer: a systematic review and meta-analysis. Sci Rep. 2020;10:2012. https://doi.org/10.1038/s41598-020-59035-x.
171. Dusingize JC, Olsen CM, Pandeya N, Thompson BS, Webb PM, Green AC, Neale RE, Whiteman DC, Skin Study Q. Smoking and cutaneous melanoma: findings from the QSkin Sun and Health Cohort Study. Cancer Epidemiol Biomark Prev. 2018;27(8):874–81. https://doi.org/10.1158/1055-9965.epi-17-1056.
172. Hughes MC, Olsen CM, Williams GM, Green AC. A prospective study of cigarette smoking and basal cell carcinoma. Arch Dermatol Res. 2014;306(9):851–6. https://doi.org/10.1007/s00403-014-1503-5.
173. Sondermeijer L, Lamboo LGE, de Waal AC, Galesloot TE, Kiemeney LALM, van Rossum M, Aben KH. Cigarette smoking and the risk of cutaneous melanoma: a case-control study. Dermatology. 2020;236(3):228–36. https://doi.org/10.1159/000502129.
174. Gęgotek A, Ambrożewicz E, Jastrząb A, Jarocka-Karpowicz I, Skrzydlewska E. Rutin and ascorbic acid cooperation in antioxidant and antiapoptotic effect on human skin keratinocytes and fibroblasts exposed to UVA and UVB radiation. Arch Dermatol Res. 2019;311(3):203–19. https://doi.org/10.1007/s00403-019-01898-w.
175. Gęgotek A, Jarocka-Karpowicz I, Skrzydlewska E. Synergistic cytoprotective effects of rutin and ascorbic acid on the proteomic profile of 3D-cultured keratinocytes exposed to UVA or UVB radiation. Nutrients. 2019;11(11):2672. https://doi.org/10.3390/nu11112672.
176. Jin GH, Liu Y, Jin SZ, Liu XD, Liu SZ. UVB induced oxidative stress in human keratinocytes and protective effect of antioxidant agents. Radiat Environ Biophys. 2007;46(1):61–8. https://doi.org/10.1007/s00411-007-0096-1.
177. Maalouf S, El-Sabban M, Darwiche N, Gali-Muhtasib H. Protective effect of vitamin E on ultraviolet B light-induced damage in keratinocytes. Mol Carcinog. 2002;34(3):121–30. https://doi.org/10.1002/mc.10055.
178. Offord EA, Gautier JC, Avanti O, Scaletta C, Runge F, Krämer K, Applegate LA. Photoprotective potential of lycopene, β-carotene, vitamin E, vitamin C and carnosic acid in UVA-irradiated human skin fibroblasts. Free Radic Biol Med. 2002;32(12):1293–303. https://doi.org/10.1016/s0891-5849(02)00831-6.
179. Caini S, Boniol M, Tosti G, Magi S, Medri M, Stanganelli I, Palli D, Assedi M, Marmol VD, Gandini S. Vitamin D and melanoma and non-melanoma skin cancer risk and prognosis: a comprehensive review and meta-analysis. Eur J Cancer. 2014;50(15):2649–58. https://doi.org/10.1016/j.ejca.2014.06.024.
180. Lim A, Shayan R, Varigos G. High serum vitamin D level correlates with better prognostic indicators in primary melanoma: a pilot study. Australas J Dermatol. 2018;59(3):182–7. https://doi.org/10.1111/ajd.12648.
181. Moreno-Arrones OM, Zegeer J, Gerbo M, Manrique-Silva E, Requena C, Traves V, Nagore E. Decreased vitamin D serum levels at melanoma diagnosis are associated with tumor ulceration and high tumor mitotic rate. Melanoma Res. 2019;29(6):664–7. https://doi.org/10.1097/cmr.0000000000000638.
182. Singh T, Katiyar SK. Green tea catechins reduce invasive potential of human melanoma cells by targeting COX-2, PGE2 receptors and epithelial-to-mesenchymal transition. PLoS One. 2011;6(10):e25224. https://doi.org/10.1371/journal.pone.0025224.
183. Vaid M, Singh T, Katiyar S. Grape seed proanthocyanidins inhibit melanoma cell invasiveness by reduction of PGE2 synthesis and reversal of epithelial-to-mesenchymal transition. PLoS One. 2011;6(6):e21539. https://doi.org/10.1371/journal.pone.0021539.
184. Singh T, Vaid M, Katiyar N, Sharma S, Katiyar SK. Berberine, an isoquinoline alkaloid, inhibits melanoma cancer cell migration by reducing the expressions of cyclooxygenase-2, prostaglandin E_2 and prostaglandin E_2 receptors. Carcinogenesis. 2011;32(1):86–92. https://doi.org/10.1093/carcin/bgq215.
185. Kim KM, Im AR, Kim SH, Hyun JW, Chae S. Timosaponin AIII inhibits melanoma cell migration by suppressing COX-2 and in vivo tumor metastasis. Cancer Sci. 2016;107(2):181–8. https://doi.org/10.1111/cas.12852.
186. Shimizu H, Ross RK, Bernstein L, Yatani R, Henderson BE, Mack TM. Cancers of the prostate and breast among Japanese and white

immigrants in Los Angeles County. Br J Cancer. 1991;63(6):963–6. https://doi.org/10.1038/bjc.1991.210.
187. Hsing AW, Tsao L, Devesa SS. International trends and patterns of prostate cancer incidence and mortality. Int J Cancer. 2000;85(1):60–7. https://doi.org/10.1002/(sici)1097-0215(20000101)85:1<60::aid-ijc11>3.0.co;2-b.
188. Hebert JR, Hurley TG, Olendzki BC, Teas J, Ma Y, Hampl JS. Nutritional and socioeconomic factors in relation to prostate cancer mortality: a cross-national study. J Natl Cancer Inst. 1998;90(21):1637–47. https://doi.org/10.1093/jnci/90.21.1637.
189. Moyad MA. Dietary fat reduction to reduce prostate cancer risk: controlled enthusiasm, learning a lesson from breast or other cancers, and the big picture. Urology. 2002;59(4):51–62. https://doi.org/10.1016/s0090-4295(01)01176-1.
190. Moyad MA. Lifestyle/dietary supplement partial androgen suppression and/or estrogen manipulation: a novel PSA reducer and preventive/treatment option for prostate cancer? Urol Clin North Am. 2002;29(1):115–24. https://doi.org/10.1016/s0094-0143(02)00037-x.
191. Bartsch H, Nair J, Owen RW. Dietary polyunsaturated fatty acids and cancers of the breast and colorectum: emerging evidence for their role as risk modifiers. Carcinogenesis. 1999;20(12):2209–18. https://doi.org/10.1093/carcin/20.12.2209.
192. Thune I, Furberg AS. Physical activity and cancer risk: dose-response and cancer, all sites and site-specific. Med Sci Sports Exerc. 2001;33(6):530–50. https://doi.org/10.1097/00005768-200106001-00025.
193. Tymchuk CN, Barnard RJ, Heber D, Aronson WJ. Evidence of an inhibitory effect of diet and exercise on prostate cancer cell growth. J Urol. 2001;166(3):1185–9.
194. Ngo TH, Barnard RJ, Leung P-S, Cohen P, Aronson WJ. Insulin-like growth factor I (IGF-I) and IGF binding protein-1 modulate prostate cancer cell growth and apoptosis: possible mediators for the effects of diet and exercise on cancer cell survival. Endocrinology. 2003;144(6):2319–24. https://doi.org/10.1210/en.2003-221028.
195. Tymchuk CN, Tessler SB, Aronson WJ, Barnard RJ. Effects of diet and exercise on insulin, sex hormone-binding globulin, and prostate-specific antigen. Nutr Cancer. 1998;31(2):127–31. https://doi.org/10.1080/01635589809514691.
196. Tymchuk CN, Barnard RJ, Ngo TH, Aronson WJ. The role of testosterone, estradiol, and insulin in diet and exercise-induced reductions in prostate cancer cell growth *in vitro*. Nutr Cancer. 2002;42(1):112–6. https://doi.org/10.1207/S15327914NC421_15.
197. Barnard RJ, Aronson WJ, Tymchuk CN, Ngo TH. Prostate cancer: another aspect of the insulin-resistance syndrome? Obes Rev. 2002;3(4):303–8. https://doi.org/10.1046/j.1467-789x.2002.00081.x.
198. LeRoith D, Roberts CT Jr. The insulin-like growth factor system and cancer. Cancer Lett. 2003;195:127–37. https://doi.org/10.1016/s0304-3835(03)00159-9.
199. Ngo TH, Barnard RJ, Tymchuk CN, Cohen P, Aronson WJ. Effect of diet and exercise on serum insulin, IGF-I, and IGFBP-1 levels and growth of LNCaP cells *in vitro*. Cancer Causes Control. 2002;13(10):929–35. https://doi.org/10.1023/a:1021911517010.
200. Gurumurthy S, Vasudevan KM, Rangnekar VM. Regulation of apoptosis in prostate cancer. Cancer Metastasis Rev. 2001;20(3-4):225–43. https://doi.org/10.1023/a:1015583310759.
201. Barnard RJ, Ngo TH, Leung P-S, Aronson WJ, Golding LA. A Low-fat diet and/or strenuous exercise alters the IGF axis in vivo and reduces prostate tumor cell growth in vitro. Prostrate. 2003;56(3):201–6. https://doi.org/10.1002/pros.10251.
202. Leung P-S, Aronson WJ, Ngo TH, Golding LA, Barnard RJ. Exercise alters the IGF axis *in vivo* and increases p53 protein in prostate tumor cells *in vitro*. J Appl Physiol. 2004;96:450–4. https://doi.org/10.1152/japplphysiol.00871.2003.
203. Ornish D, Lee KL, Fair WR, Pettengill EB, Carroll PR. Dietary trial in prostate cancer: early experience and implications for clinical trial design. Urology. 2001;57(4):200–1. https://doi.org/10.1016/s0090-4295(00)00974-2.
204. Ornish D, Weidner G, Fair WR, Marlin R, Pettengill EB, Raisin CJ, Dunn-Emke S, Crutchfield L, Jacobs FN, Barnard RJ, Aronson WJ, McCormac P, McKnight DJ, Fein JD, Dnistrian AM, Weinstein J, Ngo TH, Mendell NR, Carroll PR. Intensive lifestyle changes may affect the progression of prostate cancer. J Urol. 2005;174(3):1065–70. https://doi.org/10.1097/01.ju.0000169487.49018.73.
205. Boileau TW, Liao Z, Kim S, Lemeshow S, Erdman JW Jr, Clinton SK. Prostate carcinogenesis in N-methyl-N-nitrosourea (NMU)-testosterone-treated rats fed tomato powder, lycopene, or energy-restricted diets. J Natl Cancer Inst. 2003;95(21):1578–86. https://doi.org/10.1093/jnci/djg081.
206. Chen L, Stacewicz-Sapuntzakis M, Duncan C, Sharifi R, Ghosh L, van Breemen R, Ashton D, Bowen PE. Oxidative DNA damage in prostate cancer patients consuming tomato sauce-based entrees as a whole-food intervention. J Natl Cancer Inst. 2001;93(24):1872–9. https://doi.org/10.1093/jnci/93.24.1872.
207. Mucci LA, Tamimi R, Lagiou P, Trichopoulou A, Benetou V, Spanos E, Trichopoulos D. Are dietary influences on the risk of prostate cancer mediated through the insulin-like growth factor system? BJU Int. 2001;87(9):814–20. https://doi.org/10.1046/j.1464-410x.2001.02191.x.
208. Hsing AW, Chokkalingam AP, Gao YT, Madigan MP, Deng J, Gridley G, Fraumeni JF Jr. Allium vegetables and risk of prostate cancer: a population-based study. J Natl Cancer Inst. 2002;94(21):1648–51. https://doi.org/10.1093/jnci/94.21.1648.
209. Kristal AR, Lampe JW. Brassica vegetables and prostate cancer risk: a review of the epidemiological evidence. Nutr Cancer. 2002;42(1):1–9. https://doi.org/10.1207/S15327914NC421_1.
210. Moyad MA. The ABCs of nutrition and supplements for prostate cancer. Ann Arbor: JW Edwards Publishing; 2000.
211. Moyad MA, Carroll PR. Lifestyle recommendations to prevent prostate cancer, part II: time to redirect our attention? Urol Clin N Am. 2004;31(2):301–11. https://doi.org/10.1016/j.ucl.2004.03.007.
212. Nagata C, Inaba S, Kawakami N, Kakizoe T, Shimizu H. Inverse association of soy product intake with serum androgen and estrogen concentrations in Japanese men. Nutr Cancer. 2000;36(1):14–8. https://doi.org/10.1207/S15327914NC3601_3.
213. Dabrosin C, Chen J, Wang L, Thompson LU. Flaxseed inhibits metastasis and decreases extracellular vascular endothelial growth factor in human breast cancer xenografts. Cancer Lett. 2002;185(1):31–7. https://doi.org/10.1016/s0304-3835(02)00239-2.
214. Schuurman AG, van den Brandt PA, Dorant E, Brants HA, Goldbohm RA. Association of energy and fat intake with prostate carcinoma risk: results from The Netherlands Cohort Study. Cancer. 1999;86(6):1019–27.
215. Demark-Wahnefried W, Price DT, Polascik TJ, Robertson CN, Anderson EE, Paulson DF, Walther PJ, Gannon M, Vollmer RT. Pilot study of dietary fat restriction and flaxseed supplementation in men with prostate cancer before surgery: exploring the effects on hormonal levels, prostate-specific antigen, and histopathologic features. Urology. 2001;58(1):47–52. https://doi.org/10.1016/s0090-4295(01)01014-7.
216. Aronson WJ, Glaspy JA, Reddy ST, Reese D, Heber D, Bagga D. Modulation of omega-3/omega-6 polyunsaturated ratios with dietary fish oils in men with prostate cancer. Urology. 2001;58(2):283–8. https://doi.org/10.1016/s0090-4295(01)01116-5.
217. Chung BH, Mitchell SH, Zhang SS, Young CY. Effects of docosahexaenoic acid and eicosapentaenoic acid on androgen-mediated

cell growth and gene expression in LNCaP prostate cancer cells. Carcinogenesis. 2001;22(8):1201–6. https://doi.org/10.1093/carcin/22.8.1201.
218. Terry P, Lichtenstein P, Feychting M, Ahlbom A, Wolk A. Fatty fish consumption and risk of prostate cancer. Lancet. 2001;357(9270):1764–6. https://doi.org/10.1016/S0140-6736(00)04889-3.
219. Augustsson K, Michaud DS, Rimm EB, Leitzmann MF, Stampfer MJ, Willett WC, Giovannucci E. A prospective study of intake of fish and marine fatty acids and prostate cancer. Cancer Epidemiol Biomark Prev. 2003;12(1):64–7.
220. Albert CM, Gaziano JM, Willett WC, Manson JE. Nut consumption and decreased risk of sudden cardiac death in the physicians' health study. Arch Intern Med. 2002;162:1382–7. https://doi.org/10.1001/archinte.162.12.1382.
221. Friedenreich CM. Physical activity and cancer prevention: from observational to intervention research. Cancer Epidemiol Biomark Prev. 2001;10:287–301.
222. Babyak M, Blumenthal JA, Herman S, Khatri P, Doraiswamy M, Moore K, Craighead WE, Baldewicz TT, Krishnan KR. Exercise treatment for major depression: maintenance of therapeutic benefit at 10 months. Psychosom Med. 2000;62(5):633–8. https://doi.org/10.1097/00006842-200009000-00006.
223. Plaskon LA, Penson DF, Vaughan TL, Stanford JL. Cigarette smoking and risk of prostate cancer in middle-aged men. Cancer Epidemiol Biomark Prev. 2003;12(7):604–9.
224. Roberts WW, Platz EA, Walsh PC. Association of cigarette smoking with extraprostatic prostate cancer in young men. J Urol. 2003;169:512–6. https://doi.org/10.1097/01.ju.0000046160.80804.7f.
225. Rodriguez C, Tatham LM, Thun MJ, Calle EE, Heath CW Jr. Smoking and fatal prostate cancer in a large cohort of adult men. Am J Epidemiol. 1997;145:466–75. https://doi.org/10.1093/oxfordjournals.aje.a009129.
226. Leitzmann MF, Stampfer MJ, Wu K, Colditz GA, Willett WC, Giovannucci EL. Zinc supplement use and risk of prostate cancer. J Natl Cancer Inst. 2003;95(13):1004–7. https://doi.org/10.1093/jnci/95.13.1004.
227. Choi JW, Hua TNM. Impact of lifestyle behaviors on cancer risk and prevention. J Lifestyle Med. 2021;11(1):1–7. https://doi.org/10.15280/jlm.2021.11.1.1.
228. Porro C, La Torre ME, Tartaglia N, Benameur T, Santini M, Ambrosi A, Messina G, Cibelli G, Fiorelli A, Polito R, Messina G. The potential role of nutrition in lung cancer establishment and progression. Life. 2022;12(2):270. https://doi.org/10.3390/life12020270.
229. Halliwell B, Gutteridge J. Free radicals in biology and medicine. 4th ed. Oxford: Oxford University; 2007.
230. Afonso V, Champy R, Mitrovic D, Collin P, Lomri A. Reactive oxygen species and superoxide dismutases: role in joint diseases. J Bone Spine. 2007;74:324–9. https://doi.org/10.1016/j.jbspin.2007.02.002.
231. Filaire E, Dupuis C, Galvaing G, Aubreton S, Laurent H, Richard R, Filaire M. Lung cancer: what are the links with oxidative stress, physical activity and nutrition. Lung Cancer. 2013;82:383–9. https://doi.org/10.1016/j.lungcan.2013.09.009.
232. Weinberg RA. The biology of cancer. New York: Garland Science; 2006.
233. Hannan MA, Recio L, Deluca PP, Enoch H. Co-mutagenic effects of 2-aminoanthracene and cigarette smoke condensate on smoker's urine in the Ames Salmonella assay system. Cancer Lett. 1981;13:203–12. https://doi.org/10.1016/0304-3835(81)90019-7.
234. Lee J, Taneja V, Vassallo R. Cigarette smoking and inflammation: cellular and molecular mechanisms. J Dent Res. 2012;91:142–9. https://doi.org/10.1177/0022034511421200.
235. Møller P, Folkmann JK, Forchhammer L, Bräuner EV, Danielsen PH, Risom L, Loft S. Air pollution, oxidative damage to DNA, and carcinogenesis. Cancer Lett. 2008;266:84–97. https://doi.org/10.1016/j.canlet.2008.02.030.
236. Pryor WA. Cigarette smoke radicals and the role of free radicals in chemical carcinogenicity. Environ Health Perspect. 1997;105(4):875–82. https://doi.org/10.1289/ehp.97105s4875.
237. Sangani RG, Ghio AJ. Lung injury after cigarette smoking is particle related. Int J Chron Obstruct Pulmon Dis. 2011;6:191–8. https://doi.org/10.2147/COPD.S14911.
238. Goldkorn T, Filosto S, Chung S. Lung injury and lung cancer caused by cigarette smoke-induced oxidative stress: molecular mechanisms and therapeutic opportunities involving the ceramide-generating machinery and epidermal growth factor receptor. Antioxid Redox Signal. 2014;15:2149–74. https://doi.org/10.1089/ars.2013.5469.
239. Wang RP, Wang XH, Li ZM, Sun JR. Changes in serum inflammatory factors, adiponectin, intestinal flora and immunity in patients with non-small cell lung cancer. Eur Rev Med Pharmacol Sci. 2020;24:10566–72. https://doi.org/10.26355/eurrev_202010_23412.
240. Wei X, Zhu C, Ji M, Fan J, Xie J, Huang Y, Jiang X, Xu J, Yin R, Du L, Wang Y, Dai J, Jin G, Xu L, Hu Z, Shen H, Zhu M, Ma H. Diet and risk of incident lung cancer: a large prospective cohort study in UK Biobank. Am J Clin Nutr. 2021;114(6):2043–51. https://doi.org/10.1093/ajcn/nqab298.
241. Widiatmaja DM, Lutvyani A, Sari DR, Kurniasari H, Meiliana ID, Fasitasari M, Yamaoka Y, Rejeki PS. The effect of long-term ketogenic diet on serum adiponectin and insulin-like growth factor-1 levels in mice. J Basic Clin Physiol Pharmacol. 2021;33(5):611–8. https://doi.org/10.1515/jbcpp-2021-0287.
242. Atazadegan MA, Bagherniya M, Fakheran O, Sathyapalan T, Sahebkar A. The effect of herbal medicine and natural bioactive compounds on plasma adiponectin: a clinical review. Adv Exp Med Biol. 2021;1328:37–57. https://doi.org/10.1007/978-3-030-73234-9_4.
243. Singh RB, Niaz MA, Ghosh S, Beegum R, Bishnoi I, Agarwal P, Agarwal A. Dietary intake and plasma levels of antioxidant vitamins in health and disease. A hospital based case control study. J Nutr Environ Med. 1995;5:235–42. https://doi.org/10.3109/13590849509000224.
244. Martínez-González MA, Trichopoulou A. Observational epidemiology, lifestyle, and health: the paradigm of the mediterranean diet. Am J Health Promot. 2020;34(8):948–50. https://doi.org/10.1177/0890117120960580c.
245. Del Rio D, Rodriguez-Mateos A, Spencer JP, Tognolini M, Borges G, Crozier A. Dietary (poly)phenolics in human health: structures, bioavailability, and evidence of protective effects against chronic diseases. Antioxid Redox Signal. 2013;18(14):1818–92. https://doi.org/10.1089/ars.2012.4581.
246. Ghiselli A, Nardini M, Baldi A, Scaccini C. Antioxidant activity of different phenolic fractions separated from an Italian red wine. J Agric Food Chem. 1998;46(2):361–7. https://doi.org/10.1021/jf970486b.
247. Godos J, Sinatra D, Blanco I, Mulè S, La Verde M, Marranzano M. Association between dietary phenolic acids and hypertension in a mediterranean cohort. Nutrients. 2017;9(10):1069. https://doi.org/10.3390/nu9101069.
248. Nani A, Belarbi M, Ksouri-Megdiche W, Abdoul-Azize S, Benammar C, Ghiringhelli F, Hichami A, Khan NA. Effects of polyphenols and lipids from *Pennisetumglaucum* grains on T-cell activation: modulation of Ca(2+) and ERK1/ERK2 signaling. BMC Complement Altern Med. 2015;15:426. https://doi.org/10.1186/s12906-015-0946-3.
249. Liao BC, Hsieh CW, Liu YC, Tzeng TT, Sun YW, Wung BS. Cinnamaldehyde inhibits the tumor necrosis factor-alpha-

induced expression of cell adhesion molecules in endothelial cells by suppressing NF-kappaB activation: effects upon IkappaB and Nrf2. Toxicol Appl Pharmacol. 2008;229(2):161–71. https://doi.org/10.1016/j.taap.2008.01.021.
250. Halliwell B. Antioxidant and anti-inflammatory components of foods. Washington: ILSI International Life Sciences Institute; 2015.
251. Gorlova OY, Weng SF, Hernandez L, Spitz MR, Forman MR. Dietary patterns affect lung cancer risk in never smokers. Nutr Cancer. 2011;63(6):842–9. https://doi.org/10.1080/01635581.2011.589958.
252. Holick CN, Michaud DS, Stolzenberg-Solomon R, Mayne ST, Pietinen P, Taylor PR, Virtamo J, Albanes D. Dietary carotenoids, serum beta-carotene, and retinol and risk of lung cancer in the alpha-tocopherol, beta-carotene cohort study. Am J Epidemiol. 2002;156(6):536–47. https://doi.org/10.1093/aje/kwf072.
253. Smith-Warner SA, Spiegelman D, Yaun SS, Albanes D, Beeson WL, van den Brandt PA, Feskanich D, Folsom AR, Fraser GE, Freudenheim JL, Giovannucci E, Goldbohm RA, Graham S, Kushi LH, Miller AB, Pietinen P, Rohan TE, Speizer FE, Willett WC, Hunter DJ. Fruits, vegetables and lung cancer: a pooled analysis of cohort studies. Int J Cancer. 2003;107(6):1001–11. https://doi.org/10.1002/ijc.11490.
254. Sayin VI, Ibrahim MX, Larsson E, Nilsson JA, Lindahl P, Bergo MO. Antioxidants accelerate lung cancer progression in mice. Sci Transl Med. 2014;6(221):221ra15. https://doi.org/10.1126/scitranslmed.3007653.
255. Brennan P, Hsu CC, Moullan N, Szeszenia-Dabrowska N, Lissowska J, Zaridze D, Rudnai P, Fabianova E, Mates D, Bencko V, Foretova L, Janout V, Gemignani F, Chabrier A, Hall J, Hung RJ, Boffetta P, Canzian F. Effect of cruciferous vegetables on lung cancer in patients stratified by genetic status: a mendelian randomisation approach. Lancet. 2005;366(9496):1558–60. https://doi.org/10.1016/S0140-6736(05)67628-3.
256. Keck AS, Finley JW. Cruciferous vegetables: cancer protective mechanisms of glucosinolate hydrolysis products and selenium. Integr Cancer Ther. 2004;3(1):5–12. https://doi.org/10.1177/1534735403261831.
257. Camarena V, Wang G. The epigenetic role of vitamin C in health and disease. Cell Mol Life Sci. 2016;73(8):1645–58. https://doi.org/10.1007/s00018-016-2145-x.
258. Liu D, Perkins JT, Hennig B. EGCG prevents PCB-126-induced endothelial cell inflammation via epigenetic modifications of NF-κB target genes in human endothelial cells. J Nutr Biochem. 2016;28:164–70. https://doi.org/10.1016/j.jnutbio.2015.10.003.
259. Shankar E, Kanwal R, Candamo M, Gupta S. Dietary phytochemicals as epigenetic modifiers in cancer: promise and challenges. Semin Cancer Biol. 2016;40-41:82–99. https://doi.org/10.1016/j.semcancer.2016.04.002.
260. Zhong J, Karlsson O, Wang G, Li J, Guo Y, Lin X, Zemplenyi M, Sanchez-Guerra M, Trevisi L, Urch B, Speck M, Liang L, Coull BA, Koutrakis P, Silverman F, Gold DR, Wu T, Baccarelli AA. B vitamins attenuate the epigenetic effects of ambient fine particles in a pilot human intervention trial. Proc Natl Acad Sci U S A. 2017;114(13):3503–8. https://doi.org/10.1073/pnas.1618545114.
261. Crider KS, Yang TP, Berry RJ, Bailey LB. Folate and DNA methylation: a review of molecular mechanisms and the evidence for folate's role. Adv Nutr. 2012;3(1):21–38. https://doi.org/10.3945/an.111.000992.
262. Leng S, Picchi MA, Tesfaigzi Y, Wu G, Gauderman WJ, Xu F, Gilliland FD, Belinsky SA. Dietary nutrients associated with preservation of lung function in Hispanic and non-Hispanic white smokers from New Mexico. Int J Chron Obstruct Pulmon Dis. 2017;12:3171–81. https://doi.org/10.2147/COPD.S142237.
263. Leng S, Picchi MA, Kang H, Wu G, Filipczak PT, Juri DE, Zhang X, Gauderman WJ, Gilliland FD, Belinsky SA. Dietary nutrient intake, ethnicity, and epigenetic silencing of lung cancer genes detected in sputum in New Mexican Smokers. Cancer Prev Res. 2018;11(2):93–102. https://doi.org/10.1158/1940-6207.CAPR-17-0196.
264. Zhai T, Li S, Hu W, Li D, Leng S. Potential micronutrients and phytochemicals against the pathogenesis of chronic obstructive pulmonary disease and lung cancer. Nutrients. 2018;10(7):813. https://doi.org/10.3390/nu10070813.
265. Soeters PB, Wolfe RR, Shenkin A. Hypoalbuminemia: pathogenesis and clinical significance. JPEN J Parenter Enteral Nutr. 2019;43(2):181–93. https://doi.org/10.1002/jpen.1451.
266. Roth MJ, Wei WQ, Baer J, Abnet CC, Wang GQ, Sternberg LR, Warner AC, Johnson LL, Lu N, Giffen CA, Dawsey SM, Qiao YL, Cherry J. Aryl hydrocarbon receptor expression is associated with a family history of upper gastrointestinal tract cancer in a high-risk population exposed to aromatic hydrocarbons. Cancer Epidemiol Biomark Prev. 2009;18(9):2391–6. https://doi.org/10.1158/1055-9965.EPI-08-1098.
267. Sinha R, Kulldorff M, Curtin J, Brown CC, Alavanja MC, Swanson CA. Fried, well-done red meat and risk of lung cancer in women (United States). Cancer Causes Control. 1998;9(6):621–30. https://doi.org/10.1023/a:1008805525525.
268. Lampe JW. Dairy products and cancer. J Am Coll Nutr. 2011;30(5 Suppl 1):464S–70S. https://doi.org/10.1080/07315724.2011.10719991.
269. Thune I, Lund E. The influence of physical activity on lung-cancer risk: a prospective study of 81,516 men and women. Int J Cancer. 1997;70(1):57–62.
270. Ziegler RG, Mayne ST, Swanson CA. Nutrition and lung cancer. Cancer Causes Control. 1996;7(1):157–77. https://doi.org/10.1007/bf00115646.
271. Matthews CE, Moore SC, Arem H, Cook MB, Trabert B, Håkansson N, Larsson SC, Wolk A, Gapstur SM, Lynch BM, Milne RL, Freedman ND, Huang WY, Berrington de Gonzalez A, Kitahara CM, Linet MS, Shiroma EJ, Sandin S, Patel AV, Lee IM. Amount and intensity of leisure-time physical activity and lower cancer risk. J Clin Oncol. 2020;38(7):686–97. https://doi.org/10.1200/jco.19.02407.
272. Chao A, Connell CJ, Jacobs EJ, McCullough ML, Patel AV, Calle EE, Cokkinides VE, Thun MJ. Amount, type, and timing of recreational physical activity in relation to colon and rectal cancer in older adults: the Cancer Prevention Study II Nutrition Cohort. Cancer Epidemiol Biomark Prev. 2004;13(12):2187–95. https://pubmed.ncbi.nlm.nih.gov/15598779/.
273. Mahmood S, MacInnis RJ, English DR, Karahalios A, Lynch BM. Domain-specific physical activity and sedentary behaviour in relation to colon and rectal cancer risk: a systematic review and meta-analysis. Int J Epidemiol. 2017;46(6):1797–813. https://doi.org/10.1093/ije/dyx137.
274. Giovannucci E, Leitzmann M, Spiegelman D, Rimm EB, Colditz GA, Stampfer MJ, Willett WC. A prospective study of physical activity and prostate cancer in male health professionals. Cancer Res. 1998;58(22):5117–22. https://pubmed.ncbi.nlm.nih.gov/9823321/.
275. Giovannucci EL, Liu Y, Leitzmann MF, Stampfer MJ, Willett WC. A prospective study of physical activity and incident and fatal prostate cancer. Arch Intern Med. 2005;165(9):1005–10. https://doi.org/10.1001/archinte.165.9.1005.
276. Hamer J, Warner E. Lifestyle modifications for patients with breast cancer to improve prognosis and optimize overall health. CMAJ. 2017;189(7):E268–74. https://doi.org/10.1503/cmaj.160464.
277. Lahart IM, Metsios GS, Nevill AM, Carmichael AR. Physical activity, risk of death and recurrence in breast cancer survivors: a systematic review and meta-analysis of epidemiological studies. Acta Oncol. 2015;54(5):635–54. https://doi.org/10.3109/0284186x.2014.998275.

278. Spei ME, Samoli E, Bravi F, La Vecchia C, Bamia C, Benetou V. Physical activity in breast cancer survivors: a systematic review and meta-analysis on overall and breast cancer survival. Breast. 2019;44:144–52. https://doi.org/10.1016/j.breast.2019.02.001.

Herbal Medicines for the Treatment of COPD

Samir Ranjan Panda, Siddhi Jain, N. P. Syamprasad, Priyanka Adhikari, Meenakshi Singh, Alok Ranjan, A. Parthiban, and V. G. M. Naidu

Abstract

Numerous epidemiological studies have shown that chronic obstructive pulmonary disease (COPD) can lead to the development of lung cancer (COPD). The comorbidities are frequently linked to exposure to cigarette smoke, and patients with COPD have an eightfold increased risk of lung cancer death. Lung cancer can start in individuals with COPD due to the heterogeneous nature of the two conditions, which are defined by several sub-phenotypes brought on by telomere shortening, mitochondrial malfunction, hereditary susceptibility, failure in the body's defense system, and DNA repair processes. Emphysema, persistent inflammation, chronic bronchitis, and asthmatic bronchitis are frequently linked to COPD. Bronchodilators have been recognized as an effective therapeutic technique in reducing the symptoms associated with COPD via the evolution of research for more than 200 years. Finding medications that reduce inflammation and slow the course of COPD is still a necessity. Several inflammatory and chronic diseases have been treated with herbal remedies and medications. Many herbal medications are used to treat chronic illnesses like cancer, neurological diseases, and metabolic diseases. Along with powerful therapeutic advancements in the isolation, extraction, and therapeutic efficacy of herbal medications, there are numerous reports of decreased side effects and organ toxicity. Finding new and potent herbal medications with fewer side effects has become increasingly important due to an alarming rise in morbidity and mortality among COPD patients.

Keywords

COPD · Herbal medicines · Novel mechanism · Molecular targets

Samir Ranjan Panda and Siddhi Jain contributed equally to first author.

S. R. Panda · S. Jain · N. P. Syamprasad
Department of Pharmacology and Toxicology, National Institute of Pharmaceutical Education and Research (NIPER), Guwahati, Assam, India

P. Adhikari · M. Singh · A. Ranjan · A. Parthiban
Center for GMP Extraction Facility (Department of Biotechnology), National Institute of Pharmaceutical Education and Research (NIPER), Guwahati, Assam, India

V. G. M. Naidu (✉)
Department of Pharmacology and Toxicology, National Institute of Pharmaceutical Education and Research (NIPER), Guwahati, Assam, India

Center for GMP Extraction Facility (Department of Biotechnology), National Institute of Pharmaceutical Education and Research (NIPER), Guwahati, Assam, India
e-mail: vgmnaidu@niperguwahati.ac.in

1 Introduction

Tobacco smoking dates back to its origin in the Mayan kingdom in 600–900 AD [1]. Shamanistic rituals all over the Americas smoked tobacco and hallucinogenic drugs as early as 5000 BC [2]. Tobacco smoking is a worldwide epidemic and is the prime risk factor for chronic obstructive pulmonary disease (COPD), lung cancer, and heart disease. In the US, COPD and lung cancer represent the second and fourth leading causes of death, respectively [3]. According to the WHO, it is estimated around 3.17 million deaths occurred globally in 2015 (accounting for 5% of deaths globally in that year) due to COPD, which is currently ranked as the third leading cause of death in the world. The Global Burden of Disease Study in 2016 reported two hundred fifty-one million cases of COPD [4]. Evidence of airway obstruction with a usual decline in the functional parameters of the lung leads to aging and COPD. Progressive decline in lung function and airway obstruction causes dyspnoea (shortness of breath) and irreversible structural changes in the lung tissues [5]. Source contributing factors responsible for causing COPD has been described in Fig. 1.

A. K. Dhara, S. C. Mandal (eds.), *Role of Herbal Medicines*, https://doi.org/10.1007/978-981-99-7703-1_9

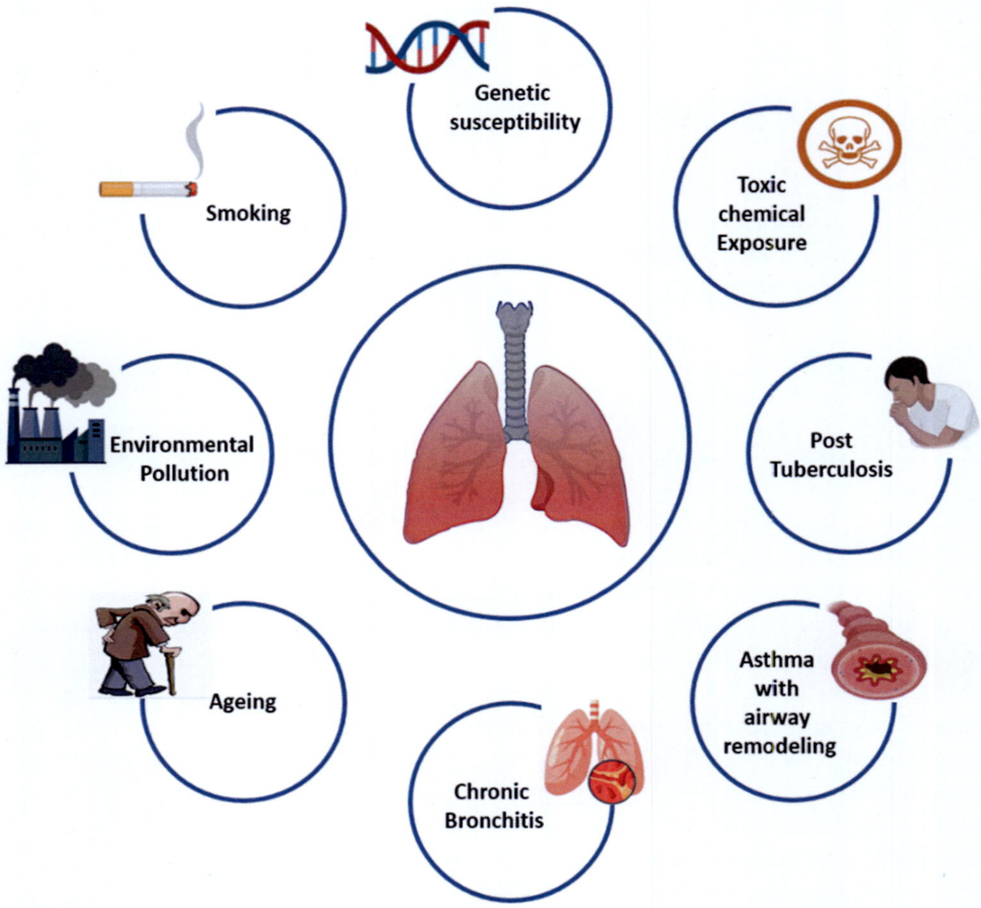

Fig. 1 Source contributing factors in causing COPD

The earliest description of COPD with the name "voluminous lungs" is given by Bonet in 1679, while Badham described the clinical understanding of chronic bronchitis and bronchiolitis of COPD in 1814. In 1846 with the invention of the spirometer by John Hutchinson, the diagnosis of COPD became easy. Christie in 1944 gives a detailed description of the individual component involved in COPD based on past research and physical examination of the patients. The physicians Barach and Beckerman 1956 wrote the book Pulmonary emphysema with a detailed description of the treatment of the disease; in this book, they also describe the expiratory airway flow volume pattern in emphysema and the use of the spirometry [6].

Herbal medicines have been used by an ancient civilization, and archaeological evidence dictates their use from the Paleolithic age. Several herbs, such as mint, clove, roses, lilies, rosemary, sage, ephedra, and fennel, have been used for their treatment against acute and chronic inflammatory diseases [7].

2 Molecular Mechanisms Involved in COPD

COPD is characterized by various processes like chronic bronchitis, emphysema, and airway remodeling, affecting all lung parts, including small and large airways and parenchyma, which contribute to chronic airway obstruction. Chronic bronchitis is a clinical feature due to a chronic increase in bronchial secretions, characterized by productive cough with submucosal glandular hypertrophy and hyperplasia, with dilated ducts in airways down to 2–4 mm in internal diameter. Emphysema is an anatomicopathological diagnosis defined by permanent destructive enlargement of airspaces distal to the terminal bronchioles, contributing to airflow limitation resulting from loss of lung elastic recoil [8]. Other features contributing to COPD are the progressive loss of the alveolar wall, vascular remodeling with pulmonary hypertension, and peribronchiolar fibrosis [9].

2.1 Role of Inflammation and Oxidative Stress

Earlier evidence confirms that biomass fuels, smoking, air pollution, and industrial dust are the principal elements of the environmental hazard causing COPD [10]. According to the American Lung Association, cigarette smoke contains several harmful constituents and is a well-recognized risk factor for about 85–90% of all COPD cases [11]. Cigarette smoke contains more than 10,000 compounds, mainly constituting nicotine and polyaromatic hydrocarbons [12]. On its continuous exposure, it activates various immune cells and airway epithelial cells by releasing various pro-inflammatory cytokines. Cigarette smoke generates lots of oxidative free radicals, which generate oxidative stress by activating various inflammatory pathways [13], as depicted in Fig. 2. Chronic exposure to cigarette smoke releases a greater amount of proinflammatory cytokines and chemokines such as tumor necrosis factor-alpha (TNFα), interferon-gamma (IFNγ), interleukins, monocyte chemoattractant protein-1 (MCP-1), matrix metalloproteinase-12 (MMP-12), and macrophage inflammatory protein-2 (MIP-2) by activating immune cells, alveolar macrophages, airway epithelial cells, smooth muscle cells, and fibroblasts [14]. The underlying cellular and molecular mechanisms involved in the onset of COPD and asthma are comparable at earlier stages. Even though both diseases include inflammation as a primary criterion for disease pathogenesis, the inflammatory signaling and therapeutic response patterns are subtler.

2.2 Role of Growth Factors

Previous studies put forward the role of growth factors like transforming growth factor beta (TGFβ), epidermal growth factor (EGF), fibroblast growth (FGF), platelet-derived growth factor (PDGF), and vascular endothelial growth factor (VEGF) in airway tissue remodeling in COPD [15]. The vascular and epithelial cells of COPD subjects showed increased expression of FGF-1, FGF-2, and FGFR1 [16]. Earlier reports also suggest that FGF–FGFR axis plays a role in tissue remodeling by increasing collagenase expression in lung fibroblasts [17]. TGFβ is another crucial growth factor involved in wound repair, overexpression of which results in epithelial to mesenchymal transition (EMT) where activated epithelial cells cause the formation of flattened mesenchymal traits leading to peribronchiolar fibrosis of small airways [18]. On the other hand, TGFβ also acts as chemotactic for cytotoxic T cells, neutrophils, and macrophages and causes activation of various proteases through EGFR [19] for release of matrix metalloproteinase (MMP-9/12), neutrophil elastase resulting in alveolar wall destruction and enhanced

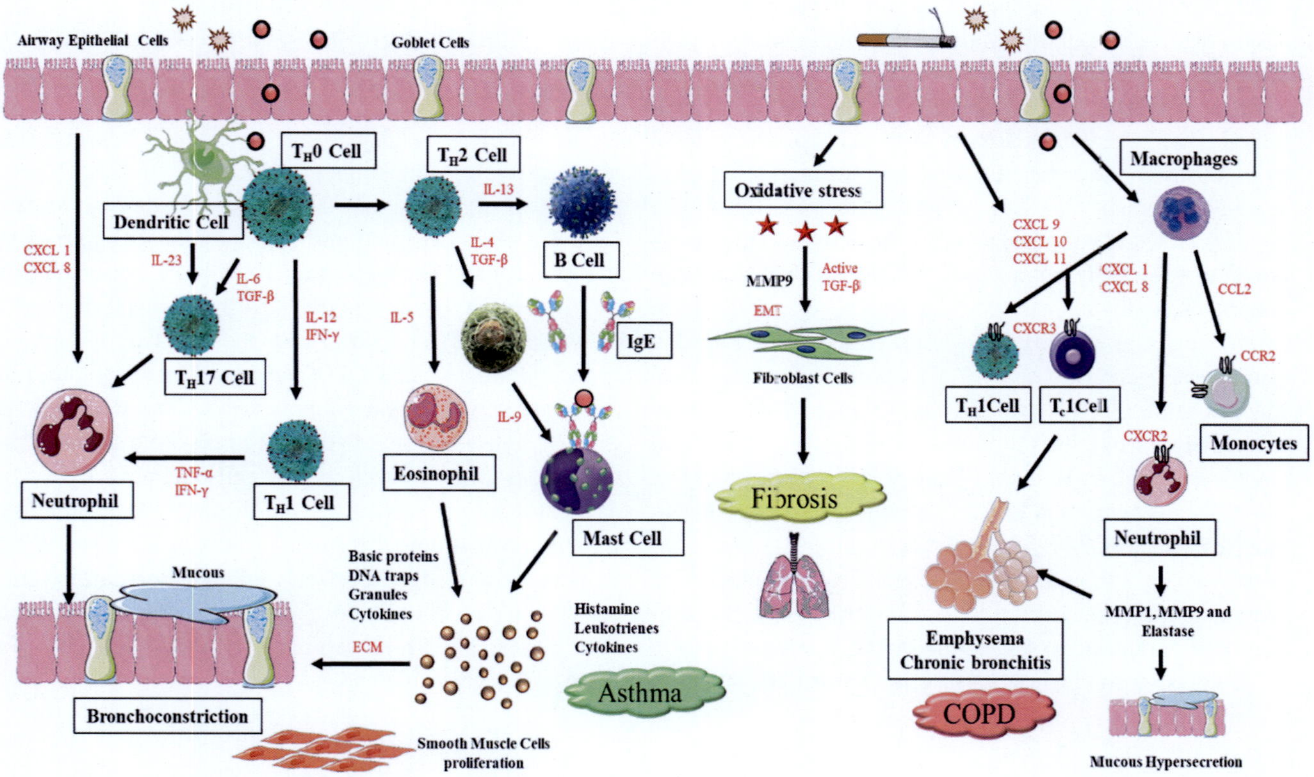

Fig. 2 Graphical representation of mechanism involved in initiation and progression of COPD

mucous secretion from goblet cells of airway tract [20]. VEGF is a chief controller of vascular growth functions involved in pulmonary vascular remodeling, which occurs due to vasoconstriction and collapse of the pulmonary circulation in severe COPD [21]. VEGF has a diverse function in chronic bronchitis and emphysema. Former evidence hypothesize the role of increased expression of VEGFR1 in emphysema and VEGFR2 in chronic bronchitis [22]. VEGF has a higher affinity for VEGFR1 than VEGFR2, resulting in vascular endothelial cell apoptosis, higher production of MMP, and alveolar and vascular destruction leading to emphysema [23]. In contrast, VEGF has a higher affinity for VEGFR2 than VEGFR1, resulting in angiogenesis and vascular remodeling leading to chronic bronchitis [24]. Hence, balance among VEGF, VEGFR1, and VEGFR2 is critical for airway disease pathogenesis.

2.3 Role of Proteases

Proteases are set of enzymes involved in inflammatory responses via the breakdown of elastin, a connective tissue network in the lung parenchyma, thereby creating an imbalance between endogenous antiproteases and proteases, leading to the development of emphysema [25]. Long ago, it was proposed that neutrophil elastase (NE), besides having strong elastolytic activity, is a potent stimulator of the MUC5AC gene, thereby resulting in excess mucus secretion in airway epithelial cells [26]. There is increasing evidence for the role of antiproteases, namely, α1-antitrypsin (α1-AT), in early-onset emphysema features [27]. α1-AT has a potent inhibitor of Proteinase 3, which is expressed on the surface of cytokine-activated neutrophils [28]. Inheritant deficiency of α1-AT or its inactivation by cigarette smoke exposure raised the possibility of emphysema development [29]. Several other cathepsins (lysosomal cysteine proteases) also show elastolytic activity, and overexpression of IFN-γ by smooth muscle cells and alveolar macrophages markedly increase the expression of cathepsins in emphysema subjects [30]. Most of the MMPs are involved in disease pathogenesis, but MMP-12 is prominent in mice, and MMP-9 is most important in humans [31]. The interest in the role of MMPs has sharply increased in the pathogenesis of emphysema and small airway fibrosis. Increased activity of MMP-1 (collagenase) in BAL concentrates and type II pneumocytes, and MMP-9 (gelatinase B) in lung parenchyma of emphysematous subjects produce chemotactic peptides which encourage macrophage recruitment in airways [32]. In addition, MMP9 can convert latent form of TGFβ to its active form by elastolysis resulting in small airway fibrosis [33].

2.4 Other Signaling Mechanisms Involved

The inflammatory changes, oxidative stress, and protease imbalance are equally responsible for impaired lung function in both smoker and non-smoker COPD subjects. The increased profile of lipid mediators like prostaglandins (PGE_2 and $PGF_{2\alpha}$) and leukotrienes (LTB_4) were found in sputum and exhaled breath condensates of COPD patients [34]. Another central molecular mechanism involved in the amplification of inflammatory responses in COPD is due to reduced HDAC activity in alveolar macrophages and peripheral lungs [35].

3 Novel Strategies and Molecular Targets in COPD

The current first-line therapeutic drugs used in treating COPD deliver bronchodilatory effects, providing only symptomatic relief, and ultimately were of no benefit. However, the underlying mechanism intricated in COPD is still not targeted due to reliance on reliever medication rather than preventer medication. Strong evidence exists for the orchestra of inflammatory response and oxidative stress in COPD. Molecules inhibiting these responses can provide a most promising approach to monitoring COPD.

3.1 Targeting NF-κB Inflammatory Signaling in Obstructive Airway Disease

NF-κB family is believed to play a cardinal role in diverse acute and chronic inflammatory diseases. In the last 20 years, NF-κB signaling has gained the focus of extensive research [36]. NF-κB is activated utilizing numerous stimuli, counting physical and chemical stress in multiple steps controlled by diverse enzymatic signaling. Further, several markers of activated NF-κB signaling were identified in sputum and bronchial biopsies of COPD patients [37]. Eucalyptol [38], Asiatic acid [39], *Zanthoxylum bungeanum* seed oil [40], and Ivy (*Hedera helix*) leaf extract [41, 42] are a few of many other plant extracts which showed anti-inflammatory activity against chronic obstructive disease via inhibiting NF-κB. PM014 herbal formula, a mixture of seven spices, exhibited more significant anti-inflammatory effects in a COPD mice model [43, 44]. Thus, anti-inflammatory molecules possessing NF-κB inhibitory activity can greatly promote lung repair.

3.2 Targeting NLRP3 Inflammasome Signaling in Obstructive Airway Disease

Recent evidence suggests the role of NLRP3 inflammasome and associated markers in airway inflammation [45]. Firstly, NLPR3 inflammasome formation is triggered under stress conditions owing to cellular and tissue damage by PAMPs, DAMPs, and toxins [46]. Secondly, the active form of caspase-1 is released in the process of inflammasome formation, which cleaves pro-IL-1β and pro-IL-18 further to orchestrate the release of mature cytokines [47]. Therefore, manipulating the inflammasome formation using NLRP3 inhibitors or blocking selective caspase-1 inhibitors can mitigate airway inflammation in COPD subjects. Melatonin [48, 49], (−)-Epicatechin [50], Silybin [51], Histidine [52], and Magnesium isoglycyrrhizinate [53] are a few natural molecules displayed prominent NLPR3 inhibitory activity in vitro and in vivo COPD models. *Hibiscus noldeae* Baker f. [54] and Berberine [55] revealed significant repressing effects on caspase-1 activities, thereby overcoming inflammation and apoptosis of airway epithelial cells. Molecules with dual NLRP3 and caspase-1 inhibitory activity could ultimately benefit managing COPD.

3.3 Targeting Cytokine Release in Obstructive Airway Disease

Mounting evidence conferred cytokines and chemokines' involvement in airway inflammation and remodeling, goblet cell hyperplasia, mucus secretion, and pyroptosis [56]. Hence, inhibiting the release of cytokines could be a promising approach to underpin progressive inflammation. Leaf extracts of *Azadirachta indica* [57], *Perilla frutescens* [58], and *Eriobotrya japonica* [59, 60] are a few examples of herbal extracts suggesting their potent use in the treatment of COPD via impeding the production of cytokines. Accordingly, this class of drugs can represent a target for novel therapies in COPD subjects.

3.4 Targeting Proteases in Obstructive Airway Disease

A line of recent studies believed that elastolytic proteases released by immune cells trigger elastin and other ECM protein destruction in the alveolar lining [61]. Anti-proteases are better therapeutic agents to block elastolytic activity. Serine protease inhibitors block the activity of neutrophil elastases and other serine proteases their reverse the impaired proteases and anti-proteases balance [62]. Circulating α1-antitrypsin is an endogenous anti-protease in the human body that blocks neutrophil elastase activity. Apocynin improves emphysema in the hamster model by selectively inhibiting NADPH oxidase and protecting the activity of endogenously released secretory leukocyte protease inhibitor (a protease inhibitor) [63]. Anemoside B4, a bioactive component of *Pulsatilla chinensis,* played a vital in restoring the protease and anti-protease imbalance in the CS-induced COPD murine model [64].

3.5 Targeting Kinases in Obstructive Airway Disease

Inflammatory and oxidative stress signaling activates the p38MAPK cascade [65]. The degree of activation of p38MAPK is directly related to the severity of lung function impairment and airway inflammation [66]. The first-line therapeutic drug, inhaled corticosteroid, requires glucocorticoid receptor (GR) to mediate its anti-inflammatory activity. Glucocorticoid resistance in COPD subjects could be due to the ability of p38MAPK to phosphorylate glucocorticoid receptor (GR), thereby restricting nuclear translocation and DNA binding of GR [67, 68]. On the other hand, phosphatidylinositol 3 kinase (PI3K) is a significant controller of diverse cellular functions, including proliferation, survival, autophagy, metabolism, and angiogenesis [69]. Activated PI3K phosphorylates phosphatidylinositol diphosphate to phosphatidylinositol triphosphate, further activating protein kinase B (AKT). Further, dysregulated PI3K/AKT signaling is involved in immune cell activation, the release of inflammatory mediators, airway remodeling, excess calcium release, and bronchoconstriction [70]. Upregulated PI3K/AKT signaling downregulates histone deacetylase HDAC2 resulting in glucocorticoid resistance [71]. Molecules targeting p38MAPK and PI3K/AKT signaling can be better therapeutic targets by impeding the inflammatory processes and overcoming glucocorticoid resistance. *Bufei Huoxue* capsule [72], Silymarin [73], and Theaflavin-3,3′-digallate [74] attenuate cigarette smoke-induced COPD via suppressing ERK/p38MAPK signaling in the in vitro and in vivo model. *Castanea crenata* inner shell extract [75], *Citrus grandis exocarpium* [76], and Glycyrrhizic acid [77] are examples shown in the current review to have an inhibitory effect on PI3K/AKT signaling to display potential benefits against cigarette smoke-induced emphysematous lesions in mice model.

3.6 Targeting Phosphodiesterase in Obstructive Airway Disease

Cyclic adenosine monophosphate (cAMP) is a secondary messenger vital in mitigating COPD's two major pathophysiological factors. An elevated level of cAMP helps relax the smooth muscle and abate inflammation. cAMP is regarded as a pompous target in treating chronic respiratory inflammatory obstructive disease [78]. Phosphodiesterases (PDE) are a superfamily of enzymes that cyclic hydrolysis nucleotides like cAMP and cGMP to their corresponding inactive derivatives AMP and GMP and help in regulating the intracellular levels of these cyclic nucleotides [79]. Altogether, PDE promotes bronchoconstriction by reducing the cellular levels of functional cAMP and elevating intracellular calcium levels. Henceforth, molecules targeting PDE enzyme or increasing cAMP levels can be beneficial in providing bronchorelaxation in COPD subjects. Commercial *Thymus vulgaris* extracts increased cAMP levels and reduced intracellular calcium levels in the reconstituted in vitro system of MucilAir 3D human COPD airway epithelium [80]. *Pistacia weinmannifolia* [81] and *Psidium guajava* [82] showed PDE inhibitory activity in lung tissue. Isoforskolin, an active component of Coleus forskohlii, is an activator of adenylyl cyclase, thereby enhancing the cellular levels of cAMP and mitigating Cigarette smoke-induced COPD in a rat model [83].

3.7 Targeting Oxidative Stress in Obstructive Airway Disease

Similar to inflammation, oxidative stress is a significant factor in the pathogenesis and exacerbation of airway disease inflammation [84]. Oxidative stress is known to intensify immune cell infiltration into the lung, the terrible release of pro-inflammatory cytokines, and the apoptosis of alveolar epithelium responsible for airway hyperresponsiveness and alveolar obliteration [85]. Repair of impaired oxidant and anti-oxidant mechanisms could mitigate oxidative stress-induced alveolar damage. Andrographolide [85], Dihydroquercetin [86], Resveratrol [87], Oroxylin A [88], Propolis [89], and Ursolic acid [90] promote lung repair via restoring anti-oxidant capacity through activation of Nrf2 pathway in COPD murine model.

4 The Natural History of Herbal Medicine

Since ancient times, to cure diseases, human beings have searched for practical solutions in Mother Nature and found most of the remedies from the plant herbs abundant in nature. In the beginning, the use of plant herbs was mainly limited to animals as the knowledge about diseases and plant herbs was limited, but gradually the use increased for diseases related to human beings as the knowledge about plant herbs increased [91]. Hence, herbal medicine has traveled a long route throughout the centuries in the form of the knowledge and skills that pass through the different generations/civilizations in the families/society for treating diseases using the medicinal plant herbs surrounding us. Traditionally, the treatment given by the use of herbal medicine is based on the holistic approach of focusing on the health rather than the disease, so the treatment methods vary according to the geographical location and environmental conditions. Also, many written historical scripts describe the use of medicinal plants for herbal medicine. For example, for details regarding Indian Vedic scripts, Pen T'Sao by China Emperor Shen Nung, Ebers Papyrus by Georg Ebers, and De Materia Medica by Dioscorides etc., refer Table 1. The oldest evidence about the plant herbs was found in the Sumerian clay slab from Nagpur, and Dioscorides, known as the father of pharmacognosy, wrote details about the plant herb's appearance, local name, collection processes, location, herbal medicine preparation, and their therapeutic effect [7, 92, 99].

Between the sixteenth and eighteenth centuries, compound drugs, which consist of plant herbs and drugs from plants and animals, increased. So, by seeing the importance of herbal medicine, Linnaeus, in 1707–1788, wrote a book called Species Plantarum in which he described the classification of the species and coined the word genus for the polynomial system. In the nineteenth century, the discovery of alkaloids, glycosides, vitamins, hormones, etc., from the plant poppy, quinine, and pomegranate, etc., initiated the scientific pharmacy for herbal medicine. With the end of the nineteenth and the start of the twentieth century, the decline in trust for herbal medicine started due to shortcomings caused by the fundamental changes in the plant herbs during the drying process. Another reason for the decline of traditional medicine is the development and mass production of chemically synthesized drugs, which give a faster therapeutic effect than herbal medicine. So, the stabilization method was developed to tackle the problem related to the imbalanced herbal medicine composition, which consists of the standardization of the cultivation and manufacturing conditions. Also, the development of the chemical and physiological studies for the numerous plant herbs along with the clinical studies to create a database of the drug from the plant-based herbs. Currently, pharmacopeias like the United State Pharmacopoeia XXXI, British Pharmacopoeia 2007, and European Pharmacopeia 6 describe the instruction for preparation, uses, and effects of herbal medicine [7, 100, 101].

Table 1 Historical evidence of plant herbs found in the ancient scripts

Place/author	Book name/year	Plants herbs/No. of plant herbs or recipes
Nagpur, Sumerian clay slab	5000-year-old	12 recipes prepared from 250 plants and herbs [92]
China, Emperor shen Nung	Pen T'Sao, 2500 BC	365 plant herbs, such as Rheirhisoma, camphor, Theae folium, Podophyllum, ginseng, and ephedra etc. [7]
India	Vedas	Turmeric, pepper, Indian sandalwood, clove, nutmeg etc. [93]
Germany, Georg Ebers	Ebers Papyrus, 1550 BC	700 plant herbs, such as pomegranate, castor oil plant, onion, fig, willow, aloe, garlic, coriander, juniper, etc. [94]
Homer's	The Iliad and the odysseys, 800 BC	63 plant herbs, such as *Inula helenium* L. Asteraceae, and *Artemisia etc.* [7]
Herodotus	500 BC	Plant herbs such as castor oil plant, Orpheus to the fragrant hellebore and garlic, and Pythagoras to the sea onion, mustard, and cabbage etc. [95]
Hippocrates	459-370 BC	300 plant herbs, Wormwood and *Centaurium umbellatum Gilib* were applied against fever; opium, henbane, fragrant hellebore, and haselwort as emetics; garlic against intestine parasites; sea onion, celery, parsley, asparagus, and garlic as diuretics; oak and pomegranate as astringents; deadly nightshade, and mandrake were used as narcotics etc. [7]
Theophrast	371–287 BC, De Causis Plantarium, Historia Plantarium	500 plant herbs such as cinnamon, iris rhizome, false and fragrant hellebore, mint, pomegranate, cardamom, and monkshood [96]
John Mesue	De re medica, 25 BC-50 AD	250 plant herbs, such as aloe, henbane, flax, poppy, pepper, cinnamon, the star gentian, cardamom, false hellebore, etc. [97]
Rome, dioscorides	De Materia Medica, 77 AD	657 plants herbs, such as fragrant and false hellebore, jimson weed, henbane, deadly nightshade, poppy, buttercup, willow, camomile, garlic, nettle, sage, common centaury, coriander, parsley, onion, marsh mallowivy, sea onion etc. [98]
Pliny, the elder	Historia naturalis, 23 AD	1000 plant herbs, including the Dioscorides work [7]
Avicenna	Canon Medicinae, 980–1037	1000 plant herbs [97]
Ibn Baitar	Liber magnae collectionis simplicum alimentorum Et medicamentorum	1000 plant herbs [97]

5 Herbal Medicines: As Affordable Precision Medicine

Despite the development and large production of chemically synthesized medicine, herbal medicine is mostly used in developing countries for primary care due to affordability, safety factor, and local people's knowledge about herbs. Now the interest shown by people from developed nations in using herbal medicine is due to the promotion of health consciousness, prominent effects in treating chronic diseases, negligible side effects compared to synthetic medicine, and preparation of the personalized herbal formulation for each patient. Recently, personalized medicine is replaced by the term precision medicine, which utilizes the patient's medical history related to genomics, metabolomics, proteomics, and environmental factors for preventing, detecting, monitoring, and curing diseases [102]. By using the omics applications tools and technologies that are focused on improving the efficacy of the herbal medicine for the new treatment option or improving the current treatment, which does not have an adequate therapeutic effect. The concept of precision medicine can be successful when the accumulation of knowledge about the methodology to analyze the parameter by advanced technology, patient clinical and demographical history, and digital medical electronic equipment's analysis data can be gathered for the precise diagnosis of the disease. Precision medicine for herbal medicine is an unexplored area and needs a more systematic research and development approach to validate the data and lead herbal medicine toward increased therapeutic efficacy [103].

Due to the recent COVID-19 pandemic, the common people are understanding the importance of herbal medicine and trying to utilize routine herbal medicine and other herbal supplements to reduce the chance of getting any disease. It motivated the researchers to explore the field of herbal medicine and take steps/initiatives toward standardizing precision medicine for herbal formulation. The diseases which use or move towards precision herbal medicine mainly include cancer therapeutics, diabetes, hypertension, and HIV/AIDS [103]. Although precision medicine is developed for COPD in the case of conventional medicine and herbal medicine is available for the treatment of COPD, not much attention has been paid to the development of precision medicine for the herbal formulation [104–106].

6 Novel and Marketed Herbal Medicine Available for the Treatment of COPD

COPD is one of the major concerns for the pharmaceutical industries; from olden times, there has been extensive use of novel and herbal mixtures preparations to extravagance COPD, predominantly in countries of Asia, i.e., China and India [107]. Herbal medicine and formulations are plant-based or preparation that practices several plant ingredients in the treatments and curations of preventive and therapeutic measures [83]. Herbal formulations comprise numerous dynamic plant-based secondary metabolites containingvarious health assistances against diverse disease symptoms [108]. Herbal cough syrup, formulated by extracts from *Hedera helix* and *Thymus vulgaris,* is one the most significant drugs on acceptance in several European countries [109]. Studies on herbal products containing *P. ginseng*, *H. helix*, *S. miltiorrhiza,* and some traditional Chinese herbal medicine decoctions generated encouraging results. However, interpretation and extrapolation of these results are difficult for a number of reasons [109, 110].

Oral injection of curcumin is also reported for pointedly reducing COPD in rats' cell by reducing the levels of IL-6, IL-8, and TNF-α after treatment. This shows that curcumin has potential value in treating and preventing COPD [111]. *Echinacea purpurea* conventionally been used to avert upper respiratory contagions [112]. One of the most studied herbal formulations for the treatment of COPD is Bufeiyishen formula I, II, and III. The main ingredients in the formulation of different medicinal plants include *Panax ginseng*, *Astragalus tibetanus*, *Cornus officinalis*, *Lyciumbarbarum*, *Schisandra arisanensis*, *Fritillaria thunbergia*, *Perilla frutescens*, *Citrus sinensis*, *Epimedium acuminatum*, *Paeonia anomala*, *Pheretima aspergillum,* and *Ardisia japonica.* Additionally, secondary metabolites were isolated from them, i.e., Ginsenoside, astragaloside IV, icariin, and nobiletin paeonol. All the authors conducted studies in rats, and they concluded that Bufeiyishen formula I, reducing airway mucus hypersecretion in COPD [113], also Bufeiyishen formula II, significantly inhibited mucus hypersecretion, which may be related to the regulation of the EGFR/PI3K/mTOR pathway [114]. Bufeiyishen formula I exerted its anti-COPD efficacy by restoring the Th17/Treg balance via activating A2aR, which may help to elucidate the underlying immunomodulatory mechanism of BYF and provide evidence for its clinical application in COPD treatment [115]. Details on different herbal formulations are given in Table 2.

In the market, numerous herbal tablets are also available as a precaution for COPD. Most of them have positive effects on lung cleansing, e.g., lung detox tablets from Trurbasics, Foresta organics, Pure nutrition (Naturals), Nature sure, Gaia herbs, and Redd remedies. These all-herbal tablets majorly contain *Glycyrrhiza glabra, Echinacea urpurea and* Gingever. The positive effect of *G. glabra,* and *E. urpurea*on COPD were previously reported by many workers [112, 120]. We found herbal tablets from the company Shaleen name Breath-easy which was particularly reported for COPD; its herbal formulation has 16 herbs or medicinal plants, i.e., *Salanumsurrsttense, Adhatodavasica, Glycyrrhiza glabra, Ocimum sanctum, Clerodendrum serratum, Aconitum ferox, Tinospora cordifolia, Zinziber officinale, Piper nigrum, Piper longum, Datura mete, Bambusa arundinacea, Myristica gragrans, Elettaria cardamomum, Syzygium aromaticum,* and *Cinnamomum zeylanicum.* A list of commonly used herbal tablets and their compositions is given in Table 3.

Table 2 List of herbal formulations studied in COPD patients or in animal models

S. No	Herbal formulations	Method	Type of study	Results	References
1	**Bufeiyishen formula III** Composition: Ginsenoside, astragaloside IV, icariin, nobiletin paeonol and N-acetylcysteine	Boiling	Rats	Formulation was reported for reducing airway mucus hypersecretion in COPD	[113]
2	**Bufeiyishen formula I** **Ingredients**: 9 g *Panax ginseng* rhizome, 15 g *Astragalus tibetanus* rhizome, 12 g *Cornus officinalis* fructus, 12 g *Lyciumbarbarum* fructus, 9 g *Schisandra arisanensis* fructus, 9 g *Fritillaria thunbergia* bulbus, 9 g *Perilla frutescens* fructus, 9 g of *Citrus sinensis* Pericarpium, 9 g of *Epimedium acuminatum* folium, 9 g *Paeonia anomala* rhizome, 12 g *Pheretima aspergillum* herba, and 15 g *Ardisia japonica* herba	Decoction and steam sterilization	Rats	Formulation exerted its anti-COPD efficacy by restoring the Th17/Treg balance via activating A2aR, which may help to elucidate the underlying immunomodulatory mechanism of BYF and provide evidence for its clinical application in COPD treatment	[115]
3	**Bufei yishen formula II** **Composition**: Ginsenoside, astragaloside, icariin, paeonol, nobiletin	Boiling	Rats	Formulation significantly inhibited mucus hypersecretion, which may be related to the regulation of the EGFR/PI3K/mTOR pathway	[116]
4	**Maxing shigan** **Ingredients**: Ephedra, almonds, gypsum, and licorice. The specific ratio of ephedra and gypsum is 1:2	Decoction	Humans	Formulation was reported to have positive influence on COPD by direct effects on inflammatory factor	[117]
5	**Xiaoqinglong** **Ingredients**: *Ephedrae sinicae* herba, *Cinnamomi ramulus*, *Pinelliae* rhizome, *Zingiberis* rhizome, *Asari Radix* rhizome, *Schisandrae Chinensis* fructus, *Paeoniae Alba* radix, *Glycyrrhiizae* radix	Decoction	Humans	Increased FEV1, FVC, and peak expiratory flow; decreased levels of serum cytokines such as IL-4, IL-8, and TNF-α	[118]
6	**Bufelikeli** **Ingredients**: 8 g/bag of *Codonopsis* rhizome, *Rehmanniae Praeparata* rhizome, Corni fructus, Ephedra Sinicaherba, and *Pericarpiumcitri* reticulatae	Boiling with water	Humans	Bufei granule can treat patients with stable COPD by lowering the frequency of acute exacerbation, improving the quality of life, and alleviating the severity of inflammation	[118]
7	**Bufeiyishen** **Ingredients**: 15 g *Astragalus tibetanus* rhizome, 15 g *Panax ginseng* rhizome, 15 g *Corni* fructus, and 9 g *Schisandrae* arisanensis Fructus	Decoction and steam sterilization	Rats	Expression of numerous interleukins, i.e., IL1*β*, IL6, IL8, and IL10 was decreased in peripheral blood and bronchoalveolar lavage fluid by formulations treatment	[116]
8	**Xiaoqinglong** **Ingredients**: *Ephedraesinicae*, *Cinnamomiramulus*, *Pinelliae* rhizome, *Zingiberis* rhizome, *Asari* rhizome, *Schisandrae chinensis* Fructus, *Paeoniae alba* radix, *Glycyrrhiizae* radix	Decoction	Humans	Decreased the total cell counts and percentage of neutrophils in induced sputum; increased expression of matrix metalloproteinase-9 (MMP-9)	[119]
9	**Qianjinweijing** **Ingredients:** *Ophiopogonis* rhizome, *Ginseng* rhizome, *Glycyrrhizae* rhizome, *Pinelliae* rhizome, *Phragmitis* rhizome, *Coicis* semen, *Benincasa* hispida, *Persicae* semen	Decoction	Humans	Formulation plays an important role in COPD treatments by restraining inflammation of the patient	[120]

(continued)

Table 2 (continued)

S. No	Herbal formulations	Method	Type of study	Results	References
10	**Maxing Shigan** **Ingredients**: *Ephedrae sinicae*, *Armeniacaeamarum*, and *glycyrrhizae*	Decoction	Rats	Decreased levels of cytokines such as IL-4, IL-8, and TNF-α; increased the level of IFN-; lowered the ratio of IFN- and IL-4 (Th1/Th2) compared to control rats in the homogenate of lung tissue from rats with COPD and syndrome of phlegm-heat obstructing lung	[121]
11	**Bushennaqihuoxue** **Ingredients**: Ginseng, clams, fairy spleen, walnut meat, psoralen, schisandra, danshen, safflower, peach and kernel	Decoction	Humans	Clinical study confirm that formulation has a good effect on COPD patients, can relieve cough, phlegm and asthma symptoms, improve lung function, and conform to the treatment principles of COPD	[122]
12	**Bakumondo** **Ingredients**: 10 g *Ophiopogonis* tuber, 5 g *Oryza sativa*, 5 g *Pinellia ternate*, 3 g *Ziziphus jujuba*, 2 g *Panax ginseng* and 2 g *Glycyrrhiza* root	Boiling of plant extract in water	Humans	Formulation did not appear to substantially affect the COPD in man	[123]

Table 3 List of commonly used herbal medicine and formulation available in market for COPD

S. No.	Name/brand	Composition	Biological applications
		Plant extracts/secondary metabolites	
1	Trurbasics/lung detox tablet	Quercetin, *Glycyrrhiza glabra*, vitamin K2, *Echinacea urpurea*, *Inula racemosa*, turmeric, vitamin C, *Stinging Nettle*, Lotus, gingever, pine bark extract, and grape seed extract	• Lung cleansing • Relieves breath illness • Boosts immune systems • Potential antioxidant activity
2	Foresta organics/lung detox tablet	Malabar nut, *Glycyrrhiza glabra*, *Echinacea*, curcumin, *Arjuna*, beetroot, *Stinging* nettle, kateri, trikuta, punarnava, ginger and piperine	• Ease lung congestion • Cleaning and purify lungs • Improves breathing • Boost immunity • Fights against lung damage
3	Himalayan organics/lung detox tablet	Vasaka leaf, arjuna bark, Beetroot, stringing nettke root, kateri, trikatu, punarnava roots, echinacea roots, curcumin, and piperine	• Relied from mucus • Stronger respiratory systems • Lessens inflammation • Healthier breathing
4	Pure nutrition (naturals)/ lung detox tablet	*Stinging* nettle extract, quercetin, *Echinacea*, lotus, ginger, licorice, *Piper nigrum*, *Inula Racemosa* root extract, vitamin C, and grape seed extracts	• Protects the lungs against the harmful effects of pollution • Helps detoxify the lungs • Improves breathing by easing congestion • Boosts respiratory health
5	Nature sure/lungs pure tablet	Vasika, guduchi, dhanyaka, kantakari, haridra, shunthi, bharangi, pipali, pine bark, oregano	• Purifies and detox lungs • Heals tissue damage due to the pollutions • Provides relief from smoker's cough • Boosts lung immunity and strength
6	Shaleen/breath-easy	*Salanumsurrsttense*, *Adhatodavasica*, *Glycyrrhiza glabra*, *Ocimum sanctum*, *Clerodendrumserratum*, *Aconitum ferox*, *Tinospora cordifolia*, *Zinziber officinale*, *Piper nigrum*, *Piper longum*, *Datura mete*, *bambusaarundinacea*, *Myristica gragrans*, *Elettaria cardamomum*, *Syzygiumaromaticum*, *Cinnamomum zeylanicum*	• Potent bronchodilator • Helpful in liquefying and relieving nasal and bronchial congestion • Its antimicrobial action combats infections caused by Gram-positive and Gram-negative bacteria • Useful in chronic obstructive pulmonary disease (COPD)
7	Gaia herbs/mighty lungs	*Olea europaea*, *Ocimum sanctum*, Mullein, *Althaea officinalis*, Plantain, *Inula helenium*, Grindelia, *Ceanothus americanus*, Hawthorn and *Schisandra chinensis*	• Support overall lung and respiratory health • Maintain healthy lung function
8	Redd remedies/lung care	*Adhatodavasica*, *Verbascum thapsus*, *Frillaria thunbergit*, *Zingiber officinale*, *Tylophora asthmatica*, *Ganoderma lucidum* and Vitamin C	• Promotes healthy lung functions • Supports clear respiratory passages • Supports healthy breathing capacity
9	Organic India/breathe free	*Inula racemose*, *Ocimum sanctum*, *Terminalia belerica*, *Piper longum*	• Relieves asthma and other respiratory complications • Naturally relieves • Shortness of breath • Relieves coughing • Improves vital capacity • Promotes expectoration
10	Planetary herbals/mullein lung complex	Platycodon, ophiopogon, long pepper fruit, black cherry, cassia, elecampane, mullein, licorice, and ginger	• Support overall lung and respiratory health • Maintain healthy lung function

7 Secondary Metabolites Involved in the Inhibition of COPD

Herbal plants are the traditional and greatest extensive system of the drug for the cure of numerous human sicknesses all over the world. Usually, the most successful sources of potential natural products drugs have been derived from the plants [124]. Plant-derived natural products are categorized by huge chemical units with molecule structure and it shows various biomedical applications [125]. Plant-derived natural products such as Resveratrol (**1**), Curcumin (**2**), Emodin (**3**), Linalool (**4**), Carvacrol (**5**), Quercetin (**6**), Forsythiaside (**7**), Berberine (**8**), Baicalin (**9**), Chrysin (**10**), Casticin (**11**), Oroxylin A (**12**), Asiatic acid (**13**), Lcariin (**14**), Andrographolide (**15**), Betulin (**16**), Eucalyptol (**17**), Epigallocatechin-3-gallate (**18**), p-Coumaric acid (**19**), Salidroside (**20**), Silymarin (**21**), Tanshinone IIA (**22**), Ursolic acid (**23**), and Phloretin (**24**) (Fig. 3) showed promising activity, particularly in the inhibition of COPD [126]. Numerous reports revealed that many of the natural products have employed efficiently for the treatment of COPD in current studiesand it also exhibit efficacy on chronic lung inflammation related to COPD as summarized in Fig. 3.

Resveratrol (**1**) (Fig. 3) is a natural polyphenol antioxidant agent and also called as 3,5,4′-trihydroxystilbene. It was mainly identified from the roots of the white hellebore *Veratrum grandiflorum* and *Polygonum japonicum* [127]. It has been used as traditional medicine such as asthma, antiseptic, wound healing for almost 2000 years [128]. It is important to consider that resveratrol have been to found to show anti-inflammatory activity is displayed in COPD [129, 130]. Curcumin (**2**) is a chief component of *Curcuma longa* (usually named as turmeric) and used as traditional medicine in the treatment of various inflammatory disorders (Fig. 3). In addition to that, it shows different pharmacological including antioxidant, antimicrobial, anti-inflammatory, anticancer, and antidiabetic properties. Curcumin might be an efficient COPD agent by preventing NF-κB signaling in CS-induced COPD mice [131].The emodin (**3**) is known as 6-methyl-1,3,8-trihydroxyanthraquinone Fig. 3, an important compound and it was isolated from the *Ventilago madraspatana*. It shows anti-inflammatory activity *via* elevation of Nrf-2 activation. Linalool (**4**) (Fig. 3) is a fragrant monoterpene alcohol, and it was isolated mainly from *Melissa officinalis* and it displayed COPD property by preventing NF-κB initiation [132]. Carvacrol (**5**) Fig. 3 is a terpenoid based compound which was identified from *Zataria multiflora* and which were used as traditional medicine such as an antiseptic and anti-inflammatory properties. Quercetin (**6**) is flavonoid-based polyphenol compound and it was obtained from *Polygonum aviculare* medicinal plants. Quercetin (Fig. 3) reduces chronic obstructive pulmonary disease by Nrf2 activation [133]. The compound forsythiaside (**7**, Fig. 3) contains 4-hydroxycinnamic acid moiety, and it was isolated from *Forsythia suspense* herbal plant. It exhibited COPD properties through inhibition of NF-κB signaling pathway [134]. It is known that the alkaloid berberine (**8**) Fig. 3, from the Mexican traditional medicine plant *Argemone ochroleuca,* in the treatment of asthma, through NF-κB activation [135, 136]. A flavonoid molecule Baicalin (**9**) (Fig. 3) has been identified from the root of *Scutellaria baicalensis* and it was exhibited promising HDAC2 activity [137]. Chrysin (**10**) Fig. 3 was obtained from medicinal plant of *Passiflora caerulea* and showed better anti-inflammatory properties *via* potency of ERK and p38 phosphorylation (Shen et al., 2015). Casticin (**11**) is flavonoid compound Fig. 3, isolated from the species of Vitex such as *Vitex rotun* difolia and *Vitex agnus-castus.* It has been found to have anti-inflammatory lung diseases, chronic obstructive pulmonary disease (COPD) [138]. The flavonoid Oroxylin A (**12**) Fig. 3 was found from medicinal plant *Scutellaria baicalensis* Fig. 3, attenuates CS-induced lung inflammation by Nrf2 activation [88]. Asiatic acid (**13**, Fig. 3), one of the bioactive natural products, isolated from *Centella asiatica*, and it possesses inhibitory property COPD through activation in lung tissue of MAPKs and NF-kB. A new flavonoid, icariin (**14**) Fig. 3, has been isolated from *Epimedium* spp and considerably showed COPD properties through the suppression of NF-kB activation. A labdane diterpenoid Andrographolide (**15**, Fig. 3) was mainly isolated from *Andrographis paniculata* herbal plant and shown anti-inflammatory properties through activation of Nrf2 signaling pathway [139]. Betulin (**16**) Fig. 3 is a rich, naturally occurring triterpene which was isolated from the bark of birch trees. It exhibited COPD with decreasing the amount of malondialdehyde (MDA) within the lung [140]. Eucalyptol (**17**) Fig. 3 is a monoterpenoid bicyclic ether and obtained from the plants of *Eucalyptus globulus*. It has been played as a potential COPD agent for the action of pulmonary inflammation [141]. Epigallocatechin-3-gallate (EGCG) (**18**) Fig. 3 is a natural polyphenol isolated from green leaf *Camellia sinensis,* and it showed promising anti-inflammatory activity through ameliorating oxidative stress. The p-Coumaric acid (**19**) Fig. 3 is an important hydroxycinnamic acid found in medicinal herbs *Bambusae Caulis* in Taeniam and inhibits NF-κB activity and for treating pulmonary inflammatory diseases. Salidroside (**20**) Fig. 3 is a glucoside of tyrosol was obtained from the medicinal plant *Rhodiola rosea* and it showed better COPD properties through MAPK/NF-κB mechanism [142]. An important flavonoid compound Silymarin (**21**), Fig. 3, has been isolated from the *Silybum marianum*, milk thistle, and it exhibited COPD on adaptable MAPK activation mechanism [143]. Tanshinone IIA (**22**, Fig. 3) is the foremost potential biomolecule which was isolated from the plant root of *Saviamiltiorrhiza*. Tanshinone IIA induced COPD

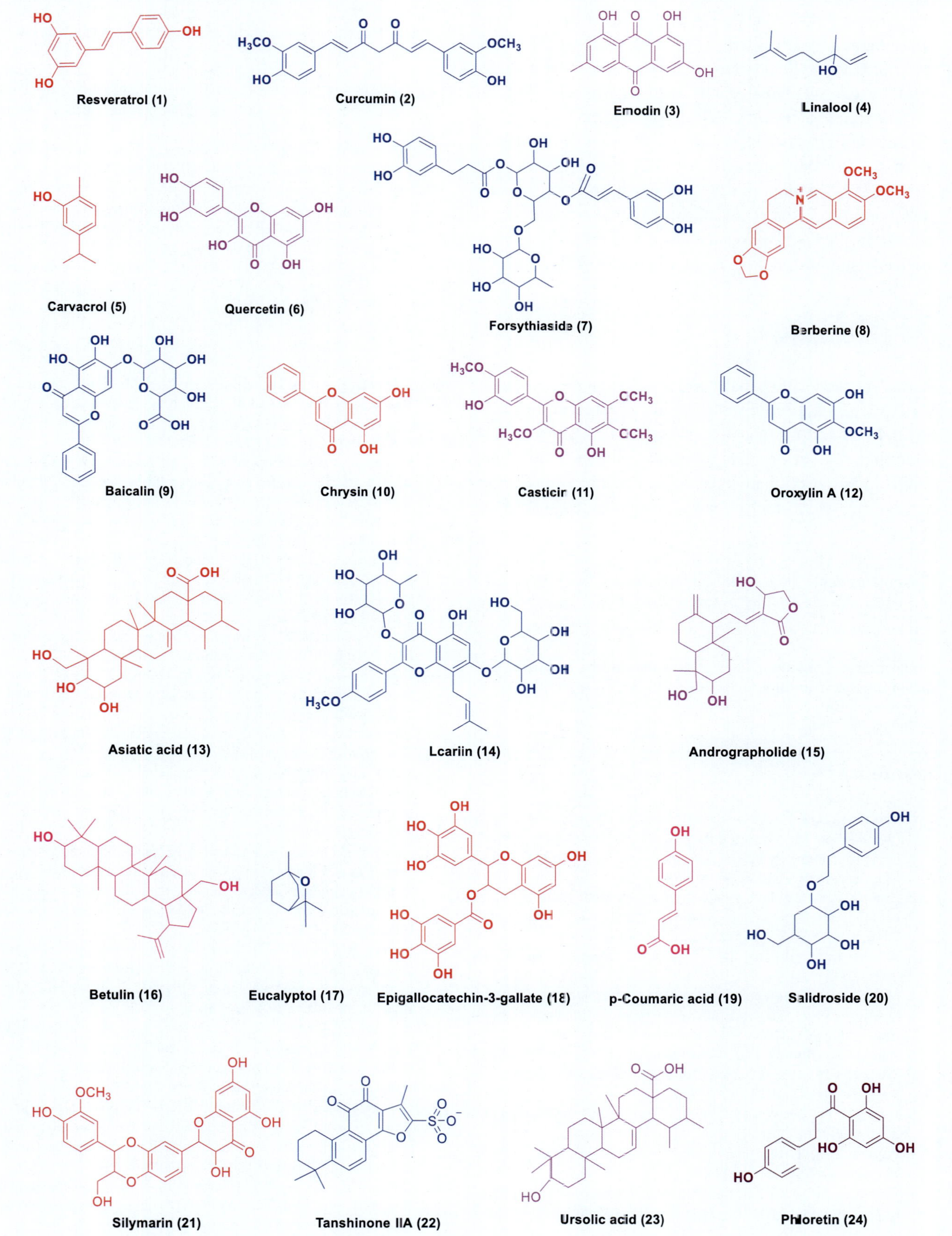

Fig. 3 Some plant-derived natural products involved in the inhibition of COPD

and down-regulation of CFTR, and it could be a potential anti-inflammatory drug for COPD. Ursolic acid (**23**), Fig. 3, is a tripterine-based terpenoid molecule, and it was obtained from various medicinal plants such as *Uncaria rhynchophylla*, *Rosmarinus officinalis*, and *Eugenia jambolana*. Ursolic acid displays excellent COPD properties through it mechanism of down-regulation of PERK and upregulation of Nrf2 [144]. Phloretin (**24**) is a natural phenol dihydrochalcone Fig. 3, and it has been identified from the apple tree leaves *Fortunella japonica* and it was pretreatment dramatically suppressed the mucins secretion [145].

8 Conclusion

Herbal medicines potentially are used in the treatment of a variety of diseases with reduced side effects and toxicities. Several herbal drugs and formulations have shown marked inhibition of acute and chronic illnesses including COPD. Thus, our chapter in detail describes the various herbal drugs and formulations with their detailed pre-clinical studies.

Acknowledgment The authors thank and acknowledge the Department of Pharmaceuticals, Ministry of chemical and fertilizers, Government of India

Declaration of Competing Interest All the authors declare no conflict of interest to publish the article.

References

1. Zagorevski DV, Loughmiller-Newman JA. The detection of nicotine in a Late Mayan period flask by gas chromatography and liquid chromatography mass spectrometry methods. Rapid Commun Mass Spectrom. 2012;26(4):403–11.
2. Braun GV. Petrography as a technique for investigating Iroquoian ceramic production and smoking rituals. J Archaeol Sci. 2012;39(1):1–10.
3. Kung H-C, Hoyert DL, Xu J, Murphy SL. Deaths: final data for 2005. 2008.
4. May SM, Li JT. Burden of chronic obstructive pulmonary disease: healthcare costs and beyond. In: Allergy and asthma proceedings. East Providence: OceanSide Publications; 2015.
5. Barnes PJ, Burney PGJ, Silverman EK, Celli BR, Vestbo J, Wedzicha JA, et al. Chronic obstructive pulmonary disease. Nat Rev Dis Primers. 2015;1(1):15076.
6. Petty TL. The history of COPD. Int J Chron Obstruct Pulmon Dis. 2006;1(1):3–14.
7. Petrovska BB. Historical review of medicinal plants' usage. Pharmacogn Rev. 2012;6(11):1.
8. Chung K, Adcock I. Multifaceted mechanisms in COPD: inflammation, immunity, and tissue repair and destruction. Eur Respir J. 2008;31(6):1334–56.
9. Milara J, Peiró T, Serrano A, Cortijo J. Epithelial to mesenchymal transition is increased in patients with COPD and induced by cigarette smoke. Thorax. 2013;68(5):410–20.
10. Zhu B, Wang Y, Ming J, Chen W, Zhang L. Disease burden of COPD in China: a systematic review. Int J Chron Obstruct Pulmon Dis. 2018;13:1353.
11. Hou W, Hu S, Li C, Ma H, Wang Q, Meng G, et al. Cigarette smoke induced lung barrier dysfunction, EMT, and tissue remodeling: a possible link between COPD and lung cancer. Biomed Res Int. 2019;2019:2025636.
12. Rodgman A, Perfetti TA. The chemical components of tobacco and tobacco smoke. Boca Raton: CRC Press; 2008.
13. John G, Kohse K, Orasche J, Reda A, Schnelle-Kreis J, Zimmermann R, et al. The composition of cigarette smoke determines inflammatory cell recruitment to the lung in COPD mouse models. Clin Sci. 2014;126(3):207–21.
14. Caramori G, Adcock IM, Di Stefano A, Chung KF. Cytokine inhibition in the treatment of COPD. Int J Chron Obstruct Pulmon Dis. 2014;9:397–412.
15. Boer WI, Alagappan VK, Sharma HS. Molecular mechanisms in chronic obstructive pulmonary disease. Cell Biochem Biophys. 2007;1(47):131–47.
16. Kranenburg AR, Willems-Widyastuti A, Mooi WJ, Saxena PR, Sterk PJ, de Boer WI, et al. Chronic obstructive pulmonary disease is associated with enhanced bronchial expression of FGF-1, FGF-2, and FGFR-1. J Pathol. 2005;206(1):28–38.
17. Kranenburg AR, De Boer WI, Van Krieken JHJ, Mooi WJ, Walters JE, Saxena PR, et al. Enhanced expression of fibroblast growth factors and receptor FGFR-1 during vascular remodeling in chronic obstructive pulmonary disease. Am J Respir Cell Mol Biol. 2002;27(5):517–25.
18. Bonniaud P, Kolb M, Galt T, Robertson J, Robbins C, Stampfli M, et al. Smad3 null mice develop airspace enlargement and are resistant to TGF-β-mediated pulmonary fibrosis. J Immunol. 2004;173(3):2099–108.
19. Fahy JV, Dickey BF. Airway mucus function and dysfunction. N Engl J Med. 2010;363(23):2233–47.
20. Takizawa H, Tanaka M, Takami K, Ohtoshi T, Ito K, Satoh M, et al. Increased expression of transforming growth factor-β 1 in small airway epithelium from tobacco smokers and patients with chronic obstructive pulmonary disease (COPD). Am J Respir Crit Care Med. 2001;163(6):1476–83.
21. Tuder RM, Yoshida T, Arap W, Pasqualini R, Petrache I. State of the art. Cellular and molecular mechanisms of alveolar destruction in emphysema: an evolutionary perspective. Proc Am Thorac Soc. 2006;3(6):503–10.
22. Santos S, Peinado VI, Ramirez J, Morales-Blanhir J, Bastos R, Roca J, et al. Enhanced expression of vascular endothelial growth factor in pulmonary arteries of smokers and patients with moderate chronic obstructive pulmonary disease. Am J Respir Crit Care Med. 2003;167(9):1250–6.
23. Rahimi N, Dayanir V, Lashkari K. Receptor chimeras indicate that the vascular endothelial growth factor receptor-1 (VEGFR-1) modulates mitogenic activity of VEGFR-2 in endothelial cells. J Biol Chem. 2000;275(22):16986–92.
24. Kearney JB, Ambler CA, Monaco K-A, Johnson N, Rapoport RG, Bautch VL. Vascular endothelial growth factor receptor Flt-1 negatively regulates developmental blood vessel formation by modulating endothelial cell division. Blood. 2002;99(7):2397–407.
25. Gaggar A, Jackson PL, Noerager BD, O'Reilly PJ, McQuaid DB, Rowe SM, et al. A novel proteolytic cascade generates an extracellular matrix-derived chemoattractant in chronic neutrophilic inflammation. J Immunol. 2008;180(8):5662–9.
26. O'Reilly P, Jackson PL, Noerager B, Parker S, Dransfield M, Gaggar A, et al. N-α-PGP and PGP, potential biomarkers and therapeutic targets for COPD. Respir Res. 2009;10:1–8.
27. Senior RM, Tegner H, Kuhn C, Ohlsson K, Starcher BC, Pierce JA. The induction of pulmonary emphysema with human leukocyte elastase. Am Rev Respir Dis. 1977;126(3):469–75.

28. Duranton J, Bieth JG. Inhibition of proteinase 3 by α1-antitrypsin in vitro predicts very fast inhibition in vivo. Am J Respir Cell Mol Biol. 2003;29(1):57–61.
29. Barnes PJ, Shapiro SD, Pauwels R. Chronic obstructive pulmonary disease: molecular and cellularmechanisms. Eur Respir J. 2003;22(4):672–88.
30. Wang Z, Zheng T, Zhu Z, Homer RJ, Riese RJ, Chapman HA Jr, et al. Interferon γ induction of pulmonary emphysema in the adult murine lung. J Exp Med. 2000;192(11):1587–600.
31. Morris DG, Huang X, Kaminski N, Wang Y, Shapiro SD, Dolganov G, et al. Loss of integrin αvβ6-mediated TGF-β activation causes Mmp12-dependent emphysema. Nature. 2003;422(6928):169–73.
32. Imai K, Dalal SS, Chen ES, Downey R, Schulman LL, Ginsburg M, et al. Human collagenase (matrix metalloproteinase-1) expression in the lungs of patients with emphysema. Am J Respir Crit Care Med. 2001;163(3):786–91.
33. Yu Q, Stamenkovic I. Cell surface-localized matrix metalloproteinase-9 proteolytically activates TGF-β and promotes tumor invasion and angiogenesis. Genes Dev. 2000;14(2):163–76.
34. Barnes PJ. Cellular and molecular mechanisms of chronic obstructive pulmonary disease. Clin Chest Med. 2014;35(1):71–86.
35. Ito K, Watanabe S, Kharitonov S, Hanazawa T, Adcock I, Barnes P. Histone deacetylase activity and gene expression in COPD patients. Eur Respir J. 2001;18:316S.
36. Edwards MR, Bartlett NW, Clarke D, Birrell M, Belvisi M, Johnston SL. Targeting the NF-κB pathway in asthma and chronic obstructive pulmonary disease. Pharmacol Therap. 2009;121(1):1–13.
37. Caramori G, Romagnoli M, Casolari P, Bellettato C, Casoni G, Boschetto P, et al. Nuclear localisation of p65 in sputum macrophages but not in sputum neutrophils during COPD exacerbations. Thorax. 2003;58(4):348–51.
38. Kennedy-Feitosa E, Cattani-Cavalieri I, Barroso MV, Romana-Souza B, Brito-Gitirana L, Valenca SS. Eucalyptol promotes lung repair in mice following cigarette smoke-induced emphysema. Phytomedicine. 2019;55:70–9.
39. Lee J-W, Park HA, Kwon O-K, Jang Y-G, Kim JY, Choi BK, et al. Asiatic acid inhibits pulmonary inflammation induced by cigarette smoke. Int Immunopharmacol. 2016;39:208–17.
40. Hou J, Wang J, Meng J, Zhang X, Niu Y, Gao J, et al. Zanthoxylum bungeanum seed oil attenuates LPS-Induced BEAS-2B cell activation and inflammation by inhibiting the tlr4/Myd88/nf-κb signaling pathway. Evid Based Complement Alternat Med. 2021;2021:1–13.
41. Sierocinski E, Holzinger F, Chenot J-F. Ivy leaf (Hedera helix) for acute upper respiratory tract infections: an updated systematic review. Eur J Clin Pharmacol. 2021;77:1113–22.
42. Guo R, Pittler M, Ernst E. Herbal medicines for the treatment of COPD: a systematic review. Eur Respir J. 2006;28(2):330–8.
43. Kim J-Y, Shin D, Lee G, Kim J-M, Kim D, An Y-M, et al. Standardized herbal formula PM014 inhibits radiation-induced pulmonary inflammation in mice. Sci Rep. 2017;7(1):45001.
44. Jasemi SV, Khazaei H, Momtaz S, Farzaei MH, Echeverría J. Natural products in the treatment of pulmonary emphysema: therapeutic effects and mechanisms of action. Phytomedicine. 2022;2022:153988.
45. Birrell MA, Eltom S. The role of the NLRP3 inflammasome in the pathogenesis of airway disease. Pharmacol Therap. 2011;130(3):364–70.
46. Zhang J, Xu Q, Sun W, Zhou X, Fu D, Mao L. New insights into the role of NLRP3 inflammasome in pathogenesis and treatment of chronic obstructive pulmonary disease. J Inflamm Res. 2021;14:4155.
47. Swanson KV, Deng M, Ting JP-Y. The NLRP3 inflammasome: molecular activation and regulation to therapeutics. Nat Rev Immunol. 2019;19(8):477–89.
48. Peng Z, Zhang W, Qiao J, He B. Melatonin attenuates airway inflammation via SIRT1 dependent inhibition of NLRP3 inflammasome and IL-1β in rats with COPD. Int Immunopharmacol. 2018;62:23–8.
49. Mahalanobish S, Dutta S, Saha S, Sil PC. Melatonin induced suppression of ER stress and mitochondrial dysfunction inhibited NLRP3 inflammasome activation in COPD mice. Food Chem Toxicol. 2020;144:111588.
50. Tian X, Xue Y, Xie G, Zhou Y, Xiao H, Ding F, et al. (−)-Epicatechin ameliorates cigarette smoke-induced lung inflammation via inhibiting ROS/NLRP3 inflammasome pathway in rats with COPD. Toxicol Appl Pharmacol. 2021;429:115674.
51. Zhang B, Wang B, Cao S, Wang Y, Wu D. Silybin attenuates LPS-induced lung injury in mice by inhibiting NF-κB signaling and NLRP3 activation. Int J Mol Med. 2017;39(5):1111–8.
52. Tian Q, Xu M, He B. Histidine ameliorates elastase-and lipopolysaccharide-induced lung inflammation by inhibiting the activation of the NLRP3 inflammasome. Acta Biochim Biophys Sin. 2021;53(8):1055–64.
53. Yang Y, Huang L, Tian C, Qian B. Magnesium isoglycyrrhizinate inhibits airway inflammation in rats with chronic obstructive pulmonary disease. BMC Pulm Med. 2021;21(1):1–11.
54. Tomani JCD, Kagisha V, Tchinda AT, Jansen O, Ledoux A, Vanhamme L, et al. The inhibition of NLRP3 inflammasome and IL-6 production by hibiscus noldeae baker f. Derived constituents provides a link to its anti-Inflammatory therapeutic potentials. Molecules. 2020;25(20):4693.
55. Moon P-D, Choi I-H, Kim H-M. Berberine inhibits the production of thymic stromal lymphopoietin by the blockade of caspase-1/NF-κB pathway in mast cells. Int Immunopharmacol. 2011;11(11):1954–9.
56. De Boer W. Cytokines and therapy in COPD: a promising combination? Chest. 2002;121(5):209S–18S.
57. Lee J-W, Ryu HW, Park S-Y, Park HA, Kwon O-K, Yuk HJ, et al. Protective effects of neem (Azadirachta indica A. Juss.) leaf extract against cigarette smoke-and lipopolysaccharide-induced pulmonary inflammation. Int J Mol Med. 2017;40(6):1932–40.
58. Yuan J, Li X, Fang N, Li P, Zhang Z, Lin M, et al. Perilla leaf extract (PLE) attenuates COPD airway inflammation via the TLR4/Syk/PKC/NF-κB pathway in vivo and in vitro. Front Pharmacol. 2022;12:763624.
59. Jian T, Ding X, Li J, Wu Y, Ren B, Li J, et al. Triterpene acids of loquat leaf improve inflammation in cigarette smoking induced COPD by regulating AMPK/Nrf2 and NFκB pathways. Nutrients. 2020;12(3):657.
60. Jian T, Chen J, Ding X, Lv H, Li J, Wu Y, et al. Flavonoids isolated from loquat (Eriobotrya japonica) leaves inhibit oxidative stress and inflammation induced by cigarette smoke in COPD mice: the role of TRPV1 signaling pathways. Food Funct. 2020;11(4):3516–26.
61. Lourenco JD, Neves LP, Olivo CR, Duran A, Almeida FM, Arantes PM, et al. A treatment with a protease inhibitor recombinant from the cattle tick (Rhipicephalus Boophilus microplus) ameliorates emphysema in mice. PLoS One. 2014;9(6):e98216.
62. Rooney CP, Taggart C, Coakley R, McElvaney NG, O'Neill SJ. Anti–proteinase 3 antibody activation of neutrophils can be inhibited by α 1-antitrypsin. Am J Respir Cell Mol Biol. 2001;24(6):747–54.
63. Stolk J, Rossie W, Dijkman JH. Apocynin improves the efficacy of secretory leukocyte protease inhibitor in experimental emphysema. Am J Respir Crit Care Med. 1994;150(6):1628–31.
64. Ma H, Zhou Z, Chen L, Wang L, Muge Q. Anemoside B4 prevents chronic obstructive pulmonary disease through alleviating cigarette smoke-induced inflammatory response and airway epithelial hyperplasia. Phytomedicine. 2022;107:154431.

65. Luo F, Shi J, Shi Q, Xu X, Xia Y, He X. Mitogen-activated protein kinases and hypoxic/ischemic nephropathy. Cell Physiol Biochem. 2016;39(3):1051–67.
66. Barnes PJ. Novel signal transduction modulators for the treatment of airway diseases. Pharmacol Therap. 2006;109(1-2):238–45.
67. Barnes PJ. Corticosteroid resistance in patients with asthma and chronic obstructive pulmonary disease. J Allergy Clin Immunol. 2013;131(3):636–45.
68. Wang C, Zhou J, Wang J, Li S, Fukunaga A, Yodoi J, et al. Progress in the mechanism and targeted drug therapy for COPD. Signal Transduct Target Ther. 2020;5(1):248.
69. Manning BD, Toker A. AKT/PKB signaling: navigating the network. Cell. 2017;169(3):381–405.
70. Jiang H, Abel PW, Toews ML, Deng C, Casale TB, Xie Y, et al. Phosphoinositide 3-kinase γ regulates airway smooth muscle contraction by modulating calcium oscillations. J Pharmacol Exp Ther. 2010;334(3):703–9.
71. Yao H, Rahman I. Current concepts on oxidative/carbonyl stress, inflammation and epigenetics in pathogenesis of chronic obstructive pulmonary disease. Toxicol Appl Pharmacol. 2011;254(2):72–85.
72. Shi L, Yan J, Meng Y, Wang G, Du J, Feng C. Systematic pharmacology-based strategy to explore the mechanism of bufei huoxue capsule in the treatment of chronic obstructive pulmonary disease. Evid Based Complement Alternat Med. 2022;2022:1129567.
73. Li D, Hu J, Wang T, Zhang X, Liu L, Wang H, et al. Silymarin attenuates cigarette smoke extract-induced inflammation via simultaneous inhibition of autophagy and ERK/p38 MAPK pathway in human bronchial epithelial cells. Sci Rep. 2016;6(1):37751.
74. Luan G, Zhu Z, Wu K, Yin S. Theaflavin-3,3′-digallate attenuates cigarette smoke extract-induced pulmonary emphysema in mice by suppressing necroptosis. Exp Ther Med. 2022;23(1):1–9.
75. Jeong J-S, Kim J-W, Kim J-H, Kim C-Y, Ko J-W, Kim T-W. Protective effects of chestnut (Castanea crenata) inner shell extract in macrophage-driven emphysematous lesion induced by cigarette smoke condensate. Nutrients. 2023;15(2):253.
76. Zhou W, Dong M, Wu H, Li H-L, Ma R-y, Su W-W, et al. Common mechanism of Citrus Grandis Exocarpium in treatment of chronic obstructive pulmonary disease and lung cancer. Chin Herb Med. 2021;13(4):525–33.
77. Kao T-C, Wu C-H, Yen G-C. Glycyrrhizic acid and 18β-glycyrrhetinic acid recover glucocorticoid resistance via PI3K-induced AP1, CRE and NFAT activation. Phytomedicine. 2013;20(3-4):295–302.
78. Pacini ES, Sanders-Silveira S, Godinho RO. The extracellular cAMP-adenosine pathway in airway smooth muscle. J Pharmacol Exp Ther. 2018;366(1):75–83.
79. Zuo H, Cattani-Cavalieri I, Musheshe N, Nikolaev VO, Schmidt M. Phosphodiesterases as therapeutic targets for respiratory diseases. Pharmacol Ther. 2019;197:225–42.
80. Nabissi M, Marinelli O, Morelli MB, Nicotra G, Iannarelli R, Amantini C, et al. Thyme extract increases mucociliary-beating frequency in primary cell lines from chronic obstructive pulmonary disease patients. Biomed Pharmacother. 2018;105:1248–53.
81. Lee JW, Ryu HW, Lee SU, Kim MG, Kwon OK, Kim MO, et al. Pistacia weinmannifolia ameliorates cigarette smoke and lipopolysaccharide-induced pulmonary inflammation by inhibiting interleukin-8 production and NF-κB activation. Int J Mol Med. 2019;44(3):949–59.
82. Tang G-H, Dong Z, Guo Y-Q, Cheng Z-B, Zhou C-J, Yin S. Psiguajadials A–K. Unusual Psidium meroterpenoids as phosphodiesterase-4 inhibitors from the leaves of Psidium guajava. Sci Rep 2017;7(1):1-15.
83. Xiao C, Cheng S, Lin H, Weng Z, Peng P, Zeng D, et al. Isoforskolin, an adenylyl cyclase activator, attenuates cigarette smoke-induced COPD in rats. Phytomedicine. 2021;91:153701.
84. Faux SP, Tai T, Thorne D, Xu Y, Breheny D, Gaca M. The role of oxidative stress in the biological responses of lung epithelial cells to cigarette smoke. Biomarkers. 2009;14(sup1):90–6.
85. Li Y-j, Yu C-h, Li J-b, Wu X-y. Andrographolide antagonizes cigarette smoke extract-induced inflammatory response and oxidative stress in human alveolar epithelial A549 cells through induction of microRNA-218. Exp Lung Res. 2013;39(10):463–71.
86. Liu X, Ma Y, Luo L, Zong D, Li H, Zeng Z, et al. Dihydroquercetin suppresses cigarette smoke induced ferroptosis in the pathogenesis of chronic obstructive pulmonary disease by activating Nrf2-mediated pathway. Phytomedicine. 2022;96:153894.
87. Ma B-n, Li X-j. Resveratrol extracted from Chinese herbal medicines: a novel therapeutic strategy for lung diseases. Chin Herb Med. 2020;12(4):349–58.
88. Li J, Tong D, Liu J, Chen F, Shen Y. Oroxylin A attenuates cigarette smoke-induced lung inflammation by activating Nrf2. Int Immunopharmacol. 2016;40:524–9.
89. Barroso MV, Cattani-Cavalieri I, de Brito-Gitirana L, Fautrel A, Lagente V, Schmidt M, et al. Propolis reversed cigarette smoke-induced emphysema through macrophage alternative activation independent of Nrf2. Bioorg Med Chem. 2017;25(20):5557–68.
90. Lin L, Yin Y, Hou G, Han D, Kang J, Wang Q. Ursolic acid attenuates cigarette smoke-induced emphysema in rats by regulating PERK and Nrf2 pathways. Pulm Pharmacol Therap. 2017;44:111–21.
91. Stojanoski NJ. Development of health culture. In: Veles and its region from the past to the end of the 20th century, vol. 13. Veles: Society of Science and Art; 1999. p. 34.
92. Kelly K. The history of medicine. New York: Facts on File Inc; 2009.
93. Tucakov J. Healing with plants–phytotherapy. JBC. 1971;180:90.
94. Glesinger L. Medicine through centuries. JZZ. 1954;1:21–38.
95. Bojadzievski P. The health services in Bitola through the centuries. Bitola: Society of Science and Art; 1992. p. 15–27.
96. Bazala V. The historical development of medicine in the Croatian lands. Zagreb: Croation Publishing Bibliographic; 1943. p. 9–20.
97. Tucakov J. Pharmacognosy. Beograd: Academic Books; 1948. p. 8–21.
98. Thorwald J. Power and knowledge of ancient physicians. Zagreb: August Cesarec; 1991. p. 10–255.
99. Benzie I, Wachtel-Galor S. Herbal medicine: an introduction to its history, usage, regulation, current trends, and research needs. Boca Raton: CRC Press; 2011. p. 464.
100. Jancic R, Lakušic BJF. Zavod za botaniku. Belgrade: Botanika Farmaceutika; 2002.
101. Mannino D, Watt G, Hole D, Gillis C, Hart C, McConnachie A, et al. The natural history of chronic obstructive pulmonary disease. Eur Respir J. 2006;27(3):627–43.
102. Collins FS, Varmus H. A new initiative on precision medicine. N Engl J Med. 2015;372(9):793–5.
103. Thomford NE, Senthebane DA, Rowe A, Munro D, Seele P, Maroyi A, et al. Natural products for drug discovery in the 21st century: innovations for novel drug discovery. Int J Mol Sci. 2018;19(6):1578.
104. Leung JM, Me O, Sadatsafavi M, Sin DD. Introduction to precision medicine in COPD. Eur Respir J. 2019;53(4):1802460.
105. Lyu YR, Lee S-W, Kim S-Y, Han H-B, Yang W-K, Kim S-H, et al. Herbal medicines for the treatment of chronic obstructive airway diseases (asthma or chronic obstructive pulmonary disease): a prospective observational study. Evid Based Complement Alternat Med. 2022;2022:3485757.

106. Sidhaye VK, Nishida K, Martinez F. Precision medicine in COPD: where are we and where do we need to go? Eur Respir Rev. 2018;27(149):180022.
107. Iwakura M, Okura K, Kubota M, Sugawara K, Kawagoshi A, Takahashi H, et al. Estimation of minimal clinically important difference for quadriceps and inspiratory muscle strength in older outpatients with chronic obstructive pulmonary disease: a prospective cohort study. Phys Ther Res. 2021;24(1):35–42.
108. Clarke R, Lundy FT, McGarvey L. Herbal treatment in asthma and COPD–current evidence. Int J Phytomed Phytother. 2015;1(1):1–7.
109. Deligiannidis KM, Freeman MP, Obstetrics RC. Complementary and alternative medicine therapies for perinatal depression. Gynaecology. 2014;28(1):85–95.
110. Guo S, Sun Z, Liu E, Feng J, Fu M, Li Y, et al. Effect of Bufei granule on stable chronic obstructive pulmonary disease: a randomized, double blinded, placebo-controlled, and multicenter clinical study. J Tradit Chin Med. 2014;34(4):437–44.
111. Zhang M, Xie Y, Yan R, Shan H, Tang J, Cai Y, et al. Curcumin ameliorates alveolar epithelial injury in a rat model of chronic obstructive pulmonary disease. Life Sci. 2016;164:1–8.
112. Karsch-Völk M, Barrett B, Kiefer D, Bauer R, Ardjomand-Woelkart K, Linde K. Echinacea for preventing and treating the common cold. Cochrane Database Syst Rev. 2014;2:CD000530.
113. Ma J, Liu X, Wei Y, Lu R, Xu K, Tian Y, et al. Effective component compatibility of Bufei Yishen formula III which regulates the mucus hypersecretion of COPD rats via the miR-146a-5p/EGFR/MEK/ERK pathway. BMC Pulm Med. 2022;2022:9423435.
114. Li J, Ma J, Tian Y, Zhao P, Liu X, Dong H, et al. Effective-component compatibility of Bufei Yishen formula II inhibits mucus hypersecretion of chronic obstructive pulmonary disease rats by regulating EGFR/PI3K/mTOR signaling. J Ethnopharmacol. 2020;257:112796.
115. Zhao P, Liu X, Dong H, Tian Y, Feng S, Zhao D, et al. Bufei Yishen formula restores Th17/Treg balance and attenuates chronic obstructive pulmonary disease via activation of the adenosine 2a receptor. Front Pharmacol. 2020;11:1212.
116. Li J, Yang L, Yao Q, Li Y, Tian Y, Li S, et al. Effects and mechanism of Bufei Yishen formula in a rat chronic obstructive pulmonary disease model. Evid Based Complement Alternat Med. 2014;2014:381976.
117. Liu S, Chen J, Zuo J, Lai J, Wu L, Guo XJB, et al. Comparative effectiveness of six Chinese herb formulas for acute exacerbation of chronic obstructive pulmonary disease: a systematic review and network meta-analysis. BMC Complement Altern Med. 2019;19:1–14.
118. Wu Y, Liu T, Liu S. Impact of Xiaoqinglong decoction on lung function and serum cytokines in patients with acute exacerbation of chronic obstructive pulmonary disease. Med Innov China. 2015;12:109–11.
119. Cao D. Effects of Xiaoqiongtang decoction on airway inflammation and airway remodeling in patients with COPD. J West China. 2009;21(4):575–7.
120. Jing X, Li X, Liu J. Effect of modified Qianjin Weijing decoction on nuclear factor-Kb of chronic obstructive pulmonary disease in acute stage. J Shandong Univ TCM. 2007;14(5):271–4.
121. Zhang W, Zhang X, Shao Y. Changes in the level of cytokine in rats with chronic obstructive pulmonary disease of phlegm heat cumber lung type after treatment of Maxing Shigan decoction. Chin J Tissue Eng Res. 2006:167–70.
122. Cai M, Chen H, Wang X-D, Zhang H-Y. Clinical observation of bushen naqi huoxue granules in treating patients with chronic obstructive lung disease. JCIM J. 2005;3(6):483–5.
123. Soraoka H, Oniki K, Matsuda K, Ono T, Taharazako K, Uchiyashiki Y, et al. The effect of Yokukansan, a traditional herbal preparation used for the behavioral and psychological symptoms of dementia, on the drug–metabolizing enzyme activities in healthy male volunteers. Biol Pharm Bull. 2016;39(9):1468–74.
124. Rafiee Z, Nejatian M, Daeihamed M, Jafari SM. Application of curcumin-loaded nanocarriers for food, drug and cosmetic purposes. Trends Food Sci Technol. 2019;88:445–58.
125. Scotti L, Ishiki H, Júnior F, Cruz R, Tavares J, Diniz M, et al. Flavonoids as multi-target compounds in drug discovery. Curr Pharm Des. 2017;14(4):333–41.
126. Bolognesi ML, Cavalli AJC. Multitarget drug discovery and polypharmacology. Hoboken: Wiley; 2016. p. 1190–2.
127. Gambini J, Inglés M, Olaso G, Lopez-Grueso R, Bonet-Costa V, Gimeno-Mallench L, et al. Properties of resveratrol: in vitro and in vivo studies about metabolism, bioavailability, and biological effects in animal models and humans. Oxidative Med Cell Longev. 2015;2015:837042.
128. Pangeni R, Sahni JK, Ali J, Sharma S, Baboota S. Resveratrol: review on therapeutic potential and recent advances in drug delivery. Expert Opin Drug Deliv. 2014;11(8):1285–98.
129. Wang X-L, Li T, Li J-H, Miao S-Y, Xiao X-Z. The effects of resveratrol on inflammation and oxidative stress in a rat model of chronic obstructive pulmonary disease. J Med. 2017;22(9):1529.
130. Harikumar KB, Kunnumakkara AB, Sethi G, Diagaradjane P, Anand P, Pandey MK, et al. Resveratrol, a multitargeted agent, can enhance antitumor activity of gemcitabine in vitro and in orthotopic mouse model of human pancreatic cancer. Int J Cancer. 2010;127(2):257–68.
131. Yuan J, Liu R, Ma Y, Zhang Z, Xie Z. Curcumin attenuates airway inflammation and airway remolding by inhibiting NF-κB signaling and COX-2 in cigarette smoke-induced COPD mice. Inflammation. 2018;41:1804–14.
132. Ma J, Xu H, Wu J, Qu C, Sun F, Xu S. Linalool inhibits cigarette smoke-induced lung inflammation by inhibiting NF-κB activation. Int Immunopharmacol. 2015;29(2):708–13.
133. Mitani A, Azam A, Vuppusetty C, Ito K, Mercado N, Barnes PJ. Quercetin restores corticosteroid sensitivity in cells from patients with chronic obstructive pulmonary disease. Exp Lung Res. 2017;43(9-10):417–25.
134. Cheng L, Li F, Ma R, Hu X. Forsythiaside inhibits cigarette smoke-induced lung inflammation by activation of Nrf2 and inhibition of NF-κB. Int Immunopharmacol. 2015;28(1):494–9.
135. Lin K, Liu S, Shen Y, Li Q. Berberine attenuates cigarette smoke-induced acute lung inflammation. Inflammation. 2013;36:1079–86.
136. Xu D, Wan C, Wang T, Tian P, Li D, Wu Y, et al. Berberine attenuates cigarette smoke-induced airway inflammation and mucus hypersecretion in mice. Int J Clin Exp Med. 2015;8(6):8641.
137. Park K, Lee JS, Choi JS, Nam YJ, Han JH, Byun HD, et al. Identification and characterization of baicalin as a phosphodiesterase 4 inhibitor. Phytother Res. 2016;30(1):144–51.
138. Lee H, Jung K-H, Lee H, Park S, Choi W, Bae H. Casticin, an active compound isolated from Vitex Fructus, ameliorates the cigarette smoke-induced acute lung inflammatory response in a murine model. Int Immunopharmacol. 2015;28(2):1097–101.
139. Liao W, Tan WD, Wong WF. Andrographolide restores steroid sensitivity to block lipopolysaccharide/IFN-γ-induced IL-27 and airway hyperresponsiveness in mice. J Immunol. 2016;196(11):4706–12.
140. Chunhua M, Long H, Zhu W, Liu Z, Jie R, Zhang Y, et al. Betulin inhibited cigarette smoke-induced COPD in mice. Biomed Pharmacother. 2017;85:679–86.

141. Yu N, Sun Y-T, Su X-M, He M, Dai B, Kang J. Treatment with eucalyptol mitigates cigarette smoke-induced lung injury through suppressing ICAM-1 gene expression. Biosci Rep. 2018;38(4):1636.
142. Luo F, Liu J, Yan T, Miao M. Salidroside alleviates cigarette smoke-induced COPD in mice. Pharmacotherapy. 2017;86:155–61.
143. Daryanto S, Wang L, Jacinthe P-A. Impacts of no-tillage management on nitrate loss from corn, soybean and wheat cultivation: a meta-analysis. Sci Rep. 2017;7(1):12117.
144. Liu W, Tan X, Shu L, Sun H, Song J, Jin P, et al. Ursolic acid inhibits cigarette smoke extract-induced human bronchial epithelial cell injury and prevents development of lung cancer. Molecules. 2012;17(8):9104–15.
145. Wang H, Yang T, Wang T, Hao N, Shen Y, Wu Y, et al. Phloretin attenuates mucus hypersecretion and airway inflammation induced by cigarette smoke. Int Immunopharmacol. 2018;55:112–9.

Herbal Medicines for the Treatment of Liver Cirrhosis

Tanmoy Banerjee, Shuvam Sar, Saptapadi Saha, Arun Baidya, Arnab Sarkar, Sanmoy Karmakar, Amit Kumar Halder, and Nilanjan Ghosh

Abstract

Liver cirrhosis is a disastrous terminal consequence of different types of chronic liver diseases (CLDs), with fibrosis serving as an initial stage of cirrhosis. Obesity, hyperlipidaemia, and type 2 diabetes mellitus (T2DM) significantly raise the possibility of liver cirrhosis and its associated consequences. Non-alcoholic steatohepatitis (NASH), a physiological condition with a 20% progression rate to cirrhosis, is brought on by the excessive lipid accumulation in non-alcoholic fatty liver disease (NAFLD) that results in inflammation and destroys the normal hepatocytes. Cirrhosis and fibrosis of the liver are complex diseases that include different inflammatory cytokines and genetic, oxidative, and endoplasmic reticulum (ER) stress factors. The prevalence of cirrhosis in humans is increasing, which is of particular concern to medical specialists. However, there are not many efficient treatments readily accessible right now. Numerous molecules from phytochemical classes, including alkaloid, flavonoid, glycoside, polyphenol, tannin, and terpenoid, have shown promising results in treating several CLDs progressing to cirrhosis. These compounds influence hepatic lipid metabolism and de novo lipogenesis (DNL) through various molecular signalling pathways. In this chapter, several natural compounds implicated in reversing liver cirrhosis development via numerous molecular pathways have been briefly described. In-depth descriptions of various phytochemicals acting on multiple cell lines, and well-established animal models of cirrhosis are discussed in detail. Therefore, with adequate scientific evidence, this chapter conveys the role of herbal medicines in altering the progression of liver cirrhosis.

Keywords

Liver cirrhosis · Herbal compounds · Chronic liver diseases · Lipid metabolism · Molecular targets

Abbreviations

ACC	Acetyl-CoA carboxylase
ALT	Alanine aminotransferase
AST	Aspartate aminotransferase
CCl4	Carbon tetrachloride
ChREBP	Carbohydrate-responsive element-binding protein
HDL-C	High-density lipoprotein cholesterol
IL-1β	Interleukin-1β
IL-6	Interleukin 6
LDL-C	Low-density lipoprotein cholesterol
MDA	Malondialdehyde
mRNA	Messenger ribonucleic acid
NLRP3	NLR family pyrin domain containing 3
Nrf2	Factor-erythroid 2 related factor 2
p53	Tumour protein P53
PPAR-α	Peroxisome proliferator-activated receptor α
PPAR-γ	Peroxisome proliferator-activated receptor γ
PPAR-δ	Peroxisome proliferator-activated receptor δ
SOD	Superoxide dismutase
SREBP-1c	Sterol regulatory element-binding protein-1c
TC	Total cholesterol

1 Introduction

Liver cirrhosis is a major cause of mortality and a serious public health concern globally. In 2016, liver cirrhosis was responsible for 2.2% of fatalities and 1.5% of patient disabil-

T. Banerjee · S. Sar · S. Saha · A. Baidya · A. Sarkar · S. Karmakar · N. Ghosh (✉)
Department of Pharmaceutical Technology, Jadavpur University, Kolkata, West Bengal, India
e-mail: nilanjanghosh.phamacy@jadavpuruniversity.in

A. K. Halder
Dr. B.C. Roy College of Pharmacy & Allied Health Sciences, Durgapur, India

A. K. Dhara, S. C. Mandal (eds.), *Role of Herbal Medicines*, https://doi.org/10.1007/978-981-99-7703-1_10

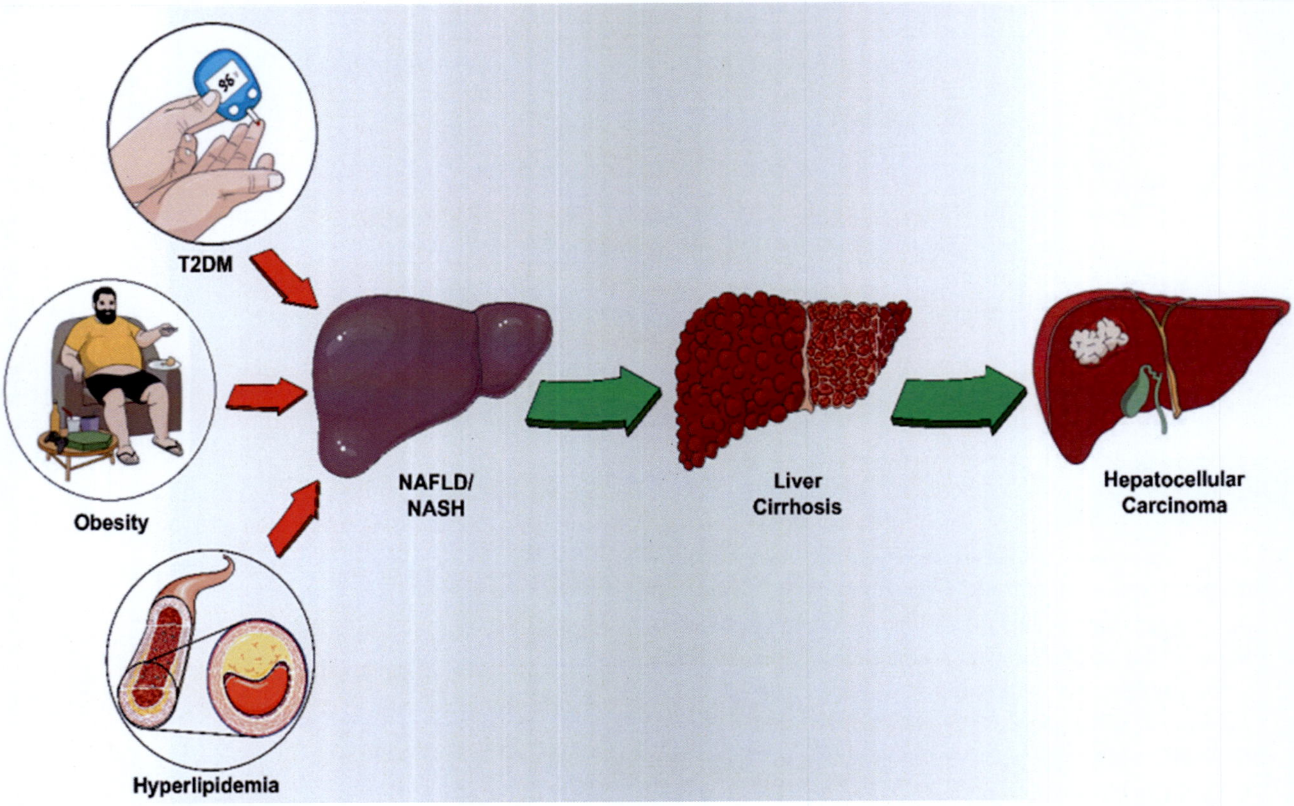

Fig. 1 Progression of liver cirrhosis from NAFLD and NASH to hepatocellular carcinoma

ity globally, making it the eleventh major death-related factor [1]. The majority of severe liver illnesses are characterized by extracellular matrix protein buildup, particularly collagen, resulting in hepatic fibrosis. Liver cirrhosis is the outcome of progressive liver fibrosis [2]. The fibrosis and regenerative nodule development can describe liver cirrhosis because chronic damage disrupts the hepatic lobular structure. Viral infections, poisons, genetic disorders, and autoimmune processes are all examples of assaults that may harm healthy hepatocytes. In prolonged damage conditions, most hepatocyte fibrosis results in liver impairment [3]. Non-alcoholic fatty liver disease (NAFLD) can be described by an accumulation of lipids in the liver that cannot be attributed to alcohol intake in individuals [4]. Non-alcoholic steatohepatitis (NASH) is characterized by a higher degree of fibrosis advancement from NAFLD [5]. NASH includes a spectrum of clinical conditions that may progress from steatosis to cirrhosis and ultimately cause liver cancer (Fig. 1) [6].

Synthetic medications traditionally used to cure liver disorders have proven fatal in several instances. Herbal medicines play a crucial role in a wide variety of hepatic disorders for having less or negligible side effects and long-term therapeutic effects, which makes them significantly popular [7]. About 65% of sufferers in Europe and the USA rely on herbal medicaments for the treatment of liver ailments. The prompting move towards natural compounds or their derivatives could be beneficial in retarding liver cirrhosis. Most drugs used to manage liver illnesses are now derived from natural sources. Furthermore, some data states that the bioactive chemicals found in botanicals can operate as hepatoprotective agents [8]. Hence, this chapter aims to conduct an in-depth evaluation of the bioactive chemicals from plant species that inhibit liver cirrhosis and associated liver ailments by regulating multiple gene and protein expressions by several molecular signalling pathways.

2 Pathophysiology

The complicated processes that may be responsible for the development of NAFLD in patients with type 2 diabetic mellitus (T2DM) have mostly been the focus of research [9]. It has been established that fatty liver disease, obesity, and insulin resistance (IR) play a significant role as co-factors in developing liver damage [10]. Absorption of free fatty acids (FFAs) and de novo lipogenesis (DNL) in the liver are the two primary contributors to NAFLD, which is the outcome of an intracellular buildup of triglyceride (TG). The hepatic damage comprises cellular destruction and inflammation. These illnesses result from a rise in mitochondrial oxidative stress on TG, forming free radicals and peroxisomes as a downstream consequence [11]. DNL regulates the balance between lipid synthesis and its breakdown, and the imbalance between these two is what causes hepatic lipid accumulation. DNL is a metabolic pathway that is complex and tightly regulated. DNL metabolizes additional carbohydrates into FFAs, which are then esterified into TGs [12]. FAS catalyses the final step in fatty acid production [13]. Liver-specific fatty acid synthase (FAS) inhibition increased hepatocyte malonyl-CoA levels that block fatty acid beta-oxidation. ACC catalyses a major rate-limiting step in fatty acid biosynthesis and controls FAO by synthesizing malonyl-CoA [14]. Two important transcription factors for fatty acid synthesis, SREBP-1c, and ChREBP are engaged in lipid synthesis and have been linked to promoting DNL in NAFLD. SREBP-1c controls lipid metabolism transcriptionally [15]. The liver X receptor alpha (LXRα) is crucial in the upregulation of SREBP-1c and ChREBP [16].

The activity of adipokines, which are cytokines generated by adipocytes, such as leptin and tumour necrosis factor-α (TNF-α), which are produced in excess, also contributes to an increase in the oxidative stress that is placed on mitochondria [17]. It is more likely for inflammatory adipokines to be active when there is a decrease in adiponectin that regulates other chemical mediators [18]. These chemical mediators, produced by inflammatory response and cell death, activate the hepatic stellate cells (HSCs) and stimulate collagen levels, ultimately promoting fibrosis and cirrhosis [19].

3 Herbal Molecules Involved in the Management of Liver Cirrhosis

3.1 Alkaloids

Alkaloids are organic compounds found in plants that are alkaline and contain nitrogen. Almost all alkaloids have cyclic structures and substantial impacts on lipid metabolism. Importantly, alkaloids act on AMPK, enhancing PPAR-α and carnitine palmitoyl transferase (CPT) 1 expression while decreasing SREBP-1c and PPAR-γ. Because phytochemicals in this class promote lipid catabolic metabolism and inhibit anabolic lipid metabolism, the availability of lipids like fatty acids may be constrained [20]. In addition, some alkaloids with potent hepatoprotective action have been discussed below as shown in Fig. 2.

3.1.1 Berberine

Berberine (BBR) is an isoquinoline alkaloid obtained from the stem of *Berberis aristate* [21]. Enhancement in insulin secretion and reversal of IR are the main functions of BBR. BBR has shown a decrease in liver lipid accumulation and prevention of hepatic steatosis, which simultaneously inhibits hepatic inflammation and oxidative stress by hampering lipogenesis [22]. Lipogenesis was lowered by the activating transcription factor 6 (ATF6) and SREBP-1c pathways. SREBP-1c is a major target protein of BBR, which may have led to the prevention of lipid deposition and endoplasmic reticulum (ER) stress-induced activation of SREBP-1c [23]. In another experiment, through the stimulation of the adenosine 5′-monophosphate activated protein kinase (AMPK), SREBP-1c pathways, BBR decreased liver TG production and mitigated hepatic steatosis. The downregulation of Sirtuin 3 (SIRT3) expression might occur in rat liver [24]. The anti-inflammatory effect of BBR was shown by downregulating the expression of inflammation markers, like TNF-α and IL-6. In 3T3-L1 adipocytes, BBR lowered leptin and glycerol secretion [25]. This alkaloid decreases liver inflammation in mice by reducing the phosphorylation state of c-Jun N-terminal protein kinase 1 (JNK1) and mRNA expression levels. BBR has been found to significantly reduce inflammation in hepatocytes in a time and dose-dependent manner and diminished the expressions of FAS and ACC by BBR [26]. BBR has been shown to inhibit hypoxia-inducible factor (HIF-2α) storage, which lowered ceramide and ameliorated IR by *HIF-2α* genes. Through the PP2A-AKT-GSK3 signalling pathway, BBR improved IR by promoting glycogen production [27].

3.1.2 Caffeine

Caffeine (CFN) is obtained from the roasted seeds of *Coffea arabica* and *Coffea canephora*. In individuals with NAFLD and hepatitis C and B virus infection, drinking coffee loosens up the liver, which might lead to reduced fibrosis and inflammation. Liver diseases have been found to occur less often in those who drink coffee. CFN has been shown to prevent liver fibrosis by inhibiting hepatic stellate cell activation by blocking A2A receptors. New research suggests that CFN may also positively affect angiogenesis and hepatic hemodynamics. The expression of connective tissue growth factor produced by transforming growth factor β (TGF-β) in hepatocytes is inhibited by CFN. CFN treatment in rats with chemically induced liver fibrosis has lowered their TGF-β levels. Hepatoprotective properties are shown by chlorogenic acid (an ester of CFN) that lowered tetrachlorobenzoquinone-induced oxidative stress and also decreased liver fibrosis and collagen I and III expressions; therefore, it must have a hepatoprotective character [28]. The metabolic actions of CFN that protected the liver were unclear. In hepatitis C virus-infected individuals, a higher CFN intake was linked to a lower level of liver fibrosis [29]. CFN and its metabolite paraxanthine reduced hepatic fibrosis by preventing the liver parenchyma from secreting connective tissue growth factor and also lowered iron buildup in the liver and ALT levels. Additionally, CFN reduced the production of alpha smooth muscle actin (α-SMA) in LX-2 cells and promoted stellate cell deactivation. Treatment with CFN resulted in a rise in apoptotic cells. The increased apoptosis rate was directly correlated to the length of exposure and the amount of CFN present. Cell migration and adhesion are all regulated by F-actin. In response to coffee administration, F-actin expression was significantly lowered. Focal adhesion

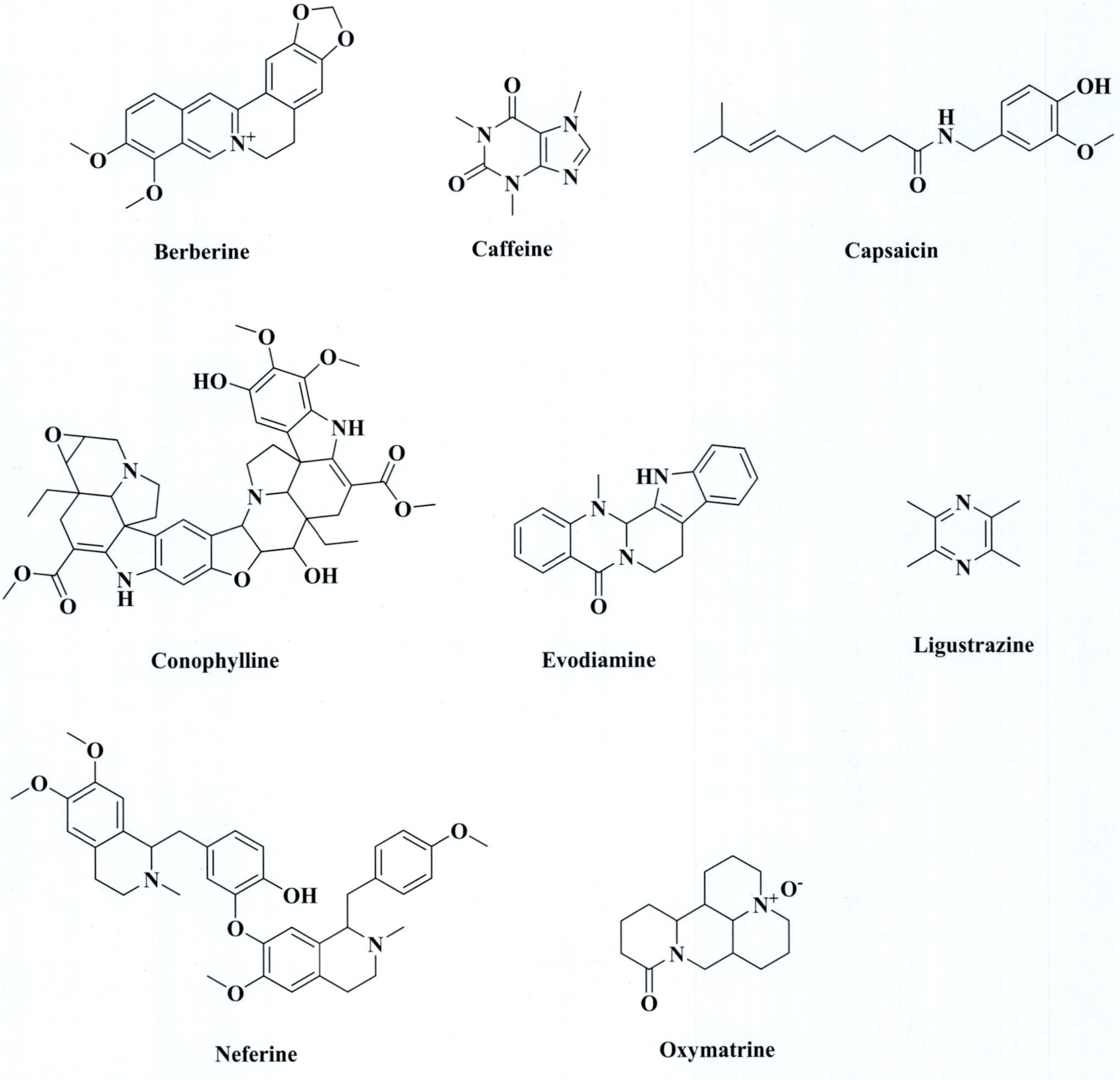

Fig. 2 Alkaloids in the treatment of liver cirrhosis

complex and integrins trigger focal adhesion kinase (FAK), which is important for ECM growth factor stimulation and cell attachment and motility. CFN downregulated the HSCs expression of FAK. CFN modulated suppressor of mothers against decapentaplegic (SMAD)-induced production of ECM components to enhance intracellular cAMP concentration [30].

3.1.3 Capsaicin

The primary ingredient in *Genus Capsicum* is capsaicin (CSC) [31]. In obese mice, CSC showed anti-steatotic action. Mice on a high-fat diet (HFD) had their liver fat decreased when CSC was applied topically. CSC decreased the expression of key enzymes involved in the synthesis of fatty acids, such as ACC and FAS, while increasing the expression of CPT1 and cluster of differentiation 36 (CD36), which are linked to oxidation and the influx of FFAs into the liver. Mice treated with CSC elevated adiponectin concentration in liver tissue. For the transient receptor potential vanilloid 1 (TRPV1) channel, CSC is a highly selective agonist, attenuating metabolic abnormalities by activating TRPV1, which stops the progression of fatty liver in vivo. In mice, dietary CSC, which activated TRPV1, enhanced PPAR-dependent autophagy and thus prevented NAFLD. Additionally, AMP-activated protein kinase (AMPK), PPAR, uncoupling protein 1 (UCP1), and glucagon-like peptide 1 (GLP-1) were all altered by the liver tissue's elevated expression of TRPV1. CSC reduced liver tissue inflammation, which may have been brought on by AMPK signalling via the TRPV1 receptor being phosphorylated. As one of the modes of action of CSC, a systemic increase in adiponectin levels facilitates hepatic AMPK activation. Through the activation of AMPK, topical CSC treatment combined with exercise reduced the symptoms of metabolic syndrome brought on by hypoestrogenism. CSC inhibited fatty liver by inhibiting lipogenesis and increased fatty acid oxidation by activating AMPK [32].

3.1.4 Conophylline

The vinca alkaloid conophylline (CNP) is taken from the leaves of the tropical plant *Tabernaemontana divaricate*. It improved hepatic fibrosis diseased mice condition and cured NASH also in mice. Additionally, the diet decreased the expression levels of hepatic PPAR-α mRNA. PPAR-α participates in β-oxidation. Hepatic PPAR-α and its target genes, such as CPT1 and CPT2, were upregulated in the liver by CNP in a dose-dependent manner. CNP elevated hepatic β-hydroxybutyrate level in a dose-dependent manner and decreased the expression of the p62 protein, which is reported to be preferentially destroyed during autophagy while enhancing the expression of the autophagosomal marker light chain 3 II (LC3-II). CNP reduced hepatic steatosis by elevating liver β-oxidation and autophagy [33].

3.1.5 Evodiamine

One of the main elements taken from *Evodia rutaecarpa* is evodiamine (EDM). ALT and histological abnormalities in liver tissues might all be improved by EDM. EDM decreased IL-6, TNF-α, collagen I, and collagen III. EDM decreased the protein expression of TGF-β1, phosphorylation of SMAD 2/3 (p-SMAD 2/3), and α-smooth muscle actin (α-SMA) in liver tissues as well as the mRNA expression of TGF-β1 and α-SMA. EDM dose-dependently lowered the cell proliferation, hydroxyproline formation, and protein expression of TGF-β1, p-SMAD 2/3, and α-SMA in HSCs. EDM also reduced collagen metabolism and HSC proliferation [34].

3.1.6 Ligustrazine

Ligustrazine (LTZ) was discovered in the Chinese plant *Ligusticum wallichii*. It can block calcium channels, cause vasodilation, decrease platelet bioactivity and aggregation, scavenge oxygen free radicals, protect the liver, and have antifibrosis properties. While lowering the expression of proteins linked to hepatic fibrosis and apoptosis, LTZ has also decreased the number of foci containing CD45-positive cells. Hepatic steatosis and hyperlipidaemia brought on by alcohol were also eliminated. LTZ enhanced the PPAR-α and CPT1 expressions, which decreased the number of lipid droplets in the liver by inhibiting the gene and protein expression of SREBP-1c and FAS. LTZ diminished the alcohol-mediated upregulation of SREBP-1c and downregulated the expression of PPAR-α, resulting in increased lipogenesis and reduced lipolysis. Furthermore, LTZ has shown antioxidant activity by increasing the production of glutathione, SOD, and catalase which may have been lowered by alcohol. These results support the potency of LTZ against liver cirrhosis [35].

3.1.7 Neferine

Nelumbo nucifera seed embryo is the source of neferine (NFR). It has anti-oxidative, anti-inflammatory, and anti-pulmonary fibrosis properties. Collagen's primary building block, hydroxyproline, forms the extracellular matrix (ECM). At the level of liver cirrhosis, ECM deposition might permanently worsen the function of the liver. NFR promoted apoptosis of cultured hepatic stellate cells (HSCs) by reducing their stimulation. The mediator causing chronic liver disease through inflammation, and fibrogenesis, is TGF-β1. To form the ECM, TGF-β1 can induce HSCs to produce and secrete collagen and fibronectin. Plasma levels of ALT and AST were dramatically reduced by NFR. Thus, NFR exhibited an antifibrosis impact on the hepatic fibrosis brought on by CCl4 in vivo [36].

3.1.8 Oxymatrine

Sophora flavescens contains significant amounts of oxymatrine (OXM). OXM lessened fat accumulation by regulating the expression of miR-182. The ratio of liver to body weight, the amount of TG, hepatic lipid buildup, and steatosis were all decreased by OXM. Following OXM intervention, the levels of SREBP-1c, ACC, and fatty acid synthase (FAS) were all dramatically reduced, while the level of CPT1A was increased in the high-fat diet (HFD)-fed rat in the experiment. In the same experiment in miR-182 knockdown rats, SREBP-1c, ACC, and FAS levels were all elevated, but the CPT1A level was decreased [37]. SOD, glutathione peroxidase (GPx), and choline acetyl-transferase (CAT) activity reduced in response to As2O3, and the decline was prevented by treatment with OXM. OXM also enhanced the expression of Nrf2 and heme oxygenase-1 (HO-1) while reducing the retention of arsenic in liver tissues. Through the Nrf2/HO-1 signalling pathway, OXM has shown protection against oxidative damage carried out by As2O3 [38]. PPAR-α, CPT1A, and microsomal TG transfer protein (MTTP) mRNA and protein levels were all significantly raised by OXM therapy. The finding of a reduction in lipid buildup in the liver served as an additional confirmation of OXM-positive effects [39]. Apart from the above discussed herbal compounds, (Table 1) enlists several phytochemicals involved in the treatment of various liver ailments by targeting lipid biomarkers.

Table 1 Names, biological sources, experiment research models, targeted biomarkers of alkaloids

Compound name	Biological source	Research model	Biomarkers	Refs.
Betaine	Vegetables like beets, spinach, etc	In vivo female Sprague-Dawley rats	Downregulated α-SMA, COL1A1, and TGF-β1 protein expressions and TIMP-1, TIMP-2, and MMP-2 mRNA expressions	[40]
Nuciferine	*Nelumbo nucifera* leaf	In vivo male C57BL/6J mice	Upregulated the expression of PPAR-α	[41]
		HepG2 cells in vitro	Lowered both TG and low-density lipoprotein metabolism	
Piperine	Black pepper (*Piper nigrum*)	In vivo AML-12 hepatocytes and human LX-2 HSCs	Downregulated TGFβ1-induced hepatocyte EMT and HSC	[42]
		In vitro TGF-β1/SMAD cascade	Upregulated Nrf2 signalling	
Trigonelline	Cantaloupe, corn, onions, peas, soybeans	In vivo male Wistar albino rats	Decreased serum AST level and inhibited the release of TNF-α	[43]
Sophocarpine	Foxtail-like sophora herb and seed	In vivo male Sprague-Dawley (SD) rats	Activated AMPK signalling pathway decreased glucose output, cholesterol, and TG synthesis	[44]
Liensinine	*Nelumbo nucifera,* such as its root, flowers, leaves, and seeds	In vivo male C57BL/6 mice	Reduced lipid accumulation and TG	[45]
		In vitro AML12 cells	Activated AMPK/ACC	
Theacrine	*Camellia assamica*	In vivo male Sprague-Dawley rats	Activated SIRT3/AMPK and inhibited SCD1 signals. Reduced the expression of α-SMA and HSCs	[46]
Kukoamine A	The root barks of *Lycium chinense*	In vivo C57 mice	Increased SREBP-1c protein, decreased FAS, and ACC	[47]
Tiliamosine	*Tiliacora racemosa*	In vivo male Wistar rats	Reduced TG	[48]
		In vitro HepG2 cells		
Koumine	*Gelsemium elegans*	In vivo male Sprague-Dawley rats	Lowered TG, TC, LDL, ALT, AST, and MDA reduced the production of IL-6, IL-1β, IFN-γ, and TNF-α	[49]

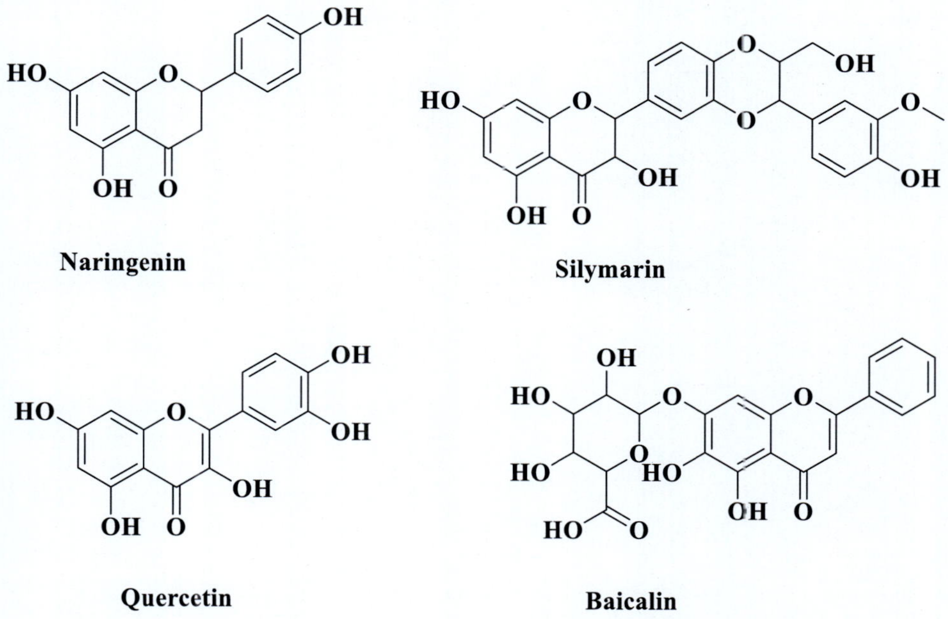

Fig. 3 Flavonoids in the treatment of liver cirrhosis

3.2 Flavonoids

Fruits, vegetables, cereals, bark, roots, stems, flowers, tea, and wine are all good sources of flavonoids, a class of natural compounds with varying phenolic structures (Fig. 3). It is hoped that isolating these natural compounds would yield their health benefits. Therefore, flavonoids are now widely recognized as an essential component in a wide range of nutraceutical, pharmacological, medical, and cosmetic products [50]. This section explores the pathological role of flavonoids (Table 2) in management of cirrhosis and other liver ailments.

3.2.1 Naringenin

Naringenin (NGN) is a flavonoid with anti-inflammatory, anti-fibrogenic, antioxidant, and anticancer effects. NGN has shown protective effects in liver illnesses by inhibiting oxidative stress through the TGF pathway and preventing the transdifferentiation of HSCs, which decreases collagen formation. Non-canonical pathways for the toll-like receptor (TLR), TGF-β, and mitogen-activated protein kinase (MAPK), which, when inhibited, further caused a significant decrease in extracellular matrix (ECM) synthesis and deposition. Through regulating lipid metabolism, NGN has demonstrated a positive effect on NAFLD and influenced the production and oxidation of lipids and cholesterol [51].

3.2.2 Silymarin

Silymarin (SMR) is the name for the flavonolignans identified in milk thistle (*Silybum marianum*). The levels of ALT and AST were somewhat decreased by SMR. In addition, both SMR and its metabolites have lowered IR and hyperlipidaemia [52].

3.2.3 Baicalin

Baicalin (BAI) is a monomeric flavonoid mainly obtained from *Scutellaria baicalensis* Georgi. This active phytochemical is used in the composition of herbal medicine for the treatment of liver disease. BAI had anti-oxidative, anti-inflammatory, and hepatoprotective activity against different types of liver injury caused by carbon tetrachloride, iron overload, and acetaminophen [53]. BAI protects the liver through its antioxidant effect by suppressing the production of reactive oxygen species, which help to protect against liver cirrhosis in ischemia/reperfusion injury in fatty liver disease. BAI reduced hepatic damage by suppressing the TLR4-mediated inflammatory signalling pathway, providing protection against post-ischemic injury in the alcoholic fatty liver [54]. BAI boosted the activity of antioxidant response kinases like SOD and GSH-PX while decreasing the yield of inflammatory factors, including TNF-α, IL-1β, and IL-6. BAI also lowered the regulations of the extracellular signal-regulated kinase (ERK), one of the primary targets. BAI also downregulates SOD and GSH-Px activity to reduce oxidative stress leading to a decrease in hepatic fibrosis. By controlling the activity of SREBP-1c, FAS, PPAR-γ, and CPT1, BAI can partially decrease the hepato-lipids accumulation triggered by an MCD diet [55].

3.2.4 Quercetin

Quercetin (QRC), a plant pigment and more precisely a flavonol, is primarily present in citrus fruits, berries, and grapes. It is an antioxidant that could prevent cellular damage caused

Table 2 Names, biological sources, experiment research models, targeted biomarkers of flavonoids

Compound name	Biological source	Research model	Biomarkers	Refs.
Nobiletin	*Citrus sinensis* and *Citrus limon*	In vivo male C57BL/6J mice	Lowered HDL-C/TC ratio. Downregulation of SREBP-2 and ACAT2	[57]
			Decreased CRP, IL-6, and IL-1β	
Dihydromyricetin	*Ampelopsis grossedentata*	In vivo mice	Decreased ALT, AST, γ-glutamyl transpeptidase, LDL-C, TNF-α, and fibroblast growth factor 21 blocked the PI3K/Akt/NF-κB signalling pathway	[58]
Rutin	Fruits, tea, and wine, *Ruta graveolens*	In vivo male C57BL/6 mice	Downregulated PPAR-α, CPT1, and CPT2 suppressed SREBP-1c, DGAT-1 and 2, ACC, TNF-α, and IL-1β. Reduced TG	[59]
Luteolin	*Dendranthema morifolium*	In vivo Wistar rats	Inhibited NF-κB and TLR4 signalling pathway	[60, 61]
Scutellarin	*Erigeron breviscapus*	In vivo male C57BL/6 mice	Suppressed IRE1α/XBP1 pathway	[62]
			Downregulated SREBP-1c, FAS, and ACC	
			Upregulated Beclin1 and Foxo1 proteins	
Genistein	Soybean products (*Cordyceps sinensis*)	In vivo male Wistar rats	Downregulated JAK2/STAT3/SOCS3 signalling pathway	[63]
			Downregulated genes MMP2/9 and TIMP1	
			Upregulated SOCS3 protein and CD163 and CD206 inhibited CD68	
Fisetin	Fruits and vegetables	In vivo male albino rats	Upregulated HNF4α/lipin1 signalling	[64]
			Inhibited ROS mediated TXNIP induction	
Quercetin3-O-β-D-glucuronide	*Polygonum perfoliatum*	In vivo male SD rats; in vitro Q3GA gene	Upregulated PPAR-α, MCAD, and CPT1 Downregulated FFA and SREBP-1c	[65]
Quercitrin	Fruits and vegetables	In vivo female C57BL/6 mice	Reduced TG, TC, LDL-C; Downregulated TNF-α, IL-6, and IL-1β	[66]
Vitexin	*Ficus deltoidei* and hawthorn	In vivo male C57BL/6J mice	Increased PPAR-γ protein; activated AMPK	[67]
Chrysin	*Passiflora caerulea* and *Oroxylum indicum*	In vivo Wistar rats	Upregulated RAS, ACE2, Ang (1–7), and Mas	[68]
Galangin	Rhizome of *Alpinia galangal*	In vivo male Wistar rats; in vitro BRL cell	Decreased ALT/AST; mediated caspase 3; inhibited BRL cell apoptosis and PI3K/AKT pathway	[69]
Hesperetin	Citrus fruits (oranges, grapefruit, and lemons)	In vivo male Wistar rats; in vitro HepG2 cells	Upregulated SOD/GPx/HO-1; triggered the PI3K/AKT-Nrf2 pathway; reduced ROS accumulation and TG, TC, LDL-C levels	[70]
Hesperidin	*Citrus genus*	In vivo HFD-fed rats	Suppressed GRP94, ATF6, PERK, IRE1α, IL-1β, IL-6, and TNF-α; activated IRE1α	[71]
		In vitro THP1 and HepG2 cells		
Bavachinin	*Fructus Psoraleae*	In vitro HepaRG cells	Inhibited FDFT1 via the AKT/mTOR/SREBP-2 pathway and cholesterol synthesis	[72]
Didymin	*Mentha spicata*	In vivo HFD-fed mice	Suppressed the TLR4/NF-κB and PI3K/Akt pathways; decreased ALT, AST; reduced the expressions of ACC, SREBP-1c, LXRα, and FAS	[73]
Ugonin J	*Helminthostachys zeylanica*	In vivo male C57BL/6J mice; in vitro, PA-induced steatotic human HuS-E/2	Upregulated pAMPK, ACC, and CPT1; downregulated SREBP-1c	[74]
Hyperoside	Hawthorn, artemisia capillaris, and *Hypericum perforatum*	In vivo Wistar male rats	Altered DNL, ApoE, and BSEP; increased NRs, FXR, LXRα	[75]
Icariin	Epimedium genus	In vivo male C57BL/6J mice	Upregulated of the cleaved caspase 3/9, SREBP1c, and DGAT2; enhanced CPT1, ACC, AMPKα1, PGC1α, and GLUT4; activated AMPKα1/PGC1α/GLTU4 pathway	[76]
Xanthohumol	*Humulus lupulus*, beer	In vivo male Wistar rats	Decreased GPx and GST; increased TBARS	[77]
Breviscapine	*Erigeron breviscapus*	In vivo male C57BL/6J mice	Inhibited TGF-β-activated kinase 1 (TAK1) phosphorylation; inhibited NF-κB signalling pathway; downregulated Col1a1, Col3a1, Ctgf, and Timp1 genes; downstream JNK/p38 activation	[78]
Myricetin	Berries, vegetables, and herbs	In vivo male BALB/c mice	Blocked phosphorylation of SMAD2/Akt/MAPK; inhibited PDGF and TGF-β1; activated JNK phosphorylation	[79]

by the toxicities of different drugs. Through several pathways, including nuclear factor kappa B (NF-κB), p53, and phosphoinositide-3-kinase-protein kinase B/Akt (PI3K/Akt) downregulation, QRC has been demonstrated to inhibit apoptosis in normal cells. The anti-inflammatory mechanisms of QRC also demonstrate that it inhibits many pro-inflammatory mediators and enzymes implicated in the processes of hepatotoxicity, including TNF-α, IL-1β, IL-6, cyclooxygenase-2 (COX-2), and nitric oxide synthase (iNOS) [56].

3.3 Polyphenols

Polyphenols are bioactive compound that play an important role in the survival of the plant against harmful environmental factors including UV light and infections. Dietary plant polyphenols have gained a lot of attention in the past decade due to the possibility that they may improve health by acting as antioxidants. Polyphenols may be found in a wide variety of whole foods, including vegetables, fruits, grains, and even certain drinks. For instance, the fresh weight of 100 grams of grapes, apples, pears, cherries, and berries might contain anywhere from 200 to 300 mg of polyphenols [80]. Bioactive polyphenols in the management of liver cirrhosis have been discussed, summarized (Table 3) and illustrated below in Fig. 4.

3.3.1 Catechin

Camellia sinensis is high in polyphenolic catechin (CAT), which directly shields progression of NAFLD. Consuming green tea is linked to decreased levels of aminotransferases, TGs, and atherogenic lipoproteins in the bloodstream, as well as a reduced risk of cardiovascular disease and all-cause mortality closely related to NAFLD as determined by human clinical studies. Green tea CATs may prevent steatosis by inhibiting intestinal lipid and carbohydrate absorption, reducing adipose lipolysis and both adipose and hepatic DNL, stimulating hepatic β-oxidation and thermogenesis, and the improvement of insulin sensitivity [81]. While raising the quantity of the antioxidant enzyme SOD, GTP lowered inflammation by lowering levels of IL-6 and TNF-α, as well as the histological signs of liver damage. The modification of AMPK activity, as GTP-enhanced AMPK phosphorylation, may be the suggested mechanism [82].

3.3.2 Resveratrol

A naturally occurring polyphenolic substance called resveratrol (RSV) may be found in jackfruit, peanuts, and grapes. In vivo RSV-mediated changes to the gut microbiota, associated gut metabolites, and intestinal environment redox status helped protect C57BL/6J male mice from developing metabolic syndrome when they were given an HFD. Two RSV gut metabolites, 4-hydroxyphenyl acetic acid and 3-hydroxyphenylpropionic acid, have demonstrated enhanced lipid metabolism in vitro [83]. This polyphenol stopped the downregulation of SREBP-1c and FAS brought on by steatosis and its effect on phosphorylated AMPK. After RSV administration, plasma ALT levels were dramatically reduced [84].

3.3.3 Curcumin

Curcumin (CUR) is primarily derived from *Curcuma longa*. CUR liposomes improved pharmacological and anticancer effects [85]. CUR imparted significant reduction of hepatic fat levels in NAFLD patients by reducing the serum lipogenic markers. The glycemic and lipid profiles, transaminase levels, and anthropometric measurements were also examined. The effectiveness of CUR in preventing liver outcomes like cirrhosis and hepatocellular cancer, as well as its anti-steatotic activities, is dose-dependent. Following treatment with CUR, significant reductions in FBS and HbA1c levels were noted [86].

3.3.4 Chlorogenic Acid

The most prevalent antioxidant in coffee (*Coffea arabica*) is chlorogenic acid (CGA), though other plants also produce it. Coffee has anti-inflammatory and antioxidant properties that help fight cancer, cirrhosis, and liver fibrosis. CFN treatment decreased lipogenesis and boosted lipid beta-oxidation in HFD rats. The PPAR-α pathway is inactivated, which has an anti-inflammatory and antioxidant impact in addition to reduced fat storage [87]. The control of glucose and lipid metabolism and associated diseases, including diabetes, cardiovascular disease, obesity, cancer, and hepatic steatosis, is thought to be impacted significantly by CGA [88].

3.3.5 Anthocyanin

In peripheral blood mononuclear cells, anthocyanin (ACY) supplementation significantly reduced plasma levels of IL-1β and IL-18 in NAFLD patients relative to controls and mRNA expression of NLRP3 inflammasome components (caspase-1, IL-1β, and IL-18). NAFLD pathophysiology is related to the activation of NLRP3 inflammasomes [89]. ACY reduced hepatic lipid buildup and mitigated oxidative stress and hepatic inflammation. Reduced lipogenesis and increased lipolysis caused by increased PPAR-*α* activity were proposed as the causes of the lower hepatic fat content. One research found enhanced AMPK pathway activity, and several studies using experimental models of NAFLD found improved hepatic antioxidant activity following exposure to ACY [90].

3.3.6 Mangiferin

A typical aqueous stem bark extract from *Mangifera indica* exhibits strong in vitro and in vivo antioxidant activity.

Table 3 Names, biological sources, experiment research models, targeted biomarkers of polyphenols

Compound name	Biological source	Research model	Biomarkers	Refs.
Epigallocatechin-3-gallate	Green tea (*Camellia sinensis)*	In vivo HFD-fed mice	Downregulated CYP450 enzymes inhibited the expression of PXR and AHR, IL-1β, IL-2, CD4+ T cell expansion, neutrophil proliferation, and cell division; suppressed immune cells and sIgA	[93]
Hydroxytyrosol	Virgin olive oil (*Olea europaea*)	In vivo male C57BL/6 mice; in vitro RAW 264.7 cells	Attenuated M1 macrophages and increased M2 macrophages after LPS stimulation; descended cytokine (TNF-α, IL-1β, IL-6, IL-10, IL-4) level; suppressed ERK pathway in macrophages	[94, 95]
Rosmarinic acid	*Rosmarinus officinalis*	In vitro HepG2 cells	Reduced ROS and p-PERK, p-IRE-1, ATF-6, p-eIF-a, and CHOP; increased activity of enzymic and non-enzymic antioxidants; alleviated lipogenesis via the AMPK pathway	[96, 97]
Ellagic acid	Various fruits, including strawberries, pomegranates, guava, etc.	In vivo male Wistar rats	Reduced hepatic levels of ROS, MDA, TNF-α, and IL-6 and the nuclear activity of NF-κB, p65; increased Nrf2 and GSH and SOD; downregulated SREBP1, SREBP2, ACC, FAS, and HMG-CoAR stimulated PPAR-α and CPT1A, CPT1A, and AMPK	[98]
Dihydrocurcumin	Turmeric (*Curcuma longa*)	In vitro human liver cell line L02 and HepG2 cell	Activated PI3K/AKT pathway; reduced NO and ROS via Nrf2 signalling pathways; inhibited SREBP-1c and PNPLA3, and upregulated PPAR-α	[99]
Chicoric acid	*Cichorium intybus*	In vivo male C57BL/6 mice	Decreased AST and ALT; activated AMPK/Nrf2 pathway; decreased P-JNK, P-ERK, and P-P38	[100]
Gallic acid	Fruits and vegetables teas, grapes, berries, pequi (*Caryocar brasiliense*), and wine; *Phyllanthus emblica*	In vivo male SWISS mice	Improved A CC and FAS genes; increased PPAR-α and CPT1 expression; decreased FAS and HMG-CoA reductase expression	[101]
Diphlorethohydroxycarmalol	*Ishigeo kamurae*	In vitro HepG2 cells; in vivo transgenic Zebrafish	SREBP-1c, C/EBP1α, and FAS IL-1β, TNF-α, and COX-2a decreased; upregulated phosphorylation of AMPK and SIRT1	[102]
CAT	*Camellia sinensis*	In vivo male Sprague-Dawley rats	Reduced ROS formation; reduced the growth of breast and prostate cancers	[103]
CGA	Coffee (*Coffea arabica*), apples, pears, berries, artichokes, and eggplant	In vivo male C57BL/6 mice; in vitro LX2 cells	Reduced ECM production, HMGB1 release, hyaluronic acid, and collagen IV; promoted mitochondrial biogenesis; decreased α-SMA, SMAD2/3, COL1A1 and COL3A1, TG, and NEFA	[104]
Salvianolic acid A	*Salvia miltiorrhiza*	In vivo Sprague-Dawley rats	Reduced TG, TC, LDL, ALT, and AST; Downregulated α-SMA and TGF-β1	[105]
6-Gingerol	*Zingiber officinale*	In vivo HFD male C57BL/6Jmice	Upregulated microRNA-467b-3p, HNF4α; decrease in glycerol-3-phosphate acyl-transferase-1; Downregulated GPAT1	[106, 107]
Methyl brevifolincarboxylate	*Phyllanthus urinaria*, *Canarium album*	In vivo male C57BL/6 mice; in vitro human SK-HEP-1 cell	Promoted PPAR-α, CPT1, and ACOX1; activated AMPK phosphorylation; dampened TNF-α, IL-6, IL-8, and IL-1β secretion, and NF-κB expression	[108]

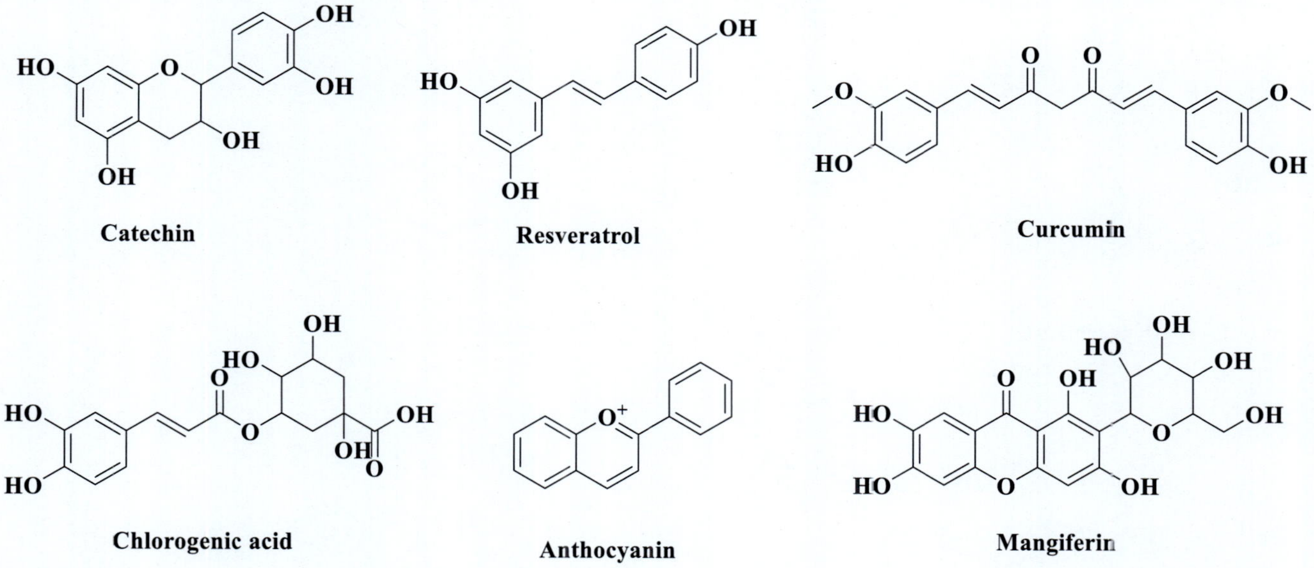

Fig. 4 Polyphenols in the treatment of liver cirrhosis

Mangiferin (MFR) is the primary polyphenolic component of Vimang. Polyphenols have a dual synergistic effect, making them potentially effective antioxidants against pathological iron overload diseases such as thalassemia, Friedreich's ataxia, and hemochromatosis. The redox cycling of the metal and the ligand, notably MFR, the primary catecholic polyphenol in the natural extract, might be used to explain the improved free radical scavenging capacity of Vimang following iron coordination. The second primary polyphenol of Vimang, MFR, is a catechol-containing molecule that likewise has significant antioxidant activity; it has been shown that iron increases its superoxide scavenging capacity [91]. Furthermore, MFR pretreatment has been demonstrated to have a strong hepatoprotective impact by reducing blood levels of the inflammatory mediator TNF-α, AST, ALT, alkaline phosphatase, and bilirubin. In addition, MFR prevented the reduction of glutathione reductase, glutathione-S-transferase, SOD, and total reduced glutathione activity and increased hepatic MDA formation in the liver of CCl4-injected male Swiss albino mice [92].

3.4 Glycosides

In glycosides, a sugar moiety is attached to an aglycone structure (Fig. 5) through an O-, S-, N-, or C-glycosidic bond, making for yet another unique family of secondary metabolites. The glycosidic residue is crucial for bioactivity in several compounds. Hydrolysis of the glycosides, however, produces an active aglycone or a cytotoxic defence chemical like hydrogen cyanide [109].

3.4.1 Glycyrrhizin

Glycyrrhizin (GLZ) is a naturally occurring glycoside isolated from the roots of the *Glycyrrhiza uralensis*. GLZ frequently employs as a flavouring and sweetening factor in sweets, drinks, chewing gum, and tobacco products. It is also recognized for its pleasant fragrant sweet taste. GLZ helps to protect animals from multiple types of experimental liver damage. This active phytochemical had anti-inflammatory, antioxidant, antimicrobial, and antiviral properties. GLZ suppressed TNF-α and CD4+ T cell-mediated cytotoxicity [110]. The deposit of bile acids caused by cholestatic hepatic injury in the liver and bloodstream typically causes severe hepatic inflammation and helps in the progression of fibrosis and cirrhosis. GLZ stimulated the activity of the pregnane X receptor (PXR) in hepatic cells leading to downregulate the cholesterol 7-alpha-monooxygenase (CYP7A1) enzyme, an essential enzyme in the production of bile acids. GLZ significantly reduced ANIT-induced acute cholangiocyte and protected the liver against cholestasis by increased ANIT clearance and decreased inflammation [111]. As a powerful antioxidant, GLZ significantly reduced caspase-3 and caspase-9 action and inhibited hepatocyte necrosis, inflammation, and fibrosis [112]. GLZ markedly reduced the production of both pro-inflammatory mediators, such as TNF-α, iNOS, and COX-2 [113].

3.4.2 Paeoniflorin

The primary active constituent, paeoniflorin (PAE), was found in the roots of *Paeonia lactiflora Pall* and *Paeonia veitchii Lynch*. It is a monoterpenoid glycoside molecule. Recent ethnopharmacological studies demonstrate the

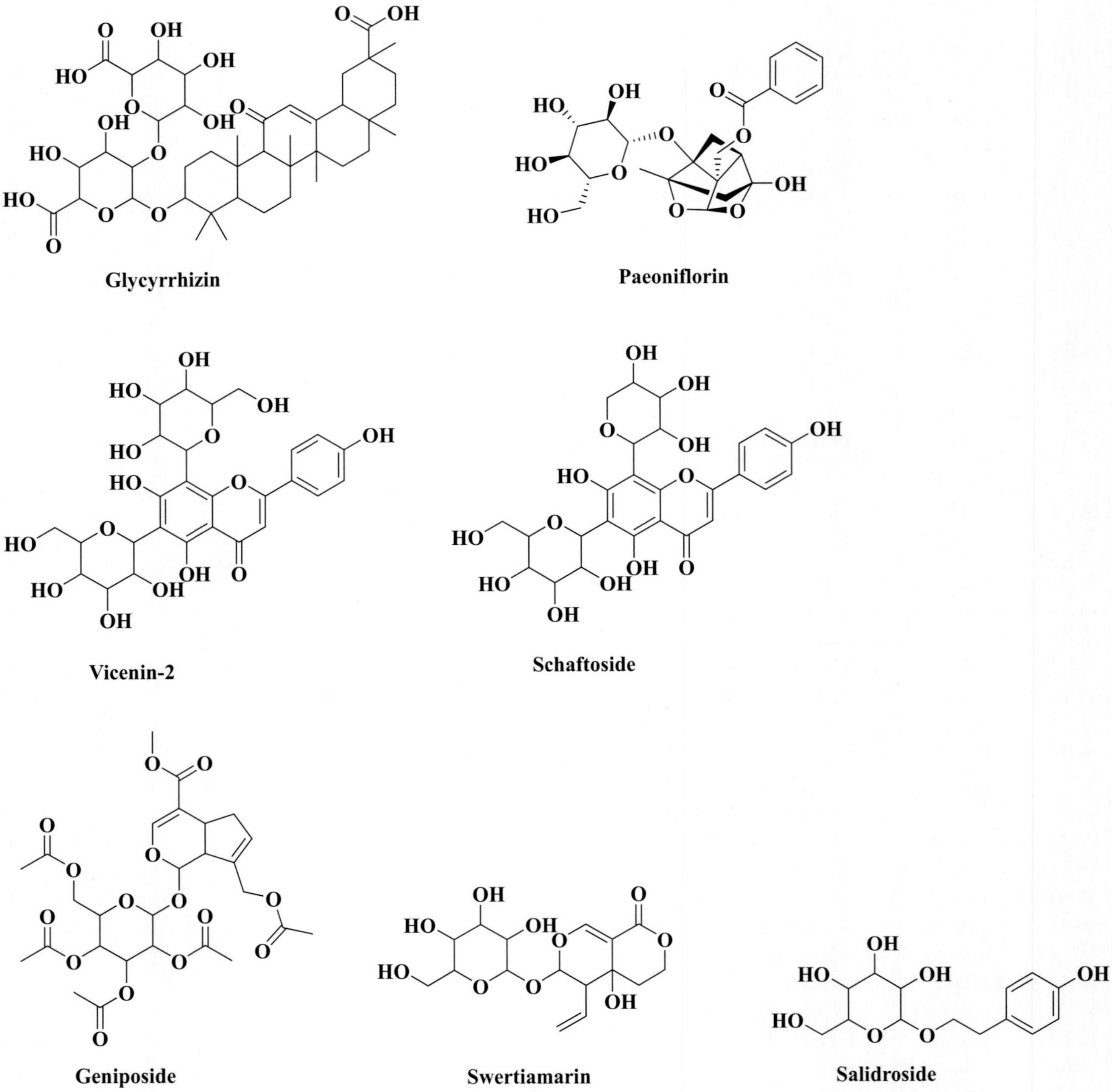

Fig. 5 Glycosides in the treatment of liver cirrhosis

hepatoprotective activity of PAE. A mechanistic study investigating the metabolic effects of PAE revealed that it primarily modulated phenylalanine, tryptophan, and tyrosine biosynthesis, linoleic acid metabolism, inositol phosphate metabolism, glycosylphosphatidylinositol-anchor biosynthesis, and primary bile acid biosynthesis, which may be responsible for its protective action in cholestatic liver injury [114]. PAE significantly reduces serum biochemical markers and prevents hepatic tissue injury, indicating that it might have effectiveness against NIT-induced cholestasis [115]. PAE reduces PPAR-γ mRNA levels. It lowers cholesterol by downregulating PPAR-γ-LXRα-ABCA-1 pathway. As a potential herbal medication, PAE has anti-NAFLD activity. Its NAFLD preventive impact was mediated by various pathways, including the de novo pathway, the lipid oxidation metabolism pathway, and the cholesterol synthesis and output pathway [116]. HIF-1α might induce liver fibrosis partially via mTOR. PAE administration potentially diminished the mTOR phosphorylation and HIF-1α expression, and thus it inhibits fibrogenesis. PAE also downregulates the COL III and α-SMA genes responsible for the fibrosis of rats [117].

3.4.3 Vicenin-2

Vicenin-2 (VCN-2), a flavonoid glycoside with the chemical name apigenin-6, is 8-di-C-D-glucopyranoside. The ethnopharmacological sources include *Cyclopia subternata* and *Perilla frutescens*. Studies have demonstrated the antidiabetic, antioxidant, antiglycation, anti-inflammatory, antithrombotic, and antiplatelet properties of VCN-2. VCN-2 post-treatment significantly reduces the activation of MAPK, such as ERK, JNK, and p38, indicating that VCN-2 prevents sepsis by preventing MAPK signalling. VCN-2 represents a possible pharmaceutical strategy to prevent liver failure by attenuating liver damage by suppressing TLR signalling [118]. By suppressing the protein tyrosine kinase Src and activating the protein tyrosine phosphatase SHP-1, VCN-2 efficiently reduced both constitutive and induced STAT3 phosphorylation. JAK protein, which controls STAT3 upstream activity, was suppressed by VCN-2 [119]. Pro-inflammatory mediators, like IL-6 and TNF-α, are downregulated by VCN-2, which directly responsible for pro-inflammatory responses, such as leukocyte adherence, permeability, and migration towards endothelial cells. VCN-2 might prevent NF-κB activation [120].

3.4.4 Schaftoside

Schaftoside (SFS) is an active phytoconstituent of the Chinese herb *Herba Desmodii Styracifolii*. SFS exhibits many potent pharmacological activities, including inflammatory inhibition, anti-hepatitis, anticancer, etc. SFS also has a vital role in regulating liver cirrhosis and lipid homeostasis [121]. SFS promoted lipolysis in Huh-7 cells, resulting in the downregulation of TG and LDL-C, which eventually decreased liver steatosis. In addition, SFS could inhibit SREBP-1c expression, reducing HFD-induced elevations in hepatic TG level, suppressing lipid droplet formation, and ameliorating HFD-induced liver injury [122]. At both the mRNA and protein levels, SFS reduced the lipopolysaccharide-enhanced upregulation of ATF4 (activating transcription factor 4), XBP-1 (X-box-binding protein-1), and CHOP (C/EBP homologous protein) [123].

3.4.5 Geniposide

Geniposide (GPS) is the major active phytochemical of *Gardenia jasminoides Ellis* and having the molecular formula of $C_{17}H_{24}O_{10}$.GPS is an example of an iridoid glycoside [124]. Numerous pharmacologic actions of this active phytochemical include anti-inflammatory analgesic, hepatoprotective, and choleretic properties. CGA and GPS treat fatty liver disease by reducing the fat content upregulated by free fatty acids (FFAs) in HepG2 cells [125]. GPS-mediated preventative action is maintained by downregulating TC, TG, and LDL-C levels and increasing HDL-C biosynthesis. GPS increased insulin levels and decreased FFA levels, suggesting that it may be a possible treatment for preventing high-fat-diet-induced hepatic steatosis. GPS upregulates PPAR-α expression and downregulates TNF-α. This indicated that GPS modulation of lipid metabolism could be associated with PPAR-α expression and adipocytokine release. GPS significantly decreased GSH-Px and SOD, reduced MDA, and lowered CYP2E1 activity. The antioxidant activity of GPS might be explained by its capacity to inhibit the production of free radicals or by its free radical scavenging activity [126].

3.4.6 Swertiamarin

Enicostemma littorale Blume, a member of the Gentianaceae family, is the major source of swertiamarin (SWT). SWT is a seco-iridoid glycoside type of phytochemical. SWT has been discovered to have beneficial effects, including analgesic, hepatoprotective, antidiabetic, and gastroprotective effects [127]. Hepatic SOD, GPx, CAT, GST, and GSH activities were all significantly reduced by CCl4 treatment. SWT significantly improved the levels of GSH and the activity of those antioxidant enzymes in the liver of CCl4-treated rats. It was proposed that SWT might reduce oxidative stress by restoring the activity of antioxidant enzymes and elevating the level of GSH in the liver, at least in part. In addition, SWT significantly lowered levels of inflammatory cytokines (IL-1β, IL-6, and TNF-α), which reduced inflammatory response. In animals with acute cirrhosis, the concentrations of both total CYP and CYP isoenzymes were shown to be lower. Due to oxidative stress, the primary liver P450 isoenzymes, CYP2C6, CYP1A2, CYP2E1, and CYP3A2, and their protein expression, drastically decreased in activity during acute liver damage brought on by CCl4 treatment [128]. Revealed SWT could accelerate the healing of CCl4-induced liver damage by recovering hepatic CYP3A and CYP2E1 expression. Treatment with SWT also restored NQO1 expression that CCl4 had downregulated. These findings suggested that the upregulation of NQO1 and HO-1 and the Nrf2 pathway activation may be related to SWT hepatoprotective properties [129]. SWT treatment successfully decreased the expression of SREBP-1, ACC, and FAS, suggesting that the drug suppressed DNL through its impact on these regulating elements. By controlling RAS, SWT has the potential to be an effective treatment for hepatic fibrosis [130].

3.4.7 Salidroside

Salidroside (SDS) is a phenolic glycoside having the chemical name [2-(4-hydroxyphenyl) ethyl-beta-D-glucoside]. SDS is commonly present in *Rhodiola sacinehalnsis*. *Ligustrum lucidum* has recently been discovered to have a significant level of SDS. SDS showed several pharmacological effects, including hepatoprotective, neuroprotection, free radical scavenging, immunological modulation, and kidney protection [131]. SDS might be used to treat NAFLD because it can inhibit the deposition of lipids in muscle and liver tissues by stimulating AMPK activity. SDS can treat NAFLD, which works by controlling the AMPK-dependent NLRP3/TXNIP pathway and oxidative stress [132]. It was found that SDS effectively lowered ALT and AST levels in the serum in CCl4-induced hepatic fibrosis mice. Still, it was also found to prevent the liver tissue, demonstrating that SDS could lessen the damage caused by inflammation in the progression of liver fibrosis. Sphingosine kinase 1 (SphK1) suppression significantly lowered the degree of liver fibrosis, whereas high levels of SphK1 expression have also been seen to promote the advancement of liver fibrosis. Conversely, levels of SphK2 are lower in alcoholic-induced cirrhosis and hepatocellular carcinoma in livers. Research showed that SDS might reverse these modifications, whether helpful or harmful [133]. Table 4 includes biological role of few glycosides involved in recovering cirrhosis and various chronic liver ailments.

3.5 Terpenoids

The two most significant categories of plant secondary metabolites are terpenes and terpenoids. They are hardly ever linked to hepatoprotective effects. Terpenoids exhibit hepatoprotection by suppressing apoptosis in liver cells by limiting the release of cytochrome c from the cytoplasm, lowering the Bax/Bcl-2 ratio, and blocking the phosphorylation of JNK and ERK. Terpenoids comprise extracts with strong radical-scavenging properties that can boost the overall antioxidant capability and lessen hepatitis-related inflammation. By modifying the activity of cytochrome P450 enzymes, they play a significant part in the biotransformation of xenobiotics. In-depth explanations are provided for other effects, including regulating TGF-β stimulated collagen type I production linked to liver fibrosis and the specific actions of certain terpenoids [138]. Some potent terpenoids have been depicted in Fig. 6 and described below in detail (Table 5).

3.5.1 Betulinic Acid

Betulinic acid (BLA) is a cyclic triterpene in several plants, including Betula. By altering the AMPK-SREBP signalling pathway, BLA significantly lowered hepatic fat buildup [139]. BLA activates calmodulin-dependent protein kinase to activate AMPK. Additionally, BLA inhibits the expression of S6 kinase-mediated SREBP1. Treatment with BLA blocks the effects of the HFD on the activation of nuclear SREBP1 and the subsequent buildup of TG in the liver [140].

3.5.2 Ursolic Acid

Ursolic acid (URA), a natural terpenoid, is the main active component of many traditional medicinal plants, including *Cornus officinalis*. URA has been implicated in various biological activities, such as lipid regulation, glucose metabolisms, and anti-atherosclerotic [141]. URA-fed mice had much smaller and lighter adipose tissues than the high-fat diet controls due to a decrease in lipid formation in adipose tissue. The main mechanisms by which URA exerts its hypolipidemic effects are the activation of PPAR and the management of its responsive genes involved in fatty acid and lipoprotein metabolism [142]. URA may have triggered endogenous ligand production or increased the availability of endogenous ligands through indirect methods to activate PPAR. After URA administration, the expression of PPAR and its responsive genes in lipid metabolism was dramatically altered in the liver by regulating the CPT1 and ACOX1 genes. In addition, SREBP1, FAS, and SCD1 gene expression levels were suppressed by URA administration, which decreased the production of hepato-fatty acids and reduced TG levels in the blood [143].

3.5.3 Artemisinin

Artemisia annua L. is the source of numerous terpenoid-based bioactive phytochemicals. The antimalarial drug artemisinin (ATM) is one of those. Hepatic ACC levels were significantly greater in *Artemisia annua* L. extract-fed animals. By administering artemisia extract, HFD-induced nuclear ChREBP and SREBP1 expression were considerably reduced. The extract dramatically reduced HMGB1 expression. RAGE expression was not significantly reduced by ATM treatment in the HFD-fed mice, though. The use of ATM significantly reduced COX-2 levels as well. The treatment of Artemisia annua extracts also reduced CTGF and TGF-β1 levels [144].

3.5.4 Oleanolic Acid

Oleanolic acid (OLA) is a triterpenoid isolated from *Olea ferruginea*, which grows in Kashmir, India. Numerous plant species worldwide also contain OLA, including *Eugenia jambolana*, *Tiarella polyphylla*, *Clerodendrum spicatum*, *Olea europaea,* etc. OLA is a very potent phytochemical that exhibits a wide range of biological actions [145]. The administration of OLA restored the levels of hepatotoxic blood marker enzymes and showed a substantial protective impact on ethanol-induced liver injury. The elevated level of CCL4-induced blood hepatic enzymes significantly decreased OLA. OLA reversed the elevated levels of MDA and reduced the activity of SOD and GPx [146]. The liver lipid levels progressively recovered to normal after OLA administration, while ALT and AST levels were markedly lowered. OLA decreased the upregulation of IL-6, IL-1β, and TNF-α and the production of pro-inflammatory factors in HFD mice. In liver tissues, OLA dramatically reduced the phosphorylation and degradation of IκB-α and p65-induced nuclear translocation [147].

Table 4 Names, biological sources, experiment research models, and targeted biomarkers of glycosides

Compound name	Biological source	Research model	Biomarkers	Refs.
Didymin	*Origanum vulgare* L.	Male C57BL/6J mice (SPF, 20 ± 2 g)	Didymin decreased ERS, ROS, Bax, GRP78, XBP1 ATF6 cytochrome c, and caspase-9 enhanced the expressions of CAT, GSH, and SOD, Bcl-2	[134]
GPS	*Gardenia jasminoides*	In vitro male Sprague-Dawley rats, 200 ± 20 g.	GPS caused increased expression of PPAR-α, SOD, and GSH-Px, HDL-C activity, and decreased MDA, CYP2E1, TG, TC, and LDL-C activity	[126]
Gentiopicroside	Roots of *Gentiana manshurica Kitagawa*	Animal experiment male C57BL/6 mice (8–10 weeks, 20–22 g)	Gentiopicroside inhibited protein and mRNA expression of α-SMA and collagen I in TGF-β-activated LX-2 cells. Decreased SREBP1 and increased AMPKα, PPAR-α and phosphorylated ACC	[135]
Diosgenin	Solanum and Dioscorea species	In vitro, the human HSC cell line, HSC-T6	Diosgenin Inhibits the expression of α-SMA and TGF-β. In addition, it inhibited the TGF-β1-induced expression of collagen I and α-SMA in HSC-T6 cells	[136]
Tectoridin	The flower of *Pueraria lobata*	In vitro male C57BL/6 mice	Tectoridin causes a reduction of serum ALT and AST levels, expression of mRNAs encoding DGATs and GPAT, and increases serum and hepatic TG levels. In addition, tectoridin significantly elevated the content of GSH and the activities of GSH-Px and SOD	[137]

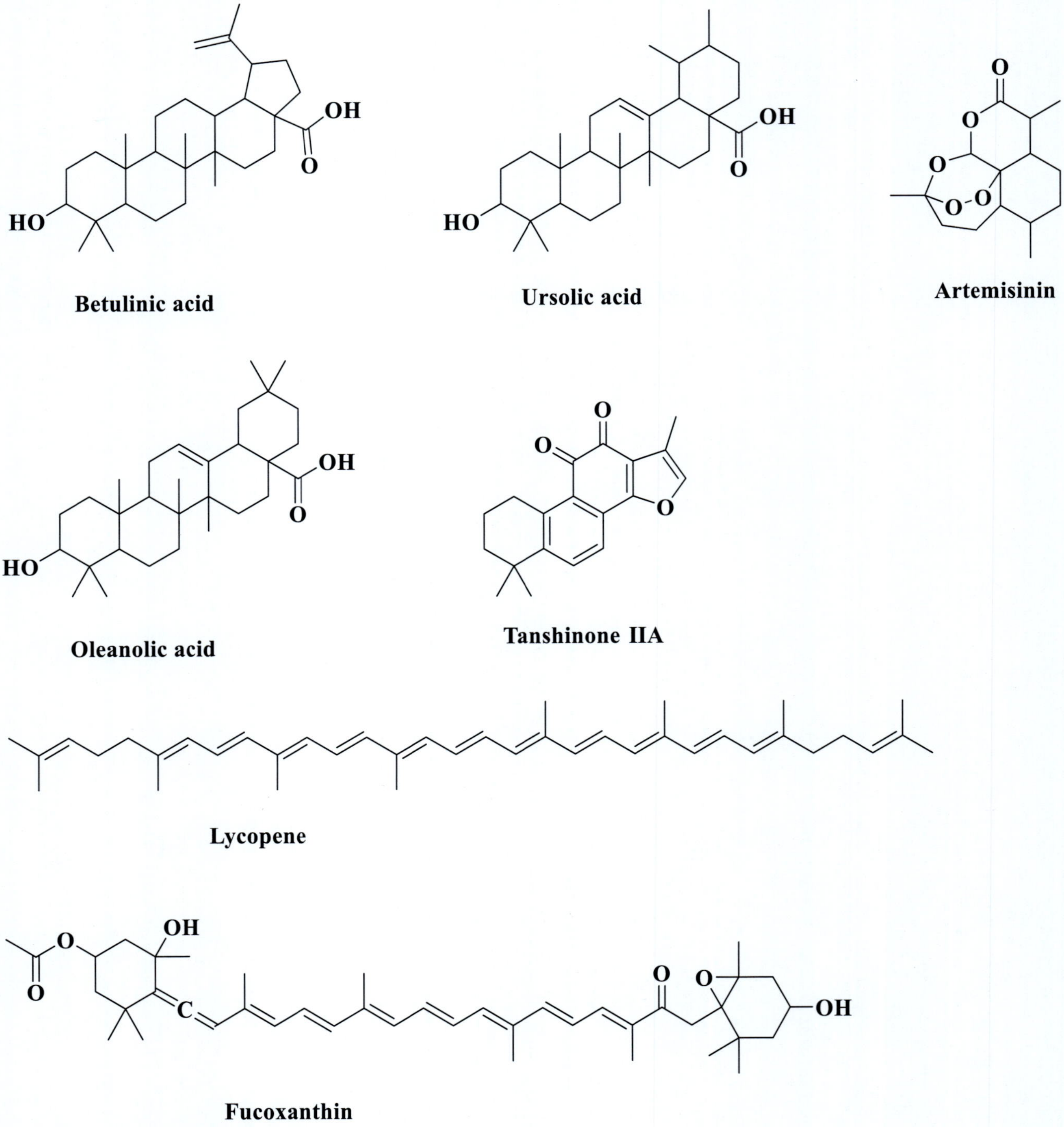

Fig. 6 Terpenoids in the treatment of liver cirrhosis

3.5.5 Lycopene

Lycopene (LYP) is present in many natural plants and food sources. LYP is primarily found in red-coloured vegetables and fruits, including tomatoes, pink guavas, watermelons, and apricots [148]. Pretreatment with LYP revealed that LYP could prevent the changes in ALT and AST, lower TC, TG, LDL-C, and FFA levels, and raise HDL-C. In addition, LYP elevated SOD and GSH and decreased MDA activities against NAFLD, indicating that the activity of antioxidants may be involved in the mechanism of its hepatoprotective effects. In contrast, animals on HFD had a substantial downregulation of CYP2E1 expression after receiving the LYP treatment [149]. LYP inhibited NF-κB activation and decreased interleukin IL-6, IL-1β, and TNF-α levels. The liver antioxidant enzymes GPx, SOD, and CAT were enhanced by LYP. LYP suppressed HSC

Table 5 Names, biological sources, experiment research models, and targeted biomarkers of terpenoid

Compound name	Biological source	Research model	Biomarkers	Refs.
Lutein	Dark green leafy vegetables, such as kale and spinach	Wistar rats weighing 150–200 g	Downregulated ALT, AST, LDH, and ALP in ALD, downregulating the inflammatory mediators like NF-κB, COX-2, iNOS and pro-inflammatory cytokines (TNF-α, IL-6, MCP-1, and IL-1β)	[159]
B-cryptoxanthin	Green and yellow vegetables and algae	In vitro human voluntary overweight/obese ($25 \leq$ BMI <40 kg/m^2) adults ($18 \leq$ age < 60 years)	B-cryptoxanthin restraining M1 or M2 activation of liver macrophages/Kupffer cells. Reductions in serum levels of MDA, hs-CRP, IL-6, and CK18-M65	[160]
Actein	*Actaea racemosa* L. syn. And *Cimicifuga racemosa* (L) Nutt.	In vitro HepG2 (p53 positive) human liver cancer cells and female Sprague-Dawley rats, 56-week old	Downregulation of CYP2C, TNF-α, SREBP-1c, IL-1β, FAS, MCP1, SCD1, α-SMA, CD36, IRS1, FOXO1, AKT, and GSK3β erythropoietin, ME1 and CCND1 and upregulating the expression of IPP, HMGCS, FDPS, CXCL1, C4BP, CYP7A1, A2M, TFRC, FLT1, (PSMB 10, NQ01) and p53 signalling gene (TNFRSF6, FAS)	[161, 162]
Andrographolide	*Andrographis paniculata*	In vitro CCl4-induced liver fibrosis male C57BL/6 mice and in vitro HepG2 cells	Downregulated α-SMA, TGF-β1, NF-κB, IL-6, IL-1β, TXNIP, GPx, while upregulated GSH, Nrf2, HO-1, and p53	[163, 164]

activation, and cellular lipid storage was altered. LYC treatment increased PPAR-α along with PPAR-β and PPAR-γ expressions of retinoid X receptor [150].

3.5.6 Fucoxanthin

A naturally occurring, low molecular weight, and chemically active carotenoid, fucoxanthin (FXN) is primarily generated from several marine algae. FXN concentration in seven microalgae isolates was measured, and the Synuroceae family had abundant FXN. The microalgae *Tisochrysis lutea* can also produce a high yield of FXN [151]. Dietary FXN significantly reduced the liver mRNA expression of fibrotic genes such as Timp1, collagen type I alpha 1 chain (Col1α1), and TGF-β1. As an indicator of activated HSCs, αSMA mRNA and protein expression were decreased by FXN. These findings imply that dietary FXN inhibits the progression of fibrosis by controlling the expression of fibrogenic factors [152]. In LX-2 cells, FXN suppressed the basal and TGF-β1-induced mRNA and protein expression of pro-fibrogenic genes by dramatically inhibiting SMAD3 activation. The TGF-1 signalling pathway, which is implicated in the onset of liver fibrosis, must include SMAD3. FXN markedly decreased both baseline and TGF-β1-induced TIMP1 expression. Its inhibitory effects on ROS buildup likely influence the anti-fibrogenic actions of FXN through decreasing NOX4 expression and on TGF-triggered suppression of PPAR-γ [153]. FXN inhibited lipid accumulation by suppressing the expression of lipogenesis and β-oxidation, including adipose TG lipase, phosphorylated hormone-sensitive lipase, PPAR, CPT1, CPT2, and upregulating the expression of genes involved in lipolysis and β-oxidation [154].

3.5.7 Tanshinone IIA

One of the main lipophilic substances obtained from *Salvia miltiorrhiza* root and the extract is tanshinone IIA (TIIA) [155].TIIA has a variety of biological actions, including the ability to protect against liver damage, hepatic steatosis, liver cirrhosis, hepatic fibrosis, and HCC [156]. The high levels of AST, ALT, ALP and TBIL in serum and of Hyp in hepatic tissue can be dramatically reduced by TIIA. TIIA can significantly lessen liver fibrosis, collagen deposition in hepatic tissue, and liver damage. TIIA successfully reduced the two markers (α-SMA and COL1A2) in vivo. MMP9, c-Jun, PI3K, c-Myc, and P38 could target genes for TIIA, and all these proteins were increased in CCl4-induced liver fibrosis. In TGF-treated LX2 cells, TIIA significantly reduced the expression of c-Jun, MMP9, c-Myc, CCND1, P-P65, PI3K, and P38. TIIA may reduce liver fibrosis through various targets and signalling pathways, and it can shed light on the pharmacological processes by which it works to treat hepatic fibrosis [157]. It is commonly accepted that the excess deposition of ECM proteins is a key component of hepatic fibrosis. The extensive synthesis and secretion of ECM are controlled by activated α-SMA and HSCs, and COL1A2 are molecular indicators of HSC activity. TIIA markedly downregulated the protein expression of these genes' products. TIIA therapy reduced ECM synthesis (COL1A2), HSC proliferation, and α-SMA expression, reducing the activation of HSC-LX2 cells [158].

3.6 Tannins

Tannins are sophisticated polyphenolic substances found in a wide variety of plants, particularly vascular plants. Tannin can capture free radicals. In addition, tannins have a hydroxyl group and a conjugated double bond (Fig. 7) that allow the electron to delocalize, making them particularly efficient as an electron/hydrogen atom donor and chelating metal [165].

3.6.1 Tannic Acid

Tannic acid (TNA) is a hydrolysable tannin polyphenol produced by plants. It is a gallic acid polymer glucoside that may be found in a variety of dietary plant products, including sorghum grain, coffee, tea, and cocoa [166]. NAFLD development is significantly influenced by ChREBP, the transcriptional activator of lipogenic genes. Through acetylation of Lys-672, p300 raises ChREBP transcriptional activity, which raises p300 occupancy on target gene promoters. Additionally, NAFLD is brought on by p300 overexpression. Histone acetyltransferases and histon deacetylases (HDACs) control the typical epigenetic process known as histone acetylation. Cancer, inflammation, and metabolic problems are linked to an imbalance between HAT and HDAC activity [167]. TNA-induced suppression of HAT activity reduces the development of NAFLD pathological characteristics. TNA prevented p300 from occupying the promoters of FAS and ATP-citrate lyase sterol regulatory element, ultimately leading to the hypoacetylation of H3K9 and H3K36. TNA also reduced the acetylation of total proteins and the histone H3 protein at lysine residues 9 and 36. Through these mechanisms, TNA suppressed the in vivo accumulation of fat by lowering the mRNA expression of genes essential for lipogenesis [168].

3.6.2 Dieckol

Dieckol (DCK) is a phlorotannin found in *Ecklonia cava*. Phlorotannin is a group of substances having polymerized phloroglucinol units. DCK from *Laminaria japonica* reduces hepatic steatosis by enhancing hepatic fatty acid β-oxidation [Liu Y et al., 2019]. The impacts of DCK on NAFLD might be made by reducing the development of the NLRP3 inflammation and pyroptosis in aHFD-induced mice NAFLD model, HFD considerably enhanced food intake and weight changes, but DCK treatment markedly reduced them. HFD also raised the blood levels of LDL and total cholesterol, which were then decreased by DCK [169]. HFD enhanced

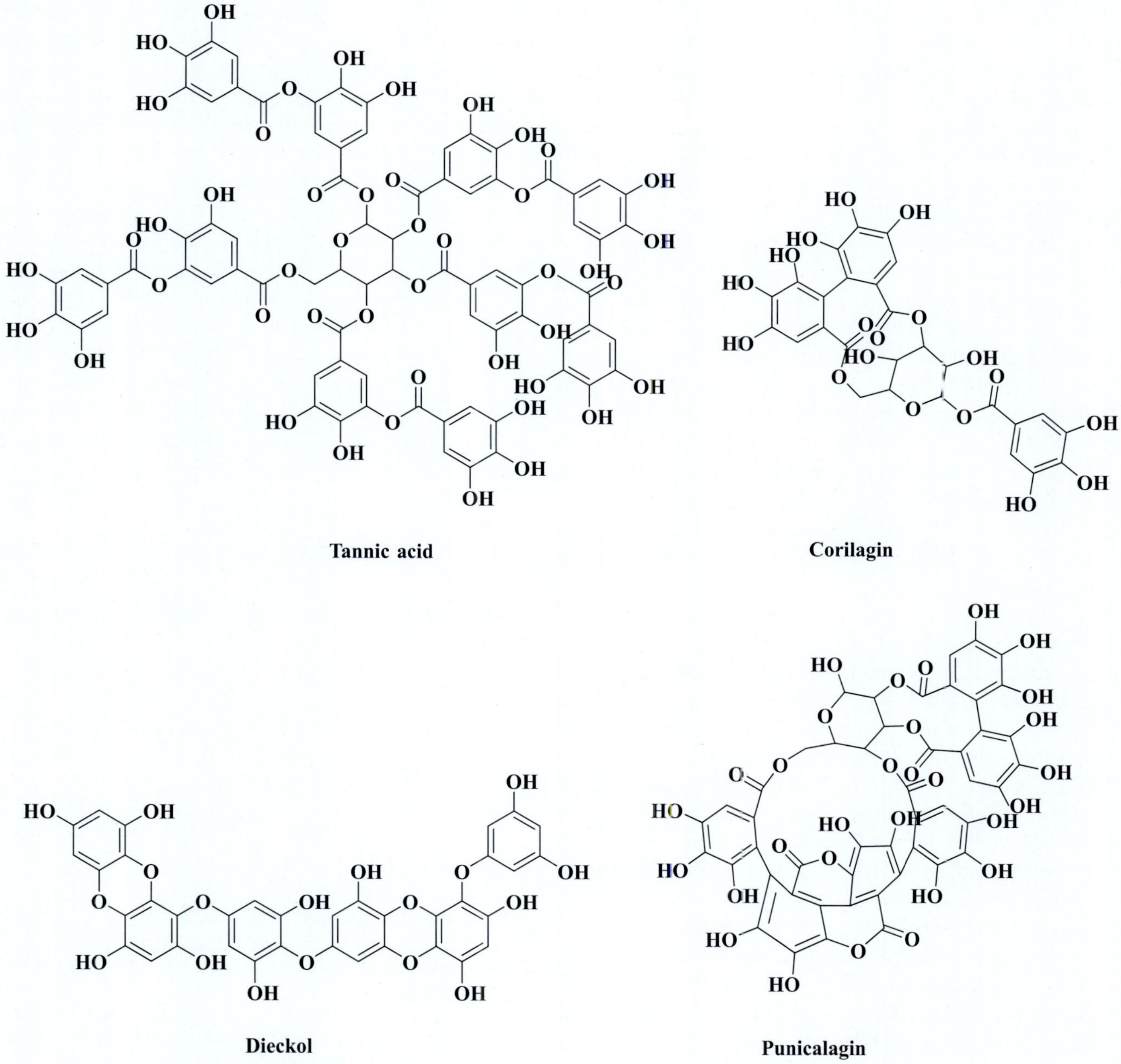

Fig. 7 Terpenoids in the treatment of liver cirrhosis

M1 infiltration but reduced M2 infiltration in the lymphatic vasculature of the liver. DCK treatment resulted in a decrease in M1 infiltration and an increase in M2 infiltration. HFD induced an uptick in IL-6 and TNF-α expression in the liver, while DCK treatment had the opposite effect. HFD lowered the expression of VEGFC and VEGFR3, whereas DCK treatment increased it. When mice were given the HFD, the expression of the lymphangiogenesis-related PI3K, pAKT, and pERK signalling pathways dropped, while DCK increased it [170]. With the administration of DCK, high-fat died induced lipogenic genes (FASN, SREBP2, and FABP4) expressions were significantly reduced. High-fat diet elevated TLR4 and NF-κB expression in the liver. DCK downregulates the overexpression. When DCK was administered, the NLRP3 inflammasome's elements, such as ASC and NLRP3, were reduced in the liver of HFD animals. In the liver, HFD raised the ratio of cleaved-caspase-1 to caspase-1, whereas DCK treatment lowered it [171].

3.6.3 Corilagin

Corilagin (CRL) is a gallotannin, one of the main active phytochemicals in various ethnopharmacological plants, including *Phyllanthus emblica, Phyllanthus reticulatus, Dimocarpus longan, and Geranium wilfordii*. According to reports, CRL exhibits many pharmacological properties, such as anti-tumour, antioxidant, hepatoprotective, and anti-inflammatory properties [172]. CRL reduced the severity of GalN/LPS-induced liver damage by regulating oxidative stress and inhibiting apoptosis. CRL prevented the development of schistosomiasis-affected hepatic fibrosis by inhibiting the miR-21/SMAD7/ERK pathway [173]. The hepatitis C virus (HCV) NS3 serine protease, a well-known target for preventing HCV replication, was demonstrated to be inhibited by CRL. CRL therapy may significantly alter the gene expression patterns in the liver brought on by a high-fat diet. High plasma ALT and AST values are two well-known biomarkers of hepatocellular damage. CRL effectively prevented the HFD-induced rise in ALT and AST levels in plasma, demonstrating its protective effects on liver function in developing NAFLD. CRL treatment in HFD-fed mice resulted in better plasma lipid profiles caused by NAFLD. It lowers levels of TG, TC, and LDL-C as well as higher levels of HDL-C. CRL downregulated Hsd3b5 expression, which may have contributed to its inhibitory effect on the development of NAFLD [174].

3.6.4 Punicalagin

Punicalagin (PCG) is a well-known phytochemical in pomegranate (*Punica granatum* L.) peel. Being used primarily in Asian nations for traditional medicine, it is currently widely grown in Southwest America, California, Mexico, and Africa. It has a variety of pharmacological properties, such as antioxidant, anti-atherosclerotic, hepatoprotective, and hypolipidemic properties, that have been reported [175]. PCG enhanced the adipokine pathway and reduced oxidative stress in eWAT (epididymal white adipose tissue), which improved lipid homeostasis in the liver and adipose tissue. When mice were given a western diet, PCG reduced the number of hepatic lipids [176]. In addition, PCG could shield rats from cyclophosphamide-induced liver toxicity. PCG reduced the levels of TNF-α, NF-κB, p65, IL-1β, inducible nitric oxide synthase, and caspases 3 and 9 to prevent cyclophosphamide-induced liver inflammation and death [177]. By decreasing the production of ROS and NO and raising the expression of SOD1, PCG prevented the oxidative stress caused by lipopolysaccharides (LPS) in macrophages. PCG activated the Nrf2-signalling pathway to significantly reduce the effects of FFA-induced lipotoxicity in HepG2 cells. PCG therapy decreased the MDA level and upregulated SOD and GPx activities against CCL4-induced liver injury [178].

4 Conclusion

Unhealthy diet and alcoholism are the two factors primarily bring on cirrhosis. DNL causes hindrance of lipid metabolism, which accelerates oxidative stress to promote lipid accumulation and inflammation in healthy hepatocytes. Naturally obtained compounds work as boons in treating these complicated diseases, while synthetic medications leave adverse imprints. Currently, herbal medicines are trying to be clinically reused like traditional medicines concomitantly with synthetic ones. The search for suitable natural products for complicated diseases stands as a gold mine for medical research. SMR, hesperidin, CUR, etc., have already shown good hepatoprotective properties and are viable therapy alternatives on their own or as a supplement when taken with other medications. Undoubtedly, there is a great potential for natural substances to be effective options for liver cirrhosis. Therefore, we expect that the study on the pathophysiology of hepatic injuries and progression to cirrhosis and future therapeutic medication development for cirrhosis and associated liver conditions may be theoretically supported by the information in this chapter.

Funding None.

Conflict of Interest The authors declare no conflicts of interest.

Acknowledgement All the authors acknowledge the Department of Pharmaceutical Technology, Jadavpur University, Kolkata, for providing research amenities.

References

1. Cheemerla S, Balakrishnan M. Global epidemiology of chronic liver disease. Clin Liver Dis (Hoboken). 2021;17:365–70.
2. Friedman SL. Liver fibrosis – from bench to bedside. J Hepatol. 2003;38:38–53.
3. Crocker TF, Brown L, Lam N, Wray F, Knapp P, Forster A. Information provision for stroke survivors and their carers. Cochrane Database Syst Rev. 2021;11:CD001919.
4. Bikbov MM, Gilmanshin TR, Zainullin RM, Kazakbaeva GM, Iakupova EM, Fakhretdinova AA, et al. Prevalence of non-alcoholic fatty liver disease in the Russian Ural Eye and Medical Study and the Ural Very Old Study. Sci Rep. 2022;12:7842.
5. Zhong F, Zhou X, Xu J, Gao L. Rodent models of nonalcoholic fatty liver disease. Digestion. 2020;101:522–35.
6. Bataller R, Brenner DA. Liver fibrosis. J Clin Invest. 2005;115:209–18.
7. Pan X, Ma X, Jiang Y, Wen J, Yang L, Chen D, et al. A comprehensive review of natural products against liver fibrosis: flavonoids, quinones, lignans, phenols, and acids. Evid Based Complement Alternat Med. 2020;2020:7171498.
8. Latief U, Ahmad R. Herbal remedies for liver fibrosis: a review on the mode of action of fifty herbs. J Tradit Complement Med. 2017;8:352–60.
9. Hickman IJ, Macdonald GA. Impact of diabetes on the severity of liver disease. Am J Med. 2007;120:829–34.
10. Tolman KG, Fonseca V, Dalpiaz A, Tan MH. Spectrum of liver disease in type 2 diabetes and management of patients with diabetes and liver disease. Diabetes Care. 2007;30:734–43.

11. Pessayre D, Fromenty B, Mansouri A. Mitochondrial injury in steatohepatitis. Eur J Gastroenterol Hepatol. 2004;16(11), 1095–1105.
12. Ameer F, Scandiuzzi L, Hasnain S, Kalbacher H, Zaidi N. De novo lipogenesis in health and disease. Metabolism. 2014;63:895–902.
13. Abu-Elheiga L, Matzuk MM, Kordari P, Oh W, Shaikenov T, Gu Z, et al. Mutant mice lacking acetyl-CoA carboxylase 1 are embryonically lethal. Proc Natl Acad Sci U S A. 2005;102:12011–6.
14. Munday MR. Regulation of mammalian acetyl-CoA carboxylase. Biochem Soc Trans. 2002;30:1059–64.
15. Clifford BL, Sedgeman LR, Williams KJ, Morand P, Cheng A, Jarrett KE, et al. FXR activation protects against NAFLD via bile-acid-dependent reductions in lipid absorption. Cell Metab. 2021;33:1671–1684.e4.
16. Bilotta MT, Petillo S, Santoni A, Cippitelli M. Liver X receptors: regulators of cholesterol metabolism, inflammation, autoimmunity, and cancer. Front Immunol. 2020;11,584303
17. Crespo J, Cayón A, Fernández-Gil P, Hernández-Guerra M, Mayorga M, Domínguez-Díez A, et al. Gene expression of tumor necrosis factor ?? And TNF-receptors, p55 and p75, in nonalcoholic steatohepatitis patients. Hepatology. 2001;34:1158–63.
18. Sanyal AJ. AGA technical review on nonalcoholic fatty liver disease. Gastroenterology. 2002;123:1705–25.
19. Bertolani C, Marra F. The role of adipokines in liver fibrosis. Pathophysiology. 2008;15:91–101.
20. Zheng X, Wu F, Lin X, Shen L, Feng Y. Developments in drug delivery of bioactive alkaloids derived from traditional Chinese medicine. Drug Deliv. 2018;25:398–416.
21. Das S, Das MK, Das R, Gehlot V, Mahant S, Mazumder PM, et al. Isolation, characterization of Berberine from Berberis aristata DC for eradication of resistant Helicobacter pylori. Biocatal Agric Biotechnol. 2020;26:101622.
22. Xu X, Yi H, Wu J, Kuang T, Zhang J, Li Q, et al. Therapeutic effect of berberine on metabolic diseases: both pharmacological data and clinical evidence. Biomed Pharmacother. 2021;133:110984.
23. Zhang Z, Li B, Meng X, Yao S, Jin L, Yang J, et al. Berberine prevents progression from hepatic steatosis to steatohepatitis and fibrosis by reducing endoplasmic reticulum stress. Sci Rep. 2016;6:20848.
24. Zhang Y, Deng Y, Tang K, Chen R, Liang S, Liang Y, et al. Berberine ameliorates high-fat diet-induced non-alcoholic fatty liver disease in rats via activation of SIRT3/AMPK/ACC pathway. Curr Med Sci. 2019;39:37–43.
25. Choi B-H, Ahn IS, Kim Y-H, Park J-W, Lee S-Y, Hyun C-K, et al. Berberine reduces the expression of adipogenic enzymes and inflammatory molecules of 3T3-L1 adipocyte. Exp Mol Med. 2007;38:599–605.
26. Guo T, Woo S-L, Guo X, Li H, Zheng J, Botchlett R, et al. Berberine ameliorates hepatic steatosis and suppresses liver and adipose tissue inflammation in mice with diet-induced obesity. Sci Rep. 2016;6:22612.
27. Xia S-F, Le G-W, Wang P, Qiu Y-Y, Jiang Y-Y, Tang X. Regressive effect of myricetin on hepatic steatosis in mice fed a high-fat diet. Nutrients. 2016;8(12), 799.
28. Nieber K. The impact of coffee on health. Planta Med. 2017;83(16), 1256–1263.
29. Oliveira K, Buss C, Tovo C. Association of caffeine intake and liver fibrosis in patients with chronic hepatitis C. Arq Gastroenterol. 2015;52:4–8.
30. Shim S, Jun DW, Kim E, Saeed W, Lee K, Lee H, et al. Caffeine attenuates liver fibrosis via defective adhesion of hepatic stellate cells in cirrhotic model. J Gastroenterol Hepatol. 2013;28(12), 1877–1884.
31. Kang C, Wang B, Kaliannan K, Wang X, Lang H, Hui S-C, et al. Gut microbiota mediates the protective effects of dietary capsaicin against chronic low-grade inflammation and associated obesity induced by high-fat diet. MBio. 2017;8:e00470–17.
32. Shin MK, Yang S-M, Han I-S. Capsaicin suppresses liver fat accumulation in high-fat diet-induced NAFLD mice. Anim Cells Syst (Seoul). 2020;24:214–9.
33. Ohashi T, Nakade Y, Ibusuki M, Kitano R, Yamauchi T, Kimoto S, et al. Conophylline inhibits high fat diet-induced non-alcoholic fatty liver disease in mice. PLoS One. 2019;14:e0210068.
34. Yang D, Li L, Qian S, Liu L. Evodiamine ameliorates liver fibrosis in rats via TGF-β1/Smad signaling pathway. J Nat Med. 2017;72(1), 145–154.
35. Lu C, Xu W, Zhang F, Jin H, Chen Q, Chen L, et al. Ligustrazine prevents alcohol-induced liver injury by attenuating hepatic steatosis and oxidative stress. Int Immunopharmacol. 2015;29:613–21.
36. Si M, Zhang J-H, Wang J-L, Lu G, Chen X-X, Xiao J-H. Antifibrotic effects of neferine on carbon tetrachloride-induced hepatic fibrosis in mice. Am J Chin Med. 2015;43:1–10.
37. Zhang H, Yang L, Wang Y, Huang W, Li Y, Chen S, et al. Oxymatrine alleviated hepatic lipid metabolism via regulating miR-182 in non-alcoholic fatty liver disease. Life Sci. 2020;257:118090.
38. Li L, Liu Q, Fan L, Xiao W, Zhao L, Wang Y, et al. Protective effects of oxymatrine against arsenic trioxide-induced liver injury. Oncotarget. 2016;8(8), 12792–12799.
39. Shi L, Shi L, Zhang H, Hu Z, Wang C, Zhang D, et al. Oxymatrine ameliorates non-alcoholic fatty liver disease in rats through peroxisome proliferator-activated receptor-α activation. Mol Med Rep. 2013;8:439–45.
40. Bingul I, Aydın A, Başaran-Küçükgergin C, Doğan Ekici AI, Çoban J, Doğru-Abbasoğlu S, et al. High-fat diet plus carbon tetrachloride-induced liver fibrosis is alleviated by betaine treatment in rats. Int Immunopharmacol. 2016;39:199–207.
41. Zhang C, Deng J, Liu D, Tuo X, Xiao L, Lai B, et al. Nuciferine ameliorates hepatic steatosis in high-fat diet/streptozocin-induced diabetic mice through a PPARα/PPARγ coactivator-1α pathway. Br J Pharmacol. 2018;175:4218–28.
42. Shu G, Yusuf A, Dai C, Sun H, Deng X. Piperine inhibits AML-12 hepatocyte EMT and LX-2 HSC activation and alleviates mouse liver fibrosis provoked by CCl4: roles in the activation of the Nrf2 cascade and subsequent suppression of the TGF-β1/Smad axis. Food Funct. 2021;12:11686–703.
43. Afifi N, Ramadan A, Erian E, Saleh D, Sedik A, Badawi M, et al. Trigonelline attenuates hepatic complications and molecular alterations in high fat high fructose-induced insulin resistance in rats. Can J Physiol Pharmacol. 2017;95:427–36.
44. Song C-Y, Shi J, Zeng X, Zhang Y, Xie W-F, Chen Y-X. Sophocarpine alleviates hepatocyte steatosis through activating AMPK signaling pathway. Toxicol In Vitro. 2013;27:1065–71.
45. Liang L, Ye S, Jiang R, Zhou X, Zhou J, Meng S. Liensinine alleviates high fat diet (HFD)-induced non-alcoholic fatty liver disease (NAFLD) through suppressing oxidative stress and inflammation via regulating TAK1/AMPK signaling. Int Immunopharmacol. 2022;104:108306.
46. Lv X-T, Wang R-H, Liu X-T, Ye Y-J, Liu X-Y, Qiao J-D, et al. Theacrine ameliorates experimental liver fibrosis in rats by lowering cholesterol storage via activation of the Sirtuin 3-farnesoid X receptor signaling pathway. Chem Biol Interact. 2022;364:110051.
47. Li G, Zhou F, Chen Y, Zhang W, Wang N. Kukoamine A attenuates insulin resistance and fatty liver through downregulation of Srebp-1c. Biomed Pharmacother. 2017;89:536–43.
48. Sylvester Darvin S, Toppo E, Esakkimuthu S, Ajeesh Krishna TP, Ceasar SA, Stalin A, et al. Hepatoprotective effect of bisbenzylisoquinoline alkaloid tiliamosine from Tiliacora racemosa in high-fat diet/diethylnitrosamine-induced non-alcoholic steatohepatitis. Biomed Pharmacother. 2018;108:963–73.
49. Yue R, Jin G, Wei S, Huang H, Su L, Zhang C, et al. Immunoregulatory effect of koumine on nonalcoholic fatty liver disease rats. J Immunol Res. 2019;2019:8325102.
50. Lee W-Y, Lee C-Y, Lee J-S, Kim C-E. Identifying candidate flavonoids for non-alcoholic fatty liver disease by network-based strategy. Front Pharmacol. 2022;13, 892559.

51. Hernández-Aquino E, Muriel P. Beneficial effects of naringenin in liver diseases: molecular mechanisms. World J Gastroenterol. 2018;24:1679–707.
52. Abenavoli L, Izzo A, Milic N, Cicala C, Santini A, Capasso R. Milk thistle (Silybum marianum): a concise overview on its chemistry, pharmacological, and nutraceutical uses in liver diseases: milk thistle and liver diseases. Phytother Res. 2018;32(11), 2202–2213.
53. Kim S-J, Moon Y-J, Lee S-M. Protective effects of baicalin against ischemia/reperfusion injury in rat liver. J Nat Prod. 2010;73:2003–8.
54. Kim S-J, Lee S-M. Effect of baicalin on toll-like receptor 4-mediated ischemia/reperfusion inflammatory responses in alcoholic fatty liver condition. Toxicol Appl Pharmacol. 2012;258:43–50.
55. Hu Q, Zhang W, Wu Z, Tian X, Xiang J, Li L, et al. Baicalin and the liver-gut system: pharmacological bases explaining its therapeutic effects. Pharmacol Res. 2021;165:105444.
56. Prabu SM, Tomczyk M, Skalicka-Woźniak K, Rastrelli L, Daglia M, Nabavi S, et al. Hepatoprotective effect of quercetin: from chemistry to medicine. Food Chem Toxicol. 2016;108(Pt B), 365–374.
57. Kim Y-J, Yoon D, Jung U. Efficacy of nobiletin in improving hypercholesterolemia and nonalcoholic fatty liver disease in high-cholesterol diet-fed mice. Nutr Res Pract. 2021;15:431.
58. Chen S, Zhao X, Wan J, Li R, Qin Y, Wang X, et al. Dihydromyricetin improves glucose and lipid metabolism and exerts anti-inflammatory effects in nonalcoholic fatty liver disease: a randomized controlled trial. Pharmacol Res. 2015;99, 74–81.
59. Lee J, Song J-H, Chung M-Y, Lee J-H, Nam TG, Park J, et al. 3,4-dihydroxytoluene, a metabolite of rutin, suppresses the progression of nonalcoholic fatty liver disease in mice by inhibiting p300 histone acetyltransferase activity. Acta Pharmacol Sin. 2020;42(9), 1449–1460.
60. Lin L-C, Pai Y-F, Tsai T-H. Isolation of luteolin and luteolin-7-O-glucoside from Dendranthema morifolium Ramat Tzvel and their pharmacokinetics in rats. J Agric Food Chem. 2015;63(35), 7700–7706.
61. Liu X, Sun R, Li Z, Xiao R, Lv P, Sun X, et al. Luteolin alleviates non-alcoholic fatty liver disease in rats via restoration of intestinal mucosal barrier damage and microbiota imbalance involving in gut-liver axis. Arch Biochem Biophys. 2021;711:109019.
62. Zhang X, Huo Z, Luan H, Huang Y, Shen Y, Sheng L, et al. Scutellarin ameliorates hepatic lipid accumulation by enhancing autophagy and suppressing IRE1α/XBP1 pathway. Phytother Res. 2022;36:433–47.
63. Xu Y, Zhang D, Yang H, Liu Y, Zhang L, Zhang C, et al. Hepatoprotective effect of genistein against dimethylnitrosamine-induced liver fibrosis in rats by regulating macrophage functional properties and inhibiting the JAK2/STAT3/SOCS3 signaling pathway. Front Biosci. 2021;26:1572–84.
64. Gaballah HH, El-Horany HE, Helal DS. Mitigative effects of the bioactive flavonol fisetin on high-fat/high-sucrose induced nonalcoholic fatty liver disease in rats. J Cell Biochem. 2019;120:12762–74.
65. Wang LL, Zhang ZC, Hassan W, Li Y, Liu J, Shang J. Amelioration of free fatty acid-induced fatty liver by quercetin-3-O-β-D--glucuronide through modulation of peroxisome proliferator-activated receptor-alpha/sterol regulatory element-binding protein-1c signaling. Hepatol Res. 2016;46:225–38.
66. Hur HJ, Jeong Y-H, Lee SH, Sung MJ. Quercitrin ameliorates hyperlipidemia and hepatic steatosis in ovariectomized mice. Life. 2020;10
67. Jiang Y, Gong Q, Gong Y, Zhuo C, Huang J, Tang Q. Vitexin attenuates non-alcoholic fatty liver disease lipid accumulation in high fat-diet fed mice by activating autophagy and reducing endoplasmic reticulum stress in liver. Biol Pharm Bull. 2022;45:260–7.
68. Attia H, Albekairi N, Albdeirat L, Soliman A, Rajab R, Ali R, et al. Chrysin attenuates fructose-induced nonalcoholic fatty liver in rats via antioxidant and anti-inflammatory effects: the role of angiotensin-converting enzyme 2/angiotensin (1-7)/Mas receptor axis. Oxidative Med Cell Longev. 2022;2022:1–14.
69. Li Y, Tong L, Zhang J, Zhang Y, Zhang F. Galangin alleviates liver ischemia-reperfusion injury in a rat model by mediating the PI3K/AKT pathway. Cell Physiol Biochem. 2018;51:1354–63.
70. Li J, Wang T, Liu P, Yang F, Wang X, Zheng W, et al. Hesperetin ameliorates hepatic oxidative stress and inflammation via the PI3K/AKT-Nrf2-ARE pathway in oleic acid-induced HepG2 cells and a rat model of high-fat diet-induced NAFLD. Food Funct. 2021;12:3898–918.
71. Xie Q, Gao S, Lei M, Li Z. Hesperidin suppresses ERS-induced inflammation in the pathogenesis of non-alcoholic fatty liver disease. Aging. 2022;14(3), 1265–1279.
72. Dong X, Zhu Y, Wang S, Luo Y, Lu S, Nan F, et al. Bavachinin inhibits cholesterol synthesis enzyme FDFT1 expression via AKT/mTOR/SREBP-2 pathway. Int Immunopharmacol. 2020;88:106865.
73. Feng Z, Pang L, Chen S, Pang X, Huang Y, Qiao Q, et al. Didymin ameliorates dexamethasone-induced non-alcoholic fatty liver disease by inhibiting TLR4/NF-κB and PI3K/Akt pathways in C57BL/6J mice. Int Immunopharmacol. 2020;88:107003.
74. Chang T-C, Chiou W-C, Lai W-H, Huang H-C, Huang Y-L, Liu H-K, et al. Ugonin J improves metabolic disorder and ameliorates nonalcoholic fatty liver disease by regulating the AMPK/AKT signaling pathway. Pharmacol Res. 2021;163:105298.
75. Wang S, Sheng F, Zou L, Xiao J, Li P. Hyperoside attenuates non-alcoholic fatty liver disease in rats via cholesterol metabolism and bile acid metabolism. J Adv Res. 2021;34:109–22.
76. Lin W, Jin Y, Hu X, Huang E, Zhu Q. AMPK/PGC-1α/GLUT4-mediated effect of icariin on hyperlipidemia-induced non-alcoholic fatty liver disease and lipid metabolism disorder in mice. Biochemistry (Mosc). 2021;86:1407–17. https://doi.org/10.1134/S0006297921110055.
77. Pinto C, Duque A, Galdón B, Cestero JJ, Macias P. Xanthohumol prevents carbon tetrachloride-induced acute liver injury in rats. Food Chem Toxicol. 2012;50:3405–12.
78. Lan T, Jiang S, Zhang J, Weng Q, Yu Y, Li H, et al. Breviscapine alleviates NASH by inhibiting TGF-β-activated kinase 1-dependent signaling. Hepatology. 2022;76:155–71.
79. Geng Y, Sun Q, Li W, Lu Z-M, Xu Z, Shi J-S, et al. The common dietary flavonoid myricetin attenuates liver fibrosis in carbon tetrachloride-treated mice. Mol Nutr Food Res. 2016;61(4), 10.1002/mnfr.201600392.
80. Ganesan K, Xu B. A critical review on polyphenols and health benefits of black soybeans. Nutrients. 2017;9(5), 455.
81. Masterjohn C, Bruno R. Therapeutic potential of green tea in non-alcoholic fatty liver disease. Nutr Rev. 2012;70:41–56.
82. Abenavoli L, Larussa T, Corea A, Procopio AC, Boccuto L, Dallio M, et al. Dietary polyphenols and non-alcoholic fatty liver disease. Nutrients. 2021;13(2), 494.
83. Wang P, Gao J, Ke W, Wang J, Li D, Liu R, et al. Resveratrol reduces obesity in high-fat diet-fed mice via modulating the structure and metabolic function of the gut microbiota. Free Radic Biol Med. 2020;156, 83–98.
84. Aguirre L, Portillo M, Hijona Muruamendiaraz E, Bujanda L. Effects of resveratrol and other polyphenols in hepatic steatosis. World J Gastroenterol. 2014;20:7366–80.
85. Feng T, Wei Y, Lee R, Zhao L. Liposomal curcumin and its application in cancer. Int J Nanomedicine. 2017;12:6027–44.
86. Rahmani S, Asgary S, Askari G, Keshvari M, Hatamipour M, Feizi A, et al. Treatment of non-alcoholic fatty liver disease with

curcumin: a randomized placebo-controlled trial: curcumin supplementation for NAFLD. Phytother Res. 2016;30(9), 1540–1548.
87. Contaldo F, Santarpia L, Pasanisi F. Chronic inflammatory liver diseases and coffee intake. Curr Opin Clin Nutr Metab Care. 2019;22(5), 389–392.
88. Tajik N, Tajik M, Mack I, Enck P. The potential effects of chlorogenic acid, the main phenolic components in coffee, on health: a comprehensive review of the literature. Eur J Nutr. 2017;56:2215–44.
89. Zhu X, Lin X, Zhang P, Liu Y, Ling W, Guo H. Upregulated NLRP3 inflammasome activation is attenuated by anthocyanins in patients with nonalcoholic fatty liver disease: a case-control and an intervention study. Clin Res Hepatol Gastroenterol. 2022;46:101843.
90. Valenti L, Riso P, Mazzocchi A, Porrini M, Fargion S, Agostoni C. Dietary anthocyanins as nutritional therapy for nonalcoholic fatty liver disease. Oxidative Med Cell Longev. 2013;2013:145421.
91. Pardo-Andreu G, Sánchez-Baldoquín C, Avila-González R, Yamamoto E, Revilla A, Uyemura S, et al. Interaction of Vimang (Mangifera indica L. extract) with Fe(III) improves its antioxidant and cytoprotecting activity. Pharmacol Res. 2006;54:389–95.
92. Rasool M, Sabina E, Mahinda P, Gnanaselvi B. Mangiferin, a natural polyphenol protects the hepatic damage in mice caused by CCl_4 intoxication. Comp Clin Pathol. 2011;21, 865-872.
93. Tang H, Hao S, Chen X, Li Y, Yin Z, Zou Y, et al. Epigallocatechin-3-gallate protects immunity and liver drug-metabolism function in mice loaded with restraint stress. Biomed Pharmacother. 2020;129:110418.
94. Fernández-Bolaños J, López Ó, Fernandez-Bolanos J, Rodríguez-Gutiérrez G. Hydroxytyrosol and derivatives: isolation, synthesis, and biological properties. Curr Organ Chem. 2008;12:442–63.
95. Yu Y-B, Zhuang H-Z, Ji X-J, Dong L, Duan M-L. Hydroxytyrosol suppresses LPS-induced intrahepatic inflammatory responses via inhibition of ERK signaling pathway activation in acute liver injury. Eur Rev Med Pharmacol Sci. 2020;24:6455–62.
96. Bekut Hitl M, Kladar N, Gavarić N, Bozin B. Rosmarinic acid–human pharmacokinetics and health benefits. Planta Med. 2020;87(4), 273–282.
97. Balachander GJ, Subramanian S, Ilango K. Rosmarinic acid attenuates hepatic steatosis by modulating ER stress and autophagy in oleic acid-induced HepG2 cells. RSC Adv. 2018;8:26656–63.
98. ALTamimi JZ, Alshammari GM, NA AF, Alagal RI, Aljabryn DH, Albekairi NA, et al. Ellagic acid protects against non-alcoholic fatty liver disease in streptozotocin-diabetic rats by activating AMPK. Pharm Biol. 2022;60:25–37.
99. Yu Q, Liu Y, Wu Y, Chen Y. Dihydrocurcumin ameliorates the lipid accumulation, oxidative stress and insulin resistance in oleic acid-induced L02 and HepG2 cells. Biomed Pharmacother. 2018;103:1327–36.
100. Li Z, Feng H, Han L, Ding L, Shen B, Tian Y, et al. Chicoric acid ameliorate inflammation and oxidative stress in lipopolysaccharide and d -galactosamine induced acute liver injury. J Cell Mol Med. 2020;24(5), 3022–3033.
101. Sousa JN, Paraíso AF, Andrade JMO, Lelis DF, Santos EM, Lima JP, et al. Oral gallic acid improve liver steatosis and metabolism modulating hepatic lipogenic markers in obese mice. Exp Gerontol. 2020;134:110881.
102. Cha S-H, Hwang Y, Heo S-J, Jun H-S. Diphlorethohydroxycarmalol attenuates palmitate-induced hepatic lipogenesis and inflammation. Mar Drugs. 2020;18(9), 475.
103. Galati G, Lin A, Sultan A, O'Brien P. Cellular and in vivo hepatoxicity caused by green tea phenolic acids and catechins. Free Radic Biol Med. 2006;40:570–80.
104. Miao H, Ouyang H, Guo Q, Wei M, Lu B, Kai G, et al. Chlorogenic acid alleviated liver fibrosis in methionine and choline deficient diet-induced nonalcoholic steatohepatitis in mice and its mechanism. J Nutr Biochem. 2022;106:109020.
105. Qiang G, Yang X, Xuan Q, Shi L, Zhang H, Chen B, et al. Salvianolic acid a prevents the pathological progression of hepatic fibrosis in high-fat diet-fed and streptozotocin-induced diabetic rats. Am J Chin Med. 2014;42:1183–98.
106. Kiyama R. Nutritional implications of ginger: chemistry, biological activities and signaling pathways. J Nutr Biochem. 2020;86:108486.
107. Ahn J, Lee H, Jung CH, Ha SY, Seo H-D, Kim YI, et al. 6-Gingerol ameliorates hepatic steatosis via HNF4α/miR-467b-3p/GPAT1 cascade. Cell Mol Gastroenterol Hepatol. 2021;12:1201–13.
108. Geethangili M, Lin C-W, Mersmann HJ, Ding S-T. Methyl brevifolin carboxylate attenuates free fatty acid-induced lipid metabolism and inflammation in hepatocytes through AMPK/NF-κB signaling pathway. Int J Mol Sci. 2021;22(18), 10062.
109. Brito-Arias M. Synthesis and characterization of glycosides. New York: Springer; 2007.
110. Dhiman R, Chawla Y. Herbal medicines for liver diseases. Dig Dis Sci. 2005;50:1807–12.
111. Li X, Sun R, Liu R. Natural products in licorice for the therapy of liver diseases: progress and future opportunities. Pharmacol Res. 2019;144:210–26.
112. Guo X-L, Liang B, Wang X-W, Fan F-G, Jin J, Lan R, et al. Glycyrrhizic acid attenuates CCl4-induced hepatocyte apoptosis in rats via a p53-mediated pathway. World J Gastroenterol. 2013;19:3781–91.
113. Lee C-H, Park S-W, Kim Y, Kang S, Kim J, Lee S, et al. Protective mechanism of glycyrrhizin on acute liver injury induced by carbon tetrachloride in mice. Biol Pharm Bull. 2007;30:1898–904.
114. Chen L, Zhao X, Wei S, Ma X, Liu H, Li J, et al. Mechanism of Paeoniflorin on ANIT-induced Cholestatic liver injury using integrated metabolomics and network pharmacology. Front Pharmacol. 2021;12, 737630.
115. Chen Z, Zhu Y, Zhao Y, Ma X, Niu M, Wang J, et al. Serum metabolomic profiling in a rat model reveals protective function of paeoniflorin against ANIT induced cholestasis. Phytother Res. 2016;30
116. Zhang L-J, Yang B, Yu B-P. Paeoniflorin protects against nonalcoholic fatty liver disease induced by a high-fat diet in mice. Biol Pharm Bull. 2015;38(7), 1005–1011.
117. Zhao Y, Ma X, Wang J, Zhu Y, Li R, Wang J, et al. Paeoniflorin alleviates liver fibrosis by inhibiting HIF-1α through mTOR-dependent pathway. Fitoterapia. 2014;99:318–27.
118. Lee I-C, Bae J-S. Hepatoprotective effects of vicenin-2 and scolymoside through the modulation of inflammatory pathways. J Nat Med. 2020;74:90–7.
119. Huang G, Li S, Zhang Y, Zhou X, Chen W. Vicenin-2 is a novel inhibitor of STAT3 signaling pathway in human hepatocellular carcinoma. Jour Funct Foods. 2020;69:103921.
120. Kang H, Ku S-K, Jung B, Bae J-S. Anti-inflammatory effects of vicenin-2 and scolymoside in vitro and in vivo. Inflamm Res. 2015;64:1005–21.
121. Liu M, Liu C, Chen H, Huang X, Zeng X, Zhou J, et al. Prevention of cholesterol gallstone disease by schaftoside in lithogenic diet-induced C57BL/6 mouse model. Eur J Pharmacol. 2017;815:1–9.
122. Liu M, Zhang G, Wu S, Song M, Wang J, Cai W, et al. Schaftoside alleviates HFD-induced hepatic lipid accumulation in mice via upregulating farnesoid X receptor. Eur J Pharmacol. 2020;255:112776.
123. Wang Y, Jiang Z-Z, Chen M, Wu M-J, Guo H-L, Sun L-X, et al. Protective effect of total flavonoid C-glycosides from Abrus mollis extract on lipopolysaccharide-induced lipotoxicity in mice. Chin. J Nat Med. 2014;12:461–8.
124. Habtemariam S, Lentini G. Plant-derived anticancer agents: lessons from the pharmacology of geniposide and its aglycone, genipin. Biomedicine. 2018;6(2), 39.

125. Peng J, Leng J, Tian H, Yang T, Fang Y, Feng Q, et al. Geniposide and chlorogenic acid combination ameliorates non-alcoholic steatohepatitis involving the protection on the gut barrier function in mouse induced by high-fat diet. Front Pharmacol. 2018;9, 1399.
126. Ma T, Huang C, Zong G, Zha D, Meng X-M, Li J, et al. Hepatoprotective effects of geniposide in a rat model of nonalcoholic steatohepatitis. J Pharm Pharmacol. 2011;63:587–93.
127. Fadzil N, Sekar M, Gan S, Bonam SR, Wu Y-S, Vaijanathappa J, et al. Chemistry, pharmacology and therapeutic potential of swertiamarin – a promising natural lead for new drug discovery and development. Drug Des Devel Ther. 2021;15:2721–46.
128. Xie Y, Hao H, Wang H, Guo C, Kang A, Wang G. Reversing effects of lignans on CCl4-induced hepatic CYP450 down regulation by attenuating oxidative stress. J Ethnopharmacol. 2014;155:213–21. https://doi.org/10.1016/j.jep.2014.05.016.
129. Wu T, Li J, Li Y, Song H. Antioxidant and hepatoprotective effect of swertiamarin on carbon tetrachloride-induced hepatotoxicity via the Nrf2/HO-1 pathway. Cell Physiol Biochem. 2017;41:2242–54.
130. Li S, Wang Q, Tao Y, Liu C. Swertiamarin attenuates experimental rat hepatic fibrosis by suppressing angiotensin II–angiotensin type 1 receptor–extracellular signal-regulated kinase signaling. J Pharmacol Exp Ther. 2016;359:247.
131. Xie H, Shen C-Y, Jiang J-G. The sources of salidroside and its targeting for multiple chronic diseases. J Funct Foods. 2020;64:103648.
132. Zheng T, Yang X, Li W, Wang Q, Chen L, Wu D, et al. Salidroside attenuates high-fat diet-induced nonalcoholic fatty liver disease via AMPK-dependent TXNIP/NLRP3 pathway. Oxidative Med Cell Longev. 2018;2018:8597897.
133. Ye Q, Zhou Y, Zhao C, Xu L, Ping J. Salidroside inhibits CCl4-induced liver fibrosis in mice by reducing activation and migration of HSC induced by liver sinusoidal endothelial cell-derived exosomal SphK1. Front Pharmacol. 2021;12, 677810.
134. Lin X, Bai F, Nie J, Lu S, Lu C, Zhu X, et al. Didymin alleviates hepatic fibrosis through inhibiting ERK and PI3K/Akt pathways via regulation of Raf kinase inhibitor protein. Cell Physiol Biochem. 2016;40:1422–32.
135. Yang H-X, Shang Y, Jin Q, Wu Y-L, Liu J, Qiao C-Y, et al. Gentiopicroside ameliorates the progression from hepatic steatosis to fibrosis induced by chronic alcohol intake. Biomol Ther (Seoul). 2020;28:320–7.
136. Xie W-L, Jiang R, Shen X-L, Chen Z-Y, Deng X-M. Diosgenin attenuates hepatic stellate cell activation through transforming growth factor-β/Smad signaling pathway. Int J Clin Exp Med. 2015;8:20323–9.
137. Xiong Y, Yang Y, Yang J, Chai H, Li Y, Yang J, et al. Tectoridin, an isoflavone glycoside from the flower of Pueraria lobata, prevents acute ethanol-induced liver steatosis in mice. Toxicology. 2010;276:64–72.
138. Chatterjee S, Patra D, Ghosh P, Prasad A, Dutta Chowdhury K. Terpenoids in treatment of liver disease. In: Terpenoids against human diseases. Boca Raton: CRC Press; 2019. p. 61–94.
139. Melo C, Queiroz M, Filho A, Rodrigues A, Sousa D, Almeida J, et al. Betulinic acid, a natural pentacyclic triterpenoid, prevents abdominal fat accumulation in mice fed a high-fat diet. J Agric Food Chem. 2009;57:8776–81.
140. Quan HY, Kim DY, Kim SJ, Jo HK, Kim GW, Chung SH. Betulinic acid alleviates non-alcoholic fatty liver by inhibiting SREBP1 activity via the AMPK–mTOR–SREBP signaling pathway. Biochem Pharmacol. 2013;85:1330–40.
141. Jang S-M, Yee S-T, Choi J, Choi M-S, Do G-M, Jeon S-M, et al. Ursolic acid enhances the cellular immune system and pancreatic β-cell function in streptozotocin-induced diabetic mice fed a high-fat diet. Int Immunopharmacol. 2009;9:113–9.
142. Jayaprakasam B, Olson L, Schutzki R, Tai M-H, Nair M. Amelioration of obesity and glucose intolerance in high-fat-fed C57BL/6 mice by anthocyanins and ursolic acid in cornelian cherry (Cornus mas). J Agric Food Chem. 2006;54:243–8.
143. Jia Y, Kim S, Kim J, Kim B, Wu C, Lee J, et al. Ursolic acid improves lipid and glucose metabolism in high-fat-fed C57BL/6J mice by activating peroxisome proliferator-activated receptor alpha and hepatic autophagy. Mol Nutr Food Res. 2015;59(2), 344–354.
144. Kim K, Ko K-H, Heo R, Yi C-O, Shin H, Kim JY, et al. Artemisia annua leaf extract attenuates hepatic steatosis and inflammation in high-fat diet-fed mice. J Med Food. 2016;19(3), 290–299.
145. Sultana N, Ata A. Oleanolic acid and related derivatives as medicinally important compounds. J Enzyme Inhib Med Chem. 2008;23(6):739–56.
146. Ayeleso TB, Matumba MG, Mukwevho E. Oleanolic acid and its derivatives: biological activities and therapeutic potential in chronic diseases. Molecules. 2017;22(11), 1915.
147. Xue C, Li Y, Lv H, Zhang L, Bi C, Dong N, et al. Oleanolic acid targets the gut–liver Axis to alleviate metabolic disorders and hepatic steatosis. J Agric Food Chem. 2021;69(28), 7884–7897.
148. Khan UM, Sevindik M, Zarrabi A, Nami M, Ozdemir B, Kaplan DN, et al. Lycopene: food sources, biological activities, and human health benefits. Oxidative Med Cell Longev. 2021;2021:2713511.
149. Jiang W, Guo M-H, Hai X. Hepatoprotective and antioxidant effects of lycopene on non-alcoholic fatty liver disease in rat. World J Gastroenterol. 2016;22:10180.
150. Mamdouh I, Althagafy H, Abd-alhameed E, Al-Thubiani WS, Hassanein E. Promising hepatoprotective effects of lycopene in different liver diseases. Life Sci. 2022;310:121131.
151. Li Y, Sun H, Wu T, Fu Y, He Y, Mao X, et al. Storage carbon metabolism of Isochrysis zhangjiangensis under different light intensities and its application for co-production of fucoxanthin and stearidonic acid. Bioresour Technol. 2019;282:94–102.
152. Takatani N, Kono Y, Beppu F, Okamatsu-Ogura Y, Yamano Y, Miyashita K, et al. Fucoxanthin inhibits hepatic oxidative stress, inflammation, and fibrosis in diet-induced nonalcoholic steatohepatitis model mice. Biochem Biophys Res Commun. 2020;528:305–10.
153. Kim M-B, Bae M, Hu S, Kang H, Park Y-K, Lee J-Y. Fucoxanthin exerts anti-fibrogenic effects in hepatic stellate cells. Biochem Biophys Res Commun. 2019;513:657–62.
154. Bae M, Kim M-B, Park Y-K, Lee J-Y. Health benefits of fucoxanthin in the prevention of chronic diseases. Biochim Biophys Acta Mol Cell Biol Lipids. 2020;1865:158618.
155. Wang X, Morris-Natschke SL, Lee K-H. New developments in the chemistry and biology of the bioactive constituents of tanshen. Med Res Rev. 2007;27:133–48.
156. Parajuli DR, Zhao Y-Z, Jin H, Chi JH, Li SY, Kim Y-C, et al. Anti-fibrotic effect of PF2401-SF, a standardized fraction of Salvia miltiorrhiza, in thioacetamide-induced experimental rats liver fibrosis. Arch Pharm Res. 2015;38:549–55.
157. Shi M-J, Yan X-L, Dong B-S, Yang W-N, Su S-B, Zhang H. A network pharmacology approach to investigating the mechanism of tanshinone IIA for the treatment of liver fibrosis. J Ethnopharmacol. 2020;253:112689.
158. Yang L, Gong Y, Yang Y, Luo S. A successful case of tanshinone II A treatment for relapsed acute promyelocytic leukemia after maintenance therapy of all-trans retinoic acid and arsenic trioxide. J Sichuan Univ Med Sci Ed. 2010;41:1065–7.
159. Du S-Y, Zhang Y-L, Bai R-X, Ai Z-L, Xie B-S, Yang H-Y. Lutein prevents alcohol-induced liver disease in rats by modulating oxidative stress and inflammation. Int J Clin Exp Med. 2015;8:8785–93.
160. Haidari F, Hojhabrimanesh A, Helli B, Seyedian S-S, Ahmadi-Angali K. An energy-restricted high-protein diet supplemented with β-cryptoxanthin alleviated oxidative stress and inflammation

in nonalcoholic fatty liver disease: a randomized controlled trial. Nutr Res. 2020;73:15–26.

161. Einbond LS, Soffritti M, Esposti DD, Park T, Cruz E, Su T, et al. Actein activates stress- and statin-associated responses and is bioavailable in Sprague-Dawley rats. Fundam Clin Pharmacol. 2009;23:311–21.
162. Chen H-J, Liu J. Actein ameliorates hepatic steatosis and fibrosis in high fat diet-induced NAFLD by regulation of insulin and leptin resistant. Biomed Pharmacother. 2018;97:1386–96.
163. Lin L, Li R, Cai M, Huang J, Huang W, Guo Y, et al. Andrographolide ameliorates liver fibrosis in mice: involvement of TLR4/NF-κB and TGF-β1/Smad2 signaling pathways. Oxidative Med Cell Longev. 2018;2018:–7808656.
164. Khole S, Mittal S, Jagadish N, Ghosh D, Gadgil V, Sinkar V, et al. Andrographolide enhances redox status of liver cells by regulating microRNA expression. Free Radic Biol Med. 2019;130:397–407.
165. Purwaningsih S, Handharyani E, Sukarno AYP. Hepatoprotective effects ethanol extract of mangrove propagule (Rhizophora mucronata) in white rat strain Sprague Dawley induced carbon tetrachloride (CCL4). KLS. 2015;141-155.
166. Martinek R, Wolman W. Xanthines tannins and sodium in coffee tea and cocoa. J Am Med Assoc. 1955;158:1031–51.
167. Lawless M, Norris S, Byrne K, Gray S. Targeting histone deacetylases for the treatment of immune, endocrine & metabolic disorders. Endocr Metab Immune Disord Drug Targets. 2009;9:84–107.
168. Chung M-Y, Song J-H, Lee J, Shin EJ, Park JH, Lee S-H, et al. Tannic acid, a novel histone acetyltransferase inhibitor, prevents non-alcoholic fatty liver disease both in vivo and in vitro model. Mol Metab. 2019;19:34–48.
169. Son M, Oh S, Choi J, Jang J, Choi C, Park KY, et al. The phlorotannin-rich fraction of Ecklonia cava extract attenuated the expressions of the markers related with inflammation and leptin resistance in adipose tissue. Int J Endocrinol. 2020;2020:1–11.
170. Byun K-A, Oh S, Son M, Park C-H, Son KH, Byun K. Dieckol decreases caloric intake and attenuates nonalcoholic fatty liver disease and hepatic lymphatic vessel dysfunction in high-fat-diet-fed mice. Mar Drugs. 2021;19(9), 495.
171. Oh S, Son M, Byun K-A, Jang J, Choi C, Son K, et al. Attenuating effects of dieckol on high-fat diet-induced nonalcoholic fatty liver disease by decreasing the NLRP3 inflammasome and pyroptosis. Mar Drugs. 2021;19:318.
172. Li X, Deng Y, Zheng Z, Huang W, Chen L, Tong Q, et al. Corilagin, a promising medicinal herbal agent. Biomed Pharmacother. 2018;99:43–50.
173. Yang F, Wang Y, Xue J, Ma Q, Zhang J, Chen Y-F, et al. Effect of Corilagin on the miR-21/smad7/ERK signaling pathway in a schistosomiasis-induced hepatic fibrosis mouse model. Parasitol Int. 2016;65:308–15.
174. Liao M, Zhang R, Wang Y, Mao Z, Wu J, Guo H, et al. Corilagin prevents non-alcoholic fatty liver disease via improving lipid metabolism and glucose homeostasis in high fat diet-fed mice. Front Nutr. 2022;9, 983450.
175. Venusová E, Kolesarova A, Horky P, Sláma P. Physiological and immune functions of punicalagin. Nutrients. 2021;13:2150.
176. Liu H, Zhan Q, Miao X, Xia X, Yang G, Peng X, et al. Punicalagin prevents hepatic steatosis through improving lipid homeostasis and inflammation in liver and adipose tissue and modulating gut microbiota in western diet-fed mice. Mol Nutr Food Res. 2021;65:2001031.
177. Fouad AA, Qutub HO, Al-Melhim WN. Punicalagin alleviates hepatotoxicity in rats challenged with cyclophosphamide. Environ Toxicol Pharmacol. 2016;45:158–62.
178. Al-khawalde AAA, Abukhalil MH, Jghef MM, Alfwuaires MA, Alaryani FS, Aladaileh SH, et al. Punicalagin protects against the development of methotrexate-induced hepatotoxicity in mice via activating Nrf2 signaling and decreasing oxidative stress, inflammation, and cell death. Int J Mol Sci. 2022 23(20), 12334.

Plants Affecting Serotonergic Neurotransmission

Keya Mallick and Sugato Banerjee

Abstract

Serotonergic disorders like depression, anxiety, or other neuropsychiatric disorders impact a person's quality of life, daily interaction, and social relationships. Several synthetic anti-depressants are available in the market today, but they have limited effects or restricted clinical applications and come with adverse drug reactions. The available medications could have negative behavioral and cognitive impacts. Anti-depressants from natural sources, such as herbal medicines that have been used traditionally, are safe for human health. Traditionally, these herbal formulations were used medicinally in different regions worldwide. Therefore, medicinal herbs, plant-based formulations, or plant extracts that have anti-depressant action have been studied in vitro or in rodent models of serotonergic disorders. These medicinal plants either prevent neurotransmitter reuptake or inhibit monoamine oxidase, which may be used to treat serotonergic disorders and replenish the neurotransmitter. A general overview of the serotonergic system, associated disorders, and natural compounds involved in the modulation of the serotonergic system with potential therapeutic applications have been reviewed in the current chapter.

Keywords

Serotonergic · Neurotransmission · Herbal medicines · Lifestyle diseases

1 Introduction

Serotonin or 5-hydroxytryptamine (5-HT) is one of the principal neurotransmitters in the central serotonergic system, which is vital for maintaining mood and controlling a variety of mental functions, such as thought and emotion [1–3]. Interestingly, this ancient molecule (5-HT) and melatonin were likely produced by the first prokaryotes on earth and served as a neuroprotective agent [4–7]. The presence of serotonin in numerous organs, including the brain, lungs, kidneys, platelets, and gastrointestinal system, was subsequently determined. Serotonin was initially examined for its involvement in platelet function, but Brodie & Shore proposed that it also served as a neurotransmitter [8, 9]. Further exploration revealed that serotonin was mainly found in the nerve ends of neurons in isolated regions in the brain of mammals [10]. These findings marked the first identification of serotonin-containing nuclei in the brain. The serotonergic system was named after these neural networks [10]. Serotonin primarily acts via 5-HT receptors. At least 16 different 5-HT receptor subtypes, seven unique families (5-HT1–5-HT7), and serotonin transporter (5-HTT) have all been discovered [11, 12]. They regulate the extracellular levels of 5-HT and facilitate its reuptake from the synaptic cleft. 5-HT1, 5-HT2, and 5-H3 are homologous [13]. Adenylyl cyclase and the 5-HT1 and 5-HT5 receptors are negatively linked since their activation leads to the downregulation of cAMP levels. The release of intracellular Ca^2+ has been associated with 5-HT2 receptor upregulation, followed by the inositol triphosphate and diacylglycerol pathways. Both 5-HT4 and 5-HT7 can increase cAMP levels. Plasma membrane depolarization is brought about by the Na+/K+ cation channel connected to 5-HT3 [8–14]. Serotonin has various actions in the central nervous system (CNS), many of which are influenced by how the serotonergic system affects the forebrain, brainstem, and cerebellum. This system's rostral nuclei project signals that control emesis, appetite, body temperature, sleep patterns, and sexual behaviour. Nociception and motor tone are both affected by projections from the caudal nuclei. The significance of serotonin in psychological illnesses in humans has been clinically relevant [8–15]. The 5-HT1A, 5-HT1B, 5-HT2A, 5-HT2C, 5-HT4, 5-HT5A, 5-HT6, 5-HT7 and 5-HTm, are found in the CNS, GI tract, and blood vessels,

K. Mallick · S. Banerjee (✉)
Department of Pharmacology and Toxicology, National Institute of Pharmaceutical Education and Research, Kolkata, India

A. K. Dhara, S. C. Mandal (eds.), *Role of Herbal Medicines*, https://doi.org/10.1007/978-981-99-7703-1_11

respectively [8–16]. The stomach fundus contains 5-HT2B, whereas the sensory enteric neurons have 5-HT3. Neuronal inhibition and behavioural consequences like anxiety, sleep, and thermoregulation are all regulated by 5-HT1A. Presynaptic inhibition, CNS behavioral effects, and pulmonary vasoconstriction were all modulated by 5-HT1B. In the CNS, 5-HTm controls movement and cerebral vasoconstriction [8–16]. However, 5-HT2A prevented smooth muscle contraction, smooth muscle vasoconstriction, platelet aggregation, neuropsychiatric alterations, and memory. CSF secretion was controlled by 5-HT2C. Emesis was regulated by 5-HT3 and 5-HT4 receptors [8–16]. Drugs like SSRIs (selective serotonin reuptake inhibitors) commonly prescribed anti-depressants have 5-HTT as their primary target [17, 18]. Each serotonin receptor subtype has unique functions and is distributed differentially across the brain. The dorsal raphe's 5-HT1A autoreceptor regulates the firing of the overall 5-HT transmission and is a postsynaptic receptor in the limbic and cortical terminal regions of the brain [17–21].

The incidence of anxiety and depression has increased in modern society which affects the quality of life of an individual [22, 23]. Serotonin is one of the neurotransmitters that profoundly contribute to regulating human mood and cognition. Several disorders, including depression, are brought on by the malfunction of the complex serotonergic system [24]. Conventional therapies to improve mood have several adverse effects like behavioural deformities and are expensive to the patients [25–27], thus highlighting the value of alternative medicines [28]. Due to fewer adverse effects and lower cost, the usage of medicinal plants is growing daily. Experimental and ethnobotanical investigations have revealed that various ailments can be prevented and treated using herbal medications [29–36].

2 Biosynthesis and Serotonin Metabolism

Serotonin functions as a non-proteinogenic amino acid that occurs naturally and functions as a neurotransmitter and a hormone in the central and peripheral nervous systems respectively [37]. Tryptophan (Trp) is the starting material for the biosynthesis of serotonin and melatonin. The principal secondary metabolites of tryptophan in plant tissues include auxin, glucosinolates, phytoalexins, alkaloids, and indoleamine, which maintain various physiological activities [38]. Tph1 and Tph2 are the two Trp genes. The majority of the serotonin in the blood is produced by Tph1, which is expressed in the enterochromaffin cells. Tph2 produces serotonin in the brain and is only expressed in serotonergic neurons of the brainstem [37]. At the same time, serotonin produced in the stomach functions as a hormone and regulates a wide range of processes, while the serotonin produced in the brain functions as a neurotransmitter. The serotonergic pathway converts about 1–2% of the consumed Trp into serotonin and melatonin [37]. Serotonin and melatonin function as inhibitors of pathogens, free radicals, abiotic, and biotic stresses in plants. Additionally, these substances promote germination, xylem sap exudation from roots, flowering, and ion permeability in plants [39]. Animals cannot produce tryptophan, which they receive from the plant source [40]. Tryptophan is currently obtained from 42 plant species belonging to 20 distinct families and is geographically scattered in leaves, stems, roots, fruits, and seeds. Two different mechanisms in plants are used for their production [41]. Tryptophan hydroxylase (TPH) initially catalyzes a hydroxylation process, which is followed by an aromatic amino acid decarboxylation (AADC) reaction, to produce the neuroactive molecule 5-hydroxytryptophan (5-HTP), which is then used to produce serotonin (5-HT) in vertebrates Fig. 1. Additionally, plants first synthesise tryptophan through the chorismate pathway, which is then decarboxylated by the enzyme tryptophan decarboxylase (TDC) to create tryptamine, a different bioactive amino acid and a typical precursor to secondary metabolites [42]. Serotonin is created from tryptamine after it has been hydroxylated by the enzyme tryptamine-5-hydroxylase (T-5-H) (Fig. 1). TDC is the rate-limiting enzyme for serotonin synthesis in plants [38–44]. Surprisingly, the Hypericaceae plant Hypericum perforatum L. produces serotonin via 5-hydroxytryptophan [45]. Even though the production and control of serotonin in vertebrates is a well-understood process, serotonin biosynthesis and control in plants is still under investigation [46]. The outer mitochondrial membrane of neurons and non-neuronal cells contain the enzyme monoamine oxidase (MAO), which has two isoforms, MAO-A and MAO-B. MAO enzymes oxidatively deaminate xenobiotics and endogenous amines [47]. Monoamine oxidase A (MAO-A) converts 5-HT into its major metabolite, 5-hydroxyindoleacetic acid (5-HIAA), which is then released from the cell [48].

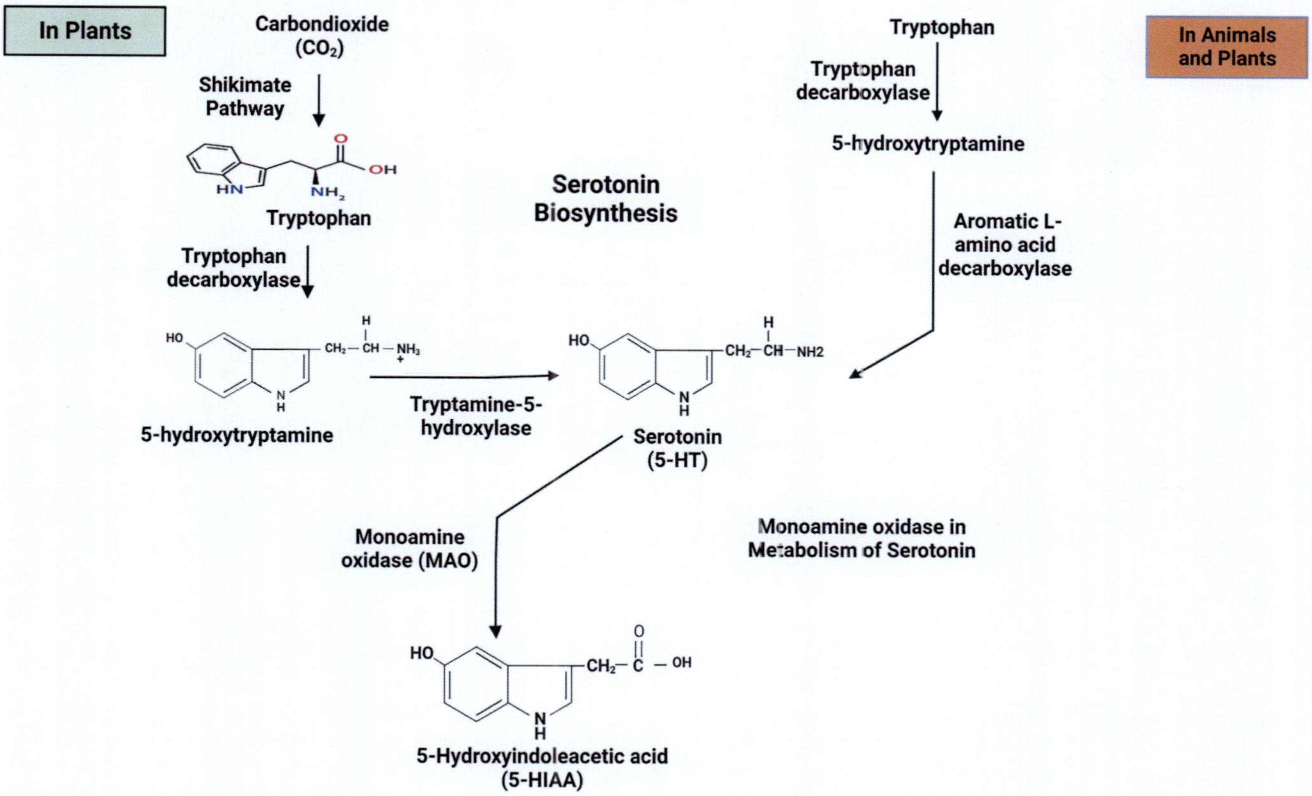

Fig. 1 Serotonin biosynthesis and breakdown

3 Complications Associated with 5-HT Dysregulation

According to the World Health Organization, more than 450 million individuals worldwide experience depression and other psychiatric disorders [49]. There have been reports of a progression of gastrointestinal disorders in 50–60% of patients. Serotonin (5-HT) receptors (5-HTRs) have been given emerging physiological roles during the past two decades, raising the possibility that these receptors could be used as a pharmaceutical target in various psychiatric and digestive disorders. More and more studies are showing that 5-HT systems impact the gut brain axis in depressed individuals [49]. Two main hypotheses based on the neurotransmitter system that mediates the development of depressive disorder have been proposed. Researchers hypothesized that serotonin (5-HT) and glutamate, two essential neurotransmitters, are involved in the development of depression (Fig. 2). Alteration in the expression levels of neurotrophic factors have also been linked to the serotonergic and glutamatergic systems (BDNF, NT-3, and NGF) [50]. The HPA axis is activated when the hypothalamic nuclei are activated, which also releases the neuroendocrine releasing factors and regulates various physiological processes for the body's maintenance and adaptation to challenging conditions [51]. Stress activates the hypothalamus, thalamic, midbrain, pons, and medullary nuclei, which due to innervation of circulatory and digestive systems change several physio-metabolic processes by cranio-spinal nerves [52]. Depressed individuals commonly report gastrointestinal issues such as stress-induced flares, edema, hyperalgesia, and alterations in gut reflexes [49] The "emotional motor system" comprising the amygdala, hypothalamus, and periaqueductal grey is linked to emotional or physical stress [53]. The sympathetic nervous system, parasympathetic nervous system, ENS, and the HPA axis are all downstream targets for these brain regions [54]. The endocrine and autonomic nervous systems mainly regulate gut function, and 5-HT plays a significant role in this process [49]. An essential component of the autonomic nervous system located inside the gut is the enteroendocrine system (ENS), which contains a substantial concentration of 5-HT with a tiny neuronal pool of 5-HT [49]. In this depot, the enzyme tryptophan hydroxylase 1 (TPH1) is in charge of producing 5-HT [55]. The rate-limiting enzyme for 5-HT synthesis in the central nervous system is TPH2 [55]. It has been previously stated that 5-HT is secreted into the intestinal mucosa by enterochromaffin cells and is involved in regulating digestive functions [56]. Peristaltic reflexes are induced by 5-HT by stimulating the mucosal processes by the submucosal primary afferent neurons. Bulbring and colleagues were initially unsure of the role enterochromaffin cells in the release of 5-HT during the onset of the peristaltic

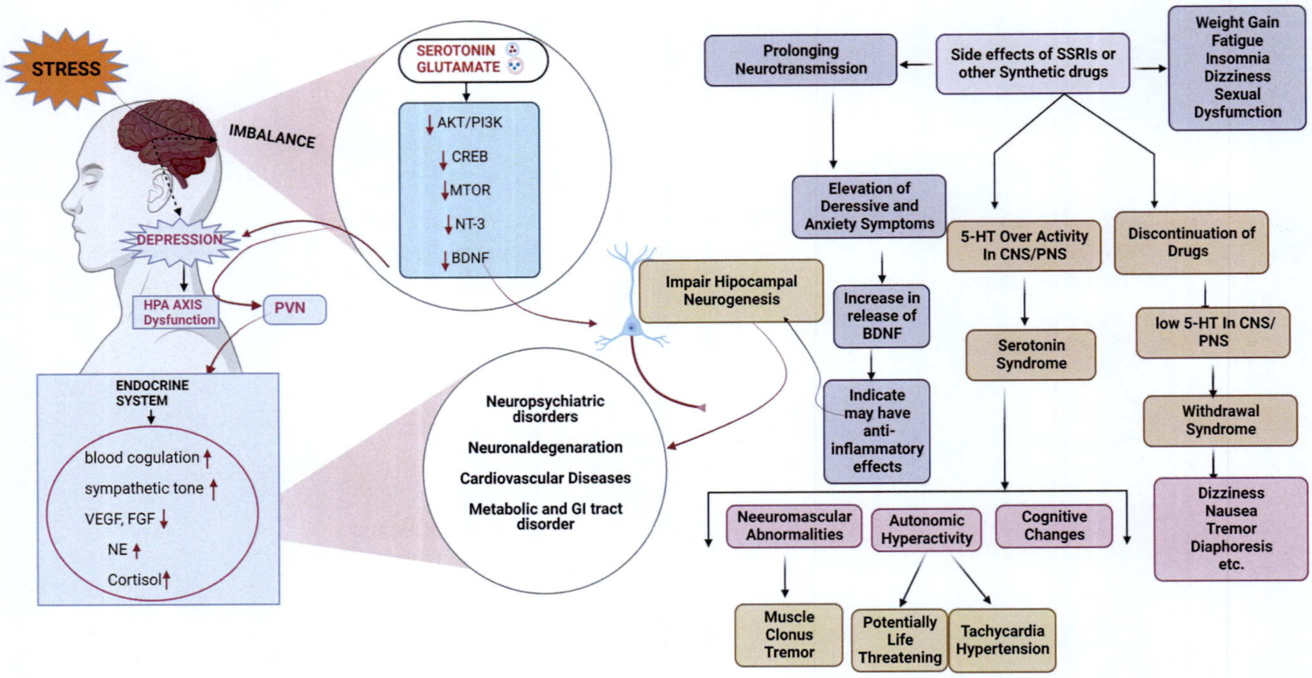

Fig. 2 Illustrative representation of the stress induced by different factors causing serotonergic disorders and affecting the physiology of various organ systems in human. (1) Stress can lead to imbalance in glutamate and serotonin levels. (2) This influences the expression of protein kinases (mTOR), transcription factors (CREB), and neurotrophic factors (BDNF, NT-3, and NGF), (3) that impact neuronal health and impair the hippocampal neurogenesis. (4) The neuroendocrine system, neuronal plasticity, gastrointestinal system, metabolic system, and cardiovascular system all undergo major alterations as a result of major/chronic depression, which increases the possibility of physiologically comorbid disorders in multiple organ systems. *AKT/PI3K* phosphoinositide 3-kinases, *BDNF* brain-derived neurotrophic factor, *HPA axis* hypothalamic-pituitary-adrenal axis, *mTOR* mammalian target of rapamycin, *FGF* fibroblast growth factor, *NGF* nerve growth factor, *NT-3* neurotrophin-3, *VEGF* vascular endothelial growth factor, *PVN* paraventricular nucleus

reflex [57]. Later studies on animals by Boullin revealed that the peristaltic responses of 5-HT-depleted animals were comparable to those of control animals. In the light of this, it was concluded that serotonin may not be necessary to trigger peristaltic reflexes but may still have an impact on them [58]. Another investigation using TPH1 selective knockouts, TPH2 selective knockouts, and TPH1 and TPH2 double knockout confirmed that the small neuronal pool of 5-HT is far more important for gastrointestinal motility than the much larger store of 5-HT in enterochromaffin cells [59]. An essential risk factor for severe unipolar depression, according to Zhang et al., is a SNP (single nucleotide polymorphism) mutation in the human TPH2 (hTHP2), a rate determining enzyme for neural serotonin production [60]. THP2 is also known to activate enteric neurons and regulate gut reflexes. The significance of neuronal 5-HT in depression and related comorbidities in the gut was later verified by a study in knock-in mice (TPH2-R439H), where levels of 5-HT were much lower in enteric neurons [61]. Lower levels of 5-HT in enteric neurons in TPH2-R439H mice resulted in anomalies in ENS and ENS-mediated gut dysfunctions like decreased motility and intestinal epithelial expansion. TPH2-R439H mice has a slower propulsive colorectal motility and total GI transit time than the control group, highlighting the significance of 5-HT as a mediator between the brain and gut axis [49]. MAOIs (monoamine oxidase inhibitors) inhibit mono amine oxidase (MAO) enzyme, thus increasing the bioavailability of 5-HT and other neurotransmitters like norepinephrine, dopamine, and epinephrine [49]. The first effective anti-depressants responsible for increasing extracellular 5-HT levels and elevate mood in depressed patients were MAOIs. However, a new class of antidepressant in the form of tricyclics were later developed for the treatment of depression due to side effects of MAOI [49].

4 5-HT-Related Abnormalities

Alec Coppen first hypothesised that depressed patients had anomalies in brain 5-HT function [62]. Determining 5-HT abnormalities in the depressed patient was difficult due to biochemical side effects of both current and previous antidepressant therapy. Reduced serotonin uptake by blood platelets [73] and lower plasma levels of tryptophan [74] provided the most reliable evidence of 5-HT abnormalities in depressed individuals. The anomalies in 5-HT in depressed individuals include (a) decreased 5-HT uptake in platelets, (b) decreased imipramine binding in platelets, (c) increased platelet binding to the 5-HT2A receptor, (d) inadequate plasma tryptophan, (e) low 5-HIAA in CSF, (f) reduced prolactin response to 5-HT-reuptake blockers, and (g) low 5-HT1A receptor binding and irregular 5-HT2A receptor and 5HTT binding in the brain [63]. Brain imaging methods have been used in more recent studies, particularly ligand imaging combined with single photon emission tomography (SPET) and positron emission tomography (PET) [64]. This makes it possible to study 5-HT receptors in the active human brain. Studies using these techniques have repeatedly and convincingly demonstrated that depressed patients have reduction in the density of 5-H1A receptors [65]. The binding appears to remain low in people who have recovered from depression, suggesting that it is not a marker of the acute depressive state [66]. Patients with panic disorder also have lower 5-HT1A-receptor binding [67]; however, these patients typically have high rates of comorbid depression [63]. An intriguing genetic vulnerability factor is allelic variation in the promoter region of the 5-HT transporter (5HTT) gene, which increases the chance of developing depression in response to stressful life events in carriers of the short allele of 5HTT (the low-activity allele). In functional imaging studies, a short allele of 5HTT also exhibits heightened neural responses to fearful facial expressions. The short allele may facilitate the processing of distressing emotional information, potentially deteriorating the psychological effects of adverse life events [63].

5 Tryptophan-Related Abnormalities

Tryptophan (Trp) is the primary starting material for 5-HT biosynthesis. Trp depletion can reduce the production of serotonin, resulting in expressive problems, depression, and mental decline. Serotonin is mostly found in the gastro intestinal tract (GIT), blood platelets, and central nervous system in mammals [37]. In mammals, the enterochromaffin cells of the GIT contain 90–95% of all serotonin. Intestinal peristalsis, nutritional absorption, vasodilatation, motility, and secretion are all encouraged by 5-HT, a key component of gastrointestinal signaling, which transmits signals from the gut to intrinsic or extrinsic neurons. Irritable bowel syndrome (IBS) and inflammatory bowel disease (IBD) are caused by the disruption of central and peripheral serotonergic signaling pathways [68]. Changes in the gut and brain serotonin levels are associated with microbiota imbalance in IBS. According to a study, short-chain fatty acids produced by bacteria can stimulate enterochromaffin cells to produce more serotonin [69]. GIT contains sites where the hormone melatonin is synthesized by the pineal gland at night. It results from the 5-HT pathway, which controls the circadian rhythm. Melatonin can also increase indoleamine 2,3-dioxygenase 1 (IDO1) activity, impacting circadian rhythms regulated by a negative feedback loop [70]. Melatonin also influences many biological processes, including oxidative stress, immunological response, apoptosis, proliferation, and angiogenesis. Since it modulates the immune function, endocrine system, and autonomic nervous system, it act as a risk factor for metabolic or cerebral disorders where sleep pattern changes is a common manifestation [37].

6 Serotonin Toxicity

The serotonin syndrome, a spectrum of serotonin toxicity, is caused by too much serotonin in the central nervous system [71]. A combination of clinical symptoms, including neuromuscular excitement, autonomic stimulation, alterations in mental states, tachycardia, and hyperreflexia, are indicative of serotonin poisoning [71, 72]. Serotonin toxicity can be mild, moderate, or severe [71]. Mild serotonin poisoning might have symptoms that the patient may or may not find concerning while major poisoning produces extreme suffering including muscle rigidity, multiple organ failure, and rapid development of severe hyperthermia like medical emergencies [71]. Serotonin poisoning [73] has emerged as an intense and frequent side effect of medications, which can be minor, serious, or even fatal. An overdose or drug-drug interaction involving serotonergic medications can all result in serotonin poisoning [71]. Selective serotonin reuptake inhibitors (SSRIs) like citalopram, paroxetine, fluvoxamine, sertraline, and fluoxetine are among the most frequently used class of antidepressants [72]. Since a growing number of people are using newer anti-depressants, it is crucial to consider the possibility of serotonin poisoning in patients receiving them [74]. The extra serotonin of the central nervous system stimulate serotonin receptors and provide toxic effects [75]. The term "Sternbach criteria" is frequently used to describe the clinical symptoms of serotonin poisoning which is also referred as the serotonin syndrome [71].

7 Common Plants Used in Serotonergic Disorders

Today, herbal therapies are regaining popularity as conventional drugs like antibiotics, which once had almost universal success against serious diseases, are losing their effectiveness [76]. About 2000 plant species are regarded as having therapeutic significance in Ayurveda (ancient Indian medicine), where the Chinese Pharmacopoeia contains about 5700 traditional medications, most of which are derived from plants [76]. Reverse pharmacology is commonly used to discover new drugs. In this method, drug candidates are first identified based on widespread population use and then verified using preclinical and clinical trials. According to experts, this method can accelerate the drug discovery process from the current 12 years to 5 years or less, and at a portion of the expense [77]. The subspecialty of psychopharmacology was subsequently created as a result of the subsequent, intense attempts to understand the structure-activity correlations and mode of action of some of first-generation psychotherapeutics [76]. The majority of currently available medications that are effective for treating mental health issues are the results of efforts to fully define, comprehend, and enhance the therapeutic potentials of drugs whose clinical efficiency was already known or by extrapolating existing knowledge on the modes of action of known agents with clearly observable effects on mood, feeling, behavior, etc. Even now, maximum attempts to find new psychoactive drugs are based solely on the information and experience obtained from such initiatives [76]. The situation has evolved a little through in the past 20 years, with the advancement of modern neurology and technologies. Several projects to discover psychoactive drugs now employ more sophisticated pharmacological models based on present knowledge of the pathological mechanisms underlying neurological problems or the various brain function-regulating mechanisms crucial to the proper operation of the central nervous system (CNS) [76]. The most widely used CNS active herbal remedies have undergone clinical and pre-clinical research and are also advised for therapeutic use by the health authorities of many Western and non-American countries. While appropriate clinical trials have repeatedly shown the clinical efficacy of several herbal extract-based remedies over the past three decades, reports of concentrated efforts to develop structurally and functionally novel psychotherapeutics based on the knowledge gained from herbal remedies are still being explored [76]. This reflects in the vast majority of psychoactive medications currently available on the market are either not natural yields or not produced from bioactive components of therapeutic plants [76]. However, in other therapeutic fields, examples of medications produced from secondary plant metabolites and their byproducts or developed based on pharmacological information obtained from investigations of herbal treatments are common [76]. A thorough market survey conducted in 1997 found that natural products or their derivatives made up 157 of the 520 medications that the Food and Drug Administration (FDA) in the USA approved between 1983 and 1994 (11 years) [78]. Additionally, this analysis showed a marked increase in success when attempts are made to find natural products for usage in clinical settings. Thus, biological compounds or their derivatives accounted for 61% of anticancer medicines licensed during the same period. Furthermore, the relatively infrequent reports on the effects of herbal extracts on neuronal function and the active ingredients that make up those extracts are generally not adequately evaluated in terms of their potential as novel CNS active drugs with novel structural or functional properties [76].

Ashwagandha in Sanskrit, *Withania somnifera* is a member of the Solanaceae family, is grown in soil unsuitable for other harvests, and needs little maintenance. The Ashwagandha plant is recognized for its variety of therapeutic applications in Indian Ayurvedic and Unani systems of medicine [79]. The rejuvenating and tonic properties of *Withania somnifera*, similar to those of *Asian ginseng*, have led to its high regard in Ayurveda. Ashwagandha has been subsequently referred as *Indian ginseng* due to these similarities. It has historically been used in Ayurveda to enhance libido, reduce exhaustion, against sickness, mental health issues, and as a Rasayana (rejuvenator). Its extract is used either alone or in conjunction with other herbal medications to treat behavioral disorders, chronic illnesses, and age-related cognitive deficiencies [80–83]. It has previously been demonstrated that the collective active principles of *W. somnifera*, which consist of equimolar quantities of withaferin A and sitoindosides VII–X [84], improve learning and memory in rats [85]. The cholinergic signal transduction cascade in the cortical and basal forebrain has been demonstrated to be preferentially affected by the *W. somnifera* extract. The cognitive-enhancing effects of *W. somnifera* extract in animals and humans can be partially explained by the *W. somnifera* induced increase in muscarinic acetylcholine receptor activity in cortical region of the brain [83]. The effectiveness of *W. somnifera* as an anti-cholinesterase and nootropic-like action has also been established [82–85]. Withaferin and glycowithanolides composed of sitoindosides VII to X are likely the active ingredients in *W. somnifera* [84]. These active ingredients have been shown to have strong anti-stress and immunomodulatory effects [82–85]. The effects of the *W. somnifera* glycowithanolides on cognitive deficits and disturbed central cholinergic indicators brought on by the neurodegeneration caused by the neurotoxins were studied in animal model of Alzheimer's disease [81]. Numerous studies confirmed the function of *W. somnifera* in neuroprotection and tardive dyskinesia [86]. An investigation revealed that *W. somnifera* glycowithanolides have anxiolytic and anti-depressant properties [87]. To support physical and mental health, Ashwagandha is frequently used [88]. Recent mouse behavioral experiments, including the tail suspension, open field, and forced swim tests (FST), have shown that *W. somnifera* has adaptogenic, antidepressant, and anxiolytic properties. It has a variety of bioactive components and is believed to operate on several sites, including the serotonergic system and others, to produce its pharmacological effects [88]. *Bacopa monnieri* or Jalanimba has been used for centuries as a medication and a meditation aid under the name "Brahmi," which also means "giving knowledge of the Supreme Reality" [89]. *B. monnieri* is highly prized in India as a reviving herb that improves memory and nervous system health. *B. monnieri* is categorized as a medhya rasayana, a substance meant to enhance memory and cognition. This herb has been utilized by Ayurvedic doctors in India for approximately 3 thousand years [89]. The Charaka Samhita (sixth century AD), suggests using *B. monnieri* in formulations for the therapy of a variety of cerebral disorders such as nervousness, impaired cognition, and absence of mind. Var-Prakarana and Bhavprakash are two early Ayurvedic monographs that reference *B. monnieri* [89]. Numerous studies on the neuropharmacological effects of *B. monnieri* extracts and isolated bacosides have established antiamnesic activity [90]. According to previous studies, bacosides cause membrane dephosphorylation and accelerate protein and RNA turnover in particular brain regions. *B. monnieri* can increase protein kinase activity in the hippocampus and show neuroprotective effects [91]. The main characteristic of Alzheimer's disease is impairment of cholinergic neuronal activity in the hippocampus [92]. The cognitive abnormalities brought on by the injection of ibotenic acid and intracerebroventricular administration of colchicines into the nucleus basalis magnocellularis were reversed by a standardized *B. monnieri* extract rich in bacosides [93]. The reduction of acetylcholine, weakening of choline acetylase action, and the decline in binding of muscarinic cholinergic receptor in the hippocampus and frontal cortex were likewise reversed by *B. monnieri*. In the rat's hippocampus, the bacopa extract has demonstrated a neuroprotective effect against oxidative damage caused by aluminium [94]. In Swiss albino mice, the *B. monnieri* extract inhibits nicotine-induced lipid peroxidation (LPO) and provides genoprotection [95]. Another study revealed that *B. monnieri* extract can treat Alzheimer's disease by lowering mouse amyloid levels [96]. Bacoside A has proven protective against the oxidative damage that prolonged cigarette smoking causes to rat brains in a recent study [97]. Additionally, *B. monnieri* extract or bacosides have been demonstrated to have anxiolytic, anti-depressant, anticonvulsant, antioxidant, anti-stress, and antiulcerogenic effects [76]. Another study concluded that GABA-ergic system mediates the effects of *B. monnieri* on the central nervous system [98].

Centella asiatica, a member of the Apiaceae (Umbelliferae) family, is referred to as mandukaparni, Indian pennywort, Jal Brahmi, and gotu kola. The plant *C. asiatica* has been utilized as a medicine from the beginning. Additionally, *C. asiatica* was mentioned in the French Pharmacopoeia in 1884, the ancient traditional Chinese Shennong Herbal about 2000 years ago, and Indian Ayurvedic medicine about 3000 years ago [99]. According to the literature, *C. asiatica* has been used to treat many illnesses, including liver, and kidney disease, periodontal disease, burn and scar treatment, against arthritis, memory improvement, improved circulation, sedation, anti-stress, anti-anxiety, anti-depressants, an aphrodisiac, anti-cancer, respiratory ailments, and as immune modulator. According to a new study, *C. asiatica* extract may help to speed up the repair of injured neurons [100]. This study showed higher axonal regeneration and speedy functional recovery, indicating that the axons developed more quickly. Because *C. asiatica* leaf extract stimulates neuronal dendritic development, it can strengthen dendrites in conditions like stress, neurodegenerative diseases, and memory difficulties [101]. An herb known as *Convolvulus pluricaulis* or as Shankhpushpi has been put to use for a very long time to treat nervous problems like insomnia, nervousness and anxiety in India. It promotes peace and relaxation and diminishes anxiety, stress, and mental exhaustion. Shankapushpi, which resembles morning glory, grows on the Indian plains. Similar to how American herbalists prescribe kava-kava and valerian, it has been frequently utilized in Ayurvedic therapy to treat nerve illnesses. Shankhpushpi has only recently been made available in American pharmacies for therapeutic purposes. According to herbalists, Shankhpushpi soothes nerves by controlling stress hormones [99]. Shankhpushpi or *C. pluricaulis* was used as anti-aging treatment known as Rasayana in Ayurvedic medicine [99]. In a study, anxiety sufferers received Shankhpushpi for 6 weeks and reported having a better sleep, greater vitality, and improved attention [102]. In one of these trials, shankhpushpi was the leading herbal supplement that was given to 28 persons with anxiety disorder. The study was published in an Indian Medical Journal in 1982. Ninety-one percent of patients reported having tremendous energy after 6 weeks of therapy, and 60–70% reported being able to sleep and focus better [102]. The use of this herb is still favored today for easing the symptoms of anxiety, panic attacks, jitters, and insomnia [102]. Amla, also known as *Emblica officinalis*, is widely distributed in China, Pakistan, Bangladesh, India, Sri Lanka, and Malaysia. In Ayurveda, the plant's fruits are utilized as a powerful Rasayana [76, 103]. The Rasayana extends life and improves health by boosting the immune system, slowing aging, and reviving the body in weakened states. In Ayurveda, the fruits of *E. officinalis* are highly regarded for their therapeutic usefulness, and Amla (*E. officinalis*) is known as a maharasayana [76]. The fruits from the primary ingredient in the Charaka Samhita-described polyherbal Ayurvedic Rasayana preparation known as chayavanprash awaleha [76]. Due to its preventative, curative, and health-restoring capabilities, this is widely used in India. Clinical research indicates that fruits have anabolic properties. Chyavanprash has been the subject of experimental studies, revealing its considerable adaptogenic, anti-stress immune potentiating, and memory-enhancing benefits [76]. A commercial herbal medication called Immuplus, which contains *Emblica officinalis* as one of its constituents, has been demonstrated to exhibit therapeutic potential effects in immune systems through cell-mediated and humoral immunity [104, 105]. Recently, the tannoid components of Amla

(*Emblica officinalis)* exhibited antioxidant properties in changes in the rat brain brought on by chronic stress [80]. A well-known ayurvedic supplement called Triphala, which contains the herbs, *Emblica officinalis, Terminalia belerica,* and *Terminalia chebula*, guards against the effects of noise stress on rats' antioxidant and cell-mediated immune systems [106]. Holi basil, or Ocimum sanctum, is also called "Tulsi" in India. It is used in Ayurveda for its healing and life-giving properties and is regarded as one of India's holiest plants. There are other Tulsi cultivars that are sometimes referred to as "Holy," but the *Ocimum sanctum* is the genuine kind [107]. Tulsi is an upright, hairy annual herb found all over India and as high as 1800 m in the Himalayas. A range of disorders can be treated using tulsi. For its impact on the central nervous system, an ethanolic extract from Ocimum sanctum leaves reduced [107]. Pentobarbital-induced reflex loss in mice was delayed, and the duration and intensity of convulsions brought on by electroshock and pentylenetetrazole were reduced. By analysing the levels of plasma corticosterone in rats exposed to 30 min of noise stress (100 dB), the ethanolic extract of leaves from Ocimum sanctum was tested for its anti-stressor effects against acute and chronic noise stress in albino rats [107]. Ethanolic extracts of Tulsi plants (Ocimum sanctum) were used to treat animals which were exposed to acute and long-term noise stress. The results showed that the plant's anti-stress qualities prevented the increase in plasma levels of corticosterone in these animals [107].

8 Plants Modulating Serotonergic Systems

To support physical and mental health, Ashwagandha (*Withania somnifera*) having adaptogenic, antidepressant, and anxiolytic properties mediated through the serotonergic system is widely used [88]. *Aloysia polystachya,* sometimes referred to as griseb, is a member of the Verbenaceae family. In female rats participating in the forced swim test, the hydroalcoholic extract of its (*A. polystachya*) leaves at varying concentrations (50, 25, and 12.5 mg/kg, i.p.) (12.5, 25, and 50 mg/kg, i.p.) demonstrated anti-depressant effects. Carvone and thujone are two bioactive compounds in *A. polystachya* with anti-depressant characteristics and properties that can treat serotonergic abnormalities [108]. *Aniba riparia* (Nees) (family, Lauraceae) contains phytoactive compounds called riparin III, showing antidepressant-like action in mice when evaluated using the Tail Suspension test (TST) and FST (25 and 50 mg/kg, intraperitoneal) [109]. Additionally, several investigations utilizing mice and rats in the behavioral study have shown that the saponins bacopasides VI-VIII, bacopaside I, bacopaside II, and bacopa saponin C. present in *B. monnieri (Brahmi)* has anti-depressant properties, regulating the serotonin concentration at doses of 20 and 40 mg/kg [88]. A member of the Theaceae family is the tea plant Camellia sinensis (L.) Kuntze. When tested in FST on male mice, the aqueous extract of its plants' leaves had potent serotonin-regulating activity (antidepressant-like effect) at the dose of 100 mg/kg [88]. A member of the Zingiberaceae family is Curcuma longa, often referred to as turmeric, haldi, or Haridra in India [88]. It includes curcumin, an active phytochemical with several pharmacological benefits and anti-inflammatory, antioxidant, and neuroprotective effects [88]. Recent research has also shown that in TST and FST, it is a more potent anti-depressant than fluoxetine [110]. The Berberidaceae family includes *Berberis aristata*, often known as Daruharidra or Daruhaldi. It has historically been used to treat inflammation, wound healing, and skin conditions. It has other pharmacological uses, including hepatoprotective, anti-diabetic, anti-inflammatory, and antioxidant. It contains berberine. This isoquinoline alkaloid is an effective anti-depressant in mice when tested using FST [88]. Root components of the Alliaceae family plant *Agapanthus campanulatus* are used to initiate conventional treatments. A type of mental disorder known as "the spirit" is reported to be treated with various parts of the plant while its extracts showed SSRI action [47]. Decoctions of the plant Boophone disticha's bulb scales are used to calm aggressive, schizophrenic patients. Bulb infusions are consumed by traditional healers and patients in South Africa as a medication to cure mental diseases and cause hallucinations for divinatory purposes [111]. Buphanidrine and buphanamine, isolated from the Amaryllidaceae plant Boophone disticha, had an affinity for the serotonin transporter (SERT) protein [147]. Powdered leaves from the *Gomphocarpus physocarpus* of the asclepiadaceae family are used as sedatives and to "strengthen the physique". *Xysmalobium undulatum* (L.) leaves extract showed SSRI action. In the Asphodelaceae family, plants like *Gasteria croucheri* were used in hysteria treatment in South Africa. Leaves of *Dioscorea species* (Dioscoreaceae family), *Millettia grandis*, *Schotia brachypetala*, *Leonotis leonurus* (Lamiaceae), *Mentha aquatica L.*, *Cinnamomum camphora (L.)*, and *Buddleja (L.) species* (Loganiaceae) have effective activity against serotonergic disorders like epilepsy, hysteria, convulsions, dispel worries and inducing sleep disorders [47].

9 Serotonin Contents in Herbal Plants extracts

In the years after its discovery, serotonin has been found in more than 37 plant families and 90 species, influenced by species, family, and the development of the maturational stage [46]. Serotonin was first discovered in plants in the legume *Mucuna pruriens* [113]. Moreover, plant serotonin has significant variations concerning environmental factors, including wavelength, light intensity, abiotic and biotic conditions, stress, and physiological conditions [46, 113, 114]. The aerial tissues of plants appear to have good serotonin levels in vegetative tissues [115–117]. Reproductive plant parts like flowers, fruits, and nuts [118–120] are generally found to have an elevated amount of serotonin than vegetative tissues, which have classes that vary from one tissue to another and with time, maturity stage, and even within the same tissue [121–123]. The edible tissues with high serotonin concentrations include black walnut (Juglans nigra L.) (4 46 g/g), butternut (Juglans cinerea L.) (398 90 g/g), Carya ovata (Mill.) K. Koch (shagbark) (143 23 g/g), and Tomato (*Solanum lycopersicum L.*) 156.1–221.9 g/g. In edible tissues, somewhat high amounts of serotonin is found in corn (Zea mays L.) (108.2 g/g) [124], cranberry (Vaccinium spp.) (105–122 g/g), banana (96.9 4.3 g/g) [119], ginger (Zingiber officinale Ros) and coffee bean (*Coffea canephora*) (63.3 μg/g) [115], spinach, Chinese cabbage, and cherry tomatoes also contain about 100 g/g of dry weight serotonin [120] (Table 1).

Table 1 The mechanisms of the effect of medicinal herbs on serotonergic transmission

Compound	Class	Pharmacological Target	Pharmacological Action	References
Anthocyanins	Flavonoid	5-HT, gut, BDNF	Anti-depressant, anti-neuroinflammatory, slows the ageing process of the brain, and controls the gut microbiota	[37, 126]
Asiaticoside	Terpenoid	BDNF, inflammatory markers,	Improves colitis, GI motility, and homeostasis while preventing neurotoxicity, lipid peroxidation, and neuroinflammation.	[37, 150]
Bacoside A	Terpenoid	BDNF, 5-HT receptors, and synaptic proteins	Anti-depressant and anti-anxiolytic	[151, 152]
Astragaloside IV	Polyphenol	Oxidative, apoptotic, and inflammatory factors, as well as 5-HT, dopamine, and MAO levels	Neuroprotection against toxicity, inflammation, oxidative stress, apoptosis, and depression	[37, 153]
Anonaine	Alkaloid	5-HT	Anti-depressants, sedatives, and anxiolytics	[154, 155]
Catechins	Flavonoid	BDNF, microbes, and MAO	Protection of the nervous system from oxidative stress, anxiety, and neurotoxicity	[37, 157]
Curcumin	Phenol	5-HT BDNF	Protection of the nervous system from oxidative stress, apoptosis, and depression	[158, 159]
Carvacrol	Phenol	5-HT and BDNF	Gastro protective and provides neuroprotection against memory degeneration, inflammation, oxidative stress, and depression	[160, 161]
Chrysin	Flavonoid	5-HT and MAO	Protection of the gut and the nervous system from oxidative damage, apoptosis, and inflammation	[37, 162]
Ferulic acid	Phenol	5-HT, MAOA, and BDNF	Increases 5-HT levels and has antidepressant and neuroprotective properties.	[37, 163]
Ginkgolides B	Terpenoid	BDNF and serotonin	Increases cognitive functions and decreases stress and anxiety	[37, 164]
Hesperidin	Flavonoid	BDNF	Controls gastrointestinal motility and prevents damage for the brain against toxicity, inflammation, and depression	[37, 165]
Hyperforin	Terpenoid	MAOA, kynurenine/Trp ratio, and 5-HT	Antidepressant	[37, 166]
Ferulic acid	Phenol	5-HT, MAO-A, and BDNF	Increases 5-HT levels and has antidepressant and neuroprotective properties	[37, 163]
Ginkgolides B	Terpenoid	BDNF and serotonin	Increases cognitive functions and decreases stress and anxiety	[37, 164]
Hesperidin	Flavonoid	BDNF	Controls gastrointestinal motility and prevents damage for the brain against toxicity, inflammation, and depression	[37, 165]
Hyperforin	Terpenoid	MAOA, kynurenine/Trp ratio, and 5-HT	Antidepressant	[37, 166]
Limonene	Terpene	Gut microbiome, BDNF, and melatonin	Neuroprotection against inflammation and oxidative stress (IBD)	[37, 165]
Ellagic acid	Phenol	5-HT and BDNF	Memory improvement and neuroprotection against oxidative stress, depression, and inflammation	[37, 167]
Ginsenoside Rg5	Terpenoid	BDNF and serotonin	Antidepressant	[37, 168]
Linalool	Terpenoid	Mucosal immunity, gut microbiota, inflammatory indicators, and 5-HT	Improves gut microbiota balance, cognitive health, and neuroprotection against oxidative stress, inflammation, and neuronal death	[37, 169]
Lycopene	Carotenoid	BDNF, serotonin, dopamine, inflammatory, and oxidative markers	Neuroprotection against oxidative stress, toxicity, inflammation, and anxiety	[37, 170]
Resveratrol	Polyphenol	Serotonin, SERT, and BDNF	Increases 5-HT levels, prevents 5-HT reuptake, and protects the brain from toxicity and oxidative damage	[37, 171]

10 Geographical Distribution of Plants Affecting Serotonergic System

In countries like India, Japan, China, Pakistan, Thailand, and Sri Lanka, traditional medicine is widely used. Medicinal plants are found in various habitats throughout India, but tropical forests in the Eastern and Western Ghats and the Himalayas are home to about 70% of the country's medicinal plants [125]. A large number of plants, derivatives or metabolites of plants, or extracts of traditional plants helpful for serotonergic transmission, are distributed worldwide according to their suitable climatic conditions (Table 2).

11 Medicinal Plants and Their Mechanism of Action on Serotonergic Transmission

Inhibition of Serotonin transporter (SERT) is the main focus because SERT transports serotonin back into the presynaptic serotonergic neuron after 5-HT release into synaptic space and reduces the serotonin level in the synaptic cleft. Various plants or their extract act as SERT inhibitors to control serotonin levels [127] (Figs. 3 and 4).

Jäger, A. K et al., 2013 has shown that many plants, like Danish folk, had more MAO-A inhibiting activity than inhibiting serotonin reuptake activity [127]. The *Borago officinalis* extract examined in a functional experiment for serotonin reuptake transporter inhibition, where the extract showed promising activity. The MAO inhibition is observed in ethanolic extracts of *Trigonella foenum-graecum* seeds (IC50 4 mg/mL), *Apium graveolens* leaves (IC50 5 mg/mL), and water extracts of aerial parts of *Calluna vulgaris* (IC50 8 mg/mL). Quercetin extracted from Calluna vulgaris has also shown MAO-A inhibition [127, 128]. Plant compounds such as phenols (like jatrorrhizine) [129, 130] and alkaloids [129–132] contain MAO-A activity. It has been demonstrated that the phenolic substance derived from the plants *Paeonia moutan* and *Primula auricular* is an MAO-A inhibitor, useful in increasing serotonin levels [129]. Monoamine oxidase A inhibitory action is present in the Paeonia seeds from *Primula veris* and *Primula elatior,* thus modulating serotonergic transmission [127]. A piperidine derivative called piperine (1-[5-(1,3-benzodioxol-5-yl)-1-oxo-2,4-pentadienyl]piperidine) preferentially inhibited MAO-A over MAO-B [129]. Eugenol inhibits human MAO-A more than MAO-B [130]. The MAO-A inhibiting activity of eugenol may be the mechanism behind its antidepressant-like effects. *Acori graminei rhizoma* is historically been used to treat depression [130]. Colossal plant kingdoms facilitate serotonergic transmission and regulate serotonin levels helping in the treatment of diseases or symptoms caused by serotonin imbalance. From various in vivo studies, it has been shown that flavonoids from *Hemerocallis citrina* affected the serotonergic and dopaminergic systems, which modulate mood [133]. In the cerebral cortex of experimental mice, ethanol extract of *Uncaria lanosa Wallich var* plants could increase 5-HT and 5-HIAA [134]., *Tagetes erecta L.* could mimic the effects of anti-depressant medications by modulating with the serotonergic, nitrergic, and sigma receptor pathways [135], *Annona cherimola* elevated the serotonin levels in the mouse brain [136], In *Curcuma longa,* curcumin exerted anti-depressant effects by increasing serotonin levels [137]. Phenols of *Rosmarinus officinalis* lead to a rise in pyruvate carboxylase and tyrosine hydroxylase [138]. *Bupleurum falcatum affected* the serotonergic systems and confers anti-depressant effects [139], via inhibiting mouse tryptophan 2, 3-dioxygenase and the expression of its genes. Extract of *H. perforatum* stimulates serotonin levels and elevates mood in mice [38, 166]. Paeonia glycosides increase the hippocampal production of serotonin (5-HT) and its metabolite, 5-hydroxy indole acetic acid [141]. Monoglycosides of *Cynanchum auriculatum Royle* inhibited serotonin reuptake in rats [142], *Hypericum perforatum* stimulate serotonin levels and elevates mood in those animals [38], *Sideritis* species as a triple monoamine reuptake inhibitor, effectively treated various illnesses, including depression and anxiety [143]. *Tagetes lucida Cav.* show anti-depressant effects via influencing the serotonergic system and serotonin production [144], *Fructus Akebiae* extract increases extracellular 5-HT and thus induce euphoria and antidepressant like effects in vivo [145], *Areca catechu nut* increases serotonin levels [146]. Active alkaloids of some plants of *the Amaryllidaceae* family has both anti-depressant and happiness-boosting properties [147], *Apium graveolens* increases serotonin through the MAO-A activation [127], *Borago officinalis* increase serotonin levels through altering the serotonin transporter [148], *Calluna vulgaris* increases serotonin levels through the MAO-A activation [148] and *Trigonella foenum-graecum* increased serotonin levels through the MAO-A activation [148] Table 2 (Figs. 3 and 4).

12 Conclusion

Numerous medicinal plants, their extract, or isolated active constituents have been shown to have pharmacological effects against serotonergic disorders like depression, anxiety, insomnia, and cognitive alterations. They have fewer side effects and are available at lower cost compared to their synthetic counterparts, thus providing promising therapeutic applications against serotonin associated psychiatric disorders.

Table 2 Serotonin contents in plants

Medicinal plants	Geographical distribution	Family	Plant parts used	Extraction Reaction	serotonin contents (µg/g Fresh Weight)	References
Spinacia oleracea	Tamil Nadu, Pondicherry, Maharashtra, Orissa, West Bengal, Bihar, Madhya	Amaranthaceae	Spinach leaves	Methanol	34.4 ± 2.4	[173]
Brassica rapa	Jammu, Kashmir, Europe, Russia, Central Asia and the Near East with Europe	Cruciferae	Chinese cabbage leaves	Methanol	110.9 ± 22.5	[174, 175]
Peganum harmala	Africa, the Middle East, India, Pakistan, South America, Mexico and several other countries	Zygophyllaceae	Leaves	Methanol	18,200	[176–178]
Lycopersicon esculentum	Madhya Pradesh, India	Solanaceae	Tomato	Methanol	221.9 ± 3.8	[179, 180]
Juglans mandshurica	Western and Central Asia, especially in China, China, India, Japan, Siberia	Juglandaceae	The embryo of the fruit	Methanol	251	[181]
Mimosa tenuiflora	Africa, South-East Asia, India	Fabaceae	Leaves	Methanol	8.3	[182–184]
Musa sapientum	All over the tropical world	Musaceae	Fruit's peel	Methanol	40–150 19–28	[187, 188]
Prunus domestica	Europe, Asia, India and Worldwide distributed	Rosaceae	Red Fruit	Methanol	10	[189, 190]

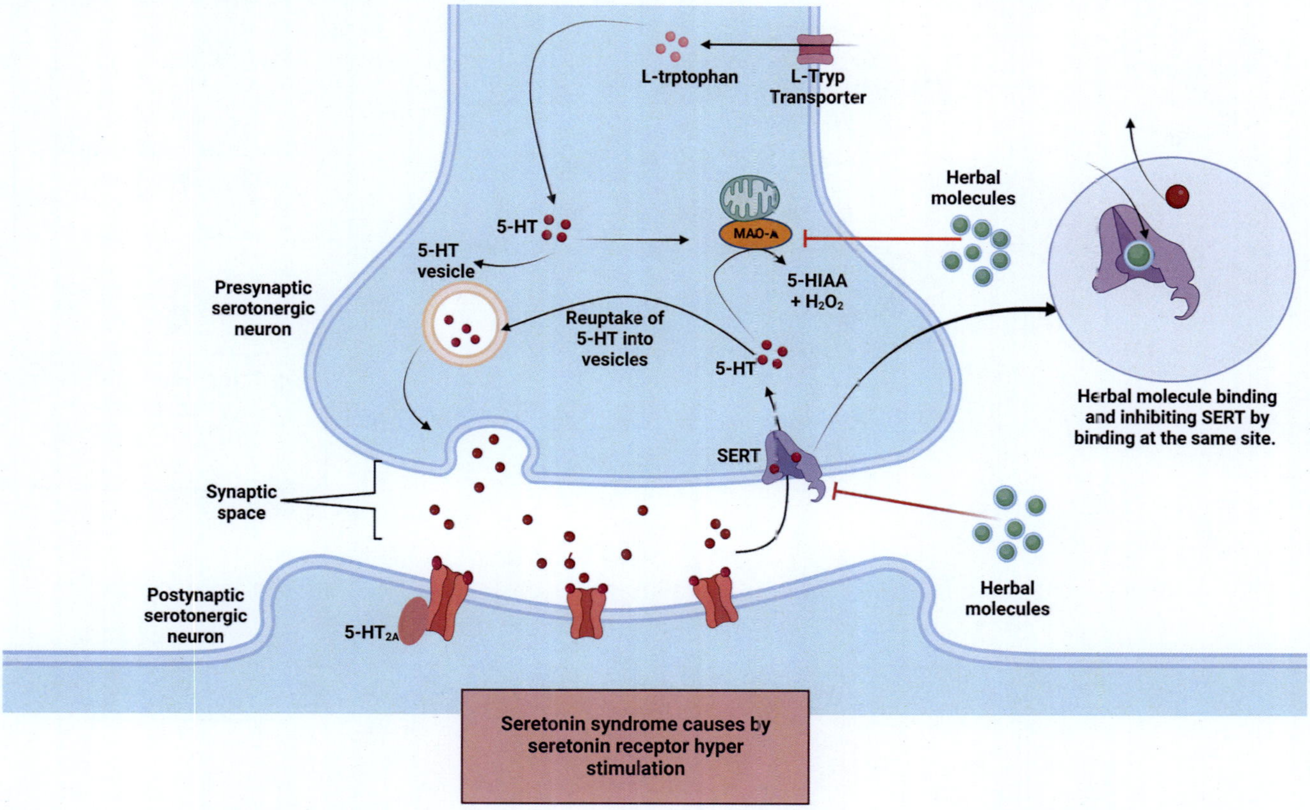

Fig. 3 Biosynthesis and metabolism of Serotonin (5-HT); reuptake of 5-HT and Serotonin syndrome caused by serotonin receptor hyper stimulation and mechanism of action of phytomolecules

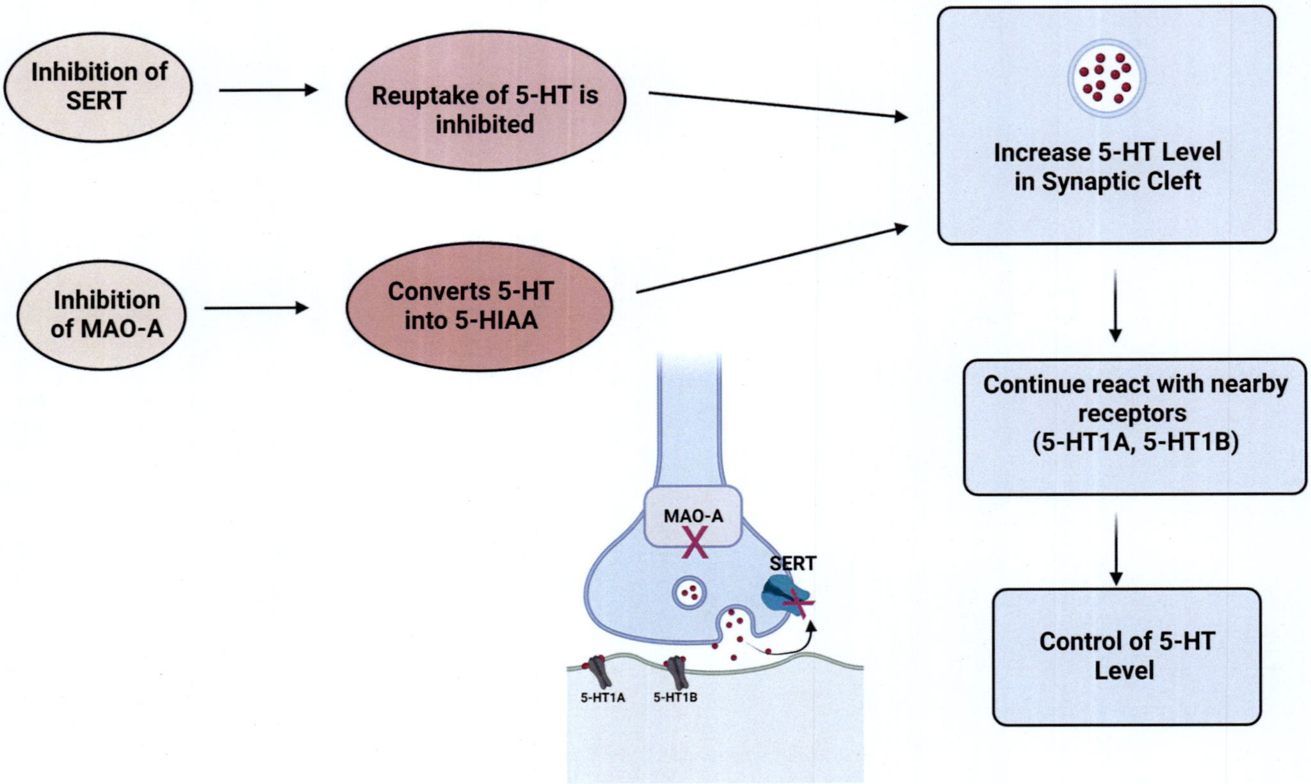

Fig. 4 Mechanism of action of Herbal Medicinal Plant to control serotonin level by inhibiting the serotonin transporter (SERT) and activity of Monoamine oxidase A enzyme which convers serotonin to HIAA

References

1. Erland LAE, Saxena PK. Beyond a neurotransmitter: the role of serotonin in plants. Neurotransmitter. 2017;4:e1538.
2. Feldman R, Myers J, Quenzer L, Cooper J, Bloom F, Roth R. The biochemical bases of neuropharma. Essent Psychopharmacol Antipsychos Mood Stab. 2002;45:133.
3. Trimble M. Molecular neuropharmacology, a foundation for clinical neuroscience. Hoboken, NJ: BMJ Publishing Group Ltd.; 2002.
4. Levy RJ. Serotonin transporter mechanisms and cardiac disease. Circulation. 2006;113(1):2–4.
5. Izzati-Zade K. The role of serotonin in the pathogenesis and clinical presentations of migraine attacks. Neurosci Behav Physiol. 2008;38(5):501–5.
6. Olivier B. Serotonin: a never-ending story. Eur J Pharmacol. 2015;753:2–18.
7. Yasin T. Parkinson's disease, subthalamic nucleus and serotonin. Front Neurosci. 2009;3:3.
8. Mohammad-Zadeh LF, Moses L, Gwaltney-Brant SM. Serotonin: a review. J Vet Pharmacol Ther. 2008;31(3):187–99.
9. Brodie BB, Shore PA. A concept for a role of serotonin and norepinephrine as chemical mediators in the brain. Ann N Y Acad Sci. 1957;66(3):631–42.
10. Johansen K. Regional distribution of circulating blood during submersion asphyxia in the duck. Acta Physiol Scand. 1964;62:1–9.
11. Hoyer D, Hannon JP, Martin GR. Molecular, pharmacological and functional diversity of 5-HT receptors. Pharmacol Biochem Behav. 2002;71(4):533–54.
12. Kitson SL. 5-hydroxytryptamine (5-HT) receptor ligands. Curr Pharm Des. 2007;13(25):2621–37.
13. Peroutka SJ, Howell TA. The molecular evolution of G protein-coupled receptors: focus on 5-hydroxytryptamine receptors. Neuropharmacology. 1994;33(3–4):319–24.
14. Derkach V, Surprenant A, North RA. 5-HT3 receptors are membrane ion channels. Nature. 1989;339(6227):706–9.
15. Lam R, Michalaak E, Swinson R. Assessment scales in depression, mania and anxiety. Boca Raton, FL: CRC; 2006.
16. Lofholm PW, Katzung BG. Rational prescribing and prescription writing. In: Basic clinical pharmacology. New York, NY: McGrew Hill; 2001. p. 1104–12.
17. Brust P, Hesse S, Muller U, Szabo Z. Neuroimaging of the serotonin transporter: possibilities and pitfalls. Curr Psychiatr Rev. 2006;2(1):111–49.
18. Meyer JH. Imaging the serotonin transporter during major depressive disorder and antidepressant treatment: 2005 CCNP young investigator award paper. J Psychiatry Neurosci. 2007;32(2):86–102.
19. Drevets WC, Thase ME, Moses-Kolko EL, Price J, Frank E, Kupfer DJ, Mathis C, biology. Serotonin-1A receptor imaging in recurrent depression: replication and literature review. Nucl Med. 2007;34(7):865–77.
20. Kumar JD, Mann JJ. PET tracers for 5-HT1A receptors and uses thereof. Drug Discov Today. 2007;12(17–18):748–56.
21. Stockmeier CA. Involvement of serotonin in depression: evidence from postmortem and imaging studies of serotonin receptors and the serotonin transporter. J Psychiatry Res. 2003;37(5):357–73.
22. Bandelow B, Michaelis S. Epidemiology of anxiety disorders in the 21st century. Dialogues Clin Neurosci. 2022;17:327.

23. Kessler RC, Bromet EJ. The epidemiology of depression across cultures. Annu Rev Public Health. 2013;34:119.
24. Jenkins TA, Nguyen JC, Polglaze KE, Bertrand PP. Influence of tryptophan and serotonin on mood and cognition with a possible role of the gut-brain axis. Nutrients. 2016;8(1):56.
25. Young SN. How to increase serotonin in the human brain without drugs. J Psychiatry Neurosci. 2007;32(6):394.
26. Evans EA, Sullivan MA. Abuse and misuse of antidepressants. Subst Abus Rehabil. 2014;5:107.
27. Zahreddine N, Richa S. Non-antidepressant treatment of generalized anxiety disorder. Curr Clin Pharmacol. 2015;10(2):86–96.
28. Shahbazi K, Solati K, Hasanpour-Dehkordi A. comparison of hypnotherapy and standard medical treatment alone on quality of life in patients with irritable Bowel syndrome: a randomized control trial. J Clin Diagn Res. 2016;10(5):Oc01–4.
29. Bahmani M, Sarrafchi A, Shirzad H, Rafieian-Kopaei M. Autism: pathophysiology and promising herbal remedies. Curr Pharm Des. 2016;22(3):277–85.
30. Asgary S, Kelishadi R, Rafieian-Kopaei M, Najafi S, Najafi M, Sahebkar AJPc. Investigation of the lipid-modifying and antiinflammatory effects of Cornus mas L. supplementation on dyslipidemic children and adolescents. Pediatr Cardiol. 2013;34(7):1729–35.
31. Asadi-Samani M, Bagheri N, Rafieian-Kopaei M, Shirzad H. Inhibition of Th1 and Th17 cells by medicinal plants and their derivatives: a systematic review. Phytother Res. 2017;31(8):1128–39.
32. Raeisi R, Heidari-Soureshjani S, Asadi-Samani M, Luther T. A systematic review of phytotherapies for newborn jaundice in Iran. Int J Pharm Sci Res. 2017;8(5):1953–8.
33. Mansouri E, Asadi-Samani M, Kooti W, Ghasemiboroon M, Ashtary-Larky D, Alamiri F, Afrisham R, Hasanzadeh NZ. Antifertility effect of hydro-alcoholic extract of fennel (Foeniculum vulgare Mill) seed in male Wistar rats. J Vet Res. 2016;60(3):357–63.
34. Kooti W, Farokhipour M, Asadzadeh Z, Ashtary-Larky D, Asadi-Samani M. The role of medicinal plants in the treatment of diabetes: a systematic review. Electron Physician. 2016;8(1):1832.
35. Asadi-Samani M, Bahmani M, Rafieian-Kopaei M. The chemical composition, botanical characteristic and biological activities of Borago officinalis: a review. Asian Pac J Trop Med. 2014;7:S22–8.
36. Shabanian S, Khalili S, Lorigooini Z, Malekpour A, Heidari-Soureshjani S. The effect of vaginal cream containing ginger in users of clotrimazole vaginal cream on vaginal candidiasis. J Adv Pharm Technol Res. 2017;8(2):80.
37. Liaqat H, Parveen A, Kim SY. Neuroprotective natural products' regulatory effects on depression via gut–brain axis targeting tryptophan. Nutrients. 2022;14(16):3270.
38. Murch S, KrishnaRaj S, Saxena P. Tryptophan is a precursor for melatonin and serotonin biosynthesis in in vitro regenerated St. John's wort (Hypericum perforatum L. cv. Anthos) plants. Plant Cell Rep. 2000;19(7):698–704.
39. Kang S, Kang K, Lee K, Back K. Characterization of tryptamine 5-hydroxylase and serotonin synthesis in rice plants. Plant Cell Rep. 2007;26(11):2009–15.
40. Iriti M, Faoro F. Grape phytochemicals: a bouquet of old and new nutraceuticals for human health. Med Hypotheses. 2006;67(4):833–8.
41. Akula R, Giridhar P, Ravishankar GA. Phytoserotonin. Plant Signal Behav. 2011;6(6):800–9.
42. Tanis JE, Moresco JJ, Lindquist RA, Koelle MR. Regulation of serotonin biosynthesis by the G proteins Gαo and Gαq controls serotonin signaling in Caenorhabditis elegans. Genetics. 2008;178(1):157–69.
43. Berlin J, Rügenhagen C, Dietze P, Fecker LF, Goddijn OJ, Hoge JHC. Increased production of serotonin by suspension and root cultures ofPeganum harmala transformed with a tryptophan decarboxylase cDNA clone from Catharanthus roseus. Transgenic Res. 1993;2(6):336–44.
44. Kang K, Kang S, Lee K, Park M, Back K. Enzymatic features of serotonin biosynthetic enzymes and serotonin biosynthesis in plants. Plant Signal Behav. 2008;3(6):389–90.
45. Khoury CK, Heider B, Castañeda-Álvarez NP, Achicanoy HA, Sosa CC, Miller RE, Scotland RW, Wood JR, Rossel G, Eserman LA. Distributions, ex situ conservation priorities, and genetic resource potential of crop wild relatives of sweetpotato [Ipomoea batatas (L.) Lam., I. series Batatas]. Front Plant Sci. 2015;6:251.
46. Erland LA, Turi CE, Saxena PK. Serotonin: an ancient molecule and an important regulator of plant processes. Biotechnol Adv. 2016;34(8):1347–61.
47. Stafford GI, Pedersen PD, Jäger AK, Van Staden J. Monoamine oxidase inhibition by southern African traditional medicinal plants. S Afr J Bot. 2007;73(3):384–90.
48. Fitzgerald LW, Kaplinsky L, Kimelberg HK. Serotonin metabolism by monoamine oxidase in rat primary astrocyte cultures. J Neurochem. 1990;55(6):2008–14.
49. Agrawal L, Korkutata M, Vimal SK, Yadav MK, Bhattacharyya S, Shiga T. Therapeutic potential of serotonin 4 receptor for chronic depression and its associated comorbidity in the gut. Neuropharmacology. 2020;166:107969.
50. Marazziti D. Understanding the role of serotonin in psychiatric diseases. F1000Res. 2017;6:1.
51. Bao A-M, Swaab DF. The human hypothalamus in mood disorders: the HPA axis in the center. IBRO Rep. 2019;6:45–53.
52. Khazaeipour Z, Taheri-Otaghsara S-M, Naghdi M. Depression following spinal cord injury: its relationship to demographic and socioeconomic indicators. Top Spinal Cord Inj Rehabil. 2015;21(2):149–55.
53. Meneses A. Chapter 9—5-HT4 Receptor. In: Meneses A, editor. The role of 5-HT Systems on memory and dysfunctional memory. San Diego: Academic; 2014. p. 39–43.
54. Furness JB. The enteric nervous system and neurogastroenterology. Nat Rev Gastroenterol Hepatol. 2012;9(5): 286–94.
55. Walther DJ, Peter JU, Bashammakh S, Hörtnagl H, Voits M, Fink H, Bader M. Synthesis of serotonin by a second tryptophan hydroxylase isoform. Science. 2003;299(5603):76.
56. Gershon MD. 5-Hydroxytryptamine (serotonin) in the gastrointestinal tract. Curr Opin Endocrinol Diabetes Obes. 2013;20(1):14–21.
57. Bülbring E, Lin RCY, Schofield G. An investigation of the peristaltic reflex in relation to anatomical observations. Q J Exp Physiol Cogn Med Sci. 1958;43(1):26–37.
58. Boullin DJ. Observations on the Significance OF 5-hydroxytryptamine in relation to the peristaltic reflex of the rat. Br J Pharmacol Chemother. 1964;23(1):14–33.
59. Li Z, Chalazonitis A, Huang YY, Mann JJ, Margolis KG, Yang QM, Kim DO, Côté F, Mallet J, Gershon MD. Essential roles of enteric neuronal serotonin in gastrointestinal motility and the development/survival of enteric dopaminergic neurons. J Neurosci. 2011;31(24):8998–9009.
60. Zhang X, Gainetdinov RR, Beaulieu J-M, Sotnikova TD, Burch LH, Williams RB, Schwartz DA, Krishnan KRR, Caron MG. Loss-of-function mutation in tryptophan hydroxylase-2 identified in unipolar major depression. Neuron. 2005;45(1):11–6.
61. Israelyan N, Del Colle A, Li Z, Park Y, Xing A, Jacobsen JPR, Luna RA, Jensen DD, Madra M, Saurman V, Rahim R, Latorre R, Law K, Carson W, Bunnett NW, Caron MG, Margolis KG. Effects of serotonin and slow-release 5-hydroxytryptophan on gastrointestinal motility in a mouse model of depression. Gastroenterology. 2019;157(2):507–521.e504.
62. Coppen A. The biochemistry of affective disorders. Br J Psychiatry. 1967;113(504):1237–64.

63. Cowen PJ. Serotonin and depression: pathophysiological mechanism or marketing myth? Trends Pharmacol Sci. 2008;29(9):433–6.
64. Grasby P. Imaging the neurochemical brain in health and disease. Clin Med. 2002;2(1):67.
65. Drevets WC, Thase ME, Moses-Kolko EL, Price J, Frank E, Kupfer DJ, Mathis C. Serotonin-1A receptor imaging in recurrent depression: replication and literature review. Nucl Med Biol. 2007;34(7):865–77.
66. Bhagwagar Z, Rabiner E, Sargent P, Grasby P, Cowen P. Persistent reduction in brain serotonin1A receptor binding in recovered depressed men measured by positron emission tomography with [11C] WAY-100635. J Mol Psychiatry. 2004;9(4):386–92.
67. Neumeister A, Bain E, Nugent AC, Carson RE, Bonne O, Luckenbaugh DA, Eckelman W, Herscovitch P, Charney DS, Drevets WC. Reduced serotonin type 1A receptor binding in panic disorder. J Neurosci. 2004;24(3):589–91.
68. Gracie DJ, Hamlin PJ, Ford AC. The influence of the brain–gut axis in inflammatory bowel disease and possible implications for treatment. Lancet Gastroenterol Hepatol. 2019;4(8):632–42.
69. Reigstad CS, Salmonson CE, Rainey JF III, Szurszewski JH, Linden DR, Sonnenburg JL, Farrugia G, Kashyap PC. Gut microbes promote colonic serotonin production through an effect of short-chain fatty acids on enterochromaffin cells. FASEB J. 2015;29(4):1395.
70. Zisapel N. New perspectives on the role of melatonin in human sleep, circadian rhythms and their regulation. Br J Pharmacol. 2018;175(16):3190–9.
71. Isbister GK, Buckley NA, Whyte IM. Serotonin toxicity: a practical approach to diagnosis and treatment. Med J Aust. 2007;187(6):361–5.
72. Dunkley E, Isbister G, Sibbritt D, Dawson A, Whyte I. The hunter serotonin toxicity criteria: simple and accurate diagnostic decision rules for serotonin toxicity. QJM. 2003;96(9):635–42.
73. Isbister GK, Buckley NA. The pathophysiology of serotonin toxicity in animals and humans: implications for diagnosis and treatment. Clin Neuropharmacol. 2005;28(5):205–14.
74. McManus P, Mant A, Mitchell PB, Montgomery WS, Marley J, Auland ME. Recent trends in the use of antidepressant drugs in Australia, 1990-1998. Med J Aust. 2000;173(9):458–61.
75. Nisijima K, Yoshino T, Yui K, Katoh S. Potent serotonin (5-HT) 2A receptor antagonists completely prevent the development of hyperthermia in an animal model of the 5-HT syndrome. Brain Res. 2001;890(1):23–31.
76. Husain G, Mishra D, Singh P, Rao CV, Kumar V. Ethnopharmacological review of native traditional medicinal plants for brain disorders. Pharmacog Rev. 2007;1:20–8.
77. Padma T. Ayurveda. Nature. 2005;436(7050):486.
78. Cragg GM, Newman DJ, Snader KM. Natural products in drug discovery and development. J Nat Prod. 1997;60(1):52–60.
79. Nadkarni K. Nadkarni's Indian materia medica. 1954.
80. Bhattacharya SK, Bhattacharya A, Chakrabarti A. Adaptogenic activity of Siotone, a polyherbal formulation of ayurvedic rasayanas. Indian J Exp Biol. 2000;38:119–28.
81. Ghayur MN, Jalil S, Riaz N, Yousuf S, Malik A. Withanolides, a new class of natural cholinesterase inhibitors with calcium antagonistic properties q. Biochem Biophys Res Commun. 2005;334:276–87.
82. Choudhary MI, Yousuf S, Nawaz SA, Ahmed S. Cholinesterase inhibiting withanolides from Withania somnifera. Chem Pharm Bull. 2004;52(11):1358–61.
83. Schliebs R, Liebmann A, Bhattacharya SK, Kumar A, Ghosal S, Bigl V. Systemic administration of defined extracts from Withania somnifera (Indian Ginseng) and Shilajit differentially affects cholinergic but not glutamatergic and GABAergic markers in rat brain. Neurochem Int. 1997;30(2):181–90.
84. Bhattacharya SK, Goel RK, Kaur R, Ghosal S. Anti-stress activity of sitoindosides VII and VIII, new acylsterylglucosides from Withania somnifera. Phytother Res. 1987;1(1):32–7.
85. Ghosal S, Lal J, Srivastava R, Bhattacharya SK, Upadhyay SN, Jaiswal AK, Chattopadhyay U. Immunomodulatory and CNS effects of sitoindosides IX and X, two new glycowithanolides from Withania somnifera. Phytother Res. 1989;3(5):201–6.
86. Jain S, Shukla SD, Sharma K, Bhatnagar M. Neuroprotective effects of Withania somnifera Dunn. in hippocampal sub-regions of female albino rat. Phytother Res. 2001;15(6):544–8.
87. Bhattacharya S, Bhattacharya A, Sairam K, Ghosal S. Anxiolytic-antidepressant activity of Withania somnifera glycowithanolides: an experimental study. Phytomedicine. 2000;7(6):463–9.
88. Kumari R, Agrawal A, Dubey G. Role of medicinal plants with antidepressant action and its mechanism: a review. Pharma Biol Eval. 2016;3(1):70–82.
89. Russo A, Borrelli F. Bacopa monniera, a reputed nootropic plant: an overview. Phytomedicine. 2005;12(4):305–17.
90. Kishore K, Singh M. Effect of bacosides, alcoholic extract of Bacopa monniera Linn. (brahmi), on experimental amnesia in mice. Indian J Exp Biol. 2005;43:640–5.
91. Singh H, Dhawan B. Neuropsychopharmacological effects of the ayurvedic nootropic Bacopa monniera Linn.(Brahmi). Indian J Pharmacol. 1997;29(5):359.
92. Enz A, Amstutz R, Boddeke H, Gmelin G, Malanowski J. Brain selective inhibition of acetylcholinesterase: a novel approach to therapy for Alzheimer's disease. Prog Brain Res. 1993;98:431–8.
93. Bhattacharya S, Kumar A, Ghosal S. Effect of Bacopa monnieri on animal models of Alzheimer's disease and perturbed central cholinergic markers of cognition in rats. Res Commun Pharmacol Toxicol. 1999;4(3/4):II–1.
94. Jyoti A, Sharma D. Neuroprotective role of Bacopa monniera extract against aluminium-induced oxidative stress in the hippocampus of rat brain. Neurotoxicology. 2006;27(4):451–7.
95. Vijayan V, Helen A. Protective activity of Bacopa monniera Linn. on nicotine-induced toxicity in mice. Phytother Res. 2007;21(4):378–81.
96. Holcomb LA, Dhanasekaran M, Hitt AR, Young KA, Riggs M, Manyam BV. Bacopa monniera extract reduces amyloid levels in PSAPP mice. J Alzheimers Dis. 2006;9(3):243–51.
97. Anbarasi K, Vani G, Balakrishna K, Devi CS. Effect of bacoside a on brain antioxidant status in cigarette smoke exposed rats. Life Sci. 2006;78(12):1378–84.
98. Singh H, Shanker G, Patnaik G. Neuropharmacological and anti-stress effects of bacosides: a memory enhancer. Indian J Pharmacol. 1996;28:47.
99. Simon JE. Herbs, spices, and medicinal plants: recent advances in botany, horticulture, and pharmacology, vol. 3. Phoenix: Oryx Press; 1998. p. 145–73.
100. Soumyanath A, Zhong YP, Yu X, Bourdette D, Koop DR, Gold SA, Gold BG. Centella asiatica accelerates nerve regeneration upon oral administration and contains multiple active fractions increasing neurite elongation in-vitro. J Pharm Pharmacol. 2005;57(9):1221–9.
101. Mohandas Rao K, Muddanna Rao S, Gurumadhva RS. Centella asiatica (L.) leaf extract treatment during the growth spurt period enhances hippocampal CA3 neuronal dendritic arborization in rats. Evid Based Complement Alternat Med. 2006;3(3):349–57.
102. Shaughnessy F. Ease anxiety with shankhpushpi: ayurvedic doctors prefer this mood-soother over kava-kava–herb brief. Nat Health Sept. 2002:1.
103. Satyavati G, Raina M, Sharma M. Medicinal plants of India, vol. 1976. New Delhi: Indian Council of Medical Research; 1976. p. 201–6.
104. Suresh K, Vasudevan DM. Augmentation of murine natural killer cell and antibody dependent cellular cytotoxicity activities by

Phyllanthus emblica, a new immunomodulator. Ethnopharmacol. 1994;44(1):55–60.
105. Dhote BS, Singh GK, Chauhan R. Effect of immuplus (an Herbal Immunomodulator) on paraspecific immune responses in chicks. J Immunol Immunopathol. 2008;10:36–41.
106. Srikumar R, Parthasarathy NJ, Manikandan S, Narayanan GS, Sheeladevi R. Effect of Triphala on oxidative stress and on cell-mediated immune response against noise stress in rats. Mol Cell Biochem. 2006;283:67–74.
107. Cohen MM. Tulsi—Ocimum sanctum: a herb for all reasons. Ayurveda Integr Med. 2014;5(4):251–9.
108. Mora S, Díaz-Véliz G, Millán R, Lungenstrass H, Quirós S, Coto-Morales T, Hellión-Ibarrola M. Anxiolytic and antidepressant-like effects of the hydroalcoholic extract from Aloysia polystachya in rats. Pharmacol Biochem Behav. 2005;82(2):373–8.
109. Sousa F, Melo C, Monteiro A, Lima V, Gutierrez S, Pereira B, Barbosa-Filho J, Vasconcelos S, Fonteles M, Viana G. Antianxiety and antidepressant effects of riparin III from Aniba riparia (Nees) Mez (Lauraceae) in mice. Pharmacol Biochem Behav. 2004;78(1):27–33.
110. Kulkarni S, Dhir A, Akula KKJT. Potentials of curcumin as an antidepressant. Sci World J. 2009;9:1233–41.
111. Van Wyk B-E, Gericke N. People's plants: a guide to useful plants of Southern Africa. Pretoria: Briza Publications; 2000.
112. Sandager M, Nielsen ND, Stafford GI, van Staden J, Jäger AK. Alkaloids from Boophane disticha with affinity to the serotonin transporter in rat brain. J Ethnopharmacol. 2005;98(3):367–70.
113. Ramakrishna A, Giridhar P, Ravishankar GA. Phytoserotonin: a review. Plant Signal Behav. 2011;6(6):800.
114. Reynolds JD, Daniel T, Weekley LB. The effect of light quality on 5-hydroxyindole metabolism in leaves of Sedum morganianum (Crassulaceae). Biochem Physiol Pflanz. 1985;180(5):345–51.
115. Engström K, Lundgren L, Samuelsson G. Bioassay-guided isolation of serotonin from fruits of Solanum tuberosum L. Acta Pharm Nord. 1992;4(2):91–2.
116. Turi CE, Murch SJ. Targeted and untargeted phytochemistry of Ligusticum canbyi: indoleamines, phthalides, antioxidant potential, and use of metabolomics as a hypothesis-generating technique for compound discovery. Planta Med. 2013;79(14):1370–9.
117. Ramakrishna A, Giridhar P, Sankar KU, Ravishankar GA. Endogenous profiles of indoleamines: serotonin and melatonin in different tissues of Coffea canephora P ex Fr. as analyzed by HPLC and LC-MS-ESI. Acta Physiol Plant. 2012;34(1):393–6.
118. Große W, Artigas F. Incorporation of 15N-ammonia into serotonin in cotyledons of maturing walnuts. Z Naturforsch. 1983;38(11–12):1057–8.
119. Lavizzari T, Veciana-Nogués MT, Bover-Cid S, Mariné-Font A, Vidal-Carou MC. Improved method for the determination of biogenic amines and polyamines in vegetable products by ion-pair high-performance liquid chromatography. J Chromatogr A. 2006;1129(1):67–72.
120. Ly D, Kang K, Choi J-Y, Ishihara A, Back K, Lee S-G. HPLC analysis of serotonin, tryptamine, tyramine, and the hydroxycinnamic acid amides of serotonin and tyramine in food vegetables. J Med Food. 2008;11(2):385–9.
121. Udenfriend S, Lovenberg W, Sjoerdsma A. Physiologically active amines in common fruits and vegetables. Arch Biochem Biophys. 1959;85(2):487–90.
122. Adão RC, Glória MBA. Bioactive amines and carbohydrate changes during ripening ofPrata'banana (Musa acuminata× M. balbisiana). Food Chem. 2005;90(4):705–11.
123. Gloria MBA, Adão RC. Effect of gamma radiation on the ripening and levels of bioactive amines in bananas cv. Prata Radiat Phys Chem. 2013;87:97–103.
124. Badria FA. Melatonin, serotonin, and tryptamine in some Egyptian food and medicinal plants. J Med Food. 2002;5(3):153–7.
125. Rasool A, Bhat KM, Sheikh AA, Jan A, Hassan S. Medicinal plants: role, distribution and future. Pharmacogn Phytochem. 2020;9(2):2111–4.
126. Wignall VR, Arscott NA, Nudds HE, Squire A, Green TO, Ratnieks FLW. Thug life: bramble (Rubus fruticosus L. agg.) is a valuable foraging resource for honeybees and diverse flower-visiting insects. Insect Conserv Divers. 2020;13(6):543–57.
127. Jäger AK, Gauguin B, Andersen J, Adsersen A, Gudiksen L. Screening of plants used in Danish folk medicine to treat depression and anxiety for affinity to the serotonin transporter and inhibition of MAO-A. Ethnopharmacol. 2013;145(3):822–5.
128. Saaby L, Rasmussen HB, Jäger AK. MAO-A inhibitory activity of quercetin from Calluna vulgaris (L.) Hull. J Ethnopharmacol. 2009;121(1):178–81.
129. Kong L, Cheng CH, Tan R. Inhibition of MAO A and B by some plant-derived alkaloids, phenols and anthraquinones. J Ethnopharmacol. 2004;91(2–3):351–5.
130. Tao G, Irie Y, Li D-J, Keung WM. Eugenol and its structural analogs inhibit monoamine oxidase A and exhibit antidepressant-like activity. Bioorg Med Chem. 2005;13(15):4777–88.
131. Lee SA, Hong SS, Han XH, Hwang JS, Oh GJ, Lee KS, Lee MK, Hwang BY, Ro JS. Piperine from the fruits of Piper longum with inhibitory effect on monoamine oxidase and antidepressant-like activity. Chem Pharm Bull. 2005;53(7):832–5.
132. Herraiz T, Chaparro C. Human monoamine oxidase enzyme inhibition by coffee and β-carbolines norharman and harman isolated from coffee. Life Sci. 2006;78(8):795–802.
133. Du B, Tang X, Liu F, Zhang C, Zhao G, Ren F, Leng X. Antidepressant-like effects of the hydroalcoholic extracts of Hemerocallis citrina and its potential active components. BMC Complement Altern Med. 2014;14(1):1–11.
134. Hsu L-C, Ko Y-J, Cheng H-Y, Chang C-W, Lin Y-C, Cheng Y-H, Hsieh M-T, Peng WH. Antidepressant-like activity of the ethanolic extract from Uncaria lanosa Wallich var. appendiculata Ridsd in the forced swimming test and in the tail suspension test in mice. Evid Based Complement Alternat Med. 2012;2012:497302.
135. Khulbe A, Pandey S, Sah SP. Antidepressant-like action of the hydromethanolic flower extract of Tagetes erecta L. in mice and its possible mechanism of action. Indian J Pharmacol. 2013;45(4):386.
136. Martínez-Vázquez M, Estrada-Reyes R, Escalona AA, Velázquez IL, Martínez-Mota L, Moreno J, Heinze G. Antidepressant-like effects of an alkaloid extract of the aerial parts of Annona cherimolia in mice. J Ethnopharmacol. 2012;139(1):164–70.
137. Kulkarni S, Akula KK, Deshpande J. Evaluation of antidepressant-like activity of novel water-soluble curcumin formulations and St. John's wort in behavioral paradigms of despair. Pharmacology. 2012;89(1–2):83–90.
138. Sasaki K, El Omri A, Kondo S, Han J, Isoda H. Rosmarinus officinalis polyphenols produce anti-depressant like effect through monoaminergic and cholinergic functions modulation. Behav Brain Res. 2013;238:86–94.
139. Kwon S, Lee B, Kim M, Lee H, Park H-J, Hahm D-H. Antidepressant-like effect of the methanolic extract from Bupleurum falcatum in the tail suspension test. Prog Neuropsychopharmacol Biol Psychiatry. 2010;34(2):265–70.
140. Bano S, Ara I, Saboohi K, Moattar T, Chaoudhry B. St. John's Wort increases brain serotonin synthesis by inhibiting hepatic tryptophan 2, 3 dioxygenase activity and its gene expression in stressed rats. Pak J Pharm Sci. 2014;27:1427.
141. Qiu F, Zhong X, Mao Q, Huang Z. The antidepressant-like effects of paeoniflorin in mouse models. Exp Ther Med. 2013;5(4):1113–6.
142. Ji C-X, Li X-Y, Jia S-B, Liu L-L, Ge Y-C, Yang Q-X, Zhang J-J. The antidepressant effect of Cynanchum auriculatum in mice. Pharm Biol. 2012;50(9):1067–72.

143. Knörle R. Extracts of Sideritis scardica as triple monoamine reuptake inhibitors. J Neural Transm. 2012;119(12):1477–82.
144. Gabriela G-C, Javier A-AF, Elisa V-A, Gonzalo V-P, Herlinda B-J. Antidepressant-like effect of tagetes lucida Cav. extract in rats: involvement of the serotonergic system. Am J Chin Med. 2012;40(04):753–68.
145. Jin Z-L, Gao N, Zhou D, Chi M-G, Yang X-M, Xu J-P. The extracts of Fructus Akebiae, a preparation containing 90% of the active ingredient hederagenin: serotonin, norepinephrine and dopamine reuptake inhibitor. Pharmacol Biochem Behav. 2012;100(3):431–9.
146. Abbas G, Naqvi S, Erum S, Ahmed S, Dar A. Potential antidepressant activity of Areca catechu nut via elevation of serotonin and noradrenaline in the hippocampus of rats. Phytother Res. 2013;27(1):39–45.
147. Bay-Smidt MG, Jäger AK, Krydsfeldt K, Meerow AW, Stafford GI, Van Staden J, Rønsted NJ. Phylogenetic selection of target species in Amaryllidaceae tribe Haemantheae for acetylcholinesterase inhibition and affinity to the serotonin reuptake transport protein. S Afr J Bot. 2011;77(1):175–83.
148. Jäger AK, Gauguin B, Andersen J, Adsersen A, Gudiksen L. creening of plants used in Danish folk medicine to treat depression and anxiety for affinity to the serotonin transporter and inhibition of MAO-A. J Ethnopharmacol. 2013;145(3):822–5.
149. Chanana P, Kumar A. Possible involvement of nitric oxide modulatory mechanisms in the neuroprotective effect of centella asiatica against sleep deprivation induced anxiety like behaviour, oxidative damage and neuroinflammation. Phytother Res. 2016;30(4):671–80.
150. Randriamampionona D, Diallo B, Rakotoniriana F, Rabemanantsoa C, Cheuk K, Corbisier A-M, Mahillon J, Ratsimamanga S, El Jaziri M. Comparative analysis of active constituents in Centella asiatica samples from Madagascar: application for ex situ conservation and clonal propagation. Fitoterapia. 2007;78(7):482–9.
151. Abdul Manap AS, Vijayabalan S, Madhavan P, Chia YY, Arya A, Wong EH, Rizwan F, Bindal U, Koshy S. Bacopa monnieri, a neuroprotective lead in Alzheimer disease: a review on its properties, mechanisms of action, and preclinical and clinical studies. Drug Target Insights. 2019;13:1177392819866412.
152. Ghosh S, Banerjee S. First report of Diaporthe phaseolorum on Bacopa monnieri in India. J Plant Pathol. 2015;97(Supplement):1004.
153. Durazzo A, Nazhand A, Lucarini M, Silva AM, Souto SB, Guerra F, Severino P, Zaccardelli M, Souto EB, Santini A. Astragalus (Astragalus membranaceus Bunge): botanical, geographical, and historical aspects to pharmaceutical components and beneficial role. Rend Lincei Sci Fis Nat. 2021;32(3):625–42.
154. Coria-Téllez AV, Montalvo-Gónzalez E, Yahia EM, Obledo-Vázquez EN. Annona muricata: a comprehensive review on its traditional medicinal uses, phytochemicals, pharmacological activities, mechanisms of action and toxicity. Arab J Chem. 2018;11(5):662–91.
155. Nayak A, Hegde K. A comprehensive review on the miracle nature of Annona muricata Linn. RGUHS J Pharm Sci. 2021;11(1):1.
156. Wee AK, Takayama K, Asakawa T, Thompson B, Sungkaew S, Tung NX, Nazre M, Soe KK, Tan HT, Watano Y. Oceanic currents, not land masses, maintain the genetic structure of the mangrove Rhizophora mucronata Lam. (Rhizophoraceae) in Southeast Asia. J Biogeogr. 2014;41(5):954–64.
157. Manilal A, Merdekios B, Idhayadhulla A, Muthukumar C, Melkie M. An in vitro antagonistic efficacy validation of Rhizophora mucronata. Asian Pac J Trop Dis. 2015;5(1):28–32.
158. Lopresti AL, Hood SD, Drummond PD. Multiple antidepressant potential modes of action of curcumin: a review of its anti-inflammatory, monoaminergic, antioxidant, immune-modulating and neuroprotective effects. J Psychopharmacol. 2012;26(12):1512–24.
159. Leong-Škorničková J, Šída O, Jarolímová V, Sabu M, Fér T, Trávníček P, Suda J. Chromosome numbers and genome size variation in Indian Species of Curcuma (Zingiberaceae). Ann Bot. 2007;100(3):505–26.
160. Vokou D, Kokkini S, Bessiere J-M. Geographic variation of Greek oregano (Origanum vulgare ssp. hirtum) essential oils. Biochem Syst Ecol. 1993;21(2):287–95.
161. Zotti M, Colaianna M, Morgese MG, Tucci P, Schiavone S, Avato P, Trabace L. Carvacrol: from ancient flavoring to neuromodulatory agent. Molecules. 2013;18(6):6161–72.
162. Temel Y, Kucukler S, Yıldırım S, Caglayan C, Kandemir FM. Protective effect of chrysin on cyclophosphamide-induced hepatotoxicity and nephrotoxicity via the inhibition of oxidative stress, inflammation, and apoptosis. Naunyn Schmiedeberg's Arch Pharmacol. 2020;393(3):325–37.
163. Karimi A, Krähmer A, Herwig N, Hadian J, Schulz H, Meiners T. Metabolomics approaches for analyzing effects of geographic and environmental factors on the variation of root essential oils of Ferula assa-foetida L. J Agric Food Chem. 2020;68(37):9940–52.
164. Sati P, Dhyani P, Bhatt ID, Pandey A. Ginkgo biloba flavonoid glycosides in antimicrobial perspective with reference to extraction method. J Tradit Complement Med. 2019;9(1):15–23.
165. Hazarika TK. Citrus genetic diversity of north-east India, their distribution, ecogeography and ecobiology. Genet Resour Crop Evol. 2012;59(6):1267–80.
166. Öztürk N, Korkmaz S, Öztürk Y. Wound-healing activity of St. John's Wort (Hypericum perforatum L.) on chicken embryonic fibroblasts. J Ethnopharmacol. 2007;111(1):33–9.
167. Chandra R, Jadhav VT, Sharma J. Global scenario of pomegranate (Punica granatum L.) culture with special reference to India. Fruit Veg Cereal Sci Biotechnol. 2010;4:7–18.
168. Sharma SK, Pandit MK. A new species of panax L.(Araliaceae) from Sikkim himalaya, India. Syst Bot. 2009;34(2):434–8.
169. Moja S, Guitton Y, Nicolè F, Legendre L, Pasquier B, Upson T, Jullien F. Genome size and plastid trnK-matK markers give new insights into the evolutionary history of the genus Lavandula L. Plant Biosyst. 2016;150(6):1216–24.
170. Nagal S, Kaur C, Choudhary H, Singh J, Bhushan Singh B, Singh K. Lycopene content, antioxidant capacity and colour attributes of selected watermelon (Citrullus lanatus (Thunb.) Mansfeld) cultivars grown in India. Int J Food Sci Nutr. 2012;63(8):996–1000.
171. Peng W, Qin R, Li X, Zhou H. Botany, phytochemistry, pharmacology, and potential application of Polygonum cuspidatum Sieb. et Zucc.: a review. J Ethnopharmacol. 2013;148(3):729–45.
172. Reddy M, Begum H, Sunil N, Pandravada S, Sivaraj N, Kumar S. Mapping the climate suitability using MaxEnt modeling approach for Ceylon spinach (Basella alba L.) cultivation in India. J Agric Sci. 2015;10:87–97.
173. Jiraungkoorskul W. Review of neuro-nutrition used as anti-alzheimer plant, spinach, Spinacia oleracea. Pharmacogn Rev. 2016;10(20):105.
174. Gazal A, Dar Z, Zafar G, Habib M. Stability analysis for yield and its contributing traits in Brown sarson (Brassica rapa L.) under Kashmir conditions in India. J Oilseed Brassica. 2016;1(1):33–8.
175. Verma D, Lakhanpal N, Singh K. Genome-wide identification and characterization of abiotic-stress responsive SOD (superoxide dismutase) gene family in Brassica juncea and B. rapa. BMC Genomics. 2019;20(1):1–18.
176. Nettleship L, Slaytor M. Limitations of feeding experiments in studying alkaloid biosynthesis in Peganum harmala callus cultures. Phytochemistry. 1974;13(4):735–42.

177. Aslam N, Wani AA, Nawchoo IA, Bhat MAJIJAR. Distribution and medicinal importance of Peganum harmala. A review. Int J Adv Res. 2014;2(2):751–5.
178. Moloudizargari M, Mikaili P, Aghajanshakeri S, Asghari MH, Shayegh J. Pharmacological and therapeutic effects of Peganum harmala and its main alkaloids. Pharmacogn Rev. 2013; 7(14):199.
179. Gupta C, Sengar S. Response of tomato (Lycopersicon esculentum Mill.) to nitrogen and potassium fertilization in acidic soil of Bastar. Veg Sci. 2000;27(1):94–5.
180. Ganal MW, Lapitan NL, Tanksley SD. A molecular and cytogenetic survey of major repeated DNA sequences in tomato (Lycopersicon esculentum). Mol Gen Genet. 1988;213(2):262–8.
181. Luan F, Wang Z, Yang Y, Ji Y, Lv H, Han K, Liu D, Shang X, He X, Zeng N. Juglans mandshurica Maxim.: a review of its traditional usages, phytochemical constituents, and pharmacological properties. Front Pharmacol. 2021;11:569800.
182. Pilar Nicasio MD, Villarreal ML, Gillet F, Bensaddek L, Fliniaux M-A. Variation in the accumulation levels of N, N-dimethyltryptamine in micro-propagated trees and in in vitro cultures of Mimosa tenuiflora. Nat Prod Res. 2005;19(1): 61–7.
183. Martínez-Higuera A, Rodríguez-Beas C, Villalobos-Noriega JMA, Arizmendi-Grijalva A, Ochoa-Sánchez C, Larios-Rodríguez E, Martínez-Soto JM, Rodríguez-León E, Ibarra-Zazueta C, Mora-Monroy R. Hydrogel with silver nanoparticles synthesized by Mimosa tenuiflora for second-degree burns treatment. Sci Rep. 2021;11(1):1–16.
184. Lima Bezerra JJ, Vieira Pinheiro AA, Barbosa LR. Phytochemistry and teratogenic potential of Mimosa tenuiflora (willd.) poir. (Fabaceae) in ruminants: a systematic review. Toxicon. 2021;195:78–85.
185. FAJ E. Melatonin, serotonin, and tryptamine in some Egyptian food and medicinal plants. J Med Food. 2002;5(3):153–7.
186. Waalkes TP, Sjoerdsma A, Creveling CR, Weissbach H, Udenfriend S. Serotonin, norepinephrine, and related compounds in bananas. Science. 1958;127(3299):648–50.
187. Imam MZ, Akter S. Musa paradisiaca L. and Musa sapientum L.: a phytochemical and pharmacological review. J Appl Pharm Sci. 2011;1(5):14–20.
188. Pari L, Maheswari JU. Hypoglycaemic effect of Musa sapientum L. in alloxan-induced diabetic rats. J Ethnopharmacol. 1999;68(1–3):321–5.
189. Marco R, Basile B, Capuozzo C, Scognamiglio P, Forlani M, Rao R, Corrado G. Molecular and phenotypic diversity of traditional European plum (Prunus domestica L.) germplasm of Southern Italy. Sustain. 2019;11(15):4112.
190. El-Beltagi HS, El-Ansary AE, Mostafa MA, Kamel TA, Safwat G. Evaluation of the phytochemical, antioxidant, antibacterial and anticancer activity of Prunus domestica fruit. Not Bot Horti Agrobot Cluj Napoca. 2019;47(2):395–404.
191. Silva LJG, Lino CM, Meisel LM, Pena A. Selective serotonin re-uptake inhibitors (SSRIs) in the aquatic environment: an ecopharmacovigilance approach. Sci Total Environ. 2012;437: 185–95.

Herbal Medicines for Management of Alzheimer's Disease

Jai Malik, Subhash C. Mandal, Sunayna Choudhary, Shweta Parihar, and Mohamed Rahamathulla

Abstract

The advancements in the therapeutic world have increased the average life expectancy of man, thereby causing an increase in the prevalence of degenerative disorders, especially Alzheimer's disease. Oxidative stress, mutation in presenilin (PSEN) genes (PSEN1 and PSEN2), inflammation, mitochondrial and neurovascular dysfunction, and calcium dysregulation are some of the other prominent causes of AD. Despite advancements in therapeutics, drug development for a safe and effective therapeutic regimen for AD has been challenging because of the high failure rate. During the last three decades, only six drugs (four acetylcholinesterase inhibitors, one NMDA antagonist-memantine, and one antibody-based drug-aducanumab) have been approved for AD by US FDA. The complex pathophysiology of this disease necessitates using medicines that can act on several targets at once. Mother Nature has proved herself from time to time incurring various illnesses, which is confirmed by the scientific literature. Various herbal drugs across the world have shown beneficial effects in AD patients either by improving the symptoms or by modifying the disease. The present chapter overviews some of the important and widely studied plants and their constituents against AD.

Keywords

Alzheimer's disease · Neurodegenerative disorders · Curcumin · Huperzine A · Ginsenosides · Resveratrol · Withanolides · Berberine

1 Introduction

The advancements in the therapeutic world have increased the average life expectancy of man, thereby causing an increase in the prevalence of degenerative disorders, especially neurodegenerative disorders. Alzheimer's disease (AD) is among the most prevalent neurological conditions in geriatric population. Despite advancements in therapeutics, the drug development for a safe and an effective therapeutic regimen for AD has been challenging because of the high (almost 100%) failure rate. During the last three decades, more than 800 drugs have been approved by USFDA for different disorders, and only six drugs (four acetylcholinesterase inhibitors, one NMDA antagonist-memantine, and one antibody-based drug-aducanumab) have been approved for AD. Very recently, a controversial accelerated FDA approval has been given to another monoclonal antibody drug lecanemab for AD. Amongst these, tacrine and memantine are no longer marketed. Furthermore, according to an estimate, there has been a staggering 150% increase in AD cases in the last three decades, and by 2050, there will be more than 130 million AD patients worldwide [1]. At present in the US, about one in nine persons of age 65 or above has AD type dementia [2]. Thus, due to very limited number of drugs available in the market and compelling increase in the AD cases across the globe, there is a dire need for newer, effective and safe drugs for AD. It is not only the number of AD patients, but also the socio-economic impact of the disease that has made AD more atrocious. The overall annual cost of dementia treatment has been estimated to be a stunning figure of USD1.3 trillion, with more than 300 billion USD spent in the US only [2].

J. Malik (✉) · S. Parihar
Department of Pharmacognosy, University Institute of Pharmaceutical Sciences—UGC Centre of Advanced Study, Panjab University, Chandigarh, India
e-mail: jmalik@pu.ac.in

S. C. Mandal
Pharmacognosy and Phytotherapy Research Laboratory, Division of Pharmacognosy, Department of Pharmaceutical Technology, Jadavpur University, Kolkata, India

S. Choudhary
Department of Pharmacognosy, M.M. College of Pharmacy, M.M. (Deemed to be University), Mullana, Ambala, Haryana, India

M. Rahamathulla
Department of Pharmaceutics, College of Pharmacy, King Khalid University, Abha, Saudi Arabia

A. K. Dhara, S. C. Mandal (eds.), *Role of Herbal Medicines*, https://doi.org/10.1007/978-981-99-7703-1_12

The work done to understand the pathophysiology of AD has shown multiple pathways involved in its development. This multi-pathway nature of the disease is also one of the major reasons for high failure rate in the drug discovery for the disease. Amongst major reasons, (1) formation and deposition of Aβ plaques, (2) neurofibrillary tangles (NFTs) formed by tau proteins hyperphosphorylation, and (3) cholinergic deficits in the memory controlling parts (cerebral cortex, hippocampus) of the brain are considered as mainstay for the disease development (Fig. 1).

Oxidative stress, mutation in presenilin (PSEN) genes (PSEN1 and PSEN2), inflammation, mitochondrial and neurovascular dysfunction, and calcium dysregulation are some of the other prominent causes of AD. Synaptic dysfunction/degeneration is also emerging as one of the prominent causes of AD development [3]. Due to better understanding of pathophysiology of AD, last two decades have observed a paradigm shift in the drug discovery for AD, from the agents giving symptomatic relief to disease modifying agents. Around 53% of the total drugs went under clinical trials for AD in 2019 were disease modifying agents [4], and this figure increased to 83% in 2021 [5] and 2022 [6]. Approval of aducanumab and lecanemab, two monoclonal antibody-based drugs, are the result of this paradigm shift. Herbal drugs and their constituents due to their multifarious effects and their effects on different molecular markers of the disease, have also emerged as promising candidates for drug development of AD. Various herbal drugs across the world have shown beneficial effects in AD patients either by improving the symptoms or by modifying the disease. It all started with physostigmine, an alkaloid from *Physostigma venenosum* (Leguminosae), that exhibited a potent acetylcholinesterase inhibitory activity which was beneficial in AD type dementia. Later in 2000, its semisynthetic and more potent derivative, Rivastigmine was approved by US-FDA for AD. In the following year, 2001, another alkaloid, Galantamine obtained from *Galanthus nivalis* (Amaryllidaceae) was also approved by FDA due to its acetylcholinesterase inhibitory activity (Fig. 2). Since then, various herbal drugs, their preparations, and isolated chemical compounds have shown their efficacy against AD. The present chapter reviews some of the important and widely studied plants along with their constituents against AD.

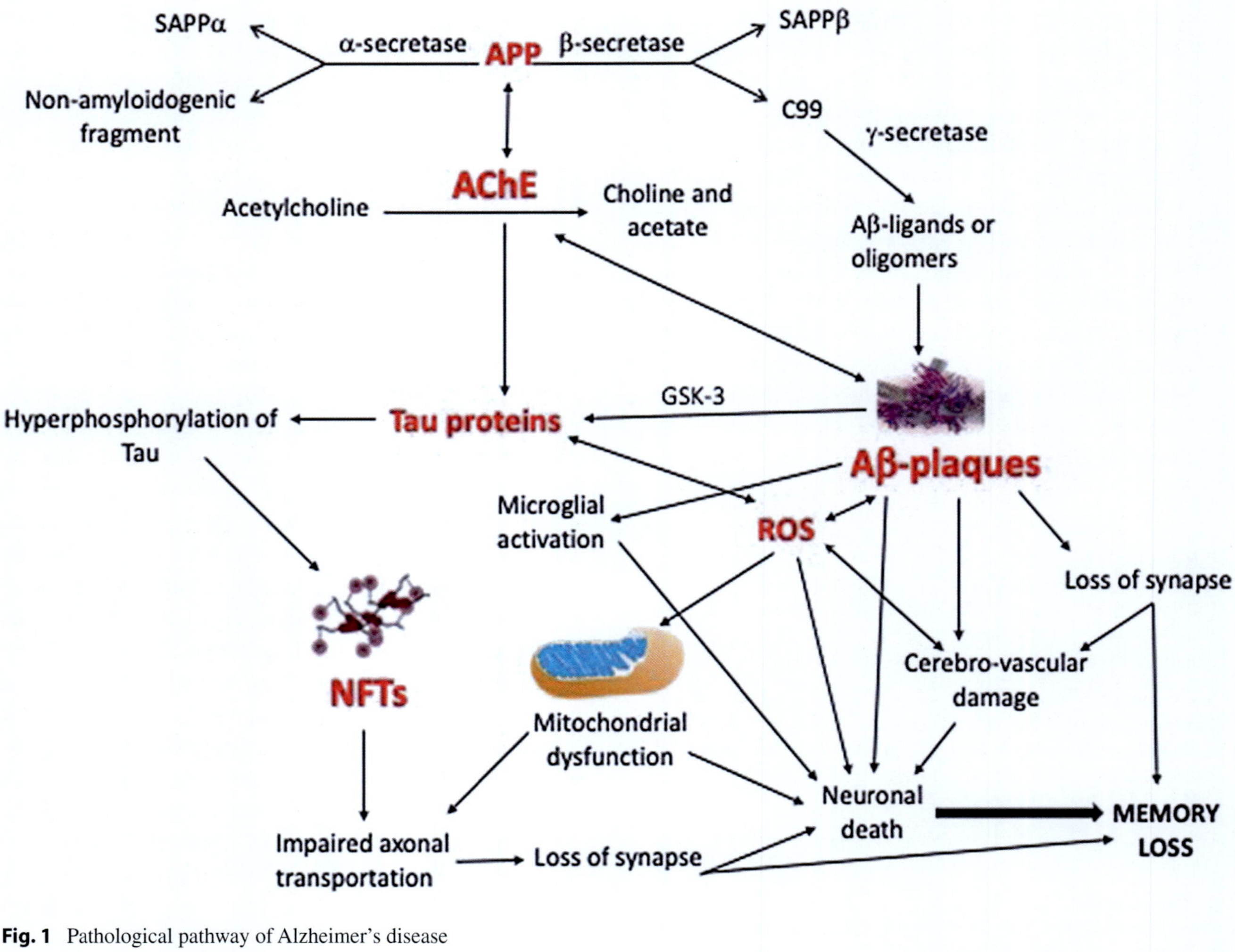

Fig. 1 Pathological pathway of Alzheimer's disease

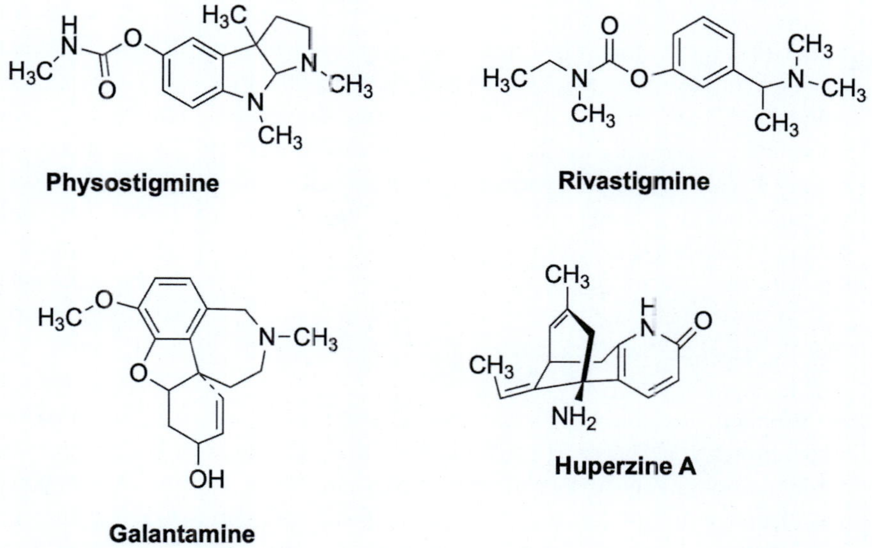

Fig. 2 Various compounds for AD from plants

2 Plants with Beneficial Effects in AD

2.1 *Huperzia serrata* (Thunb. Ex Murray) Trev

H. serrata (syn. *Lycopodium serratum* Thunb. ex Murray), belonging to the family Huperziaceae, has been used in Traditional Chinese Medicine since ages under the name 'Quian Ceng Ta', for its beneficial effects on CNS, in inflammation and pain [7]. The plant has a world-wide distribution ranging from central America to Eastern, Southern or South-east Asia. The plant is known for its alkaloids that are classified (based on their structural properties) into four categories, viz., fawcettimine-type, lycodine-type, lycopodine-type, and miscellaneous-type. Among various constituents present in *H. serrata*, Huperzine A (HupA, Fig. 2) is considered the most important constituent because majority of its biological activities have been attributed to this compound [8]. HupA was first isolated in 1986, and since then, it is amongst the most comprehensively studied natural molecules [9, 10]. At present, it has been approved for AD treatment in China and is available as a dietary supplement in the US [11].

HupA is an unsaturated lycodine type sesquiterpene alkaloid, which is optically active and occurs as (−)-HupA or L-HupA in nature. This isomer is biologically active and more potent than (+)-HupA or D-HupA, while the potency of the racemic mixture lies between the two [12]. HupA came into the picture because of its potent, specific, and reversible acetylcholinesterase inhibitory (AChEI) activity which was even more than the approved drugs (tacrine, galantamine, donepezil, and rivastigmine) [11, 13, 14]. The AChEI activity of HupA causes accumulation of acetylcholine (ACh) which acts through α7nAChRs and α4β2nAChRs receptors modulating neural stem cells function via mitogen-activated protein kinases/extracellular signal-regulated kinases (MAPK/ERK) pathway, that eventually exhibits neuroprotective and neurogenetic action [15]. It has also been observed that globular tetramer of acetylcholinesterase forms a stable complex with amyloid beta (Aβ) peptides, thereby increasing the aggregation and plaque formation in brain [16, 17]. HupAdose-dependently also prevented this aggregation by its AChEI activity when studied in various transgenic animal models [18]. Inhibition of wingless-related integration site (Wnt) signaling increases tau phosphorylation and Aβ accumulation, thereby causing cognitive decline [19]. HupA inhibits the glycogen synthase kinase-3β (GSK3β), thereby stimulating the Wnt signaling which helps in inhibiting the tau phosphorylation and Aβ accumulation by promoting non-amyloidogenic pathway [20]. Besides inhibiting the AChE activity, Aβ accumulation and tau hyperphosphorylation, HupA also promotes synaptotagmin expression causing the increased release of neurotransmitters (especially ACh) in the synaptic region from synaptic vesicles through the opening of vesicle pores [21]. This increase in ACh levels and NMDA receptor antagonistic activity also stimulates the brain-derived neurotrophic factor (BDNF)/tropomyosin receptor kinase B (TrkB) signaling in HT22 hippocampal cells. Stimulation of this pathway further stimulates the phosphatidylinositol 3-kinase (PI3K)/Aktserine-threonine kinase pathway, as well as rapamycin (mTOR) signaling route that causes suppression of neuronal apoptosis and increase the neuronal survival [22]. HupA has also exhibited beneficial effects through non-cholinergic mechanism, wherein it has shown to mitigate AD-induced mitochondrial perturbations and oxidative stress [23, 24].

Apart from numerous preclinical findings, various clinical studies have also revealed the benefits of HupA in improving cognition in AD and vascular dementia patients [25–29]. In another study, HupA, when given for 8 weeks to

AD patients, exhibited improvement in their cognition and task switching abilities when evaluated using Mini-Mental State Exam (MMSE), the AD Assessment Scale-cognitive subscale (ADASCog) [30]. Another randomized clinical trial on vascular dementia patients demonstrated similar kind of results when HupA treated patients were evaluated for their cognitive abilities using MMSE and Activities of Daily Living (ADL) tests [29].

2.2 *Ginkgo biloba* Linn

A member of family Ginkgoaceae is also known as "Maidenhair tree" due to its resemblance with "Maidenhair fern" (Adiantum) and a "living fossil" due to its presence on this planet for the last 200 million years. *G. biloba* is the only plant species of its genus. It is indigenous to East Asia (China, Japan, and Korea) but one can also find it growing in European and American continents. The plant is known for its terpene trilactone compounds, collectively known as Ginkgolides which consists of diterpene trilactones ginkgolide A, B, C, J, K, L, and M, a sesquiterpene trilactone, bilobalide, and various flavonoids - quercetin, kaempferol, myricetin, apigenin, isorhamnetin, luteolin, and their glycosides (Fig. 3). A standardized commercial extract, EGb 761 (manufactured by Beaufour-Ipsen Pharma, France, and Dr. Willmar Schwabe Pharmaceuticals, Germany), containing 5–7% ginkgolides and bilobalide (BB), along with22–24% of flavonoids and ginkgolic acids or anacardic acids (< 5 ppm, due to their cytotoxic and contact dermatitis properties) has been widely used for studying the effects of the plant [31, 32].

The efficacy of the *G. biloba* extract (GBE) against AD has been reviewed by various groups over the period [33–36]. Experts have suggested that GBE can have beneficial effects on cognition in the patients that are unresponsive to the treatment of AChEIs or NMDA receptor antagonists [37]. GBE has exhibited its beneficial effects through multifarious activities on the nervous system (Fig. 4). Various studies have shown the antioxidant potential of GBE as one of the major neuroprotective mechanisms. It has shown neuroprotective activity against oxidative stress induced by H_2O_2 [38], $FeSO_4$ [39], and Aβ [40] in brain cells. The antioxidant effect exhibited by GBE was mainly due to the upregulation of the antioxidant enzymes, including glutathione reductase, catalase, and superoxide dismutase [41]. Subsequently, its antioxidant potential has also been attributed to the modulation of certain genes [42]. The antioxidant effect of GBE has been found mainly due to flavonoids as inconsistent results have been observed with different ginkgolides [43]. The GBE, EGb761, has also shown protection against Aβ, one of the major hallmarks of AD. It prevents the toxic effects of Aβ by promoting the action of α-secretase enzyme which promotes non-amyloidogenic pathway (Fig. 4) of amyloid precursor proteins, thus inhibiting amyloidogenesis [44]. Free and high levels of cholesterol have shown to increase the formation of amyloid plaques by modulating the α-secretase activity [45]. Depletion in cholesterol levels in brain has shown to prevent the formation of Aβ [46]. EGb761 has shown to attenuate the increased free cholesterol levels, thereby preventing the formation/accumulation of Aβ plaques [47]. GBE administration has also shown to decrease levels of Aβ and tau proteins by stimulation of Heat shock factor (HSF-1) and increasing the expression of HSPs [48].

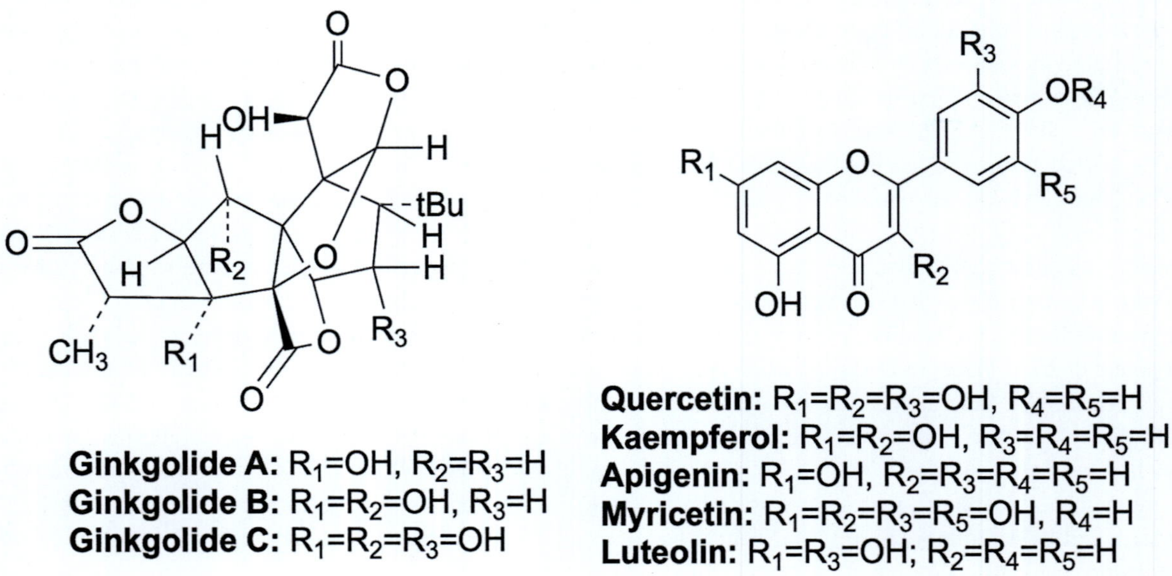

Fig. 3 Chemical constituents of *Ginkgo biloba*

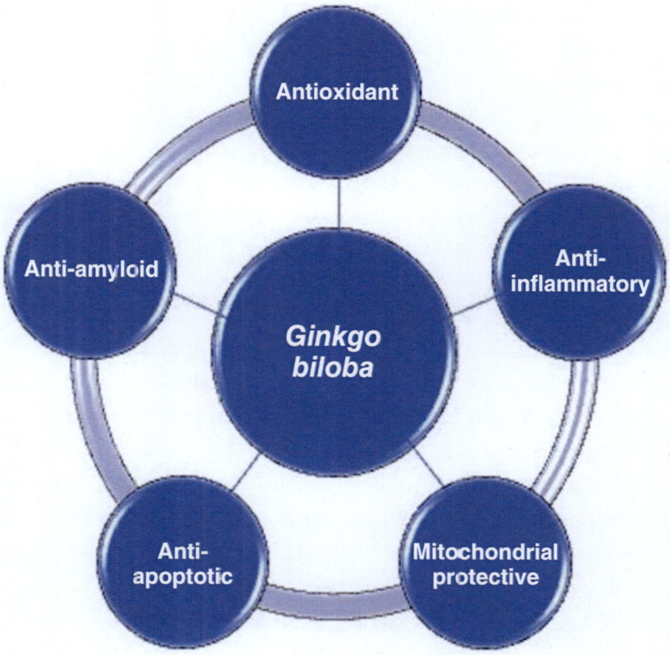

Fig. 4 Multifarious effects of *G. biloba*

Aβ also activates N-methyl-D-aspartate (NMDA) receptors leading to increased Ca^{2+} influx causing neuronal toxicity [49], and also reduces synaptic plasticity by dysregulation of α-amino-3-hydroxy-5-methyl-4-isoxazolepropionic acid (AMPA) receptors [50]. Ginkgolide A inhibits both AMPA and NMDA receptors, thereby protecting neurons from Aβ toxicity [51]. Furthermore, ginkgolide A also inhibits the Aβ induced phosphorylation of c-Jun N-terminal kinase (JNK) and prevents the neuronal apoptosis [52]. Ginkgolide A also prevents the tau protein hyperphosphorylation by activating phosphatidylinositol-3-kinase (PI3K) and kinase AKT (AKT) pathway [53], which is responsible for cell survival and downregulation of apoptotic signals [54]. In another study, ginkgolide A, bilobalide, and flavonoids promoted the tau protein degradation in the neurons by increasing the LC3B-II protein expression in tau-transgenic AD mice [55]. Inflammation and enhanced levels of platelet activating factor (PAF) have also been implicated in AD. GBE inhibits neuroinflammation by downregulating the LCN2-activated JAK2/STAT3 signaling pathway and astrocyte proliferation [56]. In another study, GBE mitigates the lead induced increase in myeloperoxidase activity, suppresses TNF-α & IL-6 secretion, and decreases the expression of NF-κB in mice hippocampus, indicating its anti-inflammatory potential [57]. The extract also attenuated the increased TNF-α and IL-1β levels and decreased the mRNA expression of NF-κB, p65 and Bax, thereby increasing the neuronal survivability in APP/PS1 double gene-transfected HEK293 (APP/PS1-HEK293) cell line model of AD [58]. According to the inflammation hypothesis of AD, inflammasome ("a cytoplasmic polymeric protein complex that functions as the platform for Caspase-1 activation and for the maturation of the pro-inflammatory cytokine interleukin-1β") of NLR family – pyrin domain containing 3 (NLRP3) inflammasome has a vital role in AD pathogenesis [59]. Ginkgolide B decreased the intracellular pro-inflammatory cytokine levels to improve cognition in SAMP8 mice. Furthermore, ginkgolide B also deactivated the NLRP3 inflammasome through ubiquitination and autophagic degradation [60].

GBE, for its efficacy in AD patients, has also been evaluated in various clinical trials that have been reviewed by various groups highlighting mixed results [43, 61–64]. But a recent report compiled various pre-clinical and clinical studies since 1980, and suggested that detailed, well planned and large-scale clinical studies are required to evaluate the benefits of long-term GBE treatment on individuals with early-stage AD [65].

2.3 *Panax ginseng* C.A. Mey

Ginseng of family Araliaceae is a popular immunomodulator drug whose roots have been used since yesteryears in China, Japan, Korea, and other East Asian countries for vitality, antiaging, as general tonic, for strengthening CNS, and other systems of the body [66–68]. The Genus name *Panax*, was given by Charles Linnaeus and is derived from Greek word 'panacea' (total healing for all), owing to its wide range of therapeutic efficacy. At present, Genus Panax contains 11 species spread across around 35 countries [69]. Among these, *Panax ginseng* (Korean and Chinese ginseng, both varieties, *white* and *red*) along with *P. quinquefolium* (American ginseng), are the most widely investigated ginseng species. Commercially, three types of ginseng are available- *raw* (extracted from the soil with at least 75% moisture content); *white* (4-year-old raw ginseng thinly peeled and sun-dried with less than 15% moisture); and *red* (raw ginseng steamed with water and then dried to get 15–18% moisture to increases the storage duration up to 15 years) [70]. Ginseng's pharmacological effects are believed to be due to its dammarane type triterpenoidal saponin compounds that are collectively known as ginsenosides. Depending upon the positioning of sugar moieties, these ginsenosides (Fig. 5) can be categorized into protopanaxadiol type (Ginsenosides Rb1, Rc, Rb2, Rd., and Rg3) and protopanaxatriol type (Ginsenosides Rg1, Re, Rg2, and Rh1). Till date, more than 100 ginsenosides have been isolated and identified from different parts of *P. ginseng*. Ginsenosides Rbl, Rb2, Rc, Rd., Rgl, Rg2, and Re are major constituents (75–80% of total ginsenosides) of fresh, white and red ginsengs, while ginsenosides Rg3, Rg5, and Rg6 are specifically found in red ginseng [67, 71].

Ginsenosides have proved themselves as potent biologically active compounds against AD by acting through multiple pathways viz., antioxidant effects [72], modulation of cholinergic pathway [73], effect on amyloid [74] and tau proteins [75], anti-inflammatory, and antiapoptotic activities [76]. Ginsenosides Rb1, Rb2, Rc, Re, Rg1, and Rg3 have significantly inhibited AChE and BChE. The order of activity of different ginsenosides was found to be Re > Rg3 > Rgl > Rb1 > Rb2 > Rc for AChE and Rg3 > Rg1 > Rb2 > Rb1 > Re > Rc for BChE [73]. It is well known that β-secretase (BACE1) is the enzyme that leads to the formation of Aβ from amyloid precursor proteins. Various ginsenosides have shown to inhibit this enzyme, thereby preventing the formation of Aβ [73]. The β-secretase inhibition by different ginsenosides was in the order of Rc > Rgl > Rb2 > Rb1 > Rg3 > Re. Another saponin, ginsenoside Rd, improved cognition by decreasing the Aβ levels via promoting non-amyloidogenic pathway, by upregulating estrogenic receptor expression, and α-secretase enzyme activity in ovariectomized rats [77]. It also decreased the expression of BACE1and showed neuroprotection by stimulating MAPK/ERK and PI3K/AKT pathways. Similar results were obtained when ginsenoside Rg1 was evaluated in ovariectomized rats and menopausal women [78, 79]. Aβ also increases the inflow of Ca^{2+} ions in the neurons causing stimulation of proteolytic enzymes, production of ROS leading to neurodegeneration [80]. Ginsenoside Rb1 [81], Rg1 [82], and Rg2 [76] have shown neuroprotective effect by mitigating the Aβ induced increase in not only Ca^{2+} ions, but also ROS and oxidative stress in the neurons. Ginsenoside Rg2 not only controlled the Ca + 2 ion concentration, but it also prevented the neuronal apoptosis by activating the PI3K/AKT pathway [76]. Ginsenoside Rg1 also reduces the Aβ levels by modulating the activities of α-, β-, and γ-secretase enzymes which are involved in the production and degradation of Aβ. It was further found that Rg1 down

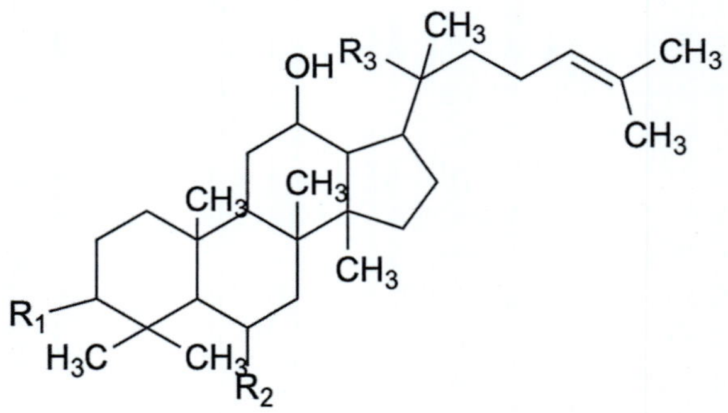

Ginsenoside Rb1: R1= -O-glc(2-1)glc, R2= H, R3= -O-glc(6-1)glc
Ginsenoside Rb2: R1= -O-glc(2-1)glc, R2= H, R3= -O-glc(6-1)ara(p)
Ginsenoside Rb3: R1= -O-glc(2-1)glc, R2=H, R3= -O-glc(6-1)xyl
Ginsenoside Rc: R1= -O-glc(2-1)glc, R2=H, R3= -O-glc(6-1)ara(f)
Ginsenoside Rd: R1= -O-glc(2-1)glc, R2=H, R3= -O-glc
Ginsenoside Rg3: R1= -O-glc(2-1)glc, R2=H, R3=OH
Ginsenoside Re: R1= OH, R2 = -O-glc(2-1)rah, R3= -O-glc
Ginsenoside Rg: R1=OH, R2= -O-glc, R3= -O-glc
Ginsenoside Rh1: R1= OH, R2= -O-glc, R3= OH

Fig. 5 Various ginsenosides from *P. ginseng*

regulates the CDK5 pathway by inhibiting the PPARγ thereby limiting the β-secretase expression and reduced the levels of Aβ in the hippocampal neurons [74]. Ginsenoside Rg1 also exhibited neuronal protection by stimulating protein kinase/c-AMP-response element-binding protein (PKA/CREB) signaling in the hippocampal region [83]. It also decreased the levels of hyperphosphorylated tau proteins in the hippocampus and cortex region of brain by upregulating PP2A expression and suppressing the GSK-3β signaling pathway [75], and by altering the gut microbiota [84]. Very recently, it has been shown that it also combats AD by restoring mitophagy [85]. Ginsenoside Rg1 has also improved AD symptoms by synaptic development. It increases the levels of synaptic formation proteins like microtubule-associated protein 2 (MAP-2) and synaptophysin [86], and also increases the expression of choline acetyltransferase, thereby improving the ACh levels in synapses [87]. Furthermore, Chen (2011) [88] improved the absorption and time of onset of action of Rg1 by preparing its nasal formulation. The results showed that there was a considerable increase in the distribution, transport efficiency, and a significant decrease in time taken by the compound to reach brain (fast onset of action). Various preclinical studies of ginsenoside Rg1 pertaining to its effect against AD have been reviewed by Liang et al. (2021), [89] while Wu et al. (2022) [90] have reviewed the mechanistic insights of Rg1 against AD. Ginsenosides Rg1 and Rg2 also decreased the deposition of Aβ and influenced the levels of biomarkers such as hypoxanthine, dihydrosphingosine, and hexadecasphinganine in the hippocampus of AD mice [91]. Studies have also shown the efficacy of ginsenoside Rg3 against AD [92, 93]. Ginsenoside Rb1 also prevented the Aβ induced tau hyperphosphorylation in cortical neurons [81]. Apart from ginsenosides, gintonin, a lysophosphatidic acid (LPA) receptor-activating ligand isolated from ginseng, has also exhibited significant improvement in cognition in preclinical [94, 95] and clinical studies [96]. Ginseng and ginsenosides have also been found effective clinically in various clinical trials [97–100], making them a strong contender for drug development for AD. However, some of the trials have shown contradictory results wherein no improvement in cognition of the subjects was recorded [101, 102].

2.4 Resveratrol (Trans-3,40,5-Trihydroxystilbene)

A natural phenol (stilbene, Fig. 6), first isolated in 1939 from *Veratrum grandiflorum,* is found in significant concentrations in the skin of red grapes and grape wines [103]. It has also been found in berries (cranberries, strawberries, blueberries, etc.), nuts, tea, and dark chocolates, though variation in the concentration is reported. It is found in two isomeric forms, *trans-* and *cis-*, and the former is predominant in nature.

It has shown protective effects against AD through multi-pathway action by acting on synaptic plasticity, Aβ peptides and tau proteins by inhibiting GSK3β [104], activating AMPK (AMP-activated protein kinase) and SIRT1 (silent information regulator 1) [105], antioxidant action and acting on PI3K/AKT [105, 106]. Chen et al., (2019) have reviewed various preclinical studies pertaining to the role of resveratrol in experimental AD models [107]. Synaptic plasticity plays a vital role in normal brain functions such as learning and memory [108]. Any alteration in the synaptic strength leads to impairment in the storage and consolidation of the information. α-amino-3-hydroxy-5-methyl-4-isoxazole-propionic acid (AMPA)-type glutamate receptors (AMPARs) also mediate in maintaining the synaptic plasticity [108]. Aβ proteins have been found to cause significant loss of AMPARs, thereby causing loss of synaptic plasticity and memory functions in AD [50]. Furthermore, AMPARs also cause the release and abnormal tau proteins phosphorylation, another hallmark of AD [109]. Thus, such evidences highlight the role of AMPARs in AD pathophysiology. Resveratrol activates AMPK and downstream PI3K/AKT signaling leading to increased AMPAR proteins expression, thereby exhibiting neuroprotective action [110, 111].

SIRT1 is a homologue of NAD-dependent deacetylase sirtuin family and has been found responsible for synaptic plasticity, dendrite branching, and axon elongation [112].

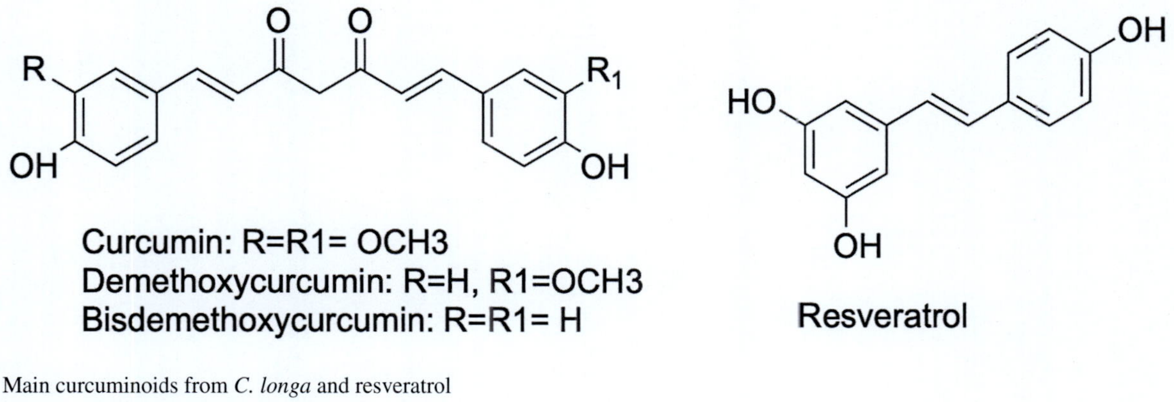

Fig. 6 Main curcuminoids from *C. longa* and resveratrol

Furthermore, SIRT1 also increases the Aβ degradation and processing of APP through α-secretase, thereby leading to decreased toxicity and production of Aβ [113]. Furthermore, resveratrol, through SIRT1, also activates AMPK-PI3K/AKT signaling pathway to exhibit neuroprotective effects [114]. Intriguingly, SIRT1 has also been to be associated with increasing the sensitivity of neurons towards oxidative damage through IGF-1/IRS-2/Ras/ERK1/2 signaling [115].

Resveratrol has also exhibited direct effects on Aβ by inhibiting its aggregation and stabilizing its structure [116], enhancing its clearance through proteasomal degradation [117]; altering the amyloid fibrillar intermediates into non-toxic fragments [118]; reducing the length and number of Aβ fibrils [119] and reducing the Aβ plaque formation [120]. It also inhibits β-secretase activity, thereby reducing the formation of pathogenic Aβ. On the contrary, reports have stated that though it can change the configuration of Aβ oligomer [121] but it cannot reduce its formation as it has no direct interaction with β-secretase [117]. It also decreased Aβ induced rise in the pro-inflammatory molecules such as monocyte chemoattractant protein-1, nitric oxide, interleukins, [122] and nuclear factor-κB [123]. Resveratrol has also exhibited its protective effects against AD by its anti-tau action. It significantly prevented the tau phosphorylation [124], reduced the levels of phosphorylated tau and inhibited its aggregation, thereby by preventing tau induced synaptic loss [125, 126]. Multifarious mechanisms of resveratrol are summarized in Table 1 below.

2.5 Curcuma longa

Curcuma longa (Fam: Zingiberaceae), has been used in India, Asia and Middle Eastern countries as an important ingredient of curries and other spicy dishes. *C. longa* like other botanical remedies, was first used as a food ingredient before it was discovered to have powerful medicinal properties. For centuries, Indian system of medicine has used it extensively as a pain relieving and anti-inflammatory agent, in the skin ailments like acne, eczema, and other skin infections [132–134]. Among various constituents present in *C. longa,* curcumin (1,7-bis [4-hydroxy-3-methoxyphenyl]-1,6-heptadiene-3,5-dione, also named diferuloyl-methane) is considered as most important because majority of its biological activities have been attributed to this compound [135]. It is a phenolic compound with two ferulic acid moieties attached through their carboxyl groups by an additional carbon (methane) inbetween (Fig. 6). Curcumin was isolated in 1815 by Vogel and Pelletier for the first time from the rhizomes of *C. longa*. Curcumin and it is derivative bisdemethoxycurcumin are considered to be the most active constituents. But majority of studies have been done keeping curcumin in focus. It exhibits its protective effects against AD through its multifarious and multitargeted effect. Cholinergic, anti-oxidative, anti-amyloid, anti-tau, anti-inflammatory and anti-apoptotic properties of curcumin have been found responsible for its beneficial effects in AD. Curcumin inhibits the AChE, maintains the acetylcholine levels and strengthens the cholinergic system. The anti-amyloid property of curcumin is an amalgamation of its ability to inhibit β-secretase enzyme, Aβ oligomerization and promote the disaggregation of the existing Aβ fibrils [136] and destabilization of the Aβ fibrils [137]. It has also been found that curcumin stimulates the estrogen receptor β (ERβ), and this estrogenic action is believed to also contribute towards its β-secretase inhibitory property [138]. Curcumin binds to the Aβ fibril at the hydrophobic sites present on the fibril and this hydrophobic interaction converts

Table 1 Neuroprotective mechanisms of resveratrol

S.no	Molecular target	Mechanism	Reference
1	SIRT 1	Stimulate its activity Increase axon length, dendrite branching and synaptic plasticity	[112]
2	Aβ proteins	Increase its degradation Prevent oligomerization, Aggregation and formation	[117] [116, 120]
3	Tau	Prevent its phosphorylation Prevent aggregation Upregulate BAG2	[124–126]
4	Oxidative markers	Antioxidant Reduce ROS Increase endogenous antioxidant levels	[127, 128]
5	Inflammatory markers	Reduce levels of interleukines, NF-κB and NO	[123]
6	Neuronal apoptotic markers	Reduce Bax/Bcl ratio Increase levels of protein kinase JNK, PKC	[129–131]
7	AMPK-PI3K/AKT	Stimulate this pathway, inhibit m-TOR, increase autophagy Neuroprotection by decreasing Aβ and tau formation	[114] [111]

these fibrils into non-toxic proteins [139]. It also prevented the synaptic, mitochondrial oxidative and inflammatory damage induced by Aβ [140]. Reddy et al., (2018) have extensively reviewed the protective effects of curcumin and its derivatives against Aβ in AD [140]. In addition to this, bisdemethoxycurcumin enhances the Aβ clearance by increased uptake by macrophages. It is believed that this compound, by enhancing the transcription of MGAT3 and TLRs, may stimulate/correct the defect in the immune system of AD patients and can be helpful in AD immunotherapy approach [141]. Apart from its action on Aβ, curcumin also reduces the load of hyperphosphorylated tau proteins by disintegrating already formed tau fibrils and inhibiting the oligomerization of new one's [142]. It has further been found that curcumin binds to the tau protein, breaks the H-bonds between the β-sheets, thereby altering its pentamer structure and causing its disintegration [143]. Curcumin could aid macrophages in removing amyloid plaques, which are involved in Alzheimer's disease. Macrophages of the AD patients treated with curcumin showed increased absorption and breakdown of Aβ plaques in comparison to control patients indicating that curcumin could facilitate the clearance of amyloid proteins via immune system. Curcumin also exhibited neuroprotective property in experimental AD by virtue of its potent anti-inflammatory activity. Curcumin inhibits Aβ-induced Egr-1 protein expression as well as Egr-1 DNA-binding activity in THP-1 monocytic cells. Egr-1 has been implicated in Aβ-induced cytochemokine gene expression in monocytes. Curcumin reduces the inflammation by inhibiting Egr-1 DNA-binding activity [144, 145]. It has been found to downregulate the gene expression for pro-inflammatory mediators such as NO, NFκB, IL-1β and TNF-α, affects mitochondrial dynamics and also bring some epigenetic changes [146]. Curcumin also inhibited the expression of COX2, and iNOS expression, NFκB translocation along with MAPK in the microglial cells [147]. Curcumin also stimulates PPARγ which binds to peroxisome proliferator response element (PPRE) and decreases the expression of NFκB that further suppress the Aβ induced microglial cells from producing pro-inflammatory cytokines, thereby exhibiting neuroprotective effects in AD [148].

2.6 *Bacopa monnieri* Linn

Bacopa monnieri (syn. *Herpestis monniera,* Fam: Scrophulariaceae), often known as Brahmi or Jalanimba, has been categorized and utilized in Ayurveda as nervine tonic [149]. Since ages, it has long been used to treat epilepsy, sleeplessness, anxiety, and mental deterioration [150]. Bacosides A and B, triterpenoid saponins of the dammarane class, are the primary components of the plant. Bacoside A is a combination of four saponins (Fig. 7), including bacoside A3, bacopaside II, jujubogenin, and bacopasaponin C [151]. In addition to these major components, the plant has flavonoids like apigenin and luteolin, alkaloids like monnierasides I-III, herpestine, brahmine, hydrocotyline, and glycosides like asiaticoside [152]. Several processes, includ-

Fig. 7 Components of Bacoside A from *B. monnieri*

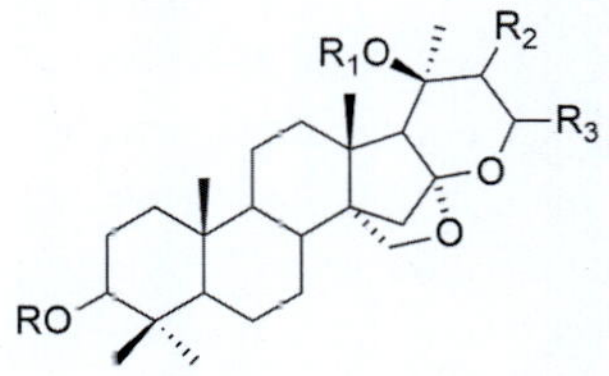

	R	R1	R2	R3
Bacoside A3	[β-D-glucopyranosyl-(1→3)]-O-{α-L-arabinofuranosyl-(1→2)}-O-(β-D-glucopyranosyl]	H	-CH=C-CH3 CH3	H
Bacopaside II	[α-L-arabinofuranosyl-(1→2)-{β-D-glucopyranosyl-(1→3)}-β-D-glucopyranosyl]	H	H	-CH=C-CH3 CH3
Jujubogenin	H	H	-CH=C-CH3 CH3	H
Bacopasaponin C	[β-D-glucopyranosyl-(1→3)]-{α-L-arabinofuranosyl-(1→2)}-α-L-arabinopyranosyl]	H	H	-CH=C-CH3 CH3

ing chelation of metal ions, scavenging of free radicals, and improved anti-oxidative system, may account for the protection of neurons and memory boosting benefits of BM [153, 154].

CDR1–08, an ethanolic extract of BM prepared by Central Drug Research Institute, Lucknow, India, has shown to significantly increase the cognitive activity in both healthy participants and older individuals with cognitive impairment [155, 156]. It is sold under the brand names KeenMind® (marketed by FlordisTM, Australia) and SynapsaTM (Soho FlordisTM, International, Australia). In scopolamine-induced amnesic mice, CDR1–08 dramatically increases the expression of the GluN2B (subunit of the NMDA receptor), reduces acetylcholinesterase activity, and improves spatial memory [157]. It boosted memory and cognition in a rat pups by improving synaptic plasticity via improved BDNF levels [158]. Bacognize®, another extract of BM, showed a substantial enhancement in memory skills and a rise in blood calcium levels (within the normal range) after 6 weeks in a clinical study [159]. Memory dysfunction and neurotoxicity caused by intracerebroventricular-streptozotocin and okadaic acid injection in male Wistar rats were alleviated by the *Bacopa monniera* standard extract (BME). In the hippocampus, it significantly decreased lipid peroxidation (LPO), boosted the levels of enzymatic antioxidants like glutathione, SOD, and catalase. It also revived the expressions of nuclear factor erythroid 2-related factor 2 (Nrf2), heme oxygenase-1 (HO1), and glutamate-cysteine ligase catalytic subunit enzyme (GCLC) [160, 161]. The BME normalizes the ATPase system and helps maintain ion gradient, membrane stability, and cellular viability [162]. It also inhibits the amyloid fibrils development in the brains of PSAPP mice [163]. Bacopa also protected rodents against diazepam and scopolamine induced amnesia by downregulating the increased expression of protein kinase C and A, MAP kinase, iNOS, and pCREB molecules [164, 165].

Bacosides A and B decreased the levels of lipofuscin (an age-related biomarker) and amyloid beta aggregation, increased acetylcholine synthesis, modulated monoaminergic neurotransmitter metabolism, and blocked LPO in the brain cortex of aged rats, and showed neuroprotective effects against Senile Dementia of Alzheimer's Type (SDAT) [166–168]. Moreover, Bacoside A improves kinase activity, neural development, and synaptic activity for a more efficient nerve conduction [169]. Colchicine, phenytoin, diazepam, scopolamine, sodium nitrite, and BN52021 (a platelet-activating factor antagonist) all caused cognitive and memory impairment, but bacoside A treatment improved these symptoms by increasing acetylcholine levels in the brain [152]. It also reduced Aβ42 cytotoxicity by preventing fibril formation and interfering with membrane interactions of the peptide. It binds to the oligomer faces through amphiphilic contacts, preventing the oligomers from causing membrane disruption or the amyloidogenic amyloid polypeptide from self-assembling [170]. In SH-SY5Y neuroblastoma cells, oxidized low-density lipoprotein (LDL) toxicity may be prevented by using bacosides such as bacoside A, bacopaside I & II, and bacoside A3. These compounds inhibit lipid peroxidation, which in turn protects cells against acetylcholinesterase activity and oxidized LDL-induced damage [171]. In addition, multiple randomized, placebo-controlled investigations have shown that BM significantly enhances learning, memory development, and rational memory [152, 172]. In addition to the protection against the neurotoxicity caused by aluminum chloride, it has also exhibited beneficial effects on learning and memory in rats when given along with other herbal medications, plant extracts, or synthetic treatments [173–175]. The anticholinesterase and memory-enhancing actions of a combination of *B. monniera* and *GBE* were effective against scopolamine-induced impairments in adult Swiss mice [176]. Brahmi's benefits extend beyond memory enhancement; it is effective in treating epilepsy [177], Parkinson's disease, and Huntington's disease [178].

The molecular processes of active plant ingredients used to treat different neurodegenerative illnesses have been better understood with modern analytical technologies like cheminformatics analysis and system pharmacology. The interactions of the compounds with the human target are assessed, and the active and new ingredients with high pharmacological activity are filtered out using cheminformatics. After identifying promising human targets, researchers may use those leads to search for databases that include information on the illnesses, pathways, and mechanisms of action of compounds that they are looking to treat [179]. Using cheminformatics and the PubChem database, 52 active molecules identified from BM were found interacting with 780 direct human targets. All the compounds were able to act on more than 15 active targets, revealing several interactions between compounds, targets, and disease. This strongly supports the multi-target effect of herbals and indicates that herbal drugs can play a significant role in controlling various disorders [179]. In another docking study, bacoside II and XII exhibited a strong interaction with hyperphosphorylated tau proteins, but had a poor score in terms of flexibility, polarity, and size thereby indicating their bioavailability issues [180].

2.7 *Withania somnifera* Dunal (Ashwagandha, Indian Ginseng)

It is one of the most popular *Rasayana* drugs in Ayurvedic medicine, belonging to family Solanaceae, that is used to promote longevity and vigour [181]. The plant has antioxidant, anti-inflammatory, immunomodulating, stress-relieving, memory-improving, and anticonvulsant effects. Ashwagandha is known for its steroidal lactone compounds collectively known as withanolides, and alkaloids (withanine, somniferine, somnine, and somniferinine). Withaferin A, withanolide A, withanone, sitoindoside, and withanamides (A–I) are some of the most significant withanolides (Fig. 8) identified from the plant [182, 183]. WS has been extensively studied plant for its neuroprotective benefits in various neurodegenerative disorders such as AD, Parkinson's disease (PD) etc. [184].

The plant stimulates many neurotransmitter receptor systems; boosts cortisol and muscular strength in stressed animals while reducing lethargy and depression [185, 186]. The acetylcholine level and choline acetyltransferase activity in the rat brain were improved by administering aqueous methanol extract of WS roots, demonstrating beneficial effects on learning and memory [185, 187, 188]. It has also exhibited neuroprotective action in AD by boosting the Aβ breakdown and improving peripheral clearance of the protein [189]. In vitro research showed the neuroprotective action of WS by restoring neuronal cell development, spine density, dendrite diameter, and spine area to normal after exposure to hazardous chemicals [190]. WS root and leaf extract exhibited a soothing effect in clinical testing on chronically stressed individuals by reducing stress hormones dose-dependently. It improves concentration, memory [191], executive function, attention, and information processing speed in a randomized, double-blind, placebo-controlled trial on 50 persons with moderate cognitive impairment [192]. Moreover, the aqueous extract of the plant also protects against neuroinflammation, motor function impairment, synaptic plasticity, and cognitive decline caused by the lipopolysaccharide [193].

Withanolides, namely, sitoindosides VII–X and withaferin A, increased endogenous SOD, catalase, and ascorbic acid levels, as well as decrease lipid peroxidation [194, 195]. These chemicals have also been shown to prevent the neurotoxicity caused by ibotenic acid [196]. The neuroprotective effects of withanolide A and withanosides IV and VI have been reported in the brain's cortex against Aβ peptide-induced neurotoxicity [197]. The root extract of WS and withanolide A prevents memory impairment induced by hypobaric hypoxia [198, 199]. Also, human neuroblastoma SH-SY5Y cell lines showed substantial neurite outgrowth activity in response to withanoside IV and VI [200]. Sominone (an active metabolite of Withanoside IV) induces regeneration and synaptic reconstruction of the cultured cortical and hippocampus neurons damaged by Aβ (25–35) [197]. It also improves spatial memory in rats and stimulates the RET-mediated neurite outgrowth [201]. Sominone also improved the Aβ induced memory impairment in experimental animals by significantly improving the synaptic, dendritic and axonal connection [202].

Withanone prevented DNA damage and oxidative stress in C6 cells caused by scopolamine [203]. Withanamides A and C protected PC-12 cells by binding to the active amyloid motif and stopped the amyloid fibrils development [204]. Furthermore, Withanolide S, isolated from WS leaf extract, exhibited antioxidant, nitric oxide scavenging, anti-inflammatory, and β-secretase and AChE inhibitory activity [205]. Molecular docking studies indicate that many other secondary metabolites in WS, including anaferine, anahygrine, cuscohygrine, and isopelletierine, behave as agonists to the nicotinic acetylcholine receptors (nAChRs) and might be promising lead compounds for the development of novel medicines for AD [206]. WS and its active constituents alter many sites connected to the processing of Aβ precursor protein to promote its removal from the brain [207]. Research on commercially available WS products has also shown its neuroprotective properties. Pre-treatment with the root extract of WS (Stresscom® capsules, Dabur India Ltd.) significantly reduced degeneration of the neuronal cells in the CA (2) and CA (3) subareas of the hippocampal region. It demonstrated neuroprotective and antistress activity in female Swiss albino rats [208]. The polyherbal medicine EuMil®, which consists of standardized extracts of WS, *Ocimum sanctum*, *Asparagus racemosus*, and *Emblica officinalis*, reduced tribulin activity in the rat brains and ameliorated the stress induced alteration in the levels of nor-adrenaline, dopamine, and 5-HT [209].

2.8 Berberine

Berberine, a bitter-tasting isoquinoline alkaloid, has been used for many centuries in traditional Chinese and Ayurvedic medicine. It has been found in plants belonging to different genera and families such as Berberis and Caulophyllum of Berberidaceae; Tinospora (Meniseprmaceae), Papaver, Argemone and Chelidonium (Papavaraceae), Xanthorhiz and Coptis (Ranunculaceae), Xanthoxyllum and Phellodendron (Rutaceae) etc. [210]. Amongst various sources, species belonging to genus Berberis namely, *B. vulgaris, B. aristae, B. lycium* and *B. aquifolium* have been given special attention w.r.t. berberine. Berberine has proved its efficacy in many disorders such as cancer, cognitive impairment, diabetes, arrhythmias, and neurodegenerative disorders (AD and PD). Cytotoxic nature of berberine, which is mainly at higher doses, is the major bottleneck in the full utilization of this potent molecule for AD and other neurodegenerative disorders [211]. At high-dose levels, berberine has shown to cause mitochondrial damage, increased oxidative stress and depletion in ATP production in neurons causing neuronal damage. But at appropriate dose levels, it has shown protective and beneficial effects in AD. Like other phytocom-

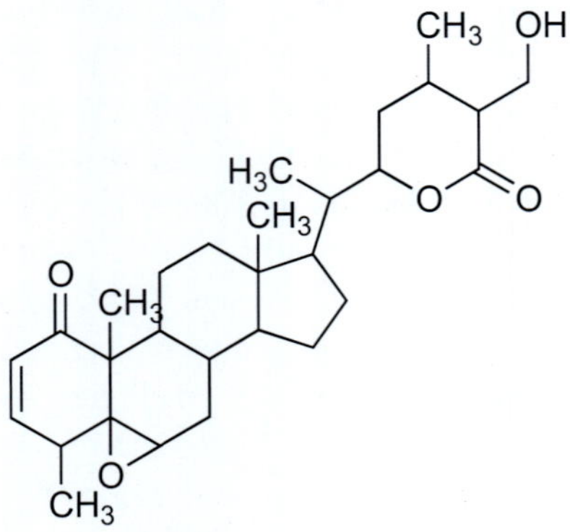

Withaferin A

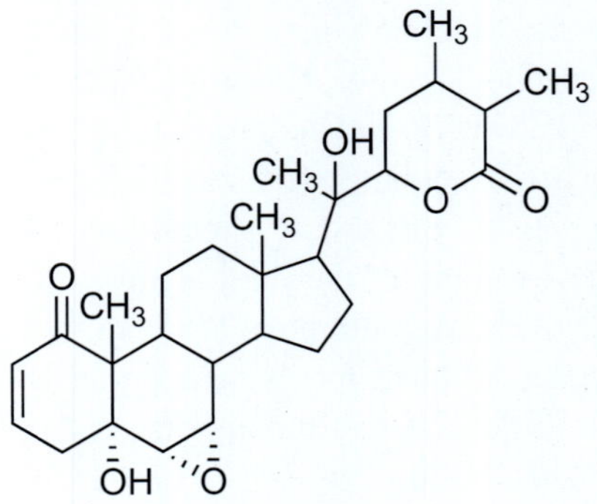

Withanolide A

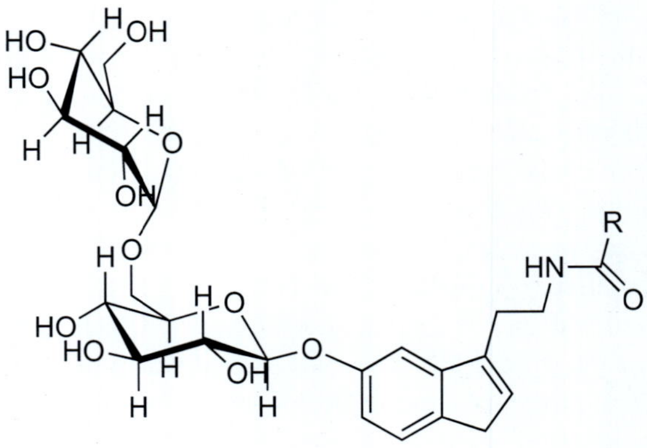

R

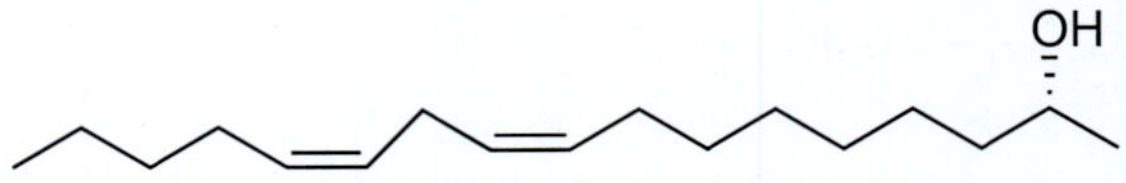

WA

R

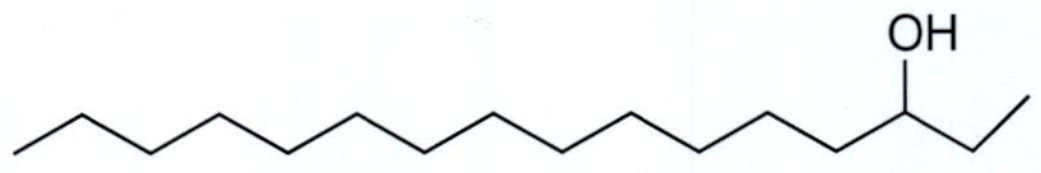

WC

R consisting of Withanamide A (WA) and Withanamide C (WC)

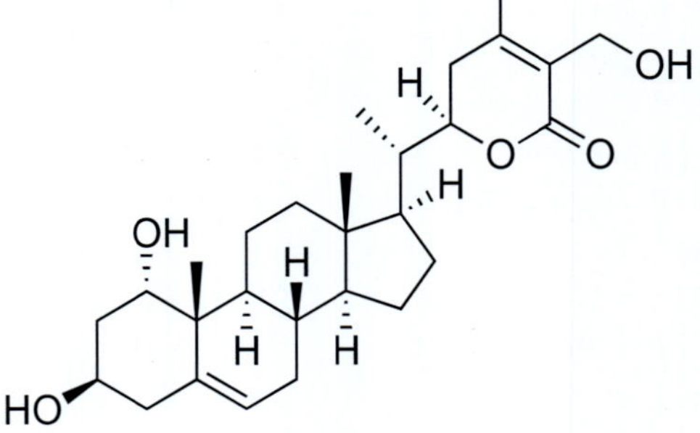

Sominone

Fig. 8 Constituents of *W. somnifera*

pounds, berberine is also considered as a multitarget agent that exhibits its action by acting on Aβ production and clearance, tau NFTs [212, 213], inflammatory markers and cholinergic system. Berberine inhibits β- and γ-secretase activity thereby inhibiting the production of pathological Aβ [214], increase activity of α-secretase leading to enhanced production of non-pathogenic Aβ [215], and also enhances the autophagic clearance of Aβ by inhibiting the mTOR/p70S6K signalling pathway [216]. A couple of other reports also showed that berberine downregulates the β-secretase expression by stimulating AMPK pathway but has no effect on mTOR [217, 218]. Berberine also decreased the Aβ peptide levels by regulating the APP processing and controlling the tau hyperphosphorylation to treat senile dementia [212]. It inhibited Aβ40/42 synthesis by blocking BACE expression through ERK1/2 pathway activation in HEK293 cells [219]. Berberine also suppressed Aβ42 production via inhibiting the PERK-eIF2α-BACE1 signalling pathway [220]. Various reports have also shown the protective effect of berberine against tau and its hyperphosphorylation which is considered as another hallmark of AD after Aβ [220–222]. Berberine not only improve the learning and memory of 3 × Tg AD mice, it also inhibited the tau hyperphosphorylation by modulating the AKT/GSK-3β and protein phosphatase 2A activity [222]. Furthermore, it also reduced the tau levels by enhancing its autophagic clearance by stimulating PI3K/beclin-1 pathway. Furthermore, berberine also exhibited protection against axonopathy and tau hyperphosphorylation by stimulating the PI3K/AKT/GKS3β pathway that eventually caused dephosphorylation of tau proteins and improved axonal transportation in T2D rats [220, 221].

Apart from its action Aβ and tau, berberine also has an impact on the inflammatory markers. It improved the memory of APP/PS1 mice by its antioxidant potential, reducing tau phosphorylation and preventing the activation of NFκB signalling pathway [223]. It also exhibited protective effects against neuroinflammation by attenuating lipopolysaccharide induced increase in NF-κB, toll-like receptor 4 (TLR4), TNFα, IL-6 levels and NO production in rat brain [224, 225]. In addition to it, berberine also reestablished the hippocampal 3-nitrotyrosine, cyclooxygenase 2, glial fibrillary acidic protein, SIRT1, and p38MAPK activity in rat brain microglia and C57BL/6 J mice [226]. Berberine also suppressed the AChE activity and influences the cholinergic system in a dose-dependent manner [227]. It also ameliorates excessive glutamate release and NMDA activity in the brain's hippocampal region. It significantly suppresses the increases in intracellular Ca2+ levels and inhibits the release of nitric oxide to prevent excitotoxicity in neurons [227]. It also improved the cell survival by decreasing pro-apoptotic caspase-3 activity and slowing the apoptosis rate through the miR-188/NOS1 pathway, suggesting that it may be a valuable medication for treating AD [228].

3 Conclusion

Alzheimer's disease eludes from its cure since last so many decades. Today, with better understanding of its pathophysiology, we are now stressing upon the search of disease-modifying agents Due to its multi-pathway pathophysiology, multitarget agents are believed to be effective in modifying the course of the disease. Colossal literature on the medicinal plants and their biological potential has strengthened my belief that Mother Nature, in her lap, has cure/treatment for every disorder. We only need an eye to identify and exploit that cure. Resurgence of interest in the herbal drugs has led to extensive work that establishes phytocompounds as multitarget agents and can be helpful in treatment of AD. Else, they can act as lead molecule for developing potent therapeutic agents for the same. Though, bioavailability and toxicity issues of some of the phytocompounds like berberine, curcumin, and bacosides, etc. act as major hurdles in their developmental paths, modern scientific approaches like novel drug delivery systems, development of pro-drugs or derivatives, have helped in overcoming such issues. With the meticulous amalgamation of the potential of phytocompounds and these modern scientific approaches, one can always hope for a new effective and a safe therapeutic agent for AD. But still ample work is warranted in this direction, and we should always remember that it is not only the drugs, but also nutritional diet, healthy lifestyle and physical exercise play a vital role in keeping these neurodegenerative disorders at bay.

References

1. Li X, Feng X, Sun X, Hou N, Han F, Liu Y. Global, regional, and national burden of Alzheimer's disease and other dementias, 1990–2019. Front Aging Neurosci. 2022;18:14.
2. Association As. 2022 Alzheimer's disease facts and figures. Alzheimers Dement. 2022;18:700–89.
3. Tzioras M, McGeachan RI, Durrant CS, Spires-Jones TL. Synaptic degeneration in Alzheimer disease. Nat Rev Neurol. 2023;9:19–38.
4. Cummings J, Lee G, Ritter A, Sabbagh M, Zhong K. Alzheimer's disease drug development pipeline: 2019. Alzheimers Dement (N Y). 2019;5:272–93.
5. Cummings J, Lee G, Zhong K, Fonseca J, Taghva K. Alzheimer's disease drug development pipeline: 2021. Alzheimers Dement (N Y). 2021;7(1):e12179.
6. Cummings J, Lee G, Nahed P, Kambar MEZN, Zhong K, Fonseca J, et al. Alzheimer's disease drug development pipeline: 2023. Alzheimers Dement (N Y). 2022;8(1):e12295.
7. Yuan J, Zhou X, Wang S. Advances in studies on chemical constituents of Huperzia serrata and their pharmacological effects. Chin Tradit Herb Drug. 2012;43:399–407.

8. Zhang H-y. New insights into huperzine A for the treatment of Alzheimer's disease. Acta Pharmacol Sin. 2012;33(9):1170–5.
9. Liu J, Yu C, Zhou Y, Han Y, Wu F, Qi B, et al. Study on the chemistry of huperzine-A and huperzine-B. Acta Chim Sin. 1986;44(10):1035–40.
10. Ma X, Tan C, Zhu D, Gang DR, Xiao P. Huperzine A from Huperzia species—an ethnopharmacolgical review. J Ethnopharmacol. 2007;113(1):15–34.
11. Orhan EI, Orhan G, Gurkas E. An overview on natural cholinesterase inhibitors-a multi-targeted drug class-and their mass production. Mini Rev Med Chem. 2011;11(10):836–42.
12. Ferreira A, Rodrigues M, Fortuna A, Falcão A, Alves G. Huperzine A from Huperzia serrata: a review of its sources, chemistry, pharmacology and toxicology. Phytochem Rev. 2016;15(1):51–85.
13. Yan Y-P, Chen J-Y, Lu J-H. Disease-modifying activity of Huperzine A on Alzheimer's disease: evidence from preclinical studies on rodent models. Int J Mol Sci. 2022;23(23):15238.
14. Tun MKM, Herzon SB. The pharmacology and therapeutic potential of (−)-huperzine A. J Exp Pharmacol. 2012;4:113.
15. Ma T, Gong K, Yan Y, Zhang L, Tang P, Zhang X, et al. Huperzine a promotes hippocampal neurogenesis in vitro and in vivo. Brain Res. 2013;1506:35–43.
16. Alvarez A, Alarcón R, Opazo C, Campos EO, Muñoz FJ, Calderón FH, et al. Stable complexes involving acetylcholinesterase and amyloid-β peptide change the biochemical properties of the enzyme and increase the neurotoxicity of Alzheimer's fibrils. J Neurosci. 1998;18(9):3213–23.
17. Bartolini M, Bertucci C, Cavrini V, Andrisano V. β-Amyloid aggregation induced by human acetylcholinesterase: inhibition studies. Biochem Pharmacol. 2003;65(3):407–16.
18. Yuan Q, Lin Z, Wu W, Albert W, Zee B. Huperzine A in treatment of amyloid-β-associated neuropathology in a mouse model of alzheimer disease: abridged secondary. Hong Kong Med J. 2020;26(6 Supplement 8):34.
19. Tapia-Rojas C, Burgos PV, Inestrosa NC. Inhibition of Wnt signaling induces amyloidogenic processing of amyloid precursor protein and the production and aggregation of amyloid-β (Aβ) 42 peptides. J Neurochem. 2016;139(6):1175–91.
20. Wang C-Y, Zheng W, Wang T, Xie J-W, Wang S-L, Zhao B-L, et al. Huperzine A activates Wnt/β-catenin signaling and enhances the nonamyloidogenic pathway in an Alzheimer transgenic mouse model. Neuropsychopharmacology. 2011;36(5):1073–89.
21. Wang Y, Tang XC, Zhang HY. Huperzine A alleviates synaptic deficits and modulates amyloidogenic and nonamyloidogenic pathways in APPswe/PS1dE9 transgenic mice. J Neurosci Res. 2012;90(2):508–17.
22. Mao X-Y, Zhou H-H, Li X, Liu Z-Q. Huperzine a alleviates oxidative glutamate toxicity in hippocampal HT22 cells via activating BDNF/TrkB-dependent PI3K/Akt/mTOR signaling pathway. Cell Mol Neurobiol. 2016;36(6):915–25.
23. Lu H, Jiang M, Lu L, Zheng G, Dong Q. Ultrastructural mitochondria changes in perihematomal brain and neuroprotective effects of Huperzine A after acute intracerebral hemorrhage. Neuropsychiatr Dis Treat. 2015;11:2649.
24. Friedli MJ, Inestrosa NC. Huperzine a and its neuroprotective molecular signaling in Alzheimer's disease. Molecules. 2021;26(21):6531.
25. Ghassab-Abdollahi N, Mobasseri K, Dehghani Ahmadabad A, Nadrian H, Mirghafourvand M. The effects of Huperzine A on dementia and mild cognitive impairment: an overview of systematic reviews. Phytother Res. 2021;35(9):4971–87.
26. Zafonte RD, Fregni F, Bergin MJ, Goldstein R, Boudreau N, Monge I, et al. Huperzine A for the treatment of cognitive, mood, and functional deficits after moderate and severe TBI (HUP-TBI): results of a phase II randomized controlled pilot study: implications for understanding the placebo effect. Brain Inj. 2020;34(1):34–41.
27. Xing S-h, Zhu C-x, Zhang R, An L. Huperzine a in the treatment of Alzheimer's disease and vascular dementia: a meta-analysis. Evid Based Complement Altern Med. 2014;2014:363985.
28. Yang G, Wang Y, Tian J, Liu J-P. Huperzine A for Alzheimer's disease: a systematic review and meta-analysis of randomized clinical trials. PLoS One. 2013;8(9):e74916.
29. Xu Z-Q, Liang X-M, Zhang Y-F, Zhu C-X, Jiang X-J. Treatment with Huperzine a improves cognition in vascular dementia patients. Cell Biochem Biophys. 2012;62(1):55–8.
30. Gul A, Bakht J, Mehmood F. Huperzine-A response to cognitive impairment and task switching deficits in patients with Alzheimer's disease. J Chin Med Assoc. 2019;82(1):40–3.
31. Nakanishi K. Terpene trilactones from Gingko biloba: from ancient times to the 21st century. Bioorg Med Chem. 2005;13(17):4987–5000.
32. Jacobs BP, Browner WS. Ginkgo biloba: a living fossil. Am J Med. 2000;108(4):341–2.
33. Niu T, Yuan B, Liu G. Ginkgolides and bilobalide for treatment of Alzheimer's disease and COVID-19: potential mechanisms of action. Eur Rev Med Pharmacol Sci. 2022;26(24):9502–10.
34. Mohanta TK, Tamboli Y, Zubaidha PK. Phytochemical and medicinal importance of *Ginkgo biloba* L. Nat Prod Res. 2014;28(10):746–52.
35. Luo Y. Ginkgo biloba neuroprotection: therapeutic implications in Alzheimer's disease. J Alzheimers Dis. 2001;3:401–7.
36. Oken BS, Storzbach DM, Kaye JA. The efficacy of Ginkgo biloba on cognitive function in Alzheimer disease. Arch Neurol. 1998;55(11):1409–15.
37. Kandiah N, Ong PA, Yuda T, Ng LL, Mamun K, Merchant RA, et al. Treatment of dementia and mild cognitive impairment with or without cerebrovascular disease: expert consensus on the use of Ginkgo biloba extract, EGb 761®. CNS Neurosci Ther. 2019;25(2):288–98.
38. Wei T, Ni Y, Hou J, Chen C, Zhao B, Xin W. Hydrogen peroxide-induced oxidative damage and apoptosis in cerebellar granule cells: protection by Ginkgo biloba extract. Pharmacol Res. 2000;41(4):427–33.
39. Bridi R, Crossetti F, Steffen VM, Henriques AT. The antioxidant activity of standardized extract of Ginkgo biloba (EGb 761) in rats. Phytother Res. 2001;15(5):449–51.
40. Wu Y, Wu Z, Butko P, Christen Y, Lambert MP, Klein WL, et al. Amyloid-β-induced pathological behaviors are suppressed by Ginkgo biloba extract EGb 761 and ginkgolides in transgenic Caenorhabditis elegans. J Neurosci. 2006;26(50):13102–13.
41. Singh SK, Srivastav S, Castellani RJ, Plascencia-Villa G, Perry G. Neuroprotective and antioxidant effect of Ginkgo biloba extract against AD and other neurological disorders. Neurotherapeutics. 2019;16(3):666–74.
42. Gohil K, Packer L. Global gene expression analysis identifies cell and tissue specific actions of Ginkgo biloba extract, EGb 761. Cell Mol Biol (Noisy-le-grand). 2002;48(6):625–31.
43. Shi C, Liu J, Wu F, Yew DT. Ginkgo biloba extract in Alzheimer's disease: from action mechanisms to medical practice. Int J Mol Sci. 2010;11(1):107–23.
44. Colciaghi F, Borroni B, Zimmermann M, Bellone C, Longhi A, Padovani A, et al. Amyloid precursor protein metabolism is regulated toward alpha-secretase pathway by Ginkgo biloba extracts. Neurobiol Dis. 2004;16(2):454–60.
45. Bodovitz S, Klein WL. Cholesterol modulates α-secretase cleavage of amyloid precursor protein. J Biol Chem. 1996;271(8):4436–40.
46. Simons M, Keller P, De Strooper B, Beyreuther K, Dotti CG, Simons K. Cholesterol depletion inhibits the generation of β-amyloid in hippocampal neurons. Proc Natl Acad Sci. 1998;95(11):6460–4.

47. Yao Z-X, Han Z, Drieu K, Papadopoulos V. Ginkgo biloba extract (Egb 761) inhibits β-amyloid production by lowering free cholesterol levels. J Nutr Biochem. 2004;15(12):749–56.
48. Verma S, Sharma S, Ranawat P, Nehru B. Modulatory effects of Ginkgo biloba against amyloid aggregation through induction of heat shock proteins in aluminium induced neurotoxicity. Neurochem Res. 2020;45(2):465–90.
49. Texidó L, Martín-Satué M, Alberdi E, Solsona C, Matute C. Amyloid β peptide oligomers directly activate NMDA receptors. Cell Calcium. 2011;49(3):184–90.
50. Zhang Y, Guo O, Huo Y, Wang G, Man H-Y. Amyloid-β induces AMPA receptor ubiquitination and degradation in primary neurons and human brains of Alzheimer's disease. J Alzheimers Dis. 2018;62(4):1789–801.
51. Kuo L-C, Song Y-Q, Yao C-A, Cheng IH, Chien C-T, Lee G-C, et al. Ginkgolide A prevents the amyloid-β-induced depolarization of cortical neurons. J Agric Food Chem. 2018;67(1):81–9.
52. Wang X, Jiang CM, Wan HY, Wu JL, Quan WQ, Wu KY, et al. Neuroprotection against permanent focal cerebral ischemia by ginkgolides A and B is associated with obstruction of the mitochondrial apoptotic pathway via inhibition of c-Jun N-terminal kinase in rats. J Neurosci Res. 2014;92(2):232–42.
53. Chen Y, Wang C, Hu M, Pan J, Chen J, Duan P, et al. Effects of ginkgolide A on okadaic acid-induced tau hyperphosphorylation and the PI3K-Akt signaling pathway in N2a cells. Planta Med. 2012;78(12):1337–41.
54. Heras-Sandoval D, Perez-Rojas J, Hernandez-Damian J, Pedraza-Chaverri J. The role of PI3K/AKT/mTOR pathway in the modulation of autophagy and the clearance of protein aggregates in neurodegeneration. Cell Signal. 2014;26(10):2694–701.
55. Qin Y, Zhang Y, Tomic I, Hao W, Menger MD, Liu C, et al. Ginkgo biloba extract EGb 761 and its specific components elicit protective protein clearance through the autophagy-lysosomal pathway in tau-transgenic mice and cultured neurons. J Alzheimers Dis. 2018;65(1):243–63.
56. Zhang Y, Liu J, Yang B, Zheng Y, Yao M, Sun M, et al. Ginkgo biloba extract inhibits astrocytic lipocalin-2 expression and alleviates neuroinflammatory injury via the JAK2/STAT3 pathway after ischemic brain stroke. Front Pharmacol. 2018;9:518.
57. Adebayo OG, Ben-Azu B, Ajayi AM, Wopara I, Aduema W, Kolawole TA, et al. Gingko biloba abrogate lead-induced neurodegeneration in mice hippocampus: involvement of NF-κB expression, myeloperoxidase activity and pro-inflammatory mediators. Biol Trace Elem Res. 2022;200(4):1736–49.
58. Niu T-T, Yin H, Xu B-L, Yang T-T, Li H-Q, Sun Y, et al. Protective effects of ginkgolide on a cellular model of Alzheimer's disease via suppression of the NF-κB signaling pathway. Appl Biochem Biotechnol. 2022;194:1–17.
59. Feng Y-S, Tan Z-X, Wu L-Y, Dong F, Zhang F. The involvement of NLRP3 inflammasome in the treatment of Alzheimer's disease. Ageing Res Rev. 2020;64:101192.
60. Shao L, Dong C, Geng D, He Q, Shi Y. Ginkgolide B inactivates the NLRP3 inflammasome by promoting autophagic degradation to improve learning and memory impairment in Alzheimer's disease. Metab Brain Dis. 2022;37(2):329–41.
61. Gauthier S, Schlaefke S. Efficacy and tolerability of Ginkgo biloba extract EGb 761® in dementia: a systematic review and meta-analysis of randomized placebo-controlled trials. Clin Interv Aging. 2014;9:2065–77.
62. Hashiguchi M, Ohta Y, Shimizu M, Maruyama J, Mochizuki M. Meta-analysis of the efficacy and safety of Ginkgo biloba extract for the treatment of dementia. J Pharm Health Care Sci. 2015;1(1):1–12.
63. Yang G, Wang Y, Sun J, Zhang K, Liu J. Ginkgo biloba for mild cognitive impairment and Alzheimer's disease: a systematic review and meta-analysis of randomized controlled trials. Curr Top Med Chem. 2016;16(5):520–8.
64. Liu H, Ye M, Guo H. An updated review of randomized clinical trials testing the improvement of cognitive function of Ginkgo biloba extract in healthy people and Alzheimer's patients. Front Pharmacol. 2020 10:1688.
65. Xie L, Zhu Q, Lu J. Can we use Ginkgo biloba extract to treat Alzheimer's disease? Lessons from preclinical and clinical studies. Cell. 2022;11(3):479.
66. Shi Z-Y, Zeng J-Z, Wong AST. Chemical structures and pharmacological profiles of ginseng saponins. Molecules. 2019;24(13):2443.
67. Shin B-K, Kwon SW, Park JH. Chemical diversity of ginseng saponins from Panax ginseng. J Ginseng Res. 2015;39(4):287–98.
68. Dela Pena I, Yoon SY, Kim HJ, Park S, Hong EY, Ryu JH, et al. Effects of ginseol k-g3, an Rg3-enriched fraction, on scopolamine-induced memory impairment and learning deficit in mice. J Ginseng Res. 2014;38(1):1–7.
69. Potenza MA, Montagnani M, Santacroce L, Charitos IA, Bottalico L. Ancient herbal therapy: a brief history of Panax ginseng. J Ginseng Res. 2022;47(3):359.
70. Lee SM, Bae B-S, Park H-W, Ahn N-G, Cho B-G, Cho Y-L, et al. Characterization of Korean Red ginseng (Panax ginseng Meyer): history, preparation method, and chemical composition. J Ginseng Res. 2015;39(4):384–91.
71. Ru W, Wang D, Xu Y, He X, Sun Y-E, Qian L, et al. Chemical constituents and bioactivities of Panax ginseng (C. A Mey). Drug Discov Ther. 2015;9(1):23–32.
72. Park SK, Hyun SH, In G, Park C-K, Kwak Y-S, Jang Y-J, et al. The antioxidant activities of Korean red ginseng (Panax ginseng) and ginsenosides: a systemic review through in vivo and clinical trials. J Ginseng Res. 2021;45(1):41–7.
73. Choi RJ, Roy A, Jung HJ, Ali MY, Min B-S, Park CH, et al. BACE1 molecular docking and anti-Alzheimer's disease activities of ginsenosides. J Ethnopharmacol. 2016;190:219–30.
74. Quan Q, Li X, Feng J, Hou J, Li M, Zhang B. Ginsenoside Rg1 reduces β-amyloid levels by inhibiting CDK5-induced PPARγ phosphorylation in a neuron model of Alzheimer's disease. Mol Med Rep. 2020;22(4):3277–88.
75. Zhang Y, Zhang Z, Wang H, Cai N, Zhou S, Zhao Y, et al. Neuroprotective effect of ginsenoside Rg1 prevents cognitive impairment induced by isoflurane anesthesia in aged rats via antioxidant, anti-inflammatory and anti-apoptotic effects mediated by the PI3K/AKT/GSK-3β pathway. Mol Med Rep. 2016;14(3):2778–84.
76. Cui J, Wang J, Zheng M, Gou D, Liu C, Zhou Y. Ginsenoside Rg2 protects PC12 cells against β-amyloid25-35-induced apoptosis via the phosphoinositide 3-kinase/Akt pathway. Chem Biol Interact. 2017;275:152–61.
77. Yan X, Hu G, Yan W, Chen T, Yang F, Zhang X, et al. Ginsenoside Rd promotes non-amyloidogenic pathway of amyloid precursor protein processing by regulating phosphorylation of estrogen receptor alpha. Life Sci. 2017;168:16–23.
78. Shi C, Zheng D-d, Fang L, Wu F, Kwong WH, Xu J. Ginsenoside Rg1 promotes nonamyloidgenic cleavage of APP via estrogen receptor signaling to MAPK/ERK and PI3K/Akt. Biochim Biophys Acta. 2012;1820(4):453–60.
79. Shi C, Na N, Zhu X, Xu J. Estrogenic effect of ginsenoside Rg1 on APP processing in post-menopausal platelets. Platelets. 2013;24(1):51–62.
80. Ekinci FJ, Malik KU, Shea TB. Activation of the L voltage-sensitive calcium channel by mitogen-activated protein (MAP) kinase following exposure of neuronal cells to β-amyloid: MAP kinase mediates β-amyloid-induced neurodegeneration. J Biol Chem. 1999;274(42):30322–7.
81. Chen X, Huang T, Zhang J, Song J, Chen L, Zhu Y. Involvement of calpain and p25 of CDK5 pathway in ginsenoside Rb1's attenu-

ation of β-amyloid peptide25–35-induced tau hyperphosphorylation in cortical neurons. Brain Res. 2008;1200:99–106.
82. Quan Q-k, Li X, Yuan H-f, Wang Y, Liu W-l. Ginsenoside Rg1 inhibits high-voltage-activated calcium channel currents in hippocampal neurons of beta-amyloid peptide-exposed rat brain slices. Chin J Integr Med. 2016:1–6.
83. Fang F, Chen X, Huang T, Lue L-F, Luddy JS, Yan SS. Multi-faced neuroprotective effects of ginsenoside Rg1 in an Alzheimer mouse model. Biochim Biophys Acta Mol Basis Dis. 2012;1822(2):286–92.
84. Wang L, Lu J, Zeng Y, Guo Y, Wu C, Zhao H, et al. Improving Alzheimer's disease by altering gut microbiota in tree shrews with ginsenoside Rg1. FEMS Microbiol Lett. 2020;367(4):fnaa011.
85. Wang N, Yang J, Chen R, Liu Y, Liu S, Pan Y, et al. Ginsenoside Rg1 ameliorates Alzheimer's disease pathology via restoring mitophagy. J Ginseng Res. 2022;47(3):448–57.
86. Li W, Chu Y, Zhang L, Yin L, Li L. Ginsenoside Rg1 attenuates tau phosphorylation in SK-N-SH induced by Aβ-stimulated THP-1 supernatant and the involvement of p38 pathway activation. Life Sci. 2012;91(15–16):809–15.
87. Lu D, Xu A, Mai H, Zhao J, Zhang C, Qi R, et al. The synergistic effects of heat shock protein 70 and ginsenoside Rg1 against tert-butyl hydroperoxide damage model in vitro. Oxid Med Cell Longev. 2015;2015:437127.
88. Chen XM. Effect of dosage form and administration route to distribution of Ginsenoside Rg1 in serum and brain of rats. Chin J Exp Tradit Med Formul. 2011;17:43–6.
89. Liang HY, Zhang PP, Zhang XL, Zheng YY, Huang YR, Zheng GQ, et al. Preclinical systematic review of ginsenoside Rg1 for cognitive impairment in Alzheimer's disease. Aging (Albany NY). 2021;13(5):7549–69.
90. Wu J-j, Yang Y, Wan Y, Xia J, Xu J-F, Zhang L, et al. New insights into the role and mechanisms of ginsenoside Rg1 in the management of Alzheimer's disease. Biomed Pharmacother. 2022;152:113207.
91. Li N, Liu Y, Li W, Zhou L, Li Q, Wang X, et al. A UPLC/MS-based metabolomics investigation of the protective effect of ginsenosides Rg1 and Rg2 in mice with Alzheimer's disease. J Ginseng Res. 2016;40(1):9–17.
92. Zhang Y, Yang X, Wang S, Song S. Ginsenoside Rg3 prevents cognitive impairment by improving mitochondrial dysfunction in the rat model of Alzheimer's disease. JAgric Food Chem. 2019;67(36):10048–58.
93. Ahn JW, Jang SK, Jo BR, Kim HS, Park JY, Park HY, et al. A therapeutic intervention for Alzheimer's disease using ginsenoside Rg3: its role in M2 microglial activation and non-amyloidogenesis. J Physiol Pharmacol 2021;72(2).
94. Hwang SH, Shin E-J, Shin T-J, Lee B-H, Choi S-H, Kang J, et al. Gintonin, a ginseng-derived lysophosphatidic acid receptor ligand, attenuates Alzheimer's disease-related neuropathies: involvement of non-amyloidogenic processing. J Alzheimers Dis. 2012;31(1):207–23.
95. Kim H-J, Kim D-J, Shin E-J, Lee B-H, Choi S-H, Hwang S-H, et al. Effects of gintonin-enriched fraction on hippocampal cell proliferation in wild-type mice and an APPswe/PSEN-1 double Tg mouse model of Alzheimer's disease. Neuro Chem Int. 2016;101:56–65.
96. Moon J, Choi S-H, Shim J-Y, Park H-J, Oh M-J, Kim M, et al. Gintonin administration is safe and potentially beneficial in cognitively impaired elderly. Alzheimer Dis Assoc Disord. 2018;32(1):85–7.
97. He Y, Yang J, Lv Y, Chen J, Yin F, Huang J, et al. A review of ginseng clinical trials registered in the WHO international clinical trials registry platform. Biomed Res Int. 2018;2018:1843142.
98. Fan S, Zhang Z, Su H, Xu P, Qi H, Zhao D, et al. Panax ginseng clinical trials: current status and future perspectives. Biomed Pharmacother. 2020;132:110832.
99. Li J, Huang Q, Chen J, Qi H, Liu J, Chen Z, et al. Neuroprotective potentials of Panax ginseng against Alzheimer's disease: a review of preclinical and clinical evidences. Front Pharmacol. 2021;12:688490.
100. Chen W, Li X, Chen Z, Hao W, Yao P, Li M, et al. A comprehensive quality analysis of randomized controlled clinical trials of Asian ginseng and American ginseng based on the CONSORT guideline. J Ginseng Res. 2022;46(1):71–8.
101. Cardinal BJ, Engels H-J. Ginseng does not enhance psychological well-being in healthy, young adults: results of a double-blind, placebo-controlled, randomized clinical trial. J Am Diet Assoc. 2001;101(6):655–60.
102. Kudoh C, Arita R, Honda M, Kishi T, Komatsu Y, Asou H, et al. Effect of ninjin'yoeito, a K ampo (traditional Japanese) medicine, on cognitive impairment and depression in patients with Alzheimer's disease: 2 years of observation. Psychogeriatrics. 2016;16(2):85–92.
103. Shaito A, Posadino AM, Younes N, Hasan H, Halabi S, Alhababi D, et al. Potential adverse effects of resveratrol: a literature review. Int J Mol Sci. 2020;21(6):2084.
104. Salehi B, Mishra AP, Nigam M, Sener B, Kilic M, Sharifi-Rad M, et al. Resveratrol: a double-edged sword in health benefits. Biomedicine. 2018;6(3):91.
105. Yan Y, Yang H, Xie Y, Ding Y, Kong D, Yu H. Research progress on Alzheimer's disease and resveratrol. Neurochem Res. 2020;45(5):989–1006.
106. Andrade S, Ramalho MJ, Pereira MC, Loureiro JA. Resveratrol brain delivery for neurological disorders prevention and treatment. Front Pharmacol. 2018;9:1261.
107. Chen J-Y, Zhu Q, Zhang S, OuYang D, Lu J-H. Resveratrol in experimental Alzheimer's disease models: a systematic review of preclinical studies. Pharmacol Res. 2019;150:104476.
108. Kulijewicz-Nawrot M, Syková E, Chvátal A, Verkhratsky A, Rodríguez JJ. Astrocytes and glutamate homoeostasis in Alzheimer's disease: a decrease in glutamine synthetase, but not in glutamate transporter-1, in the prefrontal cortex. ASN Neuro. 2013;5(4):AN20130017.
109. Tracy TE, Gan L. Acetylated tau in Alzheimer's disease: An instigator of synaptic dysfunction underlying memory loss: increased levels of acetylated tau blocks the postsynaptic signaling required for plasticity and promotes memory deficits associated with tauopathy. BioEssays. 2017;39(4):1600224.
110. Meftahi G, Ghotbedin Z, Eslamizade MJ, Hosseinmardi N, Janahmadi M. Suppressive effects of resveratrol treatment on the intrinsic evoked excitability of CA1 pyramidal neurons. Cell J (Yakhteh). 2015;17(3):532.
111. Hu W, Yang E, Ye J, Han W, Du ZL. Resveratrol protects neuronal cells from isoflurane-induced inflammation and oxidative stress-associated death by attenuating apoptosis via Akt/p38 MAPK signaling. Exp Ther Med. 2018;15(2):1568–73.
112. Sugino T, Maruyama M, Tanno M, Kuno A, Houkin K, Horio Y. Protein deacetylase SIRT1 in the cytoplasm promotes nerve growth factor-induced neurite outgrowth in PC12 cells. FEBS Lett. 2010;584(13):2821–6.
113. Li M-Z, Zheng L-J, Shen J, Li X-Y, Zhang Q, Bai X, et al. SIRT1 facilitates amyloid beta peptide degradation by upregulating lysosome number in primary astrocytes. Neural Regen Res. 2018;13(11):2005.
114. Li Y, Yang W, Quinones-Hinojosa A, Wang B, Xu S, Zhu W, et al. Interference with protease-activated receptor 1 alleviates neuronal cell death induced by lipopolysaccharide-stimulated microglial cells through the PI3K/Akt pathway. Sci Rep. 2016;6(1):1–12.

115. Li Y, Xu W, McBurney MW, Longo VD. SirT1 inhibition reduces IGF-I/IRS-2/Ras/ERK1/2 signaling and protects neurons. Cell Metab. 2008;8(1):38–48.
116. Ribeiro CA, Saraiva MJ, Cardoso I. Stability of the transthyretin molecule as a key factor in. PLOS One. 2012;7(9):e45368–e76.
117. Marambaud P, Zhao H, Davies P. Resveratrol promotes clearance of Alzheimer's disease amyloid-peptides. J Biol Chem. 2005;280(45):37377–82.
118. Ladiwala ARA, Lin JC, Bale SS, Marcelino-Cruz AM, Bhattacharya M, Dordick JS, et al. Resveratrol selectively remodels soluble oligomers and fibrils of amyloid Aβ into off-pathway conformers. J Biol Chem. 2010;285(31):24228–37.
119. He X-P, Deng Q, Cai L, Wang C-Z, Zang Y, Li J, et al. Fluorogenic resveratrol-confined graphene oxide for economic and rapid detection of Alzheimer's disease. ACS Appl Mater Interfaces. 2014;6(8):5379–82.
120. Karuppagounder SS, Pinto JT, Xu H, Chen H-L, Beal MF, Gibson GE. Dietary supplementation with resveratrol reduces plaque pathology in a transgenic model of Alzheimer's disease. Neurochem Int. 2009;54(2):111–8.
121. Feng Y, Wang X-p, Yang S-g, Wang Y-j, Zhang X, Du X-t, et al. Resveratrol inhibits beta-amyloid oligomeric cytotoxicity but does not prevent oligomer formation. Neurotoxicology. 2009;30(6):986–95.
122. Wight RD, Tull CA, Deel MW, Stroope BL, Eubanks AG, Chavis JA, et al. Resveratrol effects on astrocyte function: relevance to neurodegenerative diseases. Biochem Biophys Res Commun. 2012;426(1):112–5.
123. Yao Y, Li J, Niu Y, Yu JQ, Yan L, Miao ZH, et al. Resveratrol inhibits oligomeric Aβ-induced microglial activation via NADPH oxidase. Mol Med Rep. 2015;12(4):6133–9.
124. Schweiger S, Matthes F, Posey K, Kickstein E, Weber S, Hettich MM, et al. Resveratrol induces dephosphorylation of tau by interfering with the MID1-PP2A complex. Sci Rep. 2017;7(1):13753.
125. Sun X-Y, Dong Q-X, Zhu J, Sun X, Zhang L-F, Qiu M, et al. Resveratrol rescues tau-induced cognitive deficits and neuropathology in a mouse model of tauopathy. Curr Alzheimer Res. 2019;16(8):710–22.
126. Ashrafizadeh M, Zarrabi A, Najafi M, Samarghandian S, Mohammadinejad R, Ahn KS. Resveratrol targeting tau proteins, amyloid-beta aggregations, and their adverse effects: an updated review. Phytother Res. 2020;34(11):2867–88.
127. Ro J-H, Liu C-C, Lin M-C. Resveratrol mitigates cerebral ischemic injury by altering levels of trace elements, toxic metal, lipid peroxidation, and antioxidant activity. Biol Trace Elem Res. 2021;199(10):3718–27.
128. Chung JH, Lee J-S, Lee HG. Resveratrol-loaded chitosan–γ-poly (glutamic acid) nanoparticles: optimization, solubility, UV stability, and cellular antioxidant activity. Colloids Surf B Biointerfaces. 2020;186:110702–8.
129. Pineda-Ramírez N, Alquisiras-Burgos I, Ortiz-Plata A, Ruiz-Tachiquín M-E, Espinoza-Rojo M, Aguilera P. Resveratrol activates neuronal autophagy through AMPK in the ischemic brain. Mol Neurobiol. 2020;57(2):1055–69.
130. Tong J, Gao J, Liu Q, He C, Zhao X, Qi Y, et al. Resveratrol derivative excited postsynaptic potentiation specifically via PKCβ-NMDA receptor mediation. Pharmacol Res. 2020;152:104618.
131. Auti A, Alessio N, Ballini A, Dioguardi M, Cantore S, Scacco S, et al. Protective effect of resveratrol against hypoxia-induced neural oxidative stress. J Pers Med. 2022;12(8):1202.
132. Pawlik A, Wała M, Hać A, Felczykowska A, Herman-Antosiewicz A. Sulforaphene, an isothiocyanate present in radish plants, inhibits proliferation of human breast cancer cells. Phytomedicine. 2017;29:1–10.
133. Shishodia S, Sethi G, Aggarwal BB. Curcumin: getting back to the roots. Ann N Y Acad Sci. 2005;1056(1):206–17.
134. Hatcher H, Planalp R, Cho J, Torti FM, Torti SV. Curcumin: from ancient medicine to current clinical trials. Cell Mol Life Sci. 2008;65(11):1631–52.
135. Mishra S, Palanivelu K. The effect of curcumin (turmeric) on Alzheimer's disease: an overview. Ann Indian Acad Neurol. 2008;11(1):13.
136. Yang F, Lim GP, Begum AN, Ubeda OJ, Simmons MR, Ambegaokar SS, et al. Curcumin inhibits formation of amyloid β oligomers and fibrils, binds plaques, and reduces amyloid in vivo. J Biol Chem. 2005;280(7):5892–901.
137. Garcia-Alloza M, Borrelli L, Rozkalne A, Hyman B, Bacskai B. Curcumin labels amyloid pathology in vivo, disrupts existing plaques, and partially restores distorted neurites in an Alzheimer mouse model. J Neurochem. 2007;102(4):1095–104.
138. Huang P, Zheng N, Zhou H-b, Huang J. Curcumin inhibits BACE1 expression through the interaction between ERβ and NFκB signaling pathway in SH-SY5Y cells. Mol Cell Biochem. 2020;463(1):161–73.
139. Martin TD, Malagodi AJ, Chi EY, Evans DG. Computational study of the driving forces and dynamics of curcumin binding to amyloid-β protofibrils. J Phys Chem B. 2018;123(3):551–60.
140. Reddy PH, Manczak M, Yin X, Grady MC, Mitchell A, Tonk S, et al. Protective effects of Indian spice curcumin against amyloid-β in Alzheimer's disease. J Alzheimers Dis. 2018;61(3):843–66.
141. Fiala M, Liu PT, Espinosa-Jeffrey A, Rosenthal MJ, Bernard G, Ringman JM, et al. Innate immunity and transcription of MGAT-III and toll-like receptors in Alzheimer's disease patients are improved by bisdemethoxycurcumin. Proc Natl Acad Sci. 2007;104(31):12849–54.
142. Rane JS, Bhaumik P, Panda D. Curcumin inhibits tau aggregation and disintegrates preformed tau filaments in vitro. JAlzheimers Dis. 2017;60(3):999–1014.
143. Zou Y, Qi B, Tan J, Sun Y, Gong Y, Zhang Q. Mechanistic insight into the disruption of tau R3–R4 protofibrils by curcumin and epinephrine: an all-atom molecular dynamics study. Phys Chem Chem Phys. 2022;24(34):20454–65.
144. Giri RK, Rajagopal V, Kalra VK. Curcumin, the active constituent of turmeric, inhibits amyloid peptide-induced cytochemokine gene expression and CCR5-mediated chemotaxis of THP-1 monocytes by modulating early growth response-1 transcription factor. J Neurochem. 2004;91(5):1199–210.
145. Pendurthi UR, Rao LVM. Suppression of transcription factor Egr-1 by curcumin. Thromb Res. 2000;97(4):179–89.
146. Hatami M, Abdolahi M, Soveyd N, Djalali M, Togha M, Honarvar NM. Molecular mechanisms of curcumin in neuroinflammatory disorders: a mini review of current evidences. Endocr Metab Immune Disord Drug Targets. 2019;19(3):247–58.
147. Yu Y, Shen Q, Lai Y, Park SY, Ou X, Lin D, et al. Anti-inflammatory effects of curcumin in microglial cells. Front Pharmacol. 2018;9:386.
148. Bernardo A, Plumitallo C, De Nuccio C, Visentin S, Minghetti L. Curcumin promotes oligodendrocyte differentiation and their protection against TNF-α through the activation of the nuclear receptor PPAR-γ. Sci Rep. 2021;11(1):1–13.
149. Gohil KJ, Patel JA. A review on bacopa monniera: current research and future prospects. Int J Green Pharm. 2010;4(1):1–9.
150. Shinomol GK, Muralidhara. *Bacopa monnieri* modulates endogenous cytoplasmic and mitochondrial oxidative markers in prepubertal mice brain. Phytomedicine. 2011;18:317–26.
151. Deepak M, Amit A. The need for establishing the identities of bacoside A and B, the putative major bioactive saponins of Indian medicinal plant *Bacopa monnieri*. Phytomedicine. 2004;11:264–8.
152. Chaudhari KS, Tiwari NR, Tiwari RR, Sharma RS. Neurocognitive effect of nootropic drug Brahmi (*Bacopa monnieri*) in Alzheimer's disease. Ann Neurosci. 2017;24:111–22.

153. Russo A, Borrellib F, Campisia A, Acquavivaa R, Racitia G, Vanellaa A. Nitric oxide-related toxicity in cultured astrocytes: effect of *Bacopa monniera*. Life Sci. 2003;73(12):1517–26.
154. Tripathi YB, Chaurasia S, Tripathi E, Upadhyay A, Dubey GP. *Bacopa monnieri* Linn. As an antioxidant: mechanism of action. Indian J Exp Biol. 1996;34:523–6.
155. Barbhaiya HC, Desai RP, Saxena VS, Pravina K, Wasim P, Geetharani P, et al. Efficacy and tolerability of Bacomind® on memory improvement in elderly participants-a double blind placebo controlled study. J Pharmacol Toxicol. 2008;3(6):425–34.
156. Benson S, Downey LA, Stough C, Wetherell M, Zangara A, Scholey A. An acute, double-blind, placebo-controlled cross-over study of 320 mg and 640 mg doses of *Bacopa monnieri* (CDRI 08) on multitasking stress reactivity and mood. Phytother Res. 2014;28(4):551–9.
157. Rai R, Singh HK, Prasad S. A special extract of *Bacopa monnieri* (CDRI-08) restores learning and memory by upregulating expression of the NMDA receptor subunit GluN2B in the brain of scopolamine-induced amnesic mice. Evid Based Complement Alternat Med. 2015;2015:254303.
158. Preethi J, Singh HK, Rajan KE. Possible involvement of standardized *Bacopa monniera* extract (CDRI-08) in epigenetic regulation of reelin and brain-derived neurotrophic factor to enhance memory. Front Pharmacol. 2016;7:166.
159. Kumar N, Abichandani LG, Thawani V, Gharpure KJ, Naidu MU, Venkat RG. Efficacy of standardized extract of *Bacopa monnieri* (Bacognize®) on cognitive functions of medical students: a six-week, randomized placebo-controlled trial. Evid Based Complement Alternat Med. 2016;2016:4103423.
160. Dwivedi S, Nagarajan R, Hanif K, Siddiqui HH, Nath C, Shukla R. Standardized extract of *Bacopa monniera* attenuates okadaic acid induced memory dysfunction in rats: Effect on Nrf2 pathway. Evid Based Complement Alternat Med. 2013;2013:294501.
161. Khan MB, Ahmad M, Ahmad S, Ishrat T, Vaibhav K, Khuwaja G, et al. *Bacopa monniera* ameliorates cognitive impairment and neurodegeneration induced by intracerebroventricular-streptozotocin in rat: behavioral, biochemical, immunohistochemical and histopathological evidences. Metab Brain Dis. 2015;30(1):115–27.
162. Kunte KB, Kuna Y. Neuroprotective effect of *Bacopa monniera* on memory deficits and ATPase system in Alzheimer's disease (AD) induced mice. J Sci Innov Res. 2013;2(4):719–35.
163. Mathew M, Subramanian S. Evaluation of the anti-amyloidogenic potential of nootropic herbal extracts in vitro. Int J Pharm Sci. 2012;3(11):4276–80.
164. Saraf M, Prabhakar S, Pandhi P, Anand A. Bacopa monniera ameliorates amnesic effects of diazepam qualifying behavioral–molecular partitioning. Neuroscience. 2008;155(2):476–84.
165. Saraf MK, Anand A, Prabhakar S. Scopolamine induced amnesia is reversed by bacopa monniera through participation of kinase-CREB pathway. Neurochem Res. 2010;35:279–87.
166. Holcomb LA, Dhanasekaran M, Hitt AR, Young KA, Riggs M, Manyam BV. *Bacopa monniera* extract reduces amyloid levels in PSAPP mice. J Alzheimers Dis. 2006;9(3):243–51.
167. Limpeanchob N, Jaipan S, Rattanakaruna S, Phrompittayarat W, Ingkaninan K. Neuroprotective effect of Bacopa monnieri on beta-amyloid-induced cell death in primary cortical culture. J Ethnopharmacol. 2008;120(1):112–7.
168. Rastogi M, Ojha RP, Prabu PC, Devi BP, Agrawal A, Dubey GP. Prevention of age-associated neurodegeneration and promotion of healthy brain ageing in female Wistar rats by long term use of bacosides. Biogerontology. 2012;13(2):183–95.
169. Vishwakarma RK, Kumari U, Khan BM. Memory booster plant *Bacopa monniera* (brahmi): biotechnology and molecular aspects of bacoside biosynthesis. In: Tsay H-S, Shyur L-F, Agrawal DC, Wu Y-C, Wang S-Y, editors. Medicinal plants- recent advances in research and development. Singapore: Springer; 2016. p. 167–90.
170. Malishev R, Shaham-Niv S, Nandi S, Kolusheva S, Gazit E, Jelinek R. Bacoside-A, an Indian traditional-medicine substance, inhibits beta-amyloid cytotoxicity, fibrillation, and membrane interactions. ACS Chem Neurosci. 2017;8(4):884–91.
171. Yamchuen P, Chaiwiang N, Lapphanichayakool P, Ingkaninan K, Limpeanchob N. Neuroprotective effect of *Bacopa Monnieri* extract on oxidized low density lipoprotein-induced neurotoxicity in SH-SY5Y neuroblastoma cells. Thai J Pharmacol. 2017;39(1):5–18.
172. Calabrese C, Gregory WL, Leo M, Kraemer D, Bone K, Oken B. Effects of a standardized *Bacopa monnieri* extract on cognitive performance, anxiety, and depression in the elderly: a randomized, double-blind, placebo-controlled trial. J Altern Complement Med. 2008;14(6):707–13.
173. Cicero AF, Bove M, Colletti A, Rizzo M, Fogacci F, Giovannini M, et al. Short-term impact of a combined nutraceutical on cognitive function, perceived stress and depression in young elderly with cognitive impairment: a pilot, double-blind, randomized clinical trial. J Prev Alzheimers Dis. 2017;4(1):12–5.
174. Jyoti A, Sethi P, Sharma D. *Bacopa monniera* prevents from aluminium neurotoxicity in the cerebral cortex of rat brain. J Ethnopharmacol. 2007;111(1):56–62.
175. Thippeswamy AH, Rafiq M, Viswantha GL, Kavya KJ, Anturlikar SD, Patki PS. Evaluation of *Bacopa monniera* for its synergistic activity with rivastigmine in reversing aluminum-induced memory loss and learning deficit in rats. J Acupunct Meridian Stud. 2013;6(4):208–13.
176. Das A, Shanker G, Nath C, Pal R, Singh S, Singh H. A comparative study in rodents of standardized extracts of *Bacopa monniera* and *Ginkgo biloba*: anticholinesterase and cognitive enhancing activities. Pharmacol Biochem Behav. 2002;73(4):893–900.
177. Komali E, Venkataramaiah C, Rajendra W. Antiepileptic potential of *Bacopa monnieri* in the rat brain during PTZ-induced epilepsy with reference to cholinergic system and ATPases. J Tradit Complement Med. 2020;11:137–43.
178. Mathur D, Goyal K, Koul V, Anand A. The molecular links of re-emerging therapy: a review of evidence of Brahmi (Bacopa monniera). Front Pharmacol. 2016;7(44):1–15.
179. Jeyasri R, Muthuramalingam P, Suba V, Ramesh M, Chen J-T. *Bacopa monnieri* and their bioactive compounds inferred multi-target treatment strategy for neurological diseases: a cheminformatics and system pharmacology approach. Biomol Ther. 2020;10(4):536.
180. Dixit H, Selvaa Kumar C, Dasgupta D, Gadewal N. Molecular docking analysis of hyperphosphorylated tau protein with compounds derived from Bacopa monnieri and Withania somnifera. Bioinformation. 2021;17(9):798–804.
181. Winters M. Ancient medicine, modern use: *Withania somnifera* and its potential role in integrative oncology. Altern Med Rev. 2006;11(4):269–77.
182. Mirjalili MH, Moyano E, Bonfill M, Cusido RM, Palazon J. Steroidal lactones from *Withania somnifera*, an ancient plant for novel medicine. Molecules. 2009;14:2373–93.
183. Singh G, Sharma PK, Dudhe R, Singh S. Biological activities of *Withania somnifera*. Ann Biol Res. 2010;1(3):56–63.
184. Dar NJ, Ahmad M. Neurodegenerative diseases and Withania somnifera (L.): an update. J Ethnopharmacol. 2020;256:112769.
185. Naidu PS, Singh A, Kulkarni SK. Effect of *Withania somnifera* root extract on reserpine-induced orofacial dyskinesia and cognitive dysfunction. Phytother Res. 2006;20(2):140–6.
186. Singh B, Chandan BK, Gupta DK. Adaptogenic activity of a novel withanolide-free aqueous fraction from the roots of *Withania somnifera* dun, vol. 17. Part II: Phytother Res; 2003. p. 531–6.
187. Dhuley JN. Nootropic-like effect of ashwagandha (*Withania somnifera* L.) in mice. Phytother Res. 2001;15(6):524–8.

188. Schliebs R, Liebmann A, Bhattacharya SK, Kumar A, Ghosal S, Bigl V. Systemic administration of defined extracts from *Withania somnifera* (Indian ginseng) and Shilajit differentially affects cholinergic but not glutamatergic and GABAergic markers in rat brain. Neurochem Int. 1997;30(2):181–90.
189. Sehgal N, Gupta A, Vallia RK, Joshi SD, Mills JT, Hamel E, et al. *Withania somnifera* reverses Alzheimer's disease pathology by enhancing low-density lipoprotein receptor-related protein in liver. Proc Natl Acad Sci U S A. 2012;109(9):3510–5.
190. Kurapati KR, Atluri VS, Samikkannu T, Nair MP. Ashwagandha (*Withania somnifera*) reverses beta-amyloid1-42 induced toxicity in human neuronal cells: implications in HIV-associated neurocognitive disorders (HAND). PLoS One. 2013;8(10):e77624.
191. Auddy B, Hazra J, Mitra A. A standardized *Withania somnifera* extract significantly reduces stress-related parameters in chronically stressed humans: a double-blind, randomized, placebo-controlled study. JANA. 2008;11(1):50–6.
192. Choudhary D, Bhattacharyya S, Bose S. Efficacy and safety of ashwagandha (*Withania somnifera* (L.) Dunal) root extract in improving memory and cognitive functions. J Diet Suppl. 2017;14(6):599–612.
193. Gupta M, Kaur G. *Withania somnifera* (L.) Dunal ameliorates neurodegeneration and cognitive impairments associated with systemic inflammation. BMC Complement Altern Med. 2019;19(1):217.
194. Bhatnagar M, Sisodia SS, Bhatnagar R. Antiulcer and antioxidant activity of *Asparagus racemosa* WILLD and *Withania somnifera* DUNAL in rats. Ann N Y Acad Sci. 2005;1056:261–78.
195. Gupta SK, Dua A, Vohra BP. *Withania somnifera* (ashwagandha) attenuates antioxidant defense in aged spinal cord and inhibits copper induced lipid peroxidation and protein oxidative modifications. Drug MetabDrug Interact. 2003;19:211–22.
196. Bhattacharya SK, Kumar A, Ghosal S. Effects of glycowithanolides from Withania somnifera on an animal model of Alzheimer's disease and perturbed central cholinergic markers of cognition in rats. Phytother Res. 2006;9(2):110–3.
197. Kuboyama T, Tohda C, Komatsu K. Withanoside IV and its active metabolite, sominone, attenuate Aβ (25–35)-induced neurodegeneration. Eur J Neurosci. 2006;23:1417–26.
198. Baitharu I, Jain V, Deep SN, Hota KB, Hota SK, Prasad D, et al. *Withania somnifera* root extract ameliorates hypobaric hypoxia induced memory impairment in rats. J Ethnopharmacol. 2013;145(2):431–41.
199. Baitharu I, Jain V, Deep SN, Shroff S, Sahu JK, Naik PK, et al. Withanolide a prevents neurodegeneration by modulating hippocampal glutathione biosynthesis during hypoxia. PLoS One. 2014;9(10):e105311.
200. Zhao J, Nakamura N, Hattori M, Kuboyama T, Tohda C, Komatsu K. Withanolide derivatives from the roots of *Withania somnifera* and their neurite outgrowth activities. Chem Pharm Bull. 2002;50(6):760–5.
201. Tohda C, Joyashiki E. Sominone enhances neurite outgrowth and spatial memory mediated by the neurotrophic factor receptor. RET Br J Pharmacol. 2009;157(8):1427–40.
202. Uddin MS, Al Mamun A, Kabir MT, Jakaria M, Mathew B, Barreto GE, et al. Nootropic and anti-Alzheimer's actions of medicinal plants: molecular insight into therapeutic potential to alleviate Alzheimer's neuropathology. Mol Neurobiol. 2019;56:4925–44.
203. Kuboyama T, Tohda C, Komatsu K. Effects of ashwagandha (roots of *Withania somnifera*) on neurodegenerative diseases. Biol Pharm Bull. 2014;37(6):892–7.
204. Jayaprakasam B, Padmanabhan K, Nair MG. Withanamides in Withania somnifera fruit protect PC-12 cells from beta-amyloid responsible for Alzheimer's disease. Phytother Res. 2010;24(6):859–63.
205. Mahrous RSR, Ghareeb DA, Fathy HM, RMA EL-K, Omar AA. The protective effect of Egyptian *Withania somnifera* against Alzheimer's. Med Aromat Plants. 2017;6(2):2167.
206. Remya C, Dileep K, Variayr E, Sadasivan C. An in silico guided identification of nAChR agonists from *Withania somnifera*. Front Life Sci. 2016;9(3):201–13.
207. Patil SP, Maki S, Khedkar SA, Rigby AC, Chan C. Withanolide a and asiatic acid modulate multiple targets associated with amyloid-β precursor protein processing and amyloid-β protein clearance. J Nat Prod. 2010;73(7):1196–202.
208. Jain S, Shukla SD, Sharma K, Bhatnagar M. Neuroprotective effects of *Withania somnifera* Dunn. In hippocampal sub-regions of female albino rat. Phytother Res. 2001;15(6):544–8.
209. Bhattacharya A, Muruganandam AV, Kumar V, Bhattacharya SK. Effect of poly herbal formulation, EuMil, on neurochemical perturbations induced by chronic stress. Indian J Exp Biol. 2002;40(10):1161–3.
210. Akbar M, Shabbir A, Rehman K, Akash MSH, Shah MA. Neuroprotective potential of berberine in modulating Alzheimer's disease via multiple signaling pathways. J Food Biochem. 2021;45(10):e13936.
211. Singh N, Sharma B. Toxicological effects of berberine and sanguinarine. Front Mol Biosci. 2018;5:21.
212. Huang M, Jiang X, Liang Y, Liu Q, Chen S, Guo Y. Berberine improves cognitive impairment by promoting autophagic clearance and inhibiting production of β-amyloid in APP/tau/PS1 mouse model of Alzheimer's disease. Exp Gerontol. 2017;91:25–33.
213. Mendelson K, Evans T, Hla T. Sphingosine 1-phosphate signalling. Development. 2014;141(1):5–9.
214. Liang Y, Ye C, Chen Y, Chen Y, Diao S, Huang M. Berberine improves behavioral and cognitive deficits in a mouse model of Alzheimer's disease via regulation of β-amyloid production and endoplasmic reticulum stress. ACS Chem Neurosci. 2021;12(11):1894–904.
215. Cai Z, Wang C, He W, Chen Y. Berberine alleviates amyloid-beta pathology in the brain of APP/PS1 transgenic mice via inhibiting β/γ-secretases activity and enhancing α-secretases. Curr Alzheimer Res. 2018;15(11):1045–52.
216. Wang Y-Y, Yan Q, Huang Z-T, Zou Q, Li J, Yuan M-H, et al. Ameliorating ribosylation-induced amyloid-β pathology by berberine via inhibiting mTOR/p70S6K signaling. J Alzheimers Dis. 2021;79(2):833–44.
217. Zhang H, Zhao C, Cao G, Guo L, Zhang S, Liang Y, et al. Berberine modulates amyloid-β peptide generation by activating AMP-activated protein kinase. Neuropharmacology. 2017;125:408–17.
218. Qin S, Tang H, Li W, Gong Y, Li S, Huang J, et al. AMPK and its activator berberine in the treatment of neurodegenerative diseases. Curr Pharm Des. 2020;26(39):5054–66.
219. Zhu F, Wu F, Ma Y, Liu G, Li Z, Sun Y, et al. Decrease in the production of beta-amyloid by berberine inhibition of the expression of beta-secretase in HEK293 cells. BMC Neurosci. 2011;12(125):1–17.
220. Wu Y, Chen Q, Wen B, Wu N, He B, Chen J. Berberine reduces Aβ42 deposition and tau hyperphosphorylation via ameliorating endoplasmic reticulum stress. Front Pharmacol. 2021;12:640758.
221. Wang S, He B, Hang W, Wu N, Xia L, Wang X, et al. Berberine alleviates tau hyperphosphorylation and axonopathy-associated with diabetic encephalopathy via restoring PI3K/Akt/GSK3β pathway. J Alzheimers Dis. 2018;65(4):1385–400.
222. Chen Y, Chen Y, Liang Y, Chen H, Ji X, Huang M. Berberine mitigates cognitive decline in an Alzheimer's disease mouse model by targeting both tau hyperphosphorylation and autophagic clearance. Biomed Pharmacother. 2020;121:109670.
223. He W, Wang C, Chen Y, He Y, Cai Z. Berberine attenuates cognitive impairment and ameliorates tau hyperphosphorylation by limiting the self-perpetuating pathogenic cycle between NF-κB

signaling, oxidative stress and neuroinflammation. Pharmacol Rep. 2017;69(6):1341–8.

224. Nam KN, Kim JH, Jung HJ, Park JM, Moon SK, Kim YS, et al. Berberine inhibits inflammatory activation of rat brain microglia. Neural Regen Res. 2010;5(18):1384–90.
225. Sadraie S, Kiasalari Z, Razavian M, Azimi S, Sedighnejad L, Afshin-Majd S, et al. Berberine ameliorates lipopolysaccharide-induced learning and memory deficit in the rat: insights into underlying molecular mechanisms. Metab Brain Dis. 2019;34:245–55.
226. Chen N, Wang X-C, Fan L-L, Zhu Y-H, Wang Q, Chen Y-B. Berberine ameliorates lipopolysaccharide-induced cognitive impairment through SIRT1/NRF2/NF-κB signaling pathway in C57BL/6J mice. Rejuvenation Res. 2022;25(5):233–42.
227. Lin X, Zhang N. Berberine: pathways to protect neurons. Phytother Res. 2018;32(8):1501–10.
228. Chen M, Li L, Liu C, Song L. Berberine attenuates Aβ-induced neuronal damage through regulating miR-188/NOS1 in Alzheimer's disease. Mol Cell Biochem. 2020;474:285–94.

Herbal Medicine for the Management of Anxiety, Depression, and Insomnia

Sunday Oritsetimenyin Otimenyin

Abstract

The world has advanced in technology, health care, financial instruments and management, and other aspects of human endeavors. This advancement has impacted positively and negatively on the human race. Human demands are insatiable; hence, they crave more and more feats. This has created a lot of unnecessary pressure on man: the quest to meet targets, to achieve set goals, and to measure up with peer groups and pals. These unimaginable pressures have impacted negatively on the human and sometimes animal race. Such pressures have led to anxiety, depression, and sometimes insomnia. These have become a great economic and social burden. The need to address these burdens has increased over the years. Affected subjects sort care from orthodox and traditional medical practitioners. Traditional medical practitioner uses herbs and other natural products/methods to manage these conditions (anxiety, depression and/or insomnia). Herbs used for the management of anxiety, depression, and insomnia have been reported to be rich in alkaloids and phenolic compounds. Alkaloids with these activities often contain quercetin, rutin, apigenin, luteolin, and kaempferol. Extensive research has led to some of the findings, and revealed that these alkaloids exhibit their actions by acting on GABA receptors, serotonin receptors, monoamine oxidase (MAO), and BDNF expression. These herbs also alter endocrine, immunological, and monoamine systems, such as NE, DA, MDA formation, IL-6, BDNF, and TNF- α levels. These mechanisms of action support their claims. The herbal remedy is worth trying if orthodox medicines fail.

Keywords

Anxiety · Depression · Insomnia · Herb · Alkaloids · Quercetin · Rutin · Apigenin · Luteolin · Isorhamnetin · Kaempferol

S. O. Otimenyin (✉)
Department of Pharmacology, Faculty of Pharmaceutical Sciences, University of Jos, Jos, Nigeria

1 Introduction

Psychiatric disorders appear to be common in recent times, which may be attributed to the changes in life style over the decades. There seems to have been an increased pressure on the human race to achieve and possess. Psychiatric disorders that are commonly seen include anxiety, depression, and insomnia [1]. Depressive disorders are said to be one of the most prevalent types of mental illness. Depressive disorders have a huge influence on society, especially the plagued individual. These psychiatric disorders have been the major contributors to the incidence of suicide and ischemic heart disease. It was ranked the second leading cause of disability worldwide [2]. Depression, anxiety, and insomnia tend to occur concurrently in subjects. Their symptoms are almost similar. Major depression presents with many symptoms including insomnia and anxiety. It is also observed that drugs used in managing these disease conditions have related mechanisms of action.

Anxiety is a common emotional disorder that is intimately associated with proper fear. Anxiety may serve as a mechanism to adapt to the environment psychologically. It presents as a common symptom of many medical disorders, especially psychiatric disorders. Most people who suffer from anxiety most likely present with sleep disorder [3]. Most patients with the most common sleep disorder subjectively complain of their inability to initiate and/or maintain sleep. They may complain of poor quantity and quality of sleep, and hangovers. Anxiety, depression, and insomnia do occur in the same space of time. They are common co-morbid psychiatric disorders in the complexity of mental health disorders [4]. The link between anxiety, insomnia, and major depressive disorders impairs immune and cardiovascular systems function [4]. These psychiatric disorders (anxiety, insomnia, and major depressive disorders) affect the subject's work, daily life, quality of life, sleep, and well-being. Being a common problem, the subjects and their relatives sought cures in orthodox and herbal medicine. The first point of call in rural settings is the traditional healer (who uses herbs as their remedies).

A. K. Dhara, S. C. Mandal (eds.), *Role of Herbal Medicines*, https://doi.org/10.1007/978-981-99-7703-1_13

Mental illnesses are often attributed to negative spiritual intervention in the human system or activities. There is the common belief that orthodox remedies cannot proffer solution to the illness; hence, they turn to the spiritualist for help. These spiritualists use herbal medications in most instances. Research has documented the benefits of these herbs to mental health, and in some areas, it is being encouraged. New drug leads for use in orthodox medicine are often obtained from herbs, hence the need to document these plants for future research and drug discovery. New drug discovery is hoped to resolve the present situation seen across the population (rapid increase in the number of people with mental illness is rapidly increasing across the world).

Herbal medicines are commonly used to alleviate or cure central nervous system disorders, these herbs affect multiple target sites, and they stimulate more than one receptor in most cases. Keeping in view that central nervous system disorders, such as anxiety, depression, and insomnia, are habitually found together in a subject and share some similarities in their neurological basis, medicines for managing these disorders may have similar mechanisms of action. Drugs for managing these diseases might be interconnected with each other; they may act on the same group of receptors but elicit different effects in different regions of the brain.

2 Mechanisms of Herbs Used in the Management of Anxiety, Depression, and Insomnia

Herbal medicines used for the management of anxiety, depression, and insomnia act by altering the chemical states of the brain or by correcting hormonal imbalances.

Most of these herbs act by:

1. Increasing CNS concentrations of norepinephrine (NE), 5-hydroxytryptamine (5-HT), and dopamine (DA)
2. Enhancing the transmission of GABAergic and/or glutamatergic transmitters
3. Suppressing the reuptake of norepinephrine (NE), 5-hydroxytryptamine (5-HT), and dopamine (DA) in the brain
4. Inhibiting monoamine oxidases
5. Interferencing with the hypothalamic–pituitary–adrenal (HPA) axis
6. Depressing the release corticotropin releasing factor (CRF).
7. Down regulating the secretion of adrenocorticotropin (ACTH) from the anterior pituitary gland
8. Attenuating the production and secretion of glucocorticoids from the adrenal cortex
9. Suppressing the elevation of corticosterone concentration
10. Restoring the normal expression and function of GR
11. Decreasing brain-derived neurotrophic factor (BDNF)
12. Increasing the phosphorylation level of CREB (cAMP response element-binding protein) and the expression of BDNF in the frontal cortex and hippocampus
13. Inhibiting the p-SAPK/p-JNK signaling pathways induced by stress
14. Decreasing in expression levels of both mRNA and protein of BDNF in the hippocampus

3 Medicinal Plants Used for the Management of Depression

Central nervous system (CNS) disorders are poorly understood and present with complicated disorders. The CNS functionality has not been fully decoded; this explains the mystery that underlines its disorders. This also explains why such diseases are linked to spiritual problems. Diseases of the CNS are numerous among which are depression, anxiety, and insomnia. Depression is a syndrome characterized by some observable behaviors. The behaviors incapacitate the patients and render them unproductive or harmful to themselves. It is characterized by profound sadness and the inhibition of psychic/CNS functions. It sometimes present with neuro-vegetative symptoms. Depression has been reported to be the second leading cause of death (suicide) and disability [5]. It is foremost among CNS disorders and is a commonly diagnosed disease across the world; about 1 in 20 people suffer from depression yearly. The population of people of all ages suffering from depression is estimated to be about 280 million; of this number, 700,000 die from suicide every year [6]. More than 75% of people in low-income and middle-income countries receive some sort of treatment [7]. Depression often starts at a young age, but it cuts across all ages. Adolescents with depression present with a reduction in functional activities. Depression in adolescents often reoccurs. Reduction in productivity is one of the presentations of depression; to arrest this, research is ongoing with the aim of arriving at a break-through that is hoped to improve depression.

Across the globe, there is demand for effective management of depression and the saving of loss of man hours and reduction of suicides attributed to depression. Depression is a heterogeneous disorder that is associated with psychological, behavioral, and physiological symptoms. Observed symptoms of depression include mood changes, difficulty in thinking, and loss of interest in the environment. The patient may experience physical symptoms, such as sleep disorders, sexual problems, loss of stamina, and headaches.

3.1 Symptoms of Depression

Gateway symptoms

- Depressed mood
- Loss of interest in pleasure (anhedonia)
- Headache
- Weight loss, appetite change, and diminished interest in pleasurable activities – presenting as social withdrawal
- Insomnia and hypersomnia
- Increased somatic symptoms – fatigue, pain, and insomnia
- Psychomotor agitation or retardation (feeling of helpless/hopeless)
- General body weakness and diminished energy
- Feelings of uselessness, worthlessness, or/and guilt
- Difficulty taking decisions and paying attention to details
- Thoughts of death or/and suicide

Clinical depression shows physical, behavioral/attitude, and emotional presentation. The diagnostic effect sort include:

Physical

1. Sleep disturbances – insomnia, oversleeping, and waking up early
2. Changes in appetite or eating: much more or much less
3. Decreased energy and fatigue
4. Unexplained digestive problems, headaches, and sometimes stomachaches

Behavioral/Attitude

1. Loss of interest in pleasurable activities that were once pleasurable, such as sex, hobbies outings, sports, etc
2. Amnesia, difficulty making decisions, and concentrating
3. Looking unpresentable and neglecting responsibilities

Emotional

1. Sadist or flat mood occurring for up to two or more weeks
2. Sober or crying for no just cause
3. Feeling helpless, worthless, hopeless, or guilty
4. Feeling anxious, agitated, or irritable
5. Suicide or death thoughts

These symptoms of depression occur when there is an alternation in chemical concentration in the central nervous system. It affects all aspects of the function of the brain. There is an altered cognitive, biological, psychomotor, motivational, behavioral, and emotional process. Depression affects a patient's quality of daily life and may be the cause of suicidal death [5, 8]. Subjects are diagnosed to be depressed if they exhibit the above-listed symptoms consistently for 2 weeks in a roll, daily, and appear clinically distressed or display impairment in occupational, social, and/or other areas of daily activity. Drug-induced depression is due to the direct effect of the drug on the physiological capabilities of the patient. This state of drug-induced depression in an individual is different from a normal depressed individual. Normal depressed individuals present with complex mood disorders with many subtypes, multiple etiologies, symptomatology ranging from mild to severe with or without psychotic features, and interactions with other psychiatric and somatic disorders. Normal depression is also accompanied by: a change in appetite, loss of weight, psychomotor retardation or agitation, hypersomnia or insomnia, loss of energy or fatigue, indecisiveness or diminished ability to concentrate or think, feelings of worthlessness or excessive/inappropriate guilt, and recurrent thoughts of death and/or suicidal ideation or attempt [9, 10].

Severe depression is characterized by a depressed mood, lasting for at least 2 years, with other symptoms like fatigue, feeling hopeless, poor concentration, insomnia or hypersomnia, overeating or poor appetite, and low self-esteem [9] (Fig. 1).

4 Definition of Depression

Depression is a state of negative alteration of the whole body, mood, and thoughts. Stress is the main trigger of depression. Stress produces reactions toward stimulus (causes of stress), displayed in the form of physical, mental, and/or emotional responses. It occurs when the brain is down regulated, evident in a reduction in the levels of most neurotransmitters in the brain. This results in an alteration of the subject's personality, evident in the eating and sleeping habits of the subject, the feeling about self, and perception of things. The functionality of the whole body is affected (it is a "whole-body" illness). In depression, there are negative changes in the behavior, feelings, thinking abilities, and physical well-being of the affected patient. The subject sometimes experiences problems with decision-making, short-term memory, negative thoughts and thinking, pessimism, poor self-esteem, excessive guilt and self-criticism, and self-destructive thoughts. Subjects may be sad for no reason, not enjoying activities that were found pleasurable before depression, lack motivation, and become more apathetic, feeling "slowed down" and fatigued all the time, irritable, sometimes feeling helpless and hopeless, and experiencing difficulty in controlling temper [10]. Behavior is altered: Subjects prefer to be alone, avoid external interference, are socially withdrawn, act out their anger with temper outbursts, are not interested in sexual activities, and neglect their personal appearance and basic hygiene. Depressed subjects experience chronic fatigue, more sleeping or problems with sleeping, may stay awake during sleeping hours, and/or awaken many times at night. Some

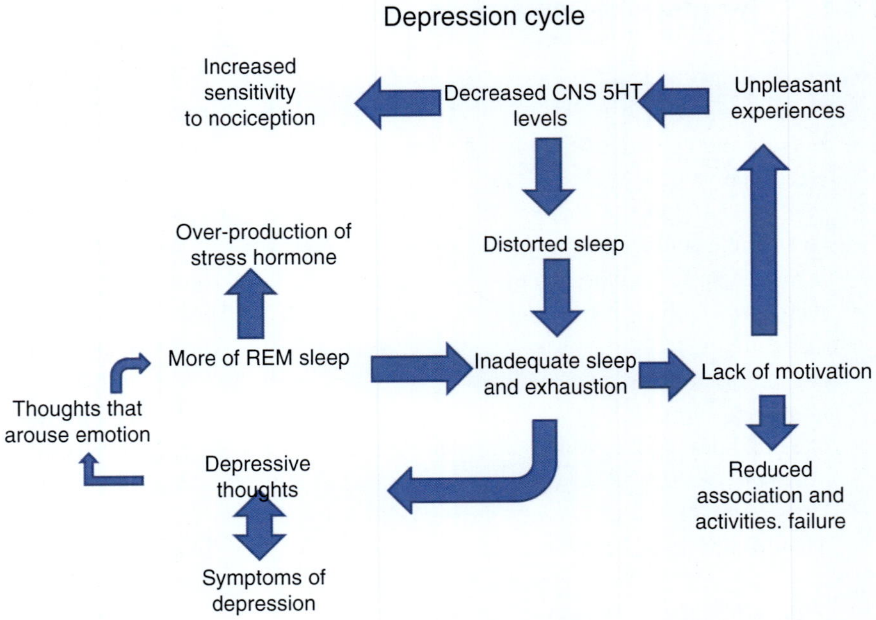

Fig. 1 Depression cycle

patients may sleep for more than the sleeping hours, sleep most of the day and wake up with fatigue, experience loss of appetite and diminished memory, weight loss/gain, suicidal or morbid thoughts, and digestive disorders [2, 5].

4.1 Types of Depression

1. Major depression
2. Dysthymic disorder
3. Unspecified depression
4. Adjustment disorder, with depression
5. Bipolar depression
6. Minor depression
7. Psychotic depression
8. Postpartum depression
9. Seasonal affective disorder (SAD)
10. Unipolar depression

4.2 Unipolar Depression

This is a condition characterized by apathy, anhedonia (loss of interest or capacity for pleasure), changes in sleep and appetite, sadness, and sometimes suicidal thoughts.

4.3 Bipolar Affective Disorder

This typically consists of both manic and depressive episodes separated by periods of normal mood. Manic episodes involve increased energy and elevated mood, resulting in over-activity, pressure of speech, and decreased need for sleep.

4.4 Stress and Depression

Stress has been shown to be the major cause of depression, especially if it is not well managed. Stress can be relieved by engaging in activities that enhance brain activity and relieve stress. Such activities include:

1. Exercising the body, brain, and mind
1. walking
2. riding bicycles
3. attending aerobic classes
4. practicing yoga
5. swimming
6. playing tennis

4.5 Depression and Diet

The brain needs nutrients to function normally. Such nutrients are found in food and some drinks. In addition, some agents may be harmful to the brain, interfering with the normal functioning of the brain. If such are found in the diet, it may spell doom for the brain and the individual. Diet plays a major role in the management of stress and in maintaining the brain in good health. It is known that food rich in sugar stimulates or prolongs the response to stress, while high-cholesterol foods adversely affect blood chemistry. Some medicinal plants and edible vegetables are rich in the

Table 1 Plants with antidepressant properties

Plant	Indication /part used	Constituents	Mechanism of action
Citrus Limon [14]	Depression Anxiety[14] Sedation Essential oil Aerial parts	Apigenin (biflavonoid)	Increases motor activity Inhibits GABAA receptor [15] Mediated via noradrenergic and serotoninergic mechanisms
Clitoria ternatea	Depression	Quercetin	Mediated via nor-adrenaline, serotonin, and acetylcholine [16]
Clitoria ternatea, Linn	Depression	Quercetin	Increase noradrenaline and serotonin levels
Coriandrum sativum	Depression	Quercetin, rutin, apigenin, and luteolin	Interaction with the adrenergic, dopamine-ergic, and GABA-ergic system
Crocus sativus	Depression Petal of Crocus sativus	Crocin 1, picrocrocin, safranal, rutin, apigenin	Increase the levels of VGF, BDNF, CREB, and P-CREB
Cuscuta planiflora	Depression	Quercetin, rutin, apigenin, luteolin, and kaempferol	
Cyperus rotundus.	Depression	Iridoids, apigenin, and luteolin	
Eschscholzia californica	Depression Anxiety Insomnia	Reticuline, aporphine, N-methyllaurotetanine, californidine, and corydine	GABAA and serotonin receptors
Apium graveolens	Depression	3, n-butylphthalide and sedanenolide	
Avena sativa	Depression, anxiety, and nervous exhaustion	Quercetin, rutin, apigenin, luteolin, isorhamnetin, and kaempferol	
Bacopa monniera	Depression Anxiety	Bacopasides I and II, Bacosides A and B, and bacopasaponin C	
Benincasa hispida	Depression Anxiety	Quercetin	Inhibits MAO-A, increases dopamine, α1-adrenaline, serotonin, and GABA
Daucus carota	Depression	Quercetin, rutin, apigenin, and luteolin,	
Eschscholtzia californica		Protopine and o-diphenol	Inhibits monoamine oxidase (MAO-B)
Foeniculum vulgare	Depression	Quercetin, rutin, apigenin, and luteolin	
Vetiveria zizanioides	Depression	Alpha tocopherol, quercetin, rutin, apigenin, and luteolin	
Gossypium herbaceum	Depression Anxiety	Quercetin, rutin, apigenin, and luteolin	Activation of AC-cAMP pathway
Helianthus annuus	Depression Anxiety	Quercetin, rutin, apigenin, and ferulic acid	
H. Tuberculatum	Depression Anxiety	Sylimarin, myrecetin, naringenin, rutin, and quercetin	
Hyoscyamus niger	Antidepressant anxiolytic	Chlorogenic acid, quercetin, rutin, and apigenin	
Bacopa monniera	Anxiolytic effects, antidepressant activity, and anticonvulsant actionantioxidant	Quercetin, rutin, apigenin, luteolin, isorhamnetin, and kaempferol	
Jasminum sambac	Antidepressant	Isoquercetin, rutin, apigenin, and luteolin	
Datura fastuosa	Depression	Scopolamine, atropine, hyoscyamine, and quercetin	
Cassia occidentalis	Antidepressant	Quercetin, rutin, apigenin, and luteolin	

precursors needed by the brain for normal activity; such plants may be beneficial in the management of CNS-related problems. Plants that have been reported to improve depression (anti-depressants) are listed in Table 1.

Stress is inevitable in life. Individuals must devise ways of managing stress either by relaxing or engaging in positive relationships. The relaxation response reverses the stress response in the human mind–body system.

4.6 Causes of Depression

Depression is caused by chemical imbalances in the brain. It may run in families or may be caused by environmental factors or individual factors such as low self-esteem and self indulgence (e.g. alcoholism).

Depression is a syndrome characterized by some observable behaviors. It is characterized by profound sadness and the inhibition of psychic functions, sometimes accompanied by neuro-vegetative symptoms. Several causative factors that have been identified to cause depression, such as:

1. genetic factors
2. changes in brain neurotransmitter concentration
3. psychosocial problems
4. psychodynamic factors
5. changes blood hormone concentration
6. some physical illnesses
7. drugs/medication
8. inadequate food and minerals
9. diet/food

Depression is caused by an unfavorable change in environmental, biochemical, genetic, and psychological factors. A number of theories have been proposed as an explanation for the patho-physiological basis of depression.

4.7 Possible Causative Pathological Factors of Depression

1. Impaired sleep (circadian rhythms)
2. Alteration in neurotransmitter levels
3. dysregulated GABA
4. glutamate signaling
5. altered hypothalamus–pituitary axis activity

Adverse life events, such as economic hardship (which often exacerbates mental health problems, incapacitates citizens, reduces office man hours, and impoverishes people), unexpected outcomes to hard work, unstable environments, and many more have recently been observed to be the main predisposition to depression. The central nervous system controls humans and animals through a complex network of neurons. Such central neurons receive signals/information from other parts of the body and the environment. Biological, psychological, social, and epigenetic factors have signaled these neurons to alert them of possible threat or sad events.

4.8 Drug Induced Depression

Medications/drugs that produce symptoms that mimic depression:

- Anabolic steroids
- Opioids
- Digitalis
- Metoclopramide
- Glucocorticoids
- H_2 blockers

Some beta-blockers ß ß

- Anti-arrythmics
- Carbidopa/levodopa
- Anti-convulsants
- Clonidine
- Barbituates
- Benzodiazepenes

4.9 Management of Depression

Depression has become a great burden in most countries, leading to loss of man hours and economic loss. Governments in those countries have in place agencies that formulate policies for the prevention and control of depression and other related diseases. The factors that predispose to depression (one of which is economic hardship) are difficult to control as they affect all nations across the world. Economic and socio-health factors are global issues that cut across almost all countries and are difficult to control. This impacts the control of depression and related mental health issues and makes it more difficult to prevent depression and access effective treatments.

The role of the monoaminergic system in depression has been well studied. Most drugs are expected to elicit their effects on this system, by increasing monoamine and their metabolites as well as transporters and precursors in the brain. This is thought to be the main mechanism of action of herbal antidepressants. Only a handful of herbal medicines act on other pathways to control depression. Patients treated with herbal remedies often regain normal mood and behaviors within a few weeks. The state of their brain may not be returned to the initial state before depression, but they can function normally and reintegrate into the work force and other societal systems.

Herbal medicines have played a major role in the control and management of depression in many countries [11–13]. The major drawback in its use is the unresearched aspect of toxicity, as they are presumed to be safe [13]. It is a general belief that herbal medicines are safer than orthodox medicines. More so orthodox medicine has not been well accepted despite the decades of use. Herbs form part of the human diet, and it is generally believed that it replaces what is lacking in the body. Taking herbs as medicines is seen as taking food rather than medicines, which makes it more acceptable.

4.10 Medicinal Plants and Depression

Medicinal plants have met human health needs for decades (even before the establishment of orthodox medicines), among such medicinal plants are the antidepressant medicinal plants. Traditional healers have used plants and other naturally occurring substances to treat/manage depression. It forms part of their traditional practice, a sizable number of these plants have and are undergoing clinical trials while some (or their active principle) have found their way into orthodox medicine. In traditional medicine practice, multiple plants are often used for a particular disease condition, with each plant eliciting the same effect or different effects, but the effects are thought to be beneficial in the management of the disease conditions concerned. Some of the plants are used extensively while others are more specific and were employed only for specific disease conditions.

4.11 Use of Herb to Manage Depression

Most herbal prescriptions do not have specified dosages. The quantity of herbal products in a cup, and the size of the cup may vary. The mechanism of action, point of action, and effects of other untargeted organs are not established in herbal medicine. It is reasonable to sound a note of warning regarding the use of herbal medicines. This implies that herbal preparation must be standardized before it is used in the human population. Flavonoids are found in plants as natural polyphenols. They have been extensively investigated for their antidepressant properties, and a number of plants used for the management of depression have been confirmed to contain active flavonoids (Table 2). Most grains, fruits, vegetables, tea, and alcohol are the rich source of flavonoids. They are known for their effects on stress (preventing or reversing stress through a number of mechanisms) (Tables 2 and 3). Some flavonoids have been shown to have antidepressant potential and their use has established this fact. In this chapter, we have attempted to compile potential molecules with beneficial anxiolytic, sedative, and/or antidepressant actions. It is hoped that this review will contribute to the development of safe and effective nutriceuticals that will reduce in incidence of anxiety, insomnia, and depression in humans.

Table 2 Plants used for the management of anxiety and insomnia

Plant name/Family	Constituents	Active biomolecule(s)	Pharmacological effects
Achillea moschata L. Asteraceae	Apigenin, luteolin, di-succinilcaffeoylquinic acid, and isorhamnetin-3-O-glucoside	Apigenin	Anxiety, insomnia, and antibacterial
Adiantum capillus-veneris L. Adiantaceae	Carvacrol, hexadecanoic acid, thymol, quercetin −3-O-glucoside, and isoadiantol	Quercetin	Anxiety, anti-inflammatory, and hypoglycemic activity
Agrimonia eupatoria L. Rosaceae	Apigenin, naringin, and gallic acid	Apigenin and naringin	Antioxidant, antimicrobial, anxiety, and insomnia
Alchemilla group alpina L. Asteraceae	Linalool, cis-3-hexenol, quercitrin, isoquercetin, and apigenin-8-C-glucoside	Apigenin and quercetin	
Alchemilla group vulgaris L. Asteraceae	Flavonoids and apigenin	Apigenin	Anti-inflammatory, anti-depressant, and anxiety
Aloysia citrodora Paláu (L'Hér.) Kuntze) Verbenaceae	Limonene, linalool, flavonoids (apigenin, petalite, and luteolin), thymol, and geranial	Flavonoids	Antioxidant, antimicrobial, anxiety, insomnia, anti-inflammatory, and hypnotic
Angelica archangelica L. Apiaceae	Flavonoids (quercetin and alkaloids) and coumarins (osthole and angelicin).	Quercetin	Antioxidant, anxiety, insomnia, anti-inflammatory, and hypnotic
Angelica sylvestris L. Apiaceae	Flavonoids (apigenin-7-glucoside)	Apigenin	Anti-cancer, antioxidant, anxiety, insomnia, anti-inflammatory, and hypnotic
Anthemis arvensis L. Asteraceae	Torreyol and apigenin	Apigenin	Antioxidant, antimicrobial, anxiety, and anti-inflammatory
Artemisia absinthium L. Asteraceae	Naringenin and flavonoid(quercetin and catechin)	Quercetin	Hypoglycemic, anti-oxidant, and anxiety
Artemisia arborescens L. Asteraceae	Chrysoeriol, apigenin, and artemetine	Apigenin	Antioxidant, antimicrobial, cytotoxic, anxiety, insomnia, anti-inflammatory, and hypnotic
Artemisia vulgaris L. Asteraceae	Eupatilin, apigenin, jaceosidin, luteolin, and galangin	Apigenin	Antioxidant, anxiety, insomnia, anti-inflammatory, and hypnotic
Arum italicum miller Araceae	Apigenin-6,8-C-pentoside-hexoside	Apigenin	Antioxidant and anxiety

(continued)

Table 2 (continued)

Plant name/Family	Constituents	Active biomolecule(s)	Pharmacological effects
Arum palaeictum L. f. Araceae			
Avena fatua L. Poaceae	Beta-carotene nd flavonoids	Flavonoids	Antioxidant and isomnia
Avena sativa L. Poaceae	Apigenin, quercetin, naringenin, and luteolin	Apigenin and quercetin	Antioxidant, anxiety, anti-inflammatory, and anti-thrombotic effect
Ballota nigra L. Solanaceae	Flavonoids (apigenin, ladanein, luteolin), diterpenes, betaines, pinene, linalool, and cadinol	Apigenin	Antioxidant, antimicrobial, anxiolytic, insomnia, anti-inflammatory, sedative-hypnotic antidepressant, anticholinesterase, and inhibit the development of cervix cancer
Borago officinalis L. Boraginaceae	Flavonoids (quercetin), gallic acid, kaempferol, and vitamin C	Quercetin	Antioxidant, anxiolytic, sedative-hypnotic antidepressant, and carcinogenic
Calendula officinalis L. Asteraceae	Quercetin, vitexin, apigenin, myricetin, and kampferol	Quercetin	Antioxidant, antimicrobial, anxiety, insomnia,, hypnotic
Centranthus ruber (L.) DC. Valerianaceae	Valtrate (valepotriates) and flavonoids	Valtrate	Antioxidant, anxiolytic, insomnia, sedative-hypnotic antidepressant
Chamaemelum nobilis all. Asteraceae	Flavonoids (apigenin, luteolin, and quercetin)	Apigenin And quercetin	Antioxidant, and antitumor, anxiolytic, insomnia, and sedative-hypnotic antidepressant
Citrus aurantium L. Rutaceae	Kaempferol, rutin, hesperidin, naringenin, quercetin, and synephrine	Quercetin	Antioxidant, anxiolytic, insomnia, anti-inflammatory, analgesic, and sedative-hypnotic antidepressant
Citrus Limon L. Rutaceae	Pinene, limonene, rutin, hesperidin, luteolin, apigenin, diosmetin, and quercetin	Apigenin	Antioxidant, antimicrobial, anxiolytic, insomnia, anti-inflammatory, sedative-hypnotic antidepressant, anticancer, and anticonvulsant
Citrus x sinensis (L.) Osbeck Rutaceae	Quercetin, rutin, kaempferol, apigenin, diosmetin, luteolin, and hesperidin	Quercetinand apigenin	Antioxidant, antimicrobial, anxiolytic, antidepressant, and anticancer
Clinopodium nepeta (L.) Kuntze (=Calamintha nepeta (L.) Savi) Lamiaceae	Catechins, luteolin, quercetin, apigenin, and myricetin	Quercetin and apigenin	Antioxidant, antimicrobial, anxiolytic, insomnia, sedative-hypnotic antidepressant, and anti cancer
Conium maculatum L. Apiaceae Leaves Infusion [62] Corydalis cava (L.	Flavonoids and coumarins	Flavonoid	Antioxidant, anti-inflammatory, analgesic, and anti cancer
Corydalis cava (L.) Schweigg. and Körte Papaveraceae	Quercetin, catechin, kaempferol, apigenin, diosmetin, luteolin, and hesperidin		Antimicrobial, anxiolytic, sedative-hypnotic
Crataegus laevigata (Poir.) DC. Rosaceae	Proanthocynidins, fisetin, quercetin, and lupeol		Lowers blood pressure, antioxidant, anxiolytic, sedative-hypnotic, and anticancer
Crataegus monogyna Jacq. Rosaceae	Quercetin, lupeol, apigenin, myricetin, and rutin		Lowers blood pressure, antioxidant, anxiolytic, and sedative-hypnotic,
Crocus sativus L. Iridaceae	Crocin, picrocrocin, crocetin, safranal, zeaxanthin, flavonoids, kaempferol, nigricin, tectoridin, and quercetin	Crocin	Antioxidant, antiviral, anti-inflammatory, hypotensive, anticonvulsant, antiAlzheimer, antidepressant, anxiolytic, sedative-hypnotic, anti-inflammatory, and cytotoxic
Cydonia oblonga mill. Rosaceae	Quinic acid, quercetin, rutin, kaempferol, epicatechin, and pectin		Antioxidant, anxiolytic, insomnia, anti-inflammatory, sedative-hypnotic antidepressant, and diuretic
Cynodon dactylon (L.) Pers Poaceae	Quercetin, rutin, kaempferol, catechin, myricetin, carotene, violaxanthin, and luteolin		Antioxidant, antimicrobial, anxiolytic, anti-inflammatory, sedative-hypnotic antidepressant, anticancer, and antipyretic
Dianthus seguieri Vill. Caryophyllaceae	Quercetin, rutin, and kaempferol		Anticancer, antiviral, and antibacterial
Ecballium elaterium (L.) A. Rich Cucurbitaceae	Rutin, quercetin, luteolin, and apigenin		Antimicrobial, antioxidant, and anticancer

Table 2 (continued)

Plant name/Family	Constituents	Active biomolecule(s)	Pharmacological effects
Foeniculum vulgare miller Apiaceae	Rutin, quercetin, apigenin, and caffeic acid		Antioxidant, antimicrobial, hypnotic, anxiolytic, antidepressant, and antiamnesic
Humulus lupulus L. Cannabaceae	Humulone, lupulon, naringenin, quercetin, kaempferol, and taxifolin		Antioxidant, anxiolytic, insomnia, antibacterial, antifungal, anti-inflammatory, and sedative-hypnotic antidepressant
Hyoscyamus niger L. Solanaceae	Chlorogenic acid, rutin, quercetin, and scopolamine		Antioxidant, antimicrobial, sedative-hypnotic, and calming agent (anxiolytic) antidepressant
Hypericum perforatum L	Hypericin, hyperforin, quercetin, kaempferol, biapigenin, and rutin		Antioxidant, antimicrobial, anxiolytic, anti-inflammatory, sedative-hypnotic antidepressant, and anti cancer
Ilex aquifolium L. Aquifoliaceae	Quercetin, rutin, kaempferol, and chlirogenin acid		Antioxidant, antimicrobial, anxiolytic, and sedative-hypnotic
Lactuca sativa L. Asteraceae	Quercetin		Antioxidant, antimicrobial, anxiolytic, sedative-hypnotic antidepressant, and analgesic
Laurus nobilis L. Lauraceae	Pinene, terpineol, quercetin, kaempferol, luteolin, and apigenin		Antioxidant, anxiolytic, anti-inflammatory, analgesic, sedative-hypnotic, and antimicrobial
Lavandula angustifolia miller Lamiaceae	Isoquercetin, quercetin, rutin, kaempferol, and apigenin		Antioxidant, anxiolytic, anti-inflammatory, and sedative-hypnotic
Lavandula officinalis Chaix Lamiaceae	Borneol, cineole, and linalool,		Antioxidant, antimicrobial, anxiolytic, anti-inflammatory, and sedative-hypnotic
Lavandula stoechas L. Lamiaceae	Lupeol, amyrin, vanillin, rutin, luteolin, and apigenin		Antioxidant, anxiolytic, anti-inflammatory, antifungal, and sedative-hypnotic
Leucanthemum alpinum lam. Asteraceae	Flavonoid		Antioxidant
Lolium multiflorum lam. Poaceae	Chorogenic acid, Isoquercetin, quercetin, rutin, kaempferol, apigenin,luteolin, and ferulic acid		Anti oxidant, anti-inflammatory, and antimicrobial
Lolium perenne L. Poaceae	Quercetin		Antioxidant,
Lotus corniculatus L. Fabaceae	Quercetin, rutin, kaempferol, isorhamnetin, luteolin, and biochanin		Antibacterial, antioxidant, anxiolytic, antidepressant, anti-inflammatory, sedative-hypnotic, anticancer,
Malva cretica Cav. Malvaceae	Isoquercetin, quercetin, rutin, kaempferol, apigenin, and luteolin		Antioxidant, and anxiolytic
Malva neglecta Wallr. Malvaceae	Isoquercetin, quercetin, rutin, kaempferol, apigenin, luteolin, chrysin, coumarin, gallic acid, and vanillin		Antioxidant, anxiolytic, antimicrobial, anticholinesterase, anti-inflammatory, and sedative-hypnotic
Malva sylvestris L. Malvaceae	Isoquercetin, quercetin, rutin, kaempferol, apigenin, and luteolin		Antioxidant, anxiolytic, anti-inflammatory, antifungal, and sedative-hypnotic
Matricaria camomilla L. (=Chamomilla recutita (L.) Rauschert) Asteraceae	Quercetin, rutin, kaempferol, apigenin, and luteolin		Antioxidant, anxiolytic, anti-inflammatory, antispasmodic, antidepressant, sedative-hypnotic, hypolipidemic, anticonvulsant, antiparasitic, an analgesic
Matricaria discoidea DC. Asteraceae	Chorogenic acid, Isoquercetin, quercetin, rutin, kaempferol, apigenin,luteolin, isorhamnetin, and cosmosin		Antioxidant, anxiolytic, anti-inflammatory, antispasmodic, antidepressant, sedative-hypnotic, hypolipidemic, anticonvulsant, antiparasitic, and analgesic
Melilotus officinalis pall. Fabaceae	Coumarin, Isoquercetin, quercetin, rutin, kaempferol, and apigenin		Antioxidant, antimicrobial, antibiofilm, anxiolytic, anti-inflammatory, anticancer, sedative-hypnotic, and analgesia

(continued)

Table 2 (continued)

Plant name/Family	Constituents	Active biomolecule(s)	Pharmacological effects
Melissa officinalis L. Lamiaceae	Isoquercetin, quercetin, rutin, kaempferol, apigenin, and rhamnocitrin		Antioxidant, anxiolytic, antibacterial, antiviral, anti-Alzheimer, anti-inflammatory, antidepressant, sedative-hypnotic, cardioprotective, antidiabetic, and analgesic
Melittis melissophyllum L. Lamiaceae	Myricetin,, quercetin, rutin, kaempferol, apigenin, and luteolin		Antioxidant, anxiolytic, anti-inflammatory, antifungal, and sedative-hypnotic
Mentha aquatica L. Lamiaceae	Isoquercetin, quercetin, rutin, kaempferol, apigenin, luteolin, and naringenin		Hepatoprotection, antioxidant, anxiolytic, anti-inflammatory, analgesic, sedative, appetite stimulant, and anticonvulsant
Mentha piperita L. Lamiaceae	Epicatechin, syringic acid, Isoquercetin, quercetin, rutin, kaempferol, apigenin, and luteolin		Antispasmodic, antioxidant, anxiolytic, anti-inflammatory, analgesic, sedative, and antidepressant
Mentha spicata L. (=M. Spicata L. subsp. glabrata (Lej. and Courtois) Lebeau) Lamiaceae	Quercetin, rutin, kaempferol, apigenin, and luteolin		Antioxidant, antimicrobial, anxiolytic, anti-inflammatory, sedative-hypnotic, antidepressant, and cytotoxic
Mentha suaveolens Ehrh. Subsp. suaveolens Lamiaceae	Isoquercetin, rutin, and apigenin		Antioxidant, anxiolytic, antibacterial, anti-inflammatory, analgesic, sedative-hypnotic
Mentha viridis L. Lamiaceae	Quercetin, rutin, and kaempferol		Antioxidant, anxiolytic, antimicrobial, anticancer, and anti-inflammatory
Myrtus communis L. Myrtaceae	Linalool, myrecitin, Isoquercetin, quercetin, rutin, kaempferol, and apigenin		Antioxidant, anxiolytic, antimicrobial, anti-inflammatory, anticancer, antidepressant, and sedative-hypnotic,
Nepeta cataria L. Lamiaceae	Chlorogenic acid, quinic acid, Isoquercetin, quercetin, myricetin, nd apigenin,		Antioxidant, anxiolytic, antimicrobial, antidepressant, anti-inflammatory, and sedative-hypnotic
Ocimum basilicum L. Lamiaceae	Isoquercetin, quercetin, rutin, kaempferol, apigenin, catechin, naringin, and genistein		Antioxidant, anxiolytic, anti-inflammatory, and sedative-hypnotic
Olea europaea L. Oleaceae [17]	Luteolin, diosmetin quercetin, rutin, kaempferol, apigenin, and oleuropein		Antioxidant, anxiolytic, neuroprotective, anti-inflammatory, analgesic, antifungal, sedative-hypnotic, and anti-tumor
Opuntia ficus-indica (L.) mill. Cactaceae	Isoquercetin, quercetin, rutin, kaempferol, apigenin, and isorhamnetin[18]		Antioxidant, anxiolytic, antibacterial, anti-inflammatory, analgesic, sedative-hypnotic, antidepressant. Anticancer, neuroprotective, and anti-viral
Origanum majorana L. Lamiaceae	Quercetin, rutin, apigenin, naringenin, luteolin, and scutellarein		Antioxidant, anxiolytic, anti-inflammatory, and sedative-hypnotic
Origanum vulgare L. subsp. viridulum (Martrin-Donos) Nyman Lamiaceae	Quercetin, rutin, apigenin, naringenin, luteolin, and scutellarein		Antioxidant, anxiolytic, antimicrobial, antidepressant, anti-inflammatory, antifungal, and sedative-hypnotic
Paeonia mascula (L.) mill. Paeoniaceae	Isoquercetin, quercetin, rutin, kaempferol, and apigenin		Antioxidant, antimicrobial, anxiolytic, analgesic, anti-inflammatory, and anticholinesterase
Papaver rhoeas L. Papaveraceae [19]	Isoquercetin, quercetin, rutin, kaempferol, and apigenin		Antioxidant, anxiolytic, anti-inflammatory, and sedative
Papaver setigerum DC. Papaveraceae [20]	Quercetin, ruti,, and apigenin		Antioxidant, anxiolytic, anti-inflammatory, antifungal, and sedative
Papaver somniferum L. Papaveraceae	Isoquercetin, quercetin, rutin, kaempferol, apigenin, and luteolin		Antioxidant, anxiolytic, anti-inflammatory, antifungal, and sedative-hypnotic
Pimpinella anisum L. Apiaceae [21]	Quercetin, rutin, kaempferol, and apigenin		Antioxidant, anxiolytic, antimicrobial, analgesic, anti-inflammatory, antispasmodic, antidepressant, sedative-hypnotic, antiviral, and anticonvulsant

Table 2 (continued)

Plant name/Family	Constituents	Active biomolecule(s)	Pharmacological effects
Polypodium vulgare L. Polypodiaceae	Catechin, kaempferol, and apigenin		Antioxidant, anxiolytic, antimicrobial, analgesic, anti-inflammatory, antidepressant, sedative-hypnotic, antiviral, and anticonvulsant
Primula veris L. Primulaceae	Quercetin, rutin, kaempferol, apigenin,		Antioxidant, anxiolytic, and anti-inflammatory
Primula vulgaris Huds. Primulaceae	Quercetin, rutin, and kaempferol		Antioxidant, anxiolytic, anti-inflammatory, and sedative-hypnotic
Prunus persica (L.) Batsch Rosaceae	Quercetin, rutin, kaempferol, apigenin, and cinnamic acid		Antioxidant, anxiolytic, antimicrobial, anti-inflammatory, antispasmodic, antidepressant, sedative-hypnotic, anticholinesterase, anti-obesity, and hepatoprotective
Robinia pseudacacia L. Fabaceae	Myricetin, chorogenic acid, isoquercetin, quercetin, rutin, kaempferol, apigenin, naringenin, chrysoeriol, and isoliquiritigenin		Antioxidant, anxiolytic, antimicrobial, anti-inflammatory, antifungal, antidepressant, inhibits interleukin-1*B*, and antiviral
Rosmarinus officinalis L. Lamiaceae	Ferulic acid, rosmarinic acid, chorogenic acid, Isoquercetin, quercetin, rutin, kaempferol, apigenin, chlorogenic acid, and rosmaric acid,		Antioxidant, anxiolytic, antimicrobial, analgesia, anti-inflammatory, neuroprotective, sedative-hypnotic, antitumor, and anticonvulsant
Ruta chalepensis L. Rutaceae	Quercetin, rutin, kaempferol, and apigenin		Antioxidant, anxiolytic, antimicrobial, antipyretic anti-inflammatory, antidepressant, sedative-hypnotic, and anticonvulsant
Salix alba L. Salicaceae	Quercetin, rutin, and salicin		Antioxidant, anxiolytic, anti-inflammatory, and analgesia
Salvia officinalis L. Lamiaceae	Quercetin, rutin, hesperidin, luteolin, apigenin, chrysin, sagerinic acid, pinocembrine, catechin, and cirsimaritin		Antioxidant, anxiolytic, and antimicrobial
Santolina insularis (Gennari ex Fiori) Arrigoni Asteraceae	, Luteolin, rhamnocitrin, andnepetin		Antioxidant, anxiolytic, and anti-inflammatory
Senecio delphinifolius Vahl. Asteraceae			Antioxidant, anxiolytic, anti-inflammatory, cytotoxic
Solanum nigrum L. Solanaceae [22]	Quercetin		Antioxidant, anxiolytic, anti-inflammatory, and antidiabetic
Sonchus oleraceus L. Asteraceae	Quercetin, rutin, apigenin, apigenin, and luteolin		Antioxidant, antimicrobial, anxiolytic, gastroprotective, anti-inflammatory, and anti-cancer
Stachys recta L. Lamiaceae	Caffeoylquinic acid, rutin, and flavonoids		Antioxidant, anxiolytic, and anti-inflammatory
Tanacetum balsamita L. (=Balsamita major Desf.) Asteraceae	Isoquercetin, nepetin, kaempferol, axillarin, quercetin, rutin, apigenin, naringenin, luteolin, scutellarein, isoramnetin, hyperoside, and syringic acid [18]		Antioxidant, anxiolytic, anti-inflammatory, and ant-spasmodic
Thalictrum aquilegiifolium L. Ranunculaceae	Thalifoline		Antioxidant, anxiolytic, and anti-fungal
Thymbra capitata (L.) Cav. Lamiaceae [23]	Rosmarinic acid, quercetin, rutin, kaempferol, and apigenin		Antioxidant, anxiolytic, and anti-inflammatory
Valeriana officinalis L. Valerianaceae [24]	Valerenic acid	Insomnia Restlessness	Insomnia Restlessness

Table 3 Mechanism of selected flavonoids found in anxiolytic and sedative/hypnotic plants

Name	Similar brain chemical	Uses/effects	Mechanism(s)
7, 8-dihydroxyflavone [25]		Anti-depressant and Parkinson's disease	It acts on activated TrkB (tropomyosin receptor kinase B) receptors and nitric oxide signaling pathway It mimics brain-derived neurotrophic factor (BDNF) and also elevates BDNF expression and levels in the hippocampus [26]
Amentoflavon [25]	Cnestis ferruginea Vahl ex DC	Neuroprotective action, antioxidant, and anti-inflammatory anxiolytic and antidepressant	Acts via interactions with serotonin (5-HT2), ionotropic GABA (gamma-aminobutyric acid), and adrenergic receptors (α1- and α2-), [26]
Apigenin [27]		Antioxidant, anti-inflammatory, anti depressant, and antitumor activities [28, 29]	Apigenin inhibits NLRP3 inflammasome expression, upregulates PPAR (peroxisome proliferator-activated receptor) expression, and inhibits IL-1 production. Up-regulates the level of BDNF in the hippocampal region of the brain and inhibits mono-amino oxidase enzyme
Baicalein (5, 6, 7-trihydroxyflavone)	Scutellaria baicalensis Georgi	Antioxidant antidepressant	Increase ERK phosphorylation and elevate hippocampal BDNF expression Increase levels of BDNF and dopamine in the hippocampus Inhibit cyclooxygenase
Chrysin	Found in honey	Antineoplastic, antioxidant, anti-inflammatory, antidepressant, and hypolipidemic	Chrysin up-regulates BDNF expression and elevates BDNF levels in the hippocampus and prefrontal cortex It also modulates 5-hydroxy-tryptamine metabolism, caspase activities, pro-inflammatory cytokines synthesis, and kyanize
Luteolin [30] (3′, 4′ 5, 7, tetrahydroxyflavone)	Cirsium japonicum Var. Maackii	Anticancer, antioxidant, anxiolytic, memory-boosting properties, and antidepressant effects	Actions are mediated via potentiation of GABA-A Decrease stress-related protein of the endoplasmic reticulum, Inhibit the MAO enzyme resulting in an increase in CNS neurotransmitter levels
Orientin	Highly abundant in herbs, fruits, and millet	Antioxidant activity and antidepressant	Increases amine neurotransmitters in the brain by inhibiting MAO and increases synaptic proteins and BDNF expression in the prefrontal cortex and hippocampus
Nobiletin (5,6,7,8,3′,4′-hexamethoxyflavone)		Neuroprotection Anti-depressant	Noboletin interaction with dopaminergic (D1 and D2), noradrenergic (α1), receptors, and serotonergic (5-HT1A and 5-HT2) receptors
Icariin		Anti-depressant [31]	Icarin inhibits inducible nitric oxide synthase enzyme activity. It also elevates hippocampal BDNF levels [32, 33]
Vitexin (is an apigenin flavone glucoside)	Passiflora incarnates L.	Antitumor, anti-inflammatory, analgesic, and antioxidant Anxiolytic and anti-depressant activities	Vitexin inhibits urease, platelet aggregation, α-glucosidase, and adipogenesis Mediated via interaction with serotonergic(5-HT1A), dopaminergic (D1, D2, and D3), and noradrenergic(α2) receptors. It also increases the synaptic concentration of this neurotransmitter
Fisetin (7, 3′, 4′-flavon-3-ol)	Fruits (especially strawberries) and vegetables are rich in fisetin	Anti-inflammatory, antioxidant, and neuroprotective antidepressant	Its action is mediated via activation of the TrkB signaling pathway and leads to an increase in phosphorylation of TrkB level without disturbing total TrkB

Table 3 (continued)

Name	Similar brain chemical	Uses/effects	Mechanism(s)
Isoquercitrin	St. John's wort	Anti-depressant Antioxidant [34]	Isoquercitin elicits these actions via the modulation of hypothalamus/pituitary axis (HPA) functions. It prevents hypersecretion of adrenocorticotropic hormone and cortisol
Kaempferitrin	Justicia spicigera Schltdl	Anti-depressant	Regulates the HPA axis and modulates the serotonergic system (mainly, presynaptic 5-HT1A receptors)
Kaempferol	Fruits and vegetables	Anti-depressant [35]	Elevates plasma β-endorphin, nor-epinephrine, dopamine, and serotonin (reduces serotonin metabolism) concentration
Myricetin	In fruits, red wine, berries, vegetables, nuts, and tea	Anti-apoptotic Anti-inflammatory, anti-depressant, antioxidant, and neuroprotective	Increases the activity of enzyhippocampus glutathione peroxidase and BDNF level
Myricitrin [36]		Antidepressant, antioxidant, anti-fibrotic, and anti-inflammatory [37]	
Quercetin [38]	Broccoli, wine, apple, onion, and plants (e.g., Green tea), [39]	Anti-depressant [40] Antiinfamatory	Inhibition of MAO enzyme
Rutin	Vegetables, citrus fruits, and plants (such as figs, green, and black tea	Anti-inflammatory, neuroprotective, anti-tumor, and antioxidant	
Flavanone		Anti-inflammatory and antioxidant	Elevates CNS monoamine level and increases hippocampal BDNF and expression
Hesperidine hesperidin		Antidepressant Antioxidant	Increase in hippocampal BDNF concentrations, activate dopaminergic systems, and enhance 5-HT1A receptors. It also inhibits monoamine oxidases and suppresses neuro-endocrine signaling
Naringin (4′-hydroxyflavanones)		Antioxidant, antidepressant, Anticarcinogenic, and blood lipid-lowering	Inhibits cytochrome P450 enzymes (CYP1A2)
Naringenin	Citrus fruits peel	Anti-inflammatory, antioxidant, neuroprotective, cognition enhancer, anti-bacterial, and wound-healing	MAO inhibitor Suppress neuro-endocrine signaling
Silibinin		Cardioprotective Anti-inflammatory Anticancer Anti-depressant [41]	Alters endocrine, immunological, and monoamines (5-HT, DA, and NE). Alters MDA formation, IL-6, TNF-α, and BDNF levels
Flavan-3-ols (−)-epicatechin and (+)-catechin	Cocoa Uncaria rhynchophylla (Miq	Anxiety and depression	Inhibiting MAO-B enzyme activity
Astilbin	Hypericum perforatum	Anti-inflammatory antioxidant, and antidepressant [42]	Up regulates neurotransmitter release and BDNF signaling pathway [43]
Dihydromyricetin	Cercidiphyllum japonicum Salix sachalinensis	Anti-depressant	Increase in BDNF expression and inhibit neuro-inflammation [44]

5 Anxiety

Anxiety is one of the most common mental illnesses plaguing one-eighth of the total population and has become of great economic burden in most countries. It is a psychiatric problem that is not easily recognized and attended to. Anxiety is a persistent feeling of apprehension, dread, and impending disaster, or uneasiness and/or tension. It is a precise class of psychopathology that presents with future-oriented apprehension and increased threat value associated with mental, physical, and social stimuli. Anxious people may commit crimes to cover up the source or cause of anxiety.

Over the years, anxiety has received great attention. This has led to the discovery of a number of drugs thought to be of help in managing anxiety. As a central nervous system (CNS) disorder, it has become a very important area of research in psychopharmacology. Stress often precedes anxiety, hence the word "'stress-induced anxiety." Anxiety is a common psychiatric problem that has been aggravated by the style of this modern world and lifestyles. Pressure from work, pairs groups, families, and self-ambition are some of the numerous predisposal to anxiety. As humans, it is in our traits to be a little stressed and anxious, this is healthy for the progress and development of the human race and for achieving great fits. Stress and anxiety could motivate an individual and spur such to be more productive. Too much of everything appears to be bad in all areas, so also in stress and anxiety. It could cripple an individual if such an individual is stressed beyond boundaries and over anxious. It could result in poor health as well as specific psychological or physical illnesses like depression and insomnia. The inability to control persistent stress can lead to anxiety and unhealthy behaviors. Anxiety is a common emotional phenomenon in humans, and it is an unpleasant emotional state that is associated with concern or fear about some defined or undefined future threat, uneasiness, and discomfort.

Anxiety, as presented in normal individuals, is characterized by

1. heart palpitations
2. nausea and fatigue
3. shortness of breath and respiratory insufficiency
4. fear of the unknown
5. over reaction to minor issues
6. and inability to make progress

5.1 Symptoms of Anxiety

1. Trembling, sweating, feelings of choking, abdominal distress, nausea, dizziness, pins and needles, feelings of impending doom, or losing control over minor issues.
2. Symptoms of anxiety include (a) physiological, (b) cognitive, and (c) behavioral symptoms.
   (a) Cognitive distortions, lack of attention, interpretation, and memory for information processing [10].
   (b) Physiological aspects of anxiety include autonomic or somatic sensations. It also includes avoidance related to nightmares, sleep, insomnia, and fear of sleeping alone. Increased heart rate, heart palpitations, shortness of breath, chest pain, nausea, and difficulty in swallowing.
   (c) Behavioral aspects of anxiety disorders include: avoidance of self-acclaimed fearful events, avoidance of specific stimuli (individual, things) that are presumed to make the subject fearful (e.g., bridges), or avoidance of situations (e.g., public speaking) that make the sufferer uncomfortable. These results in the inability to carry out or maintain daily routines in academic, family, and/or social functions.

Anxiety disorders are disturbing psychiatric disorders affecting all the population. About 30.5% of women and 19.5% of men are affected and nearly 25% of the adult population at some point in their life is affected. Young adults are more plagued with anxiety, attention must be given to this population if the nation is to have a future. Anxiety can aggravate many physical and mental ailments and also impede recovery from any other problems.

Identifying the cause of the anxiety (anxiogenic stimulus) may be the way to go in the management of anxiety. A subject may experience anxiety at a particular moment when the causative factor is present. The levels of anxiety in this case depend on the level of the anxiogenic stimulus. This can be referred to as "State anxiety", and it varies from time to time depending on the presence of the stimulus. Some subjects experience an enduring anxiety ("Trait anxiety") that cannot be tied to a particular cause. This type of anxiety does not vary from moment to moment; it endures in the subject and may become a feature of the subject.

5.2 Management of Anxiety

There are orthodox medicines known to improve the lifestyle of subjects with anxiety, but these drugs have side effects that interfere negatively with the lifestyle of the subject. Most subjects resort to psychotherapy and the use of herbal medicinal agents. Research has isolated herbal preparations that are helpful in the management of anxiety. Herbal medicines seem to work well for those subjects with mild to moderate symptoms of anxiety and sleep disorders. The use of natural or herbal remedies for the management of disease conditions by man has increasingly gained attention in recent times, partly because of the high cost of drugs and because of the side effects and non-acceptability. Some believe that evil spirits are responsible for CNS disease, and they use herbs and incantations to ward away the spirits and regain normalcy. This method works, not because spirits are driven away, but as a form of psychotherapy and distractor from the cause of anxiety or any related CNS disorders. Different types and forms of herbal medications are used for self-treatment of various stress-related afflictions. Research studies have shown that some of these plants have one or more CNS effects. Research studies into many that have not been researched have been advocated.

5.3 Types of Anxiety Disorders

Anxiety disorders encompass the following clinical conditions

1. Panic disorder
2. Social anxiety disorder and phobias
3. Generalized anxiety disorder
4. Post-traumatic stress disorder
5. Obsessive–compulsive disorder.

5.3.1 Generalized Anxiety Disorders

This disorder involves a broad presentation of anxiety. Generalized anxiety disorder presents with long-lasting anxiety that may span up to or over 6 months. Such anxiety may have more than one causative situation and may not be focused on one causative object or situation. Subjects with this type of anxiety experience persistent fear (non-specific – cannot be traced to a real fear stimulant) and worry (when there is no cause to worry) and become overly disturbed with everyday stuff like health, money, work, family, and experience these symptoms even when there are no clear signs of trouble in this aspect of their life.

5.3.2 Obsessive–Compulsive Disorder

This type of anxiety disorder presents with obsessions; recurrent thoughts of danger or threat that do not exist or may not be about real-life threat or problem, which the patient fails to suppress or ignore. Compulsions present as repetitive behaviors that the patient feels pushed to execute in response to an obsession. The compulsive behaviors are thought to be an attempt to reduce the distress from the obsessions. The behaviors may not be beneficial to the person or the community.

5.3.3 Panic Disorder

Subjects with this disorder present with apprehension and brief attacks of intense terror, which is often characterized by shaking, confusion, trembling, dizziness, nausea, and sometimes difficulty in breathing, lasting for a few minutes or longer. The patient believes that he is seriously sick or about to die. This feeling can leave the person shaken or depressed immediately after the experience and for a period of time. Panic disorder results in marked impairment in function if it continues for a sustained period of time.

5.3.4 Social Anxiety Disorder

Social anxiety disorder is a marked and persistent fear of performance or social situations. The person feels he is not measuring up with social requirements and obligations.

5.3.5 Phobias

A phobia is an exaggerated or unrealistic fear of a specific stimulus: darkness, heights, enclosed places, cooking fire, or other situations. Phobic individuals sometimes experience full panic attacks after exposure to unrealistic stimuli. Phobia seems the most common form of anxiety disorder. Panic disorders are not common and are fairly rare occurrences in the general population.

5.4 Physiology of Anxiety

The brain is a complex organ and is the center of the human nervous system. It controls the "mind" and the physical activities. Its importance explains why it is well secured and protected from the outside world and most chemical agents that may interfere with its functioning. The Locus ceruleus and the amygdala in the brain trigger a response to danger and the acquisition and expression of fear conditioning, respectively. Over activity/hyperactivity in the locus ceruleus transmits signals to the amygdala, which is sensed as danger. The resultant instinct is the instruction to run from danger and activation of the heart to beat faster, the lungs to take in more air (breathing to become rapid), and in turn, activates all the biological components of fight/flight response.

5.4.1 Symptoms of Anxiety

- Increase in breathing rate and increase in heartbeat
- Trembling/twitching
- Muscle contraction
- Headaches
- Sweating
- Difficulty in swallowing/dry mouth
- Abdominal pain

Other symptoms include:

- Dizziness/blurred vision
- Sexual problems
- Diarrhoea or urinary urgency
- Irritability/loss of temper
- Fearful dreams, nightmares, or/and sleeping difficulties
- Decreased concentration

The symptoms of anxiety are normal to every individual, and it is intended for the safety of the person. It is a series of events that alert the person to danger and proffer a solution to be safe. In most cases, it signals the person to run from the point of danger. A problem arises if these responses proceed beyond normal.

5.5 Mechanism of Anxiety

The brain performs its functions by varying the concentrations of neurotransmitters at various synapses, varying the population of receptors expressed on the recipient cells, and altering the state of the neurons in it. Anxiety is a recognized state of the brain, which is known to be one of the most important emotional processes with firm neurobiological roots. The mechanism of anxiety and its neurochemistry is not well understood. This has made it a major area of research, and the outcome of this research has led to new approaches to the management of anxiety. Anxiety is thought to occur as a result of an imbalance in the neurochemistry of the brain. Some neurochemicals are on the high side while others are very low. A number of neurotransmitters are thought to be involved in anxiety. The levels of these neurotransmitters determine the type of anxiety, and the ease to resolve and manage. Neurochemicals that may be involved are acetylcholine, adrenaline, dopamine, endorphins, serotonin, gamma amino butyric acid, glutamate, etc.

The brain is complex and difficult to isolate parts for in-depth research that could give a clue to understanding its action. Most research has been centered on the drugs that have an effect on the brain: either worsening a disease condition or alleviating the disease condition. In the case of anxiety, most information has come from studying the action of anxiety-reducing or anxiolytic drugs on the brain: their effects on the levels of the neurochemicals at known regions of the brain. The results obtained when compared with a non-anxious brain reveal that anxiety is caused by the dysfunction of one or more neurotransmitters and their receptors. The gamma amino butyric acid (GABA) mechanisms, the serotonergic system, noradrenergic mechanisms, and neuropeptides have been implicated in anxiety. Adenosine and cholecystokinin have also been implicated in anxiety.

GABA is an inhibitory neurotransmitter in the brain, and it regulates neuronal excitability, a mechanism by which it eases stress and calms the brain. The GABA receptors (especially GABAA receptors) regulate the neuronal excitability that is responsible for rapid mood changes (e.g., panic, anxiety, and stress response) via fast inhibitory synaptic transmissions. Drugs that sedate the brain (e.g., benzodiazepines, barbiturates, neurosteroids, and ethanol) are thought to act via GABAA receptors. These drugs elevate GABA levels or enhance GABA-mediated transmission and are known to be targets of sedating drugs. Action such as alteration of the influx of chloride ions within this receptor complex is associated with the development of anxiety. All commonly used anti-anxiety agents (barbiturates, benzodiazepines, and ethanol) selectively enhance only GABA-mediated transmission, exerting anti-anxiety, relaxing, and anti-convulsant effects.

Serotonin/serotonin receptors have long been implicated in anxiety in various animal models. Reduced levels of serotonin and its action on the serotonin 1A (5HT1A) receptor can produce anxiolytic effects. Serotonin 1A receptor subtype (5-HT1A) is an autoreceptor located presynaptically on serotonin neurons. When stimulated, this receptor inhibits the synthesis and release of serotonin, thereby reducing its post-synaptic activity. This action mimics the action of a serotonin antagonist (which causes sedation) in the brain. Buspirone, a 5 HT 1A agonist, exhibits anxiolytic effects in animals and is useful in the treatment of generalized anxiety disorder but not in panic disorder. The 5-HT2A, 5-HT2C, and 5-HT3 receptors are also thought to be involved in the mechanism of anxiety.

The neurotransmitter released from adrenergic nerves in the brain, noradrenaline, is helpful in situations of emergencies or fight/flight response. Noradrenaline is elevated in these emergencies, but when its levels are elevated when there is no danger or emergencies, anxiety, fear, and irritability states are produced.

6 Neurotransmitters in Anxiety

6.1 Different Neurotransmitters Involved in Anxiety

6.1.1 Gamma-Amino Butyric Acid

1. The main inhibitory neurotransmitter in the brain is the Gamma-amino butyric acid (GABA). Anxiety regulation is centered on GABA. GABA is the target of most drug therapies.
2. GABA controls excitability states in areas of the brain. The brain state at a particular point in time is determined by the equilibrium between inhibitory GABAergic activity and excitatory inputs (mostly glutamatergic).
3. Sedation often implies that the balance is swinging toward GABA. The subject may also experience ataxia and/or amnesia.
4. Arousal, anxiety, restlessness, insomnia, and exaggerated reactivity result from the mildest attenuation of the GABAergic system.

6.1.2 Glutamate

1. L-glutamate in the mammalian brain and spinal cord has been known to be the main excitatory transmitter in the brain since the 1950s.

2. Glutamate is the main excitatory neurotransmitter in the central nervous system (CNS).
3. Glutamate is distributed all around the brain (ubiquitous). It plays important roles in brain functions, including acute neurodegeneration (e.g., cerebral ischemia and traumatic brain injury), neurodevelopment (e.g., differentiation, migration, and survival), chronic neurodegeneration (e.g., Huntington's disease and Alzheimer's disease), learning (e.g., long-term potentiation and depression), and, more recently, the stress response and anxiety disorders.
4. Glutamate through its wide range of presynaptic receptors controls the synaptic release of neurotransmitters. These controls are effected via the Group II and Group III glutamate metabotropic receptors, cholinergic (nicotinic and muscarinic) receptors, adenosine (A1), kappa opioid, γ-amino butyric acid (GABA B), cholecystokinin, and neuropeptide Y (Y2) receptors.
5. The glutamatergic system plays a major role in the pathogenesis of fear conditioning and anxiety.
6. Glutamatergic neurotransmission in the limbic system plays a pivotal role in the pathogenesis of anxiety disorders.
7. Neuronal damage and/or death are often caused by severe stress exposure, which directly leads to glutamate excitotoxicity.
8. A decrease in the CNS levels of endogenously released glutamate induces anxiolysis. A decrease in brain concentration of endogenously released glutamate results in the activation of fewer NMDA receptors, but to a less extent the CNS excitation would remain at a stable stage. Such an effect can be achieved by switching on the regulatory machinery of presynaptic glutamate release.

Monosodium glutamate (MSG) [C5H8NO4NaH2O] is one of the flavor enhancers. It is sometimes added to food during cooking as an ingredient in various food products. Monosodium glutamate is a sodium salt of naturally occurring (non-essential) L-form of glutamic acid. Glutamate is linked to metabolic syndrome and obesity that is not dependent on calories ingested and physical activity. It is reported to be neurotoxic, which may be due to over excitation of the neurons.

6.1.3 Serotonin (5HT)

1. The persistence of anxiety disorders and their development is controlled by 5-HT and GABA
2. A decrease in central nervous system 5-HT concentration increases anxiety symptoms, while an increase decreases anxiety symptoms
3. Serotonergic neurons play a major role in sleep, mood, appetite, cognitive, and energy symptoms in anxiety
4. 5HT modulates the effect on the locus coeruleus and its projections to the amygdale (areas that have a prominent role in anxiety)
5. Stress and fear activate serotonergic pathways in the brain
6. Venlafaxine (an SNRI) inhibits neuronal uptake of 5-HT and resolves the patient's anxiety

6.2 Management of Anxiety

The nature or type of the anxiety disorder and the subject's personality or characteristics determine the drugs that would be used and the method of management of anxiety. Psychotherapy, the use of orthodox drugs, and the use of alternative medications are the options for the management of anxiety.

Drugs: The first line of drugs for the management of anxiety (especially generalized social phobia) are the selective serotonin reuptake inhibitors (SSRIs). This class of drugs elevates the serotonin levels in the brain. Examples include fluoxetine, sertraline, paroxetine, citalopram, etc. Benzodiazepines are also used in the management of anxiety. This class of drugs enhances the effects of GABA in the brain. Examples of benzodiazepines include diazepam, chlordiazepoxide, etc. Monoamine oxidase inhibitors (MAOIs) (phenelzine and moclobemide) prevent the breakdown of serotonin and noradrenalin. Beta-blockers like propranolol and atenolol, which reduce the ability to produce adrenaline, are other classes of drugs that are useful in the management of anxiety. The common limitations of anxiety medications or drug therapy include co-morbid psychiatric disorders and an increase in dose leading to unbearable side-effects, such as allergic reactions, nausea, fatigue, drowsiness, coordination problems, mental confusion, and addiction liability among others.

Psychotherapy: Activities that can reduce the concentration or activity of stress neurotransmitters or neurochemicals are beneficial to subjects with anxiety. Counseling, removal of causative factors of anxiety, provision of needs, etc. may alleviate anxiety states. Cognitive-behavioral therapy and exposure therapy are effective in the management of anxiety disorders. Through psychotherapy, the pattern of thinking (Cognitive therapy) of the subject can be changed; this often diverts the subject's attention from the causative thoughts. Changes in beliefs that are associated with, and trigger, anxiety could also be helpful to the subject. Behavior therapy works when the subject is exposed the reality. Subjects are trained to confront their fears to desensitize them from the anticipated dangers/fears that are triggering anxiety.

Alternative Treatments: Some exercises have a positive impact on subjects with anxiety. Exercises that remould the mind and the thinking and belief of the subjects with anxiety have been found to be beneficial. Meditation, a form of mind exercise, is beneficial to subjects with phobias and panic disorders, and it diverts their attention from the causative factor of anxiety (phobia and fear).

Hydrotherapy: Promotes general relaxation of the nervous system. Exercise engages the subjects and diverts their focus of thought. Generally, exercise is a natural stress buster and anxiety reliever, and it strengthens the nerves and makes them more resilient. Relaxation techniques (Yoga) could be employed in the management of anxiety. These techniques include progressive muscle relaxation and controlled breathing which when practised regularly relieves anxiety. Sensors that measure physiological functions like heart rate, breathing, and muscle tension give feedback (biofeedback) that helps recognize the body's anxiety response. The information generated helps stimulate subjects to use relaxation techniques. Sleep is an important endeavor of man and animals. It helps to relax the brain and protect it from the outside world. It refreshes and prepares animals for a new day and events. Most anxious subjects often present with insomnia caused largely by their fears. Efforts are made through cognitive-behavioral therapy to make subjects sleep. Induction of sleep (Hypnotherapy) involves the application of different therapeutic approaches to help the subject sleep. Sleep often causes a state of deep relaxation (mind and muscle relaxation). Medicinal plants with hypnotic activity are beneficial in the induction of sleep in anxious subjects. Chinese traditional medicine utilizes acupuncture to help alleviate anxiety. This works under the strong belief that it relaxes the mind and muscles and subsequently may induce sleep.

7 Herbal Anxiolytes

Plants are the major natural resource that are used in alternative medicine. They are used solely or in combination with other natural substances. Plant parts consist of seeds, flowers, leaves, stems, and roots. All these parts contain varying concentrations of a particular biomolecule found in that plant. Some plants concentrate some bioactive agent preferentially in their leaves, stem, or root. Medicinal plants used for the management of anxiety utilize any of the parts for the treatment of anxiety.

7.1 Isomnia

Insomnia is a state of lack of sleep or inadequate sleep. It can occur if a subject is over stressed. Insomnia is the most common sleep disorder, about one-third of the world population suffers from a long-term derangement of sleep and wakefulness [45]. Medicinal plants that have been reported to have hypnotic effects are listed in Table 3.

Abnormality in various pathways in the brain has been credited for sleep disorders. Cortisol, GABA receptor and cytokines levels, circadian rhythm (melatonin secretion and adenosine receptors), and excitatory amino acid (glutamate and aspartate) are thought to have malfunctioned/altered to cause insomnia [46].

References

1. German-Ponciano LJ, Rosas-Sánchez GU, Rivadeneyra-Domínguez E, Rodríguez-Landa JF. Advances in the preclinical study of some flavonoids as potential antidepressant agents. Scientifica. 2018;2963565:1–14.
2. American Psychiatric Association. Diagnostic and statistical manual for mental disorders, 5a ed (DSM-5). Arlington: American Psychiatric Publishing; 2013.
3. American Psychiatric Association. Diagnostic and statistical manual of mental disorders. 4th ed. Washington, DC: American Psychiatric Association; 1994.
4. Taylor DJ, Lichstein KL, Durrence HH. Insomnia as a health risk factor. Behav Sleep Med. 2003;1(4):227–47.
5. Ferrari AJ, Charlson FJ, Norman RE, Patten SB, Freedman G, Murray CJ, Vos T, Whiteford HA. Burden of depressive disorders by country, sex, age, and year: findings from the global burden of disease study. PLoS Med. 2013;10(11):e1001547.
6. Institute of Health Metrics and Evaluation. Global Health Data Exchange (GHDx). Htts://vizhubhealthdataorg/gbd-results/. Accessed 4 march 2023.
7. Evans-Lacko S, Aguilar-Gaxiola S, Al-Hamzawi A, Alonso J, Benjet C, Bruffaerts R, Chiu WT, Florescu S, de Girolamo G, Gureje O, Haro JM, He Y, Hu C, Karam EG, Kawakami N, Lee S, Lund C, Kovess-Masfety V, Levinson D, Navarro-Mateu F, Pennell BE, Sampson NA, Scott KM, Tachimori H, ten Have M, Viana MC, Williams DR, Wojtyniak BJ, Zarkov Z, Kessler RC, Chatterji S, Thornicroft G. Socio-economic variations in the mental health treatment gap for people with anxiety, mood, and substance use disorders: results from the WHO World Mental Health (WMH) surveys. Psychol Med. 2018;48(9):1560–71.
8. Firth J, Marx W, Dash S, Carney R, Teasdale SB, Solmi M, Stubbs B, Schuch FB, Carvalho AF, Jacka F, et al. The effects of dietary improvement on symptoms of depression and anxiety: a meta-analysis of randomized controlled trials. Psychosom Med. 2019;81:265–80.
9. Dwyer AV, Whitten DL, Hawrelak JA. Herbal medicines, other than St. John`s wort, in the treatment of depression. Sci Rev Altern Med. 2011;16(1):281–4.
10. Tegegne MT, Mossie TB, Awoke AA, Assaye AM, Gebrie BT, Eshetu DA. Depression and anxiety disorder among epileptic people at Amanuel specialized mental hospital, Addis Ababa, Ethiopia. BMC Psychiatry. 2015;15(1):210.
11. Kessler D, Bennewith O, Lewis G, Sharp D. Detection of depression and anxiety in primary care: follow up study. BMJ. 2002;325(7371):1016–7.
12. Unützer J, Klap R, Sturm R, Young AS, Marmon T, Shatkin J, Wells KB. Mental disorders and the use of alternative medicine: results from a national survey. Am J Psychiatry. 2000;157(11):1851–7.
13. Knaudt PR, Connor KM, Weisler RH, Churchill LE, Davidson JR. Alternative therapy use by psychiatric outpatients. J Nerv Ment Dis. 1999;187(11):692–5.
14. Khan RA, Riaz A. Behavioral effects of Citrus Limon in rats. Metab Brain Dis. 2015;30(2):589–96. https://doi.org/10.1007/s11011-014-9616-2. Epub 2014 Sep 17.

15. Viana MDM, da Silva Neto GJ, de Lima AA, Leite AB, Souza IT, Santana AEG, Campesatto EA, Moreira MSA. Citrus limon (L.) Burm f. Essential oil has anxiolytic and sedative properties by modulating GABAA-receptors human and animal health. Braz Arch Biol Technol. 2020;63:1.
16. Mittal S, Gupta P, Nigam V. Evaluation of antidepressant-like effect of Clitoria ternatea Linn. Res J Pharm Technol. 2021;14:6437–41.
17. Motti R, de Falco B. Traditional herbal remedies used for managing anxiety. Horticulturae. 2021;7:523.
18. Pannu A, Sharma PC, Thakur VK, Goyal RK. Emerging role of flavonoids as the treatment of depression. Biomol Ther. 2021;2021(11):1825.
19. Uncini Manganelli RE, Tomei PE. Ethnopharmacobotanical studies of the Tuscan archipelago. J Ethnopharmacol. 1999;65:181–202.
20. Guarino C, De Simone L, Santoro S. Ethnobotanical study of the Sannio area, Campania. South.Italy. Ethnobot Res Appl. 2008;6:255–317.
21. Tuttolomondo T, Licata M, Leto C, Bonsangue G, Letizia Gargano M, Venturella G, La Bella S. Popular uses of wild plant species for medicinal purposes in the Nebrodi Regional Park (north-eastern Sicily, Italy). J Ethnopharmacol. 2014;157:21–37.
22. Pieroni A, Nebel S, Quave C, Munz H, Heinrich M. Ethnopharmacology of liakra: traditional weedy vegetables of the Arbereshe of the vulture area in southern Italy. J Ethnopharmacol. 2002;81:165–85.
23. Licata M, Tuttolomondo T, Leto C, Virga G, Bonsangue G, Cammalleri I, Gennaro MC, La Bella S. A survey of wild plant species for food use in Sicily (Italy)–results of a 3-year study in four regional parks. J Ethnobiol Ethnomed. 2016;12:12.
24. Abourashed EA, Koetter U, Brattstrom A. In vitro binding experiments with a valerian, hops and their fixed combination extract (Ze91019) to selected central nervous system receptors. Phytomedicine. 2004;11:633–8.
25. Kim HK, Son KH, Chang HW, Kang SS, Kim HP. Amentoflavone, a plant biflavone: a new potential anti-inflammatory agent. Arch Pharm Res. 1998;21:406–10.
26. Ishola I, Chatterjee M, Tota S, Tadigopulla N, Adeyemi OO, Palit G, Shukla R. Antidepressant and anxiolytic effects of amentoflavone isolated from Cnestis ferruginea in mice. Pharmacol Biochem Behav. 2012;103:322–31.
27. Salehi B, Venditti A, Sharifi-Rad M, Kregiel D, Sharifi-Rad J, Durazzo A, Lucarini M, Santini A, Souto EB, Novellino E, et al. The therapeutic potential of Apigenin. Int J Mol Sci. 2019;20:1305.
28. Li R, Wang X, Qin T, Qu R, Ma S. Apigenin ameliorates chronic mild stress-induced depressive behavior by inhibiting interleukin-1β production and NLRP3 inflammasome activation in the rat brain. Behav Brain Res. 2016;296:318–25.
29. Weng L, Guo X, Li Y, Yang X, Han Y. Apigenin reverses depression-like behavior induced by chronic corticosterone treatment in mice. Eur J Pharm. 2016;774:50–4.
30. Ishisaka M, Kakefuda K, Yamauchi M, Tsuruma K, Shimazawa M, Tsuruta A, Hara H. Luteolin shows an antidepressantlike effect via suppressing endoplasmic reticulum stress. Biol Pharm Bull. 2011;34(9):1481–6.
31. Liu B, Xu C, Wu X, Liu F, Du Y, Sun J, Tao J, Dong J. Icariin exerts an antidepressant effect in an unpredictable chronic mild stress model of depression in rats and is associated with the regulation of hippocampal neuroinflammation. Neuroscience. 2015;294:193–205.
32. Wei K, Xu Y, Zhao Z, Wu X, Du Y, Sun J, Yi T, Dong J, Liu B. Icariin alters the expression of glucocorticoid receptor, FKBP5 and SGK1 in rat brains following exposure to chronic mild stress. Int J Mol Med. 2016;38:337–44.
33. Gong MJ, Han B, Wang SM, Liang SW, Zou ZJ. Icariin reverses corticosterone-induced depression-like behavior, decrease in hippocampal brain-derived neurotrophic factor (BDNF) andmetabolic network disturbances revealed by NMR based metabonomics in rats. J Pharm Biomed Anal. 2016;123:63–73.
34. Butterweck V, Hegger M, Winterhoff H. Flavonoids of St. John's wort reduce HPA axis function in the rat. Planta Med. 2004;70:1008–11.
35. Park S, Sim Y, Han P, Lee J, Suh H. Antidepressantlike effect of kaempferol and quercitirin, isolated from Opuntia ficus-indica var. Saboten Exp Neurobiol. 2010;19:24–30.
36. Domitrovi'C R, Rashed K, Cvijanovi'C O, Vladimir-Kneževi'C S, Škoda M, Višni'C A. Myricitrin exhibits antioxidant, antiinflammatory and antifibrotic activity in carbon tetrachloride-intoxicated mice. Chem Biol Interact. 2015;230:21–9.
37. Meyer E, Mori MA, Campos AC, Andreatini R, Guimarães FS, Milani H, de Oliveira RMW. Myricitrin induces antidepressant-like effects and facilitates adult neurogenesis in mice. Behav Brain Res. 2017;315:59–65.
38. Terao J, Yamaguchi S, Shirai M, Miyoshi M, Moon JH, Oshima S, Inakuma T, Tsushida T, Kato Y. Protection by quercetin and quercetin 3-O-beta-D-glucuronide of peroxynitrite-induced antioxidant consumption in human plasma low-density lipoprotein. Free Radic Res. 2001;35:925–31.
39. Rinwa P, Kumar A. Quercetin suppress microglial neuroinflammatory response and induce antidepressent-like effect in olfactory bulbectomized rats. Neuroscience. 2013;255:86–98.
40. Holzmann I, Silva LM, Corrêa da Silva JA, Steimbach VM, De Souza MM. Antidepressant-like effect of quercetin in bulbectomized mice and involvement of the antioxidant defenses, and the glutamatergic and oxidonitrergic pathways. Pharm Biochem Behav. 2015;136:55–63.
41. Pauff JM, Hille R. Inhibition studies of bovine xanthine oxidase by luteolin, silibinin, quercetin, and curcumin. J Nat Prod. 2009;72:725–31.
42. Petacci F, Freitas SS, Brunetti IL, Khalil NM. Inhibition of peroxidase activity and scavenging of reactive oxygen species by astilbin isolated from Dimorphandra mollis (Fabaceae, Caesalpinioideae). Biol Res. 2010;43:63–74.
43. Lv QQ, Wu WJ, Guo XL, Liu RL, Yang YP, Zhou DS, Zhang JX, Liu JY. Antidepressant activity of astilbin: involvement of monoaminergic neurotransmitters and BDNF signal pathway. Biol Pharm Bull. 2014;37:987–95.
44. Ren Z, Yan P, Zhu L, Yang H, Zhao Y, Kirby BP, Waddington JL, Zhen X. Dihydromyricetin exerts a rapid antidepressantlike effect in association with enhancement of BDNF expression and inhibition of neuroinflammation. Psychopharmacology. 2018;235:233–44.
45. Liua L, Liu C, Wang Y, Wang P, Li Y, Li B. Herbal medicine for anxiety, depression and insomnia. Curr Neuropharmacol. 2015;13:481–93.
46. Khanum F, Razack S. Anxiety-herbal treatment: a review. Res Rev Biomed Biotechnol. 2010;1(2):71–89.

Role of Herbal Medicines for the Treatment of Arthritis

Vafa Baradaran Rahimi and Vahid Reza Askari

Abstract

The use of herbal medicines has extensively propagated in almost every country worldwide and also in developed countries. Nowadays, various herbal therapies are well-studied and used to treat many chronic and acute disorders, including rheumatoid arthritis and osteoarthritis. This chapter focuses on the promising anti-arthritis effects of *Nigella sativa* and its major ingredient thymoquinone, *Zingiber officinale* and its constituents 6-gingerols and 6-shogaols, *Curcuma longa* and its main component curcumin, *Tripterygium wilfordii* Hook and its major constituent triptolide, *Punica granatum* and its major constituents ellagic acid and punicalagin, *Boswellia serrata* and its major constituent boswellic acid, *Crocus sativus* and its main constituent crocin, *Berberis vulgaris* and its main ingredient berberine, *Withania somnifera*, and *Eucommia ulmoides*. We reviewed different animal models and clinical studies of rheumatoid arthritis and osteoarthritis. The important mechanisms responsible for the anti-arthritis properties of these herbal medicines were mainly anti-oxidative, anti-inflammatory, and anti-apoptotic, regulating M1/M2 balance, T cell suppression, suppressing TLR4/NF-κB, and mTOR signaling pathways.

Keywords

Rheumatoid arthritis · Osteoarthritis · Herbal medicine · Inflammation

V. Baradaran Rahimi
Department of Cardiovascular Diseases, Faculty of Medicine, Mashhad University of Medical Sciences, Mashhad, Iran
e-mail: baradaranrv@mums.ac.ir

V. R. Askari (✉)
Applied Biomedical Research Center, Mashhad University of Medical Sciences, Mashhad, Iran

International UNESCO Center for Health-Related Basic Sciences and Human Nutrition, Mashhad University of Medical Sciences, Mashhad, Iran
e-mail: askariv@mums.ac.ir

1 Introduction

Herbal medicines with promising anti-arthritis effects are illustrated in Fig. 1. Additionally, Table 1 reviewed the protective effects of herbal medicines against animal models of arthritis. Additionally, clinical studies supporting the anti-arthritis effects of the mentioned herbal medicines are presented in Table 2.

1.1 *Nigella sativa* and Its Major Ingredient Thymoquinone

Nigella sativa (*N. sativa*), popularly named black cumin, is an annual flowering herb member of Ranunculaceae [85]. Its small and black seeds have been administered in many traditional foods and also as traditional medicine. Thymoquinone is the major active constituent of *N. sativa* seeds. Several pieces of evidence support that *N. sativa* has antitussive, anti-oxidant, anti-cancer, anti-inflammatory, analgesic, anti-anxiety, nephroprotective, as well as anti-arthritis activities [86, 87].

1.2 Animal Studies

In the arthritis model caused by complete Freund's adjuvant (CFA) injection, treatment with N. sativa oil notably decreased paw volume and arthritis score while improving spontaneous locomotor activity and anxiety-like behavior in rats [1].

1.2.1 Thymoquinone

In collagen-induced arthritis (CIA), thymoquinone notably diminished arthritis scoring, paw volume, articular elastase activity, myeloperoxidase (MPO) activity, malondialdehyde (MDA), and nitric oxide (NO) levels while enhancing glutathione (GSH) level, superoxide dismutase (SOD), and catalase (CAT) activities in rat joints. Thymoquinone also

A. K. Dhara, S. C. Mandal (eds.), *Role of Herbal Medicines*, https://doi.org/10.1007/978-981-99-7703-1_14

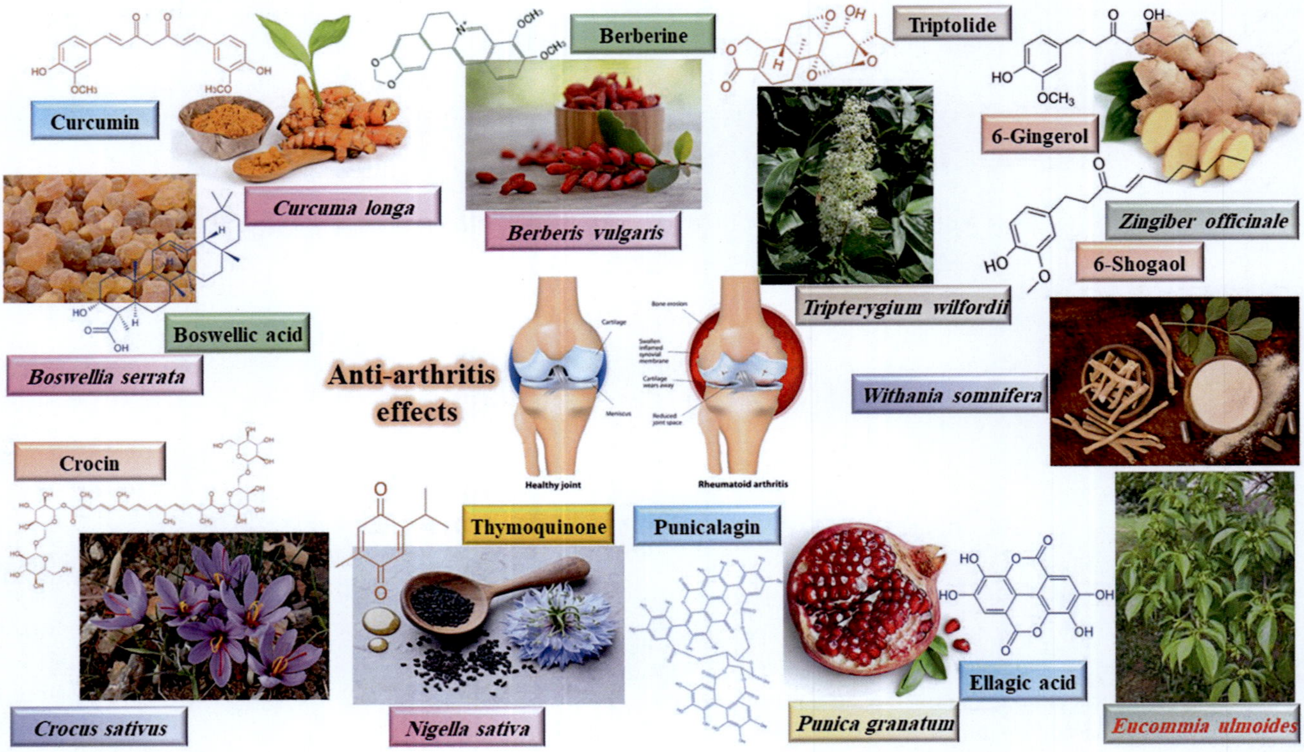

Fig. 1 Herbal medicines with promising anti-arthritis effects

diminished interleukin (IL)-1β and 6, tumor necrosis factor-alpha (TNF-α), interferon-gamma (IFN-γ), and prostaglandin E_2 (PGE2) levels while increasing IL-10 levels in joints of CIA rats [2].

In CFA-induced arthritis in rats, thymoquinone markedly attenuated arthritis scoring, histopathological scoring, C-reactive protein (CRP) level, toll-like receptor (TLR)2 and 4, IL-1, nuclear factor-kappa B (NF-κB), and TNF-αmRNA expression [3]. Similarly, thymoquinone strikingly mitigated arthritis scoring, IL-1β, TNF-α, PGE2 levels, and alkaline phosphatase activity following CFA-induced arthritis [4].

Furthermore, in pristine-induced arthritis in female rats, thymoquinone provided a significant decrement in paw weight and histopathological score [5].

1.3 Clinical Studies

N. sativa oil notably diminished disease activity score (DAS-28), swollen joints number, the duration of morning stiffness, and VAS pain score in female rheumatoid arthritis (RA) patients [61]. Similarly, *N. sativa* oil markedly reduced high sensitivity (hs)-CRP, the score of DAS-28, and swollen joint numbers in female RA patients. Moreover, N. sativa oil modulated T lymphocytes by decreasing $CD8^+$ T cell percentage and $CD4^+/CD8^+$ while increasing the percentage of $CD4^+CD25^+$ T cells in RA patients [62]. Hadi et al. supported that *N. sativa* oil meaningfully alleviates MDA and NO while elevating IL-10 in RA patients [63].

In addition, Khabbazi and co-workers conducted a meta-analysis on 19 animal, cellular, and clinical studies. They emphasized that *N. sativa* remarkably improved RA in animal and cellular studies through anti-oxidative, anti-inflammatory, and immunomodulatory effects. However, the results of five clinical studies did not show a significant improvement in RA patients [88].

Collectively, further clinical studies are necessary to explain and confirm the anti-RA activities of *N. sativa* and thymoquinone.

2 *Zingiber officinale* and Its Constituents 6-Gingerols and 6-Shogaols

Zingiber officinale (*Z. officinale*), popularly named ginger, is a perennial, herbaceous plant and member of ingiberaceae [89]. Its roots are widely used as a spice and also in folk medicine worldwide. Besides that, *Z. officinale* has anti-emetic, anti-oxidant, anti-inflammatory, antidiabetic, cardio-protective, as well as anti-arthritis activities. The phenolic constituents of *Z. officinale* are mainly responsible for its several effects, including 6-gingerol and 6-shogaol [90, 91].

Table 1 Protective effects of herbal medicines against animal models of arthritis

Extract or constituent type	Dose/Concentration	An overview of the study model	Results	Ref
NSO	1596, and 798 mg/kg/day; orally over 25 days of treatment	CFA-induced arthritis in rats	↓ paw volume and arthritis score Improved spontaneous locomotor activity and anxiety-like behavior	[1]
Thymoquinone	5 mg/kg; orally over 21 days	Arthritis following collagen injection in rats	↓ arthritis scoring, paw volume ↓articular elastase activity, MPO activity, MDA, and NO levels ↑ GSH and IL-10 level ↑ SOD and CAT activities in joints ↓ IL-1β, IL-6, TNF-α, IFN-γ, and PGE2	[2]
Thymoquinone	10 mg/kg; given I.P. During 20 days	CFA-caused arthritis in rats	↓ arthritis and histopathological ↓ CRP level ↓ TLR2 and 4, IL-1, NF-κB, and TNF-α	[3]
Thymoquinone	5 mg/kg; given orally for 28 days	CFA-caused arthritis in rats	↓ arthritis scoring ↓HNE, IL-1β, TNF-α, and PGE2 levels ↓ alkaline phosphatase activity	[4]
Thymoquinone	2 mg/kg; injected I.P. Over 15 days	Pristine-caused arthritis in female rats	↓ paw weight ↓histopathological score	[5]
Zingiber officinale extract and essential oil	26 and 28 mg/kg/day); I.P. For 28 days	Arthritis following streptococcal cell wall in female rats	↓arthritic index, joint swelling, and cartilage destruction index	[6, 7]
Zingiber officinale rhizomes powder	200 mg/kg; given orally over 28 days	CFA-caused arthritis in rats	↓ paw edema ↓ IL-1β, −6, and − 4, and TNF-α levels ↑ IL-10 level	[8]
70% Hydroalcoholic extract of rhizomes of *Zingiber officinale*	50, 100, and 200 mg/kg; injected I.P. Over 26 days	Collagen-induced arthritis in rats	↓ clinical scores, arthritic incidence, joint temperature and swelling, and cartilage destruction ↓ ESR, IL-1β, IL-6, IL-2, TNF-α, and anti-CII antibody levels	[9]
6-shogaol	30 mg/kg; injected I.P. During three continuous days	Gouty arthritis induction by monosodium urate crystal in mice	↓ paw volume ↓ lysosomal enzymes activities, lipid peroxidation, and TNF-α levels ↑ Gpx, SOD, and CAT activities	[10]
6-shogaol	6.2 mg/kg; orally over 28 days	CFA-caused arthritis in rats	↓ knees swelling and edema, leukocyte infiltration into the synovial cavity ↓ VCAM-1 level	[11]
Curcumin	100 and 200 mg/kg; given orally over18 days	Collagen-induced arthritis in rats	↓ arthritis score, histopathological changes in the joints ↓ TNF-α, IL-17, IL-1β, and TGF-β levels ↓ IκBα degradation ↓ COX-2 expression	[12]
Curcumin	100 mg/kg; administered for 22 days	Collagen-caused arthritis in mice	↓ clinical score, histology score ↓ MMP-1, 3, 9, and 13 ↓ anti-collagen type II antibodies ↓ TNF-α, IL-1β, −6, -33, -15, -16, and -33 ↓ IFN-γ, IP-10, MIP-α, and MCP-1	[13]

(continued)

Table 1 (continued)

Extract or constituent type	Dose/Concentration	An overview of the study model	Results	Ref
Curcumin	200 mg/kg; orally for 3 weeks	Collagen-induced arthritis in rats	↓ redness and paw edema, arthritis score, penetration of inflammatory cells into the synovium ↓ TNF-α, IL-1β, MMP-1, and -3 ↓ mTOR, p70S6K, and 4E-BP1 expression	[14]
Curcumin	50 mg/kg; orally for 28 days	Collagen-caused arthritis in mice	↓ arthritis scores, paw edema ↓ TNF-α, IL-6, IL-17 ↓ p-PI3K/PI3K, and p-AKT/AKT ratios	[15]
Curcumin	1 and 2 g/kg; taken orally over 14 days	CFA-induced arthritis in rats	↑ SOD, CAT, and Gpx activities ↑ MSRA gene expression and protein level	[16]
Curcumin	20 mg/kg; orally for 14 days	Monosodium iodoacetate- induced knee osteoarthritis in rats	↓ TNF-α, IL-1β, MMP-3, COMP ↑ SOD and collagen II levels ↓ TLR4, COX-2, and p-p65 expression levels	[17]
Curcumin	50 mg/kg; orally for 60 days	Zymosan-induced osteoarthritis in rats	↑ number of chondrocytes ↑ collagen 2, SOX-5, and Indian hedgehog expression	[18]
Curcumin	150 mg/kg; administered I.P.	Monosodium urate-induced acute gout in mice	↓ paw and ankle joint swelling, MPO activity ↓ p-p65 and p-p50 phosphorylation ↑ IκBα protein level ↓ COX-2, TLR4, MyD88, and NLRP3 expression	[19]
Curcumin	100, 150, and 200 mg/kg; given I.P. Over seven days	Acute gout caused by monosodium urate in rats	↓ ankle circumference, penetration of neutrophils in knee joints ↓ TNF-α, IL-1β, and elastase levels ↓ NLRP3, pro-caspase-1, caspase-1, pro- IL-1β, and IL-1β expression	[20]
Total alkaloids of *Tripterygium wilfordii*	0.25, 0.5, and 1 g/kg; orally during 28 days	Collagen-caused arthritis in rats	↓ paw swelling ↓ TNF-α, IL-6 and -8, and NF-kB	[21]
Tripterygium wilfordii glycosides	6 mg/kg; taken orally over 21 days	Collagen-caused arthritis in rats	↓ paw volume, arthritis score ↓histopathological damages ↓anti-typeIIcollagen antibodies, and IL-1β	[22]
Triptolide	45 μg/kg/d; orally for 28 days	Collagen-induced arthritis in rats	↓ paw thickness, arthritis score ↓ MPO, COX-2, iNOS ↓ oxidative stress	[23]
Triptolide	45 μg/kg; orally for 51 days	Collagen-caused arthritis in rats	↓ arthritis score ↓ TNF-α, IL-1β and -6, MMP-3 and -9, URI1, and FRAT1 mRNA expression	[24]
Triptolide	0.4 mg/kg; orally for 14 days	Adjuvant arthritis model in rats	↑ PI3K/AKT ↓ VEGFA and C1QTNF3	[25]
Triptolide	3.3, 10, and 33 μg/kg/d; for 6 weeks	TNF transgenic mice with RA	↓ arthritis score ↓ TNF-α, IL-1β, IL-α levels ↑ arthritis detumescence percentages ↑ apoptosis rates of OCP and T lymphocytes	[26]
Triptolide	10 and 33 μg/kg/day, orally for 42 days	TNF transgenic mice with RA	↓ arthritis symptoms, joint swelling ↓ OC cell levels of inflammatory joints ↓ inflammation, swelling, and bone erosion ↑ apoptosis rate of OCP cells	[27]

Extract or constituent type	Dose/Concentration	An overview of the study model	Results	Ref
P. granatum rind extract	5%, 2.5%, and 1%, w/w; topical	CFA-induced arthritis in rats	↓ paw volume, histopathological damages, MPO activity ↓ TNF-α, and IL-1β levels	[28]
P. granatum rind extract	50–200 mg/kg; taken orally over a period of 21 days	CFA-induced arthritis in rats	↓ paw volume ↓ MDA, TNF-α levels ↓ TNF-R1, TNF-α, IL-1β, IL-6, and NF-κB expression levels ↑ GSH and SOD activity	[29]
Ethanolic extract of *P. Granatum*	500–750 mg/kg; taken orally over 4 weeks	CFA-caused arthritis in rats	↓ paw volume, arthritic score ↓ histological damages ↓ MDA, RF, TNF-α, and IL-1β	[30]
Butanol fraction of *P. granatum* rind extract	50 and 75 mg/kg; taken orally for continuous 28 days	CFA-induced arthritis in rats	↓ arthritic score, paw volume, and joint diameter ↓ ESR level	[31]
Ellagic acid	700 mg/kg; given orally over 4 weeks	Adjuvant-caused arthritis in mice	↓ paw edema, histopathological damages ↓ TNF-α, IL-1β, and IL-17 levels ↑ IL-10 and IFN-γ levels	[32]
Ellagic acid	50 mg/kg; orally for 20 days	Arthritis induction by adjuvant in mice	↓ paw edema ↓ CHI3L1, NF-κB, IL-1β, MMP-9, VEGF, and Caspase-3 protein expression ↓ MDA and NO levels ↑ GSH level	[33]
Ellagic acid	25–100 mg/kg; orally during 21 days	CFA-induced arthritis in rats	↓ paw swelling, arthritis index, and pathological damages ↓ TNF-α, IL-1β, and IL-6, MTA1, and HDAC1 ↑ Nur77	[34]
Punicalagin	50 mg/kg; injected I.P. During 14 days	CIA in mice	↓ paw volume, clinical arthritis score, histology score ↓ TNF-α and IL-6 levels	[35]
Punicalagin	10–20 mg/kg; injected I.P. Over 27 days of treatment	CIA in mice	↓ paw thickness, arthritis score, and bone destruction ↓changed macrophages from the M1 to the M2 phenotype ↓ iNOS, TNF-α, IL-1β and – 6 ↑Arginase-1 and IL-10 expression ↓ NLRP3 and caspase-1 expression	[36]
Boswellia serrate gum resin extract	100 and 200 mg/kg; taken orally over 21 days	CIA in rats	↓ paw diameter, clinical severity score ↓ elastase, MPO, MDA, and NO ↑ GSH level, SOD, and CAT activity ↓ TNF-α, IL-1β and – 6, IFN-γ, and PGE2 ↑ IL-10 levels	[37]
Boswellia serrate gum resin extract	40–80 mg/kg; taken orally for 21 days	CIA in rats	↓ arthritis index, paw volume, pain, motility score, stair climbing, and histopathological damages ↓ collagen antibodies, CRP, and ESR levels in serum ↑ Hyaluronan and cartilage oligomeric matrix protein in synovial fluid	[38]

(continued)

Table 1 (continued)

Extract or constituent type	Dose/Concentration	An overview of the study model	Results	Ref
Boswellic acid	30 mg/kg; administered I.P.	Gouty arthritic in mice following monosodium urate crystal consumption	↓ paw thickness, lysosomal enzymes ↓ MDA, and TNF-α levels ↑ SOD, GPx, and CAT activities	[39]
Boswellic acid	20–40 mg/kg; administered orally over 45 days	CFA-caused arthritis in rats	↓ paw edema, ESR, and MDA levels ↑ anti-oxidant enzymes	[40]
Crocin	10 and 20 mg/kg; taken orally for 14 days	CFA-caused arthritis in rats	↓ paw swelling ↓ Exoglycosidases, Cathepsin-D, alkaline phosphatase, acid phosphatase, and tartrate-resistant acid phosphatase activities ↓ TNF-α, IL-1β, IL-6, NF-kB, COX-2, PGE2, and ROS ↑ GSH, SOD, and CAT	[41]
Crocin	6.25–25 mg/kg; given orally over 28 days	CFA-induced arthritis in rats	↓ paw swelling, arthritis score ↓ iNOS and NO levels ↓ TNF-α, IL-1β, and IL-6 levels	[42]
Crocin	10, 20, and 40 mg/kg; given orally during a period of 14 days	CIA in mice	↓ arthritis score, paw thickness, histological score ↓ MMP-1, −3, and − 13, TNF-α, IL-6, −17, and − 8, and CXCL8	[43]
Crocin	50 mg/kg; injected I.P. for 1 week	CIA in mice	↓ mean clinical score ↓ TNF-α, IL-1β, and − 6 levels	[44]
Crocin	50–100 mg/kg; injected I.P.	Adjuvant-caused arthritis in rats	↓ pain ↓ TNF-α and IL-1β levels ↓ Wnt5a/β-catenin pathway ↓ GFAP and Iba-1 levels	[45]
Crocin	30 mg/kg; taken orally over ten days	Meniscectomy surgery-induced osteoarthritis in rats	↓ joint pain ↓ ROS, lipid peroxidation, JNK ↑ GSH and GPx activities ↓ IL-6 level, citrate synthase activity, and MHC IIα expression	[46]
Berberine	75 and 150 mg/kg; taken orally for 21 days	Arthritis by CFA in rats	↓ paw swelling, arthritis index score ↓ IL-6 and -17, anti-IL-17 antibody ↑ IL-10, TGF-β, and anti-IL-10 antibody	[47]
Berberine	40, 80, and 160 mg/kg; taken orally during a period of 14 days	CFA-caused arthritis in rats	↓ paw swelling, arthritis global assessment, and histopathological damages ↓ TNF-α, IL-1β and-6, and iNOS ↑ IL-10, TGF-β1, and Arg1 ↓ Th17/Treg ratio ↓ numbers of Th17 ↑ number of Treg ↑ p-AMPK expression ↓ p-p65, p-IκBα and COX-2	[48]

Extract or constituent type	Dose/Concentration	An overview of the study model	Results	Ref
Berberine	40, 80, and 160 mg/kg; taken orally over 14 days of treatment	CFA-caused arthritis in rats	↓ TNF-α, IL-1β, -6, and -17, MCP-1, and IFN-γ ↓ proportion of M1 macrophages ↑ IL-10, IL-4, TGF-β1 ↑ proportion of M2 macrophages ↑ p-AMPK expression ↓ HIF-1α expression in synovial macrophages	[49]
Berberine	1 mg/kg/day; injected I.P. Over 28 days	CIA in mice	↓ number of CD4 + T_h, $CD4^+CXCR5^+T_{fh}$ cells ↓ CD28, CD154 expression ↑ $Foxp3^+$ Treg cells	[50]
Berberine	100–200 mg/kg; given orally over 14 days	CIA in rats	↓ paw swelling, arthritis index, histological score, joint destruction ↓ RANKL, TNF-α, IL-1β, and IL-6 ↓ proportion of Th17 cells, IL-17 levels, and phosphorylation STAT3	[51]
Berberine	200 mg/kg; orally for 42 days	CIA in rats	↓ arthritis score ↓ anti-CII total IgG, IgG1, IgG2a ↓ TNF-α, IL-1β, -6, and -17, VEGF ↓ VEGF and CD34 ↓ p-ERK, p-p38, and p-JNK	[52]
Berberine	15–60 mg/kg; orally during 12 weeks of treatment	Glucocorticoid-induced osteoporosis in rats	↑bone mineral density, trabecular number and thickness, mineral apposition rate, bone formation rate/total tissue area, and bone formation rate/trabecular bone surface ↓ trabecular separation, and osteoclast number/trabecular bone volume	[53]
Aqueous extract or roots of *Withania somnifera*	100, 200, 300 mg/kg; orally for 20 days	CIA in rats	↓ arthritic index, rheumatoid factor ↑ GSH and ferric ability of plasma ↓ anti-cyclic citrullinated peptide antibody, ANA, anti-collagen type II, CRP, MDA, and GST	[54]
Aqueous extract or roots of *Withania somnifera*	300 mg/kg; taken orally over 45 days	CIA in rats	↓ clinical severity, histological score ↓ TNF-α, IL-1β and -6, NF-kB, MMP-8, and iNOS levels ↑ IL-10 levels	[55]
70% ethanolic extract of roots of *Withania somnifera*	250–500 mg/kg/day	CFA-induced arthritis in rats	↓ paw swelling, arthritic score ↓ ESR, CRP, and rheumatoid factor ↓ IL-1β, -6, and -17a, TNF-α, and RANKL, MDA ↑ IL-4, INF-γ, and osteoprotegerin ↑ CAT and SOD activities	[56]
70% ethanolic extract of *Eucommia ulmoides*	2.7 g/kg; orally for 45 days	CIA in rats	↓ paw volume, arthritis index, pathological score ↓ Th17-positive cell numbers ↓ TNF-α, IL-1β, and IL-17 ↓ RANKL mRNA levels ↑ IL-10 and OPG mRNA level	[57]

(continued)

Table 1 (continued)

Extract or constituent type	Dose/Concentration	An overview of the study model	Results	Ref
70% ethanol extract, ethyl acetate fraction, and n-butyl alcohol fraction of *Eucommia ulmoides*	4 g/kg; orally for 4 weeks	CIA in rats	↓ ankle swelling ↓ TNF-α, IL-1β, and IL-17 levels ↓ RANKL/OPG ratio ↓ MMP-9, p65 NF-κB, and p-IKKαβexpression	[58]
70% ethanol extract of *Eucommia ulmoides*	4 g/kg; orally for 2 weeks	CIA in rats	↓ paw volume, ankle joint score ↓ TNF-α, IL-6, IL-17, IL-1β, and NO levels	[59]
Aqueous extract of *Eucommia ulmoides*	6 mL/kg/day; orally during 4 weeks	Knee osteoarthritis in rat	↓ histopathological damages ↓ MMP-1, −3, and − 13	[60]

NSO Nigella sativa oil, *CFA* complete Freund's adjuvant, *MPO* myeloperoxidase, *MDA* malondialdehyde, *NO* nitric oxide, *GSH* glutathione, *SOD* superoxide dismutase, *CAT* catalase, *IL* interleukin, *TNF-α* tumor necrosis factor alpha, *IFN-γ* interferon gamma, *PGE2* prostaglandin E_2, *CRP* C-reactive protein, TLR toll-like receptor, *NF-κB* nuclear factor-kappa B, *HNE* 4-hydroxynonenal, *ESR* erythrocyte sedimentation rate, *Gpx* glutathione peroxidase, *VCAM* vascular cell adhesion molecule, *MMP* matrix metalloproteinases, *IP-10* IFN-gamma-inducible protein 10, *MIP* macrophage inflammatory proteins, *MCP-1* -α, monocyte chemoattractant protein-1, *mTOR* mammalian target of rapamycin, *p70S6K* p70S6 kinase, *4E-BP1* eukaryotic initiation factor 4E binding protein 1, *MSRA* methionine sulfoxide reductase A, *COMP* cartilage oligomeric matrix protein, *MyD88* myeloid differentiation factor 88, *NLRP3* NLR family pyrin domain containing 3, *iNOS* inducible nitric oxide synthase, *URI1* unconventional prefoldin RPB5 interactor 1, *FRAT1* frequently rearranged in advanced T cell lymphomas-1, *VEGF* vascular endothelial growth factor, *C1QTNF3* A and C1q and tumor necrosis factor related protein 3, *OCP* osteoclast precursor cells, *OC* osteoclast, *TNF-R1* tumor necrosis factor receptor 1, *RF* rheumatoid factor, *CHI3L1* chitinase-3-like protein-1, *MTA1* metastasis-associated gene 1, *HDAC1* histone deacetylase 1, *ROS* reactive oxygen species, *GFAP* glial fibrillary acidic protein, *JNK* c-Jun N-terminal kinase, *MHC* myosin heavy chain, *HIF-1α* hypoxia inducible factor 1α, *STAT3* signal transducer and activator of transcription 3, *ANA* anti-nuclear antibody, *GST* glutathione-S-transferase activity, *AMPK* AMP-activated protein kinase, *OPG* osteoprotegerin

Table 2 Clinical studies of the anti-arthritis effects of herbal medicines

Extract type or plant part	Study design	Study model	Results	Ref
Nigella sativa oil	• 500 mg two times/daily for 30 days • Placebo	Placebo, controlled clinical trial on female RA patients	↓ disease activity score, number of swollen joints, morning stiffness duration, VAS pain score	[61]
Nigella sativa oil	• 500 mg two times/daily for 8 weeks • Placebo	A randomized, double-blinded, placebo-controlled, parallel-group clinical trial on female RA patients	↓ hs-CRP level ↓ DAS-28 score ↓swollen joint numbers ↓ $CD8^+$ T cell ↓ $CD4^+/CD8^+$ ratio ↑ $CD4^+CD25^+$ T cell	[62]
Nigella sativa oil	• 500 mg two times/daily for 8 weeks • Placebo	A randomized, double-blind, placebo-controlled clinical trial on RA patients	↓ MDA and NO ↑ IL-10 level	[63]
Zingiber officinale rhizomes powder	• 1500 mg /daily over 12 weeks of treatment • Placebo	A randomized, double-blind, placebo-controlled clinical trial on RA patients	↓ DAS-28 arthritis score ↓ T-bet, RORγt, and NF-kB gene expression ↑ FoxP3, GAT A3, and PPAR-γ gene expression	[64]
Zingiber officinale rhizomes powder	• 1000 mg /daily for 12 weeks • Placebo	A randomized placebo-controlled clinical trial on knee osteoarthritis patients	↓ VAS pain score ↑ satisfaction with treatment	[65]
95% hydroethanolic extract of *Zingiber officinale*	• 30 mg/daily over 30 days • Placebo	A randomized placebo-controlled clinical trial on osteoarthritis patients	↓ VAS and gelling pain score	[66]
Zingiber officinale rhizomes powder	• 250 mg/daily for 6 weeks • Placebo	A randomized placebo-controlled clinical trial on knee osteoarthritis patients	↓ VAS pain score, WOMAC pain score, morning stiffness, and difficulty	[67]
Curcumin	• 180 mg /daily for 8 weeks • Placebo	A randomized, double-blind,placebo-controlled prospective study on knee osteoarthritis patients	↓ VAS pain score ↓need for NSAIDs consumption	[68]
Curcumin	• 500 mg + diclofenac 50 mg twice/daily for 28 days • Diclofenac 50 mg twice/daily	A prospective, randomized, open-label parallel-group study on knee osteoarthritis patients	↓ pain, need for rescue analgesics, need for histamine 2 blockers ↓ side effects ↑ quality of life	[69]
Curcumin	• 120 mg to 1500 mg for 4–36 weeks • Placebo	A meta-analysis consisting of 20 clinical studies involving 1927 patients with osteoarthritis	↓ VAS pain score ↓ WOMAC-pain, physical function, and stiffness ↓ ESR level ↓ oxidative stress markers ↑ anti-oxidative stress markers	[70]
Curcumin	• 250 mg twice daily • 500 mg twice daily for 90 days • Placebo	A randomized, double-blind, placebo-controlled, two-dose,three-arm, and parallel-study on rheumatoid arthritis patients	↓ VAS pain score ↓ DAS 28 score ↓ CRP, rheumatoid factor, and ESR levels	[71]
Curcumin	• 120 mg to 1500 mg for 4–36 weeks • Placebo	A meta-analysis of six clinical studies involving 326 patients with rheumatoid arthritis.	↓ DAS 28 score ↓ rheumatoid factor ↓ ESR and CRP levels	[70]

(continued)

Table 2 (continued)

Extract type or plant part	Study design	Study model	Results	Ref
Curcumin	• 250 to 1500 mg/day for 8–12 weeks • Placebo	A meta-analysis of six clinical studies involving 438 patients with RA and ulcerative colitis	↓ ESR level ↓CRP level	[72]
Tripterygium wilfordii	• 60 mg three times daily • Sulfasalazine 1 g twice daily taken for 24 weeks	A randomized, controlled trial on rheumatoid arthritis patients	↑ number of patients with at least 20%, 50%, and 70% improvement in disease activity ↓ number of swollen and tender joints ↓ IL-6 and rheumatoid factor levels	[73]
Tripterygium wilfordii	• *Tripterygium wilfordii*10 mg three times daily + methotrexate 7.5 mg once a week • Methotrexate 15 mg once a week for 3 months	A randomized, controlled trial on rheumatoid arthritis patients	↓ IgG, IgA, IgM, CRP, IL-6, and TNF-α levels ↑ patients satisfaction degree ↓ lower rate of adverse effects	[74]
Tripterygium wilfordii	• T. Wilfordii + methotrexate • Methotrexate monotherapy	A meta-analysis of 40 clinical studies involving 3092 RA patients	↓ morning stiffness, tender and swollen joint count ↓ VAS pain score ↓ CRP, ESR, and rheumatoid factor levels	[75]
Punica granatum	• 250 mg twice a day over 8 weeks • Placebo	A double-blind, randomized clinical trial on RA patients	↓ DAS28, number of swollen joints, tender joints, and pain intensity ↓ HAQ score, morning stiffness ↓ ESR level ↑ GPx level	[76]
Boswellia serrata	• 333 mg three times per day over 8 weeks • Placebo	A randomized, double-blind, placebo-controlled crossover trial on patients with knee osteoarthritis	↓ knee pain, swelling ↑ knee flexion and walking distance	[77]
Boswellia serrata	• 169.33 mg two times a day for 120 days • Placebo	A pilot, randomized, double-blind, placebo-controlled trial on knee osteoarthritis patients	↓ WOMAC index, VAS pain score, European Q5D quality of life ↓hs-CRP level ↑ 6-min walk test score ↑ joint space ↓ osteophytes	[78]
Crocus sativus	• 100 mg/day (pure saffron powder) for 90 days • Placebo	A double-blind, placebo-controlled, randomized trial on RA patients	↓ DAS28ESR disease activity score	[79]
Crocus sativus	• 100 mg/day taken over 12 weeks • Placebo	A randomized, double-blind, placebo-controlled clinical trial on RA patients	↓ number of tender and swollen joints, VAS pain score, DAS28 score ↑ physician global assessment ↓hs-CRP, TNF-α, IFN-γ, and MDA levels ↑ Total anti-oxidant capacity	[80]
Crocin	• 15 mg/day for 4 months • Placebo	A randomized, double-blind, placebo-controlled clinical trial on osteoarthritis patients	↓ VAS pain score ↓IL-17 and CRP levels ↑ T reg cells ↓ T helper 17 cells	[81]

Extract type or plant part	Study design	Study model	Results	Ref
Crocin	• 15 mg/day for 4 months • Placebo	A randomized, double-blind, placebo-controlled clinical trial on osteoarthritis patients	↓ microRNA-21 gene expression levels ↑ microRNA-155 gene expression levels	[82]
Aqueous extract of roots and leaves of *Withania somnifera*	• 125 and 250 mg twice a day over 12 weeks • Placebo	A randomized, double-blind, placebo-controlled on patients with knee osteoarthritis	↓ modified WOMAC score, knee swelling index ↓ VAS pain score, stiffness, and disability	[83]
Eucommia ulmoides granules	• 12 g three times a day for 1 month + meloxicam • Meloxicam	A clinical study evaluating patients with osteoarthritis and osteonecrosis	↓ WOMAC pain, stiffness, dysfunction score ↓ BGP, S100A12, IL-17, and SIP levels	[84]

hs-CRP high sensitivity C-reactive protein, *MDA* malondialdehyde, *NO* nitric oxide, *IL* interleukin, *T-bet* T-box transcription factor TBX, *RORγt* RAR-related orphan receptor γt, *FoxP3* forkhead box P3, *GAT A3* GATA binding protein 3, *PPAR-γ* peroxisome proliferator-activated receptor gamma, *VAS* visual assessment scale, *BGP* bone-GLA protein, *S100A12* recombinant human S100 calcium-binding protein, *SIP* sphingosine 1-phosphate

2.1 Animal Studies

In streptococcal cell wall (SCW)-induced arthritis, *Z. officinale* extract and essential oil firmly decrease arthritic index, joint swelling, and cartilage destruction index in female rats [6, 7]. In addition, *Z. officinale* rhizomes powder remarkably alleviated paw edema, TNF-α, IL-1β, -6, and 4 levels while elevating IL-10 levels following CFA-induced arthritis in rats [8].

Seventy percent of hydroethanolic extract of *Z. officinale* rhizomes meaningfully reduced clinical scores, arthritic incidence, erythrocyte sedimentation rate (ESR) level, joint temperature, joint swelling, and destruction of cartilage following CIA. *Z. officinale* also mitigated the serum levels of IL-1β, -6, and -2, TNF-α, and anti-CII antibodies following CIA in rats [9].

2.1.1 6-Shogaol

6-Shogaol notably alleviated paw volume, activities of lysosomal enzymes, lipid peroxidation, as well as TNF-α level while enhancing glutathione peroxidase (Gpx), SOD, and CAT activities in gouty arthritis associated with monosodium urate crystal administration in mice [10]. Moreover, 6-shogaol significantly inhibited knee swelling and edema, leukocyte infiltration into the synovial cavity, and VCAM-1 level in CFA-induced arthritis in rats [11].

2.2 Clinical Studies

Z. officinale rhizomes powder remarkably decreased DAS-28 arthritis score, T-bet, NF-κB, and RORγt gene expression in RA patients. It also stimulated the forkhead box P3 (FoxP3), GATA binding protein 3 (GAT A3), and peroxisome proliferator-activated receptor gamma (PPAR-γ)in RA patients [64].

Z. officinale rhizomes powder strikingly mitigated visual assessment scale (VAS) pain score while improving satisfaction with treatment in patients with knee osteoarthritis [65]. Similarly, 95% hydroethanolic extract of *Z. officinale* markedly reduced VAS and gelling pain scores in patients with osteoarthritis [66]. Additionally, *Z. officinale* powder provided a significant decrement in VAS pain score, WOMAC pain score, morning stiffness, as well as difficulty in knee osteoarthritis patients [67].

3 *Curcuma longa* and Its Major Active Constituent Curcumin

Curcuma longa (*C. longa*), popularly named turmeric, is a perennial rhizomatous herb and member of Zingiberaceae. It is cultivated chiefly in Asia, India, Iran, and China. Curcumin is the major and well-studied polyphenol isolated from the rhizome of *C. longa*. Curcumin possesses plenty of promising impacts, namely anti-oxidant, anti-inflammatory, immunomodulatory, antimicrobial, anti-cancer, neuroprotective, and anti-arthritis effects [92, 93].

3.1 Animal Studies

In the CIA model, curcumin meaningfully decreased arthritis score, joint histopathological changes, levels of TNF-α, IL-17 and 1β, and TGF-β, degradation of IκBα, and COX-2 expression in the synovium of the rat joints [12]. Similarly, curcumin markedly diminished clinical score, histology score, matrix metalloproteinases (MMP)-1, 3, 9, and 13 protein expression, and anti-collagen type II antibodies following CIA. It also diminished inflammatory markers, including TNF-α, IL-1β, 6, 33, 15, and 16, IFN-γ, IFN-gamma-inducible protein 10 (IP-10), macrophage inflammatory proteins (MIP)-α, and monocyte chemoattractant protein-1 (MCP-1) in CIA mice [13].

Curcumin notably mitigated redness and paw edema, arthritis score, inflammatory cells permeation into the synovium, and TNF-α, IL-1β, MMP-1, and 3 levels in serum and synovium following CIA in rats. It also suppressed mTOR, p70S6 kinase, and eukaryotic initiation factor 4E binding protein 1 expression in the synovium of CIA rats [14]. Additionally, curcumin remarkably alleviated arthritis scores, paw edema, TNF-α, IL-6, IL-17, phospho-(phosphoinositide-3 kinase) p-PI3K/PI3K ratio, and p-AKT/AKT ration in CIA in mice [15].

In CFA-induced arthritis in rats, curcumin significantly propagated SOD, CAT, and Gpx activities, and methionine sulfoxide reductase A (MSRA)gene expression [16].

In the model of knee osteoarthritis caused by monosodium iodoacetate (MIA) in rats, curcumin strikingly mitigated TNF-α, IL-1β, MMP-3, cartilage oligomeric matrix protein (COMP) while elevated SOD, and collagen II levels. It also inhibited the TLR4/NF-κB signaling pathway by suppressing TLR4, cox-2, and p-p65 expression levels [17]. Furthermore, curcumin stimulated the number of chondrocytes, collagen 2, SOX-5, and Indian hedgehog (IHH) expression in zymosan-induced osteoarthritis in rats [18].

In monosodium urate (MSU)-induced acute gout in mice, curcumin firmly attenuated paw and ankle joint swelling, MPO activity, and p-p65 and p-p50 phosphorylation while enhancing IκBα protein levels. It also provided a significant decrement in COX-2, TLR4, myeloid differentiation factor 88, and NLRP3 expression in the footpad tissue [19]. Similarly, curcumin considerably prevented ankle circumference, penetration of neutrophils in knee joints, TNF-α, IL-1β, and elastase levels in MSU-induced acute gout in rats. In addition, curcumin suppressed the levels of NLRP3, pro-

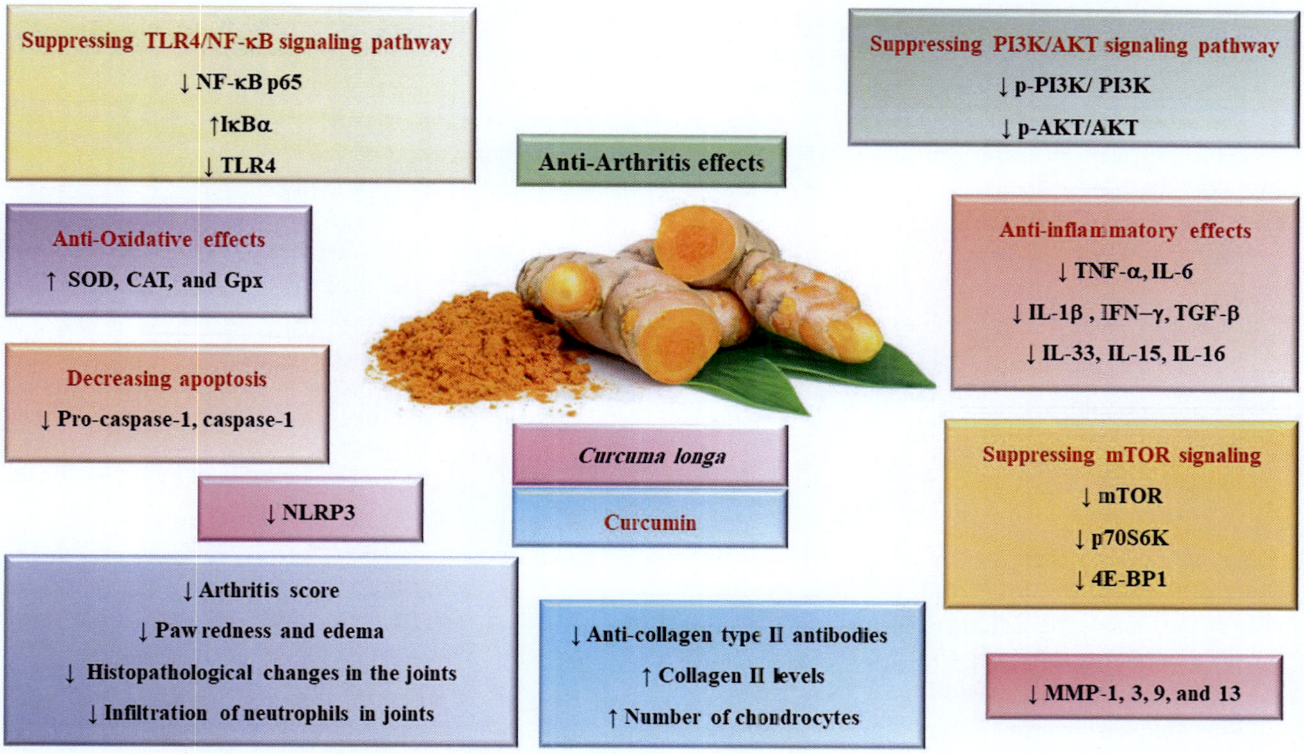

Fig. 2 The anti-arthritis mechanisms of *Curcuma longa* and its main active ingredient curcumin. *TLR4* toll-like receptor 4, NF-κB nuclear factor-κB, *SOD* superoxide dismutase, *CAT* catalase, *Gpx* glutathione peroxidase, *NLRP3* NLR family pyrin domain containing3, *MMP* matrix metallopeptidase, *mTOR* mammalian target of rapamycin, *IL* interleukin, TNF-α tumor necrosis factor-α, *TGF-β1* transforming growth factor-β1, *IFN-γ* interferon-gamma, *PI3K* phosphoinositide-3 kinase

caspase-1, caspase-1, pro-IL-1β, and IL-1β in the rat knee joints [20].

3.2 Clinical Studies

Curcumin significantly decreased VAS pain score and the need for NSAID consumption in patients with knee osteoarthritis [68]. Interestingly, curcumin indicated similar effects to diclofenac in bettering pain and quality of life in knee osteoarthritis patients. However, curcumin was better tolerated and had significantly lower adverse effects compared to diclofenac [94]. Additionally, curcumin with diclofenac remarkably diminished pain, the need for rescue analgesics, the need for histamine 2 (H2) blockers, and fewer side effects compared to diclofenac consumption in patients with knee osteoarthritis [69]. Recently, a meta-analysis of 20 clinical studies involving 1927 patients with osteoarthritis emphasized that curcumin markedly attenuated VAS pain score, WOMAC-pain, physical function and stiffness, ESR level, and oxidative stress markers while elevating anti-oxidative stress markers [70].

Curcumin also possesses anti-inflammatory and analgesic impacts in RA patients. In this regard, curcumin meaningfully alleviated VAS pain score, DAS 28 score, rheumatoid factor values, CRP, and ESR levels in RA patients [71]. Additionally, Zeng and coworkers conducted a meta-analysis of six clinical studies evaluating 326 patients with RA. They noticed that curcumin significantly mitigated the DAS 28 score, rheumatoid factor values, ESR, and CRP levels [70]. Another meta-analysis investigated six clinical studies involving 438 patients and showed that curcumin notably decreased ESR and CRP levels in rheumatoid arthritis and ulcerative colitis patients [72].

Taken together, curcumin ameliorated the severity of pain and inflammation along with safety in RA and osteoarthritis patients (Fig. 2).

4 *Tripterygium wilfordii* Hook and Its Main Component Triptolide

Tripterygium wilfordii Hook (*T. wilfordii*), commonly named as thunder god vine, is a traditional Chinese plant belonging to Celastraceae. The whole plant may cause acute toxicity, and its roots are the medicinal part of the plant. Triptolide is the major active diterpenoid isolated from the roots of *T. wilfordii*. Several pieces of evidence support that *T. wilfordii* and triptolide may be beneficial in treating inflammatory and auto-immune disorders, such as RA, multiple sclerosis, Lupus, psoriasis, diabetic nephritic syndrome, and Behcet's disease [95, 96].

4.1 Animal Studies

In CIA in rats, total alkaloids of *T. wilfordii* significantly decreased paw swelling, TNF-α, IL-6 and 8, and NF-κB levels in serum and synovial tissue [21]. Similarly, *T. wilfordii* glycosides markedly diminished paw volume, arthritis score, histopathological damages, anti-typeIIcollagen antibodies, and IL-1β levels in CIA rats [22].

4.1.1 Triptolide

In CIA in rats, triptolide notably mitigated paw thickness, arthritis score, levels of MPO, COX-2, iNOS, and oxidative stress in the articular tissue [23]. Similarly, triptolide considerably reduced arthritis score, TNF-α, IL-1β and 6, MMP-3 and -9, unconventional prefoldin RPB5 interactor 1, frequently rearranged in advanced T cell lymphomas-1 mRNA expression in synovial tissue of CIA rats [24]. Zou et al. noticed that the anti-arthritis properties of triptolide are through activating PI3K/AKT signaling cascade while inhibiting VEGF and tumor necrosis factor-related protein 3 in adjuvant arthritis model in rats [25].

In TNF transgenic mice with RA, triptolide firmly alleviated arthritis score, TNF-α, IL-1β, and IL-α levels while elevating arthritis detumescence percentages and apoptosis rates of osteoclast precursor cells (OCP) and T lymphocytes [26]. Similarly, triptolide meaningfully decreased arthritis symptoms, joint swelling, osteoclast (OC) cell levels, inflammation, swelling, and bone erosion while stimulating the rate of apoptosis in OCP cells in TNF transgenic mice [27].

4.2 Clinical Studies

T. wilfordii extract remarkably propagated the number of patients with at least 20%, 50%, and 70% improvement in disease activity compared to the sulfasalazine group in RA patients. Moreover, *T. wilfordii* extract notably decreased swollen and tender joint numbers, IL-6, and rheumatoid factor levels in RA patients [73]. In contrast, Zhou and coworkers suggested that *T. wilfordii* extract was not inferior to methotrexate treatment in RA patients [97].

Han and co-workers investigated the effects of combination therapy of total glycosides of *T. wilfordii* and methotrexate. They showed that *T. wilfordii* plus methotrexate treatment meaningfully diminished IgG, IgA, IgM, CRP, IL-6, and TNF-α levels in RA patients. In addition, *T. wilfordii* also improved the patient's satisfaction degree while having a lower rate of adverse effects than methotrexate monotherapy [74].

Additionally, several systematic reviews and meta-analysis studies determined the effects of *T. wilfordii* in RA patients. In this regard, two studies supported that *T. wilfordii* significantly improved grip strength, swelling joint count, morning stiffness, CRP, rheumatoid factor, ESR levels, and risk of adverse events than the placebo and methotrexate monotherapy [98, 99]. Similarly, a meta-analysis of 40 clinical studies involving 3092 RA patients emphasized that combination therapy of *T. wilfordii* and methotrexate notably alleviated morning stiffness, tender and swollen joint count, VAS pain grade, CRP, ESR, and rheumatoid factor levels compared to methotrexate monotherapy [75].

In general, *T. wilfordii* extract and its main constituent, triptolide, may be promising agents for treating RA patients.

5 *Punica granatum* and Its Major Constituent Ellagic Acid

Punica granatum (*P. granatum*), universally famous as pomegranate, is a small deciduous tree belonging to Punicaceae. Pomegranate is mainly cultivated in the Mediterranean countries, namely India, Iran, Turkey, and China. Multiple pharmacological effects have been emphasized for *P. granatum* fruit, namely anti-oxidant, anti-inflammatory, antidiabetic, hepatoprotective, cardioprotective, and anti-arthritis activities [100, 101]. In addition, ellagic acid (EA) and punicalagin are the major and well-studied active polyphenolic compounds found in pomegranate fruits [102, 103].

5.1 Animal Studies

In the CFA-induced arthritis model, *P. granatum* rind extract significantly decreased paw volume, histopathological damage, MPO activity, TNF-α, and IL-1β levels [28]. Similarly, *P. granatum* rind extract notably reduced paw volume, MDA, TNF-α levels, TNF receptor 1 (TNF-R1), TNF-α, IL-1β and 6, and NF-κB expression while increasing GSH levels and SOD activity in the synovial joint following CFA-caused arthritis in rats [29].

Following CFA-induced arthritis in rats, ethanolic extract of *P. granatum* markedly diminished paw volume, arthritic score, histological damages, MDA, rheumatoid factor (RF), TNF-α, and IL-1β levels [30]. Similarly, the *P. granatum* extract meaningfully alleviated the arthritic score, paw volume, joint diameter, as well as level of ESR [31].

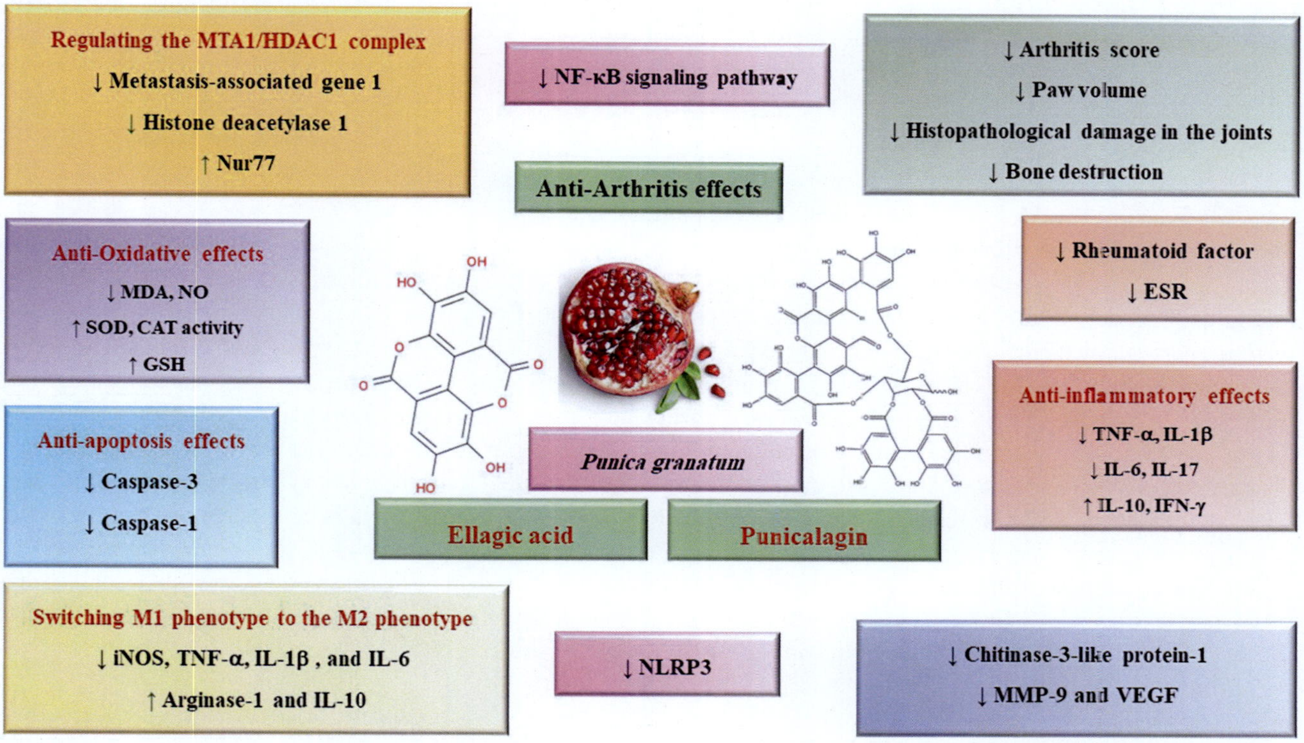

Fig. 3 The anti-arthritis mechanisms of *Punica granatum* and its main active constituents, EA and Punicalagin. *MDA* malondialdehyde, *NO* nitric oxide, *SOD* superoxide dismutase, *GSH* glutathione, *CAT* catalase, *iNOS* inducible nitric oxide synthase, *IL* interleukin, TNF-α tumor necrosis factor-α, *MMP* matrix metallopeptidase, *VEGF* vascular endothelial growth factor, *IFN-γ* interferon-gamma, *ESR* erythrocyte sedimentation rate, *NF-κB* nuclear factor-κB

5.1.1 Ellagic Acid (EA)

In arthritis induction by adjuvant in mice, EA markedly decreased paw edema, histopathological damage, TNF-α, IL-1β, and IL-17 levels while increasing IL-10 and IFN-γ levels [32]. Similarly, EA significantly mitigated paw edema, chitinase-3-like protein-1 (CHI3L1), NF-κB, IL-1β, MMP-9, VEGF, and Caspase-3 expression in rats. Moreover, EA provided a significant reduction in oxidative stress markers while enhancing GSH levels following adjuvant-caused arthritis in rats [33].

In CFA-induced arthritis, EA notably alleviated paw swelling, arthritis index, pathological damages, and synovial tissue levels of TNF-α, IL-1β , and 6. Additionally, EA down-regulated the expression of metastasis-associated gene (MTA)-1 and histone deacetylase 1 while up-regulating Nur77 in CFA-induced arthritis in rats [34].

5.1.2 Punicalagin

In CIA in mice, punicalagin markedly attenuated paw volume, clinical arthritis score, histology score, TNF-α, and IL-6 levels [35]. Similarly, punicalagin considerably decreased paw thickness, arthritis score, and bone destruction. Furthermore, punicalagin changed the phenotype of macrophages from M1 to M2 through diminishing iNOS, TNF-α, IL-1β, and IL-6 while stimulating arginase-1 (Arg-1) and IL-10 expression. Punicalagin also provided a significant decrement in the expression of NLRP3 and caspase-1 in the joint synovial membrane of CIA mice [36].

5.2 Clinical Studies

P. granatum extract significantly attenuated DAS28, swollen joint numbers, tender joints, and pain intensity in RA patients. Additionally, *P. granatum* extract meaningfully diminished HAQ score, morning stiffness, and ESR level while propagating GPx level in RA patients [76].

Collectively, *P. granatum* extract and its main constituents, EA and punicalagin, may be promising candidates for improving arthritis (Fig. 3).

6 *Boswellia serrata* and Its Major Constituent Boswellic Acid

Boswellia serrata (*B. serrata*), commonly called Indian frankincense, is a branching tree member of Burseraceae. Chiefly, *B. serrata*is native to India, and the oleo gum-resins are the most used part of the plant. *B. serrata* possesses anti-inflammatory properties and is frequently used to treat RA and osteoarthritis. In addition, boswellic acid is the major and well-studied active constituent of *B. serrata* [104, 105].

6.1 Animal Studies

In CIA model, extract of *B. serrata* gum resin notably decreased paw diameter, clinical severity score, elastase, and levels of MPO, MDA, and NO while enhancing GSH level, SOD, and CAT activities in joints. It also attenuated inflammatory markers, including TNF-α, IL-1β and 6, IFN-γ, and PGE2, while stimulating IL-10 levels in CIA in rats [37]. Similarly, *B. serrata* extracts markedly alleviated arthritis index, paw volume, pain, motility score, stair climbing, and histopathological damages in CIA rats. Moreover, *B. serrata* meaningfully attenuated collagen antibodies, CRP, and ESR levels in serum while elevating the synovial fluid levels of hyaluronan and cartilage oligomeric matrix protein (COMP) in CIA rats [38].

In CFA-induced arthritis in rats, *B. serrata* gum resin extract notably diminished ankle diameter and arthritis index while improving body weight [106].

6.1.1 Boswellic Acid

Boswellic acid significantly ameliorated carrageenan and Mycobacterium-caused paw edema and arthritis in rats. Interestingly, the promising anti-arthritis effects of boswellic acid were significantly greater than glucosamine. Additionally, combination therapy of boswellic acid and glucosamine showed synergistic effects in chronic arthritis models in rats [107]. Similarly, boswellic acid considerably alleviated paw edema, ESR, and MDA levels while increasing anti-oxidant enzymes following CFA-induced arthritis in rats [40].

In the gouty arthritic model following monosodium urate crystal administration in mice, boswellic acid remarkably reduced paw thickness, lysosomal enzymes, MDA, and TNF-α levels. Furthermore, boswellic acid also enhanced SOD, GPx, and CAT activities in gouty arthritis mice [39].

6.2 Clinical Studies

B. serrata extract significantly alleviated knee pain and swelling while elevating knee flexion and walking distance in patients with knee osteoarthritis. Additionally, no significant adverse effects were observed following *B. serrata* extract treatment, and it was well tolerated [77]. Similarly, *B. serrata* extract remarkably mitigated WOMAC index, VAS pain score, European Q5D quality of life, and hs-CRP level while promoting a 6-min walk test score in knee osteoarthritis patients. Furthermore, the beneficial effects of *B. serrata* extract have been confirmed by radiographic assessments showing an increment in joint space and a decrement in osteophytes. They also reported no serious adverse events following treatment with *B. serrata* extract [78].

Taken together, *B. serrata* and its main constituent boswellic acid may be effective in improving RA and osteoarthritis.

7 *Crocus sativus* and Its Main Constituent Crocin

Crocus sativus (*C. sativus*), universally famous as saffron, is a perennial flowering plant and member of Iridaceae. The red flower stigmas of *C. sativus* are known as saffron spice or red gold, widely used as spice and preparing food colorant in foods worldwide. Crocin is considered the main active and well-studied component of *C. sativus*. Several pieces of evidence support that *C. sativus* and crocin have anti-oxidant, anti-inflammatory, anti-depressant, analgesic, and anti-arthritis effects [108, 109].

7.1 Animal Studies

7.1.1 Crocin

In CFA-induced arthritis in rats, crocin markedly reduced paw swelling, exoglycosidases, Cathepsin-D, alkaline phosphatase, acid phosphatase, and tartrate-resistant acid phosphatase activities. It also mitigated inflammatory biomarkers, namely TNF-α, IL-1β and 6, NF-kB, cyclooxygenase-2, and PGE2. Besides that, crocin provided a meaningful suppression in reactive oxygen species (ROS) while enhancing anti-oxidative markers in CFA-induced arthritis in rats [41]. Similarly, crocin strikingly mitigated paw swelling, arthritis score, levels ofiNOS, NO, TNF-α, IL-1β, and 6 in CFA-induced arthritis in rats [42].

In CIA in mice, crocin notably alleviated arthritis score, paw thickness, histological score, MMP-1, -3, and -13, TNF-α, IL-6, -17, and -8, and CXCL8 levels [43]. Similarly, crocin significantly reduced mean clinical score, TNF-α, IL-1β, and -6 levels following CIA in mice [44].

In adjuvant-caused arthritis in rats, crocin markedly diminished pain, TNF-α and IL-1β levels. Crocin also suppressed the Wnt5a/β-Catenin Pathway and glial activation by decreasing the levels of GFAP and Iba-1 in the spinal cords of AIA rats [45].

In meniscectomy surgery-induced osteoarthritis, crocin firmly alleviated joint pain, ROS, lipid peroxidation, and JNK levels while enhancing GSH and GPx activities. Crocin also improved muscle dysfunction by reducing muscular levels of IL-6, citrate synthase activity, and myosin heavy chain IIα in osteoarthritis rats [46].

7.2 Clinical Studies

C. sativus significantly reduced DAS28ESR disease activity score in RA patients following 90 days of treatment [79]. Similarly, *C. sativus* notably diminished the number of tender and swollen joints, VAS pain grade, and DAS28 score while improving physician global assessment in RA patients. It also mitigated hs-CRP, TNF-α, IFN-γ, and MDA levels while enhancing total anti-oxidant capacity in RA patients [80].

Additionally, crocin remarkably diminished VAS pain score, CRP, and IL-17 levels in osteoarthritis patients. Crocin also elevated the percentage of T regulatory (Tregs) while decreasing T helper (Th) 17 cells in osteoarthritis patients. In fact, crocin shifted the Treg/Th17 ratio toward Treg cells in osteoarthritis patients [81]. Similarly, crocin attenuated the microRNA-21 gene expression levels while stimulating microRNA-155 gene expression levels in osteoarthritis patients [82].

In general, *C. sativus* and its main ingredient, crocin, may be a promising candidate for treating RA and osteoarthritis.

8 *Berberis vulgaris* and Its Main Constituent Berberine

Berberis vulgaris (*B. vulgaris*), commonly known as barberry, is a deciduous shrub and member of Berberidaceae. The edible part of *B. vulgaris* is its oblong red fruit with a sour flavor. Its fruits are frequently used for cooking and making jams in different countries worldwide. The main and well-studied active constituent of the plant is berberine. *B. vulgaris* and berberine showed anti-inflammatory, anti-oxidant, analgesic, and anti-arthritis properties [110, 111].

8.1 Animal Studies

8.1.1 Berberine

In CFA-induced arthritis in rats, berberine considerably reduced paw swelling, arthritis index score, IL-6, IL-17, and anti-IL-17 antibodies while elevating IL-10, TGF-β, and anti-IL-10 antibodies [47].

In CFA-caused arthritis, berberine remarkably diminished paw swelling, arthritis global assessment, and histopathological damages in rats. Berberine also regulated M1/M2 balance by attenuating the M1 markers, namely, TNF-α, iNOS, and IL-1β and -6 while elevating the M2 markers, including IL-10, TGF-β1, and Arg1. Moreover, berberine alleviated the Th17/Treg cells ratio by decreasing the number of Th17 while increasing Treg cells. In addition, berberine stimulated the p-AMPK expression while downregulated p-p65, p-IκBα, and COX-2 expression in CFA-induced arthritis in rats [48]. Similarly, berberine suppressed the inflammatory marker levels (TNF-α, IL-1β, -6, and -17, MCP-1, and IFN-γ) as well as the proportion of M1 macrophages. In contrast, it propagated the levels of anti-inflammatory indicators, such as IL-10 and -4, and TGF-β1 as well as the proportion of M2 macrophages in CFA-caused arthritis. Additionally, berberine up regulated p-AMPK expression while diminishing HIF-1α expression in synovial macrophages [49].

In CIA in mice, berberine leads to T cell suppression through attenuating the number of $CD4+T_h$, $CD4^+CXCR5^+T_{fh}$ cells, CD28, CD154 expression, and anti-CII total IgG serum levels while enhancing $Foxp3^+$ Treg cells [50]. Similarly, berberine markedly diminished paw swelling, arthritis index, histological score, joint destruction, RANKL, TNF-α, IL-1β, and IL-6 in CIA in rats. Berberine prevented the proportion of Th17 cells, IL-17 levels, and phosphorylation of STAT3 [51]. Similarly, berberine notably alleviated arthritis score, anti-CII total IgG, IgG1, IgG2a, TNF-α, IL-1β, -6, and -17, and VEGF serum levels following CIA in rats. Besides that, berberine strikingly decreased the expression of VEGF and CD34, as well as p-ERK, p-p38, and p-JNK activation in the synovium [52].

In glucocorticoid-induced osteoporosis in rats, berberine remarkably elevated bone mineral density, trabecular number and thickness, rate of mineral apposition, and bone formation rate/trabecular bone surface. Furthermore, berberine notably alleviated the trabecular separation and osteoclast/trabecular bone surface in osteoporosis rats. They supported that berberine ameliorated osteoporosis in rats by preventing bone resorption and propagating bone formation [53].

Taken together, berberine ameliorated RA through anti-inflammatory and immunomodulatory properties (Fig. 4).

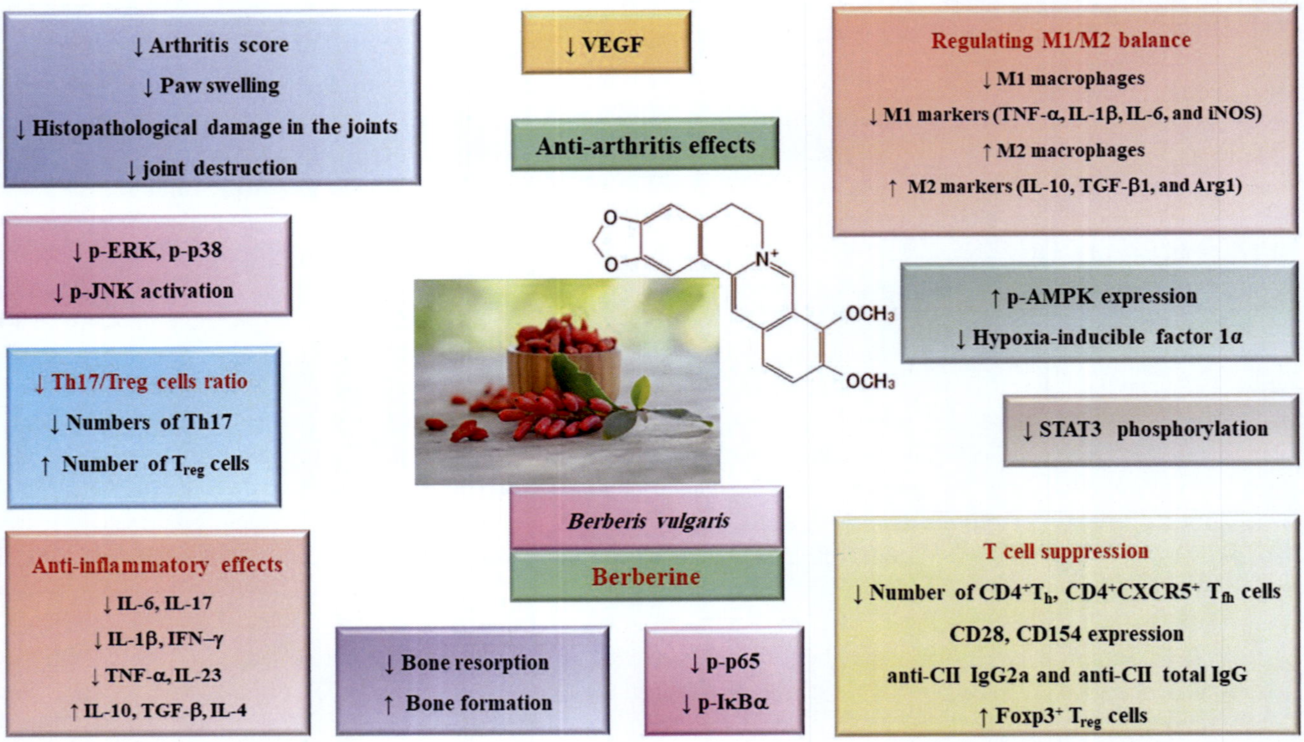

Fig. 4 The anti-arthritis mechanisms of *Berberis vulgaris* and its main active component, berberine. *JNK* c-Jun N-terminal kinase, *IL* interleukin, *TNF-α* tumor necrosis factor-α, *IFN-γ* interferon-gamma, *TGF-β1* transforming growth factor-β1, *STAT3* signal transducer and activator of transcription 3, *AMPK* AMP-activated protein kinase, *iNOS* inducible nitric oxide synthase, *Arg-1* arginase-1, *VEGF* vascular endothelial growth factor

9 Withania somnifera

Withania somnifera (*W. somnifera*), commonly named Ashwagandha, is a woody shrub belonging to Solanaceae. Its roots have been widespreadly administered in traditional Indian medicine for over 3000 years. Additionally, multiple pharmacological properties have been found for *W. somnifera*, namely anti-inflammatory, anti-oxidant, immunomodulatory, neuro-protective, as well as anti-arthritis effects [112, 113].

9.1 Animal Studies

In CIA in rats, aqueous extract of roots of *W. somnifera* notably reduced arthritic index, rheumatoid factor, anti-collagen type II, and CRP levels. *W. somnifera* also mitigated oxidative stress by decreasing MDA and glutathione-S-transferase activity (GST) while increasing GSH and ferric ability of plasma in CIA rats [54]. Similarly, aqueous extract or roots of *W. somnifera* markedly diminished clinical severity, histological score, TNF-α, IL-1β, IL-6, NF-kB, MMP-8, and iNOS while elevating IL-10 levels in CIA in rats [55]. Additionally, *W. somnifera* root powder considerably alleviated paw thickness, ankle size, arthritic score, and pain, as well as radiological score in CIA in rats [114].

In CFA-induced arthritis, 70% ethanolic extract of *W. somnifera* roots meaningfully alleviated paw swelling, arthritic score, ESR, CRP, and rheumatoid factor in rats. It also suppressed inflammation by decreasing IL-1β, -6, and -17a, TNF-α, and RANKL expression while enhancing IL-4, INF-γ, and osteoprotegerin. Furthermore, *W. somnifera* prevented oxidative stress by decreasing MDA while improving CAT and SOD activities in CFA-induced arthritis in rats [56].

9.2 Clinical Studies

Ramakanth et al. showed that Aqueous extract of *W. somnifera* roots and leaves significantly diminished modified WOMAC score, knee swelling index, VAS pain score, stiffness, and disability in patients with knee osteoarthritis. Additionally, the extract was well tolerated, and no severe adverse effects were noticed [83]. Additionally, the chondroprotective impacts of aqueous extracts of *W. somnifera* root were reported in patients with osteoarthritis [115].

10 *Eucommia ulmoides*

Eucommia ulmoides (*E. ulmoides*), commonly named Du Zhong in Chinese, is a small tree member of Eucommiaceae. Chiefly, it is native to China, and its bark, leaves, seed, and male flowers are widely administrated in traditional Chinese medicine. Additionally, *E. ulmoides* possessed multiple pharmacological activities, including anti-oxidant, anti-inflammatory, anti-allergic, anti-diabetes, anti-aging, cardio-protective, as well as anti-arthritis [116, 117].

10.1 Animal Studies

In CIA in rats, *E. ulmoides* extract markedly decreased paw volume, arthritis index, pathological score, Th17-positive cell numbers, inflammatory markers (TNF-α, IL-1β, and -17), as well as RANKL mRNA levels while enhancing IL-10 and OPG mRNA level in the joint tissue [57]. Similarly, the *E. ulmoides* extract notably decreased ankle swelling, TNF-α, IL-1β, and IL-17 levels, RANKL/OPG ratio, MMP-9, p65 NF-κB, and p-IKKαβexpression in CIA in rats [58]. In addition, *E. ulmoides* bark, leaf, and male flower extract considerably mitigated paw volume, ankle joint score, and levels of TNF-α, IL-6, -17, and -1β, and NO [59].

In the knee osteoarthritis model, *E. ulmoides* extract remarkably alleviated histopathological damages, and MMP-1, -3, and -13 levels in the rat synovial fluid and serum [60].

10.2 Clinical Studies

E. ulmoides granules significantly diminished WOMAC pain, stiffness, dysfunction score, bone-GLA protein (BGP), recombinant human S100 calcium-binding protein, IL-17, and sphingosine 1-phosphate levels in patients with osteoarthritis and osteonecrosis [84].

Acknowledgments This study was supported by the research council of Mashhad University of Medical Sciences.

References

1. Nasuti C, Fedeli D, Bordoni L, Piangerelli M, Servili M, Selvaggini R, et al. Anti-inflammatory, anti-arthritic and anti-nociceptive activities of Nigella sativa oil in a rat model of arthritis. Antioxidants (Basel). 2019;8(9):342.
2. Umar S, Zargan J, Umar K, Ahmad S, Katiyar CK, Khan HA. Modulation of the oxidative stress and inflammatory cytokine response by thymoquinone in the collagen induced arthritis in Wistar rats. Chem Biol Interact. 2012;197(1):40–6.
3. Arjumand S, Shahzad M, Shabbir A, Yousaf MZ. Thymoquinone attenuates rheumatoid arthritis by downregulating TLR2, TLR4, TNF-α, IL-1, and NFκB expression levels. Biomed Pharmacother. 2019;111:958–63.
4. Vaillancourt F, Silva P, Shi Q, Fahmi H, Fernandes JC, Benderdour M. Elucidation of molecular mechanisms underlying the protective effects of thymoquinone against rheumatoid arthritis. J Cell Biochem. 2011;112(1):107–17.
5. Faisal R, Ahmad N, Fahad YS, Chiragh S. Anti-arthritic effect of Thymoquinone in comparison with methotrexate on Pristane induced arthritis in female Sprague Dawley rats. J Ayub Med Coll Abbottabad. 2018;30(1):3–7.
6. Funk JL, Frye JB, Oyarzo JN, Timmermann BN. Comparative effects of two gingerol-containing Zingiber officinale extracts on experimental rheumatoid arthritis. J Nat Prod. 2009;72(3):403–7.
7. Funk JL, Frye JB, Oyarzo JN, Chen J, Zhang H, Timmermann BN. Anti-inflammatory effects of the essential oils of ginger (Zingiber officinale roscoe) in experimental rheumatoid arthritis. PharmaNutrition. 2016;4(3):123–31.
8. Ramadan G, Al-Kahtani MA, El-Sayed WM. Anti-inflammatory and anti-oxidant properties of Curcuma longa (turmeric) versus Zingiber officinale (ginger) rhizomes in rat adjuvant-induced arthritis. Inflammation. 2011;34(4):291–301.
9. Fouda AM, Berika MY. Evaluation of the effect of hydroalcoholic extract of Zingiber officinale rhizomes in rat collagen-induced arthritis. Basic Clin Pharmacol Toxicol. 2009;104(3):262–71.
10. Sabina EP, Rasool M, Mathew L, Ezilrani P, Indu H. 6-Shogaol inhibits monosodium urate crystal-induced inflammation—an in vivo and in vitro study. Food and Chem Toxicol. 2010;48(1):229–35.
11. Levy AS, Simon O, Shelly J, Gardener M. 6-Shogaol reduced chronic inflammatory response in the knees of rats treated with complete Freund's adjuvant. BMC Pharmacol. 2006;6:12.
12. Wang Q, Ye C, Sun S, Li R, Shi X, Wang S, et al. Curcumin attenuates collagen-induced rat arthritis via anti-inflammatory and apoptotic effects. Int Immunopharmacol. 2019;72:292–300.
13. Hemshekhar M, Anaparti V, El-Gabalawy H, Mookherjee N. A bioavailable form of curcumin, in combination with vitamin-D- and omega-3-enriched diet, modifies disease onset and outcomes in a murine model of collagen-induced arthritis. Arthritis Res Ther. 2021;23(1):39.
14. Dai Q, Zhou D, Xu L, Song X. Curcumin alleviates rheumatoid arthritis-induced inflammation and synovial hyperplasia by targeting mTOR pathway in rats. Drug Des Devel Ther. 2018;12:4095–105.
15. Xu Z, Shang W, Zhao Z, Zhang B, Liu C, Cai H. Curcumin alleviates rheumatoid arthritis progression through the phosphatidylinositol 3-kinase/protein kinase B pathway: an in vitro and in vivo study. Bioengineered. 2022;13(5):12899–911.
16. Meshkibaf MH, Maleknia M, Noroozi S. Effect of curcumin on gene expression and protein level of methionine sulfoxide reductase a (MSRA), SOD, CAT and GPx in Freund's adjuvant inflammation-induced male rats. J Inflamm Res. 2019;12:241–9.
17. Guan T, Ding LG, Lu BY, Guo JY, Wu MY, Tan ZQ, et al. Combined administration of Curcumin and chondroitin sulfate alleviates cartilage injury and inflammation via NF-κB pathway in knee osteoarthritis rats. Front Pharmacol. 2022;13:882304.
18. Nicoliche T, Maldonado DC, Faber J, Silva M. Evaluation of the articular cartilage in the knees of rats with induced arthritis treated with curcumin. PLoS One. 2020;15(3):e0230228.
19. Chen B, Li H, Ou G, Ren L, Yang X, Zeng M. Curcumin attenuates MSU crystal-induced inflammation by inhibiting the degradation of IκBα and blocking mitochondrial damage. Arthritis Res Ther. 2019;21(1):193.
20. Li X, Xu DQ, Sun DY, Zhang T, He X, Xiao DM. Curcumin ameliorates monosodium urate-induced gouty arthritis through nod-like receptor 3 inflammasome mediation via inhibiting nuclear factor-kappa B signaling. J Cell Biochem. 2019;120(4):6718–28.

21. Zhang Y, Xu W, Li H, Zhang X, Xia Y, Chu K, et al. Therapeutic effects of total alkaloids of *Tripterygium wilfordii* Hook f. on collagen-induced arthritis in rats. J Ethnopharmacol. 2013;145(3):699–705.
22. Zhu Y, Zhang L, Zhang X, Wu D, Chen L, Hu C, et al. *Tripterygium wilfordii* glycosides ameliorates collagen-induced arthritis and aberrant lipid metabolism in rats. Front Pharmacol. 2022;13:938849.
23. Yu GM, Zhou LF, Zeng BX, Huang JJ, She XJ. The anti-oxidant effect of triptolide contributes to the therapy in a collagen-induced arthritis rat model. Redox Rep. 2021;26(1):197–202.
24. Piao X, Zhou J, Xue L. Triptolide decreases rheumatoid arthritis fibroblast-like synoviocyte proliferation, invasion, inflammation and presents a therapeutic effect in collagen-induced arthritis rats via inactivating lncRNA RP11-83J16.1 mediated URI1 and β-catenin signaling. Int Immunopharmacol. 2021;99:108010.
25. Zou Y, Hu W. Investigation of gene expression profiles in a rat adjuvant arthritis model suggests an effective role of triptolide via PI3K-AKT signaling. Exp Ther Med. 2019;17(5):3999–4006.
26. Wang S, Zuo S, Liu Z, Ji X, Yao Z, Wang X. Study on the efficacy and mechanism of triptolide on treating TNF transgenic mice with rheumatoid arthritis. Biome Pharmacother. 2018;106:813–20.
27. Wang S, Liu Z, Wang J, Wang Y, Liu J, Ji X, et al. The triptolide-induced apoptosis of osteoclast precursor by degradation of cIAP2 and treatment of rheumatoid arthritis of TNF-transgenic mice. Phytother Res. 2019;33(2):342–9.
28. Mo J, Panichayupakaranant P, Kaewnopparat N, Nitiruangjaras A, Reanmongkol W. Topical anti-inflammatory and analgesic activities of standardized pomegranate rind extract in comparison with its marker compound ellagic acid in vivo. J Ethnopharmacol. 2013;148(3):901–8.
29. Karwasra R, Singh S, Sharma D, Sharma S, Sharma N, Khanna K. Pomegranate supplementation attenuates inflammation, joint dysfunction via inhibition of NF-κB signaling pathway in experimental models of rheumatoid arthritis. J Food Biochem. 2019;43(8):e12959.
30. Wang Y, He T, Li Z, Gai S. Effect of ethanol extract of Punica granatum L against Freund's complete adjuvant-induced arthritis in rats. Trop J Pharm Res. 2019;18(3):591–5.
31. Gautam RK, Sharma S, Sharma K, Gupta G. Evaluation of antiarthritic activity of butanol fraction of Punica granatum Linn Rind extract against Freund's complete adjuvant-induced arthritis in rats. J Environ Pathol Toxicol Oncol. 2018;37(1):53.
32. Allam G, Mahdi EA, Alzahrani AM, Abuelsaad AS. Ellagic acid alleviates adjuvant induced arthritis by modulation of pro- and anti-inflammatory cytokines. Cent Eur J Immunol. 2016;41(4):339–49.
33. Fikry EM, Gad AM, Eid AH, Arab HH. Caffeic acid and ellagic acid ameliorate adjuvant-induced arthritis in rats via targeting inflammatory signals, chitinase-3-like protein-1 and angiogenesis. Biomed Pharmacother. 2019;110:878–86.
34. Song H, Wu H, Dong J, Huang S, Ye J, Liu R. Ellagic acid alleviates rheumatoid arthritis in rats through inhibiting MTA1/HDAC1-mediated Nur77 deacetylation. Mediat Inflamm. 2021;2021:6359652.
35. Huang M, Wu K, Zeng S, Liu W, Cui T, Chen Z, et al. Punicalagin inhibited inflammation and migration of fibroblast-like Synoviocytes through NF-κB pathway in the experimental study of rheumatoid arthritis. J Inflamm Res. 2021;14:1901–13.
36. Ge G, Bai J, Wang Q, Liang X, Tao H, Chen H, et al. Punicalagin ameliorates collagen-induced arthritis by downregulating M1 macrophage and pyroptosis via NF-κB signaling pathway. Sci China Life Sci. 2022;65(3):588–603.
37. Umar S, Umar K, Sarwar AH, Khan A, Ahmad N, Ahmad S, et al. Boswellia serrata extract attenuates inflammatory mediators and oxidative stress in collagen induced arthritis. Phytomedicine. 2014;21(6):847–56.
38. Majeed M, Nagabhushanam K, Lawrence L, Nallathambi R, Thiyagarajan V, Mundkur L. Boswellia serrata extract containing 30% 3-Acetyl-11-keto-Boswellic acid attenuates inflammatory mediators and preserves extracellular matrix in collagen-induced arthritis. Front Physiol. 2021;12:735247.
39. Sabina EP, Indu H, Rasool M. Efficacy of boswellic acid on lysosomal acid hydrolases, lipid peroxidation and anti-oxidant status in gouty arthritic mice. Asian Pac J Trop Biomed. 2012;2(2):128–33.
40. Banji D, Banji OJF, Rashida S, Alshahrani S, Alqahtani SS. Bioavailability, anti-inflammatory and anti-arthritic effect of acetyl keto Boswellic acid and its combination with methotrexate in an arthritic animal model. J Ethnopharmacol. 2022;292:115200.
41. Hemshekhar M, Sebastin Santhosh M, Sunitha K, Thushara RM, Kemparaju K, Rangappa KS, et al. A dietary colorant crocin mitigates arthritis and associated secondary complications by modulating cartilage deteriorating enzymes, inflammatory mediators and anti-oxidant status. Biochimie. 2012;94(12):2723–33.
42. Li X, Jiang C, Zhu W. Crocin reduces the inflammation response in rheumatoid arthritis. Biosci Biotechnol Biochem. 2017;81(5):891–8.
43. Liu W, Sun Y, Cheng Z, Guo Y, Liu P, Wen Y. Crocin exerts anti-inflammatory and anti-arthritic effects on type II collagen-induced arthritis in rats. Pharm Biol. 2018;56(1):209–16.
44. Li L, Zhang H, Jin S, Liu C. Effects of crocin on inflammatory activities in human fibroblast-like synoviocytes and collagen-induced arthritis in mice. Immunol Res. 2018;66(3):406–13.
45. Wang JF, Xu HJ, He ZL, Yin Q, Cheng W. Crocin alleviates pain hyperalgesia in AIA rats by inhibiting the spinal Wnt5a/β--catenin signaling pathway and glial activation. Neural Plast. 2020;2020:4297483.
46. Lei M, Guo C, Hua L, Xue S, Yu D, Zhang C, et al. Crocin attenuates joint pain and muscle dysfunction in osteoarthritis rat. Inflammation. 2017;40(6):2086–93.
47. Wang X, He X, Zhang CF, Guo CR, Wang CZ, Yuan CS. Anti-arthritic effect of berberine on adjuvant-induced rheumatoid arthritis in rats. Biomed Pharmacother. 2017;89:887–93.
48. Zhou J, Yu Y, Yang X, Wang Y, Song Y, Wang Q, et al. Berberine attenuates arthritis in adjuvant-induced arthritic rats associated with regulating polarization of macrophages through AMPK/NF-κB pathway. Eur J Pharmacol. 2019;852:179–88.
49. Yu Y, Cai W, Zhou J, Lu H, Wang Y, Song Y, et al. Corrigendum to "Anti-arthritis effect of berberine associated with regulating energy metabolism of macrophages through AMPK/HIF-1α pathway". [Int. Immunopharmacol. 87 (2020) 106830]. Int Immunopharmacol. 2021;99:107911.
50. Vita AA, Aljobaily H, Lyons DO, Pullen NA. Berberine delays onset of collagen-induced arthritis through T cell suppression. Int J Mol Sci. 2021;22(7):3522.
51. Yue M, Xia Y, Shi C, Guan C, Li Y, Liu R, et al. Berberine ameliorates collagen-induced arthritis in rats by suppressing Th17 cell responses via inducing cortistatin in the gut. FEBS J. 2017;284(17):2786–801.
52. Wang Z, Chen Z, Yang S, Wang Y, Huang Z, Gao J, et al. Berberine ameliorates collagen-induced arthritis in rats associated with anti-inflammatory and anti-angiogenic effects. Inflammation. 2014;37(5):1789–98.
53. Xu D, Yang W, Zhou C, Liu Y, Xu B. Preventive effects of berberine on glucocorticoid-induced osteoporosis in rats. Planta Med. 2010;76(16):1809–13.
54. Khan MA, Subramaneyaan M, Arora VK, Banerjee BD, Ahmed RS. Effect of Withania somnifera (Ashwagandha) root extract on amelioration of oxidative stress and autoantibodies production in collagen-induced arthritic rats. J Complement Integr Med. 2015;12(2):117–25.
55. Khan MA, Ahmed RS, Chandra N, Arora VK, Ali A. In vivo, extract from Withania somnifera root ameliorates arthritis via

regulation of key immune mediators of inflammation in experimental model of arthritis. Antiinflamm Antiallergy Agents Med Chem. 2019;18(1):55–70.

56. Hussain A, Aslam B, Muhammad F, Faisal MN, Kousar S, Mushtaq A, et al. Anti-arthritic activity of Ricinus communis L. and Withania somnifera L. extracts in adjuvant-induced arthritic rats via modulating inflammatory mediators and subsiding oxidative stress. Iran J Basic Med Sci. 2021;24(7):951–61.
57. Wang JY, Yuan Y, Chen XJ, Fu SG, Zhang L, Hong YL, et al. Extract from Eucommia ulmoides Oliv. Ameliorates arthritis via regulation of inflammation, synoviocyte proliferation and osteoclastogenesis in vitro and in vivo. J Ethnopharmacol. 2016;194:609–16.
58. Wang JY, Chen XJ, Zhang L, Pan YY, Gu ZX, He SM, et al. Comparative studies of different extracts from Eucommia ulmoides oliv against rheumatoid arthritis in CIA rats. Evid Based Complement Alternat Med. 2018;2018:7379893.
59. Xing YY, Wang JY, Wang K, Zhang Y, Liu K, Chen XY, et al. Inhibition of rheumatoid arthritis using bark, leaf, and male flower extracts of Eucommia ulmoides. Evid Based Complement Alternat Med. 2020;2020:3260278.
60. Lu H, Jiang J, Xie G, Liu W, Yan G. Effects of an aqueous extract of Eucommia on articular cartilage in a rat model of osteoarthritis of the knee. Exp Ther Med. 2013;6(3):684–8.
61. Gheita TA, Kenawy SA. Effectiveness of Nigella sativa oil in the management of rheumatoid arthritis patients: a placebo controlled study. Phytother Res. 2012;26(8):1246–8.
62. Kheirouri S, Hadi V, Alizadeh M. Immunomodulatory effect of Nigella sativa oil on T lymphocytes in patients with rheumatoid arthritis. Immunol Investig. 2016;45(4):271–83.
63. Hadi V, Kheirouri S, Alizadeh M, Khabbazi A, Hosseini H. Effects of Nigella sativa oil extract on inflammatory cytokine response and oxidative stress status in patients with rheumatoid arthritis: a randomized, double-blind, placebo-controlled clinical trial. Avicenna J Phytomed. 2016;6(1):34–43.
64. Aryaeian N, Shahram F, Mahmoudi M, Tavakoli H, Yousefi B, Arablou T, et al. The effect of ginger supplementation on some immunity and inflammation intermediate genes expression in patients with active rheumatoid arthritis. Gene. 2019;698:179–85.
65. Alipour Z, Asadizaker M, Fayazi S, Yegane N, Kochak M, Haghighi Zadeh MH. The effect of ginger on pain and satisfaction of patients with knee osteoarthritis. J Chronic Dis Care. 2017;6(1):e34798.
66. Haghighi M, Khalvat A, Toliat T, Jallaei S. Comparing the effects of ginger (Zingiber officinale) extract and ibuprofen on patients with osteoarthritis. Arch Iran Med. 2005;8(4):267–71.
67. Zahra Z, Shahrokh I, Zohreh B, Farhang S, Behzad N, Ghasemirad M. Evaluating the effects of ginger extract on knee pain, stiffness and difficulty in patients with knee osteoarthritis. J Med Plant Res. 2011;5(15):3375–9.
68. Nakagawa Y, Mukai S, Yamada S, Matsuoka M, Tarumi E, Hashimoto T, et al. Short-term effects of highly-bioavailable curcumin for treating knee osteoarthritis: a randomized, double-blind, placebo-controlled prospective study. J Orthop Sci. 2014;19(6):933–9.
69. Shep D, Khanwelkar C, Gade P, Karad S. Efficacy and safety of combination of curcuminoid complex and diclofenac versus diclofenac in knee osteoarthritis: a randomized trial. Medicine. 2020;99(16):e19723.
70. Zeng L, Yang T, Yang K, Yu G, Li J, Xiang W, et al. Efficacy and safety of curcumin and Curcuma longa extract in the treatment of arthritis: a systematic review and meta-analysis of randomized controlled trial. Front Immunol. 2022;13:891822.
71. Amalraj A, Varma K, Jacob J, Divya C, Kunnumakkara AB, Stohs SJ, et al. A novel highly bioavailable curcumin formulation improves symptoms and diagnostic indicators in rheumatoid arthritis patients: a randomized, double-blind, placebo-controlled, two-dose, three-arm, and parallel-group study. J Med Food. 2017;20(10):1022–30.
72. Ebrahimzadeh A, Abbasi F, Ebrahimzadeh A, Jibril AT, Milajerdi A. Effects of curcumin supplementation on inflammatory biomarkers in patients with rheumatoid arthritis and ulcerative colitis: a systematic review and meta-analysis. Complement Ther Med. 2021;61:102773.
73. Goldbach-Mansky R, Wilson M, Fleischmann R, Olsen N, Silverfield J, Kempf P, et al. Comparison of Tripterygium wilfordii hook F versus sulfasalazine in the treatment of rheumatoid arthritis: a randomized trial. Ann Intern Med. 2009;151(4):229–40, w49–51.
74. Han Y, Jin J, Wu F, Wang Z. Impacts of low-dose Total glycosides of Tripterygium wilfordii plus methotrexate on immunological function and inflammation level in patients with rheumatoid arthritis. Comput Math Methods Med. 2022;2022:7523673.
75. Zheng W, Mei Y, Chen C, Cai L, Chen H. The effectiveness and safety of Tripterygium wilfordii glycosides combined with disease-modifying anti-rheumatic drugs in the treatment of rheumatoid arthritis a systematic review and meta-analysis of 40 randomized controlled trials. Phytother Res. 2021;35(6):2902–24.
76. Ghavipour M, Sotoudeh G, Tavakoli E, Mowla K, Hasanzadeh J, Mazloom Z. Pomegranate extract alleviates disease activity and some blood biomarkers of inflammation and oxidative stress in rheumatoid arthritis patients. Eur J Clin Nutr. 2017;71(1):92–6.
77. Karlapudi V, Prasad Mungara AVV, Sengupta K, Davis BA, Raychaudhuri SP. A placebo-controlled double-blind study demonstrates the clinical efficacy of a novel herbal formulation for relieving joint discomfort in human subjects with osteoarthritis of knee. J Med Food. 2018;21(5):511–20.
78. Majeed M, Majeed S, Narayanan NK, Nagabhushanam K. A pilot, randomized, double-blind, placebo-controlled trial to assess the safety and efficacy of a novel Boswellia serrata extract in the management of osteoarthritis of the knee. Phytother Res. 2019;33(5):1457–68.
79. Sahebari M, Heidari H, Nabavi S, Khodashahi M, Rezaieyazdi Z, Dadgarmoghaddam M, et al. A double-blind placebo-controlled randomized trial of oral saffron in the treatment of rheumatoid arthritis. Avicenna J Phytomed. 2021;11(4):332–42.
80. Hamidi Z, Aryaeian N, Abolghasemi J, Shirani F, Hadidi M, Fallah S, et al. The effect of saffron supplement on clinical outcomes and metabolic profiles in patients with active rheumatoid arthritis: a randomized, double-blind, placebo-controlled clinical trial. Phytother Res. 2020;34(7):1650–8.
81. Poursamimi J, Shariati-Sarabi Z, Tavakkol-Afshari J, Mohajeri SA, Ghoryani M, Mohammadi M. Immunoregulatory effects of Krocina™, a herbal medicine made of Crocin, on osteoarthritis patients: a successful clinical trial in Iran. Iran J Allergy Asthma Immunol. 2020;19(3):253–63.
82. Mohebbi M, Atabaki M, Tavakkol-Afshari J, Shariati-Sarabi Z, Poursamimi J, Mohajeri SA, et al. Significant effect of Crocin on the gene expression of MicroRNA-21 and MicroRNA-155 in patients with osteoarthritis. Iran J Allergy Asthma Immunol. 2022;21(3):322–31.
83. Ramakanth GS, Uday Kumar C, Kishan PV, Usharani P. A randomized, double blind placebo controlled study of efficacy and tolerability of Withaina somnifera extracts in knee joint pain. J Ayurveda Integr Med. 2016;7(3):151–7.
84. Hu CX, Hu KY, Wang JF. Potential role of the compound Eucommia bone tonic granules in patients with osteoarthritis and osteonecrosis: a retrospective study. World J Clin Cases. 2020;8(1):46–53.
85. Mokhtari-Zaer A, Norouzi F, Askari VR, Khazdair MR, Roshan NM, Boskabady M, et al. The protective effect of Nigella sativa

extract on lung inflammation and oxidative stress induced by lipopolysaccharide in rats. J Ethnopharmacol. 2020;253:112653.

86. Hosseini A, Baradaran Rahimi V, Rakhshandeh H, Askari VR. Nigella sativa oil reduces LPS-induced microglial inflammation: an evaluation on M (1)/M (2) balance. Evid Based Complement Alternat Med. 2022;2022:5639226.
87. Yimer EM, Tuem KB, Karim A, Ur-Rehman N, Anwar F. Nigella sativa L. (Black Cumin): a promising natural remedy for wide range of illnesses. Evid Based Complement Alternat Med. 2019;2019:1528635.
88. Khabbazi A, Javadivala Z, Seyedsadjadi N, Malek MA. A systematic review of the potential effects of Nigella sativa on rheumatoid arthritis. Planta Med. 2020;86(7):457–69.
89. Yücel Ç, Karatoprak G, Açıkara ÖB, Akkol EK, Barak TH, Sobarzo-Sánchez E, et al. Immunomodulatory and anti-inflammatory therapeutic potential of gingerols and their nanoformulations. Front Pharmacol. 2022;13:902551.
90. Yahyazadeh R, Baradaran Rahimi V, Yahyazadeh A, Mohajeri SA, Askari VR. Promising effects of gingerol against toxins: a review article. Biofactors. 2021;47(6):885–913.
91. Alipour A, Baradaran Rahimi V, Askari VR. Promising influences of gingerols against metabolic syndrome: a mechanistic review. Biofactors. 2022;48(5):993–1004.
92. Fuloria S, Mehta J, Chandel A, Sekar M, Rani N, Begum MY, et al. A comprehensive review on the therapeutic potential of Curcuma longa linn in relation to its major active constituent Curcumin. Front Pharm. 2022;13:820806.
93. Dastani M, Rahimi HR, Askari VR, Jaafari MR, Jarahi L, Yadollahi A, et al. Three months of combination therapy with nano-curcumin reduces the inflammation and lipoprotein (a) in type 2 diabetic patients with mild to moderate coronary artery disease: evidence of a randomized, double-blinded, placebo-controlled clinical trial. Biofactors. 2022;49:108.
94. Shep D, Khanwelkar C, Gade P, Karad S. Safety and efficacy of curcumin versus diclofenac in knee osteoarthritis: a randomized open-label parallel-arm study. Trials. 2019;20(1):214.
95. Li H, Hu R, Xu S, Dai Z, Wu X, Hu J, et al. Tripterygium wilfordii hook f preparations for rheumatoid arthritis: an overview of systematic reviews. Evid Based Complement Alternat Med. 2022;2022:3151936.
96. Song X, Zhang Y, Dai E. Therapeutic targets of thunder god vine (*Tripterygium wilfordii* Hook) in rheumatoid arthritis (review). Mol Med Rep. 2020;21(6):2303–10.
97. Zhou YZ, Zhao LD, Chen H, Zhang Y, Wang DF, Huang LF, et al. Comparison of the impact of Tripterygium wilfordii hook F and methotrexate treatment on radiological progression in active rheumatoid arthritis: 2-year follow up of a randomized, non-blinded, controlled study. Arthritis Res Ther. 2018;20(1):70.
98. Zhou YY, Xia X, Peng WK, Wang QH, Peng JH, Li YL, et al. The effectiveness and safety of Tripterygium wilfordii hook. F extracts in Rheumatoid arthritis: a systematic review and meta-analysis. Front Pharmacol. 2018;9:356.
99. Wang J, Chen N, Fang L, Feng Z, Li G, Mucelli A, et al. A systematic review about the efficacy and safety of Tripterygium wilfordii hook.f. preparations used for the management of Rheumatoid arthritis. Evid Based Complement Alternat Med. 2018;2018:1567463.
100. Ghadiri M, Baradaran Rahimi V, Moradi E, Hasanpour M, Clark CCT, Iranshahi M, et al. Standardised pomegranate peel extract lavage prevents postoperative peritoneal adhesion by regulating TGF-β and VEGF levels. Inflammopharmacology. 2021;29(3):855–68.
101. Rakhshandeh H, Baradaran Rahimi V, Habibi Z, Sirousi Z, Askari VR. Punica granatum seed oil detracts peritoneal adhesion: perusing anti-oxidant, anti-inflammatory, antifibrotic, and antiangiogenic impacts. Physiol Rep. 2022;10(24):e15545.
102. Rahimi VB, Askari VR, Mousavi SH. Ellagic acid dose and time-dependently abrogates d-galactose-induced animal model of aging: investigating the role of PPAR-γ. Life Sci. 2019;232:116595.
103. Baradaran Rahimi V, Ghadiri M, Ramezani M, Askari VR. Anti-inflammatory and anti-cancer activities of pomegranate and its constituent, ellagic acid: evidence from cellular, animal, and clinical studies. Phytother Res. 2020;34(4):685–720.
104. Taherzadeh D, Baradaran Rahimi V, Amiri H, Ehtiati S, Yahyazadeh R, Hashemy SI, et al. Acetyl-11-keto-β-Boswellic acid (AKBA) prevents lipopolysaccharide-induced inflammation and cytotoxicity on H9C2 cells. Evid Based Complement Alternat Med. 2022;2022:2620710.
105. Rahimi VB, Askari VR, Mehrdad A, Sadeghnia HRJAPP. Boswellia serrata has promising impact on glutamate and quinolinic acid-induced toxicity on oligodendroglia cells: in vitro study. Acta Pol Pharm. 2017;74(6):1803–11.
106. Kumar R, Singh S, Saksena AK, Pal R, Jaiswal R, Kumar R. Effect of Boswellia Serrata extract on acute inflammatory parameters and tumor necrosis factor-α in complete Freund's adjuvant-induced animal model of rheumatoid arthritis. Int J Appl Basic Med Res. 2019;9(2):100–6.
107. Singh S, Khajuria A, Taneja SC, Khajuria RK, Singh J, Qazi GN. Boswellic acids and glucosamine show synergistic effect in preclinical anti-inflammatory study in rats. Bioorg Med Chem Lett. 2007;17(13):3706–11.
108. Rahmanian-Devin P, Rakhshandeh H, Baradaran Rahimi V, Sanei-Far Z, Hasanpour M, Memarzia A, et al. Intraperitoneal lavage with Crocus sativus prevents postoperative-induced peritoneal adhesion in a rat model: evidence from animal and cellular studies. Oxidative Med Cell Longev. 2021;2021:5945101.
109. Baradaran Rahim V, Khammar MT, Rakhshandeh H, Samzadeh-Kermani A, Hosseini A, Askari VR. Crocin protects cardiomyocytes against LPS-induced inflammation. Pharmacol Rep. 2019;71(6):1228–34.
110. Kalmarzi RN, Naleini SN, Ashtary-Larky D, Peluso I, Jouybari L, Rafi A, et al. Anti-inflammatory and immunomodulatory effects of barberry (Berberis vulgaris) and its main compounds. Oxidative Med Cell Longev. 2019;2019:6183965.
111. Ghorbani A, Baradaran Rahimi V, Sadeghnia HR, Hosseini A. Effect of berberine on the viability of adipose tissue-derived mesenchymal stem cells in nutrients deficient condition. Nat Prod Res. 2018;32(5):592–5.
112. Paul S, Chakraborty S, Anand U, Dey S, Nandy S, Ghorai M, et al. Withania somnifera (L.) Dunal (Ashwagandha): a comprehensive review on ethnopharmacology, pharmacotherapeutics, biomedicinal and toxicological aspects. Biomed Pharmacother. 2021;143:112175.
113. Dar NJ, Hamid A, Ahmad M. Pharmacologic overview of Withania somnifera, the Indian ginseng. Cell Mol Life Sci. 2015;72(23):4445–60.
114. Gupta A, Singh S. Evaluation of anti-inflammatory effect of Withania somnifera root on collagen-induced arthritis in rats. Pharm Biol. 2014;52(3):308–20.
115. Sumantran VN, Kulkarni A, Boddul S, Chinchwade T, Koppikar SJ, Harsulkar A, et al. Chondroprotective potential of root extracts of Withania somnifera in osteoarthritis. J Biosci. 2007;32(2):299–307.
116. Huang L, Lyu Q, Zheng W, Yang Q, Cao G. Traditional application and modern pharmacological research of *Eucommia ulmoides* Oliv. Chin Med. 2021;16(1):73.
117. Hussain T, Tan B, Liu G, Oladele OA, Rahu N, Tossou MC, et al. Health-promoting properties of Eucommia ulmoides: a review. Evid Based Complement Alternat Med. 2016;2016:5202908.

Role of Herbal Medicines for the Treatment of Chronic Kidney Disease

Vafa Baradaran Rahimi and Vahid Reza Askari

Abstract

The use of herbal medicines has tremendously increased globally and in developed countries over the past decades. Nowadays, several herbal medicines are used to treat many chronic and acute disorders, including chronic kidney diseases (CKD). The promising effects of many of them are well investigated and documented by researchers worldwide. This chapter focuses on the promising anti-CKD effects of *Abelmoschus manihot*, *Salvia miltiorrhiza,* and their constituents, salvianolic acid-A, salvianolic acid-B, and tanshinone-IIA, berberine, *Zingiber officinale,* and its constituents 6-gingerols and 6-shogaols, *Punica granatum* and its major constituent ellagic acid, *Vitis vinifera*, *Curcuma longa,* and their major constituent curcumin, *Nigella sativa*, *Rosmarinus officinalis,* and their major constituent rosmarinic acid, in different animal and human clinical studies. CKD was mainly treated with these herbal medicines through anti-oxidative, anti-inflammatory, anti-fibrotic, and anti-apoptotic effects, improving renal function, downregulating the nuclear factor-kappa B (NF-κB) signaling pathway, and stimulating the nuclear factor erythroid 2-related factor-2 (Nrf-2) cascade.

Keywords

Chronic kidney disease · NF-κB · Inflammation · Herbal medicines · Oxidative stress

1 Introduction

Herbal medicines with promising anti-chronic kidney disease (CKD) effects are illustrated in Fig. 1. Table 1 summarizes herbal medicine's protective effects against animal models of CKD. In addition, human clinical studies supporting the anti-CKD effects of herbal medicines are shown in Table 2.

V. Baradaran Rahimi
Department of Cardiovascular Diseases, Faculty of Medicine, Mashhad University of Medical Sciences, Mashhad, Iran
e-mail: baradaranrv@mums.ac.ir

V. R. Askari (✉)
Applied Biomedical Research Center, Mashhad University of Medical Sciences, Mashhad, Iran

International UNESCO Center for Health-Related Basic Sciences and Human Nutrition, Mashhad University of Medical Sciences, Mashhad, Iran
e-mail: askariv@mums.ac.ir

A. K. Dhara, S. C. Mandal (eds.), *Role of Herbal Medicines*, https://doi.org/10.1007/978-981-99-7703-1_15

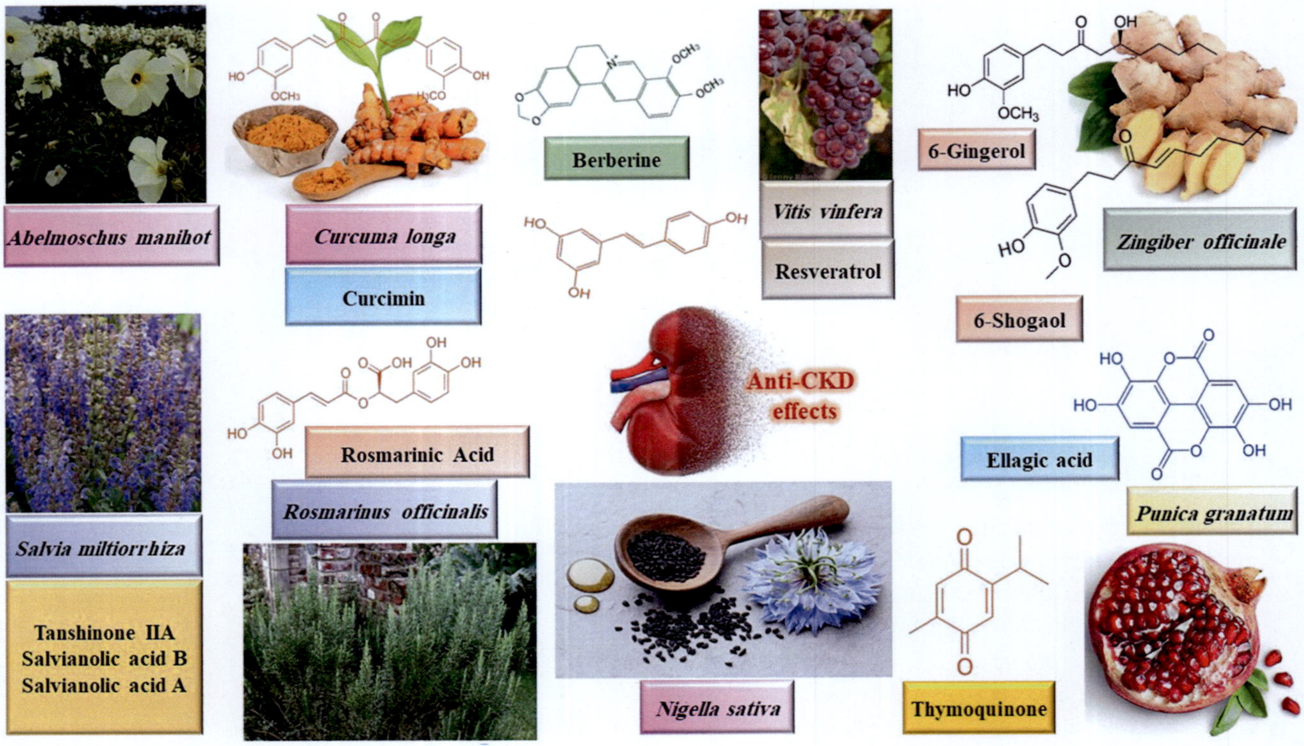

Fig. 1 Herbal medicines with promising anti-CKD effects

Table 1 Protective effects of herbal medicines against animal models of CKD

Extract or constituent type	Dose/concentration	An overview of the study model	Results	References
Abelmoschus manihot (HKC)	75, 175, and 300 mg/kg/day; given orally for 12 weeks	Unilateral nephrectomy and streptozotocin-induced DN in male Sprague-Dawley rats	↓ TGF-β and collagen type IV expression in the kidney ↑ PPAR-α, CPT-1, and ACO expression ↑ PPAR-γ, CD36, and LPL expression ↑ Serum adiponectin ↓ TNF-α, IL-6, IL-1β and IL-2	[3]
Abelmoschus manihot (HKC)	0.75 and 2.0 g/kg/day; orally for 8 weeks	Unilateral nephrectomy and streptozotocin- induced DN in male Sprague-Dawley rats	↓ BUN and UA levels ↓ Urinary albumin levels ↓ Oxidative stress parameters, including MDA, 8OhDG, and NOX4 ↑ Anti-oxidant SOD activity ↓ Phosphorylated p38MAPK, phosphorylated Akt (p-Akt), TNF-α, and TGF-β protein expressions in the kidney	[4]
Abelmoschus manihot (HKC)	0.5 and 2.0 g/kg/day; orally for 4 weeks	Unilateral nephrectomy and adriamycin-induced nephropathy in male Sprague-Dawley rats	↑ Body weight and general status ↓ Kidney/body weight ratio, urinary protein, serum BUN, and Cr levels ↓ Protein expression levels of TNF-α, IL-2, TGF-β1, and p-p38MAPK	[5]
Abelmoschus manihot (HKC)	2.0 g/kg/day; orally for 4 weeks	Unilateral nephrectomy and adriamycin-induced nephropathy in male Sprague-Dawley rats	↓ Body weight, serum albumin, Cr, and urinary albumin levels Improved glomerular pathological changes ↓ p-Akt, p-mTOR, p-p70S6K, and TGF-β1	[6]

Table 1 (continued)

Extract or constituent type	Dose/concentration	An overview of the study model	Results	References
Abelmoschus manihot (HKC)	0.75 g/kg/day; orally for 4 weeks	Induction of chronic renal failure in rats by adenine	↓ Serum Cr, BUN, and urinary protein ↓ NOX1, NOX2, NOX4, α-SMA, p-ERK expression	[7]
70% ethanolic extract of flower and leaf of *Abelmoschus manihot*	100 mg/kg/day; orally for 5 weeks	High-fat diet, unilateral nephrectomy, and streptozotocin-induced nephropathy	↓ Serum Cr, BUN, urinary albumin, glomerular, and tubular damage ↓ TNF-α, IL-6 ↑ IL-10 in kidney tissue	[8]
Ethyl acetate extract of *Salvia miltiorrhiza*	30 mg/kg/day; orally for 18 weeks	Streptozotocin-induced DN in C57BL/6 mice	↓ Urinary albumin excretion, albumin/creatinine ratio, serum Cr, BUN Improved the pathological changes in kidney tissue ↑ Nrf-2, HO-1, and NQO1 mRNA expression ↓ Keap-1 in kidney tissue	[12]
SAA	10 mg/kg; orally for 8 weeks	5/6 nephrectomized rats	↓ Kidney/body weight ratio, Scr, BUN, creatinine clearance rate, and MDA level ↑ Body weight and SOD activity ↓ TGF-β1 in kidney tissue ↑ BMP-7 and Smad6 expression	[13]
SAA	2.5, 5, and 10 mg/kg; daily, given I.P. over 28 days	5/6 nephrectomized rats	↓ Urine protein, BUN, Scr ↓ TGF-β1, α-SMA, TNF-α, IL-1β expression levels ↓ NF-κB and MAPK signaling cascade ↓ ICAM-1 VCAM-1	[14]
SAA	2.5, 5, and 10 mg/kg; daily, given I.P. over 28 days	5/6 nephrectomized rats	↑ SOD and CAT activities ↓ MDA, ROS, and NOX-4 levels ↑ p-Akt, p-Nrf-2, and HO-1 in kidney tissue	[15]
SAA	10 mg/kg; injected into the tail vein for 3 weeks	Adriamycin-induced MCD rat model	↓ Scr, BUN, urinary levels of Angptl4 ↑ PPARγ expression in the kidney tissue	[16]
SAB	3, 6.25, and 12.5 mg/kg; given I.P. for 4 weeks	Renal tubular damage by fatty acid in C57BL/6 mice	↓ Urinary output, cystatin C, Scr ↓ ICAM-1, VCAM-1, IL-1β and 6, and TNF-α mRNA expression in renal tissue ↓ Apoptotic Bax and cleaved caspase-3 expression ↑ Anti-apoptotic BCL-2 expression	[17]
SAB	50, 100, and 200 mg/kg; administered orally	Balb/c mice suffering from renal ischemia-reperfusion injury	↓ Scr, BUN ↓ MDA, IL-1β, and TNF-α levels ↓ Keap-1 ↑ Nrf-2 and HO-1 ↑ SOD and GSH activities in kidney tissue	[18]
SAB	6.25, 12.5, 25 mg/kg; given for 14 days	Kidney injury by unilateral ureteral obstruction in C57BL/6 mice	↓ BUN, Scr ↓ α-SMA, FGF-2, TGF-β1 expression ↑ E-cadherin protein expression in kidney tissue	[19]
Tanshinone IIA	10 mg/kg; oral administration over 12 weeks	Streptozotocin-induced DN	↓ Urine albumin excretion rate, CRP, and MDA levels ↑ SOD activity ↓ TGF-β1, P-selectin, and MCP-1 mRNA expression in kidney tissue	[20]

(continued)

Table 1 (continued)

Extract or constituent type	Dose/concentration	An overview of the study model	Results	References
Tanshinone IIA	I.P. doses of 2, 4, and 8 mg/kg daily; used for 42 days	Streptozotocin-induced DN	↓ TGF-β1, TSP-1, Grp78, CHOP mRNA expression in kidney tissue ↓ p-PERK, p-elf2α and ATF-4	[21]
Tanshinone IIA	25 mg/kg; S.C.	Contrast-induced nephropathy	↓ MDA, Scr, BUN ↑ Nrf-2 and HO-1 in kidney tissue	[22]
Berberine	200 mg/kg; given orally for 14 weeks	Kidney injuries in rats caused by UUO	↓ MDA level ↑ SOD and CAT activities ↓ TGF-β1, α-SMA, and pSmad3 expression in kidney tissue	[27]
Berberine	Oral dosage of 25 and 100 mg/kg; administered for 16 weeks	Streptozocin-induced DN in rats	↓ Kidney/body weight, BUN, Scr, and urinary protein levels over 24 h ↑ p-AMPK/AMPK, P-ULK/ULK, LC3-II/LC3-I, beclin-1 levels ↓ pmTOR/mTOR, TGF-β1, α-SMA protein expression	[29]
Berberine	400 mg/kg; orally for 83 days	Streptozocin-induced DN in rats	↓ Scr, BUN, protein ↓ Expression levels of TGF-β, vimentin, and α-SMA mRNA in kidney tissue	[30]
Berberine	Eight weeks of oral administration at 300 mg/kg	Male db/db diabetic mice	↓ Drp1 expression and translocation from cytoplasm to the mitochondria	[28]
Berberine	50, 100, or 200 mg/kg; given orally for 83 days	Streptozocin-induced DN in rats	↓ 24-h urinary protein, Scr, BUN ↓ IL-1β, IL-6, and MCP-1 levels in kidney tissue ↓ TLR-4 protein level, p-IκBα/IκBα ratio, and p-p65/p65 ratio	[31]
Berberine	50, 100, or 200 mg/kg; given orally for 83 days	Streptozocin-induced DN in rats	↓ MMP9, TIMP-1, TIMP-2, TGF-β1, fibronectin, and type IV collagen in kidney ↑ MMP2 expression levels	[32]
Berberine	100 mg/kg; taken for 20 weeks orally	Spontaneously hypertensive rats	↓ Hypertension ↓ Angiotensin II, aldosterone, IL-6, IL-17, IL-23, osteopontin, and KIM-1	[33]
Berberine	2 μg/h; bilateral paraventricular nucleus infusion for 28 days	2K1C renovascular hypertensive rats	↓ Mean arterial pressure ↓ NOX2, NOX4, Erk1/2, iNOS ↑ copper/zinc superoxide dismutase levels in the paraventricular nucleus	[34]
80% ethanolic extract of *Zingiber officinale*	400 and 800 mg/kg given orally for 6 weeks	Streptozocin-induced DN in rats	↓ Scr, BUN, and MDA ↑ GSH level, SOD, CAT activity ↓ TNF-α, IL-6, IL-1β, cytochrome *c*, caspase-3 in kidney tissue	[40]
70% ethanolic extract of *Zingiber officinale*	1 g/kg/day; orally for 28 days	Streptozocin-induced DN in rats	↓ MDA and TNF-α level ↑ TAC, SOD, CAT, Gpx activities ↑ Nrf-2 expression in renal tissue	[41]
20% aqueous extract of *Zingiber officinale*	125 mg/kg; orally for 3 days	Mercuric chloride-induced kidney injury	↑ GSH level, SOD and CAT, GR, GPx, and GST activities ↓ MDA level in renal tissue	[42]
Ethanolic extract of *Zingiber officinale*	150 mg/kg; orally for 3 weeks	Rat renal toxicity caused by lead	↑ GSH level, GPx, GST, and catalase activities in kidney tissue	[43]
6-S	20 mg/kg; given I.P.	Renal ischemia-reperfusion in C57BL/6 male mice	↓ Scr, BUN ↓ mRNA expression of kidney neutrophil gelatinase-associated lipocalin, IL-6, MCP-1, MIP-2, KC ↑ HO-1 mRNA expression	[44]
6-S	5 or 10 mg/kg; given I.P. for 2 weeks	Streptozocin-induced DN	↑ Nrf-2 mRNA expression	[45]
6-S	20 mg/kg; given I.P.	Cisplatin-induced renal injury in C57BL/6 male mice	↓ Scr, BUN, MDA levels ↓ KIM-1, NGAL, NOX4, TNF-α, IL-6, and MCP-1 ↑ GSH level, SOD, and catalase activities in kidney tissue	[46]

Table 1 (continued)

Extract or constituent type	Dose/concentration	An overview of the study model	Results	References
6-G	10 mg/kg; orally for 8 weeks	Streptozocin-induced DN in rats	↓ Levels of Scr, BUN, MDA, CRP, IL-6 and 1β, and TNF-α ↑ GSH level ↑ SOD and CAT activities	[47]
6-G	100 mg/kg; I.P. over 10 days	Gentamicin-induced kidney injury	↓ Scr, BUN, MDA level ↑ GSH level Improved pathological changes in kidney tissue	[48]
6-G	50, 100, and 200 mg/kg; orally over 14 continuous days	Carbendazim-induced kidney damage in rats	↓ MDA and H_2O_2 levels ↑ SOD and CAT activities ↑ GSH level in kidney tissue	[49]
6-G	50 mg/kg; orally over 3 days	Mercuric chloride-induced kidney injury	↑ GSH level, SOD, catalase, GR, GPx, and GST activities ↓ Kidney tissue MDA level	[42]
Punica granatum leaves extract	50, 100, and 200 mg/kg; orally over 28 days	Streptozocin-induced DN in rats	↓ Scr, BUN, urinary total protein, urine volume, MDA Improved histopathological changes ↑ GSH, catalase, and SOD activities in renal tissue	[55]
A methanolic extract of *Punica granatum* leaves	100, 200, and 400 mg/kg; taken orally for 8 weeks	Streptozocin-induced DN in rats	↓ Scr, BUN ↓ MDA level ↑ GSH level, CAT, and SOD activities in renal tissue	[56]
A methanolic extract of *Punica granatum* leaves	100, 200, and 400 mg/kg; taken orally for 8 weeks	Gentamicin-caused nephropathy	↓ Scr, BUN ↓ MDA and TNF-α level ↑ GSH level, catalase, and SOD activities in renal tissue	[57]
PPEE	100 mg/kg; orally over 2 weeks	Vancomycin-induced kidney injury in rats	↓ MDA, C-reactive protein ↑ GSH level, catalase, and SOD activities in the renal ↓ Caspase-3 ↑ Bcl-2 renal expression	[58]
Punica granatum juice	Administered orally for 10 weeks	CCl_4-induced nephrotoxicity in rats	↓ NO, and MDA level ↑ GSH level, CAT, SOD, GST, GPx activities in kidney tissue	[59]
EA	50, 100, and 150 mg/kg; taken orally for 4 weeks	Induction of DN in rats by streptozocin	↓ Scr, BUN, MDA, and TNF-α levels ↑ SOD activity in kidney tissue ↓ TLR-4, IRAK4, TRAF6, IKK-β, NF-κBp65, and HMGB1 protein expression in kidney tissue	[60]
EA	Dosage of 10 and 30 mg/kg; administered orally over 21 days	Rat renal injury caused by sodium arsenite	↓ Scr, BUN, MDA, and NO levels ↑ GSH level and SOD activity in renal tissue	[61]
EA	20 and 40 mg/kg daily; orally over 14 days	Chronic renal failure caused by 5/6 nephrectomy	↓ Urine volume, urine protein, BUN, Scr, MDA, TNF-α, IL-6, and ICAM-1 level ↑ GSH level and SOD activity in renal tissue ↓ miR-182, TGF-β1, fibronectin, and Bax expression levels ↑ FOXO3a and Bcl-2 expression levels	[62]
EA	50, 100, and 150 mg/kg; given orally	Renal ischemic-reperfusion injury in rats	↓ TNF-α, IL-1β, IL-6, MCP-1, MDA, LDH, iNOS, COX-2 levels ↓ Bax and caspase-3 expression ↑ GSH level, SOD activity, and Bcl-2 expression in kidney tissue ↓ p-JAK1, p-JAK2, p-STAT1, and NOX4 expression levels	[63]

(continued)

Table 1 (continued)

Extract or constituent type	Dose/concentration	An overview of the study model	Results	References
EA	10 mg/kg, given I.P., over 8 weeks	Carbon tetrachloride-caused kidney damage in rats	↓ MDA level ↓ TNF-α, NF-κB, COX-2, and VEGF ↑ GSH level, CAT activity ↑ Nrf-2 expression in kidney tissue	[64]
EA	10 mg/kg; orally taken for 7 days	Nephrotoxicity caused by amikacin in rats	↓ MDA, TNF-α, IL-6 levels, NF-κB, and Bax expression ↑ GSH level, SOD, and CAT activities	[65]
Vitis vinifera seed proanthocyanidin extracts	250 mg/kg/day; orally for 16 weeks	Streptozotocin-induced DN in rats	↓ GRP78, p-ERK and Caspase-12 in kidney tissue	[67]
Vitis vinifera seed proanthocyanidin extracts	125 and 250 mg/kg/day; GSPE over 8 weeks	Streptozotocin-induced DN in rats	↓ Scr, BUN, and MDA levels ↑ GSH level, SOD, and TAC activities ↑ Nrf-2, HO-1, GST, and NQO1 in kidney tissue	[68]
Vitis vinifera seed proanthocyanidin extracts	Oral administration of 200 mg/kg for 28 days	Carboplatin and thalidomide-induced nephrotoxicity	↓ p53, TNF-α, and IL-6 expression ↓ MDA and NO levels ↑ GSH, GST, TAC levels ↑ SOD, CAT, and GPx activities	[69]
Aqueous extracts of *Vitis vinifera* fruits	400 mg/kg; given orally over 4 weeks	Induction of DN by doxorubicin in rats	↓ Scr, BUN, and hs-CRP ↑ Nrf-2 mRNA expression in kidney tissue	[70]
V. vinifera seed extract	100 mg/kg/day; orally over 8 weeks	Amiodarone-induced nephrotoxicity in rats	↓ Scr, MDA, and IL-6 levels ↑ SOD activity	[71]
Curcuma longa	250 mg/kg/day; orally for 8 weeks	Rats with kidney damage caused by doxorubicin	↓ MCP-1 and TGF-β1 level ↓ Desmin, vimentin, and ED-1+ cells immunostaining	[75]
Curcumin	50, 100, and 200 mg/kg; taken orally over 28 days	The nephrotic syndrome produced in rats by doxorubicin	↓ Urinary protein, MDA levels ↓ NF-κB p65 and Keap-1 expression ↑ HO-1, NQO-1, Nrf-2, IκBα expression ↑ SOD activity	[76]
Curcumin	200 mg/kg; orally for 6 weeks	Sodium arsenate-induced kidney injury in mice	↓ IL-1β, IL-6, TNF-α, IFN-γ, TGF-β1 ↓ NF-κB ↓ P-JNK, P-ERK1/2, and P-p38 phosphorylation ↑ Nrf-2, NQO1, and HO-1 expression	[77]
Curcumin	100 mg/kg; given orally over 8 days	Gentamicin-induced nephrotoxicity in rats	↓ MDA and NO, IL-1β, IL-6, TNF-α ↑ SOD, CAT, and GSH-Px activities ↓ Caspase-3, and Bax ↑ Bcl-2	[78]
Curcumin	Two doses of 20 and 40 mg/kg; orally over 14 continuous days	Potassium oxonate-induced kidney inflammation in mice	↓ IL-1β, NLRP3, ASC, and caspase1 expression	[79]
Curcumin	300 mg/kg/day; orally over 30 days	Passive Heymann nephritis in rats	↓ Scr, BUN, and MDA levels ↑ GSH, SOD, CAT levels ↓ Bax, Caspase-3, p62, PI3K, p-AKT, and p-mTOR expression ↑ Bcl-2, beclin1, Nrf-2, and HO-1	[80]
Curcumin	100 mg/kg; orally for 8 days	Cisplatin-induced kidney damage in rats	↓ Scr, BUN, MDA levels ↓ ERK1/2 phosphorylation ↓ NF-κB expression ↓ Bax/Bcl-2 ratio ↓ TNF-α, IL-6, KIM-1, and NGAL ↑ GSH level ↑ IL-10 mRNA expression in kidney tissue	[81]
70% hydro-ethanolic extract of *Nigella sativa*	200 and 400 mg/kg; I.P. taken over 18 days	UUO in rats	↓ Angiotensin II, MCP-1, MDA, and TNF-α levels ↑ GSH level, SOD, and catalase activities in kidney tissue	[92]

Table 1 (continued)

Extract or constituent type	Dose/concentration	An overview of the study model	Results	References
Nigella sativa	1 g/kg/day; orally for 10 days	Radiation-induced kidney injury in rats	↓ Total oxidant status and lipid hydroperoxide levels ↑ Total anti-oxidant status ↑ Paraoxonase and ceruloplasmin activities in kidney tissue	[93]
Nigella sativa oil and thymoquinone	2 mL/kg; orally 1.5 mg/kg; orally over 34 days	Cisplatin-induced kidney injury in rats	↓ Scr, BUN, and MDA level ↑ GSH level ↑ SOD, CAT, GPx, and GR activities	[94]
Aqueous extract of *Rosmarinus officinalis*	100 mg/kg; taken orally for 14 days	Carbon tetrachloride-induced nephropathy in mice	↓ Scr, BUN, LDH, MDA ↑ GPx and GSH levels ↑ SOD and CAT activities in kidney tissue	[100]
Rosmarinus officinalis essential oil	0.02% v/w; orally for 60 days	Diethylnitrosamine-induced renal injury in rats	↓ Urea, Scr, and MDA levels ↑ Total anti-oxidant concentration, CAT, GPx activities improved histopathological changes	[101]
Rosmarinus officinalis essential oil	0.5 mL/kg; orally over 14 days	Potassium dichromate-induced kidney injury in rats	↓ Urea, creatinine, uric acid, MDA, and H_2O_2 levels ↑ GSH level, SOD, catalase, GPx, and GST activities Improved histological changes	[102]
Rosmarinic acid	5, 10, 20 mg/kg; given orally over 8 days	Kidney damage caused by cisplatin in mice	↓ MDA, NO levels ↓ IL-1β, IL-6, and TNF-α ↑ SOD, CAT activities, GSH, and TAC levels in kidney tissue ↓ Keap-1 in kidney tissue	[103]
Rosmarinic acid	25–100 mg/kg; taken orally over 28 days	Chlorpyrifos-induced kidney injury in rats	↓ Urea, Scr, KIM-1, ROS, MDA, NO, IL-1β, TNF-α, and NF-κB p65 ↑ GSH, SOD, and CAT activities ↓ Bax, caspase-3, Keap-1 ↑ Renal Bcl-2, Nrf-2, HO-1 and SIRT1	[104]
Rosmarinic acid	25 mg/kg; given orally over 60 days	Chromium-induced kidney injury in rats	↓ MDA level ↑ GSH level ↑ Nrf-2 mRNA level	[105]
Rosmarinic acid	50 mg/kg; taken orally over 14 days	Cadmium-induced nephrotoxicity in mice	↓ CRP, IL-1β, IL-6, TNF-α, ROS, NO, H_2O_2, and MDA levels ↑ GSH, GPx, and GR levels ↑ SOD and CAT activities ↓ Caspase-3, caspase-8, caspase-9, NF-κB, PKC, TNFR, TGF-β1, SMAD3, and α-SMA expression levels in kidney tissue	[106]

Huangkui capsule (HKC), diabetic nephropathy (DN), intraperitoneal (IP), salvianolic acid A (SAA), salvianolic acid B (SAB), 6-shogaol (6-S), 6-gingerol (6-G), Ellagic acid (EA), blood urea nitrogen (BUN), uric acid (UA), malondialdehyde (MDA), 8-hydroxy-2′-deoxyguanosine (8OhDG), superoxide dismutase (SOD), Interleukin (IL), creatinine (Cr), mammalian target of rapamycin (mTOR), transforming growth factor-β1 (TGF-β1), nicotinamide adenine dinucleotide phosphate oxidase 4 (NOX4), tumor necrosis factor-α (TNF-α), α-smooth muscle actin (α-SMA), nuclear factor erythroid-2-related factor 2 (Nrf-2), phosphorylation-extracellular signal-regulated kinase (p-ERK), heme oxygenase-1 (HO-1), kelch-like ECH-associated protein 1 (Keap-1), bone morphogenetic protein 7 (BMP-7), NAD(P)H quinone dehydrogenase 1 (NQO1), α-smooth muscle actin (α-SMA), nuclear factor-κB (NF-κB), glutathione peroxidase (GPx), catalase (CAT), p38 mitogen-activated protein kinase (MAPK), p-glycogen synthase kinase-3β (p-GSK-3β), intercellular adhesion molecule-1 (ICAM-1), minimal change disease (MCD), vascular cell adhesion molecule-1 (VCAM-1), angiopoietin-like 4 (Angptl4), fibroblast growth factor-2 (FGF-2), glutathione (GSH), thrombospondin-1 (TSP-1), glucose-regulated protein 78 (Grp78), activating transcription factor (ATF)-4, phosphorylated eukaryotic initiation factor 2 alpha (p-elf2α), phosphorylated Smad3 (pSmad3), dynamin-related protein 1 (Drp1), kidney-injury-molecule (KIM-1), inducible nitric oxide synthase (iNOS), glomerular filtration rate (GFR), matrix metallopeptidase (MMP), tissue inhibitor matrix metalloproteinase (TIMP), reactive oxygen species (ROS), macrophage inflammatory protein-2 (MIP-2), TNF Receptor Associated Factor 6 (TRAF6), AMP-activated protein kinase (AMPK), keratinocyte chemoattractant (KC), glutathione reductase (GR), glutathione-*S*-transferase (GST), nitric oxide (NO), Interleukin-1 receptor-associated kinase 4 (IRAK4), inhibitor of nuclear factor kappa-B kinase (IKK-β), vascular endothelial growth factor family (VEGF), high mobility group box 1 (HMGB1), Interferon gamma (IFN-γ), neutrophil gelatinase-associated lipocalin (NGAL), Sirtuin 1 (SIRT1), protein kinase C (PKC), tumor necrosis factor receptor (TNFR)

Table 2 Clinical studies of the anti-CKD effects of herbal medicines

Extract type or plant part	Study design	Study model	Results	References
Abelmoschus manihot (HKC)	• 2.5 g 3 times a day	Prospective, open-label, multicenter, randomized, controlled clinical trial on biopsy-proven primary glomerular disease patients	↓ 24-h proteinuria greater than losartan potassium	[9]
	• 50 mg losartan daily for 24 weeks			
Abelmoschus manihot (HKC)	• 2.5 g 3 times a day	Randomized, non-inferiority, double-blind, multicenter clinical trial on IgAN patients	↓ 24-h proteinuria similar to losartan potassium	[10]
	• 100 mg losartan daily for 48 weeks		No significant changes in eGFR	
Salvia miltiorrhiza	• Sulfotanshinone IIA + Western standard treatment	Meta-analysis of 21 clinical studies involving 1857 patients with CKD	↓ BUN, Scr, urinary protein level over 24 h	[23]
	• Western standard treatment		↓ Microalbuminuria, β2-macroglobulin, and cystatin C	
			↑ GFR levels	
Salvia miltiorrhiza	• Sodium tanshinone IIA sulfonate injection + ARBs	Meta-analysis of 16 clinical studies involving 1696 patients with hypertensive nephropathy	↓ 24-h urinary protein, Scr, cystatin-C	[24]
	• ARB monotherapy		↓ Urinary immunoglobulin G and transferrin	
			↓ SBP and DBP	
			↑ GFR level	
Berberine	• 0.1 g, 3 times/day for 2 years + standard medication	Randomized, controlled clinical trial evaluating hypertensive and type 2 diabetes mellitus patients	↓ Scr, urine albumin/Cr	[35]
	• Standard medication		↓ Osteopontin and KIM-1	
			↓ VCAM-1 and CRP	
			↓ MDA level	
			↑ SOD, GSH-Px, and TAC activities	
Berberine	• 0.4 g, 3 times/day over 6 months + standard medication	Controlled, randomized clinical trial investigating type 2 diabetes mellitus patients	↓ BUN, urine albumin/Cr ratio	[36]
	• Standard medication		↓ hs-CRP and cystatin C	
			↑ eGFR level	
Zingiber officinale	• *Z. officinale* (1000 mg/day) over 10 weeks	Randomized, placebo-controlled, double-blind clinical trial on patients on peritoneal dialysis	↓ Serum fasting glucose	[50]
	• Placebo			
Vitis vinifera	• *V. vinifera* seed extract (2100 mg/day) for 6 months	Randomized, placebo-controlled, double-blind clinical trial determining chronic kidney disease patients	↓ Urinary protein	[72]
	• Placebo		↓ MDA, protein carbonylation, H_2O_2	
			↓ CRP levels	
			↑ GFR, CAT, GPx, and SOD activity	
Curcuma longa	• *C. longa* containing 95% curcumin (2.5 g, 3 times per week) over 12 weeks	Randomized, placebo-controlled, double-blind clinical trial on patients undergoing hemodialysis	↓ NF-kB mRNA expression	[82]
	• Placebo		↓ hs-CRP level	
Curcumin	• Curcumin (1 g/day) over 12 continuous weeks	Double-blind, randomized, , placebo-controlled clinical trial on patients undergoing hemodialysis	↑ CAT, GPx activities	[83]
	• Placebo		No significant changes in hs-CRP, MDA levels, and GR activity	
Curcumin	• Curcumin (320 mg/day) administered over 8 weeks	Randomized, placebo-controlled, double-blind clinical trial on non-diabetic or diabetic proteinuric CKD	↓ MDA level in non-diabetic patients	[84]
	• Placebo		↑ Anti-oxidant markers in diabetic patients	
			No significant differences in the urinary protein, eGFR, and Nrf-2 activation	

Table 2 (continued)

Extract type or plant part	Study design	Study model	Results	References
Curcumin	• Curcumin (500 mg twice a day) over 6 weeks	Randomized, double-blind, placebo-controlled trial on CKD patients	↓ MDA, MCP-1, IL-4, and IFN-γ levels	[85]
	• Placebo			
Curcumin	• Curcumin	Meta-analysis of 5 clinical studies involving 290 patients with diabetic kidney disease	↓ Scr, total cholesterol, SBP, and fasting blood glucose	[86]
	• Placebo		No significant changes in BUN, urinary protein, triglyceride, and diastolic blood pressure levels	
Curcumin	• Curcumin	Meta-analysis of 10 clinical studies involving 523 CKD patients	No significant differences in IL-6, TNF-α, and hs-CRP levels	[87]
	• Placebo			
Curcumin	• Curcumin	Meta-analysis of nine clinical studies on CKD patients	↓ Total cholesterol and TNF-α levels	[88]
	• Placebo			
Nigella sativa oil	• 2.5 mL, taken orally, once daily over 12 weeks	Prospective, randomized, and open-labeled clinical study on patients with Stages 3–4 of CKD	↓ Blood urea, Scr, and 24-h urinary protein	[96]
	• Control		↑ Total urine volume over 24 h and GFR	
Nigella sativa	• *Nigella sativa*	Meta-analysis of 19 clinical studies involving 1295 participants	↓ ALP and BUN levels	[97]
	• Placebo			

Huangkui capsule (HKC), IgA nephropathy (IgAN), blood urine nitrogen (BUN), systolic blood pressure (SBP), serum creatinine (Scr), glomerular filtration rate (GFR), chronic kidney disease (CKD), and diastolic blood pressure (DBP), angiotensin receptor blockers (ARBs), kidney-injury-molecule (KIM-1), malondialdehyde (MDA), superoxide dismutase (SOD), vascular cell adhesion molecule-1 (VCAM-1), glutathione (GSH), estimated glomerular filtration rate (eGFR), high-sensitive C-reactive protein (hs-CRP)

2 *Abelmoschus manihot* L.

Abelmoschus manihot (*A. manihot*), an annual flowering herb, belongs to Malvaceae. It is widely found in eastern European countries, northern Australia, and Asian countries, such as Indonesia, India, Korea, and China [1]. *A. manihot* has been extensively administered in China as folk medicine, especially for treating CKD. Additionally, the Huangkui capsule (HKC) is a Chinese-approved drug containing the ethanolic extract of *A. manihot* flowers and is prescribed for treating kidney disorders [2].

2.1 Animal Studies

An experiment conducted by Ge et al. determined the effect of HKC in diabetic nephropathy (DN) caused by unilateral nephrectomy and intraperitoneal injection of streptozotocin in Sprague–Dawley rats. They supported that HKC significantly improved the formation of glomerulosclerosis fibrosis and mitigated transforming growth factor-β (TGF-β) and collagen type IV expression in the renal tissue of DN rats. It also propagated the peroxisome proliferator-activated receptor (PPAR)-α and its downstream target genes (CPT-1 and ACO) and PPAR-γ and its target genes (CD36 and LPL) mRNA expression in the kidney tissue. Furthermore, HKC enhanced the serum adiponectin levels while reducing inflammation-induced cytokines, namely, tumor necrosis factor-alfa (TNF-α), Interleukin (IL)-6, IL-1β, and IL-2 in the kidney tissue of DN rats. Therefore, they suggested HKC as a potential anti-DN agent through activating PPARα/γ [3]. Similarly, HKC notably decreased total body weight and kidney weight, blood urea nitrogen (BUN), uric acid (UA), and urinary albumin levels in unilateral nephrectomy and streptozotocin injection in rats. It also improved renal fibrosis by inhibiting the cell number and extracellular matrix amount in the glomerulus. In addition, HKC markedly attenuated the oxidative stress parameters, including malondialdehyde (MDA) and nicotinamide adenine dinucleotide phosphate oxidase 4 (NOX4), while elevating the activity of anti-oxidant superoxide dismutase (SOD). Furthermore, it meaningfully downregulated phosphorylated p38MAPK, phosphorylated Akt (p-Akt), TNF-α, and TGF-β protein expressions in the kidney tissue of DN rats [4].

Tu and coworkers reported that HKC remarkably enhanced body weight and general status while decreasing kidney/body weight ratio, urinary protein, serum BUN, and creatinine (Cr) levels in unilateral nephrectomy and adriamycin-induced nephropathy in male Sprague-Dawley rats. It also improved renal fibrosis by reducing the cell numbers and rate of extracellular matrix in the glomerulus and reduced the number of ED1(+) and ED3(+) macrophages infiltrating the glomeruli. In addition, HKC provided a significant decrement in TNF-α, IL-2, TGF-β1, and phosphorylated p38MAPK protein expression in the kidney of adriamycin-induced nephropathy in rats [5]. Similarly, HKC strikingly attenuated body weight, serum albumin, SCr, and urinary albumin levels, and improved glomerular pathological changes in unilateral nephrectomy and adriamycin-induced nephropathy in male Sprague-Dawley rats. Moreover, it notably downregulated the expression levels of p-Akt, p-p70S6K, p-mTOR, and TGF-β1 in the rats' renal tissue following adriamycin-induced nephropathy [6].

Cai et al. noticed that HKC firmly alleviated serum Cr, BUN, and urinary protein levels in chronic renal failure produced by adenine in male Sprague–Dawley rats. It also suppressed NOX1, NOX2, NOX4, α-SMA, and phosphorylated extracellular signal-regulated kinase (p-ERK) expression in kidney tissue [7]. Furthermore, Kim et al. showed that 70% ethanolic extract of flower and leaf of *A. manihot* markedly mitigated serum Cr, BUN, urinary albumin, glomerular and tubular damage in unilateral nephrectomy, excessive fat consumption, and streptozotocin induction in C57BL/6 mice. *A. manihot* extract also reduced the mRNA expression of TNF-α, IL-6, and p-IκBα while elevating IL-10 in kidney tissue [8]. Taken together, *A. manihot* and HKC may be effective candidates for improving CKD.

2.2 Clinical Studies

The renoprotective properties of *A. manihot* have been approved in several clinical studies. In this regard, Zhang et al. evaluated the impact of HKC (2.5 g, 3 times/day) on 417 patients with biopsy-confirmed primary glomerular disease. They revealed that HKC strikingly decreased proteinuria after 24 weeks of treatment. Surprisingly, the effect of HKC on proteinuria was more significant than losartan potassium (50 mg/day). In addition, no severe adverse effects were observed with HKC treatment [9]. Similarly, HKC showed promising results for treating IgA nephropathy (IgAN) patients. In this study, 1600 biopsy-proven cases of IgAN were randomized to receive either HKC (dosage of 2.5 g, 3 times daily) or losartan potassium (dosage of 100 mg daily) continuously over 48 weeks. Both groups notably diminished the 24-h proteinuria following 48 weeks of treatment, and HKC produced similar results as losartan. In addition, the glomerular filtration rate (eGFR) did not differ significantly in both HKC or losartan groups [10].

A meta-analysis conducted by Shi and colleagues involved 72 human clinical studies comprising 5895 patients with DN. They emphasized that the combined therapy of HKC and RAS blockers mitigated levels of urinary protein in 24 h, rate of urinary albumin excretion, and serum Cr more effectively than RAS blockers alone. However, they did not find any significant changes in eGFR levels in both groups. In addition, no severe adverse effect was reported following HKC treatment, and HKC did not elevate the adverse effects rates.

It could be concluded that *A. manihot* may be a promising and safe treatment for ameliorating the severity of proteinuria as well as renal function in CKD patients.

3 *Salvia miltiorrhiza* and Its Major Active Components: Tanshinone and Salvianolic Acid

The perennial flowering plant *Salvia miltiorrhiza (S. miltiorrhiza)* is a member of Lamiaceae. Its roots are extensively used in traditional Chinese medicine. Some major active ingredients of *S. miltiorrhiza* are salvianolic acid A and B, tanshinone I, IIA, and IIB [11].

3.1 Animal Studies

The ethyl acetate extract of roots of *S. miltiorrhiza* significantly mitigated urine level, urinary albumin excretion, albumin/Cr ratio, serum Cr, BUN, and improved the pathological changes in streptozotocin-induced DN in male C57BL/6 mice. *S. miltiorrhiza* also elevated nuclear factor erythroid-2-related factor 2 (Nrf-2), heme oxygenase-1 (HO-1), and NAD(P)H quinone dehydrogenase 1 (NQO1) mRNA expression while suppressing Keap-1 mRNA expression in the kidney of DN mice. They supported that *S. miltiorrhiza* ameliorates DN through the Nrf-2-mediated anti-oxidant system [12].

3.1.1 Salvianolic Acid A

Salvianolic acid A remarkably reduced kidney/body weight ratio, Scr, BUN, creatinine clearance rate, and MDA level while increasing body weight and SOD activity in 5/6 nephrectomized rats. It also downregulated TGF-β1 while upregulating bone morphogenetic protein 7 (BMP-7) and Smad6 protein expression levels in the kidney [13]. Furthermore, Zhang et al. found that salvianolic acid A dose-dependently alleviated TGF-β1, α-smooth muscle actin (α-SMA), TNF-α, IL-1β expression levels in 5/6 nephrectomized rats. It also suppressed nuclear factor-kappa B (NF-κB) and p38 mitogen-activated protein kinase (MAPK) signaling cascade in the kidneys of 5/6Nx rats [14]. Similarly, salvianolic acid A dose-dependently stimulated the activity of SOD, catalase (CAT), and glutathione peroxidase (GPx) while diminishing MDA, reactive oxygen species (ROS), and NOX-4 levels in 5/6 nephrectomized rats. It also propagated p-Akt, p-GSK-3β, p-Nrf-2, and HO-1 levels in the renal tissue of 5/6Nx rats [15].

In minimal change disease (MCD) caused by adriamycin in rats, salvianolic acid A firmly attenuated Scr, BUN, and urinary levels of angiopoietin-like 4 (Angptl4). It also decreased Angptl4 while increasing PPARγ mRNA expression in MCD rats [16].

3.1.2 Salvianolic Acid B

Salvianolic Acid B significantly decreased urinary output, cystatin C, Scr, intercellular adhesion molecule-1, VCAM-1, IL-1β, Interleukin-6, and TNF-αmRNA expression in the kidney tissue of renal tubular damage induced by fatty acid in C57BL/6 mice. It also diminished endoplasmic reticulum stress markers and apoptosis by decreasing apoptotic Bax and c-caspase-3 while increasing anti-apoptotic BCL-2expression [17]. In a model of renal injury caused by ischemia–reperfusion in Balb/c mice, salvianolic Acid B notably reduced Scr, BUN, MDA, IL-1β and TNF-α mRNA levels, and Keap-1 protein expression in kidney tissue. It also propagated Nrf-2 and HO-1 levels, and SOD and glutathione (GSH) activities in kidney tissue [18].

Hu and coworkers showed that salvianolic acid B markedly alleviated BUN, Scr, a-SMA, and protein expression of TGF-β and FGF-2, while increasing E-cadherin protein expression in the kidney tissue of unilateral ureteral obstruction-induced renal damage in C57BL/6 mice [19].

3.1.3 Tanshinone IIA

In streptozotocin-induced DN, Tanshinone IIA strikingly diminished urine albumin excretion rate, CRP, and MDA levels while enhancing SOD activity. It also provided a notable decrement in TGF-β1, P-selectin, and monocyte chemotactic protein-1 (MCP-1) mRNA expression levels in DN rats' renal tissue [20]. Similarly, Tanshinone IIA remarkably attenuated mRNA expression of TGF-β1, thrombospondin-1 (TSP-1), glucose-regulated protein 78 (Grp78), and CHOP in kidneys of DN rats. In addition, Tanshinone IIA significantly mitigated endoplasmic reticulum stress by decreasing the protein levels of p-PERK, p-elf2α, and activating transcription factor (ATF)-4 in the renal tissue of streptozotocin-induced DN in rats [21]. Liang et al. revealed that Tanshinone IIA ameliorated contrast-induced nephropathy through decreasing MDA and increasing Nrf-2 and HO-1 signaling pathways [22].

Taken together, *S. miltiorrhiza* and its main active ingredients may be promising candidates for modulating CKD through anti-oxidative and anti-fibrotic effects.

3.2 Clinical Studies

Zhou and coworkers conducted a meta-analysis on 21 clinical studies involving 1857 patients with CKD. They realized that *S. miltiorrhiza* significantly attenuated BUN, Scr, 24-h urine protein levels, microalbuminuria, β2-macroglobulin, and cystatin C (CysC) while improving GFR levels in CKD patients. Neither the safety nor the adverse effects were meaningfully different between the extract and control groups [23].

Similarly, Xu et al. analyzed 16 clinical studies involving 1696 cases of hypertensive nephropathy. They supported that sodium tanshinone IIA sulfonate combined with angiotensin II receptor blockers (ARBs) remarkably mitigated 24 h urinary protein, Scr, cystatin-C, urinary immunoglobulin G, urine transferrin level, systolic (SBP) and diastolic blood (DBP) pressures while propagating GFR compared to monotherapy with ARB in patients with hypertensive nephropathy [24].

It could be summarized that supplementation therapy with *S. miltiorrhiza* and its constituents may be effective and safe in treating CKD patients.

4 Berberine

Berberine is an alkaloid widely found in the barks, roots, rhizomes, and stems of species Berberis (*B. vulgaris, B. aristata, B. aquifolium*) and Coptis (*Coptis japonica, Coptis chinensis*). Berberine possesses plenty of promising pharmacological effects, including anti-inflammatory, antioxidant, anti-diabetic, anti-hypertensive, hypolipidemic, cardioprotective, and renoprotective effects [25, 26].

4.1 Animal Studies

Berberine notably decreased MDA level while enhancing SOD and CAT activities in the kidney tissue of unilateral ureteral obstruction (UUO)-induced kidney injury in Sprague-Dawley rats. It also significantly diminished TGF-β1, α-SMA, and phosphorylated Smad3 (pSmad3) expression in the kidney tissue of UUO rats [27]. Qin et al. suggested that berberine ameliorates diabetic kidney disease by decreasing the expression as well as dynamin-related protein 1 (Drp1) translocation from the cytoplasm to the mitochondria [28].

In the streptozocin-induced DN model, berberine markedly reduced kidney/body weight, BUN, Scr, 24-h urinary protein levels, and protein expression of extracellular matrix protein deposition-related markers. It also promoted the AMP-activated protein kinase (AMPK) phosphorylation and autophagy-related markers in DN rats [29]. Similarly, berberine meaningfully alleviated Scr, BUN, as well as TGF-β, vimentin, and α-SMA expression in DN rats [30]. Furthermore, Zhu and coworkers noticed that berberine strikingly mitigated kidney/body weight ratio, 24-h urinary protein, Scr, BUN, IL-1β, IL-6, and MCP-1 levels in the kidney tissue of streptozocin-induced DN rats. Berberine also suppressed the TLR-4 and NF-κB cascade through decreasing TLR-4 protein level, p-IκBα/IκBα, and p-p65/p65 in DN rats [31]. In addition, berberine provided a significant decrement in matrix metallopeptidase (MMP)-9, tissue inhibitor

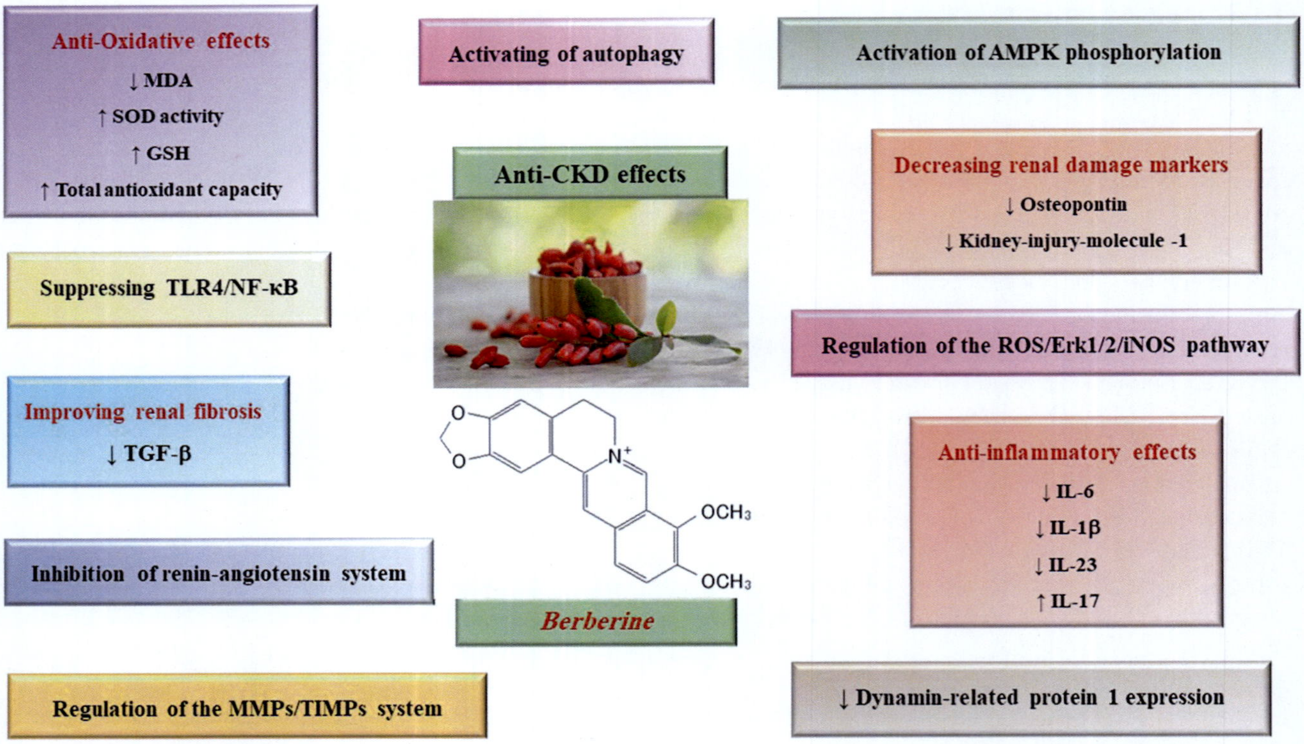

Fig. 2 The anti-CKD and reno-protective mechanisms of berberine. Malondialdehyde (MDA), reactive oxygen species (ROS), superoxide dismutase (SOD), Interleukin (IL), transforming growth factor-β1 (TGF-β1), matrix metallopeptidase (MMP), AMP-activated protein kinase (AMPK), tissue inhibitor matrix metalloproteinase (TIMP), nuclear factor-κB (NF-κB), inducible nitric oxide synthase (iNOS), toll-like receptor 4

matrix metalloproteinase (TIMP)-1 and -2, TGF-β1, fibronectin, and type IV collagen while increasing MMP2 expression levels in streptozocin-induced DN in the rat [32].

In spontaneously hypertensive rats, berberine remarkably attenuated hypertension, Ang II, aldosterone, kidney-injury molecule (KIM-1), IL-6, IL-17, IL-23, and osteopontin [33]. Tian and coworkers showed that berberine firmly alleviated mean arterial pressure, NOX2 and 4, Erk1/2, and iNOS while stimulating copper/zinc superoxide dismutase levels in hypertensive rats [34].

Taken together, berberine may effectively protect against CKD through several mechanisms, including anti-oxidative, anti-inflammatory, and anti-fibrotic impacts (Fig. 2).

4.2 Clinical Studies

Dai and coworkers evaluated the impact of berberine (0.1 g, 3 times daily) on 69 hypertensive patients with type 2 diabetes mellitus. They revealed that berberine notably decreased Scr, urine albumin-to-creatine ratio, renal damage markers including osteopontin and KIM-1, as well as inflammatory markers, including VCAM-1 and CRP. Berberine also inhibited oxidative stress through reducing MDA while stimulating superoxide dismutase, GSH-Px, and total anti-oxidant capacity (TAC) activities in hypertensive and type 2 diabetes mellitus patients [35]. Similarly, Li and coworkers determined the effects of berberine (0.4 g, 3 times daily) on 114 patients with type 2 diabetes mellitus. Berberine significantly diminished hs-CRP, BUN, urine albumin/Cr ratio, and cystatin C while improving eGFR level than the control group. Besides that, no important adverse effects were observed following treatment with berberine [36].

5 *Zingiber officinale* and Its Constituents: 6-Gingerols and 6-Shogaols

Zingiber officinale (*Z. officinale*), popularly known as ginger, is a perennial herb belonging to Zingiberaceae [37]. Its roots are widely used as a spice and also in traditional medicine worldwide. It has been emphasized that *Z. officinale* has anti-oxidant, anti-inflammatory, anti-microbial, anti-emetic, anti-diabetic, cardioprotective, and renoprotective effects. The phenolic constituents of *Z. officinale* are mainly responsible for their several effects, including 6-gingerols and 6-shogaols [38, 39].

5.1 Animal Studies

In the model of streptozocin-induced DN, *Z. officinale* extract significantly decreased Scr, BUN, and MDA while enhancing GSH level, and activities of SOD and CAT in renal tissue. In addition, *Z. officinale* markedly diminished inflammatory cytokines, namely, TNF-α, IL-6, and IL-1β, and apoptosis (cytochrome *c*, caspase-3) in the kidney tissue of streptozocin-induced DN in rats [40].

Furthermore, *Z. officinale* extract notably reduced MDA and TNF-α levels while increasing activities of TAC, SOD, catalase, Gpx, and expression of Nrf-2 in the kidney tissue of ethanol-induced kidney injury in ats [41].

In the model of mercuric chloride-induced kidney injury, *Z. officinale* extract meaningfully elevated GSH level, SOD, catalase, GPx, glutathione reductase, and glutathione-*S*-transferase (GST) activities while decreased MDA levels in kidney tissue [42]. Similarly, *Z. officinale* extract firmly stimulated GSH levels, GPx, GST, and CAT in the kidney tissue of lead-induced renal toxicity in rats [43].

5.1.1 6-Shogaol

In the renal ischemia–reperfusion (I/R) model, 6-shogaol notably diminished Scr, BUN, kidney neutrophil gelatinase-associated lipocalin (NGAL), IL-6, MCP-1, macrophage inflammatory protein-2 (MIP-2), and keratinocyte chemoattractant expression. It also stimulated HO-1 expression in I/R-induced kidney injury in mice [44]. In addition, 6-shogaol also protects against streptozocin-induced DN through propagating Nrf-2mRNA expression in C57BL/6 male mice [45].

Gwon et al. revealed that 6-shogaol remarkably mitigated Scr, BUN, MDA levels, and KIM-1, NGAL, NOX4, TNF-α, IL-6, and MCP-1 mRNA expression in acute kidney injury following cisplatin in mice. It also elevated GSH levels, SOD, and catalase activities in kidney tissue [46].

5.1.2 6-Gingerol

In streptozocin-induced DN, 6-gingerol strikingly reduced Scr, BUN, MDA, CRP, IL-6, IL-1β, and TNF-α levels. Moreover, it elevated the GSH level, and activities of SOD and catalase in the renal tissue of DN rats [47]. 6-Gingerol also alleviated Scr, BUN, and MDA levels while improving GSH levels and pathological changes in the kidney tissue of gentamicin-induced renal injury [48].

Salihu and coworkers supported that 6-gingerol markedly reduced MDA and hydrogen peroxide levels while increasing anti-oxidant activities in carbendazim-induced renal damage in rats [49].

In the mercuric chloride-induced kidney injury model, 6-gingerol firmly propagated GSH level, SOD and CAT, GR, GPx, and GST activities while reducing MDA level in kidney tissue [42].

5.2 Clinical Studies

Imani et al. investigated the effects of *Z. officinale* (1000 mg/day) on 36 patients on peritoneal dialysis. They showed that *Z. officinale* notably mitigated fasting glucose levels in serum, which is an important risk factor for DN, hyperinsulinemia, and cardiovascular disorders [50].

6 *Punica granatum* and Its Major Constituent Ellagic Acid

Punica granatum (*P. granatum*), which is popularly famous as pomegranate, is a small deciduous tree and a member of Punicaceae. It is mostly cultivated in the Mediterranean countries, including India, Iran, Turkey, and China. Several pharmacological effects have been attributed to *P. granatum* fruit, including anti-oxidant, anti-inflammatory, anti-diabetic, hepatoprotective, cardioprotective, and renoprotective properties [51, 52]. In addition, ellagic acid (EA) is one of the major and well-studied active polyphenolic compounds found in pomegranate fruits [53, 54].

6.1 Animal Studies

In the model of streptozotocin-induced DN, the *P. granatum* leaves extract containing a high amount of flavonoids (PGFF) meaningfully decreased Scr, BUN, urinary total protein, urine volume, and MDA level in rats. PGFF also improved histopathological changes and anti-oxidant markers in the kidneys of DN rats [55]. Similarly, the methanolic extract of *P. granatum* leaves (MPGL) remarkably diminished Scr, BUN, and MDA levels while promoting GSH level, CAT, and SOD activities in the kidney tissue of streptozotocin-induced DN in rats [56]. MPGL also reduced oxidative stress and TNF-α level in the rat's kidney tissue of gentamicin-induced nephropathy [57].

P. granatum peel ethanol extract (PPEE) notably mitigated oxidative stress markers, C-reactive protein, and caspase-3 while elevating Bcl-2 renal expression in vancomycin-induced kidney injury in the rat [58]. In addition, *P. granatum* juice markedly attenuated levels of nitric oxide (NO) and MDA while enhancing anti-oxidant status in the kidney tissue of carbon tetrachloride-induced nephrotoxicity in rats [59].

6.1.1 Ellagic Acid (EA)

In the model of streptozotocin-induced DN, EA firmly alleviated Scr, BUN, MDA, and TNF-α levels while increasing SOD activity in kidney tissue. EA also suppressed the HMGB1-TLR-4-NF-κB pathway by downregulating TLR-4, Interleukin-1 receptor-associated kinase 4 (IRAK4), TNF

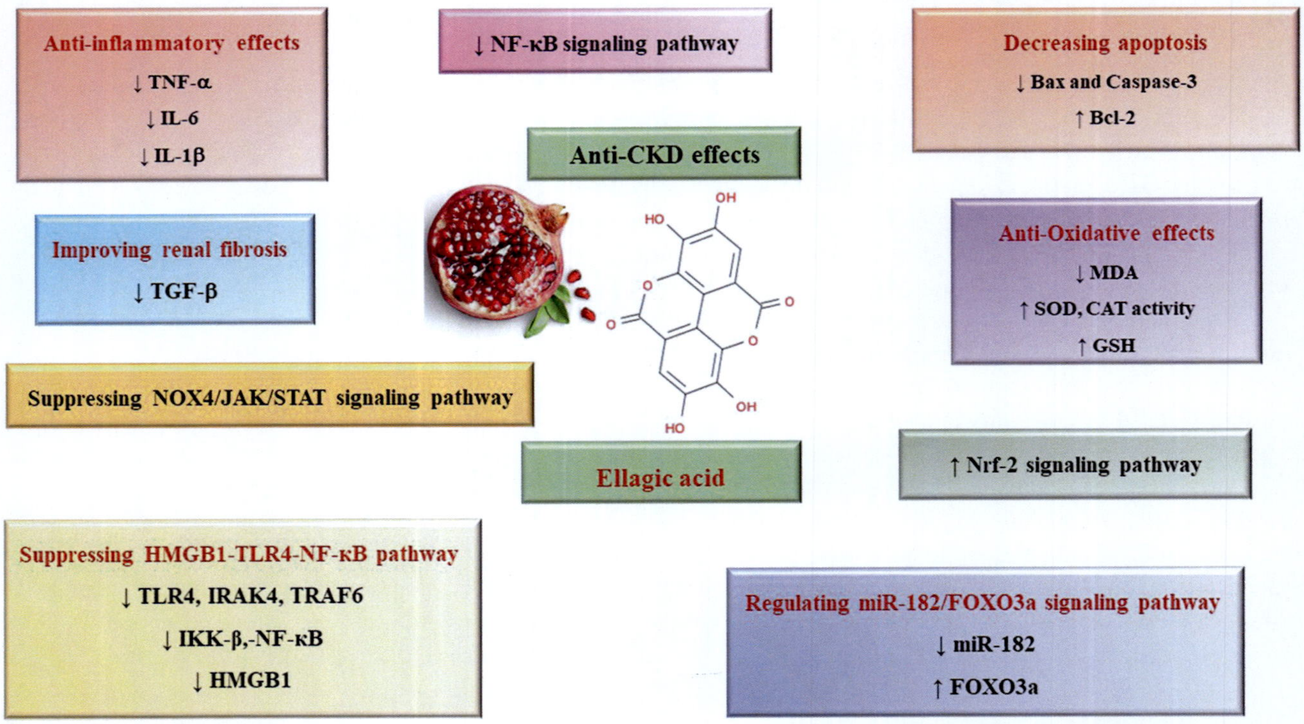

Fig. 3 The anti-CKD and reno-protective mechanisms of *Punica granatum* and its main active constituent, EA. Interleukin (IL), Interleukin-1 receptor-associated kinase 4 (IRAK4), tumor necrosis factor-α (TNF-α), TNF receptor associated factor 6 (TRAF6), transforming growth factor-β1 (TGF-β1), nicotinamide adenine dinucleotide phosphate oxidase (NOX), Toll-like receptor 4 (TLR-4), inhibitor of nuclear factor kappa-B kinase (IKK-β), nuclear factor-kappaB (NF-κB), nuclear factor erythroid-2-related factor 2 (Nrf-2), High mobility group box 1 (HMGB1), malondialdehyde (MDA), superoxide dismutase (SOD), glutathione (GSH), catalase (CAT)

receptor-associated factor 6 (TRAF6), inhibitor of NF-κB kinase (IKK-β), NF-κBp65, and HMGB1 protein expression in the kidney tissue of DN mice [60]. EA also ameliorated sodium arsenite-induced renal injury in rats by anti-oxidative effects [61].

In chronic renal failure induced by 5/6 nephrectomy, EA significantly decreased urine volume, urine protein, BUN, Scr, MDA, TNF-α, IL-6, and ICAM-1 level while increasing GSH level and superoxide dismutase activity in renal tissue. In addition, EA markedly diminished miR-182, TGF-β1, fibronectin, and Bax while stimulating FOXO3a and Bcl-2 expression levels in rats [62]. Similarly, in the renal ischemic–reperfusion injury model, EA notably alleviated TNF-α, IL-1β, IL-6, and MCP-1, MDA, LDH, iNOS, COX-2 levels, Bax and caspase-3 expression while promoted GSH level, SOD activity, and Bcl-2 expression in kidney tissue. Furthermore, EA inhibited the p-JAK1, p-JAK2, p-STAT1, and NOX4 expression levels [63].

In the carbon tetrachloride-induced kidney damage model, EA meaningfully attenuated MDA level, and COX-2, TNF-α, NF-κB, and vascular endothelial growth factor family (VEGF) expression. It also propagated GSH level, CAT activity, and Nrf-2 expression in kidney tissue [64]. Moreover, EA suppressed MDA, TNF-α, IL-6 levels, NF-κB, and Bax expression while increasing anti-oxidant markers following amikacin-induced nephrotoxicity in rats [65].

Collectively, *P. granatum* and its main active constituent, EA, may ameliorate CKD through several mechanisms, including anti-oxidative, anti-inflammatory, anti-apoptotic, and anti-fibrotic effects (Fig. 3).

7 *Vitis vinifera*

Vitis vinifera (*V. vinifera*), commonly known as grape, is a flowering plant belonging to Vitaceae. Nowadays, *V. vinifera* is cultivated in almost every country around the world. The grape juice and seeds have high levels of polyphenols and anti-oxidants and show multiple pharmacological properties, containing immunomodulatory, anti-oxidant, anti-inflammatory, anti-atherosclerotic, anti-diabetic, anti-aging, cardioprotective, and renoprotective properties [66].

7.1 Animal Studies

Grape seed proanthocyanidin extracts (GSPE) significantly ameliorated endoplasmic reticulum stress-induced apoptosis through diminishing GRP78, p-ERK, and caspase-12 expression in streptozotocin-induced DN in rats [67]. Similarly, GSPE markedly reduced levels of Scr, BUN, and MDA while increasing anti-oxidant activities in DN rats. GSPE also meaningfully propagated Nrf-2, HO-1, GST, and NQO1 renal mRNA expression [68]. Following the carboplatin and thalidomide-induced nephrotoxicity, GSPE firmly alleviated p53 expression, TNF-a, IL-6, MDA, and NO levels, while elevating GSH, GST, TAC levels, SOD, and catalase, activities in renal tissue [69].

In doxorubicin-induced DN, aqueous extracts of *V. vinifera* fruits remarkably attenuated Scr, BUN, and hs-CRP while stimulating the mRNA expression of Nrf-2 in kidney tissue [70]. In addition, *V. vinifera* seed extract notably decreased the Scr, MDA, and IL-6 levels while enhancing SOD activity in amiodarone-induced nephrotoxicity in rats [71].

7.2 Clinical Studies

Imani et al. determined the impacts of grape seed extract (GSE, 2100 mg/day) by evaluating 33 patients with chronic kidney disease. They revealed that GSE meaningfully attenuated urinary protein, MDA, protein carbonylation, H_2O_2, and CRP levels while propagating GFR, CAT, GPx levels, and SOD activity [72].

Taken together, *V. vinifera* may ameliorate renal failure and chronic kidney disease through anti-oxidant and anti-inflammatory activities.

8 *Curcuma longa* and Its Major Component: Curcumin

Curcuma longa (*C. longa*), commonly known as turmeric, is a perennial rhizomatous herb member of Zingiberaceae. It is chiefly cultivated in Asia, India, Iran, and China. Curcumin is the major and well-studied polyphenol isolated from the rhizome of *C. longa*. Curcumin has many promising pharmacological properties, including anti-oxidant, anti-inflammatory, immunomodulatory, anti-microbial, anti-cancer, neuroprotective, hepatoprotective, and renoprotective properties [73, 74].

8.1 Animal Studies

In a model of kidney injury induced by doxorubicin, powdered dried rhizomes of *C. longa* significantly decreased MCP-1 and TGF-β1 levels, pathological changes, and desmin, vimentin, and ED-1+ cells immunostaining in the rat kidney tissue [75]. Similarly, curcumin notably reduced urinary protein, MDA levels, protein expression of NF-κB p65, and Keap-1, while stimulating the protein expression of HO-1, NQO-1, Nrf-2, IκBα, and SOD activity in the doxorubicin-induced nephrotic syndrome [76].

In sodium arsenate-induced kidney injury in mice, curcumin markedly diminished IL-1β, IL-6, TNF-α, Interferon-gamma (IFN-γ), TGF-β1, and NF-κB mRNA levels and suppressed MAPK signaling through decreasing P-JNK, P-ERK1/2, and P-p38 phosphorylation in kidney tissue. Additionally, curcumin activated the Nrf-2 signaling cascade by enhancing Nrf-2, NQO1, as well as HO-1 protein expression [77]. Cao et al. also suggested that curcumin meaningfully alleviated levels of MDA and NO, IL-1β, IL-6, and TNF-α while elevating SOD, CAT, and GSH-Px activities in gentamicin-induced nephrotoxicity in rats. In addition, curcumin inhibited apoptosis by decreasing caspase-3 and Bax while enhancing Bcl-2 expression [78].

Curcumin remarkably mitigated NLRP3 inflammasome activation by decreasing IL-1β, NLRP3, ASC, and caspase1 mRNA and protein expression in hyperuricemia and kidney inflammation following potassium oxonate consumption in mice [79]. Furthermore, in passive Heymann nephritis model in rats, curcumin firmly attenuated Scr, BUN, and MDA levels while increasing GSH levels, and SOD, and catalase in kidney tissue. Besides that, curcumin suppressed Bax, Caspase-3, p62, PI3K, p-AKT, and p-mTOR while stimulating Bcl-2, Nrf-2, beclin1, and HO-1proteins expression in kidney tissue [80].

In cisplatin-induced kidney damage in rats, curcumin provided a significant decrement in Scr, BUN, MDA levels, ERK1/2 phosphorylation, NF-κB expression, Bax/Bcl-2 ratio, and TNF-α, IL-6, KIM-1, and NGAL mRNA levels. In contrast, curcumin increased GSH levels and IL-10 mRNA expression in renal tissue [81].

8.2 Clinical Studies

Alvarenga and coworkers evaluated the effects of *C. longa* containing 95% curcumin (2.5 g, 3 times/week) on 31 hemodialysis patients. They noticed that curcumin remarkably diminished the mRNA expression of NF-κB and hs-CRP levels [82]. Similarly, curcumin (1 g/day for 12 weeks) markedly propagated CAT and GPx activities in hemodialysis patients. Besides that, hs-CRP and MDA levels, and GR activity did not significantly change [83].

Osorio and coworkers determined the impacts of curcumin (320 mg/day) on non-diabetic or diabetic proteinuric CKD. They revealed that curcumin firmly alleviated MDA levels in non-diabetic patients while elevating anti-oxidant markers in diabetic proteinuric CKD patients. They reported no meaningful differences in the urinary protein, eGFR, and Nrf-2 activation [84]. Furthermore, curcumin (500 mg/tablet twice for 6 months) strikingly attenuated MDA, MCP-1, IL-4, and IFN-γ levels in CKD patients [85].

Additionally, a meta-analysis of 5 human clinical studies involving 290 diabetic kidney disease (DKD) patients

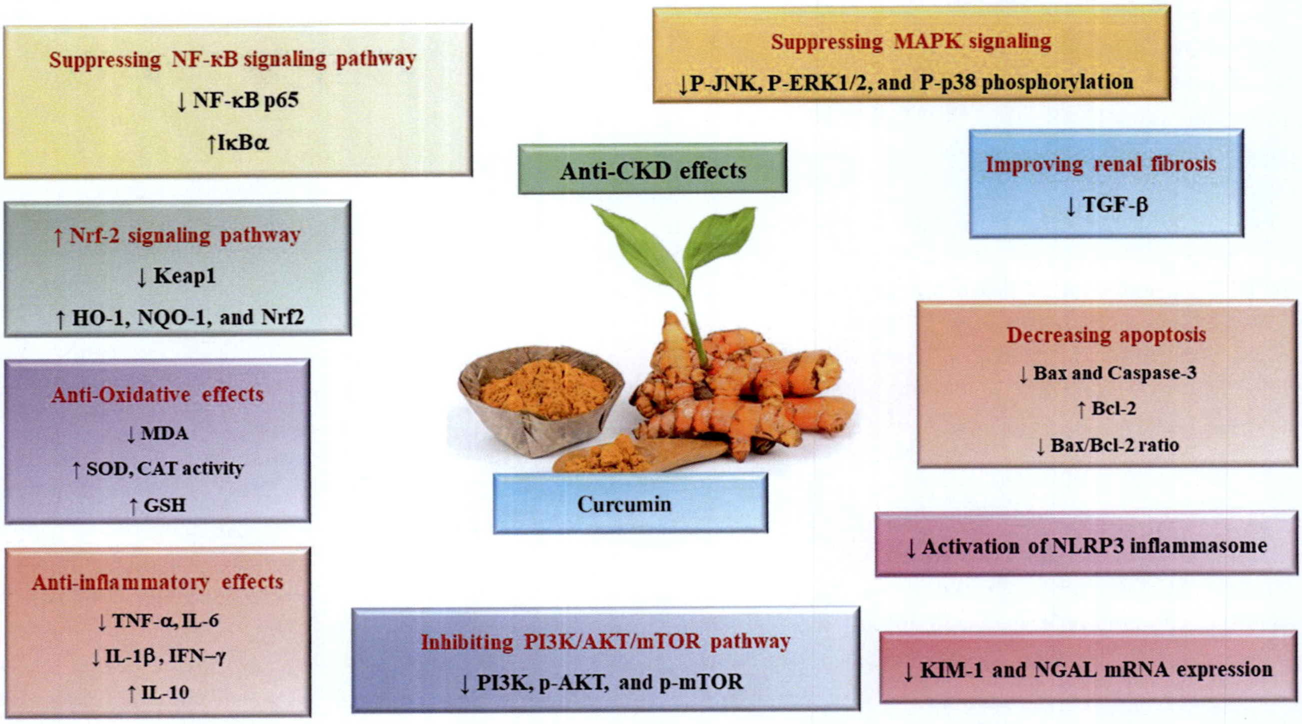

Fig. 4 The anti-CKD and reno-protective mechanisms of *Curcuma longa* and its major active constituent, curcumin. Nuclear factor-κB (NF-κB), nuclear factor erythroid-2-related factor 2 (Nrf-2), inhibitor of nuclear factor kappa-B (Iκβα), heme oxygenase-1 (HO-1), mammalian target of rapamycin (mTOR), NAD(P)H quinone dehydrogenase 1 (NQO1), phosphorylation-extracellular signal-regulated kinase (p-ERK), kelch-like ECH-associated protein 1 (Keap-1), malondialdehyde (MDA), superoxide dismutase (SOD), glutathione (GSH), catalase (CAT), Interleukin (IL), transforming growth factor-beta 1 (TGF-β1), tumor necrosis factor-α (TNF-α), NLR family pyrin domain containing 3 (NLRP3), kidney-injury-molecule (KIM-1), Interferon-gamma (IFN-γ), neutrophil gelatinase-associated lipocalin (NGAL)

emphasized that curcumin notably diminished Scr, total cholesterol, SBP, and fasting blood glucose. In contrast, curcumin showed no significant changes in BUN, urinary protein, triglyceride, and diastolic blood pressure levels [86]. Emami et al. also analyzed 10 clinical studies involving 523 CKD patients. In comparison with the control group, they found no notable changes in IL-6, TNF-α, and hs-CRP levels following curcumin consumption [87]. In contrast, a meta-analysis consisting of nine clinical studies supported that curcumin markedly mitigated total cholesterol and TNF-α in CKD patients [88].

Collectively, *C. longa* and its main active constituent, curcumin, may ameliorate CKD through several mechanisms, including anti-oxidative, anti-inflammatory effects and PI3K/AKT/mTOR, Nrf-2/HO-1, and NF-κB signaling pathways (Fig. 4).

9 *Nigella sativa*

Nigella sativa (*N. sativa*), commonly known as black cumin, is an annual herbaceous flowering plant belonging to Ranunculaceae [89]. Its small and black seeds have been used in many traditional foods and as folk medicine. Thymoquinone is the major active ingredient of seeds of *N. sativa*. Several pieces of evidence support that *N. sativa* has anti-tussive, anti-oxidant, anti-cancer, anti-inflammatory, anti-anxiety, analgesic, hepatoprotective, and nephroprotective properties [90, 91].

9.1 Animal Studies

In the UUO model in rats, 70% hydro-ethanolic extract of *N. sativa* significantly suppressed the levels of angiotensin II, MCP-1, MDA, and TNF-α in kidney tissue. It also increased GSH levels, and activities of SOD and catalase [92]. In addition, *N. sativa* oil notably diminished total oxidant status and lipid hydroperoxide levels while enhancing total anti-oxidant status, paraoxonase, and ceruloplasmin activities in radiation-induced kidney damage in rats [93].

In cisplatin-induced kidney damage, *N. sativa* oil and thymoquinone remarkably attenuated Scr, BUN, and MDA levels while stimulating the anti-oxidant status in rat kidney tissue [94]. Similarly, 70% hydro-ethanolic extract of *N. sativa* significantly improved histopathological changes and GSH levels in kidney tissue [95].

9.2 Clinical Studies

Alam and coworkers demonstrated the impact of *N. sativa* oil (2.5 mL, taken orally, once daily) on 150 patients with Stages 3 and 4 of CKD. They showed that *N. sativa* oil markedly diminished the blood urea, Scr, and 24-h urinary protein while notably increasing the 24-h total urine volume and GFR level [96]. Meanwhile, a meta-analysis of 19 clinical studies involving 1295 participants emphasized that *N. sativa* notably mitigated ALP and BUN levels [97].

10 *Rosmarinus offcinalis* and Its Major Constituent: Rosmarinic Acid

The herb *Rosmarinus officinalis,* common name rosemary, is an aromatic evergreen perennial shrub of Lamiaceae. Its dark green elongated and aromatic leaves are used to season food, as well as in folk medicine for centuries. The main active constituents of *R. officinalis* are carnosic acid, carnosol, and rosmarinic acid. Previous studies revealed that *R. officinalis* and its active components possessed anti-nociceptive, anti-oxidant, anti-inflammatory, neuroprotective, anti-depressant, and nephroprotective properties [98, 99].

10.1 Animal Studies

In carbon tetrachloride-induced nephropathy in mice, *R. officinalis* significantly reduced Scr, BUN, LDH, and MDA while increasing anti-oxidant markers activities in kidney tissue [100]. Similar to this, rosemary essential oil notably diminished urea, Scr, and MDA levels while improving histopathological changes, total anti-oxidant concentration, CAT, and GPx activities in diethylnitrosamine-induced renal injury in rats [101]. *R. officinalis* essential oil also decreased urea, creatinine, uric acid, MDA, and H_2O_2 levels while improving histological changes and anti-oxidant activities in potassium dichromate-induced kidney damage in rats [102].

10.1.1 Rosmarinic Acid

Rosmarinic acid markedly attenuated MDA, NO levels, and mRNA expression of IL-1β, IL-6, and TNF-α while elevating SOD, catalase, and TAC activities in kidney injury following cisplatin administration in mice. Rosmarinic acid also stimulated the Nrf-2 signaling cascade through stimulating Nrf-2 and HO-1 while inhibiting Keap-1 mRNA expression in kidney tissue [103]. In chlorpyrifos-induced kidney injury in rats, rosmarinic acid firmly mitigated serum urea, Scr, KIM-1, ROS, MDA, NO, IL-1β, TNF-α, and NF-κB p65 levels while promoting GSH, SOD, and CAT activities in kidney tissue. Additionally, rosmarinic acid strikingly suppressed Bax, caspase-3, and Keap-1 while stimulating the expression of Bcl-2, Nrf-2, HO-1, and Sirtuin 1 (SIRT1) in kidney tissue [104].

Rosmarinic acid also prevented MDA levels and histopathological changes while propagating GSH levels and Nrf-2 mRNA levels in chromium-induced renal injury in rats [105]. In cadmium-induced nephrotoxicity, rosmarinic acid considerably reduced CRP, IL-1β, IL-6, TNF-α, ROS, NO, H_2O_2, and MDA levels while stimulating GSH, GPx, and GR levels, as well as SOD and CAT activities. Furthermore, rosmarinic acid hampered caspase-3, 8, and 9, NF-κB, protein kinase C (PKC), tumor necrosis factor receptor, TGF-β1, SMAD3, and α-SMA expression in kidney tissue [106].

Acknowledgments This study was supported by the research council of Mashhad University of Medical Sciences.

References

1. Li J, Ye G-y, Liu H-l, Wang Z-h. New insights on Abelmoschus manihot flower development: dynamic changes of flavonoids based on a metabolomic approach. J Plant Biochem Biotechnol. 2022;31(2):351–60.
2. Luan F, Wu Q, Yang Y, Lv H, Liu D, Gan Z, et al. Traditional uses, chemical constituents, biological properties, clinical settings, and toxicities of Abelmoschus manihot L.: a comprehensive review. Front Pharmacol. 2020;11:1068.
3. Ge J, Miao JJ, Sun XY, Yu JY. Huangkui capsule, an extract from Abelmoschus manihot (L.) medic, improves diabetic nephropathy via activating peroxisome proliferator-activated receptor (PPAR)-α/γ and attenuating endoplasmic reticulum stress in rats. J Ethnopharmacol. 2016;189:238–49.
4. Mao ZM, Shen SM, Wan YG, Sun W, Chen HL, Huang MM, et al. Huangkui capsule attenuates renal fibrosis in diabetic nephropathy rats through regulating oxidative stress and p38MAPK/Akt pathways, compared to α-lipoic acid. J Ethnopharmacol. 2015;173:256–65.
5. Tu Y, Sun W, Wan YG, Che XY, Pu HP, Yin XJ, et al. Huangkui capsule, an extract from Abelmoschus manihot (L.) medic, ameliorates adriamycin-induced renal inflammation and glomerular injury via inhibiting p38MAPK signaling pathway activity in rats. J Ethnopharmacol. 2013;147(2):311–20.
6. Wu W, Hu W, Han WB, Liu YL, Tu Y, Yang HM, et al. Inhibition of Akt/mTOR/p70S6K signaling activity with Huangkui capsule alleviates the early glomerular pathological changes in diabetic nephropathy. Front Pharmacol. 2018;9:443.
7. Cai HD, Su SL, Qian DW, Guo S, Tao WW, Cong XD, et al. Renal protective effect and action mechanism of Huangkui capsule and its main five flavonoids. J Ethnopharmacol. 2017;206:152–9.
8. Kim H, Dusabimana T, Kim SR, Je J, Jeong K, Kang MC, et al. Supplementation of Abelmoschus manihot ameliorates diabetic nephropathy and hepatic steatosis by activating autophagy in mice. Nutrients. 2018;10(11):1703.
9. Zhang L, Li P, Xing CY, Zhao JY, He YN, Wang JQ, et al. Efficacy and safety of Abelmoschus manihot for primary glomerular disease: a prospective, multicenter randomized controlled clinical trial. Am J Kidney Dis. 2014;64(1):57–65.
10. Li P, Lin H, Ni Z, Zhan Y, He Y, Yang H, et al. Efficacy and safety of Abelmoschus manihot for IgA nephropathy: a multicenter randomized clinical trial. Phytomedicine. 2020;76:153231.

11. Wang L, Ma R, Liu C, Liu H, Zhu R, Guo S, et al. Salvia miltiorrhiza: a potential red light to the development of cardiovascular diseases. Curr Pharm Des. 2017;23(7):1077–97.
12. An L, Zhou M, Marikar F, Hu XW, Miao QY, Li P, et al. Salvia miltiorrhiza lipophilic fraction attenuates oxidative stress in diabetic nephropathy through activation of nuclear factor erythroid 2-related factor 2. Am J Chin Med. 2017;45(7):1441–57.
13. Zhang G, Cui G, Tong S, Cao QJACB. Salvianolic acid A alleviates the renal damage in rats with chronic renal failure. Acta Cir Bras. 2019;34(2):e201900204.
14. Zhang HF, Wang YL, Gao C, Gu YT, Huang J, Wang JH, et al. Salvianolic acid A attenuates kidney injury and inflammation by inhibiting NF-κB and p38 MAPK signaling pathways in 5/6 nephrectomized rats. Acta Pharmacol Sin. 2018;39(12):1855–64.
15. Zhang HF, Wang JH, Wang YL, Gao C, Gu YT, Huang J, et al. Salvianolic acid A protects the kidney against oxidative stress by activating the Akt/GSK-3β/Nrf-2 signaling pathway and inhibiting the NF-κB signaling pathway in 5/6 nephrectomized rats. Oxidative Med Cell Longev. 2019;2019:2853534.
16. Wang X, Qi D, Fu F, Li X, Liu Y, Ji K, et al. Therapeutic and antiproteinuric effects of salvianolic acid A in combined with low-dose prednisone in minimal change disease rats: involvement of PPARγ/Angptl4 and Nrf-2/HO-1 pathways. Eur J Pharmacol. 2019;858:172342.
17. Mai X, Yin X, Chen P, Zhang M. Salvianolic acid B protects against fatty acid-induced renal tubular injury via inhibition of endoplasmic reticulum stress. Front Pharmacol. 2020;11:574229.
18. Pang Y, Zhang PC, Lu RR, Li HL, Li JC, Fu HX, et al. Andrade-oliveira salvianolic acid B modulates caspase-1-mediated pyroptosis in renal ischemia-reperfusion injury via Nrf-2 pathway. Front Pharmacol. 2020;11:541426.
19. Hu Y, Wang M, Pan Y, Li Q, Xu L. Salvianolic acid B attenuates renal interstitial fibrosis by regulating the HPSE/SDC1 axis. Mol Med Rep. 2020;22(2):1325–34.
20. Chen X, Wu R, Kong Y, Yang Y, Gao Y, Sun D, et al. Tanshinone II attenuates renal damage in STZ-induced diabetic rats via inhibiting oxidative stress and inflammation. Oncotarget. 2017;8(19):31915.
21. Xu S, He L, Ding K, Zhang L, Xu X, Wang S, et al. Tanshinone IIA ameliorates streptozotocin-induced diabetic nephropathy, partly by attenuating PERK pathway-induced fibrosis. Drug Des Devel Ther. 2020;14:5773–82.
22. Liang R, Zhao Q, Jian G, Cheng D, Wang N, Zhang G, et al. Tanshinone IIA attenuates contrast-induced nephropathy via Nrf-2 activation in rats. Cell Physiol Biochem. 2018;46(6):2616–23.
23. Zhou Y, Jiang SM, Li L, Wang Y, Ding L, Liu CX, et al. Efficacy and safety of tanshinone for chronic kidney disease: a meta-analysis. Evid Based Complement Alternat Med. 2020;2020:3091814.
24. Xu J, Zhang C, Shi X, Li J, Liu M, Jiang W, et al. Efficacy and safety of sodium tanshinone IIA sulfonate injection on hypertensive nephropathy: a systematic review and meta-analysis. Front Pharmacol. 2019;10:1542.
25. Neag MA, Mocan A, Echeverría J, Pop RM, Bocsan CI, Crişan G, et al. Berberine: botanical occurrence, traditional uses, extraction methods, and relevance in cardiovascular, metabolic, hepatic, and renal disorders. Front Pharmacol. 2018;9:557.
26. Zhou M, Deng Y, Liu M, Liao L, Dai X, Guo C, et al. The pharmacological activity of berberine, a review for liver protection. Eur J Pharmacol. 2021;890:173655.
27. Wang FM, Yang YJ, Ma LL, Tian XJ, He YQ. Berberine ameliorates renal interstitial fibrosis induced by unilateral ureteral obstruction in rats. Nephrology (Carlton). 2014;19(9):542–51.
28. Qin X, Zhao Y, Gong J, Huang W, Su H, Yuan F, et al. Berberine protects glomerular podocytes via inhibiting Drp1-mediated mitochondrial fission and dysfunction. Theranostics. 2019;9(6):1698–713.
29. Zhang M, Zhang Y, Xiao D, Zhang J, Wang X, Guan F, et al. Highly bioavailable berberine formulation ameliorates diabetic nephropathy through the inhibition of glomerular mesangial matrix expansion and the activation of autophagy. Eur J Pharmacol. 2020;873:172955.
30. Li Z, Zhang W. Protective effect of berberine on renal fibrosis caused by diabetic nephropathy. Mol Med Rep. 2017;16(2):1055–62.
31. Zhu L, Han J, Yuan R, Xue L, Pang W. Berberine ameliorates diabetic nephropathy by inhibiting TLR-4/NF-κB pathway. Biol Res. 2018;51(1):9.
32. Ni WJ, Ding HH, Zhou H, Qiu YY, Tang LQ. Renoprotective effects of berberine through regulation of the MMPs/TIMPs system in streptozocin-induced diabetic nephropathy in rats. Eur J Pharmacol. 2015;764:448–56.
33. Guo Z, Sun H, Zhang H, Zhang Y. Anti-hypertensive and renoprotective effects of berberine in spontaneously hypertensive rats. Clin Exp Hypertens. 2015;37(4):332–9.
34. Tian H, Kang YM, Gao HL, Shi XL, Fu LY, Li Y, et al. Chronic infusion of berberine into the hypothalamic paraventricular nucleus attenuates hypertension and sympathoexcitation via the ROS/Erk1/2/iNOS pathway. Phytomedicine. 2019;52:216–24.
35. Dai P, Wang J, Lin L, Zhang Y, Wang Z. Renoprotective effects of berberine as adjuvant therapy for hypertensive patients with type 2 diabetes mellitus: evaluation via biochemical markers and color Doppler ultrasonography. Exp Ther Med. 2015;10(3):869–76.
36. Li ZY, Liu B, Zhuang XJ, Shen YD, Tian HR, Ji Y, et al. [Effects of berberine on the serum cystatin C levels and urine albumin/creatine ratio in patients with type 2 diabetes mellitus]. Zhonghua Yi Xue Za Zhi. 2018;98(46):3756–61.
37. Yücel Ç, Karatoprak G, Açıkara ÖB, Akkol EK, Barak TH, Sobarzo-Sánchez E, et al. Immunomodulatory and anti-inflammatory therapeutic potential of gingerols and their nanoformulations. Front Pharmacol. 2022;13:902551.
38. Yahyazadeh R, Baradaran Rahimi V, Yahyazadeh A, Mohajeri SA, Askari VR. Promising effects of gingerol against toxins: a review article. Biofactors. 2021;47(6):885–913.
39. Alipour A, Baradaran Rahimi V, Askari VR. Promising influences of gingerols against metabolic syndrome: a mechanistic review. Biofactors. 2022;48(5):993–1004.
40. Al Hroob AM, Abukhalil MH, Alghonmeen RD, Mahmoud AM. Ginger alleviates hyperglycemia-induced oxidative stress, inflammation and apoptosis and protects rats against diabetic nephropathy. Biomed Pharmacother. 2018;106:381–9.
41. Fathi R, Akbari A, Nasiri K, Chardahcherik M. Ginger (Zingiber officinale roscoe) extract could upregulate the renal expression of NRF-2 and TNFα and prevents ethanol-induced toxicity in rat kidney. Avicenna J Phytomed. 2021;11(2):134–45.
42. Joshi D, Srivastav SK, Belemkar S, Dixit VA. Zingiber officinale and 6-gingerol alleviate liver and kidney dysfunctions and oxidative stress induced by mercuric chloride in male rats: a protective approach. Biomed Pharmacother. 2017;91:645–55.
43. Reddy YA, Chalamaiah M, Ramesh B, Balaji G, Indira P. Ameliorating activity of ginger (Zingiber officinale) extract against lead induced renal toxicity in male rats. J Food Sci Technol. 2014;51(5):908–14.
44. Han SJ, Kim M, D'Agati VD, Lee HT. 6-Shogaol protects against ischemic acute kidney injury by modulating NF-κB and heme oxygenase-1 pathways. Am J Physiol Renal Physiol. 2019;317(3):F743–56.
45. Yi JK, Ryoo ZY, Ha JJ, Oh DY, Kim MO, Kim SH. Beneficial effects of 6-shogaol on hyperglycemia, islet morphology and apoptosis in some tissues of streptozotocin-induced diabetic mice. Diabetol Metab Syndr. 2019;11:15.
46. Gwon MG, Gu H, Leem J, Park KK. Protective effects of 6-shogaol, an active compound of ginger, in a murine model of cisplatin-induced acute kidney injury. Molecules. 2021;26(19):5931.

47. Almatroodi SA, Alnuqaydan AM, Babiker AY, Almogbel MA, Khan AA, Husain Rahmani A. 6-Gingerol, a bioactive compound of ginger attenuates renal damage in streptozotocin-induced diabetic rats by regulating the oxidative stress and inflammation. Pharmaceutics. 2021;13(3):317.
48. Hegazy AM, Mosaed MM, Elshafey SH, Bayomy NA. 6-Gingerol ameliorates gentamicin induced renal cortex oxidative stress and apoptosis in adult male albino rats. Tissue Cell. 2016;48(3):208–16.
49. Salihu M, Ajayi BO, Adedara IA, Farombi EO. 6-Gingerol-rich fraction from Zingiber officinale prevents hematotoxicity and oxidative damage in kidney and liver of rats exposed to Carbendazim. J Diet Suppl. 2016;13(4):433–48.
50. Imani H, Tabibi H, Najafi I, Atabak S, Hedayati M, Rahmani L. Effects of ginger on serum glucose, advanced glycation end products, and inflammation in peritoneal dialysis patients. Nutrition. 2015;31(5):703–7.
51. Ghadiri M, Baradaran Rahimi V, Moradi E, Hasanpour M, Clark CCT, Iranshahi M, et al. Standardised pomegranate peel extract lavage prevents postoperative peritoneal adhesion by regulating TGF-β and VEGF levels. Inflammopharmacology. 2021;29(3):855–68.
52. Rakhshandeh H, Baradaran Rahimi V, Habibi Z, Sirousi Z, Askari VR. Punica granatum seed oil detracts peritoneal adhesion: perusing anti-oxidant, anti-inflammatory, anti-fibrotic, and antiangiogenic impacts. Physiol Rep. 2022;10(24):e15545.
53. Rahimi VB, Askari VR, Mousavi SH. Ellagic acid dose and time-dependently abrogates d-galactose-induced animal model of aging: investigating the role of PPAR-γ. Life Sci. 2019;232:116595.
54. Baradaran Rahimi V, Ghadiri M, Ramezani M, Askari VR. Antiinflammatory and anticancer activities of pomegranate and its constituent, ellagic acid: evidence from cellular, animal, and clinical studies. Phytother Res. 2020;34(4):685–720.
55. Ankita P, Deepti B, Nilam M. Flavonoid rich fraction of Punica granatum improves early diabetic nephropathy by ameliorating proteinuria and disturbed glucose homeostasis in experimental animals. Pharm Biol. 2015;53(1):61–71.
56. Mestry SN, Dhodi JB, Kumbhar SB, Juvekar AR. Attenuation of diabetic nephropathy in streptozotocin-induced diabetic rats by Punica granatum Linn. leaves extract. J Tradit Complement Med. 2017;7(3):273–80.
57. Mestry SN, Gawali NB, Pai SA, Gursahani MS, Dhodi JB, Munshi R, et al. Punica granatum improves renal function in gentamicin-induced nephropathy in rats via attenuation of oxidative stress. J Ayurveda Integr Med. 2020;11(1):16–23.
58. El Bohi KM, Abdel-Motal SM, Khalil SR, Abd-Elaal MM, Metwally MMM, ELhady WM. The efficiency of pomegranate (Punica granatum) peel ethanolic extract in attenuating the vancomycin-triggered liver and kidney tissues injury in rats. Environ Sci Pollut Res Int. 2021;28(6):7134–50.
59. Abdel Moneim AE, El-Khadragy MF. The potential effects of pomegranate (Punica granatum) juice on carbon tetrachloride-induced nephrotoxicity in rats. J Physiol Biochem. 2013;69(3):359–70.
60. Zhou B, Li Q, Wang J, Chen P, Jiang S. Ellagic acid attenuates streptozocin induced diabetic nephropathy via the regulation of oxidative stress and inflammatory signaling. Food Chem Toxicol. 2019;123:16–27.
61. Mehrzadi S, Fatemi I, Malayeri AR, Khodadadi A, Mohammadi F, Mansouri E, et al. Ellagic acid mitigates sodium arsenite-induced renal and hepatic toxicity in male Wistar rats. Pharmacol Rep. 2018;70(4):712–9.
62. Chen S, Zhou M, Ying X, Zhou C. Ellagic acid protects rats from chronic renal failure via MiR-182/FOXO3a axis. Mol Immunol. 2021;138:150–60.
63. Liu Q, Liang X, Liang M, Qin R, Qin F, Wang X. Ellagic acid ameliorates renal ischemic-reperfusion injury through NOX4/JAK/STAT signaling pathway. Inflammation. 2020;43(1):298–309.
64. Aslan A, Gok O, Beyaz S, Ağca CA, Erman O, Zerek A. Ellagic acid prevents kidney injury and oxidative damage via regulation of Nrf-2/NF-κB signaling in carbon tetrachloride induced rats. Mol Biol Rep. 2020;47(10):7959–70.
65. Saeed ZM, Khattab MI, Khorshid NE, Salem AE. Ellagic acid and cilostazol ameliorate amikacin-induced nephrotoxicity in rats by downregulating oxidative stress, inflammation, and apoptosis. PLoS One. 2022;17(7):e0271591.
66. Martin ME, Grao-Cruces E, Millan-Linares MC, Montserrat-de la Paz S. Grape (Vitis vinifera L.) seed oil: a functional food from the winemaking industry. Foods. 2020;9(10):1360.
67. Gao Z, Liu G, Hu Z, Shi W, Chen B, Zou P, et al. Grape seed proanthocyanidins protect against streptozotocin-induced diabetic nephropathy by attenuating endoplasmic reticulum stress-induced apoptosis. Mol Med Rep. 2018;18(2):1447–54.
68. Ding Y, Li H, Li Y, Liu D, Zhang L, Wang T, et al. Protective effects of grape seed proanthocyanidins on the kidneys of diabetic rats through the Nrf-2 signalling pathway. Evid Based Complement Alternat Med. 2020;2020:5205903.
69. Yousef MI, Khalil D, Abdou HM. Neuro- and nephroprotective effect of grape seed proanthocyanidin extract against carboplatin and thalidomide through modulation of inflammation, tumor suppressor protein p53, neurotransmitters, oxidative stress and histology. Toxicol Rep. 2018;5:568–78.
70. Abdelsalam HM, Samak MA, Alsemeh AE. Synergistic therapeutic effects of Vitis vinifera extract and silymarin on experimentally induced cardiorenal injury: the pertinent role of Nrf-2. Biomed Pharmacother. 2019;110:37–46.
71. Eid RA, Zaki MSA, Al-Shraim M, Eldeen MA, Haidara MA. Grape seed extract protects against amiodarone-induced nephrotoxicity and ultrastructural alterations associated with the inhibition of biomarkers of inflammation and oxidative stress in rats. Ultrastruct Pathol. 2021;45(1):49–58.
72. Turki K, Charradi K, Boukhalfa H, Belhaj M, Limam F, Aouani E. Grape seed powder improves renal failure of chronic kidney disease patients. EXCLI J. 2016;15:424–33
73. Fuloria S, Mehta J, Chandel A, Sekar M, Rani N, Begum MY, et al. A comprehensive review on the therapeutic potential of Curcuma longa Linn. in relation to its major active constituent curcumin. Front Pharmacol. 2022;13:820806.
74. Dastani M, Rahimi HR, Askari VR, Jaafari MR, Jarahi L, Yadollahi A, et al. Three months of combination therapy with nano-curcumin reduces the inflammation and lipoprotein (a) in type 2 diabetic patients with mild to moderate coronary artery disease: evidence of a randomized, double-blinded, placebo-controlled clinical trial. Biofactors. 2022;49:108.
75. Russo ER, Facincani I, Nakazato KC, Coimbra TM, Crevelin EJ, Pereira AMS, et al. Oral administration of powdered dried rhizomes of Curcuma longa L. (turmeric, Zingiberaceae) is effective in the treatment of doxorubicin-induced kidney injury in rats. Phytother Res. 2018;32(12):2408–16.
76. Fan HY, Wang XK, Li X, Ji K, Du SH, Liu Y, et al. Curcumin, as a pleiotropic agent, improves doxorubicin-induced nephrotic syndrome in rats. J Ethnopharmacol. 2020;250:112502.
77. Xu G, Gu Y, Yan N, Li Y, Sun L, Li B. Curcumin functions as an anti-inflammatory and anti-oxidant agent on arsenic-induced hepatic and kidney injury by inhibiting MAPKs/NF-κB and activating Nrf-2 pathways. Environ Toxicol. 2021;36(11):2161–73.
78. Cao L, Zhi D, Han J, Kumar Sah S, Xie Y. Combinational effect of curcumin and metformin against gentamicin-induced nephrotoxicity: involvement of antioxidative, anti-inflammatory and antiapoptotic pathway. J Food Biochem. 2019;43(7):e12836.

79. Chen Y, Li C, Duan S, Yuan X, Liang J, Hou S. Curcumin attenuates potassium oxonate-induced hyperuricemia and kidney inflammation in mice. Biomed Pharmacother. 2019;118:109195.
80. Di Tu Q, Jin J, Hu X, Ren Y, Zhao L, He Q. Curcumin improves the renal autophagy in rat experimental membranous nephropathy via regulating the PI3K/AKT/mTOR and Nrf-2/HO-1 signaling pathways. Biomed Res Int. 2020;2020:7069052.
81. Soetikno V, Sari SDP, Ul Maknun L, Sumbung NK, Rahmi DNI, Pandhita BAW, et al. Pre-treatment with curcumin ameliorates cisplatin-induced kidney damage by suppressing kidney inflammation and apoptosis in rats. Drug Res (Stuttg). 2019;69(2):75–82.
82. Alvarenga L, Salarolli R, Cardozo L, Santos RS, de Brito JS, Kemp JA, et al. Impact of curcumin supplementation on expression of inflammatory transcription factors in hemodialysis patients: a pilot randomized, double-blind, controlled study. Clin Nutr. 2020;39(12):3594–600.
83. Rodrigues HCN, Martins TFP, Santana N, Braga CC, Silva MAC, Cunha LCD, et al. Anti-oxidant and anti-inflammatory response to curcumin supplementation in hemodialysis patients: a randomized, double-blind, placebo-controlled clinical trial. Clin Nutr ESPEN. 2021;44:136–42.
84. Jiménez-Osorio AS, García-Niño WR, González-Reyes S, Álvarez-Mejía AE, Guerra-León S, Salazar-Segovia J, et al. The effect of dietary supplementation with curcumin on redox status and Nrf-2 activation in patients with nondiabetic or diabetic proteinuric chronic kidney disease: a pilot study. J Ren Nutr. 2016;26(4):237–44.
85. Pivari F, Mingione A, Piazzini G, Ceccarani C, Ottaviano E, Brasacchio C, et al. Curcumin supplementation (Meriva(®)) modulates inflammation, lipid peroxidation and gut microbiota composition in chronic kidney disease. Nutrients. 2022;14(1):231.
86. Jie Z, Chao M, Jun A, Wei S, LiFeng M. Effect of curcumin on diabetic kidney disease: a systematic review and meta-analysis of randomized, double-blind, placebo-controlled clinical trials. Evid Based Complement Alternat Med. 2021;2021:6109406.
87. Emami E, Heidari-Soureshjani S, Sherwin CM. Anti-inflammatory response to curcumin supplementation in chronic kidney disease and hemodialysis patients: a systematic review and meta-analysis. Avicenna J Phytomed. 2022;12(6):576–88.
88. Futuhi F, Naghibzadeh Tahami A, Azmandian J, Saber A. The effects of curcumin-containing supplementations on inflammatory markers and lipid profiles in patients with chronic kidney diseases: a systematic review and meta-analysis of randomized controlled trials. J Complement Integr Med. 2022;19(3):531–41.
89. Mokhtari-Zaer A, Norouzi F, Askari VR, Khazdair MR, Roshan NM, Boskabady M, et al. The protective effect of Nigella sativa extract on lung inflammation and oxidative stress induced by lipopolysaccharide in rats. J Ethnopharmacol. 2020;253:112653.
90. Hosseini A, Baradaran Rahimi V, Rakhshandeh H, Askari VR. Nigella sativa oil reduces LPS-induced microglial inflammation: an evaluation on M(1)/M(2) balance. Evid Based Complement Alternat Med. 2022;2022:5639226.
91. Yimer EM, Tuem KB, Karim A, Ur-Rehman N, Anwar F. Nigella sativa L. (black cumin): a promising natural remedy for wide range of illnesses. Evid Based Complement Alternat Med. 2019;2019:1528635.
92. Hosseinian S, Ebrahimzadeh Bideskan A, Shafei MN, Sadeghnia HR, Soukhtanloo M, Shahraki S, et al. Nigella sativa extract is a potent therapeutic agent for renal inflammation, apoptosis, and oxidative stress in a rat model of unilateral ureteral obstruction. Phytother Res. 2018;32(11):2290–8.
93. Alkis H, Demir E, Taysi MR, Sagir S, Taysi S. Effects of Nigella sativa oil and thymoquinone on radiation-induced oxidative stress in kidney tissue of rats. Biomed Pharmacother. 2021;139:111540.
94. Farooqui Z, Shahid F, Khan AA, Khan F. Oral administration of Nigella sativa oil and thymoquinone attenuates long term cisplatin treatment induced toxicity and oxidative damage in rat kidney. Biomed Pharmacother. 2017;96:912–23.
95. Hosseinian S, Hadjzadeh MA, Roshan NM, Khazaei M, Shahraki S, Mohebbati R, et al. Renoprotective effect of Nigella sativa against cisplatin-induced nephrotoxicity and oxidative stress in rat. Saudi J Kidney Dis Transpl. 2018;29(1):19–29.
96. Alam MA, Nasiruddin M, Haque SF, Khan RA. Evaluation of safety and efficacy profile of Nigella sativa oil as an add-on therapy, in addition to alpha-keto analogue of essential amino acids in patients with chronic kidney disease. Saudi J Kidney Dis Transpl. 2020;31(1):21–31.
97. Razmpoosh E, Safi S, Abdollahi N, Nadjarzadeh A, Nazari M, Fallahzadeh H, et al. The effect of Nigella sativa on the measures of liver and kidney parameters: a systematic review and meta-analysis of randomized-controlled trials. Pharmacol Res. 2020;156:104767.
98. Roohbakhsh Y, Baradaran Rahimi V, Silakhori S, Rajabi H, Rahmanian-Devin P, Samzadeh-Kermani A, et al. Evaluation of the effects of peritoneal lavage with Rosmarinus officinalis extract against the prevention of postsurgical-induced peritoneal adhesion. Planta Med. 2020;86(6):405–14.
99. Farhadi F, Baradaran Rahimi V, Mohamadi N, Askari VR. Effects of rosmarinic acid, carnosic acid, rosmanol, carnosol, and ursolic acid on the pathogenesis of respiratory diseases. Biofactors. 2022;49:478.
100. Hamed H, Boulila S, Ghrab F, Kallel R, Boudawara T, El Feki A. The preventive effect of aqueous extract of Rosemary (Rosmarinus officinalis) leaves against the nephrotoxicity of carbon tetrachloride in mice. Arch Physiol Biochem. 2020;126(3):201–8.
101. Hassanen NHM, Fahmi A, Shams-Eldin E, Abdur-Rahman M. Protective effect of rosemary (Rosmarinus officinalis) against diethylnitrosamine-induced renal injury in rats. Biomarkers. 2020;25(3):281–9.
102. El-Demerdash FM, El-Sayed RA, Abdel-Daim MM. Rosmarinus officinalis essential oil modulates renal toxicity and oxidative stress induced by potassium dichromate in rats. J Trace Elem Med Biol. 2021;67:126791.
103. Xiang Y, Ji M, Wu L, Lv L, Liang Q, Deng R, et al. Rosmarinic acid prevents cisplatin-induced liver and kidney injury by inhibiting inflammatory responses and enhancing total anti-oxidant capacity, thereby activating the Nrf-2 signaling pathway. Molecules. 2022;27(22):7815.
104. Abduh MS, Alruhaimi RS, Alqhtani HA, Hussein OE, Abukhalil MH, Kamel EM, et al. Rosmarinic acid mitigates chlorpyrifos-induced oxidative stress, inflammation, and kidney injury in rats by modulating SIRT1 and Nrf-2/HO-1 signaling. Life Sci. 2023;313:121281.
105. Khalaf AA, Hassanen EI, Ibrahim MA, Tohamy AF, Aboseada MA, Hassan HM, et al. Rosmarinic acid attenuates chromium-induced hepatic and renal oxidative damage and DNA damage in rats. J Biochem Mol Toxicol. 2020;34(11):e22579.
106. Joardar S, Dewanjee S, Bhowmick S, Dua TK, Das S, Saha A, et al. Rosmarinic acid attenuates cadmium-induced nephrotoxicity via inhibition of oxidative stress, apoptosis, inflammation and fibrosis. Int J Mol Sci. 2019;20(8):2027.

Herbal Medicines for the Management of Irritable Bowel Syndrome and Constipation Problem

Eswara Rao Puppala, Neethu Prasad, Meenakshi Singh, Arun N. Prakash, Md Abubakar, Priyanka Adhikari, and V. G. M. Naidu

Abstract

Irritable bowel syndrome (IBS) is a serious digestive disorder, which is mainly of two forms ulcerative colitis (UC) and Crohn's disease (CD). IBS is characterised by abdominal pain, bloating, diarrhoea, and constipation periods. Among these constipation is accompanied by a number of bowel symptoms including difficulty passing stool, hard stool, and inadequate evacuation sensations. Although the aetiology is unknown, it is widely thought that diet and microbiota contribute to the development of IBS. By targeting the various aetiological factors, medical treatment aims to control the symptoms. Due to the multifactorial causes of these disorders, unimpressive results from conventional IBS medications, the economic burden for patients and pharmacologic effects, the treatment of IBS and constipation is challenging and the use of complementary and alternative medicines, especially herbal therapies is increasing. Complementary and alternative medicines (CAM) for IBS and related disorders include hypnosis, acupuncture, cognitive behaviour therapy, yoga, faecal microbiota transplantation, and nutritional supplementation with probiotics, prebiotics, and herbal medicine. The practice of prescribing herbal medications for a variety of gastrointestinal disorders has a long history in various parts of the world. Herbal medications have the potential to target multiple organs and have exhibited efficacy for UC and CD in experimental models and clinical trials by maintaining the epithelial integrity barrier, regulating macrophage activation, and modulating the immune response. In this chapter, the author investigates the efficacy of herbal medicines in the management of irritable bowel syndrome, constipation, and related disorders.

Keywords

Lifestyle diseases · Constipation · Herbal medicines

Eswara Rao Puppala and Neethu Prasad contributed equally with all other contributors.

E. R. Puppala · P. Adhikari
University of Maryland, School of Medicine, Marlene and Stewart Greenebaum Comprehensive Cancer Center, Baltimore, Maryland, USA

N. Prasad · A. N. Prakash · V. G. M. Naidu (✉)
Department of Pharmacology and Toxicology, National Institute of Pharmaceutical and Research (NIPER)-Guwahati, Guwahati, Assam, India
e-mail: vgmnaidu@niperguwahati.ac.in

M. Singh
Centre for GMP Extraction Facility, Sponsored by Department of Biotechnology, National Institute of Pharmaceutical Education and Research (NIPER)-Guwahati, Guwahati, Assam, India

M. Abubakar
Department of Pharmacology and Toxicology, National Institute of Pharmaceutical and Research (NIPER)-Guwahati, Guwahati, Assam, India

Department of Pharmacology and Toxicology, National Institute of Pharmaceutical Education and Research (NIPER)-Hajipur, Hajipur, Bihar, India

A. K. Dhara, S. C. Mandal (eds.), *Role of Herbal Medicines*, https://doi.org/10.1007/978-981-99-7703-1_16

Abbreviations

ATF4	Activating transcription factor 4
ATG5	Autophagy related 5
Bcl-2	B-cell lymphoma 2
CAM	Complementary and alternative medicine
CAT	Catalase
CCR2	C-C chemokine receptor type 2
CD	Crohn's disease
CD4	Clusters of differentiation 4
CDX-2	Caudal Type Homeobox 2
CHOP	C/EBP homologous protein
CINC-3	Cytokine-induced neutrophil chemoattractant 3
COX2	Cyclooxygenase-2
DAI	Disease activity index
DAPI	4′,6-Diamidino-2-phenylindole
DCFDA	2′,7′-Dichlorofluorescein diacetate
DIM	Diindolylmethane
DSS	Dextran sodium sulphate
DSS	Dextran sulfate sodium
FFAs	Free fatty acids
FOXP3	Forkhead box P3
GBF	Germinated barley foodstuff
GLUT1	Glucose transporter 1
GOS	Glucooligosaccharide
GSH	Glutathione
GSH/GSH-Px	Glutathione peroxidase
GSH-Rd	Glutathione reductase
GSK-3	Glycogen synthase kinase 3
H&E	Haematoxylin–eosin
H3K9me3	Histone 3 lysine 9 trimethylation
HO-1	Haeme oxygenase-1
HP	Haptoglobin
IBD	Inflammatory bowel disease
ICAM-1	Colonic intercellular adhesion molecule-1
IFN-γ	Interferon gamma
IgA	Immunoglobulin A
IL-1β	Interleukin 1 beta
IL-6	Interleukin 6
ISAPP	International Scientific Association for Probiotics and Prebiotics
JNK	c-Jun NH2-terminal kinase
JNKs	c-Jun N-terminal kinases
LPS	Lipopolysaccharide
MCP-1	Monocyte chemoattractant protein-1
MD	Molecular dynamics
MDA	Malondialdehyde
MIP-2	Macrophage inflammatory protein
MLNs	Mesentric lymph nodes
MMPs	Matrix metalloproteases
MPO	Myeloperoxidase
mRNA	Messenger RNA
Msi-1	Musashi RNA binding protein 1
mTOR	Mammalian target of rapamycin
MyD88	Myeloid differentiation primary response 88
NCI	National Cancer Institute
NF-κB	Nuclear factor kappa-light-chain-enhancer of activated B cells
NO	Nitric oxide
Nrf2	Nuclear factor erythroid 2-related factor 2
PAR2	Protease activated receptor 2
PERK	Protein kinase R-like endoplasmic reticulum kinase
PGE2	Prostaglandin E2
PHF	Poly herbal formulation
PI3k	Phosphatidylinositol 3-kinase
PKB	Protein kinase B
p-NF-κB	Phosphorylated nuclear factor kappa-light-chain-enhancer of activated B cells
QRZSLXF	Qing Re Zao Shi Liang Xue receipt
RIPA	Radio-immunoprecipitation assay
RORγ	RAR-related orphan receptor gamma
ROS	Reactive oxygen species
SCFA	Short chain fatty acids
SHD	Sanhuangshu'ai decoction
SIRT1	Sirtuin 1
smad	Suppressor of mothers against decapentaplegic
SOC-1	Suppressor of cytokine signalling 1
SOD	Superoxide dismutase
STAT	Signal transducer and activator of transcription
TBARS	Thiobarbituric acid reactive substances
TF	Transcription factor
Th1	Type 1 T helper
Th17	Type 17 T helper
TNBS	Trinitrobenzene sulphonic acid
TNF-α	Tumor necrosis factor alpha
UC	Ulcerative colitis
ULK	Unc-51-like kinase 1
Wnt	Wingless-related integration site
ZO-1	Zonula occludens
ZWBQD	Zhen-Wu-Bu-Qi Decoction.

1 Introduction

Inflammatory bowel syndrome (IBS) or inflammatory bowel disease (IBD) is a chronic relapsing inflammation of the gastrointestinal tract (GIT) resulting from a dysregulated immune response and was initially reported as a rare case in the Western world. Since then, their numbers and geographic distribution have progressively grown, and they are now recognised on a global scale. Wilks described ulcerative colitis

(UC) as a distinct pathological condition along with Moxon in 1875. Later, Dr. Crohn and two colleagues submitted an essay on the disorder "Terminal Ileitis" to the American Medical Association in May 1932, in which they detailed the symptoms of Crohn's disease (CD) [1]. These two main types of IBD share similar pathogenesis and both are characterised by a protracted inflammatory state. However, they differ considerably in characteristics such as location, the involvement of layers in the intestine, and clinical complications [2].

In UC, mucosal inflammation with ulceration extends continuously from the rectum to the proximal part of the colon and the main symptoms include bloody diarrhoea, abdominal pain, urgency, rectal bleeding, fever, and tenesmus [3]. The sub-acute stage of the disease is marked by mural thickening and luminal narrowing. Chronic UC is characterised by rectal constriction and broadening of the presacral region. On the other hand, any part of the digestive system can be damaged by Crohn's disease, but the proximal colon and terminal ileum are the most frequently affected [4] (see Fig. 1). Focused inflammation, aphthoid ulceration, and nearby cobble-stoning are the hallmarks of the acute Crohn's disease. The chronic phase of CD is accompanied by transmural fibrosis and cicatrisation along with perianal disease in most of the patients. Like UC, patients might go through periods of flare-ups and remission with symptoms such as abdominal pain, diarrhoea, fever, anal fissures, and rectal bleeding [5].

The intestinal epithelium functions as a barrier to prevent the inversion of antigens. The defective regulation of tight junctions increases permeability and thereby the uptake of antigens causes inflammation as an immuno-protective mechanism [6]. The aetiology of IBD is still poorly known. Many factors, including genetic, immunological, gut microbiome, and environmental including nutrition, stress, and free radicals, may contribute to the onset of IBD [7]. An irregular immune system increases the production of cytokines and chemokines that activate cellular or humoral responses and exacerbate IBD progression by disrupting homeostasis and persistently stimulating the local immune response in the intestine [8]. Generally, the gut immune system maintains a commensal relationship with the microbiome and shows tolerance to dietary antigens. However, aberrant immunological reactions to commensal non-pathogenic microorganisms lead to dysbiosis in the GIT. In IBD pathogenesis environmental factors play an important influence that is one of the least understood and most challenging to address. Based on the proportionally rising incidence of IBD in the industrialised nations and the emergence of new feeding behaviours in these regions over the past few decades, numerous studies have emphasised the probable aetiological role played by the diet [9].

IBD was previously widely reported in Europe, North America, and Australia. But now it has a much wider geographic distribution and is becoming more common even in nations that were previously thought to be low-risk zones. Two population-based studies in Haryana and Punjab of India reported a prevalence rate of 42.8/100,000 and 44.3/100,000 [10]. Additionally, hospital-based studies indicated a higher incidence of IBD in the southern states of India. Urbanisation, nutritional shifts such as the Westernisation of the Indian cuisine, improvements in environmental and hygienic conditions, as well as underlying genetic predispositions have all certainly contributed to a rise in the incidence of IBD. One modelling study from Iran predicted a 1.5-fold increase in prevalence for East Asia with 4.5 million cases of IBD and quadrupling of the occurrence for India with 2.2 million cases of IBD in 2035, as compared to 2020 [11].

Regrettably, there is little hope for a complete cure for CD and UC either through medical nor surgical treatment. Approximately 10–20% patients experience relapses and recurring symptoms. Medical treatment mostly addresses symptoms of IBD and aims to treat abdominal pain; infec-

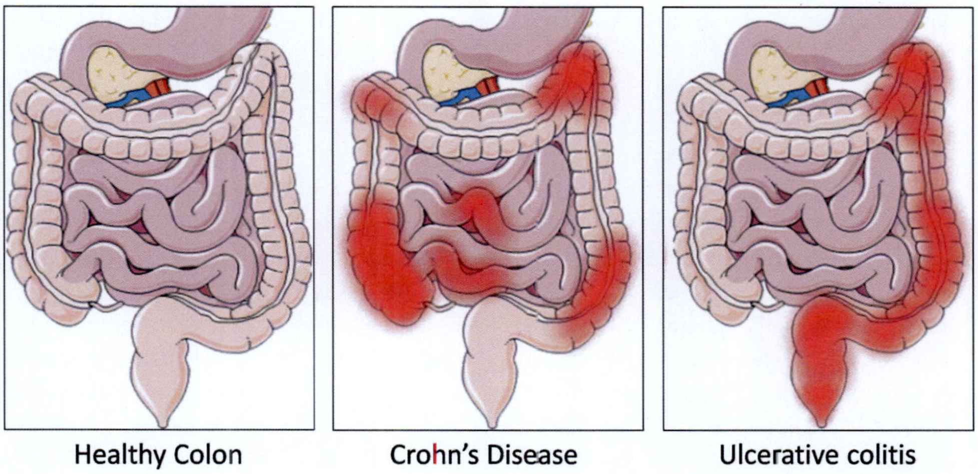

Fig. 1 Schematic representation showing the differences between the two main forms of inflammatory bowel disease (IBD), Crohn's disease, and ulcerative colitis, in terms of their location and the pattern of the affected areas (red-coloured) in the gastrointestinal tract (GIT)

tion; diarrhoea; and deficiencies in proteins, electrolytes, and vitamins. Medical treatments in allopathic medicine include intestinal antibiotics, anti-diarrhoea medications, steroids, immunomodulatory medicines, and surgery [12]. However, these treatments merely manage the disease's episodes; they do not provide long-term treatment. Additionally, a number of adverse effects are there for these modern drugs. Steroids can damage CNS and can produce diseases like cardiovascular disease, infertility, renal disease, and liver disease. Long intake of antibiotics has detrimental consequences on the digestive system. Surgery can be performed to treat complications rather than to treat the disease and relapse is still a possibility even after surgery [13].

In the modern era, Ayurveda is a good IBD therapy option. As per Ayurveda, Crohn's disease is specifically considered as "Vatikagrahani" and ulcerative colitis as "PittajaGrahani". According to Ayurveda, perfect health exists when the three basic energies (doshas: Vata, Pitta, and Kapha) are in harmony [14]. The underlying cause of IBD lies within the process of digestion, supported by a healthy lifestyle and right quantity of food [15]. Ayurveda treatment is essential and has been shown to be successful in alleviating the symptoms of IBD. Ayurveda uses both allopathic and ayurvedic criteria to assess the severity of the IBD condition. Internal medications (ShamanaChikitsa), Panchakarma therapy (ShodanaChikitsa), dietary restrictions, and lifestyle changes are all part of the treatment. Panchakarma therapy is crucial in more severe cases as it swiftly detoxifies and regenerates the digestive system, reduces inflammation, ceases rectal bleeding, and cures ulcerations [16]. Overall health and digestion of the patients are improved by several herbal treatments, which also help to keep these conditions under control. There are several plants, including Bilva (*Aegle marmelos*), Vacha (*Acorus calamus*), Giloy (*Tinospora cordifolia*), Vaividang (*Emblicaribes*), Mulethi (Liquorice root), Daruharidra (*Berberis aristata*), Kutaj (*Holarrhena antidysenterica*), and Mustak (*Cyperus rotundus*) that have anti-inflammatory, anti-microbial, and anti-oxidant properties [14]. Thus, lifestyle management through the use of mindfulness techniques promotes overall health once the disease has been put into remission by taking these herbal medicines [17].

Constipation is another very common medical condition worldwide, negatively affecting patient's quality of life and health-care system. It is accompanied by nausea, anorexia, and discomfort. Moreover, it may result in extra-intestinal symptoms such as headaches, bad breath, restlessness, and confusion. Collectively, these symptoms significantly lower the patients' quality of life. It is reported that 12–19% of Americans and 14% of Asians suffer from constipation, with symptoms varying by geographic location. Pharmacologic agents for constipation include mainly laxatives, but many patients are dissatisfied with these due to ineffectiveness, inconsistent results, and safety-related or adverse-effect concerns and are interested in other treatment options and complementary/alternative therapies. Three often used herbal remedies for achieving regular bowel movements are rhubarb, senna leaf, and aloe. In addition, rhubarb is a significant component of MaZiRenWan, a Chinese medicine formulation occasionally used in eastern nations to treat constipation.

2 Unravelling the Aetiopathogenesis of IBS and Constipation

Even though the exact pathogenesis of IBS remains obscure, it is generally believed that the disease is caused by crosstalk between the host immune system and the gut microbiota, along with genetic predisposition, environmental factors, and socio-economic development. The causes of constipation are mainly three: lifestyle related, disease related, and drug induced. A diet low in both soluble and insoluble fibre, inadequate daily intake of fluids, lack of physical activity, routine changes, age, and certain drugs are all common causes of constipation. Research on the pathogenesis of IBS and constipation has increasingly relied on the gut microbiota in the past few decades [18]. It should be noted that several IBS drugs, especially antibiotics have been suspected of affecting either the metabolic stability of intestinal cells or the metabolism of gut microbiota. Numerous investigations have revealed alterations in the gut flora in patients having CD, UC, and constipation demonstrating the momentous role of gut microbiota in intestinal homeostasis [7].

2.1 A Healthy Gut Microbiota and Its Role in Human Health

The "gut microbiota", a population of more than 100 trillion microorganisms comprising bacteria, virus, archaea, and eukaryote that inhabit the GI tract has a significant impact on the shaping and maturation of the immune system [19]. The majority of the GI tract is composed of bacteria, both useful bacteria "commensals" and opportunistic harmful bacteria "pathobionts" [20]. The prevailing consensus is that the development of the microbiota is a lifetime process that begins at birth and changes erratically with age and in response to environmental changes. Additionally, each person's pace of change in the composition of the microbial population differs and there is no one-size-fits-all recipe for a healthy microbiome [21].

Altogether, Firmicutes and Bacteroidetes make up the majority of the healthy gut microbiota, followed by the phyla Actinobacteria, Proteobacteria, Fusobacteria, and Verrucomicrobia [22]. While virome primarily resides in the form of bacteriophages, the most frequent fungal genera that cohabit with bacteria in the gut are *Aspergillus, Bullera, Candida, Galactomyces, Penicillium, Pleospora, Rhodotorula, Sclerotinia, Saccharomyces, and Trametes.* Nearly 96% of healthy people possess archaea mainly, *Methanobrevibacter smithii* in their GI microbiota, which generates methane from the hydrogen produced by bacterial catabolism [23].

The gut microbiota delivers a number of promising benefits to the host, among which the most significant roles in maintaining the integrity of the mucosal barrier, supplying nutrients like vitamins, and mitigation of infections [24]. Gut bacteria involves in the breakdown and absorption of bile acids, urea, sterols, essential vitamins, xenobiotics, plant-based polysaccharides, and some amino acids [25]. The subsequent metabolites are SCFAs like propionate, butyrate, and acetate which regulate functions including altering interstitial epithelial cell signalling, gene expression, chemotaxis, differentiation, proliferation, and apoptosis. Firmicutes are involved in butyrate production and Bacteroidetes phylum can produce propionate [26]. Butyrate, a primary energy source for colon epithelial cells and a regulator of mucin secretion, helps to maintain intestinal homeostasis, by keeping the integrity of the intestinal epithelial barrier and controlling the immunological response. Acetate and propionate, on the other hand, serve as substrates for the processes of lipogenesis and gluconeogenesis [27]. The GI microbiota is also vital for the de novo synthesis of essential vitamins which the host is unable to generate, such as riboflavin, vitamin K, biotin, nicotinic acid, pyridoxine, pantothenic acid, and thiamine [28]. Vitamin B12 is mostly produced by lactic acid bacteria, and folate, a vitamin essential for the creation and repair of host DNA, is primarily produced by bifidobacteria [19]. Gut microbiota interacts with innate immunity and adaptive immunity in an extensive bidirectional manner and the interaction site is the intestinal barrier which requires the interplay of multiple components: interstitial epithelial cells (IECs), the gut microbiota, and mucosal and submucosal immune cells [29]. IECs serve as innate immunity mediators under normal physiological circumstances. IECs express cell-surface and intracellular pattern recognition receptors such as toll-like receptors (TLRs) and NOD-like receptors, respectively, just like other cell types. Intestinal homeostasis depends on the immune response, which is activated as a result of gut microorganisms stimulating PRRs [30].

2.2 Gut Microbiota Dysbiosis in IBS and Related Disorders

Numerous studies have revealed the fact that many risk variables for IBS contribute to the host–microbe interactions that lead to the severe imbalance of gut microbiota which is a key feature of IBS (i.e. dysbiosis) [31]. Dysbiosis occurs when the microbiota is unable to provide the host with all of its advantageous traits. Usually, various factors can contribute to inflammation in IBS. The imbalance of commensal to pathogenic microbiota may cause an excessive amount of pro-inflammatory chemicals to be produced, which exacerbates intestinal inflammation [7, 32] (Fig. 2).

The gut microbiota of IBS patients exhibits decreased microbial diversity, less prevalence of the Firmicutes phylum, particularly *F. prausnitzii*, one of the most prominent bacterial species in the GI tract, and decreased abundance of Bacteroidetes phylum [7]. A decreased population of mucolytic bacteria such as *Akkermansiamuciniphila* increaseingthe abundance of mucin-degrading bacteria such as *Ruminococcusgnavus* and decreasing the presence of butyrate producers (e.g. *Blautia, Lachnospiraceae, Roseburia, and Eubacterium rectale*) were also observed in CD and UC patients [33]. Distortions in the epithelial barrier's integrity can enhance the uptake of luminal antigens, which in turn can cause inflammation. Additionally, IBS may be brought on by a decrease in mucin production driven by the destruction of goblet cells and malfunctioning of epithelial cell tight junctions [6]. This supports the key role of dysbiosis in IBS development. Recent studies have found that germ-free mice (deficient in the gut microbiota) which received intestinal microbiota transplantation from IBS individuals exhibit an exacerbation in the severity of colitis induction compared with the healthy population [34]. Even though it is proven that faecal microbiota transplantation effectively treats IBS, its safety is yet unknown. Variations in nutrient sources, oxygen levels, luminal pH, and the proliferation of different bacterial species can all be altered by inflammation and have an impact on the bacterial population's equilibrium [35]. The representative gut microbiota alterations and interactions with host cells seen in IBS patients are listed in Table 1.

2.3 Association of Pathobionts and PRRs in IBS

Pathobionts are bacteria that are typically present in the intestines but can become pathogenic when certain endogenous and exogenous triggers alter the gut microbiota's function in the emergence of pathological diseases. Virulent *E. coli* strains, *Helicobacter* species, *Klebsiella* species, *Campylobacter concisus*, and Vancomycin-resistant *Enterococcus faecalis* are some of the pathobionts. By disrupting the intestinal barrier and regulating inflammatory reactions, these bacteria promote inflammation [32].

Intestinal epithelial cells are involved in pathobiont interactions, which can be direct (via adhesion and translocation) or indirect (by the modification of epithelial cell activities). Pathobionts can also interfere with the proper growth and

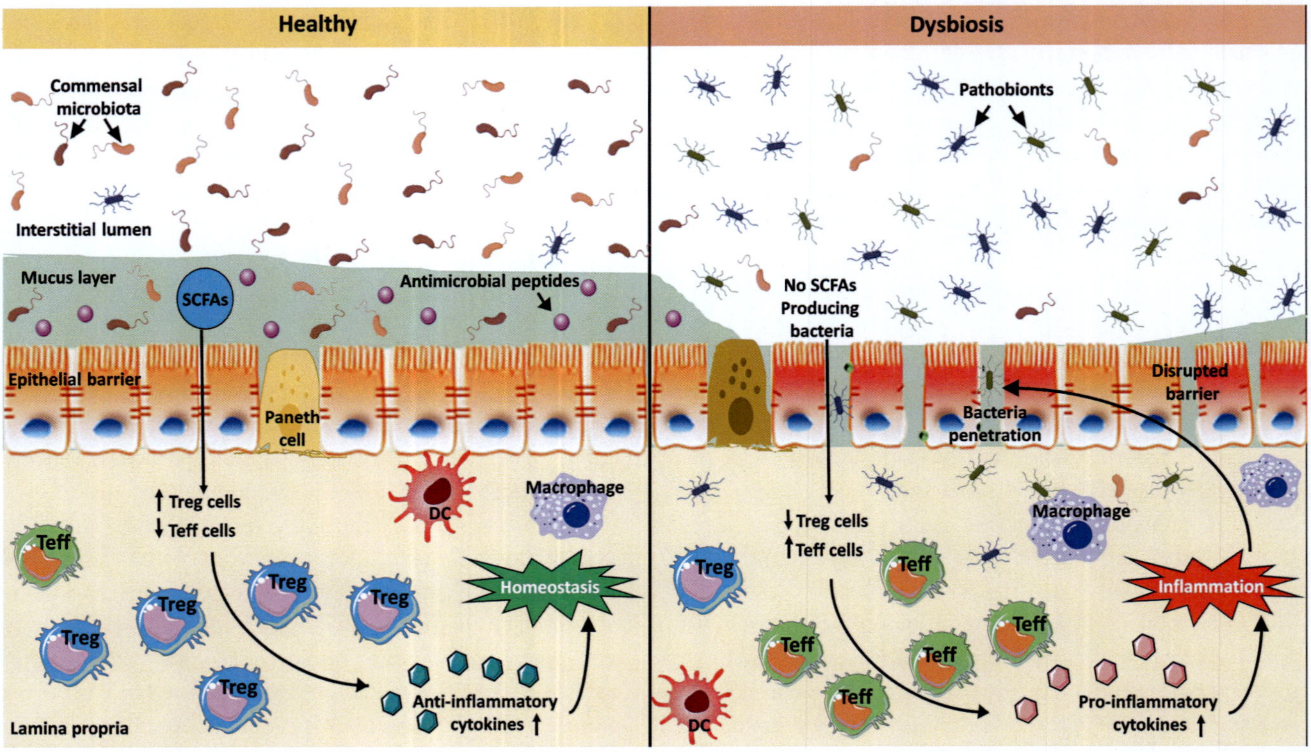

Fig. 2 Dysbiotic interaction in IBS and related disorders. In healthy state, commensal bacteria performs functions which are beneficial to the host. To maintain the homeostasis of the intestinal tract, certain bacteria produce SCFAs like butyrate. The intact mucus layer and antimicrobial peptides produced by paneth cells stop bacteria from penetrating the epithelial layer. SCFAs help in maintaining the epithelial barrier, and also promote the production of anti-inflammatory cytokines through facilitating the regulatory T cells (Treg) and effector T cells (Teff). Contrarily, in IBS patients, the decrease in commensal bacteria, including those that produce SCFAs, leads to imbalance of Treg and Teff cells which further increases the production of pro-inflammatory cytokines. As a result, the mucosal layer is disrupted resulting in the invasion of bacteria. The mucosal layer is then damaged, allowing bacteria to invade and prompting an inflammatory cascade to combat the invasive bacteria. Moreover, pathobionts and commensal bacteria that could exhibit pathogenic traits under the appropriate circumstances, arise and proliferate in inflammatory conditions. As a result, the modified pathobionts survive, further limiting the growth of commensal organisms. The reduction of commensals, the growth of pathobionts, and an imbalance in immune regulation lead to a chronic inflammatory condition

function of intestinal epithelial cells. For instance, pathobionts like *Enterobacteriaceae faecalis* can adhere to and move through monolayers of epithelial cells, while *C. concisus* can cause apoptosis in the same monolayers of epithelial cells. In addition, pathobionts may continue to trigger inflammation long after infection by increasing pro-inflammatory signalling in intestinal epithelial cells which might result in dysbiosis and colitis. *C. difficile* is one of the notable pathobionts which is associated with nosocomial infections that induce mild to severe pseudomembranous colitis. Usually, a healthy microbiome offers resistance against *C. difficile* pathogenicity [51]. However, a dysbiotic microbiota is what fuels the growth of this bacteria and ensuing disease pathology. The pathogenesis of *C. difficile* pathobionts begins with the germination of its spores into vegetative cells, which multiply in the GI tract and release toxins that break the intestinal epithelial barrier to cause severe pseudomembranous colitis [52]. IBS-associated *E. coli* is also a significant factor in the aetiology of IBS and may contribute to disease relapses in IBS patients [53]. Pathobiont bacteria suppress the expression of tight junction proteins and enhance mislocalisation, cause dysregulation of apoptosis and proliferation, and enhance ER stress and pro-inflammatory signalling. These effects result in the development of dysbiosis and facilitate inflammatory signalling following infection. In contrast, commensal bacteria provide mucosal health by enhancing mucin and IgA secretion, bolstering barrier integrity, pro-inflammatory responses and apoptosis, and facilitating commensal colonisation. These actions prolong inflammatory signalling after infection and aid in the development of symbiosis [54].

Pattern recognition receptors such as toll-like receptors (TLRs), NOD-like receptors (NLRs), and RIG-I-like receptors are capable of detecting a wide variety of microbes and their metabolites and can exert both pro- and anti-inflammatory actions. During infections, cells express these PRRs more intensely and produce more pro-inflammatory mediators. PRRs can activate specific T-cell phenotypes; for example, few Clostridia strains can activate Tregs in colonic mucosa and thereby protect from colitis examples include

Table 1 Dysbiosis and its interaction in IBS and related disorders

Gut microbiota	Changes in IBS	Mechanisms	Evidence	References
Firmicutes				
Clostridioides difficile	↑	Toxins released by this bacteria can activate pro-inflammatory cytokines and can cause epithelial barrier damage	High levels of *C. difficile* in IBS patients	[36]
Eubacterium spp.	↓	Produce butyrate and is essential in maintaining the epithelial barrier integrity	Low prevalence in CD and UC patients	[37]
Faecalibacterium prausnitzii	↓	Highly active metabolic commensal bacteria that contributes to the production of butyrate which is essential for the anti-inflammatory activity, maintenance of homeostasis of barrier function, and immune responses	Low *F. prausnitzii* levels may be indicative of CD and IBS. Surgery patients with IBS who have low *F. prausnitzii* levels are more likely to experience post-operative recurrence	[38]
Listeria monocytogenes	↑	Cause epithelial cell infection	High infection rates in IBS patients	[39]
Roseburia intestinalis	↓	Protect colonic mucosa Produce butyrate	Decreased in CD/UC patients	[40]
Ruminococcus albus	↓	Produce SCFAs and have anti-inflammatory actions	Decreased in IBS patients' samples	[41]
Ruminococcus gnavus	↓	Involved in bile and amino-acid biosynthesis pathways	Decreased in CD patient stool samples	[42]
Bacteroidetes				
Bacteroidetes spp.	↑	Production of propionate and damages the mucin layer	Increased in CD patients' samples	[43]
Verrucomicrobia				
Akkermansia muciniphila	↓	Production of SCFAs and activate the number of T cells in colon	Less prevalence in stool of both CD and UC patients	[44]
Actinobacteria				
Bifidobacterium bifidum	↓	Strengthen epithelial barrier function by enhancing the mucin production	Decreased levels in IBS patients. Also, few studies reveal that treatment with probiotics which contain this bacterium can show positive response in IBS	[45]
Eggerthella lenta	↑	Lack of possible mechanism	More prevalence in IBS patients	[46]
Mycobacterium avium	↑	Increased production of pro-inflammatory cytokines	High abundance in IBS patients	[47]
Atopobium parvulum	↑	Induction of colitis and mitochondrial dysfunction in IBS	Increased in CD patients	[48]
Proteobacteria				
Campylobacter spp.	↑	Can survive even in anaerobic conditions in IBS	Increased in IBS patients	[49]
Eikenella corrodens	↑	Involved in pro-inflammatory response	Increased in IBS patients	[42]
Escherichia coli	↑	Damage intestinal epithelial cells monolayers	High abundance in colonic and ileal mucosa of IBS patients	[50]
Haemophilus parainfluenzae	↑	Promote inflammation in IBS	Increased in stool of IBS patients	[42]
Fusobacteria				
Fusobacterium nucleatum	↑	Involved in damaging mucus layer	Increased in CD samples	[49]

MAMPs, PAMPs, and DAMPs. While NOD-like receptors (NLRs) and RIG-I-like receptors are found in the cytoplasm, TLRs and CLRs are found on the surface of immune cells, epithelial cells, and other cell types. PRRs are generally expressed at greater levels in colonic epithelia and regulating PRR responses is essential to achieving improved clinical outcomes in IBS [32].

2.4 Diet–Microbiota Interaction in IBS and Related Disorders

Diet, gut microbiota, colonocytes, and immune cells are intimately associated during both the intestinal healthy conditions and inflammatory state. The composition of microbiota is dynamic, fluctuating with age and environmental factors with food intake patterns being the most pervasive of these. The infant's immune system and intestinal homeostasis are progressively shaped by the microbiota, confirming the notion that diet in early life influences the later risk of developing immune diseases. Studies suggest that a calorie-rich diet, composed of fat and carbohydrate-based foods, leads to favourable development of Bacteroidetes phylum in adults. The overconsumption of omega-6 fatty acids is pro-inflammatory and leads to loss of homeostasis because of its metabolised product arachidonic acid, a precursor for pro-inflammatory eicosanoids. In contrast to this, a high intake of omega-3-fatty acids includes a-linolenic acid from plants and eicosapentaenoic acid and docosahexaenoic acid from fish are anti-inflammatory. Additionally, there is a link between low dietary fibre consumption and elevated risk for Crohn's disease [55].

Western diets, which typically consist of a high calorie intake with large consumption of carbohydrates, are linked to less abundance of microbial diversity, contrary to the Mediterranean diet, which focuses on fruits, vegetables, and red wine. Western diet is rich in simple carbohydrates (such as fructose and sucrose) and fat (such as long-chain fatty acids like arachidonic acid) and has little to no fibre. Moreover, it is supplemented with emulsifiers and food colourings. SCFAs the products of bacterial digestion of fibres, which are having homeostatic functions in the mucosa and anti-inflammatory properties, were shown to plummet with Western diet. Several studies reported that consuming more simple carbohydrates induced dysbiosis and intestinal inflammation. On the other hand, complex carbohydrates which are often obtained from vegetables and the bacterial metabolites they produce, have a protective effect. Many studies revealed that a high-fat diet boosts the anaerobic prevalence of Bacteroides. A high-fat diet is a key root problem of gut dysbiosis which may enhance gut inflammation. Whey consumption and pea protein extracts promote the growth of Bifidobacterium and Lactobacillus, while inhibiting the prevalence of *Bacteroides fragilis* and *Clostridium perfringens*. An essential amino acid tryptophan, which is present in dietary protein and is metabolised by the gut microbiota in the colon, regulates microbial diversity and gut immunity. Contrary to plant-based protein, animal-based protein consumption led to a rise in the number of bile-tolerant anaerobes like Bacteroides, Bilophila, and Alistipes [56].

Excessive consumption of food additives disrupts microbial composition and fosters gut inflammation. Saccharin and other artificial sweeteners induce dysbiosis in humans. Emulsifiers disturb gut microbial community structure and elevate the risk of gut inflammation. Likewise, titanium dioxide, typically used as a white powder in candies and sugar-coated chewing gums, reduces gut permeability and stimulates gut inflammation. Nutrition also promotes epigenetic alterations either directly or indirectly through the action of the gut microbiome, since some metabolites have the potential to impact gene expression, chromatin remodelling, and DNA methylation. Healthy diets with sufficient concentrations of vitamin D have been shown to be beneficial in IBS by boosting dendritic cell proliferation and reducing the production of pro-inflammatory cytokines. Zinc is also involved in DNA replication and transcription and has immune regulatory effects. Iron, folate, and vitamin B12 are critical for haemoglobin/blood cell formation. Additionally, specific carbohydrate diet (SCD, which excludes all carbohydrates except monosaccharides) and a diet low in fermentable oligo-, di-, and monosaccharides and polyols have shown favourable results in IBS. Micronutrients also have the potential to influence the microbiota. One cohort study reported that the development of IBS has a positive correlation with the iron level of drinking water. These findings provide credence to the notion that dietary manipulations might modify the gut microbiota [57].

3 Current Approaches for Managing IBS and Related Disorders

3.1 Marketed Drugs, Drawbacks of Conventional Treatment

Treatment for IBS is aimed at bringing about and keeping the disease in remission. The aims have altered in recent years from clinical remission to achieving "sustained profound remission" via mucosal healing. Nonbiological treatments that include anti-inflammatory or immunosuppressive medicines comprise aminosalicylates, steroids, and 6-mercaptopurine. These drugs alleviate UC and CD symptoms but do not alter the progression of the disease.

For instance, despite their effectiveness in reducing mild inflammation in UC patients, aminosalicylates are not very effective in treating symptoms of CD. Also, attempting to take these traditional drugs for a long time often weakens the immune system and has other serious side effects [4, 58–62]. So, there is a strong need for other treatments that can start and keep remission going. One of the ancient medical interventions is the use of therapeutic herbs. Hence, patients with inflammatory bowel disease are more likely to use complementary and alternative medicine (21–60%), especially herbal treatments, because of the positive benefits they have been documented to have on patients' health. This alternative herbal therapy may still be the sole curative option available to communities as well as ethnic groups without a formal system of health care even today, especially in developing nations. By inhibiting the synthesis of intercellular adhesion molecules including NO, iNOS, COX-2, and NF-κB, as well as lowering ROS and enhancing antioxidant effects, these natural treatments decrease the course of IBS [59, 63–69]. Tables 2, 3, and 4 illustrate the application of medicinal

Table 2 Modern evidence with in vivo studies with the efficacy of single plant-based bioactive on IBS

Type of model	Plant source	Bioactive molecules	Role in IBS	References
DSS induced murine colitis	*Sophora alopecuroides*	Sophocarpine and sophoridine	↓IL-1β, IL-6, MPO levels and ↓ICAM-1	[70, 71]
TNBS induced acute colitis TNBS and DSS induced colitis	*Sophora moorcroftiana and Sophora alopecuroides*	Matrine (stereoisomer of sophoridine)	↓TNF-α, IL-1β, and IL-6, IL-12/23p40, IFN-γ, IL-17	[71–74]
DSS induced colitis	*Sophora alopecuroides*	Aloperine	Inhibits PI3K/PKB/mTOR	[75–77]
		Oxymatrine	↓PI3K/AKT signalling, suppress Th1 and Th17 cell in colon	[78]
DSS and TNBS induced colitis	*Coptis chinensis* and *Phellodendron chinense*	Berberine	↓ZO-1, IFN-c, IL-17, IL-6, IL-1β, and TNF-a cadherin expression while ↑IL-22 levels	[79, 80]
DSS-induced chronic colitis			Initiates the AKT1/SOCS1 signalling pathway and blocks p65 NF-κB subunit phosphorylation, ↓colonic pSTAT3, ↓MPO, IL-17 while ↑IL-10 expression levels	[81, 82]
DSS-induced colitis	*Lindera aggregata*	Boldine, Norisoboldine	↓TNF-α, IL-6, and IL-17 secretion, ↓NF-κB and STAT3	[83, 84]
TNBS-induced colitis	*Isatis tinctoria*	Isatin	↓TNF-α, COX-2, PGE2 and MPO, ↑colonic IL10, SOD, GSH-Px, and GSH-Rd levels	[85–87]
DSS-induced colitis	*Polygonum tinctorium*	Tryptanthrin	↓TNF-α, NO activation by regulating NF-κB/STAT3 signalling	[88–90]
DSS and acetic acid induced colitis	*Citrus, grapefruits, and tomatoes*	Naringenin	↓ICAM-1. iNOS expression, MCP-1, TNF-α. Also ↓Cox2, MIP-2, NO, IL-1β, IL-6, ↑the colonic mucosal content	[91–93]
DSS-induced colitis	*Nicotiana tabacum*	Nicotine	Suppress leukocyte recruitment, downregulates colonic TNFα, ↓IL-6 in CD4T cells, ↓MAdCAM-1 expression, inhibits STAT3 signalling	[94–98]
DSS-induced enteritis	*Epimedium sagittatum, Epimedium pubescens, Epimedium wushanense, and Epimedium koreanum*	Icariin	Hinder p-STAT1, p-STAT3, and p-p65 expression; ↓T lymphocytic growth	[92, 99–101]
TNBS-induced colitis	*Citrus fruits*	Hesperetin	↓p-JAK2 and p-STAT3, ↑SOCS3 expression	[102]
TNBS-induced colitis	*Rhododendron*	Farrerol	↓TNF-α, IL-6, and IL-1β, NF-κB phosphorylation	[103]
TNBS and DSS induced colitis	*Citrus fruits*	Quercetin	↓Colon damage, iNOS, regulate MPO activity, MDA levels and ↑GSH expression	[104, 105]
DSS induced colitis	*Myrica rubra, Myrica cerifera*	Myricetin and kaempferol	Kaempferol-↓IL-6, IL-1 β, COX-2, TNF-α, iNOS, MPO, PGE2, NO levels in colonic mucosa Myricetin-↑TGF-β and IL-10, T regulatory cells, ↑claudin, occludins expression	[106–109]
TNBS and DSS induced colitis	*Panax quinquefolius* L. *and other species*	Ginsenoside	Anti-inflammatory, Ginsenosides Rd-downregulates infiltration of neutrophil, inhibited inflammatory cell recruitment into colonic tissue, ↓NLRP3 inflammasome Ginsenosides Rh2-↑TGF-β, Smad phosphorylation. But ↓NF-κB, MAPK Ginsenoside RK3-↓TNF-α, IL-6 and IL-1β Ginsenoside Rb1-↓IL-6, and TNF-α, IL1β, ↑IL-10 expression	[110–120]
DSS induced colitis	*Rutaceae and Meliaceae families, Chinese medicine like Evodia rutaecarpa and Coptidisrhizoma*	Limonin	↓STAT3/miR-214 signalling, IL6, TNF-α, IL-1β, IL-6, COX-2, iNOS, block PERK-ATF4-CHOP pathway, NF-κB, ↑IL-10 level	[121, 122]
TNBS induced colitis	*Cruciferous vegetables*	Indole-3-carbinol	↓IL-17 in MLNs, ↑IL-22, CD4+ Foxp3+ cells	[123]

(continued)

Table 2 (continued)

Type of model	Plant source	Bioactive molecules	Role in IBS	References
Oxazolone-induced colitis in mice	*Cruciferous vegetables*	DIM	↓Colonic RORγt mRNA expression, MPO, IL-17A, ↑Foxp3 mRNA expression in colon	[124]
DSS induced colitis	*Radix Linderae, the dry root of Lindera aggregate*	Norisoboldine	↑p-smad 2 and 3, IL-10 but ↓MPO, TNF-α, IL-17, IL-1β, SIRT1	[125, 126]
DSS induced colitis	*Indigo naturalis*	Indirubin	Ameliorates MAPK NF-κB stimulation, ↓TNF-α, IL-6, Th17 cell differentiation, p-STAT3, ↑p-GSK-3β	[127–131]
TNBS-induced colitis	*Pilocarpusmicrophyllus*	Epiisopiloturine	↓IL-1β,MDA, iNOS, NO,COX-2	[132, 133]
TNBS, acetic acid induced and DSS induced colitis	*Piper nigrum* L., *Piper longum* L.	Piperine	↓TNF-α, IL-1β, IL-6, NO, iNOS, ICAM-1, TLR4, IL-1β levels, CCR2, MCP-1, FFAs	[134–138]
TNBS-induced colitis	*Poncirus trifoliata*	Poncirin	↓COX-2, NF-κB, TNFα, iNOS, TLR4, RORγt expression; Th17 cell differentiation. Unlike ↑differentiation of Treg cell, Foxp3, ZO-1 expression	[138, 139]
TNBS-induced colitis	*Salvia miltiorrhiza*	Miltirone	↓TLR4-mRNA levels, NF-kB, IL-1β, IL-6, TNF-α IL-8, MyD88	[140]
DSS and TNBS induced colitis	*Curcuma longa*	Curcumin	↓TNF-α, IL-6, IL-17, ATG5, MPO NO, LC-3II, Beclin-1, ↑IL-10, bcl-2	[141–144]

Table 3 Preclinical polyherbal drugs formulation in mitigating IBS treatment

Model	Plant source	Role in IBS	References
Acetic acid induced model	PHF-*Eryngium foetidum, Manilkara zapota, Murrayakoenigii*	↓MPO levels, free radical scavenger, restore crypt architecture	[145]
DSS induced colitis	CM102 þ and CM132-*Astragalus root; Barley and Lycium fruit respectively*	↓IL-1β, IL-6, TNF-α	[146]
DSS induced colitis	*Red ginseng extract, Epimedium KoreanumNakai*	↓TNF-α, NO, iNOS, COX-2, IL-5, IL-6, IL-1β, IL-13, ↑CD19+ cells, Tregs cells	[147]
Acetic acid-induced colitis	*Terminalia chebula and Zingiber officinale*	↓PGE2, MPO, TNF-α	[148]
DSS induced colitis	ZenWu Decoction (MZWD*)—Poriacocos* (Schw.) Wolf, *Paeonia lactiflora, Atractylodes macrocephala* Koidz., *Zingiber officinale* Rosc., *Codonopsispilosula; Coptis chinensis*	↓NF-κB, TNF-α, IL-1β, PAR2 expression	[149]
DSS induced colitis	KM1608-*A. lappa, T. chebula and Z. officinale*	↓IL-6, MCP-1 NO, iNOS, TNF-α	[150]
TNBS induced colitis	*Triphala-Terminalia bellirica (*Gaertn.*)* Roxb., *Terminalia chebula* Retz., Roxb., *Emblica officinalis* Gaertn.	↓MDA, ↑SOD, GSH, CAT expression	[151]
DSS induced colitis	*Bibhitaki Terminalia bellirica* Roxb., *Phyllanthus emblica* Itis., *Haritaki Terminalia chebula* Retz.	↓MDA, CRP but ↑SOD, GPx level	[152]
DSS induced colitis	SHD-*Scutellariabaicalensis* Georgi, *Coptis chinensis Franch, Phellodendron chinense* Schneid, *Artemisia argyi* LévL et Vant	Inhibits NF-κB, ↓TNF-α, IL-6, MPO, IL1β, MDA	[153]
TNBS induced colitis	*Mume Fructus, Ginseng Radix et Rhizoma, Zanthoxyli Pericarpium, Asari Radix et Rhizoma, Typhonii Rhizoma, Zingiberis Rhizoma, Angelicae Sinensis Radix, Coptidis Rhizoma, Cinnamomi Ramulus, Phellodendri Chinensis Cortex*	↓NF-κBp65, STAT3, IL-6, IFN-γ while mitigates TGF-β/Smad, Wnt/β-catenin signalling	[154]
DSS induced colitis	ZWBQD-*Poriacocos* (Schw.) Wolf; *Atractylodes macrocephala* Koidz; *Zingiber officinale* Rosc; *Codonopsispilosula* (Franch.) *Nannf; Coptis chinensis* Franch	↓, p-p38 MAPK, p-ERK1/2, p-AKT, p-JNK, NF-κB expression	[155]

Table 3 (continued)

Model	Plant source	Role in IBS	References
DSS induced colitis	STW 5-*Iberis Amara, Glycyrrhiza glabra, Matricariarecutita, Silybum marianum, Angelica archangelica, Melissa officinalis, Mentha piperita, Carum carvi, Chelidonium majus*	↓MPO, TNFα, CINC-3, ↑GSH, SOD	[156]
TNBS induced colitis	AA-*Saccharum officinarum* L., *Terminalia arjuna* (Roxb.) Wight and Arn. (TA), *Woodfordiafruticosa* (L.) Kurz., *Vitis vinifera* L., *Madhuca indica* J.F.Gmel.	↓NO, MDA, MCP-1, IL-6, TNF-α, IL-1β, while ↑SOD, GSH, CAT	[157]
TNBS induced colitis	QRZSLXF-*Glycyrrhiza uralensis* Fish., *Sophora flavescensAiton, Baphicacanthuscusia* (Nees) *Bremek., Coptis chinensis France., Bletilla striata* Rchb. f.	↑FOXP3, Treg cells but ↓STAT3, Th17 cells, IL-6, RORγt expression	[158]

Table 4 Plant derived molecules under clinical trials for the management of IBS

Constituents	Sponsor	Study aim or objective/outcome measure	Current status	Clinical trials. Gov. identifier
Curcumin	Seattle Children's Hospital	To assess the curcumin tolerance in paediatric with IBS	Completed Phase I	NCT00889161
	Sheba Medical Center	To compare Curcumin with 5ASA combinational therapy to 5ASA alone in terms of efficacy in patients with mild to moderate UC	Completed Phase III	NCT01320436
Diosmin	Tanta University	Diosmin effectiveness and tolerability in the management of mild to moderately active ulcerative colitis	Completed Phase III	NCT05626166
Berberine	NCI	Assess the safety of berberine in UC participants	Completed Phase I	NCT02365480
Triptolide	Zhu Weiming	Tripterygium glycosides safety and effectiveness in the therapy of CD	Concluded Phase III	NCT02044952
Pentoxifylline	Tanta University	Pentoxifylline's potential efficacy in patients with UC treated with mesalamine	Ongoing Phase II	NCT05558761
Cannabidiol	Meir Medical Center	Cannabidiol used in IBS treatment	Completed Phase II	NCT01037322
Berberine	Xijing Hospital of Digestive Diseases	Medication with Berberine to sustain symptomatic relief in UC	Completed Phase IV	NCT02962245
Anthocyanin	University of Zurich	To investigate ACRE efficacy, safety, and tolerability in UC patients	Completed Phase II	NCT04000139
Black raspberry	UConn Health	A study of black raspberry freeze-dried for UC management	Completed Phase I	NCT02267694
Aloe vera	National Institute of Medical Sciences and Nutrition, Salvador Zubiran	Aloe Vera's impact on mild ulcerative colitis inflammatory response	Concluded Early Phase I	NCT01783119
Herbal treatment (SA100)	Stanford University	Assessing the oral dose of SA100 in patients with mild, moderate, or severe ulcerative colitis for safety and effectiveness	Completed Phase I	NCT02442960

herbs acting independently or in combinational treatment for alleviating IBS.

3.2 Complementary and Alternative Medicine (CAM) Include Prebiotics and Probiotics

CAM encompasses a broad range of curative and preventative therapies, including herbal medicines, fundamental biological therapy, physical adjustment counselling, and spirit concept therapy. Over the last several decades, CAM has grown in popularity in Western nations' IBS patients. This offers an alternative to conventional medical knowledge for individuals who have not been healed or who are reluctant to use Western medications. Complementary and alternative medicine include non-mainstream drugs that are used alongside standard medicine, while non-mainstream procedures are utilised in place of conventional therapies. Since it has a mild nature, positive therapeutic efficacy, and significantly fewer associated complications than Western medications, CAM is progressively embraced and appreciated by patients in opposition to traditional medical practices. Their failure to respond to traditional treatment with conventional medication, recognised favourable safety characteristics, and their perceptions of their capacity to exert further impact contribute to the high proportion of CAM usage among IBS patients [159–169]. Patients of all ages often use fish oil, probiotics, nutrients, and herbal products as CAMs. Acupuncture is another technique that works with the mind and body [160, 165, 170–173]. Individuals with IBS have varying microbiota in their guts than individuals in good health. With a drop in the *Firmicutes* phylum, the number of bacteria in different Clostridium classes, like *Faecalibacterium prausnitzii*, has declined significantly. *Escherichia coli*, an Enterobacteriaceae species, and *Proteobacteria*, a phylum, have both been found to show a dramatic increase. This means that CD patients have more faeces and mucosa-associated microbiota [174–182]. Moreover, by in situ fluorescence hybridisation, stool specimens from UC patients lacked anti-inflammatory *Pediococcus acidilactici and Lactobacillus*. However, these strains begin to multiply again after the restoration of UC [183, 184].

The human gut is home to 10–100 trillion beneficial bacteria that aid in the digestion and synthesis of essential vitamins. However, stress, cleanliness, and antibiotic usage cause dysbiosis, making it challenging to maintain a better and healthier microflora. Prebiotics are defined as "a substrate that is particularly utilised by host organisms conferring a health advantage" in the ISAPP consensus statement on the scope and definition of prebiotics. The term "probiotics" is derived from a Greek phrase that signifies "for life" and is employed to characterise active, non-pathogenic microorganisms with their clinical efficacy on hosts through modulating immune system responses, boosting mucosal IgA secretion, and fending off infectious agents. In comparison to synthetic medicines like Mesalazine, which are typical treatments for UC, probiotics are also involved in the maintenance of remission. Probiotic bacillus forms such as *Lactobacillus caseistrains, Lactobacillus rhamnosus, L. reuteri, bifidobacteria, Lactobacillus acidophilus-group, Bacillus coagulans, Enterococcus faecium SF68, the yeast Saccharomyces boulardii, and Escherichia coli strain Nissle 1917* are among them [182, 183, 185–190]. To maintain the protective effect, these probiotics reduced TNF-α activation, cytokine secretion, and stimulated IL10, a cytokine known to have therapeutic properties, is encouraged [191–195] (Fig. 3).

Individuals with mild or moderate UC were given salicylates, but *E. coli Nissle 1917* had similar effectiveness and safety characteristics [196–199]. The anti-inflammatory impacts of mesalazine therapy alone are not as strong as those of adjuvant with *Lactobacillus rhamnosus GG* and mesalazine, where it significantly expanded complete improvement in patients with UC. Moreover, adjuvant therapies with *Bifidobacteria* and *lactic acid bacteria* have been shown to slow the progression of the illness and maintain clinical remission in UC patients [196, 200]. Also, *Lactobacillus reuteri ATCC 55730* rectal enemas delivered as additional treatment to 5-ASA resulted in endoscopic and histological recovery as compared to the placebo group in neonates and adolescents with mild to severe UC [201]. Another commercially available and exceptionally concentrated (450 billion bacteria/sachet) probiotics supplement, *VSL#3* cocktail, has allegedly shown benefits in sustaining response to therapy in active UC patients by downregulating interferon-γ, TNF-α expression, and MMP-2 and 9 which have comparable exacerbations as 5-ASA. This is supported by meta-analysis research data that indicate markedly higher therapeutic efficacy compared to conventional therapy alone [202–207]. However the meta-analysis found that the effectiveness of *VSL#3 and E. coli Nissle 1917* in keeping UC patients in remission was only moderate. This requires more validation in terms of research findings before it can be utilised as a probiotic [196, 197]. Also, the same bacteria are present in the *VSL#3* probiotic blend, but there is no scientific evidence to support its efficacy in treating IBS patients [196, 199] (see Table 5).

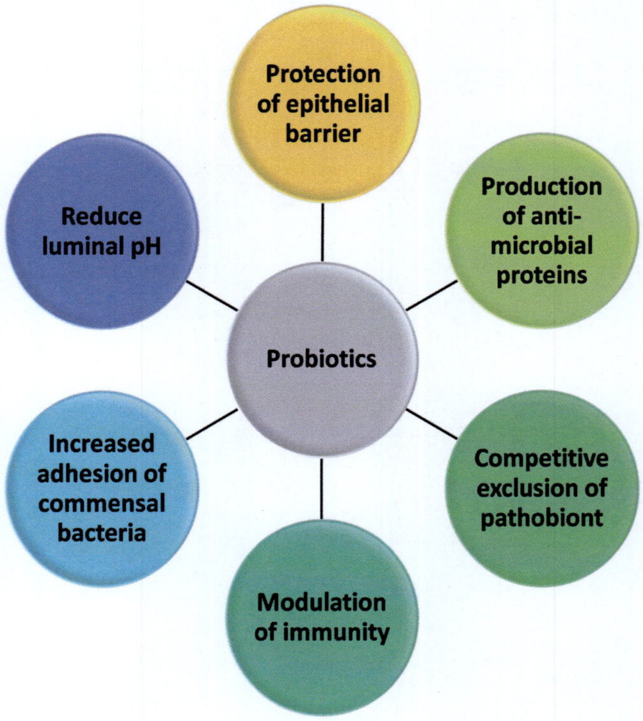

Fig. 3 Mechanisms involved in probiotic-induced protection against intestinal dysbiosis. Probiotics suppress pathogens through various actions, including lowering luminal pH, production of antimicrobial proteins, inhibition of adhesion and translocation of flora, competitive exclusion of pathogens, improvement of the epithelial barrier, enhancement of adhesion of commensal bacteria to the intestinal mucosa, and modulation of the gastrointestinal mucosal immune system

Indigestible food items called prebiotics are known to favourably encourage the growth and activity of specific bacterial species in the intestine. The host organism's well-being would be improved by this. The most often used prebiotics are inulin, lactulose, oligosaccharides such as soya-oligosaccharides, xylo-oligosaccharides, GOS, FOS, derivatives of β-glucans, and galactose [209–212]. Prebiotics are digested as fermentation products by anaerobic gut microbes, SCFA, and gas inside the colon microenvironment. This lowers the colon pH, allowing *Bifidobacteria, Lactobacilli,* and the proliferation of non-pathogenic *E. coli* while suppressing Bacteroidaceae and other possibly harmful bacteria [213–215]. Such microorganisms enhance gut barrier integrity and regulate mucosal and systemic immune function by modulating the pro-inflammatory cytokines IL-17, IL-10, TNF-α, and IL-1β [216, 217]. Following the consumption of resistant starches (RS), particularly RS1 and RS2, both studies on animals and humans demonstrate higher SCFAs with selective enrichment of *Bifidobacterium.* In individuals, RS seems to trigger primarily *Ruminococcus bromii* colonisation and boost butyrogenic species *Faecalibacterium prausnitzii* and *Eubacterium rectale* [212, 218–221]. GBF, which is high in glutamine and hemicellulose, improves DAI in patients with mild-to-moderate UC by enhancing the growth of protective microbes such as *Bifidobacterium* and *Eubacterium,* in addition to the faecal butyrate level, whilst also significantly reducing serum CRP, IL-6, and IL-8 levels, and lowering mast cells. Moreover, in TNBS-induced colitis and CD patients, FOS enhanced the beneficial bacterial growth with a significant increase in the butyrate level. In UC patients, oligo-fructose-enriched inulin reduced faecal calprotectin, implying a decline in inflammation and endoscopic histology with an elevation in butyrate and microbiota changes [222–235]. Nevertheless, prebiotics generate GI adverse effects such as nausea and stomach discomfort, leading to patient intolerance [215]. Table 6 shows several probiotic organism investigations in clinical trials for IBS treatment.

Table 5 Role of prebiotics in some colitis models with the possible outcomes

Animal model	Prebiotic component	Outcome	References
DSS-induced colitis	Dietary inulin	Reduces colitis in the distal part of the colon	[208]
HLA-B27 transgenic rat	Oligofructose-enriched inulin	Suppress the severity of colitis	
DSS and Hapten-induced colitis	Oligosaccharides derived from goat milk	Colitis reduction	
DSS induced colitis	Oligofructose	No effect on colitis	

Table 6 Clinical trials using probiotics for IBS

Prebiotics strains	UC/CD	Study	Studied group	Batch assessment	Outcome	References
Lactobacillus species like L. casei, L. plantarum, L. acidophilus, L. delbrueckii subsp. *Bulgaricus, Bifidobacterium species like B. longum, B. breve and B. infantis, Streptococcus salivarius* subsp. *Thermophils (VSL#3)*	CD	Fedorak RN. et al. Clinical Gastroenterology and Hepatology. 2014	Juvenile, adolescents	120 and 17	No endoscopic recurrence prevention benefits over placebo and compared to placebo, diminishes disease activity and enhances bodyweight and albumin thresholds respectively	[236]
		Day AS. et al. Gastroenterology 2012				
Escherichia coli Nissle 1917	CD	Malchow HA et al. J. Clin. Gastroenterol 1997	Adolescents	28	In comparison to conventional therapy alone, probiotics minimise relapse and decrease corticosteroid usage	[237]
	UC	Petersen AM et al. J Crohns Colitis. 2014	100	Adolescents	No advantage of adding probiotics to standard treatment	[238]
Saccharomyces boulardii	CD	Plein K. et al. Gastroenterol. 1993	20	Adolescents	Strong advantages in research for maintaining remission as an adjunctive treatment	[239]
	UC	Guslandi M. et al. Eur J Gastroenterol Hepatol. 2003	24	Adolescents	Higher efficacy of probiotic as add-on therapy in inducing and maintaining remission compared to mesalazin monotherapy	[240]

(continued)

Table 6 (continued)

Prebiotics strains	UC/ CD	Study	Studied group	Batch assessment	Outcome	References
Lactobacillus GG	CD	Bousvaros A. et al. Inflamm Bowel Dis. 2005	75	Juveniles	Probiotics do not prolong relapse-free time when added to traditional treatment	[241]
	CD	Gupta P. et al. JPGN 2000	4	Juveniles	Improved gut barrier function and clinical state with probiotic as an add-on medication	[242]
	UC	Zocco MA, Aliment Pharmacol Ther. 2006	36	Adolescents	Probiotic add-on treatment improves mesalazin monotherapy in extending relapse-free duration	[243]
B. breve, B. bifidum, Lactobacillus acidophilus, YIT 0168 (Bifidobacteria-Fermented Milk-BFM)	UC	Ishikawa et al. J Am Coll Nutr 2003	21	Adolescents	Greater effectiveness of probiotic combination as add-on treatment in sustaining remission and avoiding recurrence than conventional therapy alone	[244]

4 Herbal Approaches for Treating Inflammatory Bowel Disease (IBS) and Related Disorders

Before discovering the modern medical system, herbal medicine with active constituents (natural products) and their extracts were used and recorded for treating inflammatory bowel disease (IBS). The imbalance between pro-inflammatory and anti-inflammatory responses may be a fundamental cause of chronic inflammation later on accolated with inflammatory bowel disease [245]. The burden of IBS is relatively high in Asian continents and the Western world, affecting the patient's quality of life and health-care system [246]. However, conventional therapies for IBS patients are available in the market, like salicylates, steroids, immunosuppressants, and anti-TNF-α drugs [247, 248] newly approved biologicals treatments to trigger immune responses (e.g. anti-IL-6, anti-IL-12/23, anti-IL13, and integrins) [249, 250] But the long-term side effects such as quiescent illness, poor tolerability, high rate of noncompliance, perceived ineffectiveness, and variable response rates of these conventional therapies [251] were always a primary concern and have augmented the interest in complementary and herbal medicines for their safe and acceptable treatments.

Several recent reports suggested that herbal medicine is becoming more popular among IBS patients. To maintain healthy homeostasis and chronic inflammation, there is crosstalk between diet, gut microbiota, colonocytes, and immune cells [245]. Over 50% of IBS patients have used natural products, including herbs and their formulations, vitamins, probiotics, prebiotics, nano-based formulations, emulsion-based formulations, and dietary fibres [252]. These herbal formulations tend to downregulate various aberrant signalling pathways and modulate various inflammatory mediators, such as IL-1β, IL-6, IL-10, TNF-α, PGE-2, iNOS, COX-2, and IL-8. A fibre-rich diet is crucial for both preventing and treating constipation. If these preventative measures are ineffective and nutritional supplements are insufficient, patients must consider the usage of laxatives. Anthraquinone laxatives (senna, cascara, etc.) are preferred because they are easily tolerated and have a mild action in cases of constipation brought on by atonia of the colon, acute constipation, and in chronic constipation. Anthraquinone laxatives are incredibly helpful medicines when administered properly. Thus, the use of herbal medicine is an immensely vast field in IBS and constipation [253, 254].

So, in this book chapter, we have mainly focused on herbal medicine and its formulation, showing some evidence of efficacy and safety in preclinical and clinical settings with the proposed mechanisms of action. We performed a literature search in PubMed, including the terms "IBS", "colitis", "natural products", "phytochemical", "herbal", "nanocarriers", "probiotics", "gut microbiota", "dietary", "immune", "macromolecule", "colon", "constipation", and "inflammation" which resulted in around 900 references.

Different herbal medicines or derivatives have been used for IBS and related disorders in the Ayurvedic medical system. Categorially, these medicinal plants contain diverse secondary metabolites with complex functionality, such as alkaloids [71], flavonoids, phytosterols, steroids [255], terpenoids, tannin, glycosides, xanthone, polyphenolic acids, and coumarins [256] as core pharmacophores which are responsible for the treatment of IBS [257]. These herbs have various active mechanisms of action such as the downregulation of nuclear factor kappa B (NF-κB) [258], mitogen-activated protein kinase (p38-MAPK) [259], prostaglandin E2 (PGE2), laryngotracheobronchitis (LTB), nod-like receptor (NLR), jun-N-terminal kinase (JNK), extracellular signal-regulated kinase (ERK) [260], signal transducer and activator of transcription (STAT), vascular endothelial growth factor (VEGF), intercellular adhesion molecules (iCAM-1), activation of peroxisome proliferators activated receptor-gamma (PPAR-γ), and AMP-activated protein kinase pathways. Subsequently, a decrease in the expression of pro-inflammatory cytokines (IL-1β, IFN-γ, IL-1β, IL-6, IL-22, IL-17A, TNF-α, COX-2, IL-4, and IL-23), peroxidase activity (MPO and MDA), and matrix metalloproteinase (MMP-9/12/7/) [261] via these medicinal herbs is listed in Table 7.

Table 7 Herbal plants, their active constituents, possible mechanisms of action, and their targets that are used to treat inflammatory bowel syndrome and related disorders

Medicinal plants	Mechanism of action	Target/activity	References
Aloe barbadensis miller, Aloe vera active constituents: anthraquinones and their C-glycosides, anthrones, chromones, and lectins	Downregulation of PGE2 and LTB4 synthesis; increase in SOD and CAT levels	↓IL-1β, IL-8 and TNFα	[262]
Andrographis paniculate Kalmegh Ethanol extract active constituents: andrographolide, neo-andrographolide	Inhibit the proliferation of Th1 and Th17 cell lines, NF-κB, and P38-MAPK pathways	↓TNF-α, IFN-γ, IL-1β, IL-6, IL-22, IL-17A, and IL-23; ↑IL-4	[263]
Artemisia absinthium Wormwood, active constituents: α-thujone, (Z)-epoxyocimene, trans-sabinyl acetate and chrysanthenylacetate	Inhibition of NF-κB signalling; downregulation of iNOS activity	Decrease in TNF-α, IL-1β, and IL-6	[264]
Artemisia capillaris Fragrant Wormwood, active constituents: dimeric guaianolides absinthin (up to 0.28%) and anabsinthin	Downregulate the expression of ERK,JNK, and NF-κB	↓TNF-α, IL-6, IL-1β, and PGE2 Protein expression ↓COX-2	[265]
Boswellia serrata Indian olibanum active constituents: β-boswellic acid (β-BA), 11 keto-β-boswellic acid (KBA), and acetyl-11-keto-β-boswellic acid (AKBA)	Inhibition of the increase in lipid peroxidation and iNOS expression	↓TNF-α, IFN-γ, IL-1β, IL-6, and IL-12; ↑IL-4 and IL-10	[266]
Cannabis sativa Marijuana active constituents: high cannabidiol-rich extract 9-tetrahydrocannabinol (THC), cannabinoids 1 and 2	Downregulation of the p38-MAPK pathway; inhibition of the increase in iNOS expression; inhibition of ERK and NF-κB signalling, stimulation of PPARγ	↑IL-4 and IL-10 ↓TNF-α, IFN-γ, IL-1β, IL-6, and IL-10	[267]
Curcuma longa Turmeric active constituents: Curcumin-I	Downregulation of NF-κB and JAK/STAT signalling Inhibition of p38-MAPK activity, and iNOS and COX-2 expression, ↑PPAR-γ activation	↓TNF-α, IFN-γ, TGF-β, IL-1β, IL-6, IL-12, IL-15, IL-17, and IL-23 ↑IL-4 and IL-10 ↓adhesion scores, ↓colonic damage, ↓protease activity, ↓LPS activity, ↓TLR-4, ↓PG-2	[268]
Triticum aestivum Wheatgrass Juice	Downregulation of COX-2, iNOS, CCL2, and ICAM1 expression; inhibition of NLRP3 inflammasome pathways, caspase-1 and caspase-11 enzymatic activity	↓IL-1β, IL-6, and TNF-α	[269]
Ginkgo biloba Maidenhair trees active constituents: Ginkgolides, ginkgetin, bilobetin, diterpenes and its extract	Downregulation of platelet-activating factor (PAF), deceased MPO activity	↓TNF-α, ↓iNOS, and ↓Cox-2	[270]
Punica granatum Pomegranate active constituents: Tannins, ellagicacid, punicalagin, ellagitannins, proanthocyanidin	Downregulation of NF-κB, PPAR-γ, COX-2 expression, reduction of iNOS, facilitating action towards bifidobacterial, decreased MPO activity, lipid peroxidation	↓PKB/AKT activity, ↓IL-8, IL-12, IL-6, TNF-α ↓growth of pathogenic, inhibit the production of NO and ↑IL-1β stimulation	[271]
Vitis vinifera Grape active constituents: Resveratrol	Activation of PPAR, AMPK, and FoxO signalling pathways, downregulate the NLRP3 inflammasome, cytokine IL-1β	↓TNF-α, TGF-β1, IL-8, IL-6, INF-γ, mRNA downregulated the procollagen I, procollagen III, IGF-mRNAs	[272]

(continued)

Table 7 (continued)

Medicinal plants	Mechanism of action	Target/activity	References
Aconitum laciniatum Compass plant extract	Protect colonic inflammation and reduce colonic IFN-γ mRNA levels	↓TNF-α, ↓iNOS and ↓Cox-2	[273]
Plantago ovata Psyllium Husk-gel forming property	Downregulate the expression of ERK	↓protein kinase C, downregulate the expression of intercellular adhesion molecule1	[274]
Euphorbia tirucalli Milk bush active constituents: Euphol	Downregulation the expression of NOS2, VEGF, and Ki67	↓IL-1β, CXCL1/KC, MCP1, ↑MIP-2, and IL-6 TGF-β	[275]
Ganoderma lucidum Lingzhi mushroom active constituents: polysaccharides, triterpenoids, proteins, ganoderic acids C and D	Inhibition of NF-κB	↓level of IL-6, IL-8, MMP-2, MMP-9	[276]
Ipomoea asarifolia Morning-glory Decoction	Downregulation of the gene expression of JNK1, NF-κB-p65, and STAT3	↓TNF-α, IL-1β, ↑IL-10, reduce MPO activity	[277]
Styrax japonica Japanese snowbel active constituents: jegosaponins and benzofurans Styraxoside B, egonol, masutakeside, and styraxlignolid	Downregulation the expression of nitric oxide synthase (NOS) and Cyclooxygenase-2 (COX-2)	↓mRNA expression levels of NOS and COX-2, TNF-α, IL-1β; ↓DNA binding activity of NF-kB pathway	[278]
Silybum marianum Milk thistle active constituents: silybinin, silychristin, silydianin, and isosilyibinin	Reduce expression of oxidative stress, NF-κB, and proinflammatory cytokines	↓TNF- α, IL-1β, ↑IL-6	[279]
Combretum fragrans Four-leaved bushwillow extract	Reduce the expression JNK/JAK/STAT signalling pathways, upregulated the protein levels of p-AMPK, SIRT1, PCG-1α	↓IL-1β, CXCL1/KC, MCP 1, ↑MIP-2, and IL-6 TGF-β	[280]
Camellia sinensis Green tea active constituents: epigallocatechingallate (EGCG), catechin, gallocatechin, and epicatechin	Inhibit signalling pathways involved in inflammation, including nuclear factor B	↓MPO activity, ↓TNF-α, IFN-γ, NF-kB, and p65	[281]
Olea europaea Olive active constituents: hydroxytyrosol, flavonoids, secoiridoids, oleuropein	Reduce the expression of pro-inflammatory mediators (IL-1β, TNF-α and iNOS)	Improving the intestinal epithelial barrier integrity restoring the expression of ZO-1, MUC-2	[282]
Gardenia Jasminoides Cape jasmine ethanol extract	Inhibit myeloperoxidase activity	Downregulate (iNOS), COX-2, and NF-kB	[283]
Garcinia Cambogia Gambooge Aqueous extract	Inhibit expression of myeloperoxidase, COX-2, and iNOS	↓PGE2 and IL-1β	[284]
Inonotus obliquus Chaga Aqueous extract	Downregulate of the expression of TNF-α, IL-4 and STAT1, STAT6	↓TNF-α, STAT1, pSTAT1, COX-2, IFN-γ, STAT6, and pSTAT6, suppress the expression of IgE and IgA	[285]
Patrinia scabiosaefolia Golden lace Methanol extract	Downregulate the expression of nitric oxide synthase (NOS) and Cyclooxygenase-2 (COX-2)	↓TNF-α, IL-1β, and IL-6, mRNA level	[286]
Prunus mume Japanese apricot Aqueous extract	Downregulate the expression of TNF-α, COX-2, IL-4, STAT6, INF-γ, STAT1	↓IL-1β, IL-6, IL-8, inhibit myeloperoxidase (MPO) activity elevation	[287]
Commiphoramyrrha Myrrh Resin extract	Mediate TNF-α expression	↓CXCL-3, TNF-α, IL-1β, IL-6, TNF-α, PGE2, NO, ↑IL-10 and increase endogenous antioxidant activity	[288]

4.1 Nanotechnology in Inflammatory Bowel Diseases and Related Disorders

In the past decades (Fig. 4, Table 8), conventional drugs, phytoconstituents, and herbal formulations have been critical mediators fortuitously and mediated a range of signalling pathways and proteins for treating IBS and related disorders [315]. However, lack of oral bioavability, cell specificity, and rapid first pass metabolism leads to find efficient way to overcome reported drawbacks. So, with advances in science, the development of phyto-derived nanoisation and micronisation of these herbal medicine has covered a wide range and

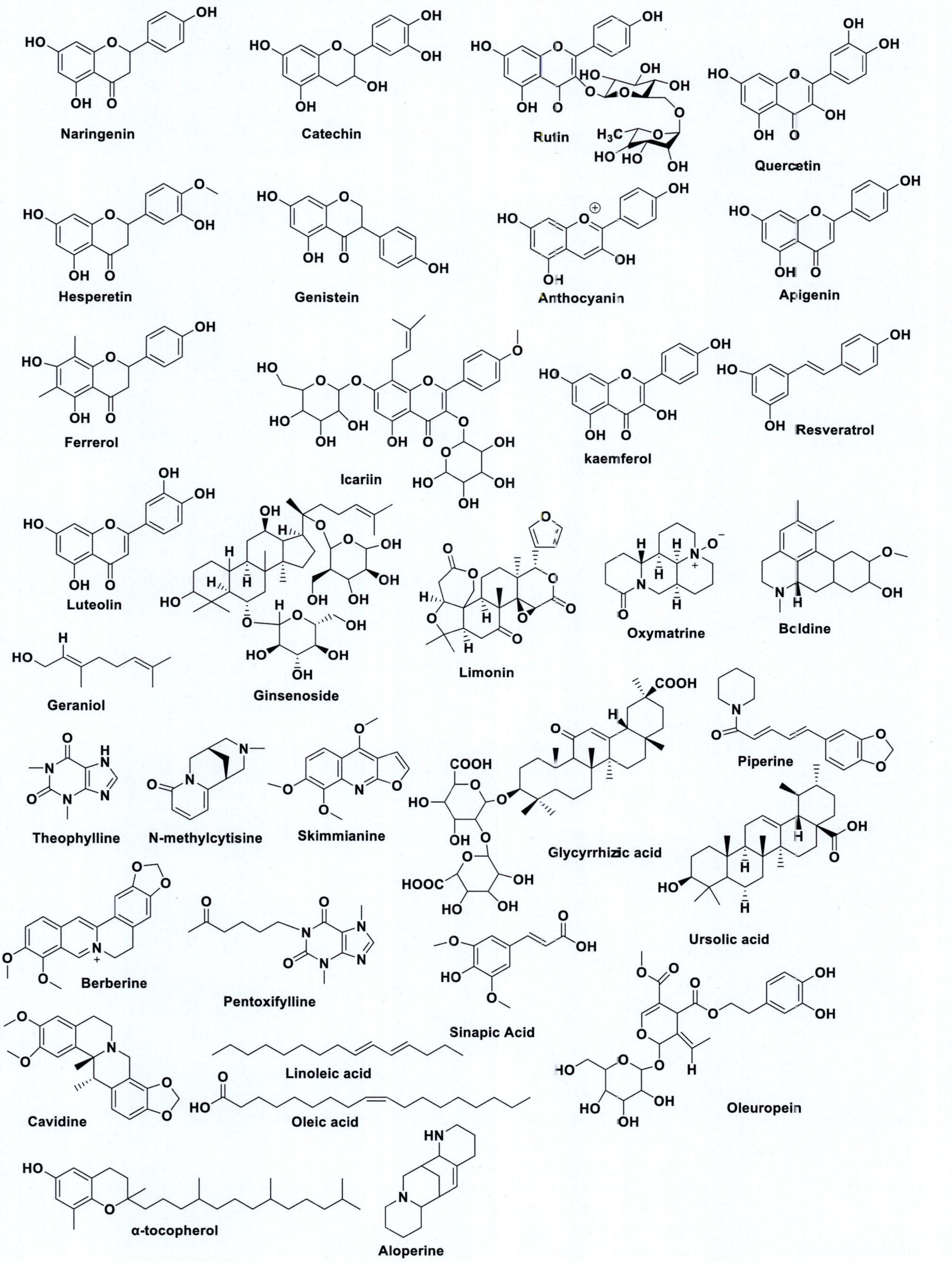

Fig. 4 Chemical structure of secondary metabolite for treatment of inflammatory bowel disease and related diseases

Table 8 Potential phytoconstituents for inflammatory bowel syndrome treatment and related diseases

Class/phytochemical constituents	Pharmacological activity	References
Flavanoids—Naringenin	Facilitate monocyte chemotaxis and colonic secretion of pro-inflammatory factors (ICAM-1, iNOS). Downregulate mRNA expression of TNF-α, Cox2, reduced the expression of NF-κB	[93]
Catechins	Improve Th1/Th2 balance and reduce TLR4/MyD88/NF-κB pathway expression. Suppress level the IL-6, MCP-1, TNF-α, reduce the MPO and LPO activity	[289]
Rutin	Decrease MPO activity, inhibit the IL-17, and NO syntheses, increase IL-4, reduce the IL-1β, and suppress the proinflammatory cytokines	[290]
Hesperetin	Decrease the protein expression of p-JAK2 and P-STAT3, reduce the MPO, MDA, and pro-inflammatory levels Increase the protein expression of SOCS3	[102]
Genistein	Facilitate the M1 macrophage to M2, reduced IL-6, TNFα, MCP-1, IL1β	[291]
Anthocyanin	Improve gut microbiota	[292]
Farrerol	Reduce the expression of TNF-α, IL-6, and IL-1β by suppressing the phosphorylation of NF-κB	[103]
Icariin	Suppress p-STAT1, p-STAT3, and p-p65 expressions, inhibit the IL-6, TNF-α	[293]
Quercetin	Inhibit the cytokine production, induce NOS through suppression of NF-κB signalling pathway, regulate MPO activity, MDA level, increase GSH content, improve intestinal oxidative stress, decrease TNF-α, IFN-γ, IL-17A, IL-6, COX-2, iNOS, IL-1β levels Activate Nrf2, upregulate the expression of Interleukin-10 and reduce gut inflammation	[294]
Myricetin and Kaempferol	Decrease pro-inflammatory factors such as IL-6, IL-1b, TNF- α, and iNOS Upregulate the expression of trefoil factor family3 (TFF3) gene, TGF-1β, and IL-10	[92]
Resveratrol	Decrease IL-1β, TNF-α, IL-1β, IL-6, MPO activity, SphK1 activity. Downregulate NLRP-3, Wnt/β-catenin pathway, SUMO1	[272]
Apigenin	Induce protective cytokines and enzyme, reduce Cox-2, MMP-3, iNOS, TNF-α and IL-1b via inflammasome pathway, decrease MPO activity	[295]
Luteolin	Suppress the JAK/STAT pathway, the activation of the Nrf2 signalling pathway, and increase the intestinal flora (Lactobacillus, Prevotella-9)	[296]
Terpenoid—Ginsenoside	Inhibit macrophage activity, modulate Th1/Treg cell differentiation, reduce the NLRP3 inflammasome via the AMPK pathway and JNK activation, decrease IL-1β, IL-6, TNF-α, inhibit LPS and TLR4, and recover the balance of Th17, Tregs. Activate the TGF-β signalling pathway, suppress the MAPK and NF-κB signalling pathway	[116]
Limonin	Reduce the disease activity index (DAI), downregulate the p-STAT3/miR-214	[121]
Geraniol	Decrease TNF- α, IL-1β, IL-6, MPO activity, iNOS and COX-2, inhibited NF-κB (p65)-DNA binding	[297]
Ursolic acid	Suppress the JNK/JAK/STAT signalling pathway, upregulate T-SOD and CAT	[298]
Glycyrrhizic acid	Reduce IL-6 production, decrease the expression level IL-1β, IL-α, increase IL-10	[299]
Polysaccharides- Astragalus polysaccharide	Regulate TNF-α, IL-1β, and NFATc4 expressions, facilitate expression of GATA-3 and T-bet, balance, reduce the NF-κB phosphorylation and downregulate IL-6, IL-1β, and TNF-α	[300]
Laminarin	Suppress the proliferation of the pathogenic bacteria, decrease ICAM-1, IL-2	[301]
Alkaloids—Oxymatrine	Facilitate the Th1/Th2 cytokines balance, reduction of colonic IL-2 expression, increase IL-10 levels, inhibition of p65 NF-kB and increase 2-β-adrenergic receptor (2BAR), decrease TNF-α and IL-6, ICAM-1	[302]
Piperine	Reduce MPO and MDA activity, upregulate PXR, Cyp3aII, MDR1a, mRNA, decrease ICAM, iNOS, IL-1β, MCP-1, IL-6, IL-1β, TNF-α, increase SOD activity	[303]
Boldine	Reduce MPO and MDA activity, decrease TNF-α, IL-6, IL-17, N and F κB expression, and increase IκB-α expression	[84]
Skimmianine	Reduce the expression levels of LBP, TNF-α, NFκB, TLR4, PGE2	[304]
Aloperine	Upregulate the Foxp3 levels, inhibit the colonic p-PI3K p85, p-mTOR, and p-Akt expressions, upregulate the protein phosphatase 2A (PP2A) expression and suppress the PI3K/Akt/mTOR signal transduction pathway	[92]
N-methylcytisine	Suppress pro-inflammatory factors via activates NF-κB	[305]

Table 8 (continued)

Class/phytochemical constituents	Pharmacological activity	References
Berberine	Suppress STAT3, NF-κB signalling, reduce IL-1, IL-1β, IL-5, IL-12, TNF-α, TGF-β, and interferon-γ, upregulate IL-4 and IL-10 expressions. Downregulate the MPO and MDA activity, decrease the IκB degradation	[306]
Theophylline	Decrease the cytokine levels (IL-6, IL-1b, and TNF-a) and MPO activity and activating the Nrf2/HO-1 signalling increase the IL-4, decrease AMCase and CHI3L1 expression	[307]
Cavidine	Reduce NF-κB, TNF-α, IL-6, MPO, and MDA activity, increase SOD activity	[308]
Pentoxifylline	Decrease the MDA levels, TGF-b1 MPO, MMP-1, and MMP-3 levels	[309]
Fatty acid—oleic acid	Decrease expression of p65, upregulate PPARγ	[310]
Linoleic acid	The histological score is improved by decreasing MPO and alkaline phosphatase activity, TNFα, IL-1β, and LC n-6PUFA	[311]
Vitamin E—α-tocopherol	Decrease MPO, MDA, and alkaline phosphatase activity; inhibit NO, PGE2, TNF-α, iNOS, COX-2	[312]
Iridoid—Secoiridoids-Oleuropein	Reduce MDA, MPO, and NO levels and elevate SOD, CAT, and GPX levels. Reduce the expression of Bax and increase the Bcl2 expression	[313]
Acids—Sinapic acid	Decrease MPO activity, facilitate SOD, GSH-Px, and catalase, and downregulate the NLRP3 inflammasome Increase the expression of ZO-1, occluding and Claudin-1	[314]

had overcome the challenges faced by pharmacokinetics and pharmacodynamics. These nanomedicines are available with unique characteristics such as a size range of ~ 100 nm, several coated delivery systems, which include transferases; liposomes; dendrimer; solid lipids; microspheres, and cellular carriers such as recombinant bacteria and macrophages [316].

4.2 Natural Products Encapsulated Drug Delivery System

Recently, alkaloid, phenolic acids, and flavonoid encapsulated nanoparticles (NPs) exhibited various immune-modulatory and anti-inflammatory effects by altering the IL-1β, TNF-α, and NF-κB signalling. Curcumin, thymoquinone quercetin, piceatannol, andrographolide, berberine, silymarin, embelin, and resveratrol are natural alkaloids; flavonoid, polyphenolic, and their acids from plants have shown various clinical pharmacological effects via reducing the production of NOD, caspase-1, NLRP3, MPO activity, IL-8, IL-6, MDA activity, Nrf2, HIF-αIL-1b, and TNF-a matrix metalloproteinase—2/3/9/13 Nrf2-, NF-κB, haeme oxygenase-1 (HO-1), and STAT3-dependent signalling pathways [259, 261, 317]. β-Lactoglobulin-nanosphere, silk and pectin NPs, gold and silver NPs, chitosan-based hydrogel nanosystem, albumin-coated polymeric, liposomes solid lipid exosome, polymeric (PVP and CMC) NPs, and polyactidecoglycolide [318] were designed for prolonged and controlled release. In addition, an amphiphilic curcumin-based polymer is reported for IBS treatment.

Sometimes single natural products are not so practical for treatment. Thus, a phyto-derived nanocarrier (size ~ 230 nm) [318] was used, including nano-size particles extracted from medicinal herbs or food. They have colossal biocompatibility and a stronghold in large industries. It is suggested that these phyto-derived nanocarriers, such as grape, ginger, blueberry, shitake mushroom exosomes, can transport across the cell membrane, can modulate intestinal tissue renewal processes, expression of inflammatory genes, accumulate in the inflamed colon, and exert potential therapeutic [319] Thus, these delivery systems can protect against liver toxicity and maintain the gut microbiota and pro-inflammatory balance.

4.3 Fermented Formulation and Dietary Supplements Used for Inflammatory Bowel Syndrome (IBS) and Related Disorders

It is noteworthy to mention that the metabolism of gut microbiota or change in the intestinal metabolic state by some of the IBS conventional medicine has been reported. These conventional medications were reported to enhance carbohydrate metabolism and citrate cycle, which leads to bacterial sensitisation to oxidative stress, and decrease bacterial colonisation and polyphosphate level [320]. Thus, fermented formulations were used to overcome these side effects. FDA and WHO defined probiotics as live microorganisms which, in adequate amounts, can be beneficial to humankind. In IBS, the primary cause of inflammation is an unbalanced gut microbiota environment which facilitates the pathogenic pathways and restricts the movement of the bowels. Therefore, increasing the beneficial bacteria subsequently decrease the inflammation associated with IBS and other

Table 9 Various fermented and dietary formulations for the management of IBS

Fermented products	Probiotic strain	Activity	References
Lactic acid bacterium	***Pediococcus acidilactici***	Induce IL-10	[322]
Dietary *Lactobacillus*	***L. plantarum***	Increase the total *Bifidobacteria* and *Lactobacilli,* decreased *Enterococci* and *C. perfringens*	[323]
Dietary probioticbacteria	***Bifidobacteria*** and ***Lactobacillus***, ***L. plantarum*** (Lp91)	Down-regulate TNF-α and COX-2 Up-regulate IL-10	[324]
Bifidobacterium strains	***B. bifidum*** Bif1, Bif2, Bif3; ***Bifidobacterium longum*** Lon4, Lon5, Lon6; ***Bifidobacterium catenulatum*** Cat7, Cat8, Cat9; ***B. breve*** Bre10, Bre11, and ***Bifidobacterium adolescentis*** Ado12	Inhibit TNF-α, IL-8 production	[325]
Fed *Saccharomyces boulardii*	***S. boulardii***	Inhibit NF-κB signalling, inhibition of CD4+ T-cell infiltration, reduce production of IFN-γ	[326]
Bifidobacterium breve strain Yakult and *Bifidobacterium bifidum* strain Yakult	***B. breve, B. bifidum***	Induce secretion of IL-10, inhibit TNF-α-induced secretion of IL-8	[327]
Administration of VSL#3	Marketed product	Decrease mucosal secretion of TNF-α, IFN-γ, prevent the epithelium from bacterial invasion	[328]
Dietary probiotic yogurt	***L. rhamnosus*** and ***L. reuteri***	Decrease IL-12, TNF-α, IL-12	[329]
Dietary probiotics powder	***Bifidobacterium bifidum***	Inhibit IFN-γ, MCP-1, increase IL-10	[330]
Administered *Lactobacillus* spp.	***Lactobacillus paracasei, Lactobacillus reuteri***	Decrease TNF-α, reduced mucosal IL-12 mRNA	[324]
Administered probiotics	***Lactobacillus casei, Lactobacillus acidophilus*** and ***Bifidobacterium lactis***	Reduce diarrhoea in *B lactis*, decrease TNF-α Production, iNOS, COX-2 expression	[324]
Administration of *Lactobacillus plantarum* 299v (*L. plantarum*)	***L. pantarum***	Decrease mucosal IL-12, IFN-γ	[324]
Dietary fibre	Soluble corn fibre, STA-LITE III polydextrose, Bio gum, Pullulan, RROMITOR-resistant starch-75, inulin, orange pulp, guar gum, a mixture of orange pulp and guar gum, germinated barley foodstuff, fermentable dextrin fibre, germinated barley foodstuff	RROMITOR-resistant starch-75 and inulin decrease IFN-γ production; STA-LITE III polydextrose, bio gum, pullulan, and RROMITOR resistant, starch-75 up-regulate colonic PPAR-γ expression; a mixture of orange pulp and guar gum increases PGE2 level	[301]
Dietary pectin	Pectin	Increase IFN-γ and IL-2, reduce TNF-α, GATA-3, IgG, and IgM	[331]

related diseases. Furthermore, these probiotics decrease visceral hypersensitivity and low-grade inflammation and increase pro-inflammatory cytokines and toll-like receptors to support the immunoregulatory effect. Pre-biotics are a substance utilised by probiotics, and postbiotics are non-living microorganisms with or without their cell components that directly give health benefits. Concerning this, the eight-strain combinations of *Lactobacillus plantarum* and *Lactobacillus paracasei* subsp. *paracasei*, *Lactobacillus acidophilus*, *Lactobacillus delbrueckii* subsp *bulgaricus, Bifidobacterium breve, Bifidobacterium longum* subsp. *longum, Bifdobacteriuminfantis*, *Bifidobacteriumlongum* subsp., and *Streptococcus salivarius* subsp. *Streptococcus thermophilus* are recommended and listed in Table 9.; Apart from these, prebiotics such as inulin, starch gum, dietary fibre, pectin, and poly oligosaccharides [318] are reported to have a potential effect in the treatment of IBS. This carbohydrate-rich diet includes long and short-chain carbohydrates (FOS and GOS) such as cellulose, hemicellulose, β-glucan, and fructans showing anti-inflammatory and immunoregulatory activities. Additionally, facilitate myenteric and colonic motility, improving the total gut movement by providing the gelling of the intestinal tract. Primary sources such as bananas, beans, onion, raw leeks, asparagus, gum acacia, barleys, husks, seaweed, oats, dandelion, wheat, garlic, and fruit vegetables rich in high fibre and carbohydrates are considered good sources of prebiotic fibres and healthy diet [321].

The incorporation of bioactive compounds via nanotechnology provides a way of developing clinical candidates for health wellness. There is an existing gap in the literature for the toxicity risk of these bioactive molecules in long term.

5 Future Prospect

In response to the rising prevalence of IBS in Westernised countries, steadily increasing costs of IBS treatments, and the ineffectiveness of current IBS drug therapies, a substantial amount of research has been done to develop alternative therapies based on natural substances that are both highly effective and safe. A focus on recognising and addressing the dietary and environmental risk factors for IBS will be a priority in the future. In the search for natural alternative treatments for IBS, the introduction of unique cell models that can imitate the GI tract is thought to be a forthcoming model of research. It is also essential to have a thorough understanding of the pathogenic mechanisms underlying IBS which will enable the researchers to develop effective treatments for this ailment employing herbal medicines.

Clinical studies indicate that herbal medicines have a promising future in IBS therapy. A few issues must first be rectified before administering the herbal medicine safely to the UC patients. There are still not enough large case-controlled studies and reliable data on the specific mechanism of use of the herbs in in vitro and in vivo studies, and only a small number of UC patients have so far taken part in clinical trials using herbal medicine treatments. This could hasten the future research and development of herbal treatments for IBS. Additionally, since herbal formulations are a mixture of a wide variety of biological compounds, it is vital to know which component in the herbs provides the precise pharmacological effects, even in certain situations where the herb mixtures exhibit clinical effects. Thus, the determination of components of herb, dosage, and course of herb treatment plays a significant role in their clinical employment. Furthermore, herbal therapy generally could carry risks and produce side effects such as liver and renal failure like other types of alternative therapy. Lead, arsenic, and mercury are just a few of the toxic ingredients found in many herb remedies. Therefore, there is a need for extensive research on the safety of herb medicines, particularly on long-term use.

6 Conclusion

From ancient times, these natural products, herbal extracts, and fermented formulations have been used to prevent and treat IBS and constipation. We now have a deeper insight into the genetics, mucosal changes, environmental triggers, and aberrant signalling that are the fundamental causes of IBS, constipation, and associated disorders. Novel therapies are being developed, yet approval of these new treatments must garner caution due to the unique and serious side effects that immunomodulator therapy entails. So, these diverse herbal formulations with complex functionality can be considered to overcome the drawbacks of conventional drugs. In addition, encapsulated bioactive molecules combination therapy is an emerging strategy for IBS treatment to synergise the protective and therapeutic potential of IBS through various signalling pathways (cellular and molecular); improving the gut microbiota and can decrease the burden on the healthcare system and improving the quality of life. Their safe nutritional amount is required for clinical trials with larger sample sizes and periodic follow-ups for future medical benefits.

Conflicts of Interest
The authors declare that there are no conflicts of interest to disclose.

References

1. Kirsner JB. Historical origins of current IBD concepts. World J Gastroenterol. 2001;7(2):175–84.
2. Hendrickson BA, Gokhale R, Cho JH. Clinical aspects and pathophysiology of inflammatory bowel disease. Clin Microbiol Rev. 2002;15(1):79–94.
3. Puppala ER, Yalamarthi SS, Aochenlar SL, Prasad N, Syamprasad NP, Singh M, et al. *Mesua assamica* (King&Prain) kosterm. Bark ethanolic extract attenuates chronic restraint stress aggravated DSS-induced ulcerative colitis in mice via inhibition of NF-κB/STAT3 and activation of HO-1/Nrf2/SIRT1 signaling pathways. J Ethnopharmacol. 2023;301:115765.
4. Ungaro R, Mehandru S, Allen PB, Peyrin-Biroulet L, Colombel JF. Ulcerative colitis. Lancet. 2017;389(10080):1756–70.
5. Safar B, Sands D. Perianal Crohn's disease. Clin Colon Rectal Surg. 2007;20(4):282–93.
6. Laukoetter MG, Nava P, Nusrat A. Role of the intestinal barrier in inflammatory bowel disease. World J Gastroenterol. 2008;14(3):401–7.
7. Lee M, Chang EB. Inflammatory bowel diseases (IBD) and the microbiome-searching the crime scene for clues. Gastroenterology. 2021;160(2):524–37.
8. Silva FAR, Rodrigues BL, Ayrizono MLS, Leal RF. The immunological basis of inflammatory bowel disease. Gastroenterol Res Pract. 2016;2016:2097274.
9. Limdi JK. Dietary practices and inflammatory bowel disease. Indian J Gastroenterol. 2018;37(4):284–92.
10. Kedia S, Ahuja V. Epidemiology of inflammatory bowel disease in india: the great shift east. Inflamm Intest Dis. 2017;2(2):102–15.
11. Park SH. Update on the epidemiology of inflammatory bowel disease in Asia: where are we now? Intest Res. 2022;20(2):159–64.
12. Srinath AI, Walter C, Newara MC, Szigethy EM. Pain management in patients with inflammatory bowel disease: insights for the clinician. Ther Adv Gastroenterol. 2012;5(5):339–57.
13. Cai Z, Wang S, Li J. Treatment of inflammatory bowel disease: a comprehensive review. Front Med. 2021;8:765474.
14. Patange G, Giri S, Kharat J. Crohn's disease: a review based on ayurveda and modern perspective. World J Pharm Res. 2019;8(10):1638–45.
15. Kalyankumarraju M, Puppala ER, Ahmed S, Jagadeesh Kumar G, Tene K, Syamprasad NP, et al. Zanthoxylum alatum Roxb. seed extract ameliorates stress aggravated DSS-induced ulcerative colitis in mice: plausible role on NF-κB signaling axis. J Ethnopharmacol. 2021;279:114385.
16. Garodia P, Ichikawa H, Malani N, Sethi G, Aggarwal BB. From ancient medicine to modern medicine: ayurvedic concepts of health and their role in inflammation and cancer. J Soc Integr Oncol. 2007;5(1):25–37.

17. Langhorst J, Schöls M, Cinar Z, Eilert R, Kofink K, Paul A, et al. Comprehensive lifestyle-modification in patients with ulcerative colitis—a randomized controlled trial. J Clin Med. 2020;9(10):3087.
18. Zheng D, Liwinski T, Elinav E. Interaction between microbiota and immunity in health and disease. Cell Res. 2020;30(6):492–506. https://doi.org/10.1038/s41422-020-0332-7.
19. Thursby E, Juge N. Introduction to the human gut microbiota. Biochem J. 2017;474(11):1823–36.
20. Pickard JM, Zeng MY, Caruso R, Núñez G. Gut microbiota: role in pathogen colonization, immune responses, and inflammatory disease. Immunol Rev. 2017;279(1):70–89.
21. Rodríguez JM, Murphy K, Stanton C, Ross RP, Kober OI, Juge N, et al. The composition of the gut microbiota throughout life, with an emphasis on early life. Microb Ecol Health Dis. 2015;26:26050.
22. Rinninella E, Raoul P, Cintoni M, Franceschi F, Miggiano GAD, Gasbarrini A, et al. What is the healthy gut microbiota composition? A changing ecosystem across age, environment, diet, and diseases. Microorganisms. 2019;7(1):14.
23. Mukhopadhya I, Segal JP, Carding SR, Hart AL, Hold GL. The gut virome: the "missing link" between gut bacteria and host immunity? Ther Adv Gastroenterol. 2019;12:1756284819836620.
24. Jandhyala SM, Talukdar R, Subramanyam C, Vuyyuru H, Sasikala M, Nageshwar Reddy D. Role of the normal gut microbiota. World J Gastroenterol. 2015;21(29):8787–803.
25. Rowland I, Gibson G, Heinken A, Scott K, Swann J, Thiele I, et al. Gut microbiota functions: metabolism of nutrients and other food components. Eur J Nutr. 2018;57(1):1–24.
26. Nogal A, Valdes AM, Menni C. The role of short-chain fatty acids in the interplay between gut microbiota and diet in cardio-metabolic health. Gut Microbes. 2021;13(1):1–24. https://doi.org/10.1080/19490976.2021.1897212.
27. Salvi PS, Cowles RA. Butyrate and the intestinal epithelium: modulation of proliferation and inflammation in homeostasis and disease. Cells. 2021;10(7):1775.
28. Hossain KS, Amarasena S, Mayengbam S. B vitamins and their roles in gut health. Microorganisms. 2022;10(6):1168.
29. Wu H-J, Wu E. The role of gut microbiota in immune homeostasis and autoimmunity. Gut Microbes. 2012;3(1):4–14.
30. Amarante-Mendes GP, Adjemian S, Branco LM, Zanetti LC, Weinlich R, Bortoluci KR. Pattern recognition receptors and the host cell death molecular machinery. Front Immunol. 2018;9:2379.
31. Alshehri D, Saadah O, Mosli M, Edris S, Alhindi R, Bahieldin A. Dysbiosis of gut microbiota in inflammatory bowel disease: current therapies and potential for microbiota-modulating therapeutic approaches. Bosn J Basic Med Sci. 2021;21(3):270–83.
32. Santana PT, Rosas SLB, Ribeiro BE, Marinho Y, de Souza HSP. Dysbiosis in inflammatory bowel disease: pathogenic role and potential therapeutic targets. Int J Mol Sci. 2022;23(7):3464.
33. Franzin M, Stefančič K, Lucafò M, Decorti G, Stocco G. Microbiota and drug response in inflammatory bowel disease. Pathogens. 2021;10(2):211.
34. Britton GJ, Contijoch EJ, Mogno I, Vennaro OH, Llewellyn SR, Ng R, et al. Microbiotas from humans with inflammatory bowel disease alter the balance of gut Th17 and RORγt(+) regulatory T cells and exacerbate colitis in mice. Immunity. 2019;50(1):212–224.e4.
35. Zeng MY, Inohara N, Nuñez G. Mechanisms of inflammation-driven bacterial dysbiosis in the gut. Mucosal Immunol. 2017;10(1):18–26.
36. Gillespie W, Marya N, Fahed J, Leslie G, Patel K, Cave DR. Clostridium difficile in inflammatory bowel disease: a retrospective study. Gastroenterol Res Pract. 2017;2017:4803262.
37. Mukherjee A, Lordan C, Ross RP, Cotter PD. Gut microbes from the phylogenetically diverse genus Eubacterium and their various contributions to gut health. Gut Microbes. 2020;12(1):1802866.
38. Miquel S, Martín R, Rossi O, Bermúdez-Humarán LG, Chatel JM, Sokol H, et al. Faecalibacterium prausnitzii and human intestinal health. Curr Opin Microbiol. 2013;16(3):255–61.
39. Miranda-Bautista J, Padilla-Suárez C, Bouza E, Muñoz P, Menchén L, Marín-Jiménez I. Listeria monocytogenes infection in inflammatory bowel disease patients: case series and review of the literature. Eur J Gastroenterol Hepatol. 2014;26(11):1247–52.
40. Zhu C, Song K, Shen Z, Quan Y, Tan B, Luo W, et al. Roseburia intestinalis inhibits interleukin-17 excretion and promotes regulatory T cells differentiation in colitis. Mol Med Rep. 2018;17(6):7567–74.
41. Kang S, Denman SE, Morrison M, Yu Z, Dore J, Leclerc M, et al. Dysbiosis of fecal microbiota in Crohn's disease patients as revealed by a custom phylogenetic microarray. Inflamm Bowel Dis. 2010;16(12):2034–42.
42. Gevers D, Kugathasan S, Denson LA, Vázquez-Baeza Y, Van Treuren W, Ren B, et al. The treatment-naive microbiome in new-onset Crohn's disease. Cell Host Microbe. 2014;15(3):382–92.
43. Zafar H, Saier MHJ. Gut bacteroides species in health and disease. Gut Microbes. 2021;13(1):1–20.
44. Zhang T, Ji X, Lu G, Zhang F. The potential of Akkermansia muciniphila in inflammatory bowel disease. Appl Microbiol Biotechnol. 2021;105(14–15):5785–94.
45. Turroni F, Duranti S, Milani C, Lugli GA, van Sinderen D. Bifidobacterium bifidum: a key member of the early human gut microbiota. Microorganisms. 2019;7:544.
46. Clooney AG, Eckenberger J, Laserna-Mendieta E, Sexton KA, Bernstein MT, Vagianos K, et al. Ranking microbiome variance in inflammatory bowel disease: a large longitudinal intercontinental study. Gut. 2021;70(3):499–510.
47. Elguezabal N, Chamorro S, Molina E, Garrido JM, Izeta A, Rodrigo L, et al. Lactase persistence, NOD2 status and Mycobacterium avium subsp. paratuberculosis infection associations to inflammatory bowel disease. Gut Pathog. 2012;4:6.
48. Mottawea W, Chiang C-K, Mühlbauer M, Starr AE, Butcher J, Abujamel T, et al. Altered intestinal microbiota-host mitochondria crosstalk in new onset Crohn's disease. Nat Commun. 2016;7:13419.
49. Brennan CA, Garrett WS. Gut microbiota, inflammation, and colorectal cancer. Annu Rev Microbiol. 2016;70:395–411.
50. Palmela C, Chevarin C, Xu Z, Torres J, Sevrin G, Hirten R, et al. Adherent-invasive Escherichia coli in inflammatory bowel disease. Gut. 2018;67(3):574–87.
51. Chandra H, Sharma KK, Tuovinen OH, Sun X, Shukla P. Pathobionts: mechanisms of survival, expansion, and interaction with host with a focus on Clostridioides difficile. Gut Microbes. 2021;13(1):1979882.
52. Zhu D, Sorg JA, Sun X. Clostridioides difficile biology: sporulation, germination, and corresponding therapies for C. difficile infection. Front Cell Infect Microbiol. 2018;8:29.
53. Mirsepasi-Lauridsen HC, Vallance BA, Krogfelt KA, Petersen AM. Escherichia coli pathobionts associated with inflammatory bowel disease. Clin Microbiol Rev. 2019;32(2):e00060-18.
54. Buttó LF, Schaubeck M, Haller D. Mechanisms of microbe-host interaction in Crohn's disease: dysbiosis vs. pathobiont selection. Front Immunol. 2015;6:555.
55. Sugihara K, Kamada N. Diet-microbiota interactions in inflammatory bowel disease. Nutrients. 2021;13(5):1533.
56. Chiba M, Nakane K, Komatsu M. Westernized diet is the most ubiquitous environmental factor in inflammatory bowel disease. Perm J. 2019;23:18–107.
57. Paul B, Barnes S, Demark-Wahnefried W, Morrow C, Salvador C, Skibola C, et al. Influences of diet and the gut microbiome on epigenetic modulation in cancer and other diseases. Clin Epigenetics. 2015;7:112.

58. Na SY, Moon W. Perspectives on current and novel treatments for inflammatory bowel disease. Gut Liver. 2019;13(6):604–16.
59. Debnath T, Kim DH, Lim BO. Natural products as a source of anti-inflammatory agents associated with inflammatory bowel disease. Molecules. 2013;18(6):7253–70.
60. Talley NJ, Abreu MT, Achkar JP, Bernstein CN, Dubinsky MC, Hanauer SB, et al. An evidence-based systematic review on medical therapies for inflammatory bowel disease. Am J Gastroenterol. 2011;106(Suppl 1):S2–31.
61. Lim WC, Wang Y, Macdonald JK, Hanauer S. Aminosalicylates for induction of remission or response in Crohn's disease. Cochrane Database Syst Rev. 2016;7(7):CD008870.
62. Li X, Lu C, Yang Y, Yu C, Rao Y. Site-specific targeted drug delivery systems for the treatment of inflammatory bowel disease. Biomed Pharmacother. 2020;129:110486.
63. Hammad MA, Sulaiman SAS, Aziz NA, Noor DAM. Prescribing statins among patients with type 2 diabetes: the clinical gap between the guidelines and practice. J Res Med Sci. 2019;24(1):15.
64. Hilsden RJ, Verhoef MJ, Rasmussen H, Porcino A, Debruyn JCC. Use of complementary and alternative medicine by patients with inflammatory bowel disease. Inflamm Bowel Dis. 2011;17(2):655–62.
65. Hung A, Kang N, Bollom A, Wolf JL, Lembo A. Complementary and alternative medicine use is prevalent among patients with gastrointestinal diseases. Dig Dis Sci. 2015;60(7):1883–8.
66. Pandurangan AK, Ismail S, Saadatdoust Z, Esa NM. Allicin alleviates dextran sodium sulfate- (DSS-)induced ulcerative colitis in BALB/c mice. Oxid Med Cell Longev. 2015;2015:605208.
67. El-Ashmawy NE, Khedr NF, El-Bahrawy HA, El-Adawy SA. Downregulation of iNOS and elevation of cAMP mediate the anti-inflammatory effect of glabridin in rats with ulcerative colitis. Inflammopharmacology. 2018;26(2):551–9.
68. Rezayat SM, Dehpour AR, Motamed SM, Yazdanparast M, Chamanara M, Sahebgharani M, et al. Foeniculum vulgare essential oil ameliorates acetic acid-induced colitis in rats through the inhibition of NF-kB pathway. Inflammopharmacology. 2018;26(3):851–9.
69. Henriques DP, de Oliveira RR, Vanni J, de Lima HP, Otiti JV, Neves FRM, et al. Complementary and alternative medicine use in Brazilian patients with inflammatory bowel disease. Arq Gastroenterol. 2022;59(3):375–82.
70. Wang XJ, Deng HZ, Jiang B, Yao H. The natural plant product sophocarpine ameliorates dextran sodium sulfate-induced colitis in mice by regulating cytokine balance. Int J Colorectal Dis. 2012;27(5):575–81.
71. Peng J, Zheng TT, Li X, Liang Y, Wang LJ, Huang YC, et al. Plant-derived alkaloids: the promising disease-modifying agents for inflammatory bowel disease. Front Pharmacol. 2019;10:1–15.
72. Cheng H, Xia B, Zhang L, Zhou F, Zhang YX, Ye M, et al. Matrine improves 2,4,6-trinitrobenzene sulfonic acid-induced colitis in mice. Pharmacol Res. 2006;53(3):202–8.
73. Xia EQ, Cui B, Xu XR, Song Y, Ai XX, Bin LH. Microwave-assisted extraction of oxymatrine from sophora flavescens. Molecules. 2011;16(9):7391–400.
74. Yao H, Shi Y, Yuan J, Sa R, Chen W, Wan X. Matrine protects against DSS-induced murine colitis by improving gut barrier integrity, inhibiting the PPAR-α signaling pathway, and modulating gut microbiota. Int Immunopharmacol. 2021;100:108091.
75. Fu X, Sun F, Wang F, Zhang J, Zheng B, Zhong J, et al. Aloperine protects mice against DSS-induced colitis by PP2A-mediated PI3K/Akt/mTOR signaling suppression. Mediators Inflamm. 2017;2017:5706152.
76. Tahir M, Ali S, Zhang W, Lv B, Qiu W, Wang J. Aloperine: a potent modulator of crucial biological mechanisms in multiple diseases. Biomedicines. 2022;10(4):1–26.
77. Jia X, Zhang H, Jiang X, Lu Y, Liu W, Yu J. Profiling and quantitation of alkaloids in different parts of Sophora alopecuroides L. extracts by high-performance liquid chromatography with electrospray ionisation ion mobility spectrometry detection. Phytochem Anal. 2021;32(6):1003–10.
78. Chen Q, Duan X, Fan H, Xu M, Tang Q, Zhang L, et al. Oxymatrine protects against DSS-induced colitis via inhibiting the PI3K/AKT signaling pathway. Int Immunopharmacol. 2017;53:149–57.
79. Zhang LC, Wang Y, Tong LC, Sun S, Liu WY, Zhang S, et al. Berberine alleviates dextran sodium sulfate-induced colitis by improving intestinal barrier function and reducing inflammation and oxidative stress. Exp Ther Med. 2017;13(6):3374–82.
80. Li C, Xi Y, Li S, Zhao Q, Cheng W, Wang Z, et al. Berberine ameliorates TNBS induced colitis by inhibiting inflammatory responses and Th1/Th17 differentiation. Mol Immunol. 2015;67(2):444–54.
81. Liu Y, Liu X, Hua W, Wei Q, Fang X, Zhao Z, et al. Berberine inhibits macrophage M1 polarization via AKT1/SOCS1/NF-κB signaling pathway to protect against DSS-induced colitis. Int Immunopharmacol. 2018;57:121–31.
82. Li YH, Xiao HT, Hu DD, Fatima S, Lin CY, Mu HX, et al. Berberine ameliorates chronic relapsing dextran sulfate sodium-induced colitis in C57BL/6 mice by suppressing Th17 responses. Pharmacol Res. 2016;110:227–39.
83. Luo Y, Liu M, Xia Y, Dai Y, Chou G, Wang Z. Therapeutic effect of norisoboldine, an alkaloid isolated from Radix Linderae, on collagen-induced arthritis in mice. Phytomedicine. 2010;17(10):726–31.
84. Pandurangan AK, Mohebali N, Hasanpourghadi M, Looi CY, Mustafa MR, Mohd EN. Boldine suppresses dextran sulfate sodium-induced mouse experimental colitis: NF-κB and IL-6/STAT3 as potential targets. BioFactors. 2016;42(3):247–58.
85. Zhou J, Qu F. Analysis of the extracts of Isatis tinctoria by new analytical approaches of HPLC, MS and NMR. Afr J Tradit Complement Altern Med. 2011;8(5 Suppl):33–45.
86. Rabelo Socca EA, Luiz-Ferreira A, De Faria FM, De Almeida AC, Dunder RJ, Manzo LP, et al. Inhibition of tumor necrosis factor-alpha and cyclooxigenase-2 by isatin: a molecular mechanism of protection against TNBS-induced colitis in rats. Chem Biol Interact. 2014;209(1):48–55.
87. Shin DW, Lim BO. Nutritional interventions using functional foods and nutraceuticals to improve inflammatory bowel disease. J Med Food. 2020;23(11):1136–45.
88. Lee S, Kim DC, Baek HY, Lee KD, Kim YC, Oh H. Anti-neuroinflammatory effects of tryptanthrin from Polygonum tinctorium Lour. in lipopolysaccharide-stimulated BV2 microglial cells. Arch Pharm Res. 2018;41(4):419–30.
89. Micallef MJ, Iwaki K, Ishihara T, Ushio S, Aga M, Kunikata T, et al. The natural plant product tryptanthrin ameliorates dextran sodium sulfate-induced colitis in mice. Int Immunopharmacol. 2002;2(4):565–78.
90. Wang Z, Wu X, Wang CL, Wang L, Sun C, Zhang DB, et al. Tryptanthrin protects mice against dextran sulfate sodium-induced colitis through inhibition of TNF-α/NF-κB and IL-6/STAT3 pathways. Molecules. 2018;23(5):1–13.
91. Dou W, Zhang J, Sun A, Zhang E, Ding L, Mukherjee S, et al. Protective effect of naringenin against experimental colitis via suppression of Toll-like receptor 4/NF-κB signalling. Br J Nutr. 2013;110(4):599–608.
92. Duan L, Cheng S, Li L, Liu Y, Wang D, Liu G. Natural anti-inflammatory compounds as drug candidates for inflammatory bowel disease. Front Pharmacol. 2021;12:1–22.
93. Al-Rejaie SS, Abuohashish HM, Al-Enazi MM, Al-Assaf AH, Parmar MY, Ahmed MM. Protective effect of naringenin on acetic acid-induced ulcerative colitis in rats. World J Gastroenterol. 2013;19(34):5633–44.

94. Ruiz Castro PA, Kogel U, Lo Sasso G, Phillips BW, Sewer A, Titz B, et al. Anatabine ameliorates intestinal inflammation and reduces the production of pro-inflammatory factors in a dextran sulfate sodium mouse model of colitis. J Inflamm (Lond). 2020;17(1):1–17.
95. AlSharari SD, Akbarali HI, Abdullah RA, Shahab O, Auttachoat W, Ferreira GA, et al. Novel insights on the effect of nicotine in a murine colitis model. J Pharmacol Exp Ther. 2013;344(1):207–17.
96. Hayashi S, Hamada T, Zaidi SF, Oshiro M, Lee J, Yamamoto T, et al. Nicotine suppresses acute colitis and colonic tumorigenesis associated with chronic colitis in mice. Am J Physiol Gastrointest Liver Physiol. 2014;307(10):G968–78.
97. Qin Z, Wan JJ, Sun Y, Wu T, Wang PY, Du P, et al. Nicotine protects against DSS colitis through regulating microRNA-124 and STAT3. J Mol Med. 2017;95(2):221–33.
98. Maruta K, Watanabe C, Hozumi H, Kurihara C, Furuhashi H, Takajo T, et al. Nicotine treatment ameliorates DSS-induced colitis by suppressing MAdCAM-1 expression and leukocyte recruitment. J Leukoc Biol. 2018;104(5):1013–22.
99. Ma H, He X, Yang Y, Li M, Hao D, Jia Z. The genus Epimedium: an ethnopharmacological and phytochemical review. J Ethnopharmacol. 2011;134(3):519–41.
100. Wang Z, Wang D, Yang D, Zhen W, Zhang J, Peng S. The effect of icariin on bone metabolism and its potential clinical application. Osteoporos Int. 2018;29(3):535–44.
101. He C, Wang Z, Shi J. Pharmacological effects of icariin. In: Advances in pharmacology, vol. 87. 1st ed. Amsterdam: Elsevier; 2020. p. 179–203.
102. Elhennawy MG, Abdelaleem EA, Zaki AA, Mohamed WR. Cinnamaldehyde and hesperetin attenuate TNBS-induced ulcerative colitis in rats through modulation of the JAk2/STAT3/SOCS3 pathway. J Biochem Mol Toxicol. 2021;35(5):1–11.
103. Ran X, Li Y, Chen G, Fu S, He D, Huang B, et al. Farrerol ameliorates TNBS-induced colonic inflammation by inhibiting ERK1/2, JNK1/2, and NF-κB signaling pathway. Int J Mol Sci. 2018;19(7):1–14.
104. Dodda D, Chhajed R, Mishra J, Padhy M. Targeting oxidative stress attenuates trinitrobenzene sulphonic acid induced inflammatory bowel disease like symptoms in rats: role of quercetin. Indian J Pharmacol. 2014;46(3):286–91.
105. Dodda D, Chhajed R, Mishra J. Protective effect of quercetin against acetic acid induced inflammatory bowel disease (IBD) like symptoms in rats: possible morphological and biochemical alterations. Pharmacol Rep. 2014;66(1):169–73.
106. Jones JR, Lebar MD, Jinwal UK, Abisambra JF, Koren J, Blair L, et al. The diarylheptanoid (+)-aR,11S-myricanol and two flavones from bayberry (Myrica cerifera) destabilize the microtubule-associated protein tau. J Nat Prod. 2011;74(1):38–44.
107. Imran M, Salehi B, Sharifi-Rad J, Gondal TA, Saeed F, Imran A, et al. Kaempferol: a key emphasis to its anticancer potential. Molecules. 2019;24(12):1–16.
108. Semwal DK, Semwal RB, Combrinck S, Viljoen A. Myricetin: a dietary molecule with diverse biological activities. Nutrients. 2016;8(2):1–31.
109. Qu X, Li Q, Song Y, Xue A, Liu Y, Qi D, et al. Potential of myricetin to restore the immune balance in dextran sulfate sodium-induced acute murine ulcerative colitis. J Pharm Pharmacol. 2020;72(1):92–100.
110. Qi LW, Wang CZ, Yuan CS. Ginsenosides from American ginseng: chemical and pharmacological diversity. Phytochemistry. 2011;72(8):689–99.
111. Yuan CS, Wei G, Dey L, Karrison T, Nahlik L, Maleckar S, et al. Brief communication: American ginseng reduces warfarin's effect in healthy patients. A randomized, controlled trial. Ann Intern Med. 2004;141(1):23–8.
112. Joh EH, Lee IA, Jung IH, Kim DH. Ginsenoside Rb1 and its metabolite compound K inhibit IRAK-1 activation—the key step of inflammation. Biochem Pharmacol. 2011;82(3):278–86.
113. Poudyal D, Le PM, Davis T, Hofseth AB, Chumanevich A, Chumanevich AA, et al. A hexane fraction of American ginseng suppresses mouse colitis and associated colon cancer: anti-inflammatory and proapoptotic mechanisms. Cancer Prev Res. 2012;5(4):685–96.
114. Dai D, Zhang CF, Williams S, Yuan CS, Wang CZ. Ginseng on cancer: potential role in modulating inflammation-mediated angiogenesis. Am J Chin Med. 2017;45(1):13–22.
115. Yang XL, Guo TK, Wang YH, Gao MT, Qin H, Wu YJ. Therapeutic effect of ginsenoside Rd in rats with TNBS-induced recurrent ulcerative colitis. Arch Pharm Res. 2012;35(7):1231–9.
116. Liu C, Wang J, Yang Y, Liu X, Zhu Y, Zou J, et al. Ginsenoside Rd ameliorates colitis by inducing p62-driven mitophagy-mediated NLRP3 inflammasome inactivation in mice. Biochem Pharmacol. 2018;155:366–79.
117. Yang XL, Guo TK, Wang YH, Huang YH, Liu X, Wang XX, et al. Ginsenoside Rd attenuates the inflammatory response via modulating p38 and JNK signaling pathways in rats with TNBS-induced relapsing colitis. Int Immunopharmacol. 2012;12(2):408–14.
118. Yang N, Liang G, Lin J, Zhang S, Lin Q, Ji X, et al. Ginsenoside Rd therapy improves histological and functional recovery in a rat model of inflammatory bowel disease. Phytother Res. 2020;34(11):3019–28.
119. Ye H, Wu Q, Zhu Y, Guo C, Zheng X. Ginsenoside Rh2 alleviates dextran sulfate sodium-induced colitis via augmenting TGFβ signaling. Mol Biol Rep. 2014;41(8):5485–90.
120. Tian M, Ma P, Zhang Y, Mi Y, Fan D. Ginsenoside Rk3 alleviated DSS-induced ulcerative colitis by protecting colon barrier and inhibiting NLRP3 inflammasome pathway. Int Immunopharmacol. 2020;85:106645.
121. Liu S, Zhang S, Lv X, Lu J, Ren C, Zeng Z, et al. Limonin ameliorates ulcerative colitis by regulating STAT3/miR-214 signaling pathway. Int Immunopharmacol. 2019;75:105768.
122. Song C, Chen J, Li X, Yang R, Cao X, Zhou L, et al. Limonin ameliorates dextran sulfate sodium-induced chronic colitis in mice by inhibiting PERK-ATF4-CHOP pathway of ER stress and NF-κB signaling. Int Immunopharmacol. 2021;90:107161.
123. Busbee PB, Menzel L, Alrafas HR, Dopkins N, Becker W, Miranda K, et al. Indole-3-carbinol prevents colitis and associated microbial dysbiosis in an IL-22-dependent manner. JCI Insight. 2020;5(1):1–18.
124. Huang Z, Jiang Y, Yang Y, Shao J, Sun X, Chen J, et al. 3,3′-Diindolylmethane alleviates oxazolone-induced colitis through Th2/Th17 suppression and Treg induction. Mol Immunol. 2013;53(4):335–44.
125. Lv Q, Qiao SM, Xia Y, Shi C, Xia YF, Chou GX, et al. Norisoboldine ameliorates DSS-induced ulcerative colitis in mice through induction of regulatory T cells in colons. Int Immunopharmacol. 2015;29(2):787–97.
126. Lv Q, Wang K, Qiao S, Yang L, Xin Y, Dai Y, et al. Norisoboldine, a natural AhR agonist, promotes Treg differentiation and attenuates colitis via targeting glycolysis and subsequent NAD+/SIRT1/SUV39H1/H3K9me3 signaling pathway article. Cell Death Dis. 2018;9(3):258.
127. Liang YN, Yu JG, Zhang DB, Zhang Z, Ren LL, Li LH, et al. Indigo naturalis ameliorates dextran sulfate sodium-induced colitis in mice by modulating the intestinal microbiota community. Molecules. 2019;24(22):4086.
128. Chang Y, Zhai L, Peng J, Wu H, Bian Z, Xiao H. Phytochemicals as regulators of Th17/Treg balance in inflammatory bowel diseases. Biomed Pharmacother. 2021;141:111931.
129. Xiao HT, Peng J, Hu DD, Lin CY, Du B, Tsang SW, et al. Qing-dai powder promotes recovery of colitis by inhibiting inflammatory

responses of colonic macrophages in dextran sulfate sodium-treated mice. Chin Med. 2015;10(1):1–13.
130. Yang L, Li X, Huang W, Rao X, Lai Y. Pharmacological properties of indirubin and its derivatives. Biomed Pharmacother. 2022;151:113112.
131. Gao W, Zhang L, Wang X, Yu L, Wang C, Gong Y. The combination of indirubin and isatin attenuates dextran sodium sulfate induced ulcerative colitis in mice. Biochem Cell Biol. 2018;96(5):636–45.
132. Sawaya ACF, Abreu IN, Andreazza NL, Eberlin MN, Mazzafera P. HPLC-ESI-MS/MS of imidazole alkaloids in Pilocarpus microphyllus. Molecules. 2008;13(7):1518–29.
133. Rodrigues de Carvalho L, de Brito TV, Simião da C, Júnior J, José Dias Júnior G, de Aguiar Magalhães D, Guimarães Sousa S, et al. Epiisopiloturine, an imidazole alkaloid, reverses inflammation and lipid peroxidation parameters in the Crohn disease model induced by trinitrobenzenosulfonic acid in Wistar rats. Biomed Pharmacother. 2018;102:278–85.
134. Guo G, Shi F, Zhu J, Shao Y, Gong W, Zhou G, et al. Piperine, a functional food alkaloid, exhibits inhibitory potential against TNBS-induced colitis via the inhibition of IκB-α/NF-κB and induces tight junction protein (claudin-1, occludin, and ZO-1) signaling pathway in experimental mice. Hum Exp Toxicol. 2020;39(4):477–91.
135. Bai R, Jie X, Yao C, Xie Y. Discovery of small-molecule candidates against inflammatory bowel disease. Eur J Med Chem. 2020;185:111805.
136. Guo BJ, Bian ZX, Qiu HC, Wang YT, Wang Y. Biological and clinical implications of herbal medicine and natural products for the treatment of inflammatory bowel disease. Ann N Y Acad Sci. 2017;1401(1):37–48.
137. Gupta RA, Motiwala MN, Dumore NG, Danao KR, Ganjare AB. Effect of piperine on inhibition of FFA induced TLR4 mediated inflammation and amelioration of acetic acid induced ulcerative colitis in mice. J Ethnopharmacol. 2015;164:239–46.
138. Dejban P, Nikravangolsefid N, Chamanara M, Dehpour A, Rashidian A. The role of medicinal products in the treatment of inflammatory bowel diseases (IBD) through inhibition of TLR4/NF-kappaB pathway. Phytother Res. 2021;35(2):835–45.
139. Kang GD, Kim DH. Poncirin and its metabolite ponciretin attenuate colitis in mice by inhibiting LPS binding on TLR4 of macrophages and correcting Th17/Treg imbalance. J Ethnopharmacol. 2016;189:175–85.
140. Wang H, Gu J, Hou X, Chen J, Yang N, Liu Y, et al. Anti-inflammatory effect of miltirone on inflammatory bowel disease via TLR4/NF-κB/IQGAP2 signaling pathway. Biomed Pharmacother. 2017;85:531–40.
141. Yue W, Liu Y, Li X, Lv L, Huang J, Liu J. Curcumin ameliorates dextran sulfate sodium-induced colitis in mice via regulation of autophagy and intestinal immunity. Turk J Gastroenterol. 2019;30(3):290–8.
142. Kunnumakkara AB, Bordoloi D, Padmavathi G, Monisha J, Roy NK, Prasad S, et al. Curcumin, the golden nutraceutical: multitargeting for multiple chronic diseases. Br J Pharmacol. 2017;174(11):1325–48.
143. Luis PB, Gordon ON, Nakashima F, Joseph AI, Shibata T, Uchida K, et al. Oxidative metabolism of curcumin-glucuronide by peroxidases and isolated human leukocytes. Biochem Pharmacol. 2017;132:143–9.
144. Mouzaoui S, Banerjee S, Djerdjouri B. Low-dose curcumin reduced TNBS-associated mucin depleted foci in mice by scavenging superoxide anion and lipid peroxides, rebalancing matrix NO synthase and aconitase activities, and recoupling mitochondria. Inflammopharmacology. 2020;28(4):949–65.
145. Dubey P, Begum RF, Chitra V, Mrinalini R, Gunasekaran H, Sumithra M. Formulation, standardization, and preclinical evaluation of polyherbal suspension against inflammatory bowel disease. J Nat Remedies. 2022;22(3):412–23.
146. Mashtoub S, Hoang BV, Vu M, Lymn KA, Feinle-Bisset C, Howarth GS. Clinical and structural effects of traditional Chinese medicine and the herbal preparation, iberogast, in a rat model of ulcerative colitis. J Evid Based Complementary Altern Med. 2014;19(1):10–9.
147. Saba E, Lee YY, Kim M, Hyun SH, Park CK, Son E, et al. A novel herbal formulation consisting of red ginseng extract and Epimedium koreanum Nakai-attenuated dextran sulfate sodium-induced colitis in mice. J Ginseng Res. 2020;44(6):833–42.
148. Gautam MK, Goel S, Ghatule RR, Singh A, Nath G, Goel RK. Curative effect of Terminalia chebula extract on acetic acid-induced experimental colitis: role of antioxidants, free radicals and acute inflammatory marker. Inflammopharmacology. 2013;21(5):377–83.
149. Tsang SW, Ip SP, Wu JCY, Ng SC, Yung KKL, Bian ZX. A Chinese medicinal formulation ameliorates dextran sulfate sodium-induced experimental colitis by suppressing the activity of nuclear factor-kappaB signaling. J Ethnopharmacol. 2015;162:20–30.
150. Shin MS, Kim SB, Lee J, Choi HS, Park J, Park JY, et al. Beneficial effect of herbal formulation KM1608 on inflammatory bowl diseases: a preliminary experimental study. Molecules. 2018;23(8):1–14.
151. Koppikar SJ, Jagtap SD, Devarshi PP, Jangle NM, Awad VB, Wele AA, et al. Triphala, an ayurvedic formulation improves the antioxidant status on TNBS induced IBD in rats. Eur J Integr Med. 2014;6(6):646–56.
152. Raju A, Rayudu V. Effect of Triphala on dextran sulphate sodium-induced colitis in rats. Ayu. 2014;35(3):333.
153. Wu Z-C, Zhao Z-L, Deng J-P, Huang J-T, Wang Y-F, Wang Z-P. Sanhuang Shu'ai decoction alleviates DSS-induced ulcerative colitis via regulation of gut microbiota, inflammatory mediators and cytokines. Biomed Pharmacother. 2020;125:109934.
154. Wu F, Shao Q, Hu M, Zhao Y, Dong R, Fang K, et al. Wu-Mei-Wan ameliorates chronic colitis-associated intestinal fibrosis through inhibiting fibroblast activation. J Ethnopharmacol. 2020;252:112580.
155. Zhai L, Peng J, Zhuang M, Chang Y-Y, Cheng KW, Ning Z-W, et al. Therapeutic effects and mechanisms of Zhen-Wu-Bu-Qi decoction on dextran sulfate sodium-induced chronic colitis in mice assessed by multi-omics approaches. Phytomedicine. 2022;99:154001.
156. Wadie W, Abdel-Aziz H, Zaki HF, Kelber O, Weiser D, Khayyal MT. STW 5 is effective in dextran sulfate sodium-induced colitis in rats. Int J Colorectal Dis. 2012;27(11):1445–53.
157. Cota D, Mishra S, Shengule S. Arjunarishta alleviates experimental colitis via suppressing proinflammatory cytokine expression, modulating gut microbiota and enhancing antioxidant effect. Mol Biol Rep. 2020;47(9):7049–59.
158. Zhang M, Fan H, Tan S, Tang Q, Liu X, Zuo D, et al. The Chinese medicinal herb decoction QRZSLXF enhances anti-inflammatory effect in TNBS-induced colitis via balancing Th17/Tregs differentiation. J Ethnopharmacol. 2020;251:112549.
159. Peregoy JA, Clarke TC, Jones LI, Stussman BJ, Nahin RL. Regional variation in use of complementary health approaches by U.S. adults. NCHS Data Brief. 2014;(146):1–8.
160. Koning M, Ailabouni R, Gearry RB, Frampton CMA, Barclay ML. Use and predictors of oral complementary and alternative medicine by patients with inflammatory bowel disease: a population-based, case-control study. Inflamm Bowel Dis. 2013;19(4):767–78.
161. Nguyen GC, Croitoru K, Silverberg MS, Steinhart AH, Weizman AV. Use of complementary and alternative medicine for inflammatory bowel disease is associated with worse adherence to con-

ventional therapy: the COMPLIANT study. Inflamm Bowel Dis. 2016;22(6):1412–7.
162. Langhorst J, Wulfert H, Lauche R, Klose P, Cramer H, Dobos GJ, et al. Systematic review of complementary and alternative medicine treatments in inflammatory bowel diseases. J Crohn's Colitis. 2015;9(1):86–106.
163. Lin SC, Cheifetz AS. The use of complementary and alternative medicine in patients with inflammatory bowel disease. Gastroenterol Hepatol. 2018;14(7):415–25.
164. Rossi M, Corradini C, Amaretti A, Nicolini M, Pompei A, Zanoni S, et al. Fermentation of fructooligosaccharides and inulin by bifidobacteria: a comparative study of pure and fecal cultures. Appl Environ Microbiol. 2005;71(10):6150–8.
165. Cheifetz AS, Gianotti R, Luber R, Gibson PR. Complementary and alternative medicines used by patients with inflammatory bowel diseases. Gastroenterology. 2017;152(2):415–429.e15.
166. Verhoef MJ, Sutherland LR, Brkich L. Use of alternative medicine by patients attending a gastroenterology clinic. CMAJ. 1990;142(2):121–5.
167. Li FX, Verhoef MJ, Best A, Otley A, Hilsden RJ. Why patients with inflammatory bowel disease use or do not use complementary and alternative medicine: a canadian national survey. Can J Gastroenterol. 2005;19(9):567–73.
168. Hilsden R. Complementary and alternative medicine use by Canadian patients with inflammatory bowel disease: results from a national survey. Am J Gastroenterol. 2003;98(7):1563–8.
169. Burgmann T, Rawsthorne P, Bernstein CN. Predictors of alternative and complementary medicine use in inflammatory bowel disease: do measures of conventional health care utilization relate to use? Am J Gastroenterol. 2004;99(5):889–93.
170. Abitbol V, Lahmek P, Buisson A, Olympie A, Poupardin C, Chaussade S, et al. Impact of complementary and alternative medicine on the quality of life in inflammatory bowel disease: results from a French national survey. Eur J Gastroenterol Hepatol. 2014;26(3):288–94.
171. Rawsthorne P, Clara I, Graff LA, Bernstein KI, Carr R, Walker JR, et al. The Manitoba inflammatory bowel disease cohort study: a prospective longitudinal evaluation of the use of complementary and alternative medicine services and products. Gut. 2012;61(4):521–7.
172. Ceballos C, Bao R, Dunkin D, Song Y, Li XM, Benkov K. Complementary and alternative medicine use at a single pediatric inflammatory bowel disease center. Gastroenterol Nurs. 2014;37(4):265–71.
173. Weizman AV, Ahn E, Thanabalan R, Leung W, Croitoru K, Silverberg MS, et al. Characterisation of complementary and alternative medicine use and its impact on medication adherence in inflammatory bowel disease. Aliment Pharmacol Ther. 2012;35(3):342–9.
174. Derikx LAAP, Dieleman LA, Hoentjen F. Probiotics and prebiotics in ulcerative colitis. Best Pract Res Clin Gastroenterol. 2016;30(1):55–71.
175. Frank DN, Amand ALS, Feldman RA, Boedeker EC, Harpaz N, Pace NR. Molecular-phylogenetic characterization of microbial community imbalances in human inflammatory bowel diseases. Proc Natl Acad Sci U S A. 2007;104(34):13780–5.
176. Jonkers D, Penders J, Masclee A, Pierik M. Probiotics in the management of inflammatory bowel disease: a systematic review of intervention studies in adult patients. Drugs. 2012;72(6):803–23.
177. Sartor RB. Microbial influences in inflammatory bowel diseases. Gastroenterology. 2008;134(2):577–94.
178. Ott SJ, Musfeldt M, Wenderoth DF, Hampe J, Brant O, Fölsch UR, et al. Reduction in diversity of the colonic mucosa associated bacterial microflora in patients with active inflammatory bowel disease. Gut. 2004;53(5):685–93.
179. Chassaing B, Darfeuillemichaud A. The commensal microbiota and enteropathogens in the pathogenesis of inflammatory bowel diseases. Gastroenterology. 2011;140(6):1720–1728.e3.
180. Veerappan GR, Betteridge J, Young PE. Probiotics for the treatment of inflammatory bowel disease. Curr Gastroenterol Rep. 2012;14(4):324–33.
181. Fedorak R, Demeria D. Probiotic bacteria in the prevention and the treatment of inflammatory bowel disease. Gastroenterol Clin North Am. 2012;41(4):821–42.
182. Khan I, Ullah N, Zha L, Bai Y, Khan A, Zhao T, et al. Alteration of gut microbiota in inflammatory bowel disease (IBD): cause or consequence? IBD treatment targeting the gut microbiome. Pathogens. 2019;8(3):1–28.
183. Astó E, Méndez I, Audivert S, Farran-Codina A, Espadaler J. The efficacy of probiotics, prebiotic inulin-type fructans, and synbiotics in human ulcerative colitis: a systematic review and meta-analysis. Nutrients. 2019;11(2):293.
184. Sasaki M, Klapproth J-MA. The role of bacteria in the pathogenesis of ulcerative colitis. J Signal Transduct. 2012;2012:704953.
185. Pandey KR, Naik SR, Vakil BV. Probiotics, prebiotics and synbiotics—a review. J Food Sci Technol. 2015;52(12):7577–87.
186. Tamaki H, Nakase H, Inoue S, Kawanami C, Itani T, Ohana M, et al. Efficacy of probiotic treatment with Bifidobacterium longum 536 for induction of remission in active ulcerative colitis: a randomized, double-blinded, placebo-controlled multicenter trial. Dig Endosc. 2016;28(1):67–74.
187. Quak SH. Role of probiotics and nutrition in the management of chronic inflammatory bowel disease in children. Singapore Med J. 2013;54(4):183–4.
188. Azad MAK, Sarker M, Li T, Yin J. Probiotic species in the modulation of gut microbiota: an overview. Biomed Res Int. 2018;2018:9478630.
189. Yoshimatsu Y, Mikami Y, Kanai T. Bacteriotherapy for inflammatory bowel disease. Inflamm Regen. 2021;41(1):3.
190. Martyniak A, Medyńska-Przęczek A, Wędrychowicz A, Skoczeń S, Tomasik PJ. Prebiotics, probiotics, synbiotics, paraprobiotics and postbiotic compounds in IBD. Biomolecules. 2021;11(12):1–24.
191. Picardo S, Altuwaijri M, Devlin SM, Seow CH. Complementary and alternative medications in the management of inflammatory bowel disease. Therap Adv Gastroenterol. 2020;13:1756284820927550.
192. Ng SC, Plamondon S, Kamm MA, Hart AL, Al-Hassi HO, Guenther T, et al. Immunosuppressive effects via human intestinal dendritic cells of probiotic bacteria and steroids in the treatment of acute ulcerative colitis. Inflamm Bowel Dis. 2010;16(8):1286–98.
193. Seth A, Yan F, Polk DB, Rao RK. Probiotics ameliorate the hydrogen peroxide-induced epithelial barrier disruption by a PKC- and MAP kinase-dependent mechanism. Am J Physiol Gastrointest Liver Physiol. 2008;294(4):1060–70.
194. Yan F, Cao H, Cover TL, Washington MK, Shi Y, Liu LS, et al. Colon-specific delivery of a probiotic-derived soluble protein ameliorates intestinal inflammation in mice through an EGFR-dependent mechanism. J Clin Invest. 2011;121(6):2242–53.
195. Borruel N, Carol M, Casellas F, Antolín M, De Lara F, Espín E, et al. Increased mucosal tumour necrosis factor α production in Crohn's disease can be downregulated ex vivo by probiotic bacteria. Gut. 2002;51(5):659–64.
196. Basso PJ, Saraiva Câmara NO, Sales-Campos H. Microbial-based therapies in the treatment of inflammatory bowel disease—an overview of human studies. Front Pharmacol. 2019;9:1571.
197. McIlroy J, Ianiro G, Mukhopadhya I, Hansen R, Hold GL. Review article: the gut microbiome in inflammatory bowel disease—avenues for microbial management. Aliment Pharmacol Ther. 2018;47(1):26–42.
198. Kruis W, Frič P, Pokrotnieks J, Lukáš M, Fixa B, Kaščák M, et al. Maintaining remission of ulcerative colitis with the probi-

otic Escherichia coli Nissle 1917 is as effective as with standard mesalazine. Gut. 2004;53(11):1617–23.
199. Bischoff SC, Escher J, Hébuterne X, Kłęk S, Krznaric Z, Schneider S, et al. ESPEN practical guideline: clinical nutrition in inflammatory bowel disease. Clin Nutr. 2020;39(3):632–53.
200. Saez-Lara MJ, Gomez-Llorente C, Plaza-Diaz J, Gil A. The role of probiotic lactic acid bacteria and bifidobacteria in the prevention and treatment of inflammatory bowel disease and other related diseases: a systematic review of randomized human clinical trials. Biomed Res Int. 2015;2015:505878.
201. Oliva S, Di Nardo G, Ferrari F, Mallardo S, Rossi P, Patrizi G, et al. Randomised clinical trial: the effectiveness of Lactobacillus reuteri ATCC 55730 rectal enema in children with active distal ulcerative colitis. Aliment Pharmacol Ther. 2012;35(3):327–34.
202. Venturi A, Gionchetti P, Rizzello F, Johansson R, Zucconi E, Brigidi P, et al. Impact on the composition of the faecal flora by a new probiotic preparation: preliminary data on maintenance treatment of patients with ulcerative colitis. Aliment Pharmacol Ther. 1999;13(8):1103–8.
203. Sood A, Midha V, Makharia GK, Ahuja V, Singal D, Goswami P, et al. The probiotic preparation, VSL#3 induces remission in patients with mild-to-moderately active ulcerative colitis. Clin Gastroenterol Hepatol. 2009;7(11):1202–1209.e1.
204. Tursi A, Brandimarte G, Papa A, Giglio A, Elisei W, Giorgetti GM, et al. Treatment of relapsing mild-to-moderate ulcerative colitis with the probiotic VSL3 as adjunctive to a standard pharmaceutical treatment: a double-blind, randomized, placebo-controlled study. Am J Gastroenterol. 2010;105(10):2218–27.
205. Mardini HE, Grigorian AY. Probiotic mix VSL#3 is effective adjunctive therapy for mild to moderately active ulcerative colitis: a meta-analysis. Inflamm Bowel Dis. 2014;20(9):1562–7.
206. Derwa Y, Gracie DJ, Hamlin PJ, Ford AC. Systematic review with meta-analysis: the efficacy of probiotics in inflammatory bowel disease. Aliment Pharmacol Ther. 2017;46(4):389–400.
207. Shen J, Zuo ZX, Mao AP. Effect of probiotics on inducing remission and maintaining therapy in ulcerative colitis, Crohn's disease, and pouchitis: meta-analysis of randomized controlled trials. Inflamm Bowel Dis. 2014;20(1):21–35.
208. Akram W, Garud N, Joshi R. Role of inulin as prebiotics on inflammatory bowel disease. Drug Discov Ther. 2019;13(1):1–8.
209. Gibson GR, Roberfroid MB. Dietary modulation of the human colonic microbiota: introducing the concept of prebiotics. J Nutr. 1995;125(6):1401–12.
210. Van Loo JAE. Prebiotics promote good health: the basis, the potential, and the emerging evidence. J Clin Gastroenterol. 2004;38(6 Suppl):S70–5.
211. Gibson GR, Hutkins R, Sanders ME, Prescott SL, Reimer RA, Salminen SJ, et al. Expert consensus document: the International Scientific Association for Probiotics and Prebiotics (ISAPP) consensus statement on the definition and scope of prebiotics. Nat Rev Gastroenterol Hepatol. 2017;14(8):491–502.
212. Basson AR, Lam M, Cominelli F. Complementary and alternative medicine strategies for therapeutic gut microbiota modulation in inflammatory bowel disease and their next-generation approaches. Gastroenterol Clin North Am. 2017;46(4):689–729.
213. Damaskos D, Kolios G. Probiotics and prebiotics in inflammatory bowel disease: microflora "on the scope". Br J Clin Pharmacol. 2008;65(4):453–67.
214. Govers MJAP, Gannon NJ, Dunshea FR, Gibson PR, Muir JG. Wheat bran affects the site of fermentation of resistant starch and luminal indexes related to colon cancer risk: a study in pigs. Gut. 1999;45(6):840–7.
215. Loganathan P, Anderson A, Hospital NM, Rivers W, Rivers CR. Nutritional management of inflammatory bowel diseases: a comprehensive guide. Berlin: Springer; 2016.
216. Nishimura T, Andoh A, Hashimoto T, Kobori A, Tsujikawa T, Fujiyama Y. Cellobiose prevents the development of dextran sulfate sodium (DSS)-induced experimental colitis. J Clin Biochem Nutr. 2010;46(2):105–10.
217. Looijer-van Langen MAC, Dieleman LA. Prebiotics in chronic intestinal inflammation. Inflamm Bowel Dis. 2009;15(3):454–62.
218. Conlon MA, Kerr CA, Mcsweeney CS, Dunne RA, Shaw JM, Kang S, et al. Resistant starches protect against colonic DNA damage and alter microbiota and gene expression in rats fed a western diet. J Nutr. 2012;142(5):832–40.
219. Wang X, Brown IL, Khaled D, Mahoney MC, Evans AJ, Conway PL. Manipulation of colonic bacteria and volatile fatty acid production by dietary high amylose maize (amylomaize) starch granules. J Appl Microbiol. 2002;93(3):390–7.
220. Leitch ECMW, Walker AW, Duncan SH, Holtrop G, Flint HJ. Selective colonization of insoluble substrates by human faecal bacteria. Environ Microbiol. 2007;9(3):667–79.
221. Ze X, Duncan SH, Louis P, Flint HJ. Ruminococcus bromii is a keystone species for the degradation of resistant starch in the human colon. ISME J. 2012;6(8):1535–43.
222. Kanauchi O, Aoata K. Protein, and dietary fiber-rich new foodstuff from brewer's spent grain increased excretion of feces and jejunum mucosal protein content in rats. Biosci Biotechnol Biochem. 1997;61(1):29–33.
223. Kanauchi O, Mitsuyama K, Homma T, Takahama K, Fujiyama Y, Andoh A, et al. Treatment of ulcerative colitis patients by long-term administration of germinated barley foodstuff: multi-center open trial. Int J Mol Med. 2003;12(5):701–4.
224. Videla S, Vilaseca J, Antolín M, García-Lafuente A, Guarner F, Crespo E, et al. Dietary inulin improves distal colitis induced by dextran sodium sulfate in the rat. Am J Gastroenterol. 2001;96(5):1486–93.
225. Schneeman BO. Nutritional and health benefits of inulin and oligofructose fiber, inulin and oligofructose: similarities and differences 1. J Nutr. 1999:1424–7.
226. Kleessen B, Hartmann L, Blaut M. Oligofructose and long-chain inulin: influence on the gut microbial ecology of rats associated with a human faecal flora. Br J Nutr. 2001;86(2):291–300.
227. Hoentjen F, Welling GW, Harmsen HJM, Zhang X, Snart J, Tannock GW, et al. Reduction of colitis by prebiotics in HLA-B27 transgenic rats is associated with microflora changes and immunomodulation. Inflamm Bowel Dis. 2005;11(11):977–85.
228. Kanauchi O, Iwanaga T, Andoh A, Araki Y, Nakamura T, Mitsuyama K, et al. Dietary fiber fraction of germinated barley foodstuff attenuated mucosal damage and diarrhea, and accelerated the repair of the colonic mucosa in an experimental colitis. J Gastroenterol Hepatol. 2001;16(2):160–8.
229. Araki Y, Andoh A, Koyama S, Fujiyama Y, Kanauchi O, Bamba T. Effects of germinated barley foodstuff on microflora and short chain fatty acid production in dextran sulfate sodium-induced colitis in rats. Biosci Biotechnol Biochem. 2000;64:1794–800.
230. Shiba T, Aiba Y, Ishikawa H, Ushiyama A, Takagi A, Mine T, et al. The suppressive effect of bifidobacteria on Bacteroides vulgatus, a putative pathogenic microbe in inflammatory bowel disease. Microbiol Immunol. 2003;47(6):371–8.
231. Gupta P, Andrew H, Kirschner BS, Guandalini S. Is Lactobacillus GG helpful in children with Crohn's disease? Results of a preliminary, open-label study. J Pediatr Gastroenterol Nutr. 2000;31(4):453–7.
232. Kanauchi O, Nakamura T, Agata K, Mitsuyama K, Iwanaga T. Effects of germinated barley foodstuff on dextran sulfate sodium-induced colitis in rats. J Gastroenterol. 1998;33(2):179–88.
233. Setoyama H, Imaoka A, Ishikawa H, Umesaki Y. Prevention of gut inflammation by Bifidobacterium in dextran sulfate-treated gnotobiotic mice associated with Bacteroides strains isolated from ulcerative colitis patients. Microbes Infect. 2003;5(2):115–22.

234. Cherbut C, Michel C, Lecannu G. The prebiotic characteristics of fructooligosaccharides are necessary for reduction of TNBS-induced colitis in rats. J Nutr. 2003;133(1):21–7.
235. Matsumoto S, Watanabe N, Imaoka A, Okabe Y. Preventive effects of Bifidobacterium- and Lactobacillus-fermented milk on the development of inflammatory bowel disease in senescence-accelerated mouse P1/Yit strain mice. Digestion. 2001;64(2):92–9.
236. Gomollón F, Dignass A, Annese V, Tilg H, Van Assche G, Lindsay JO, et al. 3rd European evidence-based consensus on the diagnosis and management of Crohn's disease 2016: part 1: diagnosis and medical management. J Crohns Colitis. 2017;11(1):3–25.
237. Schultz M, Timmer A, Herfarth HH, Sartor RB, Vanderhoof JA, Rath HC. Lactobacillus GG in inducing and maintaining remission of Crohn's disease. BMC Gastroenterol. 2004;4:3–6.
238. Petersen AM, Mirsepasi H, Halkjær SI, Mortensen EM, Nordgaard-Lassen I, Krogfelt KA. Ciprofloxacin and probiotic Escherichia coli Nissle add-on treatment in active ulcerative colitis: a double-blind randomized placebo controlled clinical trial. J Crohns Colitis. 2014;8(11):1498–505.
239. Bourreille A, Cadiot G, Le Dreau G, Laharie D, Beaugerie L, Dupas JL, et al. Saccharomyces boulardii does not prevent relapse of crohn's disease. Clin Gastroenterol Hepatol. 2013;11(8):982–7.
240. Guslandi M, Giollo P, Testoni PA. A pilot trial of Saccharomyces boulardii in ulcerative colitis. Eur J Gastroenterol Hepatol. 2003;15(6):697–8.
241. Bousvaros A, Guandalini S, Baldassano RN, Botelho C, Evans J, Ferry GD, et al. A randomized, double-blind trial of lactobacillus GG versus placebo in addition to standard maintenance therapy for children with Crohn's disease. Inflamm Bowel Dis. 2005;11(9):833–9.
242. Prantera C, Scribano ML, Falasco G, Andreoli A, Luzi C. Ineffectiveness of probiotics in preventing recurrence after curative resection for Crohn's disease: a randomised controlled trial with Lactobacillus GG. Gut. 2002;51(3):405–9.
243. Zocco MA, Dal Verme LZ, Cremonini F, Piscaglia AC, Nista EC, Candelli M, et al. Efficacy of Lactobacillus GG in maintaining remission of ulcerative colitis. Aliment Pharmacol Ther. 2006;23(11):1567–74.
244. Ishikawa H, Akedo I, Otani T, Umesaki Y, Tanaka R, Imaoka A. Randomized controlled trial of the effect of bifidobacteria-fermented milk on ulcerative colitis. J Am Coll Nutr. 2003;22(1):56–63.
245. Triantafyllidi A, Xanthos T, Papalois A, Triantafillidis JK. Herbal and plant therapy in patients with inflammatory bowel disease. Ann Gastroenterol. 2015;28:210–20.
246. Park KT, Ehrlich OG, Allen JI, Meadows P, Szigethy EM, Henrichsen K, et al. The cost of inflammatory bowel disease: an initiative from the Crohn's & Colitis Foundation. Inflamm Bowel Dis. 2020;26(1):1–10.
247. Ordás I, Eckmann L, Talamini M, Baumgart DC, Sandborn WJ. Ulcerative colitis. Lancet (London, England). 2012;380(9853):1606–19.
248. Knutson D, Greenberg G, Cronau H. Management of Crohn's disease—a practical approach. Am Fam Physician. 2003;68(4):707–14.
249. Rutgeerts P, Vermeire S, Van Assche G. Biological therapies for inflammatory bowel diseases. Gastroenterology. 2009;136(4):1182–97.
250. Weisshof R, El Jurdi K, Zmeter N, Rubin DT. Emerging therapies for inflammatory bowel disease. Adv Ther. 2018;35(11):1746–62.
251. Chudy-Onwugaje KO, Christian KE, Farraye FA, Cross RK. A state-of-the-art review of new and emerging therapies for the treatment of IBD. Inflamm Bowel Dis. 2019;25(5):820–30.
252. Currò D, Ianiro G, Pecere S, Bibbò S, Cammarota G. Probiotics, fibre and herbal medicinal products for functional and inflammatory bowel disorders. Br J Pharmacol. 2017;174(11):1426–49.
253. Hossen I, Hua W, Ting L, Mehmood A, Jingyi S, Duoxia X, et al. Phytochemicals and inflammatory bowel disease: a review. Crit Rev Food Sci Nutr. 2020;60(8):1321–45.
254. Cirillo C, Capasso R. Constipation and botanical medicines: an overview. Phytother Res. 2015;1493(10):1488–93.
255. Vezza T, Rodríguez-Nogales A, Algieri F, Utrilla MP, Rodriguez-Cabezas ME, Galvez J. Flavonoids in inflammatory bowel disease: a review. Nutrients. 2016;8(4):211.
256. Hoult JR, Payá M. Pharmacological and biochemical actions of simple coumarins: natural products with therapeutic potential. Gen Pharmacol. 1996;27(4):713–22.
257. Pithadia AB, Jain S. Treatment of inflammatory bowel disease (IBD). Pharmacol Rep. 2011;63(3):629–42.
258. Atreya I, Atreya R, Neurath MF. NF-kappaB in inflammatory bowel disease. J Intern Med. 2008;263(6):591–6.
259. Feng YJ, Li YY. The role of p38 mitogen-activated protein kinase in the pathogenesis of inflammatory bowel disease. J Dig Dis. 2011;12(5):327–32.
260. Broom OJ, Widjaya B, Troelsen J, Olsen J, Nielsen OH. Mitogen activated protein kinases: a role in inflammatory bowel disease? Clin Exp Immunol. 2009;158(3):272–80.
261. Medina C, Radomski MW. Role of matrix metalloproteinases in intestinal inflammation. J Pharmacol Exp Ther. 2006;318(3):933–8.
262. Langmead L, Makins RJ, Rampton DS. Anti-inflammatory effects of aloe vera gel in human colorectal mucosa in vitro. Aliment Pharmacol Ther. 2004;19(5):521–7.
263. Zhu Q, Zheng P, Zhou J, Chen X, Feng Y, Wang W, et al. Andrographolide affects Th1/Th2/Th17 responses of peripheral blood mononuclear cells from ulcerative colitis patients. Mol Med Rep. 2018;18(1):622–6.
264. Batiha GE-S, Olatunde A, El-Mleeh A, Hetta HF, Al-Rejaie S, Alghamdi S, et al. Bioactive compounds, pharmacological actions, and pharmacokinetics of wormwood (Artemisia absinthium). Antibiotics (Basel). 2020;9(6):353.
265. Kim H-J, Kim B, Lee M-R, Ra M, Lee Y. Bamboo shoot and artemisia capillaris extract mixture ameliorates dextran sodium sulfate-induced colitis. Curr Issues Mol Biol. 2022;44(10):5086–103.
266. Kiela PR, Midura AJ, Kuscuoglu N, Jolad SD, Sólyom AM, Besselsen DG, et al. Effects of Boswellia serrata in mouse models of chemically induced colitis. Am J Physiol Gastrointest Liver Physiol. 2005;288(4):G798–808.
267. Holleran G, Scaldaferri F, Gasbarrini A, Currò D. Herbal medicinal products for inflammatory bowel disease: a focus on those assessed in double-blind randomised controlled trials. Phytother Res. 2020;34(1):77–93.
268. Hanai H, Sugimoto K. Curcumin has bright prospects for the treatment of inflammatory bowel disease. Curr Pharm Des. 2009;15(18):2087–94.
269. Bar-Sela G, Cohen M, Ben-Arye E, Epelbaum R. The medical use of wheatgrass: review of the gap between basic and clinical applications. Mini Rev Med Chem. 2015;15(12):1002–10.
270. Zhou Y-H, Yu J-P, Liu Y-F, Teng X-J, Ming M, Lv P, et al. Effects of Ginkgo biloba extract on inflammatory mediators (SOD, MDA, TNF-alpha, NF-kappaBp65, IL-6) in TNBS-induced colitis in rats. Mediators Inflamm. 2006;2006(5):92642.
271. Viladomiu M, Hontecillas R, Lu P, Bassaganya-Riera J. Preventive and prophylactic mechanisms of action of pomegranate bioactive constituents. Evid Based Complementary Alternat Med. 2013;2013:789764.
272. Wang J, Zhang Z, Fang A, Wu K, Chen X, Wang G, et al. Resveratrol attenuates inflammatory bowel disease in mice by regulating SUMO1. Biol Pharm Bull. 2020;43(3):450–7.
273. Wangchuk P, Navarro S, Shepherd C, Keller PA, Pyne SG, Loukas A. Diterpenoid alkaloids of Aconitum laciniatum and mitigation

of inflammation by 14-O-acetylneoline in a murine model of ulcerative colitis. Sci Rep. 2015;5:12845.
274. Fernández-Bañares F, Hinojosa J, Sánchez-Lombraña JL, Navarro E, Martínez-Salmerón JF, García-Pugés A, et al. Randomized clinical trial of Plantago ovata seeds (dietary fiber) as compared with mesalamine in maintaining remission in ulcerative colitis. Spanish Group for the Study of Crohn's Disease and Ulcerative Colitis (GETECCU). Am J Gastroenterol. 1999;94(2):427–33.
275. Chen C-L, Chen Y-P, Lin M-W, Huang Y-B, Chang F-R, Duh T-H, et al. Euphol from Euphorbia tirucalli negatively modulates TGF-β responsiveness via TGF-β receptor segregation inside membrane rafts. PLoS One. 2015;10(10):e0140249.
276. Liu D-L, Li Y-J, Yang D-H, Wang C-R, Xu J, Yao N, et al. Ganoderma lucidum derived ganoderenic acid B reverses ABCB1-mediated multidrug resistance in HepG2/ADM cells. Int J Oncol. 2015;46(5):2029–38.
277. da Silva VC, de Araújo AA, de Souza Araújo DF, Souza Lima MCJ, Vasconcelos RC, de Araújo Júnior RF, et al. Intestinal anti-inflammatory activity of the aqueous extract from ipomoea asarifolia in DNBS-induced colitis in rats. Int J Mol Sci. 2018;19(12):4016.
278. Lee S-J, Lee J, Song S, Lim K-T. Glycoprotein isolated from Styrax japonica Siebold et al. Zuccarini inhibits oxidative and proinflammatory responses in HCT116 colonic epithelial cells and dextran sulfate sodium-treated ICR mice. Food Chem Toxicol. 2016;87:12–22.
279. Khare T, Palakurthi SS, Shah BM, Palakurthi S, Khare S. Natural product-based nanomedicine in treatment of inflammatory bowel disease. Int J Mol Sci. 2020;21(11):3956.
280. Marius M, Amadou D, Donatien AA, Gilbert A, William YN, Rauf K, et al. In vitro antioxidant, anti-inflammatory, and in vivo anticolitis effects of combretin A and combretin B on dextran sodium sulfate-induced ulcerative colitis in mice. Gastroenterol Res Pract. 2020;2020:4253174.
281. Rong X, Ye CX. Simultaneous analysis of purine alkaloids and catechins in Camellia sinensis, Camellia ptilophylla and Camellia assamica var. kucha by HPLC. Food Chem. 2007;100:1132–6.
282. Vezza T, Rodríguez-Nogales A, Algieri F, Garrido-Mesa J, Romero M, Sánchez M, et al. The metabolic and vascular protective effects of olive (Olea europaea L.) leaf extract in diet-induced obesity in mice are related to the amelioration of gut microbiota dysbiosis and to its immunomodulatory properties. Pharmacol Res. 2019;150:104487.
283. Debnath T, Park P, Chandra N, Nath D, Binte N, Won H, et al. Antioxidant activity of Gardenia jasminoides Ellis fruit extracts. Food Chem. 2011;128(3):697–703. https://doi.org/10.1016/j.foodchem.2011.03.090.
284. dos Reis SB, de Oliveira CC, Acedo SC, da Conceição Miranda DD, Ribeiro ML, Pedrazzoli JJ, et al. Attenuation of colitis injury in rats using Garcinia cambogia extract. Phytother Res. 2009;23(3):324–9.
285. Debnath T, Hasnat A, Pervin M, Lee SY, Park SR, Kim DH. Chaga mushroom (Inonotus obliquus) grown on germinated brown rice suppresses inflammation associated with colitis in mice. Food Sci Technol. 2012;21(5):1235–41.
286. Cho E, Shin J-S, Noh Y-S, Cho Y-W, Hong S-J, Park J-H, et al. Anti-inflammatory effects of methanol extract of Patrinia scabiosaefolia in mice with ulcerative colitis. J Ethnopharmacol. 2011;136(3):428–35.
287. Jin H, Lee B, Lim KJ, Debnath T, Shin HM, Lim BO. Anti-inflammatory effects of Prunus mume mixture in colitis induced by dextran sodium sulfate. Korean J Med Crop Sci. 2011;19:1.
288. Fatani AJ, Alrojayee FS, Parmar MY, Abuohashish HM, Ahmed MM, Al-Rejaie SS. Myrrh attenuates oxidative and inflammatory processes in acetic acid-induced ulcerative colitis. Exp Ther Med. 2016;12(2):730–8.
289. Fan F-Y, Sang L-X, Jiang M. Catechins and their therapeutic benefits to inflammatory bowel disease. Molecules. 2017;22(3):484.
290. Ye Z, Liu Z, Henderson A, Lee K, Hostetter J, Wannemuehler M, et al. Increased CYP4B1 mRNA is associated with the inhibition of dextran sulfate sodium-induced colitis by caffeic acid in mice. Exp Biol Med (Maywood). 2009;234(6):605–16.
291. Abron JD, Singh NP, Price RL, Nagarkatti M, Nagarkatti PS, Singh UP. Genistein induces macrophage polarization and systemic cytokine to ameliorate experimental colitis. PLoS One. 2018;13(7):e0199631.
292. Khan MS, Ikram M, Park JS, Park TJ, Kim MO. Gut microbiota, its role in induction of Alzheimer's disease pathology, and possible therapeutic interventions: special focus on anthocyanins. Cells. 2020;9(4):853.
293. Zhang H, Zhuo S, Song D, Wang L, Gu J, Ma J, et al. Icariin inhibits intestinal inflammation of DSS-induced colitis mice through modulating intestinal flora abundance and modulating p-p65/p65 molecule. Turk J Gastroenterol. 2021;32(4):382–92.
294. Lin R, Piao M, Song Y. Dietary quercetin increases colonic microbial diversity and attenuates colitis severity in citrobacter rodentium-infected mice. Front Microbiol. 2019;10:1092.
295. Ai X-Y, Qin Y, Liu H-J, Cui Z-H, Li M, Yang J-H, et al. Apigenin inhibits colonic inflammation and tumorigenesis by suppressing STAT3-NF-κB signaling. Oncotarget. 2017;8(59):100216–26.
296. Nunes C, Almeida L, Barbosa RM, Laranjinha J. Luteolin suppresses the JAK/STAT pathway in a cellular model of intestinal inflammation. Food Funct. 2017;8(1):387–96.
297. Medicherla K, Sahu BD, Kuncha M, Kumar JM, Sudhakar G, Sistla R. Oral administration of geraniol ameliorates acute experimental murine colitis by inhibiting pro-inflammatory cytokines and NF-κB signaling. Food Funct. 2015;6(9):2984–95.
298. El Menyiy N, El Allam A, Aboulaghras S, Jaouadi I, Bakrim S, El Omari N, et al. Inflammatory auto-immune diseases of the intestine and their management by natural bioactive compounds. Biomed Pharmacother. 2022;151:113158.
299. Kudo T, Okamura S, Zhang Y, Masuo T, Mori M. Topical application of glycyrrhizin preparation ameliorates experimentally induced colitis in rats. World J Gastroenterol. 2011;17(17):2223–8.
300. Lv J, Zhang Y, Tian Z, Liu F, Shi Y, Liu Y, et al. Astragalus polysaccharides protect against dextran sulfate sodium-induced colitis by inhibiting NF-κB activation. Int J Biol Macromol. 2017;98:723–9.
301. Nie Y, Lin Q, Luo F. Effects of non-starch polysaccharides on inflammatory bowel disease. Int J Mol Sci. 2017;18(7):1372.
302. Wen J-B, Zhu F-Q, Chen W-G, Jiang L-P, Chen J, Hu Z-P, et al. Oxymatrine improves intestinal epithelial barrier function involving NF-κB-mediated signaling pathway in CCl4-induced cirrhotic rats. PLoS One. 2014;9(8):e106082.
303. Nirmal SA, Ingale JM, Pattan SR, Bhawar SB. Amaranthus roxburghianus root extract in combination with piperine as a potential treatment of ulcerative colitis in mice. J Integr Med. 2013;11(3):206–12.
304. Zhang M, Long Y, Sun Y, Wang Y, Li Q, Wu H, et al. Evidence for the complementary and synergistic effects of the three-alkaloid combination regimen containing berberine, hypaconitine and skimmianine on the ulcerative colitis rats induced by trinitrobenzene-sulfonic acid. Eur J Pharmacol. 2011;651(1–3):187–96.
305. Jiao Y-F, Lu M, Zhao Y-P, Liu N, Niu Y-T, Niu Y, et al. N-methylcytisine ameliorates dextran-sulfate-sodium-induced colitis in mice by inhibiting the inflammatory response. Molecules. 2018;23(3):510.
306. Jiang Y, Zhao L, Chen Q, Zhou L. Exploring the mechanism of berberine intervention in ulcerative colitis from the perspective of inflammation and immunity based on systemic pharmacology. Evid Based Complement Alternat Med. 2021;2021:9970240.

307. Ghasemi-Pirbaluti M, Motaghi E, Najafi A, Hosseini MJ. The effect of theophylline on acetic acid induced ulcerative colitis in rats. Biomed Pharmacother. 2017;90:153–9.
308. Niu X, Zhang H, Li W, Wang Y, Mu Q, Wang X, et al. Protective effect of cavidine on acetic acid-induced murine colitis via regulating antioxidant, cytokine profile and NF-κB signal transduction pathways. Chem Biol Interact. 2015;239:34–45.
309. Shirakami Y, Kochi T, Kubota M, Sakai H, Ibuka T, Yoshimi K, et al. Inhibitory effects of pentoxifylline on inflammation-related tumorigenesis in rat colon. Oncotarget. 2018;9(74):33972–81.
310. Borniquel S, Jansson EA, Cole MP, Freeman BA, Lundberg JO. Nitrated oleic acid up-regulates PPARgamma and attenuates experimental inflammatory bowel disease. Free Radic Biol Med. 2010;48(4):499–505.
311. Tyagi A, Kumar U, Reddy S, Santosh VS, Mohammed SB, Ehtesham NZ, et al. Attenuation of colonic inflammation by partial replacement of dietary linoleic acid with α-linolenic acid in a rat model of inflammatory bowel disease. Br J Nutr. 2012;108(9):1612–22.
312. Sato K, Kanazawa A, Ota N, Nakamura T, Fujimoto K. Dietary supplementation of catechins and alpha-tocopherol accelerates the healing of trinitrobenzene sulfonic acid-induced ulcerative colitis in rats. J Nutr Sci Vitaminol (Tokyo). 1998;44(6):769–78.
313. Motawea MH, Abd Elmaksoud HA, Elharrif MG, Desoky AAE, Ibrahimi A. Evaluation of anti-inflammatory and antioxidant profile of oleuropein in experimentally induced ulcerative colitis. Int J Mol Cell Med. 2020;9(3):224–33.
314. Qian B, Wang C, Zeng Z, Ren Y, Li D, Song J-L. Ameliorative effect of sinapic acid on dextran sodium sulfate-(DSS-)induced ulcerative colitis in kunming (KM) mice. Oxid Med Cell Longev. 2020;2020:8393504.
315. Chen F, Liu Q, Xiong Y, Xu L. Current strategies and potential prospects of nanomedicine-mediated therapy in inflammatory bowel disease. Int J Nanomedicine. 2021;16:4225–37.
316. Pita R, Ehmann F, Papaluca M. Nanomedicines in the EU-regulatory overview. AAPS J. 2016;18(6):1576–82.
317. Rubino SJ, Selvanantham T, Girardin SE, Philpott DJ. Nod-like receptors in the control of intestinal inflammation. Curr Opin Immunol. 2012;24(4):398–404.
318. Wang L, Yu M, Yang H. Recent progress in the diagnosis and precise nanocarrier-mediated therapy of inflammatory bowel disease. J Inflamm Res. 2021;14:1701–16.
319. Yousef M, Pichyangkura R, Soodvilai S, Chatsudthipong V, Muanprasat C. Chitosan oligosaccharide as potential therapy of inflammatory bowel disease: therapeutic efficacy and possible mechanisms of action. Pharmacol Res. 2012;66(1):66–79.
320. Kanai T, Matsuoka K, Naganuma M, Hayashi A, Hisamatsu T. Diet, microbiota, and inflammatory bowel disease: lessons from Japanese foods. Korean J Intern Med. 2014;29(4):409–15.
321. Olendzki B, Bucci V, Cawley C, Maserati R, McManus M, Olednzki E, et al. Dietary manipulation of the gut microbiome in inflammatory bowel disease patients: pilot study. Gut Microbes. 2022;14(1):2046244.
322. Jang H-M, Kim J-K, Joo M-K, Shin Y-J, Lee K-E, Lee CK, et al. Enterococcus faecium and Pediococcus acidilactici deteriorate Enterobacteriaceae-induced depression and colitis in mice. Sci Rep. 2022;12(1):9389.
323. Lee SH, Kwon JE, Cho M-L. Immunological pathogenesis of inflammatory bowel disease. Intest Res. 2018;16(1):26–42.
324. Li S, Wu B, Fu W, Reddivari L. The anti-inflammatory effects of dietary anthocyanins against ulcerative colitis. Int J Mol Sci. 2019;6:1–18.
325. Leccese G, Bibi A, Mazza S, Facciotti F, Caprioli F, Landini P, et al. Probiotic Lactobacillus and Bifidobacterium strains counteract adherent-invasive Escherichia coli (AIEC) virulence and hamper IL-23/Th17 axis in ulcerative colitis, but not in Crohn's disease. Cells. 2020;9(8):1824.
326. Sivananthan K, Petersen AM. Review of Saccharomyces boulardii as a treatment option in IBD. Immunopharmacol Immunotoxicol. 2018;40(6):465–75.
327. Zuo L, Yuan K-T, Yu L, Meng Q-H, Chung PC-K, Yang D-H. Bifidobacterium infantis attenuates colitis by regulating T cell subset responses. World J Gastroenterol. 2014;20(48):18316–29.
328. Cheng F-S, Pan D, Chang B, Jiang M, Sang L-X. Probiotic mixture VSL#3: an overview of basic and clinical studies in chronic diseases. World J Clin Cases. 2020;8(8):1361–84.
329. Shadnoush M, Shaker Hosseini R, Mehrabi Y, Delpisheh A, Alipoor E, Faghfoori Z, et al. Probiotic yogurt affects pro- and anti-inflammatory factors in patients with inflammatory bowel disease. Iran J Pharm Res. 2013;12(4):929–36.
330. Zhang H, Duan Y, Cai F, Cao D, Wang L, Qiao Z, et al. Next-generation probiotics: microflora intervention to human diseases. Biomed Res Int. 2022;2022:5633403.
331. Ishisono K, Mano T, Yabe T, Kitaguchi K. Dietary fiber pectin ameliorates experimental colitis in a neutral sugar side chain-dependent manner. Front Immunol. 2019;10:2979.

Herbal Medicines for the Management of Diseases on Vitamin Deficiency

Sathi Sarkar, Anish Kumar Das, and Satheesh Kumar Nanjappan

Abstract

Vitamins are the key ingredients for the proper functioning of the body. Though it is needed in very small quantities, this does not reflect their role. Except for some vitamins, others will be obtained through diet. If the required quantities are not met by the diet for a long time, then deficiency symptoms develop. Untreated deficiency symptoms progress to deficiency disease or disorder, which can result in death or a very unpleasant life. Herbal remedies are the most ancient form of treatment. Most of the old civilizations have their own natural treatment cultures. One of the most significant advantages of these herbal remedies is their chemical and biological diversities and negligible side effects. Considering these features, there is interest in shifting from synthetic medicine to herbal or natural remedies. Many of the market surveys show the prominent trend of this shift and predict this natural product industry will be a $100 billion industry. In this context, herbal remedies are also an alternative for countering vitamin deficiency and its symptoms. This book chapter focuses on the available herbal remedies for vitamin deficiency. Along with that, we focus on the natural sources that contain the highest amount of respective vitamins, which can be used for the future development of herbal remedies, and we also raise awareness for the natural treatment of vitamin deficiency. Plants and plant-derived products are an important part of the human diet and a major source of biologically active substances such as vitamins, dietary fibre, antioxidants, and cholesterol-lowering compounds. Though there is abundant information, plant nutrition has not been defined. Trial-and-error methods were used to identify several plant nutrients and health-promoting chemicals. Chromatography, mass spectrometry, infrared spectroscopy, and nuclear magnetic resonance (NMR) allowed quantitative and qualitative analyses of plant metabolites around the turn of the century. Plant biologists developed new techniques for detecting and identifying phytochemicals. This section of the chapter discusses the qualitative and quantitative procedures of vitamin analysis. We also discuss different analytical approaches for the analysis.

Keywords

Vitamin deficiency · Lifestyle diseases · Herbal medicines

1 Introduction

Vitamins are a group of incredibly complicated organic molecules widely distributed in nature and absolutely required in the human diet. It plays a significant role in the human body's normal growth, development, and metabolism. Thirteen vitamins have been identified as essential to human nutrition and have been split into two categories based on solubility [1]. Fat-soluble vitamins comprise vitamin A, vitamin D, vitamin E, and vitamin K. Water-soluble vitamins comprise vitamin C (ascorbic acid), vitamin B complex, as well as flavonoids (vitamin P) (Fig. 1). Fat-soluble vitamins are related to fats and are absorbed with dietary lipids. The fat-soluble vitamins are absorbed in a similar way to the lipids. Water-soluble vitamins are not connected with fats and are unaffected by changes in fat absorption [2]. The biological activity of various vitamins is linked to a group of structurally related molecules known as vitamers. In most situations, vitamers with the same vitamin have similar qualitative biological attributes to one another, but their potency varies due to tiny changes in their chemical structures. Since the human body cannot synthesize vitamins, consuming them through diet is necessary. Vitamin insufficiency leads to severe or even deadly disorders. Vitamin deficiency is linked to certain diseases, such as vitamin A's link to blindness, vitamin B1's link to beriberi,

S. Sarkar · A. K. Das · S. K. Nanjappan (✉)
Department of Natural Products, National Institute of Pharmaceutical Education and Research (NIPER) - Kolkata, Kolkata, West Bengal, India
e-mail: satheesh.niperk@nic.in

A. K. Dhara, S. C. Mandal (eds.), *Role of Herbal Medicines*, https://doi.org/10.1007/978-981-99-7703-1_17

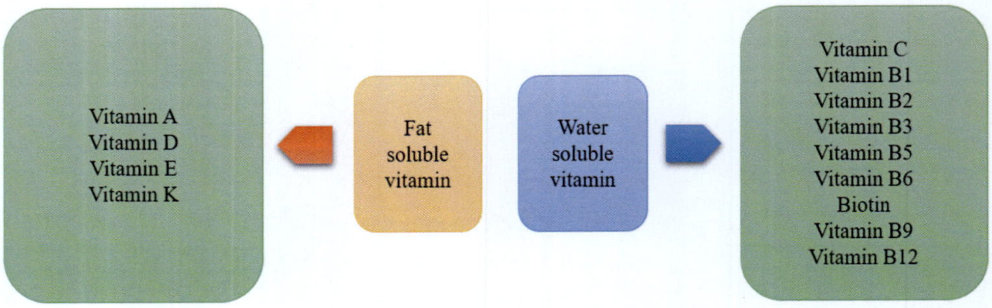

Fig. 1 Classification of vitamins

vitamin B3's link to pellagra, vitamin B6's link to anaemia, vitamin C's link to scurvy, and vitamin D's link to rickets. Thus, the deficiency of vitamins affects the health and well-being of millions worldwide. The deficit symptoms can also be treated by replenishing these nutrients [3].

Throughout the history of civilization, herbal medicines have been used to cure human ailments in every possible condition. Traditionally herbal medicine has been used in many countries against various diseases. Ancient people believed in and used plants and herbal remedies to heal illnesses. For primary health, 80% of the world's demography depends on herbal medicine [4]. Herbal medicines are natural products that are isolated from plants. Natural products have attracted interest over millions of years and are consequently being used in modern medicine. Natural products have a wide extent of chemical variety, i.e. they have multi-dimensional structures, and hence biological activities are abundant. Natural products play an increasingly important role as scientists continue to discover and refine new medications and lead compounds through various approaches. Therefore, these treatments have evolved from ancient or herbal therapies and natural resources, 90–95% of drugs used to be made from natural sources until high throughput screening was introduced to the drug discovery process [5]. Information on the source of new drugs during 1981–2007 specifies that approximately 50% of the drugs are based on natural products [6]. As herbal medicine contains naturally occurring bioactive constituents and nutritive substances, vitamin deficiency disorders can be cured using herbal medicines.

2 Role of Vitamins in the Body

Vitamins have a crucial role in maintaining health, yet only a tiny amount of them is required each day. Numerous cellular processes rely on them, including cell division and development, but their primary job is converting food into usable energy.

- Vitamin A helps teeth, bones, soft tissue, mucous membranes, and skin grow and stay healthy.
- Vitamin D is called the "sunshine vitamin" because the body makes it from sunlight. Most people at most latitudes need 10–15 min of exposure 3 times a week to synthesize enough vitamin D. Sun-deprived people may not produce enough vitamin D. Food alone cannot provide enough vitamin D. Vitamin D absorbs calcium. Calcium promotes healthy teeth and bones. It maintains calcium and phosphorus blood levels.
- Tocopherol, or vitamin E, is an antioxidant. It helps in vitamin K usage of the body and erythrocyte production.
- Blood clots cannot form without vitamin K (coagulate). There is evidence from scientific research that it benefits bone health.
- Vitamin C (Ascorbic acid) is an antioxidant that helps keep teeth and gums healthy. It enables the body to take in iron and keep its tissues healthy. It is also vital for healing wounds.
- Thiamine (vitamin B1) helps the body's cells turn carbs into energy. Getting enough carbs is very important in pregnancy or nursing conditions. It is also crucial for a healthy heart and nerve cells.
- Riboflavin (vitamin B2) helps the other B vitamins properly function in the body. It is also required for the growth and formation of red blood cells.
- Niacin is a vitamin that keeps skin and nerves healthy. At higher doses, it can also help lower cholesterol.
- Pantothenic acid (vitamin B5) is needed for the body to break down food. It also has something to do with the making of hormones and cholesterol.
- Vitamin B6 makes red blood cells and keeps the brain running well. This vitamin is also an important part of the proteins in the body, which are used in many chemical reactions.
- Folate and vitamin B12 work together to help make red blood cells. It is needed to make DNA, which controls how tissues grow and how cells work. Folate is important for any woman who is pregnant. Spina bifida and other birth defects can happen when there is not enough folate in the body. Folate is now added to a lot of foods in the form of folic acid [7].

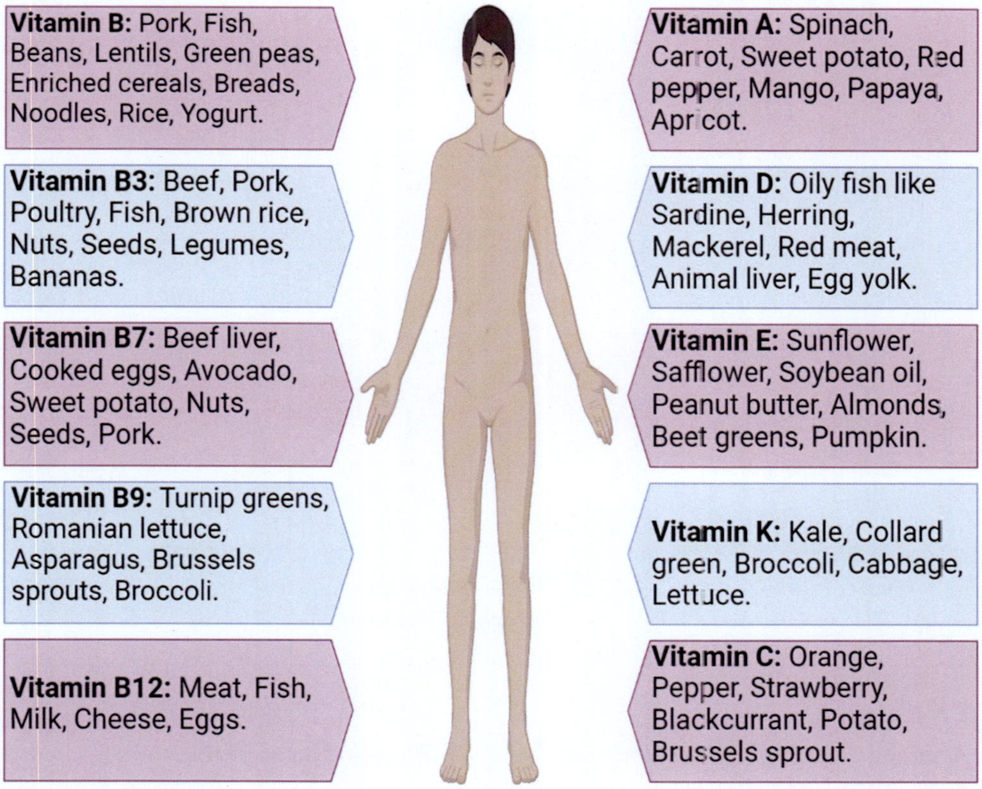

Fig. 2 Dietary sources of vitamins

3 Dietary Sources of Vitamins

Vitamins are considered "vital" since the body cannot produce them and must get them from the diet. Vitamin D boosts calcium absorption, whereas iron absorption is enhanced by vitamin C, and the B vitamins collaborate within cells. Most vitamins need to be consumed consistently rather than stored in the body, unlike vitamins A, E, and B12. Eating a variety of foods from the four main food groups such as (1) vegetables and fruits, (2) grain foods, (3) milk and milk products, and (4) legumes, nuts, seeds, fish and other seafood, eggs, poultry and red meat with the fat removed will help to achieve daily vitamin needs [8–10]. The dietary sources of vitamins are shown in Fig. 2.

Table 1 Vitamins and their required quantity

Vitamins	Required quantity	
	Men (intake per day)	Women (intake per day)
Vitamin A	700 μg	600 μg
Thiamine (B1)	1 mg	0.8 mg
Riboflavin (B2)	1.3 mg	1.1 mg
Niacin (B4)	16.5 mg	13.2 mg
Pantothenic acid (B5)	6 mg	4 mg
Pyridoxine (B6)	1.4 mg	1.2 mg
Biotin (B7)	30 μg	25 μg
Folic acid (B9)	400 μg	
Cyanocobalamin (B12)	2 μg	
Vitamin C	40 mg	
Vitamin D	10 μg	
Vitamin E	4 mg	3 mg
Vitamin K	70 μg	60 μg

4 Causes of Vitamin Deficiency

Inappropriate eating habits often cause vitamin deficiencies. Vitamins are a range of complex compounds found in various plant and animal-based foods. The daily requirement of vitamins is shown in Table 1. There are multiple food sources for each vitamin; some foods are fortified with vitamins. Milk, for instance, has vitamin D added to it and naturally includes calcium (a mineral, not a vitamin). Numerous vitamins are typically added to cereal, rice, and pasta. While proper vitamin intake from food sources is important, absorption issues can also arise from medical problems. Nutritional vitamin deficiency can be caused by any or all of

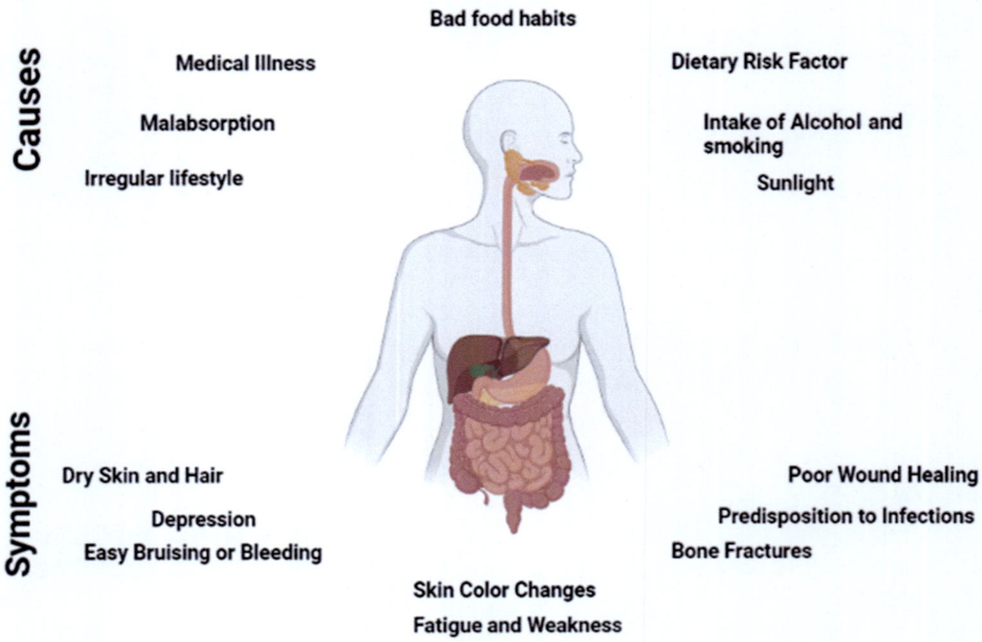

Fig. 3 Causes and symptoms of vitamin deficiency

the subsequent factors: inadequate consumption, poor absorption, inadequate use, increased need, increased excretion, and increased degradation in the body. The causes and symptoms of vitamin deficiency are shown in Fig. 3.

4.1 Dietary Risk Factors

Some eating plans are put at risk for vitamin deficiencies. Vegans and vegetarians are more likely to be deficient in vitamin B12 and biotin since these diets exclude animal sources. There is a possibility of vitamin D deficiency if dairy products are avoided. Grain products, typically enriched with vitamins in addition to their natural vitamin content, are scarce in a gluten-free diet. As a result, maintaining a gluten-free diet might result in vitamin deficiencies, notably in B vitamins such as thiamine and folate. A lack of vitamin E and K can occur on a diet high in processed foods and low in fresh fruits and vegetables [11].

4.2 Sunlight

Foods, including seafood, eggs, and dairy products, are good sources of vitamin D. Sun exposure can also provide vitamin D. Vitamin D insufficiency can be brought on by not getting enough sun. Generally, this happens in the winter in areas with a chilly environment [12].

4.3 Medical Illness

A range of diseases can hamper vitamin absorption and metabolism. Vitamin deficiencies typically result from various medical conditions, e.g. alcoholic liver disease, liver failure, kidney disease, chronic diarrhoea, malabsorption syndrome, gastric bypass, inflammatory bowel disease, Crohn's disease, irritable bowel syndrome, and pernicious anaemia [13].

5 Symptoms of Vitamin Deficiency

Symptoms of vitamin insufficiency range from lethargy to dry skin and hair to low mood to slow healing of wounds. However, many of the symptoms shared by different deficits are similar. Most people exposed to low amounts for a while will not see any visible signs until after a few months. The causes and symptoms of vitamin deficiency are shown in Fig. 3.

5.1 Fatigue and Weakness

Tiredness could be a sign of a vitamin D, B vitamin, or vitamin C deficiency. If the body does not acquire enough vitamin D, bones and muscles may suffer, as well as fatigue and weakness. Evidence links vitamin D deficiency to weariness, and studies show that supplementing with vitamin D can help treat this problem [14]. Anaemia, brought on by a lack of folate, is a common cause of tiredness and weakness. Because of a deficiency in healthy red blood cells, oxygen cannot be delivered to the body's organs and tissues. The body cannot function properly without oxygen, a vital energy source [15].

5.2 Dry Skin and Hair

Dry hair and cracked skin are common signs of vitamin A deficiency. Vitamins B, C, and D also show similar symptoms [16–18].

5.3 Depression

Vitamins B1, B3, B9, B12, vitamin C, and vitamin D deficiencies can sometimes cause clinical depression [19–21].

5.4 Easy Bruising and Bleeding

Easy bruising and bleeding can be caused by issues with blood coagulation, sluggish healing, or collagen development. (Collagen strengthens the blood vessel wall.) Significant vitamin deficiencies that can cause easy bruising or bleeding are vitamin C and vitamin K (mainly in neonates) [22, 23].

5.5 Poor Wound Healing

In case of poor wound healing, sores take longer to cure. Different vitamins, such as vitamins A, B, C, D, and K, are helpful in this healing process. Some of them promote collagen formation, while others aid in the healing of certain cell types or tissues and boost cellular health through the antioxidant action [24, 25].

5.6 Predisposition to Infections

Deficiency of some vitamins damages the system and can make it more prone to infections and infectious disorders. Examples of associated vitamins are—vitamins A, C, and D [26, 27].

5.7 Bone Fracture

Bone health and strength depend on numerous vitamins, including vitamins B6, B9, B12, C, D, and K. There is some debate as to whether or not taking nutritional supplements can reduce the incidence of fractures, even though evidence shows the deficiency of these vitamins can reduce bone density and eventually lead to fractures [28].

5.8 Skin Colour Changes

Loss of pigmentation in spots, hyperpigmentation, and overall pallor are probable outcomes of vitamin deficiency-induced alterations to the skin's hue. Possible causes of hypopigmentation and spotlightening include deficits in vitamin D (in light-skinned people).

Darker pigmentation can be due to vitamin B12 and D (in dark-skinned people). Deficiencies that can cause generally pale skin include vitamin C, vitamin B6, B9, and B12 [29]. Specific vitamin deficiency diseases for each vitamin are listed in Table 2.

Table 2 Types of vitamin deficiency diseases

Vitamin	Deficiency diseases
Vitamin A	Night blindness, xeropthalmia, keratomalacia
Vitamin D	Rickets, osteomalacia
Vitamin E	Neuromuscular, neurological disorders
Vitamin K	Reduced blood clotting leading to excessive bleeding
Vitamin B1 (Thiamin)	Beri-beri
Vitamin B2 (Riboflavin)	Retarded growth, bad skin
Vitamin B3 (Niacin)	Pellagra: dermatitis, dementia, diarrhoea
Vitamin B5 (Pantothenic acid)	Huntington's disease
Vitamin B6 (Pyridoxine)	Anaemia
Vitamin B7 (Biotin)	Alopecia and periorificial dermatitis
Folic acid	Megaloblastic anaemia
Vitamin B12 (Cyanocobalamin)	Pernicious anaemia
Vitamin C	Scurvy

6 Deficiency Disorders of Vitamins and Their Herbal Remedies

6.1 Vitamin A

Vitamin A is a lipid-soluble compound that is a necessary ingredient for human growth, survival, and the maintenance of the immune system, eyesight, and eye health [30]. It is available as a preformed vitamin (retinol) in dairy products, eggs, meat, and fish liver oil like cod liver oil [31]. Provitamin A or beta carotene is another vitamin A in dark green leafy vegetables and deep orange fruits.

6.1.1 Deficiency Disorders

Due to various causes, vitamin A deficiency and overconsumption are common in society, with both conditions resulting in poor health. However, vitamin A insufficiency is still a significant issue in South Asia and Africa, particularly among preschool-age children and women of reproductive age, most commonly at its nutritional deficiency and its adverse health consequences. The most severe clinical effects of vitamin A, like corneal xerophthalmia, severe sickness, and mortality, are most prevalent in older infants and young children of 6–59 months. In all groups with vitamin A deficiency, preventive supplementation should be given first attention [32].

A lack of vitamin A results in ocular surface (OS) abnormalities such as severe dry eye disease (DED), corneal punctate keratopathy, and goblet cell loss that progresses until they are absent. Extreme involvement causes keratinization and epithelial metaplasia, leading to xerophthalmia (OS keratinization with Bitot spots) and keratomalacia [33]. Xerophthalmia can also be directly caused by a vitamin A deficit. A condition in which the cornea swells up, the conjunctiva (the eye's covering) dries up, and the eyes get ulcerated, known as dry eye disease. Although the cause was unknown, early civilizations' records and traditions indicated that eating animals and fish livers had healing properties. Vitamin A should consequently be administered to everyone who has xerophthalmia.

6.1.2 Herbal Medicine for Vitamin A Deficiency

6.1.2.1 Antioxidants

Night blindness, cataract, and glaucoma can all be treated with herbal remedies. Since oxidative stress plays a crucial role in the pathogenesis of several common eye ailments, numerous studies have proven that herbal medications include a variety of antioxidants that may be effective for eye protection.

6.1.2.2 Carotenoids

The xanthophyll family of carotenoids, which includes lutein and zeaxanthin, is frequently found in tomatoes, Chinese wolfberries, and carrots. Additionally, lutein protects the retina from the oxidative stress brought on by diabetes [34].

6.1.2.3 Omega 3 Fatty Acid

Since oxidative stress has a major role in the pathogenesis of numerous common eye ailments, several studies have proven that herbal medications include a variety of antioxidants that may be effective for eye protection, which are essential fatty acids, crucial for maintaining good health. However since the body cannot synthesize them, they must be consumed through diet. Algal oil, flaxseed oil, and sea buckthorn seed are typical plant sources [35]. Numerous studies have been conducted on these omega-3 fatty acids' role in maintaining eye health.

6.2 Vitamin D

The fat-soluble vitamin 1a-25-dihydroxyvitamin D3 (vitamin D) is essential for human health and plays a role in bone metabolism, immunological function, and calcium homeostasis. Vitamin D is abundant in fish liver oils and several saltwater species, including herring, salmon, and sardines. The food chains from plankton are likely the source of the vitamin D found in fish liver.

6.2.1 Deficiency Disorders

Two metabolic bone illnesses, i.e. osteomalacia in adults and rickets in children, are known to be triggered by a lack of vitamin D [36]. Vitamin D insufficiency correlates to various chronic medical disorders, including osteoporosis, myopathy, lung illness, increased inflammation, and lowered immunity [37].

The lack of vitamin D has also been related to several cardiovascular risk factors [15]. Vitamin D deficiency can increase the generation of reactive oxygen species and the G protein RhoA by boosting the synthesis of renin and angiotensin II. Insulin resistance and metabolic syndrome are brought about by inhibiting intracellular glucose transporter pathways [17]. A typical ricket symptom is decreased Ca and P in the organic matrix of bone and cartilage. Since all three nutrients are essential for healthy bone production, the indications and symptoms are identical to those of a deficiency of Ca, P, or both. Osteoporosis is characterized by a loss in bone mass that can cause fractures with just minor damage. Rickets is caused by improper growth plate mineralization and is almost always found in developing youngsters. On the other hand, osteomalacia is caused by improper mineralization of the prefabricated osteoid and occurs in both adults and children in addition to rickets [36, 38]. When the bone mass is too low, skeletal integrity and mechanical support can no longer be preserved, and fractures can happen with little to no damage.

There are three different types of treatment for vitamin D deficiency: supplementation, natural ultraviolet B radiation, and sunshine. Every form of treatment has potential drawbacks and advantages [39]. Based on these findings, foods like milk and butter are now irradiated with UV light to increase their vitamin D content, and so help eliminate rickets as a major health problem around the world.

6.2.2 Herbal Medicine for Vitamin D Deficiency

For osteomalacia and rickets, Bladderwrack (*Fucus vesiculosis)* is used. Comfrey decoction and tincture Calendula (Marigold) are also used to treat osteomalacia and rickets. As an alternative to Comfrey, fenugreek seeds might be utilized. Fenugreek and St John's Wort tablets and capsules benefit from treating osteomalacia and rickets. Herbs like Chamomile, Dandelion, Clivers, Horsetail, Meadowsweet, Coltsfoot, Mistletoe, Plantain, Scarlet Pimpernel, Silverweed, Toadflax, and Shepherd's Purse can be consumed as teas, powders, tablets, or capsules. A herbal combination of Comfrey, Horsetail, Kelp, Lobelia, Marshmallow root, Oats, and Parsley root can also be used for the treatment. Along with this, exposure to sunlight is a supportive treatment option for vitamin D deficiency disease [40]. Osteomalacia is also treated using tablets or capsules prepared from Bamboo gum, Black Cohosh, Echinacea, Kelp, and Prickly Ash.

6.3 Vitamin E

Different lipid soluble components were first identified in 1922 by Evans and Bishop and were later grouped under the label "vitamin E" [41]. The vitamin E group together, known as chroman-6-ols or tocochromanols, is divided into tocopherols (α, β, γ, and δ tocopherols) and tocotrienols (α, β, γ, and δ tocopherols) that are present in food containing vitamin E as a whole [42]. Plants synthesize tocopherol and tocotrienol from homogentisic acid. Only α tocopherol satisfies the human vitamin E needs, even though these forms have antioxidant properties but cannot be transformed into one another [43]. Vegetable oils are one of the primary dietary sources of vitamin E. Vitamin E may also be found in nuts in reasonable amounts. For instance, vitamin α T is primarily present in peanuts, almonds, and sunflower seeds, whereas vitamin γ T is the main vitamin E in walnuts, pecans, pistachios, and sesame seeds [44, 45]. Compared to other oils, soybean, sunflower, maize, walnut, cottonseed, palm, and wheat germ oils have comparatively greater vitamin E concentrations (more than 50 mg of vitamin E/100 g oil) [43, 46].

For a diet high in vitamin E to work well, it must also be abundant in foods high in these other nutrients. α-Tocopherol is the predominant form of vitamin E in the circulation, although γ-tocopherol is consumed in relatively higher amounts from the diet than α-tocopherol due to the preferential binding affinity of the α-tocopherol transfer protein (α-TTP) for tocopherol. Transfer of α-tocopherol to the plasma membrane involves α-TTP [47].

In order to prevent lipid peroxidation and shield cell membranes from oxidative damage, vitamin E is the first line of defence [48]. Different experiment reports show that a combination of tocopherols had a higher inhibitory impact than alpha-tocopherol alone on lipid peroxidation in human erythrocytes. Its ability to scavenge peroxyl radicals also safeguards the polyunsaturated fatty acids in plasma lipoproteins and membrane phospholipids [49]. It has been discovered that although gamma-tocopherol captures and neutralizes the existing free radicals, the production of new free radicals is mainly prevented by alpha-tocopherol. Several ailments and diseases, including cancer, ageing, arthritis, and cataracts, have been related to oxidation. As a result, vitamin E may aid in the prevention or postponement of chronic disorders caused by reactive oxygen species molecules.

It has been discovered that elevating the level of alpha-tocopherol in endothelial cells prevents platelet aggregation and causes the endothelium to produce prostacyclin. This impact was brought on by the dysregulation of the vascular cell adhesion molecule (VCAM-1) and the intracellular cell adhesion molecule (ICAM-1), which reduced the adherence of blood cell constituents to the endothelium. Additionally, when vitamin E activates cytosolic phospholipase A2 [50] and cyclooxygenase 1 [51] in the arachidonic acid cascade, more prostacyclin is released, which acts as a significant vasodilator and inhibitor of platelet aggregation in humans [52]. Protein phosphatase 2A is activated by vitamin E, which increases PKC-dephosphorylation and reduces protein kinase C (PKC) activity. Vitamin E has also been shown to inhibit PKC in a number of different cell types, which has been shown to inhibit platelet aggregation, reduce the proliferation of monocytes, macrophages, neutrophils, and vascular smooth muscle cells, and reduce the production of superoxide in neutrophils and macrophages [42, 53].

6.3.1 Deficiency Disorders

Vitamin E insufficiency creates a specific neurological illness. It typically results from malabsorption aggravating chronic cholestasis, abetalipoproteinemia, celiac disease, or cystic fibrosis [54]. Malabsorption is a common cause of deficient states; however, they can also be brought on by a faulty ATT protein, which causes conditions like autosomal recessive ataxia with vitamin E deficiency (AVED) [54, 55]. Clinical signs of symptomatic vitamin E insufficiency comprise pigmented retinopathy, limb ataxia, hyporeflexia, loss of proprioception, loss of delicate touch, paraesthesia, and, very infrequently, myoclonic dystonia. Adults may develop symptoms after 10–20 years of malabsorption, suggesting the amount of time needed for vitamin E reserves to deplete and for clinically significant neuronal damage to accumulate [54].

6.3.2 Herbal Medicine for Vitamin E Deficiency

Treatment includes addressing the underlying cause of the deficit (fat malabsorption, abnormalities of the fat metabolism, among others), followed by oral vitamin E supplementation. Dietary changes can also help with supplementation. Increased consumption of green vegetables, whole grains, nuts, seeds, vegetable oils, and fortified cereals is strongly advised. Despite being often present in our meals, humans require 15 mg of vitamin E daily. One daily dosage of 15–25 mg/kg or 200 IU of mixed tocopherols might be utilized [56–59].

Alpha-tocopherol is advised to be consumed in the following amounts per day.

Ages 0–6 months: 3 mg;
Age 6–12 months: 4 mg;
Age 1–3 years: 6 mg; and
Age 4–10 years: 7 mg.
For adults and elderly: 10 mg.

6.4 Vitamin K

A clotting activity distinct from any known vitamin was found in an anti-haemorrhagic factor by Dam in 1935. Based on the spelling of "Koagulations" in German and Scandinavian languages, he proposed the term vitamin K. Beyond its recognized involvement in blood coagulation, the identification of additional hepatic vitamin K-dependent proteins has raised the possibility of an enlarged physiologic function for the vitamin K [60]. Vitamin K1 (phylloquinone) and vitamin K2 (menaquinone) are the two forms of natural vitamin K. Since all vitamin Ks are fat-soluble molecules with a 2-methyl-1,4-naphthoquinone nucleus in common, their 3-position side chains differ [61].

Phylloquinone is found in plants' green, leafy sections, where it serves as an electron receptor during photosynthesis. Phylloquinone is a thick, yellowish oil extracted in its purest form. In contrast to phylloquinone, menaquinone is a collection of vitamers with several isoprene units at the 3-position of the naphthoquinone ring structure [62].

Vitamin K has long been recognized for its crucial function in coagulation, but more recently, its potential significance in preserving bone health and preventing the calcification of soft tissues has come to light [62]. To activate glutamic acid (Gla) residues of vitamin K-dependent proteins, vitamin K works as a cofactor throughout this process inside the body [63]. The carboxylation process is necessary for the proteins that depend on vitamin K to bind calcium. Vitamin K's nutritional status has been evaluated using K-dependent protein carboxylation [64]. Factors II (prothrombin), VII, IX, X, and proteins C, S, and Z are the hepatic vitamin K-dependent proteins implicated in coagulation. All of these proteins require vitamin K for physiological activation [65].

Phylloquinone-based vitamin K is mainly absorbed from the proximal gut (80% absorption) after being solubilized into mixed micelles. Bile salts and pancreatic secretions must be present for vitamin K1 to be absorbed from the stomach into the lymphatic system. The vitamin then enters the circulation after being integrated into chylomicrons. Although K1 is the primary form of vitamin K that circulates in the body, K2 also has a lipoprotein distribution comparable to phylloquinone, making it available in plasma. Ninety per cent of the menaquinone and 10% of the phylloquinone are stored in the human liver. Since the liver's stocks of phylloquinone are unstable, excretion after 3 days reduces their starting levels by 25%. Menaquinones were shown to be more effective than phylloquinone in treating vitamin K insufficiency due to their longer biological half-lives and slower hepatic turnover [60, 61].

6.4.1 Deficiency Disorders of Vitamin K

Vitamin K deficiency is uncommon in adults and mainly occurs in infants and inborn. In 1894, a physician from Boston called Charles Townsend described a coagulopathy that exhibited all the signs of severe vitamin K (VK) deficiency. He termed the condition he described as haemorrhagic disease of the newborn after observing similarities between 50 cases of widespread bleeding in infants (HDN) [66]. Due to vitamin K deficiency's impact on homeostasis, insufficient generation of the functional molecules of factors II, VII, IX, and X results in a hypocoagulable state. There is a lot of room for the haemostatic system to work successfully at low-factor concentrations; nonetheless, there is a breaking point where the procoagulatory mechanisms give up, and bleeding occurs [66].

The major sign of vitamin K deficiency is bleeding (hemorrhage), which can occur in the stomach, intestines, nose, skin (producing bruises), or from a cut. Vomiting blood can occur when there is gastrointestinal bleeding. Urine or feces may have blood visible, or the stools may be tarry black. Babies can experience life-threatening bleeding in or around the brain. Because the liver is where clotting factors are produced, having a liver condition raises the risk of bleeding. Furthermore, a vitamin K deficit may weaken bones [67].

6.4.2 Herbal Medicine for Vitamin K Deficiency

Medications containing vitamin K, such phytonadione, are the standard treatment for vitamin K insufficiency. Both injectable and orally delivered, these drugs are used to rapidly increase vitamin K levels in the body.

In addition to consuming a range of vitamin K2 foods, eating sufficient plant-based foods high in vitamin K1 is essential. Foods that contain vitamin K1 include green leafy vegetables, natto (fermented soy), spring onions, brussels sprouts, cabbage, broccoli, prunes, cucumbers, and dried basil. Chives, parsley, oregano, cinnamon, shallots, tomatoes, lettuce, spring onions, spinach, peanut oil, sunflower oil, and olive oil can also be used to manage vitamin K deficiency [68].

6.5 Vitamin B

Vitamin B is a water-soluble component that includes vitamin B1 (thiamine), B2 (riboflavin), B3 (niacin), B5 (pantothenic acid), B6 (pyridoxine), B7 (biotin), B9 (folate), and B12 (cobalamin). These structurally different chemical molecules are needed for healthy cellular activities, growth, and development in all tissues [69]. Most water-soluble vitamins interact synergistically as important coenzymes in various metabolic pathways in the brain, primarily in the formation of cellular energy. They are necessary for central and peripheral nervous system development, myelination, and proper function [70, 71]. Except for niacin, the body cannot produce these vitamins; thus, they must be consumed regularly. A deficiency of a particular B vitamin may shift the rate-limiting step in cells to another pathway in the energy cycle, concealing a deficiency in another.

6.5.1 Deficiency Disorder of B Vitamins

B vitamins, such as folic acid, disguise vitamin B12 insufficiency in at-risk persons. Therefore, cell energy failure can lead to further neurological degeneration and poor clinical consequences. B vitamin deficiency causes various manifestations at different ages. Neural tube deficiency, mental retardation, language impairment, autonomic dysfunction, myelopathy, and optic neuropathy are shown in early life and neurodegeneration/cognitive decline, mood disorders in adulthood, and optic neuropathy in late life [72, 73].

Malabsorption is the most common cause of vitamin B12 deficiency, but the elderly, vegans, and ovo-lacto vegetarians with poor diets are also at risk. Vitamin B12 absorption issues in the ileum can also be brought on by a lack of IF production, atrophic gastritis, sickness, resection, bacterial overgrowth, drug–nutrient interactions, or rare genetic disorders [74]. Autoimmune gastritis causes pernicious anaemia by preventing the body from producing enough IF. Loss of IF is the root cause of vitamin B12 insufficiency, leading to megaloblastic anaemia and neurological consequences if left untreated. Removing the part of the stomach responsible for producing IF and acid also leads to vitamin B12 insufficiency. Competition for vitamin B12 in the ileum may reduce absorption in those with bacterial overgrowth or parasites. Ileal resection and inflammatory bowel diseases like Crohn's disease can also induce vitamin B12 malabsorption [74].

A deficiency of vitamin B12 also connects with cardiovascular problems [75], Osteoporosis [75], and neural tube defects [76]. There are no naturally occurring bioactive forms of vitamin B12 derived from plants. Some plant foods, such as seaweed and mushrooms, provide additional vitamin B12. Two studies, however, suggest that particular types of Japanese seaweed (nori) may have alleviated vitamin B12 insufficiency in vegans. In contrast, others, like mushrooms and seaweed, contain vitamin B12 mimics inactive in humans. There have been reports of vitamin B12 in some meals that have been tainted or fermented by bacteria, such as tempeh and Thai fish sauce, however, these foods may not have a high affinity for IF and may not be well absorbed.

6.5.2 Herbal Treatment of B Vitamin Deficiency

Researchers have discovered that wheatgrass is a good source of B vitamins, minerals, and antioxidants, making it helpful in treating B vitamin deficiencies. Since it purifies the blood and boosts its oxygen-carrying ability, it is sometimes called "green blood". Its natural green pigment is remarkably comparable to the HEME in human blood, which boosts the effectiveness of haemoglobin. It promotes digestion and helps with intestinal cleansing. They reduce acidity and cope with ulcers, hepatitis, wrinkles, and general weakness.

Andrographis paniculata, often known as KALMEGH, is regarded as the "KINGS OF BITTER" due to its intensely bitter flavour. Andrographolide, the primary active component of leaves, is responsible for the anti-inflammatory effect. Vitamin deficiency symptoms may also be alleviated by taking this supplement. In Ayurvedic scriptures, punarnava is used in a herbomineral preparation to treat anaemia, liver illness, and oedema. Punarnava is an excellent natural supplement for increasing haemoglobin levels, boosting red blood cell count, and enhancing overall blood quality. Vitamin and mineral-rich herbs include Amalaki, jeevanti, ashwagandha, pipali, shigru, and guggul. Vitamin B and other vitamin deficiencies, and the symptoms caused by such deficiencies, can be alleviated by using these plants [76].

6.6 Vitamin C

Vitamin C or ascorbic acid, is a water-soluble and simple, low-molecular-weight carbohydrate [77]. An essential element, like other nutrients, must be taken in regularly to avoid a deficiency [78]. Vitamin C consumption, effective absorption, recycling, and renal reuptake are all essential for keeping vitamin C levels in the body stable [79]. Sodium-dependent vitamin C transporters (SVCTs) strictly regulate the levels of vitamin C in the body's blood and tissues [80]. Vitamin C requirements may vary among tissues and organs. The vitamin is most concentrated in the brain, adrenal, and pituitary glands. One of vitamin C's primary roles is as a cofactor for a group of metalloenzymes involved in biosynthesis and regulation, including those involved in producing catecholamines and peptide hormones [81]. Vitamin C has been shown to play a role in epigenetic regulation, specifically through enzymes that control the methylation of DNA and histones and gene transcription. Therefore, it plays pleiotropic functions in human health and disease by regulating thousands of genes [82, 83]. A daily vitamin C consumption of 100–200 mg will keep blood concentrations at a healthy range of 50–75 mol/L [84].

Vitamin C (ascorbic acid) promotes keratinocyte development and decreases melanin formation, protecting against UV-induced photodamage. Vitamin C forms the skin barrier and collagen in the dermis, fights skin oxidation, and modulates cell signal pathways of cell proliferation and differentiation. Normal skin needs high amounts.

Vitamin C supports innate and adaptive immune system cell activities and protects against environmental oxidative stress by keeping epithelial barrier function and skin oxidant scavenging. Vitamin C increases chemotaxis, phagocytosis, reactive oxygen species, and microbial death in phagocytic cells like neutrophils. Apoptosis and macrophage clearance of wasted neutrophils reduce necrosis and tissue injury [81, 85].

6.6.1 Deficiency Disorders of Vitamin C

Atopic dermatitis (AD) and porphyria cutaneatarda can be caused by vitamin C deficiency (PCT) [86]. Maternal vitamin C deficiency may have severe consequences for the offspring, as evidenced by studies showing that the absence of vitamin C in the brain is detrimental to survival in newborn sodium-dependent vitamin C transporter (SVCT2) (/) (−/−) mice and that prenatal deficiency reduces hippocampal volume and neuron number and causes decreased spatial cognition in guinea pigs. The risk of stroke and its severity may also be affected by vitamin C deficiency with ageing [87].

From the 1930s until the 1950s, ascorbic acid's effects on gastritis and peptic ulcers and their consequences were researched. The ascorbic acid shortage was linked to all types of gastritis (autoimmune, chemical, and viral) due to insufficient intake, increased metabolic requirements, and GI tract damage. Significantly, *H. pylori* eradication reverses gastritis-associated anomalies in gastric ascorbic acid metabolism, but proton pump inhibitor therapy may aggravate them.

Diets high in naturally occurring ascorbic acid may protect the gastric corpus against atrophy and reduce stomach cancer. Ascorbic acid may reduce peptic ulcer haemorrhage. *H. pylori* eradication therapy may benefit from pharmacologic ascorbic acid dosages [88–90].

A lack of vitamin C is the root cause of scurvy. One of the ascorbic acid's primary functions is to aid in producing collagen. Type IV collagen is the primary structural protein in the skin, blood vessel walls, and the basement membrane zone that separates the epidermis and dermis. The lysyl enzyme hydroxylase is required for the hydroxylation and cross-linking of pro-collagen, and vitamin C facilitates this process. A deficiency of vitamin C inhibits the transcription of pro-collagen.

Furthermore, ascorbic acid deficiency causes epigenetic DNA hypermethylation, reducing the transcription of collagen types used in skin, blood vessels, and other tissues. Finally, haemorrhaging is a hallmark of scurvy and can affect virtually any organ. Additionally, bone production is disrupted, leading to brittle bones [91, 92].

Periodontal disease in diabetics is associated with vitamin C deficiency. Glucose may hamper, while insulin may improve ascorbate's transport across cell membranes. Sublethal amounts of endotoxin significantly speed up glucose consumption. Ascorbic acid insufficiency changes mucosal barrier function and increases tissue sensitivity to histamine in response to endotoxin [93, 94]. Vitamin C deficiency is related to rheumatoid fever and rheumatoid arthritis [95].

6.6.2 Herbal Medicine for Vitamin C Deficiency

In terms of vitamin C content, amla has the highest concentration compared to any plant, and when the whole fruit is used rather than just the active element, the body has no trouble absorbing it. Tannins stabilize amla fruit vitamin C. Thus, they cannot be degraded by exposure to heat or light [96, 97]. In addition to being a natural anti-ageing remedy, Amla Tonic helps by acting as a hematinic and lipolytic, preventing indigestion and controlling acidity.

The antioxidant capacity of *Terminalia ferdinandiana* is remarkable. The ascorbic acid content per gram of the fruit of *T. ferdinandiana* is more than 900 times higher than in blueberries. High quantities of additional antioxidants, such as phenolic compounds and anthocyanins, are also present in *T. ferdinandiana*. Many diseases, including cancer, heart disease, and neurodegenerative disorders, have been linked to oxidative stress and inflammation, but recent research suggests that antioxidants may help ward off these conditions. They help reduce obesity and have been connected to antidiabetic bioactivities. The damage induced by oxidative stress to proteins, lipids, and nucleic acids can be avoided by supplementing the diet with antioxidants, which can directly scavenge free radicals [98, 99].

Broccoli and cabbage get most of their natural antioxidant activity from phenolic compounds and vitamin C, accounting for 10–12% of the total antioxidant capacity [100]. Kale is rich in different nutrients. Bioactive substances such as phenolic compounds, vitamin C, glucosinolates, and carotenes have been identified in it. It contains between 237 and 572 mg/100 g of vitamin C [101].

Camu-Camu (*Myrciaria dubia)* is a shrub belonging to the Myrtaceae family. It is native to the Amazon basin and thrives in wet, swampy environments. The fruit ranges from pink to dark purple and has a diameter of 1–3 cm. It contains 850–5000 mg of vitamin C per 100 g of edible material, making it the best natural source of vitamin C [102].

Cayenne contains 3.82 mg of vitamin C in 1 tablespoon of it. For an adult, that is equivalent to 6% of the recommended daily allowance. In addition to being a great antioxidant source, it is also a mineral powerhouse. Cilantro, often known as coriander, is a fragrant green leafy herb. It possesses carminative, anti-oxidative, and prevents infection. The seeds from cilantro plants include healthy fats and vitamin C. Vitamin C is abundant in citrus fruits like oranges and grapefruits. The amount of vitamin C in a medium-sized orange is 70 mg. There are 93 and 70 mg of vitamin C in an 8-oz drink of orange and grape juice. Papaya has 10% of the recommended daily value of vitamin C in only one serving. Hundred grams of kiwi fruit and tomato contains 70 and 100 mg of vitamin C [103]. Herbal remedies for all vitamin deficiency are shown in Table 3

Table 3 Herbal remedies for all vitamin deficiency

Vitamins	Herbal remedies	References
Vitamin A	Carotenoids (tomatoes, Chinese wolfberries, and carrots) Omega 3 fatty acid (algal oil, flaxseed oil, and sea buckthorn seed)	[11, 12]
Vitamin D	Bladderwrack (*Fucus vesiculosis),* Comfrey decoction, tincture Calendula (Marigold), fenugreek and St John's Wort tablet, chamomile, Clivers, dandelion, horsetail, coltsfoot, meadowsweet, mistletoe, plantain, scarlet pimpernel, silverweed, Shepherd's purse, and toadflax	[17]
Vitamin E	Sunflower seeds, Almonds, wheat gram oil, and butternut squash	[33–36]
Vitamin C	Amla, *Terminalia ferdinandiana, Myrciaria dubia,* and Cilantro	[98–102]
Vitamin B	Wheat grass, Kalmegh, Amla, Jeevanti, Ashwagandha, Pipali, Shigru, and Guggul	[76]

7 Development and Analytical Approaches for Herbal Medicine (Fig. 4)

Globally, there has been a long history of change in the forms and applications of herbal therapy. Millions of people have used herbal remedies or comparable procedures during the past few decades, indicating that they have become increasingly popular globally. Herbal medications have gained growing interest due to their distinctive effects and multiple components and targets. But in today's era of evidence-based medicine, it is vital for the advancement of herbal medicines to undergo scientific review employing modern medical practice procedures. Interest in studying herbal medicine has risen due to the potential benefits of plant secondary metabolites in public health and the identification of lead compounds in medication discovery [104]. The combination of chromatographic separation techniques and spectroscopic methods significantly improved the efficiency of the isolation and identification of natural products [105].

Introducing system biology technologies and omics techniques to phytomedicine science, such as proteomics, genomics, and metabolomics, aided in creating herbal medications. However, metabolomics has proven to be faster and more accurate than any other technology for identifying patterns in biological materials. High-throughput activity screening was brought into the investigation of bioactive components of herbal medicines with the progress of contemporary biotechnology, giving a valuable tool in the research process [106]. Computational approaches, most commonly pharmacophore-based virtual screening and molecular docking, are quite well-established tools in drug discovery for identifying plant materials with high biological activity. Following almost a decade of development, chinmedomics has been successfully used for the research of hundreds of herbal formulas or herbal drugs. These successful operations have accumulated a plethora of important data for the scientific interpretation of critical challenges in the manufacture of herbal medicines, such as confirmation of therapeutic effects, investigation of pharmacodynamic component foundation, and the establishment of quality standards [107].

Because there are so many vitamin forms that occur naturally in food, liquid chromatography (LC), including high-performance liquid chromatography (HPLC), ultra-high-performance liquid chromatography (UHPLC), nano-liquid chromatography (nano-LC), and two-dimensional liquid chromatography (2D-LC) are the technique best suited for fat-soluble vitamin (FSV) analysis [108].

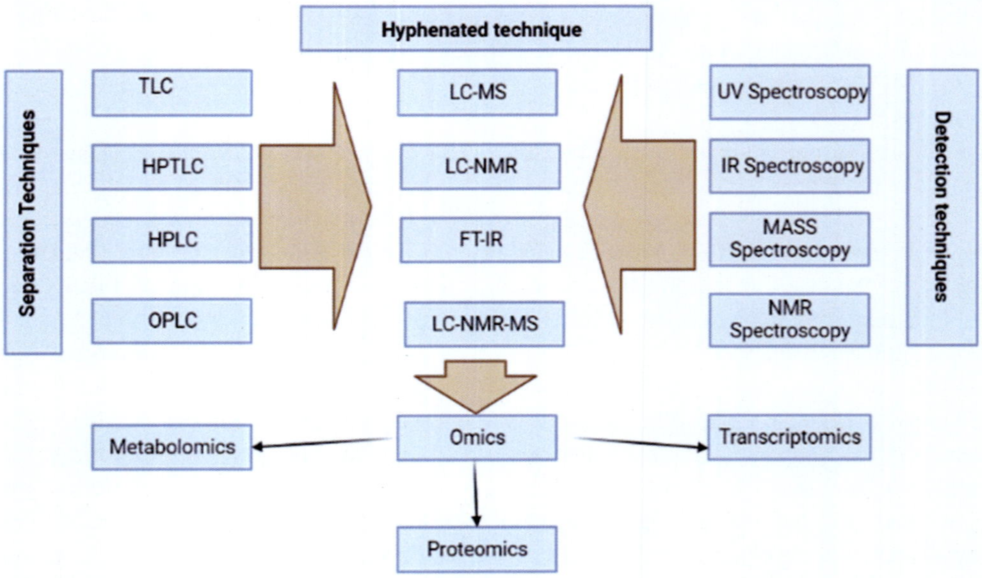

Fig. 4 Various techniques for the development and analysis of herbal medicines

FSVs and carotenoids are found in animal-based diets' lipid fraction, mostly made up of triglycerides and partially of sterols and phospholipids. The comparable solubility of these compounds makes it difficult to isolate the vitamins and creates a cause of disturbances during the LC analysis [109, 110]. Saponification is another helpful technique for eliminating undesirable lipids and chlorophylls, but it should only be used selectively, such as when analysing foods of animal origin. It is more costly to use lipase-based enzymatic digestion to extract vitamin K from milk and other dairy products [111, 112]. In this section, we have discussed different qualitative and quantitative analytical approaches for the identification of vitamin content from herbal sources.

7.1 Qualitative Test

7.1.1 Vitamin A

Dissolve 250 mg of the powder sample in 5 mL chloroform. Filter the mixture. Add 5 mL antimony trichloride solution. Instantaneously a fleeting blue colour will form.

7.1.2 Vitamin C

Add 1 drop of freshly prepared 5% w/v sodium nitroprusside solution and 2 mL of dilute sodium hydroxide solution to 1 mL of aqueous sample solution after diluting with 5 mL of water. Drop by drop, add 0.6 mL of HCl, and stir until the yellow colour changes to blue.

7.1.3 Vitamin D

Dissolve 500 mg of plant material in 10 mL of chloroform and filter. Add 10 mL antimony solution, immediately, a pinkish-red hue will arise.

7.1.4 Vitamin E

Soak 500 mg of powdered sample in 10 mL of ethanol for 5 min. Add 0.1% of ferric chloride in ethanol and 0.25 mL of 2–2 dipyridyl to 1 mL filtrate. The white background becomes red. The background turns pink [113, 114].

7.2 Quantitative Test

7.2.1 Riboflavin

Extract 5 g of plant sample with 100 mL of 50% ethanol and agitate for 1 h pipette out 10 mL of the extract into a 50 mL volumetric flask. Add 10 mL of 5% potassium permanganate and 10 mL of 30% H_2O_2 and leave for 30 min. Then add 2 mL of 40% sodium sulphate. Measure the absorbance at 510 nm [115].

7.2.2 Niacin

Homogenize fresh or dried plant samples in sodium hydroxide and distilled water. Add 17 g of ammonium sulphate after the liquid is heated for 1 h in a boiling water bath, cool the solution, and adjust the pH to 4.5. Colour will be developed from the interaction of extract with cyanogen bromide. Evaluate the absorbance at 450 nm [116].

7.2.3 Thiamin

Homogenize 0.5 g of material in 50 mL ethanolic sodium hydroxide. Add 10 mL filtrate to 10 mL potassium dichromate. Measure the absorbance at 360 nm [115].

7.2.4 Ascorbic Acid

Mix 5 g sample with 100 mL EDTA/TCA (2:1). Centrifuged this mixture at 3000 rpm for 20 min. Titrate 20 mL of this mixture with 20% $CuSO_4$ until a dark endpoint appears.

7.2.5 Vitamin A

Mix 1–5 g of plant material with 1 mL of KOH solution for saponification. Reflux the tube for 20 min at 60 °C, cool the mixture, and add 20 mL water. Extract this with 10 mL of pet. Ether twice in a separating funnel. Separate the organic layer and add sodium sulphate to remove moisture for 30–60 min. Evaporate 5 mL of ether extract at 60 °C. The residue is dissolved in chloroform. Add 2 mL of TCA solution to the chloroform mixture. Measure the absorbance at 620 nm.

Phytochemical assays are based on the following steps: separation, which includes TLC, HPLC, and HPTLC and detection techniques like UV, IR, NMR, mass spectroscopy, and FT-IR. Various techniques for the development and analysis of herbal medicines are shown in Fig. 4.

7.3 Thin Layer Chromatography (TLC)

In thin-layer chromatography, adsorption, and partition serve as the principle. The stationary phase is created as a thin layer on a flat surface, like, a plastic plate or glass surface or aluminum sheet. In this instance, a mixture is separated across a thin layer of alumina (Al_2O_3) or silica gel, where the components are absorbed by specific physical forces [117].

7.4 High-Performance Liquid Chromatography (HPLC)

High-performance liquid chromatography (HPLC) is a technique for separating chemicals based on their interaction with a packed column and a mobile phase solvent. High pressures of up to 400 bar are needed to elute the analyte via the column and detector. HPLC helps with substances that cannot be vaporized or disintegrate at high temperatures. HPLC analyses quantitatively and qualitatively in one step [118].

7.5 High-Performance Thin Layer Chromatography (HPTLC)

HPTLC is an improved version of TLC. Thin films with excellent performance planar chromatography is a technique for separating components of a sample by using high-performance layers and a sophisticated workstation. These layers are pre-coated with a 5–7 μm sorbent and 150–200 μm thick. Thinner layers and smaller particles increase plate efficiency and separation. HPTLC allows for qualitative, quantitative, and micro-preparative analyses [119].

7.6 Optimum Performance Laminar Chromatography (OPLC)

OPLC is a TLC/HPLC hybrid. The device separates 10–15 mg samples and can handle 4 or 8 samples simultaneously. In OPLC, a pump forces a liquid mobile phase through a silica or bonded-phase stationary medium. The OPLC column structure allows the use of flat planar columns as cylinders. A solvent delivery pump forces the mobile phase through a 50-bar-pressurized flat column at constant linear velocity. The workstation includes a four-channel diode array detector for monitoring the eluent, a solvent delivery system with a mobile phase degasser and pump, and an OPLC purification unit for collecting fractions in six 96-well plates [120].

7.7 UV Spectroscopy

UV and visible spectroscopy measure light attenuation after passing through a sample or reflecting off a sample surface. UV spectroscopy is used to investigate various components, including inorganic substances in solution because UV radiation can boost outer electron energies. Energy level transitions cause this. Using Beer's law, Lambert's absorbance at a specific wavelength can calculate analyte solution concentration. Absorption, transmission, and reflectivity of pigments and plant components can be characterized. This qualitative application needs to record a section of the UV–visible spectrum to describe optical or electronic materials [121].

7.8 IR Spectroscopy

The functional group of a material is determined via infrared spectroscopy. Infrared absorption spectroscopy determines the wavelength and intensity of absorption in a sample. Molecular vibrations are increased by mid-infrared light. IR spectroscopy is most useful for qualitatively studying organic and organometallic compounds. IR spectroscopy confirms a compound's identity and identifies newly produced molecules [122].

7.9 Mass Spectroscopy

Mass spectrometry identifies compounds effectively. Mass spectrometry can analyse tiny molecules and characterize proteins, making it an essential biological technique. Mass spectrometry discovers undiscovered molecules, quantifies existing compounds, and reveals molecular structure and properties. By simultaneously identifying the molecular weight and a diagnostic fragment of the molecule, this method is useful in a wide variety of situations, including the structural elucidation of organic molecules, the sequencing of peptides or oligonucleotides, and the monitoring of previously characterized compounds in complex mixtures [123].

7.10 Nuclear Magnetic Resonance (NMR) Spectroscopy

The NMR Spectroscopy reveals physical, chemical, and biological features. NMR spectroscopy is a one-dimensional technique used to examine the chemical structure. Complex molecules are represented as two-dimensional models. These methods are gradually replacing X-ray crystallography. Time domain NMR probes molecular dynamics in solutions. SS Solids' molecular structure is determined via NMR. ^{13}C-NMR identifies carbon types in a chemical. ^{1}H-NMR determines the types of hydrogen in a chemical and how they are related [124].

7.11 X-Ray Crystallography

X-ray crystallography utilizes how crystals diffract X-rays. X-rays have the right wavelength (10–8 Å) to scatter off an atom's electron cloud. The crystal's electron density can be reconstructed from its X-ray diffraction pattern. Additional phase information from diffraction data or supplemental experiments may be used to complete the reconstruction. An accurate molecular structure is produced by refining a model based on data and experimental electron density [125].

7.12 Fourier Transform Infrared (FTIR) Spectroscopy

FTIR is a recent breakthrough in collecting and translating data from an interference pattern to a spectrum. FTIR identifies organic and inorganic compounds. The FTIR technique can analyse solids, liquids, and gaseous mixtures. Computerized FTIR instruments are rapid and have greater sensitivity than dispersive ones. FTIR can identify paints, polymers, coatings, medicines, and pollutants. FTIR is best for finding chemical bonding (functional groups). The chemical bond's wavelength is shown in this annotated spectrum.

Recent advancement in analytical science makes hybrid techniques, i.e. LC-MS, LC-NMR, GC–MS, LC–PDA, LC–FTIR, LC–NMR, LC–NMR–MS, and CE–MS. Along with these inventions, several omic approaches, like metabolomics, proteomics, and transcriptomics, are widely used for analysis.

As we use phytochemicals for medications and other uses, studying the genetic basis of secondary plant metabolites helps expand their utility. The current edge in plant metabolome analysis is marginal for metabolite signals [126]. The database for metabolomic analysis needs three types of data and infrastructure to elucidate metabolite structures. They are (1) tandem mass spectra for metabolite structure elucidation, (2) a mass spectral database of phytochemicals, and (3) a way to detect false positives. Further development of metabolomics technologies and databases is complex yet necessary for studying secondary plant metabolites.

8 Conclusion

Herbal medicine has grown interest in treating various diseases due to the biodiversity of medicinal plants, low cost, synergistic effect, and negligible side effects. To date, vitamin deficiency disorders have been treated using conventional medicine, mainly the synthetic analogue of vitamins, but these medicines are costly and unaffordable for the economically weaker section of society. Herbal medicine can be a better and safer option for managing vitamin deficiency diseases irrespective of all economic classes. Secondary metabolites obtained from natural sources have an enormous biological activity to mitigate various ailments. In this context, this book chapter focuses on alternative treatment for vitamin deficiency disorders using herbal sources provided with analytical methodologies.

Acknowledgments The authors would like to thank the Department of Pharmaceuticals, Ministry of Chemicals and Fertilizers, for their support.

References

1. Ball GF. Vitamins: their role in the human body. New York: John Wiley & Sons; 2008.
2. Sinbad OO, Folorunsho AA, Olabisi OL, Ayoola OA, Temitope EJ. Vitamins as antioxidants. J Food Sci Nutr Res. 2019;2(3):214–35.
3. Maqbool MA, Aslam M, Akbar W, Iqbal Z. Biological importance of vitamins for human health: a review. J Agric Basic Sci. 2018;2(3):50–8.
4. Mathew L, Babu S. Phytotherapy in India: transition of tradition to technology. Curr Bot. 2011;2(5):26–30.
5. Newman DJ, Cragg GM. Natural products as sources of new drugs over the last 25 years. J Nat Prod. 2007;70(3):461–77.
6. Butler MS. Natural products to drugs: natural product-derived compounds in clinical trials. Nat Prod Rep. 2008;25(3):475–516.
7. https://medlineplus.gov/ency/article/002399.htm#:~:text=Vitamins%20are%20a%20group%20of,the%20body%20to%20work%20properly. Accessed 10 Jan 2023.

8. Benedik E. Sources of vitamin D for humans. Int J Vitam Nutr Res. 2022;92(2):118–25.
9. Shahidi F, Pinaffi-Langley ACC, Fuentes J, Speisky H, de Camargo AC. Vitamin E as an essential micronutrient for human health: common, novel, and unexplored dietary sources. Free Radic Biol Med. 2021;176:312–21.
10. Watanabe F. Vitamin B12 sources and bioavailability. Exp Biol Med. 2007;232(10):1266–74.
11. Vici G, Belli L, Biondi M, Polzonetti V. Gluten free diet and nutrient deficiencies: a review. Clin Nutr. 2016;35(6):1236–41.
12. Holick MF. Sunlight and vitamin D: both good for cardiovascular health. J Gen Intern Med. 2002;17(9):733.
13. Zhang FF, Barr SI, McNulty H, Li D, Blumberg JB. Health effects of vitamin and mineral supplements. BMJ. 2020;369:m2511.
14. Nowak A, Boesch L, Andres E, Battegay E, Hornemann T, Schmid C, et al. Effect of vitamin D3 on self-perceived fatigue: a double-blind randomized placebo-controlled trial. Medicine. 2016;95(52):e5353.
15. Tardy A-L, Pouteau E, Marquez D, Yilmaz C, Scholey A. Vitamins and minerals for energy, fatigue and cognition: a narrative review of the biochemical and clinical evidence. Nutrients. 2020;12(1):228.
16. Umar M, Sastry KS, Al Ali F, Al-Khulaifi M, Wang E, Chouchane AI. Vitamin D and the pathophysiology of inflammatory skin diseases. Skin Pharmacol Physiol. 2018;31(2):74–86.
17. Rembe J-D, Fromm-Dornieden C, Stuermer EK. Effects of vitamin B complex and vitamin C on human skin cells: is the perceived effect measurable? Adv Skin Wound Care. 2018;31(5):225–33.
18. Pullar JM, Carr AC, Vissers MC. The roles of vitamin C in skin health. Nutrients. 2017;9(8):866.
19. Mikkelsen K, Stojanovska L, Apostolopoulos V. The effects of vitamin B in depression. Curr Med Chem. 2016;23(38):4317–37.
20. Pullar JM, Carr AC, Bozonet SM, Vissers MC. High vitamin C status is associated with elevated mood in male tertiary students. Antioxidants. 2018;7(7):91.
21. Xie F, Huang T, Lou D, Fu R, Ni C, Hong J, et al. Effect of vitamin D supplementation on the incidence and prognosis of depression: an updated meta-analysis based on randomized controlled trials. Front Public Health. 2022;10:903547.
22. Galimberti F, Mesinkovska NA. Skin findings associated with nutritional deficiencies. Cleve Clin J Med. 2016;83(10):731–9.
23. Wolpert K, Szadkowski M, Miescier M, Hewes HA. The presentation of a fussy infant with bruising: late-onset vitamin K deficiency bleeding. Pediatr Emerg Care. 2019;35(4):e70–1.
24. Barchitta M, Maugeri A, Favara G, Magnano San Lio R, Evola G, Agodi A, et al. Nutrition and wound healing: an overview focusing on the beneficial effects of curcumin. Int J Mol Sci. 2019;20(5):1119.
25. Pazyar N, Houshmand G, Yaghoobi R, Hemmati AA, Zeineli Z, Ghorbanzadeh B. Wound healing effects of topical Vitamin K: a randomized controlled trial. Indian J Pharmacol. 2019;51(2):88.
26. Katona P, Katona-Apte J. The interaction between nutrition and infection. Clin Infect Dis. 2008;46(10):1582–8.
27. Som PM, Brandwein MS. Inflammatory diseases. In: Head and neck imaging, vol. 1. St. Louis: Mosby; 2003. p. 193–259.
28. Burt LA, Billington EO, Rose MS, Raymond DA, Hanley DA, Boyd SK. Effect of high-dose vitamin D supplementation on volumetric bone density and bone strength: a randomized clinical trial. JAMA. 2019;322(8):736–45.
29. Kristine Hoffman D. When vitamin and nutritional deficiencies cause skin and nail changes. Podiatry Today. 2014;28(1).
30. Rice AL, West KP Jr, Black RE. Vitamin A deficiency. In: Comparative quantification of health risks: global and regional burden of disease attributable to selected major risk factors, vol. 1. Geneva: WHO; 2004. p. 211–56.
31. Sommer A. Xerophthalmia and vitamin A status. Prog Retin Eye Res. 1998;17(1):9–31.
32. Ross DA. Recommendations for vitamin A supplementation. J Nutr. 2002;132(9):2902S–6S.
33. Fogagnolo P, De Cilla' S, Alkabes M, Sabella P, Rossetti L. A review of topical and systemic vitamin supplementation in ocular surface diseases. Nutrients. 2021;13(6):1998.
34. Chu K-O, Pang C-P. Herbal molecules in eye diseases. Taiwan J Ophthalmol. 2014;4(3):103–9.
35. Fatima T, Snyder CL, Schroeder WR, Cram D, Datla R, Wishart D, et al. Fatty acid composition of developing sea buckthorn (*Hippophae rhamnoides* L.) berry and the transcriptome of the mature seed. PLoS One. 2012;7(4):e34099.
36. Roth DE, Abrams SA, Aloia J, Bergeron G, Bourassa MW, Brown KH, et al. Global prevalence and disease burden of vitamin D deficiency: a roadmap for action in low- and middle-income countries. Wiley Online Library; 2018. Report no.: 0077-8923.
37. Fitzpatrick TB, Basset GJ, Borel P, Carrari F, DellaPenna D, Fraser PD, et al. Vitamin deficiencies in humans: can plant science help? Plant Cell. 2012;24(2):395–414.
38. Uday S, Högler W. Nutritional rickets & osteomalacia: a practical approach to management. Indian J Med Res. 2020;152(4):356.
39. Cannell J, Hollis B, Zasloff M, Heaney R. Diagnosis and treatment of vitamin D deficiency. Expert Opin Pharmacother. 2008;9(1):107–18.
40. Bartram T. Bartram's encyclopedia of herbal medicine. New York: Hachette UK; 2013.
41. Rizvi S, Raza ST, Ahmed F, Ahmad A, Abbas S, Mahdi F. The role of vitamin E in human health and some diseases. Sultan Qaboos Univ Med J. 2014;14(2):e157.
42. Zingg J-M. Vitamin E: a role in signal transduction. Annu Rev Nutr. 2015;35:135–73.
43. Lee GY, Han SN. The role of vitamin E in immunity. Nutrients. 2018;10(11):1614.
44. Jiang Q. Natural forms of vitamin E: metabolism, antioxidant, and anti-inflammatory activities and their role in disease prevention and therapy. Free Radic Biol Med. 2014;72:76–90.
45. Dreher ML. Pistachio nuts: composition and potential health benefits. Nutr Rev. 2012;70(4):234–40.
46. Sheppard AJ, Pennington JA, Weihrauch J. Analysis and distribution of vitamin E in vegetable oils and foods. In: Vitamin E in health and disease. London: IntechOpen; 1992. p. 9–31.
47. Traber MG. Vitamin E regulatory mechanisms. Annu Rev Nutr. 2007;27:347–62.
48. Howard AC, McNeil AK, McNeil PL. Promotion of plasma membrane repair by vitamin E. Nat Commun. 2011;2(1):1–8.
49. Tran K, Wong JT, Lee E, Chan AC, Choy PC. Vitamin E potentiates arachidonate release and phospholipase A2 activity in rat heart myoblastic cells. Biochem J. 1996;319(2):385–91.
50. Freedman JE, Farhat JH, Loscalzo J, Keaney JF. α-Tocopherol inhibits aggregation of human platelets by a protein kinase C-dependent mechanism. Circulation. 1996;94(10):2434–40.
51. Li D, Saldeen T, Romeo F, Mehta JL. Different isoforms of tocopherols enhance nitric oxide synthase phosphorylation and inhibit human platelet aggregation and lipid peroxidation: implications in therapy with vitamin E. J Cardiovasc Pharmacol Ther. 2001;6(2):155–61.
52. Brigelius-Flohé R, Traber MG. Vitamin E: function and metabolism. FASEB J. 1999;13(10):1145–55.
53. Traber MG, Atkinson J. Vitamin E, antioxidant and nothing more. Free Radic Biol Med. 2007;43(1):4–15.
54. Aslam A, Misbah S, Talbot K, Chapel H. Vitamin E deficiency induced neurological disease in common variable immunodeficiency: two cases and a review of the literature of vitamin E deficiency. Clin Immunol. 2004;112(1):24–9.
55. Aparicio JM, Bélanger-Quintana A, Suárez L, Mayo D, Benítez J, Díaz M, et al. Ataxia with isolated vitamin E deficiency: case

report and review of the literature. J Pediatr Gastroenterol Nutr. 2001;33(2):206–10.
56. Kemnic TR, Coleman M. Vitamin E deficiency. In: StatPearls. Treasure Island, FL: StatPearls Publishing; 2022.
57. Tokgöz Y, Terlemez S, Karul A. Fat soluble vitamin levels in children with newly diagnosed celiac disease, a case control study. BMC Pediatr. 2018;18(1):1–5.
58. Teriaky A, Mosli M, Chandok N, Al-Judaibi B, Marotta P, Qumosani K. Prevalence of fat-soluble vitamin (A, D, and E) and zinc deficiency in patients with cirrhosis being assessed for liver transplantation. Acta Gastroenterologica Belgica. 2017;80(2):237–41.
59. Gomez-Pomar E, Hatfield E, Garlitz K, Westgate PM, Bada HS. Vitamin E in the preterm infant: a forgotten cause of hemolytic anemia. Am J Perinatol. 2018;35(3):305–10.
60. Berenjian A, Mahanama R, Kavanagh J, Dehghani F. Vitamin K series: current status and future prospects. Crit Rev Biotechnol. 2015;35(2):199–208.
61. Shearer MJ, Newman P. Metabolism and cell biology of vitamin K. Thromb Haemost. 2008;100(10):530–47.
62. Brodie AF, Ballantine J. Oxidative phosphorylation in fractionated bacterial systems: II. The role of vitamin K. J Biol Chem. 1960;235(1):226–31.
63. Booth SL. Roles for vitamin K beyond coagulation. Annu Rev Nutr. 2009;29:89–110.
64. Geleijnse JM, Vermeer C, Grobbee DE, Schurgers LJ, Knapen MH, Van Der Meer IM, et al. Dietary intake of menaquinone is associated with a reduced risk of coronary heart disease: the Rotterdam Study. J Nutr. 2004;134(11):3100–5.
65. Truong JT, Booth SL. Emerging issues in vitamin K research. J Evid Based Complement Alternat Med. 2011;16(1):73–9.
66. Shearer MJ. Vitamin K deficiency bleeding (VKDB) in early infancy. Blood Rev. 2009;23(2):49–59.
67. https://www.msdmanuals.com/en-in/home/disorders-of-nutrition/vitamins/vitamin-k-deficiency. Accessed 29 Dec 2022.
68. https://draxe.com/nutrition/vitamin-k-deficiency/. Accessed 7 Jan 2023.
69. Said HM. Water-soluble vitamins. Nutrition for the primary care provider, vol. 111. Basel: Karger Publisher; 2015. p. 30–7.
70. Kumar N. Neurologic presentations of nutritional deficiencies. Neurol Clin. 2010;28(1):107–70.
71. Douaud G, Refsum H, de Jager CA, Jacoby RE, Nichols T, Smith SM, et al. Preventing Alzheimer's disease-related gray matter atrophy by B-vitamin treatment. Proc Natl Acad Sci. 2013;110(23):9523–8.
72. Safi J, Joyeux L, Chalouhi G. Periconceptional folate deficiency and implications in neural tube defects. J Pregnancy. 2012;2012:295083.
73. Kumar N. Neurologic aspects of cobalamin (B12) deficiency. Handb Clin Neurol. 2014;120:915–26.
74. Thiamin R. Dietary reference intakes for thiamin, riboflavin, niacin, vitamin B6, folate, vitamin B12, pantothenic acid, biotin, and choline. Washington, DC: National Academies Press; 1998.
75. Herrmann M, Schmidt JP, Umanskaya N, Wagner A, Taban-Shomal O, Widmann T, et al. Corrigendum to: the role of hyperhomocysteinemia as well as folate, vitamin B6 and B12 deficiencies in osteoporosis—a systematic review. Clin Chem Lab Med. 2021;59(2):455.
76. Liu S, West R, Randell E, Longerich L, O'Connor KS, Scott H, et al. A comprehensive evaluation of food fortification with folic acid for the primary prevention of neural tube defects. BMC Pregnancy Childbirth. 2004;4(1):1–10.
77. Michels A, Frei B, Vitamin C. Biochemical, physiological, and molecular aspects of human nutrition. Philadelphia: Saunders; 2012.
78. Carr AC, Frei B. Toward a new recommended dietary allowance for vitamin C based on antioxidant and health effects in humans. Am J Clin Nutr. 1999;69(6):1086–107.
79. Levine M, Padayatty SJ, Espey MG. Vitamin C: a concentration-function approach yields pharmacology and therapeutic discoveries. Adv Nutr. 2011;2(2):78–88.
80. Hornig D. Distribution of ascorbic acid, metabolites and analogues in man and animals. Ann N Y Acad Sci. 1975;258(1):103–18.
81. Englard S, Seifter S. The biochemical functions of ascorbic acid. Annu Rev Nutr. 1986;6(1):365–406.
82. Young JI, Züchner S, Wang G. Regulation of the epigenome by vitamin C. Annu Rev Nutr. 2015;35:545.
83. Vissers MC, Kuiper C, Dachs GU. Regulation of the 2-oxoglutarate-dependent dioxygenases and implications for cancer. Biochem Soc Trans. 2014;42(4):945–51.
84. Levine M, Conry-Cantilena C, Wang Y, Welch RW, Washko PW, Dhariwal KR, et al. Vitamin C pharmacokinetics in healthy volunteers: evidence for a recommended dietary allowance. Proc Natl Acad Sci. 1996;93(8):3704–9.
85. Kim KP, Shin K-O, Park K, Yun HJ, Mann S, Lee YM, et al. Vitamin C stimulates epidermal ceramide production by regulating its metabolic enzymes. Biomol Ther. 2015;23(6):525.
86. Butler J, Bergsten P, Welch RW, Levine M. Ascorbic acid accumulation in human skin fibroblasts. Am J Clin Nutr. 1991;54(6):1144S–6S.
87. Hansen SN, Tveden-Nyborg P, Lykkesfeldt J. Does vitamin C deficiency affect cognitive development and function? Nutrients. 2014;6(9):3818–46.
88. Waring AJ, Schorah CJ. Transport of ascorbic acid in gastric epithelial cells in vitro. Clin Chim Acta. 1998;275(2):137–49.
89. Boyer JC, Campbell CE, Sigurdson WJ, Kuo S-M. Polarized localization of vitamin C transporters, SVCT1 and SVCT2, in epithelial cells. Biochem Biophys Res Commun. 2005;334(1):150–6.
90. Ingalls TH, Warren HA. Asymptomatic scurvy: its relation to wound healing and its incidence in patients with peptic ulcer. N Engl J Med. 1937;217(11):443–6.
91. Akikusa J, Garrick D, Nash M. Scurvy: forgotten but not gone. J Paediatr Child Health. 2003;39(1):75–7.
92. Carpenter KJ. The history of scurvy and vitamin C. Cambridge: Cambridge University Press; 1988.
93. Gould BS. Collagen formation and fibrogenesis with special reference to the role of ascorbic acid. Int Rev Cytol. 1963;15:301–61.
94. Aleo JJ. Vitamin C deficiency possible relationship to periodontal disease. J Dent Res. 1980;59:386.
95. Rinehart JF. Studies relating vitamin C deficiency to rheumatic fever and rheumatoid arthritis; experimental, clinical and general considerations: II. Rheumatoid (atrophic) arthritis. Ann Intern Med. 1935;9(6):671–89.
96. Nisha P, Singhal RS, Pandit AB. A study on degradation kinetics of ascorbic acid in amla (Phyllanthus emblica L.) during cooking. Int J Food Sci Nutr. 2004;55(5):415–22.
97. Gopalan C, Rama Sastri B, Balasubramanian S. Nutritive value of Indian foods. Hyderabad: National Institute of Nutrition; 1971.
98. Woods BE. A study of the intra-specific variations and commercial potential of *Terminalia Ferdinandiana* (Exell) (the Kakadu Plum). Darwin: Charles Darwin University; 1995.
99. Mohanty S, Cock IE. The chemotherapeutic potential of Terminalia ferdinandiana: phytochemistry and bioactivity. Pharmacogn Rev. 2012;6(11):29.
100. Podsędek A. Natural antioxidants and antioxidant capacity of Brassica vegetables: a review. LWT-Food Sci Technol. 2007;40(1):1–11.
101. Becerra-Moreno A, Alanís-Garza PA, Mora-Nieves JL, Mora-Mora JP, Jacobo-Velázquez DA. Kale: an excellent source of vitamin C, pro-vitamin A, lutein and glucosinolates. CyTA J Food. 2014;12(3):298–303.

102. Rodrigues RB, De Menezes HC, Cabral LM, Dornier M, Reynes M. An Amazonian fruit with a high potential as a natural source of vitamin C: the camu-camu (Myrciaria dubia). Fruits. 2001;56(5):345–54.
103. https://www.planetayurveda.com/home-remedies-vitamin-c-deficiency. Acessed 20 Dec 2022.
104. Shyur L-F, Yang N-S. Metabolomics for phytomedicine research and drug development. Curr Opin Chem Biol. 2008;12(1):66–71.
105. Phillipson JD. Phytochemistry and medicinal plants. Phytochemistry. 2001;56(3):237–43.
106. Qv X-Y, Jiang J-G, Piao J-H. Pharmacodynamic studies of Chinese medicine at levels of whole animal, cell and molecular models. Curr Med Chem. 2010;17(36):4521–37.
107. Han Y, Sun H, Zhang A, Yan G, Wang X-j. Chinmedomics, a new strategy for evaluating the therapeutic efficacy of herbal medicines. Pharmacol Ther. 2020;216:107680.
108. Fanali CD, Orazio G, Fanali S, Gentili A. Advanced analytical techniques for fat-soluble vitamin analysis TrAC. Trends Anal Chem. 2017;87:82–97.
109. Ball GF. Vitamins in foods: analysis, bioavailability, and stability. Boca Raton: CRC Press; 2005.
110. Blake CJ. Status of methodology for the determination of fat-soluble vitamins in foods, dietary supplements, and vitamin premixes. J AOAC Int. 2007;90(4):897–910.
111. Indyk HE, Woollard DC. Vitamin K in milk and infant formulas: determination and distribution of phylloquinone and menaquinone-4. Analyst. 1997;122(5):465–9.
112. Woollard DC, Indyk HE, Fong BY, Cook KK. Determination of vitamin K1 isomers in foods by liquid chromatography with C30 bonded-phase column. J AOAC Int. 2002;85(3):682–91.
113. Balamurugan V, Fatima S, Velurajan S. A guide to phytochemical analysis. Int J Adv Res Innov Ideas Educ. 2019;5(1):236–45.
114. Pearson D. The chemical analysis of foods. 7th ed. Edinburgh: Churchill Livingstone; 1976.
115. Okwu D. Phytochemical and vitamin content of indigenous spices of South Eastern Nigeria. J Sustain Agric Environ. 2004;6:30–4.
116. Chemists AoOA, Horwitz W. Official methods of analysis. Washington, DC: Association of Official Analytical Chemists; 1975.
117. Harborne A. Phytochemical methods a guide to modern techniques of plant analysis. Berlin: Springer Science & Business Media; 1998.
118. Kazakevich YV, Lobrutto R. HPLC for pharmaceutical scientists. New York: John Wiley & Sons; 2007.
119. Srivastava M. High-performance thin-layer chromatography (HPTLC). Berlin: Springer Science & Business Media; 2010.
120. Raaman N. Phytochemical techniques. Chennai: New India Publishing; 2006.
121. Förster H. UV/Vis spectroscopy. In: Characterization I. Berlin: Springer; 2004. p. 337–426.
122. Günzler H, Gremlich H-U. IR spectroscopy: an introduction. Weinheim: Wiley; 2002.
123. Duckworth HE, Barber RC, Venkatasubramanian V. Mass spectroscopy. Cambridge: Cambridge University Press; 1986.
124. Marion D. An introduction to biological NMR spectroscopy. Mol Cell Proteomics. 2013;12(11):3006–25.
125. Smyth M, Martin J. X ray crystallography. Mol Pathol. 2000;53(1):8.
126. Iijima Y, Nakamura Y, Ogata Y, Tanaka K, Sakurai N, Suda K, et al. Metabolite annotations based on the integration of mass spectral information. Plant J. 2008;54(5):949–62.

Herbs for Autoimmune Diseases

A. Mukne, S. Dangat, P. Shirodkar, and K. Sawate

Abstract

The incidence of autoimmune diseases worldwide is on the rise, with extensive impact on lifestyle of individuals and spiralling healthcare costs. Phytotherapies are increasingly being deployed in management and treatment of various autoimmune diseases, primarily due to the restricted therapeutic benefit and serious side effects and toxicities associated with the long-term use of biologicals and other immunosuppressant drugs that currently form the mainstay of autoimmune disease management. This chapter focuses on providing an update on herbs reported in management and treatment of autoimmune diseases, with detailed, in-depth understanding of their molecular-level mechanism of action and modulation of various cell signalling pathways including NF-κB, STAT/JAK, MAPK and iNOS. Insights into the aetiology and pathogenesis of some of the major autoimmune diseases help in figuring out the arsenal of herbs that can be developed as effective therapeutic agents against these difficult-to-manage disease conditions. We take a look at phytotherapies reported for treatment and management of organ-specific and systemic autoimmune conditions including systemic lupus erythematosus, skin-related autoimmune conditions (atopic dermatitis, vitiligo and psoriasis), Addison's disease and neurodegenerative diseases (multiple sclerosis, Alzheimer's disease and Parkinson's disease).

Keywords

Autoimmune diseases · Phytotherapies · Systemic lupus erythematosus · Addison's disease · Neurodegenerative diseases · Autoimmune skin conditions

A. Mukne (✉) · S. Dangat · P. Shirodkar · K. Sawate
Department of Pharmacognosy and Phytochemistry, Bombay College of Pharmacy, Mumbai, Maharashtra, India

Abbreviations

(NF-κB)	Nuclear factor kappa light of activation B cell
11-HSD	11 beta-hydroxysteroid dehydrogenase
4-TBC	4-*tert*-Butylcatechol
4-TBP	4-*tert*-Butylphenol
6-OHDA	6-Hydroxydopamine
AAD	Autoimmune Addison's disease
ABTS	2,2′-Azino-*bis*(3-ethylbenzothiazoline-6-sulfonic acid)
ACE	Angiotensin-converting enzyme
ACTH	Adrenocorticotropic hormone
ADCC	Antibody-dependent cell-mediated cytotoxicity
ADs	Autoimmune diseases
AIRE	Autoimmune regulator
ALD	Alzheimer's disease
AP-1	Activator protein-1
APCs	Antigen-presenting cells
APECED	Autoimmune polyendocrinopathy candidiasis ectodermal dystrophy
APP	Amyloid precursor protein
APS1	Autoimmune polyglandular syndrome type 1
APS2	Autoimmune polyglandular syndrome type 2
AR	Aldose reductase
AS-IV	Astragaloside IV
B cells	Bursa-derived cell
BBB	Blood brain barrier
CAT	Catalase
CCL	Chemokine ligand
CD	Clusters of differentiation cell
CFA	Complete Ferund's adjuvant
CLDN1	Claudin-1
CNS	Central nervous system
COX-2	Cyclooxygenase-2

A. K. Dhara, S. C. Mandal (eds.), *Role of Herbal Medicines*, https://doi.org/10.1007/978-981-99-7703-1_18

CTLA-4	Cytotoxic T Lymphocyte Antigen-4
CTLs	Cytotoxic T lymphocytes
CXCL	CXC chemokine ligand
DA	Dopamine
DAD	Diallyl disulfide
DAT	Diallyl trisulfide
DCs	Dendritic cells
DJ-1	Protein deglycase
DLE	Discoid lupus erythematosus
DPPH	α,α-Diphenyl-β-picryl hydrazyl
EAE	Experimental autoimmune encephalitis
EBV	Epstein-Barr virus
EC	Epicatechin
ECG	Epicatechin gallate
EGC	Epigallocatechin
EGCG	Epigallocatechin gallate
ER	Endoplasmic reticulum
ERK	Extracellular-signal-regulated kinase
FLG	Filaggrin
Foxp3	Forkhead box protein 3 gene
GluR3	Glutamate receptor
GM-CSF	Granulocyte macrophage colony stimulating factor
GR	Glucocorticoid receptor
HBD-2	Human beta-defensin
HD	Huntington's disease
HHV6	Human herpes virus-6
HLA	Human lymphocyte antigen
HLA	Human lymphocyte antigen-27
IB kinase	I-kappa B kinase
IBD	Inflammatory bowel disease
IFNs	Type 1 interferons
IFN-β	Interferon-β
IFN-γ	Interferon-gamma
IgE	Immunoglobulin E
IL	Interleukin
ILC3	Innate lymphoid cells 3
IMQ	Imiquimod
iNOS	Inducible nitric oxide synthase
IPEX syndrome	Immune polyendocrinopathy X-linked syndrome
IRF	Interferon regulatory factor
JAK/STAT	Janus kinase/signal transducer and activators of transcription
LN	Lupus nephritis
LPS	Lipopolysaccharide
MAPK	Mitogen-activated protein kinase
MBEH	Monobenzyl ether of hydroquinone
MC1R	Melanocortin 1 receptor
MEOg	Methanol extraction of *O. gratissimum* leaves
MHC	Major histocompatibility complex
MMP	Matrix metalloproteinase
MOG	Myelin oligodendrocyte glycoprotein
MPP+	1-Methyl-4-phenylpyridinium ion
MPTP	1-Methyl-4-phenyl-1,2,3,6-tetrahydropyridine
MS	Multiple sclerosis
MSH	Melanocyte stimulating hormone
n-BuOH	*n*-Butanolic
NBUVB	Narrowband UVB
NGF	Nerve growth factor
NL	Neonatal lupus
NMDA	*N*-methyl-D-aspartate
NO	Nitric oxide
NSAID	Nonsteroidal anti-inflammatory drugs
NSO	Noni seed oil
OCA2	Oculocutaneous albinism
OVA	Ovalbumin
OXA	Oxazolone
PAG	Processed *Aloe vera* gel
PD	Parkinson's disease
PGE2	Prostaglandin E2
PHF	Pentaherbs formula
PINK1	PTEN induced putative kinase 1
PPMS	Primary progressive MS
ProFLG	Pro-filaggrin
PSEN1	Presenilin 1
PSEN2	Presenilin 2
PTS	Propyl-propane thiosulfinate
PTSO	Propyl-propane thiosulfonate
RA	Rheumatoid arthritis
ROS	Reactive oxygen species
RRMS	Relapsing remitting MS
SAC	*S*-allyl-cysteine
SLE	Systemic lupus erythematosus
SPMS	Secondary progressive MS
SPP1	Secreted phosphoprotein 1
SS	Sjogren's syndrome
STAT-3	Signal transducer and activator of transcription 3
T CELLS	T lymphocytes
TAP1	Transporters associated with antigen processing protein-1
T-bet	T-box expressed in T cells
TCR	T-cell receptor
T_{EFF}	Effector T cells
Tfh	T follicular helper
TGF-β	Transforming growth factor beta
Th	T helper cell
TH	Tyrosine hydroxylase
TLRs	Toll-like receptors
TNFATP3	TNF alpha induced protein 3

TNF-α	Tumor necrosis factor-α
T_{reg}	Regulatory T cells
TSLP	Lymphopoietin
TSN	Tanshione IIA
TwHF	*Tripterygium wilfordii hook f.*
TYR	Tyrosine
UCH-L1	Ubiquitin carboxyl-terminal hydrolase L1
UPR	Unfolded protein response
UPS	Ubiquitin-proteasome system
WS	*Withania somnifera*
ZO1	Zonula occludens-1

1 Introduction

Autoimmunity is a disease state in which the body is unable to distinguish between self and foreign antigens. At the beginning of twentieth Century, Paul Ehrlich brought in the concept of "horror autotoxicus" and emphasised on the ability of the immune system to distinguish between self and foreign antigens [1]. The immune system exhibits tolerance to the molecules identified as "self" under normal conditions; the mechanism of self-tolerance involves central and peripheral tolerance. Aetiology of autoimmune diseases is complex; it involves a combination of environmental triggers and genetic factors which lead to failure of immune tolerance mechanisms [2]. The failure of immune tolerance mechanisms is thus the characteristic feature of autoimmune diseases (ADs). It has been reported that an overall global prevalence of autoimmune diseases are increasing annually at rates of 19.1% and 12.5%, respectively [3]. Autoimmune diseases are chronic and clinically multifactorial, primarily caused by dysregulation of the cell-mediated and antibody-mediated immune response. Autoantibodies and major histocompatibility complex (MHC) are the main predictors in the development of autoimmune diseases. In humans, MHC also known as human lymphocyte antigen (HLA) accounts for almost 50% of genetic predisposition in ADs [4, 5].

Current pharmacological therapies for the management and control of ADs include calcineurin inhibitors (cyclosporine, tacrolimus), cytotoxic drugs (cyclophosphamide), glucocorticoids (prednisolone, methylprednisolone) and biological agents (rituximab, infliximab) that suppress inflammation. However, long-term use of these drugs often leads to serious drug-induced toxicity and decline of organ function [6]. Phytotherapies with major active constituents including flavonoids, terpenoids, alkaloids and phenolic acids have been reported to produce significant anti-inflammatory effect, without the attendant toxicities and side effects. Here, we look at recent advances in developing such therapies as effective medicinal agents in treatment and management of ADs. Table 1 gives an overview of various organ-specific and systemic autoimmune diseases along with common herbs used in treatment and management of these ADs.

Organ-specific autoimmune diseases: The antibodies and T lymphocytes (T cells) react with self-antigens in targeted organs.
Non-organ-specific (multiple organ)/systemic autoimmune diseases: The antibodies and T cells attack antigens spread throughout tissues and organs.

1.1 Aetiology of ADs

1.1.1 Dysregulation of Innate and Adaptive Immunity

Environmental triggers and genetic factors cause failure of immune tolerance mechanisms that leads to a disturbance in numbers and outputs of regulatory T cells (T_{reg}) and effector T cells (T_{EFF}). In order to maintain immunological homeostasis, regulatory and pathogenic effector T cell subsets must coexist in proportion [31]. Figure 1 gives a schematic illustration of overview of dysregulation of immune homeostasis.

Innate immunity has a key role in concert with adaptive immune-mediated tissue damage in aetiology of ADs [32]. Figure 2 illustrates the pathways of tissue damage and malfunctioning of innate and adaptive immunity in cell-mediated and autoantibody-mediated ADs. Antigen-presenting cells (APCs) identify antigens and trigger the activation of innate immunity cells (dendritic cells (DCs), macrophages, B cells and natural killer cells) by toll-like receptors (TLRs). Activation of innate immune cells triggers release of pro-inflammatory cytokines such as granulocyte macrophage colony stimulating factor (GM-CSF), type 1 interferons (IFNs) and interleukins((IL) like IL-12, IL-1β, IL-10, IL-6 that are associated with tissue damage in autoimmune conditions [33, 34]. Further, it stimulates the activation of adaptive immune response through maturation of T cells into clusters of differentiation cells (CD)—$CD4^+$ and $CD8^+$. Immature ($CD8^+$) and ($CD4^+$) T lymphocytes are exposed to MHC I and MHC II peptides, respectively, on the surface of the APCs. Thus, immature $CD8^+$ and $CD4^+$ T cells get activated and differentiated into effector T helper cell (Th) subsets, namely T follicular helper (Tfh), Th1, Th2, Th17 and Treg cells; each of them exhibits a distinct phenotypic profile and triggers release of unique sets of cytokines that exert different functions in the immune response. These subsets of T cells negatively control both innate and adaptive immunity and their reduction causes failure of tolerance mechanisms, as seen in various autoimmune diseases [33].

Table 1 Organ-specific and systemic ADs along with common herbs reported in their management

Types of diseases	Disease name	Epidemiology	Clinical manifestation	Herbs used for treatment	References
Non-organ specific disease	Systemic lupus erythematosus (SLE)	13–7713.5 per 100,000 individuals found to be affected globally [7]	Maculopapular, mucocutaneous, musculoskeletal, cerebrovascular, hepatic, renal, pancreatic, pulmonary, cardiopulmonary, haemolytic, ocular, oesophageal, gynaecological furthermore, multiple organ failure	*Tripterygium wilfordii, Artemisia annua, Artemisia apiaceae*	[8, 9]
Skin-related diseases	Atopic Dermatitis/ Eczema	10–20% and 1–3% of children and adults, respectively, found to be affected globally [10]	Red dry, patches of skin, rashes, swollen discoloured skin	*Aloe vera*, Pentaherbs formula (*Flos lonicerae, Herba menthae, Cortex phellodendri, Cortex moutan* and *Rhizoma atractyodis*)	[11, 12]
	Vitiligo	0.2–2% population affected worldwide [13]	Non-scaly, chalky macules	*Piper nigrum, Nigella sativa, Ammi visnaga, Picrorhiza kurroa, Angelica sinesis*	[14–17]
	Psoriasis	2% population affected worldwide [18]	Dry itchy skin, silver scaly patches, pruritus, swelling of joints, discoloration of skin	*Oryza sativa* L., *Curcuma longa*	[19–21]
Adrenal gland-related disease	Autoimmune Addison's Disease (AAD)	1/5000 to 1/7000 individuals are affected [22]	Weight loss, anorexia, nausea, vomiting, diarrhoea, salt craving and hyperpigmentation of skin and mucosal membranes	*Glycyrrhiza glabra, Curcuma longa, Emblica officinalis*	[23, 24]
Neurological autoimmune diseases	Multiple Sclerosis (MS)	2.8 million people are affected worldwide [25]	Disturbance of sensory and motor neurons	*Artemisia dracunculus, Curcuma longa*	[26]
	Alzheimer's Disease (ALD)	44 million people affected worldwide [27]	Cognitive impairment, loss of memory, dementia	*Withania somnifera, Bacopa monnieri, Ginkgo biloba, Crocus sativus*	[28]
	Parkinson's Disease (PD)	1–3% of the population over 60 years of age found to be affected globally [29]	Bradykinesia, tremor, muscle rigidity, postural instability, cognitive impairment, dementia, hallucination and autonomic dysfunction	*Citrus sinensis, Hypericum perforatum, Ginkgo biloba, Panax ginseng*	[30]

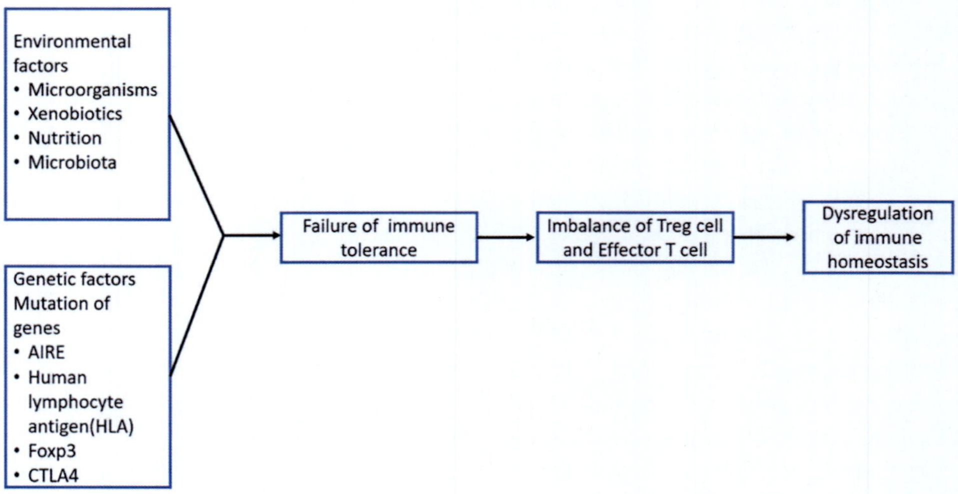

Fig. 1 Outline of dysregulation of immune homeostasis

The transcription factor T-bet (T-box expressed in T cells) and the cytokines (Type 1 IFN and IL-12) are activated to promote differentiation of CD8$^+$ and CD4$^+$ naïve helper T cells. Activated CD8$^+$ T cells are differentiated into Th1 cells, through activation of signal transducer and activator of transcription 3 (STAT-3) in T cells. Th1 cells aid in the production of cytotoxic T lymphocytes (CTLs) with release of IL-12 and IL-27. Further, autoreactive CTLs produce cytotoxic granules, activate Fas-Fas ligand and release cytokines tumor necrosis factor-α (TNF-α) and interferon-gamma

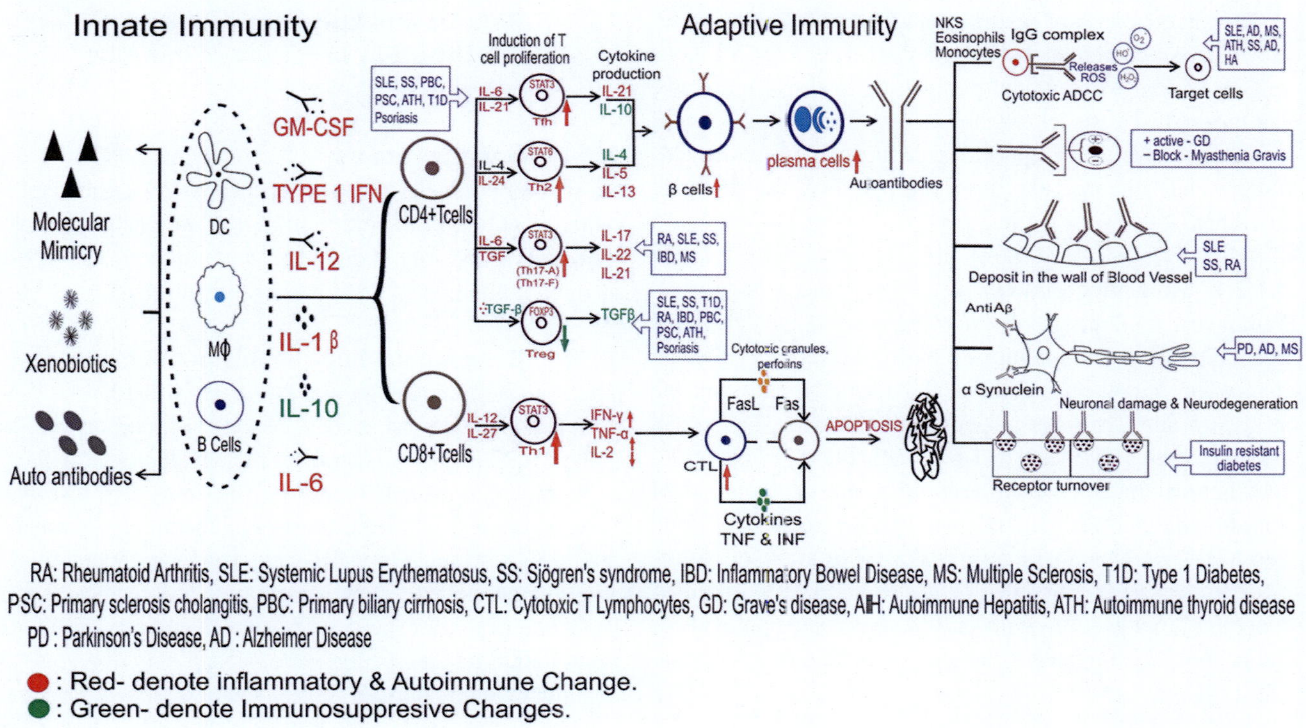

Fig. 2 Pathways of tissue damage and malfunctioning of innate and adaptive immunity in cell-mediated and autoantibody-mediated autoimmune diseases. (Reproduced with permission from Lifeng Wang, Fu-Sheng Wang, M. Eric Gershwin, Journal of Internal Medicine, Human autoimmune diseases, Published by John Wiley and Sons, Year-2015)

(IFN-γ), IL-2 which leads to apoptosis. Similarly, activation of the transcription factors STAT3 by release of cytokines IL-21, IL-1β, IL-6 are essential for the differentiation of naïve $CD4^+$ helper T cells into Th17 cells. Elevated levels of Th17 have been associated with the development of most of autoimmune diseases such as SLE, rheumatoid arthritis (RA), Sjogren's syndrome (SS), inflammatory bowel disease (IBD) and MS and crucial for both host defence against external pathogens and inflammatory ADs [35]. Th2 and Tfh cells promote the production of autoantibodies by stimulating B-cell activation, maturation, plasma cell differentiation and autoantibody production which eventually leads to tissue damage through antibody-dependent cell-mediated cytotoxicity (ADCC), release of reactive oxygen species (ROS) and binding to cell surface receptors which aid in tissue damage, as in SLE, MS and SS, in which large proportion of host antibodies are directed against own cellular structural components and receptors. A common example of immune complex impairment-mediated disease is SLE. Synovial injury in RA is mediated via rheumatoid factor-IgG complexes. Autoantibodies may also interact with cell surface receptors, which can both upregulate (antithyroid hormone for Graves' disease) and inhibit certain processes (anti-acetylcholine receptor for myasthenia gravis). The insufficient clearance of cellular debris, enriched with DNA and RNA ligands, can activate the TLR7, TLR8 and TLR9 [36], resulting in formation of immune complexes with pre-existing autoantibodies [2].

IL-1β is involved in differentiation of immature $CD4^+$ T cells into Treg cells. Treg cells express transcription factor forkhead box protein 3 gene (Foxp3) that plays a crucial role in the control and suppression of immune response by inhibiting activation of helper T and bursa-derived cell (B cells) [37, 38]. Treg cells also secrete transforming growth factor beta (TGF-β) and IL-10 that regulate immunological responses [39]. Treg cells with deficient transcription factor Foxp3 are unable to suppress production of Th1 and Th17 cells, leading to inflammation, antibody production and tissue injury [31].

1.1.2 Failure of Central and Peripheral Tolerance

Central tolerance is conferred in the thymus and bone marrow by way of negative selection wherein, in the dendritic and medullary epithelial cells, immature lymphocytes with strong affinity for auto-antigens are removed via apoptosis [40]. The lymphocytes with low affinity for auto-antigens are positively selected for survival and differentiate into mature single lymphocytes ($CD4^+$ or $CD8^+$) in the cortex region of thymus [41]. Despite the strictness of the central tolerance mechanism, few autoreactive lymphocytes ($CD4^+$ or $CD8^+$ lymphocytes) manage to escape the process [37].

Peripheral tolerance occurs in the spleen and lymph nodes and can be prevented by mechanism of energy (T cells remain unresponsive to self-peptides) and clonal ignorance (autoreactive T cells ignore the self- peptides).

Molecular mimicry, epitope spreading and bystander activation mechanisms are involved in loss of immune tolerance.

1.1.2.1 Molecular Mimicry

Pathogens including the microorganisms and chemical agents induce autoimmunity by molecular mimicry mechanism. The hypothesis of molecular mimicry relies on the structural similarity of the peptides of host and pathogen, which activates autoreactive T or B cells [42]. The other factors that may induce molecular mimicry and development of autoimmune disease include chemicals, cosmetics and food additives [43]. Rheumatic fever, a typical example of molecular mimicry, is an outcome of cross reaction between infective microorganism *Streptococcus pyogenes* and antigens of the heart of the human host leading to damage of heart muscles [44].

1.1.2.2 Epitope Spreading

Epitope spreading is the broadening of immune response induced by an antigen to T or B cells in the process of development of autoimmune disease [45]. Though epitope spreading is one of the contributing factors in the progression or onset of chronic disease, it is also protective in function [46]. The first step in epitope spreading is molecular mimicry. The diversification of immune response may take place for the same antigen or for separate antigens, which are referred to as intramolecular spreading and intermolecular spreading, respectively. Epitope spreading is triggered by interaction with T cells, somatic hypermutation and endocytic processing in B cell and can be the cause of RA, SLE and MS. Epstein-Barr virus (EBV) are reported to diversify other auto-antigens through B-cell epitope spreading [47, 48].

1.1.2.3 Bystander Activation

Bystander activation is characterised by non-specific lymphocyte activation and stimulation of T and B cell receptors regardless of their specificity. Bystander T cell activation is independent of T-cell receptor (TCR) signalling and responds rapidly to the inflammatory mediators (cytokines and TLR signalling) [49]. Bystander activation of $CD8^+$ memory T cells is mediated by cytokines; studies demonstrated that lipopolysaccharide (LPS) injection stimulates IFN-1 production, which enhances proliferation of $CD8^+$ T cells irrespective of TCR and antigen specificity [50]. Bystander activation and diverse pathogens (viruses, bacteria and parasites) are associated with onset and relapse of several autoimmune diseases including SLE, autoimmune hepatitis and type I diabetes [49, 51].

1.2 Genetic and Epigenetic Factors Contributing to Failure of Immune Tolerance

1.2.1 Genetic Factors

Most autoimmune diseases are suspected to be polygenic, in which mutation or alteration of more than one gene occurs. However, some rare monogenic autoimmune diseases are also reported.

1.2.1.1 Gene Autoimmune Regulator (AIRE)

The gene autoimmune regulator (AIRE), which encodes 545 amino acids and consists of 14 exons is found on chromosome 21q22.3, and is crucial for the development of T cells in thymus that present tissue-specific antigen. It also regulates the expression of self-antigen by medullary thymic epithelial cells, which is crucial for the development of tolerance and positive and negative selection of autoreactive T cells [52]. The failure of the tolerance mechanism is brought on by mutation or lack of the gene AIRE, which prevents the deletion of autoreactive T cells and allows them to invade the periphery. The AIRE gene has more than 100 known variants, and each mutation exhibits a particular pattern, such as missense/nonsense, splicing and deletion.

AIRE gene encodes a transcription factor that plays a role in the presentation of tissue-restricted antigens during T-cell maturation in the thymus. AIRE's participation in the negative selection of autoreactive T cells has a significant impact on the development of the T-cell repertoire. The mutation of gene AIRE was identified in patients with autoimmune polyglandular syndrome type 1 (APS1); it is also referred to as 'autoimmune polyendocrinopathy candidiasis ectodermal dystrophy' (APECED). It is a rare, multiple organ autosomal recessive autoimmune disease, characterised by variable autoimmune reactions affecting endocrine and non-endocrine organs. Two of the following diseases, hypoparathyroidism, AAD or chronic mucocutaneous candidiasis, are typically found in coexistence with APECED [53].

1.2.1.2 Forkhead Box P3 (Foxp3)

Foxp3, which is encoded with 431 amino acids on the X chromosome, is the main gene controlling the growth and function of Treg cells. Immune polyendocrinopathy X-linked syndrome (IPEX syndrome) is an example of mutation of Foxp3 with consequent immune dysregulation [54].

1.2.1.3 Cytotoxic T Lymphocyte Antigen-4 (CTLA-4)

The negative regulation of the T cell immune response and self-tolerance is facilitated by the CTLA-4. CD28 and CTLA-4, which share the ligands B7.1 (CD80) and B7.2 (CD86) on the APCs are essential for up- and down-regulation of T cell activation, respectively. By competing with CD28 receptor for binding to their common ligand B7.2, CTLA-4 inhibits T cell activation [55].

1.2.1.4 Human Lymphocyte Antigen (HLA)

HLA gene encodes for cell surface proteins and are thought to make up the MHC in humans. Mutations in HLA is strongly associated with different autoimmune conditions such asRA (RA: HLA-DR4), SLE (SLE: HLA-DR2, DR3), type 1 diabetes mellitus (T1D: HLA-II-DR3, DR4), MS (MS: HLA-DR2), celiac disease (CD: DQA1), SS (SS: HLA-DR2) and psoriasis (HLA-B-13,16,17). Whereas, individuals with HLA-B27 have greater predisposition to ankylosing spondylitis as compared to the individuals who are B27 negative. Individuals with DQ6 gene are more likely to develop MS and narcolepsy [5].

1.2.2 Epigenetic and Environmental Factors

Epigenetics is defined as the heritable and stable change in gene expression without alteration in DNA sequence, and is linked to environmental triggers [56]. The last few decades have witnessed significant increase in development of autoimmune diseases due to increase in exposure to environmental pollutants and other agents [57]. The environmental factors associated with autoimmunity include nutrition, xenobiotics, hormones, diet, stress, smoking, biological agents and infectious agents.

1.2.2.1 Stress

The immune system is dysregulated by stress-triggered neuroendocrine hormones which results in development of autoimmune diseases. Studies suggest that stress management intervention is necessary in the treatment of autoimmune diseases in order to prevent immunological imbalance triggered by stress [58]. The development of type 1 diabetes, SLE, and MS is aided by the elevated levels of IL-17, IFN-γ, and interferon-β (IFN-β), also seen in psychological stress-induced abnormalities in gastrointestinal barriers and systemic immune response [59].

1.2.2.2 Hormones

Women are more predisposed to development of autoimmune diseases as compared to men, as oestrogen is reported to be one of the more potent stimulators of autoimmunity whereas androgen is protective in function. The sex bias is mainly seen in autoimmune thyroid diseases, SLE and SS [60]. SLE is associated with early menarche where there is increased exposure to endogenous and exogenous sex hormones (contraceptive pills or hormone relapse therapy) which affects gene expression in women.

1.2.2.3 Diet/Nutrition

In vitro study by Lerner et al. (2015) demonstrated that commonly used food additives disrupt intestinal mucosal barrier function leading to leaky gut syndrome, that is, to say, the fountainhead of autoimmune conditions via molecular mimicry [61, 62]. Dietary iodine, a crucial element of thyroid hormone, is a major environmental risk factor in ADs [63].

2 Phytoconstituents/Herbs as Immunosuppressive and Immunomodulatory Agents

Numerous plant-derived phytoconstituents hold promise as immunoregulatory therapeutic options. Their principal mechanism of action in maintaining normal immune homeostasis is by either regulating the release of inflammatory cytokines and/or moderating functions of immune cells. Phytoconstituents exert immunosuppressive effects by antagonistic action on oxidative stressors. Various plants have been found to exert immunomodulatory effects, including *Acacia farnesiana, Allium sativum, Andrographis paniculata, Angelica glauca, Arundo donax, Camellia sinensis, Cymbopogon citratus, Dracocephalum rupestre and Malus domestica* [9]. Phytochemicals with anti-inflammatory and immunoregulatory effect include alkaloids (berberine), lectins (tomato lectin), flavonoids (quercetin), phenolic glycosides (curcumin), terpenoids (azadirachtin) and saponins (ginsenoside). These have been shown to suppress inflammatory cytokines such as IL-6, IL-10, IL-12, IL-17, IL-21, IL-23 and preserve immunological homeostasis and intracellular signalling. Adaptogenic herbs like amla, ashwagandha, ginseng, licorice and brahmi used in stress management have a key role to play in treatment and management of ADs [64]. Figure 3 illustrates the biomolecular mechanism of action of prominent indigenous herbs that are commonly used in Ayurveda and other traditional Indian systems of medicine for treatment and management of ADs. Overview of herbs used for management of autoimmune diseases with their molecular targets is elucidated in Table 2.

2.1 Anti-inflammatory Herbs Commonly Used in Management of Various Autoimmune Diseases

2.1.1 *Curcuma longa*

Curcumin, a naturally occurring flavonoid, is isolated from the rhizome of *Curcuma longa*, family Zingiberaceae. It exhibits strong anti-inflammatory, antioxidant and antibacterial properties. Curcumin suppresses activation of NF-κB, mitogen-activated protein kinase (MAPK), activator protein-1 (AP-1), Janus kinase/signal transducer and activators of transcription (JAK/STAT) and other signalling pathways and proinflammatory transcriptional factor that control the expression of genes involved in the early stages of inflammatory response; suppression of NF-κB results in downregulation of COX-2 and in iNOS [81], eventually leading to inhibition of pro-inflammatory cytokines IL-6, IL-23, IL-12, IL-8, IL-1, IL-17, IFN-γ, TNF-α, monocyte chemoattractant proteins (also known as CCL2) and NO release [87]. Curcumin can be used in the treatment of IBD, arthritis, psoriasis, depression and atherosclerosis and other inflamma-

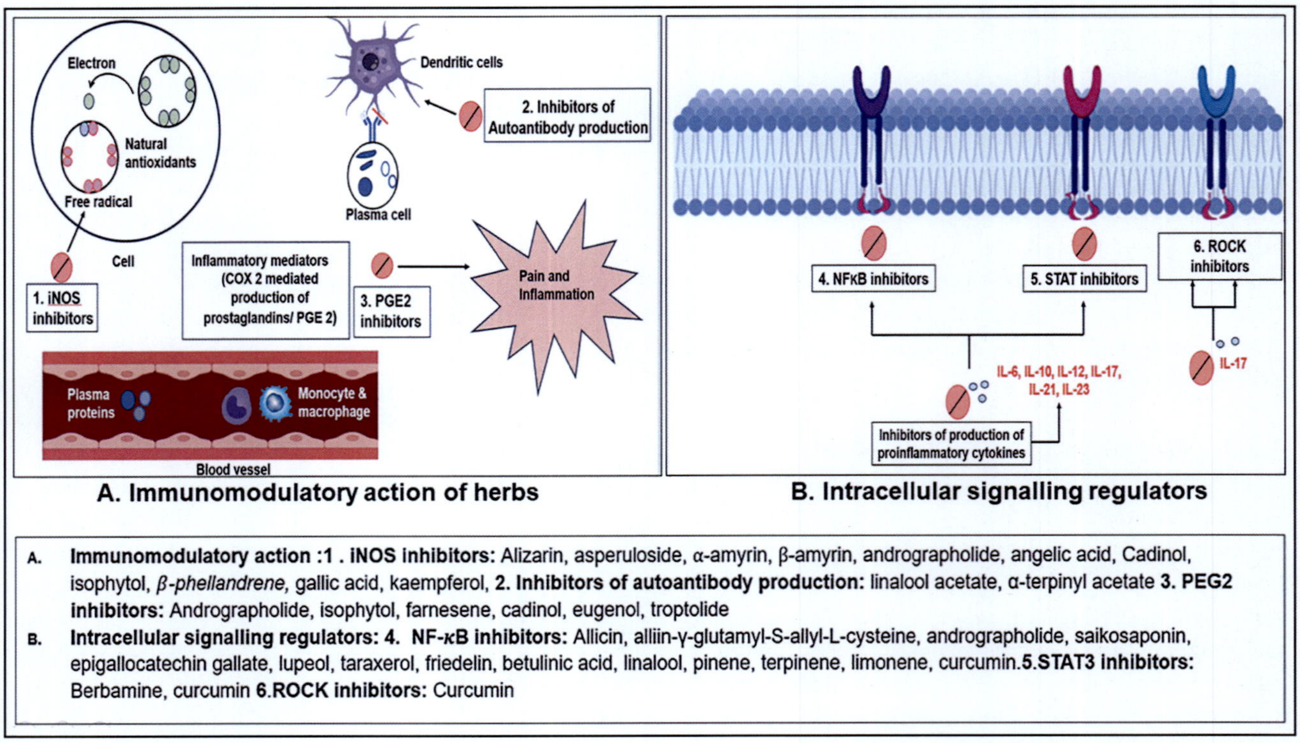

Fig. 3 Bimolecular mechanism of action of herbs used in treatment and management of ADs. (Reproduced with permission from Acharya Balkrishna, MDPI Cells 2020, 9(4) Published by MDPI, year-2020). (**a**) **Immunomodulatory action: 1. iNOS inhibitors:** Alizarin, asperuloside, α-amyrin, β-amyrin, andrographolide, angelic acid, Cadinol, isophytol, *β-phellandrene*, gallic acid, kaempferol, **2. Inhibitors of autoantibody production**: linalool acetate, α-terpinyl acetate **3. PEG2 inhibitors**: Andrographolide, isophytol, farnesene, cadinol, eugenol, troptolide. (**b**) **Intracellular signalling regulators: 4. NF-κB inhibitors**: Allicin, alliin-γ-glutamyl-S-allyl-L-cysteine, andrographolide, saikosaponin, epigallocatechin gallate, lupeol, taraxerol, friedelin, betulinic acid, linalool, pinene, terpinene, limonene, curcumin **5. STAT3 inhibitors**: Berbamine, curcumin **6. ROCK inhibitors**: Curcumin

tory diseases, in which curcumin reduces inflammatory response, providing effective symptomatic relief as well as disease treatment.

The antioxidant property of curcumin protects vital organs such as heart, liver, kidney and brain from oxidative stressors, thus protecting from inflammatory damage. Curcumin increases the activity and differentiation of oligodendrocytes, improves myelin genesis, reduces astrocyte proliferation and thus plays an important role in management of neurodegenerative conditions [88].

Curcumin is also used as a golden nutraceutical both topically and orally to alleviate and cure a variety of skin-related disorders (psoriasis, skin inflammation, acne and skin cancer) and metabolic diseases [87]. Studies have demonstrated the protective effect of curcumin against H_2O_2-induced skin injury to keratinocytes. *In vitro* studies carried out by Kang et al. on transgenic mouse model show that curcumin inhibits release of proinflammatory cytokines IL-17, IL-22, IFN-γ, IL-2, IL-8 and TNF-α in mice with psoriasis-like condition [89]. Curcumin treatment has been found to be effective in alleviating psoriasis symptoms in a number of clinical trials [87, 90].

The poor absorption, fast metabolism and rapid systemic clearance of curcumin may be the main causes of its limited bioavailability in both plasma and tissue [87]. Novel drug delivery platforms for curcuminoid phytomolecules have been developed for circumventing these issues and improving bioavailability [91, 92].

2.1.2 *Allium sativum*

Garlic (*Allium sativum* L.) is an herbaceous plant, belongs to Amarillidaceae family and contains a number of biologically active compounds, including alkaloids, phenolic compounds, saponins, polysaccharides, as well as organosulfur compounds namely—diallyl thiosulfonate (allicin), diallyl sulfide (DAS), DADS, DATS, *E*/*Z*-ajoene, *S*-allyl-cysteine (SAC), as well as *S*-allylcysteine sulfoxide (alliin) [93]. Studies conducted by Zhu et al. (2022) reported that garlic-derived preparation containing two organosulfur metabolites, propyl-propane thiosulfonate (PTSO) and propyl-propane thiosulfinate (PTS), could reduce inflammation primarily by inhibiting a number of inflammatory biomarkers, such as NO, TNF and IL-1, that regulate the expression of IL-6, IL-10, TNF-α and IFN-γ in RAW 264.7 murine macrophages that were previously treated with LPS [83]. Due to its minimal or complete lack of toxicity, garlic has a significant potential to heal a variety of inflammatory disorders.

Table 2 Molecular targets for herbs used in management of autoimmune diseases

Sr. no	Plant source with active metabolite	Mechanism of action	Therapeutic effect	References
1	*Machaerium acutifolium;* Lectin	Decrease proinflammatory cytokines, TNF-α and increase level of anti-inflammatory cytokine IL-10	Anti-inflammatory	[65]
2	*Glycine max* (Vegetable soyabean); quercetin, kaempferol	Reduced TNF-α, IL-6 and IL-1β production, inhibition of nitric oxide (NO) synthesis	Antioxidant, anti-inflammatory	[66]
3	*Andrographis paniculata;* (Kalmegh) Andrographolide (Diterpene)	Free radical scavenging activity by inhibiting formation of oxygen-derived free radicals	Anti-inflammatory, antioxidant	[67–69]
4	*Uncaria tomentosa;* Mitraphylline	Inhibit LPS stimulated inducible nitric oxide synthase (iNOS) by suppressing the action of nuclear factor kappa light of activation B cell (NF-κB), inhibit release of cytokines IL-1, IL-6, IL-8, IL-17 and TNF-α	Immunomodulatory and anti-inflammatory activity	[70–72]
5	*Phyllanthus emblica;* gallic acid Ellagic acid	Inhibit cyclooxygenase-2 (COX-2), iNOS and NF-κB	Antioxidant, anti-inflammatory	[73]
6	*Citrus sinensis;* hesperidin methyl chalcone	Inhibition of pro-inflammatory cytokines TNF-α, IL-6, IL-1β and IL-33	Antioxidant, anti-inflammatory	[74]
7	*Sarsaparilla;* Steroids: sarsapogenin similagenin, sitosterol Saponins: smilasaponin, sarsaparilloside	TNF-α-induced β activation	Anti-inflammatory	[68]
8	*Veratrum grandiflorum;* Resveratrol	Inhibition of TNF-α, IL-17, IL-6 and IL-1β and antioxidant property by neutralising ROS	Antioxidant, anti-inflammatory	[75]
9	Pentaherbs formula (PHF): *Flos lonicerae, Herba menthae, Cortex moutan, Rhizoma atractylodis* and *Cortex phellodendri* at a w/w ratio of 2:1:2:2:2 Active metabolites: gallic acid, berberine, chlorogenic acid	Suppression of pro-inflammatory cytokine IL-6 and chemokine ligand (CCL) i.e. CCL2, CCL7 and CXC chemokine ligand (CXCL) i.e. (CXCL8)	Anti-inflammatory and reduce allergic inflammation	[12]
10	*Mentha spicata;* gallic acid	Suppress inflammation, inhibit proliferation and pro-inflammatory cytokines such as TNF-α, IL-6, IL-1, IL-17 and inhibition of COX-2 and NO	Immunosuppressive, anti-inflammatory	[76]
11	*Camellia sinensis;* polyphenolic compounds: catechin-epicatechin, epigallocatechin and EGC gallate	Proliferation of proinflammatory cytokines leads to increased expression of IL-8, IL-17A and human beta-defensin (HBD-2)	Immunomodulatory, anti-inflammatory	[77]
12	*Tripterygium wilfordii hook f* (TWHF) *Active metabolites:* TWHF Triptolide, Celastrol	TWHF suppresses proliferation of inflammatory cytokines such as TNF-α, IL-6, IL-17 and downregulation of T cells and macrophages	Immunosuppressive, anti-inflammatory	[78]
13	*Artemisia annua;* artemisinin	Suppresses proinflammatory genes (TNF-gene) and stimulate anti-inflammatory and antioxidant genes	Immunosuppressive	[79]
14	*Andrographis paniculata* (Kalmegh); andrographolide (diterpene)	Inhibition of COX-2 and suppression of NF-κB	Anti-inflammatory	[80]
15	*Stephania tetrandra;* Tetrandine	Inhibition of TNF-α	Anti-inflammatory	[68]
16	*Berberis aristata;* Berberine	Inhibition of prostaglandin E2 (PGE2) and INF-γ	Anti-inflammatory	[68, 69]
17	*Curcuma longa;* Curcumin (Diferuloylmethane)	Decreases level of cytokines TNF-α, IL-1, IL-6, IL-8, IL-12, MCP-1 and IL-1β and various inflammatory enzymes and transcription factors PGE2 and INF-γ	Antioxidant, anti-inflammatory	[81]
18	*Salvia miltiorrhiza;* Tanshione IIA (TSN)	Cytokines IL-2, IL-4 and TNF-α are inhibited	Immunosuppressive	[69]
19	*Allium sativum;* polyphenols, diallyl disulfide (DAD) diallyl trisulfide (DAT) allicin, (*Z, E*)-ajoene	T cell activation is prevented by modulating the concentration of NF-κB, iNOS, COX-2, IL-6, TNF and IL-1	Immunomodulatory, antioxidant, anti-inflammatory	[82–84]
20	*Withania somnifera;* Alkaloids: Isopelletierineanaferine Steroids: Withanolides and withaferins Saponins: Sitoindoside VII, VIII and withanolides	Induction of IGg2a, downregulation of B and T cells in hyper immune states	Antioxidant, immunosuppressive	[85, 86]

2.1.3 *Zingiber officinale*

Zingiber officinale Roscoe belonging to family Zingiberaceae is native to India and Southeast Asia [94]. Both fresh and dried ginger rhizomes are used worldwide as a spice and condiment. Ginger contains gingerols, paradol and shogoal which have potent antioxidant, anti-inflammatory and anticancer properties. Previous studies reported that ginger extract was able to block the elevated expression of NF-κB and inhibit TNF-α in male Wistar rats with induced liver cancer [95]. In a study reported by Hwang et al. [96], anti-inflammatory activity of ginger extract in collagen-induced arthritic male DBA/1J mice was evaluated. Ginger was found to inhibit progression of RA by inhibiting the release of Th1/Th2 and Th17 cytokines and matrix metalloproteinase (MMPs) [96].

2.1.4 *Berberis aristata*

Berberis aristata, an Indian medicinal plant belonging to family Berberidaceae, has been traditionally used in Ayurveda for a host of medicinal uses. It is a herb used since ancient times. It is commonly known as Indian berberi "Daaru haridra" [97]. The isoquinoline alkaloid berberine found in *Berberis aristata* is reported to inhibit differentiation of proinflammatory Th1 and Th17 cells. Th cell-mediated inflammation is further downregulated by suppressing T_{regs}, DCs, macrophages and other immune cells that mediate autoreactive inflammation [98]. *In vitro* studies carried out by Yue et al. [99] on collagen-induced arthritis female Wistar rats report that by encouraging the production of cortistatin in the stomach, berberine alleviated symptoms in female Wistar rats with collagen-induced arthritis. The systemic Th17 cell response was inhibited, the upregulated cortistatin entered the bloodstream and arthritis symptoms were diminished [99].

2.1.5 *Withania somnifera*

Withania somnifera (WS), commonly referred to as Ashwagandha, Indian ginseng, member of Solanaceae family, is a widely used adaptogen, with reported use in the treatment of host of ADs including psoriasis, AAD, arthritis and rheumatism [85]. The major chemical constituents present in Ashwagandha are alkaloids (iso pelletierine anaferine), steroidal lactones (withanolides, withaferins), saponins (sitoindoside VII and VIII) and withanolides. It has been reported that WS extract inactivates NF-κB and thus suppress inflammation. Similar studies show that withaferin A, a key component of WS leaf extract, effectively inhibits NF-κB activation by inhibiting TNF-induced activation of IB kinase (I-kappa B kinase) enzyme which is responsible for antiproliferative, proapoptotic, anti-invasive, antiosteoclastogenic, antiangiogenic, antimetastatic, radio sensitising, antiarthritic and cardioprotective effects [100]. WS also helps to balance hormones in the body; it has been found to be effective in hypocortisolism (ADD) [173] and inhibiting neurodegeneration in paraquad-induced rat model of PD [22, 101].

2.1.6 *Camellia sinensis*

Commonly known as green tea is obtained from *Camellia sinensis*. Fiver major flavonoids found in green tea are the catechins, epicatechin (EC), epigallocatechin (EGC), epicatechin gallate (ECG) and epigallocatechin gallate (EGCG). It was found that green tea leaves extract inhibited the arachidonic acid-induced paw oedema in rats [102, 103]. It has been reported that anti-inflammatory and antioxidant properties of green tea have potential therapeutic benefits in xerostomia associated with SS. It has been reported that EGCG can inhibit proinflammatory IL-8 in human lung alveolar epithelial cells [104]. In addition, there are numerous applications for green tea in the treatment of diabetes, IBD, skin conditions and fitness enhancement. Green tea has a position in both the mainstream and alternative medical circles, despite the fact that there is not sufficient clinical data to support this.

2.1.7 *Uncaria tomentosa*

Uncaria tomentosa is commonly referred to as cat's claw. It is a tropical medicinal vine from the Rubiaceae family [70]. Mitraphylline is the major active constituent found in *U. tomentosa* which shows immunomodulatory and anti-inflammatory properties. Additionally, it is reported as an antioxidant, antibacterial, anti-inflammatory, and immune system booster and most effective complementary herb for treating parasites [72]. When given orally to mice at a dose of 30 mg/kg for 3 days, it was able to reduce the liberation of TNF-α by 50% and that of IL-1α, IL-1β and IL-17 by about 70% [72], demonstrating the therapeutic importance of this phytomolecule.

2.1.8 *Glycine max*

It is also known as "Vegetable soybean", mostly consumed in East Asia. It consists of a multitude of beneficial phytochemicals, including phytic acid, isoflavones and other phenolic compounds, saponins [66]. Glycine species is reported to strongly inhibit the protein expression of COX-2 and iNOS in LPS-stimulated RAW 264.7 macrophages [105] and thus shows significant anti-inflammatory and antioxidant activity. Huang et al. [105] studied *in vitro* antioxidant, anti-inflammatory and antidiabetic activities of the three Glycine species in LPS-stimulated RAW 264.7 macrophage and by using α-glucosidase and aldose reductase (AR) methods. Maximum antioxidant activity was demonstrated by aqueous extract of *Glycine tabacina* as compared to other species. Also, bioactive compounds, genistein and daidzein, were found to demonstrate antioxidant and anti-inflammatory effects with appreciable inhibition of α-glucosidase and AR [105].

2.1.9 *Andrographis paniculata*

Andrographolide, a labdane diterpene, is the main bioactive phytoconstituent in *Andrographis paniculata.* Recent clinical reports have documented the utility and potential of andrographolide as therapeutic agent in treatment of ADs like osteoarthritis, upper respiratory diseases and MS [106–109]. Methanolic and aqueous extracts of *A. paniculata* or andrographolide are reported to exhibit antioxidant and acute anti-inflammatory effect in carrageenan-induced paw oedema in rodents, with free radical scavenging activity by inhibiting formation of oxygen free radicals. Suppressing the expression of major pro-inflammatory genes like COX-2, IL-6, IL-8, IL-1β and iNOS is reported to be the underlying mechanism of action [67].

2.1.10 *Rosmarinus officinalis*

Rosmarinus officinalis, Family Lamiaceae, commonly known as Rosemary, contains important active constituents like phenolic diterpenes, triterpenes and acids. Studies have reported that essential oil of *Rosmarinus officinalis* suppresses transcription of NF-κB and the arachidonic acid pathway and thus exerts anti-inflammatory effect [110]. Phenolic compounds including carnosic acid, carnosol, ursolic acid and rosmarinic acid are shown to exhibit strong antioxidant activity [111].

2.1.11 *Coriandrum sativum*

Coriandrum sativum (Coriander) is an herbaceous annual herb belonging to the Apiaceae family. Essential oils of coriander are rich in monoterpenes, with potent anti-bacterial, antifungal, antidiabetic, antiseptic, anxiolytic and antioxidant activity. Linalool, pinene, cymene, borneol, phellandrene and geraniol are the major constituents in seed oil and decanal, decanol, cyclodecane and dodecena in the leaves [112, 113]. *In vitro* studies were carried out by Foudah et al. [114] to screen anti-inflammatory, antimicrobial and antioxidant properties of *Coriandrum sativum* leaves oil. Anti-inflammatory activity was evaluated in terms of degree of inhibition of egg albumin and trypsin-induced casein denaturation. *C. sativum* leaves oil was found to demonstrate excellent anti-inflammatory and antimicrobial activity and poor antioxidant effect [114]. The phytoconstituents from coriander suppress NF-κB and MAPKs signalling pathways with consequent reduction in levels of nitrite, ROS, IL-6 and TNF-α [112].

2.1.12 *Morinda citrifolia*

Noni (*Morinda citrifolia*), family Rubiaceae, commonly known as beach mulberry or cheese fruit, is an evergreen Polynesian herb. Noni seed oil (NSO) could be a preferable option for people with excessive cholesterol and cardiovascular disorders because of its high concentrations of polyunsaturated fatty acids, phytosterols and tocopherols [115]. Bioactive substances with anti-cancer, anti-inflammatory and antimicrobial activity have been found in noni seed extracts. Tanikawa et al. [116] have reported that *M. citrifolia* seed extract exhibits anti-inflammatory effect in LPS-induced RAW 264 cells. *M. citrifolia* seed extract was found to be effective in reducing NO generation while having no negative effect on cell viability. It was found that the *M. citrifolia* seed extract produced antioxidant effect by decreasing the expression of TNF-α and iNOS. These findings support the potential therapeutic application of *M. citrifolia* seed extract [116]. In another study, Lee et al. [117] have reported that noni fruit juice contains asperulosidic acid, rutin, nonioside A, (2*E*,4*E*,7*Z*)-deca-2,4,7-trienoate-2-*O*-β-D-glucopyranosyl-β-D-glucopyranoside and tricetin that suppress expression of NFκB p65, iNOS and COX-2 and reduce NO levels in LPS-stimulated RAW 264.7 cells [117].

2.1.13 *Cymbopogon citratus*

Cymbopogon citratus, family Gramineae, commonly known as lemongrass, contains aglycones as well as glycosides of flavonoids with anti-inflammatory and antioxidant activity [118]. *C. citratus* has potential as a therapeutic action for inflammatory disease, particularly at the gut. Suppressing expression of iNOS and NO scavenging activity are reported to be the possible mechanism of action [119]. *In vivo* studies have reported antidiabetic, anxiolytic, antibacterial and anti-hypertensive in leaf extracts of *C. citratus*. Flavonoid glycosides (isoorientin, cymaroside, cassiaoccidentalin B and kurilesin A) have been reported to inhibit α-glucosidase and exert hypoglycaemic effect [120]. Borges et al. [120] reported *in vitro* inhibition of yeast-glucosidase by various extracts and fractions of *C. citratus*. Molecular docking studies of the flavonoids provided proof-of-concept for the hypoglycaemic effect; the compounds were found to interact with the active sites of the enzyme via hydrogen bonding [120].

2.1.14 *Ocimum gratissimum*

Ocimum gratissimum L. belongs to family Lamiaceae and is a widely distributed aromatic herb in management of a range of inflammatory disorders. Ajayi et al. [121] have reported in vivo anti-inflammatory activity of various phenolic extracts and fractions of *O. gratissimum* leaves in carragenan-induced inflammation rodent models. The effect was possibly by free radical scavenging, with rutin, quercetin, caffeic acid, rosmarinic acid, circhoric acid, sitosterol, ursolic acid, salvigenin and transferulic acid being the main phytoconstituents involved [121].

2.1.15 *Tinospora cordifolia*

Tinospora cordifolia, commonly known as Guduchi, belongs to Family Menispermaceae, and is reported in Ayurveda to act as an immunomodulatory agent, improving the host ability to ward off infections. The major active constituents include alkaloids (berberine, magnoflorine), glycoside (cardifolioside A, tinocordiside), steroids (β-sitosterol) and aliphatic compounds (octacosanol, hepatocosanol) [122]. In a study by Nandan et al. [123], immunomodulatory effect and efficacy of *T. cordifolia* in Th17-linked ADs were evaluated. Aqueous extract of *T. cordifolia* was shown to suppress differentiation and proliferation of anti-CD3ε and anti-CD28-stimulated naive $CD4^+$ T cells, with reduced production of IL-17. *In silico* molecular docking studies validated the findings, with a cytokine-receptor signalling pathway reported as the possible mechanism [123].

3 Specific AD Conditions and Herbs Used in Their Management

3.1 Systemic Lupus Erythematosus

Lupus is chronic inflammatory autoimmune condition with a broad spectrum of clinical implications which eventually lead to tissue damage and multiple organ failure. Fig. 4 gives an overview of different types of lupus such as SLE, discoid lupus erythematosus (DLE), neonatal and drug-induced lupus [124].

SLE: Multiple organ failure and diverse array of clinical manifestations and production of autoantibodies against nucleic acid.
DLE: Chronic cutaneous disorder. Patients with DLE get red, raised sores on the face or scalp.
Neonatal Lupus (NL): At birth, babies have skin rashes, kidney problems or may have heart problems that slows down rhythm of heart.
Drug-induced Lupus: Some drugs (such as hydralazine, procainamide and isoniazid) may induce SLE-like condition although features go away when drug administration is stopped.

Numerous immunological and cellular regulators are deployed for managing this complicated autoimmune condition that is characterised by a dysregulated immune response. But a complete treatment plan for this disease is not yet available [124]. Also, the existing chemosynthetic medications, such as nonsteroidal anti-inflammatory drugs (NSAIDS) and corticosteroids, offer restricted therapeutic benefit with severe side effects like gastrointestinal reaction, bone marrow suppression and leukopenia with recurrence or aggravation of disease after withdrawal of therapy. Additionally, the illness heterogeneity and multiple clinical manifestations are highly extensive, making it difficult to manage with a single chemical medicine [8]. Wide range of clinical manifestations associated with this chronic autoimmune conditions are maculopapular (erythematous, elevated butterfly-shaped painful lesions on cheeks), cerebrovascular (anxiety, headache, seizures, psychosis, demyelination of nerve cells, ALD), hepatic (hepatitis, hepatomegaly, hepatic aneurysm), nephron-pathological [nephritis, proteinuria, haematuria, lupus nephritis (LN)], haematological (anaemia, leukopenia, thrombocytopenia), cardiac (pericarditis, myocardial infarction, cardiomyopathy), pulmonary (pulmonary embolism, pneumonitis, fibrosis), mucocutaneous (mouth ulcers, sores in inner cheek, dry mouth and gum disease) and musculoskeletal involvement (arthritis, osteoporosis, osteomyelitis, Raynaud's disease) [125].

Pathogenesis of SLE is complex and involves multiple cross interactions between the various elements of innate and adaptive immunity. Various aetiological factors are involved in clinical manifestation of SLE such as person's genetic predisposition [inflammation triggering genes such as interferon regulatory factor (IRF), OX40L, secreted phosphoprotein 1 (SPP1), TNF alpha induced protein 3 (TNFATP3), signalling pathway gene STAT], epigenetic (histone modification and DNA methylation), hormonal components (oestrogen, prolactin), immunological and environmental triggers [125]. Immune tolerance to self-antigens is lost in the permissive situation associated with pro-inflammatory stimuli which leads to unregulated differentiation and expression of B and

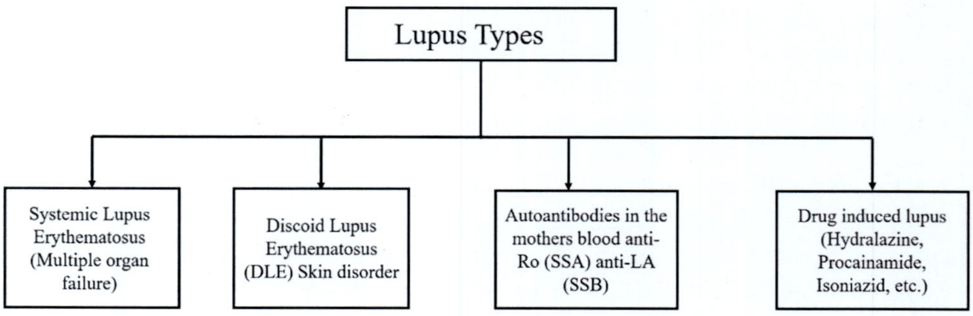

Fig. 4 Overview of different types of lupus

T cells, further inducing production of autoantibodies [126]. This overproduction of antibodies leads to DNA and ribonucleoprotein damage in healthy cells, with consequent apoptosis [124].

3.1.1 Herbs Used for Management of SLE

Immunomodulatory properties have been ascribed to a wide variety of phytoconstituents and can be deployed for treatment of autoimmune disorders such as SLE. Alkaloids (berberines), lectins (tomato lectin), flavonoids (quercetin), phenolic glycosides, terpenoids (azadirachtin) and saponins (ginsenoside) have been found to reduce inflammatory cytokines and intracellular signalling, resulting in anti-inflammatory and immunosuppressive effects [9]. Natural plant-based immunotherapies and regulators of intracellular signalling provide therapeutic alternatives with symptomatic relief by altering the cell signalling pathways and ultimately reducing release of proinflammatory cytokines production and B-T cell co-stimulation [9, 126, 127]. Figure 5 illustrates the phytoconstituents used as symptomatic relief providers in SLE.

3.1.1.1 *Tripterygium wilfordii Hook F*

Triptolide, a diterpene triepoxide, is the principal bioactive phytoconstituent in the plant TwHF, that is used in management of inflammatory and autoimmune diseases [128]. Triptolide has been reported to suppress the production of proinflammatory cytokines, including IL-2, IFN, IL-6 and TNF, as well as the upregulation of iNOS and COX-2 [128, 129].

Zhang et al. [130] reported a study evaluating the effect of (5*R*)-5-hydroxy triptolide (LLDT-8), a triptolide derivative on LN in MRL/lpr mouse model of SLE. LLDT-8 was found to have therapeutic benefit in LN by suppressing chemokine expression and inhibiting immune cell infiltration in kidneys [130].

Zhang et al. [131] reported a study evaluating the neuroprotective effect of triptolide in the treatment of SLE by using microRNA-146a in B cell TLR7 signalling pathway in mice. Triptolide produced significant reduction in levels of double-stranded DNA and IgG and also improved microRNA-146a expression, with reduced expression of TLR7 microRNA, protein levels TLR7, MyD88, p-IRAK1 and p-NF-κBp65 [131]. Similarly, Liu et al. [132] showed that the effectiveness of TwHF in combination with prednisone for the treatment of SLE was 87.5%, significantly higher than the rate for the control (prednisone in combination with Methotrexate) [132]. More extensive safety and efficacy data for triptolide in humans needs to be generated [128].

Artemisinin

Artemisinin, a sesquiterpene lactone, is isolated from *Artemisia annua* and *Artemisia apiaceae* (sweet annie, quing hao), Family Asteraceae. Both species contain active ingredient artemisinin and its congeners that are widely used as antimalarial and immunosuppressive agents [133]. It has been reported by Zamani et al. [134] that sesquiterpene extracts of *A. annua* exerted immunosuppressive effect by lowering levels of nitric oxide in LPS-primed J774A.1 macrophages [134]. The essential oil of the plant's aerial parts are reported to demonstrate appreciable antioxidant effect [135].

The use of artemisinin therapy relieved proteinuria in LN, restored blood urea nitrogen and creatinine levels, lessened pathological damage to the kidneys and decreased mortality;

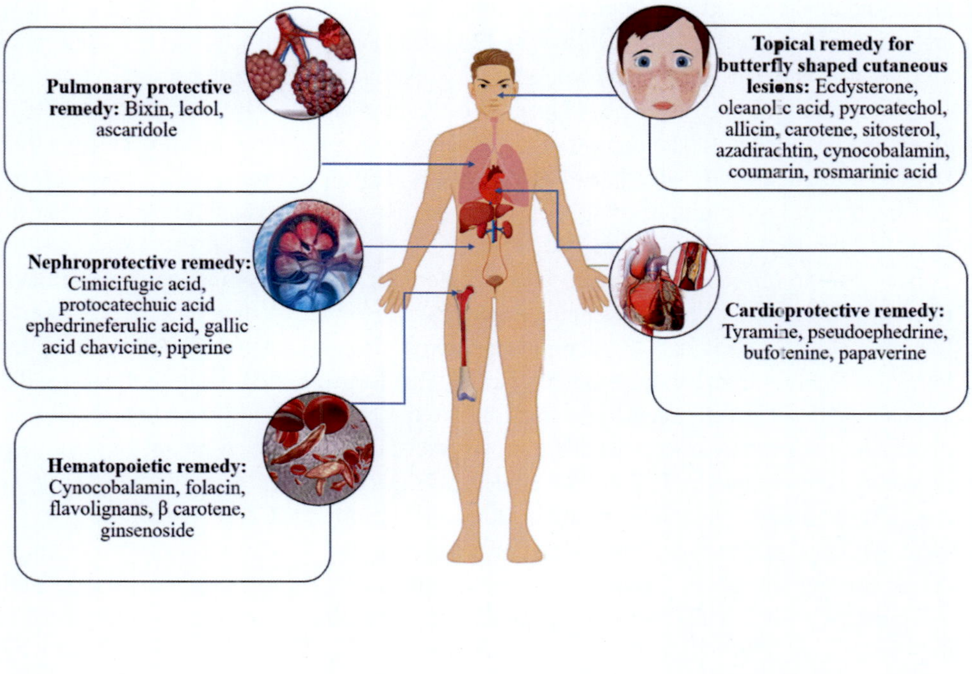

Fig. 5 Phytoconstituents used as symptomatic relief providers for SLE. (Reproduced with permission from Acharya Balkrishna, MDPI Cells 2020,9(4), Published by MDPI, Year-2020)

these effects are produced by inhibition of NF-κB and regulation of adaptive and innate immune cells, including subsets of $CD4^+$ T cells (Th1/Th2/Th9/Th17/Tfh/Tfr/Treg), $CD8^+$ memory T cells, Treg cells, B cells, dendritic cells, neutrophils, mast cells and macrophages [133].

3.2 Skin-Related Autoimmune Diseases

The body's largest organ is the skin and accounts for about 16% of the body mass. It is essential for keeping the body protected against various infectious agents and is considered as a part of the immune system. Some of the autoimmune skin disease including atopic dermatitis, vitiligo, psoriasis, pemphigus and lupus are discussed with their respective pathophysiology and herbal therapies for their management and treatment.

3.2.1 Atopic Dermatitis/Eczema

Atopic dermatitis, often known as eczema, is a chronic relapsing inflammatory skin condition characterised by impaired epidermal barrier function and hyperimmune response [136]. The complex involvement of various immunological elements, coupled with genetic and environmental factors, is linked to onset and progression of atopic dermatitis [137]. The three main stages of atopic dermatitis are infantile, childhood and adulthood [138].

Filaggrin (FLG), a crucial protein in epidermis development and claudin-1 (CLDN1) and a structural component of tight junctions in epidermis, was found to have a genetic relationship with atopic dermatitis. Pro-filaggrin (ProFLG), a protein that functions as a structural component of stratum corneum, is encoded by the filaggrin gene [139]. ProFLG polymers are cleaved and dephosphorylated by proteases to generate FLG monomers, which are linked to the accumulation of keratin filaments and the development of stratum corneum. The moisture and acidic pH of skin are both regulated by FLG metabolites, urocanic acid and pyrrolidine carboxylic acid [137]. The homozygous mutation of filaggrin gene is linked to early onset and severe atopic dermatitis. The expression of proteins filaggrin, keratin, transglutaminase and other intracellular proteins is downregulated, resulting in aberrant corneocyte development, decreased skin moisture and elevated skin pH.

The compromised skin barrier allows allergens and microbes to enter the skin and stimulate keratinocytes into releasing IL-1β, IL-25, IL-33, macrophage-derived chemokines and thymic stromal lymphopoietin (TSLP) that further activate dendritic and Langerhans cells [140]. Acute atopic dermatitis is caused by response of Th2 subset; the activated dendritic cells stimulate Th2 cells to release IL-4, IL-5, IL-13, IL-31 and IL-33 which impairs keratinocyte development, reduces antimicrobial peptide production and produces burning sensation. The chronic atopic dermatitis is associated with Th1, Th22 and Th17 responses which results in aberrant keratinocyte proliferation and thickening of epidermis [141–143].

Current therapies for management of atopic dermatitis include emollients, calcineurin inhibitors, phosphodiesterase-4 inhibitors (Crisaborole) and biological agents (dupilumab, baricitinib) [144]. Some of the plant-based therapies reported for atopic dermatitis are discussed below.

3.2.1.1 Herbs Used for Management of Atopic Dermatitis

Aloe vera

Aloe vera is a perennial succulent xerophyte with antioxidant and antibacterial properties [145]. *Aloe vera* leaf extracts with high polysaccharide content are routinely used in cosmetics and over-the-counter medications to cure sunburn, skin irritation and skin diseases like eczema. Studies were performed by Na et al. [11] to investigate the oral administration of processed *Aloe vera* gel (PAG) containing low molecular weight aloe polysaccharides in treating ovalbumin (OVA)-induced atopic dermatitis in BALB/c mice. The expression of tight junction genes, including claudin-1, claudin-8 and zonula occludens-1 (ZO-1), was downregulated in atopic dermatitis epidermis as compared to normal epidermis. It was found that PAG administration through oral route lowered the thickness of epidermis and suppressed both overall and OVA-specific immunoglobulin E (IgE) production. In addition, inflammatory cytokines, such as IFN-γ, IL-4 and IL-17A, were suppressed and restored tight junctions to protect skin barrier function [11].

3.2.1.2 Pentaherbs Formula

In traditional Chinese herbal medicine, PHF, a mixture of five herbs is used as an alternative adjunct therapy for atopic dermatitis. PHF contains *Flos lonicerae, Herba menthae, Cortex phellodendri, Cortex moutan and Rhizoma atractylodis* in a 2:1:2:2:2 w/w ratio. Tsang et al. [12] investigated the effectiveness of PHF using an in vivo mouse model with oxazolone (OXA)-mediated dermatitis. Additionally, the active components of PHF water extract were isolated and their in vitro anti-inflammatory effects were evaluated on pruritogenic cytokine IL-31 and IL-33-activated human eosinophils and dermal fibroblasts. In this study, PHF therapy either orally or topically significantly decreased ear swelling, epidermis thickness, eosinophil infiltration in epidermal and dermal layers, and the levels of serum IL-12 in mouse OXA-mediated dermatitis. The active ingredients of PHF water extract—gallic acid, berberine and chlorogenic acid—were reported to be effective in inhibiting release of pro-inflammatory cytokines (IL-6, CCL7) [12].

3.2.2 Vitiligo

Vitiligo is a chronic depigmentation disorder of skin, characterised by destruction and loss of melanocytes which results in non-scaly depigmented patches on the skin [13]. Systemic inflammation of the ears and eyes can coexist with vitiligo. The brain, eye and hairs are affected by the rarest form of vitiligo known as Vogt-Koyanagi-Harada syndrome [146]. Aetiology of vitiligo is unclear, but it is mainly due to the destruction of melanocytes; about 80% of genetic and 20% of environmental factors continue to remain the leading causes of vitiligo [13].

The inheritance of vitiligo is polygenic, including various alleles and gene loci. The HLA involved in vitiligo patients were found to be A2, DR4, DR7 and DQB1*0303. The candidate genes involved in vitiligo are transporters linked to antigen processing protein-1 (TAP1) gene, angiotensin-converting enzyme (ACE) gene, CTLA-4 and catalase (CAT) gene [147]. The innate immune response is mediated by the genes CASP7, NLRP1 and TICAM1, while the adaptive immune response is mediated by the genes CTLA4, CD80, HLA and Foxp3; majority of the genes are immune genes, that emphasise the key role of immune system in the pathogenesis of vitiligo. A small subset of risk alleles tyrosine (TYR), Oculocutaneous albinism (OCA2) and melanocortin 1 receptor(MC1R) are solely expressed in melanocytes, implying a role for melanocytes in the onset of disease [146]. The rate-limiting steps of melanin production are catalysed by the enzyme tyrosinase, which is encoded by the TYR gene [148].

Environmental risk factors associated with vitiligo include phenolic compounds that are tyrosine analogues and interfere with tyrosinase enzyme and synthesis of melanin. Chemically induced depigmentation was found in skin-lightening products, hair dyes, paints, adhesives and industrial oil which may contain the phenolic compounds such as rhododendrol, monobenzyl ether of hydroquinone (MBEH), 4-*tert*-butylcatechol (4-TBC) and 4-*tert*-butylphenol (4-TBP) [149].

Hereditary factors, autoimmune reactions, oxidative stress and production of inflammatory mediators have been linked to pathophysiology of vitiligo [148]. Melanogenesis is an energy-demanding process carried out by melanocytes that results in a pro-oxidant condition in the skin. The synthesis of melanin requires essential protein tyrosine-related protein 1. Tyrosine-related protein 1 interacts with the calnexin complex in response to oxidative stress, which reduces its stability and induces the formation of hazardous melanin intermediates. Individuals with vitiligo demonstrate modified functionality of mitochondria, that seem to be the main source of ROS [146, 148]. Overproduction of ROS triggers the unfolded protein response (UPR), which causes melanocytes to produce exosomes that carry molecular protein patterns associated with damage, as well as melanocyte-specific antigens, miRNAs and heat shock proteins [150]. Exosomes aid in their development into efficient antigen-presenting cells by transferring vitiligo target antigens to nearby dendritic cells. Stressed melanocytes release CXCL12 and CCL5, which are also implicated in T cell localisation to the skin. In vitiligo lesions, $CD8^+$ T cells release IFN-γ; IFN-γ binds to its receptor and activates the JAK/STAT pathway, as a result CXCL9 and CXCL10 are secreted in the skin. The melanocyte-specific $CD8^+$ T cells are recruited through cognate receptors CXCR3, CXCL9 into the skin and CXCL10 increases their localisation within the epidermis which results in acceleration of inflammation through a positive feedback mechanism. Melanocyte-reactive resident memory T cells are responsible for maintaining established vitiligo in skin through IL-15-dependent signalling [13, 151].

The current therapies for management of vitiligo include topical corticosteroids, mini pulse oral corticosteroids, narrow band UVB phototherapy and methotrexate, that need to be used over extended periods of time with resultant skin atrophy, acne and contact allergy. Some of the herbal therapies for management of vitiligo with least adverse events are discussed here.

3.2.2.1 Herbs Used for Management of Vitiligo

Piper nigrum

Piper nigrum fruit belongs to the family Piperaceae, with reported bioactivities including antioxidant, anti-inflammatory and immunomodulatory [15]. Melanocyte proliferation stimulants are of interest as potential treatment for vitiligo. The amides such as piperine, guineesine and pipericide found in *P. nigrum* have the ability to promote the growth of melanocytes [14, 152]. In a study by Shafiee et al. [153], patients with facial vitiligo were given topical piperine, along with narrowband UVB (NB-UVB) irradiation which appreciably increased the incidence of re-pigmentation at 1, 2 and 3 months compared to the untreated control group [153]. Another clinical trial was performed on human subjects affected by vitiligo in which ointments containing pure piperine and *P. nigrum* fruit extract were tested and found to repigment the affected skin areas. Crude extract produced results more quickly and effectively compared to pure piperine [14].

Nigella sativa

Nigella sativa, family Ranunculaceae, is commonly known as black seed or black cumin [16]. *N. sativa* contains natural bioactive substances like alkaloids, saponins, alpha-hederin and thymoquinone, which are responsible for its wide list of therapeutic activities including analgesic, anti-inflammatory, antioxidant and immunomodulatory activity. *N. sativa* inhibits the enzymes elastase and myeloperoxidase and also lipid peroxidation, raising the defence against free radicals with consequent exertion of antioxidant effect [154]. Clinical trials were carried out by Sarac et al. [17] to examine the effectiveness of topical *N. sativa* seed oil in vitiligo patients. *N. sativa*-containing cream was used topically on commonly affected areas of face, hands and genital region twice a day for 6 months. The studies reported noteworthy hyperpigmentation with no adverse effects [17].

3.2.3 Psoriasis

Psoriasis is a chronic inflammatory autoimmune skin condition that affects individuals of all age groups, involving interaction of immune cells and epidermal keratinocytes and is characterised by excessive proliferation and infiltration of immune cells [155]. The most common clinical symptoms include recurrent flare ups of thick patches of red and inflamed skin covered by flaky and silver scales on the surface of skin, dry itchy skin, pruritis, swelling of joints and discoloration of skin. Clinically, psoriasis is divided into two groups: psoriasis with pustular and non-pustular lesions. The non-pustular type of psoriasis involves plaque, arthritis, vulgaris, inverse, guttate and erythrodermic, whereas pustular psoriasis involves impetigo herpetiformis, palmoplantar and acrodermatitis continua of hallopeau [156, 157]. Aetiology of psoriasis is not clear; interplay of genetic, epigenetic and environmental factors triggers the onset of disease. With involvement of myeloid dendritic cells, cytokines stimulate aberrant keratinocyte growth and induction of various inflammatory mediators. IL-17-producing Th cells, with innate lymphoid cells 3 (ILC3), apparently induce psoriasis rashes without T-cell/antigen interaction, in response to antimicrobial peptides secreted by activated keratinocytes and inflammatory cytokines [18]. Cell signalling pathways play a major role in aetiology of psoriasis; the key regulatory pathways involve STAT, NF-κB and MAPK signalling [158].

3.2.3.1 Herbs Used for Management of Psoriasis

Number of medicinal plants have been reported in literature, for management and therapeutic interventions of psoriasis including *Silybum marianum, Wrightia tinctoria, Curcuma longa, Aloe vera, Angelica sinesis, Cassia tora, Capsicum annum, Smilax china* and many more [159–161]. Some of the more commonly used herbs are discussed with their respective molecular mechanisms of action.

Oryza sativa

Oryza sativa L. is commonly known as black rice. It has been reported by Thepthanee et al. [162] that *n*-butanolic (*n*-BuOH) extract of *Oryza sativa* enriched with total flavonoids, phenolics and anthocyanins, possesses strong antimicrobial and immunomodulatory effects. It was also shown to have strong antioxidant properties as determined by 2,2′-azino-*bis*(3-ethylbenzothiazoline-6-sulfonic acid) (ABTS) and α,α-diphenyl-β-picryl hydrazyl (DPPH) radical-scavenging tests [162]. *In vitro* and *in vivo* studies were performed by Ampawong et al. [19] on human psoriatic artificial skin and imiquimod (IMQ)-induced rat psoriasis model, respectively. Both *in vitro* and *in vivo* models have demonstrated superior activity of *Oryza sativa* extract in substantially reducing severity of psoriasis as compared to untreated skin specimens. In this study, *Oryza sativa* extract has been found to increase expression of anti-inflammatory cytokines (IL-10 and TGF-β) and downregulation of proinflammatory cytokines (IL-6, IL-8, IL-20, IL-22 and TNF-α). It was reported to decrease inflammation of epidermis and extent of apoptosis through induction of caspase-3. Anthocyanins being antioxidant and anti-inflammatory in nature, reduce keratinocyte proliferation [19].

Perilla frutescens

Various portions of *Perilla frutescens* have been reported to contain a diverse range of phytoconstituents like volatile oil, fatty acids, policosanols, tocopherols, phytosterols and hydrophilic compounds like phenols, flavonoids and triterpenes. Xu et al. [163] analysed the effectiveness of the essential oil isolated from *P. frutescens* stem and leaves using IMQ-induced psoriasis-like lesions in the BALB/c mice model. It has been observed that essential oil is able to reduce the levels of pro-inflammatory cytokines IL-6, IL-17, IL-23 and NF-B [163].

3.3 Autoimmune Addison's Disease

AAD, usually referred as primary adrenocortical insufficiency, is characterised by inefficacy of the adrenal cortex in synthesis of adequate amount of mineralocorticoids and vital steroidal hormones-glucocorticoids. Effective management of AAD requires continuous hormone replacement therapy. Depending on whether the condition is acute or chronic, several clinical manifestations may occur; the most frequent symptoms are exhaustion, weight loss, anorexia, nausea, vomiting, diarrhoea and salt craving. The most specific symptom is hyperpigmentation of the surfaces of the skin and mucosa, and is related to elevated levels of adrenocorticotropic hormone (ACTH) and melanocyte stimulating hormone (MSH) [164]. Melanin production is catalysed by the enzyme tyrosinase, that converts amino acid tyrosine into L-3,4-DOPA, which is further oxidised to form DOPA quinine and leads to production of melanin by free radical coupling pathway.

Common cause of primary adrenocortical insufficiency is destruction of adrenal tissues and failure of adrenal cortex; it may occur in combination with APS1 and autoimmune polyglandular syndrome type 2 (APS2) or isolated AADs [165]. Mutation in AIRE is thought to be the major causative factor for monogenetic autoimmune disorder APS-1 [166]. Aetiology of primary adrenocortical insufficiency includes destruction of adrenal cortex, deficiency of steroidogenic enzyme, impaired steroidogenesis pathway and infectious agents [167]. Congenital adrenal hyperplasia is the most frequent hereditary form of primary adrenocortical insufficiency caused by deficiency of steroidogenic enzyme. The immune system targets enzyme 21-hydroxylase which is located in smooth endoplasmic reticulum (ER) of adrenocortical cells, resulting in autoimmune destruction of the adrenal cortex, which is a hallmark of autoimmune primary adrenal insufficiency. The gene CYP21A2 encodes the enzyme 21-hydroxylase, which is essential for biosynthesis of the steroidal hormones cortisol and aldosterone. Mutation of gene CYP21A2 is associated with congenital adrenal hyperplasia [168]. Patients with AAD develop autoantibodies against self-antigen 21-hydroxylase, which is a key indicator in clinical diagnosis [164]. Autoantibodies against 21-hydroxylase attack the carboxy terminal of the enzyme and consist of IgG1 isotype which can cross the placenta [169]. Studies have reported that 21-hydroxylase autoantibodies inhibit the activity of 21-hydroxylase enzyme *in vitro*, but do not show inhibitory effect *in vivo*. Though these autoantibodies can cross the placenta, they do not cause hypoadrenalism in newborns. The pathophysiology of AAD may not be influenced by steroidal 21-hydroxylase autoantibodies. [170].

AAD is mediated by 21-hydroxylase-specific $CD8^+$ T cells; the cells of adrenal cortex respond to inflammatory and stressful events by releasing proinflammatory cytokines IL-6, IL-8, IL-18, IL-1 and TNF-α. The infections caused by *Mycobacterium tuberculosis* lead to cytotoxic T lymphocyte infiltration that reaches adrenal glands through blood circulation leading to initiation of AAD. The specific epitope HLA-A2 (21-hydroxylase 342–350 amino acid) binds with specific $CD8^+$ T lymphocytes and stimulates production of IFN-γ. The $CD8^+$ T lymphocytes lyse the cells that express 21-hydroxylase through the production of granzyme B [170].

Hormone replacement therapy with glucocorticoids forms the mainstay for current AAD management. However, chronic administration of this therapy is linked to severe toxicity and drug-induced autoimmunity. In AAD, hyperpigmentation of skin is associated with stimulation of melanocyte receptors by ACTH [170]. Herbs like *Curcuma longa, Glycyrrhiza glabra, Emblica officinalis, Trigonella foenum graecum* (fenugreek), *Boswellia serrata* and many more commonly used for management of AAD are discussed below.

3.3.1 Herbs Used for Management of Addison's Disease

3.3.1.1 *Glycyrrhiza glabra*

Glycyrrhiza glabra, Family Fabaceae, has been used in treatment of mild-to-moderate adrenal insufficiency. Glycyrrhizin increases the cortisol level by inhibiting the enzyme 11beta-hydroxysteroid dehydrogenase (11-HSD) which is responsible for oxidation of cortisol to its inactive form cortisone [23]. Long-term use of glycysrrhizin is associated with hyper mineralocorticoid-like effect in humans and animals. However, these effects are reversible upon withdrawal of glycyrrhizin [24]. Methlie et al. reported increased serum cortisol levels for 2.6 h after consumption of *G. glabra* in 17 AAD patients, selected from Haukeland University Hospital's outpatient clinic [171]. Kumari et al. [172] studied *in vitro* tyrosinase-inhibitory activity of different concentrations of glabridin on B16 melanoma cells and reported concentration-dependent reduction in melanin [172].

3.3.1.2 *Boswellia serrata*

Boswellia serrata, Burseraceae, commonly known as Salai/Salai guggul, has been in use for a long time in folk medicine to combat a variety of chronic inflammatory diseases as well as in ceremonial and religious events as incense. It contains oleo gum resin obtained from incision made on the trunk of Boswellia tree and consists of 30–60% resin, 5–10% essential oils, and polysaccharides. These extracts contain 4 major pentacyclic triterpenic acids—11-keto-β-boswellic acid, acetyl-β-boswellic acid, β-boswellic acid and acetyl-11-keto-β-boswellic acid—and monoterpenes, diterpenes, triterpenes and tetracyclic triterpenic acids, that are reported to be responsible for inhibition of pro-inflammatory enzymes [173, 174]. Morsy et al. [173], reported a study comparing the efficacy of a host of plant steroids in interacting with glucocorticoid receptor (GR) by *in silico* techniques and further validation of their *in vivo* anti-inflammatory activity in the classical cotton pellet-induced granuloma models in rats. Boswellic acid demonstrated superior interaction with GR, comparable to fluticasone as well as brought in a significant decrease in serum levels of IL-6 and TNF-α [173].

3.3.1.3 *Fenugreek*

Trigonella foenum graecum Linn, Fabaceae, used as dietary supplement and spice in India contains steroidal sapognins diosgenin along with lecithin, phosphates, trimethylamine, trigonelline, choline and iron and has antioxidant, hypolipidemic, hypoglycaemic and immunomodulatory activity. Its potential role in managing various neurological disorders such as cognitive impairment, depression and neurodegenerative disorders (PD, ALD and Huntington's disease (HD)) is being examined [175]. Aylanc et al. [176] evaluated *in vitro* tyrosinase-inhibitory activity of various phenolic extracts of *Trigonella spruneriana* BOISS and reported that its methanolic, ethanolic and ethyl acetate extracts showed appreciable tyrosinase-inhibitory activity when compared with kojic acid [176]. Prema et al. [177], reported that fenugreek seed extract was able to inhibit the action of AChE, a crucial enzyme involved in the progression of ALD, when tested in aluminium chloride-induced experimental AD model in male albino Wistar rats [177].

3.4 Neurological Diseases

Progressive destruction and malfunction of the neurons or nerve cells are characteristics of neurodegenerative disorders with protein deterioration, oxidative stress, inflammation, environmental factors, mitochondrial deficiencies, family history and aberrant protein build-up in neurons listed as some of the most extensively researched causes. Neurological autoimmune diseases that cause inflammation and neurodegeneration of central nervous system (CNS) include MS, ALD and PD.

3.4.1 Commonly Used Herbs for Neurodegenerative Diseases

Medicinal plants and products such as *Baccopa monnieri, Withania somnifera* (ashwagandha), curcumin, ginseng, jatamansi, Shankhpushpi, guduchi, resveratrol, *Ginkgo biloba, Centella asiatica* and Triphala have been used to prevent or treat neurological illnesses [178]. Some of these are discussed below.

3.4.1.1 *Bacopa monnieri*

Bacopa monnieri, family Scrophulariaceae, commonly known as Brahmi has been applied in Ayurveda as a nervine tonic since the sixteenth Century. Brahmi contains betullic acid, bacoside A (64.28%), bacoside B (27.11%), D-mannitol, stigmastanol and β-sitosterol. *Bacopa monnieri* contains pharmacologically active saponins (bacosides A and B, bacopide or bacopasaponins), which have sedative, anti-amnesic, anti-epileptic, anti-anxiety, anti-depressant and neuroprotective activities [179]. Its potential as anti-Parkinsonian drug is reported to be, by reducing alpha synuclein aggregation, preventing dopaminergic neurodegeneration and restoring the lipid content in nematodes on pharmacological *Caenorhabditis elegans* models of Parkinson's [180]. Kishore et al. [181] evaluated two marketed drugs of *Bacopa monnieri* for their nootropic effect in treatment of dementia in scopolamine-induced amnesic Swiss mice models. Both the products, after 14 days of administration, were able to decrease the effect of scopolamine and restore memory [181].

3.4.1.2 *Ginkgo biloba*

Ginkgo biloba, commonly referred as Living fossil [182], is well known for enhancing cerebral function [183]. Yu et al. [184] reported a study evaluating the neuroprotective effect of *Ginkgo biloba* extract and formulation in *in vitro* and *in vivo* PD models. The extract and formulations provided neuroprotective effect in 1-methyl-4-phenyl-1,2,3,6-tetrahydropyridine (MPTP)-induced damage in human neuroblastoma SH-SY5Y cells by blocking the apoptotic cycle. Similar results were observed in MPTP-induced PD mice models, with reduced neuronal damage and improved motor coordination. Improved locomotion in 6-hydroxydopamine (6-OHDA)-induced PD in zebra fish was reported, possibly due to inhibition of dopaminergic loss, as demonstrated by Tyrosine hydroxylase (TH) immunostaining [184]. In a pilot study reported by Rapp et al. [185], 189 patients suffering from ALD were divided into groups that received either *G. biloba* extracts or donepezil, over a period of 12 months. The patients who received *G. biloba* extracts showed comparable effect on cognitive symptoms, as those that were administered donepezil, but with much fewer adverse events and improved safety profile [185].

3.4.1.3 *Centella asiatica*

Centella asiatica (Gotu kola) is a conventional herb widely used in Ayurveda for memory and cognitive function improvement. Centella contains several pentacyclic triterpenoids, including asiaticoside, brahmoside and madecassic acid that shows anti-inflammatory, neuroprotective and mitoprotective effects [186–188]. Hafiz et al. [188] reported a comprehensive study evaluating the neuroprotective effect of ethanolic extract of *C. asiatica* by a series of *in vitro* and *in vivo* tests. The extract produced significant reduction in acetylcholinesterase activity in SH-SY5Y cells. A similar concentration-dependent reduction in levels of ROS and GSH and the pro-inflammatory cytokines—PGE2 and TNF-α—were observed in LPS-stimulated RAW 264.7 cells. These in vitro findings were validated in vivo in the LPS-induced Sprague Dawley rats model for neuroinflammation [188]. Fujimori et al. reported a study evaluating the neuroprotective effect of *Centella asiatica* and its component araliadiol against damage induced by ER stressors—glutamate and tunicamycin—in murine hippocampal cells (HT22). Suppression of cell death by inhibiting ROS production and phosphorylation of PERK (one of the main transducers of ER stress), respectively, was found to be the mechanism involved. They further validated this *in vivo*, in mice with scopolamine-induced amnesia, with improved cognitive function in treated mice [189]. Gray et al. [190] reported the physiological mechanism for the improved cognitive function demonstrated by *C. asiatica.* They evaluated the neuroprotective effect of *C. asiatica* on amyloid-β (Aβ) exposure in the mouse model of ALD. Water extract of *C. asiatica* increased arborisation in the dendrites and spine densities in the neurons and prevented loss of spines caused by Aβ exposure [190].

3.4.1.4 *Nardostachys jatamansi*

Nardostachys jatamansi is a blooming, rare and most ancient species within the family Valerianaceae. It has anti-inflammatory, antioxidative, cardioprotective, hepatoprotective, hypolipidemic and antifungal activity. The major chemical constituents include crystalline acid, jatamansic acid, hydrocarbons, a poly-oxygenated crystalline solid together with A-endesmol, B-eudesmol, ethanol, angelicin and 4-hydroxy thymol dimethyl ether [191]. Liu et al. [192] studied the neuroprotective effect of *N. jatamansi* root ethanolic extract in both *in vitro* cell culture system and *in vivo* Drosophila ALD model. It was observed that *N. jatamansi* extract suppressed Aβ-induced cell death in SH-SY5Y cells and also reduced Aβ42-induced cell death in the brain in flies. Intake of *N. jatamansi* extract reduced ROS level, NO level in Aβ42-expressing flies and extracellular-signal-regulated kinase (ERK) phosphorylation. It was reported by Liu et al. that *N. jatamansi* exhibits neuroprotective effect against Aβ42 neurotoxicity both *in vivo* and *in vitro*, due to its anti-inflammatory, antioxidative and ERK signalling inhibitory action [192].

Shankhpushpi

In Ayurveda, the plant Shankhpushpi *(Convolvulus pluricaulis)*, which is categorised under Medhya Rasayana, is utilised as the greatest brain tonic medication. The major bioactive components of *C. pluricaulis* are the glycosides, steroids (stigmasterol and betulinic acid), flavonoids (β-carotene and chlorogenic acid), alkaloids and coumarins (scopoletin). *C. pluricaulis* possesses anti-inflammatory, anti-oxidative, analgesic and tranquilising properties. Since ancient times, it has been used to treat chronic bronchitis, asthma, hypertension and for fever reduction [193, 194]. In Indian traditional medicine, four sources of *Shankhpushpi* reported are *Canscora decussata Schult., Clitorea ternatea Linn., Convolvulus pluricaulis and Evolvulus alsinoides Linn.* In vitro and in vivo studies on β-amyloid-induced neurotoxicity in neuroblastoma cell lines, Neuro-2a and scopolamine-induced amnesia in male Sprague Dawley rats, respectively, were conducted by Sethiya et al. [195] to examine the neuropharmacological effects of methanolic extracts of all four species of Shankhpushpi. Inhibition of the enzymes acetylcholinesterase and 5-lipoxygenase and antioxidant activity were the other *in vitro* studies conducted by them. *Evolvulus alsinoides Linn.* demonstrated superior *in vitro* and *in vivo* neuropharmacological effects in comparison to the other three varieties of Shankhpushpi suggesting that it had the potential to be developed as therapeutic agents against neurodegenerative diseases [195].

Triphala

Triphala is a widely used and extremely effective polyherbal Ayurvedic medication classified as Tridoshaic rasayana made from the fruits of the plants *Terminalia bellerica* (Bibhitaki), *Terminalia chebula* (Haritaki) and *Emblica officinalis* (Amalaki). It is a key component of digestive system and other rejuvenative therapies. In streptozotocin-induced diabetes rat models, triphala treatment was reported to suppress the progression of associated neuropathy, with significantly increased motor nerve conduction velocity and decreased thermal and mechanical hyperalgesia as well as mechanical allodynia, possibly due to elevated levels of Nerve growth factor (NGF), particularly in the sciatic nerves. TGF-1, TNF and IL-1 levels in the circulation were markedly reduced by this therapy. Histopathological investigation confirmed Triphala churna's neuroprotective properties [196]. In another study, it has been reported that triphala has potential as a neuroprotective agent, with activity against oxidative stress, both in SH-SY5Y cells and zebra fishes [197].

3.4.2 Multiple Sclerosis

MS is a chronic inflammatory disease of the CNS, categorised by demyelination, degeneration of neurons and loss of axons, giving rise to localised lesions in the grey and white matter of the brain. Aetiology of MS is not clear and includes immunological, genetic and environmental factors. The factors that trigger onset of MS include gut microbiota, smoking, deficiency of vitamin D, obesity, infections and heavy metals (cadmium, lead) in food [198]. The clinical symptoms associated with MS include dysarthria, nystagmus, tremor, disturbance of vision, sensory and motor neurons. Individuals with genetic predisposition to environmental factors and variables are at greater risk of developing MS. Infections associated with human herpes virus-6 (HHV-6), Epstein-Barr virus (EBV) and *Chlamydia pneumoniae* are responsible for increasing risk of developing MS [199]. The fundamental mechanisms behind MS remain largely unknown, which has resulted in disease classifications based on natural history and an insufficient knowledge of the processes responsible for its origin and progression. MS is classified in major types, namely relapsing remitting MS (RRMS) type in which exacerbation occurs recurrently with partial or complete recovery. About 85% of MS patients are found to be affected with RRMS [200]. Secondary progressive MS (SPMS) is the type in which there is gradual worsening of clinical disability; about 50% of patients with RRMS enter SPMS. Primary progressive MS (PPMS) is a type in which there is gradual worsening of neurologic function. Microorganisms have the ability to elicit anti-inflammatory and proinflammatory responses, activate the innate immune system and mimic myelin proteins in the CNS. Gene responsible for higher risk of developing MS is human lymphocyte antigen DRB1*15:01 allele (HLA-DRB1*15:01) [200].

Pathogenesis of MS involves interaction between immune cells and neurons. It is believed that failure of central and peripheral immune tolerance mechanisms triggers autoimmune activation of T cells [201]; the presence of T cells in CNS plays an important role in pathogenesis of MS. The activated autoreactive T cells enter CNS through impaired blood brain barrier (BBB) and create an inflammatory environment that leads to death of oligodendrocytes which ultimately leads to degeneration of myelin sheath and loss of neuron. The subsets of helper T cells, Th1 and Th17 release proinflammatory cytokines IFN-γ and IL-17, respectively. The oligodendrocytes are directly killed by cytokine IFN-γ whereas IL-17 expresses TNF-α and IL-6 [202–204]. Granzyme B can be released by $CD4^+$ and $CD8^+$ T cells that express IL-17 and destroy neurons by binding to the glutamate receptor (GluR3). The inflammatory response triggered by IL-17 and TNF-α is associated with degeneration of neurons by activation of transcriptional factors STAT-1 and NF-κB [205]. The inflammatory response is triggered when IL-17 produces inducible NO synthase, which eventually results in an increase in NO in various cells [206]. Alteration of gut microbiota, and increased exosomal protein and intracellular miRNA profile are the factors responsible for alteration in function of Treg cell in MS [207]. Activation of microglia with involvement of T and B cells, release ROS and nitrogen species which leads to mitochondrial damage and gene deletion in MS lesions with persistent demyelination or neurodegeneration [207, 208].

Currently, there are no treatments to treat MS completely; however, a number of therapeutic options are available to moderate the disease progression and provide characteristic mitigation. There are a number of phytoconstituents that are reported to decrease demyelination, these include flavonoids (quercetin, hesperidin, luteolin and icariin) [209]. Some of the herbs with potential to manage MS are discussed below.

3.4.2.1 *Artemisia dracunculus*

Artemisia dracunculus L., family Asteraceae, is an everlasting herb containing essential oils, flavonoids, coumarins, phenylpropanoids and tannins which are attributed with anti-inflammatory and antioxidant properties [26]. Since aetiology of MS is unclear, therapies focus on reducing inflammation and restoring myelin through the repair of damaged tissue. The inhibition of Th1 and Th17 cell cytokines by the aqueous extract of *A. dracunculus* was studied in the experimental autoimmune encephalitis (EAE) model by Safari et al. [210]. Immunisation with myelin oligodendrocyte glycoprotein (MOG35-55) and Complete Ferund's adjuvant (CFA) was done to induce EAE in C57BL/6 mice. *A. dracunculus* was found to markedly reduce expression of pro-inflammatory cytokines IL-1, TNF-α, IL-6 and IFN-γ through inhibition of NF-κB signalling pathway. *A. dracunculus* extracts in EAE mice appear to modify immunological responses and raise serum antioxidant levels, which reduces demyelination [210].

3.4.2.2 *Silybum marianum*

Silybum marianum belongs to the family Asteraceae and is commonly known as milk thistle. Silymarin, a complex mixture of flavonolignans, flavonoids (taxifolin, quercetin) and polyphenolic compounds, is reported to have excellent hepato-protective and antioxidant properties [211]. In vitro study performed by Shariati et al. [212] suggested that silymarin might present a potential therapeutic option for MS. In this study, the effect of silymarin on the proliferation and activation of Treg cells isolated from newly diagnosed MS patients was investigated. Expression of STAT5, JAK3 and Foxp3 genes and release of IL-10 in IFN-β treated RRMS patients were also examined. Silymarin promoted proliferation of Treg cells and the expression of genes STAT5, JAK3 and Foxp3 was improved by silymarin [212].

3.4.3 Alzheimer's Disease

ALD is the most common reason for memory loss in older adults. It is a slowly progressing, complex neurodegenerative disorder that impacts numerous regions of the cerebral cortex and hippocampus, and is characterised by neuritic plaques and neurofibrillary tangles due to accumulation of insoluble forms of amyloid-beta peptide (Aβ) in the extracellular spaces and blood vessels of the affected areas of the brain [213, 214]. The major clinical manifestations are cognitive impairment, behavioural changes, depression, hallucination and dementia. The basic aetiology of ALD is destruction of nerves, especially cholinergic nerves. The risk factors associated with ALD include genetic predisposition to environmental triggers such as smoking, infectious agents and toxins. ALD is classified into two types based on the age groups: affected-sporadic and familial ALD. Sporadic ALD affects individuals above 65 years of age and familial ALD affects individuals between 30 and 65 years age group. The risk factors associated with sporadic ALD include the presence of the epsilon-4 (ε4) allele of the APOE gene (APOEε4) and the mutations in the APP or presenilin 1 (PSEN1); presenilin 2 (PSEN2) genes are the causative factors for familial ALD [214–217].

APP exists as a transmembrane protein which helps in growth and repair of neurons, and commonly occurs in neuronal synapses. In normal non-amyloidogenic processing, APP is degraded by enzyme α-secretase followed by β-secretase that forms soluble fragments and is cleared from the brain. The enzymes α-secretase and β-secretase cleave APP during aberrant amyloidogenic processing, producing insoluble amyloid beta 42 fragments. Extracellular plaques build up as a result of impaired removal of the insoluble fragments from interstices of the brain. In neural axons, the tau protein serves as a stabilising protein for microtubules. Evidence suggests that elevated levels of ferrous (Fe^{2+}) chloride in neurons leads to aberrant phosphorylation of tau protein causing microtubule instability and formation of neurofibrillary tangles eventually inducing neurodegeneration changes [218]. Accumulation of these pathogenic proteins and free radicals causes mitochondrial damage, inflammation and oxidative stress, a deficiency of neurotransmitters mainly acetylcholine, and synaptic dysfunction, all of which ultimately result in neuronal death [213, 218].

N-methyl-D-Aspartate (NMDA)-receptor antagonist and cholinesterase inhibitors are two drug classes currently used as medication for symptomatic relief in Alzheimer's disease [219]. Some of the herbs used in management of ALD are discussed below.

3.4.3.1 Herbs Used for Management of Alzheimer's Disease

Acorus calamus

Acorus calamus rhizomes contain the phenylpropanoids α-asarone and β-asarone, which indicate anti-cancer, anti-inflammatory, anti-apoptotic, antioxidant and neuroprotective activities [220]. Various fractions of *A. calamus* extract have been studied for preventing memory loss, anxiety and oxidative stress in LPS-induced neuroinflammation rat models by Esfandiari et al. [221]. Administration of aqueous extract of *A. calamus* via oral route reduces inflammation and oxidative stress and prevents memory loss. In another study, *in vitro* effects of *A. calamus* extract and α-asarone on oxidative stress-induced and ER stress-induced cell death in hippocampal HT22 cells was investigated by Mikami et al. [222]. It was found that *A. calamus* extract and α-asarone greatly prevent cell death caused by ER stress inducer, tunicamycin, and oxidative stress inducer, L-glutamate. Asarone and *A. calamus* extract both inhibit the production of ROS, indicating that they both may be efficient therapeutic options for treating neurodegenerative disorders like ALD.

Crocus sativus

Crocus sativus has three major components—crocin (source of colouring pigment), the glycoside picrocrocin (a precursor of safranal) and safranal (the deglycosylated form of picrocrocin)—with great antioxidant potential. It has a wide spectrum of biological activities including antioxidant and anti-inflammatory activity [223]. Extensive literature was reviewed by D'Onofrio et al. [224] in which 17 *in vitro* and *in vivo* preclinical and 4 clinical studies have been reported for evaluating the effectiveness of saffron on cognitive deformity in animal models of ALD. Crocin reduces oxidative stress, modulates A tau protein aggregation and regulates glutamate levels. Saffron had a superior safety profile and its effects on cognitive impairment were similar to those brought on by donepezil and memantine.

3.4.4 Parkinson's Disease

PD is a chronic neurodegenerative disease that generally affects elderly people. Pathologically, PD is categorised by loss of dopaminergic neurons in the substantia nigra, with appearance of cytoplasmic inclusions known as Lewy bodies. Postural instability, muscle rigidity, slowness or loss of mobility and resting tremor all are common manifestations of PD. The aetiology of PD is unclear; however, it is believed to be multifactorial including both genetic and environmental variables that trigger growth of disorder [225]. Pathogenesis of PD occurs in three phases; in the first phase, the disease development is initiated by triggers (viral infections or environmental toxins), followed by spread of disease to the CNS via facilitators (genetic factors) causing systemic inflammation and mitochondrial damage. In the third phase, the aggravators (α synuclein pathology) promote inflammation leading to neurodegeneration [226]. Environmental risk factors associated with PD include pesticides (insecticide-rotenone and herbicide-paraquat), heavy metals (traffic-related air pollution as well as industrial airborne heavy metal pollution) and chlorinated solvents (drinking water) [29].

Familial forms of PD are associated with mutations in five genes that code for α-synuclein, PARKIN, ubiquitin carboxyl-terminal hydrolase L1 (UCH-L1), PTEN induced putative kinase 1 (PINK1) and Protein deglycase (DJ-1) through pathogenic pathways that may frequently evolve in imbalances in mitochondrial and ubiquitin-proteasome system (UPS) function. Oxidative stress, protein misfolding and oxidation of cytosolic dopamine (DA) contribute to the fibrillisation and aggregation of α-synuclein which impairs UPS function and may cause destruction of mitochondria. Variations in mitochondrial function trigger oxidative stress, deficiencies in ATP synthesis and α-synuclein aggregation, which may promote UPS dysfunction. Antioxidant function of DJ-1 that promotes oxidation of DA through S-nitrosylation may be affected by oxidative and nitrosative stress. Excess DA metabolism may further promote oxidative stress. DA neurons in PD eventually die due to mitochondrial and UPS malfunction, oxidative stress and synuclein aggregation [227].

3.4.4.1 Herbs Used for Management of Parkinson's Disease

Panax ginseng

The ginsenosides in *P. ginseng* possess antioxidative and immunomodulatory activities that have therapeutic action on neurodegenerative diseases [30]. Liu et al. [228] evaluated the neuroprotective effects of *P. ginseng extract* on 1-methyl-4-phenylpyridinium ion (MPPþ)-induced apoptosis in SH-SY5Y cells and reported that ginsenosides possess antioxidant properties which have protective impact on 1-methyl-4-phenylpyridinium (MPP^+) and MPTP-induced experimental PD models [228]. In another study, Zhou et al. reported neuroprotective effects of ginsenosides in both in vivo (C57BL/6J mice) and in vitro (PC12 cells) models of PD by activating the Wnt/β-catenin signalling pathway [229].

Astragalus membranaceus

One of the major constituents of *Astragalus membranaceus* aqueous extract is astragaloside IV (AS-IV), a triterpene glycoside belonging to the cycloartane category. AS-IV is known to possess anti-inflammatory, antioxidant and neuroprotective action [230]. Xu et al. [231] evaluated the neuroprotective effect of AS-IV in 6-OHDA SH-SY5Y cell model of PD through activation of JAK2/STAT3 pathways [231].

4 Conclusion

Human health has been one of the major casualties of the fast-paced modern lifestyle; autoimmune diseases form a major chunk of the difficult-to-treat and manage ailments that continue to afflict increasing numbers of populations globally. At best, what the new-age therapeutics can offer is mild-to-moderate relief in the spectrum of symptoms; not without the attendant toxicity associated with long term, chronic use of these agents. Traditional systems of medicine including Ayurveda have documented use of herbs in effectively managing the ubiquitous challenges typically associated with the poor prognosis profile in patients with autoimmune diseases. The chemically complex mix of phytomolecules in these herbs help restore the homeostasis by tempering the micromilieu within the host cells, tissues and interstitia via modulation of key signalling pathways involved in inflammation, cell necrosis and apoptosis. *In vitro* biochemical and cell-based assays and preliminary *in vivo* biological assays are being reported that provide evidence-based bioactivity validation of these herbs. The pressing need is to design clinical study protocols that can establish efficacy of phytotherapies for effective management of autoimmune diseases.

References

1. Ahsan H. Origins and history of autoimmunity—a brief review. Rheumatol Autoimmun. 2023;3:9–14.
2. Wang L, Wang FS, Gershwin ME. Human autoimmune diseases: a comprehensive update. J Intern Med. 2015;278(4):369–95.
3. Miller FW. The increasing prevalence of autoimmunity and autoimmune diseases: an urgent call to action for improved understanding, diagnosis, treatment, and prevention. Curr Opin Immunol. 2023;80:102266.
4. Muñiz-Castrillo S, Vogrig A, Honnorat J. Associations between HLA and autoimmune neurological diseases with autoantibodies. Autoimmun Highlights. 2020;11(1):1–13.
5. Matzaraki V, Kumar V, Wijmenga C, Zhernakova A. The MHC locus and genetic susceptibility to autoimmune and infectious diseases. Genome Biol. 2017;18(1):1–21.
6. Ramos-Casals M, Diaz-Lagares C, Cuadrado MJ, Khamashta MA, Group BS. Autoimmune diseases induced by biological agents: a double-edged sword? Autoimmun Rev. 2010;9(3):188–93.
7. Barber MR, Drenkard C, Falasinnu T, Hoi A, Mak A, Kow NY, et al. Global epidemiology of systemic lupus erythematosus. Nat Rev Rheumatol. 2021;17(9):515–32.
8. Fava A, Petri M. Systemic lupus erythematosus: diagnosis and clinical management. J Autoimmun. 2019;96:1–13.
9. Balkrishna A, Thakur P, Singh S, Chandra Dev SN, Varshney A. Mechanistic paradigms of natural plant metabolites as remedial candidates for systemic lupus erythematosus. Cells. 2020;9(4):1049.
10. Shin J. Epidemiology of atopic dermatitis. In: Practical insights into atopic dermatitis. Berlin: Springer; 2021. p. 11–20.
11. Na K, Lkhagva-Yondon E, Kim M, Lim YR, Shin E, Lee CK, et al. Oral treatment with Aloe polysaccharide ameliorates ovalbumin-induced atopic dermatitis by restoring tight junctions in skin. Scand J Immunol. 2020;91(3):e12856.
12. Tsang MS, Jiao D, Chan BC, Hon KL, Leung PC, Lau CB, et al. Anti-inflammatory activities of pentaherbs formula, berberine, gallic acid and chlorogenic acid in atopic dermatitis-like skin inflammation. Molecules. 2016;21(4):519.
13. Bergqvist C, Ezzedine K. Vitiligo: a focus on pathogenesis and its therapeutic implications. J Dermatol. 2021;48(3):252–70.
14. Mihăilă B, Dinică RM, Tatu AL, Buzia OD. New insights in vitiligo treatments using bioactive compounds from *Piper nigrum*. Exp Ther Med. 2019;17(2):1039–44.
15. Stojanović-Radić Z, Pejčić M, Dimitrijević M, Aleksić A, Anil Kumar NV, Salehi B, et al. Piperine—a major principle of black pepper: a review of its bioactivity and studies. Appl Sci. 2019;9(20):4270.
16. Thakur S, Kaurav H, Chaudhary G. *Nigella sativa* (Kalonji): a black seed of miracle. Int J Res Rev. 2021;8(4):342–57.
17. Sarac G, Kapicioglu Y, Sener S, Mantar I, Yologlu S, Dundar C, et al. Effectiveness of topical *Nigella sativa* for vitiligo treatment. Dermatol Ther. 2019;32(4):e12949.
18. Yamanaka K, Yamamoto O, Honda T. Pathophysiology of Psoriasis: a review. J Dermatol. 2021;48(6):722–31.
19. Ampawong S, Kengkoom K, Sukphopetch P, Aramwit P, Muangkaew W, Kanjanapruthipong T, et al. Evaluating the effect of rice (*Oryza sativa* L.: SRNC05053-6-2) crude extract on Psoriasis using in vitro and in vivo models. Sci Rep. 2020;10(1):1–14.
20. Bahraini P, Rajabi M, Mansouri P, Sarafian G, Chalangari R, Azizian Z. Turmeric tonic as a treatment in scalp Psoriasis: a randomized placebo-control clinical trial. J Cosmet Dermatol. 2018;17(3):461–6.
21. Antiga E, Bonciolini V, Volpi W, Del Bianco E, Caproni M. Oral curcumin (Meriva) is effective as an adjuvant treatment and is able to reduce IL-22 serum levels in patients with Psoriasis vulgaris. Biomed Res Int. 2015;2015:283634.
22. Saverino S, Falorni A. Autoimmune Addison's disease. Best Pract Res Clin Endocrinol Metab. 2020;34(1):101379.
23. Stansbury J, Saunders P, Winston D, Zampieron ER. Treating adrenal insufficiency and hypotension with *glycyrrhiza*. J Restor Med. 2012;1(1):102–6.
24. Pastorino G, Cornara L, Soares S, Rodrigues F, Oliveira MB. Liquorice (*Glycyrrhiza glabra*): a phytochemical and pharmacological review. Phytother Res. 2018;32(12):2323–39.
25. Walton C, King R, Rechtman L, Kaye W, Leray E, Marrie RA, et al. Rising prevalence of multiple sclerosis worldwide: insights from the Atlas of MS. Mult Scler J. 2020;26(14):1816–21.
26. Ekiert H, Świątkowska J, Knut E, Klin P, Rzepiela A, Tomczyk M, et al. *Artemisia dracunculus* (Tarragon): a review of its traditional uses, phytochemistry and pharmacology. Front Pharmacol. 2021;12:653993.
27. Dumurgier J, Sabia S. Epidemiology of Alzheimer's disease: latest trends. Rev Prat. 2020;70(2):149–51.

28. Saleem S, Muhammad G, Hussain MA, Altaf M, Bukhari SNA. *Withania somnifera* L.: insights into the phytochemical profile, therapeutic potential, clinical trials, and future prospective. Iran J Basic Med Sci. 2020;23(12):1501.
29. Ball N, Teo WP, Chandra S, Chapman J. Parkinson's disease and the environment. Front Neurol. 2019;10:218.
30. Liu H, Lu X, Hu Y, Fan X. Chemical constituents of *Panax ginseng* and Panax notoginseng explain why they differ in therapeutic efficacy. Pharmacol Res. 2020;161:105263.
31. Horwitz DA, Fahmy TM, Piccirillo CA, La Cava A. Rebalancing immune homeostasis to treat autoimmune diseases. Trends Immunol. 2019;40(10):888–908.
32. Frizinsky S, Haj-Yahia S, Machnes Maayan D, Lifshitz Y, Maoz-Segal R, Offengenden I, Kidon M, Agmon-Levin N. The innate immune perspective of autoimmune and autoinflammatory conditions. Rheumatology. 2019;58(Suppl 6):vi1–8.
33. Langan D, Rose NR, Moudgil KD. Common innate pathways to autoimmune disease. Clin Immunol. 2020;212:108361.
34. Maddur MS, Vani J, Lacroix-Desmazes S, Kaveri S, Bayry J. Autoimmunity as a predisposition for infectious diseases. PLoS Pathog. 2010;6(11):e1001077.
35. Kamali AN, Noorbakhsh SM, Hamedifar H, Jadidi-Niaragh F, Yazdani R, Bautista JM, et al. A role for Th1-like Th17 cells in the pathogenesis of inflammatory and autoimmune disorders. Mol Immunol. 2019;105:107–15.
36. El-Zayat SR, Sibaii H, Mannaa FA. Toll-like receptors activation, signaling, and targeting: an overview. Bull Natl Res Centre. 2019;43(1):1–2.
37. Xing Y, Hogquist KA. T-cell tolerance: central and peripheral. Cold Spring Harb Perspect Biol. 2012;4(6):a006957.
38. Yang L, Wang G, Xia H. Molecular mechanism for impaired suppressive function of Tregs in autoimmune diseases: a summary of cell-intrinsic and cell-extrinsic factors. J Cell Mol Med. 2020;24(19):11056–63.
39. Sehrawat S, Rouse BT. Interplay of regulatory T cell and Th17 cells during infectious diseases in humans and animals. Front Immunol. 2017;8:341.
40. Takaba H, Takayanagi H. The mechanisms of T cell selection in the thymus. Trends Immunol. 2017;38(11):805–16.
41. Khan U, Ghazanfar H. T lymphocytes and autoimmunity. Int Rev Cell Mol Biol. 2018;341:125–68.
42. Anaya JM, Restrepo Jimenez P, Rojas M, Pacheco Nieva Y, Monsalve DM, Ramirez-Santana C. Molecular mimicry and autoimmunity.
43. Rojas M, Restrepo-Jiménez P, Monsalve DM, Pacheco Y, Acosta-Ampudia Y, Ramírez-Santana C, et al. Molecular mimicry and autoimmunity. J Autoimmun. 2018;95:100–23.
44. Cunningham MW. Molecular mimicry, autoimmunity, and infection: the cross-reactive antigens of group A streptococci and their sequelae. Microbiol Spectr. 2019;7(4):7–4.
45. Cornaby C, Gibbons L, Mayhew V, Sloan CS, Welling A, Poole BD. B cell epitope spreading: mechanisms and contribution to autoimmune diseases. Immunol Lett. 2015;163(1):56–68.
46. Didona D, Di Zenzo G. Humoral epitope spreading in autoimmune bullous diseases. Front Immunol. 2018;9:779.
47. Venkatesha SH, Durai M, Moudgil KD. Epitope spreading in autoimmune diseases. In: Infection and autoimmunity. Amsterdam: Elsevier; 2015. p. 45–68.
48. Sundaresan B, Shirafkan F, Ripperger K, Rattay K. The role of viral infections in the onset of autoimmune diseases. Viruses. 2023;15(3):782.
49. Pacheco Y, Acosta-Ampudia Y, Monsalve DM, Chang C, Gershwin ME, Anaya JM. Bystander activation and autoimmunity. J Autoimmun. 2019;103:102301.
50. Lee H, Jeong S, Shin EC. Significance of bystander T cell activation in microbial infection. Nat Immunol. 2022;23(1):13–22.
51. Shim CH, Cho S, Shin YM, Choi JM. Emerging role of bystander T cell activation in autoimmune diseases. BMB Rep. 2022;55(2):57.
52. Bruserud Ø, Oftedal BE, Wolff AB, Husebye ES. AIRE-mutations and autoimmune disease. Curr Opin Immunol. 2016;43:8–15.
53. Zhao B, Chang L, Fu H, Sun G, Yang W. The role of autoimmune regulator (AIRE) in peripheral tolerance. J Immunol Res. 2018;2018:2018
54. Dominguez-Villar M, Hafler DA. Regulatory T cells in autoimmune disease. Nat Immunol. 2018;19(7):665–73.
55. Hosseini A, Gharibi T, Marofi F, Babaloo Z, Baradaran B. CTLA-4: from mechanism to autoimmune therapy. Int Immunopharmacol. 2020;80:106221.
56. Surace AE, Hedrich CM. The role of epigenetics in autoimmune/inflammatory disease. Front Immunol. 2019;10:1525.
57. Aslani S, Mahmoudi M, Karami J, Jamshidi AR, Malekshahi Z, Nicknam MH. Epigenetic alterations underlying autoimmune diseases. Autoimmunity. 2016;49(2):69–83.
58. Ilchmann-Diounou H, Menard S. Psychological stress, intestinal barrier dysfunctions, and autoimmune disorders: an overview. Front Immunol. 2020;11:1823.
59. Ilchmann-Diounou H, Ménard S. Psychological stress, intestinal barrier dysfunctions, and autoimmune disorders: an overview. Front Immunol. 2020;11:1823.
60. Ortona E, Pierdominici M, Maselli A, Veroni C, Aloisi F, Shoenfeld Y. Sex-based differences in autoimmune diseases. Ann Ist Super Sanita. 2016;52(2):205–12.
61. Mazzucca CB, Raineri D, Cappellano G, Chiocchetti A. How to tackle the relationship between autoimmune diseases and diet: well begun is half-done. Nutrients. 2021;13(11):3956.
62. Lerner A, Matthias T. Changes in intestinal tight junction permeability associated with industrial food additives explain the rising incidence of autoimmune disease. Autoimmun Rev. 2015;14(6):479–89.
63. Ferrari SM, Fallahi P, Antonelli A, Benvenga S. Environmental issues in thyroid diseases. Front Endocrinol. 2017;8:50.
64. Winston D. Adaptogens: herbs for strength, stamina, and stress relief. New York: Simon and Schuster; 2019.
65. Santos AL, Junior CP, Neto RN, Santos MH, Santos VF, Rocha BA, et al. Machaerium acutifolium lectin inhibits inflammatory responses through cytokine modulation. Process Biochem. 2020;97:149–57.
66. Lin Y, Wu S. Vegetable soybean (Glycine max (L.) Merr.) leaf extracts: functional components and antioxidant and anti-inflammatory activities. J Food Sci. 2021;86(6):2468–80.
67. Burgos RA, Alarcón P, Quiroga J, Manosalva C, Hancke J. *Andrographolide*, an anti-inflammatory multitarget drug: all roads lead to cellular metabolism. Molecules. 2020;26(1):5.
68. Beg S, Swain S, Hasan H, Barkat MA, Hussain MS. Systematic review of herbals as potential anti-inflammatory agents: recent advances, current clinical status and future perspectives. Pharmacogn Rev. 2011;5(10):120.
69. Sahoo BM, Banik BK. Medicinal plants: source for immunosuppressive agents. Immunol Curr Res. 2018;2:106.
70. Batiha GES, Magdy Beshbishy A, Wasef L, Elewa YH, Abd El-Hack ME, Taha AE, et al. *Uncaria tomentosa* (Willd. ex Schult.) DC.: a review on chemical constituents and biological activities. Appl Sci. 2020;10(8):2668.
71. Montserrat-de la Paz S, Fernandez-Arche A, De La Puerta R, Quilez AM, Muriana FJ, Garcia-Gimenez MD, et al. Mitraphylline inhibits lipopolysaccharide-mediated activation of primary human neutrophils. Phytomedicine. 2016;23(2):141–8.

72. Batiha GE, Magdy Beshbishy A, Wasef L, Elewa YH, Abd El-Hack ME, Taha AE, Al-Sagheer AA, Devkota HP, Tufarelli V. *Uncaria tomentosa* (Willd. ex Schult.) DC.: a review on chemical constituents and biological activities. Appl Sci. 2020;10(8):2668.
73. Wang HMD, Fu L, Cheng CC, Gao R, Lin MY, Su HL, et al. Inhibition of LPS-induced oxidative damages and potential anti-inflammatory effects of *Phyllanthus emblica* extract *via* down-regulating NF-κB, COX-2, and iNOS in RAW 264.7 cells. Antioxidants. 2019;8(8):270.
74. Guazelli CF, Fattori V, Ferraz CR, Borghi SM, Casagrande R, Baracat MM, et al. Antioxidant and anti-inflammatory effects of hesperidin methyl chalcone in experimental ulcerative colitis. Chem Biol Interact. 2021;333:109315.
75. de Oliveira ALB, Monteiro VVS, Navegantes-Lima KC, Reis JF, Gomes RdS, Rodrigues DVS, et al. Resveratrol role in autoimmune disease—a mini-review. Nutrients. 2017;9(12):1306.
76. Bai J, Zhang Y, Tang C, Hou Y, Ai X, Chen X, et al. Gallic acid: pharmacological activities and molecular mechanisms involved in inflammation-related diseases. Biomed Pharmacother. 2021;133:110985.
77. Rahayu RP, Prasetyo RA, Purwanto DA, Kresnoadi U, Iskandar RPD, Rubianto M. The immunomodulatory effect of green tea (*Camellia sinensis*) leaves extract on immunocompromised Wistar rats infected by Candida albicans. Vet World. 2018;11(6):765–70.
78. Yuan K, Li X, Lu Q, Zhu Q, Jiang H, Wang T, et al. Application and mechanisms of triptolide in the treatment of inflammatory diseases—a review. Front Pharmacol. 2019;10:1469.
79. Song J, Hu Y, Wang L, Ao C. Ethanol extract of Artemisia Annua prevents LPS-induced inflammation and blood–milk barrier disruption in bovine mammary epithelial cells. Animals. 2022;12(10):1228.
80. Vetvicka V, Vannucci L. Biological properties of andrographolide, an active ingredient of *Andrographis Paniculata*: a narrative review. Ann Transl Med. 2021;9(14):1186.
81. Peng Y, Ao M, Dong B, Jiang Y, Yu L, Chen Z, et al. Anti-inflammatory effects of curcumin in the inflammatory diseases: status, limitations and countermeasures. Drug Des Devel Ther. 2021:4503–25.
82. Batiha GE, Beshbishy AM, Wasef LG, Elewa YH, Al-Sagan AA, El-Hack MEA, Taha AE, Abd-Elhakim YM, Devkota HP. Chemical constituents and pharmacological activities of garlic (*Allium sativum* L.): a review. Nutrients. 2020;12(3):872.
83. Zhu L, Myhill LJ, Andersen-Civil AI, Thamsborg SM, Blanchard A, Williams AR. Garlic-derived organosulfur compounds regulate metabolic and immune pathways in macrophages and attenuate intestinal inflammation in mice. Mol Nutr Food Res. 2022;66(7):2101004.
84. Zamani Taghizadeh Rabe S, Ghazanfari T, Siadat Z, Rastin M, Zamani Taghizadeh Rabe S, Mahmoudi M. Anti-inflammatory effect of garlic 14-kDa protein on LPS-stimulated-J774A.1 macrophages. Immunopharmacol Immunotoxicol. 2015;37(2):158–64.
85. Alanazi HH, Elfaki E. The immunomodulatory role of *Withania somnifera* (L.) dunal in inflammatory diseases. Front Pharmacol. 2023;14:430.
86. Bashir A, Nabi M, Tabassum N, Afzal S, Ayoub M. An updated review on phytochemistry and molecular targets of *Withania somnifera* (L.) dunal (Ashwagandha). Front Pharmacol. 2023;14:643.
87. Kunnumakkara AB, Bordoloi D, Padmavathi G, Monisha J, Roy NK, Prasad S, et al. Curcumin, the golden nutraceutical: multitargeting for multiple chronic diseases. Br J Pharmacol. 2017;174(11):1325–48.
88. Ghanaatian N, Lashgari NA, Abdolghaffari AH, Rajaee SM, Panahi Y, Barreto GE, et al. Curcumin as a therapeutic candidate for multiple sclerosis: molecular mechanisms and targets. J Cell Physiol. 2019;234(8):12237–48.
89. Kang D, Li B, Luo L, Jiang W, Lu Q, Rong M, et al. Curcumin shows excellent therapeutic effect on psoriasis in mouse model. Biochimie. 2016;123:73–80.
90. Zhang S, Wang J, Liu L, Sun X, Zhou Y, Chen S, Lu Y, Cai X, Hu M, Yan G, Miao X. Efficacy and safety of curcumin in psoriasis: preclinical and clinical evidence and possible mechanisms. Front Pharmacol. 2022;13:903160.
91. Chopra H, Dey PS, Das D, Bhattacharya T, Shah M, Mubin S, et al. Curcumin nanoparticles as promising therapeutic agents for drug targets. Molecules. 2021;26(16):4998.
92. Bulboacă AE, Porfire AS, Tefas LR, Boarescu PM, Bolboacă SD, Stănescu IC, et al. Liposomal curcumin is better than curcumin to alleviate complications in experimental diabetic mellitus. Molecules. 2019;24(5):846.
93. Recinella L, Gorica E, Chiavaroli A, Fraschetti C, Filippi A, Cesa S, et al. Anti-inflammatory and antioxidant effects induced by *Allium sativum* L. extracts on an *ex vivo* experimental model of ulcerative colitis. Foods. 2022;11(22):3559.
94. Ezzat SM, Ezzat MI, Okba MM, Menze ET, Abdel-Naim AB. The hidden mechanism beyond ginger (*Zingiber officinale* R.) potent in vivo and in vitro anti-inflammatory activity. J Ethnopharmacol. 2018;214:113–23.
95. Habib SH, Makpol S, Hamid NA, Das S, Ngah WZ, Yusof YA. Ginger extract (*Zingiber officinale*) has anti-cancer and anti-inflammatory effects on ethionine-induced hepatoma rats. Clinics. 2008;63(6):807–13.
96. Hwang JH, Jung HW, Oh SY, Kang JS, Kim JP, Park YK. Effects of *Zingiber officinale* extract on collagen-induced arthritis in mice and IL-1β-induced inflammation in human synovial fibroblasts. Eur J Inflamm. 2017;15(3):168–78.
97. Karthikeyan V, Balakrishnan BR, Senniappan P, Janarthanan L, Venkateswarlu BS, Anandharaj G. Berbris aristata DC: pharmacognostical standardization and phytochemical studies of its leaves. J Drug Deliv Ther. 2019;9(4-A):229–36.
98. Ehteshamfar SM, Akhbari M, Afshari JT, Seyedi M, Nikfar B, Shapouri-Moghaddam A, et al. Anti-inflammatory and immune-modulatory impacts of berberine on activation of autoreactive T cells in autoimmune inflammation. J Cell Mol Med. 2020;24(23):13573–88.
99. Yue M, Xia Y, Shi C, Guan C, Li Y, Liu R, et al. Berberine ameliorates collagen-induced arthritis in rats by suppressing Th17 cell responses *via* inducing cortistatin in the gut. FEBS J. 2017;284(17):2786–801.
100. Sharifi-Rad J, Quispe C, Ayatollahi SA, Kobarfard F, Staniak M, Stępień A, Czopek K, Sen S, Acharya K, Matthews KR, Sener B. Chemical composition, biological activity, and health-promoting effects of *Withania somnifera* for pharma-food industry applications. J Food Qual. 2021;2021:8985179.
101. Fry CH, Fluck D, Han TS. Adrenal hypofunction associated with ashwagandha (*Withania somnifera*) supplementation: a case report. Toxicol Environ Health Sci. 2022;14(2):141–5.
102. Khan N, Swami DH, Gourav P, Kushwah M, Raj P. Phytochemical investigation and evaluation of anti-inflammatory activity of *Camellia sinensis* (Green Tea) leaves in experimental animals. 8(2).
103. Oliveira AP, Guimarães AL, de Oliveira-Júnior RG, Quintans JD, de Medeiros FA, Barbosa-Filho JM, Quintans-Júnior LJ, da Silva Almeida JR. *Camellia sinensis* (L.) Kuntze: a review of chemical and nutraceutical properties. Nat Prod Res Rev. 2016;4:21–62.
104. Kim J, Choi H, Choi DH, Park K, Kim HJ, Park M. Application of green tea catechins, polysaccharides, and flavonol prevent fine dust induced bronchial damage by modulating inflammation and airway cilia. Sci Rep. 2021;11(1):2232.
105. Huang SS, Su SY, Chang JS, Lin HJ, Wu WT, Deng JS, et al. Antioxidants, anti-inflammatory, and antidiabetic effects of the

aqueous extracts from glycine species and its bioactive compounds. Bot Stud. 2016;57(1):1–11.
106. Hancke JL, Srivastav S, Cáceres DD, Burgos RA. A double-blind, randomized, placebo-controlled study to assess the efficacy of Andrographis paniculata standardized extract (ParActin®) on pain reduction in subjects with knee osteoarthritis. Phytother Res. 2019;33(5):1469–79.
107. Suriyo T, Pholphana N, Ungtrakul T, Rangkadilok N, Panomvana D, Thiantanawat A, Pongpun W, Satayavivad J. Clinical parameters following multiple oral dose administration of a standardized Andrographis paniculata capsule in healthy Thai subjects. Planta Medica. 2017;83(9):778–89.
108. Ciampi E, Uribe-San-Martin R, Cárcamo C, Cruz JP, Reyes A, Reyes D, Pinto C, Vásquez M, Burgos RA, Hancke J. Efficacy of andrographolide in not active progressive multiple sclerosis: a prospective exploratory double-blind, parallel-group, randomized, placebo-controlled trial. BMC Neurol. 2020;20:173.
109. Lu J, Ma Y, Wu J, Huang H, Wang X, Chen Z, et al. A review for the neuroprotective effects of andrographolide in the central nervous system. Biomed Pharmacother. 2019;117:109078.
110. Borges RS, Ortiz BLS, Pereira ACM, Keita H, Carvalho JCT. *Rosmarinus officinalis* essential oil: a review of its phytochemistry, anti-inflammatory activity, and mechanisms of action involved. J Ethnopharmacol. 2019;229:29–45.
111. Andrade JM, Faustino C, Garcia C, Ladeiras D, Reis CP, Rijo P. *Rosmarinus officinalis* L.: an update review of its phytochemistry and biological activity. Future Sci OA. 2018;4(4):FSO283.
112. Yuan R, Liu Z, Zhao J, Wang QQ, Zuo A, Huang L, et al. Novel compounds in fruits of coriander (Coşkuner & Karababa) with anti-inflammatory activity. J Funct Foods. 2020;73:104145.
113. Chahal KK, Singh R, Kumar A, Bhardwaj U. Chemical composition and biological activity of *Coriandrum sativum* L.: a review. Indian J Nat Prod Resour. 2018;8:193–203.
114. Foudah AI, Alqarni MH, Alam A, Salkini MA, Ahmed EOI, Yusufoglu HS. Evaluation of the composition and in vitro antimicrobial, antioxidant, and anti-inflammatory activities of Cilantro (*Coriandrum sativum* L. leaves) cultivated in Saudi Arabia (Al-Kharj). Saudi J Biol Sci. 2021;28(6):3461–8.
115. Jahurul MHA, Patricia M, Shihabul A, Norazlina MR, George MR, Noorakmar AW, et al. A review on functional and nutritional properties of noni fruit seed (*Morinda citrifolia* L.) and its oil. Food Biosci. 2021;41:101000.
116. Tanikawa T, Kitamura M, Hayashi Y, Tomida N, Uwaya A, Isami F, et al. Anti-inflammatory effects of *Morinda citrifolia* extract against lipopolysaccharide-induced inflammation in RAW264 Cells. Medicines. 2021;8(8):43.
117. Lee D, Yu JS, Huang P, Qader M, Manavalan A, Wu X, et al. Identification of anti-inflammatory compounds from Hawaiian noni (*Morinda citrifolia* L.) fruit juice. Molecules. 2020;25(21):4968.
118. Salaria D, Rolta R, Sharma N, Dev K, Sourirajan A, Kumar V. *In silico* and *in vitro* evaluation of the anti-inflammatory and antioxidant potential of *Cymbopogon citratus* from North-western Himalayas. BioRxiv. 2020;2020.05.31.124982.
119. Ekpenyong CE, Akpan E, Nyoh A. Ethnopharmacology, phytochemistry, and biological activities of *Cymbopogon citratus* (DC.) Stapf extracts. Chinese J Nat Med. 2015;13(5):321–37.
120. Borges PH, Pedreiro S, Baptista SJ, Geraldes CF, Batista MT, Silva MM, et al. Inhibition of α-glucosidase by flavonoids of *Cymbopogon citratus* (DC) Stapf. J Ethnopharmacol. 2021;280:114470.
121. Ajayi AM, Ologe MO, Ben-Azu B, Okhale SE, Adzu B, Ademowo OG, Ocimum gratissimum Linn. Leaf extract inhibits free radical generation and suppressed inflammation in carrageenan-induced inflammation models in rats. J Basic Clin Physiol Pharmacol. 2017;28(6):531–41.
122. Tiwari P, Nayak P, Prusty SK, Sahu PK. Phytochemistry and pharmacology of *Tinospora cordifolia*: a review. Syst Rev Pharm. 2018;9(1):70–8.
123. Nandan A, Sharma V, Banerjee P, Sadasivam K, Venkatesan S, Prasher B. Deciphering the mechanism of *Tinospora cordifolia* extract on Th17 cells through in-depth transcriptomic profiling and in silico analysis. Front Pharmacol. 2022;13:1056677.
124. Maidhof W, Hilas O. Lupus: an overview of the disease and management options. Pharm Ther. 2012;37(4):240–9.
125. Balkrishna A, Thakur P, Singh S, Chandra Dev SN, Varshney A. Mechanistic paradigms of natural plant metabolites as remedial candidates for systemic lupus erythromatosus. Cells. 2020;9(4):1049.
126. Pan L, Lu MP, Wang JH, Xu M, Yang SR. Immunological pathogenesis and treatment of systemic lupus erythematosus. World J Pediatr. 2020;16 19–30.
127. Dörner T, Furie R. Novel paradigms in systemic lupus erythematosus. Lancet. 2019;393(10188):2344–58.
128. Song CY, Xu YG, Lu YQ. Use of Tripterygium wilfordii Hook F for immune-mediated inflammatory diseases: progress and future prospects. J Zhejiang Univ Sci B. 2020;21(4):280–90.
129. Sun Q, Liu Y, Lu D, Ji L, Wang W, Wang X, et al. Therapeutic effects of triptolide on the balance of Th17/Treg cells *via* AKT/mTOR/p70S6k signaling pathway in lupus-like mic. 2022 [cited 2022 Dec 18]. https://doi.org/10.21203/rs.3.rs-1354719/v1.
130. Zhang LY, Li H, Wu YW, Cheng L, Yan YX, Yang XQ, Zhu FH, He SJ, Tang W, Zuo JP. (5R)-5-hydroxytriptolide ameliorates lupus nephritis in MRL/lpr mice by preventing infiltration of immune cells. Am J Physiol Renal Physiol. 2017;312(4):F769–77.
131. Zhang Y, Zhang F, Gao Y, Wang M, Li H, Sun J, Wen C, Xie Z. Triptolide in the treatment of systemic lupus erythematosus-regulatory effects on miR-146a in B cell TLR7 signaling pathway in mice. Front Pharmacol. 2022;13:952775.
132. Liu W, Yan L, Zhu Q, Shao FM. Therapeutic effect of Tripterygium glycosides plus prednisone on moderate active systemic lupus erythematosus. J Chin Pract Diagn Ther. 2014 28(12):1234–5.
133. Wang YJ, Li YX, Li S, He W, Wang ZR, Zhan TP, et al. Progress in traditional Chinese medicine and natural extracts for the treatment of lupus nephritis. Biomed Pharmacother. 2022;149:112799.
134. Zamani S, Emami SA, Iranshahi M, Rabe SZ, Mahmoudi M. Sesquiterpene fractions of Artemisia plants as potent inhibitors of inducible nitric oxide synthase and cyclooxygenase-2 expression. Iran J Basic Med Sci. 2019;22(7):774.
135. Chebbac K, Benziane Ouaritini Z, El Moussaoui A, Chalkha M, Lafraxo S, Bin Jardan YA, Nafidi HA, Bourhia M, Guemmouh R. Antimicrobial and antioxidant properties of chemically analyzed essential oil of *Artemisia annua* L. (Asteraceae) native to Mediterranean Area. Life. 2023;13(3):807.
136. Avena-Woods C. Overview of atopic dermatitis. Am J Manag Care. 2017;23(8 Suppl):S115–23.
137. Otsuka A, Nomura T, Rerknimitr P, Seidel JA, Honda T, Kabashima K. The interplay between genetic and environmental factors in the pathogenesis of atopic dermatitis. Immunol Rev. 2017;278(1):246–62.
138. David Boothe W, Tarbox JA, Tarbox MB. Atopic dermatitis: pathophysiology. In: Management of atopic dermatitis: methods and challenges. Berlin: Springer; 2017. p. 21–37.
139. Sroka-Tomaszewska J, Trzeciak M. Molecular mechanisms of atopic dermatitis pathogenesis. Int J Mol Sci. 2021;22(8):4130.
140. Peng W, Novak N. Pathogenesis of atopic dermatitis. Clin Exp Allergy. 2015;45(3):566–74.
141. Kim J, Kim BE, Leung DY. Pathophysiology of atopic dermatitis: clinical implications. In: Allergy and asthma proceedings. East Providence: OceanSide Publications; 2019 p. 84.
142. Kwatra SG, Misery L, Clibborn C, Steinhoff M. Molecular and cellular mechanisms of itch and pain in atopic dermatitis

and implications for novel therapeutics. Clin Transl Immunol. 2022;11(5):e1390.
143. Fania L, Moretta G, Antonelli F, Scala E, Abeni D, Albanesi C, et al. Multiple roles for cytokines in atopic dermatitis: from pathogenic mediators to endotype-specific biomarkers to therapeutic targets. Int J Mol Sci. 2022;23(5):2684.
144. Dattola A, Bennardo L, Silvestri M, Nisticò SP. What's new in the treatment of atopic dermatitis? Dermatol Ther. 2019;32(2):e12787.
145. Maliehe TS, Nqotheni MI, Shandu JS, Selepe TN, Masoko P, Pooe OJ. Chemical profile, antioxidant and antibacterial activities, mechanisms of action of the leaf extract of *Aloe arborescens* Mill. Plants. 2023;12(4):869.
146. Rodrigues M, Ezzedine K, Hamzavi I, Pandya AG, Harris JE, Group VW. New discoveries in the pathogenesis and classification of Vitiligo. J Am Acad Dermatol. 2017;77(1):1–13.
147. Spritz RA, Santorico SA. The genetic basis of Vitiligo. J Investig Dermatol. 2021;141(2):265–73.
148. Bergqvist C, Ezzedine K. Vitiligo: a review. Dermatology. 2020;236(6):571–92.
149. Harris JE. Chemical-induced Vitiligo. Dermatol Clin. 2017;35(2):151–61.
150. Manga P, Choudhury N. The unfolded protein and integrated stress response in melanoma and Vitiligo. Pigment Cell Melanoma Res. 2021;34(2):204–11.
151. Seneschal J, Boniface K, D'Arino A, Picardo M. An update on Vitiligo pathogenesis. Pigment Cell Melanoma Res. 2021;34(2):236–43.
152. Stojanović-Radić Z, Pejčić M, Dimitrijević M, Aleksić A, Anil Kumar NV, Salehi B, Cho CW, Sharifi-Rad J. Piperine—a major principle of black pepper: a review of its bioactivity and studies. Appl Sci. 2019;9(20):4270.
153. Shafiee A, Hoormand M, Shahidi-Dadras M, Abadi A. The effect of topical piperine combined with narrowband UVB on Vitiligo treatment: a clinical trial study. Phytother Res. 2018;32(9):1812–7.
154. Hwang JR, Cartron AM, Khachemoune A. A review of Nigella sativa plant-based therapy in dermatology. Int J Dermatol. 2021;60(12):e493–9.
155. Hugh JM, Weinberg JM. Update on the pathophysiology of psoriasis. Cutis. 2018;102(5S):6–12.
156. Sarac G, Koca TT, Baglan T. A brief summary of clinical types of psoriasis. North Clin Istanb. 2016;3(1):79.
157. Yan BX, Chen XY, Ye LR, Chen JQ, Zheng M, Man XY. Cutaneous and systemic psoriasis: classifications and classification for the distinction. Front Med. 2021;8:649408.
158. Zhou X, Chen Y, Cui L, Shi Y, Guo C. Advances in the pathogenesis of psoriasis: from keratinocyte perspective. Cell Death Dis. 2022;13(1):81.
159. Herman A, Herman AP. Topically used herbal products for the treatment of psoriasis—mechanism of action, drug delivery, clinical studies. Planta Med. 2016;82(17):1447–55.
160. Rout SK, Tripathy BC, Kar BR. Natural green alternatives to psoriasis treatment—a review. Glob J Pharm Pharm Sci. 2017;4(1):1–7.
161. Aghmiuni AI, Khiavi AA. Medicinal plants to calm and treat psoriasis disease. In: Aromatic and medicinal plants—back to nature, vol. 2016. London: IntechOpen; 2017. p. 1–28.
162. Thepthanee C, Liu CC, Yu HS, Huang HS, Yen CH, Li YH, et al. Evaluation of phytochemical contents and in vitro antioxidant, anti-inflammatory, and anticancer activities of black rice leaf (*Oryza sativa* L.) extract and its fractions. Foods. 2021;10(12):2987.
163. Xu Y, Shi Y, Huang J, Gu H, Li C, Zhang L, et al. The essential oil derived from *Perilla frutescens* (L.) Britt. attenuates imiquimod-induced psoriasis-like skin lesions in BALB/c mice. Molecules. 2022;27(9):2996.
164. Hellesen A, Bratland E, Husebye ES. Autoimmune Addison's disease—an update on pathogenesis. In: Annales d'endocrinologie. Amsterdam: Elsevier; 2018. p. 157–63.
165. Bancos I, Hahner S, Tomlinson J, Arlt W. Diagnosis and management of adrenal insufficiency. Lancet Diabetes Endocrinol. 2015;3(3):216–26.
166. Perniola R, Fierabracci A, Falorni A. Autoimmune Addison's disease as part of the autoimmune polyglandular syndrome type 1: historical overview and current evidence. Front Immunol. 2021;12:606860.
167. Husebye ES, Pearce SH, Krone NP, Kämpe O. Adrenal insufficiency. Lancet. 2021;397(10274):613–29.
168. Pignatti E, Flück CE. Adrenal cortex development and related disorders leading to adrenal insufficiency. Mol Cell Endocrinol. 2021;527:111206.
169. Hellesen A, Bratland E, Husebye ES. Autoimmune Addison's disease—an update on pathogenesis. In: Annales d'endocrinologie, vol. 79, no. 3. Paris: Elsevier Masson; 2018. p. 157–63.
170. Betterle C, Presotto F, Furmaniak J. Epidemiology, pathogenesis, and diagnosis of Addison's disease in adults. J Endocrinol Invest. 2019;42(12):1407–33.
171. Methlie P, Husebye EE, Hustad S, Lien EA, Løvås K. Grapefruit juice and licorice increase cortisol availability in patients with Addison's disease. Eur J Endocrinol. 2011;165(5):761–9.
172. Kumari A, Kumar R, Sulabh G, Singh P, Kumar J, Singh VK, Ojha KK. *In silico* ADMET, molecular docking and molecular simulation-based study of glabridin's natural and semisynthetic derivatives as potential tyrosinase inhibitors. Adv Tradit Med. 2023;23:733–51.
173. Morsy MA, Patel SS, El-Sheikh AA, Savjani JK, Nair AB, Shah JN, Venugopala KN. Computational and biological comparisons of plant steroids as modulators of inflammation through interacting with glucocorticoid receptor. Mediators Inflamm. 2019;2019:3041438.
174. Boswellia Serrata, a potential antiinflammatory agent: an overview—PMC. [cited 2023 Mar 3]. https://www.ncbi.nlm.nih.gov/pmc/articles/PMC3309643/.
175. Zameer S, Najmi AK, Vohora D, Akhtar M. A review on therapeutic potentials of Trigonella foenum graecum (fenugreek) and its chemical constituents in neurological disorders: complementary roles to its hypolipidemic, hypoglycemic, and antioxidant potential. Nutr Neurosci. 2018;21(8):539–45.
176. Aylanc V, Eskin B, Zengin G, Dursun M, Cakmak YS. *In vitro* studies on different extracts of fenugreek (*Trigonella spruneriana* BOISS.): phytochemical profile, antioxidant activity, and enzyme inhibition potential. J Food Biochem. 2020;44(11):e13463.
177. Prema A, Thenmozhi AJ, Manivasagam T, Essa MM, Akbar MD, Akbar M. Fenugreek seed powder nullified aluminium chloride induced memory loss, biochemical changes, Aβ burden and apoptosis *via* regulating Akt/GSK3β signaling pathway. PLoS One. 2016;11(11):e0165955.
178. Ratheesh G, Tian L, Venugopal JR, Ezhilarasu H, Sadiq A, Fan TP, et al. Role of medicinal plants in neurodegenerative diseases. Biomanuf Rev. 2017;2(1):2.
179. Rai K, Gupta N, Dharamdasani L, Nair P, Bodhankar P. Bacopa monnieri: a wonder drug changing fortune of people. Int J Appl Sci Biotechnol. 2017;5(2):127–32.
180. Fatima U, Roy S, Ahmad S, Ali S, Elkady WM, Khan I, Alsaffar RM, Adnan M, Islam A, Hassan MI. Pharmacological attributes of *Bacopa monnieri* extract: current updates and clinical manifestation. Front Nutr. 2022;9:972379.
181. Kishore D, Babu RS, Begum A, Noor A, Farheen S, Kauser SM, Yaser AA. Evaluation of nootropic activity of two marketed drugs

of *Bacopa monnieri* in scopolamine induced amnesic models. Indian J Res Pharm Biotechnol. 2018;6:84–90.
182. Nowak A, Kojder K, Zielonka-Brzezicka J, Wróbel J, Bosiacki M, Fabiańska M, et al. The use of *Ginkgo Biloba* L. as a neuroprotective agent in the Alzheimer's Disease. Front Pharmacol. 2021;12:775034. https://www.frontiersin.org/articles/10.3389/fphar.2021.775034.
183. Das R, Lami MS, Chakraborty AJ, Mitra S, Tallei TE, Idroes R, Mohamed AA, Hossain M, Dhama K, Mostafa-Hedeab G, Emran TB. *Ginkgo biloba*: a treasure of functional phytochemicals with multimedicinal applications. Evid Based Complement Alternat Med. 2022;2022:8288818.
184. Yu D, Zhang P, Li J, Liu T, Zhang Y, Wang Q, Zhang J, Lu X, Fan X. Neuroprotective effects of *Ginkgo biloba* dropping pills in Parkinson's disease. J Pharm Anal. 2021;11(2):220–31.
185. Rapp M, Burkart M, Kohlmann T, Bohlken J. Similar treatment outcomes with *Ginkgo biloba* extract EGb 761 and donepezil in Alzheimer's dementia in very old age: a retrospective observational study. Int J Clin Pharmacol Ther. 2018;56(3):130.
186. Sun B, Wu L, Wu Y, Zhang C, Qin L, Hayashi M, et al. Therapeutic potential of *Centella asiatica* and its triterpenes: a review. Front Pharmacol. 2020;11:568032.
187. Wong JH, Barron AM, Abdullah JM. Mitoprotective effects of *Centella asiatica* (L.) Urb.: anti-inflammatory and neuroprotective opportunities in neurodegenerative disease. Front Pharmacol. 2021;12:687935.
188. Hafiz ZZ, Amin MAM, Johari James RM, Teh LK, Salleh MZ, Adenan MI. Inhibitory effects of raw-extract *Centella asiatica* (RECA) on acetylcholinesterase, inflammations, and oxidative stress activities *viain vitro* and in *vivo*. Molecules. 2020;25(4):892.
189. Fujimori H, Ohba T, Mikami M, Nakamura S, Ito K, Kojima H, Takahashi T, Iddamalgoda A, Shimazawa M, Hara H. The protective effect of *Centella asiatica* and its constituent, araliadiol on neuronal cell damage and cognitive impairment. J Pharmacol Sci. 2022;148(1):162–71.
190. Gray NE, Zweig JA, Murchison C, Caruso M, Matthews DG, Kawamoto C, Harris CJ, Quinn JF, Soumyanath A. *Centella asiatica* attenuates Aβ-induced neurodegenerative spine loss and dendritic simplification. Neurosci Lett. 2017;646:24–9.
191. Nakoti SS, Juyal D, Josh AK. A review on pharmacognostic and phytochemical study of a plant *Nardostachys Jatamansi*. Pharma Innov. 2017;6(7, Part G):936.
192. Liu QF, Jeon Y, Sung Y-W, Lee JH, Jeong H, Kim YM, et al. *Nardostachys jatamansi* ethanol extract ameliorates Aβ42 cytotoxicity. Biol Pharm Bull. 2018;41(4):470–7.
193. Kaur N, Sarkar B, Gill I, Kaur S, Mittal S, Dhiman M, et al. Indian herbs and their therapeutic potential against Alzheimer's disease and other neurological disorders. In: Neuroprotective effects of phytochemicals in neurological disorders. New York: Wiley; 2017. p. 79–112.
194. Devi P. An updated review on *Shankhpushpi*—As Medhya Rasayana. J Ayurvedic Herb Med. 2021;7(2):119–23.
195. Sethiya NK, Nahata A, Singh PK, Mishra SH. Neuropharmacological evaluation on four traditional herbs used as nervine tonic and commonly available as *Shankhpushpi* in India. J Ayurveda Integr Med. 2019;10(1):25–31.
196. Suryavanshi SV, Barve K, Addepalli V, Utpat SV, Kulkarni YA. Triphala Churna—a traditional formulation in ayurveda mitigates diabetic neuropathy in rats. Front Pharmacol. 2021;12:662000.
197. Ning W, Li S, Tsering J, Ma Y, Li H, Ma Y, et al. Protective effect of Triphala against oxidative stress-induced neurotoxicity. Biomed Res Int. 2021;2021:6674988.
198. Tarlinton RE, Khaibullin T, Granatov E, Martynova E, Rizvanov A, Khaiboullina S. The interaction between viral and environmental risk factors in the pathogenesis of multiple sclerosis. Int J Mol Sci. 2019;20(2):303.
199. Guan Y, Jakimovski D, Ramanathan M, Weinstock-Guttman B, Zivadinov R. The role of Epstein-Barr virus in multiple sclerosis: from molecular pathophysiology to in vivo imaging. Neural Regen Res. 2019;14(3):373.
200. Zéphir H. Progress in understanding the pathophysiology of multiple sclerosis. Rev Neurol (Paris). 2018;174(6):358–63.
201. Dighriri IM, Aldalbahi AA, Albeladi F, Tahiri AA, Kinani EM, Almohsen RA, Alamoudi NH, Alanazi AA, Alkhamshi SJ, Althomali NA, Alrubaiei SN. An overview of the history, pathophysiology, and pharmacological interventions of multiple sclerosis. Cureus. 2023;15(1):e33242.
202. Kaskow BJ, Baecher-Allan C. Effector T cells in multiple sclerosis. Cold Spring Harb Perspect Med. 2018;8(4):a029025.
203. Huang WJ, Chen WW, Zhang X. Multiple sclerosis: pathology, diagnosis and treatments. Exp Ther Med. 2017;13(6):3163–6.
204. Moser T, Akgün K, Proschmann U, Sellner J, Ziemssen T. The role of TH17 cells in multiple sclerosis: therapeutic implications. Autoimmun Rev. 2020;19(10):102647.
205. Qureshi M, Al-Suhaimi E, Shehzad A. Curcumin impact on multiple sclerosis. In: Curcumin for neurological and psychiatric disorders. Amsterdam: Elsevier; 2019. p. 365–80.
206. Balasa R, Barcutean L, Balasa A, Motataianu A, Roman-Filip C, Manu D. The action of TH17 cells on blood brain barrier in multiple sclerosis and experimental autoimmune encephalomyelitis. Hum Immunol. 2020;81(5):237–43.
207. Kimura K. Regulatory T cells in multiple sclerosis. Clin Exp Neuroimmunol. 2020;11(3):148–55.
208. Barcelos IP, Troxell RM, Graves JS. Mitochondrial dysfunction and multiple sclerosis. Biology. 2019;8(2):37.
209. Bayat P, Farshchi M, Yousefian M, Mahmoudi M, Yazdian-Robati R. Flavonoids, the compounds with anti-inflammatory and immunomodulatory properties, as promising tools in multiple sclerosis (MS) therapy: a systematic review of preclinical evidence. Int Immunopharmacol. 2021;95:107562.
210. Safari H, Anani Sarab G, Naseri M. Artemisia dracunculus L. modulates the immune system in a multiple sclerosis mouse model. Nutr Neurosci. 2021;24(11):843–9.
211. Adetuyi BO, Omolabi FK, Olajide PA, Oloke JK. Pharmacological, biochemical and therapeutic potential of milk thistle (silymarin): a review. World News Nat Sci. 2021;37:75–91.
212. Shariati M, Shaygannejad V, Abbasirad F, Hosseininasab F, Kazemi M, Mirmosayyeb O, et al. Silymarin restores regulatory T cells (tregs) function in multiple sclerosis (MS) patients in vitro. Inflammation. 2019;42:1203–14.
213. Eratne D, Loi SM, Farrand S, Kelso W, Velakoulis D, Looi JC. Alzheimer's disease: clinical update on epidemiology, pathophysiology and diagnosis. Australas Psychiatry. 2018;26(4):347–57.
214. Twohig D, Nielsen HM. α-Synuclein in the pathophysiology of Alzheimer's disease. Mol Neurodegener. 2019;14(1):1–19.
215. Scheltens P, Blennow K, Breteler MM, De Strooper B, Frisoni GB, Salloway S, et al. Alzheimer's disease. Lancet. 2016;388(10043):505–17.
216. Masters CL, Bateman R, Blennow K, Rowe CC, Sperling RA, Cummings JL. Alzheimer's disease. Nat Rev Dis Primer. 2015;1(1):1–18.
217. Wan W, Cao L, Kalionis B, Murthi P, Xia S, Guan Y. Iron deposition leads to hyperphosphorylation of tau and disruption of insulin signaling. Front Neurol. 2019;10:607.

218. Kumar A, Singh A. A review on Alzheimer's disease pathophysiology and its management: an update. Pharmacol Rep. 2015;67(2):195–203.
219. Khoury R, Rajamanickam J, Grossberg GT. An update on the safety of current therapies for Alzheimer's disease: focus on rivastigmine. Ther Adv Drug Saf. 2018;9(3):171–8.
220. Balakrishnan R, Cho DY, Kim IS, Seol SH, Choi DK. Molecular mechanisms and therapeutic potential of α- and β-asarone in the treatment of neurological disorders. Antioxidants. 2022;11(2):281.
221. Esfandiari E, Ghanadian M, Rashidi B, Mokhtarian A, Vatankhah AM. The effects of *Acorus calamus* L. in preventing memory loss, anxiety, and oxidative stress on lipopolysaccharide-induced neuroinflammation rat models. Int J Prev Med. 2018;9:85.
222. Mikami M, Takuya O, Yoshino Y, Nakamura S, Ito K, Kojima H, et al. *Acorus calamus* extract and its component α-asarone attenuate murine hippocampal neuronal cell death induced by l-glutamate and tunicamycin. Biosci Biotechnol Biochem. 2021;85(3):493–501.
223. Talebi M, Talebi M, Samarghandian S. Association of Crocus sativus with cognitive dysfunctions and Alzheimer's disease: a systematic review. Biointerface Res Appl Chem. 2021;11(1):7468–92.
224. D'Onofrio G, Nabavi SM, Sancarlo D, Greco A, Pieretti S, Crocus sativus L. (Saffron) in Alzheimer's disease treatment: bioactive effects on cognitive impairment. Curr Neuropharmacol. 2021;19(9):1606.
225. Bloem BR, Okun MS, Klein C. Parkinson's disease. Lancet. 2021;397(10291):2284–303.
226. Johnson ME, Stecher B, Labrie V, Brundin L, Brundin P. Triggers, facilitators, and aggravators: redefining Parkinson's disease pathogenesis. Trends Neurosci. 2019;42(1):4–13.
227. Maiti P, Manna J, Dunbar GL. Current understanding of the molecular mechanisms in Parkinson's disease: targets for potential treatments. Transl Neurodegener. 2017;6:1–35.
228. Liu Y, Zhang RY, Zhao J, Dong Z, Feng DY, Wu R, et al. Ginsenoside Rd protects SH-SY5Y cells against 1-methyl-4-phenylpyridinium induced injury. Int J Mol Sci. 2015;16(7):14395–408.
229. Zhou T, Zu G, Zhang X, Wang X, Li S, Gong X, et al. Neuroprotective effects of ginsenoside Rg1 through the Wnt/β--catenin signaling pathway in both in vivo and in vitro models of Parkinson's disease. Neuropharmacology. 2016;101:480–9.
230. Zhang J, Wu C, Gao L, Du G, Qin X. Astragaloside IV derived from *Astragalus membranaceus*: a research review on the pharmacological effects. Adv Pharmacol. 2020;87:89–112.
231. Xu Z, Yang D, Huang X, Huang H. Astragaloside IV protects 6-hydroxydopamine-induced SH-SY5Y cell model of Parkinson's disease *via* activating the JAK2/STAT3 pathway. Front Neurosci. 2021;15:631501.

Role of Herbal Medicine in Boosting Immune System

Anchal Dabas, Parul Yadav, Geetanjali, and Ram Singh

Abstract

Herbal medicines are the medicines obtained from plant sources. This is an alternate type of treatment based on traditional knowledge and scientific evidences in many cases. The utility of plants as whole or their parts like fruits, flowers, leaves, stem, and roots have been established for the treatment of many diseases. The active ingredients present in the plants either alone or in combination have improved the immune system if taken with proper care. Many studies have shown that these are better in comparison to conventional modern treatment with respect to less or no side effects. This chapter focused on the use of herbal medicines in boosting the immune system. This chapter covers why people prefer herbal medicines, role of some herbal plants in enhancing the immune system, and role of some individual natural products in boosting the immune system.

Keywords

Herbal medicine · Natural products · Secondary metabolites · Flavonoids · Polyphenols · Immunity · Immunomodulators

1 Introduction

Mother Nature has provided us with numerous ways to maintain our health. Herbal drugs are one of them that are used to treat various diseases. As per World Health Organization (WHO), herbal medicines include herbs or herbal materials that contain active ingredients present in plants. Herbal materials are plant materials taken from plants as a whole or its parts like flowers, fruits, leaves, stems, roots, and leaves. Herbal plants used to boost the immune system is a centuries-old practice (Fig. 1). These plants contain bioactive compounds that have the potential to improve the body's defense mechanism [1]. These plants are used to cure various diseases like diseases caused by bacteria, viruses, fungi, etc. [2]. Plants high in flavonoids, vitamin C, or carotenoids have been shown to improve immune function. This can stimulate interferon production, lymphocyte activity, and increase phagocytosis [3]. Traditional medicine systems around the world have relied on the therapeutic properties of various plants to boost the immune system. Scientific research has started throwing light on the potential benefits of these herbal remedies for immune support, as interest in natural remedies and holistic health has grown in recent years [2]. Naturally, there is a wide range of herbs with medicinal properties that are used to make herbal medicines. This is an excellent alternative to allopathic drugs. Allopathic drugs are available to cure diseases and thus improve immunity, but the prohibitive cost and side effects of these allopathic drugs demand the search for an alternative. The Ayurvedic medical system which utilizes herbal materials not only provides that alternative, but it also outperforms allopathic medicine in terms of the limitations that it possesses [4].

Herbal medicine is an alternative treatment that involves the therapeutic use of various plants and plant extracts [5]. Many drugs, including morphine (derived from the opium poppy), aspirin (derived from willow bark), quinine (derived from cinchona bark), and digitoxin (derived from foxglove), have herbal sources [6]. Few other examples of herbs used in the medicinal field are *Aloe barbadensis, Ocimum tenuiflorum, Echinacea purpurea, Tinospora cordifolia, Panax ginseng, Withania somnifera, Allium sativum, Emblica officinalis, Curcuma longa, Azadirachta indica, Zingiber officinale,* and many more. Around 80% of the population uses herbal remedies for primary health care in developing countries, owing to the widespread belief that herbal drugs

A. Dabas · P. Yadav · R. Singh (✉)
Department of Applied Chemistry, Delhi Technological University, Delhi, India
e-mail: ramsingh@dtu.ac.in

Geetanjali
Department of Chemistry, Kirori Mal College, University of Delhi, Delhi, India

A. K. Dhara, S. C. Mandal (eds.), *Role of Herbal Medicines*, https://doi.org/10.1007/978-981-99-7703-1_19

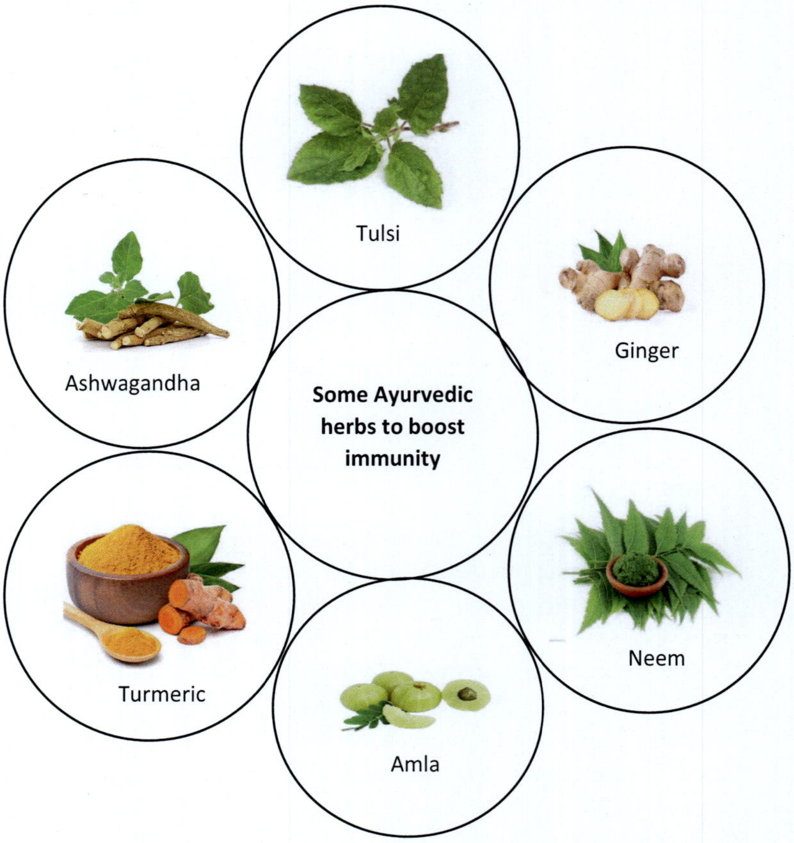

Fig. 1 Examples of herbal plants used to boost the immune system

are inexpensive, without side effects, and readily available. The earliest records of its use date back to writings written in Indian, Egyptian, Syrian, Chinese, Roman, and Greek languages. Classical Indian literature that explains the herbal medicinal system includes the Sushruta Samhita, Atharvaveda, Charak Samhita, Rigveda, and others. As a result, the traditional medicines and herbal remedies are a result of the rich cultural traditions and scientific history of earlier civilizations [7, 8]. According to a report by Spherical Insights, the Global herbal medicine market is expected to reach USD 430.05 billion in 2030. This huge market is self-explanatory toward rise in demand of herbal medicines.

Throughout the range of global public health, the immune system plays an important role to keep human healthy. As multiple-component agent, the herbal medicines control the complex immune process and did not allow infections by pathogens. These herbal medicines are excellent immunomodulators, and they are gaining popularity in today's world [9]. Furthermore, micronutrients and bioactive compounds found in functional foods, herbal remedies, and dietary supplements can help boost the immune system [9, 10, 11]. In this chapter, we will discuss about the use of herbal plants in boosting the immunity. From anti-inflammatory and antioxidant to antimicrobial and immune-modulating effects, these herbs have provided us with different ways to enhance the immunity.

2 Why People Use Herbal Medicines?

Since ancient times, the human usage of plants for healing purposes is well known [12]. People have been using herbal medication for many diseases without understanding their mechanistic pathways. This has been stated that using herbs gives one a "sense of control and a mental comfort." Also, patients seek home remedies for acute, frequently self-limiting diseases like colds, sore throats, or bee stings because its access is easy than the professional care which is uncomfortable, expensive, or sometime takes too much time [12].

Usually, people from rural places use herbal medicines. This is because they are well acquainted with traditional knowledge about the herbal medicines which they have inherited from their ancestors. This is advantageous to them because the rural area lacks access to modern medical care [13–15]. Additionally, reports of herbal toxicities are significantly less common than reports of unfavorable side effects of modern allopathic drugs. The inclusion of herbal medicines in daily routine supports a strong immune system, general health, and well-being. The likes of brewed herbal teas, soups, tinctures, or nutraceuticals enhances the immune system. In recent times, even the urban people have started using the herbal medicines and supplements to keep themselves healthy.

3 Immunity

The body's inherent defense mechanism against a wide range of illnesses and conditions is referred to as immunity [16]. The human immune system is very complex but controls many infections by generating a limitless variety of cells. There are certain medicines, natural or synthetic, that make the immune system optimum so it works effectively, are known as immunomodulators. The diseases like cancer and autoimmune, apart from other diseases have been treated by the immunomodulators. Immunosuppressants and immunostimulants are two distinct types of immunomodulators in use [17]. One of the biological research areas that is expanding quickly is immunology, which holds enormous potential for treating a wide range of illnesses and disorders. In general, human immunity is grouped into two types: innate and adaptive. A first-line defense from pathogens is known as innate immunity. The pathogens try to go inside the human body but they are prevented by protective barriers. The other system known as adaptive immunity (also called acquired immunity) recognizes the pathogens and then further regulation done by cells and organs. An adaptive immune response includes the two immune response which is called humoral immune response that involves lymphocytes and cell-mediated cytotoxic response mediated by T cells. All the immune system's cells, including their stem cells, are made in the bone marrow in a process called hematopoiesis. Either they develop into mature cells or they move to other distant areas to move [18]. Along with a wide variety of immune cells with specialized functions, certain molecules called cytokines, one of the key immune system mediators, mediate communication between the immune system's specialized cells. This fully integrates the behavior and action responses of the cells [19].

4 Herbal Medicine in Boosting the Immune System

Plants can improve immune function because they are high in flavonoids, vitamin C, or carotenoids. Herbs high in flavonoids may potentially have a slight anti-inflammatory effect. They are known to have anti-inflammatory and immune-stimulating effects. It can enhance phagocytosis, stimulate the synthesis of interferon, and stimulate lymphocyte activity. Approximately 500 medicinal plants are mentioned in ancient literature, and approximately 800 plants have been used in indigenous medical systems. India is home to a vast array of medicinal plants used in traditional medical treatments [20]. Several plant species are used to treat various ailments in indigenous systems such as Siddha, Ayurveda, Unani, and Allopathy [21]. Many scientific studies have highlighted the importance and contribution of many plant families used as medicinal plants, including Asteraceae, Liliaceae, Apocynaceae, Solanaceae, Caesalpiniaceae, Rutaceae, Piperaceae, and Sapotaceae. Medicinal plants are important in the development of new drugs. In India, natural products account for 70% of modern medicines. India has a very small (1.6%) share of this rapidly expanding global market. To compete in a growing market, it is critical to use and scientifically validate more medicinally useful plants [22].

Indian Materia Medica, which made great use of leaves, bark, flowers, roots, fruits, tubers, and fluids, was predominately composed of vegetable materials. Ayurvedic pharmacology was founded on the theories of rasa, vipaka, virya, and prabhava, which did not clearly distinguish between diet and medication because both were essential elements of treatment [23, 24]. More than 700 herbal medications were documented by Charak, Sushruta, and Vagbhata along with their characteristics and medical use. Based on clinical results, 50 different drug categories have been identified, including digestive stimulants, laxatives, anti-diarrhea, anti-hemorrhoids, anti-emetic, antipyretic, anti-inflammatory, anti-pruritic, antiasthmatic, antiepileptic, antihelminthic, hemopoietic, hemostatic, analgesics, sedative, promoter of life (Rasayana), promoter of strength [23].

The practice of Ayurveda, which experienced significant neglect at the hands of the colonial authorities, was hampered by the introduction of Western medicine in the eighteenth century. Numerous laboratories and university departments carried out a tremendous amount of characterization of medicinal plants after the discovery of reserpine, but the results were depressing due to the poorly coordinated, diffused, and unfocused work [23]. Ayurveda and contemporary medicine now have a new interface, thanks to molecular pharmacology. Numerous Ayurvedic medicine categories could produce inventive molecular probes when used using contemporary methods. It is now possible to investigate how Ayurvedic medicines work using the most recent understanding of molecular pharmacology.

Indian science has been aware of the use of herbal therapy for jaundice, most likely caused by viral hepatitis, since the Vedic era. There have been reports of about 170 phytoconstituents with liver protecting properties, identified from 110 plants from 55 families. The number of commercial herbal preparations used as hepatoprotective medications is thought to be around 6000. In the Indian market, there are roughly 40 patent polyherbal compositions made from combinations of 93 Indian herbs from 44 families [25]. However, the most effective herbal remedies for the treatment of viral hepatitis have been identified as the following four:

(a) Silymarin derived from Silibum marianum seeds.
(b) Picrorrhiza kurroa, often known as 'Kutaki' extracts.

(c) The most utilized Phyllanthus species, Phyllanthus niruri and Phyllanthus amarus, have extracts that have been employed as hepatoprotective agents.
(d) With varying degrees of success, *glycyrrhizin* preparation has been used in the past to treat liver disorders and peptic ulcers [26].

4.1 The Effects of Medicinal Plants on the Immune System

In addition to other qualities that are known to either balance or strengthen the organism's defense and response, a wide variety of plants that can be found all over the world include immunomodulatory capabilities [27]. These features regulate our system to support every part of our body. The information in this study focuses on herbal plants that have immunomodulatory qualities as well as antiviral, antioxidant, antibacterial, or anti-inflammatory capabilities. Some of the representative plants have been discussed to understand the utility of herbal medicines in protecting the immune system. A list of plants and their formulations have been listed in Tables 1 [16–18, 28–51] and 2 [52–60].

4.1.1 Turmeric

Originally from Asia, the Zingiberaceae family includes turmeric (*Curcuma longa*). The plant material obtained from its roots and rhizomes has shown potential for various medicinal applications. Due to its pharmacological characteristics, turmeric has been utilized for millennia. Curcuminoids give turmeric its yellow color, and curcumin and curcuminoids together make up most of this well-known culinary spice [61]. Historically used topically to heal wounds or treat skin ulcers [62], as well as to treat rheumatism, sinusitis, the common cold, and liver disorders [63]. According to recent investigations, curcumin is regarded as a potent immunomodulatory drug capable of altering immune cell activity and reducing proinflammatory cytokines [64], as well as exhibiting antioxidant [65] and anti-inflammatory actions [66]. Both in vitro evidence and in vivo have demonstrated its valuable therapeutic properties in a variety of illnesses, including anxiety and depression [67] as well as cancer [68], cardiometabolic disorders, and lung and liver diseases [69].

Turmeric acts as a scavenger of oxygen free radicals and it can also prevent hemoglobin from oxidation. The main curcuminoid present in turmeric is curcumin which exhibits strong antioxidant properties [70], whereas other curcumanoids present

Table 1 Herbs with their botanical name and applications in immune system

S. No.	Name of herb (botanical name) [Ref]	Type of immune system
1.	Aloe vera (*Aloe barbadensis miller*) [28–30]	Anti-inflammatory, immune-boosting, antiviral, anti-diabetic, immune-modulatory, anti-inflammatory, antioxidant, immune-stimulant, antiviral, anti-inflammatory
2.	Garlic (*Allium sativum*) [14, 15, 31, 32]	Anti-inflammatory, antioxidant, antibacterial, antiviral, anticancer, immune-stimulant, antibacterial, antiviral, anticancer, anti-diabetic
3.	Tea (*Camellia sinensis*) [33–35]	Antimicrobial, antioxidant, anti-inflammatory, antifungal, antiviral, anti-infective, antioxidant
4.	Senna [36]	Antioxidant
5.	Ashwagandha (*Withania somnifera*) [37, 38]	Antioxidant, ant-inflammatory, anti-inflammatory
6.	Tulsi (*Ocimum sanctum*) [39, 40]	Antiviral, anti-inflammatory, anti-diabetic, antimicrobial
7.	Amla (*Phyllanthus emblica*) [16]	Antioxidant, anti-diabetic
8.	Neem (*Azadirachta indica*) [17]	Antiviral, antifungal, and antibacterial
9.	*Ginger* (*Zingiber officinale*) [18, 41, 42]	Antioxidant, anti-inflammatory, anticancer, antibacterial, antiviral, antifungal, antioxidant, anti-inflammatory, and immune-booster
10.	Cinnamon (*Cinnamomum verum*) [43, 44]	Anti-diabetic, hypolipidemic, antifungal, antibacterial, anti-inflammatory
11.	Black pepper (*Piper nigrum*) [45]	Antihypertensive, anti-Alzheimer's, antidepressant, antiplatelets, anti-inflammatory, antioxidant, antipyretic, antitumor, antiasthamatic, analgesic, antimicrobial
12.	Shalaparni (*Desmodium gangeticum*) [46]	Antiviral, antiasthamatic
13.	Lasuna (*Allium sativum*) [46]	Antibacterial, antiviral
14.	Bahera (*Terminalia bellirica*) [47]	Immunostimulant
16.	Guduchi (*Tinospora cordifolia*) [48]	Immunostimulant, anti-rheumatic, antiallergic
17.	Gunja (*Abrus precatorius*) [49]	Immunostimulant
18.	Sada Bahar (*Catharanthus roseus*) [50]	Immunostimulant, induces antibody production
20.	Turmeric (*Curcuma longa*) [51]	Immunostimulatory, anti-inflammatory

Table 2 Combination of herbs with their botanical name and applications in immune system

S. No.	Formulations of herbs [Ref]	Type of immune system
1.	Ashwagandha (*Withania somnifera*) and Maitake (*Grifola frondosa*) [52]	Cellular immunity
2.	Ashwagandha (*Withania somnifera*) and Brahmi (*Bacopa monnieri*) [53]	Enhances immunoglobulin production
3.	Nyctanthes arbor-tristis L. and *Withania somnifera* [54]	Immunomodulatory, anti-inflammatory, antitumor, inhibition of oxidants
4.	Sweet basil, holy basil, Tulsi [55]	Antitumor, antioxidant, anti-inflammatory, antifungal, antiviral, antibacterial
5.	Spearmint, peppermint [56]	Antimicrobial, antispasmodic, carminative, and antiviral agents, antioxidant, antiulcer, cytoprotective, hepatoprotective, cholagogue, chemo-preventive, anti-inflammatory, antidiabetogenic
6.	Cinnamon, Ceylon, Cassia, Saigon [57]	Antibacterial, antimicrobial, antioxidant, anti-inflammatory, anti-diabetic, antifungal, antiviral
7.	Cumin, black cumin, black seed [58]	Antimicrobial, antioxidant, anticancer, antitumor, antifungal, antibacterial, anti-inflammatory
8.	Chili pepper, paprika [59]	Antimicrobial, chemo-preventative, analgesic, antioxidant, anticancer, antifungal, antiviral
9.	*Echinacea purpurea, Astragalus membranaceus, and Glycyrrhiza glabra* [60]	Immunomodulator

are demethoxycurcumin and bisdemethoxycurcumin (Fig. 2) [71], they possess radical molecules that prevent free radical formation. Curcumin has been shown to have antiallergic properties by inhibiting histamine release from mast cells [72]. It also acts as an immunomodulator by inhibiting the entry and replication of SARSCoV-2 virus into host cells [73].

4.1.2 Common fig

The Mediterranean region's tropical and temperate fig (*Ficus carica*) tree, a member of the Moraceae family, is a source of flavonoids, phenolics, and polysaccharides [74]. The primary bioactive substance derived from *F. carica* is polysaccharide, which has been shown to have high antioxidant activity [75], antitumor capabilities [76], and, most importantly, immunoregulation function [77]. However, it has previously been demonstrated in induced colitis in vivo testing [78]. This has positive effects on remodeling gut microbiome and treating colitis symptoms. More research is still required to confirm its anti-inflammatory characteristics.

4.1.3 Astragalus

Astragalus (*Astragalus membranaceus*), a member of the Fabaceae family of plants that is mostly grown in China, Mongolia, and Korea, has been the subject of numerous studies and has been shown to have several advantages [79]. Immunomodulation [80], antioxidant activity [81], anti-inflammatory effects [82], anticancer activity [83], and antiviral activity [84] are the principal uses for this herb. The amount of active chemicals in its dried roots varies depending on where they were grown and when they were initially picked [85], and it can be mixed with other herbs for enhanced performance. Enhancing the immune system, or using it as a tonic to strengthen the body's defenses [86], can help with immunodeficiency disorder therapies by decreasing the side effects brought on by conventional pharmacological treatments [87].

4.1.4 Garlic

Another amazing plant that has a significant impact on the immune system is garlic (*Allium sativum*). It comprises a bulb from the Liliaceae family. It grows from 1.5 to 2.5 cm and has traits and an Aromatic Odor. Bulb colors range from pink to white. It has been found that using garlic to improve immunity causes natural killer cells to become more active. For instance, certain studies strongly imply that garlic is a viable candidate for an immunological modulator that protects the immune system's homeostasis [34]. This is because it contains more sulfur compounds, which are what give it its medicinal properties. Other medical conditions that can be treated with garlic's chemical components include cancer, diabetes, atherosclerosis, and hyperlipidemia. Alliin, volatile oil, and allicin are present in garlic. Garlic is used to treat lung infections and is a carminative, expectorant, antimicrobial, and disinfectant [88]. The immunomodulatory effect

(i) inhibits the growth of cancer cells,
(ii) modulates activity of chemical carcinogens,
(iii) enhances capillary skin perfusion, and,
(iv) enhances macrophage (oxidative burst) and T-lymphocytes [88].

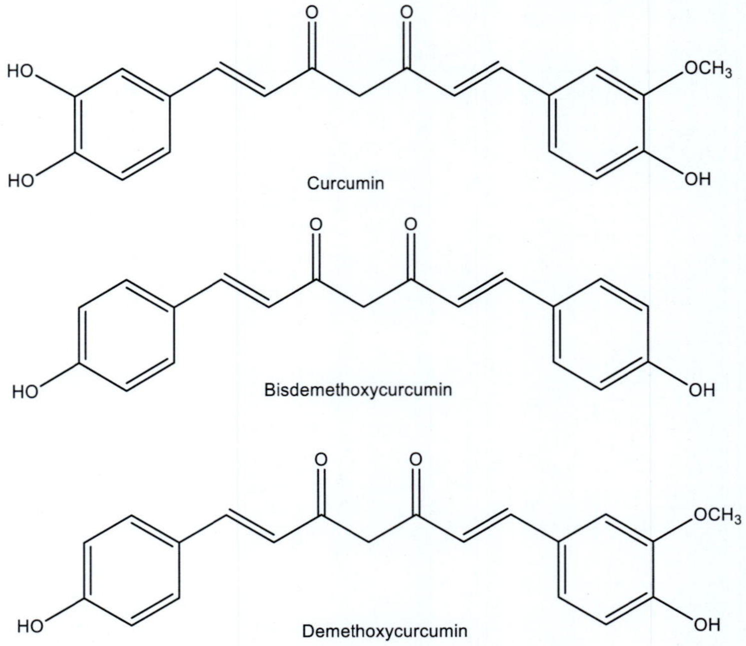

Fig. 2 Chemical structures of curcumin and its derivatives

Fig. 3 Pictures of curry leaves and guava from DTU campus

4.1.5 Curry Leaves

Curry leaves (Fig. 3), also known as *Murraya koenigii*, are a type of tree that grows in tropical and subtropical climates and belongs to the Rutaceae family. Originally from India, it is presently grown in another region of the world. The tree is tiny, at 6 m tall. Bipinnate in structure, the leaves have 9–25 or more leaflets on a short stalk [89]. Due to its distinctive aromatic quality, it is regarded as an herb and frequently employed as a flavoring agent and enhancer. Conversely, curry leaves are shown to have positive health impacts when added to food or ingested in other ways due of their nutritious qualities [90]. Curry leaves include the following bioactive substances: Murrayanine, Scopoletin, Murrayanine glycosides, Koenine, Mahanine, Girinimbine, Koenidine, Koenigine, and Isomahanine [91]. Additionally, it is abundant in minerals like calcium, iron, and phosphorus as well as vitamins like vitamins A, B (niacin, thiamine, and riboflavin), and C. Researchers have discovered that curry leaves have hepatoprotective, anti-inflammatory, anti-diabetic, antimicrobial, and anti-hypercholesterolemia characteristics. Additionally, it aids in the treatment of vomiting, diarrhea, and a calcium shortage [88]. The leaves also contain the triterpenoid alkaloids cyclomahanimbine and tetrahydromahanimbine. There are tonic, stomachic, and carminative properties in the leaves, roots, and bark. Additionally, the leaves exhibit anti-emetic qualities. The stem distillate of the leaves has proven to have purgative qualities [92]. In addition to having anti-diabetic properties, its leaf extracts also have some actions that control immunology in relation to oxidative stress metabolism. Expression of the cytokines interleukin (IL)-2, 4, 10, and tumor necrosis factor-alpha (TNF-alpha) demonstrated this immunomodulatory and anti-inflammatory action [93].

4.1.6 Ghritkumari (*Aloe Vera*)

This succulent perennial plant, a member of the genus Aloe, is native to the desert regions of the Arabian Peninsula. It has a distinctive appearance because of the gel-like substance found in its thick, fleshy leaves, which has led to its popularity as a decorative plant. Due to its extraordinary healing powers, aloe vera has been utilized for millennia in both traditional and alternative medicine. Its gel was used to treat burns, wounds, and other skin conditions. According to ancient Chinese, Indian, and Greek literature, aloe vera was highly prized for its therapeutic benefits [87]. It has the following immunomodulatory effects:

(i) Its polysaccharides exhibit adjuvant activity for the formation of antibodies and DTH, and it prevents UV-induced suppression of DTH.
(ii) Reduces inflammation.
(iii) Promotes faster healing of wounds and serves as an oxygen radical scavenger.
(iv) Cooperates with NO.
(v) Promotes tumor regression [94].

4.1.7 Guava

Guava (Fig. 3) is a tropical fruit, high in nutrients such as potassium, vitamin A, and vitamin C [95]. It also has a variety of phytochemicals and antioxidants. Because of its nutritional content and bioactive compounds, many studies suggest that guava may help boost the immune system in humans [96]. Guava is one of the richest sources of vitamin C, vitamin C acts as antioxidant and protects our immune system by neutralizing harmful free radicals which can damage cells and weaken the body's defenses [97]. It also helps to produce white blood cells, which are necessary for fighting infections. Guava contains minerals such as Mg and Cu, all of which are important for immune system health. Guava components have anti-inflammatory properties, which may help in the reduction of inflammation. Guava contains antioxidants such as carotenoids, polyphenols, and flavonoids that help in the reduction of oxidative stress in the body [98]. Oxidative stress can have a negative impact on the immune system but can be mitigated by eating antioxidant-rich foods like guava. Guava is also high in dietary fiber, which is good for gut health. A healthy gut is associated with a stronger immune system because it houses a large portion of the body's immune cells [99]. The active compounds present in guava extract such as polyphenol compounds act as an immunostimulant though these are not only compounds responsible for immunostimulatory activity but they play the major role. Phenolic compounds are mainly responsible for the high antioxidant activity in Guava [44].

4.1.8 Ginger

Ginger contains bioactive compounds such as phenolic groups, alkaloids, and steroids, all of which have medicinal properties [100]. Gingerol is the active compound that has powerful anti-inflammatory effects [100]. It also contains antioxidants, which help protect our cells from free radical oxidative stress, when our cells are protected from these damages, our immune system works more efficiently [101]. Ginger is beneficial for increasing immunity due to its high antioxidant content, lowering stress due to its influential anti-inflammatory properties, and improving blood circulation [102].

Ginger is used to boost immune system because of its antibacterial nature. Health benefits of ginger are as follows: relieves nausea, aids digestion, reduces oxidative stress, reduces the risk of heart disease, mitigates obesity [102]. The mechanisms for its anti-inflammatory properties include:

(a) inhibition of arachidonic acid-induced platelet aggregation and formation of thromboxane B,
(b) upregulation of histone H3 acetylation, and suppressed histone deacetylase (HDAC)1 expression,
(c) inhibition of IL-1, TNF-α, and IL-8 and downregulation of inflammatory inducible NO synthase (iNOS).

4.1.9 Neem

Neem plant parts have antimicrobial properties due to their inhibitory effect on microbial growth. Neem has blood purifying properties as well. Neem leaves strengthen our immune system while also cooling our body internally. Neem has antibacterial and antifungal properties, which help to keep your skin clean, radiant, and healthy. Neem extracts contained immunomodulators that stimulated cells and macrophages [103]. The active ingredient in neem is nimbin, which has antioxidant and anti-inflammatory properties [103]. Numerous biologically active substances, such as nimbin, nimbidin, nimbolide, and limonoids, are present in neem. Quercetin and β-sitosterol were the first polyphenolic flavonoids isolated from fresh neem leaves [104]. Neem has anti-inflammatory properties that can aid in the reduction of proinflammatory cytokines and mediators [104]. This action aids in the regulation of excessive immune responses, which can result in chronic inflammation and tissue damage. Macrophages are immune cells which help in engulfment of foreign substances and pathogens. Neem can increase macrophage activity, boosting the immune response to infections [105]. Neem has antimicrobial properties that can aid in the fight against pathogens such as bacteria, viruses, and fungi. Neem helps the immune system fight infections by inhibiting the growth and replication of these microbes [104].

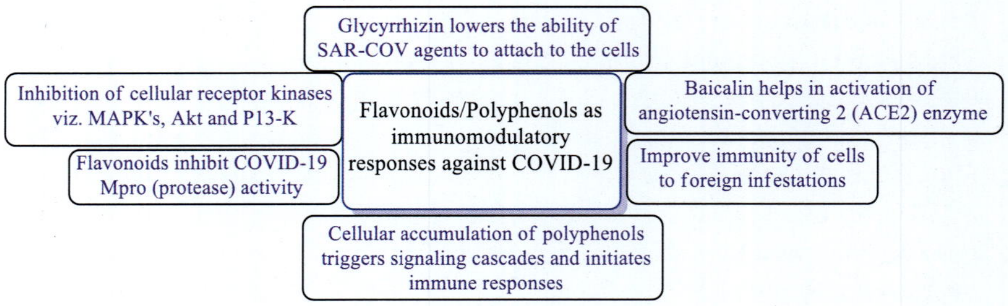

Fig. 4 Flavonoids/polyphenols as immunomodulator against COVID-19

4.1.10 Arjuna

Terminalia arjuna (Roxb.) Wight & Arn., commonly referred to as Arjuna is one of the most prevalent and useful medicinal plants in traditional medicine. It is prevalent across India's tropical and semitropical regions. Each component of *Terminalia arjuna*, including leaves, stem bark, roots, and fruit ha medicinal properties [106]. It contains analgesic, anti-inflammatory, and immunostimulatory effects and has been utilized to treat a variety of human diseases as well as to maintain good health. The bark extract has a long history of use as a cardiac stimulant for its favorable effects in angina, i.e., "hritshool," and is used in the treatment of heart failure, atherosclerosis, and hypercholesterolemia [107]. The herbal immunomodulator that contains *T. arjuna* is quite beneficial for strengthening immunity and battling against *Caecal coccidiosis* [108]. The saponin glycosides are thought to be responsible for its inotropic effects, while flavonoids/phenolics may provide antioxidant activity as well as vascular strengthening activity, confirming the medicinal plant's multifaceted role in cardio-protection [108].

Compared to their synthetic counterparts, many herbs and spices naturally have higher concentrations of bioactive chemicals. Ginger, a common spice, is used in both food and medicine since it has long been known to help with digestion. Basil, rosemary, cilantro, mint, and turmeric are just a few of the various herbs and spices that have historically been researched for their therapeutic benefits [109]. Due to their high concentration of phytonutrients and other bioactive substances, herbs and spices can offer significant benefits to overall nutrition and hence toward immunity. Individuals with current eating patterns may nevertheless be deficient in the critical nutrients needed for optimal health. The enormous array of health advantages of herbs and spices has been found to be unknown to many people [109]. Now, there is a growing interest and desire among both the general public and healthcare experts to understand more about the use of herbs and spices in health promotion. Regular use of these nutrients may help to strengthen immune defenses, which may reduce health risks and may speed up recovery. Recent studies have even concentrated on the use of herbs and spices to treat obese, metabolic syndrome-afflicted, and overweight people in order to lessen the effects of cytokine cascades of inflammation in the body, which are comparable to those seen in COVID-19 instances (Fig. 4) [110].

5 Some Specific Effect of Herbal Medicines

5.1 Modulation of Cytokines by Herbal Plant Products

Numerous in vitro and in vivo investigations to determine the impact of herbal medicine on cytokines have demonstrated that they have a significant impact on a variety of cytokines. Interleukins (ILs), interferons, chemokines, and other soluble extracellular proteins or glycoproteins known as cytokines are essential for both innate and acquired kinds of immunity. These cytokines, which are secreted by all nucleated cells in an induced response to damage, maintain physiological stability through intermolecular interactions [19]. In reality, understanding the medical literature on numerous diseases makes it clear that these disease situations are connected to cytokine releases. These cytokines play a significant part in most psychiatric disorders as well as diseases of the central nervous system, and aberrant discharges of these molecules have been shown. Under the control of cytokines, numerous neurochemicals, neuroendocrine, and neuroimmune substances have manifested [111]. Numerous behavioral changes, good and negative emotions, stress, infections, and other factors have all been shown to trigger cytokine release, and their significance in cases of depression [112], Alzheimer's disease [113], and schizophrenia [114] has been well documented.

5.2 Prebiotic Effects on Humoral Immunity in Medicinal Herbs

The gut microbiome, which is made up of more than 1000 different microbial species and is the first barrier to overcome when taking herbal medications orally, is crucial for both localized immune system training and food absorption that affects lipid metabolism, which indirectly affects immunological profile [115]. It is well known that the active components of herbs have the potential to change the dynamic of microbial composition when considering individual gut enterotypes [116]. A higher concentration in the intestinal lumen and epithelium can significantly affect the gut microbiota given the low dose of active herbal components and difficulties in being absorbed by the human body [117]. In addition to the direct impact of herbal ingredients on immune cells, lactobacilli and bifidobacteria, which control the microbiota, play a critical role in immunological maturation and, ultimately, pathogen defense [118]. Prebiotics have a significant impact on probiotic-mediated immunological improvement that is connected with modification of effector and regulatory T cell responses, according to randomized clinical trials on humans and multiple animal studies. In fact, oligosaccharides, such as inulin, fructooligosaccharides, and galactooligosaccharides, which are active components of herbs, promote the selective proliferation of these beneficial bacteria in the gut, polarizing the immune response toward Th1 or Th2, and determining the outcome of antibody production by activated B cells [119]. *Glabra Glycyrrhiza* comparing recognized plants with such efficient production, such as Lens culinaris and Camellia sinensis, Radix typically had the biggest magnitude increase in prebiotic benefits [120].

5.3 Herbal Medicine that Directly Stimulates B Cells

Only a limited portion of the medicinal component can be absorbed and get to the Peyer's patches and first-cluster gut-associated lymphoid tissue. By taking immunomodulatory herbs, it is possible to control the immunological response in Peyer's patches [121]. Thuja summitates, Echinacea purpurea radix, Echinacea pallida radix, Baptisia tinctoria radix, and Echinacea pallida radix may all be able to activate B cells within Peyer's patches cells [122]. Additionally, these herbs can enhance the activation of antibodies against other infections, such as lipopolysaccharide (a gram-negative bacterium's endotoxin from its cell wall). It is widely known that 1,3-D-glucan, which is present in *Sclerotinia sclerotiorum*, is one of the active macromolecules incorporated by M cells (transporting particles with diameter range of 100 nm to 10 m) via endocytosis to reach the Peyer's plaques [123]. However, this glucan also activates peritoneal macrophages and causes the production of cytokines that helps to modulate immunity. Additionally, *Glycyrrhiza glabra* Radix, *Withania somnifera* Radix, *Astragalus propinquus* Radix, *Oldenlandia diffusa*, and *Carthamus tinctorius* have experimental data supporting their ability to modulate B-cell activity [124].

5.4 Herbal Remedies that Indirectly Stimulate Cellular Immunity

Follicular helper T cells and dendritic cells, in particular, have the ability to drastically alter both the number and quality of the humoral response of B cells to antigens [125]. Even in immunodeficient patients with HIV infections, clinical trial studies on the ingestion of *Phyllanthus emblica*, *Echinacea purpurea*, and *Plantago asiatica* showed their herbal components serve as a potential immunostimulatory formulation with a significant increase in effector and cytotoxic T cells and indirectly B lymphocytes response [126].

5.5 Bioavailability

A factor of great significance is the herb's active ingredients' bioavailability. A substance must enter the bloodstream from the gastrointestinal tract before it can act systemically. Surprisingly little is known about herbal ingredients in this area. Goldenseal (*Hydrastis canadensis* L.), a well-known plant, contains substances like berberine and hydrastine that are virtually not absorbed after oral ingestion. Parenteral dosing of these alkaloids has been used in all animal studies demonstrating systemic effects. However, Goldenseal (*Hydrastis canadensis*) continues to be one of the most popular herbs, is heavily advertised, and is recognized as a general immunostimulant by an uninformed audience [127].

6 Representative Examples of Natural Products with Immunomodulating Properties

The active ingredients from herbal medicines are mainly responsible for their biological activities [128–130]. The natural products belonging to the groups which are also known as secondary metabolites like phenolics, saponins, terpenoids, alkaloids, fatty acids, etc. play crucial role in enhancing the human immune system. The compounds like curcumin, quercetin, luteolin, propolis, and other secondary metabolites had shown potential immunostimulatory actions [4]. Some examples of natural products have been discussed here.

A flavonoid quercetin is present in many plant species and not synthesized in human body. The studies show that this compound possesses antioxidant, antibacterial, anti-inflammatory, and antiviral properties [131]. In one of the clinical trials patients with rheumatoid arthritis were put on quercetin medication with 50 mg/day for 8 weeks. This was observed that the disease symptoms improved with reduction in pain plasma and levels of hs-TNF compared with the placebo group [132].

A dietary fiber glucan which is a polymer of D-glucose is a natural product and linked in various combination through glycosidic bond. These are present in cell walls of many fungi, algae, and bacteria. Also, these are present in wheat, rye, barley, and oats. The studies have shown that the β-glucan enhances the cytotoxic activity of macrophages and NK cells, this helps in improvement in host immunity [133].

These fibers also interact with several cell surface receptors such as lactosylceramide selected scavenger, dectin-1, and complement receptor 3 and enhances immune system [4, 133]. The mushroom reishi (*Ganoderma lucidum*) produces a glycan which enhances immunity through IgG or Iga expression in small intestinal and serum fluid [4].

Another immunoregulator natural product molecule is melatonin, an indole derivative hormone. This molecule has been obtained from grapes, polives, maize, cherries, black pepper, seeds of *Coffea arabica*, *Curcuma longa*, *Hypericum perforatum*, etc. This molecule lowers the proinflammatory cytokines and inhibits the immunity response [134].

7 Conclusions

The immune system plays a pivotal role to maintain human health by protecting the body from pathogens. The adaptive and innate types of the immune system encounter viruses, bacteria, or fungi to eliminate them and guard the body. However, the over expressive immune system is a problem to human beings as this leads to autoimmune diseases and allergies. The dietary balance, adequate food, minerals, and vitamins help in developing strong immunity systems. The traditional knowledge along with modern scientific tools has established the role of herbal medicines along with different natural ingredients in enhancing the immunity. The presence of bioactive phytochemicals like flavonoids, terpenoids, polyphenols, vitamins, etc. in herbal materials showed immunomodulatory properties. The bioactive phytochemicals control the optimum functioning of immune cells, including basophils, NK cells, neutrophils, lymphocytes, macrophages, and mast cells, thereby enhancing the human immunity.

The role of herbal medicines in health-promoting effects including immunity still has to go a long way because there is lack of scientific evidences and clinical trials of various herbal materials and formulations. Another challenge is the quantification of active ingredients which keep changing due to change in altitude, ecology, and environments. Future research in this field requires proper quantification of phytochemicals, efficacy of the extracts, proper dose range, etc.

Acknowledgments The authors Anchal Dabas, Parul Yadav, and Ram Singh are thankful to DTU, and Geetanjali to Kirori Mal College for infrastructural support.

References

1. Tan BK, Vanitha J. Immunomodulatory and antimicrobial effects of some traditional Chinese medicinal herbs: a review. Curr Med Chem. 2004;11(11):1423–30.
2. Sharma P, Kumar P, Sharma R, Gupta G, Chaudhary A. Immunomodulators: role of medicinal plants in immune system. Natl J Physiol Pharm Pharmacol. 2017;7(6):552.
3. Debnath S, Chakravorty R, Devi D. A review on role of medicinal plants in immune system. Asian J Pharm Technol. 2020;10(4):273–7.
4. Gasmi A, Shanaida M, Oleshchuk O, Semenova Y, Mujawdiya PK, Ivankiv Y, Pokryshko O, Noor S, et al. Natural ingredients to improve immunity. Pharmaceuticals. 2023;16:528.
5. Maqbool M, Dar MA, Gani I, Mir SA, Khan M. Herbal medicines as an alternative source of therapy: a review. World J Pharm Pharm Sci. 2019;3:374–80.
6. Butler MS. The role of natural product chemistry in drug discovery. J Nat Prod. 2004;67(12):2141–53.
7. Kew J, Morris C, Aihie A, Fysh R, Jones S, Brooks D. Arsenic and mercury intoxication due to Indian ethnic remedies. Br Med J. 1993;306(6876):506.
8. Pal SK, Shukla Y. Herbal medicine: current status and the future. Asian Pac J Cancer Prev. 2003;4(4):281–8.
9. Das S, Bordoloi R, Newar N. A review on immune modulatory effect of some traditional medicinal herbs. J Pharm Chem Biol Sci. 2014;2(1):33–42.
10. Frazzoli C, Grasso G, Husaini DC, Ajibo DN, Orish FC, Orisakwe OE. Immune system and epidemics: the role of African indigenous bioactive substances. Nutrients. 2023;15(2):273.
11. Panyod S, Ho CT, Sheen LY. Dietary therapy and herbal medicine for COVID-19 prevention: A review and perspective. J Tradit Complement Med. 2020;10(4):420–7.
12. Winslow LC, Kroll DJ. Herbs as medicines. Arch Intern Med. 1998;158(20):2192–9.
13. Miller LG. Herbal medicinals: selected clinical considerations focusing on known or potential drug-herb interactions. Arch Intern Med. 1998;158(20):2200–11.
14. Tsai CW, Chen HW, Sheen LY, Lii CK. Garlic: health benefits and actions. Biomedicine. 2012;2(1):17–29.
15. Gebreyohannes G, Gebreyohannes M. Medicinal values of garlic: A review. Int J Med Med Sci. 2013;5(9):401–8.
16. Zheng D, Liwinski T, Elinav E. Interaction between microbiota and immunity in health and disease. Cell Res. 2020;30:492–506.
17. Bhowmik D, Chiranjib YJ, Tripathi KK, Kumar KS. Herbal remedies of Azadirachta indica and its medicinal application. J Chem Pharm Res. 2010;2(1):62–72.
18. Sahoo JP, Behera L, Praveena J, Sawant S, Mishra A, Sharma SS, Ghosh L, Mishra AP, Sahoo AR, Pradhan P, Sahu S. The golden spice turmeric (Curcuma longa) and its feasible benefits in prospering human health—a review. Am J Plant Sci. 2021;12(3):455–75.
19. Sahoo JP, Mohapatra U, Mishra AP, Samal K. Turmeric (Haldi)-A strapping strategy for enhancing the immune system to reduce the effect of SARS-CoV-2. Food Sci Rep. 2020;1(8):10.
20. Chopra RN, Nayar SL. Glossary of Indian medicinal plants. Council of Scientific and Industrial Research; 1956.
21. Rabe T, Van Staden J. Antibacterial activity of South African plants used for medicinal purposes. J Ethnopharmacol. 1997;56(1):81–7.
22. Verma S, Singh SP. Current and future status of herbal medicines. Vet World. 2008;1(11):347.
23. Lele RD. Ayurveda (ancient Indian system of medicine) and modern molecular medicine. J Assoc Physicians India. 1999;47(6):625–8.
24. Valiathan MS. Healing plants. Curr Sci. 1998;75(11):1122–7.
25. Bhatt AD, Bhatt NS. Indigenous drugs and liver disease. Indian J Gastroenterol. 1996;15(2):63–7.
26. Patil US, Jaydeokar AV, Bandawane DD. Immunomodulators: A pharmacological review. Int J Pharm Pharm Sci. 2012;4(1):30–6.
27. Falzon CC, Balabanova A. Phytotherapy: an introduction to herbal medicine. Prim Care. 2017;44(2):217–27.

28. Rajeswari R, Umadevi M, Rahale CS, Pushpa R, Selvavenkadesh S, Kumar KS, Bhowmik D. Aloe vera: the miracle plant its medicinal and traditional uses in India. J Pharmacogn Phytochem. 2012;1(4):118–24.
29. Hamman JH. Composition and applications of aloe vera leaf gel. Molecules. 2008;13(8):1599–616.
30. Qadir MI. Composition and applications of aloe vera leaf gel. Int J Nat Sci. 2009;2:21–6.
31. Singh UP, Prithiviraj B, Sarma BK, Singh M, Ray AB. Role of garlic (Allium sativum L.) in human and plant diseases. Indian J Exp Biol. 2001;39(4):310–22.
32. Butt MS, Sultan MT, Butt MS, Iqbal J. Garlic: nature's protection against physiological threats. Crit Rev Food Sci Nutr. 2009;49(6):538–51.
33. Sun Y, Xun K, Wang Y, Chen X. A systematic review of the anticancer properties of berberine, a natural product from Chinese herbs. Anti-Cancer Drugs. 2009;20(9):757–69.
34. Wang S, Li Z, Ma Y, Liu Y, Lin CC, Li S, Zhan J, Ho CT. Immunomodulatory effects of green tea polyphenols. Molecules. 2021;26(12):3755.
35. Steinmann J, Buer J, Pietschmann T, Steinmann E. Anti-infective properties of epigallocatechin-3-gallate (EGCG), a component of green tea. Br J Pharmacol. 2013;168(5):1059–73.
36. Ikram A, Khalid W, Saeed F, Arshad MS, Afzaal M, Arshad MU. Senna: as immunity boosting herb against COVID-19 and several other diseases. J Herb Med. 2023;10:100626.
37. Gupta S, Bansal RN, Sodhi SP, Brar GK, Molhotra M. Ashwagandha (Withania somnifera)–a herb with versatile medicinal properties empowering human physical and mental health. J Pre-Clin Clin Res. 2021;15(3):129–33.
38. Gayen KC, Jana P, Giri A. Use of ashwagandha to boost immunity to combat COVID-19. In: Immunity boosting functional foods to combat COVID-19, vol. 14. CRC Press; 2021. p. 47–52.
39. Palai SS, Giri A. Tulsi-A potential immune boosting functional food ingredient to combat COVID-19. In: Immunity boosting functional foods to combat COVID-19, vol. 14. CRC Press; 2021. p. 27–36.
40. Srivastava AK, Singh VK. Tulsi (*Ocimum sanctum*): A potent adaptogen. Clin Res Notes. 2021;2(2):01–5.
41. Priya FF, Islam MS. Phyllanthus emblica Linn. (Amla)—a natural gift to humans: an overview. J Dis Med Plants. 2019;5:1–9.
42. Abinaya S, Gayatri Devi R, Lakshmanan G. Knowledge and awareness about ginger and turmeric as a herbal cure for COVID-19. Int J Pharm Res. 2020:768–77.
43. Singh S, Paul V, Singh R. Cinnamon powder and its importance for immunity and human health. BMC Complement Altern Med. 13:275Y284.
44. Dutta A, Chakraborty A. Cinnamon in anticancer armamentarium: a molecular approach. J Toxicol. 2018;29:2018.
45. Singh NA, Kumar P, Jyoti KN. Spices and herbs: potential antiviral preventives and immunity boosters during COVID-19. Phytother Res. 2021;35(5):2745–57.
46. Singh S, Parmar N, Patel B. A review on Shalparni (*Desmodium gangeticum* DC.) and Desmodium species (*Desmodium triflorum* DC. & *Desmodium laxiflorum* DC.)–ethnomedicinal perspectives. J Med Plants. 2015;3(4):38–43.
47. Sabu MC, Kuttan R. Antidiabetic and antioxidant activity of Terminalia bellirica. Roxb. Indian J Exp Biol. 2009;47(4):270–5.
48. Saha S, Ghosh S. Tinospora cordifolia: one plant, many roles. Anc Sci Life. 2012;31(4):151.
49. Bhakta S, Das SK. The medicinal values of Abrus precatorius: a review study. J Adv Biotechnol Exp Ther. 2020;3(2):84–91.
50. Pattanayak S. Use of succulent bio-medicines to control COVID-19. Sikdar Bagan Street: Calcutta Block and Print; 2020.
51. Allam G. Immunomodulatory effects of curcumin treatment on murine schistosomiasis mansoni. Immunobiology. 2009;214(8):712–27.
52. Vetvicka V, Vetvickova J. Immune enhancing effects of WB365, a novel combination of Ashwagandha (*Withania somnifera*) and Maitake (*Grifola frondosa*) extracts. N Am J Med Sci. 2011;3(7):320.
53. Yamada K, Hung P, Park TK, Park PJ, Lim BO. A comparison of the immunostimulatory effects of the medicinal herbs Echinacea, Ashwagandha and Brahmi. J Ethnopharmacol. 2011;137(1):231–5.
54. Padhy M. A review on medicinal plants *Withania somnifera* and *Nyctanthes arbortristis*: boosting of immune system during SARS-CoV-2. Lett Appl NanoBioScience. 2020;9:1538–46.
55. Singh P, Chakraborty P, He DH, Mergia A. Extract prepared from the leaves of *Ocimum basilicum* inhibits the entry of Zika virus. Acta Virol. 2019;1:63.
56. McKay DL, Blumberg JB. A review of the bioactivity and potential health benefits of chamomile tea (*Matricaria recutita* L.). Phytother Res. 2006;20(7):519–30.
57. Marissal-Arvy N, Batandier C, Dallennes J, Canini F, Poulet L, Couturier K, Hininger-Favier I, Moisan MP, Roussel AM, Mormede P. Effect of a high-fat–high-fructose diet, stress, and cinnamon on central expression of genes related to immune system, hypothalamic–pituitary–adrenocortical axis function and cerebral plasticity in rats. Br J Nutr. 2014;111(7):1190–201.
58. Randhawa MA, Alghamdi MS. Anticancer activity of Nigella sativa (black seed)—a review. Am J Chin Med. 2011;39(06):1075–91.
59. Sun T, Xu Z, Wu CT, Janes M, Prinyawiwatkul W, No HK. Antioxidant activities of different colored sweet bell peppers (*Capsicum annuum* L.). J Food Sci. 2007;72(2):S98–102.
60. Brush J, Mendenhall E, Guggenheim A, Chan T, Connelly E, Soumyanath A, Buresh R, Barrett R, Zwickey H. The effect of *Echinacea purpurea*, *Astragalus membranaceus* and *Glycyrrhiza glabra* on CD69 expression and immune cell activation in humans. Phytother Res. 2006;20(8):687–95.
61. Kocaadam B, Şanlier N. Curcumin, an active component of turmeric (*Curcuma longa*), and its effects on health. Crit Rev Food Sci Nutr. 2017;57(13):2889–95.
62. Kunnumakkara AB, Bordoloi D, Padmavathi G, Monisha J, Roy NK, Prasad S, Aggarwal BB. Curcumin, the golden nutraceutical: multitargeting for multiple chronic diseases. Br J Pharmacol. 2017;174(11):1325–48.
63. Kant V, Gopal A, Pathak NN, Kumar P, Tandan SK, Kumar D. Antioxidant and anti-inflammatory potential of curcumin accelerated the cutaneous wound healing in streptozotocin-induced diabetic rats. Int Immunopharmacol. 2014;20(2):322–30.
64. Ganjali S, Sahebkar A, Mahdipour E, Jamialahmadi K, Torabi S, Akhlaghi S, Ferns G, Parizadeh SM, Ghayour-Mobarhan M. Investigation of the effects of curcumin on serum cytokines in obese individuals: a randomized controlled trial. Sci World J. 2014;2014
65. Sahebkar A. Dual effect of curcumin in preventing atherosclerosis: the potential role of pro-oxidant–antioxidant mechanisms. Nat Prod Res. 2015;29(6):491–2.
66. Panahi Y, Sahebkar A, Parvin S, Saadat A. A randomized controlled trial on the anti-inflammatory effects of curcumin in patients with chronic sulphur mustard-induced cutaneous complications. Ann Clin Biochem. 2012;49(6):580–8.
67. Momtazi AA, Shahabipour F, Khatibi S, Johnston TP, Pirro M, Sahebkar A. Curcumin as a microRNA regulator in cancer: a review. Rev Physiol Biochem Pharmacol. 2016;171:1–38.
68. Momtazi AA, Derosa G, Maffioli P, Banach M, Sahebkar A. Role of microRNAs in the therapeutic effects of curcumin in non-cancer diseases. Mol Diagn Ther. 2016;20:335–45.
69. Esmaily H, Sahebkar A, Iranshahi M, Ganjali S, Mohammadi A, Ferns G, Ghayour-Mobarhan M. An investigation of the effects of curcumin on anxiety and depression in obese individuals: A randomized controlled trial. Chin J Integr Med. 2015;21:332–8.

70. Rathaur P, Raja W, Ramteke PW, John SA. Turmeric: the golden spice of life. Int J Pharm Sci Res. 2012;1;3(7):1987.
71. Gautam SC, Gao X, Dulchavsky S. Immunomodulation by curcumin. In: The molecular targets and therapeutic uses of curcumin in health and disease, vol. 1. Boston, MA: Springer; 2007. p. 321–41.
72. Kurup VP, Barrios CS. Immunomodulatory effects of curcumin in allergy. Mol Nutr Food Res. 2008;52(9):1031–9.
73. Irfandi R, Ilham M, Erwing R, Arafah M, Rompegading AB, Putri SE, Sartika SD, Fauziah S, Agustina AS, Akbar H. Review on curcumin compounds in turmeric plants for the treatment of COVID-19. Int J Des Nat Ecodyn. 2022;17:957–65.
74. Gibbons A. Archaeology. Ancient figs push back origin of plant cultivation. Science. 2006;312(5778):1292.
75. Yang XM, Yu W, Ou ZP, Ma HL, Liu WM, Ji XL. Antioxidant and immunity activity of water extract and crude polysaccharide from Ficus carica L. fruit. Plant Foods Hum Nutr. 2009;64(2):167–73.
76. Chen R, Li H, Li S, Jin C, Lu J. Extraction optimization, preliminary characterization and immunological activity of polysaccharides from figs. Int J Biol Macromol. 2015;72:185–94.
77. Du J, Li J, Zhu J, Huang C, Bi S, Song L, Hu X, Yu R. Structural characterization, and immunomodulatory activity of a novel polysaccharide from *Ficus carica*. Food Funct. 2018;9(7):3930–43.
78. Zou Q, Zhang X, Liu X, Li Y, Tan Q, Dan Q, Yuan T, Liu X, Liu RH, Liu Z. *Ficus carica* polysaccharide attenuates DSS induced ulcerative colitis in C57BL/6 mice. Food Funct. 2020;11(7):6666–79.
79. Zhang J, Wu C, Gao L, Du G, Qin X. Astragaloside IV derived from Astragalus membranaceus: A research review on the pharmacological effects. Adv Pharmacol. 2020;87:89–112.
80. Yang ZG, Sun HX, Fang WH. Haemolytic activities and adjuvant effect of Astragalus membranaceus saponins (AMS) on the immune responses to ovalbumin in mice. Vaccine. 2005;23(44):5196–203.
81. De-Hong YU, Yong-Ming BA, Li-Jia AN, Ming YA. Protection of PC12 cells against superoxide-induced damage by isoflavonoids from *Astragalus mongholicus*. Biomed Environ Sci. 2009;22(1):50–4.
82. Qin Q, Niu J, Wang Z, Xu W, Qiao Z, Gu Y. Astragalus membranaceus inhibits inflammation via phospho-P38 mitogen-activated protein kinase (MAPK) and nuclear factor (NF)-κB pathways in advanced glycation end product-stimulated macrophages. Int J Mol Sci. 2012;13(7):8379–87.
83. Shen HH, Wang K, Li W, Ying YH, Gao GX, Li XB, Huang HQ. *Astragalus Membranaceus* prevents airway hyperreactivity in mice related to Th2 response inhibition. J Ethnopharmacol. 2008;116(2):363–9.
84. Zhu H, Zhang Y, Ye G, Li Z, Zhou P, Huang C. In vivo and in vitro antiviral activities of calycosin-7-O-β-D-glucopyranoside against coxsackie virus B3. Biol Pharm Bull. 2009;32(1):68–73.
85. Fu J, Wang Z, Huang L, Zheng S, Wang D, Chen S, Zhang H, Yang S. Review of the botanical characteristics, phytochemistry, and pharmacology of *Astragalus membranaceus* (Huangqi). Phytother Res. 2014;28(9):1275–83.
86. Auyeung KK, Han QB, Ko JK. *Astragalus membranaceus*: a review of its protection against inflammation and gastrointestinal cancers. Am J Chin Med. 2016;44(01):1–22.
87. Yin X, Zhang Y, Wu H, Zhu X, Zheng X, Jiang S, Zhuo H, Shen J, Li L, Qiu J. Protective effects of Astragalus saponin I on early stage of diabetic nephropathy in rats. J Pharmacol Sci. 2004;95(2):256–66.
88. Ramalingum N, Mahomoodally MF. The therapeutic potential of medicinal foods. Adv Pharmacol Pharm Sci. 2014;2014:354264.
89. Bhandari PR. Curry leaf (*Murraya koenigii*) or cure leaf: review of its curative properties. J Med Nutr Nutraceut. 2012;1(2):92.
90. Lal M, Kaur N. Effect of processing on nutritional and antinutritional composition of curry leaves (*Murraya koenigii*). Int J Home Sci. 2019;5(3):186–90.
91. Saini SC, Reddy GB. A review on curry leaves (*Murraya koenigii*): versatile multi-potential medicinal plant. Am J Phytomed Clin Ther. 2015;3(04):363–8.
92. Kaur P, Davar V. Exploring the nutraceutical and therapeutic potential of commonly used Indian spices. Int J Physiol Nutr Phys Educ. 2018;3:410–9.
93. Paul S, Bandyopadhyay TK, Bhattacharyya A. Immunomodulatory effect of leaf extract of *Murraya koenigii* in diabetic mice. Immunopharmacol Immunotoxicol. 2011;33(4):691–9.
94. Bhargava K, Singh DK. Aloe vera medical plant. Eur Chem Bull. 2023:1755–64.
95. Haque MN, Saha BK, Karim MR, Bhuiyan MN. Evaluation of nutritional and physico-chemical properties of several selected fruits in Bangladesh. Bangladesh J Sci Ind Res. 2009;44(3):353–8.
96. Vijaya Anand A, Velayuthaprabhu S, Rengarajan RL, Sampathkumar P, Radhakrishnan R. Bioactive compounds of guava (*Psidium guajava* L.). Bioactive compounds in underutilized fruits and nuts 2020:503–527.
97. Mumtaz S, Ali S, Tahir HM, Kazmi SA, Shakir HA, Mughal TA, Mumtaz S, Summer M, Farooq MA. Aging and its treatment with vitamin C: a comprehensive mechanistic review. Mol Biol Rep. 2021;1:1–3.
98. Naseer S, Hussain S, Naeem N, Pervaiz M, Rahman M. The phytochemistry and medicinal value of Psidium guajava (guava). Clin Phytosci. 2018;4(1):1–8.
99. Jiménez-Escrig A, Rincón M, Pulido R, Saura-Calixto F. Guava fruit (Psidium guajava L.) as a new source of antioxidant dietary fiber. J Agric Food Chem. 2001;49(11):5489–93.
100. Laily N, Kusumaningtyas RW, Sukarti I, Rini MRDK. The potency of guava Psidium Guajava (L.) leaves as a functional immunostimulatory ingredient. Procedia Chem. 2015;14:301–7.
101. Lashgari NA, Momeni Roudsari N, Khayatan D, Shayan M, Momtaz S, Roufogalis BD, Abdolghaffari AH, Sahebkar A. Ginger and its constituents: role in treatment of inflammatory bowel disease. Biofactors. 2022;48(1):7–21.
102. Ahmed RS, Suke SG, Seth V, Chakraborti A, Tripathi AK, Banerjee BD. Protective effects of dietary ginger (Zingiber officinale Rosc.) on lindane-induced oxidative stress in rats. Phytother Res. 2008;22(7):902–6.
103. Al-Hashemi ZSS, Hossain MA. Biological activities of different neem leaf crude extracts used locally in ayurvedic medicine. Pac Sci Rev A Nat Sci Eng. 2016;18(2):128–31.
104. Patel MJ, Tripathy S, Mukhopadhyay KD, Wangjam T, Cabang AB, Morris J, Wargovich MJ. A supercritical CO_2 extract of neem leaf (*A. indica*) and its bioactive liminoid, nimbolide, suppresses colon cancer in preclinical models by modulating pro-inflammatory pathways. Mol Carcinog. 2018;57(9):1156–65.
105. Sharma SK, Ilavazhagan G, Kumar D, Selvamurthy W. Immunomodulatory effects of NIM-76, a volatile fraction from neem oil. J Ethnopharmacol. 1997;55(2):133–9.
106. Kapoor D, Vijayvergiya R, Dhawan V. *Terminalia arjuna* in coronary artery disease: ethnopharmacology, pre-clinical, clinical & safety evaluation. J Ethnopharmacol. 2014;155(2):1029–45.
107. Maulik SK, Talwar KK. Therapeutic potential of *Terminalia arjuna* in cardiovascular disorders. Am J Cardiovasc Drugs. 2012;12:157–63.
108. Gaherwal S, Anare R, Nageshwar W, Prakash MM. Immunomodulatory efficacy of *Terminalia arjuna* against *Aspiculuris tetraptera* in mice. Int J Microbiol Res. 2014;5(1):23–9.
109. Atanasov AG, Waltenberger B, Pferschy-Wenzig EM, Linder T, Wawrosch C, Uhrin P, Temml V, Wang L, Schwaiger S, Heiss EH, Rollinger JM. Discovery and resupply of pharmacologically active plant-derived natural products: A review. Biotechnol Adv. 2015;33(8):1582–614.

110. Low Z, Lani R, Tiong V, Poh C, AbuBakar S, Hassandarvish P. COVID-19 therapeutic potential of natural products. Int J Mole Sci. 2023;24(11):9589.
111. Fredrickson BL. The role of positive emotions in positive psychology: the broaden-and-build theory of positive emotions. Am Psychol. 2001;56(3):218.
112. Wilson DR, Warise L. Cytokines and their role in depression. Perspect Psychiatr Care. 2008;44(4):285–9.
113. Rubio-Perez JM, Morillas-Ruiz JM. A review: inflammatory process in Alzheimer's disease, role of cytokines. Sci World J. 2012;2012
114. Mansur RB, Zugman A, Asevedo ED, da Cunha GR, Bressan RA, Brietzke E. Cytokines in schizophrenia: possible role of anti-inflammatory medications in clinical and preclinical stages. Psychiatry Clin Neurosci. 2012;66(4):247–60.
115. Arrieta MC, Finlay BB. The commensal microbiota drives immune homeostasis. Front Immunol. 2012;3:33.
116. Li M, Wang B, Zhang M, Rantalainen M, Wang S, Zhou H, Zhang Y, Shen J, Pang X, Zhang M, Wei H. Symbiotic gut microbes modulate human metabolic phenotypes. Proc Natl Acad Sci. 2008;105(6):2117–22.
117. Sun S, Wang Y, Wu A, Ding Z, Liu X. Influence factors of the pharmacokinetics of herbal resourced compounds in clinical practice. Evid Based Complement Alternat Med. 2019;2019
118. Vlasova AN, Kandasamy S, Chattha KS, Rajashekara G, Saif LJ. Comparison of probiotic lactobacilli and bifidobacteria effects, immune responses and rotavirus vaccines and infection in different host species. Vet Immunol Immunopathol. 2016;172:72–84.
119. Ruiz L, Delgado S, Ruas-Madiedo P, Sánchez B, Margolles A. Bifidobacteria and their molecular communication with the immune system. Front Microbiol. 2017;8:2345.
120. Peterson CT, Sharma V, Uchitel S, Denniston K, Chopra D, Mills PJ, Peterson SN. Prebiotic potential of herbal medicines used in digestive health and disease. J Altern Complement Med. 2018;24(7):656–65.
121. Koboziev I, Karlsson F, Grisham MB. Gut-associated lymphoid tissue, T cell trafficking, and chronic intestinal inflammation. Ann N Y Acad Sci. 2010;1207:86–93.
122. Classen B, Thude S, Blaschek W, Wack M, Bodinet C. Immunomodulatory effects of arabinogalactan-proteins from Baptisia and Echinacea. Phytomedicine. 2006;13(9–10):688–94.
123. Batbayar S, Lee DH, Kim HW. Immunomodulation of fungal β-glucan in host defense signaling by dectin-1. Biomol Ther. 2012;20(5):433.
124. Li ZX, Zhao GD, Xiong W, Linghu KG, Ma QS, Cheang WS, Yu H, Wang Y. Immunomodulatory effects of a new whole ingredients extract from Astragalus: a combined evaluation on chemistry and pharmacology. Chin Med. 2019;14:1–10.
125. Du X, Zhao B, Li J, Cao X, Diao M, Feng H, Chen X, Chen Z, Zeng X. Astragalus polysaccharides enhance immune responses of HBV DNA vaccination via promoting the dendritic cell maturation and suppressing Treg frequency in mice. Int Immunopharmacol. 2012;14(4):463–70.
126. Phetkate P, Kummalue T, Kietinun S. Significant increase in cytotoxic T lymphocytes and natural killer cells by triphala: A clinical phase I study. Evid Based Complement Alternat Med 2012;2012.
127. Tyler VE. Phytomedicines: back to the future. J Nat Prod. 1999;62(11):1589–92.
128. Rathore D, Srivastava R, Geetanjali SR. Phytochemical screening, antioxidant and antimicrobial activities of Bombax ceiba flower. Alger J Nat Prod. 2019;7(1):651–6.
129. Rathore D, Srivastava R, Geetanjali SR. Phytochemistry and pharmacology of genus Bombax. Nat Prod J. 2019;9(3):184–96.
130. Singh R, Geetanjali. *Asparagus racemosus*: a review on its phytochemical and therapeutic potential. Nat Prod Res. 2016;30(17):1896–908.
131. Gasmi A, Mujawdiya PK, Lysiuk R, Shanaida M, Peana M, Gasmi Benahmed A, Beley N, Kovalska N, Bjorklund G. Quercetin in the prevention and treatment of coronavirus infections: a focus on SARS-CoV-2. Pharmaceuticals. 2022;15:1049.
132. Javadi F, Ahmadzadeh A, Eghtesadi S, Aryaeian N, Zabihiyeganeh M, Rahimi Foroushani A, Jazayeri S. The effect of quercetin on inflammatory factors and clinical symptoms in women with rheumatoid arthritis: a double-blind, randomized controlled trial. J Am Coll Nutr. 2017;36:9–15.
133. Akramiene D, Kondrotas A, Didziapetriene J, Kevelaitis E. Effects of beta-glucans on the immune system. Medicina. 2007;43:597–606.
134. Farnoosh G, Akbariqomi M, Badri T, Bagheri M, Izadi M, Saeedi-Boroujeni A, Rezaei E, Ghaleh H, Aghamollaei H, Fasihi-Ramandi M, et al. Efficacy of a low dose of melatonin as an adjunctive therapy in hospitalized patients with COVID-19: a randomized, double-blind clinical trial. Arch Med Res. 2020;53:79–85.

Medicinal Plants, Antioxidant Potential, and Applications to Aging

Aabir Pramanik, Sonali Maheshwari, and Niyati Acharya

Abstract

Aging is a progressive process in the body where wrinkling, fine lines of the skin, graying and thinning of hair, brittle and pale nails, and fragility in bones are obvious due to the breakdown of collagens, elastin, reduced production of melanin, and hormonal changes. DNA damage, telomerase shortening, alteration in specific genes, increase in the reactive oxygen species level, and mitochondrial dysfunction are the leading causes of aging. These internal factors are aggravated by some external factors like chronic sun exposure specifically UVA and UVB, smoking, alcohol consumption, radiation, pollution, exposure to metal ions and toxic chemicals, and nutrient deficiencies. To correct these signs of aging, antioxidants are widely used to treat and prevent the early signs of aging. Many medicinal plants, which have been used for thousands of years, are found in Rasayana category, a category of the Indian traditional health care system (Ayurveda) advocated for its intriguing antioxidant effects. Some of the medicinal plants utilized in Ayurveda as Rasayana for their therapeutic activity have been well-researched for its antiaging properties. This chapter focuses on some important plant material and isolated compounds that are reported to have ability to slow down the rate of aging and used worldwide. Majority of them exhibited anti-aging effects through their antioxidants potential.

Keywords

Aging · Antioxidant · Anti-aging · Medicinal plants · Phytoconstituents

A. Pramanik · S. Maheshwari
Institute of Pharmacy, Nirma University, Ahmedabad, India

N. Acharya (✉)
Department of Pharmacognosy, Institute of Pharmacy, Nirma University, Ahmedabad, Gujarat, India
e-mail: niyati.acharya@nirmauni.ac.in

1 Introduction

Aging is a biological process that gradually influences the normal processes in the cells and the tissues. Since the body is unable to generate new body cells during this phase, there is no elimination of dead or defective cells from multiple parts of the body [1]. Internal and environmental variables are thought to play significant roles in the aging process. Internal factor involves normal biological processes of the cell. Environmental variables include ultraviolet (UV) irradiation, prolonged sun exposure, nutritional deficiencies, hormone imbalance, pollution, and other factors like smoking, fast food, and even lifestyle [2, 3]. Wrinkling is related to skin aging due to loose elasticity, whereas graying and hair thinning are associated with hair aging due to reduced melanin production. These can be avoided or slowed by adopting preventative steps like taking antioxidant-rich supplements, eating a balanced diet, and adopting good skincare habits [4]. These changes lead to restrained free radicals which are the main cause of aging [5].

Aging refers to an inevitable yet natural and progressive process characterized by specific physiological changes occurring in living organisms with key features, such as senescence, telomere shortening, and altered gene expression at the cellular and molecular level which altogether leads to increased susceptibility to age-related disorders [6]. Chronological aging and photoaging are the two types of aging. Chronological aging is a natural and inevitable process, characterized by internal factors such as a reduction in the contents of collagen and elastin and reduced epidermal hydration leading to fine lines, thinning of skin and hairs, and reduction in elasticity. On the other hand, premature aging of the skin due to environmental aggressors, such as solar and ultraviolet radiations leading to leathery texture, wrinkles, and sunspots are clubbed under photoaging [7]. Recently, an interdisciplinary field of research called Geroscience has been introduced for gaining insights into the fundamental processes of aging to identify potential interventions and strategies to promote healthy aging without chronic disability [8].

A. K. Dhara, S. C. Mandal (eds.), *Role of Herbal Medicines*, https://doi.org/10.1007/978-981-99-7703-1_20

1.1 Manifestations

Wrinkling, pigmentation heterogeneity, circulatory disorders, and ptosis are specific indications of aging. Wrinkling along with yellowing, coarseness, atrophy, and others is caused due to loss or reduction in the levels of collagen, elastin, and hyaluronic acid [9]. College and elastin are responsible for providing elasticity and strength to the epidermal and dermal layers of skin, while hyaluronic acid is essential for cutaneous hydration [10]. Commonly observed in females, fragility in nail plates is caused due to abnormalities in the maturation and dedifferentiation of keratinocytes in the matrix of the nails [11]. Another indication of aging is graying of hairs caused due to impaired melanin pigmentation. A study to indicate the elevation of hair graying due to smoking in young people also demonstrated the damage to melanocytes due to the generation of reactive oxidative species (ROS) because of smoking [12, 13].

Changes at the cellular and molecular levels such as elevation in oxidative stress, DNA mutation, reduction in mitochondrial biogenesis, and decrease in mitophagy have all been linked to aging and aging-related disorders such as osteoarthritis, osteoporosis, Parkinson's disease, chronic obstructive pulmonary disease (COPD), atherosclerosis, and others [14]. Imbalances in the hypothalamic-pituitary-gonadal axis are responsible for menopause in aged females due to elevation in estradiol and reduction in progesterone [15].

1.2 Mechanism of Action

Although aging theories and concepts have not been fully elucidated, various mechanisms such as DNA damage, protein aggregation, and misfolding, electricity, Ca2+ signaling, mammalian target of rapamycin, the roles of non-coding RNAs, oxidative stress, apoptosis, alteration in mitochondrial permeability, etc. have been reported as major underlying mechanisms. Understanding these mechanisms could provide valuable insights for the development of new targets aimed at designing anti-aging drugs [16].

Various factors such as solar radiation, ultraviolet radiation, smoking, mitochondrial, metabolism, chemical toxins release several free radicals such as hydroxyl radical •OH, superoxide radical •O2-, nitric oxide radical •NO, peroxide radical ROO•, singlet oxygen 1O2, and others produce oxidative stress, causing DNA damage leading to misfolding and degradation of proteins which synergistically cause aging.

Damage to the DNA, impaired repairing of DNA, activation of an oncogene, certain epigenetic alterations, oxidative stress, and shortening of telomeres are the key internal factors governing the progressive process of aging, Fig. 1 [17]. Conversely, pollution, smoking, alcohol consumption, certain chemical compounds such as toxins, inflammatory markers, ultraviolet radiation, and others are the group of external factors that produce cumulative detrimental changes in the morphology and physiology of skin cells leading to aging. Being the largest organ of the human body, from working as a shield to protect against pathogens and pollutants to regulating the body temperature, multiple approaches have been reported in the journals to prevent and treat senescence [18].

The initial theories of the by-products of metabolism and free radicals generated by mitochondria attacking the cell and its constituents suggest the production of highly reactive and unstable oxygen that oxidizes to form reactive oxygen species (ROS). The free radical theory of aging, first proposed by Denham Harman, helps in understanding the mechanism of action [8]. The generated ROS species have the potential to damage the cells by lipid peroxidation, protein modifications, and damage to the DNA. Elevated levels of ROS cause a reduction in the concentration of non-enzymatic proteins and hamper the antioxidant of the defense system, leading to an imbalance in homeostasis. ROS are known to attack the nitrogenous bases and sugar-phosphate backbone of DNA, leading to premature aging [19, 20].

Telomere shortening or replicative senescence during replication is an important process that governs senescence that leads to cellular senescence and mitochondrial dysfunction. Telomere shortening is also connected to mitochondrial malfunctioning via the peroxisome proliferator-activated receptor gamma coactivator 1a/b which regulates the biogenesis and role of mitochondria. Activation of p53 and DNA damage response pathways due to the shortening of telomeres suppresses the peroxisome proliferator-activated receptor gamma coactivator 1a/b and activation of p53 (a tumor protein) causing aging at the molecular level [21, 22].

The exposure and accumulation of ultraviolet radiation have been linked to the loss of structural integrity of the skin's dermal connective tissue leading to the development of signs of skin aging, such as wrinkles, fine lines, and sagging. A well-known treatment for immunosuppression, exposure to solar and UV radiation at different wavelengths, goes in the process of aging skin cells. UV radiation, particularly UVB (constituting 5% of total UV radiation), can activate immunosuppressive cells like Tregs, myeloid-derived suppressor cells (MDSCs), and regulatory dendritic cells (DCreg). This activation has the potential to harm DNA and proteins by causing the formation of pyrimidine dimers and initiating a sequence of oxidative processes, ultimately resulting in oxidative stress [23–25].

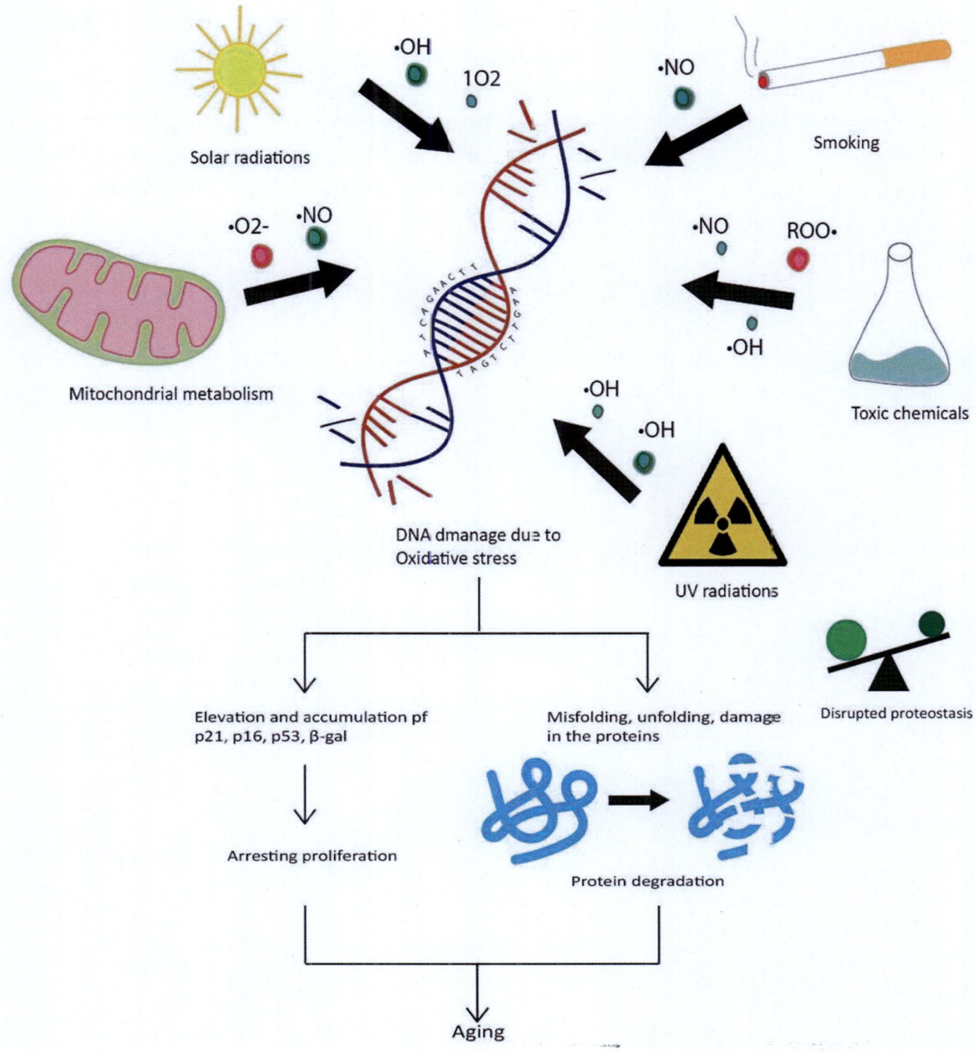

Fig. 1 Mechanism of various factors inducing oxidative stress leading to aging

1.3 Conventional Anti-aging Methods

To reduce the signs of aging and promote overall skin health, several conventional anti-aging modalities are available in the market. They include topical skincare products with retinoids, hyaluronic acid, alpha hydroxy acids, and idebenone which not only provide moisture to the epidermal and dermal layers but also promote the production of collagen. Hormone replacement therapy, laser-in-light therapy, dermal fillers, microdermabrasion, botulinum toxin injections, and chemical peels are available in the market for dermatological rejuvenation. Various pharmacological treatments have also been explored for anti-aging potential such as senolytic drugs, telomerase activators, and epigenetic drugs [25].

1.4 Potential of Medicinal Plants as Anti-aging Agents

While the mentioned treatment options are expensive and raise concerns about safety, researchers from all around the globe are trying to exploit medicinal plants as anti-aging agents. The concept of using medicinal plants as anti-aging agents goes back to the ancient Ayurveda literature categorized as "*Vayasthapana*," meaning to arrest aging or maintain youthfulness, portraying the potential of the herb *Centella asiatica* having anti-aging properties [26]. Various studies on Withanolide, the steroidal lactones extracted from the roots of Ashwagandha, showed a 20% extension of the lifestyle by acting as agents that protect telomeres and repair DNA [27].

Different parts of four South African Plants *C. glabrum, S. brachypetala, P. africanum, and P. capensis* were evaluated for anti-aging potential in an in vitro study based on their activity against elastases, hyaluronidase, collagenase. The results revealed their ability to scavenge free radical and enzyme inhibitory, thereby helping in restoration of skin elasticity, and delaying the wrinkling [10]. *Lycium barbarum*, a medicinal plant of Chinese origin, comprising of betaine, cerebroside, β-sitosterol, flavonoids, riboflavin, and other phytoconstituents, has been reported to decrease the oxidative stress while simultaneously boosting the efficacy of antioxidant enzymes. Furthermore, it inhibits cell apoptosis and reduces DNA damage, thereby acting as an anti-aging plant [28]. Certain Malaysian plants *Cnestis palala, Urceola micrantha, Marantodespumilum,* and *Microporus xanthopus* have reported to have anti-collagenase activities, highest anti-tyrosinase, acetyl- and butyryl-cholinesterase activity, all of which indicating anti-aging potential [29] (Fig. 2).

Resveratrol, a phytoalexin stilbenoid, is a well-known fighter constituent from grapevine. It is a potent antioxidant that acts as a free radical scavenger, thereby inhibiting the DNA damage and protein degradation and protecting epidermal and dermal cells from oxidative stress. Simultaneously, it also activates SIRT1 which aids in regulating the cellular processes which collectively support cellular health and longevity.

Considering the current literature, this chapter shall focus on the mechanism and regulation of aging along with the effect of several medicinal plants acting as antioxidants to be further developed as anti-aging agents and restrict senescence. The purpose of this chapter is to indicate the enormous potential of medicinal plants and their phytoconstituents across the globe that can be built for dermatological characterization and upliftment. However, it should be noted that dietary antioxidants play a crucial role in preventing oxidative damage in the body, as the endogenous antioxidant defense mechanism alone may not be fully efficient.

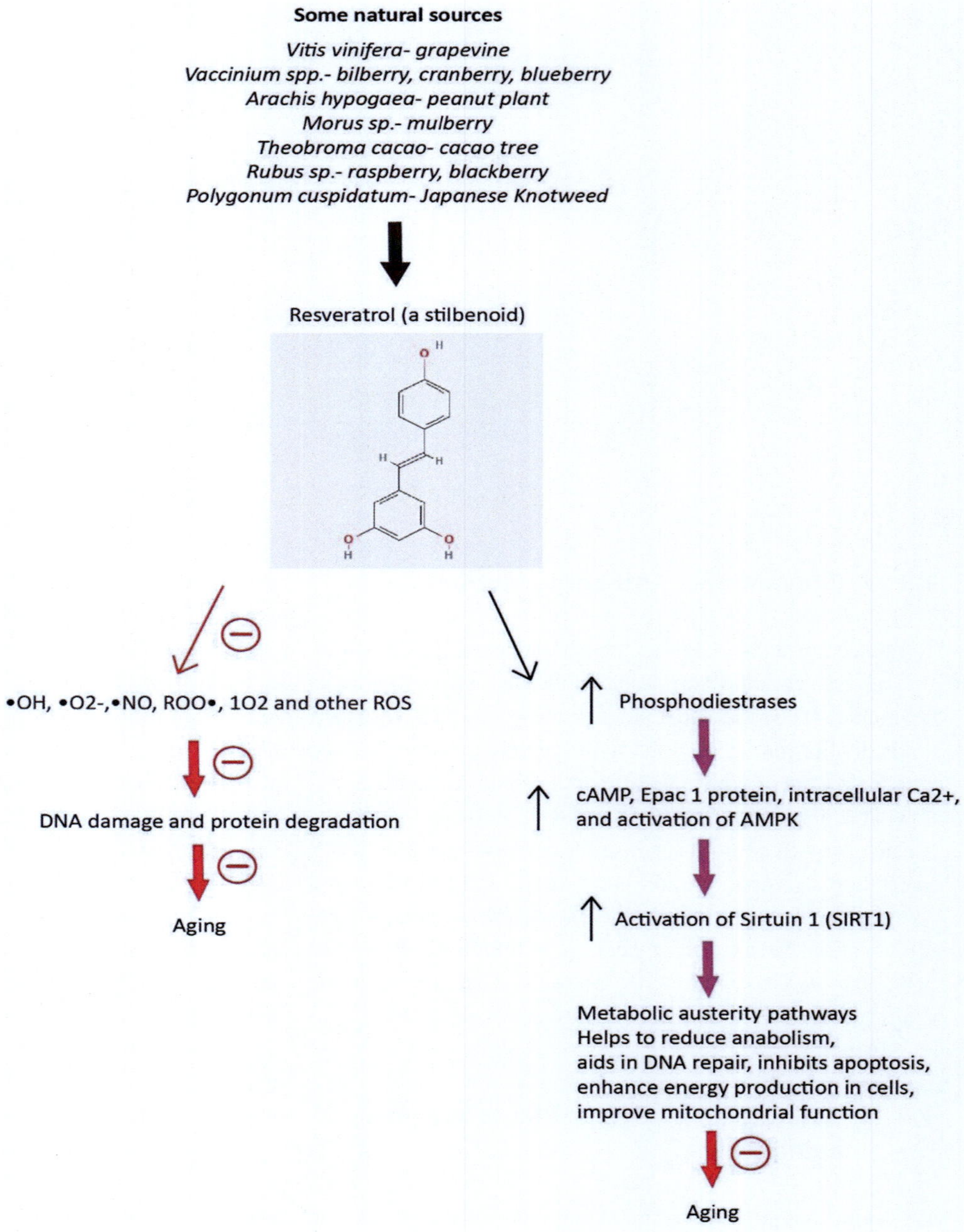

Fig. 2 Mechanism of resveratrol acting as a well-known anti-aging agent

2 Plant Sources with Anti-aging Potential

Plants are the source of many medicines to treat almost all kinds of diseases. Plants and their secondary metabolites have both enzymatic and non-enzymatic antioxidant defense mechanisms by which they can prevent the toxic effects of reactive oxygen species (ROS) like superoxide anion ($O_2^{\bullet-}$), hydrogen peroxide (H_2O_2), hydroperoxyl radical ($HO_2^{\bullet}$), hydroxyl radical (OH), hydroperoxyl radical ($^{\bullet}OH$), peroxynitrite (ONOO-), lipid peroxyl radical ($LOO^{\bullet}$), and nitric oxide ($NO^{\bullet}$). These free radicals attack DNA, protein, and lipids. Various plant metabolites like polyphenols, vitamins, carotenoids, and flavonoids are reported as potent antioxidants [30]. Some plants that show anti-aging properties have been described in Table 1 with the probable mechanism of action.

Table 1 Plant sources having an anti-aging effect

Plant	Plant part(s) used	Active phytoconstituent(s)	Model	Dose	MOA	References
Bacopa monnieri (Brahmi)	Whole plant	Bacosides, Brahmine, and Herpestine	In vitro (human fibroblast cells)	[a]NP	In a dose-dependent way, it boosts the activity of ROS-scavenging enzymes such as catalase (CAT), glutathione peroxidase (GPX), and superoxide dismutase (SOD)	[31–34]
Curcuma longa (turmeric)	Rhizome	Curcumin	In vivo (rat)	200 mg and 400 mg of Curcumin/kg body weight for 6 months	It assists in the lowering of ROS generation, the scavenging of free oxygen radicals, and the blocking of lipid peroxidation. In human fibroblasts, curcumin promotes a cellular stress response via redox signaling through the phosphatidylinositol 3-kinase/Akt (protein kinase B; PKB) pathway. It has the likelihood to avert cellular senescence. It also helps in C-reactive protein (CRP) reduction, an anti-aging inflammatory marker	[35–39]
Emblica officinalis L. (amla)	Fruit	Tannoids (ascorbic acid, gallic acid, elaeocarpusin)	In vitro (NB1RGB human skin fibroblasts)	0–40 g/mL for 48 h	It inhibits the activities of GABA and monoamine oxidase-A (MAO-A). It also inhibited type I collagen collagenase and increased TIMP-1 levels; it inhibited cellular proliferation and protected procollagen 1 from UVB-induced depletion by inhibiting UVB-induced MMP-1	[40–42]
Ginkgo biloba (gingko)	Leaves	Ascorbic acid, catechin, shikimic acid, lactone derivatives (ginkgolides), isorhamnetin, kaempferol 3-O-β-D-glucopyranoside, isorhamnetin-3-O-glucoside, myricetin, ginkgolide A, bilobalide	In vitro (human dermal fibroblasts)	0.1, 0.2 mg/mL for 24 h	It decreased the levels of liver metalloproteinase and malondialdehyde while enhancing SOD activity to prevent oxidative stress. It inhibits production of ROS and the breakdown of MMP-1 in human dermal fibroblasts	[43–45]
Glycyrrhiza glabra (licorice)	Rhizome, root	Glycyrrhizin, glabridin, and isoliquiritigenin	In vitro assay	[a]NP	It acts by chelating with metal ions and scavenging free radicals. It also has tyrosinase and elastase inhibitory activity	[46, 47]

(continued)

Table 1 (continued)

Plant	Plant part(s) used	Active phytoconstituent(s)	Model	Dose	MOA	References
Panax ginseng (Asian ginseng)	Root	Ginsenoside	In vitro and in vivo (human volunteer)	0.05% PGLE for 8 weeks	It can reduce lipid peroxidation. It is linked to the activation of the foxO3a gene, often known as the longevity gene. It stimulates transformation of growth factor-β (TGF-β) in human skin fibroblast cells that promotes collagen production	[48–51]
Citrus sinensis L. (sweet orange)	Fruit	Anthocyanins, flavanones, hydroxycinnamic acid, and ascorbic acid	In vitro human keratinocytes (HaCaT cell line)	15 and 30 μg/mL for 7 h	It works by translocating NF-B and AP-1 and cleaving procaspase-3	[52]
Litchi chinensis (litchi)*; Nephelium lappaceum L.* (rambutan)*; Tamarindus indica* (tamarind)	Fruit	Ferulic acid, gallic acid, epigallocatechin	In vitro (human skin fibroblasts)	0.05, 0.01, and 0.007 mg/mL for 72 h	It inhibits melanin synthesis in B16F10 melanoma cells by inhibiting tyrosinase and TRP-2; it is also efficient at inhibiting elastase and collagenase	[53]
Citrus reticulata Blanco (mandarin orange)	Fruit	D-Limonene, n-hexadecanoic acid	In vitro assay	[a]NP	It inhibits collagenase and elastase and has anti-enzymatic action	[54]
Salacca zalacca (Gaert.) Voss (snake fruit)	Fruit	Chlorogenic acid	*In silico*	[a]NP	It acts by MMP-1 inhibition	[55]
Citrus sunki Hort. ex Tanaka, Citrus unshiu Marcov, Citrus sinensis Osbeck, Citrus reticulata Blanco, and Vitis vinifera L. (mandarin, grapes)	Fruit	Narirutin, hesperidin, ascorbic acid	*In vitro* and in vivo (cell culture and mice)	33, 100, 300 mg/kg for 10 weeks	Increased antioxidant enzyme expression levels; skin thickness and wrinkle development are reduced, while collagen levels are increased in a hairless mouse model exposed to UVB	[56]
Daucus carota L. (carrot)	Root	Carrot glycoprotein	In vitro (cell culture)	0.3, 0.5, 1 mg/mL	It neutralizes ROS and takes part in cell membrane protection	[57]
Carthamus tinctorius (safflower)	Seed oil	Phenol	In vitro assay	[a]NP	It has the capability to inhibit collagenase assay and elastase assay	[58]
Chaenomeles sinensis (Chinese quince)	Fruit	β-1,4-xyloglucan	In vitro assay	[a]NP	It has the capability of inhibiting the activity of dermal extracellular matrix proteases: elastase and collagenase	[59]
Prunus dulcis (almonds)	Seed	α-tocopherol	Observational study (observational study)	340 kcal/day of almonds (58.9 g) for 16 weeks	It works on wrinkle, decreases its severity in postmenopausal females	[60]
Panax ginseng Meyer and Crataegus pinnatifida (Korean ginseng, mountain hawthorn)	Root, fruit	Ginsenoside	In vitro and observational study (human dermal fibroblasts, healthy human skin)	100 μg/mL for 12 weeks	It prevents photoaging of the skin caused by UVB-exposure by regulating procollagen type 1 and MMP-1 expression in NHDFs	[61]

Table 1 (continued)

Plant	Plant part(s) used	Active phytoconstituent(s)	Model	Dose	MOA	References
Eleutherococcus senticosus (Siberian ginseng)	Root	Phlorizin	In vitro (human keratinocytes)	[a]NP	It works via miR135b suppression that leads to improvement of the microenvironment and increases the proliferative potential of basal epidermal cells	[62]
Citrus limon (lemon)	Fruit	Eriocitrin	In vivo (mice)	4 mL and 6 mL/day	It increases in aging-related scores (e.g., periophthalmic lesions) and delay in locomotor atrophy	[63]
Sclerocarya birrea (marula)	Fruit, oil	Quinic acid, catechin, epigallocatechin gallate, and epicatechin gallate	In vitro assay	100, 200 μg/mL	It blocks the activity of collagenase	[64]
Zizania aquatica (black rice)	Grain	Cyanidin -3-O-glucoside	In vivo (mice)	15, 30, and 60 mg/kg	It increases superoxide dismutase (SOD) and catalase (CAT) while decreases MDA and the activity of monoamine oxidase (MAO)	[65]
Camellia sinensis L. (black and green tea)	Leaves	Epigallocatechin gallate, epigallocatechin-3-gallate	In vitro assay, in vivo (Caenorhabditis elegans)	50–300 μM for 6 days	Black tea inhibits elastase activity, green tea helps in extension of life span through mitohormesis	[66, 67]
Musa sapientum (banana)	Fruit	Corosolic acid	In vitro assay	[a]NP	It has inhibitory effects on MMPs activities	[68]
Oryza sativa (Rice)	Grain	Vanillin and coumaric acid	In vitro assay	[a]NP	It has elastase inhibitory activity	[69]
Withania somnifera (Ashwagandha)	Root	Withanolide	In vitro assay (HeLa cell lines), in vivo	10–50 μg	It resulted in an increase in telomerase activity. It also shown anti-genotoxic properties in human peripheral blood cells against H_2O_2-induced DNA damage	[70, 71]
Tinospora cordifolia (Guduchi)	Stem	NS	In vitro assay (PC12 cell lines)	10 ng, 100 ng, or 1 μg	It protects against cytotoxicity and DNA damage caused by radiation	[72]
Ocimum basilicum L. (Tulsi)	Essential oil	NS	In vitro assay (K562 cells)	100 μL	Essential oil raises the apparent telomeres length and downregulates the telomeric repeat binding factor 1 (TERF 1)	[73]
Centella asiatica L. (Mandukaparni)	Whole extract	NS	In vivo (human PBMCs)	0.02, 0.2, 2 μg/ml	The extract has the capability to increase telomerase activity almost nine-fold	[74]
Aloe barbadensis miller (aloe vera)	Gel	Acemannan	In vivo (human volunteer)	1200 mg, 3600 mg	Gel increases collagen production by decreasing collagen-degrading MMP-1 gene expression	[75]
Rosa gallica (French rose)	Petal	NS	In vitro assay (B16F10 melanoma cells)	[a]NP	It decreases MMP-1 gene expression	[76]
Hibiscus rosa sinensis, Hibiscus roseus	Leaves	Ferulic acid, *p*-coumaric acid derivatives	In vivo (mouse skin)	[a]NP	It inhibits collagenase and elastase. It also blocks UV radiation in the skin	[77]

[a]*NP* Not precise, *NS* Not specified

3 Phytoconstituents Having Potent Anti-aging Property

Plants are supreme producers of secondary metabolites like alkaloids, glycosides, tannins, resin, volatile oil, phenolic and polyphenolic compounds, and flavonoid compounds which are widely available in fruits, flowers, seeds, vegetables, grains, and cereals. Out of these secondary metabolites phenolic, polyphenolic, and flavonoid compounds possess very potent antioxidant anti-aging properties. Due to the hunger for an extended lifespan, people are increasingly drawn to natural antioxidants capable of promoting the generation of new cells, preventing DNA damage, addressing mitochondrial dysfunction, inhibiting ROS production, and safeguarding against UV radiation without causing any harmful effects. Some active phytoconstituents that have been reported to possess very distinct activities to prevent aging are listed in Table 2 (Fig. 3).

Table 2 Phytochemicals with anti-aging effects

Phytoconstituent	Model	Dose	MOA	References
Ascorbic acid (vitamin C)	In vitro (HS68 fibroblast cell line)	10–40 μg/mL	It promotes procollagen content and inhibits matrix metalloproteinase levels in skin fibroblasts. It also blocks hyaluronidase, elastase, and MMP-1	[78, 79]
α-Tocopherol (vitamin E)	In vitro	[a]NP	It blocks the lipid peroxidation chain reaction	[80]
Anthocyanins	In vivo (fruit flies)	30 mg/dL	It downregulates Mth gene and upregulates CAT, Rpn11, SOD1, and SOD2 gene expression	[81]
Lycopene	In vitro (A549 cell line)	75 mg/day. 1 nM and 10 nM	It reduces MMP-1 activity. It induces repair of base excision	[82, 83]
Resveratrol	In vivo (gray mouse lemur)	200 mg/kg	It stimulates the activity of nuclear factor erythroid-2 related factor 2 (Nrf2) and activates SOD, CAT to inhibit ROS. It activates anti-aging factor sirtuin1 (Sirt 1). It inhibits ROS by downregulating Akt/mTOR pathway	[84]

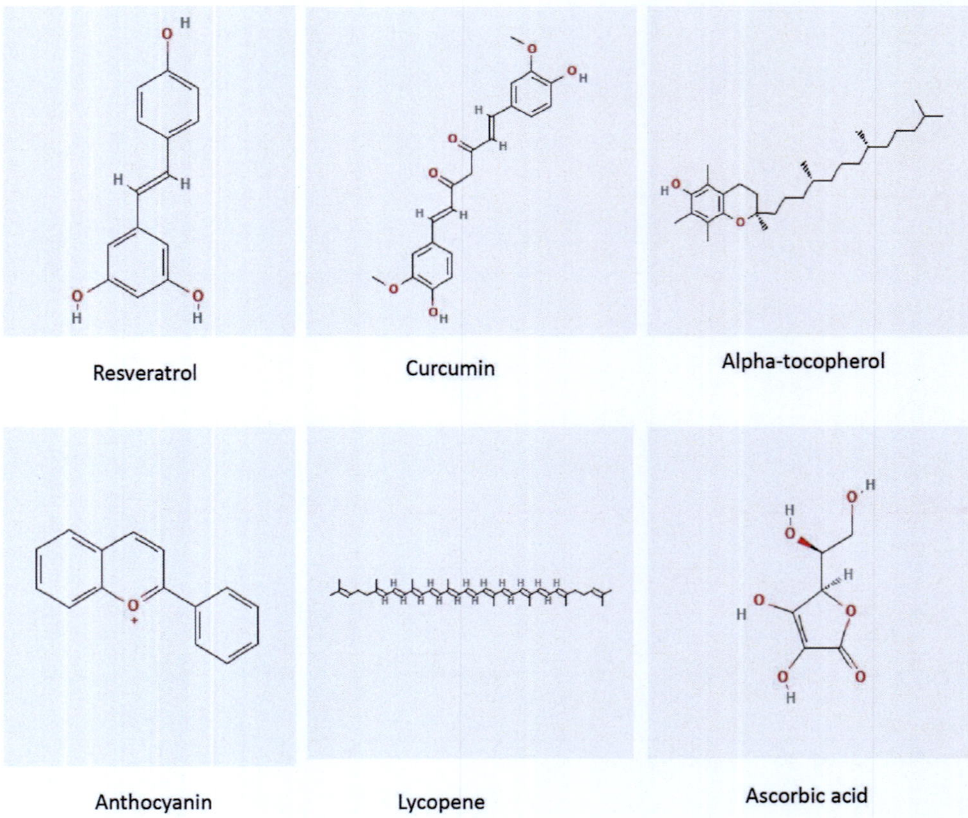

Fig. 3 Chemical structures of phytoconstituents used as antioxidant and anti-aging agents

4 Conclusion

The purpose of our chapter is to indicate the enormous potential of medicinal plants and their phytoconstituents that can be built for dermatological characterization and upliftment. While various plants and their phytoconstituents have been studied exhaustively for their antioxidant potential to further develop as anti-aging agents, additional investigations are required to develop optimum extraction methods, to elucidate the molecular mechanisms of action of medicinal plants having antioxidant potential. A holistic approach to healthy aging, including proper nutrition, sun protection, regular exercise, and other lifestyle factors, remains essential for maintaining overall health and well-being as we age. While the above-mentioned list of phytoconstituents such as polyphenols, flavonoids, carotenoids, and tocopherols, among others, has shown promising free radical scavenging properties in various studies, it should be noted that omega-3 fatty acids such as eicosapentaenoic acid (EPA) and docosahexaenoic acid (DHA) also have the antioxidant potential but have not been exhaustively characterized yet. Retinol, hyaluronic acid, niacinamide, and ceramides are some of the versatile synthetic compounds that are well incorporated to address skin concerns such as signs of aging, dryness, uneven skin tone, and sensitivity. There are numerous unknown medicinal plants with ROS-scavenging potential that can be further expanded as anti-aging agents with comprehensive research. The major classics of Indian traditional medicine such as Charak Samhita and Sushruta Samhita mentioning over 700 herbs have elucidated the role of over 200 herbs aiding in skin wellness depicting promising prospects as anti-aging agents. A detailed insight into those unexplored sources may open a new path to the development of novel anti-aging therapeutics.

References

1. Muravchick S. Theories of aging. Geriatr Anesthesiol. 2008;(July):29–37.
2. Zhang S, Duan E. Fighting against skin aging: the way from bench to bedside. Cell Transplant. 2018;27(5):729–38.
3. Bocheva G, Slominski RM, Slominski AT. Neuroendocrine aspects of skin aging. Int J Mol Sci. 2019;20(11):1–19.
4. Schagen SK, Zampeli VA, Makrantonaki E, Zouboulis CC. Discovering the link between nutrition and skin aging. Dermatoendocrinol. 2012;4(3):298–307.
5. Lobo V, Patil A, Phatak A, Chandra N. Free radicals, antioxidants and functional foods: impact on human health. Pharmacogn Rev. 2010;4(8):118–26.
6. Corrêa RCG, Peralta RM, Haminiuk CWI, Maciel GM, Bracht A, Ferreira ICFR. New phytochemicals as potential human anti-aging compounds: reality, promise, and challenges. Crit Rev Food Sci Nutr. 2018;58(6):942–57.
7. Mukherjee PK, Maity N, Nema NK, Sarkar BK. Bioactive compounds from natural resources against skin aging. Phytomedicine [Internet]. 2011;19(1):64–73. Available from: https://doi.org/10.1016/j.phymed.2011.10.003.
8. Vaiserman AM, Lushchak OV, Koliada AK. Anti-aging pharmacology: promises and pitfalls. Ageing Res Rev [Internet]. 2016;31:9–35. Available from: https://doi.org/10.1016/j.arr.2016.08.004.
9. Flament F, Bazin R, Laquieze S, Rubert V, Simonpietri E, Piot B. Effect of the sun on visible clinical signs of aging in Caucasian skin. Clin Cosmet Investig Dermatol. 2013;6:221–32.
10. Ndlovu G, Fouche G, Tselanyane M, Cordier W, Steenkamp V. In vitro determination of the anti-aging potential of four southern African medicinal plants. BMC Complement Altern Med. 2013;13
11. Chessa MA, Iorizzo M, Richert B, López-Estebaranz JL, Rigopoulos D, Tosti A, et al. Pathogenesis, clinical signs and treatment recommendations in brittle nails: a review. Dermatol Ther (Heidelb). 2020;10(1):15–27.
12. Zayed A, Shahait A, Ayoub M, Yousef AM. Smokers' hair: does smoking cause premature hair graying? Indian Dermatol Online J. 2013;4(2):90.
13. Nishimura EK, Granter SR, Fisher DE. Mechanisms of hair graying: incomplete melanocyte stem cell maintenance in the niche. Science. 2005;307(5710):720–4.
14. Guo J, et al. Aging and aging-related diseases: from molecular mechanisms to interventions and treatments. Signal Transduct Target Ther. 2022;7:391. https://doi.org/10.1038/s41392-022-01251-0.
15. Wu JM, Zelinski MB, Ingram DK, Ottinger MA. Ovarian aging and menopause: current theories, hypotheses, and research models. Exp Biol Med (Maywood). 2005;230:818–28.
16. Maldonado E, Morales-Pison S, Urbina F, Solari A. Aging hallmarks and the role of oxidative stress. Antioxidants. 2023;12(3):1–37.
17. Schumacher B, Pothof J, Vijg J, Hoeijmakers JHJ. The central role of DNA damage in the ageing process. Nature [Internet]. 2021;592(7856):695–703. Available from: https://doi.org/10.1038/s41586-021-03307-7.
18. Costa EF, Magalhães WV, Di Stasi LC. Recent advances in herbal-derived products with skin anti-aging properties and cosmetic applications. Molecules. 2022;27(21):7518. https://doi.org/10.3390/molecules27217518.
19. Tsang Y-L, Kao C-L, Lin S-CA, Li C-J. Mitochondrial dysfunction and oxidative stress in aging and disease. Biomedicines. 2022;10:2872. https://doi.org/10.3390/biomedicines10112872.
20. Hajam YA, et al. Oxidative stress in human pathology and aging: molecular mechanisms and perspectives. Cells. 2022;11(3):552. https://doi.org/10.3390/cells11030552.
21. Shchelkunov SN, Totmenin AV, Safronov PF, Mikheev MV, Gutorov VV, Ryazankina OI, et al. Analysis of the monkeypox virus genome. Virology. 2002;297(2):172–94
22. Dabrowska A, Venero JL, RyotaIwasawa M-kH, Rahman S, Boobis A, et al. PGC-1α controls mitochondrial biogenesis and dynamics in lead-induced neurotoxicity. Aging (Albany NY). 2015;7(9):629–43.
23. Kligman LH. Photoaging. Manifestations, prevention, and treatment. Dermatol Clin. 1986;4(3):517–28.
24. Salminen A, Kaarniranta K, Kauppinen A. Photoaging: UV radiation-induced inflammation and immunosuppression accelerate the aging process in the skin. Inflamm Res [Internet]. 2022;71(7–8):817–831. Available from: https://doi.org/10.1007/s00011-022-01598-8.
25. Salminen A, Suuronen T, Huuskonen J, Kaarniranta K. NEMO shuttle: a link between DNA damage and NF-κB activation in progeroid syndromes? Biochem Biophys Res Commun. 2008;367(4):715–8.
26. Bowler PJ. Impact on facial rejuvenation with dermatological preparations. Clin Interv Aging. 2009;4:81–90.
27. Datta HS, Paramesh R. Trends in aging and skin care: ayurvedic concepts. J Ayurveda Integr Med. 2010;1(2):110–3.

28. Gao Y, Wei Y, Wang Y, Gao F, Chen Z. Lycium barbarum: a traditional Chinese herb and a promising anti-aging agent. Aging Dis. 2017;8(6):778–91.
29. Ranneh Y, Abu Bakar MF, Ismail NA, Kormin F, Mohamed M, Md Akim A, et al. Anti-aging and antioxidant of four traditional Malaysian plants using simplex centroid mixture design approach. Saudi J Biol Sci [Internet]. 2021;28(12):6711–6720. Available from: https://doi.org/10.1016/j.sjbs.2021.07.048.
30. Kasote DM, Katyare SS, Hegde MV, Bae H. Significance of antioxidant potential of plants and its relevance to therapeutic applications. Int J Biol Sci. 2015;11:982–91.
31. Aguiar S, Borowski T. Neuropharmacological review of the nootropic herb Bacopa monnieri. Rejuvenation Res. 2013;16(4):313–26.
32. Jain P, Das D. Pharmacognostic comparison of Bacopa monnieri, Cyperus rotundus, and Emblica officinalis. Int J Pharm Pharm Sci. 2016;4:16–26.
33. Sinensis C. Synergistic antioxidant activity of three medicinal plants Hypericum perforatum, Bacopa monnieri, and Camellia Sinensis. Indo Am J Pharm Res. 2014;4(January):2563–8.
34. Bhattacharya SK, Bhattacharya A, Kumar A, Ghosal S. Antioxidant activity of Bacopa monniera in rat frontal cortex, striatum and hippocampus. Phytother Res. 2000;14(3):174–9.
35. Kocaadam B, Şanlier N. Curcumin, an active component of turmeric (Curcuma longa), and its effects on health. Crit Rev Food Sci Nutr. 2017;57:8398.
36. Engwa GA. Free radicals and the role of plant phytochemicals as antioxidants against oxidative stress-related diseases. In: Phytochemicals - source of antioxidants and role in disease prevention. InTech; 2018. https://doi.org/10.5772/intechopen.76719.
37. Lima CF, Pereira-wilson C, Rattan SIS. Curcumin induces heme oxygenase-1 in normal human skin fibroblasts through redox signaling: relevance for anti-aging intervention. Mol Nutr Food Res. 2011;55:430–42.
38. Hamidie RDR, Yamada T, Ishizawa R, Saito Y, Masuda K. Curcumin treatment enhances the effect of exercise on mitochondrial biogenesis in skeletal muscle by increasing cAMP levels. Metabolism [Internet]. 2015; Available from: https://doi.org/10.1016/j.metabol.2015.07.010.
39. Shailaja M, Damodara Gowda KM, Vishakh K, Suchetha Kumari N. Anti-aging role of curcumin by modulating the inflammatory markers in albino Wistar rats. J Natl Med Assoc [Internet]. 2017;109(1):9–13. Available from: https://doi.org/10.1016/j.jnma.2017.01.005.
40. Bhattacharya A, Ghosal S. Antioxidant activity of tannoid principles of Emblica officinalis (arnla) in chronic stress induced changes in rat brain. Indian J Exp Biol. 2000;38(September):877–80.
41. Dhingra D, Joshi P, Gupta A, Chhillar R. Possible involvement of monoaminergic neurotransmission in antidepressant-like activity of Emblica officinalis fruits in mice. CNS Neurosci Ther:1–7.
42. Fujii T, Wakaizumi M, Ikami T, Saito M. Amla (Emblica officinalis Gaertn.) extract promotes procollagen production and inhibits matrix metalloproteinase-1 in human skin fibroblasts. J Ethnopharmacol. 2008;119(1):53–7.
43. Van Beek TA. Chemical analysis of Ginkgo biloba leaves and extracts. J Chromatogr A. 2002;967(1):21–55.
44. Huang SZ, Luo YJ, Wang L, Cai KY. Effect of ginkgo biloba extract on livers in aged rats. World J Gastroenterol. 2005;11(1):132–5.
45. Wang L, Cui J, Jin B, Zhao J, Xu H, Lu Z, et al. Multifeature analyses of vascular cambial cells reveal longevity mechanisms in old Ginkgo biloba trees. Proc Natl Acad Sci U S A. 2020;117(4):2201–10.
46. Grodzicki W, Dziendzikowska K. The role of selected bioactive compounds in the prevention of Alzheimer's disease. [El papel de los compuestosbioactivosseleccionadosen la prevención de la enfermedad de Alzheimer]. Antioxidants. 2020;9(3):1–18.
47. Ciganović P, et al. Glycerolic licorice extracts as active cosmeceutical ingredients: extraction optimization, chemical characterization, and biological activity. Antioxidants (Basel, Switzerland). 2019;8(10):445. https://doi.org/10.3390/antiox8100445.
48. Yu H, Zhao J, You J, Li J, Ma H, Chen X. Factors influencing cultivated ginseng (Panax ginseng C. A. Meyer) bioactive compounds. PLoS One. 2019;14(10):1–17.
49. Lee YM, Yoon H, Park HM, Song BC, Yeum KJ. Implications of red Panax ginseng in oxidative stress associated chronic diseases. J Ginseng Res [Internet]. 2017;41(2):113–119. Available from: https://doi.org/10.1016/j.jgr.2016.03.003.
50. Kim J, Cho SY, Kim SH, Cho D, Kim S, Park CW, et al. Effects of Korean ginseng berry on skin antipigmentation and antiaging via FoxO3a activation. J Ginseng Res [Internet]. 2017;41(3):277–283. Available from: https://doi.org/10.1016/j.jgr.2016.05.005.
51. Shin S, Lee JA, Son D, Park D, Jung E. Anti-skin-aging activity of a standardized extract from panax ginseng leaves in vitro and in human volunteer. Cosmetics. 2017;4(2)
52. Cimino F, Cristani M, Saija A, Paolo F. Protective effects of a red orange extract on UVB-induced damage in human keratinocytes. Biofactors. 2007;30:129–38.
53. Lourith N, Kanlayavattanakul M, Chaikul P, Chansriniyom C, Bunwatcharaphansakun P. In vitro and cellular activities of the selected fruits residues for skin aging treatment. An Acad Bras Cienc. 2017;89(1 suppl):577–89. https://doi.org/10.1590/0001-3765201720160849.
54. Apraj VD, Pandita NS. Evaluation of skin anti-aging potential of Citrus reticulata Blanco peel. Pharmacognosy Res. 2016;8:160–8.
55. FAO. The State of Food and Agriculture 2019. Moving forward on food loss and waste reduction. Rome. Licence: CC BY-NC-SA 3.0 IGO; 2019.
56. Kim D, Shin G, Kim J, Kim Y, Lee J, Seok J, et al. Antioxidant and anti-ageing activities of citrus-based juice mixture. FOOD Chem [Internet]. 2016;194:920–927. Available from: https://doi.org/10.1016/j.foodchem.2015.08.094.
57. Lee M, Jeong N, Jang B. Antioxidative activity and antiaging effect of carrot glycoprotein. J Ind Eng Chem [Internet]. 2014; Available from: https://doi.org/10.1016/j.jiec.2014.10.037.
58. Zemour K, Labdelli A, Adda A, Dellal A, Talou T, Merah O. Phenol content and antioxidant and antiaging activity of safflower seed oil (Carthamus Tinctorius L.). Cosmetics. 2019;6:55.
59. Itoh S, Yamaguchi M, Shigeyama K, Sakaguchi I. The anti-aging potential of extracts from Chaenomeles sinensis. Cosmetics. 2019;6(1):21. https://doi.org/10.3390/cosmetics6010021.
60. Foolad N, Vaughn AR, Chodur GM, Newman JW, Steinberg FM, Sivamani RK. Prospective randomized controlled pilot study on the effects of almond consumption on skin lipids and wrinkles. Phytother Res. 2019;(August):3212–7.
61. Hwang E, Park S, Yin CS, Kim H, Kim YM, Yi TH. Anti-aging effects of the mixture of Panax ginseng and Crataegus pinnatifida in human dermal fibroblasts and healthy human skin. J Ginseng Res [Internet]. 2016; Available from: https://doi.org/10.1016/j.jgr.2016.01.001.
62. Choi H, Nam K, Lee H, Yang S, Kim Y, Lee J, et al. Phlorizin, an active ingredient of Eleutherococcus senticosus, increases proliferative potential of keratinocytes with inhibition of MiR135b and increased expression of type IV collagen. Oxid Med Cell Longev. 2016;2016:3859721.
63. Shimizu C, Wakita Y, Inoue T, Hiramitsu M, Okada M, Mitani Y, Segawa S, Tsuchiya Y, Nabeshima T. Effects of lifelong intake of lemon polyphenols on aging and intestinal microbiome in the senescence-accelerated mouse prone 1 (SAMP1). Sci Rep. 2019;9(1):3671. https://doi.org/10.1038/s41598-019-40253-x. PMID: 30842523; PMCID: PMC6403313
64. Shoko T, Maharaj VJ, Naidoo D, Tselanyane M, Nthambeleni R. Anti-aging potential of extracts from Sclerocarya birrea

(A. Rich.) Hochst and its chemical profiling by UPLC-Q-TOF-MS. BMC Complement Altern Med. 2018:1–14.
65. Wei TT, Li MJ, Guo L, Xie YD, Chen WH, Sun Y, Liu GH, Ding Y, Chai YR. Resveratrol ameliorates thymus senescence changes in D-galactose induced mice. Microbiol Immunol. 2020;64(9):620–9. https://doi.org/10.1111/1348-0421.12833. Epub 2020 Aug 27. PMID: 32691886
66. Xiong L, Chen Y, Tong J, Gong Y, Huang J. Epigallocatechin-3-gallate promotes healthy lifespan through mitohormesis during early-to-mid adulthood in Caenorhabditis elegans. Redox Biol. 2018;14(September 2017):305–15.
67. Ratnasooriya WD, Abeysekera WKSM, Bs T. In vitro antiglycation and cross-link breaking activities of Sri Lankan low-grown orthodox orange pekoe grade black tea (Camellia sinensis L). Trop J Pharm Res. 2014;13(April):567–71.
68. Pereira A, Maraschin M. Author' s accepted manuscript. J Ethnopharmacol [Internet]. 2014; Available from: https://doi.org/10.1016/j.jep.2014.11.008.
69. Widowati W, Herdiman H, Afni M, Afifah E, Kusuma H, Nufus H, Arumwardana S, Rihibiha D. Antioxidant and anti aging assays of Oryza sativa extracts, Vanillin and Coumaric acid. J Nat Remedies. 2016;16:88. https://doi.org/10.18311/jnr/2016/7220.
70. Raguraman V, Subramaniam JR. Withania somnifera root extract enhances telomerase activity in the human HeLa cell line. Adv Biosci Biotechnol. 2016;7(April):199–204.
71. Kumar N, Yadav A, Gupta R, Aggarwal N. Antigenotoxic effect of Withania somnifera (Ashwagandha) extract against DNA damage induced by hydrogen peroxide in cultured human peripheral blood lymphocytes. Int J Curr Microbiol Appl Sci. 2016;5(4):713–9.
72. Taylor P, Masuma R, Okuno T, Shahabuddin M, Choudhuri K, Saito T. Effect of Tinospora cordifolia on the reduction of ultraviolet radiation-induced cytotoxicity and DNA damage in PC12 cells. J Environ Sci Health B. 2014;(December):37–41.
73. Plant J. Effects of essential oils on telomere length in human cells. Med Aromat Plants. 2016;5:230. 1-6. https://doi.org/10.4172/2167-0412.1000230.
74. Tsoukalas D, Fragkiadaki P, Docea AO, Alegakis AK, Sarandi E, Thanasoula M, et al. Discovery of potent telomerase activators: Unfolding new therapeutic and anti-aging perspectives. Mol Med Rep. 2019;20:3701–8.
75. Cho S, Ph D, Lee S, Lee M, Lee DH. Dietary aloe vera supplementation improves facial wrinkles and elasticity and it increases the type I procollagen gene expression in human skin in vivo. Ann Dermatol. 2009;21(1):6–11.
76. Ju E, Han SA, Song MLY, Lim T, Lee M, Lee S. Extraction conditions for Rosa gallica petal extracts with anti-skin aging activities. Food Sci Biotechnol [Internet]. 2019;28(5):1439–1446. Available from: https://doi.org/10.1007/s10068-019-00596-7.
77. Dos Santos Nascimento LB, Gori A, Raffaelli A, Ferrini F, Brunetti C. Phenolic compounds from leaves and flowers of Hibiscus roseus: potential skin cosmetic applications of an under-investigated species. Plants (Basel). 2021;10(3):522. https://doi.org/10.3390/plants10030522. PMID: 33802222; PMCID: PMC8000889
78. Adil MD, Kaiser P, Satti NK, Zargar AM, Vishwakarma RA, Tasduq SA. Effect of Emblica officinalis (fruit) against UVB-induced photo-aging in human skin fibroblasts. J Ethnopharmacol [Internet]. 2010;132(1):109–114. Available from: https://doi.org/10.1016/j.jep.2010.07.047.
79. Nema NK, Maity N, Sarkar B, Mukherjee PK. Cucumis sativus fruit-potential antioxidant, anti-hyaluronidase, and anti-elastase agent. Arch Dermatol Res. 2011;303(4):247–52. https://doi.org/10.1007/s00403-010-1103-y. Epub 2010 Dec 14. PMID: 21153830
80. Traber MG, Bruno RS. Chapter 7 - Vitamin E [Internet]. In: Present knowledge in nutrition, vol. 1. Elsevier Inc.; 2020. p. 115–136. Available from: https://doi.org/10.1016/B978-0-323-66162-1.00007-X.
81. Manuscript A. Food & Function. 2011;(207890).
82. Przybylska S. (2019). Lycopene – a bioactive carotenoid offering multiple health benefits: a review. Int J Food Sci Technol. 2019;55. https://doi.org/10.1111/ijfs.14260.
83. Cheng J, Miller B, Balbuena E, Eroglu A. Lycopene protects against smoking-induced lung cancer by Inducing Base excision repair. Antioxidants (Basel). 2020:1–14.
84. Zhou D, Luo M, Huang S, Saimaiti A, Shang A, Gan R, et al. Review article: Effects and mechanisms of resveratrol on aging and age-related diseases. Oxid Med Cell Longev. 2021;2021

Role of Natural Polysaccharides in the Management of Lifestyle Diseases

Shuvam Sar, Tanmoy Banerjee, Ankit Kumar, Arun Baidya, Saptapadi Saha, Jayashree Mondal, Rituparna Chaki, Amit Kumar Halder, and Nilanjan Ghosh

Abstract

Metabolic syndrome (MetS) affects 20–25% of adults and as many as 19.5–20% of adolescents. MetS negatively impacts multiple physiological processes. Natural compounds have an exceptionally bright aspect with a promising alternative to synthetic medications. Plant polysaccharides (PPSs), which are made up of at least ten monosaccharides, exhibit excellent safety and a range of pharmacological actions, such as anti-apoptotic, antihyperglycemic, antihyperlipidemic, antioxidant, and antifibrotic benefits. However, the benefits of numerous PPS for metabolic disorders, including gastroesophageal reflux disease, cardiovascular disease, chronic renal disease, and pulmonary diseases, are often brought on by a malfunction in the metabolism of lipids, glucose, and peptides. Therefore, the quantity of articles focusing on antioxidative PPSs have steadily increased over the past decade. This chapter outlines numerous attributes and cellular mechanisms of several PPS in metabolic disorders. In addition, the interaction of PPS with various genes and proteins involved in molecular signaling pathways is briefly elucidated.

Keywords

Metabolic disease · Plant polysaccharides · Inflammatory mediators · Molecular signaling · mRNA expressions

S. Sar · T. Banerjee · A. Kumar · A. Baidya · S. Saha · J. Mondal · N. Ghosh (✉)
Department of Pharmaceutical Technology, Jadavpur University, Kolkata, West Bengal, India
e-mail: nilanjanghosh.phamacy@jadavpuruniversity.in

R. Chaki · A. K. Halder
Dr. B.C. Roy College of Pharmacy and Allied Health Sciences, Durgapur, India

1 Introduction

Lifestyle diseases occur due to the unhealthy habits of people which detract them from the regular activities and lead them to a sedentary routine causing various health problems that can be even to some extent lethal. In a broader sense, non-communicable or non-infectious diseases are generally caused due to malfunctioning of metabolism regulator enzymes, eventual accumulation of primary and secondary products of metabolism, and deficiency of desired molecules for biochemical reactions. These diseases are known as metabolic disorders/diseases (MDs) or non-communicable diseases (NCDs). So, it can be said that all non-communicable diseases in a sense are metabolic disorders. This chapter focuses on metabolic disorders arising from lifestyle changes in modern days. There might be various forms of metabolic diseases, but the primary causing factor lies between the junction of hyperglycemia and hyperlipidemia. Disturbances in the homeostasis of blood glucose level and free fatty acid (FFAs) level rise as the initiation point of all cardiovascular, hepatic, and nephrological diseases. Diabetes works longer and in a deeper way than hyperlipidemia. Prolonged insulinemia and hyperglycemia gradually accumulate fat globules in adipose tissues of liver, gastrointestinal viscera, and other parts of body. Concomitantly prolonged deposition of lipid in the abdominal area becomes the factory of glucose through neoglucogenesis and other pathways. Chronic kidney/renal disease usually occurs as a consequence of diabetic toxicity leading to death. Similarly, cardiovascular and hepatic metabolic ailment may serve lethal for patients. MDs kill about 40 million people each year, which is nearly 70% of all death around the globe. Thus, meeting this crisis through correct treatment approach is of current interest to researchers [1].

The formation and progression of MDs, such as diabetes, obesity, cardiovascular diseases (CVDs), and non-alcoholic steatohepatitis, are caused by biomolecules such as glucose, protein, fat, vitamins, and minerals in the body that are not metabolized effectively. These MDs are the result of a com-

A. K. Dhara, S. C. Mandal (eds.), *Role of Herbal Medicines*, https://doi.org/10.1007/978-981-99-7703-1_21

plicated interaction between diet, susceptibility, genetics, and epigenetics. MDs are caused, to a massive extent, by social and behavioral risk factors, like smoking, unhealthy diet, less physical activity, reckless alcoholism, and unsurpassable stress in work–life balance. Obesity or rising blood lipids, elevated blood glucose, and raised high pressure are the root causes of numerous deadly problems in the cardiovascular, respiratory, and renal systems. Many of the risk factors are common for a number of MDs. According to the WHO, smoking causes the deaths of more than seven million people each year, and that figure is expected to rise. Excessive consumption of dietary sodium kills 4.1 million people annually, while alcohol consumption leads to approximately 1.65 million deaths from NCDs. Simple lack of exercise kills 1.6 million people a year [2].

2 General Pathophysiology of Metabolic Diseases

MDs have a continuously progressive pathophysiological state comprising several interdependent phenomena like inflammation, ectopic (especially liver and muscle) and visceral obesity, insulin resistance. The inflammatory and prethrombotic markers, C-reactive protein (CRP), synthesized and secreted by hepatocytes, insulin levels, plasminogen activator inhibitors, leptin, adiponectin, resistin, tumor necrosis factor-α (TNF-α), leptin, monocyte chemoattractant protein-1 (MCP-1), interleukin-6 (IL-6), uric acid (UA) levels, and microalbuminuria are considered to better understand the molecular scenario of MDs [3]. Obesity, abdominal adiposity, hypertension, and hyperglycemia are common signs of various metabolic diseases. IkB kinase and c-Jun terminal kinase (JNK), two pro-inflammatory serine kinases that are activated by FFA buildup, help to release IL-6 from adipose tissue. Additionally, elevated FFA encourages hepatic protein kinase activation, which results in gluconeogenesis and lipogenesis. These lead to components of MDs such as IR, lipotoxicity, altered glucose metabolism, and atherosclerosis [4]. The evolution of CVDs such as atherosclerosis, hypertension, heart failure, and peripheral artery disease can be sped up by IR because it promotes the generation of reactive oxygen species (ROS) and causes an imbalance in the enzyme endothelial nitric oxide synthase (eNOS) [5].

3 Plant Polysaccharides and Its Composition

The most prevalent naturally occurring macromolecular polymer may be found in polysaccharides, which are produced by bacteria, fungus, plants, and even algae [6]. A variety of monosaccharides are combined to form polysaccharide structures, which are joined together by glycosidic linkages. Numerous research suggest that plant polysaccharides (PPSs) exhibit a variety of biological activities including immunomodulation, antineoplasia, anti-inflammation, antiviral, antioxidative, osteogenesis, hypolipidemic and hypoglycemic effects. Despite PPSs' potency, there are still significant drawbacks, including poor targeting, unstable, quick blood clearance rate, and unfocused scope of action, which substantially impair bioavailability and clinical use.

Recent studies have found that PPSs molecular weights can be up to thousands or even millions. Hydrophilic groups such as OH, COOH, and NH2 groups can form bioadhesive layers with epithelial and mucosal tissues [7]. The most common component of polysaccharides is d-glucose. However, d-fructose, d-galactose, l-galactose, d-mannose, l-arabinose, rhamnose, fucose, uronic acid, and d-xylose also can be found as depicted in Fig. 1 [8]. Amino sugars are one type of monosaccharide derivative found in polysaccharides. Some plants contain their derivatives (N-acetylneuraminic acid and N-acetylmuramic acid) as well as the monosaccharide acids (glucuronic acid and iduronic acid) [9].

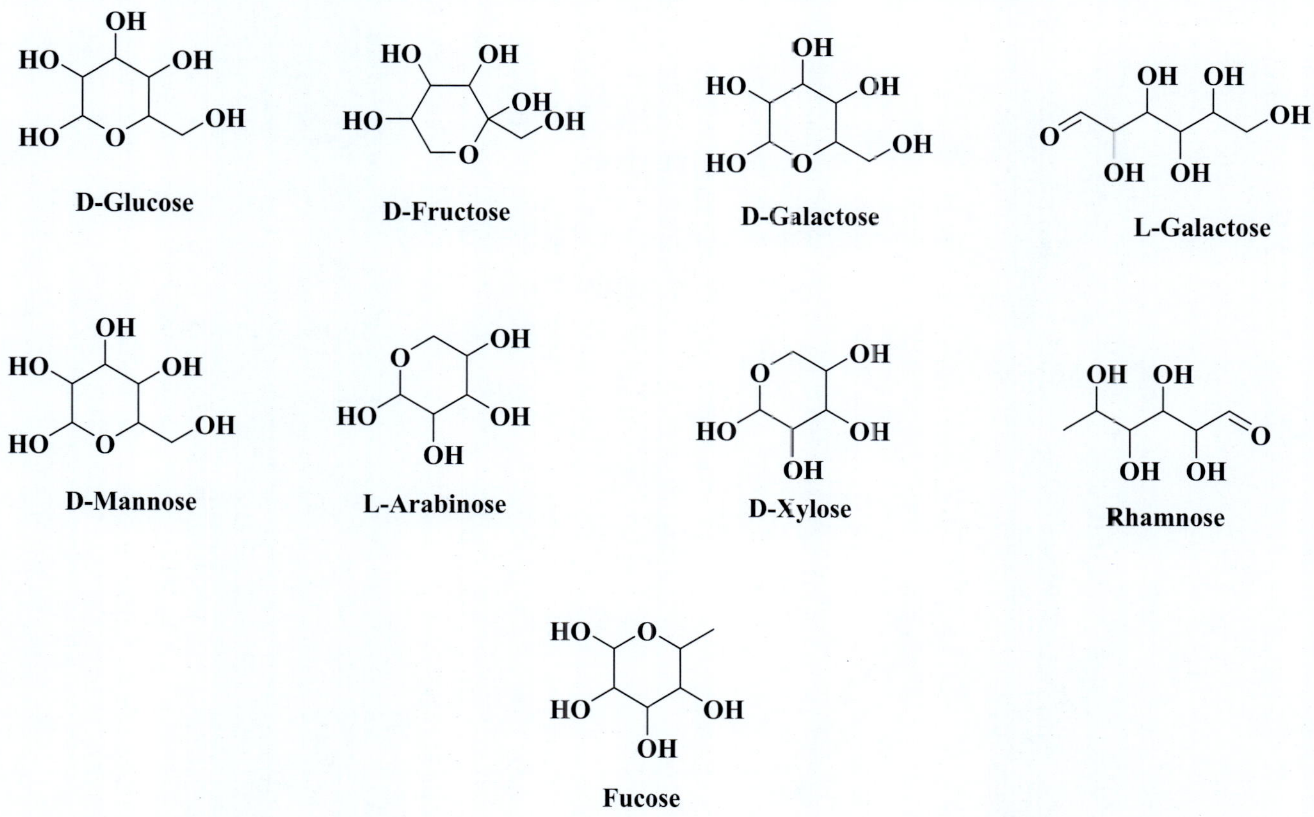

Fig. 1 Structures of components of polysaccharides

4 Management of Lifestyle Disorders Using Natural Polysaccharides

4.1 Hyperglycemia

Hyperglycemia is the name of the condition or symptom caused by diabetes mellitus, which is characterized by insufficient insulin secretion, IR, or reduced sensitivity of the target organs to insulin when it is secreted normally, followed by consequences of MDs. β cells of the pancreas secrete insulin, but due to the lack of sensitivity of target organ cells, insulin cannot bind to the corresponding receptors upon the cells, and as a result, the secreted insulin stays in the blood for long, causing hyperinsulinemia. However, when the body cannot maintain normal blood sugar level through the feedback mechanism in insulin-glucagon concentration, they develop type 2 diabetes mellitus (T2DM). Hyperglycemia is the main among MDs that in turn gives rise to several other organ system disorders, like chronic renal failure (CRF), cardiac diseases, hyperlipidemia, nervous system diseases, etc. That is why hyperglycemia is known as the silent killer. One of the main stimulators of T2DM is obesity alone, particularly abdominal adiposity. Consumption of high-fat-containing foods, fast foods made with highly saturated oils, lack of exercise, and an overall sedentary lifestyle are the underlying causes of abdominal accumulation of fats. In case of abdominal fat, lipid is deposited within adipocyte cells of adipose tissue, in the abdominal area which works as source of calorie supply in fasting [10]. But when adipocyte cells accumulate lipids more than their normal capacity, due to high-fat diet intake and sedentary lifestyle, they lead to hepatic and systemic insulin resistance through interlukin-6 (IL-6) and other adipose-specific cytokines [11]. This is very clear that no other metabolic diseases are so directly linked to lifestyle than hyperglycemia. Not only fatty foods, fast food consumption, and irregularity in having breakfast, lunch, and dinner also lead to the development of hyperglycemia. That is why modern views portray hyperglycemia and hyperlipidemia as linked to each other, as their further complications also progress simultaneously (summarized in Fig. 2).

Polysaccharides acting against hyperglycemia with their sources and probable mechanisms of action are hereby summarized in Table 1.

4.2 Hyperlipidemia

Hyperlipidemia is the second-most prevalent metabolic disorder that has been occurring from the dawn of modern days,

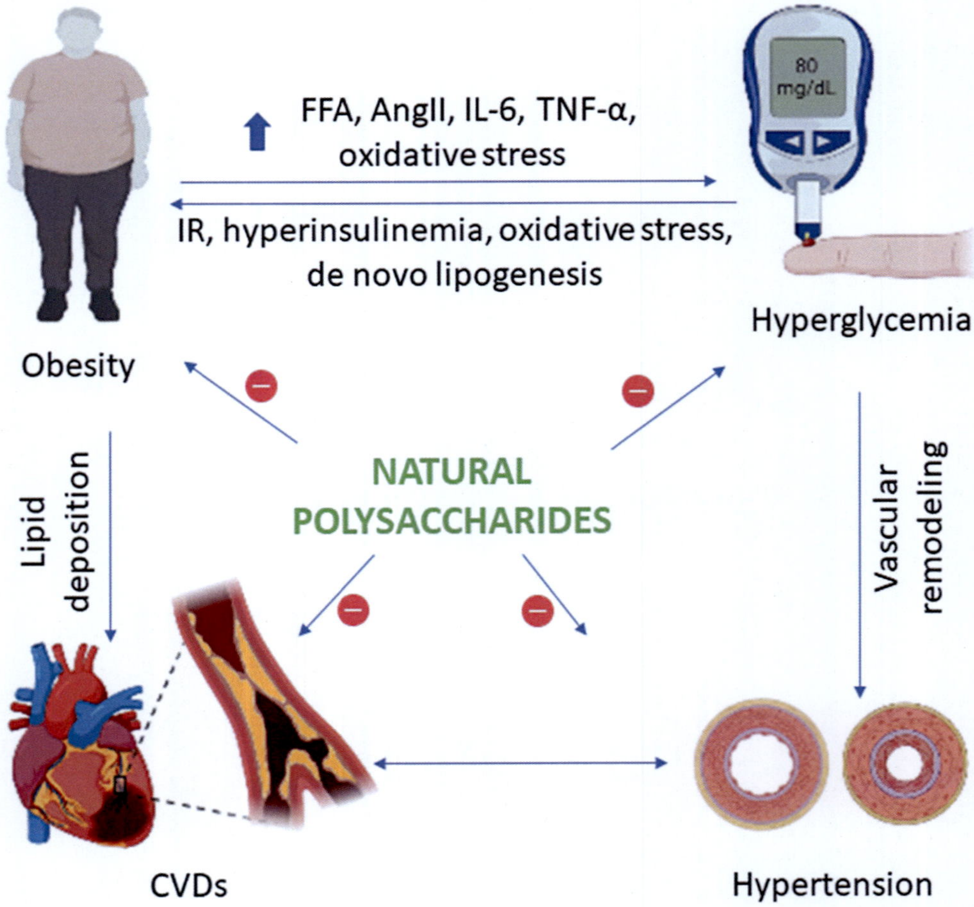

Fig. 2 Interlink between hyperglycemia, obesity, hypertension, and CVDs and action of natural polysaccharides on them

mostly due to the high intake of fatty foods and consequently the gradual rise in FFAs, triglycerides (TGs), cholesterol, and low-density lipoprotein (LDL-C) levels and decreasing high-density lipoprotein (HDL-C) cholesterol levels. This occurs due to unhealthy lifestyles including high-fat diet and other lifestyle behavioral factors. Evidence indicates that smoking can disrupt lipid metabolism. Surprisingly, eating betel nuts increases the risk of hyperlipidemia as well as increasing the likelihood of obesity because it increases hunger by blocking GABA receptors. Cardiovascular disorders including atherosclerosis and coronary heart disease are caused by persistent hyperlipidemia [27]. With the constant changes in lifestyle, environment, and food habits, more and more people suffer from this and its consequences. Therefore, downregulating serum lipid levels is the ultimate way to prevent the risk of hyperlipidemia. Additionally, aberrant adipokine synthesis, including that of adiponectin and leptin, is linked to hyperlipidemia [28]. Cholesterol, after being absorbed from the intestine, is transported through the chylomicron to the liver. The cholesterol absorption route is crucial for treating hyperlipidemia because of the intimate connection between lipid metabolism and cholesterol absorption. Thus, lowering the intestinal absorption of cholesterol is another way to avoid hyperlipidemia [29].

4.2.1 Polysaccharides Against Hyperlipidemia

4.2.1.1 Cyclocarya Polysaccharide

The "sweet tea tree" is another name for the species *Cyclocarya paliurus* (Batal) Iljinskaja. By slowing fatty acid production and degrading fatty acids, *Cyclocarya paliurus* polysaccharide (CPP) has been found to reduce FFAs in KKAy mice. In addition, there is evidence that CPP causes an increase in the expression of adipose triglyceride lipase (ATGL) and peroxisome proliferator-activated receptor (PPAR) and a relative decrease in FFAs and hepatic 3-hydroxy-3-methyl glutaryl-Coenzyme-A (HMG-CoA) mRNA. Downregulating the expression of HMG-CoA reductase prevents the liver's production of new cholesterol, hence lowering blood cholesterol levels [30].

4.2.1.2 Pleurotus Mycelia Polysaccharides

The polysaccharides are extracted from mycelia of *Pleurotus eryngii var. tuoliensis*, a medicinal mushroom that belongs to the family Pleurotaceae of the phylum Basidiomycota. It has

Table 1 Polysaccharides acting against hyperglycemia

Plant polysaccharide	Biological source	Research model	Action on parameters	Remarks	References
Codium fragile sulfated polysaccharide (CFSP)	*Codium fragile* alga	Alloxan-induced diabetic rats	Decrease in blood glucose and intestinal amylase levels Enhanced antioxidant enzymes activity	Increased glucose tolerance and lowered α-amylase activity Improved lipid profile	[12]
Alhagi polysaccharide (neutral Aps-1, acidic Aps-2 and 3)	*Alhagi pseudalhagi*	In vitro	Antioxidant effects in DPPH and ABTS assays α-Glucosidase inhibition and ABTS scavenging activity by APs-3	Hypoglycemic activity through free radical scavenging action	[13]
Polysaccharides from Adlay seeds (PAS)	*Coix lacryma jobi L., commonly known as adlay*	8 weeks old ICR male mice	Decreased blood glucose and insulin levels Reduced plasma levels of amyloid β42 and glycated hemoglobin (HbA1c) Increased expression of glucagon-like peptide-1 (GLP-1) and superoxide dismutase (SOD) activity	Not a very potent anti-diabetic Good antioxidant Significant hypolipidemic action	[14]
Acanthopanax polysaccharide (ASP)	*Acanthopanax senticosus*	Alloxan-induced diabetes mice	Lowered fasting blood glucose level and blood insulin level		[15]
Grifola polysaccharides	*Grifola frondosa*	KK-ay mice Hep G2 cells	Lowered blood glucose and reversed insulin resistance Enhanced glucose uptake	Hypoglycemic action by α-glucosidase inhibition and insulin sensitization Increased phosphorylated-AktSer473 Synthesis of intracellular glycogen through the Akt/GSK-3 pathway	[16]
Astragalus polysaccharide	*Dried roots of Astragalus membranaceus*	Fat-fed streptozotocin-treated T2DM rats	Decreased plasma glucose level	Normalized insulin-induced PKB-Ser473 phosphorylation and GLUT-4 translocation in skeletal muscle Reduced PTP1B protein level and activity in the muscle, but not in the liver along with NF-kB activation	[17]
Cordyceps polysaccharide	*Fruiting bodies of Cordyceps sinensis*	Streptozotocin and nicotinamide induced diabetic male Wistar rats	Improved response in OGTT and lowered post-prandial blood glucose level		[18]
Fructus polysaccharide	*Fructus corni*	Streptozotocin-induced diabetic rats RIN-m5f cells (in vitro)	Lowered blood glucose level and reversed insulin resistance	α-Amylase and α-glucosidase inhibition Increased insulin secretion and promoted pancreatic β-cell proliferation	[19]
Tea polysaccharide	*Camellia sinensis L.*	High-fat diet and streptozotocin-induced male Kunming mice	Hypoglycemic action	Hypoglycemic action through PI3Kp85/p-Akt/GLUT4 signaling pathway	[20]
The protein-bound polysaccharide from pumpkin fruits (PBPP)	*Pumpkin fruits*	Alloxan-induced Sprague-Dawley rats	Hypoglycemic action		[21]

(continued)

Table 1 (continued)

Plant polysaccharide	Biological source	Research model	Action on parameters	Remarks	References
Nigella seed polysaccharide	*Nigella sativa* seed	High-fat diet and streptozotocin-induced Kunming mice	Lowered fasting blood glucose (FBG), glycosylated serum protein (GSP), TG, TC, LDL-C, Malondialdehyde (MDA), TNF-α, IL-6, and IL-1β Insulinemia. Raised HDL-C, total antioxidant capacity (T-AOC), SOD, catalase (CAT), and p-AKT and GLUT4 expression	Hypoglycemic action was done by PI3Kp85/p-Akt/GLUT4 pathway	[22]
Polysaccharides from Hovenia (Guaizao)	*Hovenia dulcis*	Streptozotocin-induced diabetic male Sprague-Dawley rat	Lowered FBG. Higher body weight, and liver glycogen levels Insulinemia	Hypoglycemic mechanisms may include balancing the regeneration and apoptosis of β-cells and activating liver glycometabolism-linked Signaling pathways in diabetic rats	[23]
Sargassum polysaccharides	*Sargassum fusiforme*	High glucose and high-fat diets fed T2DM mice	Improved insulin resistance. Fasting blood glucose (FPG) and fasting insulin (FINS) were significantly decreased Lowered insulin resistance (IR) Expression of Glut-2 in pancreatic tissue is increased		[24]
Suillellus polysaccharides	*Suillellus luridus*	Streptozotocin-induced diabetic male ICR mice	Blood glucose level reduced to normal Decline in insulin level	Increased insulin sensitivity	[25]
AAMP-N	*Armillaria mellea fruiting body*	Db/db mice Hepa 1–6 cells	Significantly reduced blood glucose level Antagonized insulin resistance	Phosphorylation of insulin receptor and AKT	[26]

been proved that zinc-containing polysaccharides show antioxidant properties. Additionally, it has been demonstrated that natural alkali-extractable (AlMZPS) and acidic (AcMZPS) compounds both dramatically raise HDL-C levels and lower blood levels of LDL-C, VLDL-C, TC, and TG, therefore establishing their hyperlipidemic effects [31].

4.2.1.3 Ulvan

Ulvan polysaccharide (PU3) has been isolated from *Ulvan pertusa*, a green alga. The administration of PU3 has shown inhibition of lipid peroxidation in hyperlipidemic mice by increasing and activating antioxidant enzymes. PU3 lessen risk factors rising from interrelation between oxidative stress and hyperlipidemia [32].

4.2.1.4 Morchella Polysaccharides

Morchella esculenta polysaccharides (EnMPS and MPS fractions) from Morchella mushroom is a large complex carbohydrate. Following administration of both fractions, higher HDL-C and lower LDL-C values were seen, indicating a possible inhibition against hyperlipidemia-derived liver damage. These 2 fractions also have protective action against oxidative stress-induced hyperlipidemic complications. EnMPS has been found to enhance the antioxidant enzyme activities and reduced the myeloperoxidase (MPO) content, both in vivo and in vitro experiments [33].

4.2.1.5 TDGP-3 Polysaccharides

Traditional Chinese medicine uses a kind of Vitaceae plant called *Tetrastigma hemsleyanum Diels et Gilg* (TDGP3). A lipid metabolic pathway that is out of balance might result from too many free radicals and the subsequent oxidative stress, which can result in hyperlipidemia. In vivo, TDGP-3 has demonstrated strong antioxidant activity, lowering LDL-C, TC, and TG levels while raising HDL-C levels [34].

4.2.1.6 Pine Needle Polysaccharide (PNP)

Pine needle polysaccharide (PNP), extracted from the *Pinus massoniana* tree has shown hypolipidemic and antioxidant effects. High-fat diets can increase the production of free radicals and weaken the antioxidant defense system. Because oxidized-LDL-C, which is intimately linked to the development of hyperlipidemia, is produced when reactive oxygen species (ROS) oxidize LDL-C. PNP increases antioxidant enzymes and non-enzymatic antioxidants, reducing oxidative damage in mice subjected to HFD [35].

4.2.1.7 Selenium Modified Lachnum Polysaccharide

Lachnum is a saprophytic fungus with various biological actions. SeLEP-1b, an organically modified selenium containing Lachnum polysaccharide, effectively reduced lipids and protected the liver. SeLEP-1b had shown hypolipidemic action probably by reducing the expression of the genes for fatty acid synthase (FAS) and adipocyte fatty acid binding protein (aP2), which are involved in TG production and lipid storage in adipocytes. Another possibility is that SeLEP-1b, which downregulates the liver x receptor ligands known as the major regulators of lipid metabolism, increased lipid metabolism and broke down the fat and cholesterol in the liver. The HMG-CoA reductase enzyme in the liver may be inhibited to reduce cholesterol production, which would be the third possible cause [36].

4.2.1.8 Prunella Polysaccharides

The *Prunella vulgaris* L. (Labiatae family) plant is the source of the *Prunella vulgaris* polysaccharides, which are also referred to as "self-heal" in Western herbal therapy. PVPs markedly decrease plasma TC, TGs, and LDL-C levels and increased HDL-C in high-fat/high-cholesterol (HFHC) diet-fed mice. PVPs also lowered MDA and TNF-α levels in hyperlipidemic rats and stimulated glutathione peroxidase (GSH-Px), indicating its antioxidant property [37].

4.2.1.9 Apostichopus Polysaccharides

Apostichopus japonicus polysaccharide (AJP) is obtained from sea cucumber *Apostichopus japonicus Selenka* (also identified as *Stichopus japonicus Selenka*). HDL-C levels were raised, while TC, TG, and LDL-C levels significantly decreased, following AJP administration, which proves the biosynthesis and uptake of cholesterol. To modulate cholesterol metabolism, AJP interacts with lipids and functions as a carrier, improving the transit and excretion of serum lipids. AJP can also act as bile acid synthesis stimulators. Synthesis of bile acids depletes hepatic cholesterol [38].

4.2.1.10 Termitomyces Polysaccharides

Two fractions, EIPS and AIPS are hydrolysis products of intracellular polysaccharide from the fruiting bodies of *Termitomyces albuminosus*, also known as "Chicken Mushroom" or "Jizong" in China. Treatment EIPS lowered the serum lipids and restored lipid metabolic disturbances induced by high-fat emulsion [39].

4.2.1.11 Helvella Polysaccharides

The treatment of high doses of Helvella polysaccharides (HLP), obtained from *Helvella leucopus*, also named Bachu mushroom, partially delayed the uprising serum LDL-C level, and HDL-C decrease. The effect was comparable to atorvastatin treatment, correlated with the reduction in serum TGs [40].

4.2.1.12 Ganoderma Polysaccharides

The primary active ingredients of *Ganoderma lucidum* (GL), a Chinese herbal remedy that has been used for millennia in Asia, are Ganoderma lucidum polysaccharides (GLP), which, depending on the dose, show immunoregulatory, antioxidant, and hypolipidemic effects [41].

4.2.1.13 Quinoa Polysaccharide

The Incas is formerly referred to the Andean pseudo-cereal quinoa (Chenopodium quinoa Willd., family Amaranthaceae) as the "mother of all grains." Quinoa seed eating has been linked to reduce blood TG and LDL-C levels in vivo [42].

4.2.1.14 Holothuria Polysaccharides

Holothuria leucospilota polysaccharides (HLP), obtained from dry black sea cucumbers. Alleviated high-fat diet induced hyperlipidemia. HLP is difficult to digest in the upper gastrointestinal tract, so in the large intestine it is converted to short-chain fatty acids (SCFAs) by gut microbiota and regulates lipid metabolism. HLP has been reported to suppress acetyl-CoA carboxylase (ACC) and fatty acid translocase CD36 expressions to inhibit lipid accumulation in liver cells, thereby secretion of excess leptin and adiponectin from adipocytes was inhibited. Leptin and adiponectin stimulate inflammatory factors and insulin secretion is elevated, leading to IR. HLP also downregulated the expressions of TNF-α and NF-kB, which directly alleviated hyperlipidemia [43].

4.2.1.15 Rice Bran Polysaccharide

Rice bran polysaccharide (RBP) is got from the rice (*Oryza sativa L.*) milling. RBP-mediated modulation of expressions of PPARs, ACC, FAS, sterol regulatory element-binding protein 1 (SREBP-1C), sirtuin 1(SIRT1) in HFD-induced mouse model implies that RBP might be candidate for adjuvant therapies for hyperlipidemia [44].

4.2.1.16 Pleurotus Polysaccharides

In a zebrafish hyperlipidemia model, it has been demonstrated that polysaccharides from the fruiting bodies (PEPE) of *Pleurotus eryngii* (king oyster mushroom) may reduce the lipid level during the lipid absorption phase. Although it has considerable toxicity, it can significantly reduce the amount of oxidized LDL-induced foam-cell production in murine macrophage cells (RAW264.7) [45].

4.3 Hypertension

Long-term hypertension results in ultimate organ failure and is defined as an increase in blood pressure (BP) that is more than systolic 130–139/diastolic 85–89 mmHg. Arterial hypertension is a major risk factor because of its consequences like coronary heart disease, left ventricular hypertrophy, and renal diseases. The involvement of target organs (i.e., heart, brain, and kidneys) implies the severity and fatality of the outcome. In addition to mediating vasodilation, the endothelium releases nitric oxide (NO), which controls leukocyte adherence to endothelial cells and lowers the production of peptides that constrict blood vessels and raise blood pressure. Hypertension results from reactive oxidative stress signaling, which also causes vascular inflammation and persistently elevated vascular tone [46].

Hyperglycemia and hyperlipidemia directly or indirectly cause hypertension. Although most of the causal factors lie in lifestyle habits. A sedentary lifestyle, lack of physical exercise, consumption of food with high salt, sugar, and oil content, smoking, alcoholism, regular stress at work and social life, and lack of deep enough sleep widen the scope for hypertension to occur [47].

4.3.1 Polysaccharides Used Against Hypertension

4.3.1.1 Cordyceps Polysaccharides

Caterpillars are parasitized by a fungus called *Cordyceps sinensis*. Systolic and diastolic blood pressure (SBP and DBP) were demonstrated to be susceptible to reduction by *C. sinensis* polysaccharides (CSP1) in spontaneously hypertensive rats. Additionally, it increased NO, lowered the levels of epinephrine, noradrenaline, endothelin-1 (ET-1), and angiotensin II (Ang-II), and repressed the release of TGF-I and CRP [48].

4.3.1.2 Watermelon Rinds Polysaccharide

Watermelon polysaccharides (WMRP) are extracted from *Citrullus lanatus* rinds by hot water. WMRP exhibited ACE-inhibitory action in a dose-dependent fashion probably by the galacturonic acid present in the extract. The acidic pH might have denatured the ACE enzyme also [49].

4.3.1.3 White Mulberry Fruit Polysaccharides

In Sprague-Dawley rats, white mulberry fruit polysaccharides (WMFPs) from the mulberry (*Morus alba*) of the family Moraceae have been demonstrated to boost NO production and relax blood vessels. After receiving WMFP treatment, rats' mean artery pressure (MAP) decreased by more than 10%, establishing a NO dependent pathway [50].

4.3.1.4 Althea Polysaccharides

Water-soluble acidic bioactive polysaccharides are present in the whole plant *Althea rosea* is also known as "saze posh." The dose-dependent ACE-inhibitory effect of Althea polysaccharides has been seen in studies. Plant mucilages can reduce blood pressure by lowering endothelial dysfunction and decreasing inflammation in the vasculature, through the blockage of HMGCoA formation [51].

4.3.1.5 Sodium Alginate Oligosaccharides

A fiber-rich polysaccharide called alginate is obtained from seaweeds like kombu and giant kelp. Dahl S rats treated with sodium alginate oligosaccharide demonstrated a dose-dependent attenuation against salt-induced hypertension. The BP significantly declined after alginate administration and was maintained throughout the experiment [52].

4.3.1.6 Undaria Polysaccharide

An *Undaria pinnatifida* (Qundaicai) brown alga fucoidan-rich sulfated polysaccharide showed significant anti-hypertensive effects on vascular relaxation both in vivo and in vitro. Along with reducing blood pressure in chronically hypertensive rats receiving L-NAME treatment, this medication also protects the thoracic aorta's histological structure through activating the Akt/eNOS pathway. Fucoidan reduced the elevated inducible nitric oxide synthase production and systemic inflammation in those animals [53].

4.3.1.7 Other Polysaccharides

Polysaccharide from Xin-Ji-Er-Kang (XJEK), the Chinese medicine, exerts protection against L-NAME-induced hypertensive mice and Ang II-induced damage in human umbilical vein endothelial cells (HUVECs) and it might be done by improving endothelial function and the inflammation in blood vessels [54]. *Cystoseira crinita* sulfated polysaccharide shows anti-hypertensive action through ACE inhibition [55]. By triggering the eNOS/NO signaling pathway, astragalus polysaccharides decreased the MPA pressure in rats with monocrotaline-induced pulmonary arterial hypertension [56]. Acidic polysaccharides from Gastrodia rhizome, Chickpea water-soluble polysaccharide, *Cymodocea nodosa* sulfated polysaccharide, *Momordica charantia* polysaccharide, polysaccharides from *Ephedra alata*, Almond and Pistachio not only alleviate vasoconstriction by downregulating angiotensin II, but also decreased vasodilator bradykinin metabolism by blocking kininase II [57]. Polysaccharide as the active constituent of *Dendrobium officinale* can reverse metabolic hypertension in rats by activating the intestinal SCFAs-GPCR43/41 pathway [58]. Acidic polysaccharides from *Gastrodia elata* Blume rhizome have shown a reduction in hypertension and improved serum lipid levels [59].

The water-soluble polysaccharides AWSP isolated from almond (*Prunus amygdalus var. amara*) and PWSP from pistachio (*Pistacia vera*) juice processing by-products, both have shown ACE-inhibitory activity in concentration-dependent way [60]. Polysaccharide fraction of *Lycium barbarum L.* (Gouqi) efficiently lowers nitric oxide or endothelium-derived relaxation factor (EDRF) or both SBP and DBP in 2K1C-renovascular hypertensive rats. *Dendrobium officinale* (Tiepishihu) has recently shown lowering of BP in hypertensive rats. Through ACE-inhibitory activity, a homogeneous polysaccharide known as CGP, isolated from the enzyme-assisted extraction of Crassostrea gigas, has demonstrated similar hypotensive effects to captopril in both SBP and DBP [61].

4.4 Cardiovascular Diseases

In addition to hypertension, coronary heart disease, cerebrovascular disease, peripheral arterial disease, rheumatic heart disease, congenital heart disease, deep vein thrombosis, and pulmonary embolism, a group of heart and blood vessel disorders known as CVDs are the leading cause of death worldwide. Hyperglycemia and hyperlipidemia pose the highest risk to cardiovascular health. Hyperglycemia indirectly weakens heart muscles and perfusion. Hyperlipidemia directly deposits lipids in blood vessels, causing serious issues like atherosclerosis, coronary thrombosis, and deep vein thrombosis. All the lifestyle factors that cause hyperglycemia and hyperlipidemia, like smoking, a sedentary lifestyle, alcoholism, lack of physical exercise, less sleep, and huge stress eventually become reasons for CVDs [62].

4.4.1 Polysaccharides Acting Against CVDs

All polysaccharides that act against hyperlipidemia and hyperglycemia usually have cardioprotective action. Antihypertensive polysaccharides have already been discussed. So, this portion will try to discuss basically on the polysaccharides acting against other CVDs.

4.4.1.1 Auricularia Polysaccharides

Polysaccharides from *Auricularia auricula* have been found to possess antioxidant activity and anticoagulant activity. In mice, polysaccharide administration dramatically increased myocardial SOD and GSH-Px levels and decreased myocardial MDA levels, demonstrating antioxidant activity [63].

4.4.1.2 Astragalus Polysaccharide

Astragalus membranaceus, commonly known as Huang Qi in China, produces astragalus polysaccharide (APS), which is derived from the dried roots, has shown dose-dependent anti-hypertrophic effect in isoproterenol (ISO)-induced cardiomyocyte hypertrophy. The production of fetal genes such as atrial natriuretic peptide (ANP) and B-type natriuretic peptide (BNP), which were downregulated by APS, is increased in cardiomyocytes during hypertrophy [64].

4.4.1.3 Red Microalga Polysaccharides

Red microalga polysaccharides (PSs) are produced from *Porphyridium* sp. (Rhodophyta), algal capsule. The PS positively regulates NO formation and endothelium-dependent vessel relaxation in different models and/or inactivates ROS [65].

4.4.1.4 Dendrobium Polysaccharides

Dendrobium officinale polysaccharide stimulated SOD, meis1 expression, and downregulated cardiomyocyte apoptosis, thereby improved myocardial ischemia in mice with ligated coronary arteries. Effects of *Dendrobium officinale* on the management of oxidative stress and apoptosis have been observed. Additionally, H_9C_2 cells injured by hydrogen peroxide (H_2O_2) have demonstrated polysaccharides on cardiomyocytes [57].

4.4.1.5 Ophiopogon Polysaccharide

Through the stimulation of the SPHK/S1P/bFGF/AKT/ERK and eNOS/NO signaling pathways, *Ophiopogon japonicus* polysaccharide has demonstrated increased angiogenesis in myocardial ischemic tissue and reduced infarct size in acute myocardial ischemia rats. Additionally, it increased endogenous antioxidants while preserving the ionic environment in rats with ISO-induced myocardial ischemia and sustaining Na + -K + -ATPase and Ca_2 + -Mg_2 + -ATPase activity [66].

4.4.1.6 Zhiqiao

Using the PI3K/Akt and ERK1/2 pathways, the acidic heteropolysaccharide CALB-3, obtained from *Citrus aurantium* Linn, has been demonstrated to have cardioprotective properties against the cytotoxicity in ISO-treated rats. In vitro study using H9c2 cells has further confirmed myocardial protection activity of CALB-3. Mechanistically, through the stimulation of the Akt signal pathway, CALB-3 partly prevented cardiomyocyte death [67].

4.4.1.7 Padina Tetrastromatica

The mRNA expressions of the anti-inflammatory cytokines IL-10 and NF-κB, which are produced by cardiomyocytes, were significantly upregulated by bioactive sulfated polysaccharides (PSPS) from the brown algae species *Padina tetrastromatica*. Hence PSPS can control inflammation linked to CVDs [68].

4.4.1.8 Opuntia Polysaccharides

Water-soluble polysaccharide opuntia polysaccharides (ODP-Ia), purified from *Opuntia dillenii* has shown action against inflammation, hyperglycemia, and hyperlipidemia. ODP-Ia inhibits the expression of hepatic apolipoprotein B and hepatic diglyceride acyltransferase proteins, which prevents lipid accumulation and inflammatory cell infiltration in hyperlipidemic, high-calcium diet-fed rats. ODP-Ia also modulates the enzymatic activity involved in cholesterol metabolism. Suggesting ODP-Ia has a good preventive effect on atherosclerosis [69].

4.4.1.9 Plantago Polysaccharides

P. asiatica L. polysaccharide (PLP), a dietary supplement having anti-inflammatory and anti-atherosclerosis effects, is derived from *Plantago asiatica* L. In rats with hyperlipidemia brought on by a high-fat diet (HFD), PLP decreased TC, TG, and non-esterified fatty acid levels while increasing HDL-C, may be by regulating the gut microbiota and increasing SCFA levels. By lowering lipid peroxidation, boosting the generation of NO, and inhibiting the mRNA expression of monocyte chemotactic protein-1 and c-myc, PLP has demonstrated inhibitory impact on the proliferation of smooth muscle cells produced by oxidized low-density lipoprotein (Ox-LDL) [70].

4.4.1.10 Polysaccharide Krestin

In numerous studies, the protein-bound polysaccharide Krestin (PSK), which is derived from the *Coriolus versicolor* plant, has demonstrated atherosclerosis prevention by preventing the buildup of lipid peroxides and the development of lipid-containing macrophages (foam cells) by Ox-LDL [71].

4.4.1.11 Poria Polysaccharides

The Polyporaceae family of medicinal fungi *Poria cocos* produces Poria polysaccharide (PCP), which has a number of biological properties, including anti-inflammatory, antioxidant, and immunity-boosting properties. PCP can reduce the proliferation and migration of vascular smooth muscle cells (VSMCs) brought on by Ox-LDL by blocking the TLR4/NF-B p65 signaling pathway. By modulating the PPAR-LXR-ABCA1 pathway in ApoE-knockout mice, PCP has been discovered to enhance HDL function and boost cholesterol reverse transport, which together contribute to its anti-atherosclerotic action [72].

4.4.1.12 Dictyophora Polysaccharides

Two distinct polysaccharide fractions, Al-DPS and En-DPS, were produced by alkali extraction and enzymatic hydrolysis, respectively, from *Dictyophora indusiate*, which is a saprophytic fungus having culinary and therapeutic significance. Al-DPS and En-DPS can both lower the risks of hyperlipidemia, according to in vivo testing, suggesting a potential to lessen atherosclerosis [73].

4.4.1.13 Momordica Polysaccharide

As evidenced by decreased serum levels of the biomarker enzymes aspartate transaminase (AST), creatine kinase (CK) isoenzyme MB (CK-MB), and lactate dehydrogenase, *Momordica charantia* polysaccharide (MCP), derived from *Momordica charantia* (Kugua), has shown protection from ISO-induced myocardial infarction in vivo rodent models. MCP also upregulated antioxidant enzymes SOD and CAT, anti-apoptotic protein Bcl-2 expression, and lowered nuclear factor kappa B (NF-κB) and pro-apoptotic proteins (Bax and caspase-3) expressions [74].

4.4.1.14 Cornus Polysaccharide

In a left anterior descending coronary artery ligated MI rat model, *Cornus officinalis* polysaccharide (COP) from *Cornus officinalis* (Shanzhuyu) enhanced left ventricular contractile performance and infarct size. The underlying mechanisms of cardioprotective activity of COP may be the upregulation of mRNA expression of peroxisome proliferator-activated receptor-gamma coactivator (PGC)-1, 1 and nuclear respiratory factor 1 (NRF-1) and downregulation of glycogen synthase kinase-3 [75].

4.4.1.15 Tamarindus Xyloglucan

Tamarind xyloglucan (TXG), a branched heteropolysaccharide derived from seeds of *Tamarindus indica* Linn., has demonstrated reduction of phosphorylation of mitogen-activated protein kinases (MAPKs) p38 and c-Jun N-terminal kinase (JNK), which suggests its activity to reduce ROS production. These indicate a significant cardioprotective effect [76].

4.4.1.16 Panax Ginseng

H9c2 cells pre-treated with AP1, an acidic polysaccharide extract of *Panax ginseng* exhibited the cardioprotective effect by partially suppressing the intrinsic apoptotic pathway through mitochondrial malfunction and ROS-mediated JNK/p38-MAPK activation [61].

4.5 Asthma and COPD

Asthma and COPD cannot be called metabolic disorders. Still, these 2 pulmonary disorders may be brought on by hyperlipidemia and other causes, even though chronic obstructive pulmonary disease (COPD) is more common among doctors. Airway hyperresponsiveness, congestion, excessive mucus production, and thickening and rigidity of the airway wall are all symptoms of bronchial asthma, an allergic inflammatory disease of the airways, which can lead to chronic progression of the condition or COPD to some extent. Conversely, COPD is characterized by a weakly reversible, frequently increasing airflow limitation and linked to an aberrant inflammatory response of the lung. There are two key factors contributing to the decreased flow seen in COPD, namely increased airflow resistance by airway blockage and reduction in lung elastic recoil tension by parenchymal deterioration [77].

The main lifestyle risk factors that directly cause asthma are poor diet quality, a sedentary lifestyle, smoking, and obesity. According to several population-based surveys, people with asthma engage in less physical exercise than people without asthma [78]. Childhood asthma symptoms are significantly correlated with a sedentary lifestyle. When physical exercises cause breathlessness, a phobia or unwillingness may develop to avoid doing it, which makes muscles more oxygen-seeking, which causes further breathlessness. Thus, in a cyclic way asthma or COPD may develop. Consumption of fast food, salt, and trans fatty acids is positively correlated with asthma [79]. Chronic stress exposures like domestic stress, community stressors, high crime rates, and poverty are linked to an elevated inflammatory profile, which is linked to an increase in asthma symptoms [80]. More than 90% of people with COPD presently attribute their condition to cigarette smoking. The development of airflow restriction in smokers with chronic bronchitis is linked to a further rise in macrophages and T lymphocytes in the subepithelium. Smokers suffering from COPD have more inflammatory cells present in the submucosa than in the adventitia [81].

4.5.1 Polysaccharides Acting Against COPD and Asthma

4.5.1.1 Chitin

Fish scales, arthropod exoskeletons, fungal cell walls, and yeast cell walls are sources of chitin, which has a variety of uses in respiratory illnesses. Chitin functions as a powerful T and B-cell adjuvant and resembles fungal spores in the form of phagocytosable microparticles. Chitin microparticles (CMP) can induce immune responses through T helper cell 1 (Th1) maturation that can be beneficial for asthma. In a study of ragweed induced allergic mouse model, oral and intranasal administration of CMP successfully decreased serum IgE, airway hyperresponsiveness, lung inflammation, and lung eosinophilia. They also increased the secretion of IL-12, IFN-γ, and TNF-a while decreasing the production of IL-4 [82]. Asthma developed in mice appears to be prevented by microgram dosages of CMP administered intranasally during the neonatal period [83].

4.5.1.2 Chitosan

A polysaccharide called chitosan is created when chitin is deacetylated. Chitosan is aβ-(1, 4)-linked-D-N-acetyl glucosamine polysaccharide that is a very significant carrier to deliver a huge number of medicines, proteins, and biologics, and it may be easily associated with negatively charged pharmaceuticals via electrostatic binding contact. Human airway and alveolar region cellular linings have shown that chitosan-based nanoparticles have very low cytotoxicity. Chitosan-loaded medications have been shown to have an improved therapeutic effect because they are mucoadhesive and create gels, so they can stay in the respiratory system for longer. Ovalbumin-induced asthma mice models' IL-4, TNF-α, IL-5, and IL-13 levels and their mRNA expressions have decreased as a result of this [84]. Rat mast cells interact with chitosan-hyaluronic acid nanoparticles, which has a positive impact by decreasing inflammation and airtrack remodeling [85].

4.5.1.3 Astragalus and Codonopsis Polysaccharides

Polysaccharides extracted from *Astragalus membranaceus* and *Codonopsis pilosula* have combinatorial effects on asthma. The benefits of combining Astragalus polysaccharide (APS) with budesonide in the treatment of asthma may be due to the modulation of dendritic cells (DCs), natural killer cells (NK), T-regulatory cells (TREG), IL-4, and IL-10. This new method of treating asthma may be due to these factors. In mice treated for asthma, APS plus budesonide significantly reduced DCs compared to budesonide alone, indicating a reversible immunological response. In COPD mouse models, administration of Astragalus polysaccharide and/or Codonopsis pilosula polysaccharides (CPPs) significantly increased alveolar macrophages (AM) phagocytosis and decreased the inflammatory mediators in the lung and circulation. In COPD mice, the effects of Astragalus polysaccharide and/or CPPs on AM phagocytosis and regional or systemic inflammation have not entirely returned to baseline. These results suggest that CPPs and Astragalus polysaccharide may offer some protection against COPD [86].

4.5.1.4 Antrodia Polysaccharides

In Asian traditional medicine, the fungus *Antrodia camphorata* is frequently employed. In this study, the capacity of dendritic cells (DCs) to suppress OVA-induced asthma and the immunomodulatory effects of A. camphorata polysaccharides (GF2) on DCs were investigated. GF2 treatment on LPS-activated DCs suppressed CD4+ T-cell proliferation and T helper 2 (Th2) cell polarization with generation of IL-10 in an allogeneic mixed lymphocyte response. GF2 injection in an allergen-induced asthma mouse model dose-dependently reduced Th2 responses, eosinophilia in the airways, and airway hyperresponsiveness [37]. In that model, the production of anti-OVA IgG2a and IFN-c was is induced, which further triggered Th1 responses. It has recently been demonstrated that asthma is prevented by GF2 through an inducible Th1-like TREG that also produces IL-10 and IFN-c [88].

4.5.1.5 Portulaca Polysaccharides

The plant *Portulaca oleracea* is the source of portulaca polysaccharides (PoPs), which demonstrates potent anti-asthmatic effects by downregulating NO, inflammatory cytokines, and chemokines, decreasing oxidative stress while enhancing antioxidant markers, and also by having potent relaxant effects on tracheal smooth muscles via stimulation of the adrenoceptor [89].

4.5.1.6 Ganoderma Polysaccharides

A polysaccharide fraction, PS-F2, isolated from the submerged culture fluid of *G. formosanum*, stimulates macrophage activation by activating Toll-like receptor 4 (TLR4), Dectin-1, and complement receptor 3.These findings imply that PS-F2 may be developed into a treatment for allergic asthma [90].

4.5.1.7 Boletus Edulis Polysaccharide

The polysaccharide extract from the edible fungus Boletus edulis (BEP) significantly reduced airway resistance. The ability of BEP to diminish inflammatory cell infiltration was equivalent to that of conventional dexamethasone therapy. Furthermore, after receiving Boletus edulis polysaccharide treatment, mice had less muco substance accumulation in their lung tissues. In youngsters with asthma, IL-4 levels rise, while IFN-levels fall. BEP-treated mice's lung tissues had higher IL-4 and lower IFN-γ levels. The fraction of anti-inflammatory CD4 + CD25 + FOXP3+ TREG cells considerably increased (*P* 0.05) when compared to untreated asthma animals, whereas the expression of cyclophilin A significantly reduced. BEP therapy outcome observed in induced and airway resistance is found to be nearly similar to dexamethasone therapy [91].

4.6 Chronic Kidney Disease

Chronic kidney disease (CKD) refers to the clinical condition of slow gradual loss of renal function to filter blood through them, caused by different metabolic factors, leading to complete loss of filtration capacity. CKD causes further consequences such as anemia, stroke, cardiovascular diseases, hypocalcemia, hypokalemia, depression, frequent infections, etc. Renal fibrosis and inflammation are frequent side effects of CKD development and progression, which can ultimately lead to end-stage renal failure. The potential apoptotic mechanism of glomerular sclerosis in CKD may be controlled by the pro-apoptotic factor Bax and the anti-apoptotic factor Bcl-2 (B-cell lymphoma/leukemia-2). Therefore, it appears that it is required to block the inflammatory and apoptotic processes in order to treat and prevent CKD [92]. As people age, the prevalence of CKD grows steadily, as do other risk factors such as arterial hypertension, metabolic syndrome, diabetes, obesity, and indiscriminate use of nephrotoxic medicines [93].

In all affluent and many developing nations, diabetes, hyperlipidemia, and hypertension are the main causes of CKD, but glomerulonephritis and other unidentified causes are more prevalent in Asia and sub-Saharan Africa. The burden of CKD in underdeveloped nations is exacerbated by insufficient sanitation, a lack of access to clean water, environmental pollution, pesticide usage, painkiller misuse, and the use of unrestricted food additives [94]. Consuming foods high in sodium causes whole-body volume overload and hypertension, which suggests CKD development and cardiovascular remodeling. Even when kidney function is normal, eating too much protein may not be a good idea. A high amino acid load leads to afferent arteriole expansion, an increase in intraglomerular pressure, and glomerular hyperfiltration. In contrast, a low-protein diet reduces high glomerular filtration load and improves afferent arteriole growth. Protein restriction has positive effects on the development of CKD, blood pressure, and proteinuria, according to many meta-analyses. Nevertheless, there is still no established protein consumption guideline for preventing CKD. High-calorie diets generate overweight and obesity, and obesity accelerates the onset and progression of CKD. In CKD, visceral fat and abdominal obesity are implicated in the regulation of many cytokines and adipokines that are linked to insulin resistance and hyperglycemia. As a result of poor kidney function and a steady drop-in glomerular filtration rate brought on by hypertension, BP regulation is further compromised. Endothelial dysfunction and modification of the renal microvasculature are results of arterial hypertension. Kidney disease and CKD are caused by endothelial dysfunction. Blood flow to the glomeruli declines with an increase in RAAS activity, further deteriorating kidney function. When they block the renin-angiotensin-aldosterone system (RAAS), angiotensin-converting enzyme inhibitors and angiotensin receptor blockers reduce blood pressure and protect the kidneys and the body's microcirculation [95].

4.6.1 Polysaccharides Used in CKD

4.6.1.1 Soybean Soluble Polysaccharides

Seeds of the Glycine sp. plant are used to make soybean soluble polysaccharide (SSPS). In CRF mice, the renal pathological damage was markedly reversed by the SSPS therapy. High-dose SSPS was able to minimize the amount of renal fibrosis, according to Masson histological examination, suggesting that it was successful in reducing kidney damage in CRF animals.TGF-1 production has been identified as a critical mediator of renal fibrosis, and SSPS has drastically decreased its mRNA expression [96].

4.6.1.2 Pleurotus Mycelia Polysaccharides

The genus Pleurotus of the Pleurotaceae family contains *Pleurotus djamor*, and its polysaccharides display a wide range of biological activity, including anti-inflammatory and antioxidant properties. It is possible to quantify TNF-α and IL-6 in blood and tissue, two crucial inflammatory factors in the inflammatory response. A prominent reduction in IL-6 and TNF-α levels and relative mRNA expression following PMPS application to adenine-induced CRF mice suggests that PMPS has potent anti-inflammatory, anti-fibrosis, and renal protective effects [97].

4.6.1.3 Ophiocordyceps Polysaccharides

The eastern Himalayan region is home to the fungus *Ophiocordyceps lanpingensis*. According to studies, *O. lanpingensis* reduces oxidative stress, which helps to improve renal dysfunction in glycerol-induced acute renal failure. By lowering the level of ROS in CKD mouse kidneys, ophiocordyceps polysaccharides (OLP) reduced oxidative stress. The MAPK pathway, which is implicated in renal interstitial fibrosis and renal cell death, was activated by OLP, whereas the pro-inflammatory cytokines were downregulated [98].

4.6.1.4 Auricularia Polysaccharides

It has been proven that A. polytricha's fruiting body contains bioactive polysaccharides, which have historically been utilized as both a food and a medication. Enzymatic-hydrolysis APS (EnAPS) and A. polytricha polysaccharide (APS) show antioxidant activity in their protective effects. The oxidative stress induces DNA damage, which leads to apoptosis in the kidney. Caspase-3, the primary downstream effector of apoptosis among the caspases, may contribute to cell death via mitochondria-dependent (Bax/Bcl-2)-induced apoptosis. On the one hand, caspase-3 activation may control the equilibrium between the anti-apoptotic protein Bcl-2 and the pro-apoptotic protein Bax, whose imbalance results in apoptosis. Renal fibrosis is significantly influenced by apoptosis. By lowering Bax expression, blocking caspase-3 activations, and raising Bcl-2 expression, APS and EnAPS demonstrated potential anti-apoptotic actions. EnAPS's low molecular weights resulted in somewhat better effects. According to the research now being conducted, APS and its derivatives may be created as novel functional dietary supplements to help prevent and reduce CKD and associated problems [99].

4.6.1.5 Cordyceps Polysaccharides

The fungus *Cordyceps militaris* (L.), a member of the Ascomycetes class, utilized in traditional Chinese medicine (TCM) in East Asia, is used as a safe source of pertinent bioactive compounds. During the course of CRF, nitric oxide (NO) has a protective effect for the kidney, inhibiting the mesangial cell proliferation and decreasing the synthesis of the mesangial matrix. The buildup of P^{3+} and excretion of Ca^{2+} can be facilitated by the reduced glomerular filtration rate following the development of CRF. CRF may also lower the sources of estrogen and prevent the production of LH and E2, which can result in an endocrine problem. According to our findings, treatment with C. militaris, either alone or in combination with H. brevicor, considerably reduced the lowered concentrations of NO, Ca^{2+}, and LH and the rise in P^{3+} caused by adenine. Accordingly, the progression of CRF was either delayed or stopped, damage to the renal tubular epithelial cells was decreased, renal tubular function was restored, functional protein synthesis was increased, the negative nitrogen balance was addressed, and so overall renal function was improved [100].

4.6.1.6 Chitosan

Chitosan is a non-toxic polysaccharide. Industrial production of it involves deacetylating chitin from crab and prawn shells. Chitosan interacts with several acidic compounds thought to be uremia toxins, causing the body to expel them and improving any residual renal function. Because blood levels of urea and creatinine are crucial markers of renal function, a decrease in these values denotes an improvement in renal function. The considerable rise in serum hemoglobin levels without the use of human erythropoietin in the therapy group was another indication of improved renal function. A decrease in the serum hemoglobin levels, or anemia, is frequently linked to the nephrotic syndrome. Curing this type of anemia is challenging. Human erythropoietin is typically used to treat anemia nephrotic syndrome patients receiving hemodialysis, although this treatment must be long term and comes with a number of negative effects. Therefore, people with renal insufficiency may benefit from chitosan as a dietary supplement [101].

4.6.1.7 Fucoidan

Fucoidan, sulfated polysaccharides derived from the brown alga Laminaria japonica, includes significant amounts of the fucose and sulfate groups as well as negligible amounts of the following sugars: mannose, glucuronic acid, glucose, xylose, arabinose, and rhamnose. By controlling HO-1 and SOD-1 gene expression through the Nrf2 signaling pathway in HaCaT cells, fucoidan reduced oxidative stress. Fucoidan can enhance SOD and GSH-Px stimulation in the kidney and hippocampus and lower the level of MDA, according to the results of the current study. Additionally, the findings revealed that fucoidan protected tissue from adenine-induced harm by downregulating GSK-3 and activating the antioxidant responses mediated by the Nrf2-HO-1 signaling pathway. According to another investigation, downregulation of the MAPK and NF-κB signaling pathways and the subsequent decrease in pro-inflammatory cytokines are two probable anti-inflammatory mechanisms of fucoidan. TNF-α, IL-1, and iNOS mRNA expression in M1 microglia was significantly reduced by fucoidan, but IL4, CD206, Arg1, and TGF- mRNA expression in M2 microglia was significantly increased. Consequently, fucoidan has antioxidant and anti-inflammatory properties [102].

4.6.1.8 *Cordyceps sinensis*

Since more than a 1000 years ago, cordyceps, one of the most prized traditional Chinese medicines, has been widely utilized to treat and prevent many human illnesses. It is made up of the dried *Cordyceps sinensis* fungus, which has immune-boosting and antioxidative qualities and on which caterpillar larvae develop. Through the transforming growth factor 1 (TGF 1) pathway, cordyceps polysaccharides (CPS) may lessen the production of ROS and lessen damage. TGF-1 and extracellular matrix (ECM) levels in patients may return to normal under the influence of CPS. CPS improved CRF considerably and in a dose-dependent way [103].

4.6.1.9 Chinese Chive Polysaccharides

Chinese chive, also known as *Allium tuberosum Rottl. Ex Spreng*, is a well-known culinary herb used in traditional Chinese medicine to protect the kidneys. The Chinese chive polysaccharides (CCP) were obtained through hot water extraction. In a dose-dependent way, CCP greatly inhibited the rise in TGF-levels in CRF animals, and TGF-mRNA expression similarly decreased. According to this finding, CCP may help those with chronic renal insufficiency have better renal function. The antioxidant, anti-inflammatory, and anti-fibrotic properties of CCP may potentially be involved in the protective processes [104].

4.7 Inflammatory Bowel Syndrome

According to the location and intensity of the lesion, ulcerative colitis (UC) or Crohn's disease can be classified as an inflammatory bowel syndrome (IBS) or irritable bowel disease (IBD). IBD symptoms include diarrhea, stomach discomfort, distension, and sometimes bloody stools. The two regions with the highest rates of IBD are North America and Europe, where it is estimated that 1.4 and 2.2 million people, respectively, have CD and UC, and that more than 0.3% of the population live in several European countries [105].

Numerous lifestyle choices, such as smoking, obesity, stress, sleep disruption, and a lack of physical activity, may increase the chance of developing IBS. Obesity increases the likelihood of recurrence and is associated with higher levels of anxiety, hopelessness, tiredness, and pain as well as higher healthcare expenses. Additionally, obesity increases the chance of therapy failure and adversely affects the pharmacokinetics of biological medicines. Independent of disease activity, sleep disruption is quite common in IBD patients and raises the chance of recurrence. Similarly, stress, especially perceived stress as opposed to significant life events, may cause a clinical flare in IBD patients; however, it is unknown how stress affects inflammation. Despite some research suggesting that smoking may be associated with a decreased incidence of flare-ups, smoking did not significantly alter outcomes in people with ulcerative colitis. Smoking increases the likelihood of corticosteroid dependency, accelerates the course of Crohn's disease in patients, and may even need surgery [106]. According to some research, cannabis may help IBD patients with chronic pain without having a major impact on biological remission. However, the effects of alcohol and cannabis usage on IBD patients are inconsistent [107]. Although these lifestyle variables may be changed, few interventional studies have been carried out. Although a healthy lifestyle may help, studies of organized exercise, psychological treatment, mindfulness-based therapies like yoga and meditation, and IBD patients' endoscopic or clinical disease activity have not always improved with gut-directed hypnotherapy [108].

4.7.1 Polysaccharides Against IBS

4.7.1.1 Oat Bran Glucan

Oat (*Avena sativa*) bran, whose primary components are beta 1,3- and beta 1,4-glucan, significantly reduced the dextran sulfate sodium-induced (DSS) colitis mice model's clinical symptoms, weight loss, diarrhea, the infiltration of inflammatory cells, and increased colon length. In addition to lowering the levels of MPO, NO, and MDA, beta-glucan administration also downregulates the expression of pro-inflammatory markers in the colonic tissues [109].

4.7.1.2 Mushroom Glucan

Mushrooms are used as food as well as a form of alternative medicine and to reduce inflammation since the dawn of time. Patients with IBD have showed improvement in their inflammatory symptoms after taking AndoSanTM, a mixed extract from basidiomycetes mushrooms. The study also found that calprotectin, an important indicator of inflammation found in the feces of UC patients, had reduced [110]. Another study revealed that Chaga mushroom extract can shield IBD patients' lymphocytes' DNA from oxidative degradation. Another randomized placebo-controlled research found that AndoSanTM had an anti-inflammatory impact on individuals with both UC and CD. After being administered orally to mice, Lentinus edodes beta-glucan had an anti-inflammatory effect. It also reduced the MDA and MPO levels in colonic tissues and lowered the production of iNOS, TNF-α, IL-1, and IL-6. Glucan decreased pro-inflammatory factor synthesis, PPAR phosphorylation at Ser112 and Elk-1 phosphorylation at Ser84 in a colitis model generated by DSS and in a RAW264.7 cell model caused by LPS. This shows that a sophisticated signaling network was used by beta-glucan to contribute in the anti-inflammatory activity. Other researchers have shown that giving P. pulmonarius glucan to animals with colitis decreased their intestinal inflammation and related symptoms [111].

4.7.1.3 Seaweed-Derived β-Glucan

A research demonstrates that β-glucan derived from Laminaria hyperborea and Laminaria digitata may both dramatically reduce the production of Th17-associated cytokines (IL-17a, IL-17F, and IL-22), as well as receptor IL23R and IL-6, with no change to the TREG related targets. An important focus for treating IBD's underlying pathophysiology is the Th17 inflammatory response. When administered to pigs with DSS, the algal polysaccharide laminarin (containing β-glucan) alleviated clinical symptoms, body weight loss, colonic Enterobacteriaceae, and a decrease in colonic IL-6 mRNA expression level. But there has not yet been a human clinical trial [111].

4.7.1.4 Fucoidan

Fucoidans are acidic, sulfated macromolecules consisting of L-fucose and a variety of other oligosaccharides including mannose, galactose, and xylose that are frequently recovered from brown algae such as Fucus vesiculosus and the sporophyll of Undaria pinnatifida. The oral treatment of fucoidan considerably reduced the symptoms of colitis, and a histological investigation revealed a clear decrease in crypt architecture and goblet cells as well as immune cell infiltration and edema. These protective effects were verified by reducing the expression of 15 pro-inflammatory cytokines in the intestinal tissues. In pigs with DSS-induced colitis, a diet high in fucoidan lowered the expression of IL-6 and improved diarrhea and pathology scores. By lowering the expression of the aP2 and PPAR genes, which in turn reduced the expression of genes associated with inflammation, fucoidan also suppressed fat storage. These data imply that fucoidan may help reduce IBD-related symptoms [112].

4.7.1.5 Plantago Seeds Polysaccharides

Plantago ovata seeds and mesalazine (5-aminosalicylic acid) were used in a clinical experiment on UC patients, and the results were same for both. It improved intestinal cytoarchitecture in HLA-B27 transgenic rats given 5% Plantago ovata seeds, and it reduced the production of pro-inflammatory cytokines such as NO, leukotriene B4, and TNF-α. It also reduced colonic inflammation. Compared to the control group, rats receiving supplementation generated more SCFAs. On colitis brought on by trinitrobenzenesulfonic acid (TNBS), Plantago polysaccharides also show a comparable protective effect [113].

4.7.1.6 Wheat Bran and Barley

Whole grain cereals store a lot of arabinoxylans, and it has been suggested that supplementing with them has biological benefits for IBS patients. In a clinical experiment involving UC patients in remission, germinated barley (GB) was employed, and it dramatically improved the disease index and decreased the recurrence rate despite having no adverse effects when compared to the control group [114].

4.7.1.7 Gums

Partially hydrolyzed guar gum (PHGG), an enzyme-produced water-soluble polysaccharide synthesized from guar gum beans, decreased MPO activity, TNF-protein, and mRNA expression in the colonic mucosa, and improved colonic damage in mice with TNBS-induced colitis. PHGG-fed mice showed considerably greater amounts of Clostridium coccoides, Clostridium leptum, and Bacteroides fragilis, according to an intestinal microbiota research [115]. Similar to this, another study shown that giving animals with DSS-induced colitis a gum supplement—guar gum and PHGG—significantly decreased their clinical score. In a research involving volunteers, the PHGG and fructooligosaccharides (FOS) in the biscuits reduced the irritable bowel syndrome symptoms of abdominal discomfort and loose stool [116]. According to these investigations, consuming PHGG may be beneficial for the creation of IBD treatments.

4.7.1.8 Pectins

A study looked at how wild jujube (Zizyphus spinosus Hu) pulp polysaccharides (WJPs) affected an experimental IBD model that was brought on by an intrarectal injection of TNBS. WJPs are acidic heteropolysaccharides made up of the remaining 60% of galactose and the remaining 40% of arabinose glucose. WJPs were shown to dramatically improve the condition in colitis rats by reducing weight loss, disease activity index scores, and mucosal damage. Additionally, WJPs downregulated the production of TNF-α, IL-1, and IL-6 as well as MPO activity, which partially suppressed the inflammatory response. In Caco-2 cells treated with TNF, WJPs' effect on TER and FITC-conjugated dextran permeability suggested that its protective effects on the colon in colitis were connected to barrier function through the activation of AMP-activated kinase (AMPK) activity. 67.3% anhydrogalacturonic acid and 38 mg/g calcium make up calcium pectate (CP), which has been researched for its gastroprotective properties. Results showed that giving mice CP (39.3 kDa) made from citrus at dosages of 25 and 50 mg/kg considerably reduced the mean area of lesions and ulcers. This preventive effect was even more effective than common ulcer-treating drugs like famotidine [117].

4.8 Gastric and Peptic Ulcer

Due to the oversecretion of pepsin or stomach acid, an ulcer forms on the inner lining of the gastrointestinal tract (GIT). This is the most directly lifestyle-related disorder that also has impression on metabolism. Very indisciplined food habit is the main cause behind not only peptic ulcer, even gastric ulcer too. *Helicobacter pylori* bacteria is the second cause of ulceration but that is not linked to lifestyle habits. Fasting for more than 12 h after dinner and being very callous about having early breakfast initiate ulceration day by day. Continuous acid and enzyme secretion in an empty stomach make the mucosal layer of GIT thinner each day and at a certain point, the serous lining under the mucosa becomes disclosed and very low pH gastric content makes wound on it, which can occur both in the stomach and any other area of the GIT [118]. Although ulcers can also be caused by dietary components, a hypersecretory acidic environment, a stressful environment, and other lifestyle factors like tobacco smoking, alcohol consumption, etc. [119]. The most common symptoms of peptic ulcer are acute upper abdominal pain while having food. Sometimes the upper abdominal pain is relieved with eating, due to concomitant mucus secretion. But the discomfort is frequently characterized as a burning or aching sensation. Other symptoms include vomiting, weight loss, and a lack of appetite. In certain cases, these symptoms may be stomach bleeding, perforation, or obstruction. Obesity, smoking, and alcoholism also play vital roles in the progression of ulcer. Obese people have an excess of visceral adipose tissues that are metabolically active, which may worsen the development of GIT inflammation and release of more pro-inflammatory cytokines. As a result, it has been hypothesized that obesity may affect peptic ulcer disease [120]. The biggest risk factor that can be changed is smoking. There is growing evidence connecting cigarette smoking to gastrointestinal disorders, including chronic inflammation, GI tract cancers, and peptic ulcers [106]. Smoking decreases the level of epithelial growth factor (EGF), leading to the prevention of mucosal cell proliferation. Smoking enhances gastric acid secretion and downregulates the production of bicarbonate anions. It causes pyloric ineffectiveness and increases biliary reflux, which allows bile salts to damage the stomach mucosa, as well as reduces the renewal of the epithelial cell line in GIT. Alcoholism drastically reduces the activity of SOD, CAT, and GPx. Additionally, it results in microcirculatory disruption, harms the vascular endothelial cells in the stomach mucosa, and causes hypoxia, which is brought on by an excess of oxygen radicals [121].

4.8.1 Polysaccharides Against Ulcer

4.8.1.1 Sodium Alginate

Sodium alginate polysaccharides are hydrophilic, found in marine algae and brown seaweed. It is biocompatible and non-toxic. It has mucoadhesive properties and favorable biological properties also. The drug entrapment effectiveness of floating microparticles is reported to be improved by sodium alginate. By keeping the drug in the stomach for a long time after administration, sodium alginate microbeads ensured a totally aqueous environment, eliminating the need of organic solvent in the preparation for the successful treatment of peptic ulcer [122].

4.8.1.2 Ganoderma Lucidum Polysaccharide (GLP)

One of the primary sources of its pharmacological activity is Ganoderma lucidum polysaccharide (GLP). Research on the biological effects of GLP has focused on its anti-inflammatory, anti-tumor, antioxidant, anti-diabetic, and immunomodulatory properties [41]. Antioxidant enzymes, including SOD, GPX, and MPO, may be controlled in alcohol-induced gastric damage treated by GLPs in rats, preserving the stomach tissue. In addition, GLPs can maintain steady levels of EGF and signal molecule NO, protecting the stomach mucosa from harm. TNF-α, IL-1, and IL-18 may be inhibited by GLP, which would prevent the deterioration of gastric damage. With a specific dose-dependent impact, all three GLPs can dramatically lessen ethanol-induced acute stomach damage in rats [123].

4.8.1.3 Pectic Polysaccharide

The natural food source of pectic polysaccharide (PP) is the root of Decalepis *hamiltonii*, where PP is an arabinogalactan-type polysaccharide. By enhancing gastric mucin formation and providing cytoprotection to mucosal cells via a PGE2-mediated route, PP dramatically increases endogenous PGE2 levels in the stomach. The exact pathway through which this occurs is yet unknown though [124].

4.8.1.4 Fucoidan

A sulfated polysaccharide called fucoidan is produced from the body walls of sea cucumbers and brown algae. By preventing changes in alanine aminotransferase, aspartate aminotransferase, IL-10, and IL-6, fucoidan can help prevent stomach ulcers caused by aspirin. This suggests that fucoidan helps prevent inflammatory cytokine-mediated oxidative damage to the gastric mucosa [125].

4.8.1.5 Bletilla Polysaccharide

Bletilla striata, also known as Baiji in Chinese, is a significant astringent hemostatic medicinal plant that has been used for thousands of years to treat gastrointestinal ulcers, skin and mucosal wounds, severe bleeding, hematemesis, and hemoptysis [126]. Polysaccharides have the easy capacity to gel at low pH levels, making them potentially effective anti-gastric ulcer medications. In gastric tissue, Bletilla striata polysaccharide (BSP) may lower MDA and MPO levels while increasing SOD activity. Through the antioxidant system, studies have shown that BSP has gastroprotective qualities that may reduce neutrophil hyperplasia and lipid peroxidation brought on by high ROS levels. In ethanol-induced gastric ulcers, BSP has shown anti-inflammatory effects by inhibiting inflammatory cytokine levels (TNF-α, IL-1, IL-6, and IL-18) through MAPK/NF-κB pathway. BSP pretreatment significantly raised PGE2 levels. These suggest that BSP may disrupt the control of stomach pH, mucus production, and gastric mucosal integrity [127].

4.8.1.6 Arabinogalactan

A polymer called arabinogalactan is found in the cell walls of higher plants. Additionally, it is a part of exudates and gums. Fruits like the mango (Mangifera indica L) and Lycium ruthenicum Murr also contain arabinogalactan. By lowering acid and pepsin secretion, increasing mucous synthesis, and coating the mucosal membranes, arabinogalactan can protect stomach tissue. Activating the TGF-β, the arabinogalactan protein from Jatropha curcas seed endosperm was shown to stimulate cellular differentiation and mucosal layer renewal and to promote gastric ulcer repair [128].

4.8.1.7 Rhamnogalacturonan

The polysaccharide named rhamnogalacturonan (RG) is present in the leaves of the Acmella oleracea (L.) shrub, known to the locals as jambu [129]. Using ethanol-induced gastric lesions, the anti-ulcer efficacy of RG has been investigated. Through Toll-like receptor 4, RG stimulates dendritic cells, maturing their phenotype and activating the mitogen-activated protein kinase (MAPK) signaling cascade.

4.9 Gastroesophageal Reflux Disease

The widespread medical illness called gastroesophageal reflux disease (GERD) is brought on by the irritating effect of hydrochloric acid and other substances (such as pepsin) found in stomach secretions that cause damage to the esophageal mucosa. The esophageal epithelial lining cells act as buffer against he stomach secretions. Heartburn, a burning sensation in the chest that typically occurs after eating, is the predominant symptom of GERD [130]. GERD is not a direct lifestyle metabolic disease, but disoriented lifestyle management may cause it sometimes. Obesity is sometimes linked to overeating, which results in stomach distension and causes brief relaxations of the lower esophageal sphincter (LES). The distal esophageal epithelium of obese people shows increased permeability, indicating a disturbed epithelial barrier, even in the absence of pathologic reflux. Metabolic consequences of central obesity may also be involved [131]. Other potential risk factors for GERD include dietary/lifestyle choices, alcohol consumption, and cigarette smoking [132].

4.9.1 Polysaccharides Against GERD

4.9.1.1 Chitosan

Chitosan is a natural polysaccharide, mainly derived from chitin. Chitosan-made microspheres showed extended drug release and stayed buoyant for more than 11 h [133].

4.9.1.2 Carbopol

Carbopol is a polymer that is pH dependent. As there is a change in pH, a change in the state of Carbopol is observed. Solution form can be noticed at acidic pH and at alkaline pH, the solution gets changed into low viscosity gel. There are some water-soluble polymers of Carbopol system [134].

4.9.1.3 Pectin

Pectin is the second-most used polymer in raft formation. Pectin is typically extracted from plant cell walls and is also an anionic polysaccharide. Divalent cations are responsible for the conversion of pectin into the gel state in the stomach. Calcium ions are also required to make the formulation that forms a gel [135].

4.9.1.4 Starch

Jackfruit seeds are the source of the most prevalent polymer, starch. Both carbs and protein are abundant in these seeds. Two polysaccharides called amylose and amylopectin make up the majority of starch. Utilizing an emulsion cross-linking process, the starch microspheres can be made [136].

4.9.1.5 Alginates

Brown algae and certain bacteria produce alginate, which is a phycocolloid and a microbial polysaccharide. Na alginate can be used alone or in mixtures with magnesium (Mg) alginate, mannitol, sodium bicarbonate, potassium bicarbonate, calcium carbonate, and other alginate preparations. Mg alginate is sometimes used in place of Na alginate in formulations. Without HCO_3, alginate preparations prevent reflux by thickening the stomach's contents, whereas with HCO_3, alginate preparations neutralize stomach acid and create a "foam raft" that floats on top of the stomach's contents and prevents reflux [137].

4.9.1.6 Cashew Gum

The Anacardium occidentale L. tree yields cashew gum, a biopolymer. Cashew gum can reduce the degradation of barrier function and inflammatory episodes brought on by refluxate contents into the esophageal mucosa in both humans and animals with experimentally produced GERD, and no hazardous indications were noted [138].

4.9.1.7 Pectin

After alginates, pectin is the biopolymer that has received the most attention for its potential to produce rafts. Commercial production uses citrus peels and apple pomace. The pectin-containing anti-reflux formulation has effectively reduced the moderate to severe heartburn and erosive esophagitis returning with GERD and can be used as a maintenance therapy in patients with healed esophagitis. Additionally, it was discovered that pectin reduced symptoms like nausea, nausea, and regurgitation [139].

4.9.1.8 Guar Gum: Seed Gum

Guar gum is a polymer that is made from the *Cyamopsis tetragonolobus* bean plant's seed, gets dissolved in cold water, and hydrates readily to form very viscous solutions at low concentrations. These compounds are real emulsifiers since they demonstrate interfacial binding. When guar gum and xanthan gum are mixed, viscosity synergism is demonstrated. Guar gum was utilized in an antacid preparation that combined its raft-forming function with well-known antacids to create a protective coating [140].

4.9.1.9 Carrageenans: Seaweed Gum

As the carrageenan molecule contains up to 1000 galactose residues, it is linear sulfated galactan derived from red seaweeds (Rhodophyceae). The three primary types of these are typically defined as kappa, iota, or lambda. Additionally, the fractions μ, η, and xi have been discovered. These types exhibit different gelling properties despite being stable across a wide pH range. Iota carrageenan gels are weaker and less likely to synergize than kappa carrageenan gels, particularly when potassium ions are present. Lambda carrageenans bind strongly with proteins to create a pseudoplastic thickening rather than gelling in water [141].

4.9.1.10 Plantago Polysaccharide

Isapgol (*Plantago ovata*) is a naturally occurring dietary fiber that is accessible locally and recognized by the Indian Pharmacopeia. It is composed of polysaccharides and has long been used as a bulk laxative. Its cytoprotective activity has been supported by recent investigations. In an acidic media, a viscous dispersion of isapgol husk in water transforms into a swelling gel-like substance. Isapgol was examined for its suitability as a raft-forming agent and was discovered to be a strong contender for usage in the creation of raft-forming antacid suspensions. However, substantial clinical trials are required to determine this product's in vivo effectiveness in the management of GERD [142].

The above polysaccharides are used in peptic ulcers and other GIT disorders by the formation of Raft, a novel approach to treating the problem of acidity and GERD (shown in Fig. 3).

4.10 Constipation

A typical gastrointestinal ailment known as constipation is characterized by irregular bowel motions, or peristalsis (less

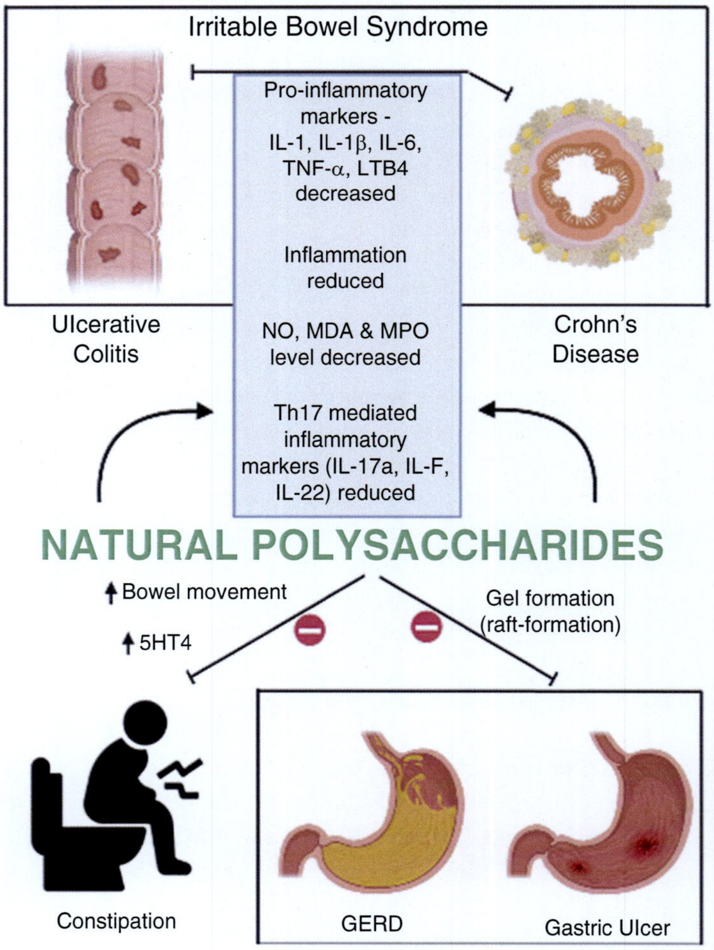

Fig. 3 Action of natural polysaccharides on gastrointestinal diseases

than three times per week), and challenging stool transit. A sense of incomplete defecation, stomach pain, bloating, excessive straining and a sensation of anorectal obstruction during stool transit, and occasionally needing manual assistance to evacuate the stool are additional symptoms that might vary from patient to patient [143]. Diet and lifestyle modifications, medication, and surgery are the cornerstones of treatment. According to epidemiological research, a high frequency of chronic constipation is linked to aging. Pathogenesis of constipation is multifactorial and primary or idiopathic subtypes of causes are mostly related to lifestyle like overuse of laxatives, low-fiber diet, sedentary lifestyle, ignoring the urge to defecation, inadequate fluid intake, haphazard changes in daily routine, lack of regular exercise, work stress, along with genetic predisposition and pharmaceutical factors. Along with these, other social variables that affect constipation include lower socioeconomic position, less parental education, physical activity at work, medicines, depression, physical and sexual abuse, and other traumatic occurrences. Constipation can be brought on by a lack of a diet high in vegetables and a low fluid intake. The risk has been shown to rise with certain drugs and physiological situations (such as pregnancy and aging). Actually, the regular busy schedules of people all over the world force to keep them off from early morning rising, missing the urge to defecate, and lack of physical exercise. However, the young generation is getting addicted to fast food, and funky foods with lesser roughage content, and they are also less interested in having green leafy vegetables and homemade bread, which as a whole lead to a consequence of developing constipation [144].

4.10.1 Plant Polysaccharides for Constipation

The anti-constipation efficacy of Spirulina Polysaccharide, isolated from Spirulina platensis, was shown to be comparable to or superior to that of phenolphthalein in mice with constipation symptoms [145]. Chronic constipation was alleviated by the polysaccharide fractions extracted from *Chimonobambusa quadrangularis* using microwave-assisted extraction (MAE) in the presence of fake gastric juice [146]. The kiwi berry (Actinidia arguta) can help with constipation, enhancing intestinal motility, adjusting neurotransmitter levels, and controlling expression of the SCF/c-kit signaling pathway [147]. Psyllium seeds, also known commercially as Indian Psyllium or Ispaghula, are a naturally occurring polysaccharide that is used extensively and most often around the world to cure diseases like constipation. The modified polysaccharide (PhPPS) can be classified as a hydrogel due to its high swelling index [148]. Galactomannan, a complex polysaccharide made of d-galactose and d-mannose and found in the powdered endosperm of Cyamopsis tetragonolobus (a leguminous crop) seeds, is used to treat constipation [149]. ALP-2, a polysaccharide derived from the roots of Arctium lappa L., may be used as a medicinal agent in the treatment of constipation or as an active ingredient in functional foods. In mice treated with ALP-2 for constipation, the weight of feces and the pace of small intestine movement were dramatically increased [150]. Acid and enzymatic hydrolysis partly destroyed the pectin from Arctium. 2 pieces, ALP2-A and ALP2-E, enhanced intestinal peristalsis, and mice in the ALP2-A group had a higher rate of intestinal motility than mice in the ALP2-E group did.

Enteromorpha Polysaccharide (EP) is a common marine alga-derived polysaccharide with stimulated intestinal motility function and has alleviated constipation-derived intestinal inflammation in loperamide-induced constipated mice model. EP drastically lowered serum NO concentration, downregulated VIPR1 expression, and upregulated 5-HT4 expression in the distal colon. After EP therapy, the intestinal microecological alterations brought on by constipation healed, according to genomic stool DNA MiSeq sequencing analysis [151].

Durio zibethinus Murr rind polysaccharide (DZMP) inhibited constipation in loperamide hydrochloride induced constipation Sprague-Dawley (SD) rats. DZMP increased the intestinal transit rate, motilin, gastrin, substance P levels and concentration of SCFAs, reduced the somatostatin levels, and improved the gastrointestinal peristalsis of rats [152].

Konjac glucomannan (KGM) polysaccharide is extracted from *Konjac glucomannan* (KGM) tubers. KGM may accelerate small intestine growth and stomach emptying in mice with constipation. To examine the impact on constipation, Bifidobacterium F1–7 with KGM and an aqueous extract of Prunus persica (PP) were combined. The outcomes demonstrated that KGM might encourage F1–7 in vitro development. F1–7-KGM reduced constipation and increased intestinal motility in mice. Additionally, the combination enhanced the water content of the mice's stools by downregulating the expression of AQP3 and upregulating the expression of 5-HT4GPCR/MUC-2 in the humoral transport pathway [153]. In addition, this combination effectively increased the level of tryptamine and content of SCFAs such as acetic acid, propionic acid, butyric acid, and valeric acid. This combination effectively promoted the humoral transport process in constipated mice. Not only KGM, konjac oligosaccharides (KOS) also upregulated the protein expression of stem cell factors (SCF)/c-kit and significantly promoted the secretion of mucus. Compared to KGM, KOS had a conspicuous laxative effect at a lower dosage implying higher potency [154].

***Nostoc sphaeroides Kutz* Polysaccharide (NSKP)** was used in a loperamide-induced slow transit constipation (STC) model. Nitric oxide (NO) and vasoactive intestinal peptide (VIP) levels were noticeably decreased, while 5-HT levels rose. The c-kit/SCF signaling pathway increased the number of interstitial cells of Cajal (ICC), moistened the colon, and alleviated constipation [155].

Inula japonica **polysaccharides (IJP)** from *Inula japonica*, in spite of having hypoglycemic and hypolipidemic effects, significantly improved constipation symptoms in induced constipated mice. It was found to be effective in stimulating intestinal motility and improving constipation in elderly diabetic patients [156].

Galacto-oligosaccharide (GOS) has demonstrated the ability to relieve constipation and facilitate stool defecation. GOS-1000-S significantly enhanced small bowel propulsion in mice with constipation, decreased the time before the first feces, and increased the weight and quantity of feces. This study provided a foundation for the creation of GOS as a dietary supplement for those with constipation [157].

4.11 Osteoporosis

Osteoporosis, which literally translates to "pore development on bones," is a skeletal illness characterized by weakened bones that put a person at higher risk for fractures. When compared to normal controls, osteoporotic bones provide evidence of an overall loss in trabecular bone volume and substantial variations in microstructure [158]. Osteoporosis is also considered a silent disease, as there are commonly no symptoms until the first fracture occurs. Globally, females are more susceptible to osteoporosis than men due to genetic and epigenetic factors such as bone mineral density (BMD), maternal body build, lifestyle, and 25(OH)-vitamin D status. Bone minerals are accumulated for the maintenance of bone shape and strength by osteoblasts, whereas bone minerals are removed by osteoclasts for supporting other body functions using the minerals in the whole circulation. Osteoporosis is frequently caused by an imbalance in the equilibrium between osteoblasts (bone-forming cells) and osteoclasts (bone-resorbing cells). The genetics and the environment present at different periods of life might be seen as contributing factors to the osteoporotic phenotype. Early research has demonstrated that circulating microRNAs (miRNAs) are related to fragility fracture risk, at least in postmenopausal women with type 2 diabetes mellitus. Specific gene loci are being searched to study the possibility for epigenetic pathways to alter BMD [159]. Among common metabolic diseases, diabetes helps osteoclast activation through the AGE-RAGE pathway. There are no studies that compare physical activity's efficacy to that of other therapies [160]. The Fracture Risk Assessment Tool or FRAX includes smoking status, glucocorticoid use, and units of alcohol per day among lifestyle factors for assessment of osteoporosis because modifiable lifestyle factors include not only these 3 but others like inadequate calcium and vitamin D intake also [161].

4.11.1 Plant Polysaccharides Against Osteoporosis

Polysaccharides obtained from plant, fungal, and algal sources, which have shown inhibitory effects in osteoporosis or helped in bone formation and maintenance, in vivo rat or mice models or in vitro cell lines, are depicted in the table with their probable mechanistic pathways. Other polysaccharides obtained from different other sources are also discussed followed by table. But very little information is available on them and their mechanisms.

4.11.1.1 Polysaccharide-Based Iron Oxide Nanoparticles (NPs)

By scavenging reactive oxygen species, PSC-Fe_2O_3 polysaccharide-based iron oxide nanoparticles reduced osteoporosis caused by iron buildup [170].

4.11.1.2 ABPB-4

The heteropolysaccharide ABPB-4, which comes from the rhizome of Achyranthes bidentata, has osteogenic activity and greatly encourages MC3T3-E1 cell proliferation, differentiation, and mineralization. Runx2 mRNA expression is noticeably increased in cells treated with ABPB-4. The relative fluorescence intensity of the skull bone mass was dramatically raised by ABPB-3 in a concentration-dependent manner, showing that it encouraged bone-forming activity. So, ABPB and ABPB-3 may 1 day be used as anti-osteoporosis medications [171].

4.11.1.3 Arabinogalactan

Phellodendron chinense Schneid's arabinogalactan, known as PPCP1, could be employed as a food supplement for bone health and/or in the development of drugs to treat diabetic osteoporosis [172].

4.11.1.4 Lycium Ruthenicum Polysaccharide (LRPS)

The anti-osteoporosis mechanisms of LRPS may involve a number of transcription factors as well as signaling pathways like Wnt/−catenin, BMP/SMAD/RUNX2, OPG/RANKL/RANK, and apoptosis-mediated pathways (Table 2) [173].

4.11.1.5 Spirulina Fusiformis

Against osteoporosis brought on by Rosiglitazone in insulin-resistant rats, *Spirulina fusiformis* polysaccharide extract has preventive effects [174].

4.11.1.6 Cistanche

In SAMP6 mice, *Cistanche deserticola* polysaccharides prevent osteoporosis by triggering the Wnt/−catenin (Fig. 4) signaling pathway [175]. Polysaccharides of *C. orchioides* may also be effective for the prevention and treatment of osteoporosis [176].

Table 2 Polysaccharides acting against osteoporosis

Plant polysaccharide	Biological source	Research model	Action on parameters	Remarks	References
Polygonati rhizoma polysaccharide (PRP)	Dried rhizome of *Polygonatum sibiricum* red (PS), *Polygonatum cyrtonema* Hua (PC), and *Polygonatum kingianum* Coll. Et Hemsl. (PK)	In vivo LPS-induced C57BL/J6 mice; in vitro BMMs from C57BL/6 mice	Osteogenic differentiation of mouse BMSCs; inhibition of the RANKL-induced osteoclastogenesis. Protection against LPS-induced osteolysis Reversed bone loss, and prevented osteoporosis in ovariectomized rats	miR-1224/hippo signaling pathway	[162]
Morinda officinalis polysaccharide (MOP)	*Morinda officinalis roots, Guangdong, Guangxi, Fujian, and Hainan*	In vivo female Wistar rats; MC3T3-E1 cells in vitro	Enhanced whole femoral bone mineral density (BMD); decreased IL-6 and TNFα; enhanced the concentrations of ca, P, mg, Zn, Mn, cu, Fe; upregulated Bmp2, Runx2, and alp gene expression		[163]
Cashew gum polysaccharide (CG-P)	*Anacardium occidentale*	In vivo female Wistar rats (Rattus norvegicus)	Decreased TNF-α and IL-1β; inhibited bone loss; lowered MPO activity in the periodontal tissue of rats	RANK/RANKL/OPG pathway	[164]
Chitosan (CS)	*Deacetylation of chitin*	In vivo	Osteoblast attachment, proliferation, and differentiation		[165]
BLE0 polysaccharide	*Young barley leaves Hordeum vulgare*	In vivo ovariectomized ICR mice	Bone loss inhibition, inhibited expressions of c-Fos and NFATc1	MAP kinases/NF-κB/NFATc1 pathways, RANKL signaling pathways	[166]
Dipsacus asper polysaccharide	*Roots of Dipsacus asper wall*	In vivo SPF female rats	Prevented bone loss, increased U-ca/Cr, U-P/Cr, ALP, TRAP, OC, and DPD/Cr levels	PI3K/Akt/eNOS signaling pathway, RANKL/RANK/OPG/VEGF pathway	[167]
Cibotium barometz polysaccharide	*Root of Cibotium barometz*	In vitro U87 human glioblastoma cell	Enhanced apoptosis, cell cycle arrest, metabolic changes, GSH depletion, and ROS accumulation	Downregulating MGMT gene expression	[168]
ASP2–1	*Root of Acorus tatarinowii Schott*	In vitro BMMs from C57BL/6 mice	Inhibited osteoclastogenesis induced by RANKL, suppressed osteoclast differentiation and bone resorption induced by RANKL, alleviated bone loss	Suppressed Blimp1 signaling and activation of the PLCγ2-Ca^{2+} – calcineurin signaling axis	[169]

Via reducing the expression of c-fos and NFATc1, **lotus leaf polysaccharides** (LLEP) exert anti-osteoporotic actions by decreasing osteoclastogenesis [177].

4.11.1.7 Polygonatum Sibiricum

By raising the expression of Limd1, the miRNA-1224 target gene, in the *Polygonatum sibiricum* Polysaccharide (PSP), osteoclast-derived macrophages are converted to osteoclasts [178]. PSP might prevent in vivo osteolysis caused by LPS and inhibit the receptor activation of nuclear factor-KB ligand (RANKL)-induced osteoclastogenesis [179]. Polygonatum sibiricum polysaccharide inhibits osteoporosis by promoting osteoblast formation and blocking osteoclastogenesis through the Wnt/βcatenin signaling pathway [162].

4.11.1.8 Angelica Polysaccharide (AP)

In marrow stromal cells, AP dramatically boosted cell viability, elevated cyclin D1, and raised RUNX2, OCN, ALP, and BMP-2 protein levels (MSCs) (Fig. 4). MSCs' PI3K/AKT and Wnt/−catenin signaling pathways are likely activated by AP. Additionally, AP dose-dependently increased the relative expression level of H19 [180].

4.11.1.9 Cuscutae Semen

When estrogen insufficiency causes osteoporosis, *Cuscutae semen* polysaccharide (CSP) aids in bone growth and prevents bone resorption [181].

Polysaccharides from persimmon leaves (PLE0) and tea have shown anti-osteoporotic effects in vivo model of ovariectomy (OVX)-induced bone loss and an in vitro system of receptor activator of nuclear factor-κB ligand (RANKL)-induced osteoclast development [182, 183].

Fig. 4 Natural polysaccharides acting against osteoporosis

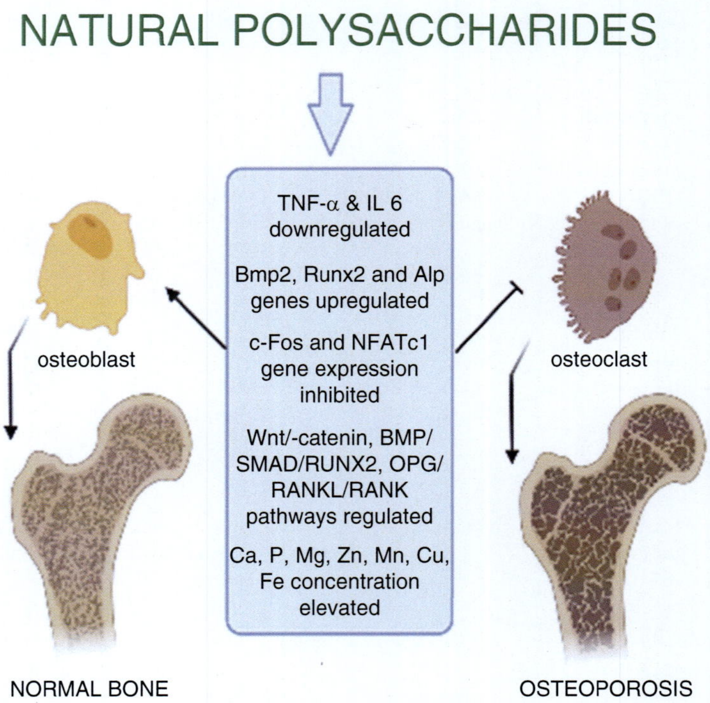

5 Conclusion

Consumption of fatty foods, especially fast foods, and red meat, along with psychological lethargy in doing regular physical exercise are the major causes that lead to hyperglycemia and hyperlipidemia. Uprising demands, consumerism, and professional competition compels people to stay involved in demanding work pressure that causes health being less prioritized than being wealthy. Natural remedies along with healthy food habits and regular physical exercise are the need of the hour to deal with the above crisis. Through this chapter, diseases such as diabetes and hyperlipidemia are highlighted and further research and findings are needed for critical and comparatively newer areas like NAFLD & NASH. Some other disorders have become daily companions for females, e.g., polycystic ovarian syndrome, and osteoporosis, that also rise directly or indirectly from lifestyle factors. Unfortunately, there are no significant natural polysaccharides or other plant-derived remedies that can be used or further developed for the betterment of female health which can also be a relevant topic of discussion and to be worked on.

Acknowledgments All the authors acknowledge the Department of Pharmaceutical Technology, Jadavpur University, Kolkata, for providing research amenities.

Conflict of InterestThe authors declare no conflicts of interest.

Funding None.

References

1. Tabish SA. Lifestyle diseases: consequences, characteristics, causes and control. J Cardiol Curr Res. 2017;9:9.
2. Hsu K-Y, Tsai Y-F, Huang C-C, Yeh W-L, Chang K-P, Lin C-C, et al. Tobacco-smoking, alcohol-drinking, and betel-quid-chewing behaviors: development and use of a web-based survey system. JMIR Mhealth Uhealth. 2018;6:e142.
3. Al-Hamodi Z, Al-Habori M, Al-Meeri A, Saif-Ali R. Association of adipokines, leptin/adiponectin ratio and C-reactive protein with obesity and type 2 diabetes mellitus. Diabetol Metab Syndr. 2014;6:99.
4. Yung JHM, Giacca A. Role of c-Jun N-terminal kinase (JNK) in obesity and type 2 diabetes. Cell. 2020;9:706.
5. Sugamura K, Keaney JF. Reactive oxygen species in cardiovascular disease. Free Radic Biol Med. 2011;51:978–92.
6. Darge HF, Andrgie AT, Tsai H-C, Lai J-Y. Polysaccharide and polypeptide based injectable thermo-sensitive hydrogels for local biomedical applications. Int J Biol Macromol. 2019;133:545–63.
7. Ahmadi F, Oveisi Z, Samani SM, Amoozgar Z. Chitosan based hydrogels: characteristics and pharmaceutical applications. Res Pharm Sci. 2015;10:1–16.
8. Nazari M, Riebeling S, Banfield CC, Akale A, Crosta M, Mason-Jones K, et al. Mucilage polysaccharide composition and exudation in maize from contrasting climatic regions. Front Plant Sci. 2020;11:11.
9. Mohammed ASA, Naveed M, Jost N. Polysaccharides; classification, chemical properties, and future perspective applications in fields of pharmacology and biological medicine (a review of current applications and upcoming potentialities). Polym Environ. 2021;29:2359–71.
10. Freemantle N, Holmes J, Hockey A, Kumar S. How strong is the association between abdominal obesity and the incidence of type 2 diabetes? Int J Clin Pract. 2008;62:1391–6.
11. Kim J, Bachmann RA, Chen J. Chapter 21 Interleukin-6 and insulin resistance. In: Vitamins & hormones, vol. 80. Academic Press; 2009. p. 613–33.

12. Kolsi RBA, Fakhfakh J, Sassi S, Elleuch M, Gargouri L. Physico-chemical characterization and beneficial effects of seaweed sulfated polysaccharide against oxydatif and cellular damages caused by alloxan in diabetic rats. Int J Biol Macromol. 2018;117:407–17.
13. Mutailifu P, Nuerxiati R, Lu C, Huojiaaihemaiti H, Abuduwaili A, Yili A. Extraction, purification, and characterization of polysaccharides from Alhagi pseudalhagi with antioxidant and hypoglycemic activities. Proc Biochem. 2022;121:121.
14. Chen L-C, Jiang B-K, Zheng W-H, Zhang S-Y, Li J-J, Fan Z-Y. Preparation, characterization and anti-diabetic activity of polysaccharides from adlay seed. Int J Biol Macromol. 2019;139:605–13.
15. Fu J, Fu J, Liu Y, Li R, Gao B, Zhang N, et al. Modulatory effects of one polysaccharide from Acanthopanax senticosus in alloxan-induced diabetic mice. Carbohydr Polym. 2012;87:2327–31.
16. He X, Wang X, Fang J, Chang Y, Ning N, Guo H, et al. Polysaccharides in Grifola frondosa mushroom and their health promoting properties: a review. Int J Biol Macromol. 2017;101:910–21.
17. Jin M, Zhao K, Huang Q, Shang P. Structural features and biological activities of the polysaccharides from Astragalus membranaceus. Int J Biol Macromol. 2014;64:257–66.
18. Lo H-C, Hsu T-H, Tu S-T, Lin K-C. Anti-hyperglycemic activity of natural and fermented Cordyceps sinensis in rats with diabetes induced by Nicotinamide and Streptozotocin. Am J Chin Med. 2006;34:819–32.
19. Wang D, Li C, Fan W, Yi T, Wei A, Ma Y. Hypoglycemic and hypolipidemic effects of a polysaccharide from Fructus corni in streptozotocin-induced diabetic rats. Int J Biol Macromol. 2019;133:420–7.
20. Wu J, Shi S, Wang H, Wang S. Mechanisms underlying the effect of polysaccharides in the treatment of type 2 diabetes: a review. Carbohydr Polym. 2016;144:474–94.
21. Li Q, Caili F, Yukui R, Guanghui H, Tongyi C. Effects of protein-bound polysaccharide isolated from pumpkin on insulin in diabetic rats. Plant Foods Hum Nutr. 2005;60:13–6.
22. Wang P-C, Zhao S, Yang B-Y, Wang Q-H, Kuang H. Anti-diabetic polysaccharides from natural sources: a review. Carbohydr Polym. 2016;148:148.
23. Yang B, Luo Y, Wei X, Kan J. Polysaccharide from Hovenia dulcis (Guaizao) improves pancreatic injury and regulates liver Glycometabolism to alleviate STZ-induced type 1 diabetes mellitus in rats. Int J Biol Macromol. 2022;214:655–63.
24. Zhang R, Zhang X, Tang Y, Mao J. Composition, isolation, purification and biological activities of Sargassum fusiforme polysaccharides: a review. Carbohydr Polym. 2020;228:115381.
25. Zhang L, Liu Y, Ke Y, Liu Y, Luo X, Li C, et al. Antidiabetic activity of polysaccharides from Suillellus luridus in streptozotocin-induced diabetic mice. Int J Biol Macromol. 2018;119:134–40.
26. Yang S, Yan J, Yang L, Meng Y, Wang N, He C, et al. Alkali-soluble polysaccharides from mushroom fruiting bodies improve insulin resistance. Int J Biol Macromol. 2019;126:466–74.
27. Lin W-Y, Chiu T-Y, Lee L-T, Lin C-C, Huang C-Y, Huang K-C. Betel nut chewing is associated with increased risk of cardiovascular disease and all-cause mortality in Taiwanese men. Am J Clin Nutr. 2008;87:1204–11.
28. Jung UJ, Choi M-S. Obesity and its metabolic complications: the role of Adipokines and the relationship between obesity, inflammation, insulin resistance, dyslipidemia and nonalcoholic fatty liver disease. Int J Mol Sci. 2014;15:6184–223.
29. Trapani L, Segatto M, Pallottini V. Regulation and deregulation of cholesterol homeostasis: the liver as a metabolic "power station". World J Hepatol. 2012;4:184–90.
30. Kurihara H, Asami S, Shibata H, Fukami H, Tanaka T. Hypolipemic effect of Cyclocarya paliurus (Batal) Iljinskaja in lipid-loaded mice. Biol Pharm Bull. 2003;26:383–5.
31. Zhang B, Li Y, Zhang F, Linhardt RJ, Zeng G, Zhang A. Extraction, structure and bioactivities of the polysaccharides from Pleurotus eryngii: a review. Int J Biol Macromol. 2020;150:1342–7.
32. Cindana F, Wilar G, Devkota H, Wathoni N. Ulvan, a polysaccharide from Macroalga ulva sp.: a review of chemistry, biological activities and potential for food and biomedical applications. Appl Sci. 2020;10:5488.
33. Badshah S, Riaz A, Muhammad A, Tel-Çayan G, Çayan F, Duru M, et al. Isolation, characterization, and medicinal potential of polysaccharides of Morchella esculenta. Molecules. 2021;26:1459.
34. Ji T, Ji WW, Wang J, Chen HJ, Peng X, Cheng KJ, et al. A comprehensive review on traditional uses, chemical compositions, pharmacology properties and toxicology of Tetrastigma hemsleyanum. J Ethnopharmacol. 2021;264:113247.
35. Chu L, Yang L, Lin L, Wei J, Wang N, Xu M, et al. Chemical composition, antioxidant activities of polysaccharide from pine needle (Pinus massoniana) and hypolipidemic effect in high-fat diet-induced mice. Int J Biol Macromol. 2019;125:445–52.
36. Surhio MM, Wang Y, Xu P, Shah F, Li J, Ye M. Antihyperlipidemic and hepatoprotective properties of selenium modified polysaccharide from Lachnum sp. Int J Biol Macromol. 2017;99:88–95.
37. Pan J, Wang H, Chen Y. Prunella vulgaris L.—a review of its ethnopharmacology, phytochemistry, quality control and pharmacological effects. Front Pharmacol. 2022;13:13.
38. Liu X, Sun Z, Zhang M, Meng X, Xia X, Yuan W, et al. Antioxidant and antihyperlipidemic activities of polysaccharides from sea cucumber Apostichopus japonicus. Carbohydr Polym. 2012;90:1664–70.
39. Nhi NTN, Khang DT, Dung TN. Termitomyces mushroom extracts and its biological activities. Food Sci Tech. 2022;42:42.
40. Ge Y, Qiu H, Zheng J. Physicochemical characteristics and anti-hyperlipidemic effect of polysaccharide from BaChu mushroom (Helvella leucopus). Food Chem. 2022;15:100443.
41. Ahmad MF. Ganoderma lucidum: persuasive biologically active constituents and their health endorsement. Biomed Pharmacother. 2018;107:507–19.
42. Cao Y, Zou L, Li W, Song Y, Zhao G, Hu Y. Dietary quinoa (Chenopodium quinoa Willd.) polysaccharides ameliorate high-fat diet-induced hyperlipidemia and modulate gut microbiota. Int J Biol Macromol. 2020;163:55–65.
43. Yuan Y, Liu Q, Zhao F, Cao J, Shen X, Li C. Holothuria Leucospilota polysaccharides ameliorate hyperlipidemia in high-fat diet-induced rats via short-chain fatty acids production and lipid metabolism regulation. Int J Mol Sci. 2019;20:4738.
44. Duansak N, Schmid-Schönbein GW, Srisawat U. Anti-obesity effect of rice bran extract on high-fat diet-induced obese mice. Prev Nutr Food Sci. 2022;27:172–9.
45. Wei H, Yue S, Zhang S, Lu L. Lipid-lowering effect of the Pleurotus eryngii (king oyster mushroom) polysaccharide from solid-state fermentation on both macrophage-derived foam cells and zebrafish models. Polymers. 2018;10:492.
46. Sandoo A, Veldhuijzen van Zanten J, Metsios G, Carroll D, Kitas G. The endothelium and its role in regulating vascular tone. Open Cardiovasc Med J. 2010;4:302–12.
47. Foex P, Sear J. Hypertension: pathophysiology and treatment. Continuing Edu Anaest Crit Care Pain. 2004;4:4.
48. Yuan Q, Xie F, Tan J, Yuan Y, Mei H, Zheng Y, et al. Extraction, structure and pharmacological effects of the polysaccharides from Cordyceps sinensis: a review. J Funct Foods. 2022;89:104909.
49. Romdhane M, Haddar A, Ghazala I, Jeddou K, Claire B, Chaabouni ES. Optimization of polysaccharides extraction from watermelon rinds: structure, functional and biological activities. Food Chem. 2017;216:355.
50. Zhang H, Ma ZF, Luo X, Li X. Effects of Mulberry fruit (Morus alba L.) consumption on health outcomes: a mini-review. Antioxidants. 2018;7:7.

51. Hassan I, Gani A, Ahmad M, Banday J. Extraction of polysaccharide from Althea rosea and its physicochemical, anti-diabetic, antihypertensive and antioxidant properties. Sci Rep. 2022;12:17116.
52. Ueno M, Tamura Y, Toda N, Yoshinaga M, Terakado S, Otsuka K, et al. Sodium alginate oligosaccharides attenuate hypertension in spontaneously hypertensive rats fed a low-salt diet. Clin Exp Hypertens. 2012;34:305–10.
53. Wang Y, Xing M, Cao Q, Ji A, Liang H, Song S. Biological activities of fucoidan and the factors mediating its therapeutic effects: a review of recent studies. Mar Drugs. 2019;17:183.
54. Ding L, Cheng P, Wang L, Hu J, Zhang Y-X, Cai G-W, et al. The protective effects of polysaccharide extract from Xin-Ji-Er-Kang formula on Ang II-induced HUVECs injury, L-NAME-induced hypertension and cardiovascular remodeling in mice. BMC Complement Altern Med. 2019;19:127.
55. Ben Gara A, Ben Abdallah Kolsi R, Jardak N, Chaaben R, El-Feki A, Fki L, et al. Inhibitory activities of Cystoseira crinita sulfated polysaccharide on key enzymes related to diabetes and hypertension: in vitro and animal study. Arch Physiol Biochem. 2017;123:31–42.
56. Yuan L-B, Hua C-Y, Gao S, Yin Y-L, Dai M, Meng H-Y, et al. Astragalus polysaccharides attenuate monocrotaline-induced pulmonary arterial hypertension in rats. Am J Chin Med. 2017;45:773–89.
57. Dong X, Zhou M, Li Y, Li Y, Ji H, Hu Q. Cardiovascular protective effects of plant polysaccharides: a review. Front Pharmacol. 2021;12:783641.
58. Li B, Wang H-Y, Huang J-H, Xu W-F, Feng X-J, Xiong Z-P, et al. Polysaccharide, the active component of Dendrobium officinale, ameliorates metabolic hypertension in rats via regulating intestinal Flora-SCFAs-vascular axis. Front Pharmacol. 2022;13:935714.
59. Lee O-H, Kim K-I, Han C-K, Kim Y-C, Hong H-D. Effects of acidic polysaccharides from Gastrodia rhizome on systolic blood pressure and serum lipid concentrations in spontaneously hypertensive rats fed a high-fat diet. Int J Mol Sci. 2012;13:698–709.
60. Sila A, Bayar N, Ghazala I, Bougatef A, Ellouz-Ghorbel R, Ellouz-Chaabouni S. Water-soluble polysaccharides from agro-industrial by-products: functional and biological properties. Int J Biol Macromol. 2014;69:236–43.
61. Xie M, Tao W, Wu F, Wu K, Huang X, Ling G, et al. Anti-hypertensive and cardioprotective activities of traditional Chinese medicine-derived polysaccharides: a review. Int J Biol Macromol. 2021;185:917–34.
62. Marc T. Cardiovascular disease: an introduction; 2018. p. 1–90. https://doi.org/10.1007/978-3-319-89315-0_1.
63. Chen G, Luo Y-C, Ji B-P, Li B, Su W, Xiao Z-L, et al. Hypocholesterolemic effects of Auricularia auricula ethanol extract in ICR mice fed a cholesterol-enriched diet. J Food Sci Technol. 2011;48:692–8.
64. Zheng Y, Ren W, Zhang L, Zhang Y, Liu D, Liu Y. A review of the pharmacological action of Astragalus polysaccharide. Front Pharmacol. 2020;11:349.
65. Tannin-Spitz T, Bergman M, van-Moppes D, Grossman S, Arad S. Antioxidant activity of the polysaccharide of the red microalga Porphyridium sp. J Appl Phycol. 2005;17:215–22.
66. Wang S, Lin X, Wang L-Y, Ruan K-F, Feng Y, Li X-Y. A polysaccharides MDG-1 augments survival in the ischemic heart by inducing S1P release and S1P1 expression. Int J Biol Macromol. 2012;50:734–40.
67. Gao T, Jiang M, Deng B, Zhang Z, Fu Q, Fu C. Aurantii fructus: a systematic review of ethnopharmacology, phytochemistry and pharmacology. Phytochem Rev. 2021;20:1–36.
68. Lekshmi VS, Rauf AA, Kurup GM. Sulfated polysaccharides from the edible marine algae Padina tetrastromatica attenuates isoproterenol-induced oxidative damage via activation of PI3K/Akt/Nrf2 signaling pathway—an in vitro and in vivo approach. Chem Biol Interact. 2019;308:258–68.
69. Gao J, Han Y-L, Jin Z-Y, Xu X-M, Zha X-Q, Chen H-Q, et al. Protective effect of polysaccharides from Opuntia dillenii haw. Fruits on streptozotocin-induced diabetic rats. Carbohydr Polym. 2015;124:25–34.
70. Nie Q, Hu J, Gao H, Fan L, Chen H, Nie S. Polysaccharide from Plantago asiatica L. attenuates hyperglycemia, hyperlipidemia and affects colon microbiota in type 2 diabetic rats. Food Hydrocoll. 2019;86:34–42.
71. Yuan C, Mei Z, Shangxi L, Yi L. PSK protects macrophages from Lipoperoxide accumulation and foam cell formation caused by oxidatively modified low-density lipoprotein. Atherosclerosis. 1996;124:171–81.
72. Zhao J, Niu X, Yu J, Xiao X, Li W, Zang L, et al. Poria cocos polysaccharides attenuated ox-LDL-induced inflammation and oxidative stress via ERK activated Nrf2/HO-1 signaling pathway and inhibited foam cell formation in VSMCs. Int Immunopharmacol. 2020;80:106173.
73. Jiang C, Li X, Shen W, Pan L, Li Q, Luo J, et al. Bioactive polysaccharides and their potential health benefits in reducing the risks of atherosclerosis: a review. J Food Biochem. 2022;46:46.
74. Raish M, Ahmad A, Ansari MA, Alkharfy KM, Aljenoobi FI, Jan BL, et al. Momordica charantia polysaccharides ameliorate oxidative stress, inflammation, and apoptosis in ethanol-induced gastritis in mucosa through NF-kB signaling pathway inhibition. Int J Biol Macromol. 2018;111:193–9.
75. Gao X, Liu Y, An Z, Ni J. Active components and pharmacological effects of Cornus officinalis: literature review. Front Pharmacol. 2021;12:633447.
76. Mishra A, Malhotra A. Tamarind xyloglucan: a polysaccharide with versatile application potential. J Mater Chem. 2009;19:19.
77. Papandrinopoulou D, Tzouda V, Tsoukalas G. Lung compliance and chronic obstructive pulmonary disease. Pulm Med. 2012;2012:542769.
78. Bos ER, Jamison DT, Bainga F, Feacham RGA, Makgoba M, Hofman KJ, et al. Disease and mortality in sub-Saharan Africa. Washington, DC: World Bank; 2006.
79. Nkosi V, Rathogwa-Takalani F, Voyi K. The frequency of fast food consumption in relation to wheeze and asthma among adolescents in Gauteng and North West provinces, South Africa. Int J Environ Res Public Health. 2020;17:1994.
80. Wright RJ. Epidemiology of stress and asthma: from constricting communities and fragile families to epigenetics. Immunol Allergy Clin N Am. 2011;31:19–39.
81. Saetta M, Di SA, Turato G, Facchini FM, Corbino L, Mapp CE, et al. CD8+ T-lymphocytes in peripheral airways of smokers with chronic obstructive pulmonary disease. Am J Respir Crit Care Med. 1998;157:822–6.
82. Zuoren Z, Pinhua P, Hongyi T, Yongbin H, Chengping H. Effects of chitin micro-particles on airway inflammation in a mouse neutrophilic asthmatic model. Zhonghua Jie He He Hu Xi Za Zhi. 2015;38:185–90.
83. Aun MV, Bonamichi-Santos R, Arantes-Costa FM, Kalil J, Giavina-Bianchi P. Animal models of asthma: utility and limitations. J Asthma Allergy. 2017;10:293–301.
84. Mikušová V, Mikuš P. Advances in chitosan-based nanoparticles for drug delivery. Int J Mol Sci. 2021;22:9652.
85. Oyarzun-Ampuero F, Brea J, Loza MI, Torres D, Alonso M. Chitosan-hyaluronic acid nanoparticles loaded with heparin for the treatment of asthma. Int J Pharm. 2009;381:122–9.
86. Chu X, Liu X-J, Qiu J-M, Zeng X-L, Bao H-R, Shu J. Effects of Astragalus and Codonopsis pilosula polysaccharides on alveolar macrophage phagocytosis and inflammation in chronic obstructive pulmonary disease mice exposed to PM2.5. Environ Toxicol Pharmacol. 2016;48:48.

87. Liu K-J, Leu S-J, Su C-H, Chiang B-L, Chen Y-L, Lee Y-L. Administration of polysaccharides from Antrodia camphorata modulates dendritic cell function and alleviates allergen-induced T helper type 2 responses in a mouse model of asthma. Immunology. 2010;129:351–62.
88. Martín-Orozco E, Norte-Muñoz M, Martínez-García J. Regulatory T cells in allergy and asthma. Front Pediatr. 2017;5:117.
89. Khazdair MR, Anaeigoudari A, Kianmehr M. Anti-asthmatic effects of Portulaca oleracea and its constituents, a review. J Pharmacopunct. 2019;22:122–30.
90. Wang C-L, Lu C-Y, Pi C-C, Zhuang Y-J, Chu C-L, Liu W-H, et al. Extracellular polysaccharides produced by Ganoderma formosanum stimulate macrophage activation via multiple pattern-recognition receptors. BMC Complement Altern Med. 2012;12:119.
91. Wu S, Wang G, Yang R, Cui Y. Anti-inflammatory effects of boletus edulis polysaccharide on asthma pathology. Am J Transl Res. 2016;8:4478–89.
92. Rapa SF, Di Iorio BR, Campiglia P, Heidland A, Marzocco S. Inflammation and oxidative stress in chronic kidney disease-potential therapeutic role of minerals, vitamins and plant-derived metabolites. Int J Mol Sci. 2019;21:263.
93. Kazancioğlu R. Risk factors for chronic kidney disease: an update. Kidney Int Suppl. 2013;3:368–71.
94. Prüss-Ustün A, Bartram J, Clasen T, Colford JM Jr, Cumming O, Curtis V, et al. Burden of disease from inadequate water, sanitation and hygiene in low- and middle-income settings: a retrospective analysis of data from 145 countries. Tropical Med Int Health. 2014;19:894–905.
95. Remuzzi G, Perico N, Macia M, Ruggenenti P. The role of renin-angiotensin-aldosterone system in the progression of chronic kidney disease. Kidney Int Suppl. 2005;68:S57–65. https://doi.org/10.1111/j.1523-1755.2005.09911.x.
96. Bashi D, Tafaghodi M, Fazly Bazzaz BS, Jafari B. Characterization of soluble soybean (SSPS) polysaccharide and development of eco-friendly SSPS/TiO2 nanoparticle bionanocomposites. Int J Biol Macromol. 2018;112:852.
97. Li H, Feng Y, Sun W, Kong Y, Jia L. Antioxidation, anti-inflammation and anti-fibrosis effect of phosphorylated polysaccharides from Pleurotus djamor mycelia on adenine-induced chronic renal failure mice. Int J Biol Macromol. 2021;170:652–63.
98. Zhou S, He Y, Zhang W, Xiong Y, Jiang L, Wang J, et al. Ophiocordyceps lanpingensis polysaccharides alleviate chronic kidney disease through MAPK/NF-κB pathway. J Ethnopharmacol. 2021;276:114189.
99. Song X, Pang H, Cui W, Zhang J, Li J, Jia L. Renoprotective effects of enzyme-hydrolyzed polysaccharides from Auricularia polytricha on adenine-induced chronic kidney diseases in mice. Biomed Pharmacother. 2021;135:111004.
100. Ma C, Wang X, Shao W, Zhao Q, Wei J, Liu Z, et al. Effects of a Cordyceps militaris with Herba epimedii complex on chronic renal failure induced by Adenine in vivo. Nat Pro Commun. 2022;17:1934578X221105373.
101. JING S, Li L, Ji D, Takiguchi Y, Yamaguchi T. Effect of chitosan on renal function in patients with chronic renal failure. J Pharm Pharmacol. 1997;49:721–3.
102. Ryu M, Chung H. Fucoidan reduces oxidative stress by regulating the gene expression of HO-1 and SOD-1 through the Nrf2/ERK signaling pathway in HaCaT cells. Mol Med Rep. 2016;14:3255.
103. Wang Y, Yin H, Lv X, Wang Y, Gao H, Wang M. Protection of chronic renal failure by a polysaccharide from Cordyceps sinensis. Fitoterapia. 2010;81:397–402.
104. Hu G, Lu Y, Wei D. Chemical characterization of Chinese chive seed (Allium tuberosum Rottl.). Food Chem. 2006;99:693–7.
105. Burisch J, Jess T, Martinato M, Lakatos PL, On behalf of ECCO -EpiCom. The burden of inflammatory bowel disease in Europe. J Crohns Colitis. 2013;7:322–37.
106. Berkowitz L, Schultz BM, Salazar GA, Pardo-Roa C, Sebastián VP, Álvarez-Lobos MM, et al. Impact of cigarette smoking on the gastrointestinal tract inflammation: opposing effects in Crohn's disease and ulcerative colitis. Front Immunol. 2018;9:74.
107. Martinho-Grueber M, Kapoglou I, Bravo F, Sarraj R, Benz E, Restellini S, et al. Alcohol and cannabis consumption in patients with inflammatory bowel disease: prevalence, pattern of consumption and impact on the disease. Eur J Gastroenterol Hepatol. 2022;35:21; Publish Ahead of Print
108. Hoekman D, Vlieger A, Stokkers P, Mahhmod N, Rietdijk S, Boer N, et al. Hypnotherapy for irritable bowel syndrome-type symptoms in patients with quiescent inflammatory bowel disease: a randomized, controlled trial. J Crohns Colitis. 2020;15:1106.
109. Liu B, Lin Q, Yang T, Zeng L, Shi L, Chen Y, et al. Oat β-glucan ameliorates Dextran Sulfate Sodium (DSS)-induced ulcerative colitis in mice. Food Funct. 2015;6:3454.
110. Førland DT, Johnson E, Saetre L, Lyberg T, Lygren I, Hetland G. Effect of an extract based on the medicinal mushroom Agaricus blazei Murill on expression of cytokines and calprotectin in patients with ulcerative colitis and Crohn's disease. Scand J Immunol. 2011;73:66–75.
111. Nie Y, Lin Q, Luo F. Effects of non-starch polysaccharides on inflammatory bowel disease. Int J Mol Sci. 2017;18:1372.
112. Wei Q, Xing M, Wang K, Yang Q, Zhao J, Wang Y, et al. Fucoidan is not completely dependent on degradation to Fucose to relieve Ulcerative colitis. Pharmaceuticals. 2022;15:430.
113. Fernández-Bañares F, Hinojosa J, Sánchez-Lombraña JL, Navarro E, Martínez-Salmerón JF, García-Pugés A, et al. Randomized clinical trial of Plantago Ovata seeds (dietary fiber) as compared with Mesalamine in maintaining remission in ulcerative colitis. Am J Gastroenterol. 1999;94:427.
114. Hanai H, Kanauchi O, Mitsuyama K, Andoh A, Takeuchi K, Takayuki I, et al. Germinated barley foodstuff prolongs remission in patients with ulcerative colitis. Int J Mol Med. 2004;13:643–7.
115. Takagi T, Naito Y, Higashimura Y, Ushiroda C, Mizushima K, Ohashi Y, et al. Partially hydrolysed guar gum ameliorates murine intestinal inflammation in association with modulating luminal microbiota and SCFA. Br J Nutr. 2016;116 1199–205.
116. Paul SP, Barnard P, Edate S, Candy DCA. Stool consistency and abdominal pain in irritable bowel syndrome may be improved by partially hydrolysed guar gum. J Pediatr Gastroenterol Nutr. 2011;53:582.
117. Yue Y, Wu S, Li Z, Li J, Li X, Xiang J, et al. Wild jujube polysaccharides protect against experimental inflammatory bowel disease by enabling enhanced intestinal barrier function. Food Funct. 2015;6:2568–77.
118. Woolf A, Rehman RB, Rose R. Gastric ulcer. Treasure Island (FL): StatPearls Publishing; 2022.
119. Lee YB, Yu J, Choi HH, Jeon BS, Kim H-K, Kim S-W, et al. The association between peptic ulcer diseases and mental health problems: a population-based study: a STROBE compliant article. Medicine. 2017;96:e7828.
120. Emerenziani S, Guarino MPL, Trillo Asensio LM, Altomare A, Ribolsi M, Balestrieri P, et al. Role of overweight and obesity in gastrointestinal disease. Nutrients. 2019;12:111.
121. Tsermpini EE, Plemenitaš Ilješ A, Dolžan V. Alcohol-induced oxidative stress and the role of antioxidants in alcohol use disorder: a systematic review. Antioxidants. 2022;11:1374.
122. Bilal M, Iqbal HMN. Marine seaweed polysaccharides-based engineered cues for the modern biomedical sector. Mar Drugs. 2020;18:7.
123. Tian B, Zhao Q, Xing H, Xu J, Li Z, Zhu H, et al. Gastroprotective effects of Ganoderma lucidum polysaccharides with different

molecular weights on ethanol-induced acute gastric injury in rats. Nutrients. 2022;14:1476.
124. Srikanta BM, Sathisha UV, Dharmesh SM. Alterations of matrix metalloproteinases, gastric mucin and prostaglandin E2 levels by pectic polysaccharide of swallow root (Decalepis hamiltonii) during ulcer healing. Biochimie. 2010;92:194–203.
125. Choi J, Raghavendran HRB, Sung N-Y, Kim J-H, Chun BS, Ahn DH, et al. Effect of fucoidan on aspirin-induced stomach ulceration in rats. Chem Biol Interact. 2010;183:249–54.
126. He X, Wang X, Fang J, Zhao Z, Huang L, Guo H, et al. Bletilla striata: medicinal uses, phytochemistry and pharmacological activities. J Ethnopharmacol. 2017;195:20–38.
127. Bang COJ, Ham I, Choi H. Anti-inflammatory effect of Bletilla striata through the inhibition of inflammatory mediators expression in raw 264.7 cells. Planta Med. 2008;74:PA260.
128. Mostofa R, Ahmed S, Begum MM, Sohanur Rahman M, Begum T, Ahmed SU, et al. Evaluation of anti-inflammatory and gastric anti-ulcer activity of Phyllanthus niruri L. (Euphorbiaceae) leaves in experimental rats. BMC Complement Altern Med. 2017;17:267.
129. Maria-Ferreira D, da Silva LM, Mendes DAGB, de Cabrini DA, Nascimento AM, Iacomini M, et al. Rhamnogalacturonan from Acmella oleracea (L.) R.K. Jansen: gastroprotective and ulcer healing properties in rats. PLoS One. 2014;9:e84762.
130. Antunes C, Aleem A, Curtis SA. LVHN Scholarly Work.
131. Wu JC, Mui L, Cheung CM, Chan Y, Sung JJ. Obesity is associated with increased transient lower esophageal sphincter relaxation. Gastroenterology. 2007;132:883–9.
132. Zhang M, Hou Z-K, Huang Z-B, Chen X-L, Liu F-B. Dietary and lifestyle factors related to gastroesophageal reflux disease: a systematic review. Ther Clin Risk Manag. 2021;17:305–23.
133. El-Nahas HM, Hosny KM. Chitosan-based floating microspheres of Trimetazidine Dihydrochloride; preparation and in vitro characterization. Indian J Pharm Sci. 2011;73:397–403.
134. GUPTA S, VYAS SP. Carbopol/chitosan based pH triggered in situ gelling system for ocular delivery of Timolol maleate. Sci Pharm. 2010;78:959–76.
135. Gawkowska D, Cybulska J, Zdunek A. Structure-related gelling of pectins and linking with other natural compounds: a review. Polymers. 2018;10:762.
136. Lovegrove A, Edwards CH, De Noni I, Patel H, El SN, Grassby T, et al. Role of polysaccharides in food, digestion, and health. Crit Rev Food Sci Nutr. 2017;57:237–53.
137. Bor S, Kalkan İH, Çelebi A, Dinçer D, Akyüz F, Dettmar P, et al. Alginates: from the ocean to gastroesophageal reflux disease treatment. Turk J Gastroenterol. 2019;30:109–36.
138. Nicolau LAD, Batista-Lima FJ, Santana APM, Sales TMAL, Carmo-Neto JP, Freitas GB, et al. Cashew gum, a biopolymer, topically protects Oesophageal mucosa in non erosive reflux disease: a promising translational study. Carbohydr Polym. 2019;226:115205.
139. Havelund T, Aalykke C, Rasmussen L. Efficacy of a pectin-based anti-reflux agent on acid reflux and recurrence of symptoms and oesophagitis in gastro-oesophageal reflux disease. Eur J Gastroenterol Hepatol. 1997;9:509–14.
140. Kapadia CJ, Mane VB. Raft-forming agents: Antireflux formulations. Drug Dev Ind Pharm. 2007;33:1350–61.
141. Cosenza VA, Navarro DA, Stortz CA, Rojas AM. Rheology of partially and totally oxidized red seaweed galactans. Carbohydr Polym. 2020;230:115653.
142. Moazen M, Shafaghi A, Ebrahimi-Najafabadi H, Ghasemi S, Ashoobi MT, Manoochehri S. Optimization of pH-sensitive ingredients and characterization of raft-forming alginate-based oral suspensions as reflux suppressant. J Drug Delivery Sci Tech. 2022;68:103124.
143. Steele SR, Mellgren A. Constipation and obstructed defecation. Clin Colon Rectal Surg. 2007;20:110–7.
144. Fuhrman J. The hidden dangers of fast and processed food. Am J Lifestyle Med. 2018;12:375–81.
145. Ma H, Xiong H, Zhu X, Ji C, Xue J, Li R, et al. Polysaccharide from Spirulina platensis ameliorates diphenoxylate-induced constipation symptoms in mice. Int J Biol Macromol. 2019;133:1090–101.
146. Li G, Zou X, Kuang G, Ren Y, Deng C, Lin Q, et al. Preventative effects of fermented Chimonobambusa quadrangularis shoot on activated carbon-induced constipation. Exp Ther Med. 2017;13:1093–100.
147. Drummond L, Gearry RB. Chapter twelve—kiwifruit modulation of gastrointestinal motility. In: Boland M, Moughan PJ, editors. Advances in food and nutrition research, vol. 68; 2013. p. 219–32.
148. Rao MRP, Warrier DU, Gaikwad SR, Shevate PM. Phosphorylation of psyllium seed polysaccharide and its characterization. Int J Biol Macromol. 2016;85:317–26.
149. Polymeros D, Beintaris I, Gaglia A, Karamanolis G, Papanikolaou IS, Dimitriadis G, et al. Partially hydrolyzed guar gum accelerates colonic transit time and improves symptoms in adults with chronic constipation. Dig Dis Sci. 2014;59:2207–14.
150. Jiang Y, Yu J, Li Y, Wang L, Hu L, Zhang L, et al. Extraction and antioxidant activities of polysaccharides from roots of Arctium lappa L. Int J Biol Macromol. 2019;123:531–8.
151. Ren X, Liu L, Gamallat Y, Zhang B, Xin Y. Enteromorpha and polysaccharides from Enteromorpha ameliorate loperamide-induced constipation in mice. Biomed Pharmacother. 2017;96:1075–81.
152. Jiang H, Dong J, Jiang S, Liang Q, Zhang Y, Liu Z, et al. Effect of Durio zibethinus rind polysaccharide on functional constipation and intestinal microbiota in rats. Food Res Int. 2020;136:109316.
153. Lu Y, Yu Z, Zhang Z, Liang X, Gong P, Yi H, et al. Bifidobacterium animalis F1-7 in combination with Konjac glucomannan improves constipation in mice via humoral transport. Food Funct. 2021;12:791–801.
154. Hu M, Fang C, Liu Y, Gao M, Zhang D, Shi G, et al. Comparative study of the laxative effects of Konjac oligosaccharides and Konjac glucomannan on loperamide-induced constipation in rats. Food Funct. 2021;12:7709–17.
155. Liu Y, Yang L, Bi C, Tang K, Zhang B. Nostoc sphaeroides Kütz polysaccharide improved constipation and promoted intestinal motility in rats. Evid Based Complement Alternat Med. 2021;2021:5596531.
156. Shan JJ, Zhang Y, Diao YL, Qu WS, Zhao XN. Effect of an antidiabetic polysaccharide from Inula japonica on constipation in normal and two models of experimental constipated mice. Phytother Res. 2010;24:1734–8.
157. Niittynen L, Kajander K, Korpela R. Galacto-oligosaccharides and bowel function. Scand J Food Nutr. 2007;51:62–6.
158. Föger-Samwald U, Dovjak P, Azizi-Semrad U, Kerschan-Schindl K, Pietschmann P. Osteoporosis: pathophysiology and therapeutic options. EXCLI J. 2020;19:1017–37.
159. Bocheva G, Boyadjieva N. Epigenetic regulation of fetal bone development and placental transfer of nutrients: progress for osteoporosis. Interdiscip Toxicol. 2012;4:167–72.
160. Asadipooya K, Uy EM. Advanced glycation end products (AGEs), receptor for AGEs, diabetes, and bone: review of the literature. J Endocr Soc. 2019;3:1799–818.
161. Unnanuntana A, Gladnick BP, Donnelly E, Lane JM. The assessment of fracture risk. J Bone Joint Surg Am. 2010;92:743–53.
162. Du L, Nong M-N, Zhao J-M, Peng X-M, Zong S-H, Zeng G-F. Polygonatum sibiricum polysaccharide inhibits osteoporosis by promoting osteoblast formation and blocking osteoclastogenesis through Wnt/β-catenin signalling pathway. Sci Rep. 2016;6:32261.
163. Wu P, Jiao F, Huang H, Liu D, Tang W, Liang J, et al. Morinda officinalis polysaccharide enable suppression of osteoclastic differentiation by exosomes derived from rat mesenchymal stem cells. Pharm Biol. 2022;60:1303–16.

164. Souza Filho MD, Medeiros JVR, Vasconcelos DFP, Silva DA, Leódido ACM, Fernandes HF, et al. Orabase formulation with cashew gum polysaccharide decreases inflammatory and bone loss hallmarks in experimental periodontitis. Int J Biol Macromol. 2018;107:1093–101.
165. Rochet N, Balaguer T, Boukhechba F, Laugier J-P, Quincey D, Goncalves S, et al. Differentiation and activity of human pre-osteoclasts on chitosan enriched calcium phosphate cement. Biomaterials. 2009;30:4260–7.
166. Choi S-W, Kim S-H, Lee K-S, Kang HJ, Lee MJ, Park K-I, et al. Barley seedling extracts inhibit RANKL-induced differentiation, fusion, and maturation of osteoclasts in the early-to-late stages of osteoclastogenesis. Evid Based Complement Alternat Med. 2017;2017:6072573.
167. Sun X, Wei B, Peng Z, Fu Q, Wang C, Zhen J, et al. Protective effects of Dipsacus asper polysaccharide on osteoporosis in vivo by regulating RANKL/RANK/OPG/VEGF and PI3K/Akt/eNOS pathway. Int J Biol Macromol. 2019;129:579–87.
168. Zhao X, Wu Z-X, Zhang Y, Yan Y-B, He Q, Cao P-C, et al. Anti-osteoporosis activity of Cibotium barometz extract on ovariectomy-induced bone loss in rats. J Ethnopharmacol. 2011;137:1083–8.
169. Wang J, Zhang Y, Xu X, Jin W, Jing C, Leng X, et al. ASP2-1, a polysaccharide from Acorus tatarinowii Schott, inhibits osteoclastogenesis via modulation of NFATc1 and attenuates LPS-induced bone loss in mice. Int J Biol Macromol. 2020;165:2219–30.
170. Yu P, Zheng L, Wang P, Chai S, Zhang Y, Shi T, et al. Development of a novel polysaccharide-based iron oxide nanoparticle to prevent iron accumulation-related osteoporosis by scavenging reactive oxygen species. Int J Biol Macromol. 2020;165:1634–45.
171. Lin Z, Li T, Yu Q, Chen H, Zhou D, Li N, et al. Structural characterization and in vitro osteogenic activity of ABPB-4, a heteropolysaccharide from the rhizome of Achyranthes bidentata. Carbohydr Polym. 2021;259:117553.
172. Wang N, Xu P, Yao W, Zhang J, Liu S, Wang Y, et al. Structural elucidation and anti-diabetic osteoporotic activity of an arabinogalactan from Phellodendron chinense Schneid. Carbohydr Polym. 2021;271:118438.
173. Shan LS, Su J, Zhang Y, Wen HX, Ping WX, Cong HM, et al. Benefits and mechanisms of polysaccharides from Chinese medicinal herbs for anti-osteoporosis therapy: a review. Int J Biol Macromol. 2021;193:1996–2005.
174. Gupta S, Hrishikeshvan HJ, Sehajpal PK. Spirulina protects against rosiglitazone induced osteoporosis in insulin resistance rats. Diabetes Res Clin Pract. 2010;87:38–43.
175. Wang F, Tu P, Zeng K, Jiang Y. Total glycosides and polysaccharides of Cistanche deserticola prevent osteoporosis by activating Wnt/β-catenin signaling pathway in SAMP6 mice. J Ethnopharmacol. 2021;271:113899.
176. Wang X, Zhang M, Zhang D, Wang X, Cao H, Zhang Q, et al. Structural elucidation and anti-osteoporosis activities of polysaccharides obtained from Curculigo orchioides. Carbohydr Polym. 2019;203:292–301.
177. Hwang Y-H, Jang S-A, Lee A, Cho C-W, Song Y-R, Hong H-D, et al. Polysaccharides isolated from lotus leaves (LLEP) exert anti-osteoporotic effects by inhibiting osteoclastogenesis. Int J Biol Macromol. 2020;161:449–56.
178. Li B, Wu P, Fu W, Xiong Y, Zhang L, Gao Y, et al. The role and mechanism of miRNA-1224 in the Polygonatum sibiricum polysaccharide regulation of bone marrow-derived macrophages to osteoclast differentiation. Rejuvenation Res. 2019;22:420–30.
179. Qu Z, Zhang B, Kong L, Gong Y, Feng M, Gao X, et al. Receptor activator of nuclear factor-κB ligand-mediated osteoclastogenesis signaling pathway and related therapeutic natural compounds. Front Pharmacol. 2022;13:1043975.
180. Xie X, Liu M, Meng Q. Angelica polysaccharide promotes proliferation and osteoblast differentiation of mesenchymal stem cells by regulation of long non-coding RNA H19. Bone Joint Res. 2019;8:323–32. https://doi.org/10.1302/2046-3758.87.BJR-2018-0223.R2.
181. Liu H, Zhang H, Fan H, Tang S, Weng J. The preventive effect of Cuscutae semen polysaccharide on bone loss in the ovariectomized rat model Biomed Pharmacother. 2020;130:110613.
182. Hwang Y-H, Ha H, Kim R, Cho C-W, Song Y-R, Hong H-D, et al. Anti-osteoporotic effects of polysaccharides isolated from persimmon leaves via osteoclastogenesis inhibition. Nutrients. 2018;10
183. Xu H, Yin D, Liu T, Chen F, Chen Y, Wang X, et al. Tea polysaccharide inhibits RANKL-induced osteoclastogenesis in RAW264.7 cells and ameliorates ovariectomy-induced osteoporosis in rats. Biomed Pharmacother. 2018;102:539–48.

Role of Antioxidant Vitamins and Minerals from Herbal Source in the Management of Lifestyle Diseases

S. Ganguly and J. Kumar

Abstract

In the race of development, lifestyle of humans has become hectic and stressful. The negligence of healthy diet, over pressure of work and devoid of time has made humans prone to diseases due to lifestyle disorder. The unsystematic lifestyle paves a pathway to several diseases like diabetes, cardiac problems, hypertension, neurodegenerative disorders, and cancer. These diseases generate reactive oxygen species which produces oxidative stress in the biological system. The continuous exposure to pro-oxidants affects the DNA of mitochondria and causes cellular deformities, chromosomal aberrations, and mutation of DNA. Due to lack of time, people are prone to fast foods and chemically processed foods which trigger the problem to thousand times more in synthesizing excess amount of oxidative stress. The antioxidants are the agents which neutralize this oxidative stress by scavenging the free radicals from the biological system. Use of herbal sources of antioxidant has gained popularity because of it is cost-effective, easily available, and cheaper in price. Several fruits and vegetables are reported to have preventive efficacies in oxidative stress and have antioxidant properties. In this chapter, we have discussed about the oxidative stress, its generation in different pathological condition of diseases, and their mechanism in promoting the lifestyle diseases. Finally, concluding with the remedies and list of antioxidant vitamins and minerals along with its mode of antioxidant action which are essential to alleviate the lifestyle disease and lead a disease-free life.

Keywords

Lifestyle diseases · Oxidative stress · Antioxidant vitamins · Antioxidant minerals · Herbal sources

S. Ganguly (✉) · J. Kumar
University Department of Botany, Ranchi University, Ranchi, Jharkhand, India
e-mail: gangulysharmistha23@gmail.com

1 Introduction

In the charismatic world of science and technology, the lifestyles of humans have become very hectic and preoccupied with work. The heavy work load and short deadlines at workplaces have triggered the situation in such a manner that the compensation is paid off by the health of mankind. No matter what, in earlier days the schedule of work was fixed during the daytime, whereas, nowadays the work has been seen in different shifts including the night which even hampers the sleeping pattern and the biological rhythm of the humans. The lifestyle disorder has paved a pathway to several diseases into their lives which were not even seen in their ancestors in traces. The food habit has changed drastically and has landed up to junk foods and fast foods due to scarcity of time management. Stress is a word which has qualified a prominent place in almost everyone's life today due to lifestyle disorder and overburden of work for earning money. Lifestyle diseases are those which are chronic noncommunicable in nature and have a long-term life-threatening effects in humans. It depends of the day-to-day lifestyle management that leads to sedentary daily life. Several lifestyle diseases are cardiovascular diseases like heart failure, arrhythmias, cardiomyopathy, coronary artery disease, heart valve disease; obesity; type 2 diabetes; hypertension; Chronic Obstructive Pulmonary Diseases (COPD); asthma; osteoporosis; and Polycystic Ovarian Disease (PCOD). The only way to combat the lifestyle diseases is to incorporate a healthy lifestyle and food habits enriched with antioxidants and minerals.

Several metabolic processes in our body such as breathing, digestion, conversion of fats produce free radicals which have corrosive action in the biological system of humans. They actually hamper the natural antioxidant system of the body by damaging the cell membrane, inhibiting the enzymatic actions, distorting cell division process, destroying DNA, and also restrict generation of energy [1]. It has been observed in the last 50 decades that free radicals are responsible for development of various diseases and role of antioxi-

A. K. Dhara, S. C. Mandal (eds.), *Role of Herbal Medicines*, https://doi.org/10.1007/978-981-99-7703-1_22

dants in combating it [2]. Antioxidants are badged as bona fide warriors of the defense mechanism against free radicals which are achieved by maintenance of balance between antioxidants and oxidants [3]. Recent studies have proved that food supplements and foods with antioxidant property play a major role in maintaining the health and compensate with the required supplements for anabolic and catabolic processes of the body which ensure immunity in the body to withstand ailments. There is a huge cornucopia of antioxidant vitamins and minerals from herbal sources which promotes normal functioning of the body whose usage in day-to-day life not only normalizes the metabolism but also prevents the disease by boosting the immune system of the body and also keeping the body away from lifestyle diseases [4]. Many plants having compounds as secondary metabolites which are proved clinically having several properties like antioxidizing, analgesic, anti-inflammatory, diuretic, antipyretic, carminative, and also help in releasing stress and blood pressure as well as asthma [5]. This review is a summarization of the causes of lifestyle diseases and its management by using antioxidant vitamins and minerals from plant sources which helps in maintaining a healthy and disease-free life.

2 Role of Oxidative Stress in Lifestyle Diseases

Oxidative stress is caused by the harmful reactive oxygen species (ROS) which are produced during different metabolic activities of the cell which include processes of transcription, signal transduction, and responses of immune system [6]. Some common reactive oxygen species of the cells are hydroxyl radical (OH^-), superoxides (O_2^-), and hydrogen peroxide (H_2O_2) [7, 8]. These reactive oxygen species cause oxidative damage to different biomolecules when produced in excess and harm structural skeletons of the cell like lipids, amino acids, and DNA which gradually lead to the development of diseases like cancer, cardiovascular, respiratory, neurodegenerative, digestive, hypertension, diabetes, and aging [6]. Oxidative stress, as described is caused by the reactive oxygen species which eventually lead to apoptosis [9]. Antioxidants are patented to have free radical scavenging potential and as the reactive oxygen species are nothing but free radicals, it scavenges the ROS and reduces the oxidative stress faced by the cell [6]. Increase in oxidative stress causes several chronic diseases due to low supply or production of antioxidants in the body which becomes a flag bearer of ailments like hypertension, obesity, atherosclerosis, chronic heart failure, risk of stroke, cardiovascular diseases, cancer, and neurological disorders [3]. The intake of antioxidant-rich foods and beverages helps in neutralizing the oxidative stress produced by the harmful reactive oxygen species. Hence, malnutrition and deficiency of vitamins and minerals rich in antioxidants are directly correlated to the chronic diseases due to lifestyle disorder like diabetes, hypertension, cardiovascular, and respiratory diseases [6]. A distorted oxidation reduction balance and nutritional deficiency are noticed in the patients suffering from oxidative damage. The deficiency of antioxidants promotes the risk of disease occurrence which is due to the malnutrition [10–12]. In this review, we have tried to put collective information regarding the antioxidant vitamins and minerals which can help to regulate, cure, and keep immuned from diseases related to reactive oxygen species-mediated oxidative stress diseases.

3 Some Lifestyle Diseases Suffered by Humans

The world has been undergoing through a massive paradigm shift of technology, advancement, and urbanization which have contributed to induction of several chronic diseases caused by a lacuna in lifestyle management. Pattern of chronic diseases circles around the lifestyle disorders which are generally noncommunicable diseases like respiratory, cardiovascular, neurodegenerative, polycystic ovarian disease, cancer, digestive diseases, aging, diabetes, and hypertension (Fig. 1). The risk and detriments of these chronic diseases are a collective form of malnutritive diet, unhealthy lifestyle, smoking and drinking habits, obesity, exposure to pollutants, lack of exercise, and stress [13] and are mainly affected in the middle-aged people [14].

3.1 Respiratory Diseases

From recent studies, it is revealed that respiratory diseases like chronic obstructive pulmonary disease and asthma are major chronic health issues which are increasing the rate of mortality all over the globe [15, 16]. Exposure to environmental pollutants in the air and smoking of cigarettes promotes these diseases among urban humans. The drastic noxiousness of the reactive oxygen species damages the lungs by affecting the alveoli and connective tissues which thereby paves a pathway to chronic obstructive pulmonary disease [17]. The pathophysiology of asthma also indicates increased oxidative stress [18] which can cause inflammation and leads to progression of asthma [19]. The transcription factors like activator protein-1 (AP-1), mitogen-activated protein kinase (MAPK), nuclear factor-kappa B (NF-κB), and proinflammatory mediators get activated and cause inflammation in the airways due to increased oxidative stress during bronchial asthma [6] and promote secretion of mucin

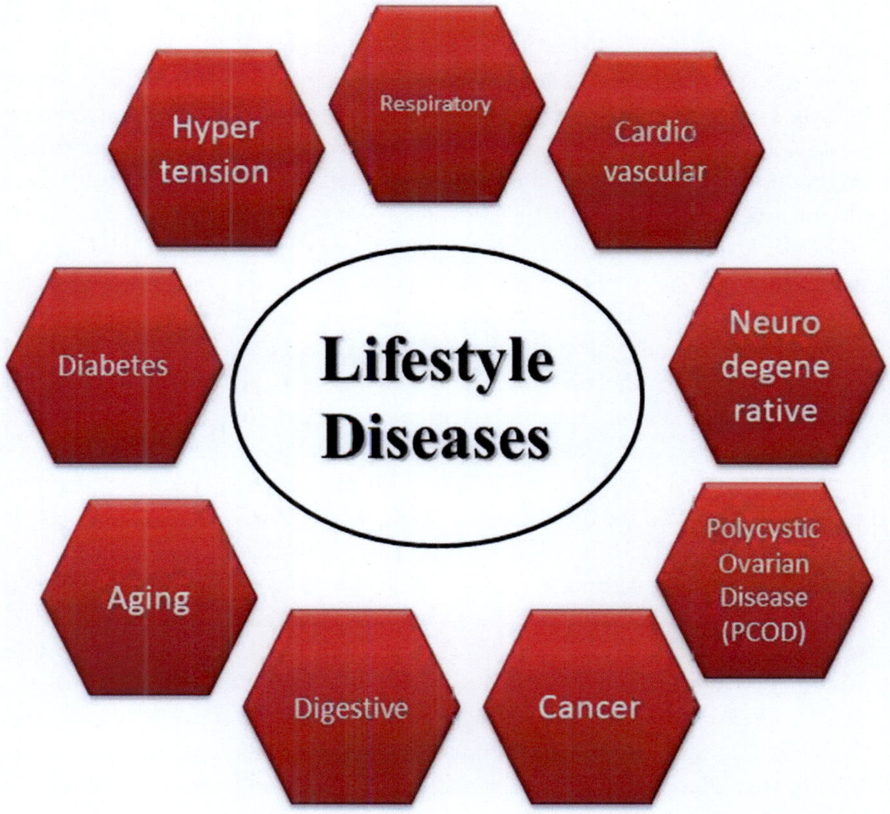

Fig. 1 Diagrammatic representation of some noncommunicable diseases suffered by humans

related to asthma [19–21]. The heavy formation of oxidative stress destroys the respiratory system which also increases the production of reactive oxygen species due to dismantled defense mechanism of antioxidant [22]. Malnutrition and stumpy body mass index are correlated to the austerity of chronic obstructive pulmonary disease [23]. The chronic obstructive pulmonary disease patients those who are underweight are more familiar to pulmonary impairment, intolerant to exercise, and have high rates of mortality than people with standard weights [23, 24]. Deficiency in dietary antioxidants like vitamin E, vitamin C, polyphenols, carotenoids, etc. deteriorates the antioxidant defense mechanism and aggravates the development of the disease [23, 25]. The diet rich in fruits, green vegetables, whole grains, and fish has been found to be related to enhancement in the function of pulmonary tissues which lowers the risk of chronic obstructive pulmonary disease [26]. The supplements has proved to improve the function of respiratory tissues in patients with poor diet plans in patients suffering from chronic obstructive pulmonary disease [6]. Vitamin D supplements help to strengthen the muscles responsible for respiratory function and increase their capacity to perform exercise in individuals suffering from chronic obstructive pulmonary disease [27]. Vitamin C helps in reduction of oxidative stress and pulmonary inflammation not only by scavenging the reactive oxygen species but also activating the NF-κB pathway [28]. Melatonin is another antioxidant which regulates the circadian biological clock and also diminishes lung damage related to oxidative stress [29]. A study has proved that intake of vitamin A and vitamin C is inversely proportional to the occurrence of asthma in an individual [30]. The juxtaposition of Vitamin C along with Vitamin E helps in stimulating the process of regeneration of α-tocopherol from its state of oxidization [31]. The carotenoids which are rich sources of antioxidants have shown promising outcomes in asthma and healthy functioning of lungs [32]. The natural antioxidant plays an important role between oxidative stress and incidence of respiratory disorders [33]. Though intake of antioxidants helps in attenuating the respiratory disorders, but these supplements when consumed above the recommended dose can act as oxidative stress promoters and prooxidants [34]. These studies help in supporting the fact that natural antioxidants have the potential to alleviate respiratory diseases like chronic obstructive pulmonary disease and asthma.

3.2 Cardiovascular Diseases

In USA, cardiovascular diseases are responsible for nearly more than a million of deaths every year [6]. In majority of the cases of cardiovascular disorder, development of atherosclerosis caused by oxidative stress plays a pivotal role [35].

In a vascular cell, excess amount of reactive oxygen species can be produced from NAD(P) H oxidase, uncoupling of nitric oxide synthases, plus mitochondria, which are responsible for modification in lipoprotein of low density by oxidation [35–37]. These oxidized form of low-density lipoproteins are transported by the arterial lumens and persuade the process of apoptosis in the cells of endothelium and smooth muscles. After phagocytosis of oxidized low-density lipoproteins, the macrophages get transformed into foam cells which discharge some growth mediators for attracting the smooth muscle cells inside the intima. The extracellular matrix secreted from the smooth muscle cells forms a cap which is thin and fibrous and surrounds the fatty streak [6, 35–37]. The fatty streaks get converted to fibrous plaque of advanced level by constant proliferation of monocytes, smooth muscles cells, and macrophages which thereby paves a pathway to occlusion of vessels [35]. Oxidative stress has also been reported to subsidize cardiac failure by development of cardian hypertrophy, apoptosis of myocytes, and ischemic-reperfusion injuries [35, 38]. Several studies have been focused of evaluation of natural antioxidants in cardiovascular diseases implicated by reactive oxygen species [6]. The intake of green vegetables and fruits has found to elevate the antioxidant levels in the blood which include vitamin C and carotene and also helps to reduce oxidation of cholesterol [6]. In a study where meta-analysis was done, about 16 reported cases showed the inverse ratio of cardiovascular diseases and consumption of green vegetables and fruits [39]. Natural α-tocopherol has been reported to reduce cardiovascular-related mortality in patients with coronary atherosclerosis [6]. Antioxidants like polyphenolic compounds and lycopene can play a major role in protecting cardiovascular diseases [40]. RRR-AT which is the natural form of antioxidant vitamin E has shown positive response against this disease [35, 41]. The prooxidant activity of vitamin C at elevated doses may cause production of reactive oxygen species which hinders the protective nature of the antioxidant supplements. Lycopene that is present in tomatoes is a source of natural antioxidants which was found to reduce the cardiovascular diseases in a study [6, 42, 43]. Lycopene inhibits the synthesis of cholesterol by hindering the oxidation of low-density lipoproteins and further degrades it [40]. In another study, it was revealed that lycopene has substantial shielding role in patients of myocardial infarction. So, it likely supported the intake of vegetables rich in lycopene [44]. Almost all fruits, green vegetables, beverages, and cereals contain ample amount of polyphenolic compounds which significantly reduce the threat for cardiovascular diseases [45]. Flavonoids are found to mend better functioning of endothelium which plays a prominent role in cardiovascular diseases [6]. The polyphenols have intrinsic anti-inflammatory potential which helps in vasodilation regulation and endothelial cell apoptosis [46].

3.3 Neurodegenerative Diseases

Neurons contain high amount of polyunsaturated fatty acids containing cell membranes and have elevated mandate for consuming oxygen which makes them vulnerable to oxidative stress-associated destruction [47]. Several studies revealed that the reactive oxygen species triggers the incidence of neurodegenerative disorders like Huntington's disease, Alzheimer's disease, Parkinson's disease, amyotrophic lateral sclerosis, and spinocerebellar ataxia [8, 48–52]. Huntington's disease is caused by expansion of CAG repeat after mutation in exon 1 of the gene and produces a mutated protein "huntingtin" (mHtt). It is an autosomal dominant disease. The oxidative stress is a characteristic of Huntington's disease which increases the oxidation on DNA in the patient's brain. The dysfunctional pattern of mitochondria is induced by oxidative stress which intensifies the formation of reactive oxygen species. An enzyme of Kreb's cycle, mitochondrial aconitase, is damaged in this disease by reactive oxygen species-induced oxidation of Fe-S cluster present in the enzyme [53]. The expression of glucose transporter (GLUT)-3 decreases due to oxidative stress which leads to glucose uptake restrictions and excessive lactate accumulation [52]. Alzheimer's causes dementia in old-aged people as aging triggers the production of oxidative stress which thereby leads to lesion formation in Alzheimer's disease [54, 55]. Before the occurrence of plaque pathology and accumulation of amyloid-β (Aβ), the events of oxidation initiates which paves a pathway for oxidative stress-induced damages in Alzheimer's disease [56, 57]. The JNK/p38 MAPK pathways get modulated by oxidative stress which leads to Aβ accumulation and tau proteins hyper-phosphorylation [50]. Excessive production of reactive oxygen species plays a major role in the mechanism related to progression of Parkinson's disease. The antioxidant defense mechanism is damaged and presence of glutathione is noted at the initial stages of the disease. The proteins, lipids, and DNA get oxidized at an alarming rate which produces toxic end products and causes death of the cell [58]. The complex I of the mitochondria responsible for respiration show retarded activity which induces the production of reactive oxygen species responsible for dopaminergic neuron apoptosis [6]. Mutation of superoxide dismutase (SOD) 1 and degeneration of mitochondria are the two major reasons responsible for amyotrophic lateral sclerosis [59]. A substantial degeneration of the vacuoles in mitochondria were noticed in the neurons of SOD1 mutant mice which depicts that malfunctioning of mitochondria is the initiation point for amyotrophic lateral sclerosis. The abnormal interaction between mitochondria and mutant SOD1 leads to release of cytochrome C and apoptosis activation [56]. The SOD1 mutation causes damage in the antioxidant mechanism which leads to degeneration of motor neurons [8]. The excessive amount of oxidative stress inhibits the neuroprotective IGF-I/AKT pathway which results in dysfunctional neuron cells [48]. Again, alterations in the mitochondria instigated by oxidative stress are found to be responsible for spinocerebellar ataxia development [60]. The optimization of reactive oxygen species is considered to be a prospective method for treating and mitigating neurodegenerative diseases [61]. There are several clinical evidences which indicate that proper consumption of natural antioxidants in the diet can ameliorate neurodegenerative disorders [62]. Vitamin D is a potent antioxidant which helps to regulate the excite-toxicity of the neurons which is calcium-mediated and induces the neurotransmitters and structural proteins of the synapse. Its deficiency can become a major subsidizing aspect which leads to neurological aberration [63, 64]. Deficiency of Vitamin D in blood serum is correlated to dopaminergic neurons loss in the brains of patient suffering from Parkinson's disease [64]. Vitamin B is another major antioxidant which when deficient causes impairment of neurons related to Parkinson's disease and Alzheimer's disease [65]. In both in vitro and in vivo studies, it is revealed that resveratrol, rutin, and vitamin E, improve the neurodegenerative conditions caused by reactive oxygen species associated cascades like NF-κB and JNK [8]. Vitamin E works more actively in modulation of oxidative stress than Vitamin A and Vitamin C in rodent brains [66]. In a cohort study, it was found that vitamin E administration at a dose of 2000 IU per day showed positive response in Alzheimer's disease [67]. The amalgamation of coenzyme Q10 and Vitamin E helps to improve generation of energy in Friedreich ataxia by mitigating oxidative stress and functional restoration of mitochondria [68]. Several phytochemicals, phenolic components, and flavonoids present in green vegetables and fruits have higher antioxidizing potential which can be effective in Parkinson's disease [69]. Strawberries contain anthocyanins which have antioxidant, antiapoptotic, and anti-inflammatory properties. They help in amelioration of astrogliosis and preservation of neuromuscular junctions and functions of muscles in amyotrophic lateral sclerosis [70]. Resveratrol protects the neurons from Aβ and oxidative stress-induced toxicity in Alzheimer's disease [71, 72]. Lipoic acid enhances the generation of glutathione and helps in depletion of lipid peroxide which protects the proper functioning of mitochondria in the neurons from oxidative stress [73, 74]. MitoQ is another antioxidant which resorts the function of mitochondria in Purkinje cells and helps to alleviate motor incoordination in spinocerebellar ataxia [60]. Neurodegenerative disorders target the antioxidant defense system as a whole, so consumption of single antioxidant would not enough to revive the cellular damages [75].

3.4 Cancer

The initiations, promotion, and proliferation are the three stages of cancer growth which involves association of reactive oxygen species [76–78]. Firstly, during the stages of initiation, accumulation of mutated DNA occurs in the cancerous cells which are induced by reactive oxygen species [79]. The over production of reactive oxygen species leads to oncogenic DNA mutation which contributes to cancer development [80]. Due to alteration in the metabolic system and high demand of energy, excess synthesis of reactive oxygen species is a characteristic feature of every cancer cell when compared to normal cells [81]. The reactive oxygen species triggers cell proliferation by enhancing cellular signaling, resistance of tumors during therapies, supply of blood to tumors, and accelerating metastasis [82]. They also promote cellular expansion by modification of genes relation to transcription factors, apoptosis, and cell proliferation [83]. It causes an upregulation of antiapoptotic genes and downregulation of proapoptotic proteins by PI3K/AKT and ERK/MEK pathways [84]. It upregulates the metalloproteinases of the matrix, inhibits the antiproteases action and angiogenesis which leads to metastasis [85–87]. The endogenous antioxidants get depleted which disrupts the redox equilibrium which triggers development of cancer. Several studies proved that modification in the diet helps to prevent cancer by 35% [88]. The green vegetables and fruits containing antioxidants help to protect from the risk of cancer development [89, 90]. The polyphenolic components from plant sources have proved to show antioxidant activity [91–94]. These antioxidants have anticancerous potential which helps them to combat against breast, lung, larynx, tongue, gastric, prostate, and colon cancers [95, 96]. Fruits which possess elevated amount of phenolic compounds have the potency to substitute OH^- group in their aromatic ring [97, 98]. They can persuade cancer cell apoptosis, inhibition of mutated cell proliferation, reduction of cyclooxygenase-2 (COX-2) production, and downregulation of gene expression related to cancer [99–102]. Vitamins and minerals show antioxidizing potency which inhibits cell proliferation, maintains methylation of DNA, and promotes arresting of cell cycle [88, 103]. Consumption of antioxidant-rich diets modifies the biological markers related to progression of cancer [104]. Food from natural plant sources reduces the risk of breast cancer [105, 106], neck and head cancer [107], and colorectal cancer [108]. Oral cancer can be inhibited by Vitamin A and E [109]. Vitamin D and some minerals like selenium and carnitine show promising results in cancer treatment [110, 111]. Effective consumption of green vegetables, fruits, and whole grains with lower intake of red meat and alcohol helps to maintain a healthy lifestyle and decreases the probability of cancer threat [112].

3.5 Digestive Diseases

Several studies have established that the oxidative stress related to inflammation in the intestine plays a pivotal role in gastrointestinal diseases and inflammatory bowel syndrome [113]. The pathological symptoms showing the etiology of inflammatory bowel disease clearly signify reactive oxygen species to be responsible [114, 115]. Imbalance in the intestinal microflora, presence of pathogens, and food particles in the gastrointestinal tract trigger the excessive production of reactive oxygen species and hamper the endogenous mechanism of antioxidant defense [116]. Epithelium barrier of the intestine gets disrupted by oxidative stress which elevates the permeability of the intestine thereby causing inflammation. Inflammatory bowel disease comprises of ulcerative colitis which is defined by chronic inflammation related to oxidative stress in the gastrointestinal tract [113]. The patients diagnosed by inflammatory bowel syndrome show increased level of proinflammatory mediators like leukotriene B4 (LTB4) and platelet-activating factor in the mucous sample which triggers the reactive oxygen species by excessive stimulation of phagocytes and these reactive oxygen species are cytotoxic in nature [117–119]. Throughout the massive infiltration of neutrophils which are polymorphonuclear and macrophages, myeloperoxidases get secreted which produce hypochlorous acid by hydrogen peroxide metabolism. Different enzymes like 5-lipoxygenase, cyclooxygenase, and xanthine oxidase are the foundation of reactive oxygen species in the intestinal mucosa. Some deficiency in antioxidants in diet also subsidizes oxidative stress development [120]. In patients with Crohn's disease shows deficiency of antioxidant vitamins and minerals which is due to poor nutritional diet [120, 121]. Vitamin A concentration is greatly influenced in malnourished patients of inflammatory bowel syndrome due to lower consumption of green vegetable and fruits [113]. The mucosal linings are protected and repaired by Vitamin C which is deteriorated in gastritis and peptic ulcers [122]. Deficiency of Vitamin D causes morbidity of the skeletons in patients of Crohn's disease [123, 124]. Excessive oxidative stress damages the permeability of epithelium in gastrointestinal tracts through lipid peroxidation and distortion of tight junction. This hampers the microflora of the gastrointestinal tract and inhibits its colonization which promotes the occurrence of pathogenic bacteria [123, 124]. These pathogenic bacteria cause infections and trigger the synthesis of reactive oxygen species and lead to inflammation which increases the threat of inflammatory bowel syndrome [114]. The consumption of antioxidants from natural sources help to maintain the redox potential which helps to mitigate the damages caused to the intestine by reactive oxygen species and leads to healthy gastrointestinal tract [115]. Consumption of antioxidants showed fruitful results in Crohn's disease with upliftment in serum antioxidants. It also attenuated inflammation in Crohn's disease when consumed in combination of $n - 3$ fatty acids [125]. Intake of over dose of antioxidant supplements has detrimental effect by removing reactive oxygen species from biological metabolism. So it is highly recommended to consume antioxidants from natural sources [126, 127]. α-tocopherol, which is a component of vitamin E and some minerals has the potential to alleviate reactive oxygen species like selenium interacts with GPx and zinc interacts with superoxide dismutase to fight the oxidative stress. Selenium and Vitamin E in combination protect the colon from oxidative damage in rats [128].

3.6 Aging

During the process of aging, accretion of oxidative destruction inside the cells is unalterable and advanced [129]. The redox balance gets shifted toward oxidized state in the aging cells which is designated by low GSH/GSSG ratio. This distorts the capacity of the cell to duffer reactive oxygen species in normal physiological conditions and during exterior stress [130]. The accumulation of reactive oxygen species damages the proteins, DNA, lipids which dismantle the normal function of the cell [8]. Mitochondrial DNA gets mutated easily due to oxidative stress and hence triggers the process of aging which was supported by the study done in mice which showed signs of aging like loss of hair, osteoporosis, short life span with mutated mitochondrial DNA [131]. Elevated levels of reactive oxygen species cause shortening of telomere which accelerates senescence of cells [130]. The oxidative stress-associated aging causes inflammation by activating the NF-κB, which are regulatory factors for interleukin (IL)-1β, IL-6, and tumor necrosis (TNF-α). The NF-κB signaling has short span which is induced by oxidative stress and triggers aging. A continuous and slight inflammation can become the reason for age-related diseases like cancer, dementia, and atherosclerosis [132]. The malnutrition becomes a reason for deteriorated rate of metabolism and digestion in aged person [133]. Deficiency of Vitamin D results in loss of bone density, prone to fracture, and hyperparathyroidism. Natural sources of antioxidants can reduce these threats of elderly patients [134]. The antioxidants suppress the production of free radicals which retard the process of aging [135]. Studies show that consumption of green leafy vegetable of the family Brassicaceae shows decline in the cognitive function [136]. Increase in life span was seen in *Caenorhabditiselegans* and fruit flies where blueberry extract which is diet-rich antioxidants were provided [137, 138]. The activity of catalase and superoxide dismutase was increased which showed the fruitful effects of blueberry extract by booting the antioxidant system of the body.

Polyphenols from apple, anthocyanin from black rice, theaflavins from black tea showed elevated antioxidant potential in different animal models [138].

3.7 Hypertension

Hypertension is the major reason for cardiovascular diseases. Several studies has established the fact that excess production of reactive oxygen species causes oxidative stress which plays a pivotal role in hypertension pathogenesis. The vasomotor system gets modulated by reactive oxygen species which influences endothelin-1, angiotensin II, and urotensin-II leading to vasoconstriction. The redox system also regulates the nitric oxide which is a vasodilator. During excessive presence of reactive oxygen species, oxidative damage is caused which contributes to vascular dysfunction. The hydrogen peroxide and superoxide anions increase during pathophysiological hypertension which decreases the bioavailability of nitric oxide and antioxidant system gets damaged. In this scenario, the consumption of antioxidants neutralizes the system as they are reducing agents. The intake of Vitamin C and Vitamin E has been proved to show promising results as its helps to protect the vascular epithelial injury [139]. The hypertension caused by the reactive oxygen species triggers renal disorders by altering glomerulus by mediating glomerular lesions which has inflammations and also glomerulopathy [140]. Not only the kidney, the central nervous system is also affected by pathogenesis of hypertension which increases the central sympathetic outflow which is caused by the reactive oxygen species generated in the brainstem cells due to hypertension [141–143]. Some endogenous antioxidants of the cells are catalase, superoxide dismutase, and glutathione peroxidase which neutralize the oxidative stress. Some exogenous antioxidants have proved to sow wonders in this condition like, Vitamin C, Vitamin E, polyphenols, selenium, and N-acetylcysteine. Vitamin C upregulates the eNOS and declines the level of NADPH oxidase. The plasma ascorbate thus has proved to have inversely proportional to blood pressure. The consumption of antioxidants reduces blood pressure and stimulates vascular functioning vigorously. It acts on nitric oxide and promotes vascular dilation [139]. Increased consumption of Vitamin E regulates the generation of hydrogen peroxide and superoxides of the mitochondria via its antioxidizing potential thereby reducing the threat of cardiovascular diseases [144]. Vitamin C and Vitamin E in combination regulate the α-tocopheroxyl radical which has an antihypertensive effect [145].

3.8 Diabetes

The complication of diabetes is also triggered by the oxidative stress in both microvascular and cardiovascular diseases. The superoxides of the mitochondria are produced in excess in the endothelial cells of the vessels and myocardium during the diabetes abnormality in the metabolism. The overproduction of superoxides promotes five important pathways like elevated synthesis of advanced glycation end products, polyol pathway flux, overexpression of the advanced glycation end products receptor and its ligands activation, protein kinase C (PKC) isoforms activation, and excess activity of the hexosamine pathway. It also nullifies the activity of the enzymes, namely, eNOS and prostacyclin synthase which are antiatherosclerotic enzymes. The elevated intracellular reactive oxygen species harms the ischemia and causes defective angiogenesis. It also turns on several proinflammatory pathways and leads to enduring epigenetical changes thereby continuing the proinflammatory gene expression when diabetes becomes normal. In type 2 diabetes, cardiomyopathy and atherosclerosis are seen which lead to resistance of insulin thereby elevating the mitochondrial reactive oxygen species from fatty acids which causes antiatherosclerosis enzymes inactivation. Superoxide dismutase is the savior for cardiomyopathy, nephropathy, and diabetic retinopathy [146]. Consumption of Vitamin C and α-lipoic acid (ALA) from natural sources helps to prevent diabetes. α-lipoic acid reduces blood glucose by reducing oxidative stress which leads to lowering the insulin resistance. Antioxidant minerals like zinc and magnesium also help in triggering the glucose metabolism which leads to lowering the threat of diabetes. Vitamin A (retinol) is an antioxidant which helps to improve the function of pancreatic cells by regulating stress leading to normal synthesis of insulin [147].

3.9 Polycystic Ovarian Disease

Polycystic ovarian disease is caused in women during the reproductive stage which results in anovulation and infertility. This disease is mainly caused due to lifestyle disorder by intake of junk food and fast foods which triggers excess amount of reactive oxygen species in the body thereby leading to oxidative stress. The oxidative stress is the main reason for polycystic ovarian disease where the symptoms are obesity, hirsutism, acne, and irregular menstrual cycle. This chronic disorder leads to hypothyroidism, insulin resistance, hyperandrogenemia, and inflammation. The reactive oxygen species attacks the DNA and mutates them leading to gene aberration and mutation which suppress the tumor suppressor genes and trigger the cell proliferation which contributes to incidence of gynecological cancers in patients suffering from polycystic ovarian disease. About 45% of women suffering polycystic ovarian disease face the problem of obesity as it creates a metabolic disorder which leads to elevated levels of malondialdehyde, oxidized low-density lipoprotein, thiobarbituric reactive substances, advanced oxidation protein products, body mass index, and oxidative stress. Insulin resistance is another factor which is triggered by the high concentration of oxidative stress due to high levels of free fatty acid and low glucose metabolism [148]. Antioxidants have a promising effect in patients suffering from polycystic ovarian disease. Both enzymatic antioxidants like catalase, superoxide dismutase, glutathione reductase (GR), and glutathione peroxidase (GPx), and nonenzymatic antioxidants like α-tocopherol (vitamin E), glutathione (GSH), β-carotene and ascorbate (vitamin C), alpha lipoic acid, coenzyme Q, N-acetylcysteine, resveratrol, melatonin, and carnitine show positive results in polycystic ovarian disease [149, 150]. The antioxidants improve the environment of the ovary, reduce the level of androgen, and promote maturation of follicles. It also improves vascular epithelial function and triggers lipid metabolism which prevents deposition of lipids in obese patients [151]. Some minerals like selenium [152], magnesium, zinc [153], and chromium [154] also have antioxidant activity which helps in polycystic ovary.

4 Antioxidant Vitamins from Herbal Sources to Combat Lifestyle Diseases

Antioxidant vitamins play a major role in combating the oxidative stress caused by reactive oxygen species which are generated in different metabolism of the body leading to dreadful life-threatening diseases and are mainly caused by lifestyle disorder. Consumption of sufficient amount of vitamins helps to neutralize the oxidative condition of the cells which leads to healthy disease-free life. Each of the antioxidant vitamins, its mechanism of regulating oxidative stress, and natural sources are elaborated in the sections below.

4.1 Vitamin C

Vitamin C is a strong antioxidant as it is an electron donor and is also known as ascorbic acid. Being a donor of electron, it is hydrophilic in nature. It is an electron donor and donates two electrons from the double bond which is between third and second carbon of six carbon structure which prevents oxidization of other compounds. After donation of electron-free radical, ascorbyl is formed which is stable in nature; therefore, ascorbate is used as a strong antioxidant which reacts with harmful free radicals and for a lesser harmful ascorbyl radical which is less reactive which proves its radical scavenging potential [155]. Vitamin C acts on lipid, proteins, and DNA which are found in different envelopes of cell. In lipids, the membrane lipids when interact with reactive oxygen species undergo lipid peroxidation which produces free radicals which will lead to oxidative stress. Vitamin C helps to combat the free radicals thus produced by inhibiting the process of lipid peroxidation [156]. Low-density lipoproteins are the major carriers of triglycerides and cholesterol and also infilter arteries. This when oxidized, it causes activation of adhesion factors in the endothelial cells which lead to adherence of monocytes and macrophages are differentiated. These macrophages accumulate the low-density lipoproteins which are oxidized and stay in vascular wall forming foam cells which develop lesions in atherosclerosis. Ascorbate decreases the process of oxidation of low-density lipoproteins and reduces the initiation of oxidants [157–159]. Again oxidation of protein occurs when oxidants cleave the chain of peptides of single amino acid. Methionine and cysteine are two oxidant prone amino acids. By reducing the induction of oxidants, ascorbate helps to reduce the oxidation of proteins and amino acids. It also acted as a substrate for several end products of glycation [155]. DNA can also be attacked by the oxidative stress which includes oxidation of protein where alteration of DNA repair genes and DNA polymerases occurred. Again when lipids were attacked by oxidants, it caused lipid peroxidation which triggered mutation of DNA [160]. In DNA, guanine is more susceptible to oxidant attacks 8 hydroxyguanine and 8-hydroxy-2-deoxyguanosine [161]. When the attackers are nitrogen species, nitrosamines are produced as derivatives [155]. There are several natural sources which are rich in Vitamin C like fruits and green vegetables. The fruits include kiwi, grapefruit, honeydew, cantaloupe, orange, mango, strawberries, papaya, tangerine, tangelo, and watermelon. The vegetables include brussels sprouts, broccoli, asparagus, cauliflower, cabbage, mustard greens, kale, pepper (red or green), potatoes, plantains, sweet potatoes, tomatoes, and snow peas [155] (Table 1).

4.2 Vitamin E

The basic component of Vitamin E is α-tocopherol which is actually responsible for its antioxidant potential. They scavenge the oxidative species and nitrogen species and also restrict the 5-lipoxygenase-catalyzed eicosanoids and cyclooxygenase and suppress the signaling of inflammation like STAT3/6 and NF-κB. The eight stereoisomers of Vitamin E are α, β, γ, δ tocopherol and α, β, γ, δ tocotrienol. Dextrorotary d-α-tocopherol is hyperactive [162]. As is it fat soluble in nature, it protects the cell membranes from oxidative damage [34]. The unsubstituted fifth position γT attacks electrophiles and hence nitrogen species are trapped. The tocotrienol and tocopherol from δT have radical scavenging potential as compared to αT [163]. Vitamin E helps to maintain the homeostasis of skeletal muscles and promotes the repairing of plasma membrane. It also prevents aggregation of platelets by inhibiting protein kinase C (PKC). It also helps to inhibit the process of cholesterol biosynthesis by suppression of suppressing 3-hydroxy-3-methylglutaryl-CoA

Table 1 Showing the herbal sources of antioxidant vitamins and minerals to alleviate lifestyle diseases

Sl. no.	Name of sources	Antioxidants vitamins and minerals	References
1	Cereals	Vitamin E, zinc	[194–199]
2	Oil seeds	Vitamin E, zinc	[163, 199, 200]
3	Nuts	Vitamin E, selenium, zinc	[199–202]
4	Broccoli	Vitamin E, iodine	[197, 200]
5	Cooking oil	Vitamin E	[163, 200]
6	Brussels sprouts	Vitamin E, zinc	[199]
7	Cauliflower	Vitamin E, vitamin C, vitamin A, iodine	[203]
8	Citrus fruits	Vitamin C	[204]
9	Guava	Vitamin C, vitamin A	[205]
10	Tomatoes	Vitamin C, vitamin E, vitamin A	[203]
11	Goose berry	Vitamin C	[204]
12	Sprouted pulses	Vitamin C, zinc	[199]
13	Green leafy vegetables	Vitamin A, zinc	[199]
14	Papaya	Vitamin A	[206]
15	Musk melon	Vitamin A	[207]
16	Mango	Vitamin A	[204]
17	Pumpkin	Vitamin A, zinc	[199]
18	Carrots	Vitamin A, vitamin C, vitamin E, selenium	[203, 208, 209]
19	Mushrooms	Vitamin B, vitamin E, vitamin C, copper, selenium	[206]
20	Whole grains	Selenium, zinc, iodine	[201, 202]
21	Seafood	Vitamin E, selenium	[201, 208–210]
22	Asparagus	Vitamin C	[211]
23	Pink grape fruit	Vitamin C	[212]
24	Beet root	Vitamin C	[203, 213, 214]
25	Avocado	Vitamin E, vitamin C, vitamin A, magnesium	[215]
26	Pears	Vitamin C, vitamin A	[216]
27	Pomegranate	Vitamin C	[217]
28	Peas	Vitamin A, vitamin C, vitamin E, selenium, iodine	[203, 204, 208, 209]
29	Spinach	Vitamin A, vitamin C, vitamin E, selenium	[200, 202–204]
30	White onion	Vitamin A, vitamin C, vitamin E	[203, 204]
31	Green tea	Vitamin C, vitamin E, vitamin A, vitamin B, manganese, zinc, copper	[199, 204, 218]
32	Oatmeal	Selenium, zinc, manganese	[199, 219]
33	Banana	Selenium, iodine	[220]
34	Peach	Selenium	[221]
35	Lettuce	Selenium	[202]
36	Potatoes	Selenium	[208, 209]
37	Soybean	Zinc, vitamin E, iodine	[163, 199]
38	Flax seed	Zinc	[199]
39	Pepper	Vitamin C	[204]
40	Cabbage	Vitamins C, vitamin E, vitamin A, iodine	[203]
41	Kiwi	Vitamin E	[200]
42	Aloe	Vitamin E, vitamin C, vitamin A	[222]
43	Apple	Vitamin C, vitamin E, zinc	[223]
44	Parsley	Vitamin A, vitamin C, magnesium	[4, 224]
45	Peppermint	Vitamin E, vitamin C, vitamin A, magnesium	[4, 225]
46	Lavender	Vitamin C	[4, 226]
47	Drumstick	Vitamin C, vitamin A	[227]
48	Red pepper	Vitamin A, vitamin E	[206]
49	Squash	Vitamin A	[228]
50	Chia seed	Selenium	[183]
51	Pineapples	Manganese, iodine	[189]
52	Strawberries	Iodine	[229]
53	Cranberries	Iodine	[229]
54	Green beans	Iodine	[229]
55	Cassava	Iodine	[230]

(HMGCoA) reductase in liver cells. Vitamin E also stimulates wild-type p53 tumor suppressor gene, downregulates p53 proteins, upregulates heat shock proteins which help to prevent cancer. Γ-tocopherol prevents cell proliferation in cancer by scavenging the free radicals which causes mutation of DNA and promotes malignancy in the cells. It also arrests cell cycle by suppressing Cyclin D and promotes the process of apoptosis. The γ and δ tocopherol inhibits the sphingolipids in prostate cancer cells and induces apoptosis. It actually promotes cytochrome c synthesis, caspase-9 and caspase-3 are activated and polyadenosine diphosphate (ADP)-ribose polymerase (PARP) cleaving. The process of cytotoxicity due to production of hydrogen peroxide by oxidation of protein and peroxidation of lipids in Alzheimer's diseases are blocked by Vitamin E [164]. The consumption of food containing Vitamin E can be helpful in alleviating the diseases related to lifestyle disorders. The diet rich in Vitamin E containing herbal sources like cereals, oil seeds including sunflower seed, pumpkin seed, cooking oils including coconut oil, palm oil, corn oil, olive oil, peanut oil, soybean oil, wheatgerm oil, safflower oil, and sunflower oil, almonds, hazelnut, peanut, spinach, broccoli, kiwi, mango, tomato, brussel sprouts, carrots, mushrooms, peas, white onion, green tea, cabbage, and aloe can reduce the lifestyle diseases by promoting the radical scavenging property of Vitamin E containing foods (Table 1). Vitamin E is predominantly dependent on Vitamin C, selenium, and glutathione. A combination of Vitamin E with Vitamin C works as a more potent antioxidizing agent by enhancing the antioxidant defense system of the cell [164].

4.3 Vitamin A

Vitamin A is the transcriptional regulator which regulates the expression of hundreds of genes. It has an indirect antioxidant potential by regulation of various transcriptional genes in response to which the free radicals are scavenged. The antioxidant potential of Vitamin A convenes to polyene chains which are hydrophobic in nature and has the capacity to scavenge free radicals of oxygen and neutralizes thiyl radicals which lead to stabilization of peroxyl radical. It has auto-oxidation capacity which helps to decrease the oxygen tension during physiological process in a cell. Food rich in antioxidant Vitamin A helps to reduce coronary heart failure and diabetic cardiomyopathy by reduction of oxidative stress [165]. β-carotene is the precursor of Vitamin A, and is fat soluble in nature. This gets converted to retinol which has prominent antioxidizing potential [34]. The consumption of foods rich in Vitamin A are cauliflower, guava, tomatoes, papaya, musk melon, mango, pumpkin, carrots, pears peas, spinach, white onion, green tea, cabbage, ale, parsley, peppermint, avocado, and drumstick (Table 1) which can reduce the risk of cardiovascular diseases, cancer, and cellular function of the pancreas thereby preventing diabetes [166].

5 Antioxidant Minerals from Herbal Sources to Combat Lifestyle Diseases

Minerals are required in negligible amount in the body and its deficiency can cause several lifestyle diseases. They help in scavenging the free radicals from the biological system due to normal activity of the cell. There are some minerals required in large amount known and macronutrients and some are required in trace amount known as micronutrients. The antioxidant minerals and its herbal sources are discussed in the following section.

5.1 Selenium

Selenium is one of the most essential antioxidant microminerals which has tremendous power of antioxidant in the form of selenoprotein which has the potential to scavenge the free radicals in oxidative stress. This selenoprotein is synthesized by incorporation of selenocysteine into protein which is directed by a termination codon UGA. In selenoproteins, selenium can be present in both organic and inorganic forms. Out of all the forms, organic form of selenium as selenomethionine is predominantly present in the foods consumed by humans. The selenoproteins reduce the synthesis of oxidized low-density lipoprotein which lowers the risk of cardiac disorders [167]. TrxR is an essential selenoenzyme which reduces oxidative stress by increasing the bioavailability of nitric oxide which leads to prevention of atherosclerosis [168, 169]. The consumption of selenium-rich diet reduces reperfusion and in deficiency can cause intrinsic myocardial tolerance to ischemic insult [170–172]. Ebselen is another organoselenium compound which has activity like glutathione peroxidase which provides protection from myocardial ischemia-reperfusion injury. This compound induces stress protein HSP27 and preserves glutathione [173]. It also has neuroprotective role when administered with thrombolytic tissue plasminogen activator [174]. Selenium also has evidence to reduce the progress of cancer in men [175]. Selenium in combination with Vitamin E shows inverse relation with incidence of prostate cancer [176, 177]. The mineral reduces oxidative stress and prevents damage of DNA [88, 178–181]. It promotes the arrest of cell cycle, triggers the process of apoptosis, and inhibits cell proliferation and protein kinase C activity [178, 181–183]. The consumption of selenium-rich diet helps to reduce the risk of lifestyle diseases. The foods from herbal sources are Brazilian nuts, brown rice, chia seed, flax seed, whole wheat breads, baked beans, oatmeal, lentils, spinach, cashew nuts, green peas, bananas, potato, peach, carrot, and lettuce [184].

5.2 Zinc

Zinc has a prominent antioxidizing potential and is maximum found as cofactors in enzymes involved in antioxidant defense system. It acts as a membrane stabilizer and NADPH-Oxidase inhibitor. During diabetes, cancer, and obesity, the harmful reactive oxygen species are scavenged by metallothionein proteins whose synthesis is triggered by zinc. It regulates glutathione peroxidase and forms the structural component of superoxide dismutase of cellular cytoplasm. It regulates the glutamate-cysteine ligase which is responsible for de novo synthesis of glutathione. It regulates transcription of NF-κB by protein A20 which is anti-inflammatory and also regulates the receptor signaling pathways [185]. Deficiency of zinc creates dysfunctioning of epithelium which induces responses of inflammation mediated by a process related to elevated production of oxidative stress. It also protects the vascular system by regulating a transcription factor Nrf2, which promotes encoding of gene related to antioxidant enzymes and metallothionein [186–188]. The foods from herbal sources which are rich in zinc are cereal, oil seed, nuts, brussel sprouts, sprouted pulses, leafy vegetables, pumpkin, whole grain, green tea, oatmeal, soybean, flax seed, and apple (Table 1).

5.3 Manganese

Manganese is required in trace amount and found in livers, bones, pancreas, and kidneys. It helps in formation of connective tissues, blood clotting factor, hormones, and bones. It regulates the metabolism of carbohydrates, fats, and blood sugar. It also helps normal functioning of nerve and brain. It as an important component of superoxide dismutase enzyme which has free radical scavenging property. The free radical damages the DNA and cell membranes. It also has role in the process of aging, cardiovascular diseases, and cancer. The superoxide dismutase neutralizes the reactive oxygen species and prevents in cellular damage. Deficiency of this micronutrient causes infertility, malformation of bone, and weakness. The diet rich in nuts, seeds, and whole grains promotes the concentration of manganese in the system [189].

5.4 Iodine

Iodine is another micronutrient which is responsible for healthy functioning of thyroid. It has free radical scavenging potential. It depolarizes the hyaluronic acids, increases the antioxidant potential of human blood, promotes enzyme activities, and decreases peroxidases and malondialdehydes. It protects from the risk of cardiac diseases, and respiratory diseases which produce oxidative stress. The iodide ion attacks the lipid peroxidases and hydrogen peroxidases. A therapy known as balneotherapy is being used in patients with high oxidative stress. It is proven to have antiproliferative effect in mastalgia [190]. Its been used in therapy of prostate and breast cancer [191]. In cardiovascular diseases, the increased level of thyroid hormones is favorable. It decreases the triglyceride and cholesterol levels. It also helps in ischemia reperfusion injuries [192]. Reactive oxygen species produced by respiratory disorders can be alleviated by iodide by augmenting antiviral innate immunity [193]. Food rich in iodine like Broccoli, cauliflower, whole grain, peas, banana, soybean, cabbage, pineapples, strawberries, cranberries, green bean, and cassava should be consumed in daily diet to keep away the risk of cardiovascular and thyroid-related diseases (Table 1).

6 Summarizing Remarks and Prospects

A balanced lifestyle with healthy diet can show positive results by reducing oxidative stress. Consumption of antioxidants is the best source to alleviate oxidative stress as it damages the DNA and causes cellular damage and plays a key role in incidence of several dreadful diseases. Chromosomal aberrations and structural deformity are promoted by oxidative stress. In spite of the production of reactive oxygen species in several biological metabolism, it can be concluded that consumption of antioxidant-rich diet and maintaining a healthy lifestyle plays a pivotal role in oxidative stress control. In the recent year, herbal natural antioxidants have gained popularity due to its lesser side effect and cheaper price which has shown promising results in having therapeutic potential in not only treating the disease but also preventing it [231]. Regular consumption of these herbal sources also shows chemoprotective activity. The plant-based antioxidant vitamins and minerals are also effective against cardiac problems, neurodegenerative disorders, cancer, polycystic ovarian disease, diabetes, aging, respiratory disease, hypertension, and digestive disorders. Further, isolation of chemical compound with antioxidant potency can be achieved in future for safer and wider acceptability of these herbal sources. It can be concluded that oxidative stress is directly linked to lifestyle disorder and can be managed by consumption of antioxidant-rich diet which will help to keep away from the risk of disease development and maintain a healthy life.

References

1. Kurutas EB. The importance of antioxidants which play the role in cellular response against oxidative/nitrosative stress: current state. Nutr J. 2016;15(1):71. https://doi.org/10.1186/s12937-016-0186-5.
2. Liu ZQ. Bridging free radical chemistry with drug discovery: a promising way for finding novel drugs efficiently. Eur J Med Chem. 2020;189:112020. https://doi.org/10.1016/j.ejmech.2019.112020.
3. Sharifi-Rad M, Anil Kumar NV, Zucca P, Varoni EM, Dini L, Panzarini E, Rajkovic J, Tsouh Fokou PV, Azzini E, Peluso I, Prakash Mishra A, Nigam M, El Rayess Y, Beyrouthy ME, Polito L, Iriti M, Martins N, Martorell M, Docea AO, Setzer WN, Calina D, Cho WC, Sharifi-Rad J. Lifestyle, oxidative stress, and antioxidants: back and forth in the pathophysiology of chronic diseases. Front Physiol. 2020;11:694. https://doi.org/10.3389/fphys.2020.00694.
4. Chanda S, Tiwari RK, Kumar A, Singh K. Nutraceuticals inspiring the current therapy for lifestyle diseases. Adv Pharmacol Sci. 2019;2019:6908716. https://doi.org/10.1155/2019/6908716.
5. Bergamin A, Mantzioris E, Cross G, Deo P, Garg S, Hill AM. Nutraceuticals: reviewing their role in chronic disease prevention and management. Pharm Med. 2019;33(4):291–309. https://doi.org/10.1007/s40290-019-00289-w.
6. Liu Z, Ren Z, Zhang J, Chuang CC, Kandaswamy E, Zhou T, Zuo L. Role of ROS and nutritional antioxidants in human diseases. Front Physiol. 2018;9:477. https://doi.org/10.3389/fphys.2018.00477.
7. Rogers NM, Seeger F, Garcin ED, Roberts DD, Isenberg JS. Regulation of soluble guanylate cyclase by matricellular thrombospondins: implications for blood flow. Front Physiol. 2014;5:134. https://doi.org/10.3389/fphys.2014.00134.
8. Zuo L, Zhou T, Pannell BK, Ziegler AC, Best TM. Biological and physiological role of reactive oxygen species—the good, the bad and the ugly. Acta Physiol (Oxf). 2015;214(3):329–48. https://doi.org/10.1111/apha.12515.

9. Kannan K, Jain SK. Oxidative stress and apoptosis. Pathophysiology. 2000;7(3):153–63. https://doi.org/10.1016/s0928-4680(00)00053-5.
10. Ames BN, Wakimoto P. Are vitamin and mineral deficiencies a major cancer risk? Nat Rev Cancer. 2002;2(9):694–704. https://doi.org/10.1038/nrc886.
11. Evatt ML, Delong MR, Khazai N, Rosen A, Triche S, Tangpricha V. Prevalence of vitamin d insufficiency in patients with Parkinson disease and Alzheimer disease. Arch Neurol. 2008;65(10):1348–52. https://doi.org/10.1001/archneur.65.10.1348.
12. Schols AM, Ferreira IM, Franssen FM, Gosker HR, Janssens W, Muscaritoli M, Pison C, Rutten-van Mölken M, Slinde F, Steiner MC, Tkacova R, Singh SJ. Nutritional assessment and therapy in COPD: a European Respiratory Society statement. Eur Respir J. 2014;44(6):1504–20. https://doi.org/10.1183/09031936.00070914.
13. Mathur P, Mascarenhas L. Life style diseases: *keeping fit for a better tomorrow*. Indian J Med Res. 2019;149(Suppl):S129–35. https://doi.org/10.4103/0971-5916.251669.
14. Steyn K, Damasceno A. Lifestyle and related risk factors for chronic diseases. In: Jamison DT, Feachem RG, Makgoba MW, Bos ER, Baingana FK, Hofman KJ, Rogo KO, editors. Disease and mortality in sub-Saharan Africa. 2nd ed. Washington (DC): The International Bank for Reconstruction and Development/The World Bank; 2006. Chapter 18.
15. Masoli M, Fabian D, Holt S, Beasley R, Global Initiative for Asthma (GINA) Program. The global burden of asthma: executive summary of the GINA dissemination committee report. Allergy. 2004;59(5):469–78. https://doi.org/10.1111/j.1398-9995.2004.00526.x.
16. Pauwels RA, Rabe KF. Burden and clinical features of chronic obstructive pulmonary disease (COPD). Lancet. 2004;364(9434):613–20. https://doi.org/10.1016/S0140-6736(04)16855-4.
17. van Eeden SF, Sin DD. Oxidative stress in chronic obstructive pulmonary disease: a lung and systemic process. Can Respir J. 2013;20(1):27–9. https://doi.org/10.1155/2013/509130.
18. Comhair SA, Erzurum SC. Redox control of asthma: molecular mechanisms and therapeutic opportunities. Antioxid Redox Signal. 2010;12(1):93–124. https://doi.org/10.1089/ars.2008.2425.
19. Cho YS, Moon HB. The role of oxidative stress in the pathogenesis of asthma. Allergy Asthma Immunol Res. 2010;2(3):183–7. https://doi.org/10.4168/aair.2010.2.3.183.
20. Fitzpatrick AM, Teague WG, Holguin F, Yeh M, Brown LA, Severe Asthma Research Program. Airway glutathione homeostasis is altered in children with severe asthma: evidence for oxidant stress. J Allergy Clin Immunol. 2009;123(1):146–152.e8. https://doi.org/10.1016/j.jaci.2008.10.047.
21. Zuo L, Pannell BK, Liu Z. Characterization and redox mechanism of asthma in the elderly. Oncotarget. 2016;7(18):25010–21. https://doi.org/10.18632/oncotarget.7075.
22. Jiang L, Diaz PT, Best TM, Stimpfl JN, He F, Zuo L. Molecular characterization of redox mechanisms in allergic asthma. Ann Allergy Asthma Immunol. 2014;113(2):137–42. https://doi.org/10.1016/j.anai.2014.05.030.
23. King DA, Cordova F, Scharf SM. Nutritional aspects of chronic obstructive pulmonary disease. Proc Am Thorac Soc. 2008;5(4):519–23. https://doi.org/10.1513/pats.200707-092ET.
24. Ferreira IM, Brooks D, White J, Goldstein R. Nutritional supplementation for stable chronic obstructive pulmonary disease. Cochrane Database Syst Rev. 2012;12:CD000998. https://doi.org/10.1002/14651858.CD000998.pub3.
25. Sies H, Stahl W, Sevanian A. Nutritional, dietary and postprandial oxidative stress. J Nutr. 2005;135(5):969–72. https://doi.org/10.1093/jn/135.5.969.
26. Varraso R, Chiuve SE, Fung TT, Barr RG, Hu FB, Willett WC, Camargo CA. Alternate healthy eating index 2010 and risk of chronic obstructive pulmonary disease among US women and men: prospective study. BMJ. 2015;350:h286. https://doi.org/10.1136/bmj.h286.
27. Hornikx M, Van Remoortel H, Lehouck A, Mathieu C, Maes K, Gayan-Ramirez G, Decramer M, Troosters T, Janssens W. Vitamin D supplementation during rehabilitation in COPD: a secondary analysis of a randomized trial. Respir Res. 2012;13(1):84. https://doi.org/10.1186/1465-9921-13-84.
28. Tecklenburg SL, Mickleborough TD, Fly AD, Bai Y, Stager JM. Ascorbic acid supplementation attenuates exercise-induced bronchoconstriction in patients with asthma. Respir Med. 2007;101(8):1770–8. https://doi.org/10.1016/j.rmed.2007.02.014.
29. Gumral N, Naziroglu M, Ongel K, Beydilli ED, Ozguner F, Sutcu R, Caliskan S, Akkaya A. Antioxidant enzymes and melatonin levels in patients with bronchial asthma and chronic obstructive pulmonary disease during stable and exacerbation periods. Cell Biochem Funct. 2009;27(5):276–83. https://doi.org/10.1002/cbf.1569.
30. Allen S, Britton JR, Leonardi-Bee JA. Association between antioxidant vitamins and asthma outcome measures: systematic review and meta-analysis. Thorax. 2009;64(7):610–9. https://doi.org/10.1136/thx.2008.101469.
31. Moreno-Macias H, Romieu I. Effects of antioxidant supplements and nutrients on patients with asthma and allergies. J Allergy Clin Immunol. 2014;133(5):1237–44. https://doi.org/10.1016/j.jaci.2014.03.020; quiz 1245.
32. Wood LG, Garg ML, Smart JM, Scott HA, Barker D, Gibson PG. Manipulating antioxidant intake in asthma: a randomized controlled trial. Am J Clin Nutr. 2012;96(3):534–43. https://doi.org/10.3945/ajcn.111.032623.
33. Romieu I. Nutrition and lung health. Int J Tuberc Lung Dis. 2005;9(4):362–74.
34. Pham-Huy LA, He H, Pham-Huy C. Free radicals, antioxidants in disease and health. Int J Biomed Sci. 2008;4(2):89–96.
35. Madamanchi NR, Vendrov A, Runge MS. Oxidative stress and vascular disease. Arterioscler Thromb Vasc Biol. 2005;25(1):29–38. https://doi.org/10.1161/01.ATV.0000150649.39934.13.
36. Azumi H, Inoue N, Ohashi Y, Terashima M, Mori T, Fujita H, Awano K, Kobayashi K, Maeda K, Hata K, Shinke T, Kobayashi S, Hirata K, Kawashima S, Itabe H, Hayashi Y, Imajoh-Ohmi S, Itoh H, Yokoyama M. Superoxide generation in directional coronary atherectomy specimens of patients with angina pectoris: important role of NAD(P)H oxidase. Arterioscler Thromb Vasc Biol. 2002;22(11):1838–44. https://doi.org/10.1161/01.atv.0000037101.40667.62.
37. Ambrose JA, Barua RS. The pathophysiology of cigarette smoking and cardiovascular disease: an update. J Am Coll Cardiol. 2004;43(10):1731–7. https://doi.org/10.1016/j.jacc.2003.12.047.
38. Zhou T, Prather ER, Garrison DE, Zuo L. Interplay between ROS and antioxidants during ischemia-reperfusion injuries in cardiac and skeletal muscle. Int J Mol Sci. 2018;19(2):417. https://doi.org/10.3390/ijms19020417.
39. Wang X, Ouyang Y, Liu J, Zhu M, Zhao G, Bao W, Hu FB. Fruit and vegetable consumption and mortality from all causes, cardiovascular disease, and cancer: systematic review and dose-response meta-analysis of prospective cohort studies. BMJ. 2014;349:g4490. https://doi.org/10.1136/bmj.g4490.
40. Ignarro LJ, Balestrieri ML, Napoli C. Nutrition, physical activity, and cardiovascular disease: an update. Cardiovasc Res. 2007;73(2):326–40. https://doi.org/10.1016/j.cardiores.2006.06.030.
41. Hoppe PP, Krennrich G. Bioavailability and potency of natural-source and all-racemic alpha-tocopherol in the human: a

dispute. Eur J Nutr. 2000;39(5):183–93. https://doi.org/10.1007/s003940070010.
42. Arab L, Steck S. Lycopene and cardiovascular disease. Am J Clin Nutr. 2000;71(6 Suppl):1691S–5S. https://doi.org/10.1093/ajcn/71.6.1691S; discussion 1696S–7S.
43. Rao AV, Agarwal S. Role of antioxidant lycopene in cancer and heart disease. J Am Coll Nutr. 2000;19(5):563–9. https://doi.org/10.1080/07315724.2000.10718953.
44. Kohlmeier L, Kark JD, Gomez-Gracia E, Martin BC, Steck SE, Kardinaal AF, Ringstad J, Thamm M, Masaev V, Riemersma R, Martin-Moreno JM, Huttunen JK, Kok FJ. Lycopene and myocardial infarction risk in the EURAMIC study. Am J Epidemiol. 1997;146(8):618–26. https://doi.org/10.1093/oxfordjournals.aje.a009327.
45. Vita JA. Polyphenols and cardiovascular disease: effects on endothelial and platelet function. Am J Clin Nutr. 2005;81(1 Suppl):292S–7S. https://doi.org/10.1093/ajcn/81.1.292S.
46. Quiñones M, Miguel M, Aleixandre A. Beneficial effects of polyphenols on cardiovascular disease. Pharmacol Res. 2013;68(1):125–31. https://doi.org/10.1016/j.phrs.2012.10.018.
47. Rego AC, Oliveira CR. Mitochondrial dysfunction and reactive oxygen species in excitotoxicity and apoptosis: implications for the pathogenesis of neurodegenerative diseases. Neurochem Res. 2003;28(10):1563–74. https://doi.org/10.1023/a:1025682611389.
48. Dávila D, Torres-Aleman I. Neuronal death by oxidative stress involves activation of FOXO3 through a two-arm pathway that activates stress kinases and attenuates insulin-like growth factor I signaling. Mol Biol Cell. 2008;19(5):2014–25. https://doi.org/10.1091/mbc.e07-08-0811.
49. Hakonen AH, Goffart S, Marjavaara S, Paetau A, Cooper H, Mattila K, Lampinen M, Sajantila A, Lönnqvist T, Spelbrink JN, Suomalainen A. Infantile-onset spinocerebellar ataxia and mitochondrial recessive ataxia syndrome are associated with neuronal complex I defect and mtDNA depletion. Hum Mol Genet. 2008;17(23):3822–35. https://doi.org/10.1093/hmg/ddn280.
50. Patten DA, Germain M, Kelly MA, Slack RS. Reactive oxygen species: stuck in the middle of neurodegeneration. J Alzheimers Dis. 2010;20(Suppl 2):S357–67. https://doi.org/10.3233/JAD-2010-100498.
51. Blesa J, Trigo-Damas I, Quiroga-Varela A, Jackson-Lewis VR. Oxidative stress and Parkinson's disease. Front Neuroanat. 2015;9:91. https://doi.org/10.3389/fnana.2015.00091.
52. Covarrubias-Pinto A, Moll P, Solís-Maldonado M, Acuña AI, Riveros A, Miró MP, Papic E, Beltrán FA, Cepeda C, Concha II, Brauchi S, Castro MA. Beyond the redox imbalance: oxidative stress contributes to an impaired GLUT3 modulation in Huntington's disease. Free Radic Biol Med. 2015;89:1085–96. https://doi.org/10.1016/j.freeradbiomed.2015.09.024.
53. Browne SE, Beal MF. Oxidative damage in Huntington's disease pathogenesis. Antioxid Redox Signal. 2006;8(11–12):2061–73. https://doi.org/10.1089/ars.2006.8.2061.
54. Harman D. Alzheimer's disease pathogenesis: role of aging. Ann N Y Acad Sci. 2006;1067:454–60. https://doi.org/10.1196/annals.1354.065.
55. Pimplikar SW, Nixon RA, Robakis NK, Shen J, Tsai LH. Amyloid-independent mechanisms in Alzheimer's disease pathogenesis. J Neurosci. 2010;30(45):14946–54. https://doi.org/10.1523/JNEUROSCI.4305-10.2010.
56. Lin MT, Beal MF. Mitochondrial dysfunction and oxidative stress in neurodegenerative diseases. Nature. 2006;443(7113):787–95. https://doi.org/10.1038/nature05292.
57. Wang X, Wang W, Li L, Perry G, Lee HG, Zhu X. Oxidative stress and mitochondrial dysfunction in Alzheimer's disease. Biochim Biophys Acta. 2014;1842(8):1240–7. https://doi.org/10.1016/j.bbadis.2013.10.015.
58. Jenner P. Oxidative stress in Parkinson's disease. Ann Neurol. 2003;53(Suppl 3):S26–36. https://doi.org/10.1002/ana; discussion S36–8.
59. Rotunno MS, Bosco DA. An emerging role for misfolded wild-type SOD1 in sporadic ALS pathogenesis Front Cell Neurosci. 2013;7:253. https://doi.org/10.3389/fncel.2013.00253.
60. Stucki DM, Ruegsegger C, Steiner S, Radecke J, Murphy MP, Zuber B, Saxena S. Mitochondrial impairments contribute to spinocerebellar ataxia type 1 progression and can be ameliorated by the mitochondria-targeted antioxidant MitoQ. Free Radic Biol Med. 2016;97:427–40. https://doi.org/10.1016/j.freeradbiomed.2016.07.005.
61. Uttara B, Singh AV, Zamboni P, Mahajan RT. Oxidative stress and neurodegenerative diseases: a review of upstream and downstream antioxidant therapeutic options. Curr Neuropharmacol. 2009;7(1):65–74. https://doi.org/10.2174/157015909787602823.
62. Zandi PP, Anthony JC, Khachaturian AS, Stone SV, Gustafson D, Tschanz JT, Norton MC, Welsh-Bohmer KA, Breitner JC, Cache County Study Group. Reduced risk of Alzheimer disease in users of antioxidant vitamin supplements: the Cache County Study. Arch Neurol. 2004;61(1):82–8. https://doi.org/10.1001/archneur.61.1.82.
63. Mpandzou G, Aït Ben Haddou E, Regragui W, Benomar A, Yahyaoui M. Vitamin D deficiency and its role in neurological conditions: a review. Rev Neurol (Paris). 2016;172(2):109–22. https://doi.org/10.1016/j.neurol.2015.11.005.
64. Wang J, Yang D, Yu Y, Shao G, Wang Q. Vitamin D and sunlight exposure in newly-diagnosed Parkinson's disease. Nutrients. 2016;8(3):142. https://doi.org/10.3390/nu8030142.
65. Sechi G, Sechi E, Fois C, Kumar N. Advances in clinical determinants and neurological manifestations of B vitamin deficiency in adults. Nutr Rev. 2016;74(5):281–300. https://doi.org/10.1093/nutrit/nuv107.
66. Zaidi SM, Banu N. Antioxidant potential of vitamins A, E and C in modulating oxidative stress in rat brain. Clin Chim Acta. 2004;340(1–2):229–33. https://doi.org/10.1016/j.cccn.2003.11.003.
67. Sano M, Ernesto C, Thomas RG, Klauber MR, Schafer K, Grundman M, Woodbury P, Growdon J, Cotman CW, Pfeiffer E, Schneider LS, Thal LJ. A controlled trial of selegiline, alpha-tocopherol, or both as treatment for Alzheimer's disease. The Alzheimer's disease cooperative study. N Engl J Med. 1997;336(17):1216–22. https://doi.org/10.1056/NEJM199704243361704.
68. Lodi R, Hart PE, Rajagopalan B, Taylor DJ, Crilley JG, Bradley JL, Blamire AM, Manners D, Styles P, Schapira AH, Cooper JM. Antioxidant treatment improves in vivo cardiac and skeletal muscle bioenergetics in patients with Friedreich's ataxia. Ann Neurol. 2001;49(5):590–6. https://doi.org/10.1002/ana.1001.
69. Mazo NA, Echeverria V, Cabezas R, Avila-Rodriguez M, Tarasov VV, Yarla NS, Aliev G, Barreto GE. Medicinal plants as protective strategies against Parkinson's disease. Curr Pharm Des. 2017;23(28):4180–8. https://doi.org/10.2174/1381612823666170316142803.
70. Winter AN, Ross EK, Wilkins HM, Stankiewicz TR, Wallace T, Miller K, Linseman DA. An anthocyanin-enriched extract from strawberries delays disease onset and extends survival in the hSOD1^{G93A} mouse model of amyotrophic lateral sclerosis. Nutr Neurosci. 2018;21(6):414–26. https://doi.org/10.1080/1028415X.2017.1297023.
71. Anekonda TS. Resveratrol—a boon for treating Alzheimer's disease? Brain Res Rev. 2006;52(2):316–26. https://doi.org/10.1016/j.brainresrev.2006.04.004.
72. Bellaver B, Souza DG, Souza DO, Quincozes-Santos A. Resveratrol increases antioxidant defenses and decreases

proinflammatory cytokines in hippocampal astrocyte cultures from newborn, adult and aged Wistar rats. Toxicol In Vitro. 2014;28(4):479–84. https://doi.org/10.1016/j.tiv.2014.01.006.
73. Moreira PI, Zhu X, Wang X, Lee HG, Nunomura A, Petersen RB, Perry G, Smith MA. Mitochondria: a therapeutic target in neurodegeneration. Biochim Biophys Acta. 2010;1802(1):212–20. https://doi.org/10.1016/j.bbadis.2009.10.007.
74. Zuo L, Motherwell MS. The impact of reactive oxygen species and genetic mitochondrial mutations in Parkinson's disease. Gene. 2013;532(1):18–23. https://doi.org/10.1016/j.gene.2013.07.085.
75. Murphy MP. Antioxidants as therapies: can we improve on nature? Free Radic Biol Med. 2014;66:20–3. https://doi.org/10.1016/j.freeradbiomed.2013.04.010.
76. Khandrika L, Kumar B, Koul S, Maroni P, Koul HK. Oxidative stress in prostate cancer. Cancer Lett. 2009;282(2):125–36. https://doi.org/10.1016/j.canlet.2008.12.011.
77. Wells PG, McCallum GP, Chen CS, Henderson JT, Lee CJ, Perstin J, Preston TJ, Wiley MJ, Wong AW. Oxidative stress in developmental origins of disease: teratogenesis, neurodevelopmental deficits, and cancer. Toxicol Sci. 2009;108(1):4–18. https://doi.org/10.1093/toxsci/kfn263.
78. Katakwar P, Metgud R, Naik S, Mittal R. Oxidative stress marker in oral cancer: a review. J Cancer Res Ther. 2016;12(2):438–46. https://doi.org/10.4103/0973-1482.151935.
79. Poulsen HE, Prieme H, Loft S. Role of oxidative DNA damage in cancer initiation and promotion. Eur J Cancer Prev. 1998;7(1):9–16.
80. Valko M, Rhodes CJ, Moncol J, Izakovic M, Mazur M. Free radicals, metals and antioxidants in oxidative stress-induced cancer. Chem Biol Interact. 2006;160(1):1–40. https://doi.org/10.1016/j.cbi.2005.12.009.
81. Sosa V, Moliné T, Somoza R, Paciucci R, Kondoh H, LLeonart ME. Oxidative stress and cancer: an overview. Ageing Res Rev. 2013;12(1):376–90. https://doi.org/10.1016/j.arr.2012.10.004.
82. Brown NS, Bicknell R. Hypoxia and oxidative stress in breast cancer. Oxidative stress: its effects on the growth, metastatic potential and response to therapy of breast cancer. Breast Cancer Res. 2001;3(5):323–7. https://doi.org/10.1186/bcr315.
83. Trueba GP, Sánchez GM, Giuliani A. Oxygen free radical and antioxidant defense mechanism in cancer. Front Biosci. 2004;9:2029–44. https://doi.org/10.2741/1335.
84. McCubrey JA, Steelman LS, Chappell WH, Abrams SL, Wong EW, Chang F, Lehmann B, Terrian DM, Milella M, Tafuri A, Stivala F, Libra M, Basecke J, Evangelisti C, Martelli AM, Franklin RA. Roles of the Raf/MEK/ERK pathway in cell growth, malignant transformation and drug resistance. Biochim Biophys Acta. 2007;1773(8):1263–84. https://doi.org/10.1016/j.bbamcr.2006.10.001.
85. Maulik N. Redox signaling of angiogenesis. Antioxid Redox Signal. 2002;4(5):805–15. https://doi.org/10.1089/152308602760598963.
86. Mori K, Shibanuma M, Nose K. Invasive potential induced under long-term oxidative stress in mammary epithelial cells. Cancer Res. 2004;64(20):7464–72. https://doi.org/10.1158/0008-5472.CAN-04-1725.
87. Shinohara M, Adachi Y, Mitsushita J, Kuwabara M, Nagasawa A, Harada S, Furuta S, Zhang Y, Seheli K, Miyazaki H, Kamata T. Reactive oxygen generated by NADPH oxidase 1 (Nox1) contributes to cell invasion by regulating matrix metalloprotease-9 production and cell migration. J Biol Chem. 2010;285(7):4481–8. https://doi.org/10.1074/jbc.M109.071779.
88. Rayman MP. Selenium in cancer prevention: a review of the evidence and mechanism of action. Proc Nutr Soc. 2005;64(4):527–42. https://doi.org/10.1079/pns2005467.
89. Soerjomataram I, Oomen D, Lemmens V, Oenema A, Benetou V, Trichopoulou A, Coebergh JW, Barendregt J, de Vries E. Increased consumption of fruit and vegetables and future cancer incidence in selected European countries. Eur J Cancer. 2010;46(14):2563–80. https://doi.org/10.1016/j.ejca.2010.07.026.
90. Turati F, Rossi M, Pelucchi C, Levi F, La Vecchia C. Fruit and vegetables and cancer risk: a review of Southern European studies. Br J Nutr. 2015;113(Suppl 2):S102–10. https://doi.org/10.1017/S0007114515000148.
91. Barrajón-Catalán E, Fernández-Arroyo S, Saura D, Guillén E, Fernández-Gutiérrez A, Segura-Carretero A, Micol V. Cistaceae aqueous extracts containing ellagitannins show antioxidant and antimicrobial capacity, and cytotoxic activity against human cancer cells. Food Chem Toxicol. 2010;48(8–9):2273–82. https://doi.org/10.1016/j.fct.2010.05.060.
92. Paredes-López O, Cervantes-Ceja ML, Vigna-Pérez M, Hernández-Pérez T. Berries: improving human health and healthy aging, and promoting quality life—a review. Plant Foods Hum Nutr. 2010;65(3):299–308. https://doi.org/10.1007/s11130-010-0177-1.
93. Sreelatha S, Dinesh E, Uma C. Antioxidant properties of Rajgira (*Amaranthuspaniculatus*) leaves and potential synergy in chemoprevention. Asian Pac J Cancer Prev. 2012;13(6):2775–80. https://doi.org/10.7314/apjcp.2012.13.6.2775.
94. Cordero-Herrera I, Martín MA, Bravo L, Goya L, Ramos S. Epicatechin gallate induces cell death via p53 activation and stimulation of p38 and JNK in human colon cancer SW480 cells. Nutr Cancer. 2013;65(5):718–28. https://doi.org/10.1080/01635581.2013.795981.
95. Manikandan R, Beulaja M, Arulvasu C, Sellamuthu S, Dinesh D, Prabhu D, Babu G, Vaseeharan B, Prabhu NM. Synergistic anticancer activity of curcumin and catechin: an in vitro study using human cancer cell lines. Microsc Res Tech. 2012;75(2):112–6. https://doi.org/10.1002/jemt.21032.
96. Sak K. Site-specific anticancer effects of dietary flavonoid quercetin. Nutr Cancer. 2014;66(2):177–93. https://doi.org/10.1080/01635581.2014.864418.
97. Sun J, Chu YF, Wu X, Liu RH. Antioxidant and antiproliferative activities of common fruits. J Agric Food Chem. 2002;50(25):7449–54. https://doi.org/10.1021/jf0207530.
98. Rokayya S, Li CJ, Zhao Y, Li Y, Sun CH. Cabbage (Brassica oleracea L. var. Capitata) phytochemicals with antioxidant and anti-inflammatory potential. Asian Pac J Cancer Prev. 2014;14(11):6657–62. https://doi.org/10.7314/apjcp.2013.14.11.6657.
99. Gloria NF, Soares N, Brand C, Oliveira FL, Borojevic R, Teodoro AJ. Lycopene and beta-carotene induce cell-cycle arrest and apoptosis in human breast cancer cell lines. Anticancer Res. 2014;34(3):1377–86.
100. Li Y, Ma C, Qian M, Wen Z, Jing H, Qian D. Butein induces cell apoptosis and inhibition of cyclooxygenase-2 expression in A549 lung cancer cells. Mol Med Rep. 2014;9(2):763–7. https://doi.org/10.3892/mmr.2013.1850.
101. Xie Q, Bai Q, Zou LY, Zhang QY, Zhou Y, Chang H, Yi L, Zhu JD, Mi MT. Genistein inhibits DNA methylation and increases expression of tumor suppressor genes in human breast cancer cells. Genes Chromosomes Cancer. 2014;53(5):422–31. https://doi.org/10.1002/gcc.22154.
102. Zhang YJ, Gan RY, Li S, Zhou Y, Li AN, Xu DP, Li HB. Antioxidant phytochemicals for the prevention and treatment of chronic diseases. Molecules. 2015;20(12):21138–56. https://doi.org/10.3390/molecules201219753.
103. Pathak SK, Sharma RA, Mellon JK. Chemoprevention of prostate cancer by diet-derived antioxidant agents and hormonal manipulation (review). Int J Oncol. 2003;22(1):5–13. https://doi.org/10.3892/ijo.22.1.5.
104. Jones LW, Demark-Wahnefried W. Diet, exercise, and complementary therapies after primary treatment for cancer.

Lancet Oncol. 2006;7(12):1017–26. https://doi.org/10.1016/S1470-2045(06)70976-7.
105. Vrieling A, Buck K, Seibold P, Heinz J, Obi N, Flesch-Janys D, Chang-Claude J. Dietary patterns and survival in German postmenopausal breast cancer survivors. Br J Cancer. 2013;108(1):188–92. https://doi.org/10.1038/bjc.2012.521.
106. George SM, Ballard-Barbash R, Shikany JM, Caan BJ, Freudenheim JL, Kroenke CH, Vitolins MZ, Beresford SA, Neuhouser ML. Better postdiagnosis diet quality is associated with reduced risk of death among postmenopausal women with invasive breast cancer in the women's health initiative. Cancer Epidemiol Biomark Prev. 2014;23(4):575–83. https://doi.org/10.1158/1055-9965.
107. Arthur AE, Peterson KE, Rozek LS, Taylor JM, Light E, Chepeha DB, Hébert JR, Terrell JE, Wolf GT, Duffy SA, UM Head and Neck SPORE Program. Pretreatment dietary patterns, weight status, and head and neck squamous cell carcinoma prognosis. Am J Clin Nutr. 2013;97(2):360–8. https://doi.org/10.3945/ajcn.112.044859.
108. Pelser C, Arem H, Pfeiffer RM, Elena JW, Alfano CM, Hollenbeck AR, Park Y. Prediagnostic lifestyle factors and survival after colon and rectal cancer diagnosis in the National Institutes of Health (NIH)-AARP diet and health study. Cancer. 2014;120(10):1540–7. https://doi.org/10.1002/cncr.28573.
109. Garewal H. Antioxidants in oral cancer prevention. Am J Clin Nutr. 1995;62(6 Suppl):1410S–6S. https://doi.org/10.1093/ajcn/62.6.1410S.
110. Block KI, Koch AC, Mead MN, Tothy PK, Newman RA, Gyllenhaal C. Impact of antioxidant supplementation on chemotherapeutic toxicity: a systematic review of the evidence from randomized controlled trials. Int J Cancer. 2008;123(6):1227–39. https://doi.org/10.1002/ijc.23754.
111. Gröber U, Kisters K, Adamietz IA. Vitamin D in der Onkologie—Komplementäre Medizin: Update 2015 [Vitamin D in oncology: Update 2015]. Med Monatsschr Pharm. 2015;38(12):512–6.
112. Norat T, Scoccianti C, Boutron-Ruault MC, Anderson A, Berrino F, Cecchini M, Espina C, Key T, Leitzmann M, Powers H, Wiseman M, Romieu I. European code against cancer 4th edition: diet and cancer. Cancer Epidemiol. 2015;39(Suppl 1):S56–66. https://doi.org/10.1016/j.canep.2014.12.016.
113. Balmus IM, Ciobica A, Trifan A, Stanciu C. The implications of oxidative stress and antioxidant therapies in inflammatory bowel disease: clinical aspects and animal models. Saudi J Gastroenterol. 2016;22(1):3–17. https://doi.org/10.4103/1319-3767.173753.
114. Zhu H, Li YR. Oxidative stress and redox signaling mechanisms of inflammatory bowel disease: updated experimental and clinical evidence. Exp Biol Med (Maywood). 2012;237(5):474–80. https://doi.org/10.1258/ebm.2011.011358.
115. Bhattacharyya A, Chattopadhyay R, Mitra S, Crowe SE. Oxidative stress: an essential factor in the pathogenesis of gastrointestinal mucosal diseases. Physiol Rev. 2014;94(2):329–54. https://doi.org/10.1152/physrev.00040.2012.
116. Moura FA, de Andrade KQ, Dos Santos JCF, Araújo ORP, Goulart MOF. Antioxidant therapy for treatment of inflammatory bowel disease: does it work? Redox Biol. 2015;6:617–39. https://doi.org/10.1016/j.redox.2015.10.006.
117. Ingraham LM, Coates TD, Allen JM, Higgins CP, Baehner RL, Boxer LA. Metabolic, membrane, and functional responses of human polymorphonuclear leukocytes to platelet-activating factor. Blood. 1982;59(6):1259–66.
118. Sharon P, Stenson WF. Enhanced synthesis of leukotriene B4 by colonic mucosa in inflammatory bowel disease. Gastroenterology. 1984;86(3):453–60.
119. Wallace JL, Chin BC. Inflammatory mediators in gastrointestinal defense and injury. Proc Soc Exp Biol Med. 1997;214(3):192–203. https://doi.org/10.3181/00379727-214-44087.
120. Alzoghaibi MA. Concepts of oxidative stress and antioxidant defense in Crohn's disease. World J Gastroenterol. 2013;19(39):6540–7. https://doi.org/10.3748/wjg.v19.i39.6540.
121. Buffinton GD, Doe WF. Altered ascorbic acid status in the mucosa from inflammatory bowel disease patients. Free Radic Res. 1995;22(2):131–43. https://doi.org/10.3109/10715769509147535.
122. Aditi A, Graham DY. Vitamin C, gastritis, and gastric disease: a historical review and update. Dig Dis Sci. 2012;57(10):2504–15. https://doi.org/10.1007/s10620-012-2203-7
123. van Hogezand RA, Hamdy NA. Skeletal morbidity in inflammatory bowel disease. Scand J Gastroenterol Suppl. 2006;243:59–64. https://doi.org/10.1080/00365520600664276.
124. Alastair F, Emma G, Emma P. Nutrition in inflammatory bowel disease. JPEN J Parenter Enteral Nutr. 2011;35(5):571–80. https://doi.org/10.1177/0148607111413599.
125. Geerling BJ, Badart-Smook A, van Deursen C, van Houwelingen AC, Russel MG, Stockbrügger RW, Brummer RJ. Nutritional supplementation with N-3 fatty acids and antioxidants in patients with Crohn's disease in remission: effects on antioxidant status and fatty acid profile. Inflamm Bowel Dis. 2000;6(2):77–84. https://doi.org/10.1097/00054725-200005000-00002.
126. Bjelakovic G, Nikolova D, Simonetti RG, Gluud C. Antioxidant supplements for prevention of gastrointestinal cancers: a systematic review and meta-analysis. Lancet. 2004;364(9441):1219–28. https://doi.org/10.1016/S0140-6736(04)17138-9.
127. Poljsak B, Šuput D, Milisav I. Achieving the balance between ROS and antioxidants: when to use the synthetic antioxidants. Oxidative Med Cell Longev. 2013;2013:956792. https://doi.org/10.1155/2013/956792.
128. Bitiren M, Karakilcik AZ, Zerin M, Ozardali I, Selek S, Nazligül Y, Ozgonul A, Musa D, Uzunkoy A. Protective effects of selenium and vitamin E combination on experimental colitis in blood plasma and colon of rats. Biol Trace Elem Res. 2010;136(1):87–95. https://doi.org/10.1007/s12011-009-8518-3.
129. Mariani E, Polidori MC, Cherubini A, Mecocci P. Oxidative stress in brain aging, neurodegenerative and vascular diseases: an overview. J Chromatogr B Analyt Technol Biomed Life Sci. 2005;827(1):65–75. https://doi.org/10.1016/j.jchromb.2005.04.023.
130. Kregel KC, Zhang HJ. An integrated view of oxidative stress in aging: basic mechanisms, functional effects, and pathological considerations. Am J Physiol Regul Integr Comp Physiol. 2007;292(1):R18–36. https://doi.org/10.1152/ajpregu.00327.2006.
131. Trifunovic A, Wredenberg A, Falkenberg M, Spelbrink JN, Rovio AT, Bruder CE, Bohlooly-Y M, Gidlöf S, Oldfors A, Wibom R, Törnell J, Jacobs HT, Larsson NG. Premature ageing in mice expressing defective mitochondrial DNA polymerase. Nature. 2004;429(6990):417–23. https://doi.org/10.1038/nature02517.
132. Chung HY, Cesari M, Anton S, Marzetti E, Giovannini S, Seo AY, Carter C, Yu BP, Leeuwenburgh C. Molecular inflammation: underpinnings of aging and age-related diseases. Ageing Res Rev. 2009;8(1):18–30. https://doi.org/10.1016/j.arr.2008.07.002.
133. Brownie S. Why are elderly individuals at risk of nutritional deficiency? Int J Nurs Pract. 2006;12(2):110–8. https://doi.org/10.1111/j.1440-172X.2006.00557.x.
134. Lips P. Vitamin D deficiency and secondary hyperparathyroidism in the elderly: consequences for bone loss and fractures and therapeutic implications. Endocr Rev. 2001;22(4):477–501. https://doi.org/10.1210/edrv.22.4.0437.
135. Kandola K, Bowman A, Birch-Machin MA. Oxidative stress—a key emerging impact factor in health, ageing, lifestyle and aesthetics. Int J Cosmet Sci. 2015;37(Suppl 2):1–8. https://doi.org/10.1111/ics.12287.
136. Kang JH, Ascherio A, Grodstein F. Fruit and vegetable consumption and cognitive decline in aging women. Ann Neurol. 2005;57(5):713–20. https://doi.org/10.1002/ana.20476.

137. Wilson MA, Shukitt-Hale B, Kalt W, Ingram DK, Joseph JA, Wolkow CA. Blueberry polyphenols increase lifespan and thermotolerance in Caenorhabditis elegans. Aging Cell. 2006;5(1):59–68. https://doi.org/10.1111/j.1474-9726.2006.00192.x.
138. Peng C, Wang X, Chen J, Jiao R, Wang L, Li YM, Zuo Y, Liu Y, Lei L, Ma KY, Huang Y, Chen ZY. Biology of ageing and role of dietary antioxidants. Biomed Res Int. 2014;2014:831841. https://doi.org/10.1155/2014/831841.
139. Rodrigo R, González J, Paoletto F. The role of oxidative stress in the pathophysiology of hypertension. Hypertens Res. 2011;34(4):431–40. https://doi.org/10.1038/hr.2010.264.
140. Rodrigo R, Rivera G. Renal damage mediated by oxidative stress: a hypothesis of protective effects of red wine. Free Radic Biol Med. 2002;33(3):409–22. https://doi.org/10.1016/s0891-5849(02)00908-5.
141. Grassi G. Assessment of sympathetic cardiovascular drive in human hypertension: achievements and perspectives. Hypertension. 2009;54(4):690–7. https://doi.org/10.1161/HYPERTENSIONAHA.108.119883.
142. Guyenet PG. The sympathetic control of blood pressure. Nat Rev Neurosci. 2006;7(5):335–46. https://doi.org/10.1038/nrn1902.
143. Kishi T, Hirooka Y, Kimura Y, Ito K, Shimokawa H, Takeshita A. Increased reactive oxygen species in rostral ventrolateral medulla contribute to neural mechanisms of hypertension in stroke-prone spontaneously hypertensive rats. Circulation. 2004;109(19):2357–62. https://doi.org/10.1161/01.CIR.0000128695.49900.12.
144. Malyszko J. Mechanism of endothelial dysfunction in chronic kidney disease. Clin Chim Acta. 2010;411(19–20):1412–20. https://doi.org/10.1016/j.cca.2010.06.019.
145. Bilodeau JF, Hubel CA. Current concepts in the use of antioxidants for the treatment of preeclampsia. J Obstet Gynaecol Can. 2003;25(9):742–50. https://doi.org/10.1016/s1701-2163(16)31003-9.
146. Giacco F, Brownlee M. Oxidative stress and diabetic complications. Circ Res. 2010;107(9):1058–70. https://doi.org/10.1161/CIRCRESAHA.110.223545.
147. Zou L. Antioxidants, minerals and vitamin supplements in diabetes. Pers Oxid Antioxid Med Sci. 2022;11(8):01–2.
148. Zuo T, Zhu M, Xu W. Roles of oxidative stress in polycystic ovary syndrome and cancers. Oxidative Med Cell Longev. 2016;2016:8589318. https://doi.org/10.1155/2016/8589318.
149. Fang YZ, Yang S, Wu G. Free radicals, antioxidants, and nutrition. Nutrition. 2002;18(10):872–9. https://doi.org/10.1016/s0899-9007(02)00916-4.
150. Sulaiman MA, Al-Farsi YM, Al-Khaduri MM, Saleh J, Waly MI. Polycystic ovarian syndrome is linked to increased oxidative stress in Omani women. Int J Women's Health. 2018;10:763–71. https://doi.org/10.2147/IJWH.S166461.
151. Cheng X, He B. Clinical and biochemical potential of antioxidants in treating polycystic ovary syndrome. Int J Women's Health. 2022;14:467–79. https://doi.org/10.2147/IJWH.S345853.
152. Razavi M, Jamilian M, Kashan ZF, Heidar Z, Mohseni M, Ghandi Y, Bagherian T, Asemi Z. Selenium supplementation and the effects on reproductive outcomes, biomarkers of inflammation, and oxidative stress in women with polycystic ovary syndrome. Horm Metab Res. 2016;48(3):185–90. https://doi.org/10.1055/s-0035-1559604.
153. Afshar Ebrahimi F, Foroozanfard F, Aghadavod E, Bahmani F, Asemi Z. The effects of magnesium and zinc co-supplementation on biomarkers of inflammation and oxidative stress, and gene expression related to inflammation in polycystic ovary syndrome: a randomized controlled clinical trial. Biol Trace Elem Res. 2018;184(2):300–7. https://doi.org/10.1007/s12011-017-1198-5.
154. Jamilian M, Zadeh Modarres S, Amiri Siavashani M, Karimi M, Mafi A, Ostadmohammadi V, Asemi Z. The influences of chromium supplementation on Glycemic control, markers of cardio-metabolic risk, and oxidative stress in infertile polycystic ovary syndrome women candidate for in vitro fertilization: a randomized, double-blind, placebo-controlled trial. Biol Trace Elem Res. 2018;185(1):48–55. https://doi.org/10.1007/s12011-017-1236-3.
155. Padayatty SJ, Katz A, Wang Y, Eck P, Kwon O, Lee JH, Chen S, Corpe C, Dutta A, Dutta SK, Levine M. Vitamin C as an antioxidant: evaluation of its role in disease prevention. J Am Coll Nutr. 2003;22(1):18–35. https://doi.org/10.1080/07315724.2003.10719272.
156. Morrow JD. The isoprostanes: their quantification as an index of oxidant stress status in vivo. Drug Metab Rev. 2000;32(3–4):377–85. https://doi.org/10.1081/dmr-100102340.
157. Steinberg D. Low density lipoprotein oxidation and its pathobiological significance. J Biol Chem. 1997;272(34):20963–6. https://doi.org/10.1074/jbc.272.34.20963.
158. Heinecke JW. Oxidants and antioxidants in the pathogenesis of atherosclerosis: implications for the oxidized low density lipoprotein hypothesis. Atherosclerosis. 1998;141(1):1–15. https://doi.org/10.1016/s0021-9150(98)00173-7.
159. Witztum JL, Steinberg D. The oxidative modification hypothesis of atherosclerosis: does it hold for humans? Trends Cardiovasc Med. 2001;11(3–4):93–102. https://doi.org/10.1016/s1050-1738(01)00111-6.
160. Lee SH, Oe T, Blair IA. Vitamin C-induced decomposition of lipid hydroperoxides to endogenous genotoxins. Science. 2001;292(5524):2083–6. https://doi.org/10.1126/science.1059501.
161. Halliwell B. Why and how should we measure oxidative DNA damage in nutritional studies? How far have we come? Am J Clin Nutr. 2000;72(5):1082–7. https://doi.org/10.1093/ajcn/72.5.1082.
162. Nguyen LA, He H, Pham-Huy C. Chiral drugs: an overview. Int J Biomed Sci. 2006;2(2):85–100.
163. Jiang Q. Natural forms of vitamin E: metabolism, antioxidant, and anti-inflammatory activities and their role in disease prevention and therapy. Free Radic Biol Med. 2014;72:76–90. https://doi.org/10.1016/j.freeradbiomed.2014.03.035.
164. Clarke MW, Burnett JR, Croft KD. Vitamin E in human health and disease. Crit Rev Clin Lab Sci. 2008;45(5):417–50. https://doi.org/10.1080/10408360802118625.
165. Palace VP, Khaper N, Qin Q, Singal PK. Antioxidant potentials of vitamin A and carotenoids and their relevance to heart disease. Free Radic Biol Med. 1999;26(5–6):746–61. https://doi.org/10.1016/s0891-5849(98)00266-4.
166. Song Y, Xu Q, Park Y, Hollenbeck A, Schatzkin A, Chen H. Multivitamins, individual vitamin and mineral supplements, and risk of diabetes among older U.S. adults. Diabetes Care. 2011;34(1):108–14. https://doi.org/10.2337/dc10-1260.
167. Furman C, Rundlöf AK, Larigauderie G, Jaye M, Bricca G, Copin C, Kandoussi AM, Fruchart JC, Arnér ES, Rouis M. Thioredoxin reductase 1 is upregulated in atherosclerotic plaques: specific induction of the promoter in human macrophages by oxidized low-density lipoproteins. Free Radic Biol Med. 2004;37(1):71–85. https://doi.org/10.1016/j.freeradbiomed.2004.04.016.
168. Zhang H, Luo Y, Zhang W, He Y, Dai S, Zhang R, Huang Y, Bernatchez P, Giordano FJ, Shadel G, Sessa WC, Min W. Endothelial-specific expression of mitochondrial thioredoxin improves endothelial cell function and reduces atherosclerotic lesions. Am J Pathol. 2007;170(3):1108–20. https://doi.org/10.2353/ajpath.2007.060960.
169. World CJ, Yamawaki H, Berk BC. Thioredoxin in the cardiovascular system. J Mol Med (Berl). 2006;84(12):997–1003. https://doi.org/10.1007/s00109-006-0109-6.
170. Tanguy S, Toufektsian MC, Besse S, Ducros V, De Leiris J, Boucher F. Dietary selenium intake affects cardiac susceptibility to ischaemia/reperfusion in male senescent rats. Age Ageing. 2003;32(3):273–8. https://doi.org/10.1093/ageing/32.3.273.

171. Venardos K, Harrison G, Headrick J, Perkins A. Effects of dietary selenium on glutathione peroxidase and thioredoxin reductase activity and recovery from cardiac ischemia-reperfusion. J Trace Elem Med Biol. 2004;18(1):81–8. https://doi.org/10.1016/j.jtemb.2004.01.001.
172. Venardos K, Ashton K, Headrick J, Perkins A. Effects of dietary selenium on post-ischemic expression of antioxidant mRNA. Mol Cell Biochem. 2005;270(1–2):131–8. https://doi.org/10.1007/s11010-005-5279-y.
173. Baljinnyam E, Hasebe N, Morihira M, Sumitomo K, Matsusaka T, Fujino T, Fukuzawa J, Ushikubi F, Kikuchi K. Oral pretreatment with ebselen enhances heat shock protein 72 expression and reduces myocardial infarct size. Hypertens Res. 2006;29(11):905–13. https://doi.org/10.1291/hypres.29.905.
174. Lapchak PA, Zivin JA. Ebselen, a seleno-organic antioxidant, is neuroprotective after embolic strokes in rabbits: synergism with low-dose tissue plasminogen activator. Stroke. 2003;34(8):2013–8. https://doi.org/10.1161/01.STR.0000081223.74129.04.
175. Takasago T, Peters EE, Graham DI, Masayasu H, Macrae IM. Neuroprotective efficacy of ebselen, an anti-oxidant with anti-inflammatory actions, in a rodent model of permanent middle cerebral artery occlusion. Br J Pharmacol. 1997;122(6):1251–6. https://doi.org/10.1038/sj.bjp.0701426.
176. Lippman SM, Goodman PJ, Klein EA, Parnes HL, Thompson IM Jr, Kristal AR, Santella RM, Probstfield JL, Moinpour CM, Albanes D, Taylor PR, Minasian LM, Hoque A, Thomas SM, Crowley JJ, Gaziano JM, Stanford JL, Cook ED, Fleshner NE, Lieber MM, Walther PJ, Khuri FR, Karp DD, Schwartz GG, Ford LG, Coltman CA Jr. Designing the selenium and vitamin E cancer prevention trial (SELECT). J Natl Cancer Inst. 2005;97(2):94–102. https://doi.org/10.1093/jnci/dji009.
177. Klein EA, Thompson IM, Lippman SM, Goodman PJ, Albanes D, Taylor PR, Coltman C. SELECT: the selenium and vitamin E cancer prevention trial: rationale and design. Prostate Cancer Prostatic Dis. 2000;3(3):145–51. https://doi.org/10.1038/sj.pcan.4500412.
178. McKenzie RC, Arthur JR, Beckett GJ. Selenium and the regulation of cell signaling, growth, and survival: molecular and mechanistic aspects. Antioxid Redox Signal. 2002;4(2):339–51. https://doi.org/10.1089/152308602753666398.
179. Whanger PD. Selenium and its relationship to cancer: an update. Br J Nutr. 2004;91(1):11–28. https://doi.org/10.1079/bjn20031015.
180. El-Bayoumy K. The protective role of selenium on genetic damage and on cancer. Mutat Res. 2001;475(1–2):123–39. https://doi.org/10.1016/s0027-5107(01)00075-6.
181. Lü J, Jiang C. Selenium and cancer chemoprevention: hypotheses integrating the actions of selenoproteins and selenium metabolites in epithelial and non-epithelial target cells. Antioxid Redox Signal. 2005;7(11–12):1715–27. https://doi.org/10.1089/ars.2005.7.1715.
182. Papp LV, Lu J, Holmgren A, Khanna KK. From selenium to selenoproteins: synthesis, identity, and their role in human health. Antioxid Redox Signal. 2007;9(7):775–806. https://doi.org/10.1089/ars.2007.1528.
183. Tinggi U. Selenium: its role as antioxidant in human health. Environ Health Prev Med. 2008;13(2):102–8. https://doi.org/10.1007/s12199-007-0019-4.
184. Internet: Selenium: Fact Sheet for Health Professionals. https://ods.od.nih.gov/factsheets/Selenium-HealthProfessional/.
185. Marreiro DD, Cruz KJ, Morais JB, Beserra JB, Severo JS, de Oliveira AR. Zinc and oxidative stress: current mechanisms. Antioxidants (Basel). 2017;6(2):24. https://doi.org/10.3390/antiox6020024.
186. Biagiotti S, Menotta M, Orazi S, Spapperi C, Brundu S, Fraternale A, Bianchi M, Rossi L, Chessa L, Magnani M. Dexamethasone improves redox state in ataxia telangiectasia cells by promoting an NRF2-mediated antioxidant response. FEBS J. 2016;283(21):3962–78. https://doi.org/10.1111/febs.13901.
187. Jenner A, Ren M, Rajendran R, Ning P, Huat BT, Watt F, Halliwell B. Zinc supplementation inhibits lipid peroxidation and the development of atherosclerosis in rabbits fed a high cholesterol diet. Free Radic Biol Med. 2007;42(4):559–66. https://doi.org/10.1016/j.freeradbiomed.2006.11.024.
188. Miao X, Wang Y, Sun J, Sun W, Tan Y, Cai L, Zheng Y, Su G, Liu Q, Wang Y. Zinc protects against diabetes-induced pathogenic changes in the aorta: roles of metallothionein and nuclear factor (erythroid-derived 2)-like 2. Cardiovasc Diabetol. 2013;12:54. https://doi.org/10.1186/1475-2840-12-54.
189. https://www.mountsinai.org/health-library/supplement/manganese#:~:text=Manganese%20is%20a%20component%20of,including%20heart%20disease%20and%20cancer.
190. Kessler J. Are there side effects when using supraphysiologic levels of iodine treatment regiments? In: Preedy VR, Burrow GN, Watson RR, editors. Comprehensive handbook of iodine: nutritional, endocrine and pathological aspects. San Diego, CA: Academic Press; 2009. p. 801–10. https://doi.org/10.1016/B978-0-12-374135-6.00082-0.
191. Aceves C, Anguiano B. Is iodine an antioxidant and antiproliferative agent for the mammary and prostate glands? In: Preedy VR, Burrow GN, Watson RR, editors. Comprehensive handbook of iodine: nutritional, endocrine and pathological aspects. Academic Press: San Diego, CA; 2009. p. 249–57.
192. Iwata A, Morrison ML, Roth MB. Iodide protects heart tissue from reperfusion injury. PLoS One. 2014;9(11):e112458. https://doi.org/10.1371/journal.pone.0112458.
193. Winkler R. Iodine—a potential antioxidant and the role of iodine/iodide in health and disease. Nat Sci. 2015;7:548–57. https://doi.org/10.4236/ns.2015.712055.
194. Okarter N, Liu RH. Health benefits of whole grain phytochemicals. Crit Rev Food Sci Nutr. 2010;50(3):193–208. https://doi.org/10.1080/10408390802248734.
195. Durazzo A, Casale G, Melini V, Maiani G, Acquistucci R. Evaluation of antioxidant properties in cereals: study of some traditional Italian wheats. Foods. 2015;4(3):391–9. https://doi.org/10.3390/foods4030391.
196. Žilić S, Šukalović V, Dodig D, Maksimović V, Maksimovic M, Basić Z. Antioxidant activity of small grain cereals caused by phenolics and lipid soluble antioxidants. J Cereal Sci. 2011;54(3):417–24. https://doi.org/10.1016/j.jcs.2011.08.005.
197. Sreeramulu D, Reddy CV, Raghunath M. Antioxidant activity of commonly consumed cereals, millets, pulses and legumes in India. Indian J Biochem Biophys. 2009;46(1):112–5.
198. Djordjevic TM, Šiler-Marinkovic SS, DimitrijevicBrankovic SI. Antioxidant activity and total phenolic content in some cereals and legumes. Int J Food Prop. 2011;14(1):175–84. https://doi.org/10.1080/10942910903160364.
199. Saunders AV, Craig WJ, Baines SK. Zinc and vegetarian diets. Med J Aust. 2013;199(S4):S17–21. https://doi.org/10.5694/mja11.11493.
200. Rizvi S, Raza ST, Ahmed F, Ahmad A, Abbas S, Mahdi F. The role of vitamin e in human health and some diseases. Sultan Qaboos Univ Med J. 2014;14(2):e157–65.
201. Shargorodsky M, Debby O, Matas Z, Zimlichman R. Effect of long-term treatment with antioxidants (vitamin C, vitamin E, coenzyme Q10 and selenium) on arterial compliance, humoral factors and inflammatory markers in patients with multiple cardiovascular risk factors. Nutr Metab (Lond). 2010;7:55. https://doi.org/10.1186/1743-7075-7-55.
202. Chun OK, Floegel A, Chung SJ, Chung CE, Song WO, Koo SI. Estimation of antioxidant intakes from diet and supplements in U.S. adults. J Nutr. 2010;140(2):317–24. https://doi.org/10.3945/jn.109.114413.

203. Ou B, Huang D, Hampsch-Woodill M, Flanagan JA, Deemer EK. Analysis of antioxidant activities of common vegetables employing Oxygen Radical Absorbance Capacity (ORAC) and Ferric Reducing Antioxidant Power (FRAP) assays: a comparative study. J Agric Food Chem. 2002;50(11):3122–8. https://doi.org/10.1021/jf0116606.
204. Lykkesfeldt J, Michels AJ, Frei B. Vitamin C. Adv Nutr. 2014;5(1):16–8. https://doi.org/10.3945/an.113.005157.
205. Nantitanon W, Yotsawimonwat S, Okonogi S. Factors influencing antioxidant activities and total phenolic content of guava leaf extract. LWT- Food Sci Technol. 2010;43(7):1095–103. https://doi.org/10.1016/j.lwt.2010.02.015.
206. Anwar H, Hussain G, Mustafa I. Antioxidants from natural sources. In: Antioxidants in foods and its applications, vol. Ch.1; 2018. p. 3–28. https://doi.org/10.5772/intechopen.75961.
207. Edwards AJ, Vinyard BT, Wiley ER, Brown ED, Collins JK, Perkins-Veazie P, Baker RA, Clevidence BA. Consumption of watermelon juice increases plasma concentrations of lycopene and beta-carotene in humans. J Nutr. 2003;133(4):1043–50. https://doi.org/10.1093/jn/133.4.1043.
208. Rayman MP. Food-chain selenium and human health: emphasis on intake. Br J Nutr. 2008;100(2):254–68. https://doi.org/10.1017/S0007114508939830.
209. Rayman MP. Selenium and human health. Lancet. 2012;379(9822):1256–68. https://doi.org/10.1016/S0140-6736(11)61452-9.
210. Grażyna C, Hanna C, Adam A, Bielecka MM. Natural antioxidants in milk and dairy products. Int J Dairy Technol. 2017;70(2):165–78. https://doi.org/10.1111/1471-0307.12359.
211. Goyal S, Kaur TJ. Antioxidants: dietary scavengers in lifestyle diseases. J Pharmacogn Phytochem. 2019;8(4):96–102.
212. Delia-Gabriela D, Nicoleta-Gabriela H, Moldovan C, Diana-Nicoleta R, Mirela Viorica P, Rădoi B. Antioxidant activity of some fresh vegetables and fruits juices. J Agroaliment Process Technol. 2011;17(2):163–8.
213. Čanadanović-Brunet JM, Savatović SS, Ćetković GS, Vulić JJ, Djilas SM, Markov SL, Cvetković DD. Antioxidant and antimicrobial activities of beet root pomace extracts. Czech J Food Sci. 2011;29(6):575–85.
214. Georgiev VG, Weber J, Kneschke EM, Denev PN, Bley T, Pavlov AI. Antioxidant activity and phenolic content of betalain extracts from intact plants and hairy root cultures of the red beetroot Beta vulgaris cv. Detroit dark red. Plant Foods Hum Nutr. 2010;65(2):105–11. https://doi.org/10.1007/s11130-010-0156-6.
215. Dreher ML, Davenport AJ. Hass avocado composition and potential health effects. Crit Rev Food Sci Nutr. 2013;53(7):738–50. https://doi.org/10.1080/10408398.2011.556759.
216. Fernández-López JA, Almela L, Obón JM, Castellar R. Determination of antioxidant constituents in cactus pear fruits. Plant Foods Hum Nutr. 2010;65(3):253–9. https://doi.org/10.1007/s11130-010-0189-x.
217. Yasoubi P, Barzegar M, Sahari M, Azizi M. Total phenolic contents and antioxidant activity of pomegranate (*Punicagranatum* L.) peel extracts. J Agric Sci Technol. 2010;9:35–42.
218. Kusmita L, Puspitaningrum I, Limantara L. Identification, isolation and antioxidant activity of pheophytin from green tea (*Camelliasinensis* (L.) Kuntze). Procedia Chem. 2015;14:232–8. https://doi.org/10.1016/j.proche.2015.03.033.
219. Sunde RA. Selenium. In: Coates PM, Betz JM, Blackman MR, et al., editors. Encyclopedia of dietary supplements. 2nd ed. London and New York: Informa Healthcare; 2010. p. 711–8.
220. Terry EN, Diamond AM. Selenium. In: Erdman JW, Macdonald IA, Zeisel SH, editors. Present knowledge in nutrition. 10th ed. Washington, DC: Wiley-Blackwell; 2012. p. 568–87.
221. Institute of Medicine, Food and Nutrition Board. Dietary reference intakes: vitamin C, vitamin E, selenium, and carotenoids. Washington, DC: National Academy Press; 2000. https://doi.org/10.17226/9810.
222. Surjushe A, Vasani R, Saple DG. Aloe vera: a short review. Indian J Dermatol. 2008;53(4):163–6. https://doi.org/10.4103/0019-5154.44785.
223. Oyenihi AB, Belay ZA, Mditshwa A, Caleb OJ. "An apple a day keeps the doctor away": the potentials of apple bioactive constituents for chronic disease prevention. J Food Sci. 2022;87(6):2291–309. https://doi.org/10.1111/1750-3841.16155.
224. National Institute of Child Health and Human Development. Drugs and lactation database (LactMed®). Bethesda (MD): National Institute of Child Health and Human Development; 2006; Parsley 2021 Feb 15. https://www.ncbi.nlm.nih.gov/books/NBK501922/.
225. Uribe E, Marín D, Vega-Gálvez A, Quispe-Fuentes I, Rodríguez A. Assessment of vacuum-dried peppermint (Mentha piperita L.) as a source of natural antioxidants. Food Chem. 2016;190:559–65. https://doi.org/10.1016/j.foodchem.2015.05.108.
226. León-Ruiz V, Vera S, González-Porto AV, San Andrés MP. Vitamin C and sugar levels as simple markers for discriminating Spanish honey sources. J Food Sci. 2011;76(3):C356–61. https://doi.org/10.1111/j.1750-3841.2011.02041.x.
227. Abdull Razis AF, Ibrahim MD, Kntayya SB. Health benefits of Moringa oleifera. Asian Pac J Cancer Prev. 2014;15(20):8571–6. https://doi.org/10.7314/apjcp.2014.15.20.8571.
228. Willcox JK, Ash SL, Catignani GL. Antioxidants and prevention of chronic disease. Crit Rev Food Sci Nutr. 2004;44(4):275–95. https://doi.org/10.1080/10408690490468489.
229. https://www.parashospitals.com/blogs/food-items-that-are-good-in-iodine.
230. https://ods.od.nih.gov/factsheets/Iodine-HealthProfessional/.
231. Bouri T, Ganguly S. Documentation of traditional knowledge on edible wild plants of forests in Burdwan district, West Bengal (India). J Med Plants Study. 2016;4(5):35–8; https://www.semanticscholar.org/paper/Documentation-of-traditional-knowledge-on-edible-of-Bouri-Ganguly/0712c4ae55e4cf9bb90bd1dbc10c0dbc11b24c14.

Role of Nutraceuticals in the Management of Lifestyle Diseases

Vimalavathini Ramesh, Kayalvizhie Subbarayan, Sittarthan Viswanathan, and Kavimani Subramanian

Abstract

Lifestyle disorders have become a global health burden in terms of money, morbidity, and mortality. Since time immemorial, nutraceuticals had been a part of human life for its preventive and curative properties. The effectiveness of modern drug therapy has improved with the use of nutraceuticals, thus reducing its dosage and side effects. It includes vitamins, Prebiotics, probiotics, dietary fibers, polyunsaturated fatty acids, minerals, proteins, herbs, amino acids, and phytoconstituents. They help in preventing lifestyle diseases such as hypertension, diabetes mellitus, cancer, Alzheimer's, Parkinson's, obesity, and osteoporosis owing to their antiproliferative, anti-inflammatory, antiapoptotic, antioxidant, antimetastatic, free radical scavenging, and hepatoprotective effects. The following chapter discusses different types of nutraceuticals and its significance in the management of various lifestyle disorders.

Keywords

Lifestyle diseases · Nutraceuticals · Disease prevention · Dietary fibers · Herbs

1 Introduction

World Health Organization (WHO) states that lifestyle disorders are the leading cause of morbidity and mortality and the most significant barrier to global development. Moreover, lifestyle diseases are continuously rising throughout the third-world countries [1]. Human lifestyles have changed dramatically due to industrialization, untimely and hectic work schedules, faster living conditions, and associated psychological stress, resulting in a high prevalence of diabetes mellitus, obesity, hypertension, osteoporosis, cancer, and neurological disorders such as Alzheimer's and Parkinson's. Controlling the associated risk factors is a crucial component in the management of lifestyle diseases. Nutritional therapy is a comprehensive approach for disease prevention and treatment that employs complementary therapeutics or nutraceuticals as dietary therapies. The effectiveness of nutraceuticals as natural nutritional supplements aids in holistic detoxification, prevents mineral and vitamins insufficiency, and reestablishes more biofriendly dietary habits to go a long way to maintain health, improve disease prognosis, and abate adverse effects of modern medicines taken concurrently in these disorders. An added advantage is that they are readily available and cost-effective [2]. Nutraceuticals are essential when it comes to warding off and treating various lifestyle diseases. The following chapter defines different nutraceuticals and their prominence in alleviating various lifestyle diseases.

2 Types of Nutraceuticals

Nutraceuticals classification includes nontraditional, traditional, and commercial nutraceuticals. These include dietary fibers, prebiotics, probiotics, amino acids, minerals, fatty acids, enzymes, vitamins, proteins, herbs, and phytochemicals (Fig. 1).

2.1 Dietary Fibers

Dietary fibers are defined as food material that can be digested by gut microflora but cannot be digested by digestive enzymes in the gastrointestinal tract. Dietary fibers are mainly divided into soluble and insoluble fibers. Soluble dietary fiber includes β-glucans, pectins, gums, mucilage, and hemicelluloses fermented in the colon. Research has revealed that certain soluble fibers enhance immunity in humans. Insoluble dietary fiber includes hemicelluloses, celluloses, and lignans, which are fermented to a limited extent in the colon. For example,

V. Ramesh · K. Subbarayan · S. Viswanathan
K. Subramanian (✉)
Department of Pharmacology, College of Pharmacy, Mother Theresa Post Graduate and Research Institute of Health Sciences, Puducherry, India

A. K. Dhara, S. C. Mandal (eds.), *Role of Herbal Medicines*, https://doi.org/10.1007/978-981-99-7703-1_23

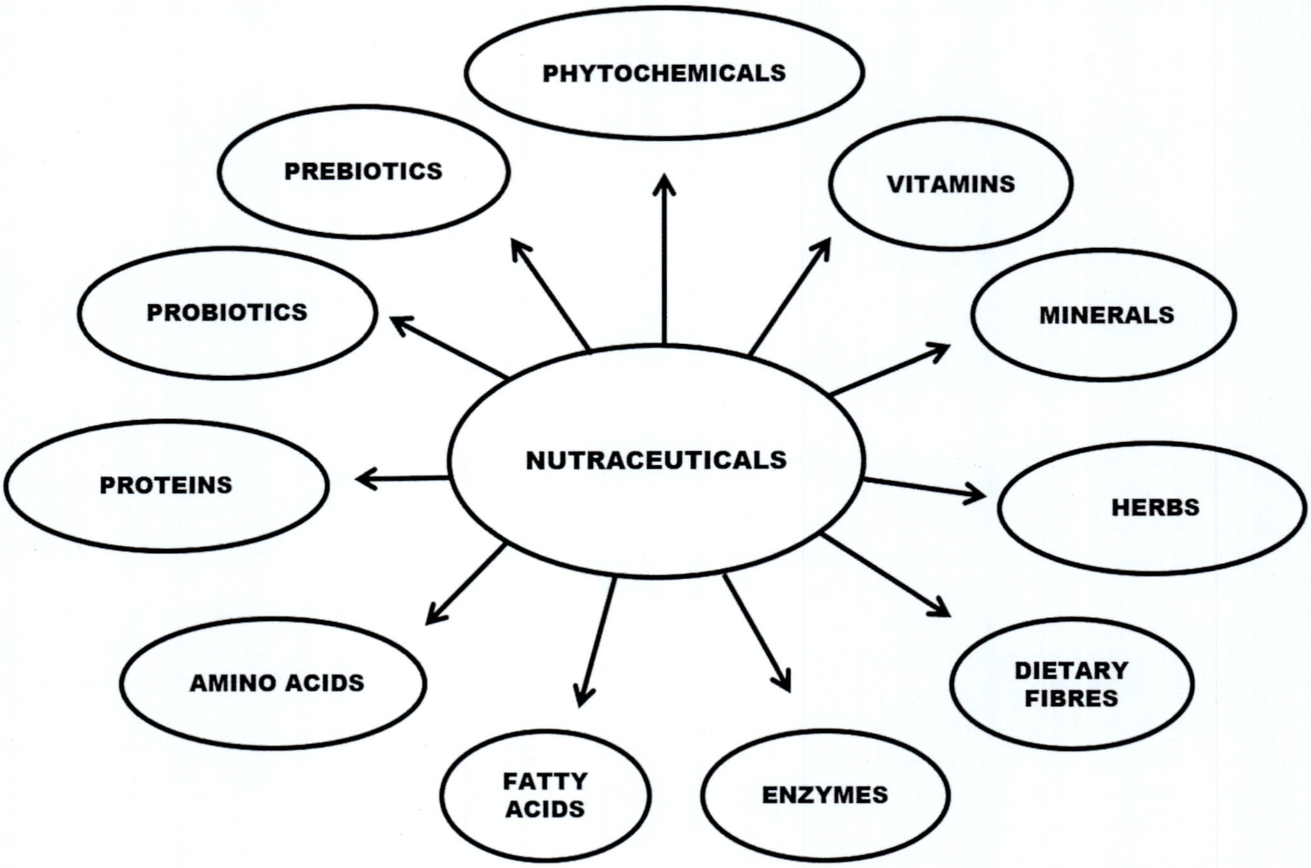

Fig. 1 Types of nutraceuticals

fructooligosaccharide is a dietary fiber obtained from tomatoes, bananas, chicory roots, and alliums that promote the growth of *Bifidobacteria* in the gut and reduce the risk of gastrointestinal disorders, hypertension, stroke, diabetes mellitus, hypercholesterolemia, and obesity [3].

2.2 Prebiotics

Prebiotics are substances that are undigested by humans during consumption and act as nutrient source for probiotic bacteria encouraging its growth which in turn prevents the growth of harmful microorganisms in the gastrointestinal tract. They also offer antitumor properties, neutralization of toxins, enhancement of the intestinal immune system, and lactose tolerance, along with reduction of constipation and blood cholesterol levels. Prebiotic, in addition, promotes the growth of symbiotic bacteria such as *Bifidobacteria* and *Lactobacillus* in the gut, thus enabling biotransformation. For example, inulin is obtained from the roots of plants such as dandelions, Jerusalem artichoke, and chicory [3, 4].

2.3 Probiotics

According to WHO "probiotics are live strains of strictly selected microorganisms that, when administered in adequate quantities, confer health benefit on the host" [5]. Endogenously, probiotics are nonpathogenic, gastric acid-resistant, antimicrobial-producing microbes present in gut epithelium [6]. Studies have shown that the administration of probiotics decrease the risk of systemic conditions such as cancer, asthma, allergy, ear, and urinary tract infections. It forms a symbiotic relationship with the human gastrointestinal tract and has been part of our food culture in the form of curd, gruel, and batter. Probiotics include *Bifidobacteria* such as *Bifidobacterium adolescentis, Bifidobacterium bifidum, Bifidobacterium thermophilum, Bifidobacterium longum,* and *Bifidobacterium infantis;* Gram-positive cocci such as *Streptococcus salivarius subsp. Thermophilus* and *Enterococcus faecium, Lactococcus lactis; Lactobacilli* such as *Lactobacillus cellobiosus, Lactobacillus casei, Lactobacillus brevis, Lactobacillus delbrueckii, and Lactobacillus acidophilus* [3].

2.4 Vitamins

Both fat and water-soluble vitamins are used as dietary supplements owing to their antioxidant properties, which are useful in preventing lifestyle diseases, such as diabetes mellitus, cancer, osteoporosis, cardiovascular diseases, and neurological disorders [5, 6]. According to research studies, high intake of tocopherol prevents PD and dehydroascorbic acid crosses the blood-brain barrier, acts as antioxidants in the central nervous system abetting in the treatment of Alzheimer's disease. Vitamins C and E are useful in reducing lipid peroxidation and atherosclerosis [7]. Seaweed contains fat and water-soluble-like vitamins A, B, C, D, E, and K, which exerts anticancer, anti-obesity, antidiabetic, antihypertensive, and antioxidant properties [8].

2.5 Minerals

Natural products and minerals obtained from plants, animals, and dairy products are effective in managing disorders like osteogenesis imperfecta and anemia, which strengthen bones and muscles, sustain rhythmicity of heart muscles, and aid in neurotransmission [5, 6]. Genetically predisposed individuals with pancreatic cancer have low-serum selenium levels, and can be prevented by selenium supplementation. Zinc is essential for digestion, metabolism, and wound healing [7]. Seaweeds contain minerals, such as fluoride, potassium, iodine, zinc, magnesium, iron, manganese, phosphorus, copper, selenium, and calcium which can be helpful in reducing the incidence of various lifestyle disorders [8].

2.6 Fatty Acids

Monounsaturated fatty acids (MUFA) and polyunsaturated fatty acids (PUFAs) are "essential fatty acids" consumed in diet. PUFAs are two types, omega-3fatty acids and omega-6fatty acids. Main omega-3-fatty acids are eicosapentaenoic acid (EPA), α-linolenic acid, and docosahexaenoic acid (DHA); Omega-6- fatty acids mainly consist of arachidonic acid, linoleic acid, and γ-linolenic acid. Studies have suggested that omega-3 fatty acids have significant effects in the therapy of arrhythmia, thrombosis, and diabetes mellitus [3].

2.7 Proteins and Amino Acids

Legumes and pulses are important ingredients of food and are main source of minerals and proteins (calcium, zinc, iron), with a low amount of lipids. Legumes are also rich in vitamins like folic acid, riboflavin, niacin, biotin, and thiamine with low concentration of saturated fats [9]. Methionine, arginine, histidine, phenylalanine, leucine, isoleucine, threonine, valine, tryptophan, and lysine are the essential amino acids used as nutraceuticals [10]. Seaweed contains essential amino acids like threonine, valine, lysine, and leucine [8]. L-arginine is a semi-essential amino acid and aids in production of nitric oxide which can be helpful in ameliorating neurodegenerative disorders and cardiovascular diseases [7].

2.8 Enzymes

Nutraceutical enzymes derived from plant, animal, and microbial sources include, lactase, papain, xylanase, casein, and coenzyme Q10 (CoQ10). They are indicated in management of medical conditions such as digestive problems, obesity, diabetes mellitus, gastroesophageal reflux disease, constipation, diarrhea, and ulcerative colitis [7].

2.9 Herbs

Herbs provide a complete storehouse of remedies to improve health, to prevent and cure acute and chronic diseases [7]. Examples include willow bark with the active component salicin, which has antiarthritic activity; lavender contains tannin, which helps cure depression, hypertension, and stress, cranberries contain proanthocyanidin which is found to be ulcer and anticancer therapy [6]. Condiments such as garlic, liquorice, ginger, buckwheat proteins, turmeric, thyme, and peppermint are effective in curbing diseases like hypertension, ulcer, hypercholesterolemia, diabetes, obesity, cancer, abdominal distention, and gastrointestinal complaints [4]. Spirulina and chlorella are used as food supplements for their anticholesteremic, anticancer, antihypertensive, and antidiabetic activities [5].

2.10 Phytochemicals

Phytochemicals are plant nutrients which help in the maintaining healthy life, preventing, and treating various harmful diseases in humans. Phenolic compounds possess hepatoprotective, antiallergic, antioxidant, antithrombotic, antiviral, anti-inflammatory, and anticarcinogenic effects [7]. Flavonoids, exert antioxidant, anticancer (breast and prostate cancer), and antidiabetic properties and also help in curing heart diseases and kidney problems. Nonflavonoid polyphenolics exhibit strong anti-inflammatory, antioxidant, blood thinning effect and reduce cholesterol [5]. Carotenoids have anticarcinogenic and hypocholesterolemic properties [4, 6].

3 Role of Nutraceuticals in Lifestyle Diseases

Lifestyle changes have culminated in chronic illness resulting in major financial burden and psychological burden on entire population [1]. The forthcoming paragraphs aim to summarize the effect of selected nutraceuticals against lifestyle disorders such as hypertension, obesity, diabetes mellitus, Parkinson's, Alzheimer's, cancer, and osteoporosis.

3.1 Neurological Disorders

The main neurodegenerative disorders (NDDs), such as Alzheimer's disease (AD), Parkinson's disease (PD), motor neuron disease, Huntington's disease, and prion disease are primarily characterized by decrease in neurotransmitters, which ultimately cause a loss of motor, sensory, and cognitive function leading to decrement in the neuronal numbers or integrity, obstructing neuronal communication. PD and AD can cause dementia and depression, leading to an increase in mortality and morbidity, which is a significant financial burden on society as a lifestyle-related disorder. By the year 2050, it is expected that approximately 47 million people worldwide would have dementia (cognitive decline) and other NDDs [11].

AD is a chronic, irreversible neurodegenerative disorder and amyloid-(β) peptide-containing extracellular amyloid plaques and hyperphosphorylated tau (p-tau) protein-containing intracellular neurofibrillary tangles (NFT) are the two major hallmarks of the disease [12]. Together with AD, PD accounts for major NDDs in the population. PD-affected brain is distinguished by presence of parkin, α-synuclein, and other proteins which are collectively known as Lewy bodies present in the cytoplasm of dead neurons located in various regions of the brain [13].

Nutraceuticals acts at various target to modulate protein synthesis and protect against dysfunctional mitochondria, endoplasmic reticulum stress. They also exert antioxidant, anti-inflammatory, antihyperlipidemic, and neuroprotective effects, resulting in the prevention of neurodegenerative disorders. They decrease β-site amyloid precursor protein-cleaving enzyme 1 (BACE-1), Aβ aggregation, and deposition, enhance neuroplasticity, inhibit acetylcholinesterase (AChE), downregulate amyloid precursor protein (APP) gene expression, and inhibit glycogen synthase kinase (GSK) signaling. Nutraceuticals such as vitamins, dietary fibers, CoQ10, creatine, lipoic acid, herbs, and phytoconstituents are easily available, inexpensive with minimal side effects, and offer an alternative method for prophylactic and therapeutic management of neurological disorders [14].

3.1.1 Role of Nutraceuticals in Parkinson's Disease

Several nutraceuticals such as vitamins (Vitamin A, C, D, and E), fatty acid (MUFA and PUFA), herbs (*Plantago ovata*, *Mucuna pruriens*, *Scutellaria baicalensis)*, phytoconstituents (curcumin, resveratrol, epigallocatechin-3-gallate, genistein, lycopene, and vincamine), and miscellaneous (CoQ10, Mito Q, creatine, lipoic acid, and N-acetyl cysteine) play a significant part in the management of Alzheimer's disease [11, 13–16].

3.1.1.1 Vitamins

Vitamin supplements such as vitamins A, C, D, and E are typical type of nutraceuticals utilized in the management of Parkinson's disease. Vitamins C and E work synergistically to prevent oxidation and restore antioxidative function in PD patients. Levodopa (L-Dopa) is more readily absorbed by older PD patients when taken along with vitamin C. Vitamin D regulates dopamine levels, controls calcium homeostasis, increases glutathione levels, reduces nitric oxide synthase (NOS) levels, and exhibits antiapoptotic activity. β-carotene, a carotenoid found in vitamin A, prevents the buildup of intracellular α-synuclein and also acts as an antioxidant. It decreases C/EBP homologous protein caspase-12 and binding immunoglobulin protein (BIP)/Grp78 expression [14, 15].

3.1.1.2 Fatty Acids

Lack of PUFAand MUFA, such as alpha-linolenic acid, linoleic acid, and docosahexaenoic acid, are strongly linked to impaired brain function. Omega-3 fatty acids from fish reduce reactive oxygen species (ROS) formation and enhance free radical scavenging properties. They also reduce neutrophil and monocyte chemotaxis, minimize the production of proinflammatory cytokines, and exert a neuroprotective effect in dopamine deficiency. MUFA can lower plasma cholesterol and triglycerides [11].

3.1.1.3 Herbs

Plantago ovata is used along with Levodopa/carbidopa to improve L-Dopa profiles and also alleviate constipation in PD patients. The seed powder of *Mucuna pruriens* is a natural source of L-Dopa and contains two neuroprotective agents, namely CoQ 10 and Nicotine Adenine Dinucleotide (NADH). NADH increases dopamine levels by enhancing the expression of tyrosine hydrolase and mitochondrial complex I. *Scutellaria baicalensis* contains baicalein that inhibits apoptosis, activates autophagy, improves antioxidant, restores dopamine levels, and reduces inflammation by suppressing inflammatory cytokines. It ultimately reduces neurotoxicity and fibrillations by the formation of an oligomer of α-synuclein [16].

3.1.1.4 Phytoconstituents

Phytoconstituents such as polyphenols have significant beneficial effects in PD by enhancing motor and gait abnormalities, preserving dopaminergic neurons, and preventive free radicals. Curcumin and curcuminoid, owing to their powerful antioxidant properties, protect neurons from protein oxidation and maintain mitochondrial complex I activity by suppressing α-synuclein aggregation. Resveratrol inhibits nicotinamide adenine dinucleotide phosphate (NADPH) oxidase and suppress the expression of inflammatory mediators like tumor necrosis factor-alpha (TNF-α) and interleukin 1-α, (IL-1). It also regulates the expression of Bcl-2-associated X protein (BAX) and B-cell lymphoma protein 2 (Bcl-2) at both mRNA and protein levels, contributing to its antiapoptotic activity. Oxyresveratrol exhibits potent free radical scavenging activity by suppressing intracellular ROS in neuronal cells [16]. Epigallocatechin-3-gallate reduces neurotoxicity by increasing dopamine levels, improving motor function, regulating protein kinase C activity, and inducing endogenous antioxidant defense systems. They also modulate the ROS-NO pathway to protect the substantia nigra dopaminergic neurons against oxidative stress caused by iron, which can lead to neuronal death [13].

Estrogen has neuroprotective effects in both PD and postmenopausal women. Estrogen-deficient individuals are more prone to get PD, and phytoestrogens supplement such as soy products, nuts, and grains which are structurally similar to estradiol, confer protection against PD. Genistein is a rich phytoestrogen and acts as a transcription activator to upregulate antioxidative and antiapoptotic genes. Ginseng contains ginsenosides Rg1 which exhibits antiapoptotic effect, cytochrome c release inhibition and stimulates PI3k/Akt/mTOR pathway. Its antioxidant activity regulates glutathione levels and oxidative stress-mediated nuclear factor kappa B (NF-κB) pathway, maintains mitochondrial function, and reduces iron levels by regulating the expression of iron transport proteins such as ferroportin 1 and divalent metal transport [16].

Vincamine helps to increase dopamine production, reduces neuronal damage, decreases the production of ROS and iron-chelating molecules, increases vasodilation activity, relaxes the muscles surrounding the capillaries in neurons, and increases the flow of nutrients and glucose to the brain by producing more adenosine triphosphate (ATP) via the Kreb's cycle. Lycopene exhibits neuroprotective action by increasing dopamine levels in the striatum. Its antioxidant effect is accompanied by an increase in the activity of superoxide dismutase (SOD) and NADH dehydrogenase, as well as a reduction in glutathione and malondialdehyde concentrations [14].

3.1.1.5 Miscellaneous

CoQ10 or Ubiquinone exhibits antioxidant and free radical scavenging activity, suppresses lipid peroxidation, and acts as a redox component of the mitochondrial electron transport chain in neuronal cells. Creatine has neuroprotective properties as it counteracts ATP depletion by increasing intracellular phosphocreatine levels and acts as an antioxidant [13, 16]. Mito Q (Triphenylphosphine) has antioxidant properties and maintains respiratory chain function [14]. Lipoic acid is a powerful antioxidant and acts as an anti-inflammatory agent by inhibiting TNF-α and NF-κB. Its neuroprotection is due to increase in intracellular cysteine levels, thereby increasing glutathione levels. N-acetylcysteine acts as a direct glutathione precursor and antioxidant that stimulates mitochondrial complex I and IV activities in neuronal cells [13].

3.1.2 Role of Nutraceuticals in Alzheimer's Disease

Several nutraceuticals such as vitamins (Vitamin B complex, C, D, and E) phytoconstituents (isoflavones, anthocyanidins, and flavones), and miscellaneous (CoQ10, Omega-3 fatty acids, vayacog, souvenaid, cerefolin NAC, and axona) play a significant part in the management of Alzheimer's disease [11, 14–18].

3.1.2.1 Vitamins

The antioxidative properties of vitamins C and E help to protect the aging brain from pathologies associated with AD. The cognitive status of older adults is improved by 25-hydroxy vitamin D3 and B vitamins maintain the integrity of the nervous and hematopoietic systems. Pyridoxin, folate, and cobalamin maintain normal brain function, energy production, and the repair of damaged cells. High levels of homocysteine are neurotoxic and supplementing with B vitamins, including vitamin B6, helps to lower blood homocysteine levels [15].

3.1.2.2 Phytoconstituents

Flavonoids are polyphenolic compounds classified into six subgroups: flavonols, flavanonols, flavones, isoflavones, flavanones, and anthocyanidins. Several neurological processes are influenced by flavonoids and their metabolites. Flavonoids alter cerebral blood flow, upregulate synaptic plasticity-related proteins and antioxidant enzymes, along with suppression of AD neuropathological process. Grape juice is a flavonol which reduces glutamate toxicity, platelet aggregation, lipid peroxidation and improves endothelial function. The intake of these supplements improves cerebral hemodynamic as well as visual and cognitive performance [15].

Oxyresveratrol found in grapes inhibits BACE-1and reduces beta-amyloid production [16]. Garlic (*Allium sativum*) has neuroprotective and antioxidant properties. Curcumin owing to its anti-inflammatory and antioxidant properties, lowers Aβ plaque development by blocking the BACE-1 enzyme [17].

Isoflavones (Genistein, Daidzein, Glycitin) enhance cholinergic function by increasing choline intake and stimulating acetylcholine release by potassium. It activates estrogen receptor beta in the brain to enhance visual memory, spatial memory, and learning ability. Anthocyanidins (Pelargonidin, Cyanidin, Malvidin) are found in blueberries, bilberries, cranberries, elderberries, raspberry seeds, and strawberries exhibit anti-inflammatory, antioxidant, improve brain function, and protect genomic DNA integrity, and increase structural and synaptic plasticity. Flavones (Luteolin, Apigenin) are found in rosemary, celery, and parsley and enhance the GABAergic and glutamatergic transmission, improve neuronal viability, protect neurons from beta-mediated toxicity caused by copper, and also have anti-inflammatory properties. Bioflavonoids, found in capers, apples, onions, and green tea, protect endothelial cells from oxidative stress, exhibit cytoprotective properties, and improve hippocampal synaptic plasticity and memory. Crocin is the main chemical compound identified in saffron, which is involved in improving learning and memory in AD, and has antioxidant activity [14, 15].

3.1.2.3 Miscellaneous

CoQ10 supplements reduce AD-related oxidative stress, inflammation, and AChE activity. Alpha-lipoic acid is a powerful antioxidant that helps glucose metabolism and utilization in the brain of AD patients. Omega-3 fatty acids enhance nootropic activity and stop brain atrophy in AD-related conditions. They also exhibit anti-inflammatory effects by suppressing microglia and astrocytes through the peroxisome proliferator-activated receptor delta (PPAR-δ) and Jun N-terminal kinase (JNK) signaling pathways, inhibiting the β-secretase enzyme, increasing neurotransmitter synthesis, and improves neurogenesis by increasing neurotrophic growth factors. Neurocognitive function in patients with AD is improved by taking neutral amino acid supplements. Memory, brain wave activity, and brain metabolism were all improved by phosphatidyl serine [11, 15].

The following marketed products, such as vayacog, souvenaid, cerefolin NAC, and axona are used in treating AD. Vayacog is combined with omega-3 fatty acids and phosphatidylserine, to supplement with lipids necessary for neuronal cell membrane fluidity and integrity. Omega-3 fatty acids, uridine monophosphate, choline, and B vitamins in souvenaid improve synaptic function. Vitamin B2, B6, B12, and L-methyl folate present in cerefolin NAC are used to the prevent neuroinflammation, oxidative stress, and cognitive impairments causing vitamin B12 and folate deficiencies. Axona, a coconut oil fraction contains caprylic triglyceride, provides energy and nootropic effects in the brain of AD patients [18].

3.2 Role of Nutraceuticals in Hypertension

Cardiovascular disease is the leading cause of mortality worldwide and is predicted to increase rapidly in the next two decades. The most significant pathophysiologic contributor to cardiovascular disease is the development of hypertension, and it is predicted that nearly 90% of people will develop hypertension in their lifetime. Many nutraceuticals such as probiotics, CoQ10, melatonin, vitamins C and D, potassium, cocoa flavonoids, L-arginine, beetroot juice, flaxseed, aged garlic extract, green tea, and resveratrol have proved to be effective in lowering the blood pressure [19–23] (Table 1).

3.2.1 Dietary Fibers

Soluble fiber consumption improves endothelium-mediated vasodilation, viability of coronary artery endothelial cells and reduces intracellular free radical generation. Soluble fibers and gut microbiota activate membrane-type receptor for bile acids (TGR5) and farnesoid X receptor (FXR), enhancing glucose, cholesterol, lipid, and energy metabolism by increasing bile acid excretion. Trimethylamine produced by gut microbiota is converted into toxic trimethylamine-N-oxide (TMAO) by flavin monooxygenase and excretes toxic metabolites through the kidney. Accumulating a high level of TMAO in serum leads to an increase in the risk of developing cardiovascular disease like stroke and myocardial infarction. Therefore, high-dietary fibers are recommended to minimize serum TMAO levels. Fibers like arabinoxylans and pectins increase the synthesis of key antioxidant enzymes by stimulating the transcription factor Nrf2, causing a decrease in lipid peroxidation and free radial generation, inhibiting the glutathione level, enhancing the inducible nitric oxide synthase (iNOS) activity in the cardiac cell leading to chronic blood pressure reduction. Secoisolariciresinol diglucoside found mainly in flaxseed, is an essential lignan, with rich fiber content and acts as an angiotensin-converting enzyme (ACE) inhibitor [20, 22].

3.2.2 Prebiotics and Probiotics

Prebiotics exert antihypertensive activity by enhancing hepatic insulin sensitivity and reducing the production of cholesterol, fatty acid, and triacylglycerol. A decrease in SCFAs lead to hypertension onset which can be rectified by supplementation with dietary fiber producing SCFAs, thus activating the G protein-coupled receptor 41 and olfactory receptor 78 present in blood vessels, through modulation of

Table 1 Role of nutraceuticals for the management of hypertension [19–23]

Sl. no	Nutraceuticals	Mechanism of action
1	Prebiotics	Reduce the synthesis of triacylglycerol, fatty acid, and cholesterol
		SCFAs activate the Olfr78 and Gpr41 receptor present in blood vessels, through modulation of renin release
2	Ascorbic acid	Improve in NO
		Enhancement endothelial function and arterial compliance
3	Vitamin D supplementation	Modulates vascular tone
		Regulates blood pressure
		Anti-inflammatory
4	Magnesium	Increase prostaglandin E production
		Induces vasodilation
5	PUFAs	Suppress thromboxane A2 synthesis
		Exert vasodilatory effect
		PGH2 and TXA2 antagonist
		Release NO
		Antioxidant activity
		Vasorelaxation via activation of K+ channels and inhibition of Ca2+ channels
6	Sour tea (*Hibiscus sabdariffa*)	Antihyperlipidemic activity
		Anti-inflammatory property
		Antioxidant activity
		NOS enzyme stimulation to cause endothelial relaxation
7	Saffron	Inhibit COX 1 and 2, prostaglandin E2 (PGE2), and TNF-α
		Inhibit calcium channel
		Antioxidant
		Downregulate ACE
		Increase eNOS activity and NO serum levels
8	Ginger	Blocks voltage-dependent calcium channels
		Promotes the production of NO
		Inhibit ACE activity
		Reduce ROS formation
		Suppress the inflammatory mediators via downregulation of intracellular adhesion molecule (ICAM-1)
9	Cinnamon, turmeric, paprika, saffron, black and red pepper	Increase the regulation of detoxification enzymes
		Enhancement DNA repair
		Enhance the antioxidant activity
		Reduce inflammation
10	Extra virgin olive oil	Antihypertension action by eNOS
		Decrease endogenously produced asymmetric dimethylarginine by inhibiting NO synthesis
11	Grape seed and red wine extracts	Inhibit platelet aggregation
		Antioxidant
12	Lycopene	Powerful antioxidant and important in preventing cardiovascular disease
		Decrease blood pressure by improving endothelial function
		Decrease lipid peroxidation
		Enhanced SOD activity
		Increase in NO levels
13	Quercetin	Potent antioxidant
		Antiatherosclerotic action
		Vasodilator
		Antiplatelet activity
		Anti-inflammatory
14	Catechin	Decrease blood pressure
		Platelet activation
		Decrease in vascular oxidative stress
		Anti-inflammatory
15	Soy isoflavones	Promotes vasodilation
		Improves endothelial function
16	Cocoa flavonols	Antihypertensive effect via increasing NO synthesis

(continued)

Table 1 (continued)

Sl. no	Nutraceuticals	Mechanism of action
17	Epicatechin and catechin	Suppress glycoprotein IIb/IIIa expression- antiplatelet action
18	Garlic (H_2S, allicin diallydisulfide diallyltrisulfide methylallyltrisulfide)	Cardioprotective signaling molecule
		Protect the cell against apoptosis, oxidative stress and necrosis
		Inhibit platelet aggregation
		Enhances vasodilatation and fibrinolysis
19	CoQ10	Act as mitochondrial electron transporter
		Antioxidant activity
		Decrease proinflammatory cytokines
20	Valyl-prolyl-l-proline and Isoleucyl-prolyl-proline	ACE inhibitor
21	α -linolenic acid	Antihypertensive action by suppressing soluble epoxide hydrolase

renin release. For example, fermentation of insoluble fiber oligosaccharides, is prebiotic that helps in the synthesis of human colonic bacteria and short-chain fatty acids (SCFAs) like butyrate, propionate, and acetate. Probiotics reduce total and low-density lipoprotein cholesterol, insulin resistance, glucose and regulates ACE, thereby reducing blood pressure [20, 22].

3.2.3 Vitamins

Vitamins A, C, D, and E are majorly involved in regulating BP. Vitamin C is a potent antioxidant and displays a synergistic effect with antihypertensive drugs, such as amlodipine. It reduces blood pressure by improving endothelial function which decreases the binding affinity to angiotensin II type 1 receptor and increases nitric oxide (NO) activity. Vitamin E has, anti-inflammatory antioxidant, cardioprotective effect, and antihypertensive activity by increasing endothelial nitric oxide synthase (eNOS), and decreasing NADPH levels. Vitamin D exerts cardiovascular protective effects, by improving vascular tone, modulating blood pressure, increasing high-density lipoprotein (HDL) levels, and exhibiting anti-inflammatory activity [21, 22].

3.2.4 Minerals

The main minerals involved in the regulation of BP are calcium, magnesium, and potassium. Potassium exerts antihypertensive effect by increasing baroreflex sensitivity, natriuresis, modulates sodium-potassium ATPase in the vascular smooth muscle cells, reduces sensitivity to angiotensin II and catecholamines, and decreases NADPH oxidase. Other mechanisms include enhancing antioxidant and anti-inflammatory, suppressing insulin resistance, decreasing asymmetric dimethylarginine, reducing intracellular sodium, and producing TNF-β. High levels of magnesium prevent calcification of atherosclerotic plaques, induce vasodilation, and also act as a natural calcium channel blocker, reducing intracellular sodium and calcium contents thus decreasing blood pressure. Calcium supplementation reduces the frequency occurrence of preeclampsia in pregnant women [21, 22].

3.2.5 Fatty Acids

PUFAs namely as omega-3 fatty acid and omega-6 fatty acid are found in cottonseed, hemp seed, salmon, safflower, soybean, flax oil, mustard oil, corn oil, and evening prime rose oil. Flaxseed is rich in omega-3 fatty acids and acts as cardioprotective factor by blocking prostaglandin H2 receptors (PGH2) and decreasing the synthesis of thromboxane A2 (TXA2). It also exerts antioxidant and free radicals scavenging activity, producing vasodilation by activating potassium channels, and inhibiting the calcium channel and ACE [22].

3.2.6 Herbs

Herbs like saffron, ginger, cinnamon, curcumin, paprika, turmeric, and pepper act by increasing eNOS activity, modulating detoxification enzymes and exert antioxidant, antihypertensive, along with calcium channel-blocking effects. Flavonoids such as resveratrol, epicatechin, and quercetin possess potent antioxidant, anti-inflammatory, antiatherosclerotic, antiproliferative, vasodilator, and antiplatelet aggregation, as well as blood pressure lowering properties [22].

3.2.7 Phytoconstituents

Polyphenols, anthocyanins, alkaloids, pectin, and wax are involved in the management of blood pressure. *Hibiscus sabdariffa* (Polyphenols and anthocyanins) exerts ACE inhibitor and has cardioprotective, vasodilator effect through the endothelium-derived nitric oxide-cyclic guanosine monophosphate (cGMP) relaxant pathway. Cocoa beans (*Theobroma cacao*) commonly used in beverages and chocolate, contain methyl-xanthine like theophylline, theobromine, and caffeine. Flavonoids such as epicatechin, catechin, procyanidins, quercetin, and isoquercitrin promote NO synthesis and encourage flow-mediated dilatation along with antiplatelet activity by downregulating glycoprotein IIb/IIIa expression and normalizing the lipid profile, and glucose level [19].

Garlic (*Allium sativum*) contains hydrogen sulfide, diallyl disulfide, diallyl trisulfide, and methylallyl trisulfide which inhibit platelet aggregation, improve vasodilation and fibrinolysis, and protect the cell from oxidative stress, apoptosis, and necrosis. Aged dry garlic extract contains sulfur compounds, flavonoids, and allicin which exerts calcium channel-blocking and ACE inhibitory activities. They enhance arterial compliance by decreasing sensitivity to catecholamines and increasing the synthesis of NO and bradykinin [19].

Betalain owing to its electron-donating properties have the ability to quench oxidative stress-induced hypertension. Lycopene, a hydrophobic carotenoid constituent of tomato, exerts antihypertensive properties by improving endothelial function, increasing NO levels, and acts as an antioxidant. Olive leaves contain triterpenoids, oleanolic acid, hydroxytyrosol, and consumption of olive oil helps in lowering blood pressure. Polyphenols increase plasma nitrite/nitrate levels and decrease serum asymmetric dimethylarginine levels, thereby increasing NO levels and decreasing blood pressure [19].

3.2.8 Miscellaneous

Nonnutrient supplements, such as CoQ10, pycnogenol, and melatonin, are not necessarily taken with food. CoQ10 has antioxidant properties, mitochondrial electron transporter, and coenzyme for mitochondrial enzymes and its deficiency is implicated in hypertension and cardiovascular disease. It suppresses production and action of proinflammatory cytokines such as C-reactive protein, TNF-α and IL-6. *Pinus pinaster*, marketed as Pycnogenol functions as a natural ACE inhibitor, protects against oxidative stress on cell membranes, boosts NO, enhances endothelial function and blood flow in the renal cortex, and has a positive impact on hypertension. Melatonin, a dark hormone secreted from the pineal gland, appears to ameliorate hypertension through central and peripheral mechanisms by improving endothelial function, NO metabolism, and antioxidant activity [19].

ALA lowers blood pressure by altering circulating oxylipin by inhibiting soluble epoxide hydrolase. L-arginine, a basic amino acid, and substrate for NOS participates in the synthesis of endothelium-derived relaxing factor. Milk-derived tripeptide has beneficial effects on hypertension because of its peptide sequences, high potassium, and calcium content. Valyl-prolyl-proline and isoleucyl-prolyl-proline act as ACE inhibitor and attenuate the development of hypertension.

3.3 Role of Nutraceuticals in Obesity

Obesity is a chronic disease that has a wide range of complications and can affect different physiological functions. Obesity is increasing dramatically in both men and women of all ages. Excessive consumption of foods with high fat and calories, along with sedentary lifestyles, is one of the factors that contribute to obesity. Consequently, a low-calorie diet and appropriate exercise help reduce obesity [24].

Several nutraceuticals such as dietary fibers (glucomannan and psyllium fiber), Prebiotics, probiotics, minerals (calcium), fatty acids (PUFA and DHA), herbs (*Murrayakoenigii, Momordica charantia, Caralluma fimbriata, Paulliniacupana, Panax ginseng, Ginkgo biloba, Hieracium* sp., *Morus nigra, Prunus armeniaca, Rhus verniciflua, Taraxacum officinale, Garcinia mangostana, Cordyceps militaris, Sorghum bicolor*, fenugreek, black gram, chili pepper, ginger, turmeric, bottle gourd, amla, green tea, apple, and annurca flesh) and phytoconstituents (capsaicin, curcumin, polyphenols, gingerol, shogaol, and baicalin) show a significant part in the management of obesity. Nutraceuticals act against obesity by the following mechanisms; increasing the bile acid synthesis and low-density lipoprotein (LDL) clearance, decreasing the absorption of fat, upregulation of LDL receptors, inhibiting insulin resistance and α-amylase, regulating lipase activity, modulating AMP-activated protein kinase (AMPK), peroxisome proliferator-activated receptor gamma (PPAR-γ), and mitogen-activated protein kinases (MAPK) pathways along with anti-inflammatory and antioxidant activity [3, 24, 25] (Fig. 2).

3.3.1 Dietary Fibers

Dietary fibers modify food structure, digestive processes, and appetite cues, improving energy balance and body weight management in humans. Dietary fibers bind with bile acids and catalyze an increase in bile acid genesis, a decrease in hepatic cholesterol, upregulation of the LDL receptor, and a rise in the clearance of LDL. Fiber increases intraluminal viscosity and inhibits the assimilation of macronutrients, which traps cholesterol in the small intestine and makes people feel fuller for longer, resulting in lower calorie consumption [25].

Dietary fiber is easily digested by the local microbiota in the colon, leading to bacterial proliferation and SCFA such as acetate and propionate, in which propionate decreases the hepatic cholesterol synthesis. SCFA may play an important part in energy balance, and insulin resistance through the stimulation of the G-protein-coupled free fatty acid receptor (FFA). In particular, FFA2 and FFA3 stimulation in the colon increases the synthesis of anorectic gut peptides like Peptide YY (PYY) and Glucagon-like peptide-1 (GLP-1) with an appetite suppressant effect by reducing food intake. FFA3 activation in adipocytes also boosts leptin expression.

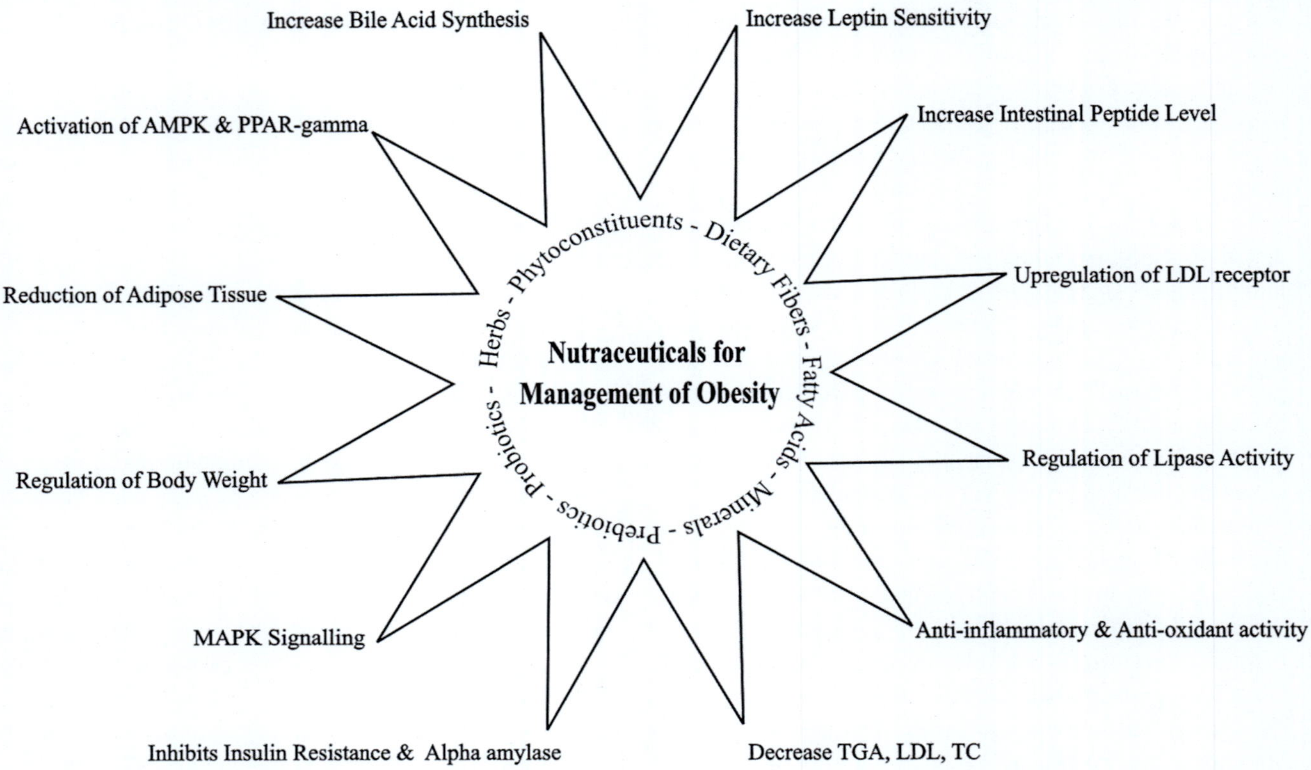

Fig. 2 Role of nutraceuticals for the management of obesity

Prebiotics change the gut microbiota composition and increase endogenous GLP-1, L-cell number, and leptin sensitivity, which manage energy homeostasis. Glucomannan is a hydrosoluble fiber, with a high-density structure that exerts hypocholesterolemic activity. Psyllium fiber attenuates low-density lipoprotein levels. β-glucans increase HDL cholesterol and decrease LDL cholesterol levels [25].

3.3.2 Minerals and Fatty Acids

Calcium-rich foods (milk, yogurt, and cheese) help in the control of body weight and reduction of adipose tissue. Fish contains arginine, which increases the endogenous synthesis of NO exerts a vasodilatory effect, thus causing a reduction in blood pressure. PUFAs reduce hyperglycemia, hyperinsulinemia, and have anti-obese action by downregulating PPAR-α. DHA acts as an anti-inflammatory factor that inhibits obesity. MUFA can lower plasma cholesterol and triglyceride levels. In order to treat fatty liver and a disturbed fat metabolism, choline lipotropic agents are employed [3, 24].

3.3.3 Phytoconstituents

Flavonoids exert cardioprotective action by reducing NO, ROS, platelet aggregation, lipid peroxidation, and proinflammatory cytokines. Fenugreek (steroidal sapogenins, galactomannans, and isoleucine), ginger (6-gingerol, 6-shogaol), bottle gourd, and amla exhibit antihyperglycemic, antihyperlipidemic, immunomodulatory, antioxidant, and anti-inflammatory activity by decreasing the serum triglycerides, total cholesterol, and LDL level, increasing HDL level, and activating AMPK signaling pathway. *Momordica charantia*, *Cordycepsmilitaris* (cordycepin), green tea (epigallocatechin-3-gallate), *Caralluma fimbriata* (pregnane glycosides, flavone glycosides, and flavonoids), and *Sorghum bicolor* decrease the adipose tissue deposition, risk of hypertension, and cardiovascular disorders, increase insulin sensitivity, and modulate the gene expression via MAPK, PI3K-Akt, and insulin signaling pathway. *Paulliniacupana*, ginger, tiliroside, baicalin, and baicalein reduce obesity by inducing expression of uncoupling protein 1 (UCP-1) and adiponectin signaling via the activation of AMPK and PPARγ. *Panax ginseng* (Ginsenoside Rg1) induces UCP-1 expression by modulating the transcription factor PR-domain containing 16 (PRDM16) and peroxisome proliferator-activated receptor gamma coactivator 1-alpha (PGC1α) [24, 26].

Flavonoids such as capsaicin, curcumin, curcuminoid, cranberry, and rose have adipocyte browning property, reduce inflammation, and modulate cytokines like TNF-α, cyclooxygenase-2 (COX-2), and interleukins 1 and 6 (IL-1&6) which constraint the activity of UDP-glucuronosyl transferase and CYP-450. *Murrayakoenigii*, black gram, and chili pepper (capsaicin) exhibit antihypertensive, anti-inflammatory, antioxidant, and antihyperglycemic properties by elevating the hepatic glycogen level, reducing the

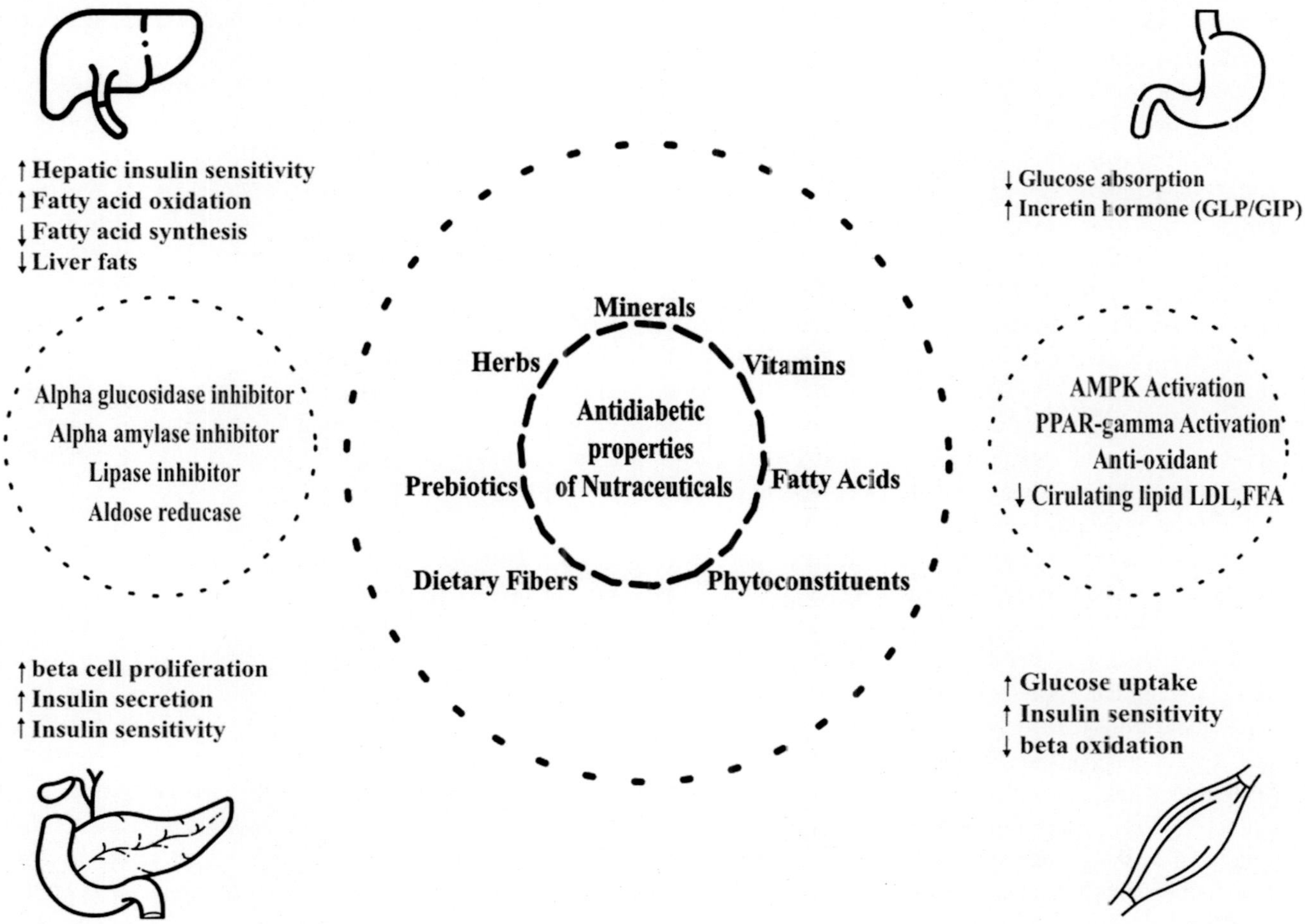

Fig. 3 Role of nutraceuticals in diabetes mellitus with their possible mechanism of action

gluconeogenic enzymatic activity, inhibiting insulin resistance and α-amylase thus counteracting obesity. *Ginkgo biloba*, *Hieracium* sp., *Morus nigra*, *Prunus armeniaca*, *Rhus verniciflua*, *Taraxacum officinale,* and annurca flesh (polyphenols) inhibit lipase activity by reducing the adipocyte volume, fat accumulation, fat mass, and lipid peroxidation thus preventing obesity [3, 21, 26, 27]. Apple (oligomeric procyanidin) acts effectively in the cholesterol uptake process [28].

3.4 Role of Nutraceuticals in Diabetes Mellitus

All over the world, diabetes mellitus is becoming more common. As a result of increasing population, aging, urbanization, and related lifestyle changes, it is predicted that high percentage of the world's adult community is prone to have diabetes mellitus. However, diabetes itself is a multimorbid metabolic disorder that can damage and impair various organs, ensuing in a poor quality of life and a higher death rate which would significantly affect the financial state of global health care [29].

Phytochemicals, dietary supplements, functional foods, bioactive components, lipids, protein, vitamins, and minerals are significant nutraceuticals that have been proved to be effective in management and prevention of diabetes mellitus [27, 30–33]. In diabetes mellitus, nutraceuticals mainly target α-amylase, α-glucosidase, dipeptidyl peptidase-IV (DPP-IV), aldose reductase, PPAR-γ, AMPK, and glucose transporter protein type-4 (GLUT4) to reduce hyperglycemic condition (Fig. 3) [34].

3.4.1 Dietary Fibers and Prebiotics

Insoluble and soluble fibers are two different types of dietary fibers and have unique role in maintaining blood glucose levels. Soluble fibers are β-glucans, inulin, and pectin. β-glucans are abundant in the cell walls of fungi, yeast, oats, and barley which reduce postprandial glycemia and delay glucose absorption and stomach emptying by modulating PI3K/AKT pathway. It also decreases LDL level and increases HDL level, thus curbing the risk of CVD in type 2 diabetes mellitus (T2DM). It also acts as a prebiotic and generates SCFA primarily propionate and acetate. Propionate promotes gluconeogenesis, inhibits hepatic cholesterol synthesis, and is associated with insulin secretion [30].

Inulin and fructooligosaccharides occur in plants and fruits like bananas, chicory root, garlic, wheat, and onions. Inulin consumption improves blood glucose management and has antioxidant properties. A unique colonic fermentation property of high-performance inulin can alter the gut microbiota composition favoring growth of *Bifidobacteria*. After fermentation, it tends to support propionate production, which reduces acetate-to-propionate, resulting in lowering the risk factors related to T2DM by reducing LDL and total serum cholesterol. Dietary fibers (psyllium) exert hypoglycemic effect in diabetic patients and lower lipid levels in hyperlipidemic conditions. Nopal (*Opuntia ficus-indica*) regulates postprandial blood glucose peaks in T2DM. Arabinoxylan, a key component of whole grains exerts diabetic-related anti-inflammatory properties and antidiabetic property by decreasing the absorption of glucose and total cholesterol. Pectins are linear polymer of galacturonic acid that reduces fasting glucose and cholesterol concentrations [30].

Insoluble dietary fibers include lignin, cellulose, xyloglucans, xylans, mannans, and glucomannans and act as physical barriers, thus modulating the motility of digestive products in GIT. Intake of high-insoluble dietary fibers helps in modulating the glycated hemoglobin A1c (HbA1c) levels thereby minimizing the risk of development of T2DM [30].

All prebiotics are fibers and get fermented in colon and stimulate the growth of *Lactobacilli*, *Bifidobacteriaceae,* and *Faecalibacterium* family, which exerts anti-inflammatory property. A high level of *Akkermansia muciniphila* increases number of L-cells synthesis of GLP-1 and GLP-2 which is associated with healthy glucose metabolism and gut barrier function [30].

3.4.2 Vitamins

Vitamin D and calcium supplementation aid in improving insulin sensitivity and preventing diabetes by suppressing parathyroid hormone (PTH) secretion. Vitamin H (chromium picolinate and biotin) may improve blood glucose management by enhancing insulin activity and glucokinase enzyme [31, 32].

3.4.3 Minerals

Minerals such as chromium, selenoprotein, zinc, magnesium, and vanadium, are used to manage diabetes mellitus. Chromium enhances both β-cell and tissue sensitivity by upregulation of insulin receptor and improving insulin binding. Selenoprotein prevents the likelihood of developing metabolic diseases, through its antioxidant and cytoprotective properties. Zinc partially exhibits an antioxidant property and is beneficial for antiaging and antidiabetic activity. Before the discovery of insulin in 1922, vanadium was used to enhance insulin sensitivity to control blood glucose levels. Oral magnesium supplements help to restore magnesium deficiencies and improve oxidative stress, inflammation, and insulin resistance in T2DM [32, 33].

3.4.4 Fatty Acids

Regular consumption of fish provides DHA and EPA fatty acids that regulate satiety and appetite by counteracting insulin resistance and plasma leptin level. Omega-6 fatty acids overcome insulin sensitivity syndrome by modulating genes related to lipid, glucose, and insulin homeostasis through PPARs [33]. Olive oil exerts antidiabetic, anti-inflammatory, antioxidative properties and improves mitochondrial function [35].

3.4.5 Herbs

Botanicals such as *Biophytum sensitivum, Caesalpinia bonducella, Catharanthus roseus, Eugenia jambolana, Citrullus colocynthis, Helicteres isora, Scoparia dulcis, Tinospora cardifolia, Cucurbita maxima,* and *Acacia arabiaca* exhibit antidiabetic activity through stimulation of pancreatic β-cell to release insulin. *Punica granatum and Cassia auriculata* inhibit intestinal α-glucosidase activity leading to antihyperglycemic effects. *Ocimum sanctum* exhibits antidiabetic activity by inhibiting cortisol levels. *Murraya koenigii, Casearia esculenta, Coccinia indica, Cassia auriculata,* and *Brassica juncea* suppress glycogen phosphorylase activity, leading to a suppression of glycogenolysis and neoglycogenesis, establishing hypoglycemic activity. *Ipomoea batatas* reduces insulin resistance, and inhibits maltase. *Cucurbita maxima* exerts hypoglycemic action and regenerates damaged pancreatic cells [36, 37].

3.4.6 Phytoconstituents

Activation of AMPK suppresses the synthesis of both fatty acid and cholesterol in the liver. AMPK activation also reduces lipogenesis and lipolysis in adipocytes. Berberine, epigallocatechin gallate, and caffeine reduce body weight, plasma triglyceride by suppressing lipogenesis-related genes and promoting gene associated with ATP production in muscle tissue and adipocytes [34].

PPAR-γ modulates gene associated with inflammation, lipid metabolism, and glucose homeostasis. Activation of PPAR-γ enhances the expression and mobilization of the glucose transporters GLUT-1 and GLUT-4 to the cell surface, thus increasing glucose uptake into liver and muscle cells. PPAR-γ agonists suppress TNF-α and enhance adiponectin levels thus improving insulin sensitivity. Natural products such as rutin of Noni fruit (*Morindacitrifolia* L.), narcissoside, nicotifloroside, honokiol, and catechin upregulate expression of adiponectin and PPAR-γ. Green tea consumption counteracts T2DM risk by promoting insulin sensitivity and PPAR-γ activation. Sargahydroquinoic acid, sargaquinoic acid, and amorfrutins are novel PPAR-α/γ dual agonists. Quercetin and kaempferol from *Euonymus alatus*, improve insulin-mediated glucose uptake [35].

The DPP-IV enzyme causes degradation and inactivation of many glucose-regulating incretin hormones, including GLP-1 and GIP, enhances insulin production in pancreatic β-cells. The phytoconstituents such as resveratrol, luteolin, apigenin, flavone, myrcenol, linalool, α-elemol, and β-eudesmol have significant interactions with PPAR-γ and DPP-IV targets [34].

Apple extracts contain phlorizin from peels and flesh, which acts as a hypoglycemic factor [28]. The cardioprotective and antidiabetic effects of soy proteins such as solcon S and solgen 40 reduce insulin resistance, and dyslipidemia in diabetic postmenopausal women. The main polyphenolic compound in coffee and chlorogenic acid is similar to acarbose and acts by inhibiting intestinal glucose transport and glucose-6-phosphate translocase, an enzyme involved in gluconeogenesis and glycogenolysis, by reducing the absorption of carbohydrates. Robinetinidol and fisetinidol present in *Acacia mearnsii*, inhibit lipase and exert antidiabetic activity. The hydroxy chalcone compound in cinnamon inhibits an enzyme called tyrosine phosphatase that targets the insulin receptor, and has the potential to enhance insulin signaling [36, 38].

3.4.7 Miscellaneous

ALA is a PPAR-γ agonist, antioxidant, and improves insulin sensitivity by tissue uptake of glucose via GLUT-4. It also reduces microvascular diabetic complications such as cataract formation, vascular damage, and polyneuropathy. Carnitine (L-Carnitine, Acetyl L-Carnitine) aids in the breakdown of fatty acids and the binding of acyl residues in the body and is useful in preventing diabetic ketoacidosis. N-Acetyl Cysteine enhances glutathione synthesis, decreases the progression of vascular damage, maintains RBC intact form, diminishes structural changes in T2DM, and exhibits antiglycation activity. Consumption of cow milk and donkey milk helps to maintain normoglycemia. Trypsin, pepsin, and pancreatin present in the milk improves damaged β-cell viability. Alcalase and pepsin found in eggs inhibit α-glucosidase and DPP-IV [36, 39].

Watermelon, cucumbers, pumpkins, and muskmelons contain L-citrulline which is crucial for urea cycle by recoupling uncoupled e-NOS. Nutraceuticals such as zinc, glycine, lipoic acid, broccoli sprouts, taurine, and N-acetylcysteine reduce dicarbonyl stress by promoting Nrf2 activation. Inositol has a role in reverting of diabetes-induced neuropathy and maintaining the normal blood cholesterol level and integrity of cell membrane. Spirulina (phycocyanin) has numerous health advantages especially antioxidant and antidiabetic activity by an increase in insulin production and reduction in the glucose absorption and HbA1c levels [39, 40].

3.5 Role of Nutraceuticals in Cancer

Nutraceuticals are efficient in the prevention of cancer rather than treatment and may be used as an adjuvant in cancer treatment by preventing metastasis, invasion, and drug resistance. They are also known to cause proteasome and epigenetic alterations that prevent and treat cancer. Common cancer molecular targets that nutraceuticals can affect are the growth factor receptors (Fig. 4), mainly protein kinases such as phosphoinositide 3-kinase (PI3K), mammalian target of rapamycin (mTOR), AMPK, breakpoint cluster region-Abelson murine leukemia viral oncogene (BCR-ABL); pro-inflammatory mediators namely interleukins, TNF-α, COX-2, and 5-lipoxygenase (5-LOX); epidermal growth factor receptor (EGFR) like vascular endothelial growth factor receptor (VEGFR) and insulin-like growth factor 1 receptor (IGF1-R). It also includes transcription factors like signal transducer and activator of transcription 3 (STAT3), nuclear factor erythroid 2-related factor 2 (Nrf2), activator protein 1 (AP-1), PPAR-γ, β-catenin, NF-κB, sonic hedgehog along with dietary supplements (vitamin E, L-glutamine, and omega-3 fatty acid), fibers, and phytoconstituents, (turmeric, saffron, black pepper) [41].

3.5.1 Dietary Fibers

Dietary fibers, namely vegetables, fruits, whole grains, legumes, nuts, seeds, wholemeal, brown rice, high-fiber cereals, muesli, and oats exhibit anticancer activity. Dietary fiber has an essential role in antineoplastic functions due to physical, prebiotic, and metabolic effects. It reduces the intestinal transit time by increasing the stool bulk and binds the carcinogens to the stool, which favors the elimination of fecal carcinogens. Colonic bacteria convert fiber into SCFA, which has anti-inflammatory, antioxidant, and anticancer properties. SCFAs with cholestyramine inhibit bile acid reabsorption and prevent liver cancer. Dietary fibers alleviate breast cancer by preventing intestinal reabsorption of estrogens, inhibiting β-d-glucuronidase activity by hydrolysis of conjugated estrogen, increasing fecal excretion, elevating the sex hormone-binding globulin levels, modulating the insulin production and glucagon-like peptide, then lowering the level of leptin. Phytoestrogens like genistein, resveratrol, enterolactone, and enterodiol, act as a competitive inhibitor of estrogen receptor, induce cancer cells differentiation, and reduce angiogenesis. Probiotics prevent cancer cell proliferation by improving host immunity and modulating apoptosis and cell differentiation [42].

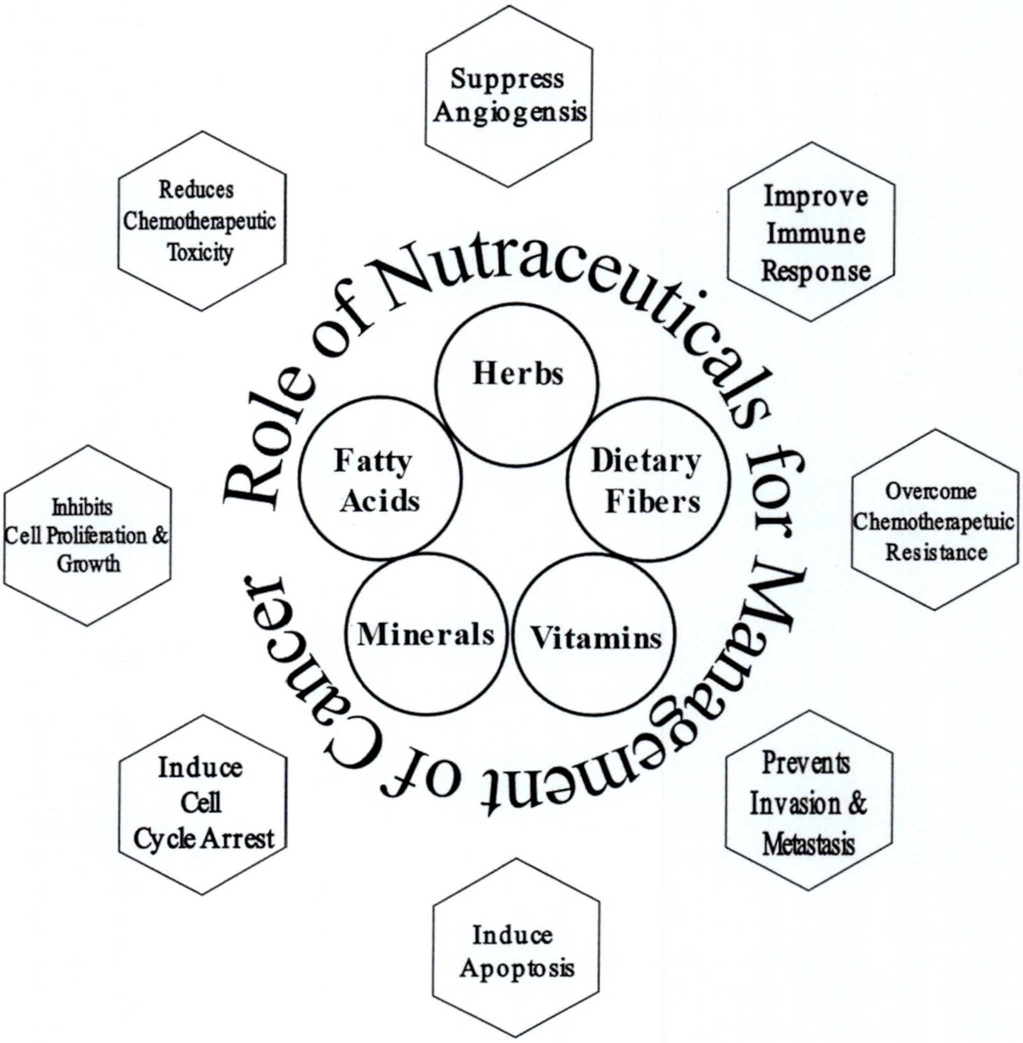

Fig. 4 Role of nutraceuticals for the management of cancer

3.5.2 Minerals and Vitamins

Minerals such as selenoprotein, zinc, germanium, magnesium, and vanadium potentially act against cancer. Selenoprotein, represses p53, acts as thioredoxin reductase by reducing NADPH of thioredoxin, decreasing the nucleotides in DNA synthesis, regenerating the antioxidant system, and regulating gene expression. Zinc exerts prophylactic anticancer activity by lower HDL and inhibiting free radical-induced DNA damage in cancer cells [43].

Organogermanium compounds like germanium sesquioxide and spirogermanium are effective in treating cancer. Germanium acts as a powerful antioxidant and anticancer agent by inhibiting the tumors' growth, metastasis, and immunomodulating interferon-α and natural killer cell. Magnesium decreases the risk of colorectal cancer. Vanadium shows antitumor effect by stimulating tyrosine phosphorylases and/or blocking the cellular tyrosine phosphatases which trigger the signal transduction pathways leading to activation of tumor suppressor gene and apoptosis. Lipoperoxidation, DNA fragmentation, and cleavage lead to cell arrest and also regulate cellular adhesive molecules by exhibiting antimetastatic action and revert chemotherapeutic drug resistance. Vitamin E, L-glutamine, and omega-3 may help to prevent chemotherapy-induced peripheral neuropathy [44].

3.5.3 Herbs and Phytoconstituents

Herbs such as turmeric, black cumin, ginger, garlic, saffron, red chili pepper, rosemary, clove, galangal, black pepper, cinnamon, and oregano exhibit anticancer activity. Turmeric (curcumin) targets the androgen receptor with minimizes cytotoxicity on normal cell and induces apoptosis, inhibits growth and proliferation, cell invasion, and angiogenesis in breast and liver cancer cells. It also inhibits G1/S phase transition, by suppressing the action of cyclin-D1 and p21-activated kinase. Black cumin (thymoquinone) induces apoptosis, suppresses growth and proliferation, migration, and cancer cell invasion. Ginger (6-shogaol) decreases tumorigenesis, proliferation, metastasis, and cell viability by stimulating apoptosis. Garlic (S-allyl mercaptocysteine, diallyl disulfide, triacremonone, diallyl sulfide, diallyl trisulfide) inhibits tumor growth, metastasis, proliferation, DNA damage, and estrogen receptor-α activity by causing G2/M arrest, induces immunomodulation, promotes apoptosis, and suppresses xenograft tumors [45].

Saffron (Crocetin and crocin) induces cell death and apoptosis, inhibits cell invasion and migration, and exhibits antioxidant and antiproliferative activity. Red chili pepper (Capsaicin), a radio-sensitizing agent, induces apoptosis, oxidative DNA damage, and cell death, inhibits proliferation, migration, invasion, angiogenesis, and metastasis. Rosemary (carnosic acid) modulates apoptosis and is autophagic by degradation of human epidermal growth factor receptor 2 and estrogen receptor-α, and also downregulates androgen receptor, suppresses cell migration and tumor growth size. Eugenol in clove suppresses cell growth and proliferation, enhances cell cycle arrest, anti-inflammatory, apoptosis, cytochrome P450 drug metabolism enzymes activity, and improves synergistic activity with gemcitabine. Galangal (Galangin) blocks the cancer cells proliferation and stimulates cell death. Black pepper (Piperine) suppresses cell growth, proliferation, and motility, produces apoptosis, enhances the efficacy of tumor necrosis factor-related apoptosis-inducing ligand TRAIL, and reduces androgen-dependent and independent tumor growth. Cinnamon (Cinnamaldehyde) regulates drug-metabolizing genes, thus decreasing chemotherapeutic toxicity. Carvacrol in oregano induces apoptosis and prevents proliferation in cancer cell. *Boswellia serrata* (boswellic acid) exhibits anticancer by enhancing G1 phase cell cycle arrest in colon cancer cells [45].

Phytochemicals like alkaloids, polyphenols (flavonols, flavones, isoflavones, stilbenes, ellagic acid, and anthocyanins), and terpenoids play an essential role as nutraceuticals in the alleviation of cancer. Resveratrol, phytoalexin, and polyphenol stimulate proapoptotic effect and reduce NF-κB activation in breast and pancreatic cancer cells. Caffeic acid phenethyl ester acts as chemotherapeutic adjuvant and upregulates sodium channels, inhibits cell proliferation, invasiveness, viability, and metastasis in colorectal cancer. Moreover, it blocks binding between DNA and p65–p50 complex, which suppresses the transcription of target genes that are regulated by NF-B. Limonene exhibits synergistic anticancer activity with docetaxel in prostate cancer. Carotenoids have oxygen quenching, antioxidant, and immunomodulatory activity and prevent like the onset of several types of cancer [46].

3.6 Role of Nutraceuticals in Osteoporosis

Osteoporosis is a bone disorder, characterized by increase in bone loss causing bone fragility and increase in fractures risk. The etiological factors include age, race, height, gender, genetics, hormonal status, alcohol consumption, diet, smoking, and insufficient physical activity. Nutraceuticals such as prebiotics, probiotics, vitamins, minerals, dairy products, herbs, and phytoconstituents are beneficial for general bone health and the treatment of osteoporosis [47] (Table 2).

Table 2 Role of nutraceuticals for the management of osteoporosis [47–49]

Sl. no	Nutraceuticals	Mechanism of action
1	Prebiotics (nondigestible oligosaccharides, galacto-oligosaccharides)	Improves mineral absorption to promote bone mineralization
		Enhances magnesium and calcium assimilation in long bone
2	Probiotics (*Lactobacillus, Bifidobacteria*)	Formation of a vitamins C, D, K, and folate, which become crucial for bone development and related to calcium metabolism
		SCFAs reduce PTH which increases mineral absorption
		Antioxidant activity thus decreasing osteoclast differentiation
		Increase calcium absorption
3	Calcium	Reduce bone loss
		Modulate turnover rate in bones
		Mobilized by PTH-mediated bone breakdown if loss of calcium from the extracellular fluid exceeds its intake via the gut
4	Magnesium	Improves ATP production and acts as enzyme cofactor
5	Boron	Enhance half-life of estrogen and vitamin D
6	Copper	Increase cross-linking of lysine in elastin and collagen by activating lysyl oxidase
7	PUFA	Increase calcium absorption
		Decrease inflammatory cytokines
		Activates PPAR-δ
8	*Herbaepimedii* (icariin)	Blocks COX-2, IL-6, TNF-α, and PGE2
		Suppress inflammatory bone loss

(continued)

Table 2 (continued)

Sl. no	Nutraceuticals	Mechanism of action
9	Horse tail *Equisetum arvense* (silica)	Collagen formation promotes bone mineralization
10	Quercetin, rutin, organosulfur flavonoids, and limonoids	Antioxidant activity
		Decreases osteoclast activity
11	Alfalfa (Ipriflavone)	Enhancement of type 1 collagen and alkaline phosphatase by enhancing osteoblastogenesis
		Modulates secretion of β-glucuronidase and cathepsin K and L
		Inhibits osteoclastogenesis
12	Red clover (Genistein, daidzein)	Inhibits topoisomerase II and tyrosine kinase
		Suppress osteoclast activity
13	Green tea (epigallocatechin-3-gallate)	Enhances osteoblast synthesis and alkaline phosphatase
		Inhibit TNF-α
		Activates caspase-3
		Suppress matrix metalloproteinase-9
		Suppress IL-6 activity and RANKL expression
		Block osteoclast synthesis and differentiation
14	Resveratrol	Reduce oxidative stress
		Reduce inflammatory cytokines
		Modulates osteoblast and enhances osteoblastogenesis
		Change the expression of osteoclast genes to inhibit osteoclastogenesis
15	Berries, plums, black cohosh (polyphenols, 25-acetylcimigenol xylopyranoside)	Reduce osteoclastogenesis by inhibiting RANKL and TNF-α
16	Kaempferol and quercetin	Suppress inflammatory pathway
		Osteoclast-mediated bone loss
17	Basic milk protein	Increases osteoblast differentiation and proliferation
		Suppress osteoclast genesis leading to bone loss
18	Cystatin C	Suppress cysteine protease
		Inhibits prostaglandin E2, interleukin-1, and thrombin-dependent release of calcium from bones
19	Lactoferrin (glycoprotein)	Prevent osteoblast apoptosis
		Decreases osteoclast genesis
		Enhance bone formation
		Stimulates MAPK to confer rapid cell mitogenesis
20	Melatonin	Antioxidant activity
		Increase osteoclast metabolism
		Increase serum corticosteroid levels
		Suppress osteoclast differentiation by enhancing osteoprotegerin

Probiotics such as *Lactobacillus* and *Bifidobacteria* promote the production of vitamins C, D, K, and folic acid, which boost calcium metabolism, lower PTH by increasing mineral and calcium absorption in the bone, and decrease in osteoclast differentiation. Prebiotics such as indigestible oligosaccharides and galactooligosaccharides improve magnesium absorption and promote bone mineralization. Omega-3 fatty acids affect bone remodeling via calcium uptake, inflammatory cytokines, and PPAR-γ. Melatonin increases serum corticosteroid levels and procollagen type I c-peptide. Minerals such as calcium, magnesium, boron, and copper protect the neuromuscular system, stabilize the half-life of estrogen and vitamin D by triggering lysyl oxidase, which causes the cross-linking of lysine into elastin. Dairy products such as milk protein, cystatin C, and lactoferrin promote osteoblast proliferation and differentiation, prevent osteoclast formation and bone resorption, and stimulate MAPK to cause rapid cell mitogenesis. Flavonoids such as ipriflavone, genistein, daidzein, quercetin, kaempferol, luteolin, rutin, epigallocatechin-3-gallate, icariin, anthocyanidin, polyphenols, 25-acetylcimigenol, xylopyranoside, resveratrol, and limonoids exhibit antiosteoporotic activity. They increase alkaline phosphatase, inhibit tyrosine kinase, topoisomerase II, β-glucuronidase, and cathepsin (K and L), suppress oxidative stress, inflammation, caspase-3, bone marrow mononuclear cell differentiation, and downregulate receptor activator of nuclear factor kappa beta ligand (RANKL) and TNF-α [48, 49].

4 Conclusion

In conclusion, nutraceutical is an effective way to prevent lifestyle diseases and improve overall health profile of an individual. Nutraceuticals are the fastest growing industry which have a bright future as therapeutic agents. Many of the nutraceuticals discussed in this chapter have been shown to be not only preventive but also therapeutic for numerous lifestyle diseases. Nutraceuticals frequently have an advantage over synthetic medications in that they have the potential to alleviate and prevent lifestyle disorders. Phytochemicals like flavonoids, alkaloids, glycosides, saponins and tannins exhibit plethora of pharmacological activities such as hepatoprotective, neuroprotective, hypotensive, antidiabetic, anti-inflammatory, antioxidant, and antiosteoporosis. They exhibit synergistic action with clinical drug therapy and act at multiple targets and bring about a holistic approach rather than targeted treatment thus minimizing dosage and adverse effects contemporary drug. However, in current scenario the role of nutraceuticals is undermined due to insufficient data on various pharmaceutical aspects. There is a need to establish a nutraceutical research community to transform natural nutraceuticals into authenticated ones by addressing their enormous benefits to humans as their prominence in the nutraceutical industry grows. Extensive and elaborate research on most of the nutraceuticals has mandatory to establish its effectiveness as novel treatment and prophylactic strategies to alleviate various lifestyle disorders.

References

1. Balwan WK, Kour S. Lifestyle diseases: the link between modern lifestyle and threat to public health. Saudi J Med Pharm Sci. 2021;7(4):179–84.
2. Chauhan B, Kumar G, Kalam N, Ansari SH. Current concepts and prospects of herbal nutraceutical: a review. J Adv Pharm Technol Res. 2013;4(1):4.
3. Das L, Bhaumik E, Raychaudhuri U, Chakraborty R. Role of nutraceuticals in human health. J Food Sci Technol. 2012;49: 173–83.
4. Ruby S, Prakash S, Pradeep, et al. A comprehensive review on nutraceuticals. Int J Pharm Sci Rev Res. 2021;68(2):136–48.
5. Borkar N, Saurabh SS, Rathore KS, Pandit A, Khandelwal KR. An insight on nutraceuticals. Pharma Tutor. 2015;3(8):13–23.
6. Chanda S, Tiwari RK, Kumar A, Singh K. Nutraceuticals inspiring the current therapy for lifestyle diseases. Adv Pharmacol Sci. 2019;2019:6908716.
7. Srividya AR, Venkatesh N, Vishnuvarthan VJ. Nutraceutical as medicine. Int J Adv Pharm Sci. 2010;1(2):132–3.
8. Shannon E, Abu-Ghannam N. Seaweeds as nutraceuticals for health and nutrition. Phycologia. 2019;58(5):563–77.
9. Carbonaro M, Nucara A. Legume proteins and peptides as compounds in nutraceuticals: a structural basis for dietary health effects. Nutrients. 2022;14(6):1188.
10. Alagawany M, Elnesr SS, Farag MR, Tiwari R, Yatoo MI, Karthik K, Michalak I, Dhama K. Nutritional significance of amino acids, vitamins, and minerals as nutraceuticals in poultry production and health–a comprehensive review. Vet Q. 2021;41(1):1–29.
11. Chiu HF, Venkatakrishnan K, Wang CK. The role of nutraceuticals as a complementary therapy against various neurodegenerative diseases: a mini-review. J Tradit Complement Med. 2020;10(5):434–9.
12. Sadhukhan P, Saha S, Dutta S, Mahalanobish S, Sil PC. Nutraceuticals: an emerging therapeutic approach against the pathogenesis of Alzheimer's disease. Pharmacol Res. 2018;129:100–14.
13. Ciulla M, Marinelli L, Cacciatore I, Stefano AD. Role of dietary supplements in the management of Parkinson's disease. Biomol Ther. 2019;9(7):271.
14. Makkar R, Behl T, Bungau S, Zengin G, Mehta V, Kumar A, Uddin MS, Ashraf GM, Abdel-Daim MM, Arora S, Oancea R. Nutraceuticals in neurological disorders. Int J Mol Sci. 2020;21(12):4424
15. Josmi PJ, Divya PD, Rosemol JM. Role of nutraceuticals in Alzheimer's disease. Pharm Innov J. 2019;8(4):1129–32.
16. Chao J, Leung Y, Wang M, Chang RC. Nutraceuticals and their preventive or potential therapeutic value in Parkinson's disease. Nutr Rev. 2012;70(7):373–86.
17. Kumar A, Ali R, Kumar R, Chawla PA. Exploring the potential of neuronutraceuticals as alternatives in the prevention of Alzheimer's disease. Pharmaspire. 2022;14:47.
18. Masters CL, Bateman R, Blennow K, Rowe CC, Sperling RA, Cummings JL. Alzheimer's disease. Nat Rev Dis Primers. 2015;1(1):1–8.
19. Sirtori CR, Arnoldi A, Cicero AF. Nutraceuticals for blood pressure control. Ann Med. 2015;47(6):447–56.
20. Tovar AR, Guevara-Cruz M, SerraldeZúñiga AE, Torres N. Dietary fiber and hyperlipidemia and cardiovascular disease. In: Science and technology of fibers in food systems; 2020. p. 219–39.
21. Chiu HF, Venkatakrishnan K, Golovinskaia O, Wang CK. Impact of micronutrients on hypertension: evidence from clinical trials with a special focus on meta-analysis. Nutrients. 2021;13(2):588.
22. Ghaffari S, Roshanravan N. The role of nutraceuticals in prevention and treatment of hypertension: an updated review of the literature. Food Res Int. 2020;128:108749.
23. Alves QL, Camargo SB, Silva DF. Role of nutraceuticals in the prevention and treatment of hypertension and cardiovascular diseases. J Hypertens Manag. 2019;5(1):1G10.
24. Nijhawan P, Behl T. Nutraceuticals in the management of obesity. Obes Med. 2020;17:100168.
25. Vitaglione P, Mennella I. Dietary fiber and obesity. In: Science and technology of fibers in food systems; 2020. p. 187–99.
26. Ammendola S, Scotto d'A A. Nutraceuticals and the network of obesity modulators. Nutrients. 2022;14(23):5099.
27. Gautam SP. Nutraceuticals: a boom to medical industry. Res Gate Pharm Sci. 2013;2:1–8.
28. Santini A, Novellino E. Nutraceuticals: beyond the diet before the drugs. Curr Bioact Compd. 2014;10(1):1–2.
29. Shi J, Hu H, Harnett J, Zheng X, Liang Z, Wang YT, Ung CO. An evaluation of randomized controlled trials on nutraceuticals containing traditional Chinese medicines for diabetes management: a systematic review. Chin Med. 2019;14(1):1–20.
30. Torres N, Avila-Nava A, Medina-Vera I, Tovar AR. Dietary fiber and diabetes. In: Science and technology of fibers in food systems; 2020. p. 201–18.

31. Valdés-Ramos R, Ana Laura GL, Beatriz Elina MC, Alejandra Donaji BA. Vitamins and type 2 diabetes mellitus. Endocr Metab Immune Disord Drug Targets. 2015;15(1):54–63.
32. Nimesh S, Ashwlayan VD. Nutraceuticals in the management of diabetes mellitus. Pharm Pharmacol Int J. 2018;6(2):114–20.
33. Dubey P, Thakur V, Chattopadhyay M. Role of minerals and trace elements in diabetes and insulin resistance. Nutrients. 2020;12(6):1864.
34. Dehdashtian E, Pourhanifeh MH, Hemati K, Mehrzadi S, Hosseinzadeh A. Therapeutic application of nutraceuticals in diabetic nephropathy: current evidence and future implications. Diabetes Metab Res Rev. 2020;36(8):e3336.
35. Alkhatib A, Tsang C, Tuomilehto J. Olive oil nutraceuticals in the prevention and management of diabetes: from molecules to lifestyle. Int J Mol Sci. 2018;19(7):2024.
36. Baldi A, Choudhary N, Kumar S. Nutraceuticals as therapeutic agents for holistic treatment of diabetes. Int J Green Pharm. 2013;7(4):278.
37. Venkatakrishnan K, Chiu HF, Wang CK. Popular functional foods and herbs for the management of type-2-diabetes mellitus: a comprehensive review with special reference to clinical trials and its proposed mechanism. J Funct Foods. 2019;57: 425–38.
38. Cicero FA, Tartagni E, Ertek S. Nutraceuticals for metabolic syndrome management: from laboratory to benchside. Curr Vasc Pharmacol. 2014;12(4):565–71.
39. Antony P, Vijayan R. Bioactive peptides as potential nutraceuticals for diabetes therapy: a comprehensive review. Int J Mol Sci. 2021;22(16):9059.
40. Lympaki F, Giannoglou M, Magriplis E, Bothou DL, Andreou V, Dimitriadis GD, Markou G, Zampelas A, Theodorou G, Katsaros G, Papakonstantinou E. Short-term effects of spirulina consumption on glycemic responses and blood pressure in healthy young adults: results from two randomized clinical trials. Meta. 2022;12(12):1180.
41. Gupta SC, Sharma A, Mishra S, Awasthee N. Nutraceuticals for the prevention and cure of cancer. In: Nutraceuticals in veterinary medicine; 2019. p. 603–10.
42. Senés-Guerrero C, Gradilla-Hernández MS, García-Gamboa R, García-Cayuela T. Dietary fiber and gut microbiota. In: Science and technology of fibers in food systems; 2020. p. 277–98.
43. Venturelli S, Leischner C, Helling T, Renner O, Burkard M, Marongiu L. Minerals and cancer: overview of the possible diagnostic value. Cancers. 2022;14(5):1256.
44. Mankaran S, Dinesh K, Gurmeet S, et al. Natural minerals and cancer. J Appl Pharm Sci. 2012;2(04):158–65.
45. Zheng J, Zhou Y, Li Y, Xu DP, Li S, Li HB. Spices for prevention and treatment of cancers. Nutrients. 2016;8(8):495.
46. Ganapathy M, Bhunia S. Nutraceuticals: the new generation therapeutics. Adv Tech Biol Med. 2016;4:1–5.
47. Rajput R, Wairkar S, Gaud R. Nutraceuticals for better management of osteoporosis: an overview. J Funct Foods. 2018;47:480–90.
48. Pandey MK, Gupta SC, Karelia D, Gilhooley PJ, Shakibaei M, Aggarwal BB. Dietary nutraceuticals as backbone for bone health. Biotechnol Adv. 2018;36(6):1633–48.
49. Jalgaonkarn K, Kumar M, Bibwe B, et al. Nutraceuticals and functional foods. In: Trends & prospects in processing of horticultural crops; 2019. p. 231–50.

Eating and Healing: Traditional Food as Medicine

Sparsha Suresh Bandekar and Ajay Gajanan Namdeo

Abstract

Food being the basic building block of biology, consists of vital micronutrients that furnish energy for different life processes. In addition, foods also contain phytochemicals responsible for their medicinal use. The integrative nature of foods as medicine has been explored and utilized in their indigenous habitats. Food when consumed adequately and essentially can serve as powerful medicine promoting health and preventing diseases through good nutrition. But the consumption of medicinal plants or traditional foods cannot be generalized because a plant showing beneficial effects can also contain toxic phytochemicals and the amount of food should be considered for optimal effects. Also, this should not affect the sustainability of wild plants by overextraction. Contributions of Ayurveda and traditional systems of medicine to Food & Nutrition can help in exploring traditional foods with medicinal benefits. The food resources providing nutrients as well as phytochemicals namely Spices, Condiments, Grains, Pulses, Vegetables, Herbs, Fruits, Berries, Teas, and Extracts can be explored with a literature-based approach to bring out the facts about the traditional foods with medicinal benefits. This chapter presents a narrative about such facts supporting the theory of eating and healing with traditional foods.

Keywords

Traditional food · Phytochemicals · Health benefits

S. S. Bandekar
Satara College of Pharmacy, Satara, Maharashtra, India

A. G. Namdeo (✉)
Department of Pharmaceutical Sciences, Hemvati Nandan Bahuguna Garhwal University (A Central University), Srinagar, Uttarakhand, India

1 Introduction

Eating could be considered one of the most significant actions of one's everyday routine. Apart from the two other basic needs of life, food can be unarguably regarded as the top need for survival. As it is evident that deprival of food or starvation has led to health conditions, food is the first contributor to health. With the advent of science, many medicines have been discovered that aided in health and cure for sickness. Over the past years of research in the field of science and medicine, we have a pool of chemicals used in the mitigation or cure of various diseases and ailments.

As said earlier that food is the first contributor to health, and ensuring the consumption of food that provides sufficient nutritional value has proved to prevent illness in the first place. Healthy eating is an alternative, but there are foods that provide medicinal value and have been consumed during health conditions to get rid of the disease. Such traditional foods are a treasure for the health care sector as it leads to healing from eating.

What is food? Food can be considered the basic building block of biology consisting of vital micronutrients in the form of carbohydrates, proteins, fats, minerals, and vitamins that furnish energy for different life processes. It can be added now that food also consists of phytochemicals (Phyto—meaning from plant origin, chemical—here means medicinal component from the plant) that contribute to mechanisms that are responsible for the cure of diseases [1, 2].

Integrative tools based on scientific evidence such as food pyramids have been suggestive of the relative consumption patterns for various foods based on the bodily requirements of humans. As this approach expands to healing foods, such tools can be suggestive of the foods essential for the human body and their serving sizes [3]. But a question that arises here is whether a single pyramid tool is suggestive of foods for all people across the globe. The Answer could be "NO" since the food availability varies across the geographical regions as well as the seasons. And especially while consid-

A. K. Dhara, S. C. Mandal (eds.), *Role of Herbal Medicines*, https://doi.org/10.1007/978-981-99-7703-1_24

ering healing foods, should be more integrative to incorporate other considerable factors like the time of consumption and the incompatibility concerns.

The indigenous people around the world do not discriminate between food & medicine. There is a continuum in the use of certain foods as medicine, especially the wild and semi-domestic ones. The integrative nature of food as medicine is evident in the plant as well as the animal kingdom. Though the availability of dietary supplements and nutraceuticals, the wholesome approach of utilizing the miraculous foods gifted to us by nature is extremely beneficial [4].

Food can serve as a powerful medicine if consumed adequately and essentially, thus promoting health and preventing diseases through good nutrition. Proper nutrition is of utmost necessity especially when a patient is sick or recovering from a disease, thus food plays its role here by contributing its bit in treating a disease. The lacuna over here is the lack of inclusion of nutrition as a prime topic in the training and practice of doctors and nurses. A sound healthcare system can support the impact of food as medicine on proper healthcare management [5].

The interventions of food as medicines can be integrated with the healthcare systems through the active participation of healthcare providers in considering the impact of food recommendations to the patients and concepts like medically tailored meals and medically tailored groceries. This would lead to increased awareness among society as doctors are most trusted by the community among healthcare providers [6].

Nutritionally adequate diet plans have supported the healthcare systems for better outcomes for their patients and this has resulted in the acceleration of the "Food and Medicine" movement across the world. But there came some limitations to these initiatives due to food insecurities which have also led to the consideration of food as the basic right of humans [7].

The consumption of medicinal plants or traditional foods cannot be generalized, as certain points are necessary to be considered before their public announcement:

1. A plant showing beneficial effects can also contain toxic phytochemicals.
2. There are cultural prescriptions that need to be considered for optimal effects.
3. Promoting wild foods can affect their sustainability by overextraction [4].

Traditional and modern eating is comprised of different facets based on which an eating habit can be regarded as traditional or modern. Some of the important facets include the food itself, and food composition. The secondary facets include methods of preparation, the origin of food, etc. Other facets may be mealtime and coadministered foods which can affect the overall health of the human [8]. Overall if we consider, which food is eaten and what it is composed of has the most alarming implication on health.

This chapter presents a wholistic scientific literature-based approach to bring out the facts about the traditional foods with medicinal benefits which can be consumed by humans in their routine diet or otherwise under certain manageable circumstances. Such foods will help a healthy society that would reduce the burden on the medical sector.

2 Contributions of Ayurveda to Food and Nutrition

As per the concept of Ayurvedic nutrition, all bodily constituents like *dosha*, *dhatu,* and *mala* are formed in balance with an appropriate and wholesome diet as widely said, "we are what we eat." As per these theories, nutrition is considered to be essential for the mind as well. Ayurveda considers the consumption of three types of foods for the mind functions and also prescribes different diets (*ahara*) for the dosha constitution like *Vatta*, *Pitta*, and *Kapha*. There is an emphasis on the consideration of sleep (*nidra*) for human well-being and the concept of *Dinacharya* is important for understanding the best routine for health as per Ayurveda. Ayurveda evolved as an ancient art of healing and living a healthy life from Indian Vedic Literature and has been believed in, today over the world. Indian Traditional foods have been considered to contain the necessary functional components and healing phytochemicals that translate their health benefits upon consumption. Foods are classified based on their nature as cereals, pulses, fruits, vegetables, milk, alcoholic beverages, and animal products. For effective consumption of food, the seasonal study also makes a difference [9].

3 Traditional Systems of Medicines and Food

From the time of history, natural sources of medicine are prevalent in various systems of medicine. Plant- and animal-based products have been used to treat or prevent diseases. Systems such as Ayurveda, traditional Chinese medicine, Unani, traditional Japanese medicine, Homeopathy, etc., commonly used nature-based healing. Such other traditional systems are seen in countries like Korea, Australia, Africa, and Russia. Medicines from natural sources have been researched over time and it is evident that different medicinal plants such as turmeric also have a place in the kitchen and can be administered as food without the need fora prescription from a doctor. The phytochemicals from medicinal foods and herbs are responsible for the healing properties of medicinal foods since human history. There is a huge development in concepts to develop drug discovery from such medicinal foods in recent years and there is the success of the researchers. Although the fact medicinal food itself is an alternative, medicine method further supports the treatments by modern medicine with a complementary approach [10, 11].

4 Medicinal Foods

Due to the use of certain foods with medicinal properties by communities for ages, it is considered safe and alongside has also led to growth in the use of functional foods and phytopharmaceuticals. Due to lifestyle changes and perspectives toward food processing and diet, it is necessary to collaboratively understand the use of such medicinal foods in today's era. Intake of food as medicine can also have the consequences like the interaction of the medicinal components of food and medicinal agents. Risks of food-drug and nutrient-drug interaction need to be anticipated and studied by experts before considering the generous use of such foods in patients [12].

5 Food, Nutrition, Energy, and Health

The basic requirement of well-being is the fulfillment of the bodily needs of energy through food. The secondary requirement is to get our body ready to fight against bodily ailments through preventive medicine that ensures well-being. Tertiary to this is the cure of any disease or ailment contracted, with the use of appropriate medicine.

It is proven that the energy generated and spent in our body is associated with food intake and nutrition therein. The balance in energy in the human body is diet, nutrition, and other human factors. Foods are not just responsible for energy generation but also energy expenditure. The energy balance of foods can be in the form of different mechanisms. Foods like turmeric, ginger, green tea, cinnamon, and chili pepper are responsible for enhanced thermogenesis. Ginger, green tea, chili pepper, nuts, and cinnamon increase energy expenditure. Others like Green tea, bitter orange, and cocoa enhance oxidative metabolism. The human gut also plays an important role in the overall process of metabolism and energy generation [13].

The energy needs of the human body are incurred by the different types of foods that humans consume. Over a while, there is very little evolution or change in the foods which can suffice the needs of energy and nutrition for the human body. A study showed that of ancient diets showed that the maximum energy needs of humans are obtained from Fats (35%), Carbohydrates (35%), and Proteins (30%). The maximum fat intake and energy supply were from poly-unsaturated fatty acids, followed by saturated fats and negligible amounts of trans-fatty acids. Carbohydrate energy supply came more from uncultivated (naturally grown/organic) fruits and vegetables (>50% carbs) and minimal amounts of grains and dairy consumption. Sugar intake was from honey or jaggery ranging to 2%. High-fiber intake was there but was kept to a bare minimum of the total food. Vitamins, Minerals, and Phytochemicals intake was as high as eight times than today's human consumption [14]. Today the scenario of the human diet has changed to a drastic level with the consumption of processed foods, junk, alcohol, imbalance in diet, and lifestyle changes. Minute things like freezing food items or artificial methods of ripening or cultivational changes affect the overall nutrition that human receives through foods.

In addition to the physical health of humans, it is evident that food plays a vital role in the mental, emotional, and psychological well-being of the human. Nutrition has shown positive implications in the elevation of behavior and mood, thus in the mitigation and treatment of mental disorders. In a conference proceeding the topic of "Diet, nutrition, mental health, and well-being" was reviewed to unwind the association of food with the mental health of humans. The universality of food has been considered as a modifiable risk factor with which even a minute modification in the nutritional environment can be translated into larger gains in positive outcomes for the mental health and well-being of society [15]. It has been the truth of the past and from time immemorial, food was associated with the mental well-being of humans and thus there were etiquettes maintained in food regimes from traditional times to ensure healthy mental well-being.

6 The Food Nutrients

Major energy sources or nutrients are carbohydrates, proteins, and fats which are obtained from grains and pulses, meat, fish, eggs, etc. These are required for the major energy supply of the human body and to perform the vital life process within the human body.

Micronutrients like vitamins, amino acids, and minerals are obtained from various plant, animal, and mineral sources and are responsible to support the vital process of life as supplementary needs. They are required in smaller amounts and affect the overall functionality of the human body.

Apart from these, the foods provide phytochemicals that aid in the prevention and cure of diseases in humans and play a role in the maintenance of overall health. Certain classes of phytochemicals that are obtained from food resources and play a major role in the health of humans can be named Polyphenols, Phenolic acids, Flavonoids, Tannins, Aromatic Compounds, lower Terpenoids, certain Alkaloids, and

Essential Fatty acids. Other medicinally important phytochemicals, like steroids, higher terpenoids, saponins, many alkaloids, glycosides, etc. are not obtained from food resources by the human body. They need to be administered in the form of medicines under the guidance of medical professionals.

The food resources can be regarded as sources of nutrients as well as phytochemicals can be classified into the following:

1. Spices and Condiments
2. Grains and Pulses
3. Vegetables and Herbs
4. Fruits and Berries
5. Teas and Extracts

The major source of phytochemicals can be attributed to spices and condiments followed by vegetables, herbs, fruits, berries, teas, and extracts. Since spices, condiments, and certain herbs are used in minor quantities in foods and are found to be an ingredient of different medicinal preparations, this has been discussed separately with detailed emphasis.

Furthermore, the phytochemical classes that have majorly been obtained from foods and are responsible for medicinal activities are flavonoids and polyphenols. Flavonoid-containing foods come from all the abovementioned resources and this again has been discussed separately.

7 Spices, Condiments, and Herbs with Medicinal Value

Spices are used in cuisines in small quantities for flavor, color, or as preservatives. They have a place in traditional medicine and are historically proven to be used for their medicinal benefits. Spices are native to various places of the globe and are now available worldwide due to better logistics and globalization [16].

The Spices have a major place in curing and medicinal benefits incurred upon them. Traditions have shown the use of spices in controlled quantities in certain foods thus obtaining maximum medicinal benefits from them. These are among those foods which contain phytochemicals other than the basic nutrients which are responsible for medical benefits and not the best nutritional benefits.

Turmeric is a yellow color spicy rhizome used traditionally in cookery and for the cure and mitigation of various ailments. It is also known as Indian saffron and is an alternative to expensive saffron. Pepper is known as the king of spices, has originated in South India, and gets its name from the Sanskrit word pipalli (for long pepper). Pepper powders can be obtained from black pepper, white pepper, red pepper, and green pepper. Cinnamon powder in water is applied to alleviate headaches and neuralgia. Cinnamon is often combined with ginger to stimulate circulation and digestion. It has been regarded as a folk remedy for indurations (of the spleen, breast, uterus, liver, and stomach) and tumors (especially of the abdomen, liver, and sinews) [16].

Clove got its name from the term nail "clou" due to its shape. The spice is used in Ayurveda, Chinese medicine, and Western herbalism and dentistry, where the essential oil is used as an anodyne (painkiller) for dental emergencies. Cardamom mixed with neem and camphor, is used as a nasal preparation to treat colds. An infusion of cardamom can be used as a gargle to relieve sore throats, which has led to its use in cough sweets. Cardamom pods, fried and mixed with mastic and milk, are used for bladder problems. Coriander, Fenugreek, Ginger, Garlic, and Chili are some of the traditional medicines that are used for their beneficial effects on health. Mint and rosemary are among the herbs and kokum fruit has also shown use traditionally [16].

Many spices and condiments have shown to provide different medicinal benefits from the foods in which they are used. Some of the important ones along with their biological source, usability in foods, phytochemical composition, and medicinal use have been provided in Table 1.

Apart from Spices and Condiments, which form a major source of distinct variety of phytochemicals, the other foods also contain bioactive phytochemicals which give beneficial health effects in humans. These are

1. Grains and Pulses
2. Vegetables and Herbs
3. Fruits and Berries
4. Teas and Extracts

One of the phytochemical classes majorly responsible for biomedical effects is Polyphenolic compounds like Flavonoids. Different flavonoids and its derivatives are present in each of these food categories.

8 Foods with Flavonoids Responsible for Health Benefits

The one prominent activity that every researcher would carry out on a new plant drug or any natural or semisynthetic drug while researching is the antioxidant activity. The in vitro antioxidant activities are carried out to test every herbal drug sample or formulation since the antioxidant potential of a drug can be a guide that the sample might show other beneficial biological activities.

What accounts for the antioxidant activity of any plant? The major credit for the antioxidant activity in plants can be due to the polyphenolics and the flavonoids or some plant pigments. Flavonoids are among those polyphenolics which

Table 1 Spices and condiments with medicinal benefits

Sr. no.	Name	Biological source	Usability as food	Type of phytochemical	Medicinal use	Reference
1	Turmeric	Rhizomes of *Curcuma longa* (Zingiberaceae)	In powdered form added to Indian cuisines as a condiment, and to sweet preparations as a coloring agent	Curcuminoids	Antioxidant, anticancer, antibacterial, anti-inflammatory, neuroprotective, hepatoprotective, cardioprotective	[17, 18]
2	Pepper	Dried fruits of *Piper nigrum* (Piperaceae)	In powdered form or as a whole added to Indian cuisines as a spice, and also used as a flavor enhancer	Piperine	Antiproliferative, antitumor, antioxidant, antidiabetic, anti-obesity, cardioprotective, antipyretic and anti-inflammatory, antimicrobial, antiaging, and immunomodulatory effects	[18, 19]
3	Long pepper	Dried fruits of *Piper longum* (Piperaceae)	In powdered form or as a whole added to Indian cuisines as a spice, and also used as a flavor enhancer	Piperine, piperidine, caffeic acid, quercetin, kaempferol	Digestive, appetizing agent, stomachic, antidiarrheal, in the lung, and live problems	[16, 20]
4	Clove	Flower buds of *Eugenia caryophyllus* (Myrtaceae)	In powdered form or as a whole added to Indian cuisines as a spice, and also used for flavoring	Phenolic compounds like eugenol, eugenol acetate, β-caryophyllene, α-humulene, gallic acid	Antimicrobial, larvicidal, antiviral, anesthetic, insecticidal, antioxidant, and anticancer	[18, 21]
5	Cinnamon	Bark of *Cinnamomum verum, C cassia, C zeylanicum, C loureirii* (Lauraceae)	In powdered form or as a whole added to Indian cuisines as a spice, and also used for flavoring	Cinnamaldehyde, cinnamate, cinnamyl acetate, b-caryophyllene, and even some amount of eugenol	Antimicrobial, antioxidant, anti-inflammatory, blood purifying, and blood thinning properties	[18, 22]
6	Green cardamom	It consists of dried pods of *Elettaria cardamomum* (Zingiberaceae)	It is used as a spice powder in south Indian foods	Essential oils and monoterpenes	Carminative, diuretic, stomachic and digestive, gastroprotective, anti-inflammatory, antibacterial, aphrodisiac	[16]
7	Black cardamom	It consists of dried pods of *Amomum subulactum* (Zingiberaceae)	It is used in spice powders in Bengal foods	Essential oils and monoterpenes	Anti-inflammatory, anti-asthmatic, digestive, aphrodisiac	[16]
8	Ginger	Rhizomes of *Zingiber officinalis* (Zingiberaceae)	Used in paste form or crushed form for flavoring foods and cuisines, and also used in flavored teas and sweets	Gingerol, zingiberene, geranial, ar-curcumene, and b-bisabolene starch, lipids, proteins, and inorganic compounds	Aphrodisiac, digestive, antibacterial, antifungal, in nausea and vomiting, metabolic syndrome, and pain	[18, 23]
9	Garlic	Bulbs of *Allium sativum* (Liliaceae)	As a whole or in paste or the crushed form used in seasoning	Allicin, allyl-alcohols, phenolics	Antibacterial, antifungal, anticancer, antioxidant	[16, 18]
10	Coriander	Dried ripe fruits and leaves of *Coriandrum sativum* (Umbelliferae)	Used as a condiment in Indian foods	Linalool	Antimicrobial	[18]
11	Cumin	Dried ripe fruits of *Syzygium cumuni* (Umbelliferae)	Indian spice	Ellagic acid, gallic acid, cuminaldehyde Quercetin, Kaempferol	Antioxidant, antimicrobial	[18, 20]

(continued)

Table 1 (continued)

Sr. no.	Name	Biological source	Usability as food	Type of phytochemical	Medicinal use	Reference
12	Caraway	Seeds of *Carum carvi* (Umbelliferae)	Indian spice	Volatile oils	Antioxidant activity	[24]
13	Black cumin	Seeds of *Nigella sativa* (Ranunculaceae)	Spice in food dishes	Nigellone, thymoquinones, hydrothymoquinones, stable oil, proteins, alkaloids, saponins, and essential oils	Antibacterial, antifungal, anticancer	[18]
14	Black mustard	Seeds of *Brassica sinapis alba* (Brassicaceae)	In seasoning for food dishes	Sinapic acid and sinapoyl conjugates	Antimicrobial	[18]
15	Fenugreek	Seeds of *Trigonella foenum graecum* (Leguminosae)	In seasoning for food dishes	Steroidal saponins (diosgenin and trigogenin), free amino acids (hydroxy isoleucine and lysine), and flavonoids and alkaloids (gentianine and trigonelline)	Improve the function of insulin, lower blood sugar levels	[24, 25]
16	Dill	Fruits and leaves of *Anethum graveolens* (Umbellifieae)	As a vegetable and flavoring agent	Carvone, α-phellandrene, limonene, dill ether, myristicin	Carminative, stomachic, and diuretic	[26]
17	Chili pepper	Fruits of *Capsicum annuum* (Solanaceae)	As spice for various foods	Capsaicinoids, capsaicin, an alkaloid	Antioxidant, anti-inflammatory, cardiovascular benefits, antimicrobial	[25]
18	Nutmeg	Dried seeds of *Myristica fragrans* (Myristicaceae)	Food spice, flavoring agent	Argenteane (lignans)	Antimicrobial, antipyretic, asthma, heart disorder, and bad breath	[18]
19	Mace	The outer covering of seeds of *Myristica fragrans* (Myristicaceae)	Whole spice and powdered spice mix	Lignans	Antimicrobial, anticaryogenic activity stomach aches	[18]
20	Star-anise	Dried fruits of *Illicium verum* (Schisandraceae)	Food spice, a major ingredient in Chinese and Vietnamese cuisines	Anethole, α-pinene, phellandrene, p-cymene, limonene, and d-terpineol	Carminative, antifungal, antibacterial, analgesic, sedative, anticarcinogenic, and antioxidant	[18]
21	Asafoetida	Exudates of rhizome or tap roots *Ferula asafoetida* (Apidae)	Used as a condiment in foods, as a digestive in food, and flavoring agent in pickles	E-1-propenyl sec-butyl disulfide (40.0%) and germacrene B (7.8%)	Sedative, expectorant, analgesic, carminative, stimulant, antiperiodic, antidiabetic, antispasmodic, emmenagogue, vermifuge, laxative, anti-inflammatory, contraceptive, and antiepileptic effects	[27]
22	Peppermint	Leaves of Mentha piperita (Lamiaceae)	Used as a condiment in food dishes	Volatile oil, monoterpenes	Antiviral, antibacterial, anti-inflammatory, antifungal, anti-asthmatic, allopathic, spasmolytic, antiheadache, antiseptic	[28]
23	Holy basil	Whole plant of *Ocimum tenuiflorum* (Lamiaceae)	Used in chutneys	Alkaloids, flavonoids, tannins, leucoanthocyanins, terpenes, phenols, Saponins	Diabetes management, and as an immunomodulatory agent	[29, 30]
24	Rosemary	Leaves of *Rosmarinus officinalis* (Lamiaceae)	A flavoring agent with an earthy, sweet-yet-savory flavor	Phenolic acids and diterpenes including carnosic acid, carnosol, caffeic acid, rosmarinic acid, flavonoids, tannins, and volatile oils	Antimicrobial, strong antioxidant, chemopreventive, anticarcinogenic, hepatoprotective	[24, 25]
25	Oregano	Leaves and flowering tops of *Origanum vulgare* (Lamiaceae)	In the crushed form used as a flavoring agent in Italian cuisines	Polyphenols and flavones	Antioxidant	[22]

Table 1 (continued)

Sr. no.	Name	Biological source	Usability as food	Type of phytochemical	Medicinal use	Reference
26	Thyme	Leaves and tops of *Thymus vulgaris* (Lamiaceae)	For flavoring different foods	Flavonoids and phenolic acids, natural terpenoid thymol, and phenol isomer carvacrol	Antioxidant, antimicrobial, antitussive, antispasmodic, and expectorant effects	[27]
27	Parsley	Leaves of *Petroselinum crispum* (Apiaceae)	For flavoring different foods	Flavones: Apigenin, chrysin, luteolin, glycosides of apigenin, luteolin, and diosmetin	Antioxidant, anti-inflammatory	[31, 32]
28	Celery	Leaves of *Apium graveolens* (Apiaceae)	For flavoring different foods	Flavones and their glycosides: apiin and apigenin 7-O-glycoside, glycosides of apigenin, luteolin, and diosmetin	Antioxidant	[32, 33]
29	Saffron	The dried red stigma of *Crocus sativus* (Iridaceas)	Coloring and flavoring agents in sweets	Crocin, crocetin, picrocrocin and safranal, carotenoids: (glycosides of crocetin)	Antioxidant, anti-inflammatory, neuroprotective	[27]

due to their diversity in structures account for various activities such as antioxidant, anti-inflammatory, anticarcinogenic, and free radical scavenging activities. Flavonoids are believed to reduce oxidative damage thus showing the potential for protection from cancer. Flavonoids are present in teas, wines, fruit juices, vegetables, and fruits and are shown to reduce the risks of mortality in patients with cancer and cardiovascular disorders. Recent studies on Flavonoids are focused on exploring and quantifying the health benefits of dietary intake of flavonoids. Even though research in this area is gaining pace, the historical evidence of the well-being of humans with a balanced intake of colorful seasonal foods is proof of the health benefits of flavonoids [31, 34].

Flavonoids can be classified into:

1. Flavonols
2. Flavones
3. Flavanols
4. Proanthocyanidins
5. Flavanones
6. Anthocyanins

Flavonols are found in fruits, vegetables (especially green leafy vegetables), and grains. Flavanol aglycons include kaempferol, quercetin, myricetin, isorhamnetin, and rhamnetin. In plants, flavonols are found as free aglycones and their glycosides. The glucosides show antimutagenic effects. The Free flavonol aglycones show anticancer potential. They also possess antihypertensive, antibacterial, and antiviral properties [33, 35].

Flavones are confined to limited edible plants and contain free aglycones and glycosides of apigenin, luteolin, tangeretin, and diosmetin. It is found in vegetables like hot pepper fresh parsley and celery. It is also found in spices and herbs like oregano, rosemary, and thyme [33, 35].

Flavanols (catechins) are found in high concentrations in Tea and to some extent in fruits and spices. They are of two types namely catechol (catechins) or pyrogallol type (gallocatechins).They majorly include Catechin, epicatechin gallate, and epigallocatechin gallate. They majorly show antihypertensive, antibacterial, and antiviral properties [33, 35].

Proanthocyanidins are polymers bound with two or more flavanols which constitute most of the tannins present in plants [33].

Flavanones are present in almost all citrus fruits but absent in most vegetables. They include aglycones and glycosides of hesperetin, eriodictyol, and naringenin [33, 35].

Isoflavones form the major component of soybean and are majorly found in Legume plants. Daidzin, genistin, glycitin, and also malonylated isoflavones are known as soy isoflavones. They are found in traces in other legumes like chickpeas and black turtle beans [33].

Anthocyanins are found in many edible plants like fruits and beans. Some Anthocyanins include Cyanidin, delphinidin, malvidin, peonidin, and pelargonidin. They are of numerous types and are present in colorful plants. They show improvement in liver health, visual function improvement, and reduce oxidative damage [33].

Flavonoids as a whole are found to possess profound antioxidant effects. They are also shown to reduce the risks of central nervous system problems. They are found to possess enzyme-level effects and thus show antidiabetic effects [33]. Flavonoids from Tea, Green Tea, Fruits, and Berries containing Anthocyanin, and Flavanones have shown positive health outcomes in cardiovascular disease-containing patients. Foods with Flavanols and Isoflavones have also shown an effect on cholesterol through different mechanisms. In addition, flavonoids have shown, anti-inflammatory effects and effect on platelet aggregation. Thus, concluding that flavonoid-rich foods have the potential to reduce cardiovascular risks [36].

Different foods containing flavonoids, the flavonoid composition, and their beneficial effects on health are shown in Table 2.

Table 2 Flavonoids containing foods showing medicinal benefits

Sr. o.	Name	Origin	Biological source	Flavonoid content	Use	Reference
Cereals and pulses						
1	Tartary buckwheat	East Asia	Seeds of *Fagopyrum tataricum* (Polygonaceae)	Proteins, polyphenols, D-chiro-inositol, flavonoids, resistant starch, dietary fiber, minerals, and vitamins	Prevention of AIDS, and prevention and treatment of cancer, and cardiovascular and cerebrovascular diseases	[37]
2	Buckwheat	Asia	Seeds of *Fagopyrum esculentum* (Polygonaceae)	Gallic acid; Flavonol glycosides: quercetin, rutin	Antioxidant, antihypertensive, antibacterial, and antiviral properties	[22, 33]
3	Black rice/ purple rice	India & China	Grains of *Oryza sativa* (Poaceae)	Anthocyanins	Antioxidant	[22]
4	Red rice	Africa and Madagascar	Grains of *Oryza punctata* (Poaceae)	Anthocyanins	Antioxidant, free radical scavenging activity	[22]
5	Sorghum	Africa	Grains of *Sorghum bicolor* (Poaceae)	Glycosides of flavones: apigenin, luteolin, and diosmetin	Antioxidant	[33]
6	Soybean	China	Beans of *Glycine max* (Leguminosae)	Caffeic acid, ferulic acid, gallic acid, isoflavones: daidzein Daidzein, genistin, glycitin, formononetin	Antioxidant activity	[20]
7	Black bean	America	Beans of *Phaseolus vulgaris* (Fabaceae)	Isoflavones	Antioxidant	[22]
8	Purple corn	South America	Beans of Zea mays (Poaceae)	Anthocyanins, cyanidin 3-O-glucoside	Antioxidant	[33]
9	Broad beans	Persia	Beans of Faba sativa (Fabaceae)	Flavanols: epicatechin, epigallocatechin, and catechin	Antioxidant	[33]
10	Moong beans	Indian subcontinent	Beans of *Vigna radiate* (Fabaceae)	Flavanols: epigallocatechin and epigallocatechin gallate	Antioxidant	[33]
11	Cowpea	Africa	Seeds of *Vigna sinensis* (Fabaceae)	Flavanols: epigallocatechin and epigallocatechin gallate	Antioxidant	[33]
12	Chickpea	Turkey	Seeds of *Cicer arientum* (Fabaceae)	Flavonol: quercetin	Antioxidant	[20]
Vegetables and herbs						
13	Oregano	Leaves and flowering tops of *Origanum vulgare* (Lamiaceae)	In the crushed form used as a flavoring agent in Italian cuisines	Polyphenols and flavones like apigenin and luteolin	Antioxidant	[22, 32]
14	Rosemary	Mediterranean, Portugal, and North-Western Spain	Leaves of *Rosmarinus officinalis* (Lamiaceae)	Phenolic acids; flavones like apigenin and luteolin	Antioxidant, chemopreventive, and anticarcinogenic	[24, 25]
15	Thyme	Southern Europe from Spain to Italy	Leaves and tops of *Thymus vulgaris* (Lamiaceae)	Phenolic acids; flavones like apigenin and luteolin	Antioxidant, antimicrobial	[27, 32]
16	Parsley	Greece and Morocco	Leaves of *Petroselinum crispum* (Apiaceae)	Flavones: apigenin, chrysin, luteolin, glycosides of apigenin, luteolin, and diosmetin	Antioxidant, anti-inflammatory	[31, 32]
17	Celery	Mediterranean region	Leaves of *Apium graveolens* (Apiaceae)	Flavones and their glycosides: apiin and apigenin 7-O-glycoside, glycosides of apigenin, luteolin, and diosmetin	Antioxidant	[32, 33]
18	Onions	Asia	Bulbs of *Allium cepa* (Amaryllidaceae)	4′-O-glucosides and 3,4′-di-O-glucosides of isorhamnetin and quercetin	Antioxidant, anticancer	[33]

Table 2 (continued)

Sr. o.	Name	Origin	Biological source	Flavonoid content	Use	Reference
19	Shallot	Asia	Bulbs of *Allium ascalonicum* (Amaryllidaceae)	4′-O-glucosides and 3,4′-di-O-glucosides of isorhamnetin and quercetin	Antioxidant, anticancer	[33]
20	Kale	Mediterranean and Anatolia	Leaves of *Brassica oleracea* (Brassicaceae)	Flavonol glycosides: quercetin 3-O-glycosides like rutin, isoquercitrin, and quercitrin; kaempferol 3-O-glycosides	Antimutagenic effects, antihypertensive, antibacterial, and antiviral properties	[33]
21	Lettuce	Egypt	Leaves of *Lactuca sativa* (Asteraceae)	Flavonol glycosides: quercetin 3-O-glycosides like rutin, isoquercitrin, and quercitrin	Antimutagenic effects, antihypertensive, antibacterial, and antiviral properties	[33]
22	Spinach	Persia	Leaves of *Spinacia oleracea* (Amaranthaceae)	Flavonol glycosides: rutin, spinacetin glycosides, and patuletin	Antihypertensive, antibacterial, and antiviral properties	[33]
23	Broccoli	Mediterranean	Flowering head and stalks of *Brassica oleracea* (Brassicaceae)	Flavonol glycoside: Kaempferol 3-O-glycosides	Antioxidant	[33]
24	Endive	Egypt and Indonesia	Leaves of Cichoriumendive (Asteraceae)	Flavonol glycoside: Kaempferol 3-O-glycosides	Antioxidant	[33]
25	Sweet potato	Central and South America	Tubers of *ipomoea Batatas* (Convolvulaceae)	Flavonols: quercetin 3-O-glycosides like rutin, isoquercitrin, and quercitrin	Antimutagenic effects, antihypertensive, antibacterial, and antiviral properties	[33]
26	Tomato	America	Fruits of *Solanum lycopersicum* (Solanaceae)	Flavanones like hesperetin, eriodictyol, and naringenin; flavonols like quercetin, kaempferol, and myricetin	Antioxidant	[32]
27	Eggplant	Northern Africa	Fruit of solanum Melongena (Solanaceae)	Anthocyanins (nasunin and malvin)	Antioxidant	[22]
28	Drumstick leaves	Indian subcontinent	Leaves of *Moringa oleifera* (Moringaceae)	Phenolic compounds: caffeic acid, ferulic acid; flavonols: quercetin, kaempferol, rutin	Antioxidant activity	[20]
29	Red chili Pepper	Bolivia	Fruits of *Capsicum annuum* (Solanaceae)	Flavonols quercetin 3-O-glycosides like rutin, isoquercitrin, and quercitrin	Antimutagenic effects, antihypertensive, antibacterial, and antiviral properties	[33]
Fruits & berries						
30	Apple	Central Asia	Fruits of malus Sylvestris (Rosaceae)	Flavonols like quercetin 3-O-glycosides like rutin, isoquercitrin, and quercitrin Flavanols like catechin, epicatechin, epigallocatechin Proanthocyanidins, anthocyanins	Antimutagenic effects, antihypertensive, antibacterial, and antiviral properties	[33]
31	Peach	Eastern China	Fruits of *Prunus persica* (Rosaceae)	Flavanols like catechin, epicatechin, epigallocatechin; anthocyanins	Antihypertensive, antibacterial, and antiviral properties	[33]
32	Pear	Europe, North Africa, and Asia	Fruits of *Pyrus pyrifolia* (Rosaceae)	Flavanols like catechin, epicatechin, epigallocatechin	Antihypertensive, antibacterial, and antiviral properties	[33]
33	Apricot	China	Fruits of *Prunus armeniaca* (Rosaceae)	Flavonols like quercetin, kaempferol, and myricetin, and flavanols like catechin, epicatechin, epigallocatechin	Antioxidant and anti-inflammatory	[36]

(continued)

Table 2 (continued)

Sr. o.	Name	Origin	Biological source	Flavonoid content	Use	Reference
34	Date fruit	Egypt and Mesopotamia	Fruits of *Phoenix dactylifera* (Arecaceae)	Carbohydrates and several essential nutrients	Potential nutritional therapeutic diet for patients with metabolic syndrome	[38]
35	Amla	India, Pakistan, and Bangladesh	Fruits of *Phyllanthus emblica* (Phyllanthaceae)	Ellagic acid, gallic acid	Antioxidant	[20, 22]
36	Lemon	Assam, Myanmar, and China	Fruits of *Citrus limoni* (Rutaceae)	Hesperidin (flavanone), taxifolin (flavanonol)	Antioxidant	[31]
37	Lime	Indonesian archipelago and the mainland of Asia	Fruits of *Citrus latifolia* (Rutaceae)	Hesperidin (flavanone), taxifolin (flavanonol) Naringenin Hesperetin	Antioxidant	[31]
38	Kokum	Tropical rain forests of Western Ghats	Fruits of Garcinia indica (Clusiaceae)	Anthocyanin, garcinol, cynidine	Antioxidant, acidulant, and appetizer	[16]
39	Orange	Southern China and Northeast India	Fruits of *Citrussinensis* (Rutaceae)	Hesperidin, naringenin, eriodictyol (flavanone), taxifolin (flavanonol)	Antioxidant	[31]
40	Mandarin	North China	Fruits of Citrus reticulata (Rutaceae)	Isoflavones: Hesperetin, Hesperidin, hesperetin 7-O-rutinoside	Antioxidant	[33]
41	Nectarines	Eastern China	Fruits of *Prunus Persica var. nucipersica* (Rosaceae)	Flavones: catechin, epicatechin, epigallocatechin	Antihypertensive, antibacterial, and antiviral properties	[33]
42	Grapefruit	Barbados	Fruits of *Citrus paradise* (Rutaceae)	Isoflavones: naringenin, naringin, and naringenin 7-O-neohesperidoside	Antioxidant	[33]
43	Mango	Southern Asia	Fruits of Mangifera indica (Anacardiaceae)	Flavones: catechin, epicatechin, epigallocatechin	Antihypertensive, antibacterial, and antiviral properties	[33]
44	Fig	Greece	Fruit of *Ficus carica* (Moraceae)	Flavonol glycoside: rutin	Antihypertensive, antibacterial, and antiviral properties	[33]
45	Plum	China	Fruits of *Prunus salicina* (Rosaceae)	Flavones: catechin, epicatechin, epigallocatechin	Antihypertensive, antibacterial, and antiviral properties	[33]
46	Grape	Middle east	Fruits of *Vitis* genus (Vitaceae)	Anthocyanins in the peel, Proanthocyanidins in pulp, Proanthocyanidins	Antioxidant	[22]
47	Black grape and raisins	Europe and Asia	Fruits of *Vitis vinifera* (Vitaceae)	Anthocyanidins	Antioxidant	[36]
48	Black currents	Europe and northern Asia	Fruits of *Ribes nigrum* (Grossulariaceae)	Flavonols: quercetin, kaempferol, and myricetin	Antioxidant	[36]
49	Cranberry	Britain	Fruits of *Vaccinium oxycoccus* (Ericaceae)	Flavonols (quercetin 3-O-glycosides like rutin, isoquercitrin, and quercitrin); quercetin and proanthocyanidins	Anticancer immunological and cardiovascular benefits, antimutagenic effects, anticancer	[33]
		North America	Fruits of *Vaccinium macrocarpon* (Ericaceae)			
50	Raspberry	Asia minor and America	Fruits of *Rubus idaeus* (Rosaceae)	Flavones: catechin, epicatechin, epigallocatechin	Antihypertensive, antibacterial, and antiviral properties	[33]
51	Mulberry	Temperate Asia and North America	Fruits of *Morus nigra* (Moraceae)	Anthocyanins (nasunin and malvin)	Antioxidant	[33]
52	Blueberry	North America	Fruits of *Vaccinium cyanococcus* (Ericaceae)	Flavones: Catechin, epicatechin, epigallocatechin	Antioxidant	[33]
53	Bilberry	Northern Europe	Berries of *Vaccinium myrtillus* (Ericaceae)	Flavones: catechin, epicatechin, epigallocatechin	Antioxidant	[33]

Table 2 (continued)

Sr. o.	Name	Origin	Biological source	Flavonoid content	Use	Reference
54	Cherry	Asia minor	Fruits of Prunus avium (Rosaceae)	Anthocyanins (nasunin and malvin)	Antioxidant	[33]
55	Strawberry	Brittany, France	Fruits of *Fragaria ananassa* (Rosaceae)	Catechin, anthocyanins	Antioxidant	[33]
56	Blackberry	Asia, Europe, and north and South America	Edible fruits of *Rubus* species (Rosaceae)	Anthocyanins and other phenolic compounds, mainly flavonols and ellagitannins	Antioxidant	[36]
57	Elderberry	Europe and Asia	Fruits of Sambucus nigra (Adoxaceae)	Anthocyanidins and flavonols	Antioxidant	[36]
Teas						
58	Tea	East Asia	Leaves of Thea sinensis (Theaceae)	Flavanols: gallocatechin, epigallocatechin, epicatechin gallate, epigallocatechin gallate, catechin and epicatechin	Antioxidant, anti-inflammatory	[31]
59	Green tea	East Asia	Leaves of *Camelia sinensis* (Theaceae)	Green tea catechin and polyphenols	Antioxidant, anti-inflammatory, radical scavenging activity	[31]

Food is the source of numerous medicinal phytochemicals that benefits the mankind in the mitigation and prevention of health issues if adequately consumed. But the absorption of the components of the food may not always be optimal in all individuals. For proper absorption, there is the role of the gut microbiota, which is responsible for assimilation of the food we consume. Here plays a role, the prebiotics and probiotics.

9 Probiotics and Prebiotics

For optimum absorption and best effects of all foods or medicines that are ingested, there needs to be proper gut microbiota which plays an important role in shaping human health. These are the microorganisms categorized for better understanding as prebiotics, probiotics, synbiotics, and postbiotics, and today, functional products are obtained with these as additives. But the resources providing these in human diet through food have been evident since, traditions. Probiotics can be termed as the live microorganisms ingested in small quantities that show positive health effects on humans and prebiotics are the substrates utilized by the host microorganisms, thus giving the health benefits. Synbiotics use the combination of prebiotics and postbiotics to confer health effects, as may be called microbiota with their substrate, administered in adequate amounts for good health effects in humans. Postbiotics are nonviable microbial products or metabolites that show effects on the health of an individual. Today there is an evolution of these to the concepts of parabiotics, oncobiotics, psychobiotic, and pharmacobiotics [39].

Simply, we can consider probiotics being nothing but foods that are substrates for microorganisms and prebiotics are preparations containing good microorganisms in live form. The concept of synbiotics and postbiotics are mere modifications and subsets of the other two. Since traditional times, the ingestion of fermented products is prevalent, one such example being yogurt which contains an important strain of probiotic called Lactobacillus. On the other hand, the foods consumed traditionally are the source of the prebiotics [40]. There are traditional beverages like lassi, buttermilk, or yogurt itself which will give us these beneficial effects without modern alterations to our traditional foods.

For the optimum results from the food and medicines that human beings consume, there are other underlying factors which can affect the overall process. These factors include the quality and the type of food, the environmental factors, the intrinsic and extrinsic factors affecting human life, and the overall code of conduct by humans.

10 Factors Affecting the Health Benefits of Foods

Apart from the food resources and the phytochemical composition, other factors that affect the overall health benefits obtained from food, diet, and nutrition include:

1. Food Preparations and Storage
2. Food Combinations
3. Seasonal Considerations (Ritucharya)
4. Daily Routine (Dinacharya)
5. Habits and Physical activity

10.1 Food Preparations and Storage

Traditionally, cooking has been done using simple techniques and cookware, usually, open or closed cooking of foods in earthen pots or vessels made up of iron, cast iron, copper, wood, or alloys like bell metal (Kansa). Even the storage or eating was done in vessels made up of the same, along with this silver was also used. These metals formed the sources of elements. Today with the modern era, there is a change in the material of construction of the cookware. There is the use of materials ranging from steel and aluminum to glass or even plastic. There is today, evidence of the harmful effects of these modern transitions. Adopting the use of traditional cookware will serve the best outcomes from the foods [41].

Apart from this, the methods used in cooking varies drastically from traditional to today's modern era. From the use of firewood for cooking to the use of pressurized cooking techniques or microwave techniques to the engineered versions of cooking like gastronomy, cooking has evolved to a great extent. In traditional cooking techniques, the impact of external factors on the foods is minimal, thus making it a minimalistic way of cooking. Even though there is so much advancement in the field of cooking, the health-conscious subset of people today is moving toward the traditional way of cooking.

10.2 Food Combinations

The effect of combinations of various foods needs to be understood. Some foods enhance the potential of others when ingested in combination and some foods can diminish the health benefits of foods that they are combined with. In Ayurveda, the other one is called Virudha Ahara and is important as it can give a concern if not considered wisely. According to Ayurveda, foods that support the three qualities of mind are classified into Sattvic, Rajasic, and Tamasic foods [42].

Sattvic means pure essence, and sattvic diet is for a consciously spiritual and healthy life. It is considered to nourish the body and maintains it in a peaceful state. It is excellent for those individuals who desire to live a quiet, peaceful, and meditative life. Sattvic foods include sprouted whole grains, fresh fruit, land and sea vegetables, pure fruit juices, nut and seed milk and cheese, legumes, nuts, seeds, sprouted seeds, honey, and herbal teas [42]. The effect of Sattvic food on human body is explained as per Bhagwad Gita Shloka in Fig. 1.

Rajasic foods aggravate pitta and vata doshas which in turn increase anger and restlessness. They stimulate more fire, outward motion, creativity, aggression, and passion. Such foods are too much spicy, salty, and sour such as pickles, tea, coffee, alcohol, and vegetables like onion, garlic, etc. The mind-body equilibrium can be destroyed if rajasic foods to greater extent [42]. The effect of Rajasic food on human body is explained as per Bhagwad Gita Shloka in Fig. 2.

Tamasic foods increase the inner darkness and confusion and include fried and are good for slowing us down, numbing, depressing, and enhancing inertia in humans. Such foods include frozen foods, fast foods, microwaved foods, processed foods, left overnight foods, meat, fish, eggs, onion,

Bhagwad Gita. 17.8

आयु:सत्त्वबलारोग्यसुखप्रीतिविवर्धना: ।
रस्या: स्निग्धा: स्थिरा हृद्या आहारा: सात्त्विकप्रिया: ॥ ८ ॥

āyuḥ-sattva-balārogya-
sukha-prīti-vivardhanāḥ
rasyāḥ snigdhāḥ sthirā hṛdyā
āhārāḥ sāttvika-priyāḥ

Translation

Foods dear to those in the mode of goodness increase the duration of life, purify one's existence and give strength, health, happiness and satisfaction. Such foods are juicy, fatty, wholesome, and pleasing to the heart.

Fig. 1 Bhagwad Gita Shloka relating to Sattvic Ahara

alcohol, etc. Tamasic food is the unhealthiest food [42]. The effect of Tamasic food on human body is explained as per Bhagwad Gita Shloka in Figs. 2 and 3.

Ayurvedic food habits help us to regularize absorption, assimilation, and elimination thus balancing the doshas in human body. Foods prescribed as per doshas and prakriti of human help in avoiding the foods that should not be taken together and also guide in appropriate combinations of the foods. The food as per doshas includes diet for vata, pitta, and kapha prakriti respectively. The functions of vata dosha

Bhagwad Gita. 17.9

कट्वम्ललवणात्युष्णतीक्ष्णरूक्षविदाहिनः ।
आहारा राजसस्येष्टा दुःखशोकामयप्रदाः ॥ ९ ॥

kaṭv-amla-lavaṇāty-uṣṇa-
tīkṣṇa-rūkṣa-vidāhinaḥ
āhārā rājasasyeṣṭā
duḥkha-śokāmaya-pradāḥ

Translation

Foods that are too bitter, too sour, salty, hot, pungent, dry and burning are dear to those in the mode of passion. Such foods cause distress, misery and disease.

Fig. 2 Bhagwad Gita Shloka relating to Rajasic Ahara

Bhagwad Gita. 17.10

यातयामं गतरसं पूति पर्युषितं च यत् ।
उच्छिष्टमपि चामेध्यं भोजनं तामसप्रियम् ॥ १० ॥

yāta-yāmaṁ gata-rasaṁ
pūti paryuṣitaṁ ca yat
ucchiṣṭam api cāmedhyaṁ
bhojanaṁ tāmasa-priyam

Translation

Food prepared more than three hours before being eaten, food that is tasteless, decomposed and putrid, and food consisting of remnants and untouchable things is dear to those in the mode of darkness.

Fig. 3 Bhagwad Gita Shloka relating to Tamasic Ahara

control the movements in body, process of elimination, the nervous system, etc. Pitta persons have hot, penetrating or sharp, flowing liquid, slightly oily, and light qualities in them. The kapha dosha in the body protects, governs the structure of the body, provides immunity, holds the cells together, forms the muscle and bone, etc. [42].

Vata people should take fresh, warm food with sweet, sour, and salty tastes. They should avoid too much fasting and also the use of spicy food to regulate their digestion. Diet for vata constitution includes warm drinks, sweet fruits, raw nuts, milk, butter, cream, hot cereals, etc. Pitta persons should take Cool foods and drinks, especially in hot weather and also eat a mostly vegetarian diet which suits the constitution, are preferred. Adequate intake of raw food and juices are indicated. Cool water should be taken and hot beverages and foods should be avoided. Kapha people should prefer low carbohydrate, low-fat diets without sugars, and avoid frequent eating, frozen edibles, and cold water [42].

10.3 Seasonal Considerations

It is considered as best for each seasonal food (the ones growing in the particular season) and also the place of growing of foods can play an important role (local foods). This help in keeping the natural harmony in the diet. In Ayurveda, the concept of Ritucharya explains its impact of it on our bodies. There is an impact of the seasonal changes on our Prakriti and hence the choice of foods should be in line with the balancing of the doshas. There are six Rutus namely Shishira—Late Winter season (Mid-Jan to Mid-March),Vasantha—Spring season (Mid-March to Mid-May), Greeshma—Summer season (Mid-May to Mid-July), Varsha—Rainy season (Mid-July to Mid-Sept), Sharat—Autumn season (Mid-Sept to Mid Nov), and Hemantha—Winter season (Mid-Nov to Mid-Jan) according to the Gregorian Calendar [43]. There are external environmental changes that can affect internal bodily changes in humans (Tridoshas). Hence, each season demands a change in the diet as per the need of the body according to the environmental impact during that season. The changes in the dietary consumption of food according to seasonal variation is called "Ritu-Bhukta," which means, diet according to season [44].

A summary of the dietary preferences and the lifestyle preferences during the different seasons (Rutus) is given in Fig. 4.

10.4 Daily Routine (Dinacharya)

It has been discussed in a national conference for public health promotion in 2018, that it is important to consider the impact of the daily routine of patients incurring of the illness. This conference highlighted the importance of the daily regime "DINACHARYA" and the seasonal regime "RITUCHARYA" to get ideal lifestyle benefits. It was considered a way to control the hazardous impact of lifestyle disorders at the community level by following healthy regimes as per Ayurveda [44]. A systematic routine helps in synchronizing biological activities to gain benefits in health. The biological clock plays an important role in maintaining life processes including metabolism, diseases, sleep, and psychiatric disorders. The chronology of the processes and certain good practices in the daily regime should be followed. Activities other than diet include wake-up time, gut hygiene, oral hygiene, eye care, skin care, exercises, bath, and sleep. The time of activities, especially the time of waking up, taking meals, routine activities, and sleep maintains equilibrium in the body and hence positively impacts longevity and health [45].

A chart showing the preferred daily regime as per the ayurvedic protocols that can benefit the human health to maximum by regulating the circadian rhythm of the body is given in Fig. 5.

10.5 Habits and Physical Activity

As rightly said in Bhagwad Gita, Fig. 6, the human habits along with physical activity can regulate the health to a greater extent. Apart from being a part of the daily routine, the physical activity and overall habits of human play a vital role in overall regulation of health.

Fig. 4 Diet and lifestyle for six seasons (Rutus)

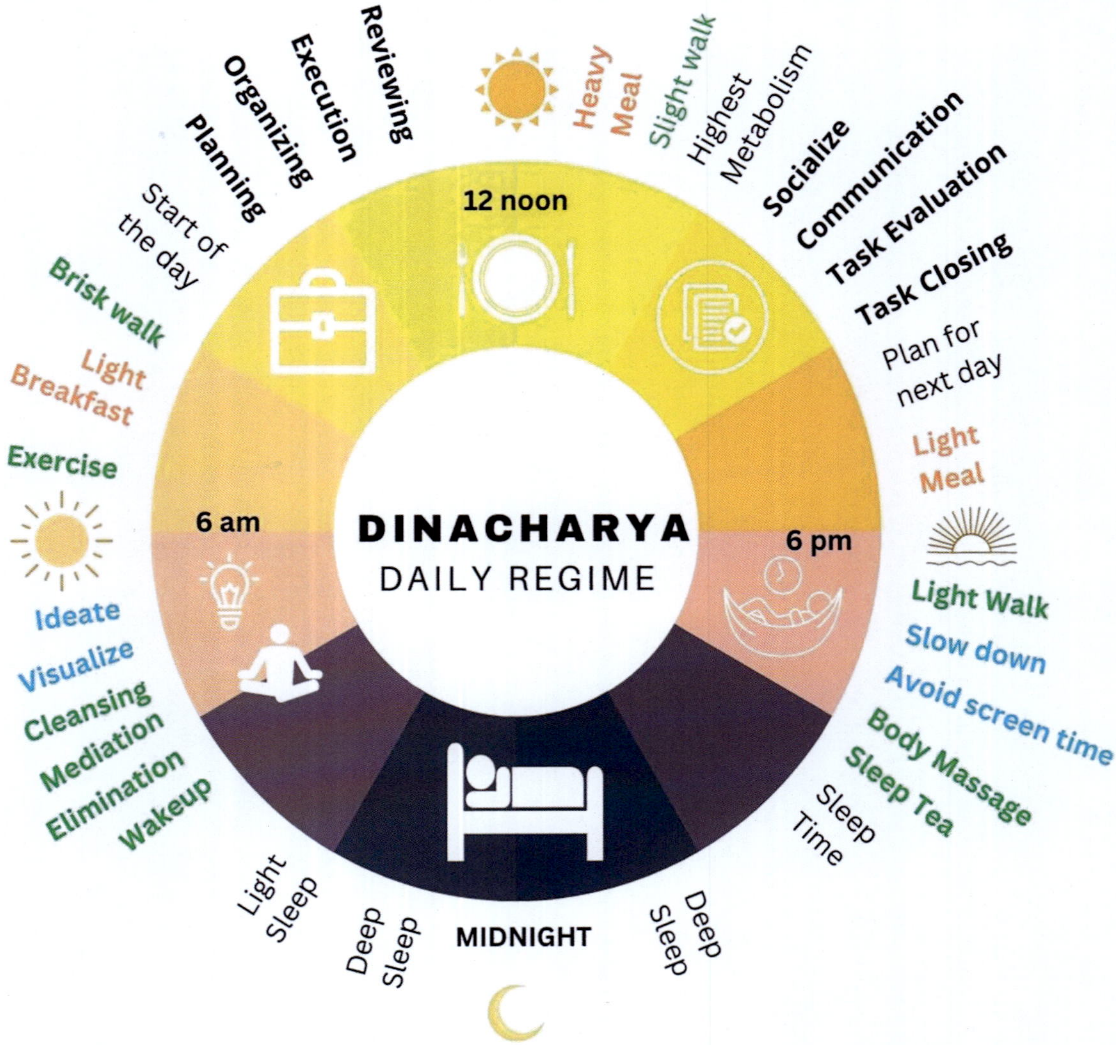

Fig. 5 Dinacharya (daily regime) as per Ayurvedic principles

Fig. 6 Bhagwad Gita Shloka relating to benefits of good habits and yoga

Bhagwad Gita. 6.17

युक्ताहारविहारस्य युक्तचेष्टस्य कर्मसु ।
युक्तस्वप्नावबोधस्य योगो भवति दु:खहा ॥ १७ ॥

yuktāhāra-vihārasya
yukta-ceṣṭasya karmasu
yukta-svapnāvabodhasya
yogo bhavati duḥkha-hā

Translation

He who is regulated in his habits of eating, sleeping, recreation and work can mitigate all material pains by practicing the yoga system.

References

1. Hyman M, Bradley E. Food, medicine, and function: food is medicine part 1. Phys Med Rehabil Clin N Am. 2022;33:553.
2. Hyman M, Bradley E. Food, medicine, and function: food is medicine part 2. Phys Med Rehabil Clin N Am. 2022;33:553.
3. Myklebust M. The healing foods pyramid: an integrative nutrition tool. Explore (NY). 2006;2:352.
4. Reyes-García V. Eating and healing. Traditional food as medicine. Econ Bot. 2006;60:60.
5. Lucan SC. When food isn't medicine—a challenge for physicians and health systems. Prev Med Rep. 2018;10:62.
6. Downer S, Berkowitz SA, Berkowitz SA, Harlan TS, Olstad DL, Mozaffarian D. Food is medicine: actions to integrate food and nutrition into healthcare. BMJ. 2020;369:m2482.
7. Barnidge EK, Stenmark SH, DeBor M, Seligman HK. The right to food: building upon "Food Is Medicine". Am J Prev Med. 2020;59:611.
8. Sproesser G, Ruby MB, Arbit N, Akotia CS, Alvarenga MDS, Bhangaokar R, et al. Understanding traditional and modern eating: the TEP10 framework. BMC Public Health. 2019;19:1606.
9. Sarkar P, Lohith KDH, Dhumal C, Panigrahi SS, Choudhary R. Traditional and ayurvedic foods of Indian origin. J Ethn Foods. 2015;2:2.
10. Yuan H, Ma Q, Ye L, Piao G. The traditional medicine and modern medicine from natural products. Molecules. 2016;21:559.
11. Ozdal T, Tomas M, Toydemir G, Kamiloglu S, Capanoglu E. Introduction to nutraceuticals, medicinal foods, and herbs. In: Aromatic herbs in food: bioactive compounds, processing, and applications; 2021.
12. Ramalingum N, Mahomoodally MF. The therapeutic potential of medicinal foods. Adv Pharmacol Sci. 2014;2014:1.
13. Bo S, Fadda M, Fedele D, Pellegrini M, Ghigo E, Pellegrini N. A critical review on the role of food and nutrition in the energy balance. Nutrients. 2020;12:1161.
14. Eaton SB. The ancestral human diet: what was it and should it be a paradigm for contemporary nutrition? Proc Nutr Soc. 2006;65:1.
15. Owen L, Corfe B. The role of diet and nutrition on mental health and wellbeing. Proc Nutr Soc. 2017;76:425.
16. Kunnumakkara AB, Koca C, Dey S, Gehlot P, Yodkeeree S, Danda D, et al. Traditional uses of spices: an overview. In: Molecular targets and therapeutic uses of spices: modern uses for ancient medicine; 2009.
17. Sharifi-Rad J, El RY, Rizk AA, Sadaka C, Zgheib R, Zam W, et al. Turmeric and its major compound curcumin on health: bioactive effects and safety profiles for food, pharmaceutical, biotechnological and medicinal applications. Front Pharmacol. 2020;11:01021.
18. D'Souza SP, Chavannavar SV, Kanchanashri B, Niveditha SB. Pharmaceutical perspectives of spices and condiments as alternative antimicrobial remedy. J Evid Based Complementary Altern Med. 2017;22:1002.
19. Haq IU, Imran M, Nadeem M, Tufail T, Gondal TA, Mubarak MS. Piperine: a review of its biological effects. Phytother Res. 2021;35:680.
20. Bajpai M, Pande A, Tewari SK, Prakash D. Phenolic contents and antioxidant activity of some food and medicinal plants. Int J Food Sci Nutr. 2005;56:287.
21. Rani R, Jena MK. Clove (SYZYGIUM AROMATICUM): beneficial effects on human health: a review. Plant Arch. 2021;21:1967.
22. Xu DP, Li Y, Meng X, Zhou T, Zhou Y, Zheng J, et al. Natural antioxidants in foods and medicinal plants: extraction, assessment and resources. Int J Mol Sci. 2017;18:96.
23. Li H, Liu Y, Luo D, Ma Y, Zhang J, Li M, et al. Ginger for health care: an overview of systematic reviews. Complement Ther Med. 2019;45:114.
24. El-Sayed SM, Youssef AM. Potential application of herbs and spices and their effects in functional dairy products. Heliyon. 2019;5:e01989.
25. Jiang TA. Health benefits of culinary herbs and spices. J AOAC Int. 2019;102:395.
26. Jana S, Shekhawat G. Anethum graveolens: an Indian traditional medicinal herb and spice. Pharmacogn Rev. 2010;4(8):179.
27. Khazdair MR, Anaeigoudari A, Hashemzehi M, Mohebbati R. Neuroprotective potency of some spice herbs, a literature review. J Tradit Complement Med. 2019;9:98.
28. Nayak P, Kumar T, Gupta AK, Joshi NU. Peppermint a medicinal herb and treasure of health: a review. Phytopathology. 2020;9:9.
29. Mahomoodally MF, Subratty AH, Gurib-Fakim A, Choudhary MI, Nahar KS. Traditional medicinal herbs and food plants have the potential to inhibit key carbohydrate hydrolyzing enzymes in vitro and reduce postprandial blood glucose peaks in vivo. Sci World J. 2012;2012:1.
30. Mahomoodally F, Mesaik A, Choudhary IM, Subratty AH, Gurib-Fakim A. In vitro modulation of oxidative burst via release of reactive oxygen species from immune cells by extracts of selected tropical medicinal herbs and food plants. Asian Pac J Trop Med. 2012;5:5.
31. Yao LH, Jiang YM, Shi J, Tomás-Barberán FA, Datta N, Singanusong R, et al. Flavonoids in food and their health benefits. Plant Foods Hum Nutr. 2004;59:113.
32. Lotito SB, Frei B. Consumption of flavonoid-rich foods and increased plasma antioxidant capacity in humans: cause, consequence, or epiphenomenon? Free Radic Biol Med. 2006;41:1727.
33. Terahara N. Flavonoids in foods: a review, vol. 10. Nat Prod Commun; 2015. p. 1934578X1501000.
34. Ivey KL, Jensen MK, Hodgson JM, Eliassen AH, Cassidy A, Rimm EB. Association of flavonoid-rich foods and flavonoids with risk of all-cause mortality. Br J Nutr. 2017;117:1470.
35. Gildawie KR, Galli RL, Shukitt-Hale B, Carey AN. Protective effects of foods containing flavonoids on age-related cognitive decline. Curr Nutr Rep. 2018;7:39.
36. Hooper L, Kroon PA, Rimm EB, Cohn JS, Harvey I, Le Cornu KA, et al. Flavonoids, flavonoid-rich foods, and cardiovascular risk: a meta-analysis of randomized controlled trials. Am J Clin Nutr. 2008;88:38.
37. Ruan J, Zhou Y, Yan J, Zhou M, Woo SH, Weng W, et al. Tartary buckwheat: an under-utilized edible and medicinal herb for food and nutritional security. Food Rev Intl. 2022;38:440.
38. Vayalil PK. Date fruits (phoenix dactylifera Linn): an emerging medicinal food. Crit Rev Food Sci Nutr. 2012;52:249.
39. Nataraj BH, Shivanna SK, Rao P, Nagpal R, Behare PV. Evolutionary concepts in the functional biotics arena: a mini-review. Food Sci Biotechnol. 2021;30:487.
40. Zawistowska-Rojek A, Tyski S. How to improve health with biological agents—narrative review. Nutrients. 2022;14:1700.
41. Bansal C, Zaina S, Parihar V. REVIEW ON HEALTH IMPACT OF HAZARDOUS AND SAFEST TRADITIONAL COOKWARE WITH AYURVEDIC APPROACH. AYUSHDHARA. 2020:2559.
42. Dhanya S, Ramesh NV, Mishra A. Traditional methods of food habits and dietary preparations in Ayurveda—the Indian system of medicine. J Ethn Foods. 2019;6:1.
43. Yashas U, Reddy PS, Beena MD. Concept of Sarat Ritucharya- a practical approach. Int J Ayurveda Pharm Res. 2022:64.
44. Rao MV, Harti S, Ghildiyal S, Rai S. AYUSHCHARYA 2018—'A National Conference on Dinacharya and Ritucharya for public health promotion'. J Ayurveda Integr Med. 2019;10:230.
45. Montaruli A, Castelli L, Mulè A, Scurati R, Esposito F, Galasso L, et al. Biological rhythm and chronotype: new perspectives in health. Biomol Ther. 2021;11:1.

Nanotechnology in Delivery and Targeting of Phytochemicals for Lifestyle Diseases

Milan Milivojević, Ivana Pajic-Lijaković, Zora Dajić, Amal Kumar Dhara, Amit Kumar Nayak, and Md Saquib Hasnain

Abstract

In many cases, conventional therapeutics are not able to cure diseases in a proper and safe manner. Conversely, plants and their derivatives have been successfully used for the treatment of numerous diseases and disorders for centuries and have proven themselves as efficient, inexpensive, environmentally friendly, faster, and less toxic. However, their use is sometimes connected to some issues, mainly due to inefficient systemic delivery and bioavailability, which prevent to translate their promising in vitro and in vivo effects into clinical use. During the past decade, nanotechnology has made big progress in developing different nanocarriers for drug delivery formulations. In the case of phytopharmaceuticals their encapsulation in nanoparticles can help in improving drug solubility, physical and chemical stability, pharmacological activity, reducing its toxicity and side effects, and providing targeted and sustained delivery which all leads to better bioavailability. The current review highlights the main problems connected to the delivery of phytopharmaceuticals, the main properties of different types of plant drugs, and gives a critical review of properties and limitations connected to the application of numerous lipid, polymer, and inorganic nanoparticles as phytodrug carriers. It also provides information about up-to-date investigated combinations of phytopharmaceuticals and nanocarriers.

Keywords

Review · Nanocarriers · Nanoparticles · Phytopharmaceuticals · Phytochemicals

Herbal medicine has been for centuries one of the sciences with the longest tradition, but in the twentieth century it appeared that it has lost its primacy in medicine due to the high expansion of the synthetic pharmaceutical industry. However, in recent decades, phytomedicine is again gaining increased interest due to many factors. Some of those factors may be scientifically still inexplicable, and perhaps unjustified, but others have good scientific and economic base. The herbal medicine is generally safe, inexpensive, universally obtainable, and efficient in preventing and healing a wide spectrum of diseases and disorders. In addition, it has deeply rooted trust due to a centuries-long tradition and positive therapeutic experience as well as smaller side effects in comparison to the many modern synthetic drugs [1–3]. The phytopharmacy origins date thousands of years back in time when people first time discovered that specific herbs, plants or their parts, have evident positive effects in treating different diseases. According to some estimations between 50,000 and 80,000 plant species have been currently considered to have medically active compounds [4]. At the beginning the remedy mechanisms were not known but traditional medicine, through the many centuries of knowledge accumulation, managed to group, organize, and categorize phytopharmaceuticals, while modern chemistry and biomedicine in most cases accomplished to detect the main active ingredients of these herbal remedies and to identify the mechanism of their action [5]. Therefore, it is not surprising that today about 80% of the developing countries population

M. Milivojević (✉) · I. Pajic-Lijaković
Faculty of Technology and Metallurgy, University of Belgrade, Belgrade, Serbia
e-mail: mmilan@tmf.bg.ac.rs

Z. Dajić
Faculty of Agriculture, University of Belgrade, Belgrade, Serbia

A. K. Dhara
Department of Pharmacy, Contai Polytechnic, Purba Medinipur, West Bengal, India

A. K. Nayak
Department of Pharmaceutics, School of Pharmaceutical Sciences, Siksha 'O' Anusandhan (Deemed to be University), Bhubaneswar, Odisha, India

M. S. Hasnain
Department of Pharmacy, Palamau Institute of Pharmacy, Chianki, Daltonganj, Jharkhand, India

A. K. Dhara, S. C. Mandal (eds.), *Role of Herbal Medicines*, https://doi.org/10.1007/978-981-99-7703-1_25

use herbal drugs as the first-choice remedy, and even in the developed countries this percent is over 25% [6]. All these, in combination to the fact that the main sources of many drugs are still of herbal origin, have led to the growth in the demand for phytochemicals in many areas of medicine. However, it needs to be stressed that phytopharmaceuticals extracted from plants are commonly composed of hundreds of different components and this overabundance of natural compounds may activate and modulate many different systems in an organism, in contrast to synthetic drugs which in most cases represent only one active component [7–9]. However, phytodrug action may vary in dependence on plant extract composition which is governed by many factors. Therefore, phytopharmaceuticals that are used for different drug formulations are usually either isolated or chemically synthetized and commonly represent single, dominant pure component of the plant extract or standardized extracts [10]. This is one of the most important differences between phytopharmaceuticals used in modern medicine and traditional herbal medicine. In addition, phytopharmaceuticals in most cases have very good safety and tolerability, although they, like all drugs, may also have some undesired side effects, drug interactions, or contraindications. But generally, they are well tolerated, with a very wide therapeutic margin and minimal interaction potential, which all make them especially suitable for patients with multiple medications [9].

The use of phytopharmaceuticals, however, has some major limitations, namely, most of the biologically active phytochemicals like flavonoids, phenolics, tannins, alkaloids, terpenoids, steroids, phenylpropanoids, etc., are secondary metabolites with large molecular size. Those secondary metabolites have proven their pharmacological effects not only in treating acute and "simple" diseases like bacterial and fungal infections, but also for chronic, systemic, and degenerative diseases, like cancer and diabetes [11]. Although those phytopharmaceuticals show magnificent in vitro results, their activity is usually much less in vivo [12]. This is not surprising since secondary metabolites are categorized as organic compounds and, as such, they have low water or lipid solubility, inadequate bioavailability, poor stability in GIT, reduced intestinal absorption after oral administration due to large MW, short biological half-life, narrow therapeutic index, poor active efflux mechanism, rapid elimination due to first-pass metabolic elimination, poor penetration into cells, rapid uptake by normal tissues (compared to cancer cells), and, in many cases, they are toxic [3, 5, 13–15]. The proper delivery of phytopharmaceuticals as it may be concluded is a challengeable task connected to many restrains.

In order to overcome those physicochemical, pharmacokinetic, and physiological limitations of phytochemicals many different techniques have been utilized in the past to improve drug solubility, bioavailability, sustainability, GIT permeability, reduce side effects, etc. However, those techniques may also have their own disadvantages. As example, in order to increase solubility organic solvents can be used or drug can be converted in its salt form, but those non-polar solvents may be toxic, while phytopharmaceutical in its salt form may have lower activity [5]. Therefore, novel drug delivery system (DDS) must accomplish few prerequisites related to drug regulatory standards (quality and safety) as well as efficacy [1]. The efficient DDS should help drug to retain it activity in therapeutic levels while at the same time improving its physical properties in desirable direction [5]. Ideally, DDS should enable controlled drug delivery rate, during desired time period and at specified body organ or system [16].

In the recent decades, nanotechnology gained rising abilities and therefore the increased attention toward its application in many different fields. This interest also had its great part in the research and developing many different nano-based drug delivery dosage forms and following 1980 there was a significant increase in the number of approved nanoformulations [3]. Nowadays, nanotechnology is widely investigated as a useful tool for reduction and elimination of aforementioned drawbacks of phytopharmaceuticals. Up to now nanodosage forms have proven their efficacy in many phytopharmaceutical formulations where they were used to enhance phytochemicals solubility, bioavailability, pharmacological activity and stability, to improve specific cell/tissue targeting and sustained delivery, to protect it from physical and/or chemical degradation, reduce toxicity and therapeutic doses, and to diminish side effects [3, 17]. All this enables effective transfer of even a poorly soluble phytopharmaceutical to the targeted site with improved therapeutic index and pharmacokinetic/bioavailability profile [18, 19].

In such manner, not only that the use of many phytopharmaceuticals was enhanced, but it also opened a possibility to reconsider use of wide range of different phytocompounds that were overlooked in the past due to the above-mentioned limitations [5]. The possibilities of nanotechnology are very extensive since nanoparticles that may be used as drug nanocarriers may be produced in many different dosage forms (polymeric nanoparticles and nanocapsules, liposomes, solid lipid nanoparticles, phytosomes, nanoemulsion, dendrimers, quantum dots, gels, etc.), shapes (spheres, cubes, rods, etc.), compositions (inorganic, organic, or hybrid), sizes, surface chemical/physical modifications, and/or functionalizations (charge, functional groups, PEGylation, targeting moieties, etc.) [13, 17]. As it can be concluded, all this offers a wide area for very fine tuning of the physiochemical properties of nanocarriers in order to more precisely obtain desired DDS properties according to the characteristics of both the drug and the targeted organ. In that way, nanophytopharmaceutical DDS offers possibility to treat effectively various diseases and disorders with the lowest side effects [5, 13] (Fig. 1).

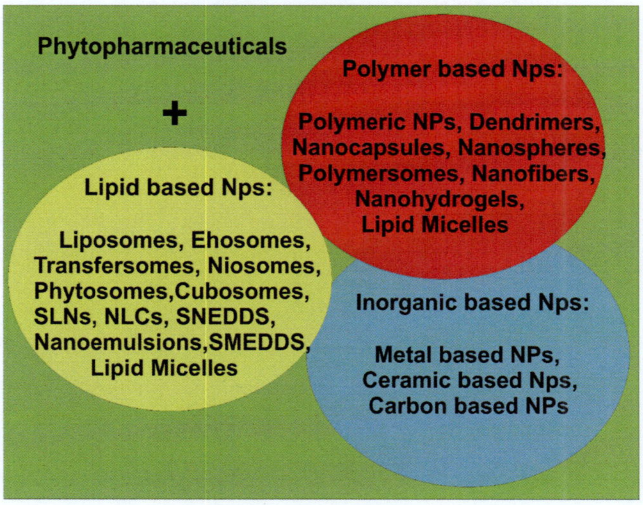

Fig. 1 Possible combinations of phytopharmaceuticals and nanoparticles

1 Phytochemicals

Phytochemicals are bioactive plant compounds present in various plant parts (like flowers, stems, leaves, barks, fruits, seeds, nuts, and roots) which in the human body generally are not used for growth and metabolism but rather for health and well-being [20, 21]. Within the plants phytochemicals are produced as secondary metabolites as a defensive response in order to protect them from a wide range of environmental biotic or abiotic stresses (pollutants, sun, fungal and microbial infections, pests, etc.) [22]. Those active ingredients may have many different beneficial properties like anti-inflammatory, antiproliferative, anticancer, antioxidant, antiaging, astringent, anti-irritant, photoprotectant, moisturizing, antimicrobial, immunosuppressive, hepatoprotective, and many others [5, 23, 24]. Therefore, they have been widely used for treating many traditional and modern diseases and disorders (like cancer, diabetes, viral infections, Alzheimer, obesity, autoimmune and cardiovascular disorders, etc.), multidrug-resistant microbial infections, diabetes, and obesity disorders [25]. However, their actions may be also harmful, especially when used in inappropriate concentrations. In addition, despite their chemical and structural diversity there are some common problems that usually emerge when those components are used as drugs. Namely, although phytochemicals are effective, their direct administration is usually associated with some limitations of physicochemical, physiological, and pharmacokinetic nature. First there is reduced stability during and after extraction from the plants. The extracted lipophilic components have low water solubility and after ingestion they are poorly dissolved in gastrointestinal fluids. In addition, even those low concentrations are further subjected to aggressive actions of wide pH range within GIT and of enzymes present not only in GIT, but in liver, kidneys, and other tissues [16, 26]. Conversely, many phytoconstituents that are soluble in water are either poorly absorbed in GIT due to their large MW which limits their passive diffusion or cannot pass across cell lipid membranes due to low lipid solubility [17]. The phytochemicals that manage to reach circulation system are further submitted to methylation, sulfation, and glucuronidation which additionally reduces already low blood concentrations [23, 27]. And above all this, there is lack of target specificity which means that those agents may bind not only to targeted site but also to other molecules and tissues [26]. Therefore, phytochemical drug levels at targeted site are commonly below the therapeutic concentrations and those are some of the main reasons why although herbal drugs have enormous therapeutic potential in vitro they in vivo show much lower or no therapeutic effects and/or undesired side effects if drug concentrations are high at undesired site [17, 28]. It must be also mentioned that phytopharmaceuticals are very diverse in chemical composition so the delivery of each in some aspects is nonspecific although there are some general aspects [29].

In spite of all mentioned limitations phytopharmaceuticals still have advantages compared to the modern medicine, not only because chemical drugs also have most of the mentioned limitations, but also due to their thousand years long application experience and confidence, as well as the higher therapeutic values and lower adverse side effects [26].

There is a tremendous number of different phytochemicals and they are classified within many distinctive groups of compounds. The most numerous and for phytomedicine most important groups are *phenolic compounds*, *alkaloids*, and *terpenoids*, while some others are *steroids*, *organosulfur compounds*, *saponins*, *anthranoids*, etc. Of course, many complex compounds may be classified within different groups based on the parts of their molecules.

Phenolic compounds are the most numerous and most important group of phytopharmaceuticals and almost all of them are *polyphenolic compounds*. Those aromatics have over 8000 up to date determined different compounds which are further classified within flavonoids, lignans, tannins, quinones, xanthones, coumarins, and other subgroups [30]. The classification is managed according to the structure of basic chemical structure, while differences between compounds within the same groups are mainly based on the degrees of oxidation, hydroxylation, methylation, glycosylation, and connections to other molecules [31]. They are extensively present in the plants providing them the characteristic color, flavor, and taste. Polyphenols generally have many preferable medical actions due to the presence of phenol rings, hydroxyl groups, and conjugated double bonds, which promote metal-chelating, fibril-destabilizing, estrogen-like, enzyme-binding, and antioxidative properties [32].

Unfortunately, they also exhibit a very low bioavailability when orally administrated due to the poor solubility in water (aromatic nature), low GIT absorption, and high metabolic elimination [5]. The main efforts of nanotechnology based administration of phenolic compounds are directed to overcome or diminish those limitations.

Among natural polyphenolic compounds *flavonoids* are not only the largest (>6000) but also the most important group since they possess potent and wide range of biological activities (antioxidant, anti-inflammatory, antimicrobial, antiviral, anti-allergic, antitumor, neuroprotective, cardiovascular protective, hepatotoxic, anti-ulcer, and many other effects). Flavonoids core structure consists of a flavan nucleus in which 15 carbon atoms are arranged in three rings (A, B, and C) [33]. They may be further divided into many subclasses (and their main representatives) according to the oxidation level of the central ring (ring C), as given below.

Phenolic Compounds

1. *Phenolic acids*: *Hydroxycinnamic acids* (caffeic acid, caftaric acid, chlorogenic acid, cinnamic acid, coumaric acid, curcumin, ferulic acid, γ-coumaric acid, p-coumaric acid, quinic acid, sinapic acid), *hydroxybenzoic acids* (p-hydroxybenzoic acid, ellagic acid, gallic acid, protocatechuic acid, salicylic acid, syringic acid, vanillic acid);
2. *Flavonoids*: *Flavonols* (fisetin, galangin, isohanmetin, kaempferol, myricetin, mirol, rutin, quercetin), *flavanols* (or catechins) (catechin, epicatechin, epigallocatechin, gallocatechin, and their gallate derivatives, thearubigins), *flavones* (apigenin, baicalein, baicalin, chrysin, luteolin, nobiletin, rutin), *flavanones* (eriodictyol, hesperetin, hespiridin, naringenin, naringin, narirutin), *flavononol* (taxifolin (or dihydroquercetin)), *isoflavones* (biochanin A, curcumin, daidzein, daidzin, equol, formononetin, genistein, genistin, glycitein, tetrahydro-curcuminoids), *flavonolignans* (silibinin, silymarin), *anthocyanins* (cyanidin, delphinidin, malvidin, pelargonidin, peonidin, petunidin), *chalcones* (phlorizin);
3. *Stilbenes* (combrastatin-A, isorhapontigenin, piceatannol, pinosylvin, pterostilbene, resveratrol, rhapontigenin, tetrahydroxystilbene glucoside);
4. *Coumarins* (cajanuslactone, esculetin (aesculetin), imperatorin, herniarin, methoxyseselin, psoralen, scopoletin, umbelliferone, warfarin);
5. *Lignans* (bicyclol, clemastanin B, diphyllin, enterodiol, enterolacton, hinokinin, honokiol, isatindolignanoside A, lariciresinol, matairesinol, niranthin, pinoresinol, phillygenin, podophyllotoxin, rubrifloralignan A, secoisolariciresinol, sesamin, sesamol, silymarin, steganacin, wikstromal, yatein);
6. *Tannins* (proanthocyanidins and hydrolysable tannins) (afzelechin, catechin, ellagic acid, ellagitannin, gallic acid, gallotannin, geranin, gllocatechin, hexahydroxydiphenic acid, procyanidins (A2 and B2), tannic acid);
7. *Quinones* (*benzoquinones, naphthoquinones, anthraquinones, and polycyclic quinones*) (β-lapachone, arnebifuranone, arnebinone, atovaquone, daunorubicin, diospyrin, emodin, isodiospyrin, lapachol, juglone, plumbagin, rhein, saprorthoquinone, sennosides, tanshinone);
8. *Xanthones* (α-mangostin, cadensin D, caloxanthone, calozeyloxantone, gentisine, mangiferin, phomoxanthone A, rutinosylxanthone, swertipunicoside, transkielcorin, vinaxanthone, xanthofulvin);
9. *Terpenes* (camosic acid, camosol, carotenoids, glycyrrhizic acid).

Alkaloids are the second largest class of plant secondary metabolites with over 5000 up to date identified phytopharmaceuticals. They are usually not present in lower plants. Their structures include rings are nitrogen atom (in most cases within a heterocyclic ring), but are very diverse in structure ranging from simple to extremely complicated [5]. Their classification and definition are problematical since they are very inhomogeneous group of compounds not only from chemical and biochemical perspective but also from physiological viewpoint. Only thing that is general for all alkaloids is that they are all nitrogen-containing compounds. Alkaloids have many different and strong biological activities in human organisms and are therefore the most important group of phytopharmaceuticals. Alkaloids may have many different pharmacological actions like analgesia, local anesthesia, cardiac stimulation, respiratory stimulation and relaxation, vasoconstriction, muscle relaxation and toxicity, as well as antineoplastic, hypertensive, and hypotensive properties [34]. Their action in most cases is connected to the presence of heterocyclic tertiary nitrogen. Some of the alkaloids within human body can interact with neurotransmitters (acetylcholine, dopamine, noradrenaline, serotonin) producing sedative, psychotropic, or analgesic action and therefore it is not surprising that many of them are doping compounds, drugs of abuse, or poisons. The fact that alkaloids are water soluble at acidic pH, while at neutral and basic pH they are lipid soluble, enables their transport in the protonated form, and pass through cell lipid membranes in the neutral form [35]. The application of nanotechnology in the case of alkaloids is mainly aimed at reducing their toxicity by enabling local and targeted applications, which greatly improves their bioavailability and facilitates medicinal use.

Alkaloids may be classified based on their botanical origin or based on amino acids from which they can be biosynthetically derived, but most common classification is based on their chemical structure. Two most important and numerous groups are indole and isoquinoline alkaloids, while other also important but less numerous groups are tropane, steroidal alkaloids pyridine and pyrrolizidine alkaloids, as well as pyrrole, piperidine, quinolone, purine, imidazole, acridones, aromatics, carbolines, ephedras, ergots, bisindoles, indolizidines, manzamines, oxindoles, quinolines, quinozolines, phenylisoquinolines, phenylethylamines, pyrrolidines, pyrroloindoles, and simple tetrahydroisoquinolines, etc. [36].

Alkaloids: Based on their chemical structures, they are classified into many groups as: *isoquinoline* (*with benzylisoquinoline and aporphine*) (largest group) (apomorphine, aporphine, berberine, chelerythrine, codeine, emetine, jatrorrhizine, morphine, muraricine, noscapine, palmatine, papaverine, sanguinarine, tetrandrine, thebaine, tubocurarine), *indole* and *terpene indole* (second largest group) (ajmaline, catharanthine, ellipticine, ergocornine, ergocristine, ergocryptine, ergonovine, ergosine, ergotamine, harmane, physostigmine, psilocybine, rescinnamine, serotonin, strictosidine, strychnine, toxiferine, tryptamine, villalstonine, vinblastine, vincamine, vincristine, vinorelbine, voacamine, yohimbine), *pyrrolizidine* (retrorsine, senecionine, senkirkine), *pyrrolidine* (aegyptolidine, bgugaine, nicotine), *steroidal alkaloids* (chaconine, cyclopamine, jervine, solanidine, solanine, solasodine, tomatidine, tomatine, veratridine), *tropane* (atropine, cocaine, hyoscyamine, scopolamine), *pyridine* (anabasine, nicotine, nornicotine, ricinine, trigonelline), *pyrrole* (keramadine, solsodomine, stevensine), *quinoline* (camptothecin, cinchonidine, quinidine, quinine, strychnine), *piperidine* (coniine, coniceine, lobeline, pinidine, piperine, sedamin, (S)-solenopsin, veratrine), *tropolone* (allantoin, colchicine, demecolcine, thiocolchicoside), *purine* (caffeine, methylxanthosine, theacrine, theobromine, theophylline), *imidazole* (pilocarpine), *phenyl* (*capsaicinoids*) (capsaicin), *amino* (ephedrine), *diterpene* (paclitaxel), *sesquiterpene* (cassinine, celapanin, evonoline, evorine) *terpenoid* (β-skytanthine, aconitine, actinidine, ajmalicine, gentianine), *phenethylamine* (mescaline), *tryptamine* (psilocin, psilocybin), etc.

Terpenoids are also a large group of phytopharmaceuticals with more than 40,000 identified compounds. They are unsaturated cyclic hydrocarbons whose classification is based on various number of isoprene units (3-methyl-1,3-butadiene) from which they are built. There is a very wide range of different terpenoids that possess pharmacological and/or preventive action in treatment of many disorders and ailments. They have been proven to possess effective anticancer, cytotoxic, antimicrobial, antifungal, antiparasitic, antiviral, anti-allergenic, anti-angiogenic, antioxidant, antispasmodic, anti-ulcer, anti-inflammatory, antihyperglycemic, antihyperlipidemic, antipyretic, analgesic, hepatoprotective, neuroprotective, and immunomodulatory activities [5, 37].

Hemiterpenoids consist of a single isoprene unit. *Monoterpenoids* consist of two isoprene units and they are naturally found in aromatic plants where they mostly contribute to their organoleptic properties as floral scents, essential oils, and defensive resins (either constitutive or induced) [38]. Monoterpenes may be further classified as unsaturated hydrocarbons, alcohols, alcohol esters, aldehydes, and ketones. They may have anti-irritant, analgesic, anti-itching, anthelmintic, vasodilatant, and other actions [34]. *Sesquiterpenoids* have three isoprene units which may be combined in linear (acyclic), monocyclic, bicyclic, and tricyclic form. Sesquiterpene lactones are a special group of sesquiterpenoids from which they chemically distinct from due to the presence of a γ-lactone system. They mainly possess antibacterial, antifungal, antiprotozoan, cardiotonic, diuretic, analgesic and anti-inflammatory activities. *Diterpenoids* are a chemically very heterogenous group of phytopharmaceuticals (over 90 different triterpenoidal carbon skeletons) that are composed of four isoprene units arranged in linear, bicyclic, tricyclic, tetracyclic, pentacyclic, or macrocyclic form. Their most important pharmacological properties are analgesic, anti-inflammatory, antibacterial, antifungal, antiprotozoal, and antineoplastic. Terpenoids also include *sesterterpenoids*, *triterpenoids*, *sesquiterpenoids*, *tetraterpenoids* (or carotenoids), as well as *polyterpenoids* and *norisoprenoids*.

The main limitations with administration of those compounds are that their oral bioavailability is greatly limited by very low water solubility, which in combination to extremely high lipophilicity and instability (easy heat and light degradation) restricts their use in food, health, and cosmetic products.

Terpenoids: *Hemiterpenoids* (isoprene, angelic acid, isovaleric acid), *monoterpenoids* (α-pinene, β-phellandrene, β-pinene, ascaridole, auraptene, borneol, camphor, carvacrol, carvone, cineol, citral, citronellal, D-limonene, eucalyptol, linalool, linalyl acetate, menthol, myrcene, perillic acid, thujone, thymol, thymoquinone), *sesquiterpenoids* (*and sesquiterpene lactones*) (α-humulene, α-zingiberene, β-bisabolene, β-caryophyllene, β-cedrol, β-elemene, β-nerolidol, β-santalol, δ-cadinene, alantolactone, ambrosin, artabsin, artemisinin, artesunate, bisabolol, caryophyllene, chamazulene, coronopilin, costunolide, dehydrocostuslactone, farnesene, farnesol, helenalin, inunolide, khushimol, matricin, nootkatone, patchoulol, parthenolide, santonin, thapsigargin, thujopsene, zerumbone), *diterpenoids* (9-geranyl-α-terpineol, abietic acid, andrographolide, carnosic acid, casbene, crocetin, gibberellin A1, ginkgolide A, ginkgolide B, marrubiin, phytol, retinal, retinol, salvinorin A, sclareol, steviol, tanshinone I, taxol, trans-retonic acid, triptolide), *sesterterpenoids* (heliocide H1, hippolide E, leucosceptrine, manoalide, nitiol, sesterstatin 7), *triterpenoids* (α-amyrin, α-spinasterol, β-amyrin, asiatic acid, betulinic acid, campesterol, cucurbitacin B, isoliquiritigenin, lupeol, oleanolic acid, stigmasterol, sitosterol, squalene, ursolic acid), *tetraterpenoids (or carotenoids)* (α-carotene, β-carotene, β-cryptoxanthin, γ-carotene, astaxanthin, crocetin, crocin, lutein, lycopene, zeaxanthin)).

Glycosides are composed of aglycone (or genin) and sugar part (or glycone) which are interlinked by the glycosidic linkage. The pharmacological action is dependent on aglycone part, while sugar part is responsible for glycosides pharmacokinetic properties (like solubility, cell permeability, etc.). They commonly have good water solubility. Glycosides are known for their cardiotonic, general tonic, bitter tonic, purgative, laxative, antileprotic, antiarrhythmia, coronary vasodilator, expectorant, anti-inflammatory, antipyretic, analgesic, antiseptic, antirheumatic, antispasmodics, sedative, hypnotic, and many other actions. Main classes of *glycosides* are: cardiac (bufalin, digitoxin, oleandrin, ouabain), quinone (anthraquinone (aloin, emodin, sennoside A, B, C, and D), naphtaquinone (β-lapachone), benzoquinone (thymoquinone)), xanthone (gambogic acid), benzofuran (isopsoralen, psoralen), cyanogenic (amygdalin), saponin (ginsenoside, glycyrrhizin), isothiocyanate, cyanophore, alcoholic, thioglycosides, phenolic, anthocyanins, coumarin, chromone, lavonoid, steviol, etc.

Saponins are group of phytochemicals that possess a polycyclic aglycone part (also called genin, or sapogenin, that represents a water non-soluble part) with either a steroid (steroidal saponins) or triterpenoid (triterpenoidal saponins) attached to a carbohydrate unit (a monosaccharide or oligosaccharide chain composed variously of pentoses, hexoses, or uronic acids, as water soluble part). As they possess hydrophobic/hydrophilic asymmetry they are lowering surface tension and behave like soap. They are dominantly concentrated in the plant roots although they may be present in all parts of plants. Although their complexity obstructs better comprehension between the chemical structure and their pharmaceutical action it has been shown that they have numerous pharmacological properties like antitumor, piscicidal, molluscicidal, spermicidal, sedative, expectorant, antitussive, analgesic, and anti-inflammatory [34]. Depending on the type of present sapogenin (aglycone counterparts), the saponins can be classified into three major groups [39]: (1) triterpenoid saponins with 30 C-atoms (most widely distributed in the plants), which according to the number of attached sugar moieties can be further classified as monodesmosidic and didesmosidic (another subclassification is onto oleanane, ursolic acid, and dammarane saponins); (2) steroid saponins with 27 C-atoms (mainly used as precursors for the partial synthesis of sex hormones) which may be also further divided into furostanol and spirostanol type; and (3) alkaloid saponins.

Organosulfur compounds have sulfur molecule included in the structure and they are extensively known for their health benefits as they have shown strong antimicrobial, cardioprotective, anticarcinogenic, antioxidant, anti-inflammatory, anticarcinogenic, and anti-angiogenic effects. Main organosulfur compounds include isothiocyanates, indoles, allylic sulfur compounds, and sulforaphane.

Phytosterols are a group of more than 200 different compounds that have cholesterol in their structure but may be found only in plants. Although they are structurally similar to cholesterol, they are much less absorbed in organism, but they may significantly reduce cholesterol absorption. In addition, they possess anticancer properties. Main phytosterols are sitosterol and campesterol, while some other examples are β-sitosterol, brassicasterol, campestanol, cycloartenol, diosgenin, ergosterol, guggulsterone, stigmastanol, stigmasterol.

Most important or most used phytochemicals in developing nanomedicines are curcumin, paclitaxel, quercetin, catechins, berberine, camptothecin, and resveratrol [40].

2 Nanotechnology

2.1 General

Nanotechnology as an emerging scientific and engineering discipline is a complex mix of materials science, physics, chemistry, and biology sciences which aim to develop, synthesize, characterize, and apply nanoparticles in many different fields [41]. Although any particle with size less than 1 μm may be considered as nanoparticle, for practical applications those particles need to be less than 100 nm in at least one dimension [42]. Namely at those size levels those nanostructures start to present magnetic, mechanical, electrical, optical, thermal, and other characteristics that are different to the macromolecular forms of those materials [43]. Due to the modified and/or increased characteristics of those nanoparticles encapsulated bioactive may exhibit increased drug solubility, permeability, stability, bioactivity, target specificity, safety, as well as reduced drug doses, side effects, invasiveness, and toxicity. Most of those characteristics beside the type of used bioactive compound are also dependent on the nanoparticle characteristics, like its size, surface activity, production method, and toxicity [21].

In the past two decades many different nano-based drug carriers have been produced (i.e. liposomes, dendrimers, nanomicelles, niosomes, metallic NPs, etc.), but only some of them were clinically authorized, although they have shown great potential in improving drug target specificity, biopharmaceutical and pharmacokinetic characteristics. Nevertheless, it is expected that increased understanding of the NPs characteristics and advances in nanotechnology will boost activities and offer new opportunities for development of NPs based on nanoherbal systems for different therapeutic applications. However, in order to translate nanomedicine from preclinical trials to its clinical application some challenging tasks must be addressed. Major challenges are connected to effective passing through different biological barriers to achieve specific targeting, successful manufacturing and characterization of nanoformulations, precise evaluation of NPs toxicity, and finally, determining main regulatory guidelines for the examination of nanomedicine therapeutics [41].

2.2 Advantages and Limitations of NPs

Different types of nanoparticles have different advantages and limitations and beside the type of particles this also may depend on loaded drug or more particularly on its molecular weight, water/lipid solubility, and permeability. However, some general observations may be drawn.

2.2.1 Advantages of Nanophytopharmaceuticals

Through the last few decades, different nanoformulations have shown promising potential to overcome main limitations of conventional phytopharmaceuticals (low solubility, hindered membrane permeability, and poor bioavailability) by offering numerous advantages for increasing their therapeutic effectiveness. Some of the most important advantages that may be obtained by use of nanotechnology are given below [5, 24, 40, 44]:

1. *Increased Solubility*: The main reason why nanotechnology can improve solubility is that decreasing particle size to nanolevel greatly increases surface area. However, it must be kept in mind that herbal drugs are comprised of thousands of different types of molecules which greatly vary in their lipid and water solubilities, and therefore it may be very challenging task to formulate them in a single uniform dosage form.
2. *Improved Permeability*: In order to be delivered to the action site phytobased drugs often need to cross different membranes. The main problems are that along this path to the action site some membranes are hydrophilic, while others are lipophilic, and that within the human organism there are different efflux mechanisms (i.e. P-glycoprotein efflux pump) whose primary role is to extrude xenobiotics and toxic substances from the body. The permeability is influenced by size, charge, and hydrophobic/hydrophilic properties of the formulated phytodrug. The maximum drug release inside the cell membrane may be obtained when applied nanoparticles have size range in the range of 100–200 nm, while the charge and hydrophylic properties influence depend on the type of phytopharmaceutical and on the site of action.
3. *Reduced Side Effects and Toxicity*: The nanoformulated drugs may increase drug solubility and improve permeability and consequently may produce similar or even better therapeutic effects with smaller drug doses. This leads to reduction of dose related adverse or toxic effects to a greater extent than for any other dosage form. In addition, toxicity is further reduced due to the absence of toxic non-polar solvents that are commonly used to increase drug solubility and improve traditional formulation characteristics.
4. *Targeted Delivery*: Targeted drug delivery is one of the most desirable but at the same time most challenging task, which hardly can be met by the conventional formulations. The use of nanoformulations for phytopharmaceutics has shown promising potential to fulfill this task which can further enhance drug bioavailability and reduce required doses.
5. *Controlled/Sustained Delivery*: Nanocarriers can be formulated to release drugs in a controlled manner but it must be kept in mind that reducing the size of the drug particles to nanolevel in most cases leads to poor drug encapsulation or loss of control for sustained drug release. Therefore, for controlled drug delivery nanoparticles of appropriate type, size, and surface characteristics must be used. Generally, 50–200 nm particles are considered as optimal for sustained release, while use of some biodegradable polymers or PEGylation may also help in extending the drug release time.
6. *Enhanced Stability*: Most of the herbal drugs are prone not only to deterioration during its storage but also due to the harsh chemical, physical, and environmental factors along their way to their site of action within the body. Entrapment within the nanocarriers may protect active phytomolecules as well as reduce unwanted reactions, thus minimizing loss of activity or production of toxic metabolites.
7. *Better Compliance*: In many cases, not only due to the mentioned advantages, nanocarriers offer a more convenient patient compliance during the treatment.

All the mentioned advantages have led to augmented bioavailability, improved therapeutic index, enhanced drug efficacy of phytopharmaceuticals, thus adjusting pharmacodynamics, pharmacokinetics, and immunogenicity of applied phytodrugs [45]. However, there are also many drawbacks connected with the use of nanoformulations and therefore one must be very careful when using those dosage form having in mind that they are relatively new and that some undesired effects may be observed only after enough time period.

2.2.2 Limitations of Nanophytopharmaceuticals

Different types of nanoparticles have different advantages and limitations and beside the type of particles this also may depend on loaded drug or more particularly on its molecular weight, water/lipid solubility, and permeability. However, some general observations may be drawn [3, 5, 24, 40–44, 46, 47]:

1. *Toxicity*: Nanosize of NPs is not only their main advantage but may also be their main drawback. Namely, due to the nanoscale dimensions they can easily penetrate many bodies barriers (skin, lungs, blood vessel wall, blood–brain barrier, body tissues, and organs), after which they can simulate the intracellular organelles or biomolecules (parts of NPs, like nucleic acids, antibody fragments, peptides, and proteins, can function as antigens). This can lead to detrimental biological interactions and toxicity. Many different unwanted consequences of nanoformulations have been observed depending on their composition like pseudo-allergy reactions (non-Ig E-mediated hypersensitivity), protein denaturation, mitochondrial disconcertion, generation of reactive oxygen species (ROS), oxidative cell damage, and perturbation of phagocytic functions. In addition, there are also toxic effects connected to chronic exposure to nanostructures since they may be uptake by nucleus, neuronal tissue, the reticuloendothelial system (RES), as well as connected to the generation of neoantigens that may provoke organ enlargement and dysfunction. Therefore, biocompatibility and potential cytotoxicity, immunotoxicity, and genotoxicity of the used nanomaterials must be thoroughly evaluated, especially long-term and in vivo toxicity, even before starting with preclinical and clinical studies. However, this is a difficult task since NPs are relatively new in the medical market and currently neither good in vivo models nor list of required guidelines and standardized safety test variables established for NP toxicity and adverse effects are established. Some of the main reasons for this are that classical drug toxicity assays are probably inadequate for NPs, while predicting of potential hazards for particles of such small diameters is difficult as their properties are different from those of bulk scale materials.
2. *Complex Composition and Chemical Nature of Herbal Extracts*: While in conventional drug delivery there is only one active component that needs to be incorporated within the nanocarrier, herbal medicine pharmaceutical action is commonly caused by synergistic action of a mixture of numbers of phytoconstituents with different molecular weights, solubilities, therapeutic potentials, and chemical classes. This brings big challenges in choosing and optimizing nanoformulation strategy. This partially may be overcome by fractionation of herbal extracts.
3. *Cost*: Formulating, optimizing, and characterizing of NPs is an expensive and complicated process, which requires many special ingredients, certain complex and expensive instruments and optimal conditions, which significantly increase the cost of nanoformulations. Since one of the primary advantages of phytopharmaceuticals is their low cost, this can greatly limit the number of competitive nanoformulations in the market unless optimal cost/benefit ratio is reached.
4. *Stability*: Although NPs may improve stability of phytopharmaceuticals in some cases they may suffer from inherent instability if applied type of NPs is prone to aggregation or swelling.
5. *Low Drug-Loading Capacity*: In most cases drug-loading capacity of the nanocarriers is low, which although may be compensated by targeted delivery can lead to overexposing to large amounts of different chemicals used in NPs and therefore to adverse effects.
6. *Dispersity*: NPs characteristics may greatly vary due to manufacturing method if produced NPs do not have the same diameter since many secondary properties such as targeting properties, drug release rate, biocompatibility, toxicity, and in vivo behaviors are affected by NP size. In some cases, even the particles with same average size but different polydispersity index may differ in secondary NP properties.
7. *Target-Specific Drug Delivery*: Although integration of target ligands on the surface of NPs can enhance targeted delivery of encapsulated phytochemicals, this is not at all easy to achieve. During the path from the application site to the action site nanocarriers need to cross several physical and biological barriers, which are usually of different types which demand to use multiple advanced nanoformulation strategies in NPs production.
8. *Characterization*: Since properties of NPs may widely vary not only due to the manufacturing procedure and batch-to-batch basis but also within the same batch, it is very difficult to characterize them. In addition, methods and instruments for their characterization are complex and expensive in most cases.
9. *Manufacturing*: The manufacturing of NPs demands not only appropriate and usually expensive equipment but also efficient controlling physicochemical properties on a batch-to-batch basis and keeping them. For NPs this is of the utmost importance since at the nanoscale any polydispersity of nanomaterials may lead to more significant variations in physical and chemical properties than for bulk materials. Yet, it is very difficult to achieve such control.
10. *Biodegradability*: Although biodegradability is dependent on the type of NPs, problems can exist for both types of behavior. In the case of nonbiodegradable NPs there is concern when lack of cellular uptake ability, and/or uncertain ligand-tagging ability can lead to the accumulation of NPs within the body. Conversely, when NPs are biodegradable and biocompatible, they can be

easily digested, or their characteristics may be altered or degraded within the body.

11. *Ethical/Regulative Concerns*: Nanomedicine as a relatively new field of science carries on considerable uncertainty and difficulty in regulating nano-based treatment and its applications, as well as some ethical concerns and questions which need to be cleared and regulated in order to expand nanotechnology's application in medicine.

2.3 Production Methods

2.3.1 Production Methods

NPs may be produced by many different methods among which some are classical, based on physical and chemical principles of synthesis, while nowadays there are also some biological methods (the so-called green synthesis) which are considered to be cheaper, faster produced, less toxic, more stable, and environmentally friendly compared to classical ones [15, 48]. Since there are many production methods careful choice must be done between them as well as careful design and optimization of formulation process in order to obtain NPs with desired properties. Generally production technologies may be classified according to the approach in two groups, as top-down and bottom-up techniques. However, in the recent years there is a new approach aimed to obtain NPs by sustainable, pollution-free method of green synthesis.

Top-Down Approach: This group of methods is based on size reduction of primary material to nanosize by different external mechanical and/or chemical processes. Those processes are generally consuming more energy, while the size and other properties of obtained NPs are hard to regulate [48]. The most used techniques in this group are grinding, milling, and laser shredding [15].

Bottom-Up Approach: The essence of those group of methods is application of thermodynamically controlled chemical reactions to build up NP structure from atoms or molecules [49]. Those methods are somewhat easier to regulate the properties of obtained NPs. Some of the most important methods based on this approach are chemical vapor condensation, chemical vapor coating, sol-gel, and spray pyrolysis [48].

Green Synthesis: Green nanotechnology is based on the fact that in nature there are numerous materials that are produced on nanoscale. This approach is based on the production of nanoparticles from living cells, plant cells, bacteria, cyanobacteria, seaweeds, fungi, etc. [48]. The main advantages of green synthesis are simplicity, cost-efficiency, low energy requirement, faster and large-scale synthesis, minimal use of hazardous substances, prolonged stability [48, 50]. Some of the green synthesis methods are irradiation, bio-assisted synthesis methods, mixed-valence polyoxometallates, polysaccharides, tollens [48].

However, it must be noted that not all methods used for production of NPs are good for nanophytomedicine formulation. Best methods are high-pressure homogenization, supercritical fluid, coprecipitation, self-assembly, complex coacervation, solvent-emulsification-diffusion, solvent displacement, and salting-out methods [3].

2.4 Characteristics and Characterization of NPs

Naturally, basic characteristics of the NPs are determined by their type. However, there are some general characteristics that define overall behavior independently of NP type. Those main parameters, that have crucial influence on NPs characteristics, are their size, size homogeneity (polydispersity index), surface charge (zeta potential), hydrophobicity or hydrophilicity, chemical and physical stabilities, drug encapsulation efficiency, and drug-loading capacity, as well as their targeting ability [51]. Those characteristics are responsible for NPs bioavailability, biodistribution, stability, and metabolism. The size of NPs will determine their ability to enter into different cells and immune system response as well as their pharmacokinetics [50]. Surface charge determines the NPs cellular absorption and cytotoxicity [52]. The hydrophilicity or hydrophobicity of NPs is controlling various biological responses like cellular absorption, interactions with plasma proteins, particle elimination, including phagocytosis, as well as immunological responses [53]. For example, smaller NPs have faster release of bioactive components, while NPs coated with hydrophilic compounds are protected against phagocytosis [21]. Chemical and physical stability are not only important during the formulation and storage processes but even more in order to provide that encapsulated drug could reach the site of release by overcoming different barriers at this trip [54]. The entrapment efficiency and loading capacity beside of the type of NPs and encapsulated drug hydrophilicity/hydrophobicity are also governed by the MW of loaded active compound. Generally, compounds with lower MW have better entrapment efficiencies compared to those with higher ones [35].

Other, also very important properties of NPs are their targeting ability and toxicity. The targeted drug delivery is among the most important properties of NPs since one of the advantages of medical nanotechnology is minimizing side and toxic effects. Having in mind that drug encapsulation of NPs is relatively low targeted delivery becomes even more important factor. There are two strategies for targeting: passive and active [55]. In passive targeting nanodelivery is attained by modifying NPs physiochemical characteristics (their hydrophobicity and pH) and due to the enhanced retention and permeability, while drug release from NPs is accomplished either by diffusion from NPs or by their erosion [22, 56]. The active delivery strategy applies different biomarkers (i.e. RNAs, proteins, lipids, carbohydrates, small metabolite molecules) to attach loaded NPs at specific sites [57]. Another mechanism is stimuli-responsive active targeting where some materials that may be activated by certain external stimuli are used (pH, pI, magnetic fields, heat, ultrasound, light, enzymes, etc.) [58].

Toxicity of the NPs is a crucial parameter for their application in medicine. The main forms of NPs toxicity are mutagenicity, immunotoxicity, neurotoxicity, hepatotoxicity, nephrotoxicity, reproductive, GIT, and pulmonary toxicity [43]. Therefore, it is important to precisely determine by which factors it is governed, and how to diminish those effects.

First reason for NPs toxicity is their chemical composition which in case of nonbiodegradable or slowly biodegradable components may cause their accumulation within the body and therefore lead to toxicity. Similarly, in the case of the metal NPs those made of soluble forms are more toxic than insoluble ones [59, 60]. Even the crystalline form has effect on NPs toxicity [43].

The size of NPs is second important parameter that may induce its toxicity but it in addition may also have an effect on stability, solubility, mucoadhesion, permeability, and cellular uptake of NPs, as well as on loading efficiency, release profile, and biodistribution of encapsulated phytodrug [40]. With size decrease there is large increase in surface area, and therefore particle reactiveness, which may affect NP distribution, absorption, and elimination [18, 43]. There are reports that optimal size of NPs is in the range of 100–200 nm, since smaller particles can cause cell toxicity, while larger ones can be eliminated from the systemic circulation by the in situ P-gp efflux pump [61]. In addition, NPs with diameters less than 50 nm can enter almost all body tissues, while those with larger size are mainly toxic to the reticuloendothelial system (RES) (liver and spleen) [62]. Furthermore, production of NPs with smaller size usually needs to use a higher quantity of surfactants, which may increase NPs toxicity [40]. Another parameter, polydispersity Index (PDI) which represents the population's size distribution, is also connected to the NPs size and may also influence their toxicity. The numerical value of PDI may vary from 0.0 to 1, for uniform and highly disperse particle size, respectively. In general, acceptable value of PDI is equal or lower than 0.3 [61].

However, not only the size but also the shape of NPs affects their in vivo toxicity since it can change their optical, electromagnetic, catalytic, or flow properties, as well as their endocytosis or phagocytosis. Generally, spherical NPs are less toxic than other shapes of the same NP type [18, 43]. Surface charge of NPs is important as it determines the interactions with the biological membranes which are all electronegative [63]. Therefore, positively charged NPs have higher toxicity, tendency to agglomerate and permeability [43]. Finally, the surface coatings and targeting ligands, mainly done for NPs functionalization and targeting delivery, can also be cause of toxicity, but in turn sometimes surface coatings can stabilize NPs and diminish their toxicity [42, 64].

In order to determine the main characteristics of the NPs appropriate characterization methods must be used [40, 42, 65].

The NPs size can be measured by dynamic light scattering (DLS) method, scanning electron microscope (SEM), and transmission electron microscope (TEM) analysis. The PDI can be determined by the dynamic light scattering (DLS). The surface properties of NPs can be determined by different methods: morphology analysis by SEM and TEM as well as by more powerful atomic force microscopy (AFM), specific surface area by Brunauer-Emmett-Teller (BET) analyzer, while surface charge by zeta potential analyzer. Zeta potential may be used to determine NPs stability.

There are also some other methods used for characterization of different types of NPs like Fourier transform infrared (FTIR) spectroscopy, nuclear magnetic resonance (NMR) spectroscopy, UV–Vis-spectroscopic analysis, infrared (IR) spectroscopy, differential scanning calorimetry (DSC), thermogravimetric analysis, solubility studies, in vivo drug release studies, X-ray diffraction analysis, etc.

3 Types of Nanoparticles

There are many different types of NPs produced and investigated as delivery vehicles for phytopharmaceuticals. Their classification may be done according to various parameters.

The first way to classify NPs may be done *according to their state*, into two main types: *liquid* (i.e. nanoliposomes, nanoemulsions, nanopolymersomes) and *solid* (i.e. nanocrystals, lipid NPs, polymeric NPs) [22].

Second way to classify NPs is *according to their nature* on [13]:

1. *Organic NPs* (lipid and polymer-based NPs: solid lipid nanoparticle (SLN), nanostructured lipid carrier (NLC), nanoemulsion (NE), nanocapsule (NC), lipid drug conjugate (LDC) or polymer drug conjugate (PDC), liposome, transferosome, niosome, ethosome, dendrimer, micelle, nanosphere (NS), nanocrystals, phytosome or herbosome, Self-nanoemulsifying drug delivery system (SNEDDS), self-microemulsifying drug delivery system (SMEDDS), nanofiber, polymersome (PS), cubosome, biopolymer-based nanocarrier (BBN), pure biopolymer nanoparticles (PBN), biopolymer-based hydrogels (BBH), biopolymer drug conjugate (BDC));

2. *Inorganic NPs* (metal nanoparticle (MN), mesoporous silica nanoparticle (MSN), magnetic nanoparticle (MNP), calcium carbonate ($CaCO_3$) nanoparticle, nanotubes (halloysite clay nanotubes (HNT), carbon nanotubes (CNT)));
3. *Hybrid NPs*; and
4. *Biological NPs*.

According to the material used for manufacturing the core of a nanocarrier, NPs may be classified as [24, 29, 66–68]:

1. *Non-organic NPs* which may be further classified as:
   (a) *Metal based*:
      – Pure metal (gold, silver, iron, zinc, copper, etc.)
      – Metal oxide based (iron oxides, etc.)
      – Metal salts based (quantum dots)
   (b) *Carbon based* (nanotubes, fullerene, etc.)
   (c) *Non-metal based* (CaCO3, halloysite clay nanotubes, mesoporous silica NPs, etc.)
2. *Organic NPs* which may be further classified as:
   (a) *Lipid based*:
      – Vesicular (liposomes, niosomes, transfersomes, ethosomes, cubosomes, etc.)
      – Non-vesicular (SLNs, NLCs, liquid crystals, etc.)
      – Emulsions (nanoemulsions, SMEEDS, SNEEDS, etc.)
   (b) *Polymer based* (nanospheres, nanocapsules, polymeric NPs, dendrimers, micelles, polymersomes, nanofibers, biopolymer-based hydrogel NP, etc.)

According to *biodegradability* NPs may be classified into two groups [47]:

1. *Biodegradable* (i.e. liposomes, SLN, NLC, niosomes, nanoemulsions, nanocrystals, etc.)
2. *Nonbiodegradable and partially biodegradable* (polymeric NPs, polymersomes, carbon nanotubes, magnetic nanoparticles, silica NPs, etc.)

As it can be seen there are many different ways to classify NPs and one NPs can be classified according to different parameters. However, more important is that NPs are effective and safe for use in nanodelivery of phytopharmaceuticals. Up to now, only few of the investigated NPs have proven itself as safe, biocompatible, biodegradable, and effective and those are mainly lipid and polymeric based NPs, such as liposomes, micelles, nanoemulsions, solid NPs, transferosomes, ethosomes, niosomes, NLCs, SLNs, and flexible nanovesicles, poly(lactic-co-glycolic acid) (PLGA) NPs [2, 42, 69].

The most investigated nanocarriers for phytopharmaceuticals are PLGA NPs, nanoemulsions, phytosomes, liposomes, SLNs, NLCs, nanosuspensions, nanofibres, ethosomes, SNEEDS, nanomicelles, CNT, etc. [40].

3.1 Lipid-Based Nanoparticles

Generally, lipid-based NPs are most promising type of nanocarriers since they are compatible with natural body lipids and because lipids in the digestive tract act as drug penetration enhancer since they support drug solubilization in the stomach and thus circumventing first-pass metabolism by diffusion of the drug through the lymphatic circulatory system [13, 70]. Main types of lipid-based NPs are liposomes, phytosomes, ethosomes, niosomes, transfersomes, cubosomes, solid lipid nanocarriers, nanostructured lipid carriers, nanoemulsions, SNEDDS, SMEDDS, and lipid micelles.

3.1.1 Liposomes

Liposomes are spherical self-assembled vesicles consisting of one or more phospholipid polar bilayer membranes formed around encapsulated amount of solvent in the core region. They can have one, several, or multiple concentric membranes which are composed of cholesterol and phospholipids (either synthetic or natural), although in some cases other constituents may be added to modify their chemical and physical properties, while encapsulated solvent is commonly water. Since polar lipids have lipophilic and hydrophilic groups on the same molecules, both hydrophilic and hydrophobic (lipophilic) drugs may be encapsulated, hydrophilic inside the aqueous compartment, and lipophilic in the membrane [3]. Their size is from 20 nm to a few micrometers, but nanoliposomes generally have diameters up to 450 nm [71]. Difference between liposomes and micelles is in the number of layers in the lipid membrane since the micelles are monolayer lipid vesicles.

There are a few different types of liposomes, depending on the preparation method (reverse-phase evaporation vesicles or vesicle-extruded technique), size (small, intermediate, or large), the number of membranes (or lamellarity) (uni-, oligo-, and multilamellar vesicles), and surface charge [41]. The main types are small unilamellar vesicles (SUV), large unilamellar vesicles (LUV), multilamellar vesicles (MLV) (multiple concentric membranes with structure like onion), and multivesicle vesicles (MVV) (with several liposomes inside larger liposome), as well as cochleate vesicles (Fig. 2) [5]. The size of SUVs is 20–100 nm, LUVs are >100 nm, while MLV are ranged from 500 nm to 5 μm [5, 22]. The multivesicular liposomes are mainly used for sustained drug release after parenteral administration, unlike the lamellar liposome technology [72].

The type of formed liposome and its size is dependent on production conditions. Thus, unilamellar vesicles are usually produced in dilute solutions of surfactants, while multilamellar ones are formed in highly concentrated systems [3]. Liposomes can be very easily produced by disturbance of the lipid film in aqueous medium which may be caused by large shear force, produced via different techniques (i.e. sonica-

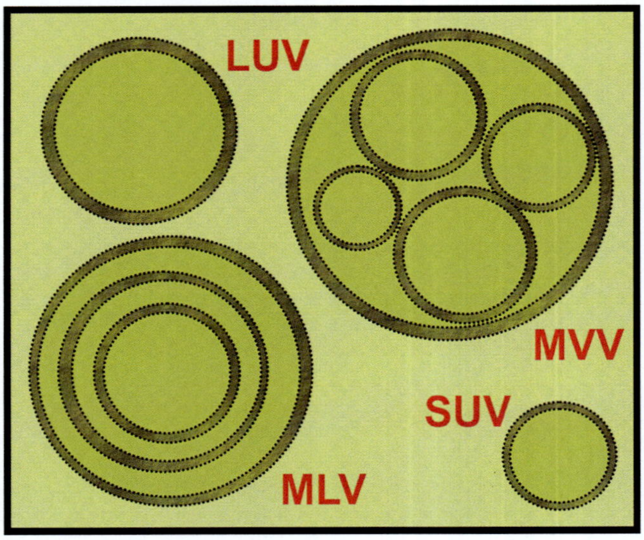

Fig. 2 Main types of liposomes

tion, stirring, etc.). They may be also produced by several freeze-thaw processes.

Advantages of nanoliposomes are numerous making them to be one of the most used nanocarriers. First of all, they are biodegradable, nontoxic, nonimmunogenic, and highly biocompatible which all reduce the unwanted side effects connected to use of nanocarries, and classify them as safe nanocarriers [73]. Second, due to their dual structure drug solubility of both lipophylic and hydrophylic components is increased, with possibility to encapsulate multiple phytochemicals into the same vesicle [23]. They protect encapsulated drug from harsh external environment, like enzymatic degradation, and chemical or immunologic inactivation, preventing a drug to be metabolized before it reaches the target tissues, which enhances drug stability and delivery efficacy [41]. Due to the structural similarity of liposome membrane to the cellular membranes they can transport any medication via endocytosis and are increasing absorption of the drug within a cell, so they are usually used to encapsulate compounds that cannot pass biological membranes [1]. Encapsulated drugs can be distributed throughout the body and released at desired site in a controlled manner [24]. They provide relatively low particle size of NPs in a cost-effective manner, good in vitro and in vivo stability, and long shelf-life [5, 17, 69]. In addition, they have relatively easy controllable properties that may be optimized according to application by changing the composition of the bilayer components, or by size and charge of liposome as well as by functionalizing and modifying the surface through addition of different functional groups, antibodies, or membrane proteins [5, 18]. There are many possible ways to modify the surface of liposomes [72], to enable targeted delivery which may significantly change/improve pharmacokinetic of drugs and increase their therapeutic index.

Disadvantages of liposomes are not so numerous and there are several ways to overcome some of them. Liposome NPs have short half-life in the circulatory system, either because they are recognized as foreign particles and rapidly absorbed in liver and taken up by macrophages, or due to their disintegration by electrostatic, hydrophobic, and van der Waals forces, which all reduce their efficacy [5, 42]. Fortunately, this problem may be diminished by coating liposomes with inert polymers like poly(ethylene glycol) (PEG), monosialoganglioside, or poloxamer, which induce steric stabilization (making them "stealth") and therefore reduce their clearance by the reticuloendothelial system (RES) and fagocitosis [1, 66]. Other way to increase their half-life is modifying liposome membrane composition by incorporating cholesterol, whose concentration is a main factor that regulates the membrane properties, or other chemicals (i.e. polyvinylpyrrolidone polyacrylamide lipids, glucuronic acid lipids, or phospholipid distearoylphosphatidylcholine (DSPC)) and by better control of their size [5, 66]. Next problem that may be encountered with liposomes is lower loading efficiency for hydrophobic phytochemicals although they are to be suitable for such application [23]. Namely, those non-polar drugs may affect the integrity of these nanovesicles [5]. The nanoliposomes may have higher production cost (especially those with targeting abilities) and there are also some problems connected to targeted delivery for chemotherapy applications [72, 74]. Finally, liposomes have low stability and quick leakage of the hydrophilic molecules [73].

Major *synthesis methods* for liposome NPs formulations are layer-by-layer electrostatic deposition technique, gentle hydration, sonication, extrusion, freeze-thawing, ether injection, microemulsification, and microfluidization [23, 56]. Layer-by-layer electrostatic deposition technique applies adding of charged polymers to a solution with a charged template structure (which may be either liposomes composed of charged polar lipids, emulsion droplets stabilized by ionic emulsifiers, or hydrogel particles composed of charged biopolymers). Liposomes may be produced by hydration of dried lipid films in aqueous solution (gentle hydration) [56]. During extrusion nanoliposomes may be formed if high power, long period, and small pore size extruder filtration are applied. However, this method and sonication are mainly for laboratory scale [56]. If liposomes are to be produced at industrial scale, microfluidization method can be applied since it has high reproducibility for large-scale preparation in addition to advantages like adjustable size of liposomes and avoiding their use of toxic solvent in preparation process. It involves a microfluidizer with fine orifices through which the stream is forced to flow by high pressure [42].

In conclusion, liposomes are generally considered as good candidates for phytopharmaceutical delivery especially for herbal drugs with poor lipid solubility and high permea-

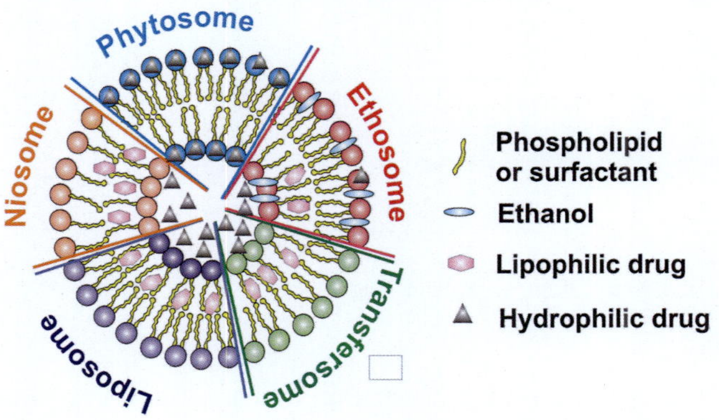

Fig. 3 Schematic representation of lipid-based NPs

bility [75]. The main delivery routes of nanoliposomes are oral, nasal, topical, and parenteral [42]. Schematic representation of liposomes is given in Fig. 3, while some of their main applications in nanoherbal drug delivery are given in Table 1.

3.1.2 Phytosome or Herbosome

Phytosomes may be considered as type of liposomes, since they are structurally almost identical, except for the entrapment mechanism of the phytopharmaceuticals. While in liposomes they are dissolved in the medium within the core, in phytosomes the active material is a vital part of the membrane where it is covalently attached by hydrogen bonds with

Table 1 Types of novel lipid-based nanocarriers and encapsulated phytopharmaceuticals

Nanocarriers	Phytopharmaceuticals	Ref
Liposomes	α-Tocopherol, avicequinone-B, actein, ampelopsin, artemisinin, artesunate, ascorbic acid, baicalein, baicalin, berberine, betulinic acid, breviscapin, brucine, camptothecin, capsaicin, carotenoids, catechins, colchicine, combretastatin A4, curcumin, cytarabine, daunorubicin, dihydroartemisinin, doxorubicin, E-resveratrol, epicatechin, epigallocatechin gallate, fisetin, gambogenic acid, garlicin, ginsenoside, glycyrrhizic acid, imperatorin, irinotecan, letrozole, magnolol, morphine sulfate, muramyl tripeptide phosphatidylethanolamine, oleanolic acid, naringenin, palmatine, polydatin, paclitaxel, polydatin, propolis flavonoids, psoralen, quercetin, resveratrol, rutin, shikonin, silibinin, silymarin, sterols, strychnine, thymoquinone, topotecan, ursolic acid, usnic acid, vinblastine, vincristine, vinorelbine, vomicine, wogonin	[12, 13, 17, 40, 47, 72, 83–85]
Phytosomes	18-β glycyrrhetinic acid, apigenin, berberine, bilobalide, boswellic acid, catechin, celastrol, chrysin, curcumin, dimeric flavonoids, epicatechin, epigallocatechin, epigallocatechin gallate, escin ß-sitosterol, flavonoids, gallic acid, ginkgo flavone glucosides, ginkgolides, gingerol, ginsenosides, isorhamnetin, kaempferol, lawsone, luteolin, morin, naringenin, persimmon, polyphenols, procyanidins, quercetin, resveratrol, rutin, silybin, silymarin, sinigrin, triterpenes, umbelliferone, vinblastine, visnadine, vitexin-2″-O-rhamnoside	[12, 13, 17, 40, 47, 72, 73, 85, 86]
Ethosomes	Aloperine, ammonium glycyrrhizinate, apigenin, capsaicin, colchicine, cryptotanshinone, curcumin, curcuminoids, diterpene triepoxide, fisetin, ginsenoside Rhl, glycyrrhizic acid, lemannine, leucocyanidin, ligustrazine, matrine, methylnicotinate, piperine, podophyllotoxin, proanthocyanidin A2, psoralen, rosmarinic acid, sericoside, silybin, silymarin, sophocarpine, sophoridine, tetrandrine, thymoquinone	[12, 13, 17, 40, 44, 47, 85]
Niosomes	β-Sitosterol, ammonium glycyrrhizinate, borreline, curcumin, embelin, gallic acid, gymnemic acid, isorhmnatin, lawsone, lycopene, papain, quercetin rosmarinic acid, ursolic acid, visnadin	[12, 13, 47, 72, 81, 87]
Transfersomes	Apigenin, capsaicin, colchicine, curcumin, emodin, paeonol, quercetin, ruscogenin, rutin, silybin, vincristine	[13, 17, 88]
Polymersomes	Bacosides, doxorubicin, lactoferrin, oxymatrine, paclitaxel	[13]
Cubosomes	Capsaicin, curcumin, hinokitiol, piperine	[13, 47]
SLNs	Andrographolide, astaxanthin, bacoside, berberine, bixin, caffeic acid, camptothecin, cantharidin, carvacrol, coumarin, cryptotanshinone, cucurbitacin B, curcumin, curcuminoids, ellagitannins, emodin, epigallocatechin, epigallocatechin-3-gallate, ferulic acid, genistein, hesperidin, hydroxycitric acid, indirubin, letrozole, luteolin, morin, myricetin, myricitrin, narngenin, noscapine, oridonin, paclitaxel, piperine, podophyllotoxin, puerarin, quercetin, resveratrol, rosmarinic acid, sesamol, silibinin (silybin), silymarin, tetrandrine, thymoquinone, transresveratrol, tryptanthrin, tripterygium, triptolide, umbelliferone, witepsol	[12, 13, 20, 22, 35, 40, 44, 47, 69, 72, 81, 85, 86, 89]
NLCs	β-Elemene, baicalein, baicalin, berberine, capsaicin, cinnamaldehyde, citral, curcumin, diosgenin, docetaxel, ferulic acid, hesperidin, lycopene, oleuropein, naringenin, nicotinamide, noscapine, quercetin, resveratrol, silymarin, thymoquinone, triptolide, ursolic acid, vincristine, zerumbone	[12, 13, 20, 44, 47, 76, 81, 89]
Nanoemulsions	5-Hydroxytryptophan, β-carotene, andrographolide, anthocyanin, apolipoprotein E, berberine, betulinic acid, baicalin, capsaicin, carotenoids, carvacrol, catechin, coenzyme Q10, coixenolide, colchicine, curcumin, docetaxel, emodin, epicatechin, ethoxyquin, eugenol, ferulic acid, flavonoids, furocoumarin, glutathione, glycyrrhizin, huperzine, hydroxysafflor yellow A, hydroxytyrosol, lapachol, lignans, lipoic acid, luteolin, lycopene, melatonin, naringenin, oleanolic acid, ursolic acid, paclitaxel, piplartine, psoralen, pterostilbene, quercetin, resveratrol, salicylic acid, silybin, silymarin, squalene, tarragon, triptolide, uric acid, vitamin A, vitamin E, vitamin K	[1, 12, 13, 40, 47, 74, 76, 79, 89, 90]

(continued)

Table 1 (continued)

Nanocarriers	Phytopharmaceuticals	Ref
SNEDDS	β-Artemether, akebia saponin, apigenin, berberine, baicalin, camptothecin, cyclovirobuxine-D, curcumin, daidzein, ellagic acid, embelin, ginkgolide, hesperetin, kaempferol, lutein, morin, myricetin, naringenin, oleanolic acid, oridonin, puerarin, quercetin, resveratrol, rutin, silymarin, thymoquinone, withaferin A, wurenchun	[12, 13, 21, 40, 69, 76]
SMEDDS	9-Nitrocamptothecin, β-artemether, baicalein, camptothecin, curcumin, lutein, pueraria flavone, silymarin	[12, 13, 72]
Nanomicelles Lipid	Apigenin, circinal–icaritin, curcumin gambogic acid, glycyrrhizin, naringenin, quercetin	[3, 26, 40, 85]

the polar head of phospholipids [5, 40, 47]. Phytosomes are developed and patented by Indena, Italian pharmaceutical and nutraceutical company, to increase the bioavailability of water soluble and/or large MW phytopharmaceuticals (like flavonoids, polyphenols, tannins, terpenoids, xanthones, etc.) by their complexation with phospholipids. Those molecules have limited ability to be absorbed by passive diffusion or to pass across the lipid biological membranes which cause their poor bioavailability [13]. Within the phospholipid complexes, phytopharmaceuticals chemically interact with the polar head of used phospholipid (any phospholipid may be used like phosphatidylcholine, phosphatidylserine, phosphatidylethanolamine, etc.). The most preferred phospholipid is phosphatidylcholine due to its low toxicity and good biocompatibility and it is commonly complexed with active component in 1:1 or 2:1 ratio depending on complexed component [17, 76]. Phytosomes may be administered orally or topically, while drug release is by bilayer diffusion, surface desorption, or degradation.

Advantages of phytosomes are better absorption of topically or orally applied phytopharmaceuticals, their better transition across the enterocyte cell membrane (from a water-soluble to a lipid-soluble state), as well as their protection from stomach fluids and gut bacteria, which all leads to a better pharmacokinetic and pharmacodynamic profile [13, 77]. In addition, lipid-insoluble phytochemicals are easily administered with enhanced bioavailability [73].

Disadvantages include somewhat tedious and time-consuming production, low encapsulation efficiency, leaching of the phytoconstitute from lipid core, and relatively lower stability and shelf-life [73].

Common *synthesis method* for phytosome NPs formulation is thin-film hydration [47]. Schematic representation of phytosomes is given in Fig. 3, while some of their main applications in nanoherbal drug delivery are given in Table 1.

3.1.3 Ethosomes

Ethosomes may be also considered as a novel type of liposomes with difference that they contain relatively high concentration of alcohol (30–45%) (ethanol and isopropyl alcohol) beside water within the core region. Their membrane is composed of elastic phospholipids like phosphatidylcholine, phosphatidylserine, phosphatidylethanolamine, or phosphatidylglycerol [13]. This combination of phospholipids and a high concentration of ethanol provides higher deformability of vesicles (they are softer), while ethanol promotes the fluidization of intercellular lipid bilayers making them preferable for topical, transdermal, and systemic applications [13, 76].

Ethosomes *advantages* include, beside the mentioned higher deformability, enhanced transdermal transport efficiency enabling phytochemicals to reach deeper layers of the skin and even systemic circulation. They have ability to improve entrapment efficiency of both, hydrophilic and lipophilic phytopharmaceuticals. In addition, they are non-invasive delivery carriers, easy to prepare, safe, and effective, with prolonged physical stability compared to liposomes, and relatively smaller size than other vesicles [47, 76, 78].

Disadvantages of ethosomes include low stability due to ethanol evaporation and leakage of loaded active compounds from ethosome core [73]. Those problems may be diminished by combining ethanol with trehalose and propylene glycol [13]. In addition, ethosomes have low yield which highly diminish their economic, and slow, rather sustained drug delivery so they are not appropriate for drugs that require high blood levels [78].

Preparation methods for ethosomes may be classified as cold and hot which are simple and do not require sophisticated equipment [78]. Schematic representation of ethosomes is given in Fig. 3, while some of their main applications in nanoherbal drug delivery are given in Table 1.

3.1.4 Niosomes

Niosomes are yet another NP vesicle with a form similar to liposomes. Their bilayer structure instead of phospholipids is made up of alkyl or dialkyl polyglycerol-based nonionic surfactants (amphiphilic lipids like alkyl amides, crown ester, poly-oxyethylene alkyl ether, spans, and tweens). Vesicles are either unilamellar or multilamellar and their size is 10–100 nm for SUV, 100–3000 nm for LUV, while other types (MLV and GLV) have sizes larger than 5 μm and cannot be considered as NPs [24]. They can also bind both hydrophobic and hydrophilic compounds and compared to liposomes they have enhanced delivery. Therefore, niosomes are usually used as replacement for liposomes since they overcome some of liposome NPs main limitations like stability problems, oxidation susceptibility, and high price.

Their *advantages* include flexibility, relatively low cost, increased drug solubility and protection, ability for controlled release of active compounds, restriction of drugs action to specific locations, low toxicity, and higher stability [13, 21].

Disadvantages of niosomes are few, like quite negligible escape of the drug from their aqueous core during dispersion and aggregation, which has made them one of the most widely studied nanocarrier delivery systems [72].

Production methods include sonication, thin-film hydration, microfluidization, multiple-membrane extrusion, remote loading, pronisome pre-formulation technique, reverse phase evaporation technique, and bubble method [47].

The main delivery routes of nanoliposomes are topical, transdermal, nasal, intravenous, ocular, and pulmonary [1]. Schematic representation of niosomes is given in Fig. 3, while some of their main applications in nanoherbal drug delivery are given in Table 1.

3.1.5 Transferosomes

Transferosomes are relatively new classes of liposomes developed to enhance their properties. They can penetrate stratum corneum with significantly increased efficiency compared to liposomes or ethosomes due to their ultra-flexible lipid-based elastic vehicle with highly deformable membranes [13]. Transferosome vesicles high elasticity and deformability are caused by their composition based on phospholipid and a single chain surfactant which endows them not only with flexibility but also with ability to entrap hydrophilic, lipophilic, and amphiphilic active components. Therefore, they are mostly utilized as transdermal nanocarriers for small molecules, and phytopharmaceuticals with a wide range of solubility [17]. Moreover, transferosomes production process is simple, easy to scale-up, and without harmful additives. However, they are not stable (predisposed to oxidative degradation) and have problems with the purity of used natural phospholipids for production [13]. Schematic representation of liposomes is given in Fig. 3, while some of their main applications in nanoherbal drug delivery are given in Table 1.

3.1.6 Cubosomes

Cubosomes are cubic like, viscous liquid vesicles, made of amphiphilic lipids (unsaturated monoglycerides) and liquid crystalline surfactants (i.e. poloxamers). Their size is in the range of 10–1000 nm, while their 3D bi-continuous structure (two distinct regions separated by surfactant bilayers) provides them with both, hydrophilic and hydrophobic domains, so they can encapsulate lipophilic, amphiphilic, and hydrophilic drugs [24]. Cubosomes have high internal surface area per unit volume (~400 m^2/g) which enable sustained drug delivery [47]. Their advantages include simple preparation, lipid biodegradability, bioadhesiveness, and controlled and targeted release [13]. Both approaches, the top-down and bottom-up may be applied for cubosomes production but most common method is dispersion or fragmentation of the cubic phases of gels in a liquid phase [47]. Due to their almost structural identity to stratum corneum they are good candidates for topical delivery of phytopharmaceuticals [13]. Some of the main applications of cubosomes in nanoherbal drug delivery are given in Table 1.

3.1.7 Lipid Nanoparticles (Solid Lipid Nanoparticles (SLNs) and Nanostructured Lipid Carriers (NLCs))

Lipid NPs are quite similar to nanoemulsions, with the exception that nanoemulsions are consisted only of liquid lipids while those are consisted of both, solid and liquid lipids. Lipid NPs can be generally categorized into two categories: solid lipid nanoparticles (SLNs) and nanostructured lipid carriers (NLCs). The main difference between those two types is their core structure, which further determines their loading capacities. Since they all have a lipid core, lipid NPs are preferable for encapsulating hydrophobic phytopharmaceuticals.

3.1.7.1 Solid Lipid Nanoparticles (SLNs)

The *SLNs* are commonly spherical particles dispersed in water or in aqueous surfactant solutions, composed of solid lipid core and surfactant single layer shell. In solid lipid core different lipids may be used such as mono-, di-, triglycerides (e.g. glyceryl monostearate, glyceryl dilaurate hydrine, tristearine), waxes (e.g. cetylpalmitate, cetyl alcohol), fatty acids (stearic acid), lecithin, and steroids, or a mixture of those, while as surfactants/emulsifiers some biocompatible stabilizers are employed (i.e. polysorbate, poloxamer sodium glycocholate, bile salts, phospholipids, sorbitan esters, fatty acid ethoxylates, etc., or a mixture of these components) [40, 42]. Each type of lipids as well as surfactants has its own advantages and limits. Used lipids, either natural or synthetic, are commonly biodegradable with melting temperatures above body temperature, while surfactants need to be biocompatible. The size of SLN is between 10 nm and 1 μm.

SLNs like many other types of lipids based nanocarriers share some of their common *advantages* like: suitability for encapsulation of hydrophobic insoluble drugs within its core, protecting the drug from harsh environment, increased drug stability with reduced toxicity, ability for surface functionalization in order to obtain targeting delivery, sustained drug release due to their highly lipophilic core. However, SLNs have some additional advantages over other types of colloidal NPs (emulsions, liposomes, polymeric NPs), since they were developed in order to overcome some of their limitations. Therefore, SLNs have easier and cheaper production process with higher reproducibility (even for different preparation methods) and easier scale-up [5, 42]. The low or no organic solvents usage in their preparation, along with use of physiological, highly degradable lipids for solid lipid core, give them lower chronic and acute toxicity, combined with higher biodegradability and biocompatibility [35, 79]. A solid lipid core significantly reduces the mobility of encapsulated drugs which greatly enhances the sustained release for a prolonged time period [21, 66]. While lipophilic drugs can be integrated in solid lipid core hydrophilic may be attached on the surface of SLNs [56]. They have ability to bypass the spleen or liver filtration for particle size in the range 120–200 nm, and easy modulation of the drug release profile, as well as easy mass-scale sterilization [35, 79]. All those advantages have made SLNs, promising candidates for the encapsulating different plant bioactive compounds such as antioxidants, antiviral, and anticancer therapeutic agents [76].

Normally, although SLNs are considered as one of the most useful lipid-based NPs, they also have some *drawbacks* like low lipophilic compounds loading capacity and drug leakage during storage, as well as burst release of the phytopharmaceuticals in the stomach at a lower pH [35, 80]. In order to increase the loading capacity, complex lipid mixtures may be employed for production of solid lipid core (i.e. mixture of monoglyceride, diglyceride, or triglyceride with fatty acids of different chain length forms) [42]. In order to prevent burst release SLNs may be coated and surface modified by heparin, albumin, PEG, and some polysaccharides. In addition, properties of SLNs are critically dependable on formulation composition and production method [35].

Production methods for SLNs are numerous but they also depend on the type of used lipids, surfactants, phytopharmaceuticals, cosurfactants, and cryoprotectants. Conversely, production method affects the stability, loading capacity, and target reachability of the loaded phytopharmaceuticals [35]. In order to produce SLNs core and disperse lipophilic compound within it, appropriate solvent and high-energy methods need to be employed like supercritical fluid extraction, nanoprecipitation, solvent injection, emulsification/solvent evaporation, double emulsification/solvent evaporation, emulsification/solvent diffusion, high-pressure homogenization, ultrasonication, melt homogenization, and microfluidization [47, 56, 72]. However, due to the lower degradation of sensitive phytopharmaceutical compounds, lower toxicity, improved stability, and higher production capacity, usually microfluidization and high-pressure homogenization techniques are employed for this type of NPs [35].

Preferable delivery routes of SLNs are oral (most preferable), dermal, and parenteral, but they may be also administered by rectal, nasal, ocular, and pulmonary routes [24, 40, 42, 69, 70].

3.1.7.2 Nanostructured Lipid Carriers (NLCs)

NLCs are second generation lipid NPs since they are developed to overcome main limitations of NLCs, low loading capacity for lipophilic drugs, and drug release during storage. Those problems arise due to structure of single solid lipid, since during SLNs formulation process, crystallization occur which leads to formation of the "brick wall" structure in the hydrophobic SLNs core. In order to produce NPs with less ordered crystalline structure which will enable to increase loading capacity of hydrophobic phytochemicals, NLCs were developed which use a mixture of solid lipids (SLs) and liquid lipids (LLs) (oils) in their hydrophobic cores. Therefore NLCs are composed by combining different SLs (hydrogenated palm oil, glyceryl monostearate, stearic acid, and cetyl alcohol glyceryl monostearate and glyceryl tripalmitate), LLs (olive oil, mustard oil, castor oil, cod liver oil, oleic acid, caprylic triglycerides, cupric triglycerides, and squalene), surfactants (lecithin and Tween 80), and cosurfactants (stabilizers) (thiomersal) [2, 47]. The type of lipid controls the stability of loaded compound, while the concentration of LLs influences the size of NPs. Smaller NPs are obtained with higher LLs concentrations but also higher is the active compound release rate [21].

Advantages of NLCs include those of SLNs but they additionally offer lower particle size, better regulation of drug release, enhanced drug-loading capacity, increased physiochemical stability, lower drug loss during encapsulation, and minimized drug ejection during storage [47, 69]. Main drawbacks of NLCs are their thermodynamical instability, as well as aggregation and sedimentation tendencies [42].

Main production methods include hot/cold/high-pressure homogenization, microemulsification, solvent emulsification evaporation, solvent displacement, solvent diffusion, solvent evaporation, phase inversion, melt emulsification, ultrasonication, lyophilization, spray drying [3, 47]. For large-scale production most effective is high-pressure homogenization, while in laboratory scale most used are emulsification and ultrasonication based methods [42]. The most preferable administration route for NLCs is oral but they also may be delivered by dermal, nasal, ocular parenteral, rectal, and pulmonary routes [13, 42]. Some of the main applications of SLNs and NLCs in nanoherbal drug delivery are given in Table 1.

3.1.8 Nanoemulsions

Nanoemulsions (NEs) are thermodynamically stable liquid dispersions of oil phase, or water phase and appropriate amphiphilic surfactants and cosurfactants (emulsifying/stabilizing agents like phospholipids, alkyl polyglycosides, PEGylated fatty acid ester, fatty alcohol) in the form of monolayer around the core. They can be either oil-in-water (O/W) (mostly used type), or water-in-oil (W/O) dispersion of two immiscible liquids, but there are also possibilities of bi-continuous W/O/W and O/W/O NEs. The volume of oil in O/W type of NEs is between 5 and 20% [81]. The size range of NE particles is between 10 and 1000 nm, but commonly it is 20–200 nm. The size of the NE particle influences its target distribution [17]. NEs with sizes lower than 100 nm require larger amounts of surfactants or cosurfactants and higher energy inputs in order to diminish the surface tension to the levels that will enable formation of small and thermodynamically stable NPs [42]. Although they are oldest delivery form and are considered as ideal carrier for lipophilic drugs, they have been improved during year in order to reduce droplet size, increase their uniformity and stability. NE are considered as a suitable vehicle for the delivery of drugs and essential oils, as well as hydrophobic and hydrophilic phytopharmaceuticals [13]. While the lipophilic drugs may be entrapped within the oil phase of O/W and O/W/O NEs, hydrophilic drugs are encapsulated within aqueous phase of W/O and W/O/W NEs [47]. Thus, encapsulated lipophilic drugs are phagocytosed by macrophage and are concentrated within the liver, spleen, and kidneys, while hydrophilic drugs are concentrated within the lymphatic system [17].

Advantages of NEs are increased stability of entrapped phytopharmaceuticals, possibility of sustained and targeted release, increased drug permeability through the skin and mucous membranes, simple production methods, lower contamination possibilities, less allergic response [47]. Their properties can be fine tuned in accordance with the entrapped phytopharmaceuticals by optimizing compositions and types of surfactants/cosurfactants, as well as the used oil [5]. The loading capacity of NEs is high, while fine tunable rheology offers possibility for extremely low viscosity [23]. They can deliver lipophilic components more easily than liposomes, and their possible delivery routes are more numerous compared to other nanocarriers [40]. In addition, they can provide higher skin permeability of lipophilic, and higher subcutaneous or intramuscular permeability of hydrophilic components [1]. Due to the dual nature (presence of polar and non-polar components) they are good solubilizing agents for components of different lipophilicities [5]. Drawbacks of NEs are not so numerous but may be serious, and they include relatively higher manufacturing cost and low thermodynamic stability due to easy aggregation and creaming [23, 40]. In addition, since surfactants are used in large quantities, they need to be nontoxic [24].

The NEs synthesis methods are generally carried out using either high-energy approaches (mechanical methods) like: ultrasonication, microfluidization, high-pressure homogenization, or low energy approaches: (chemical methods) like: spontaneous emulsification phase inversion temperature, emulsion inversion point, membrane emulsification [47]. While in mechanical methods large droplets are broken down into smaller ones by certain physical force, chemical methods are based on spontaneous formation of NEs due to hydrophobic effects of lipophilic molecules which take place in the presence of emulsifiers [22]. Homogenization and ultrasonication are mostly used in a laboratory scale, while high-pressure homogenization and microfluidization are most suitable for large amount production [23].

Administration routes of NEs include oral, transdermal, ocular, parenteral, intravenous, intralipid, intramuscular, or subcutaneous [40, 42]. Some of the main applications of nanoemulsions in nanoherbal drug delivery are given in Table 1.

3.1.9 Self-Emulsifying Drug Delivery System (SEDDS)

Self-emulsifying drug delivery systems (SEDDS) are relatively new delivery systems developed as a result of investigating self-dispersing lipid formulations and efforts to make emulsion isotropic and to increase their thermodynamic stability [72]. SEDDS are isotropic mixtures of oils and surfactants, commonly combined with one or more of cosurfactants/coemulsifiers and loaded with drugs. Concentration of oil in SEDDS is between 40–80% and of surfactant is in the range 30–40%. Those mixtures may be in solid, semiliquid, or liquid form and after oral administration they form fine O/W emulsion in contact with aqueous media like GIT fluids with particle size >300 nm [82]. There are also two nanosized types of SEDDS, self-microemulsifying drug delivery systems (SMEDDS) and self nanoemulsifying drug delivery systems (SNEDDS). Some of the main applications of nanoemulsions in nanoherbal drug delivery are given in Table 1.

3.1.9.1 Self-Nanoemulsifying Drug Delivery Systems (SNEDDS)

SNEDDS may be considered as anhydrous form of nanoemulsions since they are lipid-based anhydrous isotropic mixtures of lipid/oil (i.e. Capmul PG12, Captex 170, Captex 300, and Capmul MCM EP), surfactants (i.e. Tween-20, Tween-80, Span-80, Labrasol, and Chremophor EL), and hydrophilic cosurfactant (i.e. PEG-200, PEG-400, Labrafil CS 1499, and Transcutol) [40]. Those mixtures in aqueous media easily undergo in situ solubilization and form nanoemulsions with droplets size between 20 and 200 nm [13]. Due to their significantly smaller size they have larger concentrations of surfactant than SEDDS (40–80%) and therefore lower concentrations of oil (>20%). Used surfactants also need to have hydrophilic–lipophilic balance (HLB) value >12. SNEDDS have received significant attention due to their advantages and proved themselves as effective nanocarriers for oral delivery of highly hydrophobic compounds. Main advantages over the conventional phytomedicine drug formulations and other lipid-based NPs are improved dissolution rates and oral drug bioavailability that results in more reproducible blood–time profiles and better therapeutic ability, controlled and sustained drug release, formulation stability during storage, easy production (even in a large scale) with reduced energy requirement, global acceptance, and the regulatory approvals [13, 72, 76]. Conversely, there are some limitations connected to ocular SNEDDS delivery [40].

3.1.9.2 Self-Microemulsifying Drug Delivery Systems (SMEDDS)

SMEDDS are also lipid-based NPs composed of oil, surfactant (i.e. Cremophor RH40, Cremophor EL, or Polysorbate 80) and cosurfactant which produce microemulsions, nanoemulsions, or emulsions when injected into aqueous media with mild agitation. Only difference between SMEDDS and SNEDDS are in size of the particles which is larger for SMEDDS.

Some of the main applications of SNEDDS and SMEDDS in nanoherbal drug delivery are given in Table 1.

3.1.10 Micelles

Micelles are self-assembling colloidal dispersion of amphiphilic molecules (such as surfactants). They spontaneously self-assemble when surfactant concentrations exceed their critical micelle concentration (CMC) and temperatures reach the critical micellization temperature [18, 42]. The amphiphilic molecules may be phospholipids or some block copolymers with amphiphilic character. If lipids are used traditional micelles are formed and if block copolymers are used polymeric micelles are produced. The amphiphilic lipid molecules in aqueous solution spontaneously assemble into regular micelles in which the hydrophilic molecule part (head) is arranged outside, while the hydrophobic part (fatty acid tail) assembles in the inner core. Micelles with opposite orientation are known as reverse micelles and they may be formed in hydrophobic solutions [40]. Micelles are somewhat similar to the liposomes but differ in the number of layers in the lipid membrane since the liposomes are bilayer lipid vesicles. The size of micelles is smaller compared to other NPs and it is in the common range 10–100 nm (although sizes may vary between 2 and 400 nm), generally with spherical structure. Regular micelles may accommodate hydrophobic phytochemicals within the core region. Micelles are commonly first choice for increasing drug solubility in aqueous media and may be considered as perfect carriers since they have been used more and longer than any other NPs [5].

Advantages of micelles are enhanced hydrophobic drugs solubility (i.e. terpenoids, steroids, etc.), extended blood circulation time, increased tissue permeability and cellular uptake, low toxicity [23, 40]. In addition, micelles are simple to produce, biocompatible, and highly soluble in water [79]. While inner lipophilic core of micelles enables intracellular drug accumulation and protects it from degradation, external hydrophilic surface layer provides micelles with biocompatibility and offers possibility for grafting specific targeting ligands and other chemical moieties for active site-specific cell targeting, as well as obtaining additional thermal, pH, and photosensitivity properties [13]. The main *disadvantages* of micelles are their high production cost and instability, low encapsulation efficiency and loading capacity, and high surfactant content [23, 74]. In addition, they cannot solve the problems if natural active compound is toxic and/or instable since micelles are very dynamic systems and suffer from fast body clearance rate and low stability. However, some of these disadvantages may be to some extent reduced by the use of polymeric micelles [5].

The main micelles *production methods* are solvent evaporation, oil-in-water emulsion, dialysis, filtration, and solid dispersion [23]. Micelles may be administered by oral, nasal, ocular topical, and parenteral [42]. Some of the main applications of nanomicelles in nanoherbal drug delivery are given in Table 1.

3.2 Polymeric Nanocarriers

Polymeric nanocarriers are a large group of nanosized solid particles or dispersions, formulated as capsules, spheres, branched, core–shell, and other structures. Among them most important examples are polymeric NPs, polymeric micelles, dendrimers, nanospheres, nanocapsules, nanofibers, polymersomes, hydro- or nanogels, vesicles, prodrugs, and polyplexes. Due to broad variety of monomers, with wide variation in their structural and physicochemical properties, they offer some features that lipid-based NPs do not for production of polymeric nanocarriers with personalized and on-demand treatment possibilities [84]. Used polymers may be either of synthetic or natural base and according to this polymeric nanocarriers may be classified as polymeric nanoparticles (PNPs) or biopolymeric nanoparticles (BPNPs).

3.2.1 Polymeric Nanoparticles (PNPs)

Polymeric NPs are colloidal solid nanocarriers with size range between 10 and 100 nm which are produced from different biodegradable and biocompatible polymers. Some of the most commonly used synthetic polymers for PNPs production in phytopharmaceutical applications include poly-D, L-lactic-co-glycolic acid (PLGA), poly-D, L-lactic acid (PLA), polyglycolic acid (PGA), poly (ε-caprolactone) (PCL), and poly(ethylene glycol) (PEG) [40]. The most used one, PLGA, has both hydrophilic and hydrophobic properties, it is synthesized from glycolic acid to lactic acid, and different types of PLGA may be produced by changing their ratio. Therefore, PLGA based PNPs are hydrolyzed into lactic and glycolic acids, which are readily excreted from the body, causing minimal systemic toxicity, which facilitated FDA to approve PLGA [21, 23, 42, 85]. Another commonly used polymer as the nanoparticle's hydrophilic shell is PEG, since it is highly hydrophilic, and has low immunogenicity. PEG based PNPs are used to prolong NPs circulation time and increase their stability [23].

There are two main types of PNPs, nanospheres and nanocapsules. In the case of nanospheres drugs are either uniformly embedded (entrapped) within the NP or coupled to its surface (physically adsorbed or chemically linked), while in the case of nanocapsules drugs are confined within the inner oily or aqueous cavities and surrounded by tiny polymeric membranes [22].

PNPs have some of the *advantages* that are common with other nanocarriers such as high drug-loading capacity, stabilization of entrapped molecules, long shelf-life, sustained drug release, and biodegradability and biocompatibility [74]. They are structurally highly stable, nonimmunogenic, nonthrombogenic, nontoxic, avoid the reticuloendothelial system, have minimum leakage of the drug, and do not activate neutrophils when used in blood [3, 73]. Their production methods are easy and suitable for scale-up [5]. Additional advantages of PNPs are possibility to use "smart" polymers that respond to external physical (temperature, light, electricity, ultrasound, or mechanical stress), chemical (pH or ionic strength), or biological stimuli signals (enzymes or biomolecules), which enables obtained NPs to deliver drugs in more accurate and programmable manner [23].

The main *limitation* of PNPs is that their metabolism and long-term safety are still not adequately tested, which is of primary interest especially for treatment of chronicle diseases, and it requires further investigations, as well as their possible high toxicity after intravenous injection (depending on polymer) [23, 73].

Production methods for formulation of herbal PNPs are numerous since different polymers may be used, but some most common include solvent evaporation, solvent displacement, and solvent emulsification-diffusion methods, nanoprecipitation of preformed polymers, complex coacervation, polymeric dispersion, coprecipitation, salting-out, dialysis, and supercritical fluid technology. PNPs may be also directly synthesized by the monomers polymerization by mini- and microemulsion, interfacial polymerization, dispersion of preformed polymers, and controlled polymerization methods [22, 40, 56]. Production of nanospheres is mainly performed by solvent evaporation, solvent emulsification-diffusion, nanoprecipitation, and salting-out methods, while nanocapsules are created mainly by nanoprecipitation [1]. Methods used to synthesize PLGA NPs are solvent evaporation, solvent emulsification–diffusion, and the nanoprecipitation of preformed polymers [42].

3.2.2 Biopolymeric Nanoparticles (BPNPs)

Biopolymeric nanoparticles (BPNPs) generally share most of the properties attributed to the PNPs. Only difference is that they are produced from natural polymers and therefore are less toxic, more bioactive, biocompatible, and biodegradable than PNPs, and easier for gel formation [1]. Most used natural polymers include proteins (collagen, albumin, gelatin, etc.), polysaccharides (alginate, chitosan, cellulose, hyaluronan, and many others), as well as their modifications, derivatives, or their combinations [13]. Some of the numerous methods used for BPNPs production are coacervation, spray drying, electrospinning, electrospray, supercritical fluid, emulsion–diffusion, reverse micelle, emulsion-droplet coalescence, emulsification/solvent evaporation, salting-out, ultrasonication, and high-pressure homogenization [1, 13]. Some of the main applications of (bio)polymeric NPs in nanoherbal drug delivery are given in Table 2.

Table 2 Types of novel polymer-based nanocarriers and encapsulated phytopharmaceuticals

Nanomicelles Polymeric	3-Acetyl-11-keto-beta-boswellic acid, 5-fluorouracil, 7-ethyl-10-hydroxycamptothecin, 9-nitrocamptothecin, 10-hydroxycamptothecin, β-lapachone, adriamycin, alantolactone, artemisinin, berberine, celastrol, chrysin, cisplatin, combretastatin A4, cucurbitacin, curcumin, dihydroartemisinin, diosmin, docetaxel, doxorubicin, embelin, estradiol, galangin, gambogic acid, genistein, ginsenoside Rg3, harmine, honokiol, hydroxycamptothecin, indomethacin, luteolin, mangiferin, morin, nanoplatin, nobiletin, paclitaxel, puerarin, quercetin, resveratrol, shikonin, silibinin, silymarin, thymoquinone, tilianin, triptolide, triptorelin, vinblastine, vincristine, xanthone, wogonin	[1, 12, 13, 20, 26, 40, 42, 66, 69, 83–86, 89, 90]
Dendrimers	3, 4-Difluorobenzylidenecurcumin, anthocyanin, berberine, celastrol, chrysin, curcumin, gallic acid, liquiritin, malloapelta B, paclitaxel, puerarin, quercetin, silybin	[12, 13, 20, 40, 85, 89]
Nanospheres	Curcumin, diosmin, menthol, nerolidol, oridonin, papain, silymarin	[13, 74, 86, 87]
Nanocapsules	Artemisinin, capsaicin, catechin, curcumin, honokiol, Hydrocotyle sibthorpioides, paeonol, prednisolone, quercetin, resveratrol, tetrahydrofuran neolignane grandisin	[12, 13, 41, 74, 86]
(Bio)Polymeric NPs	9-Nitrocamptothecin, 11-keto-beta-boswellic acid (KBA), (2S)-5,7-dihydroxy-6-prenylflavanonederivatives, aclitaxel, bacoside-A, berberine, betulinic acid, bufalin, camptothecin, catechin, chrysin, cinnamaldehyde, curcumin, daunorubicin, doxorubicin, ellagic acid, embelin, emodin, epigallocatechin, epigallocatechin-3-gallate, etoposide, eugenol, ferulic acid, fisetin, fucoidan, gallate, genistein, geraniol, glycyrrhizin, hypericin, iperine, juglone, luteolin, morin, naringenin, naringin, oleanolic acid, paclitaxel, piperine, pterostilbene, quercetin, resveratrol, rutin, silbinin, silymarin, sulforaphane, tartary buckwheat flavonoids, triptolide, ursolic acid, vinblastine, vitamin E	[1, 12, 13, 20, 40, 41, 70, 73, 85, 86, 90, 96]
Nanofiber	β-Carotene, curcumin, quercetin, resveratrol, rosmarinic acid, thymol	[40, 81]
Nanogels	Acitretin, apigenin, curcumin, emodin, myricetin, puerarin, quercetin, sacchachitin	[12, 76, 81]

3.2.3 Dendrimers

Dendrimers are micellar, tree-like macromolecules with spherical, unimolecular structure, composed of single, polyfunctional, central core, and multibranched monomers uniformly branching of from this central core, and, if needed, terminal functional groups.

They are formed by engrafting branches onto the core, in a stepwise manner, until uniform three-dimensional morphology is obtained with size between 1 and 10 nm [15, 24]. Mostly used macromolecules for producing dendrimers are polyamidoamine, as well as sugars, nucleotides, and amino acids [13, 70]. Due to their hydrophobic core and hydrophilic surface they frequently form micelles [85].

Some of the most important *advantages* of dendrimers are connected with their high surface modification ability due to the large number of surface functional groups. This enables easier regulation of drug release by changing the number (density) and/or rigidity of branches, and beside this gives possibility to make dendrimers water soluble by functionalizing their outer shell with charged or other hydrophilic functional groups [66, 85]. Large number of surface groups also enables easy attachment of very specific targeting ligands (i.e. PEGylation, acetylation, glycosylation, different amino acids, etc.) and this makes dendrimers very attractive as antitumor and CNS drug nanocarriers [70]. In addition, they are biodegradable, biocompatible, have low side effects, and low polydispersity index (PDI) (are monodisperse) [76]. Other also important features of dendrimers are that they can act as a particulate system while keeping the polymer properties, as well as their small size which make their clearance from the body fast by the renal system and ease their escape from the reticuloendothelial system [3, 13]. Drug delivery by dendrimers may be accomplished by two methods. Their large internal cavity like structure enables high encapsulation ability of different hydrophobic drugs, increasing their water solubility and enabling their sustained and controlled release. On the other side, drugs may be also embedded on the outer surface by covalent or ionic interactions [3, 15]. Due to the ability to precise control dendrimers size, shape, number of branches, and attached functional groups they offer good potential for very different applications as well as for further expanding [84].

Disadvantages that limit dendrimers application are mostly connected to their inherent toxicity which may be hemolytic, hematological, or cytotoxicity. The toxicity is mainly attributed to the in vivo interactions of the positively charged dendrimer branches with the negatively charged biological membranes which may cause the membrane disruption [84]. Some additional limitations include high production cost, failure to reach the targeted site, hard scale-up production, and some safety issues [74].

Methods for dendrimers synthesis may be either divergent or convergent. While in divergent technique synthesis starts from the inner core toward the outer layers, in the convergent approach synthesis has opposite direction [15].

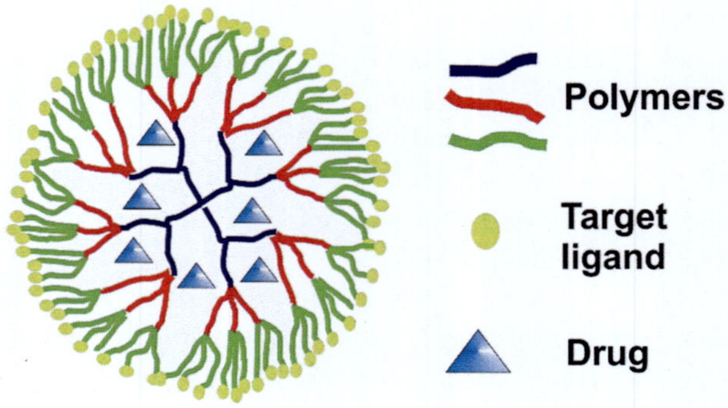

Fig. 4 Dendrimer structure

All these distinctive features, altogether with shortcomings of other nanocarriers, are making these emerging nanocarriers one of the most promising NPs for targeted delivery of phytopharmaceuticals. Schematic representation of dendrimer structure is shown in Fig. 4, while some of the main applications of dendrimers in nanoherbal drug delivery are given in Table 2.

3.2.4 Polymeric Micelles

Polymeric micelles are developed in order to reduce some disadvantages of lipid-based micelles and they are composed of synthetic amphiphilic di- or tri-block copolymers instead of phospholipids [5]. Used block copolymers are composed of two monomers with different hydrophobicities (one hydrophobic and other hydrophilic), which enable them to form a core–shell micelle structure [42]. The most preferred hydrophilic monomer is polyethylene glycol (PEG), while core-forming blocks are used different polymers like poly(caprolactone) (PCL), poly(N-vinyl-2-pyrrolidone), poly(D, L-lactic acid), poly(L-aspartic acid), poly(propyleneoxide), poly(c-benzyl-L-glutamate), etc. [13, 23, 42]. Among them combination of PEG-PCL is the mostly used due to its desirable characteristics like biodegradability, safety, and high loading capacity for hydrophobic components [13]. In addition, polymeric micelles, compared to the traditional micelles, have increased stability, longer shelf-life, and reduced level of clearance from the body. Polymeric micelles are gaining increased attention in the recent years due to their ability to accumulate in the tissue of solid cancers. Namely, the MWs of polymeric micelles are usually high and they enter tumor tissue more easily than other tissues, either via passive or active targeting mechanisms [13, 66]. However, while the preparation of polymeric micelles is simple, controlling their drug release rate from them is problematic and they commonly must be surface modified by chemical bonds to lower release of toxic drugs within the blood and increase it inside the tumor tissue [66].

Some of the main applications of lipid and polymeric micelles in nanoherbal drug delivery are given in Table 2.

3.2.5 Polymersomes

Polymersomes have some structure similarities to liposomes. Polymersomes are stable artificial nanovesicles composed of self-assembling synthetic amphiphilic block copolymers which may be hydrophilic (i.e. PEG) or hydrophobic (i.e. polybutylene oxide) [13]. Their size is from tens of nm up to 1 µm, and their membranes may be relatively very thick (up to 40 nm) compared to other types of vesicles. Their overall properties may be easily and effectively tuned by using appropriate biodegradable and/or stimuli-responsive block copolymers [56]. In such manner their drug encapsulation capabilities for hydrophilic and hydrophobic drugs either in a membrane or in core region, as well as release properties capability may be tuned. Moreover, their capability to bind with biologically active ligands and antibodies makes them very applicable for targeted and control release delivery [13]. Their advantages beside those already mentioned include increased stability, increased drug-loading capacity and lower permeability, controlled release, versatility, tunable degradation properties [22, 56]. Methods for polymersomes production are similar to those for polymeric NPs. Some of the main applications of polymersomes in nanoherbal drug delivery are given in Table 2.

3.2.6 Hydrogel Nanoparticles

Hydrogels are flexible cross-linked networks of hydrophilic polymers which can absorb and retain large amounts of water (more than 95%) without disturbing its 3D structure [79]. Those structural crosslinks are formed by different bonds like ionic, hydrogen, covalent, Van der Waals interactions, or physical entangling, hydrophobic interactions [72, 91]. For production of hydrogels many different natural and synthetic polymers may be used like alginate, gellan gum, chitosan, collagen, PLA, PLGA, PVA, and many others, but main parameters for their selection are biodegradability, biocompatibility, nontoxicity, nonimmunogenicity, and noninflammatory, beside the ability to encapsulate and controllably release drugs [70, 92]. Therefore, they may be classified as hydrogels based on natural or synthetic polymers, as well as hybrid produced by their combination. Hydrogels have large potential for various biomedical applications and offer possibility for slow release of drugs, thixotropic and sol-gel interconvertible nature, tunable stiffness, high porosity and water content, restricted penetration of various degrading enzymes, high hydrophilicity but without possibility of dissolving [91, 93, 94]. However, although hydrogel NPs are a promising drug delivery approach for improving the pharmacokinetics and pharmacodynamic prospects of phytopharmaceuticals and they can be further modified or combined with the magnetic nanoparticles to achieve targeting delivery, they are still not commercially applied. The main problems connected to hydrogel NPs are the same as problems with classic gels, significant batch-to-batch variability and potential immunogenicity due to the presence of some foreign substances that are used for the gelation process [13, 93]. In order to overcome this limitation polymers with easy ionic gelling like alginate and gellan gum may be used [93, 95]. Some of the main applications of nanohydrogels in nanoherbal drug delivery are given in Table 2.

3.2.7 Nanofibers (NFs)

Nanofibers (NFs) are relatively new nanocarriers produced from solid polymers with diameters in the range 0–1000 nm with high length diameter ratio. This ratio gives them their outstanding properties as large surface-to-volume ratio, high porosity, low density, and excellent mechanical characteristics [70]. Generally, they are produced by electrospinning method from biodegradable and biocompatible water-soluble polymers like chitosan, polyvinyl alcohol, poly-ε-caprolactone, polyvinylpyrrolidone, etc. [13, 40]. Although NFs currently have relatively small role in drug delivery their properties such as high loading capacity and encapsulation efficacy, easy and cost-effective production, possibility for simultaneous delivery of various therapeutics, production from different polymers with different physical properties and application potentials, strong adhesiveness, make them very suitable for phytopharmaceutical delivery. In addition, use of stimuli-responsive polymers (pH, temperature, electricity, magnet, light, etc.) for NFs production makes them suitable for targeted delivery [13, 70]. Some of the main applications of nanofibers in nanoherbal drug delivery are given in Table 2.

3.2.8 Nanospheres (NSs)

Nanospheres are spherical-shaped particles with matrix type structure in which drugs are dispersed all over the particles. Their size may be from 1 to 200 nm, but commonly it is between 10 and 100 nm [13]. Depending on the materials they are produced from, their nature may be either amorphous or crystalline, and they may also be regarded as biodegradable and nonbiodegradable. NSs produced from some natural polymers like gelatin, albumin, and starch, as well as from synthetic like polypropylene dextran, and polylactic acid, polyglycolic acid, and polylactide-coglycolide are biodegradable [1]. Type of polymer also determines the rate of drug release [13]. Within the NS matrix drug may be entrapped, enclosed, dissolved, or attached in order to be protected from the external environment, but it may be also adsorbed on the NS surface. They may be used for phytopharmaceuticals with low absorptivity and increased biodegradability [24]. This type of NPs may be also combined with magnetic NPs and antibodies in order to increase targeted delivery [13].

NSs have several advantages which make them attractive as nanocarriers. These are possibility of complete drug entrapments, prolonged drug release, increased drug stability and its protection from chemical and enzymatic degradation, but they also minimize drug toxicity. In addition, NSs improve drug bioavailability and make its concentrations in plasma steady. All these properties are mostly connected to hydrophobic nature of NSs surface, but this hydrophobicity is also reason for the main limitations of NSs, susceptibility for opsonization, and fast clearance by the reticuloendothelial system [1, 13].

The main techniques for NSs production are emulsion evaporation and preformed polymer nanoprecipitation, while main delivery routes are oral, but they can be also administered locally and systematically [13]. Some of the main applications of nanospheres in nanoherbal drug delivery are given in Table 2.

3.2.9 Nanocapsules (NCs)

Nanocapsules are nanovesicular colloidal dispersion systems with the core-shell structure and diameters between 10 and 1000 nm. Active ingredients are confined within the core which may be liquid, solid, or molecular, while shell is mostly made of polymers. Polymers used as shell materials are of essential importance, since they have major impact not only on protection and release profile of encapsulated drugs, but also on NCs stability, encapsulation efficiency, and biodistribution within the body. Depending on the used material and production technique they may be both lipophilic or hydrophobic, but NCs are mainly used to improve bioavailability of lipophilic drugs [13].

The main advantages of this type of NPs are high drug encapsulation efficiency, relatively lower polymer content, controlled drug release, improved drug protection [1, 76]. Preparation methods for NCs are numerous due to large variability in core-shell components and some of the most used are nanoprecipitation, emulsion–diffusion, emulsion–evaporation emulsion-coacervation, double emulsification, polymer-coating, arc discharge method, interfacial polymerization, and layer-by-layer methods [1, 13]. Some of the main applications of nanocapsules in nanoherbal drug delivery are given in Table 2.

3.3 Inorganic Nanoparticles

Currently many different types of inorganic nanocarriers are being investigated and developed for phytopharmaceutical drug delivery. Inorganic NPs may be metal based (mostly transition metals), ceramic based, and carbon based [5].

3.3.1 Metal-Based Nanoparticles (MBNPs)

Metal based or metallic NPs (MBNPs) are type of nanocarriers that are increasingly gaining attention in the recent years due to their specific potentials. They can be either magnetic or nonmagnetic and produced of metals or metal oxides. Some of the most used metals for MBNPs production are gold, silver, nickel, copper, iron, platinum, selenium, gadolinium, palladium, while most used metal oxides are iron, zinc, titanium, magnesium, aluminum, yttrium, and cerium [13, 56]. There are also some modified MBNPs like nanowires, nanofilms and quantum dots [70]. The size of MBNPs is between 1 and 100 nm, and their properties largely depend on type of used metal, but also on their size. MBNPs have some specific properties that must be kept in mind, regardless of their drug loading. As first, many transition metals when excited by external radiation release energy mostly in the form of heat and can therefore cause cell death which may be useful in cancer treatment [5]. In addition, they can be used for theranostics applications. Some of those metals may be minute essential components included in redox reactions in our body [70]. However, most of those metallic NPs are highly poisonous to our body and have increased reactivity due to their small size and increased surface ratio [84].

In order to reduce their high toxicity and reactivity MBNPs are commonly encapsulated or coated with organic or inorganic shell [1]. They can be synthesized or modified with appropriate chemical functional groups to help them conjugate with various drugs and/or to enable targeting delivery [56]. Most advantageous properties that make metallic NPs attractive are high encapsulation efficiency due to large surface/volume ratio, biocompatibility, easy chemical modification and tunable biophysical properties due to large surface energy, possibility for site-specific drug delivery, high stability, and relatively simple synthesis [22, 73]. However, many toxicological and environmental studies in the recent years have shown certain negative aspects of MBNPs like inflammation due to allergic and/or autoimmune response, oxidative stress caused by the formation of reactive oxygen species, genotoxic behavior, and accumulation within specific and sensitive organs and tissues like liver, spleen, brain, spinal cord, and heart [84]. They can be toxic, instable, and have impurities [73]. In addition, some of the MBNPs are not economical for production due to their high cost, like gold, silver, platinum etc. [24]. One of the most promising methods for MBNPs production is biological synthesis by bioreduction of metal salt solutions mixed with aqueous plant extracts [1]. Some of the main applications of MBNPs in nanoherbal drug delivery are given in Table 3.

Table 3 Types of novel inorganic based nanocarriers and encapsulated phytopharmaceuticals

CNTs	Apigenin, betulinic acid, camptothecin, carvedilol-A, Curcumin, doxorubicin, ellagic acid, folic acid, gallic acid, gambogic acid, ginsenosides Rb1 and Rg1, paclitaxel, pirarubicin, podophyllotoxin, quercetin, rutin, vinblastine	[20, 40, 70, 81, 84, 85 90]
GO NPs	Gallic acid	[89]
Metallic NPs	Aloin, apigenin, berberine, curcumin, guar gum, ellagic acids, epigallocatechin-3-gallate, garlic, gallic acid, hesperidin, licochalcone A, naringenin, opium alkaloids, quercetin, resveratrol, paclitaxel, silymarin, ursolic acid	[1, 13, 20, 44, 70, 96]
Quantum Dots	Curcumin, quercetin	[44, 99]
Magnetic NPs	Camptothecin, celastrol, curcumin, dihydroartemisinin, eugenol, gallic acid, gambogic acid, paclitaxel, quercetin, sulfasalazine	[12, 13, 26, 41, 70, 85, 90]
HCNTs	Curcumin, silibinin	[13]
MSNPs	Axitinib, celastrol, curcumin, ellagic acid, guar gum, myricetin, paclitaxel, silibinin, ursolic acid	[13, 70, 89]

3.3.2 Inorganic Based Nanoparticles (IBNPs)

There are a few types of nonmetallic inorganic based NPs that may be used for delivery of phytopharmaceuticals. These are clay NPs (mostly halloysite clay nanotubes, ceramic NPs (mostly mesoporous silica NPs) and calcium carbonate NPs.

3.3.2.1 Halloysite Clay Nanotubes (HCNTs)

There are different classes of clay and the major groups are: kaolinite (halloysite, zeolite), montmorillonite/smectite, illite, and chlorite. The main difference between clay and nanoclay is that nanoclay has at least one dimension of nanometer range and high aspect ratio, as well as improved surface integrity, thermal and mechanical characteristics [84]. Halloysite clay nanotubes (HCNTs) are made of aluminosilicate clay with tetrahedral and octahedral structural symmetries. HCNTs are natural green hollow tubes with nanocavities and high surface area and loading capacity [13]. In addition, they have low cost, large availability and biocompatibility, reduced side effects, and increased formulations' shelf-life [70]. They can be used to load either positively charged drugs on its surface, or negatively charged within the inner core since HCNTs have both positive and negative charges on the surface and drug molecules are connected to the HCNTs surface by hydrogen bonding, ion exchange, and hydrophilic/hydrophobic interactions [70, 84]. However, HCNTs are not only inert fillers but they also offer comprehensive opportunity for further surface functionalization/coating for controlled and targeted drug release. Therefore, HCNTs represent competitive alternative to CNTs and polymer NPs for controlled release of drug compounds [84]. Some of the main applications of HCNTs in nanoherbal drug delivery are given in Table 3.

3.3.2.2 Ceramic NPs

Although clay and ceramic have some similarities there are some fundamental differences. The key difference is in composition, while clay contains moistened minerals like aluminum silicates and crystalline silica, ceramic is a mixture that beside clay (in most cases only silica is used) may contain metal oxides (i.e. zirconium oxide, silica oxide), nitrides, and carbides (i.e. silica carbide), and various additives [5]. Ceramic may be made of clay but some ceramics are not made of clay, and while clay is unfired ceramic is fired.

Ceramic NPs that are used in drug delivery are mostly in the form of *mesoporous silica nanoparticles* (*MSNPs*) with diameters in the range between 2 and 50 nm [15]. MSNPs have highly porous honeycomb-like structure made of silica and can be generally classified in three groups: ordered, hollow (or rattle), and core/shell types of MSNPs [97].

Advantages of MSNPs are numerous so it is not surprising they are gaining increased attention in recent years. Their highly porous structure offers large surface area and significantly increased loading capacities. They are biocompatible and highly stable. In addition, pore size and surface morphology are adjustable, and functionalization of MSNPs surface by conjugation is easy [13]. Due to surface functionalization MSNPs may obtain increased circulation half-life (by PEGylation), better drug release control (by grafting appropriate polymers), and active targeting (by conjugation with certain ligands) [97].

MSNPs are advantageous type of nanocarriers for transporting phytopharmaceuticals that are susceptible to pH variation since they are not sensitive to this change [15]. MSNPs may be applied to improve bioavailability for poorly soluble drugs, reduce drug toxicity, and targeted delivery [13]. However, like the metal-based NPs, MSNPs are not biodegradable, which may cause undesirable effects [15]. The hydrophilic drugs are adsorbed on MSNPs negatively charged pores and surface, while the hydrophobic drugs are either attached to the MSNPs by hydrophobic interactions, or are dissolved in the organic solvent, mixed with MSNs solution and then vacuum dried to remove the solvent [97]. Some of the main applications of ceramic NPs in nanoherbal drug delivery are given in Table 3.

3.3.2.3 Calcium Carbonate Nanoparticles (CCNPs)

Calcium carbonate nanoparticles (CCNPs) are another promising type on nanocarriers for phytopharmaceutical delivery that need to be further studied. Main advantages of those NPs are biocompatibility, nontoxicity, slow biodegradability, pH-sensitivity, high surface-to-volume ratio, robust nature, simple synthesis (even in a bulk scale) and surface functionalization, and ability to exist in a variety of morphologies and polymorphs [13, 98]. Main methods for their synthesis are precipitation, slow carbonation, emulsion, polymer-mediated techniques (i.e. in situ polymerization), mechanochemical, microwave-assisted, and biological methods [98]. Some of the main applications of CCNPs in nanoherbal drug delivery are given in Table 3.

3.3.3 Carbon-Based Nanoparticles

There are a few types of carbon-based NPs with interior structure made up of graphite sheets. These are carbon nanotubes (CNTs), fullerenes (C60), graphene and graphene oxide (GO), nanodiamonds (NDs), and carbon dots (CDs) with small diameters in the range of 3–10 nm [1]. They are gaining increased interest due to some of their advantages but the major limitation of the application of carbon-based NPs is their high cytotoxicity [15]. Carbon-based NPs may be produced by laser ablation, arc discharge, or chemical vapor deposition methods [43]. Most important type of carbon-based NPs for phytopharmaceutical delivery are CNTs.

3.3.3.1 Carbon Nanotubes (CNTs)

Carbon nanotubes (CNTs) are allotropes of carbon with cylindrical hollow nanostructures and since they are produced from graphene sheets that have honeycomb-like hexagonal carbon structure. CNTs are usually formed by rolling, at specified and discrete chiral angles, a single or multiple layers of graphene sheets and depending on the number of sheets they are named single (SWCNTs), double (DWCNTs), and multi-walled CNTs (MWCNTs) [24]. Those nanocylinders may be opened or closed. While SWCNTs have dimensions 0.7 x 50 nm (diameter x length) and are naturally organizing themselves into strands bound by π-stacking, MWCNTs are in the range 20–30 x1000 nm and regular spacing between layers of 0.34–0.39 nm, and they are capped (closed) ends [24, 66, 85].

The CNTs are very promising nanocarriers and have some *advantages* compared to other NPs. Due to its large surface area they can conjugate or adsorb many different drugs from small up to large macromolecules and have high encapsulation abilities due to their hollow structure, possibility for controlled and prolonged drug release while preserving drug intact during the transport in the body to action site. They have low weight, high chemical stability and tensile strength, flexibility to external stimuli due to rich electronic polyaromatic structure, especially to the near infrared waves. Due to their small dimensions they can easily cross cell membrane and across blood–brain barrier (BBB). Possibility for functionalization enables biological applications and targeting which make them especially attractive for tumor targeting [40, 66, 70, 73].

Main *limitations* are their high toxicity in vivo due to the presence of carboxyl, hydroxyl, and carbonyl functional groups on the surface of CNTs, possibility of dissociation in biological fluids, inhibition of cell proliferation, problems with handling due to very small size, problems with elimination from the body, triggering immune response and cell apoptosis, high production cost, and polydispersity [15, 73, 74, 84]. Some of the main applications of CNTs in nanoherbal drug delivery are given in Table 3.

3.4 Polymer(Biopolymer)/Lipid/Antibody Drug Conjugates ((B)PDC, LDC, ADC)

Union of drugs with the (bio)polymer/lipid/antibody is a relatively new approach in drug delivery and it is used to modify drug properties and change its pharmacokinetics. In those conjugates one or several drugs are in most cases covalently, but sometimes noncovalently, linked (by spacer or linker linear arm) to (bio)polymer/lipid/antibody. Polymer conjugated drugs are sometimes called prodrugs, while those conjugated with lipids (especially phospholipids) are also known as pharmacosomes. For the PDCs production mainly PEG, PLA, PLGA polymers are used, BPDCs are made of thermally sensitive biopolymers, LDCs are produced from diglycerides, phosphoglycerides, and fatty acids, while in the case of ADCs they, beside antibodies, may be also produced from peptides or proteins [13, 15]. Their size is lower than those of micelles and liposomes and ADCs have size between 10 and 20 nm, while PDCs and LDCs sizes are below 10 nm [18]. Drug conjugates are mostly used for delivery of anticancer drugs as cytotoxins and small hydrophilic molecules which are then turned into water insoluble. Main advantages of drug conjugates are prolonged circulation, drug targeting (most important fact for cytotoxic anticancer drugs) and controlled release, increase in bioavailability and GIT permeability [13, 85]. Therefore, they are ideal nanocarriers for phytochemicals which cannot be used as free drugs either due to their toxicity and/or inadequate therapeutic index. Those drugs when conjugated couple high potency and enlarged therapeutic window [18]. Therefore, it is not surprising that LDCs are considered as most acceptable lipid-based NPs [13]. However, the main problem in drug conjugate design which still limits their wider application is that drugs, polymer/lipid/antibody and linkers/spacers on one side and targeted cells on the other must be optimally combined, which is a very challenging task [18]. Some of the

Table 4 Types of other novel based nanocarriers and encapsulated phytopharmaceuticals

Drug conjugates	Camptothecin, celastrol, curcumin, decitabine, docetaxel, doxorubicin, epigallocatechin-3-gallate, ginsenoside, oridonin, periplocymarin, plumbagin, quercetin, resveratrol, theaflavin, transresveratrol, vitamin E	[12, 13, 26]
Nanocrystals (nanosuspensions)	Andrographolide, apigenin, baicalein, berberine, curcumin, embelin, etoposide gambogenic acid, ginkgo lactone, ginkgolides A and B, hesperetin, hesperidin, myricetin, nabilone, naringenin, oleanolic acid, oridonin, piperine, puerarin, quercetin, resveratrol, rutin, silymarin, ursolic acid, zerumbone	[12, 13, 26, 47, 69, 70, 89]

main applications of nanodrug conjugates in nanoherbal drug delivery are given in Table 4.

3.5 Nanocrystals (NCs)

Nanocrystals (NCs) are pure drug (carrier free) nanosized particles commonly in the form of colloidal dispersion (nanosuspensions). The size of those self-carries may be within the range 10–1000 nm, but they are commonly smaller than 100 nm. Although they are primarily 100% composed of drug substance some stabilizers or surfactants are commonly used, while as a dispersing medium water, liquid PEG600, oil, or any other aqueous or non-aqueous media is employed [47]. NCs are widely applied in delivery of hydrophobic drugs with low dissolution rates (high logP values), since increased surface area of nanosized drug increases its solubility and dissolution rate, which results in higher plasma drug concentrations [15]. Therefore, NCs could specifically suitable for delivery of polyphenols and other phytochemicals that have low stability and inadequate pharmacokinetics [15]. Since NCs are self-carriers the toxic effects caused by encapsulating/solubilizing excipients are eliminated [13]. Other benefits of using NCs are high drug-loading capacity and physical stability [22, 79]. The distribution of NCs within the body is influenced by particle size and morphology, while it could be further affected by surface modifications (like ligand and/or stimuli-responsive polymers conjugation) [13]. However, NCs can be recognized by mononuclear phagocytic system as an exogenous material and remove them from blood stream which leads to their passive accumulation in liver, spleen, and lung [13].

Methods for NCs production may be either by mechanical or chemical, using bottom-up or top-down approaches, or their combination, but in most cases top-down techniques are preferable since they are simple and easy to scale-up [47, 56]. Top-down methods are wet or dry milling methods (bead milling or high-pressure homogenization), while bottom-up approach methods include different precipitation techniques (precipitation, high gravity-controlled precipitation, restricted impinging liquid jet precipitation), sono-crystallization, and multi-inlet vortex mixing [47, 69]. Among mentioned methods, milling, precipitation, and high-pressure homogenization are most commonly used. The main routes for NCs application are oral, nasal, ocular, dermal, pulmonary, and parenteral [13, 40]. Some of the main applications of nanodrug conjugates in nanoherbal drug delivery are given in Table 4.

4 Conclusions

The herbal medicine has been used for centuries in preventing and healing many different diseases and disorders. In the recent decades it is regaining increased interest either due to some limitations and drawbacks of synthetic drugs or due to increased awareness of phytopharmaceuticals advantages. However, they also have some constraints and one of the ways to overcome some of them is their encapsulation in different nanocarriers. However, since nanotechnology is relatively new science, especially in the field of medicine, there are numerous questions that need to be answered and problems that need to be solved. In this review, main problems connected to plant based drugs are stressed, as well as role of nanotechnology in their overcoming. In addition, specific limitations connected to some group of phytodrugs are emphasized, and main properties of many different types of nanocarriers are given. This chapter also provides an insight to specific applications of some phytopharmaceuticals that have been already encapsulated in different nanoparticle formulations. According to the presented data it is obvious that nanobased carriers have very promising potential for development of many different, functional phytopharmaceutical formulations. However, big care must be taken since nanocarriers are relatively new type of drug carriers and they already have some big limitations especially due to their toxicity, but they also lack investigations about long-term use effects caused by relatively short term applications. Therefore, nanophytopharmaceuticals dosage forms will represent one very wide field for further investigations, especially having in mind tremendous number of different plant drugs and a wide range of possible nanocarriers for their encapsulation. This review is aimed to help researchers to determine optimal choice for each phytodrug and to design novel, improved nanobased phytopharmaceutical DDS.

Acknowledgments This work was supported by the Ministry of Science, Technological Development and by the Science Fund of the Republic of Serbia, #GRANT No 7750160 [1], R-SPECT.

References

1. Chavda VP, Patel AB, Mistry KJ, Suthar SF, Wu ZX, Chen ZS, et al. Nano-drug delivery systems entrapping natural bioactive compounds for cancer: recent Progress and future challenges. Front Oncol. 2022;12:867655.
2. Rajagopal M, Paul AK, Lee MT, Joykin AR, Por CS, Mahboob T, et al. Phytochemicals and nano-phytopharmaceuticals use in skin, urogenital and locomotor disorders: are we there? Plants (Basel). 2022;11(9):1265.
3. Vimala K, Kannan S. Phyto-drug conjugated nanomaterials enhance apoptotic activity in cancer. Adv Protein Chem Struct Biol. 2021;125:275–305.
4. Chen SL, Yu H, Luo HM, Wu Q, Li CF, Steinmetz A. Conservation and sustainable use of medicinal plants: problems, progress, and prospects. Chin Med. 2016;11:37.
5. Odeh F, Al-Jaber H, Khater D. Nanoflora—how nanotechnology enhanced the use of active phytochemicals. In: Application of nanotechnology in drug delivery. London: InTech; 2014.
6. Ekor M. The growing use of herbal medicines: issues relating to adverse reactions and challenges in monitoring safety. Front Pharmacol. 2014;10(4):177.
7. Vriens J, Nilius B, Vennekens R. Herbal compounds and toxins modulating TRP channels. Curr Neuropharmacol. 2008;6(1):79–96.
8. Atanasov AG, Zotchev SB, Dirsch VM, Supuran CT. Natural products in drug discovery: advances and opportunities. Nat Rev Drug Discov. 2021;20:200–16.
9. Bhusnure OG, Shinde MC, Vijayendra SSM, Gholve SB, Giram PS, Birajdar MJ. Phytopharmaceuticals: an emerging platform for innovation and development of new drugs from botanicals. J Drug Deliv Ther. 2019;9:1046–57.
10. Sasidharan S, Chen Y, Saravanan D, Sundram KM, Yoga LL. Extraction, isolation and characterization of bioactive compounds from plants' extracts. Afr J Tradit Complement Altern Med. 2011;8(1):1–10.
11. Dewi MK, Chaerunisaa AY, Muhaimin M, Joni IM. Improved activity of herbal medicines through nanotechnology. Nanomaterials (Basel). 2022;12(22):4073.
12. Beg S, Barkat MA, Ahmad FJ. Nanophytomedicine. Singapore: Springer, Singapore; 2020.
13. Jain T, Srivastava K, Singh R, Ansari MJ, Singh S, Dutta PK. Role of nanotechnology in the field of phytopharmaceuticals for the delivery of herbal drugs. In: Singh S, Datta R, Johri P, Trivedi M, editors. Phytopharmaceuticals and biotechnology of herbal plants. 1st ed. Boca Raton: CRC Press; 2022.
14. Zhao Q, Luan X, Zheng M, Tian X-H, Zhao J, Zhang W-D, et al. Synergistic mechanisms of constituents in herbal extracts during intestinal absorption: focus on natural occurring nanoparticles. Pharmaceutics. 2020;12:128.
15. Manickam V, Sundar V, Panchangam RL, Tamizhselvi R. Nanotechnology in delivery and targeting of phytochemicals. In: Nanopharmaceuticals: principles and applications, vol. 2. Cham: Springer; 2021.
16. Aqil F, Munagala R, Jeyabalan J, Vadhanam MV. Bioavailability of phytochemicals and its enhancement by drug delivery systems. Cancer Lett. 2013;334:133–41.
17. Ajazuddin SS. Applications of novel drug delivery system for herbal formulations. Fitoterapia. 2010;81(7):680–9.
18. Xie J, Yang Z, Zhou C, Zhu J, Lee RJ, Teng L. Nanotechnology for the delivery of phytochemicals in cancer therapy. Biotechnol Adv. 2016;34(4):343–53.
19. Tung BT, Hoa TTT. Nanotherapeutics of phytoantioxidants for diabetes mellitus. In: Rudrapal M, editor. Phytoantioxidants and nanotherapeutics. Hoboken: Wiley; 2022. p. 433–47.
20. Rizwanullah M, Amin S, Mir SR, Fakhri KU, Rizvi MMA. Phytochemical based nanomedicines against cancer: current status and future prospects. J Drug Target. 2018;26(9):731–52.
21. Ganesan P, Ko HM, Kim IS, Choi DK. Recent trends in the development of nanophytobioactive compounds and delivery systems for their possible role in reducing oxidative stress in Parkinson's disease models. Int J Nanomedicine. 2015;10:6757–72.
22. Lushchak O, Strilbytska O, Koliada A, Zayachkivska A, Burdyliuk N, Yurkevych I, et al. Nanodelivery of phytobioactive compounds for treating aging-associated disorders. Geroscience. 2020;42(1):117–39.
23. Goktas Z, Zu Y, Abbasi M, Galyean S, Wu D, Fan Z, et al. Recent advances in nano-encapsulation of phytochemicals to combat obesity and its comorbidities. J Agric Food Chem. 2020;68(31):8119–31.
24. Vaishampayan P, Rane MM. Herbal nanocosmceuticals: a review on cosmeceutical innovation. J Cosmet Dermatol. 2022;21(11):5464–83.
25. Atanasov AG, Waltenberger B, Pferschy-Wenzig E-M, Linderd T, Wawroscha C, Uhrine P, et al. Discovery and resupply of pharmacologically active plant derived natural products: a review. Biotechnol Adv. 2015;33:1582–614.
26. Anwar DM, El-Sayed M, Reda A, Fang JY, Khattab SN, Elzoghby AO. Recent advances in herbal combination nanomedicine for cancer: delivery technology and therapeutic outcomes. Expert Opin Drug Deliv. 2021;18(11):1609–25.
27. Selby-Pham SNB, Miller RB, Howell K, Dunshea F, Bennett LE. Physicochemical properties of dietary phytochemicals can predict their passive absorption in the human small intestine. Sci Rep. 2017;7(1):1931.
28. Khan J, Alexander A, Saraf S, Saraf S. Recent advances and future prospects of phyto-phospholipid complexation technique for improving pharmacokinetic profile of plant actives. J Control Release. 2013;168(1):50–60.
29. Rudrapal M. Phytoantioxidants and nanotherapeutics. Hoboken: Wiley; 2022.
30. Mann J, Davidson RS, Hobbs JB, Banthorpe DV, Harborne JB. Natural products: their chemistry and biological significance. 1st ed. Harlow: Longman Group UK Limited; 1994.
31. Munin A, Edwards-Lévy F. Encapsulation of natural polyphenolic compounds; a review. Pharmaceutics. 2011;3:793–829.
32. Renaud J, Martinoli MG. Considerations for the use of polyphenols as therapies in neurodegenerative diseases. Int J Mol Sci. 2019;20(8):1883.
33. Pietta P-G. Flavonoids as antioxidants. J Nat Prod. 2000;63(7):1035–42.
34. Hussein R, El-Anssary AA. Plants secondary metabolites: the key drivers of the pharmacological actions of medicinal plants. In: Builders P, editor. Herbal medicine. London: Intech Open; 2018. p. 11–30.
35. Ganesan P, Ramalingam P, Karthivashan G, Ko YT, Choi DK. Recent developments in solid lipid nanoparticle and surface-modified solid lipid nanoparticle delivery systems for oral delivery of phyto-bioactive compounds in various chronic diseases. Int J Nanomedicine. 2018;13:1569–83.
36. Tadeusz A. Alkaloids: chemistry, biology, ecology, and applications. 2nd ed. Amsterdam: Elsevier; 2015.
37. Ludwiczuk A, Skalicka-Woźniak K, Georgiev MI. Terpenoids. In: Pharmacognosy. Boston: Academic Press; 2017. p. 233–66.
38. Rabi T, Bishayee A. Terpenoids and breast cancer chemoprevention. Breast Cancer Res Treat. 2009;115(2):223–39.
39. Ashour AS, El Aziz MMA, Gomha Melad AS. A review on saponins from medicinal plants: chemistry, isolation, and determination. J Nanomed Res. 2019;7:282–8.
40. Teja PK, Mithiya J, Kate AS, Bairwa K, Chauthe SK. Herbal nanomedicines: recent advancements, challenges, opportunities and regulatory overview. Phytomedicine. 2022;96:153890.
41. Upadhyay S, Bhushan R, Rudrapal M. Nanophytomedicine in disease and therapy. In: Rudrapal M, editor. Phytoantioxidants and nanotherapeutics. Hoboken: Wiley; 2022. p. 173–201.

42. Wang S, Su R, Nie S, Sun M, Zhang J, Wu D, et al. Application of nanotechnology in improving bioavailability and bioactivity of diet-derived phytochemicals. J Nutr Biochem. 2014;25(4):363–76.
43. Chakraborty T, Sarma A. Toxicity of nanostructures and nanodrugs. In: Rudrapal M, editor. Phytoantioxidants and nanotherapeutics. Hoboken: Wiley; 2022. p. 267–87.
44. Ng PQ, Ling LSC, Chellian J, Madheswaran T, Panneerselvam J, Kunnath AP, et al. Applications of nanocarriers as drug delivery vehicles for active phytoconstituents. Curr Pharm Des. 2020;26(36):4580–90.
45. Rahman HS, Othman HH, Hammadi NI, Yeap SK, Amin KM, Abdul Samad N, et al. Novel drug delivery systems for loading of natural plant extracts and their biomedical applications. Int J Nanomedicine. 2020;15:2439–83.
46. Shevalkar GB, Prajapati MK, Bhairav BA, Pardeshi CV. Nanotherapeutics of phytoantioxidants for aging and neurological disorders. In: Rudrapal M, editor. Phytoantioxidants and nanotherapeutics. Hoboken: Wiley; 2022. p. 477–93.
47. Kumari S, Goyal A, Sönmez Gürer E, Algın Yapar E, Garg M, Sood M, et al. Bioactive loaded novel nano-formulations for targeted drug delivery and their therapeutic potential. Pharmaceutics. 2022;14(5):1091.
48. Kaya G, Keskin M, Rudrapal M. Green approaches for synthesis of nanophytopharmaceuticals / nanophytodelivery systems. In: Rudrapal M, editor. Phytoantioxidants and nanotherapeutics. Hoboken: Wiley; 2022. p. 231–8.
49. Manickam V, Velusamy RK, Lochana R, Amiti RB, Tamizhselvi R. Applications and genotoxicity of nanomaterials in the food industry. Environ Chem Lett. 2017;15(3):399–412.
50. Hoshyar N, Gray S, Han H, Bao G. The effect of nanoparticle size on in vivo pharmacokinetics and cellular interaction. Nanomedicine (Lond). 2016;11:673–92.
51. Blanco E, Shen H, Ferrari M. Principles of nanoparticle design for overcoming biological barriers to drug delivery. Nat Biotechnol. 2015;33(9):941.
52. Fröhlich E. The role of surface charge in cellular uptake and cytotoxicity of medical nanoparticles. Int J Nanomedicine. 2012;7:5577–91.
53. Ajdary M, Moosavi MA, Rahmati M, Falahati M, Mahboubi M, Mandegary A, et al. Health concerns of various nanoparticles: a review of their in vitro and in vivo toxicity. Nanomaterials (Basel). 2018;8:E634.
54. Moss DM, Curley P, Kinvig H, Hoskins C, Owen A. The biological challenges and pharmacological opportunities of orally administered nanomedicine delivery. Expert Rev Gastroenterol Hepatol. 2018;12:223–36.
55. Kydd J, Jadia R, Velpurisiva P, Gad A, Paliwal S, Rai P. Targeting strategies for the combination treatment of cancer using drug delivery systems. Pharmaceutics. 2017;9:46.
56. Vaiserman A, Koliada A, Zayachkivska A, Lushchak O. Nanodelivery of natural antioxidants: an anti-aging perspective. Front Bioeng Biotechnol. 2020;7:447.
57. Conte R, Marturano V, Peluso G, Calarco A, Cerruti P. Recent advances in nanoparticle-mediated delivery of anti-inflammatory phytocompounds. Int J Mol Sci. 2017;18:709.
58. Gu M, Wang X, Toh TB, Chow EK-H. Applications of stimuli-responsive nanoscale drug delivery systems in translational research. Drug Discov Today. 2018;23:1043–52.
59. Gurr JR, Wang ASS, Chen CH, Jan KY. Ultrafine titanium dioxide particles in the absence of photoactivation can induce oxidative damage to human bronchial epithelial cells. Toxicology. 2005;213(1–2):66–73.
60. Griffitt RJ, Luo J, Gao J, Bonzongo JC, Barber DS. Effects of particle composition and species on toxicity of metallic nanomaterials in aquatic organisms. Environ Toxicol Chem. 2008;27(9):1972–8.
61. Danaei M, Dehghankhold M, Ataei S, Hasanzadeh Davarani F, Javanmard R, Dokhani A, et al. Impact of particle size and polydispersity index on the clinical applications of lipidic nanocarrier systems. Pharmaceutics. 2018;10:1–17.
62. De Jong WH, Hagens WI, Krystek P, Burger MC, Sips AJ, Geertsma RE. Particle size-dependent organ distribution of gold nanoparticles after intravenous administration. Biomaterials. 2008;29(12):1912–9.
63. Sahay G, Alakhova DY, Kabanov AV. Endocytosis of nanomedicines. J Control Release. 2010;145:182–95.
64. Gupta AK, Gupta M. Cytotoxicity suppression and cellular uptake enhancement of surface modified magnetic nanoparticles. Biomaterials. 2005;26(13):1565–73.
65. Mishra AK, Sahoo PK. Characterization of nanophytopharmaceuticals. In: Rudrapal M, editor. Phytoantioxidants and nanotherapeutics. Hoboken: Wiley; 2022. p. 239–65.
66. Khan T, Gurav P. PhytoNanotechnology: enhancing delivery of plant based anti-cancer drugs. Front Pharmacol. 2018;8:1002.
67. Sandhiya V, Ubaidulla U. A review on herbal drug loaded into pharmaceutical carrier techniques and its evaluation process. Future J Pharm Sci. 2020;6:1–16.
68. Mukherjee PK, Harwansh RK, Bhattacharyya S. Bioavailability of herbal products: approach toward improved pharmacokinetics. In: Evidence-based validation of herbal medicine; Elsevier: Amsterdam, 2015.
69. Ganesan P, Arulselvan P, Choi DK. Phytobioactive compound-based nanodelivery systems for the treatment of type 2 diabetes mellitus - current status. Int J Nanomedicine. 2017;12:1097–111.
70. Sarwa KK, Patel D, Uraon D, Jangde MS, Suryawanshi VK, Saraf S. Nanoparticulate delivery systems for phytoconstituents. In: Rudrapal M, editor. Phytoantioxidants and nanotherapeutics. Hoboken: Wiley; 2022. p. 121–38.
71. Bozzuto G, Molinari A. Liposomes as nanomedical devices. Int J Nanomedicine. 2015;10:975–99.
72. Alexander A, Ajazuddin PRJ, Saraf S, Saraf S. Recent expansion of pharmaceutical nanotechnologies and targeting strategies in the field of phytopharmaceuticals for the delivery of herbal extracts and bioactives. J Control Release. 2016;241:110–24.
73. Palai S. Nanotherapeutics of phytoantioxidants for inflammatory disorders. In: Rudrapal M, editor. Phytoantioxidants and nanotherapeutics. Hoboken: Wiley; 2022. p. 377–404.
74. Paul A, Guha N, Zaman MK. Nanotherapeutics of phytoantioxidants for CNS disorders. In: Rudrapal M, editor. Phytoantioxidants and nanotherapeutics. Hoboken: Wiley; 2022. p. 449–76.
75. Poovi G, Damodharan N. Lipid nanoparticles: a challenging approach for oral delivery of BCS Class-II drugs. Future J Pharm Sci. 2018;4:191–205.
76. Sagar R, Pandey SP, Sah AK. Nanotherapeutics of phytoantioxidants for viral infections. In: Rudrapal M, editor. Phytoantioxidants and nanotherapeutics. Hoboken: Wiley; 2022. p. 317–50.
77. Sarma H, Kashyap P, Zothantluanga JH, Devi R. Nanotherapeutics of phytoantioxidants for cardiovascular diseases. In: Rudrapal M, editor. Phytoantioxidants and nanotherapeutics. Hoboken: Wiley; 2022. p. 405–31.
78. Razavi H, Janfaza S. Ethosome: a nanocarrier for transdermal drug delivery. J Paramed Sci. 2015;6:38–43.
79. Punia S, Sandhu KS, Kaur M, Siroha AK. Nanotechnology: a successful approach to improve nutraceutical bioavailability. In: Nanobiotechnology bioformulation. Cham: Springer; 2019. p. 119–33.
80. Ghasemiyeh P, Mohammadi-Samani S. Solid lipid nanoparticles and nanostructured lipid carriers as novel drug delivery systems: applications, advantages and disadvantages. Res Pharm Sci. 2018;13(4):288–303.
81. Prajapati SK, Jain D, Parveen S, Maji S, Deb PK. Nanodelivery of antioxidant herbal extracts, spices, and dietary constituents.

In: Rudrapal M, editor. Phytoantioxidants and nanotherapeutics. Hoboken: Wiley; 2022. p. 145–71.
82. Gursoy RN, Benita S. Self-emulsifying drug delivery systems (SEDDS) for improved oral delivery of lipophilic drugs. Biomed Pharmacother. 2004;58:173–82.
83. Lim CL, Raju CS, Mahboob T, Kayesth S, Gupta KK, Jain GK, et al. Precision and advanced nano-phytopharmaceuticals for therapeutic applications. Nanomaterials (Basel). 2022;12(2):238.
84. Yata VK, Ranjan S, Dasgupta N, Lichtfouse E. Nanopharmaceuticals: principles and applications. New York: Springer International; 2020.
85. Kumar V, Garg V, Dureja H. Nanomedicine-based approaches for delivery of herbal compounds. Tradit Med Res. 2022;7(5):48.
86. Hafez DA, Elkhodairy KA, Teleb M, Elzoghby AO. Nanomedicine-based approaches for improved delivery of phyto-therapeutics for cancer therapy. Expert Opin Drug Deliv. 2020;17(3):279–85.
87. Saraf S, Gupta A, Alexander A, Khan J, Jangde M, Saraf S. Advancements and avenues in nanophytomedicines for better pharmacological responses. J Nanosci Nanotechnol. 2015;15(6):4070–9.
88. Alharbi WS, Almughem FA, Almehmady AM, Jarallah SJ, Alsharif WK, Alzahrani NM, et al. Phytosomes as an emerging nanotechnology platform for the topical delivery of bioactive phytochemicals. Pharmaceutics. 2021;13(9):1475.
89. Zothantluanga JH, Das P, Sarma H, Umar AK. Nanotherapeutics of phytoantioxidants for parasitic diseases and neglected tropical diseases. In: Rudrapal M, editor. Phytoantioxidants and nanotherapeutics. Hoboken: Wiley; 2022. p. 351–76.
90. Chakraborty P, Chakraborty DD, Chakraborty M. Biochemical and therapeutic targets for nanophytomedicines. In: Rudrapal M, editor. Phytoantioxidants and nanotherapeutics. Hoboken: Wiley; 2022. p. 203–29.
91. Milivojevic M, Pajic-Lijakovic I, Levic S, Nedovic V, Bugarski B. Alginic acid: sources, modifications and main applications. In: Moore A, editor. Alginic acid - chemical structure, uses and health benefits. New York: Nova Science Publishers, Inc.; 2015. p. 45–88.
92. Milivojevic M, Pajic-Lijakovic I, Bugarski B. Biological macromolecules in cell encapsulation. In: Nayak AK, Dhara AK, Pal D, editors. Biological macromolecules. Cambridge, MA: Academic Press, Elsevier; 2021. p. 491–528.
93. Taha M, Alhakamy NA, Md S, Ahmad MZ, Rizwanullah M, Fatima S, et al. Nanogels as potential delivery vehicles in improving the therapeutic efficacy of phytopharmaceuticals. Polymers (Basel). 2022;14(19):4141.
94. Milivojevic M, Pajic-Lijakovic I, Bugarski B, Hasnain M, Nayak AK. Gellan gum in drug delivery applications. In: Nayak AK, Hasnain M, editors. Natural polysaccharides in drug delivery and biomedical applications. Cambridge, MA: Academic Press, Elsevier; 2019. p. 145–86.
95. Milivojevic M, Pajic-Lijakovic I, Bugarski B. Recent advances of alginates as material for biomedical applications. In: Hasnain M, Nayak AK, editors. Alginates: versatile polymers in biomedical applications and therapeutics. New York: Apple Academic Press, CRC Ress, Taylor & Francis Group; 2019. p. 25–87. Chapter 2.
96. Chen F, Liu Q. Demystifying phytoconstituent-derived nanomedicines in their immunoregulatory and therapeutic roles in inflammatory diseases. Adv Drug Deliv Rev. 2022;186:114317.
97. Kleinstreuer C, Childress EM, Kennedy AS. Targeted drug delivery: multifunctional nanoparticles and direct micro-drug delivery. In: Becker S, Kuznetsov A, editors. Transport in biological media. London: Elsevier; 2012. Chap. 10.
98. Fadia P, Tyagi S, Bhagat S, Nair A, Panchal P, Dave H, et al. Calcium carbonate nano- and microparticles: synthesis methods and biological applications. 3 Biotech. 2021;11(11):457.
99. Bhushan R, Upadhyay S, Rudrapal M. Nanotherapeutics of phytoantioxidants for microbial infections. In: Rudrapal M, editor. Phytoantioxidants and nanotherapeutics. Hoboken: Wiley; 2022. p. 289–316.

Pharmacovigilance of Herbal Medicines for Lifestyle Diseases

Shuvam Sar, Tanmoy Banerjee, Arun Baidya, Saptapadi Saha, Ankit Kumar, Amit Kumar Halder, Madhurima Ghosh, Vilas D. Nasare, and Nilanjan Ghosh

Abstract

Herbal medicines contain chemical compounds which can potentially forestall illnesses and afflictions. Unfortunately, the increasing popularity of herbal drugs worldwide led to the most pervasive misconception that they are perfectly safe and are recommended by patients as self-medication, which frequently brings about unfavourable outcomes and adverse effects. To combat and create awareness of these adverse events, establishing pharmacovigilance (PV) procedures for herbal medications is becoming more widely recognized on several levels. Different ways of using herbal treatments, which might present unique toxicological issues, whether taken separately or in conjunction with other pharmaceuticals, might represent contributing factors to the unpleasant drug response in lifestyle disorders. The chapter contributes to delivering basic knowledge of PV and adverse drug reactions, along with the safe dosage range of phytochemicals in lifestyle diseases, viz. obesity, diabetes, and atherosclerosis. The discussed toxicity and phytovigilance profile of multitudinous natural remedies in enormous lifestyle disorders makes the work unique.

S. Sar · T. Banerjee · A. Baidya · S. Saha · A. Kumar
N. Ghosh (✉)
Department of Pharmaceutical Technology, Jadavpur University, Kolkata, West Bengal, India
e-mail: nilanjanghosh.phamacy@jadavpuruniversity.in

A. K. Halder
Dr. B. C. Roy College of Pharmacy and Allied Health Sciences, Dr. Meghnad Saha Sarani, Bidhannagar, Durgapur, West Bengal, India

M. Ghosh · V. D. Nasare
Department of Pathology and Cancer Screening, Chittaranjan National Cancer Institute, Kolkata, India

Keywords

Pharmacovigilance · Herbal medicine · Lifestyle diseases · Adverse drug reaction · Toxicity

1 Introduction

Pharmacovigilance (PV) is the term used to describe the awareness of processes involved in the identification, assessment, comprehension, and avoidance of adverse drug reactions (ADRs). By informing patients about the hazards and advantages of medications, PV seeks to improve patient health and safety [1]. PV can scientifically be defined by gathering, supervising, investigating, analysing, and examining data from healthcare professionals and patients on the adverse effects of pharmaceuticals, folklore medicines, biological products, blood products, herbal remedies, vaccines, and medical devices with the goal of discovering novel facts concerning the risks related to the safety of patients. Maximizing medication safety while sustaining public trust has grown to be a complicated problem. Therefore, throughout a product's lifespan, from discovery to post-market, pharmaceutical and biotechnological organizations should track drug hazards and deliberately assess and manage the risk exposures [2]. ADRs are a serious clinical problem on a worldwide scale, leading to considerable morbidity and death. ADRs are interpreted as the undesirable or detrimental responses produced by an intervention associated with the application of a medicinal product [3]. Adverse effects typically forecast risk from subsequent usage and indicate prevention, treatment, dosage adjustment, and therapy termination. A thorough medical record might help a prescriber comprehend the patient's prior drug treatment circumstances, especially in recognizing prior ADRs that can prevent further exposure to the drug. The crucial aspect of PV is voluntary reporting regarding suspicions of ADRs, yet generally, ADRs are drastically understated in every public health sector [4].

A. K. Dhara, S. C. Mandal (eds.), *Role of Herbal Medicines*, https://doi.org/10.1007/978-981-99-7703-1_26

Threats to public health and worries about safety are becoming more widely acknowledged as the usage of herbal medicines throughout the world increases and a large number of novel medications are released into the market. Many herbal medicines are still untested, despite the fact that they are frequently used, and certain varieties have intriguing potential to create ADRs. This is also well known that the majority of herbal medications are jeopardized due to poor quality controls and lack of patient knowledge [5]. Hence, it is crucial to provide adequate public information in order to enable an improved awareness of potential hazards and ADRs arising from the utilization of herbal products. The material covered in this chapter is restricted to significant PV monitoring, toxicity, and ADR-related difficulties resulting from the utilization of herbal medications in several lifestyle diseases like diabetes, obesity, and atherosclerosis.

2 Pharmacovigilance of Herbal Drugs for Lifestyle Diseases

According to the World Health Organization (WHO), traditional medicine is the knowledge, skills, and practices utilized for a long time to maintain health and prevent, diagnose, improve, or treat physical and mental illness. It is based on theories, beliefs, and experiences unique to many cultures. Regardless of culture, a holistic way of living, balancing the mind and body with the environment is the common goal. Generally, instead of concentrating on the specific condition or sickness that the patient is dealing with, the emphasis is typically placed on the person's general wellness. The use of herbs is a core part of all systems of traditional medicine. Traditional Chinese Medicine (TCM) and Ayurveda are still being used by people worldwide, mostly involving herbs. That is why traditional medicine is now regarded as equivalent to herbal medicine [6]. Three of the most popular botanical items, including *Panax ginseng*, *Allium sativum*, and *Ginkgo biloba*, have their immense contribution to TCM and are still used to treat a variety of illnesses even today. Ayurveda and TCM share some familiar biological sources as effective therapeutical remedy. Turmeric, amla, aloe vera, tulsi, etc., are the most common plants that are used for a huge variety of diseases in Ayurveda. However, herbal medicine is accepted by most of the conscious people of the world because of its low price and fewer side effects. Herbal compounds are the primary sources of approved drugs to treat various lifestyle diseases. Eventually, these must have some limits in use from the safety point of view. Whenever the dose is out of the therapeutic range, it will not suffice for the specific disease it is used for. Even when taken in high doses, the consequences may be very hazardous, even near death to some extent through ADRs [7]. Nowadays, herbal formulations are widely used for lifestyle diseases, specifically metabolic diseases like diabetes, obesity, and cardiovascular diseases. The misconception that herbal medicines have no side effects, so they are entirely safe, compels most consumers to buy and use them without a doctor's prescription. This malpractice leads to severe adverse reactions, which can also be fatal. Most adverse events (AEs) are caused due to poor product quality or improper use of herbal products. In addition, improper regulatory measures, weak quality control, and uncontrolled distribution channels contribute to the occurrence of such events. Other elements that may contribute to the toxicity of herbal medicine products include adulteration, toxic plant components or metabolites, environmental pollutants (heavy metals, pesticides), and microbial contamination (toxigenic fungi). For the patient's safety, a proper evaluation is crucial. However, the awareness of the need to develop PV practices for herbal medicines is rising at several levels. Several models of PV and its associated tools developed for synthetic drugs are currently being applied for the safety monitoring of herbal medicines [8, 9]. Strengthening national regulation, registration, quality control, and assurance of herbal medicines is the responsibility of member states of the World Health Organization (WHO) [10].

In 2000, in the United States, when herbal medicine was found to be used by a vast population, it denoted the high rise in the global use of herbal medicine as well as its acceptance. It so did the incorporation of safety management [11]. National PV centres of the international drug monitoring programmes asked for urgent assistance from WHO in 2000 and 2001 to form nation-based herbal medicine safety vigilance systems. Following that, in May 2003, the 56th World Health Assembly adopted resolution WHA56.31 on traditional medicine, asking Member States to create new safety monitoring systems or to enhance the existing ones to keep tracking herbal treatments and other traditional practices. In February 2004, Health Canada served as the venue for the WHO Consultation on Safety Monitoring of Herbal Medicines. The WHO focused on the importance of including traditional medicines in surveillance systems and issued guidelines on safety monitoring [12].

2.1 The Importance of Safety Monitoring of Herbal Medicine

Safety is always a key consideration in medical practice for any treatment. Due to the widespread use of natural remedies, the safety of traditional medicines has been a top concern for member nations. However, contrary to popular belief, misusing traditional medicine or combining it with other drugs can occasionally be hazardous to a person's health because of the synergistic effect. Drug interactions may occur when two or more medications are taken at the same time, and there is a chance that one of them might hamper the other effects. This activity could be antagonistic when a drug's effect is reduced, synergistic when the effect is boosted, or a novel effect can take place that none of the medications can produce independently [13].

2.2 Adverse Drug Reactions Caused by Herb-Drug Interaction

Interactions between herbal medicine and synthetic or semi-synthetic drugs are becoming common due to the recent tendency in medicated treatment regimens to combine traditional medicine with modern medicine. Specific complications arising from combining conventional remedies and herbal medicines affect specific patient populations, such as cancer patients, where the incidence and risk of interactions can be fatal. Ultimately, it is a major concern and should be reported promptly. It has been noticed that the group of patients who take drugs getting metabolized by cytochrome P450 enzyme families or where bioavailability is affected by P glycoprotein are at high risk of experiencing herb-drug interactions [14]. Careless monitoring of ADRs can lead to treatment failure, on-treatment AEs, and even impact clinical trial data. UMC recently conducted a study demonstrating that pharmacokinetic and pharmacodynamic drug interactions can be captured in a database. Hopefully, the herb-drug interactions might also be investigated [15]. Some common examples of herb-drug interactions have been discussed below.

Example 1: When antineoplastics like imatinib and irinotecan, antiretrovirals like indinavir and nevirapine, immunosuppressants like ciclosporin, inotropics like digoxin and warfarin are used with St. John's wort (*Hypericum perforatum* L.) and may all result in lowered plasma concentrations and reduced efficacy.

Example 2: Another example of a herb-drug interaction involves green tea (*Camellia sinensis* (L.) Kuntze) used with folic acid, which results in decreased blood levels of folate [16].

Example 3: An elevated anticoagulant activity has been noticed while Goji berry (*Lycium barbarum* L.) and warfarin are administered together [17].

2.3 Methods to Report ADR in Herbal Pharmacovigilance

A PV system network involves all stakeholders, including regional and national pharmacovigilance centres, private and public healthcare facilities, drugstores/pharmacies, patients/consumers, and producers, suggesting that everyone involved in the network system should participate in the reporting. As herbal medications are frequently purchased over-the-counter from retail pharmacies with prior post-marketing surveillance, nurses and neighbourhood chemists must report ADRs. The safety of conventional medicine can be monitored with the aid of a nationwide PV system. Three different monitoring channels are available: the spontaneous/voluntary reporting system, the stimulated reporting system, and the active and passive surveillance system [18]. However, aside from SRS, alternative monitoring systems are less established Compared to the monitoring systems used for conventional medications. In addition, prescription-event monitoring and ADR reporting forms are currently being evaluated as substitutes for ADR reporting in herbal pharmacovigilance [19].

2.3.1 Spontaneous Reporting System

Healthcare professionals must identify and report any suspected ADR to their national pharmacovigilance centres or the manufacturer to conduct pharmacovigilance globally. This kind of reporting system is known as spontaneous reporting, a vital component of worldwide pharmacovigilance exertion [20]. After the marketing of medicines, spontaneous reports are crucial to ensure their safety after marketing. Rare AEs that are overlooked while conducting clinical trials or other pre-marketing investigations could be exposed by spontaneous reports after marketing, which is why it is crucial for safety management. Furthermore, it offers vital subtle elements on at-risk socioeconomics, chance variables, and clinical characteristics of recognized noteworthy ADRs. However, an enormous downside is that, despite the seriousness of the side effects, overburdened medical staff cannot manage to continuously report the ADRs of a specific medicine [21].

2.3.2 Stimulated Reporting System

Stimulated reporting refers to the public exposure of a safety concern by publishing any national or international pharmacovigilance agency alert that causes noticeably higher AE reporting rates for the medication and the particular AE listed in the alert. It is used to encourage and enhance the reporting of adverse events by healthcare professionals based on pre-designed methodologies in specific circumstances, even in hospital settings, for new goods or for brief periods of time. Stimulated collects AE reports from consumer-based 'support group' activities and litigation-related reporting activity. The use of post-marketing AE data is significantly impacted by stimulated reporting since alert-driven changes in reporting may affect the validity of comparative studies and associated analytical techniques like disproportionality analysis. If this reporting system is appropriately used, companies can alert consumers about medicines' AEs before they are hugely marketed [22].

2.3.3 Active and Passive Surveillance System

In an active surveillance system, medical data of healthcare practitioners and patients are evaluated during visits to healthcare institutions to spot a particular illness or condition. Although less time- and money-consuming to operate, passive monitoring techniques face the risk of underreporting some illnesses. Active surveillance is a programmed monitoring technique that employs an ongoing, pre-planned strategy to properly determine the number of unfavourable events patients see using a particular medicine. Patients are asked to take a brief survey and provide their permission for more communication [23].

2.4 Different Pharmacovigilance Databases for ADR Reporting

The investigation of herbal medication safety is one of the main obstacles in pharmacovigilance. Information indicating a previously unknown relationship between herbal remedies and ADRs or certain association features is a safety signal. Signals can come from various sources, most notably active monitoring systems, a review of spontaneous case reports, and research findings. The development of computer technology in PV has a significant positive impact on extracting precise data. Each component of the pharmacovigilance system, including risk-tracking signals and case processing, may be further streamlined using AI technology and machine-learning techniques, which will shorten the processing time overall [24]. Some widely used ADR reporting PV databases have been discussed below.

2.4.1 AB Cube

It is a multi-vigilance compliance database with a fully integrated user-friendly platform with tools and features complied with modern industry standards and statutory regulations for medication safety, allowing the monitoring of drug safety profiles. This is run as a cloud-based software-as-a-service (SaaS) solution. The distinguishing features of AB Cube Safety Easy are the user interface (UI), synchronized performance, and ease of entering and removing data from the database [25].

2.4.2 ARISg

It is the most frequently used pharmacovigilance database, used to maintain critical drug safety data internationally and speed up the management of ADR. This database supports all pharmacovigilance methods, including automatically creating reports on AEs ready for resignation and case entry. In addition, modern automation and specialized workflow technologies help the purpose to be served [26].

2.4.3 Clinevo Safety

This complete pharmacovigilance safety database platform includes features including signalling management, a case interpreting, PV Intake, artificial intelligence (AI) tools, and safety data analysis. More than 100 end customers utilize it, and it presently handles 100,000 cases yearly, notifying users when cases have due dates approaching. The database is built on a simplified technical stack. Thus, there are numerous configuration possibilities and few changes. Additionally, it provides straightforward pricing that is simplified without compromising quality. It aims to minimize end-user effort by continually automating processes and increasing AI capabilities [27].

2.4.4 EVWEB (EudraVigilance Web)

This European data processing and management system makes this European data processing, and management system makes a protected possible. It also facilitates effectively creating safety and acknowledgement messages and reports for pharmaceutical products.

2.4.5 FAERS Public Database

The FAERS Public Dashboard is a freely accessible web-based service that includes mandated data reports from pharmaceutical companies and voluntary ADR reports from patients and healthcare professionals, mostly from the United States (MedWatch, ADR reporting programmes). The inquiries may be viewed as beginning in 1968, even though public access started in September 2017 [28].

2.4.6 Oracle Argus Safety

It is a sizable platform utilized to keep track of local data items and was first developed to meet the rigorous pharmacovigilance requirements of the life science industry. It provides electronic signatures, data entry via electronic means, and the capacity to search for counterfeit goods. Additionally, it makes safer conclusions when determining local labelling options based on local datasheets or licenses and ensures compliance with international requirements [29].

2.4.7 PvNET

It is one of the most effective applications for managing AE, ADR information, and ICSR submission. It is a comprehensive drug safety solution. Incorporating safety information is made possible by the MedDRA dictionary management system, supporting users in making critical judgements. Automated story writing with multilingual text support for serious AE [30].

2.4.8 VigiAccess

It is a commonly used program by which users can browse information regarding harmful medication side effects from VigiBase.

2.4.9 VigiBase

With over 30 million records, it is the most extensive database of probable pharmaceutical side effects.

2.4.10 VigiFlow

It is a web-based ICSR management system for global drug surveillance that the WHO and MedDRA endorse. It gathers, processes, and exchanges data to allow effective data analysis.

2.4.11 VigiLyze

It is a website that provides a brief but comprehensive analysis of VigiBase and may be used for online research.

2.4.12 VigiMatch

It is a system that searches through data to locate comparable individual case reports using probabilistic pattern matching.

2.4.13 VigiRank

It is an advanced technique for figuring out statistical signals that consider disproportionate reporting patterns in combination with the accuracy, rapidity, and geographic distribution of individual case monitoring [25].

2.5 Challenges in Safety Monitoring of Herbal Medicines and Their Management

2.5.1 Regulation

There are many laws in various nations. In some countries, herbal medicines can be classified as prescription or over-the-counter or other medicines. Typically, qualified suppliers and distributors of the relevant products also belong to the national regulatory framework. Consequently, their regulatory status determines the distribution method or access to these items [7].

2.5.2 Quality Assurance and Control

At a 2001 WHO meeting, the quality monitoring of finished herbal products was first briefly discussed. Although every nation that monitors herbal medicines must have quality assurance and control strategies in place, including national quality standards and requirements for herbal materials, good manufacturing practices (GMPs) for herbal medicines, labelling, and licensing structures for production, purchases, and marketing in general. As inadequate supervision and quality management may result in significant adverse responses, these processes are crucial for ensuring the security and efficacy of herbal remedies. Particularly herbal medications, which might have been contaminated with potentially hazardous components or adulterated with unknown potent substances and residues, may be turned out to be a high issue of concern. Because of this, the specifications and procedures for quality control of completed herbal products—especially for herbal combination products—are far more complicated than those for conventional medications. The calibre of the raw materials utilized impacts the quality of such items. To provide quality control and assurance, it is crucial to use proper agricultural and collecting procedures for using medicinal plants, such as plant cultivation and selection [31].

2.5.3 Safety Monitoring of Herbal Medicines

Other than doctors, dentists, chemists, and nurses, herbal medicine providers are frequently left out of ADR and AE reporting systems in some countries. National reporting methods should be remodelled to include all providers of herbal medicines (prescribers and dispensers) and providers of traditional, complementary, and alternative medicine if adequate coverage of herbal medicines is to be attained. Herbal medications make up a significant percentage of the non-prescription drug industry, and many of them enter this market without undergoing after-sale safety surveillance, unlike prescription drugs. Therefore, it is crucial to take action to strengthen pharmacovigilance measures within the context of over-the-counter medications. Despite the fact that many of these items are available outside of pharmacies, community pharmacists and nurses may contribute to ensuring the safety of over-the-counter drugs. Consumer complaints about side effects should be treated as a vital data source that might help researchers spot warning signs of potential consequences of herbal medications [10].

2.5.4 Documentation and Coding of Herbal Medicines

The coding for adverse events and reactions linked to herbal medication should be similar to that used for other drugs. Adopting a globally implemented classification and recognition system when sending data to the UMC is better. The WHO Drug Dictionary (WHO-DD), developed to arrange and categorize information on the names of herbal products and their ingredients in the same way that comparable information on other pharmaceuticals is preserved, was promoted by the UMC. The UMC proposed the herbal anatomic-therapeutic-chemical (HATC) classification system for the therapeutic categorization of herbal products. This combinatorial system used as a component of the global WHO database manages all levels of data input and is structurally identical to the anatomic-therapeutic-chemical (ATC) classification used for chemical ingredients in other therapies [32]. A system checklist for cross-referencing botanical and common names used as ingredient names was also recommended by the UMC. The WHO-DD, the HATC classification, and the checklist, according to the UMC, could provide helpful resources for the national pharmacovigilance centres when the reporter asks questions to improve the precision and clarity of reports. Herbal medicines usually contain multiple constituents, and it is not always possible to distinguish all of them. In such cases, the product name should be recorded and referred to the UMC, which will assist with identification. A national inventory or catalogue of medicinal plants may be useful as a resource of medicinal plants and their use in the community. If any issue arises with the finished herbal medicine or any of its raw ingredients were imported from another country, the drug regulatory agency of the exporting country may be able to help. The corresponding medicinal plants' precise Latin binomial botanical names should be included, along with information on the plant parts utilized, the extraction process, and the preparation steps taken. By collaborating with university pharmacognosy departments, botanists, zoologists, and personnel of botanical gardens, it is possible to avoid delays or to cancel the submission of a report by utilizing a common vernacular phrase.

2.5.5 WHO Database Analysis

The planned 'Database management and classification for coding of herbal medicines', of which the previously mentioned HATC is one aspect, is being used to strengthen the WHOWHO database handled by the UMC. Data-mining techniques that have been effective on many reports for other pharmaceuticals will be used to identify signals on reports for herbal treatments. Programs that offer regular analysis and data generation suited for local requirements should be developed. The effectiveness of these measures depends on the volume and quality of the data provided by the national Pharmacovigilance Centres [10].

3 Physiological Effects of Herbal Compounds on Lifestyle Diseases

3.1 Obesity

Obesity is an extremely prevalent worldwide health issue. Over the past few years, the incidence of obesity has been rising rapidly. An excessive deposition of body fat caused by an imbalance between energy consumption and utilization of energy characterizes this metabolic disease. More precisely, the accumulation of adipose tissue impacts the disease progression. A typical indicator for identifying obesity and overweight is body mass index (BMI), calculated as weight/(height)^2. According to WHO, overweight and obesity refer to BMI $\geq$ 25 and BMI $\geq$ 30, respectively. Obesity is a significant risk factor for a variety of diseases, such as diabetes, heart disease, endocrine and metabolic disorders, musculoskeletal problems, in cancers [33]. The best course of therapy for obesity is loss of weight, which should involve following a suitable diet plan that reduces caloric intake and changing one's lifestyle to incorporate more physical activity but maintaining a long-term lifestyle change is difficult. For this reason, sometimes drug treatment is necessary [34] but medicines used to treat obesity can have a variety of adverse effects, particularly those that affect monoamine neurotransmitters and lead to drug abuse or dependence [35]. In addition, some surgical procedures also have complications, including deep vein thrombosis, postoperative fistula, and long-term complications such as malnutrition and anaemia [36]. Several herbal medicines beneficial in treating obesity have been summarized in Table 1. Structures of most common and potent phytomolecules with action on obesity has been shown in Fig. 1.

3.2 Diabetes

Hyperglycaemia or high blood glucose level characterizes diabetes. The two most prevalent types of diabetes are type 1, in which the autoimmune loss of pancreatic β cells results in impaired insulin production, and type 2, in which insulin resistance or diminished secretion results in hyperglycaemia [51]. The most significant risk factor for type 2 diabetes mellitus (T2DM) is obesity, and the huge rise in childhood overweight and obesity appears to be the cause [52]. Diabetes can be managed nowadays with various potent synthetic medications, but the contribution of traditional natural remedies (Table 2) is unforgettable and is also used till date emerging effective therapies. Beyond obesity as a risk factor, it is recognized that a kind of 'lean diabetes mellitus' describes a condition in which fundamental inadequacies in insulin production, caused by malfunction of the pancreatic β cells, serve as the primary trigger for the onset of diabetes [53]. In the United States, the lifetime chance of developing diabetes is 40%, with 29.1 million Americans (9.3%) estimated to have the disease. In the United States, 86.1 million persons are thought to have prediabetes, in addition to those diagnosed with the disease [54]. Nearly every organ and tissue in the body is affected by the complications of diabetes, which are also the main cause of cardiovascular morbidity and death, blindness, renal failure, and amputations. A more severe form of T2DM with the early onset of major consequences has also been associated with type 2 diabetes in adolescents and young adults (up to age 40) [55]. From the dawn of civilization, people consciously or unknowingly have been consuming different plants and their different parts, that are mentioned in traditional herbal medicines and are currently being worked on to establish their roles and to assess the mechanism by which they control blood sugar level. The most common plants include turmeric, amla, bael, ginger, etc (Fig. 2) which are used by people all over the world to control diabetes even without knowing their proper dose and dosage.

Table 1 Safety dose range of herbal medicines in controlling obesity

Sl. No.	Name	Biological source	Family	Physiological effects	Dosage	Refs
1	Berberine	*Rhizoma coptidis*	Ranunculaceae	Decreased visceral adipose tissue, blood glucose, and lipid levels	1.6 g daily for human	[37]
2	Ginsenoside	*Panax ginseng*	Araliaceae	Decreased overweight, food consumption, hyperglycaemia, TC, and TG	200 mg to 3 g by mouth daily for up to 12 weeks for human	[38]
3	Astragalosides I and II	*Astragalus membranaceus* (Fisch.) Bunge	Fabaceae	Decreased body weight and food intake; increased HDL-C	Radix Astragali (2 g/kg/day)/oral gavage/12 weeks (experimental rat)	[39]
4	Catechin	*Camellia sinensis*	Theaceae	Decreased insulin resistance, TG, TC, and FFA	400 to 800 mg/ per day aqueous extract	[40]
5	Ephedra	*Ephedra sinica* Stapf	Ephedraceae	Decreased fasting glucose and increased HDL-C	The maximum safe dose of ephedra is 32 mg/day for human	[41]
6	Capsaicin	*Capsicum frutescens*	Solanaceae	Decreased heat-shock protein-27 and STEAP3 protein; increased fatty acid oxidation	2 to 6 mg per day	[36]
7	Curcumin	*Curcuma longa*	Zingiberaceae	Increase hepatic acyl-CoA oxidase activity and lower the lipid levels	Effective dose is 500 to 2000 mg per day	[42]
8	Capsicoside-G	*Capsicum annuum*	Solanaceae	Increase body weight and food efficiency ratio Decrease epididymal adipose tissue weight and adipocyte hypertrophy Increase receptor γ, CCAAT	1.08 mg/kg for human	[43]
9	Ginkgo	*Ginkgo biloba*	Ginkgoaceae	Decrease food/energy intake, body weight gain; increased IRS-1 and AdipoR1 Decrease PTP1B, TNF-α levels, and NFκB-p65	120 to 240 mg daily	[44]
10	6-gingerol	*Zingiber officinale*	Zingiberaceae	Ginger acts as a potential body cleanser which removes the food logged in the digestive system and avoids fat storage 45.7% reduction in pancreatic lipase activity	0.5 to 3 g daily for up to 12 weeks	[45]
11	Anthocyanins	Blueberry (*Vaccinium ashei*)	Ericaceae	7.30%–9.81% reduction in body weight; reduction in WAT adipocyte sizes	160 mg twice a day for 2 months for human	[46]
12	Cinnamon	*Cinnamomum verum*	Lauraceae	Significantly reduce body weight and BMI Promoting weight loss by suppressing pancreatic amylase, lowering intestinal glucose absorption, increasing cellular glucose uptake and glycogen synthesis, blocking gluconeogenesis, and activating insulin receptors	Dosages of more than 2 g/day, administer for 12 weeks	[47]
13	Soybean	*Glycine max*	Fabaceae	18.9% reduction in body weight gain; 26.6% reduction in body fat; higher thermogenin expression	FDA has determined 25 g/ day can safely be consumed	[48]
14	Coffee	*Coffea arabica*	Rubiaceae	Reduction in body weight; liver fat 16.2% reduction in organ WAT weight	400 mg/day for adults and 2.5 mg/kg/day for children and adolescents	[49]
15	Caralluma	*Caralluma fimbriata*	Apocynaceae	Decrease in hunger levels, body weight, BMI, body fat and energy intake	1 g/60 days	[50]

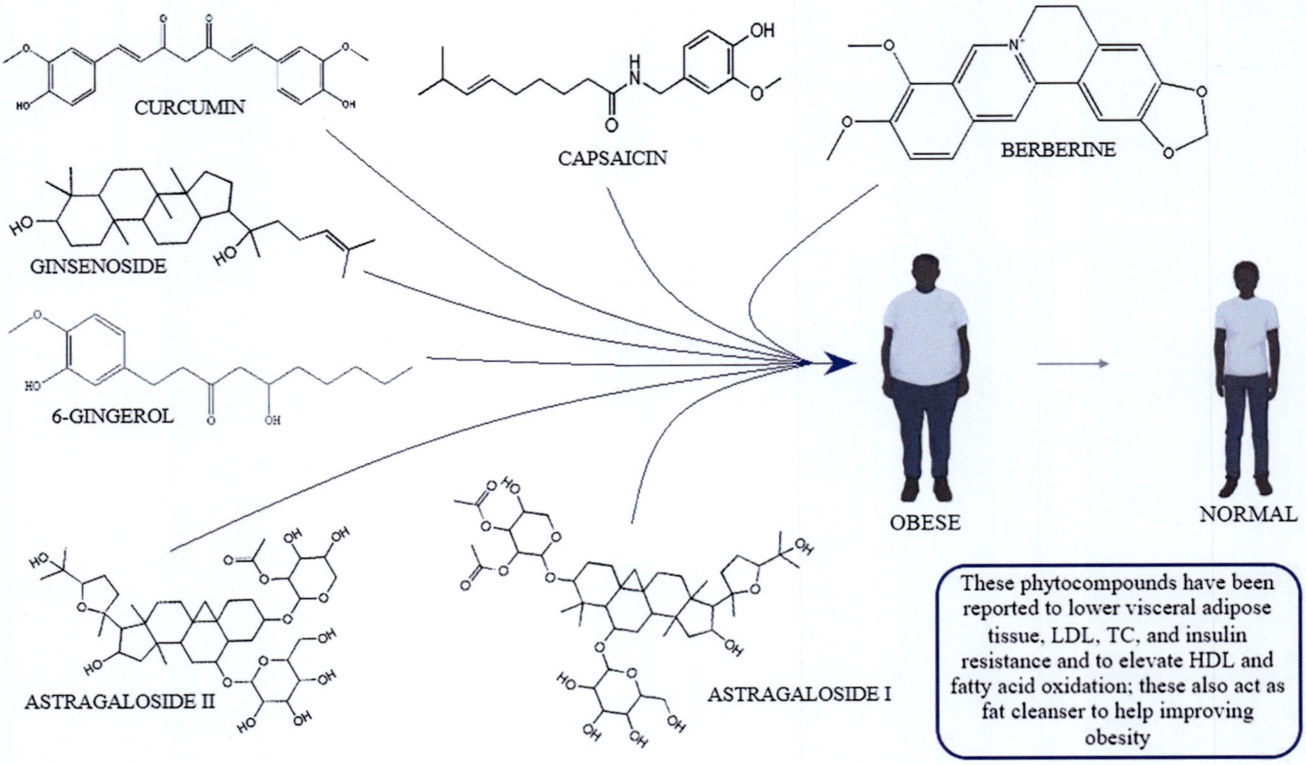

Fig. 1 Phytomolecules with reported action on obesity

3.3 Atherosclerosis

Atherosclerosis is a term of Greek origin that signifies fat accumulation and thickening of the intima of arteries. It consists of two parts: atherosis refers to the deposition of fat cells accompanied by many macrophages, while sclerosis refers to the fibrosis layer composed of smooth muscle cells, leukocytes, and connective tissue [74]. Atherosclerosis is a chronic inflammatory condition characterized by the accumulation of lipids and inflammatory cells in the walls of medium and large arteries. Atherosclerosis is significantly triggered by oxidative stress and inflammation. Excess ROS production and LDL oxidation are risk factors for atherosclerosis. Ox-LDL acting on several cell types promotes atherogenesis (Fig. 3) [75]. Increased oxidative damage is caused by hyperlipidaemia and hyperglycaemia, which affects antioxidant status and lipoprotein levels [74]. Atherosclerosis has become more common in older people and continues to be the most significant cause of death, which can be managed efficiently with herbal medicines with proper dosage (Table 3). The monolayer of endothelial cells (ECs) that lines the internal vessel wall is covered by an intimal gap that accumulates and transforms lipids, inflammatory cells, smooth muscle cells, and necrotic cell debris during a lifespan. Typically, lesion development can reduce blood flow in the lumen by more than 50%, leading to angina, especially during stress or physical activity [76]. The risk factors of atherosclerosis are unhealthy blood cholesterol and lipoproteins levels, high blood pressure, smoking, insulin resistance, diabetes, overweight or obesity, family history of early heart disease, inflammation, stress, and alcohol [74].

4 Phytovigilance of Various Herbal Medicines Across Multiple Lifestyle Diseases

It is very clear from various studies performed at different parts of the world that the whole plant at any dose can't be beneficial at all for health. Even it might be lethal to some extent. Some herbal medicines whether it is the plant extract or any isolated compound have very serious long-lasting effects if taken at wrong dosage or wrong time like pregnancy or infancy. Figure 4 depicts the reported toxicity profile of some common herbal medicines showing their range of adverse effects.

4.1 Cinnamon

Cinnamomum verum J. Presl is the potential source of cinnamon and belongs to the Lauraceae family. Cinnamon is commonly used worldwide as a cooking and colouring agent in many industries. It is a very popular traditional medicine effectively used for diabetes treatment. It also has antimicrobial and antioxidant activity [100]. Although cinnamon is a

Table 2 Safety dose range of herbal medicines in controlling diabetes

Sl. No.	Name	Biological source	Family	Physiological effects	Dosage	Refs
1.	Turmeric	*Curcuma longa*	Zingiberaceae	Inhibited ROS production and SREBP1c gene expression up-regulated PPAR-γ via AMPK activation and CPT1 and ACAT enzymes	12 g per day	[56]
2.	Onion	*Allium cepa*	Liliaceae	Boosted insulin secretion and adiponectin reduced lipid peroxidation, α-glucosidase and increased adiponectin		[57]
3.	Garlic	*Allium sativum*	Alliaceae	Antagonized LXR*α* enhanced LXR expression in the gut and control cholesterol, triglycerides, and glucose homeostasis	10 mg to 10 g	[58]
4.	Bael	*Aegiemarmelos*	Rutaceae	Lowered the concentration of thiobarbituric acid reactive substances, hydroperoxide, lipid peroxidation increased glutathione level	250 mg/kg	[59]
5.	Amla	*Emblica officinalis*	Phyllanthaceae	Lowered FBS and PPBS, HbA1c, lipid levels, SBP	1 g and 2 g per day	[60]
6.	Nayantara	*Catharanthus roseus* (vindoline)	Apocynaceae	Stimulated β cell, stopped caspase 3, caspase 9 and overexpressed Lowered SOD and lipid peroxidation	20 mg/kg body weight	[61]
7.	Neem	*Azadirachta indica*	Meliaceae	Free radical scavenging activity inhibitedα-glucosidase and α-amylase enzymes, oxidative stress, PPBS, FBS, HbA1c, and hsCRP levels Sitosterol boosted AMPKα phosphorylation and glucose absorption	50 and 100 mg/kg	[62–64]
8.	Cinnamon	*Cinnamomum zeylanicum*	Lauraceae	Upsurge insulin levels, insulin receptor signalling, HbA1c reduced intestinal α-glucosidase activity and FBGL	2 g daily	[65]
9.	Epigallocatechin gallate	*Camellia sinensis* L.	Theaceae	Up-regulated tyrosine phosphorylation of the insulin receptor, RBP-4 reduced gene expression of the phosphoenolpyruvate carboxykinase enzyme and proliferation of adipocytes	75 mg/kg; 50 mg (0.7 mg/kg)	[66]
10.	Chirayata	*Swertia chirayita*	Gentianaceae	Increased insulin secretion, expression of PPAR-γ, GLUT-4, and adiponectin	3.72 g/mL	[67]
11.	Gymnema	*Gymnema sylvestre*	Asclepiadaceae	Postponed intestine's blood glucose absorption Increased insulin release	60 days of taking 500 mg per day	[68]
12.	Aloe vera	*Aloe barbadensis*	Xanthorrhoeaceae	Enhanced insulin secretion and metabolism Lowered body weight	300 mg/kg daily	[69]
13.	Ginseng	*Panax ginseng*	Araliaceae	Improved insulin sensitivity Lowered insulin resistance Activated of IRS-1 and AKT	8 g day ginseng root or 250 to 500 mg/day ginsenoside	[70]
14.	Guava	*Psidium guajava*	Myrtaceae	Regulated glucose homeostasis and insulin sensitization Inhibited DPP4, hepatic gluconeogenesis Raised the incretin hormone half-life, insulin secretion		[71, 72]
15.	Zinger	*Zingiber officinale*	Zingiberaceae	Reduced FBS, HbA1c raised insulin resistance	1.2 g	[73]

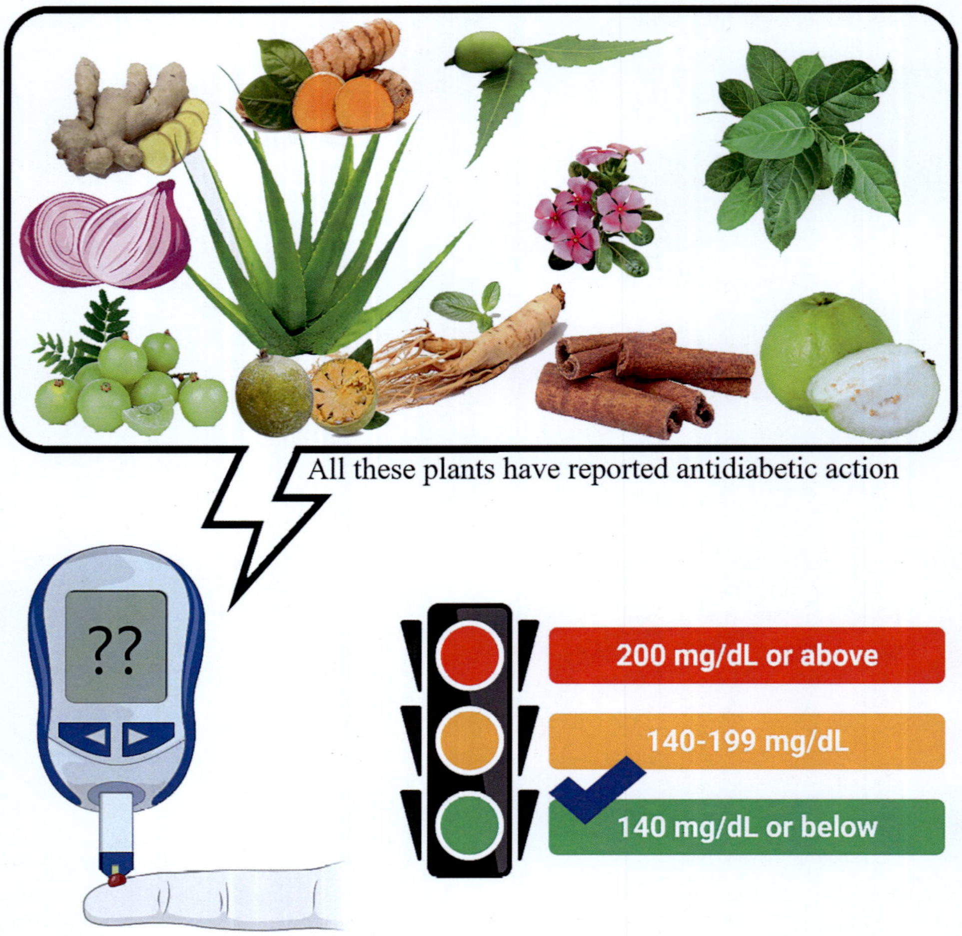

Fig. 2 Plants and plant parts commonly used as medication to control the blood glucose level

medicinal herb, there is evidence suggesting it may have side effects. Coumarin, a major substance of Cinnamon, has been found to have carcinogenic and hepatotoxic [101]. Skin sensitization may be triggered by cinnamaldehyde. To assure safe usage, the tolerated daily dosage of cinnamon was established to be 0.1 mg/kg/day [102]. Stomatitis, gingivitis, perioral dermatitis, contact dermatitis, and other hypersensitive responses can be caused by cinnamic acid exposure. Allyl cinnamyl ester can lead to skin irritation [103].

4.2 Ginkgo

Ginkgo biloba, sometimes known as ginkgo, has been used as herbal supplement for more than a millennium. Various pharmaceutical characteristics and clinical applications have been examined in more than 400 clinical trials. Antioxidant properties of ginkgo leaf extract have been employed in medicine. Ginkgo has been used to treat intestinal tract worm infections, pyogenic skin infections, enuresis, cough, and asthma. Additionally, the antioxidant properties of ginkgo leaf extract are exhibited [104]. For most people, ginkgo leaf extract is probably safe. It may result in a few mild adverse effects, including nausea, headaches, light-headedness, and allergic skin responses. Additionally, the ginkgo extract may lead to bleeding and bruising as well as cause arrhythmia. When consumed orally, the crude plant and roasted seed may not be safe. More than ten roasted seeds consumed every day might have severe negative effects, including convulsions. Consuming seeds might be fatal. They are regarded as harmful and venomous. Ginkgo may be unsafe during pregnancy if taken orally. If taken, it can result in early labour or excessive bleeding during delivery [105].

4.3 Ginsenoside

Panax ginseng, one of the most widely used traditional herbs, has been shown to have a variety of pharmacological uses. The primary active components of ginseng, known as ginsenosides, are responsible for its biological and pharmacological properties, including anti-inflammation,

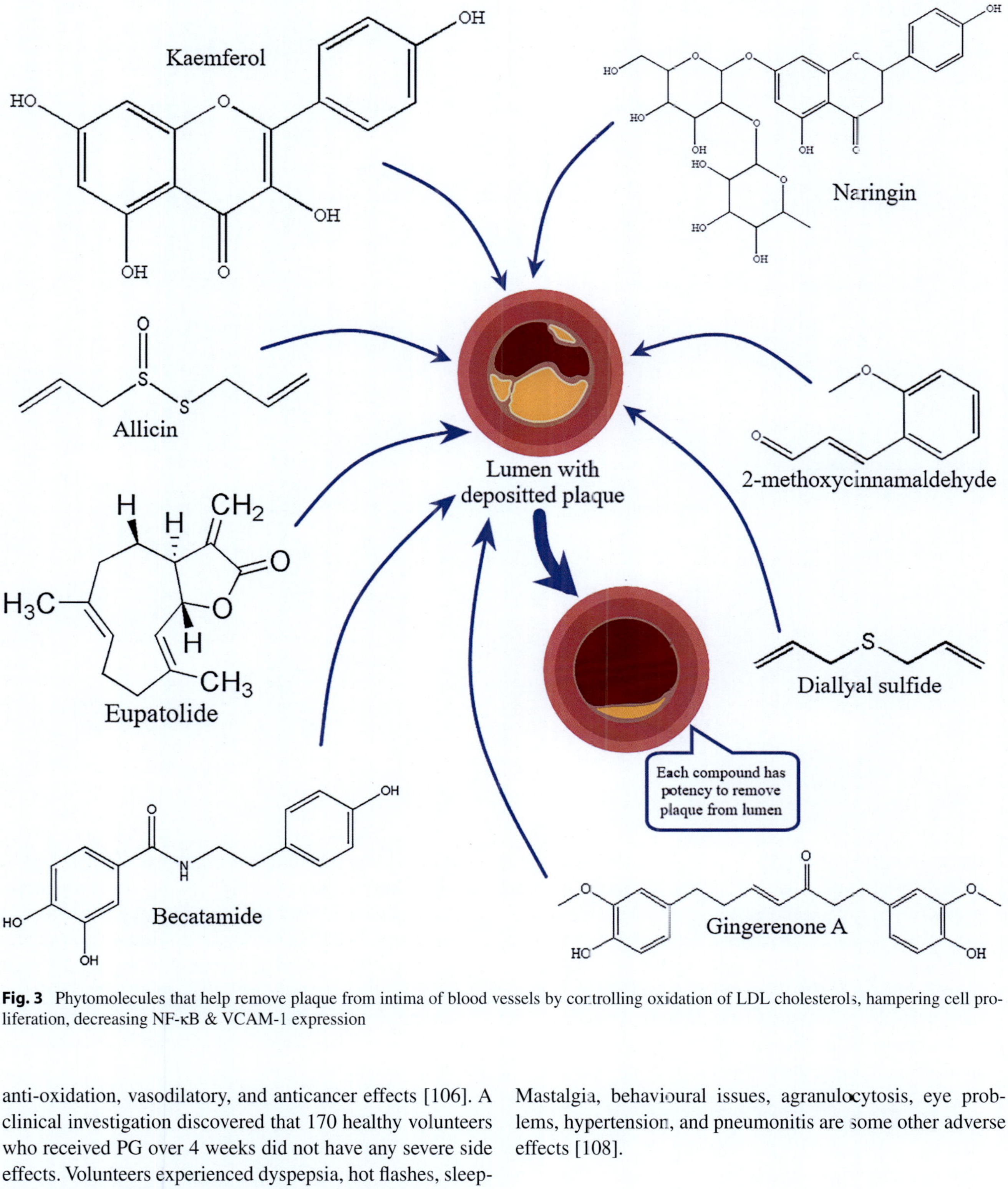

Fig. 3 Phytomolecules that help remove plaque from intima of blood vessels by controlling oxidation of LDL cholesterols, hampering cell proliferation, decreasing NF-κB & VCAM-1 expression

anti-oxidation, vasodilatory, and anticancer effects [106]. A clinical investigation discovered that 170 healthy volunteers who received PG over 4 weeks did not have any severe side effects. Volunteers experienced dyspepsia, hot flashes, sleeplessness, constipation, and other adverse effects, most of which were moderate in intensity [107]. The most usual adverse effect of *Panax ginseng* is anxiety and excitability brought on by an overdose. Once the initial few days pass, these side effects typically diminish over time. Insomnia and elevated blood pressure are the most frequent negative effects of improper applications. Stevens-Johnson syndrome, Mastalgia, behavioural issues, agranulocytosis, eye problems, hypertension, and pneumonitis are some other adverse effects [108].

4.4 Caffeine

The main active component of *Coffea arabica* is caffeine, a methylxanthine (1,3,7-trimethylxanthine), the most widely used psychoactive drug in the world. Its pharmacological impacts include activation of the central nervous system

Table 3 Safety dose range of herbal medicines in controlling atherosclerosis

Sl. No.	Name	Biological source	Family	Physiological effects	Dosage	Refs
1	Allicin	*Allium sativum*	Alliaceae	Reduction of ox-LDL	150 μg/mL	[77]
2	Diallyl sulphide	*Allium sativum*	Alliaceae	Reduction of ox-LDL	1280 mg per kg	[78]
3	Polyphenols	*Camellia sinensis*	Theaceae	Reduction of ox-LDL	1 g	[79]
4	Kaempferol	*Glycine max*	Fabaceae	Reduction of ox-LDL	50 or 100 μM	[80]
5	Avenanthramides	*Avena sativa*	Poaceae	Decreased cell proliferation	80 μM	[81]
6	2- Methoxycinnamaldehyde	*Cinnamomum cassia*	Lauraceae	Decreased cell proliferation	50 μM	[82]
7	Eupatolide	*Inula britannica*	Asteraceae	Decreased cell proliferation	10 μM	[83]
8	Flaxseed oil	*Linum usitatissimum*	Linaceae	Decreased cell proliferation	Maximum 24 g per day up to 7 weeks	[84]
9	Seed oil	*Schisandra semen*	Schisandraceae	Decreased NF-κB	100 mg twice daily	[85]
10	Gingerenone A	*Zingiber officinale*	Zingiberaceae	Decreased NF-κB	20 μM	[86]
11	Ethanol extract	*Glycine max*	Fabaceae	Decreased VCAM-1 expression	35–120 mg per day	[87]
12	Polyphenol	*Theobroma cacao*	Malvaceae	Decreased VCAM-1 expression	400 mg per kg body weight	[88]
13	Cysteinyl	*Allium sativum*	Alliaceae	Decreased platelet activation	400 mg oral dose twice per day	[89, 90]
14	Becatamide	*Houttuynia cordata*	Saururaceae	Decreased platelet activation	200–400 mg per kg is effective dose for rat	[91]
15	Naringin	*Lycopersicon esculentum*	Solanaceae	Decreased platelet activation	150–900 mg per day	[92, 93]
16	Polysaccharide-polyphenolic conjugates	*Matricaria chamomilla*	Asteraceae	Decreased platelet activation	25–2000 mg per day	[94, 95]
17	Saponin	*Camellia sinensis*	Theaceae	Decreased lipid level	338 mg EGCG per day (3 to 5 cups per day)	[96, 97]
18	Isoflavonoid	*Pueraria lobata*	Fabaceae	Decreased lipid level	20 mg toth 2.4 g per day	[98, 99]

(CNS), relaxation of smooth muscle and stimulation of cardiac muscle, and diuresis [109]. There are several negative expressions of caffeine on the digestive system. It can cause gastro-oesophageal reflux disease by loosening the lower oesophageal sphincter. Additionally, it leads to gastric hypersecretion, which increases the risk of developing an ulcer. The acidic stomach contents move into the duodenum more quickly due to the increased gastric emptying rate, which can cause inflammatory responses of the duodenal mucosa [110]. Frequent and high consumption can cause migraines in individuals. Frequent use of caffeine with analgesics may cause analgesic nephropathy. Additionally, caffeine may be harmful during pregnancy; daily intakes above 300 mg could lead to miscarriage and low baby weight [111].

4.5 Ephedra

Ephedra sinica Stapf, commonly known as ephedra, has long been used to treat bronchial colds, flu, asthma, and allergies. The oral dose of Ephedra shows help with weight loss in the short term. But it is a high-risk medicinal plant, it exhibits serious adverse effects including strokes, high blood pressure, seizures, heart attacks, an irregular heartbeat, and even death [112]. Ephedra can cause serious negative effects when used with other stimulant substances like coffee. Ephedra in high amounts may exacerbate anxiety. Ephedra has a stimulant effect on the heart, which exacerbates chest discomfort and irregular pulse. Ephedra may exacerbate high blood pressure [113].

4.6 Turmeric

The Zingiberaceae family includes turmeric (*Curcuma longa* Linn.) [114]. By triggering the AMPK system in gestational diabetic mice, 100 mg/kg of curcumin supplementation reduced glucose and insulin intolerance. For 9 months, using 250 mg of curcumin daily resulted in minimal effects of vertigo, constipation, and itching. 8 weeks of taking 300 mg of curcuminoids resulted in mild diarrhoea. With the addition of curcumin, the supplementation group had significant improvements in lipo-

CINNAMON	COUMARIN (carcinogenic, hepatotoxic); CINNAMALDEHYDE (skin sensitization); CINNAMIC ACID (stomatitis, gingivitis, perioral dermatitis, contact dermatitis, and other hypersensitive responses); ALLYL CINNAMYL ESTER (skin irritation);
GINKGO	Nausea, headaches, light-headedness, and allergic skin responses; Ginkgo extract may lead to bleeding and bruising as well as cause arrhythmia; More than ten roasted seeds consumed every day might have severe negative effects, including convulsions; Ginkgo may be unsafe during pregnancy if taken orally can result in early labour or excessive bleeding during delivery;
GINSENG	Moderate dyspepsia, hot flushes, sleeplessness, constipation; Anxiety and excitability by overdose; Insomnia, elevated blood pressure, Stevens-Johnson syndrome, Mastalgia, behavioral issues, agranulocytosis, eye problems, hypertension, and pneumonitis are among the common adverse effects;
CAFFEINE	Gastro-oesophageal reflux disease Additionally, gastric hypersecretion, ulcer; Frequent and high consumption can cause migraines in individuals; Frequent use of Caffeine with analgesics may cause analgesic nephropathy; During pregnancy; daily intake of above 300 mg may lead to miscarriage and low baby weight ;
EPHEDRA	Strokes, High blood pressure, seizures, heart attacks, an irregular heartbeat, and even death
CURCUMIN	Vertigo, constipation, diarrhoea and itching at extremely high doses
ALOE VERA	Teratogenicity and toxicological consequences on the embryo or fetus when consumed during pregnancy; Increased mortality, inflammation, and reproductive damage at high doses in rats; Diarrhoea, electrolyte imbalance from prolonged use; Allergic reactions for anthraquinone content;
NEEM	Neem oil - nausea, sluggishness, tachypnea, and recurrent generalized seizures; Toxic encephalopathy in newborns & young children; leucocytosis and metabolic acidosis; Nimbolide & Nimbic acid - lethal at high dose, due to vital organs' failure;
GINGER	Dehydration, allergic response, arrhythmia, and unexpected miscarriage are the most common adverse effects; Abdominal pain, headaches, sleepiness, belching, diarrhoea, and other minor side effects were also recorded across the studies
GARLIC	Bad breath, nausea, vomiting, stinky body odour, spinal epidural hematoma, extended bleeding time, platelet impairment, abdominal bloating & intestinal flora alteration
DANSHEN	Maculopapular rash, dizziness, nausea, vomiting, and musculoskeletal pain
HAWTHORN	Rash, sweating, headache, dizziness, vertigo, palpitation, agitation and sleepiness within therapeutic window; Sedation, hypotension, and arrhythmia at higher doses;
RED YEAST RICE	Myopathy, peripheral neuropathy, symptomatic hepatitis, erectile dysfunction, and allergic reactions.
GREEN TEA	Mild & infrequent bloating, gastric upset, dizziness, skin rash, and constipation.

Fig. 4 Adverse effects of commonly used herbal medicines

protein lipase activity, serum triglycerides, total free fatty acids (FFAs), fasting glycemia, insulin resistance, and HbA1c. Especially a reduction in serum FFA, which might result from boosting fatty acid oxidation and utilization, contributed to curcumin's ability to lower blood sugar levels. Curcumin is safe and non-toxic at doses of up to 12 g daily [56].

4.7 Aloe Vera

Aloe barbadensis Miller is the most popular species of Aloe. The concentration of the phenolic chemicals in Aloe vera whole leaf extract and latex has to be closely monitored due to the cytotoxicity, mutagenicity, and carcinogenicity of anthraquinones. The highest permissible aloin level in aloe-derived material, according to the International Aloe Science Council guideline, was less than 10 ppm; for non-medical usage, the recommended limit is 50 ppm or below [115]. Aloe vera ethanol extract, administered orally in drinking water at a dose of 100 mg/kg, increased mortality, inflammation, and reproductive damage in rats over control levels. The use of aloe vera during pregnancy has been linked to possible teratogenic and toxicological consequences on the embryo or foetus [116]. Aloe vera latex ingestion for a lengthy period may result in watery diarrhoea, which can induce electrolyte imbalance and hypokalaemia. Therefore, it is advised that people with a history of renal or cardiac issues refrain from consuming aloe vera latex [117]. Most allergic responses associated with aloe vera appear to have been caused by the products' anthraquinone content [118].

4.8 Neem

Azadirachta indica and *Azadirachta azedarach* are two species of Neem. Particularly in newborns and young children, neem oil causes toxic encephalopathy. The typical symptoms include nausea, sluggishness, tachypnoea, and recurrent generalized seizures. Important test results include leucocytosis and metabolic acidosis. The CNS and the lungs were the target organs for the harmful effects of neem oil [119]. After 6 weeks of daily dosing of 25, 50, or 100 mg/kg of nimbidin, adult rats developed subacute toxicity. When administered intravenously or intraperitoneally to mice, nimbic acid was discovered to be hazardous. Most animals died from the fatal dosage of nimbolide and nimbic acid due to liver, kidney, and small intestine failure, as well as a sharp and rapid decrease in arterial blood pressure [120]. Since the fatal range of a single dosage began at 8.7 mL/kg, the anti-fertility effects of neem oil in rats were found at 2 mL/kg. Liver toxicity is indicated by elevated blood total bilirubin and SGOT [121]. After being exposed for 24 h, human lymphocytes' cellular viability was reduced by neem [122].

4.9 Ginger

Zingiber officinale roots are the source of ginger, which is typically used in cooking and as a medicine due to its antioxidant properties. Ginger has relatively low toxicity; in rats and mice, its LD_{50} is greater than 5 g/kg body weight. Between 170 and 250 g/person was the lethal oral intake of ginger [123]. Ginger did not raise the possibility of sleepiness or heartburn as a side effect. The frequency of vomiting episodes during nausea and vomiting during pregnancy did not get considerably reduced by ginger. Dehydration, an allergic response, an arrhythmia, and an unexpected miscarriage were the most common occurrences noticed across studies. Pain in the abdomen, headaches, sleepiness, belching, diarrhoea, and other minor side effects were also recorded across the studies [124].

4.10 Garlic

The plant family Alliaceae includes *Allium sativum* L., also known as garlic. Garlic consumption can have certain negative effects on people, such as bad breath, nausea, vomiting, and stinky body odour. In addition to extending the bleeding duration and having been connected to a spontaneous spinal epidural hematoma, eating a lot of garlic on an empty stomach can cause gas, stomach discomfort, and changes in the intestinal flora. The daily recommended intake of garlic for adults is 4 g [125]. An 87-year-old man with platelet impairment consumed too much garlic [126].

4.11 Amla

Indian gooseberry, also known as Amla, is the scientific name for *Emblica officinalis* Gaertn. or *Phyllanthus emblica* Linn. (Euphorbiaceae) [127]. More than 1000 mg/kg is the LD_{50} for *E. officinalis* [128]. Due to their great safety and adequate nutritional content, *Phyllanthi fructus* (PF), the edible fruit of *Phyllanthus emblica*, served as an essential source for several health goods, cuisines, and medications. According to the findings of the safety assessment, the acute oral maximum tolerated dosage of PF was larger than 20 g/kg body weight, which is essentially non-toxic according to the acute toxicity classification. Rats were administered up to 2 g/kg body weight of PF ethanol extract orally for 14 days in the acute toxicity trial and exhibited no behavioural abnormalities or harmful effects. Somatic or germ cells were not genotoxic to PF. Haematological testing, behavioural observation, and biochemical indicators all showed that PF had no deleterious side effects, indicating that PF may be used without risk [129].

4.12 Danshen

Danshen is the dried root and rhizome of *Salvia miltiorrhiza* Bunge, which is used to cure coronary vasodilatation, inhibit platelet adhesion and aggregation, suppress the formation of thromboxane, and protect against myocardial ischemia [130]. Adverse effects of Danshen injectable were reported, such as rash, maculopapular rash, dizziness, nausea, vomiting, and musculoskeletal pain in a few per cent of patients. According to the study, rats administered an intravenous dose of 1500 mg/kg of Danshensu (the main active ingredient in Danshen) did not exhibit any adverse reactions. Acute toxicity was seen in rats when an intravenous dose of 64 g/kg body weight (equivalent to 441 mg/kg of sodium danshensu) was administered [131].

4.13 Hawthorn

Hawthorn fruits belong to the Rosaceae family. It is bright red berries of the *Crataegus* species. Hawthorn was used to treat myocardial dysfunction, hypertension, and atherosclerosis [132]. The adverse effects of hawthorn extract within the therapeutic window are rash, sweating, headache, dizziness, vertigo, palpitation, agitation, sleepiness, and gastrointestinal symptoms. Sedation, hypotension, and arrhythmia were observed to be brought on by higher doses in animal tests. Patients receiving doses of up to 1800 mg per day in a clinical study evaluating both the efficacy and safety of hawthorn extract in NYHA functional class III heart failure rarely experienced mild adverse events (AEs) like dizziness and vertigo (1.4%), which is statistically comparable to patients receiving doses of 900 mg per day (4.3%) and even less than patients receiving a placebo (10%) [133].

4.14 Quercetin

A large family of molecules known as flavonoids, including quercetin, share a common hydroxylated 3-ringed structure linked with hydroxyl groups. In general, quercetin is well tolerated. In long-term supplementation at 1000 mg/day, some mild side effects, such as a mild headache, nausea, and tingling in the extremities, were reported. Quercetin was not found to have any significant adverse effects on foetal growth in rats, although there are no human studies [134].

4.15 Red Yeast Rice

Red yeast rice (RYR) is a traditional Chinese recipe that uses rice that has been fermented with *Monascus purpureus* as a therapeutic agent. Starch, phytocholesterols, isoflavones, monounsaturated fatty acids, and polyketides known as monacolins are all present in red yeast rice extracts. The advantages of RYR extract, which includes Xuezhikang and Zhibituo in the treatment of individuals with hyperlipidaemia. To treat CVD cases, Xuezhikang has been used in clinical contexts [135]. RYR does have adverse effects due to MK and citrinin have been associated with a higher risk of myopathy, peripheral neuropathy, symptomatic hepatitis, erectile dysfunction, and allergic reactions. From April 2002 to September 2015, myopathy and hepatitis accounted for 52.73% of total AEs of RYR dietary supplements [136].

4.16 Camellia Sinensis

It is extracted from the Theaceae family member tea plant *Camellia sinensis*. *Camellia sinensis* is a 'non-fermented' tea; it contains more catechins than black or oolong tea. Catechins are potent antioxidants both in vivo and in vitro [137]. Adverse effects of Camellia sinensis were reported, i.e. mild, infrequent, and transient bloating, gastric upset, dizziness, skin rash, and constipation [138].

5 Conclusion

In medical platforms worldwide, herbal medicines have been crucial in treating illnesses and as a potential resource to sustain good health. The herbal product sector has the capability to influence the world economy profoundly. Future international labelling guidelines should tackle quality issues, given the rising use of natural remedies. The problems offered by the ever-growing variety and effectiveness of herbal drugs, each of which has an unavoidable and often unanticipated tendency to cause damage, continue to be met in large part through PV. Therefore, it is crucial to document the ADR and toxicity of examined natural remedies and their importance adequately conveyed to the public. To comprehend the utilization of herbal medicines, standardization of methods along with quality control information on safety and effectiveness are required. Quality, safety, and efficacy are used sensibly to assure medications of excellent reliability. These must be considered when treatment choices are made to reduce the adverse effects.

References

1. Abdulsalim S, Farooqui M, Alshammari MS, Alotaibi M, Alhazmi A, Alqasomi A, et al. Evaluation of knowledge, attitudes, and practices about pharmacovigilance among community pharmacists in Qassim, Saudi Arabia. Int J Environ Res Public Health. 2023;20:20. https://doi.org/10.3390/ijerph20043548.
2. Suke D, Kosta P, Negi H. Role of pharmacovigilance in India: an overview role of pharmacovigilance in India: an overview. Online J Public Health Inform. 2015;7:1–34. https://doi.org/10.5210/ojphi.v7i2.5595.
3. Lavan AH, Gallagher P. Predicting risk of adverse drug reactions in older adults. Ther Adv Drug Saf. 2016;7:11–22. https://doi.org/10.1177/2042098615615472.
4. Coleman J, Pontefract S. Adverse drug reactions. Clin Med (Lond). 2016;16:481–5. https://doi.org/10.7861/clinmedicine.16-5-481.
5. Raynor DK, Dickinson R, Knapp P, Long AF, Nicolson DJ. Buyer beware? Does the information provided with herbal products available over the counter enable safe use? BMC Med. 2011;9:94. https://doi.org/10.1186/1741-7015-9-94.
6. Jaiswal Y, Liang Z, Zhao Z. Botanical drugs in ayurveda and traditional Chinese medicine. J Ethnopharmacol. 2016;194:245–59. https://doi.org/10.1016/j.jep.2016.06.052.
7. Wachtel Galor S, Benzie I. Herbal medicine: an introduction to its history, usage, regulation, current trends, and research needs. In: Herbal medicine: biomolecular and clinical aspects. 2nd ed; 2011.
8. Wal A, Wal P, Gupta S, Sharma G, Rai AK. Pharmacovigilance of herbal products in India. J Young Pharm. 2011;3:256–8. https://doi.org/10.4103/0975-1483.83780.
9. Jitareanu A, Trifan A, Vieriu M, Caba I-C, Martu I, Agoroaei L. Current trends in toxicity assessment of herbal medicines: a narrative review. PRO. 2022;11:83. https://doi.org/10.3390/pr11010083.
10. Shetti S, Kumar CD, Sriwastava NK, Sharma IP. Pharmacovigilance of herbal medicines: current state and future directions. Pharmacogn Mag. 2011;7:69–73. https://doi.org/10.4103/0973-1296.75905.
11. Hepner DL, Harnett M, Segal S, Camann W, Bader AM, Tsen LC. Herbal medicine use in parturients. Anesth Analg. 2002;94:690.
12. Gromek K, Drumond N, Simas P. Pharmacovigilance of herbal medicines. Int J Risk Saf Med. 2015;27:55–65. https://doi.org/10.3233/JRS-150643.

13. Cascorbi I. Drug interactions-principles, examples and clinical consequences reply. Dtsch Arztebl Int. 2012;109:546–56. https://doi.org/10.3238/arztebl.2012.0546.
14. Wanwimolruk S, Prachayasittikul V. Cytochrome P450 enzyme mediated herbal drug interactions (part 1); 2014. https://doi.org/10.17877/DE290R-15628.
15. Strandell J, Wahlin S. Pharmacodynamic and pharmacokinetic drug interactions reported to VigiBase, the WHO global individual case safety report database. Eur J Clin Pharmacol. 2011;67:633–41. https://doi.org/10.1007/s00228-010-0979-y.
16. Colalto C. Herbal interactions on absorption of drugs: mechanisms of action and clinical risk assessment. Pharmacol Res. 2010;62:207–27. https://doi.org/10.1016/j.phrs.2010.04.001.
17. Rivera C, Ferro C, Bursua A, Gerber B. Probable interaction between Lycium barbarum (goji) and warfarin. Pharmacotherapy. 2012;32:e50–3. https://doi.org/10.1002/j.1875-9114.2012.01018.x.
18. Gogtay N, Bhatt HA, Dalvi S, Kshirsagar NA. The use and safety of non-allopathic Indian medicines. Drug Saf. 2002;25:1005–19. https://doi.org/10.2165/00002018-200225140-00003.
19. Walji R, Boon H, Barnes J, Austin Z, Baker G, Welsh S. Adverse event reporting for herbal medicines: a result of market forces. Healthc Policy. 2009;4:77–90. https://doi.org/10.12927/hcpol.2009.20820.
20. Olsson S, Pal S, Stergachis A, Couper M. Pharmacovigilance activities in 55 low- and middle-income countries A questionnaire-based analysis. Drug Saf. 2010;33:689–703.
21. Mclernon D, Harrild K, Bergh C, Davies M, De Neubourg D, Dumoulin J, et al. Clinical effectiveness of elective single versus double embryo transfer: meta-analysis of individual patient data from randomised trials. BMJ. 2010;341:c6945. https://doi.org/10.1136/bmj.c6945.
22. Hoffman KB, Demakas AR, Dimbil M, Tatonetti NP, Erdman CB. Stimulated reporting: the impact of US Food and Drug Administration-issued alerts on the adverse event reporting system (FAERS). Drug Saf. 2014;37:971–80. https://doi.org/10.1007/s40264-014-0225-0.
23. Nsubuga P, White M, Thacker S, Anderson M, Blount S, Broome C, et al. Public health surveillance: a tool for targeting and monitoring interventions. Disease control priorities in developing countries, 2006.
24. Murali K, Kaur S, Prakash A, Medhi B. Artificial intelligence in pharmacovigilance: practical utility. Indian J Pharmacol. 2019;51:373. https://doi.org/10.4103/ijp.IJP_814_19.
25. Choudhury A, Singh PA, Bajwa N, Dash S, Bisht P. Pharmacovigilance of herbal medicines: concerns and future prospects. J Ethnopharmacol. 2023;309:116383. https://doi.org/10.1016/j.jep.2023.116383.
26. Candore G, Juhlin K, Manlik K, Thakrar B, Quarcoo N, Seabroke S, et al. Comparison of statistical signal detection methods within and across spontaneous reporting databases. Drug Saf. 2015;38:38. https://doi.org/10.1007/s40264-015-0289-5.
27. Girard N, Audigier Valette C, Cadranel J, Monnet I, Hureaux J, Hilgers W, et al. IFCT-1502 CLINIVO: real-life experience with nivolumab in 600 patients (pts) with advanced non-small cell lung cancer (NSCLC): efficacy and safety of nivolumab and post-nivolumab treatment in the French expanded access program (EAP). Ann Oncol. 2017;28:v463. https://doi.org/10.1093/annonc/mdx380.005.
28. Pozsgai K, Szűcs G, Konig-Peter A, Balázs O, Vajda P, Botz L, et al. Analysis of pharmacovigilance databases for spontaneous reports of adverse drug reactions related to substandard and falsified medical products: a descriptive study. Front Pharmacol. 2022;13:964399. https://doi.org/10.3389/fphar.2022.964399.
29. Kumar A, Khan H. Signal detection and their assessment in pharmacovigilance. Open Pharm Sci J. 2015;2:66–73.
30. Kimura T, Matsushita Y, Kao Yang Y-H, Choi N-K, Park B-J. Pharmacovigilance systems and databases in Korea, Japan, and Taiwan. Pharmacoepidemiol Drug Saf. 2011;20:1237–45. https://doi.org/10.1002/pds.2244.
31. Ekor M. The growing use of herbal medicines: issues relating to adverse reactions and challenges in monitoring safety. Front Pharmacol. 2014;4:177. https://doi.org/10.3389/fphar.2013.00177.
32. Chen L, Zeng W-M, Cai Y-D, Feng K-Y, Chou K-C. Predicting anatomical therapeutic chemical (ATC) classification of drugs by integrating chemical-chemical interactions and similarities. PLoS One. 2012;7:e35254. https://doi.org/10.1371/journal.pone.0035254.
33. Aumeeruddy MZ. Traditional herbal medicines used in obesity management: a systematic review of ethnomedicinal surveys. J Herb Med. 2021;28:100435. https://doi.org/10.1016/j.hermed.2021.100435.
34. Thomas JG, Bond DS, Phelan S, Hill JO, Wing RR. Weight-loss maintenance for 10 years in the National Weight Control Registry. Am J Prev Med. 2014;46:17–23. https://doi.org/10.1016/j.amepre.2013.08.019.
35. Dietrich MO, Horvath TL. Limitations in anti-obesity drug development: the critical role of hunger-promoting neurons. Nat Rev Drug Discov. 2012;11:675–91. https://doi.org/10.1038/nrd3739.
36. Liu J, Lee J, Salazar Hernandez MA, Mazitschek R, Ozcan U. Treatment of obesity with celastrol. Cell. 2015;161:999–1011. https://doi.org/10.1016/j.cell.2015.05.011.
37. Xie W, Gu D, Li J, Cui K, Zhang Y. Effects and action mechanisms of berberine and rhizoma coptidis on gut microbes and obesity in high-fat diet-fed C57BL/6J mice. PLoS One. 2011;6:e24520. https://doi.org/10.1371/journal.pone.0024520.
38. Christensen LP. Chapter 1 ginsenosides: chemistry, biosynthesis, analysis, and potential health effects. In: Advances in food and nutrition research, vol. 55. Academic Press; 2008. p. 1–99. https://doi.org/10.1016/S1043-4526(08)00401-4.
39. Xu A, Wang H, Hoo RLC, Sweeney G, VanhouttePaulM WY, et al. Selective elevation of adiponectin production by the natural compounds derived from a medicinal herb alleviates insulin resistance and glucose intolerance in obese mice. Endocrinology. 2009;150:625–33. https://doi.org/10.1210/en.2008-0999.
40. Wang H, Wen Y, Du Y, Yan X, Guo H, Rycroft JA, et al. Effects of catechin enriched green tea on body composition. Obesity. 2010;18:773–9. https://doi.org/10.1038/oby.2009.256.
41. Wang J-H, Kim B-S, Han K, Kim H. Ephedra-treated donor-derived gut microbiota transplantation ameliorates high fat diet-induced obesity in rats. Int J Environ Res Public Health. 2017;14:14. https://doi.org/10.3390/ijerph14060555.
42. Asai A, Miyazawa T. Dietary curcuminoids prevent high-fat diet–induced lipid accumulation in rat liver and epididymal adipose tissue. J Nutr. 2001;131:2932–5. https://doi.org/10.1093/jn/131.11.2932.
43. Sung J, Jeong HS, Lee J. Effect of the Capsicoside G-rich fraction from pepper (Capsicum annuum L.) seeds on high-fat diet-induced obesity in mice. Phytother Res. 2016;30:1848–55. https://doi.org/10.1002/ptr.5692.
44. Hirata BKS, Cruz MM, de Sá RDCC, Farias TSM, Machado MMF, Bueno AA, et al. Potential anti-obesogenic effects of Ginkgo biloba observed in epididymal white adipose tissue of obese rats. Front Endocrinol. 2019;10:1.
45. Mahmoud R, Elnour W. Comparative evaluation of the efficacy of ginger and orlistat on obesity management, pancreatic lipase and liver peroxisomal catalase enzyme in male albino rats. Eur Rev Med Pharmacol Sci. 2013;17:75–83.
46. Wu T, Tang Q, Gao Z, Yu Z, Song H, Zheng X, et al. Blueberry and mulberry juice prevent obesity development in C57BL/6 mice. PLoS One. 2013;8:e77585.

47. Mousavi SM, Rahmani J, Kord-Varkaneh H, Sheikhi A, Larijani B, Esmaillzadeh A. Cinnamon supplementation positively affects obesity: a systematic review and dose-response meta-analysis of randomized controlled trials. Clin Nutr. 2020;39:123–33.
48. Torre-Villalvazo I, Tovar AR, Ramos-Barragán VE, Cerbón-Cervantes MA, Torres N. Soy protein ameliorates metabolic abnormalities in liver and adipose tissue of rats fed a high fat diet. J Nutr. 2008;138:462–8.
49. Murase T, Misawa K, Minegishi Y, Aoki M, Ominami H, Suzuki Y, et al. Coffee polyphenols suppress diet-induced body fat accumulation by downregulating SREBP-1c and related molecules in C57BL/6J mice. Am J Physiol Endocrinol Metab. 2011;300:E122–33.
50. Kuriyan R, Raj T, Srinivas S, Vaz M, Rajendran R, Kurpad AV. Effect of Caralluma fimbriata extract on appetite, food intake and anthropometry in adult Indian men and women. Appetite. 2007;48:338–44.
51. American Diabetes Association. (2) Classification and diagnosis of diabetes. Diabetes Care. 2014;38:S8–16. https://doi.org/10.2337/dc15-S005.
52. Todd JN, Srinivasan S, Pollin TI. Advances in the genetics of youth-onset type 2 diabetes. Curr Diab Rep. 2018;18:57. https://doi.org/10.1007/s11892-018-1025-1.
53. George A, Jacob A, Fogelfeld L. Lean diabetes mellitus: an emerging entity in the era of obesity. World J Diabetes. 2015;6:613–20. https://doi.org/10.4239/wjd.v6.i4.613.
54. Schmidt A. Highlighting diabetes mellitus: the epidemic continues. Arterioscler Thromb Vasc Biol. 2018;38:e1–8. https://doi.org/10.1161/ATVBAHA.117.310221.
55. Lascar N, Brown J, Pattison H, Barnett A, Bailey C, Bellary S. Type 2 diabetes in adolescents and young adults. Lancet Diabetes Endocrinol. 2017;6:6. https://doi.org/10.1016/S2213-8587(17)30186-9.
56. Pivari F, Mingione A, Brasacchio C, Soldati L. Curcumin and type 2 diabetes mellitus: prevention and treatment. Nutrients. 2019;11:1837. https://doi.org/10.3390/nu11081837.
57. Akash MSH, Rehman K, shuqingchen. Spice plant Allium cepa: dietary supplement for treatment of type 2 diabetes mellitus. Nutrition. 2014;30:30. https://doi.org/10.1016/j.nut.2014.02.011.
58. Ríos JL, Flavio F, Schinella GR. Natural products for the treatment of type 2 diabetes mellitus. Planta Med. 2015;81:975–94. https://doi.org/10.1055/s-0035-1546131.
59. Kamalakkannan N, Prince P. Effect of Aegle marmelos Correa. (Bael) fruit extract on tissue antioxidants in streptozotocin diabetic rats. Indian J Exp Biol. 2003;41:1285–8.
60. Majeed M, Mundkur L, Paulose S, Nagabhushanam K. Novel Emblica officinalis extract containing β-Glucogallin vs. metformin: a randomized, open-label, comparative efficacy study in newly diagnosed type 2 diabetes mellitus patients with dyslipidemia. Food Funct. 2022;13:13. https://doi.org/10.1039/D2FO01862D.
61. Oguntibeju O, Aboua Y, Goboza M. Vindoline—a natural product from Catharanthus roseus reduces hyperlipidemia and renal pathophysiology in experimental type 2 diabetes. Biomedicine. 2019;7:59. https://doi.org/10.3390/biomedicines7030059.
62. Sanni O, Erukainure OL, Chukwuma CI, Koorbanally NA, Ibeji CU, Islam MS. Azadirachta indica inhibits key enzyme linked to type 2 diabetes in vitro, abates oxidative hepatic injury and enhances muscle glucose uptake ex vivo. Biomed Pharmacother. 2019;109:734–43. https://doi.org/10.1016/j.biopha.2018.10.171.
63. Pingali U, Ali M, Gundagani S, Nutalapati C. Evaluation of the effect of an aqueous extract of Azadirachta indica (neem) leaves and twigs on glycemic control, endothelial dysfunction and systemic inflammation in subjects with type 2 diabetes mellitus—a randomized, double-blind, placebo-controlled clinical study. Diabetes Metab Syndr Obes. 2020;13:4401–12. https://doi.org/10.2147/DMSO.S274378.
64. Perez-Gutierrez R, Damian-Guzman M. Meliacinolin: a potent α-glucosidase and α-amylase inhibitor isolated from Azadirachta indica leaves and in vivo antidiabetic property in Streptozotocin-nicotinamide-induced type 2 diabetes in mice. Biol Pharm Bull. 2012;35:1516–24. https://doi.org/10.1248/bpb.b12-00246.
65. Leach M, Kumar S. Cinnamon for diabetes mellitus. Cochrane Database Syst Rev. 2012;9:CD007170. https://doi.org/10.1002/14651858.CD007170.pub2.
66. Chowdhury A, Sarkar J, Chakraborti T, Pramanik P, Chakraborti S. Protective role of epigallocatechin-3-gallate in health and disease: a perspective. Biomed Pharmacother. 2016;78:50–9. https://doi.org/10.1016/j.biopha.2015.12.013.
67. Dey P, Singh J, Jagadish Kumar S, Dilip K, Nayak J. Utilization of Swertia chirayita plant extracts for management of diabetes and associated disorders: present status, future prospects and limitations. Nat Prod Bioprospect. 2020;10:431–43. https://doi.org/10.1007/s13659-020-00277-7.
68. Pothuraju R, Kumar SR, Chagalamarri J, Jangra S, Kumar P. A systematic review on Gymnema sylvestre in the obesity and diabetes management. J Sci Food Agric. 2014;94:94. https://doi.org/10.1002/jsfa.6458.
69. Haghani F, Arabnezhad M, Mohammadi S, Ghaffarian-Bahraman A. Aloe vera and Streptozotocin-induced diabetes mellitus. Rev Bras Farm. 2022;32:32. https://doi.org/10.1007/s43450-022-00231-3.
70. Chen W, Balan P, Popovich D. Review of ginseng anti-diabetic studies. Molecules. 2019;24:4501. https://doi.org/10.3390/molecules24244501.
71. Diaz de Cerio E, Girón F, Pérez-Garrido A, Pereira A, Gabaldon J, Verardo V, et al. Fishing the targets of bioactive compounds from Psidium guajava L. leaves in the context of diabetes. Int J Mol Sci. 2023;24:5761. https://doi.org/10.3390/ijms24065761.
72. Choi E, Baek S, Baek K, Kim H-K. Psidium guajava L. leaf extract inhibits adipocyte differentiation and improves insulin sensitivity in 3T3-L1 cells. Nutr Res Pract. 2021;15:568–78.
73. Nunes Carvalho G, Garcia Lira Neto JC, Araújo MF, Freitas R, Lúcia Zanetti M, Damasceno M. Effectiveness of ginger in reducing metabolic levels in people with diabetes: a randomized clinical trial*. Rev Lat Am Enfermagem. 2020;28:28. https://doi.org/10.1590/1518-8345.3870.3369.
74. Rafieian-Kopaei M, Setorki M, Doudi M, Baradaran A, Nasri H. Atherosclerosis: process, indicators, risk factors and new hopes. Int J Prev Med. 2014;5:927.
75. Kattoor AJ, Pothineni NVK, Palagiri D, Mehta JL. Oxidative stress in atherosclerosis. Curr Atheroscler Rep. 2017;19:1–11.
76. Björkegren JL, Lusis AJ. Atherosclerosis: recent developments. Cell. 2022;185:1630.
77. Chen X, Pang S, Lin J, Xia J, Wang Y. Allicin prevents oxidized low-density lipoprotein-induced endothelial cell injury by inhibiting apoptosis and oxidative stress pathway. BMC Complement Altern Med. 2016;16:1–6.
78. Ou C, Tsao S, Lin M, Yin M. Protective action on human LDL against oxidation and glycation by four organosulfur compounds derived from garlic. Lipids. 2003;38:219–24.
79. Suzuki-Sugihara N, Kishimoto Y, Saita E, Taguchi C, Kobayashi M, Ichitani M, et al. Green tea catechins prevent low-density lipoprotein oxidation via their accumulation in low-density lipoprotein particles in humans. Nutr Res. 2016;36:16–23.
80. Li H, Ji H-S, Kang J-H, Shin D-H, Park H-Y, Choi M-S, et al. Soy leaf extract containing kaempferol glycosides and pheophorbides improves glucose homeostasis by enhancing pancreatic β-cell function and suppressing hepatic lipid accumulation in db/db mice. J Agric Food Chem. 2015;63:7198–210.

81. Nie L, Wise M, Peterson D, Meydani M. Mechanism by which avenanthramide-c, a polyphenol of oats, blocks cell cycle progression in vascular smooth muscle cells. Free Radic Biol Med. 2006;41:702–8.
82. Jin Y-H, Kim S-A. 2-Methoxycinnamaldehyde inhibits the TNF-α-induced proliferation and migration of human aortic smooth muscle cells. Int J Mol Med. 2017;39:191–8.
83. Kim N, Hwangbo C, Lee S, Lee J. Eupatolide inhibits PDGF-induced proliferation and migration of aortic smooth muscle cells through ROS-dependent heme oxygenase-1 induction. Phytother Res. 2013;27:1700–7.
84. Francis AA, Deniset JF, Austria JA, LaValleé RK, Maddaford GG, Hedley TE, et al. Effects of dietary flaxseed on atherosclerotic plaque regression. Am J Phys Heart Circ Phys. 2013;304:H1743–51.
85. Jeong J-W, Kim JW, Ku SK, Kim SG, Kim KY, Kim G-Y, et al. Essential oils purified from Schisandrae semen inhibits tumor necrosis factor-α-induced matrix metalloproteinase-9 activation and migration of human aortic smooth muscle cells. BMC Complement Altern Med. 2015;15:1–13.
86. Kim HJ, Son JE, Kim JH, Lee CC, Yang H, Yaghmoor SS, et al. Gingerenone A attenuates monocyte-endothelial adhesion via suppression of I kappa B kinase phosphorylation. J Cell Biochem. 2018;119:260–8.
87. Han J-M, Li H, Cho M-H, Baek S-H, Lee C-H, Park H-Y, et al. Soy-leaf extract exerts atheroprotective effects via modulation of Krüppel-like factor 2 and adhesion molecules. Int J Mol Sci. 2017;18:373.
88. Bao L, Zhang Y, Wei G, Wang Y, Ma R, Cheng R, et al. The anti-atherosclerotic effects of puerarin on induced-atherosclerosis in rabbits. Biomed Pap Med Fac Univ Palacky Olomouc Czech Repub. 2015;159:53–9.
89. Morihara N, Hino A. Aged garlic extract suppresses platelet aggregation by changing the functional property of platelets. J Nat Med. 2017;71:249–56.
90. Bayan L, Koulivand PH, Gorji A. Garlic: a review of potential therapeutic effects. Avicenna J Phytomed. 2014;4:1–14.
91. Park JB. Becatamide found in Houttuynia cordata suppresses P-selectin expression via inhibiting COX enzyme, not increasing cAMP in platelets. Phytother Res. 2015;29:1381–7.
92. Rebello CJ, Beyl RA, Lertora JJL, Greenway FL, Ravussin E, Ribnicky DM, et al. Safety and pharmacokinetics of naringenin: a randomized, controlled, single-ascending-dose clinical trial. Diabetes Obes Metab. 2020;22:91–8. https://doi.org/10.1111/dom.13868.
93. Xiao Y, Li L, Wang Y, Guo J, Xu W, Wang Y. Naringin administration inhibits platelet aggregation and release by reducing blood cholesterol levels and the cytosolic free calcium concentration in hyperlipidemic rabbits. Exp Ther Med. 2014;8:968–72.
94. Bijak M, Saluk J, Tsirigotis-Maniecka M, Komorowska H, Wachowicz B, Zaczyńska E, et al. The influence of conjugates isolated from Matricaria chamomilla L. on platelets activity and cytotoxicity. Int J Biol Macromol. 2013;61:218–29.
95. Kolanos R, Stice SA. Chapter 44- German chamomile. In: Gupta RC, Lall R, Srivastava A, editors. Nutraceuticals. 2nd ed. Academic Press; 2021. p. 757–72. https://doi.org/10.1016/B978-0-12-821038-3.00044-6.
96. Song C, Yu Q, Li X, Jin S, Li S, Zhang Y, et al. The hypolipidemic effect of total saponins from Kuding tea in high-fat diet-induced hyperlipidemic mice and its composition characterized by UPLC-QTOF-MS/MS. J Food Sci. 2016;81:H1313–9.
97. Hu J, Webster D, Cao J, Shao A. The safety of green tea and green tea extract consumption in adults—results of a systematic review. Regul Toxicol Pharmacol. 2018;95:412–33. https://doi.org/10.1016/j.yrtph.2018.03.019.
98. Gao Y, Wang X, He C. An isoflavonoid-enriched extract from Pueraria lobata (kudzu) root protects human umbilical vein endothelial cells against oxidative stress induced apoptosis. J Ethnopharmacol. 2016;193:524–30.
99. Bihlet AR, Byrjalsen I, Andersen JR, Simonsen SF, Mundbjerg K, Helmer B, et al. The efficacy and safety of multiple dose regimens of kudzu (Pueraria lobata) root extract on bone and cartilage turnover and menopausal symptoms. Front Pharmacol. 2021;12:760629.
100. Santos HO, da Silva GA. To what extent does cinnamon administration improve the glycemic and lipid profiles? Clin Nutr ESPEN. 2018;27:1–9.
101. Brancheau D, Patel B, Zughaib M. Do cinnamon supplements cause acute hepatitis? Am J Case Rep. 2015;16:250.
102. Abraham K, Wöhrlin F, Lindtner O, Heinemeyer G, Lampen A. Toxicology and risk assessment of coumarin: focus on human data. Mol Nutr Food Res. 2010;54:228–39. https://doi.org/10.1002/mnfr.200900281.
103. Hossein N, Zahra Z, Abolfazl M, Mahdi S, Ali K. Effect of Cinnamon zeylanicum essence and distillate on the clotting time. J Med Plants Res. 2013;7:1339–43.
104. van Beek TA, Montoro P. Chemical analysis and quality control of Ginkgo biloba leaves, extracts, and phytopharmaceuticals. J Chromatogr A. 2009;1216:2002–32.
105. Mei N, Guo X, Ren Z, Kobayashi D, Wada K, Guo L. Review of Ginkgo biloba-induced toxicity, from experimental studies to human case reports. J Environ Sci Health C. 2017;35:1–28.
106. Yu SE, Mwesige B, Yi Y-S, Yoo BC. Ginsenosides: the need to move forward from bench to clinical trials. J Ginseng Res. 2019;43:361–7.
107. Coon JT, Ernst E. Panax ginseng. Drug Saf. 2002;25:323–44.
108. Lee N-H, Yoo S-R, Kim H-G, Cho J-H, Son CG. Safety and tolerability of Panax ginseng root extract: a randomized, placebo-controlled, clinical trial in healthy Korean volunteers. J Altern Complement Med. 2012;18:1061–9.
109. Baumann TW, Gabriel H. Metabolism and excretion of caffeine during germination of Coffea arabica L. Plant Cell Physiol. 1984;25:1431–6.
110. Boekema PJ, Samsom M, van Berge Henegouwen GP, Smout AJ. Coffee and gastrointestinal function: facts and fiction: a review. Scand J Gastroenterol. 1999;34:35–9.
111. Parazzini F, Chiaffarino F, Chatenoud L, Tozzi L, Cipriani S, Chiantera V, et al. Maternal coffee drinking in pregnancy and risk of small for gestational age birth. Eur J Clin Nutr. 2005;59:299–301.
112. Andraws R, Chawla P, Brown DL. Cardiovascular effects of ephedra alkaloids: a comprehensive review. Prog Cardiovasc Dis. 2005;47:217–25.
113. Kalman D, Incledon T, Gaunaurd I, Schwartz H, Krieger D. An acute clinical trial evaluating the cardiovascular effects of an herbal ephedra–caffeine weight loss product in healthy overweight adults. Int J Obes. 2002;26:1363–6.
114. Amalraj A, Pius A, Gopi S, Gopi S. Biological activities of curcuminoids, other biomolecules from turmeric and their derivatives—a review. J Tradit Complement Med. 2016;7:7. https://doi.org/10.1016/j.jtcme.2016.05.005.
115. Guo X, Mei N. Aloe vera—a review of toxicity and adverse clinical effects. J Environ Sci Health C. 2016;34:77–96. https://doi.org/10.1080/10590501.2016.1166826.
116. Ulbricht C, Armstrong J, Basch E, Basch S, Bent S, Dacey C, et al. An evidence-based systematic review of aloe vera by the natural standard research collaboration. J Herb Pharmacother. 2007;7:279–323.
117. Boudreau M, Beland F. An evaluation of the biological and toxicological properties of aloe Barbadensis (miller), aloe vera. J Environ

Sci Health C Environ Carcinog Ecotoxicol Rev. 2006;24:103–54. https://doi.org/10.1080/10590500600614303.

118. Bottenberg M, Wall G, Harvey R, Habib S. Oral aloe vera-induced hepatitis. Ann Pharmacother. 2007;41:1740–3. https://doi.org/10.1345/aph.1K132.
119. Brahmachari G. Neem—an omnipotent plant: a retrospection. Chembiochem. 2004;5:408–21. https://doi.org/10.1002/cbic.200300749.
120. Biswas K, Chattopadhyay I, Banerjee R, Bandyopadhyay U. Biological activities and medicinal properties of neem (Azadirachta indica). Curr Sci. 2001;82:1336.
121. Gandhi M, Lal R, Sankaranarayanan A, Banerjee CK, Sharma PL. Acute toxicity study of the oil from Azadirachta indica seed (neem oil). J Ethnopharmacol. 1988;23:39–51. https://doi.org/10.1016/0378-8741(88)90113-4.
122. Jerobin J, Makwana P, Suresh Kumar RS, Sundaramoorthy R, Mukherjee A, Chandrasekaran N. Antibacterial activity of neem nanoemulsion and its toxicity assessment on human lymphocytes in vitro. Int J Nanomedicine. 2015;10 Suppl 1:77. https://doi.org/10.2147/IJN.S79983.
123. Shalaby MA, Hamowieh AR. Safety and efficacy of Zingiber officinale roots on fertility of male diabetic rats. Food Chem Toxicol. 2010;48:2920–4. https://doi.org/10.1016/j.fct.2010.07.028.
124. Viljoen E, Visser J, Koen N, Musekiwa A. A systematic review and meta-analysis of the effect and safety of ginger in the treatment of pregnancy-associated nausea and vomiting. Nutr J. 2014;13:20. https://doi.org/10.1186/1475-2891-13-20.
125. Rauf A, Abu-Izneid T, Thiruvengadam M, Imran M, Olatunde A, Shariati AM, et al. Garlic (Allium sativum L.): its chemistry, nutritional composition, toxicity, and anticancer properties. Curr Top Med Chem. 2022;22:957–72. https://doi.org/10.2174/1568026621666211105094939.
126. Al-Qattan K, Thomson M, Al-Mutawa'a S, Al-Hajeri D, Drobiova H, Ali M. Nitric oxide mediates the blood-pressure lowering effect of garlic in the rat two-kidney, one-clip model of hypertension. J Nutr. 2006;136:774S–6S. https://doi.org/10.1093/jn/136.3.774S.
127. Variya B, Bakrania A, Patel S. Emblica officinalis (Amla): a review for its phytochemistry, ethnomedicinal uses and medicinal potentials with respect to molecular mechanisms. Pharmacol Res. 2016;111:111. https://doi.org/10.1016/j.phrs.2016.06.013.
128. Pandey G, Pandey S. Phytochemical and toxicity study of Emblica officinalis (Amla). Int Res J Pharm. 2011;2:270–2.
129. Yan X, Li Q, Jing L, Wu S, Duan W, Chen Y, et al. Current advances on the phytochemical composition, pharmacologic effects, toxicology, and product development of Phyllanthi fructus. Front Pharmacol. 2022;13:13. https://doi.org/10.3389/fphar.2022.1017268.
130. Ong CE, Pan Y. Drug-herb interactions: mechanisms involved and clinical implications of five commonly and traditionally used herbs. CELLMED. 2014;4:17–1.
131. Wang M, Liu J, Zhou B, Xu R, Tao L, Ji M, et al. Acute and sub-chronic toxicity studies of Danshen injection in Sprague-Dawley rats. J Ethnopharmacol. 2012;141:96–103.
132. Dong P, Pan L, Zhang X, Zhang W, Wang X, Jiang M, et al. Hawthorn (Crataegus pinnatifida Bunge) leave flavonoids attenuate atherosclerosis development in apoE knock-out mice. J Ethnopharmacol. 2017;198:479–88.
133. Chang W-T, Dao J, Shao Z-H. Hawthorn: potential roles in cardiovascular disease. Am J Chin Med. 2005;33:1–10.
134. Dagher O, Mury P, Thorin-Trescases N, Noly PE, Thorin E, Carrier M. Therapeutic potential of quercetin to alleviate endothelial dysfunction in age-related cardiovascular diseases. Front Cardiovasc Med. 2021;8:658400.
135. Xie X, Wang Y, Zhang S, Zhang G, Xu Y, Bi H, et al. Chinese red yeast rice attenuates the development of angiotensin II-induced abdominal aortic aneurysm and atherosclerosis. J Nutr Biochem. 2012;23:549–56.
136. Zhu B, Qi F, Wu J, Yin G, Hua J, Zhang Q, et al. Red yeast rice: a systematic review of the traditional uses, chemistry, pharmacology, and quality control of an important Chinese folk medicine. Front Pharmacol. 2019;10:1449.
137. Namita P, Mukesh R, Vijay KJ. Camellia sinensis (green tea): a review. Glob J Pharm. 2012;6:52–9.
138. Nantz MP, Rowe CA, Bukowski JF, Percival SS. Standardized capsule of Camellia sinensis lowers cardiovascular risk factors in a randomized, double-blind, placebo-controlled study. Nutrition. 2009;25:147–54.

Standardization of Herbal Medicines for Lifestyle Diseases

Tiyasa Paul and K. Jayaram Kumar

Abstract

The standardization of herbal medicines is crucial for managing lifestyle diseases due to the growing interest in natural remedies. Lifestyle diseases like obesity, diabetes, cardiovascular conditions, and hypertension are a global health concern due to unhealthy lifestyles. In response to the limitations of conventional pharmaceutical treatments, people are turning to herbal medicine as an alternative approach. The present chapter highlights the significance of herbal medicine standardization in ensuring consistent quality, safety, and efficacy of herbal products. Qualitative and quantitative methods, along with fingerprint profiling, are used to characterize herbal samples and identify key bioactive compounds. Recent years have seen a notable shift toward reliance on herbal medicine for lifestyle diseases, driven by concerns over synthetic drug side effects and a desire for sustainable and personalized healthcare solutions. Advancements in technology and research have validated herbal medicine's therapeutic potential, boosting its credibility in mainstream healthcare practices. Accurate identification and quantification of active components through analytical techniques like chromatography and spectroscopy are essential. The increasing reliance on herbal remedies reflects a growing recognition of their efficacy and safety. Traditional medicine, including Chinese medicine, utilizes unique diagnostic methods that must be carefully considered during the standardization process. Recently, digital intelligence has proven its efficacy in addressing a wide range of diverse and complex challenges, and among the successful approaches in decision-making, fuzzy logic stands out prominently. Fuzzy logic is an artificially intelligent tool used to identify various diseases using the Medical Diagnostic System (MDS). It is considered one of the most effective qualitative computational techniques. Coupling human cognitive techniques with modern analysis is seen as a comprehensive and effective solution for individuals seeking natural therapies.

Keywords

Herbal medicines · Standardization · Quality control · Human cognitive techniques

1 Introduction

Herbal medicine, has a long history in human healthcare, tracing its roots back to ancient civilizations and indigenous healing traditions. In ancient India, the medical system of Ayurveda heavily relied on herbal medicine for healing. However, with the rise of modern medicine and pharmaceutical innovations in the nineteenth and twentieth centuries, herbal medicine lost prominence in some regions, becoming associated with traditional and alternative healing. India boasts an impressive abundance of medicinal plant diversity, with around 20,000 recorded species. Over 500 indigenous communities in the country use approximately 800 plant species to treat various ailments [1]. In the contemporary era, the global market for health-related products remains active, with standardization playing a crucial role in ensuring consistent availability of uniform products worldwide. There are varying responses to conventional, supplementary, and alternative therapies, with conventional medicine still widely used in developing nations and alternative and complementary medicine gaining popularity in developed countries. Policymakers, healthcare experts, and the public are debating issues related to the safety, effectiveness, access, and regulation of these medical approaches worldwide. The World Health Organization (WHO) works on developing strategies to address concerns with regulations, safety, effectiveness, access, and accountable use of conventional, complementary, and alternative medicine. WHO collaborates with health ministries to incorporate indigenous plant medi-

T. Paul · K. J. Kumar (✉)
Department of Pharmaceutical Sciences and Technology, Birla Institute of Technology, Mesra, Ranchi, Jharkhand, India

A. K. Dhara, S. C. Mandal (eds.), *Role of Herbal Medicines*, https://doi.org/10.1007/978-981-99-7703-1_27

cines into primary healthcare programs, evaluating their safety and effectiveness and implementing quality control procedures for raw and processed materials.

1.1 Herbal Drug Standardization

Herbal medicine, owing to its wide acceptance, enhanced compatibility with the human body, and minimal side effects, remains a predominant therapeutic approach for 75–80% of the global population, particularly in economically challenged nations. Notably, the herbal pharmaceutical market in India is valued at approximately $1 billion, with crude plant-based drugs accounting for export worth nearly $80 million [2]. Ayurvedic, Unani, and Siddha therapies are integral components of the over-the-counter, ethical, and traditional herbal medicine formulations in India. Despite their cultural significance, these therapies contribute a meager $80 million to the export market. Notably, 80% of the exports to industrialized nations comprise raw drugs rather than complete formulations, leading to limited revenue generation for the nation [3]. Standardization is a critical process that ensures that each dose of a herbal product contains a predetermined and consistent quantity of bioactive ingredients, maintaining a desired level of quality and therapeutic efficacy. The scientific authenticity of a herbal medication hinges on its authentication and thorough characterization, guaranteeing reproducibility in the manufacturing process. The therapeutic potential of an herbal formulation is strongly influenced by its phytochemical components. Significant hurdles exist in the development of trustworthy analytical methods for precisely characterizing the phytochemical composition, including quantitative evaluations of marker molecules and other crucial ingredients [4]. Authenticating herbal pharmaceuticals and distinguishing them from adulterated products are of utmost importance to pharmaceutical companies, public health, and ensuring the reproducible quality of herbal medicines.

The standardization process for herbal medicines involves several important stages to ensure their quality, safety, and efficacy:

- Authentication: This stage involves verifying the botanical identity and purity of the collected plant material. It includes detailed examination of the plant's phytomorphology, microscopic features, and taxonomical identity to confirm its authenticity and regional status.
- Foreign Matter: The collected herbs must be free from any foreign matter such as soil, insect parts, or animal excreta to maintain their purity and safety.
- Organoleptic Evaluation: Herbal materials are evaluated for sensory characters like taste, appearance, odor, and texture, providing valuable information about their overall quality.
- Microscopic Analysis: Microscopic and histological analysis of the drug powder allows for the identification and presence of specific tissues, providing additional evidence of the plant's authenticity.
- Ash and Extractive Values: Determination of ash and extractive values helps assess the purity and quality of the herbal material.
- Volatile Matter and Moisture Content: Evaluating the volatile matter and moisture content ensures the stability and shelf-life of the herbal product.
- Spectroscopic and Chromatographic Evaluation: Qualitative and semi-quantitative information regarding the primary active constituents in the herbal material is provided by techniques including HPTLC, TLC, and HPLC. Chromatographic fingerprints serve as chemical markers for quality control.
- Heavy Metal and Pesticide Residue Analysis: Testing for heavy metals and pesticide residues ensures that the herbal medicine meets safety standards and is free from harmful contaminants.
- Microbial Contamination: The presence of bacteria and molds in medicinal plants can be assessed through the analysis of E. coli and mold limits, providing insights into harvesting and production practices.
- Radioactive Contamination: If herbs are treated with irradiation to prevent microbial growth, potential radioactivity hazards should be monitored in accordance with international guidelines.

By implementing these standardized evaluation methods, herbal medicines can be reliably characterized, ensuring their safety, efficacy, and consistency. Quality assurance measures are essential to gaining acceptance in modern healthcare systems and providing effective treatments to patients. Standardization of bioactive extracts in herbal medicines should be conducted based on the identification and quantification of active principles or major compounds present. This process should be complemented by the establishment of chromatographic fingerprints using techniques such as high-performance thin-layer chromatography (HPTLC), thin-layer chromatography (TLC), gas chromatography (GC), and high-performance liquid chromatography (HPLC). These analytical methods allow for the characterization and quantification of specific chemical constituents, ensuring consistency and quality control of the herbal extracts [5].

Herbal formulation standardization involves a systematic approach, including the use of raw ingredients from various sources. Comparative studies of chemical efficacy among different batches of formulations are conducted to identify preparations with superior clinical efficacy. Regular assessments of physiological, chemical, and pharmacological aspects are performed to ensure consistency and reliability, guiding the selection of the final preparation and validating the entire production and manufacturing process. The standardization of herbal preparations is an important step for ensuring the reliability and efficacy of medications. This method involves determining the concentration of their active ingredients and conforming to physical, chemical, and physiochemical standards, as well as in vitro and in vivo parameters [6]. The World Health Organization (WHO) emphasizes the significance of qualitative and quantitative processes to characterize herbal samples, quantifying biomarkers, and creating a fingerprint profile [7]. In cases where active components contribute toward biological effects are identified, herbal products should be standardized by these components. Alternatively, in case of unknown active ingredients, specific marker substances are chosen for analytical purposes, specific to the botanical plants under investigation.

2 Role of Herbal Medicine in Management of Lifestyle Disease

Herbal medicine plays a significant role in management of lifestyle diseases, which are chronic conditions often linked to connected to harmful lifestyle decisions such a poor diet, inactivity, smoking, and binge drinking, and is greatly aided by herbal medicine. These diseases, including diabetes, obesity, hypertension, and cardiovascular diseases metabolic syndrome, have become a global health burden. Herbal medicine offers a holistic and natural approach to complement conventional treatments in managing lifestyle diseases. It ensures the authenticity of plant sources, quantification of bioactive compounds, and adherence to regulatory guidelines. Standardized dosage guidelines and careful consideration of herb–drug interactions further enhance the safety and efficacy of herbal treatments. This approach bridges traditional knowledge with modern science, offering promising and reliable options for alleviating the burden of lifestyle diseases on global health [8].

3 Need for Standardization

Quality control has been a focal point in almost every traditional medical system, necessitating the development of modern and objective standards to assess the efficacy, quality, and safety of these treatments [9]. Various environment-related factors, such as rain, temperature, conditions, sunlight, storage practices, altitude, harvesting techniques, soil, timing, and collection methods, as well as manufacturing processes like selection, drying, purification, and extraction, all exert significant influences on product quality and the concentrations of plant chemicals. Additionally, genetic variability plays a vital role in determining the chemical composition of plants. Furthermore, environmental factors like insect feeding and microbial infections can impact secondary metabolites, thereby affecting the overall chemical composition of the plant material. In the standardization of herbal products, two types of extracts are employed: the extract of active constituents and the extract of markers. Active constituent extracts are utilized when the therapeutic benefits of known biochemical principles are established. On the other hand, marker extracts are used when the active ingredient is still unknown and a particular compound is chosen as a marker to show that the herbal formulation contains additional medicinal biochemical compounds [10].

Techniques that incorporate both conventional and modern methods for evaluation need to be developed for traditional medicine. The quality of herbal agents will be improved by combining methodologies like fuzzy logic, neural networks, and accurate linguistic models with modern analytical methods of evaluation. This will also encourage traditional medicine practitioners to participate more in standardization [11].

4 Quality Evaluation of Herbal Drug [12]

4.1 Organoleptic or Macroscopic Evaluation

The organic drug evaluation involves qualitative assessment through the application of the five senses (nose, skin, ear, tongue, and eye) or microscopic examination, considering factors such as taste, color, size, odor, shape, and unique characteristics like touch and texture. This method focuses on the morphological and sensory features of complete pharmaceuticals. For instance, important properties of cascara barks, cinchona, and quillaia, as well as quassia wood, are identified through examination of their fragmented surfaces. Specific examples of this evaluation method include recognizing the sweet flavor of liquorice and the aromatic odor of umbelliferous fruits, where the presence of volatile oils influences the overall odor of the medications.

4.2 Microscopic Evaluation

Organoleptic evaluation is a comprehensive examination method used to identify herbal drugs based on their recognized histological characteristics. Microscopic analysis is commonly employed to assess the quality of complete and potent herbal pharmaceuticals. Important features such as trichomes, stomata, starch granules, calcium oxalate crystals, and aleuronic grains are crucial in the identification of specific herbal drugs during this evaluation process.

4.3 Chemical Evaluation

Most pharmaceuticals contain well-defined chemical constituents responsible for their biological or pharmacological activity. Qualitative chemical evaluations are employed to identify specific drugs and ensure their purity. The evaluation process involves isolation, purification, and identification of active constituents using chemical methods. For resins, the evaluation includes tests such as acid value and sulfated ash. Balsams are evaluated through tests like acid value, saponification value, and bester values. Volatile oils undergo evaluation using acetyl and ester values. Qualitative chemical testing aids in the identification of chemical components and helps detect adulteration in pharmaceutical products.

4.3.1 Finger Printing of Herbal Medicine

Herbal medicines are formulated using plant-derived ingredients, making them susceptible to contamination, degradation, and changes in composition. Ensuring quality control of herbal medications poses several challenges. Markers are chemical components present in plant material, serving as distinctive identifiers or fingerprints. These markers can be visually detected and quantified using analytical techniques such as thin-layer chromatography (TLC), high-performance liquid chromatography (HPLC), and high-performance thin-layer chromatography (HPTLC). Conversely, biomarkers are specific chemical components unique to the plant material and are also associated with its biological effectiveness [13]. In the absence of standardized reference materials for accurate identification, variations may occur from the initial raw material collection to storage and subsequent processing. Hence, it becomes essential to establish criteria for selecting and incorporating the appropriate material into the formulation. To address this, the concept of using multiple markers and biomarkers to ensure the identification of the correct material is introduced [14].

4.3.2 Chemical Marker for Herbal Standardization

Fingerprinting in the context of herbal medicines involves generating a unique chemical profile for the plant material, its specific cut, fraction, or extract. This chemical pattern serves as a distinctive identifier, enabling the accurate characterization and authentication of the herbal product. Various analytical techniques, such as chromatography, spectroscopy, and mass spectrometry, are utilized to create the fingerprint, which aids in quality control, batch-to-batch consistency, and identification of potential adulterants or contaminants [15]. It is critical to comprehend that a plant extract is made up of known chemical compounds from several types. These consist of inorganic salts, metals, and secondary and primary metabolites. Primary metabolites encompass essential compounds like carbohydrates, proteins, and lipids, which play critical roles in fundamental plant physiological processes. On the other hand, secondary metabolites are substances produced as by-products in metabolic pathways that may not be essential for the basic physiological functions of plants. These secondary metabolites often contribute to specific ecological interactions, defense mechanisms, and responses to environmental stimuli. While primary metabolites are typically present in all plant tissues, secondary metabolites tend to be more specific to certain plant parts or stages of growth. Alkaloids, flavonoids, coumarins, terpenoids, and anthocyanins are just a few of the fascinating and useful families of molecules that make up those secondary metabolites. As our understanding of chemistry has advanced, we can now use these secondary metabolites to identify plant material. Planar chromatography techniques, such as thin-layer chromatography (TLC), and its advanced version, high-performance thin-layer chromatography (HPTLC), are employed to analyze and separate various constituents in herbal extracts. HPTLC offers enhanced resolution and sensitivity compared to TLC, allowing for more accurate identification and quantification of components [16]. Liquid chromatography methods, such as high-performance liquid chromatography (HPLC), are particularly useful when a marker compound is known and can serve as a reference standard. HPLC is employed to determine the concentration of specific compounds in the herbal extract. TLC plate and developed using appropriate solvent systems. This process aids in identifying any variations in the composition of the herbal product, facilitating quality control and standardization. To create a more effective TLC pattern, we can employ various solvent solutions. This generated plate is then examined in the scanner using various circumstances.

Table 1 Physical evaluation in the standardization of herbal medicine

Physical evaluation	Purpose	Method of determination	Advantages	Disadvantages	Reference
Organoleptic evaluation	Sensory assessment of appearance, color, odor, taste, texture, etc.	Visual observation and subjective assessment	Quick and simple to perform	Subjective and may vary among evaluators	[18]
Macroscopical evaluation	Examines external features (size, shape, texture)	Naked eye, magnifying glass	Nondestructive and useful for crude drugs	Limited to external features, may not detect internal issues	[19]
Microscopical evaluation	Studies cellular and tissue structure and detects contaminants	Compound microscope	Allows identification of plant parts	Requires expertise in microscopy and interpretation	[20]
Ash value determination	Identifies inorganic content and assesses purity	Incineration of sample	Useful for detecting adulterants	Organic matter may also burn, leading to overestimation	[21]
Moisture content determination	Assesses stability and quality	Heating sample to measure weight loss	Important for formulation and storage	Requires precise temperature control	[22]
Extractive value determination	Measures extractable compounds	Extraction with a suitable solvent	Provides information on solubility	Solvent selection may influence results	[23]
Swelling index	Assesses swelling behavior of herbal material	Measures volume increase after water absorption	Useful for certain plant parts like gums	Limited applicability to all herbal materials	[24]
Foreign matter determination	Detects extraneous matter or contaminants	Visual inspection and separation	Ensures purity and quality	May miss contaminants that resemble herbal material	[25]
Particle size analysis	Assesses particle size distribution	Sieving, laser diffraction, or microscopy	Impacts dissolution and bioavailability	Requires specialized equipment and expertise	[26]
Density determination	Measures bulk density and tapped density	Volume and weight measurements	Useful for formulation and dosage calculation	Influenced by particle size, shape, and packing behavior	[27]
pH determination	Assesses acidity or alkalinity of extracts/ preparations	pH meter or pH indicator paper	Important for stability and formulation	May require calibration and proper handling of pH meter	[28]

Various analytical techniques can be employed to detect spots on chromatography plates, such as ultraviolet (UV) detection at different wavelengths [17]. Each type of plant material has a specific graphic pattern and extracts from several batches typically produce graphic patterns that overlap. Therefore, we may basically optimize this pattern to provide a unique HPTLC profile.

4.4 Physical Evaluation

The evaluation of various medications often involves the analysis of physical constants. These constants encompass properties such as moisture content, specific gravity, optical rotation, refractive index, melting point, viscosity, and solubility in various solvents. These physical characteristics play a significant role in the identification and characterization of plant components present in the medications (Table 1).

4.5 Biological Evaluation

Specific drugs exhibit distinct biological and pharmacological activities that are utilized for their evaluation. Actually, a particular type of ingredient found in the plant extract is the generating this activity. The evaluations were carried out using both intact and isolated organs of living animals for evaluation. Bioassays can be used to assess the drug's potency during production (Fig. 1).

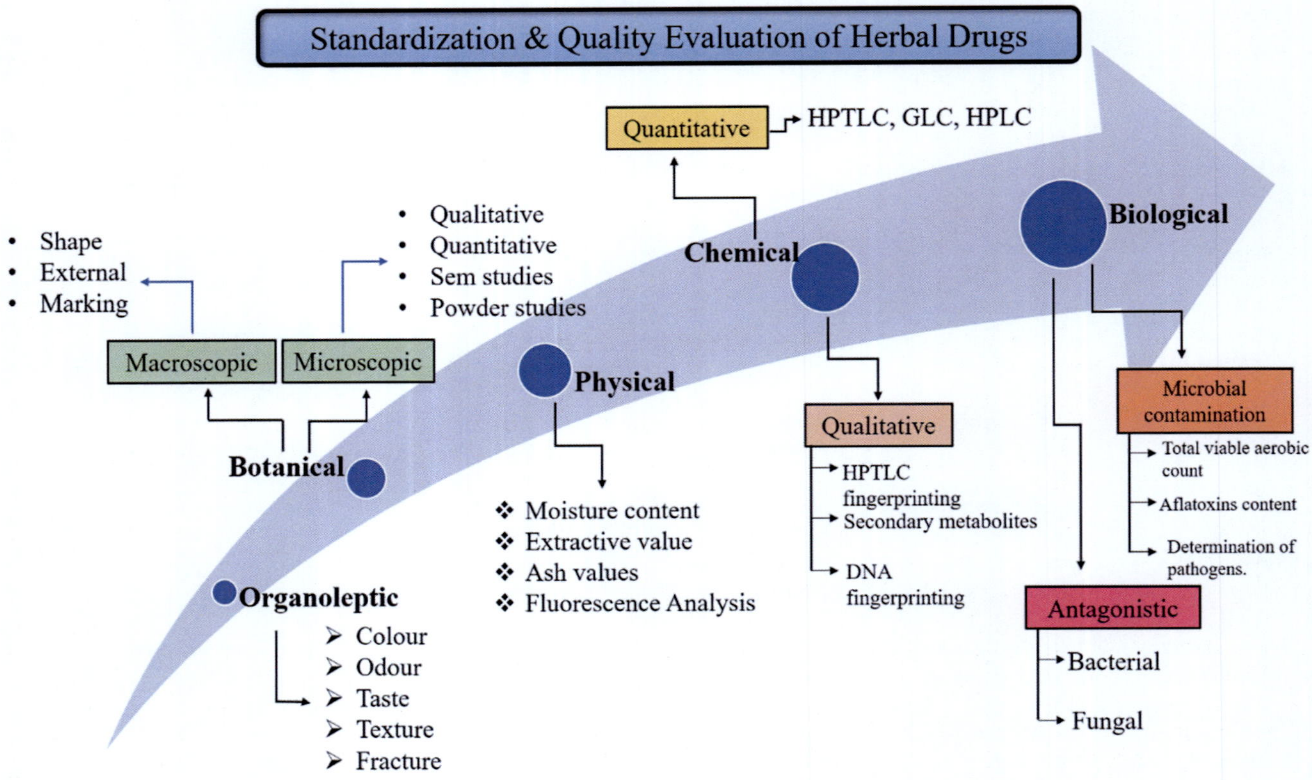

Fig. 1 A systematic diagram of herbal drug standardization

5 Techniques in Herbal Drug Identification and Characterization

The accurate taxonomic identification of plants, determination of the appropriate season and geographical location for collection, development of efficient extraction and purification methodologies for the herbs, and rationalization of the combination in the case of polyherbal formulations are all crucial factors in achieving high-quality herbal products [29]. The standardization of crude drug materials involves a series of essential steps [30] including:

1. Authentication: Verifying the identity of the medicinal plants through taxonomic classification and morphological characterization to avoid misidentification and ensure the use of correct plant species.
2. Collection and Harvesting: Identifying the optimal season and location for the collection of herbal materials, considering factors such as climate, soil conditions, and growth stage to obtain the highest quality plant parts.
3. Extraction and Purification: Implementing appropriate extraction methods, such as maceration, percolation, or Soxhlet extraction, to obtain the active constituents effectively. Subsequent purification steps are performed to eliminate undesirable compounds.
4. Phytochemical Analysis: Conducting comprehensive phytochemical analyses to quantify and identify the major chemical constituents present in the herbal material. Techniques like chromatography, mass spectrometry, and spectroscopy are utilized for this purpose.
5. Quantitative Determination: Determining the levels of active compounds in the herbal extracts to ensure consistency in therapeutic effects and potency.
6. Formulation Rationalization: For polyherbal formulations, rationalizing the combination of herbs based on traditional knowledge, pharmacological synergy, and modern scientific evidence.

The following are the physical, chemical, and microbiological stability characteristics of herbal formulations.

Friability, pH, disintegration time, hardness, flowability, viscosity, moisture content flocculation, sedimentation, settling rate, and ash values are examples of physical characteristics. Physical characteristics such as viscosity, transparency, appearance, color, and odor are essential qualities in herbal product evaluation. Chemical parameters encompass a wide range of assays, tests, and limit tests [31]. Various chromatographic techniques, such as fluorimetry, gas chromatography (GC), ultraviolet–visible spectroscopy (UV), gas chromatography–mass spectrometry (GC-MS), thin-layer

Table 2 Instrumentation method used for standardization of herbal drugs

Instrumental method	Purpose of use	Parameters assessed	Advantages	Disadvantages	Reference
High-performance Liquid chromatography	Quantitative analysis of active compounds marker compounds	Content of phytochemicals, such as alkaloids Flavonoids	High sensitivity and resolution. Fast analysis with short run times	Expensive equipment and maintenance costs	[32]
Gas chromatography	Identification and quantification of volatile components	Volatile compounds, essential oils, terpenoids, etc.	Excellent for analyzing volatile compounds High sensitivity and selectivity	Limited to volatile or semi-volatile compounds Requires derivatization for some nonvolatile compounds	[33]
Mass spectrometry	Structural analysis and identification of phytochemical constituents	Molecular weight, fragmentation patterns, etc.	High specificity and sensitivity Can provide structural information on compounds	Expensive equipment and skilled operators needed Sample preparation and ionization methods can affect the results	[34]
Infrared spectroscopy (IR)	Qualitative analysis and identification of chemical compounds	Functional groups, molecular structure, etc.	Rapid and nondestructive analysis Can be used for both liquid and solid samples	Limited sensitivity for trace analysis Overlapping bands can make interpretation complex	[33]
Nuclear magnetic resonance (NMR)	Structural elucidation and quantitative analysis of complex mixtures	Chemical shifts, molecular structure, purity, etc.	Provides detailed structural information Nondestructive and quantitative analysis	Expensive equipment and requires specialized expertise Requires samples to be dissolved in suitable solvents	[35]
Ultraviolet–visible spectroscopy (UV-vis)	Quantification of specific compounds, such as polyphenols	Absorbance of UV/vis light by chromophores, concentration determination	Simple and rapid analysis Nondestructive and can be used for liquid samples	Limited specificity for complex mixtures Requires standard compounds for accurate quantification	[36]
Fourier transform infrared (FT-IR)	Structural analysis and identification of chemical constituents	Molecular structure, functional groups, etc.	High-resolution spectra with improved sensitivity Minimal sample preparation required	Expensive equipment and skilled operators needed Overlapping bands can make interpretation complex	[37]
Thin-layer chromatography (TLC)	Qualitative analysis and fingerprinting of herbal extracts	Presence of specific compounds and chemical profiling	Simple and cost-effective Can analyze multiple samples simultaneously	Semi-quantitative results Subjective interpretation of chromatograms	[38]

chromatography (TLC), high-performance liquid chromatography (HPLC), and high-pressure thin-layer chromatography (HPTLC), are employed for the comprehensive analysis of herbal substances. Limit tests are quantitative or semi-quantitative methods utilized to monitor and control impurities, such as chemicals used in the extraction process, contaminants from production vessels, and residual solvents. By conducting rigorous analyses based on these scientific parameters, manufacturers can ensure the adherence to quality standards, guarantee product safety, and provide consumers with herbal medicines of high-quality and purity [30] (Table 2).

6 Advancements in Herbal Medicine

Advancements in herbal medicine standardization have emerged as a pivotal area of research and development, intended to improve the effectiveness, safety, and integrity of herbal goods. As the demand for natural remedies grows globally, the need for evidence-based practices becomes increasingly apparent. Advancements in herbal medicine standardization encompass a wide array of cutting-edge technologies and approaches. Metabolomics and genomics provide comprehensive insights into the chemical and genetic composition of herbal materials, while nanotechnology enhances delivery and bioavailability of bioactive compounds [39]. Biotechnological production offers sustainable solutions for obtaining specific phytochemicals, reducing environmental impact, and promoting conservation. However, amid these advancements, the integration of traditional knowledge and modern scientific findings remains pivotal. By synergizing traditional wisdom with technological innovations, herbal medicine standardization is poised to shape a new era of evidence-based, safe, and effective herbal treatments, expanding the role of herbal medicine in modern healthcare practices [40].

In herbal medicine standardization, focusing on innovative approaches that utilize modern technologies and techniques. From metabolomics and genomics to nanotechnology and biotechnological production, these advancements have the potential to revolutionize the field of herbal medicine and bridge the gap between traditional knowledge and scientific validation.

6.1 Quality Control and Safety Concerns

It involves the interactions between herbal medication, toxicity, and their safety:

6.1.1 Safety in Herbal Drugs

Significant discrepancies in the methods used to assess quality, safety, and efficacy could impede the free flow of herbal medicinal products and perhaps put consumers in danger. The intricacy of herbal medical products, specialized knowledge, and expertise are required for the optimal management of the analysis of research information on both safety and effectiveness, reflecting the experience collected in continuous use [41]. Pharmaceutical considerations, including manufacturing methodologies and the specification of herbal extracts, directly affect complex biological products, including herbal medications, in terms of their safety and effectiveness. Addressing these challenges with scientific rigor and adherence to established standards is vital to guarantee the reliable and safe use of herbal medicinal products in the market [42]. To ensure the quality evaluation of herbal medicines, a deeper understanding of agricultural elements related to the production of these medicinal plants becomes imperative. Selecting appropriate seeds, optimizing culture conditions, and implementing precise harvesting practices significantly contribute to achieving consistent quality in herbal medicines. Staying abreast of the latest developments and discussions on good agricultural practice (GAP) specific to medicinal plants is essential for maintaining high standards in the cultivation and sourcing of these botanical materials. By incorporating GAP principles, herbal medicine producers can enhance the quality and efficacy of their products while promoting sustainable and responsible practices in the herbal medicine industry [43].

6.1.2 Toxicity in Herbals and their Interactions

The FDA is responsible for assessing both the safety and efficacy of drug product, and not all botanicals and herbs are safe. The incident that occurred in Brussels, Belgium, between 1991 and 1992 is mentioned in this context. It is possible to consider a case in which 30 women who had treatment with a Chinese herbal slimming product died from kidney failure caused by the amount of aristolochic acid in the product. The classification of one of the herbs as non-toxic was inaccurate [44]. Therefore, it is crucial to supervise the accurate identification of herbal medicines from the very beginning. Along with the issue of inaccurate plant identification, some mixes could be poisonous, especially if they are utilized improperly. Continuous monitoring and aggressively seeking information should be prioritized.

7 WHO Guidelines for Assessment of Herbal Medicines

According to WHO guidelines, standardization of every herbal product is necessary to establish mechanisms for integrating traditional Phytomedicines into mainstream healthcare programs, through analyzing their efficacy as well as safety and assuring the sufficient supply and rigorous quality control of raw and processed materials. WHO recommends employing chemical analyses, less stringent screening, regulatory procedures, and clinical trials, for herbal medicines, but the quality control process for pure phytochemicals should be on par with that of synthetic drugs [45]. The World Health Organization (WHO) defines traditional medicinal products as encompassing therapeutic practices that predate the development and widespread use of modern medicine and continue to be employed today, often spanning centuries. These traditional preparations comprise organic matter, minerals, or medicinal herbs. Crucial parameters include safety assessment, stability testing investigation of specific therapeutic activities, and determination of the active components in plant raw materials and final products. Adherence to these parameters ensures the standardized quality and safety of herbal medicines in line with scientific principles. The WHO recommendations are aimed to establish fundamental guidelines for the comprehensive evaluation of the quality, safety, and efficacy of herbal medicinal products. These guidelines are intended to assist national regulatory agencies, scientific organizations, and manufacturers in their evidence-based assessments [46]. Detailed descriptions of the production method, formulation, and inclusion of excipients should be provided. Additionally, a robust technique for the detection and, whenever feasible, to quantify the plant material in the final product must be established. In cases where the identification of a specific active principle is not achievable, it is essential to identify a distinct ingredient or a combination of compounds (e.g., through "chromatographic fingerprinting") to ensure consistent quality of the herbal product. By adhering to these scientific protocols, herbal medicine manufacturers can ensure the reliability and safety of their products, facilitating their acceptance and integration into mainstream healthcare systems [47]. According to the defined scope, "herbal medicines" refer to labeled medicinal products containing aerial or underground parts of plants or other plant materials, either in their crude form or as plant preparations. This category includes substances such as gums, juices, essential oils, fatty acids, and other plant-derived components Excipients may also be present in herbal medicines, in addition to the active ingredients. However, formulations combining chemically identified active ingredients with plant material, particularly when isolated or chemically identified plant elements are present, do not fall under the classification of herbal medicines [48]. There is an exception where herbal medications may contain natural, organic, or inorganic active components, which are not derived from plants, following established practices or customs. Adherence to these precise definitions helps in distinguishing and categorizing herbal medicines appropriately within the broader spectrum of medicinal products [49]. Recent advancements in scientific techniques such as DNA fingerprinting, liquid chromatography, mass spectrometry, and high-pressure thin-layer chromatography (HPTLC) have enabled the standardization of complex multicomponent botanical compositions. When utilizing botanicals in their natural forms, the necessity for animal studies to assess safety and toxicity is diminished [50]. Nevertheless, adherence to essential pharmacopeial tests, including those for microbial, pesticide, and heavy metal contamination, must comply with global regulations. Additionally, the production of all Ayurvedic medicines should adhere to the current good manufacturing practices (cGMP) specific to herbal products. By incorporating these modern scientific methodologies and adhering to stringent quality control measures, the standardization and safety of herbal medicines can be ensured for the benefit of consumers and to meet international standards [51].

8 Application of Human Cognitive Techniques

Traditional system of medicine is ancient healing practice shaped by indigenous knowledge and cultural beliefs. Ayurveda exemplifies one such system with a profound historical legacy. According to Ayurveda, there are three humors, together referred to as "Tridoshas," (Vata, Pitta, and Kapha dosha), which control essential physiological functions. However, these traditional methods have been in practice for centuries and follow specific procedures for preparation and standardization. Additionally, the cultural elements of herbalism in various countries and societies significantly contribute to the diverse global utilization of herbal remedies. The abundance of modern technology promises to expand the opportunities for personalized treatment and disease prevention including fuzzy logic, neural networks, expert system, and digital intelligence, among which fuzzy logic is one of major tool applied for humanity to gain greater insights into lifestyle diseases [52].

To diagnose various diseases using the proposed the Medical Diagnostic System (MDS) is one of the most effective qualitative computational techniques, fuzzy logic served as the artificial intelligence tool used in the suggested methodology [53]. However, it is a fundamental part of medicine that disease diagnosis entails a wide variety of ambiguity and imprecision. Depending on the patient and with varying degrees of severity, a single disease may present itself quite differently. Additionally, there are other diseases that may possess the same symptoms. On the other hand, a patient who has multiple diseases may have interactions between them that prohibit a typical description of any one of the disorders from being accurate. Communication is imprecise and unresolved, yet it is the greatest and most accurate way to convey disease entities. Additionally, the traditional ideas of health and disease are incompatible and contrary to one another. Both ideas are implemented as alternative methods on the same continuum in certain recent approaches [54].

Digital intelligence has been employed in recent years to develop intelligent systems which have been used to address a variety of complicated challenges. In addition, for decision-making systems like expert systems and pattern classification systems, fuzzy logic has demonstrated to be a potent tool. Moreover, some medical expert systems have already used the principle of fuzzy sets [55]. However, fuzzy logic is more like human cognitive process and provides meaning to the sense like "often," "smaller," and "higher" as compared to computer software that simply understands binary operations or actual numbers like 1.8 and 2.6. Fuzzy logic involves the complexity and uncertainty of the real world; nothing can have absolute values or match a linear formula [56].

8.1 Applications of Fuzzy Logic

Fuzzy set theory can be construed as an adequate approach to deal with the imprecision inherent to many biomedical and bioinformatics challenges since fuzzy logic takes into account such uncertainty. Uncertainty is inherent in domains like medicine and enormous data in bioinformatics. Moreover, in the field of medicine, fuzzy logic is a technique for making imprecise information exact. However, there are several instances and illustrations of fuzzy logic classified below.

8.1.1 Fuzzy Logic in Smart Health

Medical diagnosis is a challenging task when maintaining the level of ambiguity and uncertainty, specifically if the disease involves multiple signs and symptoms. The fuzzy-analytic hierarchy process (fuzzy-AHP) technique has been utilized by a lot of researchers to handle ambiguous data in medical diagnosis and therapy. While the AHP may perform pairwise comparisons of choice elements to assess their relative importance in the decision process, fuzzy logic can deal with ambiguity in the process of making decisions [57]. In smart cities, the use of home-based healthcare is in increasing popularity. When making decisions about a diagnosis and necessary therapy in today's world, timely information about the health state of people with chronic heart disease can be extremely valuable. The development of intelligent decision-making systems utilizing the knowledge and observations of healthcare practitioners has proven to be possible with the help of fuzzy logic, which has proven to be a great tool. Therefore, Hussain et al. suggested a fuzzy logic-based personal healthcare system for chronic heart disease patients (under stable conditions) for follow-up and monitoring outside of hospitals. The proposed system might be an inventive, accessible resource and alternative to the current healthcare systems, assisting practitioners in providing effective treatment to cardiovascular patients who lived single at home [58].

The role of smart healthcare has become crucial in the current period of a severe public health destruction spurred on by the spread of the novel coronavirus (2019-nCoV), resulting in millions of deaths and adversely affected a large number of people globally. Medical personnel could benefit greatly from any contemporary technique that permits quick and effective monitoring of patients with COVID-19. Worldwide interest in a number of Internet of Medical Things (IoMT)-based smart healthcare systems has grown as a result of their increasing technical support for health services, particularly in the detection, prevention, and remote monitoring of the majority of infectious diseases. The role of smart healthcare has become crucial in the current period of a severe public health destruction spurred on by the spread of the novel coronavirus (2019-nCoV), resulting in millions of deaths and adversely affected a large number of people globally. Medical personnel could benefit greatly from any contemporary technique that permits quick and effective monitoring of patients with COVID-19. Worldwide interest in a number of Internet of Medical Things (IoMT)-based smart healthcare systems has grown as a result of their increasing technical support for health services, particularly in the detection, prevention, and remote monitoring of the majority of infectious diseases [59].

8.1.2 Fuzzy Logic in Bioinformatics

The study of biological data by computers gets information in the field of bioinformatics. In addition to the information encoded in the genome, these data may also include patient statistics, scientific literature, and experimental results from multiple sources with inherent uncertainty, and experimental results from other sources. The study of molecular computing, genetic engineering, and the creation of training tools for biocomputing systems are all part of the field of bioinformatics, which combines computer science, biology, physical science, and chemical principles with tools for large-scale biological data analysis and modeling [60].

8.1.3 Fuzzy Logic in Genomics

The study of genome structure and evolution is known as genomics, and it also serves as a tool to comprehend the functions of genes. Genomic research is concerned with gathering and analyzing information on all genes in a living being. It has grown in significance for biologists as a means of identifying patterns in data that can then be transformed into biological knowledge and facts. In the year of 2005, [61] suggested that the quality of genomic sequence is becoming more and more important in analyses of genetic variation, study of pathogenic alterations, structure of proteins estimation, relative analysis of sequences, and sequence annotation. An effective tool in this approach is a level of confidence in the DNA sequencing analysis process. However, Woolf et al. created a unique algorithm to analyze data on gene expression. Therefore, with the help of fuzzy logic, this program converts values for expressions into qualitative identifiers that can be assessed using an array of heuristic criteria.

8.2 Advantages of Comprehensive Standardization of Herbal Medicine

8.2.1 Enhancing Herbal Agent Quality

The integration of advanced techniques, such as fuzzy logic, neural networks, and accurate linguistic models, will lead to a significant improvement in the quality of herbal drugs. These sophisticated methodologies can analyze and evaluate the properties and efficacy of herbal remedies with greater precision, ensuring consistent and reliable products [62].

8.2.2 Fuzzy Logic for Uncertain Data Handling

By incorporating fuzzy logic into the analysis of herbal agents, we can effectively deal with the inherent uncertainty and imprecision often associated with traditional medicine data. Fuzzy logic allows for flexible and nuanced decision-making, enabling a more comprehensive evaluation of herbal properties and their potential health benefits [63].

8.2.3 Empowering Traditional Medicine Practitioners

Through the adoption of modern analytical methods in the standardization process, traditional medicine practitioners will gain access to evidence-based data supporting the effectiveness of herbal treatments. This scientific validation can boost their confidence in traditional medicine and motivate them to actively participate in the standardization efforts [64].

8.2.4 Bridging Ancient Knowledge and Modern Science

By combining traditional knowledge with cutting-edge analytical techniques, the standardization process creates a bridge between the wisdom of traditional medicine and the rigor of modern science. This synergy fosters a deeper understanding and appreciation of the potential benefits of herbal agents within both traditional and contemporary healthcare practices [65].

8.2.5 Wider Acceptance and Recognition

The improved quality and scientific validation of standardized herbal agents can lead to greater acceptance and recognition of traditional medicine within the broader medical community. As traditional remedies demonstrate their efficacy backed by data-driven evaluations, they can be more readily integrated into mainstream healthcare practices, offering patients additional choices for their well-being [66] (Fig. 2).

9 Conclusion

Herbal medicine offers a valuable adjunctive approach to the management of lifestyle diseases. The diverse array of bioactive compounds present in medicinal plants provides a broad spectrum of therapeutic benefits, ranging from antioxidant and anti-inflammatory effects to glycemic control, cardiovascular protection, weight management, and stress reduction. Integrating herbal medicine into lifestyle disease management can complement conventional treatments, reduce medication side effects, and improve overall patient outcomes. However, it is essential to seek guidance from qualified healthcare professionals and ensure proper dosing and safety considerations while incorporating herbal remedies into treatment regimens. Further research and clinical trials are warranted to strengthen the evidence base for the role of herbal medicine in managing lifestyle diseases and to pave the way for its integration into mainstream healthcare practices. The advancement of techniques for evaluating traditional medicine demands the integration of both conventional and contemporary methodologies. Coupling traditional evaluation methods with modern analytical approaches, such as fuzzy logic, neural networks, and precise linguistic models, can significantly enhance the quality of herbal drugs. This integration promotes the active participation of traditional medicine practitioners in the standardization process, fostering a more comprehensive and reliable evaluation of medicinal preparations.

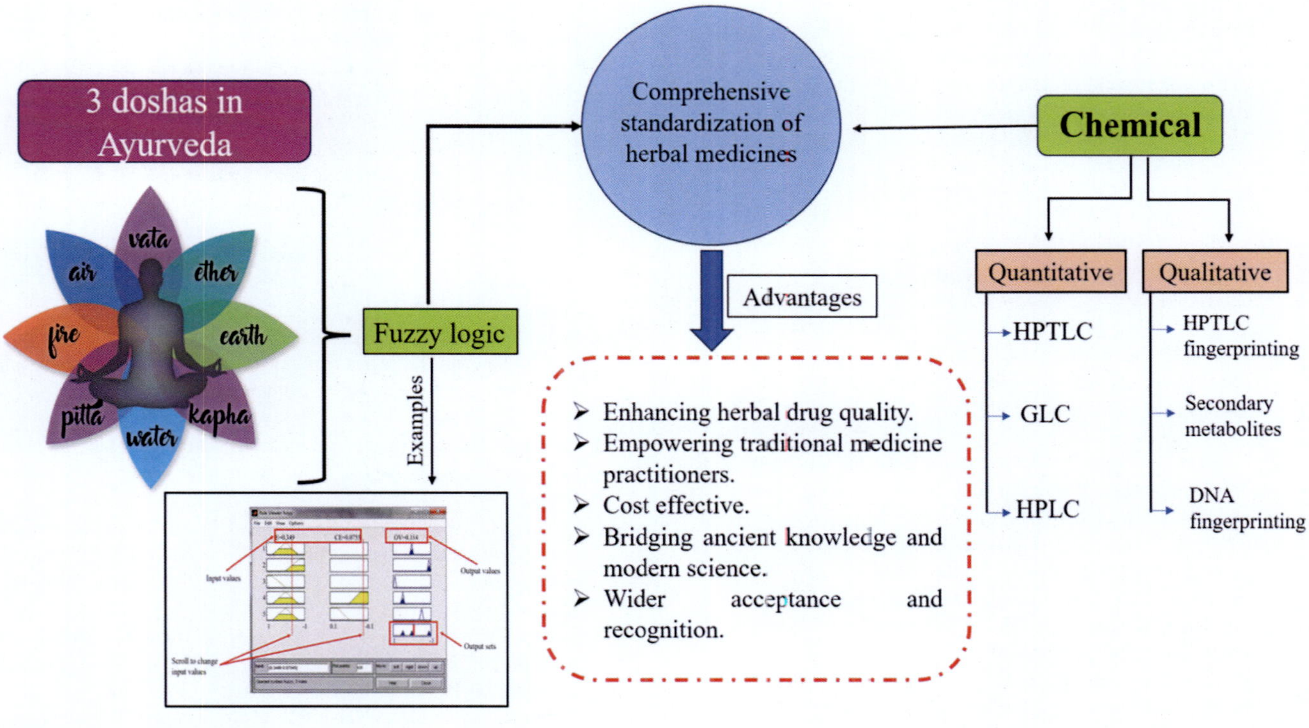

Fig. 2 Comprehensive standardization of herbal medicines

References

1. Verma S, Singh SP. Current, and future status of herbal medicines. Veterinary world. 2008;1(11):347.
2. Shrikumar S, Ravi TK. Approaches towards development and promotion of herbal drugs. Pharmacog Rev. 2007;1(1):180–4.
3. Verma NK. History of plants and animal products in the treatment of human disease-a. Res J Phytomed. 2015;1:02.
4. Calixto JB. Efficacy, safety, quality control, marketing, and regulatory guidelines for herbal medicines. Braz J Med Biol Res. 2000;33:179–89.
5. Zhang Z, Liang Y, Xie P, Chau F, Chan K. Chromatographic fingerprinting and chemometric techniques for quality control of herb medicines. In: Data analytics for traditional Chinese medicine research; 2014. p. 133–53.
6. Madhav NS, Upadhyaya K, Bisht AS. Phytochemical screening and standardization of polyherbal formulation for dyslipidemia. Int J Pharm Sci. 2011;3(3):235–8.
7. World Health Organization. Quality control methods for medicinal plant materials. World Health Organization; 1998.
8. Sumi Jain DK. Ayurvedic principles to prevent & management of life style disorders. J Sci Innov Res. 2019;8(1):24–8.
9. Barnett RN. Importance of quality control in the medical laboratory. Ann N Y Acad Sci. 1969 Sep;161(2):477–83.
10. Mukherjee PK, Harwansh RK, Bahadur S, Biswas S, Kuchibhatla LN, Tetali SD, Raghavendra AS. Metabolomics of medicinal plants–a versatile tool for standardization of herbal products and quality evaluation of ayurvedic formulations. Curr Sci. 2016 Nov;25:1624–30.
11. Kumar KJ, Bhatnagar SP. Risking alternative and complementary medicine in the name of standardization. J Alternat Complement Med. 2006;12(5):419–20.
12. Ong ES. Extraction methods and chemical standardization of botanicals and herbal preparations. J Chromatogr B. 2004;812(1–2):23–33.
13. Alam G. Mishra AK. Traditional and modern approaches for standardization of herbal drugs: a review: Acta Biomed Sci e-ISSN; 2017. p. 234–68.
14. Govindaraghavan S, Sucher NJ. Quality assessment of medicinal herbs and their extracts: criteria and prerequisites for consistent safety and efficacy of herbal medicines. Epilepsy Behav. 2015;1(52):363–71.
15. Mohammed Abubakar B, Mohd Salleh F, Shamsir Omar MS, Wagiran A. DNA barcoding and chromatography fingerprints for the authentication of botanicals in herbal medicinal products. Evid Based Complement Alternat Med. 2017;27:2017.
16. Sudberg S, Sudberg EM, Terrazas J, Sudberg S, Patel K, Pineda J, Fine B. Fingerprint analysis and the application of HPTLC to the determination of identity and quality of botanicals, from an industry perspective. J AOAC Int. 2010;93(5):1367–75.
17. Wolfender JL, Rodriguez S, Hostettmann K. Liquid chromatography coupled to mass spectrometry and nuclear magnetic resonance spectroscopy for the screening of plant constituents. J Chromatogr A. 1998;794(1–2):299–16.
18. Gautam A, Kashyap SJ, Sharma PK, Garg VK, Visht S, Kumar N. Identification, evaluation, and standardization of herbal drugs: a review. Pharm Lett. 2010;2(6):302–15.
19. Sahil K, Sudeep B, Akanksha M. Standardization of medicinal plant materials. Int J Res Ayurveda Pharmacy. 2011;2(4):1100–9.
20. Patwekar SL, Suryawanshi AB, Gaikwad MS, Pedewad SR, Potulwar AP. Standardization of herbal drugs: an overview. Pharma Innov. 2016;5(4, Part B):100.
21. Fernandes FH, Boylan F, Salgado HR. Quality standardization of herbal medicines of Spondias dulcis Parkinson using analytical and microbiological analysis. J Therm Anal Calorim. 2018;134: 1923–8.
22. Mandal M, Misra D, Ghosh NN, Mandal V. Physicochemical and elemental studies of Hydrocotyle javanica Thunb. For standardization as herbal drug. Asian Pac J Trop Biomed. 2017;7(11):979–86.
23. Alam G, Mishra AK. Traditional and Modern Approaches for Standardization of Herbal Drugs: A Review. Acta Biomed Sci. 2017:234–68.
24. Ekka NR, Namdeo KP, Samal PK. Standardization strategies for herbal drugs-an overview. Res J Pharm Technol. 2008;1(4):310–2.
25. Shivatare RS, Nagore DH, Nipanikar SU. ‘HPTLC’ an important tool in standardization of herbal medical product: a review. J Sci Innov Res. 2013;2(6):1086–96.
26. Pattanayak P, Kumar Hardel D, Mohapatra P. Standardization of Vaisvanara Churna: a poly-herbal formulation. Pharm J. 2010;2(5)
27. Prakash KD, Shivaji GS, Rohidas KP, Patel V, Bhalke R. Current quality control methods for standardization of herbal drugs. Int J Pharmaceut Drug Analysis. 2017;5(3):82–95.
28. Indrayanto G. Recent development of quality control methods for herbal derived drug preparations. Nat Prod Commun. 2018;13(12):457.
29. Wei XC, Cao B, Luo CH, Huang HZ, Tan P, Xu XR, Xu RC, Yang M, Zhang Y, Han L, Zhang DK. Recent advances of novel technologies for quality consistency assessment of natural herbal medicines and preparations. Chin Med. 2020;15:1–24.
30. Pandey A, Tripathi S. Concept of standardization, extraction and pre phytochemical screening strategies for herbal drug. J Pharmacognosy Phytochem. 2014;2(5):115–9.
31. Yu BS, Lai SG, Tan QL. Simultaneous determination of cinnamaldehyde, eugenol and paeonol in traditional Chinese medicinal preparations by capillary GC-FID. Chem Pharm Bull. 2006;54(1):114–6.
32. Kuzmenko AN. Application of gas-liquid chromatography for standardization of herbal raw materials and herbal drugs. Russ J Gen Chem. 2012;82:595–601.
33. Patil PS, Shettigar R. An advancement of analytical techniques in herbal research. J Adv Sci Res. 2010;1(01):8–14.
34. Li W, Cheng Z, Wang Y, Qu H. Quality control of Lonicerae Japonicae Flos using near infrared spectroscopy and chemometrics. J Pharm Biomed Anal. 2013;72:33–9.
35. Bokov DO. Standardization of snowdrop (Galanthus L.) herbal pharmaceutical substances by ultraviolet-spectrophotometry. Asian J Pharm Clin Res. 2018;11(10):207–11.
36. Singh SK, Jha SK, Chaudhary A, Yadava RD, Rai SB. Quality control of herbal medicines by using spectroscopic techniques and multivariate statistical analysis. Pharm Biol. 2010;48(2):134–41.
37. Ganguly P, Chowdhury S, Paul N, Haldar S, Singh R, Katiyar CK. Role of TLC to ensure manifestation of individual ingredients in herbal formulation. JPC–J Planar Chromatogr–Modern TLC. 2020;33:281–91.
38. Bhairam M, Roy A, Bahadur S, Banafar A, Turkane D. Standardization of herbal medicines–an overview. J Appl Pharmaceut Res. 2013;1(1):14–21.
39. Ruiz GG, Nelson EO, Kozin AF, Turner TC, Waters RF, Langland JO. A lack of bioactive predictability for marker compounds commonly used for herbal medicine standardization. PLos One. 2016;11(7):598–57.
40. Leonti M, Casu L. Traditional medicines and globalization: current and future perspectives in ethnopharmacology. Front Pharmacol. 2013;4:92.
41. Lutoti S, Okwany P, Ajayi CO, Oloro J. Formulation and standardization of herbal medicinal products: a review of the formulation considerations, quality control and safety of herbal products. J Pharm Drug Res. 2020;3(3):373–81.
42. Indrayanto G. Recent development of quality control methods for herbal derived drug preparations. Nat Prod Commun. 2018;13(12):193–8.
43. Zhang R, Zhang MX, Chen Y, Wang CC, Zhang CH, Heuberger H, Li HT, Li MH. Future development of good agricultural practice

in China under globalization of traditional herbal medicine trade. Chinese Herbal Med. 2021;13(4):472–9.
44. Vanherweghem JL, Tielemans C, Abramowicz D, Depierreux M, Vanhaelen-Fastre R, Vanhaelen M, Dratwa M, Richard C, Vandervelde D, Verbeelen D, Jadoul M. Rapidly progressive interstitial renal fibrosis in young women: association with slimming regimen including Chinese herbs. Lancet. 1993;341(8842):387–91.
45. World Health Organization. WHO global report on traditional and complementary medicine 2019. World Health Organization; 2019.
46. World Health Organization. Ethics and governance of artificial intelligence for health: WHO guidance. 2021
47. Kopp B. Herbal medicinal products and phytotherapy—a thematic issue. Wien Med Wochenschr. 2017;167:145–6.
48. Heinrich M. Quality and safety of herbal medical products: regulation and the need for quality assurance along the value chains. Br J Clin Pharmacol. 2015;80(1):62–6.
49. Bent S. Herbal medicine in the United States: review of efficacy, safety, and regulation: grand rounds at University of California, San Francisco medical Center. J Gen Intern Med. 2008;23:854–9.
50. Nicoletti M, Petitto V, Gallo FR, Multari G, Federici E, Palazzino G. The modern analytical determination of botanicals and similar novel natural products by the HPTLC fingerprint approach. Stud Nat Prod Chem. 2012;37:217–58.
51. Vaidya VN, Tatiya AU, Elango A, Kukkupuni SK, Vishnuprasad CN. Need for comprehensive standardization strategies for marketed ayurveda formulations. J Ayurveda Integrative Med. 2018;9(4):312–5.
52. Yuan H, Ma Q, Ye L, Piao G. The traditional medicine and modern medicine from natural products. Molecules. 2016;21(5):559.
53. Awotunde JB, Matiluko OE, Fatai OW. Medical diagnosis system using fuzzy logic. Afr J Comput ICT. 2014;7(2):99–106.
54. Torres A, Nieto JJ. Fuzzy logic in medicine and bioinformatics. Biomed Res Int. 2006:2006.
55. Phuong NH, Kreinovich V. Fuzzy logic and its applications in medicine. Int J Med Inform. 2001;62(2–3):165–73.
56. Godil SS, Shamim MS, Enam SA, Qidwai U. Fuzzy logic: a "simple" solution for complexities in neurosciences? Surg Neurol Int. 2011:2.
57. Uzoka F-ME, Obot O, Barker K, Osuji J. An experimental comparison of fuzzy logic and analytic hierarchy process for medical decision support systems. Comput Methods Prog Biomed. 2011;103(1):10–27.
58. Hussain A, Wenbi R, Xiaosong Z, Hongyang W, Da Silva AL. Personal home healthcare system for the cardiac patient of smart city using fuzzy logic. J Adv Inform Technol. 2016;7(1)
59. Panja S, Chattopadhyay AK, Nag A, Singh JP. Fuzzy-logic-based IoMT framework for COVID19 patient monitoring. Comput Ind Eng. 2023;176:108–41.
60. Woolf PJ, Wang Y. A fuzzy logic approach to analyzing gene expression data. Physiol Genomics. 2000;3(1):9–15.
61. Ressom H, Natarajan P, Varghese RS, Musavi MT. Applications of fuzzy logic in genomics. Fuzzy Sets Syst. 2005;152(1):125–38.
62. Che C-T, Wang ZJ, Chow MSS, Lam CWK. Herb-herb combination for therapeutic enhancement and advancement: theory, practice and future perspectives. Molecules. 2013;18(5):5125–41.
63. Louw G. A political-historical literature review of the statutory impact of the traditional health practitioners act (no 22, 2007) and the traditional health practitioner on the empowerment of the present and future south African healthcare establishment: North-West University; 2018.
64. Patwardhan B. Bridging ayurveda with evidence-based scientific approaches in medicine. EPMA Journal. 2014;5:1–7.
65. Shi J, Cao B, Wang X-W, Aa J-Y, Duan J-A, Zhu X-X, et al. Metabolomics and its application to the evaluation of the efficacy and toxicity of traditional Chinese herb medicines. J Chromatogr B. 2016;1026:204–16.
66. Klein-Junior LC, de Souza MR, Viaene J, Bresolin TM, de Gasper AL, Henriques AT, Vander HY. Quality control of herbal medicines: from traditional techniques to state-of-the-art approaches. Planta Med. 2021;87(12/13):964–88.

Intellectual Property Rights, Naturally Derived Bioactive Compounds, and Resource Conservation

Raju Wadekar, Subhash C. Mandal, and Kalpana Patil

Abstract

On October 20 to 22, 1994, an interim annual meeting was hosted by the American Society of Pharmacognosy in San Jose, Costa Rica. During the symposium, subject experts from all over the nation shared their insights on issues adjoining intellectual property rights in kith and kin of drug development from naturally derived compounds, plant conservation, and international conventions on bioprospecting. To guarantee that the genomic assets of all nations will be developed for the benefit of human race and observant of recent current stratagems for the benefits, scientific collaboration will be warranted. The core issue of the conferences, the fair and justifiable sharing of the compensations of bioprospecting, was articulated from a variety of opposing standpoints, as can be seen from the summaries of presentations. These divergent belvederes are the consequence of various explanations. Different viewpoints to the larger society are formed by the multiplicity midst the world, particularly those in developing nations. Additionally, a lot of the discussion and negotiations around access to resources wherever they are discovered are clouded and perplexed by the excessive emphasis on the unearthing, expansion, and business of medications as the inimitable outcome of the exploitation of genetic resources. It should be underlined that academic and research institutions would fail, due to a successful outcome from bioprospecting product. Instead, it is critical to understand that other vital goals may also be achieved, particularly in the fields of technology transfer and capacity building. The symposiums help advance research on intellectual property rights (IPR), naturally occurring bioactive substances, and resource conservation for the moral use of conventional medicines, as well as for the development of a country's economy and healthcare system. In light of the foregoing information, an effort has been made in this chapter to disseminate recent scientific advancement and to comprehend the interrelationship between IPR, bioactive of natural origin, and conservation of resources in order to establish a smooth avenue for traditional drugs and medicine thereof.

Keywords

Intellectual property rights · Bioprospecting · Genomic · Conservation · Bioactive

Abbreviations

BMBs	Bioactive-made biologics
CDSCO	The Central Drugs Standards Control Organization
EGCG	Epigallocatechin gallate
GAP	Good Agricultural Practices
IMS	Indian Medical Systems
IPR	Intellectual property rights
NGOs	Nongovernmental organizations
PPD	Phytopharmaceutical drugs
TEK	Traditional ecological knowledge
TRIPS	Trade-Related Intellectual Property Rights
WIPO	World Intellectual Property Organization

R. Wadekar (✉)
Department of Pharmacognosy and Phytochemistry, Shri Vile ParleKelavani Mandal's Institute of Pharmacy, Dhule, Maharashtra, India
e-mail: raju.wadekar@svkm.ac.in

S. C. Mandal
Division of Pharmacognosy, Department of Pharmaceutical Technology, Jadavpur University, Kolkata, India

K. Patil
Department of Pharmacognosy, K.L.E.'s College of Pharmacy, Nehru Nagar, Belagavi, Karnataka, India
e-mail: kalpanapatil@klepharm.edu

A. K. Dhara, S. C. Mandal (eds.), *Role of Herbal Medicines*, https://doi.org/10.1007/978-981-99-7703-1_28

1 Introduction

According to one definition of traditionally derived medicine, it is "the sum and set of the understanding, expertise, and practices based on the principles, beliefs, and proficiencies native to diverse cultures, whether justifiable or not, used in the preservation of health as well as in the prevention, diagnosis, improvement, or treatment of physical and mental illnesses" [1]. The term "traditional ecological knowledge (TEK)" also applies to traditional communities. As a whole, traditional knowledge (TEK) is a component of ecology science and is crucial to the development of traditional medicine and healthcare systems [2]. In contrast to the Western systems, TEK in traditional societies is carried out by "collective memory" [3]. The amount of knowledge available to humankind about the universe he lives in has increased exponentially. The scientific revolution had begun during the Renaissance and had grown throughout time as information was amassed at an ever-increasing rate. The twentieth century was an era of great change in many facets of life, mostly brought on by advances in science and technology. Although each discovery builds on the one before it, the "individual innovation" is the primary driver of this Western-originated knowledge accumulation. The society gave being "inventive" a lot of support, and it eventually developed ways to defend itself. The primary system for safeguarding innovation is a "patent." The world receives thousands of new patent applications each year. In the history of scientific creation, inventors have on occasion decided to reveal their discoveries without any financial incentive or with the hope of profiting from patents. In the current international economy, patenting is advantageous, and it positively influences invention. However, as the scope and accessibility of the global economy expand to every corner and area of the earth, some new issues surface that are difficult to address using the traditional private rights approach. A type of patenting in the area of intellectual invention is known as an intellectual property right (IPR). Intellectual property also encompasses copyright, trade names, and industrial designs, but, in this article, patents are solely discussed in relation to conventional medicine. The outcomes of Western science are essentially recalled when innovations are discussed, but there are also genuine inventions that are protected by IPR since they were the fruit of thousands of years of knowledge accumulation based on the experiences of traditional societies. One of the areas of research that has generated the most interest and questions concerning the locals' intellectual property rights (IPR) is the treatment of diseases that are unknown to contemporary medicine. A number of perspectives on the problem are presented by the conceptual framework on the domains and relationships of property and related rights [4–6]. Biopiracy, which is the act of using information without providing any profit to the owner of the entity or information, poses a concern for the future but is already a reality today and carries with it a number of legal repercussions [7]. IPR has evolved into a metaphor for indigenous ownership of all traditional knowledge, opening up possibilities for contractual procedures to guarantee benefits go back to the original cultures and nations. The IPR discussions shed light on how important biodiversity is to human health. The basis of pharmacy has been plants from antiquity to the present, but, as time goes on, the pace of extinction of species and cultures increases, and the health of humans continues to decline due to diseases for which there is now no solution. As a result, the significance of plants as medicines is now highly acknowledged, and the IPR related to their usage has been contested globally [8]. Also, efforts have also been made to utilize naturally occurring bioactive substances as delivery agents, on the other hand. Such bioactive, naturally occurring compound-derived nanocarriers, nanovesicles, phytosomes, microemulsions, etc., are anticipated to display intrinsic therapeutic efficacy in addition to having greater biocompatibility than drug delivery systems produced from synthetic materials. The active components of green tea, epigallocatechin gallate (EGCG), curcumin, and resveratrol, are typical naturally occurring bioactive chemicals that have been employed as carriers in biomedical applications. The ability of these delivery carriers to treat diseases is illustrated by a number of examples, and their potential for biological applications is evaluated [9]. 90% of the above 1400 medicinal plants used in Europe come from wild sources; in the USA, about 120 of the top 165 prescription medications are derived from natural origin [10]. According to a very conservative estimate, in every two-year earth is trailing substantial of each significant drug candidate and the rate of plant extinction is currently between 100 and 1000 times higher than what is expected by natural extinction [11]. Globally, 50,000 plus and 80,000 kinds of flowering plants are used as medicines, as per the International Union for Conservation of Nature and the World Wildlife Fund. Among them, 15,000 species are in danger of going extinct due to habitat loss and overfishing [12], and 20% of their wild resources have already been almost completely depleted due to the growing human population and plant consumption [13]. The risk of extinction of medicinal plants has increased globally, despite the fact that this apprehension has been known for years, particularly in China [14], India [15], Kenya [15], Nepal [15], Tanzania [16], and Uganda [16]. Numerous studies have been conducted on the preservation and sustainable usage of medicinal plants [17, 18]. A number of sets of recommendations have been made about their conservation, including the necessity for integrated conservation approaches based on both in situ and ex situ tactics [19] and the construction of

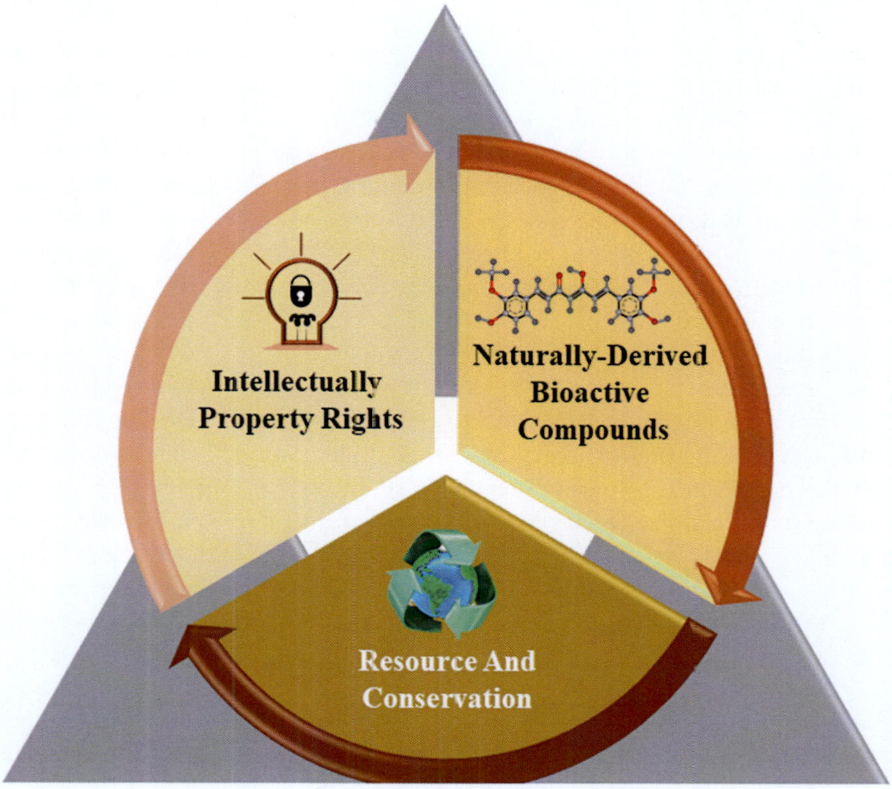

Fig. 1 Golden triangle and interrelationship between IPR for traditional knowledge of medicines/drugs

systems for species inventorying and status monitoring. Sustainable use of wild resources can be a successful conservation solution for medicinal plants with diminishing stocks. Due to the enormous needs of sizable populations, the situation is particularly dire in China and South Africa. Therefore, it is warranted to understand and apprehend the recent knowledge regarding the appropriate use of the golden triangle (Fig. 1) and interrelationship between IPR for traditional knowledge of medicines/drugs, naturally occurring bioactive compounds, and its exploitation by modern science in drug delivery and its resource and conservation in nation economy building and to provide better healthcare system to the world.

2 Intellectually Property Rights (IPR)

IPR is a weapon used by the law to safeguard industrial innovation in the contemporary era, as well as to encourage biodiversity preservation and ensure just and equitable benefit distribution among indigenous custodians [20, 21]. The majority of patented inventions are built on existing information. One benefit of the patent system is that it encourages new ideas by utilizing publicly available knowledge on the basis of nation-to-nation cultural heritage and beliefs [22]. Patent laws protect many herbal products that are derived from conventional medicine. In terms of copyrights, patents, trademarks for pharmaceutical preparations, and registered designs, phytochemicals have a wide range of intellectual property rights (IPR) [23]. Trade secrets and trademarks are the two most common types of IP protection for herbal medicines [24]. However, patenting plant-derived pharmaceutical goods and methods based on conventional wisdom has grown to be a significant intellectual property rights issue [25]. Indigenous knowledge and intellectual property rights have resulted in numerous difficult legal issues for the modern world. Indigenous knowledge preservation concerns extend beyond the realm of the law. Furthermore, communities all across the world can directly experience the issues raised by intellectual property legislation in traditional knowledge systems [26]. The word "biopiracy" is now frequently used by activists and some specialized nongovernmental organizations (NGOs) to refer to traditional knowledge regarding the incorrect or illegal usage of biological materials [27]. Additionally, the rising patent count reflects increased research into biopiracy. Indian neem tree cases have demonstrated the crucial role patents play in biopiracy. These international consultants are hesitant about the overuse of the term "biopiracy" to define particular instances of unfair claims over biological resources and traditional knowledge, despite the fact that there has been much international discussion on disputes pertaining to intellectual property and assets, traditional knowledge, and heritage. The word "bioprospecting" is relatively new and was created to define the responsible use of natural resources, respect for indigenous peoples' rights, and identification and commercialization of bio-products [22, 27–29]. The rights and interests of the traditional knowledge owners who give the "source" of these patents (such as disclosure of sources and profit sharing) must also be honored while we preserve these leading compounds and secure private rights under the current patent law system. Otherwise, it violates fundamental intellectual property system principles and ideas like the interest balance [30, 31]. A significant agreement that supports the harmonization of intellectual property rights (TRIPS) [32]. New contexts for the protection of ethnic interests have been reputable with the help of imperative organizations like the World Intellectual Property Organization (WIPO). In addition, nations and regions are mostly concerned with safeguarding traditional resources [33].

2.1 Traditional Knowledge and Intellectual Property Rights

Traditional agricultural, medicines, and ecological knowledge (TAK, TMK, and TEK) are three types of "traditional knowledge." Native American tribes, peoples, and countries have and employ indigenous knowledge, which is a subset of the outmoded knowledge category. Traditional medical knowledge (TMK) is used by indigenous people, particularly in rural communities, to preserve their healthcare systems [34, 35]. The preservation of traditional knowledge (TK), which is fundamental to the identity of most local communities, is crucial for the social and physical environment of those communities. They have a connection to the local biodiversity, which includes endemic plants, animals, fungus, and other biological materials. A large range of medical plants, healthful herbal remedies, and agricultural and forestry products that are sold internationally and have significant economic worth were discovered, developed, and preserved by traditional tribes and communities. TK thus contributes significantly to the world economy. Pharmaceutical companies have expressed interest in creating traditional medicines from a variety of plant species, which can serve as vital starting points for the development of novel pharmaceuticals. As an example, the indigenous Madagascar *Catharanthus roseus* (Apocynaceae) plant was used to create the anticancer medications vincristine and vinblastine. Patent, trademark, geographical indication, trade secret, or confidential information protection may be available for innovations based on TK. Traditional intellectual property (IP) regimes, on the other hand, do not protect traditional knowledge, which has ancient heritages and is frequently transmitted from person to person orally. By altering well-known qualities, biological resources and/or traditional knowledge (TK) have been claimed as intellectual property (IP) rights [36–38]. However, there are some concerns that need also be addressed with regard to preserving traditional knowledge and cultural manifestations [34, 35]. It is disturbing to learn that researchers from universities and major corporations have a keen interest in aboriginal plant use today for the purpose of obtaining patents to profit financially, putting the lives of millions of people living in tribal communities around the globe in danger [34, 35, 39]. We must reverence the rights and interests of owners of traditional knowledge who provide a "source" for these patents while using the current patent law framework to protect these prominent amalgams and safeguard private rights. Additionally, the possibility of patenting traditional medicine-derived products offers a significant encouragement for pharmaceutical syndicates because it produces potential profits that could accrue over-time ,and, eventually, partaking in such information would value humanity [31, 40]. Attempts to patent indigenous plant knowledge in some areas where people have been using it for centuries have frequently prompted harsh criticism. Patents issued in relation to the neem tree are an example of this. By revoking a contentious patent on the antifungal medicines derived from the Azadirachta (neem) tree, the European Patent Office (EPO) has raised awareness of biopiracy around the world [41, 42]. Traditional knowledge resources may be embezzled if they are used for commercial or industrial purposes, and this can sway the rightful custodians' interests. TK must be protected and nurtured for long-term development in order to serve the interests of those who have traditional knowledge in the face of such risks. For developing nations in particular, it is crucial to preserve, defend, and promote local communities' inventions and practices that are founded on traditional knowledge [43]. As per the summary of IPBES Global Assessment on Biodiversity and Ecosystem Services 2019 for Policymakers: The quality of life for indigenous peoples and local communities and the conservation, restoration, and sustainable use of nature, which is crucial for larger society, are frequently improved by acknowledging their literature, practices, institutions, and values. Legislation and governance rules and regulations can be an effective way to protect nature and its welfare to people by integrating locally adapted management systems and ethnic knowledge. Aboriginal peoples and indigenous communities can benefit from national recognition of land tenure, access, and resource rights in agreement with national legislation, the use of free, prior, and informed consent, amended alliance of benefits resulting from the use, and co-management agreements with local communities [41, 44].

2.2 Protection of Traditional Knowledge

The draft of an international legal instrument created by the World Intellectual Property Organization (WIPO) for the protection of TK licenses to those outside the nation or community of its traditional holders. This effort is being done through the Inter-Governmental Committee on Genetic Resources, traditional knowledge, and Folklore (IGC-GRTF). Traditional knowledge is being sought after for two different types of intellectual property protection [39, 45]. Defensive protection: As an illustration, India created the Traditional Knowledge Digital Library (TKDL), a searchable database that serves as proof that certain treatments are already used in indigenous systems of medicine. This database also works to prevent patents from being granted, which prevents biopiracy, which was made famous when the US Patent and Trademark Office granted a patent for turmeric. Defensive protection is intended to stop infringement and the use of IP in new inventions. A good example of a defensive protection measure is TKDL [38, 39, 45, 46]. It gives communities more power to advance their traditional knowledge, regulate its applications, and profit from its economic exploitation. A handful of nations have also created particular legislation, and some uses of traditional knowledge may be protected by the current intellectual property system [39, 45]. The other paradigms of IPR for the protection of traditional knowledge and resource are summarized in Table 1.

Table 1 Paradigms of IPR for the protection of traditional knowledge and resources

Sr. no	National legislation and policies	Description
01	India's Biological Diversity Act, 2002	India has taken lead in inculcating substantive legislation that is in line with the CBD's goals. It sets rules for the preservation of biological diversity, sustainable usage, and fair distribution of the rewards from the exploitation of biological resources and knowledge [47, 48]. It is forbidden to file a patent in India without first receiving prior approval from the National Biodiversity Authority (NBA) along with the consent of competent authorities for the admission of TK and the completion of IPR applications for goods or inventions using TK [49–51]
02	The Patents Act, 1970	According to the Patents Act of 1970, an innovation is "a new product or a new process for industrial application." The act states that "an invention" is defined in substance as ethnic knowledge. Applications relating to TK and/or biological material must also pass a critical evaluation in accordance with sect. 10(4)(a) and (b) of the Patents Act requirements for full and particular disclosure of the invention [50]
03	Protection of Plant Variety and Farmers' Rights Act, 2001	The Ministry of Agriculture is responsible for administering India's unique legislation, the Protection of Plant Variations and Farmers' Rights Act, which was passed in 2001 and protects farmers' rights and plant variations. The act establishes a framework to safeguard plant varieties and farmers' and plant breeders' rights, particularly those relating to their assistance to the accessibility and preservation of plant genetic resources for new plant varieties [52, 53]
04	The National Green Tribunal Act, 2010 (NGT Act, 2010)	For the instant tenacity of civil matters involving environmental protection, forest conservation, and other natural resource preservation. The tribunal contributes significantly to environmental development that is sustainable [53]
05	Forest Rights Act, 2006	The act protects individual rights to the community to owe the cultivation process, other husbandry processes such as grazing, fishing, and access to waterbodies in woods, habitat moralities for predominantly susceptible tribal groups, traditional seasonal resource access for pastoral and itinerant communities, access to biodiversity, community privileges to intellectual property and TK, acknowledgment of traditional accustomed rights, and the right to shield, reinforce, or conserve or maintain natural resources [52, 54]
06	Geographical Indication of Goods (Registration and Protection) Act (1999)	The act was passed by India as part of its compliance with TRIPS obligations. According to the act, a "geographical indication" is a signal that places agronomic, natural, or factory-made goods as coming from or being made in a country's terrain, or a particular province or locality within that region, where a particular superiority, repute, or other distinctive of those goods is primarily attributable to that locality, as well as in the case of engineering. Geographical indications in India include tea from Darjeeling, silk from Kanchipuram, and sandalwood oil from Mysore [55, 56]
07	Traditional Knowledge Digital Library (TKDL)	In addition to developing a databank on the plant's intricate and their therapeutic applications, tangible steps have been taken to build a program aimed at documenting the knowledge and evidence found in the ancient books of ASU. More than 2.90 million Ayurveda, Unani, and Siddha medical formulas have been translated by TKDL into five internationally renowned. A communal tool for managing intellectual property rights is TKDL [57, 58]
08	International Forums: The Nagoya Protocol	It is a brand-new international agreement that strengthens and aids in the application of the CBD, particularly one of its three goals: the equitable and fair distribution of gains from the use of genetic resources. A significant accord in the regulation of biodiversity on a global scale, the Nagoya Protocol, is important for a number of business and non-commercial sectors engaged in the use and exchange of genetic resources [59]
09	Convention on Biological Diversity (CBD)	International agreement was created to embolden the conservation and fair and equal dissemination of the welfare associated with the use of genetic resources as well as the sustainable development of biological diversity. It is aimed at upshot in a justifiable future. The Convention on Biological Diversity protects inherited properties, classes, and environments at all levels of biodiversity [60]

3 Naturally Derived Bioactive Compounds

There has been a tremendous amount of awareness of bioactive substances derived from natural sources, particularly secondary metabolites such as terpenes, flavonoids, alkaloids, coumarins, and stilbenes with anti-inflammatory, immunomodulatory, antioxidant, and other activities. It is crucial to investigate the action mechanisms thereof. It is noteworthy that the impacts of individual biologically active chemicals' antiviral, antibacterial, or other capabilities are more difficult to identify when plant extracts are used. The solute chemicals in extracts may also have complementary or antagonistic effects [61]. At the same time, it is widely understood how plant biodiversity affects the variety of natural chemicals found in plants. Moreover, natural product chemists are ready to avoid continually ascertaining chemicals and identify a wide range of novel secondary plant metabolites [62]. Zhao et al. (2022) identified naturally occurring micro/nanoparticles Naoluoxintong, which are mostly made of carbohydrates, sarsaponins, and proteins and have a typical size of 200 nm. It has been demonstrated that these organic nanoparticles can enhance brain cell viability, lessen oxidative stress, and enhance nerve function. Research on brain protection can be developed using the findings of the study that has been presented [63]. Terpinyl acetate chemotypefrom the genus *Thymus* was examined in the study by Vaiciulyte et al. (2021) to ascertain its phytotoxic and antibacterial properties as a source of natural terpinyl acetate. The findings revealed that just 35% of the *Thymus pulegioides* habitats studied contained terpinyl acetate, a chemical that is extremely rare in essential oils. On fungi and dermatophytes, terpinyl acetate essential oil had a strong antibacterial impact, but not as much on bacteria and Candida yeasts [64]. It was discovered that specimens found near the southernmost point of the species' geographic range in Europe have different chemical properties and antioxidant activity. The concentration of secondary metabolites and the total phenolic contents varied widely measured extract of bearberry plants that grew in two diverse habitat types, heathlands and pine woods [65]. In addition to arbutin, the enriched source of polyphenolic compounds from bearberry is proven for its antioxidant qualities. Bearberry leaves have the potential to be used as a robust natural source of antioxidants in herbal remedies due to their high phenol content and high values for antioxidant indices [65]. Fruit and vegetable bioactive components interact to affect each other's antioxidant activity. The interactions between citric acid, quercetin, and caffeine may be affected by the absence of iron during thermal processing [66]. According to the findings of Engelhardt et al. (2021), the presence of minerals, temperature, and interactions with other molecules can all affect a compound's ability to behave as an antioxidant [66]. A genus known as Carissa is indigenous to tropical and subtropical regions of Asia, Australia, and Africa. Dhatwalia et al. (2021) examined 155 research papers as part of an updated review. The carissa fruit's abundance in dietary fiber, lipids, proteins, carbs, vitamin C, and other macro- and micronutrients has been verified [67]. Novel tactics are being developed to meet the ongoing demand for new natural antibiotics in order to counteract the rising levels of antibiotic resistance. By utilizing the cutting-edge MS/MS-based molecular networking approach, Wang et al.'s study from 2021 sought to examine amicoumacin antibiotics from the desert-derived *Bacillus subtilis* PJS [68]. Methicillin-sensitive and resistant Staphylococcus epidermidis showed high antibacterial activity against hetiamacin E at 2–4 g/mL, while Staphylococcus aureus showed strong antibacterial activity against hetiamacin E at 8–16 g/mL. At 32 g/mL, hetiamacin F showed some antibacterial activity against Staphylococcus species. According to a double-fluorescent protein reporter system [68], both substances were inhibitors of protein production [68]. The division of naturally occurring bioactive substances into tiny biomolecules, biologics produced by plants, and a newly developed third category known as phytopharmaceutical medications is noteworthy. The ethno-pharmacological methodology, which is based on the traditional medical system, is the foundation for the development of phytopharmaceutical drugs. The idea of "one disease, one target drug" is losing favor, and the usage of plant fractions, extracts, and molecules is the new paradigm that offers hope for developing effective medications. This led to the development of a novel idea known as polypharmacology, in which natural compounds with origins ranging widely can interact with numerous human physiological targets [69]. Numerous naturally occurring bioactive chemicals have been identified by researchers from plants, microbes, and animals. Numerous of these substances, including resveratrol, curcumin, bilirubin, and epigallocatechin gallate (EGCG), have pharmacological or biological properties that make them potent anti-disease agents [70]. The limited bioavailability of these natural chemicals, which has made them useless in clinical trials, is one of the main obstacles to their use [71]. Curcumin, for instance, has a limited range of applications due to its low solubility in water (0.0004 mg mL-1 at pH 7.3), and it also tends to deteriorate at physiological or higher pH levels [72]. As a result, curcumin's insolubility, instability, and low bioavailability have limited its use in clinical settings. The biological transport of natural chemicals with proven therapeutic potential can be enhanced by the introduction of innovative drug delivery methods like nanotechnology. For instance, adding natural substances to nanoparticles can improve their bioavailability, better target them to specific tissues or receptors, and enable the regulated release of substances over extended periods of time. Natural substances can be delivered to tumor locations with greater therapeutic effectiveness by being encapsulated in nanoparticles, as has been adequately demonstrated in the literature for curcumin [73], flavonoids [74], and traditional Chinese medicines [75]. Long-term application of these nanocarriers is linked to concerns regarding their degradation, elimination, and toxicity in humans [76] because the majority of reported nanocarriers are synthetic (i.e., polymers, lipids, and inorganic metals) and lack intrinsic therapeutic efficacy. Additionally, the drug accumulation in the tumor is severely reduced by nanocarriers' low drug-loading capabilities (below 10% for the majority of anticancer drugs), which affects their therapeutic efficiency [77]. The new paradigms of naturally derived bioactive compounds are summarized in Table 2.

4 Resource Conservation

India has long been regarded as a treasure house of plant genetic resources or as the botanical garden due to its inimitable agro-climatic conditions and regional geography. India is observed as one of the top 12 countries with the most miscellany in the world. More than 8000 different kinds of plants are included in our herbal paragon, which accounts for over 50% of all higher-blooming plant species in India. The tropical forests of the Western Ghats are home to about 70% of the nation's medicinal plants. However, research now available indicates about 1800 species are employed in classical Indian medical systems (IMS). The growing market for herbal products has enormous potential for the economic growth of the Indian region. The use of herbs in IMS as a source of food, medicine, fragrance, flavor, colors, and other products is on the rise. 95% of the therapeutic plants utilized in the Indian herbal business today are thought to

Table 2 New Paradigms of Naturally Derived Bioactive Compounds

Paradigms	Description	Name of crude drug	Bioactive compounds	Therapeutic applications
Single molecules	Plant biosynthesizes signaling molecules such as growth regulators. They also produce alkaloids, terpenoids, and phenylpropanoids, which are important in a variety of developmental and defense processes. These molecules, which are frequently referred to as tiny molecules, are extremely important in controlling the life cycle of plants. These tiny chemicals are secreted when a plant is under stress to defend it from infections, colds, or UV rays [78]	*Hyperian perfotum*	Hypericin	Immunogenic cell death inducer [79]
		Lithospermum erythrorhizon	Shikonin	Immunogenic cell death inducer [80]
		Scutellaria baicalensis	Wogonin	Immunogenic cell death inducer [81]
Naturally derived bioactive-made biologics	In contrast to conventional methods based on in vitro models, which are reliant on exclusive facilities, these bioactive-made biologics (BMBs) are an effective, safer, and more affordable platform to generate therapeutic proteins. The patient's ability to obtain medications has also been aided by PMBs. These plant-produced proteins can be used to create a variety of life-saving medications [82]	Genetically engineered carrot Cells produce enzyme Taliglucerase Alfa	–	Gaucher's disease [83]
Phytopharmaceutical drugs	PPD extracts were enriched with polyphenolic compounds that exhibited specific pharmacological properties against a wide range of human health issues, including allergy, inflammation, and diabetes. In order to treat and prevent disease, the manufacture of herbal drugs frequently lacks adequate control and regulation [84]	*Artemisia annua*	Artemisinin	Type I diabetes and cancer [85]
		Coleus forskohlii	Forskolin	Antiglaucoma drug [86]
		Genista tinctoria L.	Genistein	Anticancer, Alzheimer's disease [87]
		Trigonella foenum-graceum L., trigonelline	Diosgenin	Antidiabetic, anti-conception [88]
		Vitis vinifera L.	Resveratrol	Chemotherapeutic, antidiabetic, antioxidant [89]
Novel drug delivery	The utilization of bioactive, molecules from a diversity of ecology to combat metabolic disorders and other medical ailments. Bioactive, natural compound-derived nanocarriers are anticipated to display intrinsic therapeutic efficacy and greater biocompatibility than drug delivery systems developed from synthetic materials [90]	Indian spice turmeric (Curcuma longa)	Curcumin nanocarriers	Anti-inflammation Anticancer Anti-angiogenic [91]

have been foraged from the wild. Every year, 1.65 lakh acres of forest are cleared and 500,000 tons of dry material are indiscriminately harvested using damaging methods. There is a significant depletion of plant genetic resource wealth, with many of them going extinct day by day, as a result of the rise in population, rapid expansion of land used for food and commercial crops, deforestation, expansion of urban areas, establishment of industries in rural areas, etc. [92, 93]. In reality, during the last few decades, the demand for wild resources has increased by 8–15% yearly [12, 13]. A species' capacity to reproduce is permanently reduced below a certain threshold [94, 95]. There are numerous unique sets of guidelines for the preservation of medicinal plants, including offering both in situ and ex situ conservation. Innate reserves and desolate are typical examples of preserving the efficacy of plants, whereas national parks and seed gene banks are major paradigms for ex situ conservation and future replanting [96, 97].

4.1 Endangered Medicinal Plants

Due to overuse and habitat destruction, many significant and beneficial plant species are on the verge of extinction. A lot of the medicinal plants have become endangered in their natural habitats, and more than 95% of them are gathered from the wild. It is necessary to promote the growth and cultivation of these plants [98].

4.2 Factors Responsible for Exploitation of Medicinal Plants and Resources Thereof

4.2.1 Environmental Factors

1. Annual rainfall has reduced over the past few years, which has affected the health of many herbaceous plants during the summer.
2. Over the past 20 years, deforestation has been documented. The main causes of the decline in the area covered by important woods include the expansion of agriculture, logging, firewood collecting, heavy wood collection, heavy grazing, etc. Due to deforestation efforts, numerous valuable wild medicinal plant species are eliminated or reduced every year.
3. In forests, siltation of water bodies has reduced their ability to hold water, which has led to the depletion of subsurface water.
4. In woods and other regions, the number of honey bee colonies has decreased by 50–60%. Reduced seed set and seed dissemination have been seen as a result of pollinator loss [99].

4.2.2 Activities for Modernization and Development

1. Due to submersion, it has been observed that many species of medicinal plants are disappearing from forests. For instance, the Maradavally forest serves as the catchment area for the Linganamakki Dam, the primary reservoir used in Karnataka for irrigation and power production. Nearly 10 sq. km. of forest were submerged during the monsoon, which led to the extinction of important medicinal plant species.
2. The Devarayanadurga forest in Karnataka has suffered significant harm as a result of road expansion, power line installation, and building construction.
3. Monoculture plantations completely undermine the stability and complexity of the natural world, which leads to a loss of medicinal plants. For instance, many woodlands contain eucalyptus and acacia species.
4. Forestland encroachments have reached troublesome heights. The agricultural methods utilized on highly sloppy lands have also resulted in soil erosion and a loss in the riches of medicinal plants, in addition to the fall of trees and clearance of vegetation.
5. There is a widespread gathering of medicinal plants from the forests. Unorganized forest collectors were responsible for the collection, and they afterward sold the product to a contractor at a price set by the latter. However, illegal gathering has now been somewhat controlled as a result of the awareness raised by the "Local Traditional Medical Practitioners Association" members.

4.3 Recent Updates on Strategies for Conservation of Medicinal and Aromatic Plants

World Conservation Strategy's main objectives for biodiversity conservation:

1. Preservation of vital ecological systems and processes that are necessary for human survival and economic activity.
2. The protection of genetic variety and species, as well as the sustainable use of ecosystems and species that support millions of rural communities and important businesses.

Scientific methods and social initiatives can both be used to address the conservation of wild medicinal plants or any other threatened species [100]. The three basic scientific conservation techniques are i. legislation, ii. in situ conservation, and iii. ex situ conservation (Fig. 2).

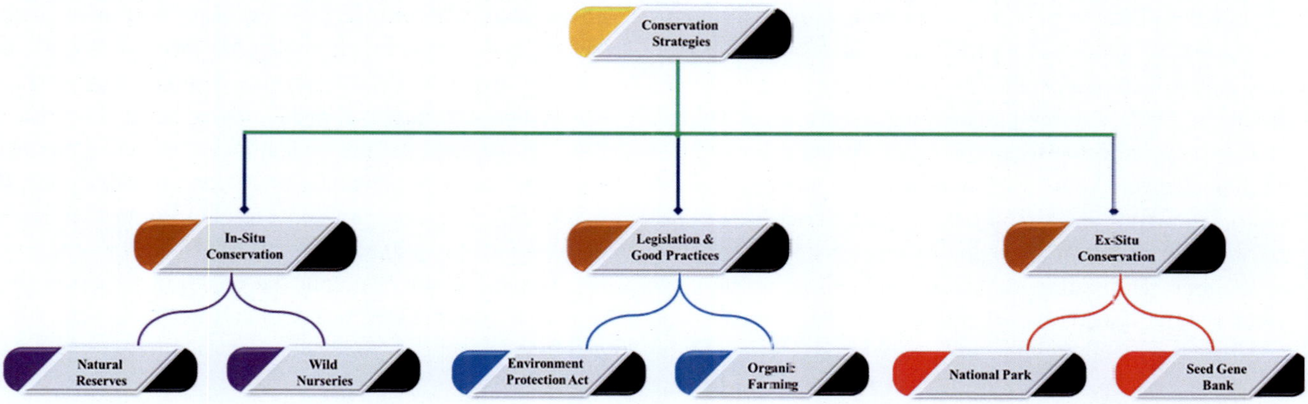

Fig. 2 Resource and conservation strategies

4.3.1 In Situ Conservation

The majority of medicinal plants are endemic species and their therapeutic benefits mostly result from the existence of secondary metabolites, which are compounds that react to stimuli in their natural settings but may not express themselves in cultured environments [97, 101]. We can preserve natural communities and indigenous species while preserving their rich web of relationships thanks to in situ conservation of complete communities [102]. Furthermore, in situ conservation reinforces the connection between resource protection and sustainable use [103] and expands the amount of diversity that can be preserved [104]. Around the world, in situ conservation initiatives have concentrated on creating protected areas and using an ecosystem-oriented strategy rather than a species-oriented one [105]. Rules, regulations, and potential compliance of medicinal plants within growth habitats are necessary for successful in situ conservation [106]. The loss of medicinal plant resources is largely due to habitat deterioration and destruction [107]. To maintain and restore biodiversity, natural reserves were established as protected areas for significant wild resources [108, 109]. More than 12,700 protected areas have been established globally, covering 13.2 million km2, or 8.81% of the land area of the planet [110]. It is necessary to evaluate the contributions and ecosystem functions of each habitat if we want to conserve medicinal plants through the protection of important natural habitats [111]. However, due to financial constraints and competing land uses, it is impractical to designate every natural wild plant habitat as a protected area [112]. In a protected region, a natural habitat, or a location that is only a short distance from where the plants naturally grow, a wild nursery is built for the species-oriented cultivation and domestication of endangered medicinal plants [113]. Although overexploitation, habitat destruction, and invasive species have severely harmed the populations of many wild species, wild nurseries can offer a practical solution for the in situ conservation of medicinal plants that are endemic, threatened, and in demand [114].

4.3.2 Legislation and Good Practices

In India, there are no specific laws or rules protecting the medicinal plants that thrive in forests. There are forestry-related regulations in place that address conservation. The regulations namely (i) Forest Act, 1927, (ii) Wildlife (Protection) Act 1972 and Wildlife (Protection) Amendment Act 1991, (iii) Forest (Conservation) Act, 1980, (iv) Environment Protection Act, 1986, (v) National Forest Policy, 1988, (vi) National Biodiversity Act, 2002, and (vii) The Scheduled Tribes and Other Traditional Forest Dwellers Act, 2006 were created by the Indian government to safeguard forests and, in some cases, the wild herbal flora [98].

In order to control production, guarantee quality, and make it easier to standardize herbal medicines, good agricultural practices (GAP) for medicinal plants have been developed [115]. By using the available knowledge to address numerous issues, a GAP strategy ensures high quality and safety [116]. GAP cover a wide range of topics, including the manufacturing facilities, storage facility, quality characteristics of weed and undesired material detection, morpho-anatomical authentication, identification of bioactive chemicals, and inspection of metal elements [117]. The adoption of GAP is actively supported by many nations. In places where these medicinal plants are customarily grown, for instance, Chinese authorities have supported GAP for the growth of regularly used herbal medications [118]. Similarly, organic farming methods have also drawn more attention due to their capacity to develop integrated, compassionate, and economically and environmentally sustainable agricultural systems. The growth of medicinal plants and the production of important compounds are substantially impacted by the application of organic fertilizers, which continuously supply soil nutrients and enhance soil stability. For instance, when organic fertilizers were used, Chrysanthemum balsamita's biomass yield increased and its essential oil concentration was high in comparison with plants that did not get organic fertilizers [119]. In reality, damaging harvesting frequently causes resource exhaustion and even species extinction for medicinal plants with low abundance and sluggish growth [120]. Therefore, it is important to think about how to use medicinal herbs sustainably and to develop good harvesting procedures. Harvesting the roots and entire plants does greater harm to medicinal plants (such as herbs, shrubs, and trees) than only taking their leaves, flowers, or buds. Using the leaves of the plants as a medicine can be a safe substitute for herbal medications comprised of whole plants or roots. For instance, Wang et al. [121] found that ginseng leaf stem and root extracts have comparable pharmacological activity, but the leaf stem has the advantage of being a more dependable supply of the herb. Furthermore, advances in tissue culture and the fermentation of medicinal plants have created new opportunities for the large-scale and extremely efficient production of desirable bioactive compounds. Genetic engineering has also made it feasible to synthesize natural products on a large scale. A promising alternative for the generation of uncommon and highly valuable secondary metabolites that are significant from a medicinal standpoint is tissue culture, which includes plant cell and transgenic hairy root culture [122]. Higher regeneration rates can be encouraged by micropropagation using tissue-encapsulated propagules, in addition to storage and transportation conveniences [120]. Synthetic seed technology, which is described as artificially encapsulated somatic embryos (or other tissues) that might be utilized for cultivation in vitro or ex vitro, is a workable option when the quantities of natural seeds are insufficient for propagation [123, 124]. The time needed for breeding may then be greatly reduced by adopting molecular marker-based methodologies applied at the genetic level to carry out breeding improvements [120, 122, 123].

4.3.3 Ex Situ Conservation

Ex situ conservation, which is frequently an immediate action done to conserve medicinal plant resources, aims to nurture and naturalize threatened species to ensure their continuous survival and occasionally to create huge quantities of planting material required in the development of medications [125, 126]. In addition to maintaining high potency when grown in gardens far from their natural habitats, many species of formerly wild medicinal plants can also have their reproductive components carefully chosen and saved in seed banks for future replanting [19]. However, the idea of a national park is crucial for ex situ preservation [127], and they can preserve ecosystems to help rare and endangered plant species survive [110]. Despite the fact that living collections typically only contain a small number of specimens of each species and are therefore ineffective for genetic conservation [128], arboreta has a number of distinctive qualities. They often have taxonomically and ecologically diverse flora and involve a large range of plant species that are grown together under similar conditions [129]. via the creation of propagation and culture guidelines, as well as via the implementation of domestication and variety breeding programs, arboreta can play a further role in the conservation of medicinal plants [130]. However, seed banks are advised to help maintain the ecological species since they provide a superior means of storing the genetic diversity [131, 132].

5 Conclusion

The recent development, implementation, applications, and interrelationship of intellectual property rights over naturally derived bioactive compounds and resource conservation thereof have been discussed and explored in the present chapter. Appropriate examples have also been used to support the topic's relevance in each segment. In conclusion, it is warranted for all researchers, academicians, and businesspeople to resurrect and promote the study of natural remedies for health care, as well as to compensate the traditional resources by abiding by strict rules, regulations, laws, and policies, and to periodically update the strategies for conserving the nation's herbal wealth. The awareness and comprehension of IPR, naturally occurring substances, and resource conservation will undoubtedly pave the road forward for global health.

References

1. WHO, General Guidelines for Methodologies on Research and Evaluation of Traditional Medicines, WHO/EDM/TRM/2000.1, Geneva, 2000.
2. Suzuki D, Knudtson P. Wisdom of the elders: honouring sacred visions of nature. Bantam Press; 1992.
3. Posey DA. Commodification of the sacred through intellectual property rights. J Ethno. 2002;83:3–12.
4. Gupta AK. Getting creative individuals and communities their due: Framework for operationalizing. Article 8j and 10c. paper prepared for the Secretariat of the CBD. 1996.
5. Gupta, AK. Accessing biological diversity and associative knowledge system: Can ethics influence equity?" Working paper no. 1340. International Information Management Association, 1996.
6. Dutfield G. Intellectual property rights, trade and biodiversity. London: Earthscan Publications Ltd; 2000.
7. Gollin MA. Legal and practical consequences of biopiracy. Diversity. 1999;15:7–9.
8. Moran K. Health: Indigenous Knowledge, Equitable Benefits World Bank, Indigenous Knowledge (IK) Notes, No.15, December 1998.
9. Phei ES, Soyoung L, Sangyong J. Naturally occurring bioactive compound-derived nanoparticles for biomedical applications. Drug Delivery Adv Therap. 2019;2. 1800146 C _ 1800146 (1 of 15)

10. Balunas MJ, Kinghorn AD. Drug discovery from medicinal plants. Life Sci. 2005;78:431–41.
11. Pimm S, Russell G, Gittleman J, Brooks T. The future of biodiversity. Science. 1995;269:347.
12. Bentley R, editor. Medicinal plants. London: Domville-Fife Press; 2010. p. 23–46p.
13. Ross IA, editor. Medicinal plants of the world (volume 3): chemical constituents, traditional and modern medicinal uses. New Jersey: Humana Press Inc; 2005. p. 110–32p.
14. Heywood VH, Iriondo JM. Plant conservation: old problems, new perspectives. Biol Conserv. 2003;113:321–35.
15. Hamilton AC. Medicinal plants in conservation and development: case studies and lessons learned. In: Kala CP, editor. Medicinal plants in conservation and development. Salisbury: Plantlife International Publisher; 2008. p. 1–43.
16. Zerabruk S, Yirga G. Traditional knowledge of medicinal plants in Gindeberet district. Western Ethiopia S Afr J Bot. 2012;78:165–9.
17. Larsen HO, Olsen CS. Unsustainable collection and unfair trade? Uncovering and assessing assumptions regarding central Himalayan medicinal plant conservation. Biodivers Conserv. 2007;16:1679–97.
18. Uprety Y, Asselin H, Dhakal A, Julien N. Traditional use of medicinal plants in the boreal forest of Canada: review and perspectives. J Ethnobiol Ethnomed. 2012;8:1–14.
19. Hamilton AC. Medicinal plants, conservation and livelihoods. Biodivers Conserv. 2004;13:1477–517.
20. WIPO 2016. Understanding Industrial Property Available from https://www.wipo.int/edocs/pubdocs/en/wipo_pub_895_2016.pdf.
21. Javed G, Priya R. Protection of traditional health knowledge: international negotiations, National Priorities and knowledge commons. Society and Culture in South Asia. 2020;6(1):98–120.
22. Brody BA. Traditional knowledge and intellectual property. Kennedy Inst Ethics J. 2010;20(3):231–49.
23. Singh MK, Singh SK, Singh AV, HariomVerma H, Singh PP, Kumar A. 12 – phytochemicals: intellectual property rights, editor(s): Bhanu Prakash. Functional and Preservative Properties of Phytochemicals: Academic Press; 2020. p. 363–375p.
24. Albert WC, Jason CL. Intellectual property protection of natural products. Asia Pacific Biotech News. 8(10):540–5.
25. Shiva V. Biopiracy: the plunder of nature and knowledge. 2016th ed. Berkeley, CA: North Atlantic Books; 1997. p. 2016.
26. Anderson JE. Indigenous knowledge and intellectual property rights [internet], International Encyclopedia of the Social &Behavioral Sciences, vol. 11. 2nd ed. Elsevier; 2015. p. 769–78.
27. López MS, Páramo IF. The identification of biopiracy in patents. World Patent Inf. 2016;47:67–74.
28. Jain A, Hallihosur S, Rangan L. Dynamics of nanotechnology patenting: An Indian scenario. Technol Soc. 2011;33(1–2):137–44.
29. Efferth T. Biopiracy of medicinal plants: finding fair solutions for the use of natural resources. Phytomedicine. 2019;53:294–5.
30. Caplanova A. Intellectual property. In: Pacheco-Torgal F, Rasmussen E, Granqvist C-G, Ivanov V, Kaklauskas A, Makonin S, editors. Woodhead Publishing Series in Civil and Structural Engineering, Start-Up Creation. 2nd ed. Woodhead Publishing; 2020. p. 81–105.
31. Xiaoting S. New problems of intellectual property during innovation of traditional Chinese medicine. World Sci Technol. 2011;13(3):466–9.
32. Bijoy CR. Access and Benefit Sharing from the Indigenous Peoples' Perspective The TBGRI-Kani 'Model.' Law 2007;3(1):1–19.
33. Gupta V. An approach for establishing a traditional knowledge digital library. J Intellect Prop Rights. 2000;05(6):307–19.
34. Ahoyo CC, Houéhanou TD, Yaoïtcha AS, Prinz K, GlèlèKakaï R, Sinsin BA, et al. Traditional medicinal knowledge of woody species across climatic zones in Benin (West Africa). J Ethnopharmacol. 2021;265
35. Overwalle V. Protecting and sharing biodiversity and traditional knowledge: holder and user tools. Ecol Econ. 2005;53:585–607.
36. WIPO 2000. Traditional knowledge. Available from: https://www.wipo.int/tk/en/tk/.
37. Banerjee M. Biopiracy in India: seed diversity and the scramble for knowledge. Phytomedicine. 2019;53(7):296–301.
38. Sahoo N, Manchikanti P, Dey SH. Herbal drug patenting in India: IP potential. J Ethnopharmacol. 2011;137(1):289–97.
39. Tedlock B. Indigenous heritage and biopiracy in the age of intellectual property rights. Explor J Sci Heal. 2006;2(3):256–9.
40. Timmermans K. Intellectual property rights and traditional medicine: policy dilemmas at the interface. Soc Sci Med. 2003;57(4):745–56.
41. IPBES. In: Brondizio ES, Settele J, Díaz S, Ngo HT, editors. Global assessment report on biodiversity and ecosystem services of the intergovernmental science-policy platform on biodiversity and ecosystem services. Bonn: IPBES secretariat; 2019.
42. Hellerer U, Jarayaman KS. Greens persuade Europe to revoke patent on neem tree. Nature. 2000;405(6784):266–7.
43. WIPO. Traditional Knowledge and Intellectual Property – Background Brief available from: https://www.wipoint/pressroom/en/briefs/tk_ip.html.
44. Forest Peoples Programme, International Indigenous Forum on Biodiversity, Indigenous Women's Biodiversity Network, Centres of Distinction on Indigenous and Local Knowledge, Secretariat of the Convention on Biological Diversity. Local Biodiversity Outlooks 2: The contributions of Indigenous Peoples and local communities to the implementation of the Strategic Plan for Biodiversity 2011–2020.
45. UNEP. The Clearing-House Mechanism of the Convention on Biological Diversity: 6th National Report-India. 2018;1–270.
46. Patents. Available from. https://ipindia.gov.in/vision-patent.htm.
47. National-IPR-Policy. 2016. Available from https://dipp.gov.in/sites/default/ files/national-IPR-Policy20114October 2020.pd.
48. Government of India. The biological diversity act. India Code Digital Repository of All Central and State Acts. 2002;2002:1–20.
49. Guidelines for processing of patent applications relating to traditional Knowledge and biological material. 2005. available from http://www. ipindia.gov.in/write read data/Portal/ IPO Guidelines Manuals/1_39_1_5-tk guidelines.pdf.
50. NBAP. India's National Biodiversity Action Plan An Overview. 2019.2019:198.
51. Protection of plant varieties and farmers' rights authority, India, available from http://www.plantauthority.gov.in/aboutauthority.
52. The Gazette of India PART II – Section 1. The protection of plant varieties and farmers' rights act, 2001.2001;(DI):1–50.
53. Government of India. The National Green Tribunal Act. 2010;2014:1–19.
54. Forest Rights Act. Govt of India. available from https://tribal.nic.in/FRA.aspx.
55. WIPO. World intellectual property indicators 2019. Geneva: World Intellectual Property Organization; 2019.
56. Sharma A. India Geographical Indication Tagging Of Agricultural Products and Foodstuffs. Available from https://www.mondaq.com/india/ trademark/898690/geographical indication- tagging-of-agricultural products.
57. Verma D, Governance G. National Intellectual Property Rights Policy of India – a review awareness about intellectual editors Dr Anita Moudgil DAV College for Girls Yamuna Nagar National Press Associates. New Delhi; 2020. p. 7.
58. Gaudillière JP. An Indian path to biocapital? The traditional knowledge digital library, drug patents, and the reformulation regime of contemporary ayurveda. East Asian SciTechnol Soc. 2014;8(4):391–415.

59. Factsheets. Introduction to access introduction to access and benefit sharing. CBD Convention on Biological Diversity. 2011:1–6.
60. The Convention on Biological Diversity, available from https://www. cbd.int/intro/.
61. Williams DH, Stone MJ, Hauck PR, Rahman SK. Why are secondary metabolites (natural products) biosynthesized? J Nat Prod. 1989;52:1189–208.
62. Kim HW, Choi SY, Jang HS, Ryu B, Sung SH, Yang H. Exploring novel secondary metabolites from natural products using preprocessed mass spectral data. Sci Rep. 2019;9:17430.
63. Zhao G, Hong L, Liu M, Jiang H, Peng D, He L, Chen W. Isolation and characterization of natural nanoparticles in NaoluoXintong decoction and their brain protection research. Molecules. 2022;27:1511.
64. Vaiˇciulyt V, Ložien K, Švedien J, Raudonien V, Paškeviˇcius A. Terpinyl acetate: occurrence in essential oils bearing *Thymus pulegioides*, phytotoxicity, and antimicrobial effects. Molecules. 2021;26:1065.
65. Sugier P, Seczyk Ł, Sugier D, Krawczyk R, Wójcik M, Czarnecka J, Oko´ n, S, Plak A. Chemical Characteristics and Antioxidant Activity of Arctostaphylos uva-ursi L. Spreng. At the southern border of the geographical range of the species in Europe. Molecules. 2021;26:7692.
66. Engelhardt L, Pöhnl T, Neugart S. Interactions of ascorbic acid, 5-caffeoylquinic acid, and Quercetin-3-Rutinoside in the presence and absence of iron during thermal processing and the influence on antioxidant activity. Molecules. 2021;26:7698.
67. Dhatwalia J, Kumari A, VermaR UN, Guleria I, Lal S, Thakur S, Gudeta K, Kumar V, Chao JC, et al. Phytochemistry, pharmacology, and nutraceutical profile of Carissa species: An updated review. Molecules. 2021;26:7010.
68. Wang T, Lu Q, Sun C, Lukianov D, Osterman IA, Sergiev PV, Dontsova OA, Hu X, You X, Liu S, et al. Hetiamacin E and F, new Amicoumacin antibiotics from Bacillus subtilis PJS using MS/MS-based molecular networking. Molecules. 2020;25:4446.
69. Nasim N, Sandeep IS, Mohanty S. Plant-derived natural products for drug discovery: current approaches and prospects. Nucleus. 2022;65:399–411.
70. Morley N, Clifford T, Salter L, Campbell S, Gould D, Curnow A. Photodermatol Photoimmunol Photomed. 2005;21(15)
71. Bonifacio BV, Silva PB, Ramos MA, Negri KM. Nanomedicine. 2014;9:1.
72. Tonnesen HH, Masson M, Loftsson T. Int J Pharm. 2002;244:127.
73. Naksuriya O, Okonogi S, Schiffelers RM, Hennink WE. Biomaterials. 2014;35:3365.
74. Leonarduzzi G, Testa G, Sottero B, Gamba P, Poli G. Curr Med Chem. 2010;17:74.
75. Liu Y, Feng N. Adv Colloid Interf Sci. 2015;221:60.
76. Zhang J, Li S, Li F, An J. Nanoscale. 2015;7:13503.
77. Huang P, Wang D, Su Y, Huang W, et al. J Am Chem Soc. 2014;136:11748.
78. Zheng XS, Chan TF, Zhou HH. Genetic and genomic approaches to identify and study the targets of bioactive small molecules. Chem Biol. 2004;11(5):609–18.
79. Krysko DV, Garg AD, Kaczmarek A, Krysko O, Agostinis P, Vandenabeele P. Immunogenic cell death and DAMPs in cancer therapy. Nat Rev Cancer. 2012;12(12):860–75.
80. Yin SY, Efferth T, Jian FY, Chen YH, Liu CI, Wang AH, Chen YR, Hsiao PW, Yang NS. Immunogenicity of mammary tumor cells can be induced by shikonin via direct binding-interference with hnRNPA1. Oncotarget. 2016;7(28):43629.
81. Li X, Dong W, Nalin AP, Wang Y, Pan P, Xu B, Zhang Y, Tun S, Zhang J, Wang LS, He X. The natural product chitosan enhances the anti-tumor activity of natural killer cells by activating dendritic cells. Onco Targets Ther. 2018;7(6):e1431085.
82. Chen Q, Davis KR. The potential of plants as a system for the development and production of human biologics. F1000Res. 2016;5. 8010:1
83. Fox JL. First plant-made biologic approved. Nat Biotechnol. 2012;30(6):472–3.
84. Nooreen Z, Rai VK, Yadav NP. Phytopharmaceuticals: a new class of drug in India. Ann Phytomed. 2018;7(1):27–37.
85. Lai HC, Singh NP, Sasaki T. Development of artemisinin compounds for cancer treatment. Investig New Drugs. 2013;31(1):230–46.
86. Wagh VD, Patil PN, Surana SJ, Wagh KV. Forskolin: upcoming antiglaucoma molecule. J Postgrad Med. 2012;58(3):199.
87. Tuli HS, Tuorkey MJ, Thakral F, Sak K, Kumar M, Sharma AK, Sharma U, Jain A, Aggarwal V, Bishayee A. Molecular mechanisms of action of genistein in cancer: recent advances. Front Pharmacol. 2019;10:1336.
88. Bahmani M, Shirzad H, Mirhosseini M, Mesripour A, Rafieian-KM. A review on ethnobotanical and therapeutic uses of fenugreek (Trigonella foenum-graceum L.). J Evid Based Complement Altern Med. 2016;21(1):53–62.
89. Ferraz da Costa DC, Pereira Rangel L, Martins-Dinis MM, Ferretti GD, Ferreira VF, Silva JL. Anticancer potential of resveratrol, β-lapachone and their analogues. Molecules. 2020;25(4):893.
90. Saw PE, Lee S, Jon S. Naturally occurring bioactive compound-derived nanoparticles for biomedical applications. Adv Therap. 2019;2:1800146 (1-15).
91. Gutierres VO, Campos ML, Arcaro CA, et al. Evid Based Complement Alternat Med. 2015;
92. Vijayalatha SJ. An ornamental garden with medicinal plants an indirect approach for conservation of medicinal plants. Indian J Arecanut Spices and Medicinal Plants. 2004;6(3):98–107.
93. Singh S. Medicinal plants –a natural gift to mankind. Agriculture Today. 2005;3(3):58–60.
94. Soule ME, Estes JA, Miller B, Honnold DL. Strongly interacting species: conservation policy, management, and ethics. Bioscience. 2005;55:168–76.
95. Semwal DP, Saradhi PP, Nautiyal BP, Bhatt AB. Current status, distribution and conservation of rare and endangered medicinal plants of Kedarnath wildlife sanctuary, Central Himalayas. India Curr Sci India. 2007;92:1733–8.
96. Sheikh K, Ahmad T, Khan MA. Use, exploitation and prospects for conservation: people and plant biodiversity of Naltar Valley, Northwestern Karakorums, Pakistan. Biodivers Conserv. 2002;11:715–42.
97. Coley PD, Heller MV, Aizprua R, Arauz B, Flores N, Correa M, Gupta M, Solis PN, Ortega-Barría E, Romero LI, Gómez B, Ramos M, Cubilla-Rios L, Capson TL, Kursar TA. Using ecological criteria to design plant collection strategies for drug discovery. Front Ecol Environ. 2003;1:421–8.
98. Lakshman Chandra De. Bio-diversity and conservation of medicinal and aromatic plants. Adv Plants Agric Res. 2016;5(4):561–6.
99. Harrington JF. Seed and pollen storage for conservation of plant genetic resources. In: Frankel OH, et al., editors. *Genetic resources in plants*. Their exploration and conservation. Oxford: Blackwell; 1970. p. 501–21.
100. Roberts EH. Problems of long–term storage and seed and pollen of genetic resources conservation. In: Frankel OH, et al., editors. Crop genetic resources for today and tomorrow. Cambridge: Cambridge University Press; 1975. p. 226–96.
101. Figueiredo MSL, Grelle CEV. Predicting global abundance of a threatened species from its occurrence: implications for conservation planning. Divers Distrib. 2009;15:117–21.
102. Gepts P. Plant genetic resources conservation and utilization: the accomplishments and future of a societal insurance policy. Crop Sci. 2006;46:2278–92.

103. Long CL, Li H, Ouyang ZQ, Yang XY, Li Q, Trangmar B. Strategies for agrobiodiversity conservation and promotion: a case from Yunnan, China. Biodivers Conserv. 2003;12:1145–56.
104. Forest F, Grenyer R, Rouget M, Davies TJ, Cowling RM, Faith DP, Balmford A, Manning JC, Proche S, Bank M, Reeves G, Terry AJ, Savolainen V. Preserving the evolutionary potential of floras in biodiversity hotspots. Nature. 2007;445:757–60.
105. Ma J, Rong K, Cheng K. Research and practice on biodiversity in situ conservation in China: progress and prospect. Sheng Wu Duo Yang Xing. 2012;20:551–8.
106. Volis S, Blecher M. Quasi in situ: a bridge between ex situ and in situ conservation of plants. Biodivers Conserv. 2010;19:2441–54.
107. Camm J, Norman S, Polasky S, Solow A. Nature reserve site selection to maximize expected species covered. Oper Res. 2002;50:946–55.
108. Rodriguez JP, Brotons L, Bustamante J, Seoane J. The application of predictive modelling of species distribution to biodiversity conservation. Divers Distrib. 2007;13:24351.
109. Chiarucci A, Maccherini S, De Dominicis V. Evaluation and monitoring of the flora in a nature reserve by estimation methods. Biol Conserv. 2001;101:305–14.
110. Huang H, Han X, Kang L, Raven P, Jackson PW, Chen Y. Conserving native plants in China. Science. 2002;297:935.
111. Liu J, Linderman M, Ouyang Z, An L, Yang J, Zhang H. Ecological degradation in protected areas: the case of Wolong nature reserve for giant pandas. Science. 2001;292(98):101.
112. Kramer AT, Havens K. Plant conservation genetics in a changing world. Trends Plant Sci. 2009;14:599–607.
113. Strandby U, Olsen CS. The importance of understanding trade when designing effective conservation policy: the case of the vulnerable Abies guatemalensis Rehder. Biol Conserv. 2008;141:2959–68.
114. Li XW, Chen SL. Conspectus of ecophysiological study on medicinal plant in wild nursery. Zhongguo Zhong yao za zhi= Zhongguo zhongyao zazhi= China Journal of Chinese Materia Medica. 2007;32:1388–92.
115. Chan K, Shaw D, Simmonds MS, Leon CJ, Xu Q, Lu A, Sutherland I, Ignatova S, Zhu YP, Verpoorte R, Williamson EM, Duezk P. Good practice in reviewing and publishing studies on herbal medicine, with special emphasis on traditional Chinese medicine and Chinese materia medica. J Ethnopharmacol. 2012;140:469–75.
116. Muchugi A, Muluvi GM, Kindt R, Kadu CAC, Simons AJ, Jamnadass RH. Genetic structuring of important medicinal species of genus Warburgia as revealed by AFLP analysis. Tree Genet Genome. 2008;4:787–95.
117. Makunga NP, Philander LE, Smith M. Current perspectives on an emerging formal natural products sector in South Africa. J Ethnopharmacol. 2008;119:365–75.
118. Liu C, Yu H, Chen SL. Framework for sustainable use of medicinal plants in China. Zhi Wu Fen Lei Yu Zi Yuan XueBao. 2011;33:65–8.
119. Suresh B. Organic farming: status, issues and prospects—a review. Agr Econ Res Rev. 2010;23:343–58.
120. Baker DD, Chu M, Oza U, Rajgarhia V. The value of natural products to future pharmaceutical discovery. Nat Prod Rep. 2007;24:1225–44.
121. Wang HW, Peng DC, Xie JT. Ginseng leaf-stem: bioactive constituents and pharmacological functions. Chin Med. 2009;4:20.
122. Rao SR, Ravishankar GA. Plant cell cultures: chemical factories of secondary metabolites. Biotechnol Adv. 2002;20:101–53.
123. Lata H, Chandra S, Khan IA, Elsohly MA. Propagation of Cannabis sativa L using synthetic seed technology. Planta Med. 2008;74:328.
124. Zych M, Furmanowa M, Krajewska-Patan A, Lowicka A, Dreger M, Mendlewska S. Micropropagation of Rhodiola kirilowii plants using encapsulated axillary buds and callus. Acta Biol Cracov Ser Bot. 2005;47:83–7.
125. Swarts ND, Dixon KW. Terrestrial orchid conservation in the age of extinction. Ann Bot. 2009;104:543–56.
126. Pulliam HR. On the relationship between niche and distribution. Ecol Lett. 2000;3:349–61.
127. Havens K, Vitt P, Maunder M, Guerrant EO, Dixon K. Ex situ plant conservation and beyond. Bioscience. 2006;56:525–31.
128. Yuan QJ, Zhang ZY, Hu JA, Guo LP, Shao AJ, Huang LQ. Impacts of recent cultivation on genetic diversity pattern of a medicinal plant, *Scutellaria baicalensis*(Lamiaceae). BMC Genet. 2010;11:52–9.
129. Primack RB, Miller-Rushing AJ. The role of botanical gardens in climate change research. New Phytol. 2009;182:303–13.
130. Maunder M, Higgens S, Culham A. The effectiveness of botanic garden collections in supporting plant conservation: a European case study. Biodivers Conserv. 2001;10:383–401.
131. Li DZ, Pritchard HW. The science and economics of ex situ plant conservation. Trends Plant Sci. 2009;14:614–21.
132. Schoen DJ, Brown AHD. The conservation of wild plant species in seed banks. Bioscience. 2001;51:960–6.

Clinical Trial on and Regulatory Considerations for Herbal Medicines for Lifestyle Diseases

Samir Bhargava, Amal Kumar Dhara, Manmohan Singhal, Neeraj Kr. Sethiya, Bhavna, Satish Shilpi, Vijay Singh Rana, Amit Kr. Dubey, Ashok Behera, and Anuj Nautiyal

Abstract

Lifestyle disorders are types of noncommunicable diseases. A poor and unhealthy lifestyle may initiate a series of interactive disorders that could become fatal (if they go unrecognised and untreated). Several traditional treatment approaches have long been available to treat such diseases. These traditional medical treatments completely rely on herbal medicines. But because of a lack of scientific evidence, these traditional approaches are not recognised by regulatory bodies. Currently, several clinical trials are in progress to look for evidence that might increase their acceptability.

In this chapter, we rely on and examine PubMed sources when discussing the current and potential herbal medications that have recently been or are currently under trial (mostly 2016–2023). The chapter clearly focuses on regulations imposed by the United States' Federal Drug Administration (FDA). The FDA has currently received more than 800 applications for acceptance, but because these applications lack scientific evidence, most of them are rejected.

Keywords

Lifestyle disease · Lifestyle disorder · Nutrition · Exercise · Diet

S. Bhargava (✉) · M. Singhal · N. K. Sethiya · Bhavna · S. Shilpi · V. S. Rana · A. K. Dubey · A. Behera
Faculty of Pharmacy, School of Pharmaceutical and Populations Health Informatics, DIT University, Dehradun, India
e-mail: samir.bhargava@dituniversity.edu.in

A. K. Dhara
Department of Pharmacy, Contai Polytechnic, Government of West Bengal, Kolkatta, West Bengal, India

A. Nautiyal
Department of Pharmacy, SGRR University, Dehradun, India

Abbreviations

AARP	American Association of Retired Persons
AIS	Athens insomnia scale
ARIA	Amyloid-related imaging abnormalities
BRT	Botanical review team
CBT	Cognitive behavioural therapy
CDER	Center for Drug Evaluation and Research
cGMP	Current good manufacturing practices
CHF	Congestive heart failure
CNS	Central nervous system
COPD	Chronic obstructive pulmonary disease
CRS	Cytokine release syndrome
DASS	Depression anxiety stress scale
ECG	Echocardiograph
eNOS	Endothelial nitrous oxide synthase
ESS	Epworth Sleepiness Scale
EWDE	Edelweiss (*Leontopodium alpinum*, var. Helvetia) extract
FMT	Faecal microbiota transplantation
FTND	Fagerstrom test for nicotine dependence
GGt	Galgeun-tang
GGT	Gamma Glutamyl transferase
GLP	Glucagon-like peptide
GLP-1	Glucagon-like peptide 1
HbA1c	Glycated haemoglobin
HFC	Hepatic fat content
HFO	Herbal formulation for obesity
HOMA	Homeostasis model assessment
HOMA-IR	Homeostasis model assessment–insulin resistance
i.v.	Intravenous
ICD	International classification of diseases
ICSD	International classification of sleep disorders
Inj	Injection
IR	Insulin resistance
ISI	Insomnia severity index
KGM	Konjac glucomannan
LSEQ	Leeds sleep evaluation questionnaire

A. K. Dhara, S. C. Mandal (eds.), *Role of Herbal Medicines*, https://doi.org/10.1007/978-981-99-7703-1_29

Mab	Monoclonal antibody
MRS	Magnetic resonance spectra
NAFLD	Nonfatty liver alcoholic disease
NnEx	*Nelumbo nucifera* leaf extract
NSCLC	Metastatic nonsmall-cell lung cancer
NT ProBNP	N-terminal pro-B-type natriuretic peptide
PSQI	Pittsburgh sleep-quality index
Q3GA	Quercetin-3-glucuronide
QoL	Quality of life
RCT	Randomised control trial
RQ	Respiratory quotient
s.c.	subcutaneous
SCLC	Small-cell lung cancer
SDS	Self-rating depression scale
SF	Short form
SF-36	Short-form health
SGOT	Serum glutamate oxaloacetic transaminase
SGPT	Serum glutamic pyruvate transaminase
STEP	Semaglutide treatment effect in people with obesity
TCM	Traditional Chinese medicine
TDT	Transfusion-dependent thalassemia
TFM	Total fat mass
TLM	Total lean mass
UACR	Urine albumin-to-creatinine ratio
UAER	Urinary albumin excretion rate
uHCC	Unresectable hepatocellular carcinoma
WHOQOL	World Health Organization Quality of Life-BREF Questionnaire

1 Introduction

Lifestyle disorders are types of noncommunicable diseases. The long-lasting effects of such diseases on a person's health may impose economic burdens on their family and the healthcare system. Herbal medicines can be used instead of other treatment systems for their holistic approaches and wide availability and to reduce some adverse effects of other medicines. Lifestyle disorders are regulated by multiple mechanisms. Many treatment strategies utilising multiple mechanisms are being developed for the treatment of lifestyle disorders. Treatment through herbal medicines can be compatible with patient needs and can be easily administered through their diets. Hence, herbal medicines, such as supplements, can be introduced into diets to treat some lifestyle disorders and reduce the adverse effects of treatment.

2 Botanical Drugs and Dietary Supplements

As per the US Federal Drug Administration (FDA), a botanical drug is one that aims to help diagnose, cure, mitigate the symptoms of, treat or prevent one or more diseases in humans. It may be a single ingredient or a complex of ingredients (made from plants, algae or fungi). To date, only two botanical products have been promoted as prescription drugs: Veregen and Mytesi. For a drug to be sold over the counter (OTC), it must have sufficient clinical data to provide evidence of its safety. A guidance for industry was published in 2004, clearly laying out the requirements for approval, specifically in response to the complexity of botanical drugs. The guidance also includes specifications for clinical studies [1]. Dietary ingredients are used in the preparation of dietary supplements and are swallowed orally. Supplements may be administered in one or more of various dosages, and they come in various forms, such as tablets, capsules, liquids and powders. However, they cannot be claimed as food items and are not drugs. They are covered under the Dietary Supplement Health and Education Act (DSHEA) of 1994. However, in contrast to botanical drugs, dietary supplements do not claim to diagnose, cure, mitigate, treat or prevent diseases [2, 3].

2.1 Botanical Review Team

The botanical review team (BRT), of CDER, is a subset of the Office of Pharmaceutical Quality. BRT have a range of experts, who can understand the complexities of natural drugs and their development. The team also ensures that the Botanical Drug Development Guidance for Industry is interpreted consistently. BRT contributes by investigating the decision-making processes of all botanical-related applications: Investigational New Drug (IND), Biological License Application (BLA), new drug application (NDA) and abbreviated new drug application (ANDA). BRT, as an expert resource, also contributes to these decisions and applications by responding to queries and promoting the development of botanicals at all levels [4]. The Manual of Policy and Procedures (MAPP 5210.9 Rev. 1) defines their activities, which BRT members must adhere to in order to ensure the quality and reliability of botanical products. The guidance has helped solve most of the challenges faced during the various phases of botanical-drug development [5].

3 Classification of Lifestyle Diseases

There are many prevalent classifications that organise the various lifestyle diseases in humans. These lifestyle diseases (Fig. 1) are generally noncommunicable but are also often hereditary. According to several leading World Health Organisation documents, they have been the leading causes of death in recent decades. These lifestyle diseases affect populations worldwide [6]. More recently, a variety of treatment strategies has become available to prevent the incidence of such diseases. Many such strategies are new and under trial. Such trials require the oversight of international regulators for the implementation of harmonised decisions. A similar report

Fig. 1 Classification of lifestyle diseases according to the ICD system (WHO)

Classification of life style diseases as per ICD system – Including the causative agents (9)	
Major Cause of Lifestyle disorder	**Code (ICD) - Disorder**
EATING / NUTRITIONAL DISORDERS	**5B81** – Obesity; **5B81.0** - Obesity due to energy imbalance; **5B81.1** - Drug-induced obesity; **8C03.3** - Polyneuropathy in nutritional deficiency; **9C40.Y** - Other specified disorder of the optic nerve (Includes Nutritional Amblyopia; Nutritional optic neuropathy); **8C03.3** - Polyneuropathy in nutritional deficiency; **3A02.1** - Folate deficiency anaemia due to low intake; **DB94.3** - Alcoholic cirrhosis of liver without hepatitis; **MB29.2** - Eating of non-nutritive substances; **8D40.3** - Intellectual developmental disorder due to nutritional deficiency; **8D4Z** - Nutritional or toxic disorders of the nervous system, unspecified; **3A00.Y** - Other specified iron deficiency anaemia; **5B6Z** - Sequelae of malnutrition or certain specified nutritional deficiencies, unspecified; **QF10** - Limited function or disability of body organ or system; **5A61.Y** - Other specified hypofunction or disorders of pituitary gland;; **8D42** - Neurological disorders due to overweight or obesity in adults or children; **5C53.22** - Coenzyme Q10 deficiency;
EXERSICE (SUBSTANTIAL)	**4A84.30** - Exercise-induced anaphylaxis; **CA23.21** - Exercise-induced bronchospasm; **EB02.0** - Cholinergic urticaria; **NF06.3** - Exercise muscle cramp; **QB95.1** - Physical rehabilitation; **8A83** - Other primary headache disorder; **5C53.2Y** - Other specified disorders of mitochondrial oxidative phosphorylation; **NF06.Z** - Effects of strenuous physical exercise, unspecified; NF06.1 - Post exercise postural hypotension; **NF06.Y** - Other specified effects of strenuous physical exercise; **5C51.3** - Glycogen storage disease; **PK9A.1** - Physical medicine devices associated with injury or harm, therapeutic, nonsurgical or rehabilitative devices **QE20** - Lack of physical exercise (Exception from all above)
SLEEP DISORDERS	**7A0Z** - Insomnia disorders, unspecified; **7B00.1** - Sleepwalking disorder; **7A4Z** - Sleep-related breathing disorders, unspecified; **7A41** - Obstructive sleep apnoea; **MG41** - Sleep disturbance, not elsewhere classified; **7B02.1** - Sleep-related hallucinations; **7A26** - Insufficient sleep syndrome; **NE60** - Harmful effects of drugs, medicaments or biological substances, not elsewhere classified; **QE11.2** - Hazardous use of sedatives, hypnotics or anxiolytics; **7A82** - Sleep-related leg cramps; **7A60** - Delayed sleep-wake phase disorder;
SMOKING	**QE13** - Tobacco use; **QD70.5** - Problems associated with exposure to tobacco smoke; **PE7Y&XE3NR** - Assault by smoke, fire or flames; **KD37** Exposure to tobacco smoke in the perinatal period
ALCOHOL DEPENDENCE	**6C40.5** - Alcohol-induced delirium, **6D84.0** - Dementia due to use of alcohol; **8D44.1** - Alcoholic myopathy; **DA42.80** - Alcoholic gastritis; **DA51.50** - Alcoholic duodenitis; **DB94.1Z** - Alcoholic hepatitis, unspecified; **DB94.3**- Alcoholic cirrhosis of liver without hepatitis; **6C40.42** - Alcohol withdrawal with seizures; **DB94.Z** - Alcoholic liver disease, unspecified; **DB94.0** - Alcoholic fatty liver; **DB94.10** - Alcoholic hepatitis with cirrhosis; **DB94.2** - Alcoholic liver fibrosis; **5A70.2** - Cushing pseudosyndrome due to alcohol; **6C40.70** - Alcohol-induced mood disorder; **6C40.71** - Alcohol-induced anxiety disorder; **8D44.Z** - Alcohol-related neurological disorders, unspecified; **DA60.60** - Alcohol-induced gastric ulcer; **DA63.5Y** - Alcohol-induced duodenal ulcer; DC31.1 - Acute alcohol-induced pancreatitis; **DC32.3** - Chronic alcohol-induced pancreatitis; **FB81.2** - Drug-induced osteonecrosis
POLLUTION	**QD70.Z** - Problems associated with the natural environment or human-made changes to the environment, unspecified; **QD70.1** - Problems associated with exposure to air pollution; **QD70.2** - Problems associated with exposure to water pollution; **QD70.3** - Problems associated with exposure to soil pollution
SUNLIGHT	**EB01.Y** - Sunlight-induced urticaria; **PB52** - Unintentional exposure to sunlight;

appeared in the UN Chronicle stating that these diseases are diseases of industrialised countries. They are the results of some unhealthy lifestyle practices, such as smoking, unhealthy dietary choices and other behaviours. These diseases progress from mild symptoms into diseases like cancer [7]. Hence, one lifestyle disorder may be the cause of another. Some of the most common lifestyle disorders are as follows: atherosclerosis, mental health, obesity, depression, alcoholism, anxiety, several cancers (colorectal, cervical, oral and breast), type2 diabetes, COPD, hypertension, and sleep deprivation [8].

4 US FDA and Lifestyle Diseases

A regulatory agency like the US FDA has as its primary role the regulation of medicines used to treat lifestyle disorders, assessing these medicines and either certifying them or not certifying them. The US FDA strictly follows cGMP practices to educate the public with correct information. The taskforce of the FDA has proposed methods to educate customers on lifestyle diseases like obesity through nutrition labelling. These methods also follow the recommendations of the AARP. The claims made in FDA rulings must factor in the following considerations: the upper limits of nutrition; nutrition labels to help guide the choices of at-risk groups; nutrition information; and potential health risks [9]. The FDA regulates lifestyle disorders through various mechanisms:

1. Nutrition claims on labels—the FDA regulates the information printed on food labels. It ensures that such information is accurate and standardised, because it is meant to educate people on eating balanced diets. Fig. 2 shows the various claims (as per FDA guidelines) made for conventional foods and supplements.

Fig. 2 Nutrition- and health-based claims, from the US FDA

Nutrition and Health based Claims (Through Label) for Conventional Foods and Dietary Supplements (11)	
Health Claims	• The 2 crucial elements of such statements are – name of substance or dietary ingredient: condition defining disease. It states that claims must be truthful. • Health claims by FDA are made through – o 1990 Nutrition Labelling and Education Act (NLEA) o 1997 Food and Drug Administration Modernization Act (FDAMA) o Interim Procedures for Qualified Health Claims in the Labelling of Conventional Human Food and Human Dietary Supplements **Health Claims Based on Authoritative Statements.** – Using FDAMA, a health claim can be made by authoritative science agency of US Government (12). FDAMA does not use dietary supplements for making health claims. **Qualified Health Claims** - Petitions are considered by FDA for a claim related to food. The petitions are supported by scientific evidence.
Nutrient Content Claims	As per NLEA, these label claims define the level of nutrients in food. But these nutrients must be authorised by FDA. The nutrient content compares level of nutrients from other foods.
Structure/Function Claims	• Claims defining the structure or function of an ingredient. • Dietary Supplement Health and Education Act of 1994 (DSHEA) – defined structural / functional labelling claims as a case of general well – being and a nutritional deficiency. Manufacturer must have evidence related to his claims and must submit the statements related to claims made to FDA (Within 30 days) after marketing his product. The claims must be supported by a "Specification" - FDA has not assessed claim. The supplement is not made to diagnose, treat or cure the disease.

2. Tobacco product requirements—the FDA regulates the production and the supply and selling of tobacco and tobacco products. Its right to regulate tobacco products is authorised by the Tobacco Control Act (2009). The FDA has the right to stop the sale of tobacco products to minors, limit the size of tobacco products within single packets, restrict the claims made by tobacco companies and halt the distribution of tobacco products as samples for promotion [10]. Even in recent times, the FDA has proposed new requirements to reduce the production and supply of tobacco products contaminated with substances like metal and plastics [11]. Hence, as per such regulations, manufacturers must comply with the following: restrictions on tobacco product designs; certifications that tobacco products meet specifications; reducing the contaminants in tobacco products; and establishing measures for identifying the remnants of contaminants. Because tobacco use is harmful to human health, the FDA supports several research programmes for reducing the risks related to tobacco use [12].
3. Drug reviews—the FDA has the prime responsibility of approving medications for several diseases (including lifestyle disorders). The FDA has to approve the drug after reviewing evidence, which includes verification data supported through conducting clinical trials.

5 Clinical Trials

Clinical trials are a subset of research studies which are initiated for developing a drug. The research involved in a trial is completed by assessing the safety and side effects of drugs on animal and human models. The to-be-tested substances include newly developed interventional molecules, drugs or medical devices. During the trials, interventional tests may be conducted on subjects from various age groups, various ethnicities and various genders to reduce the level and number of biases in outcomes. Clinical trials are completed regulated by authorities like the FDA, with informed consent from participants. Trials may be single blind, double blind or open label, depending on whether participants and/or researchers are informed of who is undergoing the therapy. Regulating authorities ensure the ethical conduct of trials. The outcomes from trials provide evidence for or against medical interventions, supporting healthcare providers in their reaching conclusions on treatment strategies.

5.1 Clinical Trials and the FDA

The FDA regulates the ethics and interests of participants involved in clinical trials through its guidance documents [13]. The trials start on laboratory animals to and may proceed to human subjects. Some risks are inevitable because of the potential risks of developing new products [14]. Clinical trials are piloted according to protocol, which establishes the following: the patients (fixed) for study; the duration of tests; the timing of dose administration; the time of study; the sampling; the dose to be given; the drugs to be administered; and the proposed outcomes.

Clinical trials are necessary parts of research, but they come with risks. Participants have the right to know what these risks are, and because they are always at risk, they can leave of their own accord after re-evaluating the risks. Although every effort must be made to control the risks, they are sometimes completely unavoidable [15, 16]. Several approvals and trials are ongoing related to treatment of life style disorders (Table 1, 2).

6 Clinical Trials and FDA-Based Recommendations for Drugs Used to Treat Nutritional Disorders

Nutrition is a significant part of our healthcare system. Better nutrition plays a significant role in reducing health risks. Intaking a healthy diet starting from a young age prevents severe lifestyle disorders. The effects of improper nutrition are broad, ranging from complex cellular mechanisms to improper physical structures.

One such lifestyle disorder is obesity, whose complications are complex. Malnutrition is also considered a nutrition-based lifestyle disorder by the WHO because a diet denied the necessary quantities of nutrients essential for biological functioning also causes diseases.

The strategy to combat nutritional disorders includes several FDA initiatives to help accelerate efforts to empower consumers with information and create a healthier food supply. Such efforts include developing an updated definition and a voluntary symbol for the "healthy" nutrient content claims, front-of-package labelling, dietary guidance statements on food labels and gathering input for nutrition labelling for groceries sold online. The strategy also reflects the need to encourage the production of healthier foods by including targets to reduce sodium levels in foods and providing greater flexibility to industry around the use of salt in standardised foods. The FDA also will explore additional ways to reduce added sugars in foods to complement its requirement that added sugars be included on the Nutrition Facts label.

6.1 Obesity

Obesity is caused by multiple factors, such as sedentary lifestyles and lifestyles where people intake processed foods with enhanced fats and low water content. Such processed foods are low in nutrients and contains excess fats. Excess fats accumulate, influencing the hormones needed to regulate glucose metabolism and the cellular-ageing process. Cellular senescence due to diet further triggers a series of inflammatory disorders in human organs, like the liver, the CNS and the gastrointestinal tract [17–19]. All people with obesity are at risk of significant social discrimination, and women with obesity are affected by menstrual abnormalities and fertility-related issues, leading to persistent stress. Obesity is now more prevalent in the urban locations of middle-income countries [20]. According to the FDA, obesity is the unnecessary deposition of fats and is linked to several other diseases, such as diabetes, cardiovascular diseases and breathing issues. The FDA crudely defines a person with a BMI $\geq$ 30 kg/m^2 as obese, while a person with a BMI between 25 and 29.9 kg/m^2 is so defined as overweight. The FDA has approved several devices for the management of weight, such as gastric bands, gastric balloons systems and endoscopic systems. The FDA advises that before starting any therapy or device for weight management, people should rely on a better diet regime, nutrition supervision and physical activity [21]. Nonetheless, several antiobesity medications targeting endogenous hormones have been certified, and more are underway [22]. Such medications may be an adjunct to the exercise and diet practices needed to regulate obesity. The majority of obesity cases are due to energy imbalances.

Lingguizhugan decoction is a Chinese herbal preparation composed of four herbs. The effect of this complex preparation for controlling lipid metabolism has been validated by an RCT. In it, 243 participants were included, and its objective was to identify insulin resistance in patients who were overweight. According to the HOMA-IR, the results showed a reduction in insulin resistance specifically in patients with obesity, but not in patients without obesity. Doses of the herbal extract resulted in improvements in insulin resistance in people with obesity and NAFLD [23].

Similarly, in another trial, a Korean herbal medicinal formula composed of seven herbs has been used for reducing weight in women, especially those with metabolic diseases. The trial was conducted on 160 participants for 12 weeks, and it has a 12-week of follow-up period. The GGt pills were administered orally three times a day. The investigated outcomes revealed the safety of the administered herbal regimen in women with or without metabolic indications. The study also had drawbacks: The included participants were kept on a reduced-calorie diet [24]. Another Korean herbal preparation (Hanslim, from Wolbigachul-tang) was administered, according to the traditional system, to patients with oedema. Here, 165 participants were allowed to take two tablets of Hanslim per day for 12 weeks. This was continued with a clinical visit every 4 weeks. The results from phase IIb trial provided evidence for using Hanslim as a regimen for obesity. The authors supported the need for a larger phase III trial, on dosage estimation [25].

Utilising the Indian system of traditional medicine, Baidyanath formulated an antiobesity tablet called HFO-2. The herbs involved in its formulation followed Indian Ayurveda systems. The participants were advised to take two tablets (each of 500 mg) per day for 3 months, with a 30-day follow-up period. HFO-2 reduced some of the characteristics in people with overweight. There was no lessening in glycaemic parameters or lipid reports. The conclusion indicated that HFO-2 redistributes fat breakages, which needs further evidence. However, the low sample size was not efficient for drawing many conclusions [26].

The US FDA could allow labelling to include caloric intake in the nutrition information, which could be used to educate people with obesity [9]. Warnings have been issued by the FDA to several manufacturers for not following cGMP guidelines (in relation to nutrition labelling). In addition to herbal drugs, the regulatory body has approved Wegovy for injection use (2.4 mg/week) for weight control in adults with obesity (along with other weight-related conditions). The drug acts by imitating hormone GLP-1, which regulates hunger and food intake. The drug is not for people with certain allergies [27]. Another drug, Saxenda, has been approved in paediatrics for children 12 years or older (on the basis of their BMI). A lower-calorie diet is required as an additional condition of using Saxenda. Saxenda is not administered with GLP-1 receptor agonists [28].

6.2 Anaemia

Anaemia is not necessarily a lifestyle disorder. It is due to inadequate nourishment and insufficient food and is a worldwide concern [29]. However, several types of anaemia can be due to several lifestyle choices, such as iron-deficiency anaemia, but also B12-deficiency anaemia and anaemia arising from chronic diseases. A Chinese herbal medicine Hongyibuxue oral solution (containing the extracts of four medicinal herbs) was administered to 50 patients with iron-deficient anaemic for 4 weeks. Ferrous sulphate was administered to the control group. The results, in terms of iron improvement (effectiveness) and dose tolerance, were found faster in the group administered with the Hongyibuxue oral solution [30]. In another study, on a traditional Chinese medicine, Shengxuening was used on patients with iron-deficient anaemia. Just 0.1 g of Shengxuening was administered three times/day for 30 days to 50 patients, and the controls were given ferrous gluconate (0.1 g three times/day for the same duration). The outcomes were assessed by measuring blood levels of transferrin, ferritin and iron-binding capacity. Iron metabolism significantly improved: a 60% higher efficacy compared to the control group, according to the increased red blood cell (RBC) levels [31].

Madhulai manappagu, a traditional Siddha herbal syrup, has been administered to patients with iron-deficient anaemia. A scientific validation of this formulation was carried out to determine its safety in animal models. For acute toxicity studies, female rats were administered with 5000 mg/kg of body weight and observed for 13 consecutive days. In a subacute toxicity study, three oral doses—500, 750 and 1000 g/kg of body weight—were administered once per day for 28 days). The results showed the safety of taking madhulai manappagu [32]. In another research study, the effectiveness of Kampo medicine (a traditional Sino-Japanese medicine) was prepared with iron for patients with hypochromic anaemia connected with uterine myoma. Uterine myoma deteriorates QoL as the symptoms related to dysmenorrhoea become severe. Hence, 10/25 patients who had haemoglobin levels of 8–12 g/dL were administered with 2.5 g of granules three times/day before meals. The remaining 15/25 patients were administered sodium ferrous citrate: 1–2 tablets/day after meals. At the end of 4 and 8 months, no side effects were reported in the group that took Kampo medicine, while side effects were seen in ferrous citrate group. Hence, the results after the administration of Kampo medicine show that it may improve the QoL of such patients [33].

6.3 Diabetes

Type 2 diabetes is a genetic metabolic disorder exacerbated by lifestyle choices and eating habits. A combination of a healthy lifestyle with appropriate medications may alleviate and manage the symptoms of diabetes. The application of traditional knowledge systems may be a treatment for diabetes. Several plants have proven hypoglycaemic effects on human and animal models. A powder (composed of three herbs as once-a-day tablet), used in traditional Persian medicine, has been evaluated for its hyperglycaemic possibilities in type 2 diabetes. Of the 93 patients, 60 were included in the interventional study, mainly those who had been taking glyburide (10 mg/day) and metformin (1 g/day) for the preceding 3 months. Fasting glucose was measured at several intervals, up to 90 days. Significant reductions in fasting glucose and HbA1c were observed, along with serum triglycerides. However, the usage of herbal drugs as adjuvants to routine clinical treatments using oral hypoglycaemic agents was considered a limitation of the study [34].

In one of the trials, herbal tea was evaluated for treating type 2 diabetes, defined as those with HbA1c levels greater than 120% of normal. After a run-in period, the patients were administered 250 mL of herbal tea per day for 10 days. A follow-up was set at up to 4 months. The analysis was based on changes of 10% in blood glucose levels, changes in HbA1c levels and changes in fructosamine levels. The levels of fructosamine were reduced after the administration of tea in groups. The data revealed only short-term benefits from the tea without leading to significant glycaemic control [35].

Diabetes comes with severe complications like diabetic nephropathy, leading to renal damage. Suggested treatments include lifestyle modifications. In a trial, Chinese herbal tablets of Shenyankangfu (composed primarily of 13 medicinal herbs) has been used with irbesartan (to reduce blood pressure) for the treatment of diabetic nephropathy (and the associated qi-yin-deficiency syndrome). The symptoms of diabetic nephropathy may lead to a future diagnosis of qi-yin-deficiency syndrome. Participants were divided into intervention and control groups. The intervention group received Shenyankangfu tablets (2.4 g three time per day) + irbesartan (0.15 g/day) for 16 weeks, while the control group received irbesartan (0.15 g/day) and a placebo for 16 weeks. Assessments were taken for UACR, UAER, changes to TCM and adverse effects [36]. The Indian system of medicine claims that Ayurvedic medicines are less toxic and more effective. A polyherbal formulation takes eight herbs to form tablets of Dianex. Furthermore, 26/37 patients who were given Dianex were started at a dose of one tablet/day, which gradually increased to two or three tablets/day (depending on sugar content) for 12 weeks. Follow-up continued for the next 12 weeks. The outcomes suggested reductions in glycosylated haemoglobin, insulin resistance and beta cell functioning. Also, there was an elevation of total cholesterol and no reduction in blood glucose. Overall, the drug was considered only as a suitable add-on therapy for type 2 diabetes. The lack of randomisation was considered a limitation of the study [37].

In one of its reports, the FDA stated that insufficient evidence is available to support using psyllium husk for reducing risks of type 2 diabetes [38]. Also, according to one study by the FDA, quinoa sprouts reduced the incidence of type 2 diabetes when administered to rats [39]. However, any drug approval from the FDA requires evidence supporting the drug's claims. Warnings have been issued by the FDA to several manufacturers for not following cGMP guidelines.

In addition to botanicals, in 2020, the FDA issued a new guidance requiring that trials include more subjects with type II diabetes. The unignorable risks are cardiovascular diseases, but scientists must look beyond diabetes to find better treatments [40]. In one of the FDA's reports on tirzepatide, it stated that the drug has more potential to cure diabetes. Weight losses of up to 15 pounds were observed in patients with obesity who received 15 mg of tirzepatide. But it could not be used in patients with histories of medullary thyroid cancer, because it caused such cancer in rats during safety trials and in patients with type 1 diabetes [41]. Even in one of the reports in June 2023, the FDA approved Empagliflozin and Empagliflozin + metformin hydrochloride with a proper dietary regime and exercise in children (10 years and younger) with type 2 diabetes. These drugs have received priority review status for approval by the FDA [42]. The FDA has also issued draft guidance for clinical trials on the medical devices needed to control or monitor type 2 diabetes. These are unenforceable guidelines, and they have been issued by the FDA for the inclusion/exclusion of patients, patient safety and protocols for patient treatment [43, 44]. Recently, Revita for duodenal mucosal resurfacing (DMR) has become a technical therapy involving thermal procedures for removing excess lining from the duodenum. The new lining generates inside the duodenum within days. The therapy has been utilised for the treatment of type 2 diabetes and is undergoing clinical trials [45].

6.4 Cardiovascular Diseases

Cardiovascular diseases, includes a cluster of diseases which may impact heart and its functioning. Many clinical trials are under progress to uproot cause of these diseases, for better care and patient comfort. Since last decade some herbal formulations have undergone through trial-based stages, to prove their efficacy. Heart failure is a final condition of heart diseases, due to poor treatments. As per current content, by FDA also it is clearly revealed that heart diseases are clearly diagnosed through symptoms. Heart diseases have many contributing factors, like diabetes, diet, alcohol, low physical activity and obesity [46].

In one of the trials, shenfuqiangxin pills (composed of phytoconstituents from seven plants and in doses of 5.4 g/day for 12 weeks) were clinically assessed for the treatment of chronic heart failure. In total, 352 recruits were equally divided into an intervention group (shenfuqiangxin pills + standard medicine) and a control group. The inclusion criteria were kept for those with history of CHF. The outcomes were at levels of NTPro-BNP (a biomarker for assessing cardiac functions) and reached several echocardiography limits. These trials were needed to support the present systems of medicine [47].

Hypertension is a heart disease that can be fatal in people with chronic conditions. It can be treated with lifestyle modifications, in addition to specific regimens. In one trial, *Apium graveolens* (celery) seed extract capsules were administered as supplements to 51 hypertensive patients (1.34 g/day—distributed four times/day) for 4 weeks. The analytical studies revealed the presence of 5.23 mg of 3-n-butylphthalide/capsule. It can be concluded from these outcomes that celery seeds may be a promising supplement for reducing blood pressure while improving kidney and liver functions [48].

Other heart diseases can also easily occur due to lifestyle modifications. Reasons like smoking, poor diet, stress and cholesterol consumption (through diet) are highly prevalent among individuals. In one trial on unstable angina, which is highly dangerous and can happen to individuals even at rest, Luofengning granule, a Chinese herbal composition of six herbs, was given to 30 patients diagnosed with angina as supplements in addition to Western medicine. Two sachets of 25.05 g were dissolved in 100 mL of water and given to these patient for 4 weeks. The outcomes were measured by estimating Angina Scale Scores, ECG images and serum factors. The trial was an open-label trial aiming to determine the effects of Luofengning granule because such effects lacked evidence [49].

In one study, the FDA stated that EPA and DHA (from several marine and plant sources) have the potential to reduce hypertension [50]. Several cautions have been issued by the FDA to several producers for not following cGMP guidelines. FDA requires production of drugs should follow cGMP guidelines.

7 Clinical Trials and FDA-Based Recommendations for Drugs Used for Exercising

Many herbs and herbal supplements are used as ergogenic aids. Herbs that have been parts of many traditional medicines could be beneficial as complementary treatments. But their use as ergogenic aids has less scientific evidence and is complicated by variations among individuals, variations among ethnicities and its lacking standardisation parameters.

Several studies in the literature have shown the efficiency of herbal supplements. But before recommending such supplements, dieticians and other specialists must consider factors such as the study test period, the number of patients included in trials, the dose-response curves, the outcomes of the treatment, the adverse effects of phytoconstituents, the widespread distribution of unfocused results and the lack of evaluations on unpatented herbs. Thus, the usage of herbs for enhancing physical performance must not be influenced by traditional use, encouragement from others or others' using them [51]. For example, even in sports, the usage of herbal supplements with ephedrine (without appropriate consultation) may be disastrous. Ephedrine (a nervine stimulant) has been shown to not enhance running, and its derivatives are considered as doping substances [52].

In one clinical trial, turmeric-based supplements (NCT04765527) were given to participants [53] for 2 weeks. Participants were encouraged to engage in exercises involving muscular movements, including weightlifting, for 1.5 h only. This was followed up with fitness tests and blood isolation for the analysis of the biomarkers (serum creatinine kinase) involved in muscular damage. The aim of the trial was to identify any nutritional based damage to muscles before and after the exercises [53]. In another trial, administrations of *Rhodiola*- and *Cordyceps*-based supplements (NCT03929575) were given to participants with BMIs between 18.5 and 35 kg/m^2. Participants were administered with either 250 mg of calcium as a placebo or 250 mg (*Rhodiola*)/(250 mg *Rhodiola* + 225 mg of *Cordyceps*). The outcomes were based on measurements of gastrointestinal distress and blood glucose levels. The aim of the trial was to identify the effects of administering herbs for distress on V02 max [54].

8 Clinical Trials and FDA-Based Recommendations for Drugs Used for Sleep Disorders

Insomnia is one of the major sleep disorders impacted by lifestyle, and the development of insomnia is closely associated with risks for other lifestyle disorder [55]. Several lifestyle factors, such as an irregular sleep schedule, the excess intake of stimulants (before sleep), reduced physical activity and stress (including anxiety) are the major causes of insomnia. Numerous clinical trials have been conducted or are underway with significant outcomes for treating insomnia (Table 3).

One such trial, based on customary Persian medicine, has been used for developing an intranasal formulation (involving oils from extracts), with sesame oil as a diluent. Patients with insomnia were evaluated according to the ICSD system. For their treatment, they were intranasally administered two drops of medicinal extract every day at noon for 8 weeks. To obtain higher-quality sleep, a reduced ISI score was desired. ISI and PSQI scores were calculated by using questionnaires filled out by patients at follow-ups: 1, 4 and 8 weeks after the initiation of treatment. During the study, even a reduction in scores (ISI and PSQI) in placebo subjects was considered because of the oleic acid and linoleic acid in sesame oil. The mechanism suggested for the action of the developed formulation was to activate GABA receptors by binding it at its benzodiazepine binding location. This mechanism for sleep was induced 4 weeks after the initiation of treatment. The limitation of study was its exclusion of polysomnography (a more sensitive method of assessment) [56].

In another trial, Zao Ren An Shen (a Chinese herbal medicine) was given to 45 volunteers as a placebo. Zao Ren An Shen, composed of three herbal Chinese compositions, was administered to volunteers as three capsules/day 1 h before sleep (amounting to 2.28 g/day) for 4 weeks. The outcomes were measured through ISI (at 4 weeks) and DASS scores (at 8 weeks). The sleep parameters were measured with sleep diaries and actigraphs. The suggested hypothesis was in favour of Zao Ren An Shen, suggesting its potential application for insomnia. The adverse outcomes of using Zao Ren An Shen were significantly lower than those of using benzodiazepine-based agonists [57].

Table 1 New drug applications approved by the Food and Drug Administration for lifestyle diseases (2017–2023)

Molecule Application Number	Usage	Dosage Form	Applicant	Recommended Population	Risks	Reference
DAPRODUSTAT; 216,951	Anaemia due to chronic kidney disease	Tablets of 1 mg, 2 mg, 4 mg, 6 mg and 8 mg, with or without food	GLAXOSMITHKLINE	For adults who have been undergoing dialysis for at least 4 months If prehaemoglobin level is < 9 g/dL, oral starting dose is 4 mg; if prehaemoglobin level is 9–10 g/dL, oral starting dose is 2 mg; if prehaemoglobin level is > 10 g/dL, oral starting dose is 1 mg	Hypertension; gastrointestinal erosion resulting in further bleeding; and abdominal pain	[65]
BEXAGLIFLOZIN; 214,373	Additive to improve to improve glycaemic control	Tablets of 20 mg	THERACOSBIO	In adults with type 2 diabetes mellitus, 20 mg with or without food	Volume depletion; urinary tract infections like urosepsis and pyelonephritis; serious hypoglycaemic conditions; and genital mycotic infections	[66]
LECANEMAB-IRMB; 761,269	Antibody for treatment of Alzheimer disease	Injection as i.v. infusion (100 mg/mL)-single dose vial	EISAI INC	Given to population with mild cognitive impairment or mild dementia stage The dosage is kept at 10 mg/kg, with dilution using 0.9% sodium chloride injection MRI is required at 5th, 7th and 14th days of administration	ARIA; infusion-related problems; and headache	[67]
MIRVETUXIMAB SORAVTANSINE-GYNX; 761,310	Administered to ovarian, fallopian tube cancer	Injection (5 mg/mL)-single dose vial	IMMUNOGEN INC	The requisite dose is administered with dextrose solution Affected patients are eligible if they have received 1–3 systemic treatments Patients are selected on basis of FDA-approved tests Dose of 6 mg/kg/3 weeks till toxicity has been reached	Effects on vision; elevated SGOT, SGPT; reduced white blood count (WBC); reduced haemoglobin; pneumonitis; and peripheral neuropathy	[68]
TECLISTAMAB CQYV; 761,291	It is for treatment of adult patients with relapsed or refractory multiple myeloma	Injection for s.c. administration (10 mg/mL); (90 mg/mL)—Single dose vial	JANSSEN BIOTECH	Patients must have received at least 4 previous therapies, including proteasome inhibitor, an immunomodulatory agent and an anti-CD38 monoclonal antibody Also, step-up dosing schedule is followed from 0.5–1.5 mg/kg Hospitalisation is required for 48 h, after administration	Lowering of WBC; pyrexia; hypersensitivity; CRS; infection in upper respiratory tract; hepatotoxicity; and embryofoetal toxicity	[69]

(continued)

Table 1 (continued)

Molecule Application Number	Usage	Dosage Form	Applicant	Recommended Population	Risks	Reference
TREMELIMUMAB-ACTL; 761,270, 761,289	For treatment of nonremovable hepatocellular carcinoma	Injection as i.v. infusion (20 mg/mL)	ASTRAZENECA AB	For uHCC, give to adults in combination with durvalumab: for adults who weigh > 30 kg, 300 mg single dose with durvalumab (1500 mg at day 1, with repeated administration every 4 weeks); for adults who weigh < 30 kg, 4 mg/kg as single dose with durvalumab (20 mg/kg—day 1, with repeated administration every 4 weeks) For NSCLC, give to adults in combination with durvalumab: for adults who weigh > 30 kg, 75 mg/3 weeks with durvalumab therapy at week 1 (1500 mg + chemotherapy—4 cycles), then durvalumab (1500 mg + histology-based pemetrexed therapy)/4 weeks; fifth dose of drug, 75 mg + sixth dose of durvalumab at week 16; for adults who weigh < 30 kg, start therapy with 1 mg/kg/3 weeks + durvalumab (20 mg/kg + chemotherapy for 4 cycles), then durvalumab (20 mg/kg + histology-based pemetrexed therapy)/4 weeks; fifth dose is 1 mg/kg + sixth dose of durvalumab at week 16	Infusion-related problems; abdominal pain, increased SGPT; and embryofoetal toxicity	[70]
OMIDENEPAG ISOPROPYL; 215,092	For reduction of elevated pressure in eyes, as in ocular hypertension	Ophthalmic solution as drops	SANTEN	As 0.02 mg/mL	Blurred vision; ocular pain; hyperaemia; keratitis; and vision impairment	[71]
TIRZEPATIDE; 215,866	As additive for diet and exercise	Injection for s.c. administration, either 2.5; 5; 7.5; 10; 12.5; or 15 mg/0.5 mL in single dose pen	ELI LILLY AND CO	Dose is started from 2.5 mg/week After 4 weeks, dose is increased in increments of 2.5 mg This is done progressively every 4 weeks till maximum dose has been reached at 15 mg/week	Pancreatitis; hypersensitivity; abdominal pain; hypoglycaemia; and reduced appetite	[72]
DARIDOREXANT HYDROCHLORIDE; 214,985	Given for insomnia	Tablets of 25 or 50 mg	IDORSIA	Administered, within the 30 min before bedtime	Induces depression; negotiated respiration-related effects; fatigue; and headache	[73]
TEBENTAFUSP-TEBN; 761,228	Administered for melanoma (metastatic/nonremovable)	100 mcg/0.5 mL for i.v. route	IMMUNOCORE LTD	Dosage is 20 μg (i.v.) on day 1; 30 μg (i.v.) on day 8; 68 μg (i.v.) on day 15; and 68 μg (i.v.)/week thereafter	CRS; increased liver-based enzymes; embryofoetal toxicity; pyrexia; vomiting; hypotension; and increased glucose content	[74]

MAVACAMTEN; 214,998	Drug is given to adults for obstructive hypertrophic cardiomyopathy	Capsules are available in 2.5; 5; 10; and 15 mg doses	BRISTOL	Dose to adults is based on clinical symptoms and ECG based assessments	Dizziness; syncope; heart failure; and embryofoetal toxicity	[75]
ADUCANUMAB-AVWA; 761,178	For Alzheimer disease	Injection (i.v.) use as 100 mg/mL vials	BIOGEN INC	Recommended dosage is 10 mg/kg (i.v.), administered for 1 h/4 weeks Brain MRI is a prerequisite for treatment initiation Similar MRIs are required after the seventh and 12th infusions Critical analysis for administration is required after a period of time The dosage is diluted with saline solution and administered via a 0.2- or 0.22-micron filter	ARIA and hypersensitivity	[76]
DASIGLUCAGON HYDROCHLORIDE; 214,231	Treatment of severe hypoglycaemia	Injection of (0.06 mg/mL) single dose as prefilled syringe/autoinjector	ZEALAND PHARMA	Dose for adults, paediatrics ≥ 6 years is 0.06 mg The dosage is injected via s.c. route in specific locations	Hypersensitivity; hypoglycaemia; nausea; vomiting; pain at injection site; and headache	[77]
VERICIGUAT; 214,377	Treatment to reduce death rate due to cardiovascular failure	Tablets of 2.5; 5; and 10 mg	MERCK SHARP DOHME	Dosage is 2.5 mg/day without food The dose is doubled every 2 weeks until target dose of 10 mg has been reached	Hypotension and anaemia	[78]
TRILACICLIB DIHYDROCHLORIDE; 214,200	Therapy for widespread SCLC	Injection of 300 mg lyophilised cake	G1 THERAP	Dosage is 240 mg/m^3/30 min 4 h before chemotherapy	Injection-related reactions; hypersensitivity; embryofoetal toxicity; hypocalcaemia; hypokalaemia; and pneumonia	[79]
BELANTAMAB MAFODOTIN-BLMF; 761,158	For relapsed myeloma, who received 4 therapies, including anti-CD38 MAb, a proteasome inhibitor and an immunomodulatory agent	Injection of 100 mg lyophilised powder for reconstitution	GLAXOSMITHKLINE	Dosage is 2.5 mg/kg/30 min every 3 weeks	Thrombocytopenia; drug infusion–related reactions; embryofoetal toxicity; keratopathy; and grade 3–4 blood-related abnormalities	[80]
LURBINECTEDIN; 213,702	Usually for SCLC	Injection—4 mg lyophilised powder	JAZZ	Dose is 3.2 mg/m^2/21 days. The dose is given as i.v. for 60 min	Myelosuppression; hepatotoxicity; embryofoetal toxicity; increased SGOT and SGPT; and reduced magnesium	[81]

(continued)

Table 1 (continued)

Molecule Application Number	Usage	Dosage Form	Applicant	Recommended Population	Risks	Reference
FLORTAUCIPIR F-18; 212,123	Radioactive diagnostic agent for Alzheimer's disease	Injection of 300–1900 MBq/mL via multiple-dose vial	AVID RADIOPHARMS INC	Dose is administered as intrabolus injection, with imaging started after 80 min	Pain at injection site and risks of misdiagnosis and radiations	[82]
VIBEGRON; 213,006	Used for overactive bladder, with increased urinary frequency in adults	Tablets of 75 mg	UROVANT	Recommended dose is 75 mg/day, with water	Urinary retention; upper respiratory tract infections; and nasopharyngitis	[83]
UBROGEPANT; 211,765	Treatment of acute migraine with or without aura in adults	Tablets of 50; 100 mg	ABBVIE	Recommended dose is 50 or 100 mg Second dose can be given after first dose, with a maximum dose of 200 mg/day	Nausea and somnolence	[84]
LEMBOREXANT; 212,028	Recommended for patients experiencing difficulty in sleep onset and maintenance	Tablets of 5 and 10 mg	EISAI INC	Recommended dose of 5 mg/day before sleeping and 7 h later, after waking In hepatic impairment, maximum dose is 5 mg	Narcolepsy, CNS depression; sleep paralysis; respiratory dysfunctions; and composite sleep behaviours	[85]
FERRIC MALTOL; 212,320	Iron replacement for deficiency conditions in adults	Capsules of 30 mg	SHIELD TX	Taken 30 mg (twice)/day	Overloading iron; staining of feces; abdominal pain; and uneasiness	[86]
ALPELISIB; 212,526	For metastatic breast cancer, approved by FDA-based tests	Tablets of 50; 100; and 200 mg	NOVARTIS	Tablets given as 150 mg (twice)/day, with food	Hypersensitivity; skin-related problems; hyperglycaemia; diarrhoea; embryofoetal toxicity; increase creatinine, lipase, and ALT; and reduced weight, calcium, and haemoglobin levels	[87]
ISTRADEFYLLINE; 022075	For Parkinson disease	Tablets of 20 and 40 mg	KYOWA KIRIN	Suggested dose is 20 mg/day, which can be increased to 40 mg The dosage is restricted to patients with hepatic impairments and who smoke	Dyskinesis; hallucinations; constipation; and nausea	[88]
NETARSUDIL MESYLATE; 208,254	For ocular hypertension	Ophthalmic solution of 0.2 mg/mL	AERIE PHARMS INC	Daily dose in the evening	Hyperaemia; pain at site; and haemorrhage	[89]
ANGIOTENSIN II ACETATE; 209,360	Used as vasoconstrictor for increasing blood pressure	Injection of 2.5 mg/mL in vial	LA JOLLA PHARMA	The dose is diluted with 0.9% sodium chloride The dose is started at 20 ng/kg/min The dose is not increased beyond 80 ng/kg/min	Thromboembolism and its related effects	[90]

ERTUGLIFLOZIN; 209,803	Used as additive to control glucose level in type II diabetes mellitus	Tablets of 5 and 15 mg	MSD SUB MERCK	Dose of 5 mg (can be increased to 15 mg), with or without food Thus, monitoring of estimated glomerular filtration rate is required	Mycotic infections; kidney injury; lowering of blood pressure and glucose; and ketoacidosis	[91]
SEMAGLUTIDE; 209,637	Used as additive to control glucose level in type II diabetes mellitus	Injection of 1.34 mg/mL as single patient use pen, delivering 0.25 to 0.1 mg/mL	NOVO	Dose is started at 0.25 mg/week for 4 weeks, which can be increased to 0.5 mg/week The dose can further increased to 1 mg/week (as needed)	Pancreatitis; lowering of glucose levels; hypersensitivity; abdominal pain; and constipation	[92]
LATANOPROSTENE BUNOD; 207,795	For ocular hypertension	Eye drops of 0.24 mg/mL	BAUSCH AND LOMB	Daily dose/day in evening	Hyperaemia; pain at site; and pigmentation	[93]
SAFINAMIDE MESYLATE; 207,145	Parkinson disease (off episodes)	Tablets of 50 and 100 mg	MDD US	Not effective as single therapy The dose is initiated with 50 mg/day, which increases till 100 mg is reached, depending on need and admissibility	Dyskinesia; hallucinations; insomnia; and may increase hypertension	[94]
ERENUMAB-AOOE; 761,077	Prevention of migraine in adults	Injection of 70 mg/mL (through autoinjector; single dose syringe)	AMGEN INC	Via s.c. route (abdomen; thigh; and upper arm) at dose of 70 mg/mL per month, which may increase up to 140 mg/mL per month	Problems at injection site; and constipation	[95]
REVEFENACIN; 210,598	For treatment of COPD	Inhalation solution of 175 μg/3 mL	MYLAN IRELAND LTD	Via oral route, using jet nebuliser	Cough; upper respiratory tract infections; severity in narrow angle glaucoma and urinary retention; and hypersensitivity	[96]
FREMANEZUMAB-VFRM; 761,089	Prevention of migraine in adults	Injection of 225 mg/1.5 mL (single dose)	TEVA PHARMS USA	For s.c. route (in abdomen; thigh; or upper arm), with dosing as 225 mg/month or 675 mg/3 months	Hypersensitivity and reactions at site	[97]
SETMELANOTIDE ACETATE; 213,793	Obesity, hunger control due to opiomelanocortin deficiency	Injection via s.c. route of 18 mg in 3 mL solution	RHYTHM	1 injection/day, at recommended dose of 3 mg/day Start at 0.6 mg/day for 1 week, increasing till dose reaches 3 mg/day Dose increments in paediatrics may take 8 weeks Dose can be up to 2.4 mg/day	Skin hyperpigmentation; depression; erectile dysfunction	[98]

Table 2 Herbal formulations under trial for lifestyle diseases (Pubmed.com; 2022–2023 1 year)

Herbal Formulation	Eligible Participants	Dosage Used	Parameters for Outcomes	Outcomes	References
Lingguizhugan decoction Composed of Poria (Fuling); RamulusCinnamomi (Guizhi); RhizomaAtractylodisMacrocephalae (Baizhu); Radix Glycyrrhizae (Gancao)	243 with NAFLD, along with spleen—yang deficit outline	Treatment initiated with few lifestyle changes Standard dose: 12 g of Fuling; 9 g of Guizhi; 6 g of Baizhu; 6 g of Gancao each day for 12 weeks Low dose: 6 g of Fuling; 4.5 g of Guizhi; 3 g of Baizhu; 3 g of Gancao each day for 12 weeks Placebo: starch; colourant; and standard decoction (10%)	Primary outcome: 1 unit reduction in IR-HOMA following treatment Secondary outcome: enzymatic and inflammatory markers	The low dose of decoction proved better than placebo for NAFLD It functions by regulating the DNA 6 mA of several guiding enzymes and channels for people with obesity	[23]
Tea or coffee	48 children (19 males; 29 females) 18—Coffee 17—Green tea 13—placebo	Through coffee: 2 cups/day (160 mg of caffeine/day) Through green tea (catechin/caffeine): 3 cups/day (252 mg of catechin; 96 mg caffeine/day)	Primary outcome: reduced BMI; BMI percent of 95 percentile; and reduced fat at 3 and 6 months of study period	Caffeine and catechins, enhance induction of fat breakdown causing release of more energy and satiety However, consumption of coffee showed better effects, due to larger concentration of caffeine	[99]
Nelumbo nucifera leaf extract (NnEx), containing quercetin-3-glucuronide (Q3GA)	60 participants: 42 male; 18 female 22 in control group; 22 in low-dose group; 16 in high-dose group	Low-dose group: 1 g NnEx-based extract/day High-dose group: 2 g NnEx-based extract/day	Reduction in body weight; visceral fat; and waist circumference, measured at 6 and 12 weeks	Extract under study resulted in more reduction of body weight in both men and women However, reductions in visceral fat and BMI were only in men	[100]
Lecithin-based tablet containing curcumin	37 participants who met defined criteria	Administered as 1000 mg (2 times/day) equivalent to 400 mg curcuminoids Placebo administered was devoid of curcuminoids Compliance: ingestion of > 75% dosage	Significant changes in liver fats; anthropometry-based estimations	No effect on MRS-based HFC analysis, while there was significant reduction in GGT, indicating hepatoprotective effect of curcumin	[101]
EWDE batch CM-19.20a, from leaves and flowers of the plant	2 women (51 and 40 years old); 1 man (34 years old)	0.001% extract till day 6, (replacement of culture media at day 3 and day 5)	Score was applied as per the developed method, and hair follicle advancement in anagen and catagen	The identified extract reduced hair loss, improved hair density by 22 hairs/cm^2, when compared to minoxidil-based formulation, which improved to 24 hairs/cm^2	[102]

Table 2 (continued)

Herbal Formulation	Eligible Participants	Dosage Used	Parameters for Outcomes	Outcomes	References
Euphytose extract: a combination of *V. officinalis* L. (from the roots), *P. incarnate* L. (aerial parts), *Crataegus* sp. (from the leaf and flower) and *B. nigra* L. (from the flowering tops)	27/31 (19 female; 12 male), after appropriate assessments	Dose: 1 tablet, combination of 50 mg of *V. officinalis* L. (from the roots), 40 mg of *P. incarnate* L. (aerial parts), 10 mg of *Crataegus* sp. (from the leaf and flower) and 10 mg of *B. nigra* L. (from the flowering tops) 2 tablets with breakfast, lunch and dinner (14 days), followed by 28-day wash-out phase	Scores were assessed using reported scales to measure the progress of stress and anxiety	Significant reduction in stress was found, after 14-day administration of extract There was lowering of dermal conductance	[103]
(W-LHIT) is composed of herbal medicines like *Ganoderma lucidum*; *Coptis chinensis*; *Astragalus membranaceus*; *Nelumbo nucifera gaertn*; and *Fructus aurantii*	37/40 (14 female; 26 male) were distributed in 2 groups: aged 18–60 (Wei Fn hospital)	Dose as per body weight, with berberine as key component 9 capsules (daily crude herbal medicine dosage) of 75 kg Hence 9–15 W-LHIT capsules/day (3–5 capsules/meal—3 times/day) for 2 months For placebo, starch was used	Outcomes were assessed through several parameters, where key measurements included weight loss and BMI	W-LHIT significantly showed reduction in body weight, along with problems associated with obesity The administration of W-LHIT normalised the gut flora	[104]
Aframomummelegueta seed based extract as AfperFit	60 (completed)/70 (included)/90 (eligible) subjects in study 35 participants were distributed to each group: aged 20–50 years	AfperFit—Estimated to contain ≤ 2% 6-paradol (one of the active constituents) 5 mg of 6-paradol/250 mg of capsule Participants received 2 oral doses of 250 mg capsules/day for 84 days	Endpoint analysis included determination of energy spending; fat dissemination at 84 days The outcomes were statistically analysed	AfperFit represented its thermogenic effect, via activation of brown adipose tissue through sympathetic pathways	[105]
TOTUM-63, composed of 5 plants: *Chrysanthellum indicum* subsp. *Afroamericanum B.L. turner*; *Cynara scolymus* L.; *Vaccinium myrtillus* L.; *Olea europaea* L.; and *Piper nigrum* L.	51 (completed)/66 (included)/157 (eligible) patients were distributed among 2 groups: aged 35–75 years 51 participants in TOTUM-63 group and 15 in placebo	Capsules as food additives contained 623.75 mg of TOTUM-63 Dose/day—8 capsules of placebo and TOTUM-63 were administered for 6 months	Alteration in fasting glucose was determined	The formulated composition was safe and lowered glucose levels This was assessed through tolerance tests for glucose Administration of TOTUM-63 proved its effectiveness by lowering the associated parameters with diabetes	[106]
Greenselect Phytosomes, composed of catechins, gallates and caffeine	28 women; mean age of 58	2 doses of phytosomes (150 mg)/day	Reductions in RQ values and fractions of carbohydrates; fat concentrations Reductions in body constitution–based determinations	The administration of extract, altered the lipolysis pathway by resisting disturbances in adipose functioning	[107]

Table 3 Herbal formulations under trial for insomnia (Pubmed.com; mostly over past 5 years)

Herbs/Composition	Eligible Participants	Dose in Trial	Parameter for Outcomes	Outcomes	References
Ashwagandha root extract (capsule KSM 66, a commercial product)	80/116 participants: 40 in arm A and arm B each Arm A: 20 ashwagandha group; 20 placebo group (healthy) Arm B: 20 ashwagandha group; 20 placebo group (people with insomnia)	2 × one capsule/day with milk/water for 8 weeks, 1 capsule weighing 300 mg	Sleep parameters: week 4 and week 8 PSQI score; Hamilton anxiety scale-A measurement; sleep-quality assessments; and mental alertness after waking up	There were significant reductions in scores compared to placebo in subjects with insomnia, indicative of improvements Even methanolic extract of ashwagandha is a potent inhibitor of GABA receptor antagonists	[108]
Saffron extract in capsule (0.9 mg of crocin and 0.7 mg of Safranal)	66/98 participants: 32 in placebo group and 34 in saffron group Insomnia severity index: 7–21 Stress scale: 6–29	Trained to have 1 glass of water (in evening for 6 weeks) 1 capsule of 15.5 mg of saffron + 259.5 mg of maltodextrin Interventions were at days 6, 35 and 42	LSEQ score (days 7 and 42); PSQI score; SF 36 (assessing QoL); and actigraphy (to determine sleep quality)	The results reveal significant improvements: Time in bed increased in saffron group than in placebo group (according to actigraphy analysis) A desirable score improvement was favoured for LSEQ and SF-36, while a decline in PSQI score was achieved Safranal from saffron stimulated sleep-enhancing neurons and inhibit wake-causing neurons in brain	[109]
Cheonwangbosimdan, a Korean medicinal syrup, composed of 13 herbs	22/24 patients, out of which 11 allocated to Cheonwang-bosimdan and another 11 allocated to CBT-1 Specific inclusion criteria: tumour with treatment (for 0–5 years); ISI: at least 8	Cheonwangbosimdan, administered orally (at meals, for 4 weeks) CBT I/week for 4 weeks over 5 different involvements (related to sleep)	ISI score/4 weeks, at least 2 times: for improving QoL Other scores required: insomnia-dependent measures (including PSQI; ESS; among 5 different subscores completed at weeks 1 and 5); daily practice-based scores; and tracker-based scores	Study focused on insomnia in cancer patients The herbal composition of Cheonwangbosimdan caused reductions in anxiety-related to heart tasks in such patients However, differences in scores were not significant, indicating importance of both types of treatments	[110]
Tian Wang Bu Xin granule, composed of 13 herbs	13 perimenopausal women with insomnia (treatment group) + their 13 male counterparts (control group)	Tian Wang Bu Xin (83 g)/bag, in morning and evening for 4 weeks (measurement of PSQI score was before treatment and on 29th day)	Outcomes were based on PSQI scores, sleep monitoring and sleep patterns Stool samples were further collected to identify potential microbes	PSQI scores declined after therapy, followed by regulation of stool-based microbial flora	[111]
Lettuce seed (capsule)	Analysed 70/100 (pregnant women) gestational age of 12–36 months 35/50 in lettuce group; 15/50 in placebo group PSQI score > 5	1 capsule/night for 2 weeks 1 capsule with 1000 mg of seeds Placebo: starch (capsules)	Analysis of PSQI scores	Lettuce seeds have significant effects on insomnia during second and third trimesters	[112]

Table 3 (continued)

Herbs/Composition	Eligible Participants	Dose in Trial	Parameter for Outcomes	Outcomes	References
Polygonatumsibiricum seeds as tablets	75/131 patients with mild insomnia: 40 in treatment group; 39 in placebo group	Rhizome extract (500 mg) in tablets (800 mg)	AIS score; actigraphy; and sleep patterns, followed by an MRI-based analysis	The treatment group showed improved sleep and reduced instability The linoleate-based phytoconstituent in rhizome is a GABA agonist, where mRNA levels in receptors are modified after administration of rhizome extract	[113]
Saffron from stigmas of *Crocus sativus* L., further standardised for a strength of 3.5% of Lepticrosalides	63/118 participants with ISI scores of 8–21 and BMIs of 18–30 kg/m^2	Each tablet had 14 mg of saffron Administration was 1 tablet 2 times/day for 28 days	ISI scores and assessments of sleep rebuilding, awakening pattern, depression and anxiety scores through several questionnaires	Significant improvement in sleep evident after 28 days of intervention, according to reductions in ISI scores, whereas DASS scores remained constant	[114]
Passion flower extracts, composed of 80% leaves and 20% fruit (air dried)	110/130 participants were included in study, where the distribution to treatment and placebo group was in 1:1 Recruitment followed diagnostic-statistical-manual 5 criteria	Received 60 mg of extract/day for 2 weeks	Estimation of sleep parameters through polysomnography; ISI and PSQI scores; assessments of DASS scores	The evaluation and outcome parameters showed improvements in sleep, but a comparison with marketed drugs is essential to analysis	[115]
Ziziphus jujuba mill of herbal seeds (powdered and granulated)	12/18 participants were involved in study Mean ISI score: 17.42 All 12 received treatment and placebo	(2 g granules/10 g of dry seed) 4 × 1 capsule (500 mg granules), 1 h before sleep for 4 weeks	Assessment was through PSQI scores; QoL questionnaires (WHOQOL); evaluations of sleep issues; assessments of DASS scores Some parameters were verified through actigraphy	Trial proved the appropriateness for insomnia CBT, as an additional strategy was recommended with treatment for more beneficial effects	[116]
Aerial parts of *Melissa officinalis* (lemon balm)	80/452 patients with BMIs of 18.5–40	3 × 1 capsule (3 g)/day—For 8 weeks	PSQI scores and DASS scores were assessed and their reliability tested	Significant improvements in depression and anxiety were found the phytoconstituents act via multiple mechanisms, such as the inhibition of GABA transaminase and serotonergic action	[117]

The outcomes of these studies have shown that in potent doses, herbal drugs have better therapeutic efficacy than some other systems of medicine. However, the treatment using herbal drugs was assessed by a set of questionnaires (ISI; PSQI), sleep charts/diaries, actigraphs or scales (Table 3). The reliability of these assessments must be validated appropriately.

9 Clinical Trials and FDA-Based Recommendations for Drugs Used Against Smoking

Smoking is the customary usage of tobacco products. It is harmful and leads to addiction, causing health risks. Smoking, as a lifestyle disorder, may lead to other health consequences and even serious environmental issues. In many cases, people experience difficulty quitting smoking. Stopping smoking or its tendency requires strong willpower and counselling. Some herbal medicines may help people stop smoking. However, the initiation of such treatments may require a consultation.

The habit of smoking arises from stress, ongoing depression, anxiety and/or social influences. Smoking has a significant impact on a person's heart and several other vital organs in that it increases their heart rate and blood pressure, causes the formation of blood clots, damages vessels, reduces oxygen supply, reduces the oxygen-carrying capacity of blood vessels and increases vascular obstructions, leading to arrythmia, myocardial infarction and stroke. Any amoung of smoking in any form, even by smoking ecigarettes, is harmful to human health [58]. Multiple clinical trials are underway to treat such complications from smoking.

Smoking may damage vascular dilation inside the vascular endothelium, causing changes to arterial stiffness. This increases blood pressure. In a trial on phytoconstituents, Sanghuang–Danshen was used for its synergistic effects. Sanghuang is an edible mushroom, and Danshen (Rhizome) contains several bioactive agents; together, they were studied in a trial on arterial stiffness due to smoking. The trial included healthier smokers (> 10 cigarettes/day) who were further divided into placebo, low-dose and high-dose groups. The dose of Sanghuang–Danshen was administered as 600 mg/day and 900 mg/day in single capsules. The volunteers in the trial were checked for compliance and the regularity of their dose administration. At end of 4 weeks, measurements of their blood pressure and brachial and ankle pulse wave velocity values revealed significant improvements in systolic blood pressure after the 900 mg dose. An increase in eNOS phosphorylation was also detectable at high-dose levels. With Western blotting, it was revealed that bioactive agents like ellagic acid, caffeic acid, protocatechuic acid, cryptotanshinone, tanshinone I and tanshinone II affected the change in vascular dilation in prolonged smokers [59].

Smoking is heavily related to stress and depression. Smoking, which contains nicotine, causes the downregulation of nicotinic receptors. This causes reduced serotonin liberation, increasing depressive tendencies in individuals. Yokukansan (a composite of seven herbs) has been useful for treating neurosis and insomnia because it reduces the uptake of glutamic acid and activates serotonin receptors. For the test, around 108 patients (who wanted to quit smoking) were each administered a dose of 7.5 g/day for 12 weeks. Older, smaller patients (age ≥ 75; weight ≤ 50 kg) were each given a dose of 5 g/day. The inclusion criteria prioritised SDS and FTND test scores. The outcomes were further assessed on the basis of any improvement in SDS test scores (for depression) and POMS test scores (for mood elevation) at the end of the 12 weeks. Yokukansan led to even less smoking stoppage, but there were clear improvements in the patients' psychological states [60].

10 Clinical Trials and FDA-Based Recommendations for Drugs Used Against Alcoholism

Alcoholism can have severe consequences on the human body. It is a chronic disorder where an individual becomes dependent on alcohol usage despite their knowing its negative effects on their health. Alcoholism can be treated with medical interventions and counselling. Herbal medicines are considered as alternatives for preventing alcoholism because the interventions of herbal medicines may be useful for relieving anxiety and stress, promoting healthier lifestyles and acting as immunomodulators. In one trial, kudzu root (composed of isoflavones) was utilised to reduce alcohol consumption. A highly concentrated extract was utilised this time because another extract with a smaller concentration of ingredients failed to produce the desired effects in previous studies. In this kudzu root study, 14 people with alcohol dependency (with negative family histories) and with BMIs between 19 and 24 kg/m^2 were included. The subjects took a kudzu capsule (2×500 mg) three times per day. The initial assessment was made by using the variables defining the desire to drink, while the outcomes were assessed according to the Alcohol Urge Questionnaire (AUQ). The results showed that cravings were reduced in the subjects, while drinking capacity did not decline. The drinking time taken to finish one glass of alcohol increased. However, the lack of subjects who had undergone therapy was a major drawback of the study [61].

The effects of heavy drinking include later hangover symptoms. During a hangover, the onset of acetate may lead to fatigue, an increase in systolic blood pressure and other symptoms, like sweating. In a trial, red ginseng was used to relieve the hangover symptoms of 25 participants (who had BMIs between 18.5 and 29.9 kg/m^2). The red ginseng syrup contained 0.321 mg/mL of extract, administered to participants at their scheduled visit (0, 1, 2 week) 5 min after they had consumed alcohol. The results revealed a decline in plasma alcohol and expiratory alcohol 30 min after the administration of alcohol to the treatment group. Hence, the Symptom Index values were reduced in the red ginseng group, but symptoms of anxiety and depression were higher in the red ginseng group. The researchers noted that the levels of alcohol, acetaldehyde and acetaldehyde dehydrogenase needed to be detected more vigorously during the study, which was therefore a limitation of the study [62]. In another similar trial, *Hovenia dulcis* Thunberg fruit extract (2460 mg/day) was used to reduce the symptoms of hangovers in 26 participants (heterozygous aldehyde dehydrogenase). The findings suggest that inflammatory molecules IL-6 and IL-10 (regulated by the phytoconstituents of the extract) help reduce hangover symptoms [63].

Similarly, in another trial, an Unani-based medicine, Sharbat-e-Afsanteen (composed of *Artemisia absinthium* L. syrup), was assessed for its ability to modify the liver enzymes affected by alcohol intake. Thus, for convenience, 20/30 patients (20 patients tested; ten controls) were administered *Artemisia absinthium* L. syrup (2 × 25 mL)/day for 28 days. It has been suggested that the phenolics and flavones in *Artemisia absinthium* L. prevent damage to liver enzymes by inhibiting the effect of inflammatory mediators. The test formulation showed a significant improvement in the concentration of liver enzymes, indicating reduced liver dysfunction [64].

The major limitations found in several studies were insufficient recruitment, lack of female participants and patients' follow-up compliance.

11 Conclusions

Developing drugs for a complex lifestyle disease is complicated by the difficulty of ascertaining its exact cause. Herbal drugs have heterogenous compositions containing several phytochemicals, so ensuring the consistency of their formulation and products is also difficult. This sets them up to fail during development [118]. Hence, manufacturers are required to follow the guidelines for scientific research and evidence established by regulators for the production of drugs.

References

1. US FDA. Botanical drug development guidance for industry. US FDA. [cited 2023 Jul 2]. https://www.fda.gov/media/93113/download.
2. US FDA. What is a botanical drug? US FDA. 2022 [cited 2023 Jul 2]. https://www.fda.gov/about-fda/center-drug-evaluation-and-research-cder/what-botanical-drug#:~:text=A botanical drug%27s special features require consideration and,facilitate development of new therapies from botanical sources.
3. US FDA. Dietary supplement ingredient directory. US FDA 2023 [cited 2023 Jul 2]. https://www.fda.gov/food/dietary-supplements/dietary-supplement-ingredient-directory.
4. US FDA. Botanical Review Team (BRT). US FDA. 2022 [cited 2023 Jul 3]. https://www.fda.gov/about-fda/center-drug-evaluation-and-research-cder/botanical-review-team-brt.
5. US FDA. OFFICE OF PHARMACEUTICAL QUALITY Review of Botanical Drug Products. 2016 [cited 2023 Jul 3]. https://www.fda.gov/media/72454/download.
6. Budreviciute A, Damiati S, Sabir DK, Onder K, Schuller-Goetzburg P, Plakys G, et al. Management and prevention strategies for non-communicable diseases (NCDs) and their risk factors. Front Public Heal. 2020;8:574111; https://www.frontiersin.org/articles/10.3389/fpubh.2020.574111/full.
7. United Nations. LIFESTYLE DISEASES: an economic burden on the health services. UN Chron 2021 [cited 2023 Feb 11]. https://www.un.org/en/chronicle/article/lifestyle-diseases-economic-burden-health-services.
8. Zahid. Health lifestyle. National Health Portal 2019 [cited 2023 Feb 11]. https://www.nhp.gov.in/Health-Lifestyle_pg.
9. US FDA. Consumer health information for better nutrition initiative—attachment I—public docket comments. US FDA. 2022 [cited 2023 Jul 2]. https://www.fda.gov/food/food-labeling-nutrition/consumer-health-information-better-nutrition-initiative-attachment-i-public-docket-comments.
10. US FDA. Family smoking prevention and tobacco control act—an overview. US FDA. 2022 [cited 2023 Jul 2]. https://www.fda.gov/tobacco-products/rules-regulations-and-guidance/family-smoking-prevention-and-tobacco-control-act-overview.
11. US FDA. FDA proposes new requirements for tobacco product manufacturing practices. US FDA. 2023 [cited 2023 Jul 2]. https://www.fda.gov/news-events/press-announcements/fda-proposes-new-requirements-tobacco-product-manufacturing-practices.
12. US FDA. Tobacco science & research. US FDA. 2023 [cited 2023 Jul 2]. https://www.fda.gov/tobacco-products/tobacco-science-research.
13. US FDA. Clinical trials guidance documents. US FDA. 2023 [cited 2023 Jul 2]. https://www.fda.gov/regulatory-information/search-fda-guidance-documents/clinical-trials-guidance-documents.
14. US FDA. Conducting clinical trials. US FDA. 2020 [cited 2023 Jul 2]. https://www.fda.gov/drugs/development-approval-process-drugs/conducting-clinical-trials.
15. US FDA. Basics about clinical trials. US FDA. 2023 [cited 2023 Jul 2]. https://www.fda.gov/patients/clinical-trials-what-patients-need-know/basics-about-clinical-trials.
16. FDA. Clinical trials and human subject protection|FDA. US FDA. 2023 [cited 2023 Jul 2]. https://www.fda.gov/science-research/science-and-research-special-topics/clinical-trials-and-human-subject-protection.
17. Rubio-Tomás T, Rueda-Robles A, Plaza-Díaz J, Álvarez-Mercado AI. Nutrition and cellular senescence in obesity-related disorders. J Nutr Biochem. 2022;99:108861; https://linkinghub.elsevier.com/retrieve/pii/S0955286321002813.
18. Tuncay C, Ergoren MC. A systematic review of precision nutrition and mediterranean diet: a personalized nutrition approaches for prevention and management of obesity related disorders. Clin Nutr ESPEN. 2020;38:61–4; https://linkinghub.elsevier.com/retrieve/pii/S2405457720300759.
19. Hampsey J, Karnsakul W. Liver disorders: nutritional management. In: Encyclopedia of human nutrition. San Diego, CA: Elsevier; 2013. p. 87–99; https://linkinghub.elsevier.com/retrieve/pii/B9780123750839001720.
20. Yuan I, Sinha AC. Obesity and nutrition disorders. In: Anesthesia and uncommon diseases. Philadelphia, PA: Elsevier; 2012. p. 215–24; https://linkinghub.elsevier.com/retrieve/pii/B9781437727876000061.
21. US FDA. Weight-loss and weight-management devices. US FDA. 2022 [cited 2023 Jul 3]. https://www.fda.gov/medical-devices/products-and-medical-procedures/weight-loss-and-weight-management-devices.
22. Jastreboff AM, Aronne LJ, Ahmad NN, Wharton S, Connery L, Alves B, et al. Tirzepatide once weekly for the treatment of obesity. N Engl J Med. 2022;387(3):205–16; http://www.nejm.org/doi/10.1056/NEJMoa2206038.
23. Dai L, Xu J, Liu B, Dang Y, Wang R, Zhuang L, et al. Lingguizhugan decoction, a Chinese herbal formula, improves insulin resistance

in overweight/obese subjects with non-alcoholic fatty liver disease: a translational approach. Front Med. 2022;16(5):745–59. https://doi.org/10.1007/s11684-021-0880-3.

24. Ko Y, Kim HJ, Kim H, Choi JB, Kwon YD, Jung WS, et al. Exploring the efficacy and safety of herbal medicine on Korean obese women with or without metabolic syndrome risk factors: a study protocol for a double-blind, randomized, multicenter, placebo-controlled clinical trial. Medicine (Baltimore). 2020;99(28):e21153.
25. Lee S, Lee H, Cho Y, Kim J, Kang JW, Seo B-K, et al. The efficacy and safety of Hanslim for obese patients. Medicine (Baltimore). 2018;97(38):e12440; https://journals.lww.com/00005792-201809210-00075.
26. Gupte P, Harke S, Deo V, Bhushan Shrikhande B, Mahajan M, Bhalerao S. A clinical study to evaluate the efficacy of herbal formulation for obesity (HFO-02) in overweight individuals. J Ayurveda Integr Med. 2020;11(2):159–62.
27. US FDA. FDA approves new drug treatment for chronic weight management, first since 2014. US FDA. 2021 [cited 2023 Jul 2]. https://www.fda.gov/news-events/press-announcements/fda-approves-new-drug-treatment-chronic-weight-management-first-2014.
28. US FDA. FDA approves weight management drug for patients aged 12 and older. US FDA. 2021 [cited 2023 Jul 2]. https://www.fda.gov/drugs/news-events-human-drugs/fda-approves-weight-management-drug-patients-aged-12-and-older.
29. WHO. Anaemia. [cited 2023 Jun 10]. https://www.who.int/health-topics/anaemia#tab=tab_2.
30. Li Y, Mai L, Chang H, Yu B, Huang Z. Efficacy of complex Hongyibuxue oral solution and ferrous sulfate for iron deficiency anemia. Di Yi Jun Yi Da Xue Xue Bao. 2005;25(6):732–3; http://www.ncbi.nlm.nih.gov/pubmed/15958324.
31. Ke Y, Wei K, Zheng J, Pu J, Zhu Y, Liang W. Study on treatment of iron-deficiency anemia by shengxuening. Zhongguo Zhong Xi Yi Jie He Za Zhi. 2004;24(10):893–6; http://www.ncbi.nlm.nih.gov/pubmed/15553821.
32. Malayappan MS, Natarajan G, Mockaiyathevar L, Ramasamy M. Acute and subacute toxicity assessment of madhulai manappagu (siddha herbal syrup formulation) in animal model. J Complement Integr Med. 2021;19(1):9–18; http://www.ncbi.nlm.nih.gov/pubmed/33818028.
33. Akase T, Akase T, Onodera S, Jobo T, Matsushita R, Kaneko M, et al. A comparative study of the usefulness of Toki-shakuyaku-san and an oral iron preparation in the treatment of hypochromic anemia in cases of uterine myoma. Yakugaku Zasshi. 2003;123(9):817–24; http://www.ncbi.nlm.nih.gov/pubmed/14513774.
34. Khalili N, Fereydoonzadeh R, Mohtashami R, Mehrzadi S, Heydari M, Huseini HF. Silymarin, olibanum, and nettle, a mixed herbal formulation in the treatment of type II diabetes: a randomized, double-blind, placebo-controlled, clinical trial. J Evid Based Complementary Altern Med. 2017;22(4):603–8; http://www.ncbi.nlm.nih.gov/pubmed/29228792.
35. Ryan EA, Imes S, Wallace C, Jones S. Herbal tea in the treatment of diabetes mellitus. Clin Invest Med. 2000;23(5):311–7; http://www.ncbi.nlm.nih.gov/pubmed/11055324.
36. Wang H, Mu W, Zhai J, Xing D, Miao S, Wang J, et al. The key role of Shenyan Kangfu tablets, a Chinese patent medicine for diabetic nephropathy: study protocol for a randomized, double-blind and placebo-controlled clinical trial. Trials. 2013;14:165; http://www.ncbi.nlm.nih.gov/pubmed/23738508.
37. Sudha V, Bairy KL, Shashikiran U, Sachidananda A, Jayaprakash B, Shalini S. Efficacy and tolerability of Dianex in type 2 diabetes mellitus: a non randomized, open label non-comparative study. Med J Malaysia. 2005;60(2):204–11; http://www.ncbi.nlm.nih.gov/pubmed/16114162.
38. US FDA. Petition for a qualified health claim for psyllium husk to reduce the risk of type 2 diabetes mellitus (Docket No. FDA-2013-Q-0167). US FDA. 2014 [cited 2023 Jul 2]. http://wayback.archive-it.org/7993/20171115122026/https://www.fda.gov/downloads/Food/IngredientsPackagingLabeling/LabelingNutrition/UCM403090.pdf.
39. US FDA. GRAS notification for quinoa sprouts rich in botanical B vitamins. US FDA. 2017 [cited 2023 Jul 2]. https://www.fda.gov/media/104075/download.
40. US FDA. FDA proposes broad approach for conducting safety trials for type 2 diabetes medications. US FDA. 2020 [cited 2023 Jul 2]. https://www.fda.gov/news-events/press-announcements/fda-proposes-broad-approach-conducting-safety-trials-type-2-diabetes-medications#:~:text=The U.S. Food and Drug Administration today issued,may be more vulnerable to drug-related side effects.
41. US FDA. FDA approves novel, dual-targeted treatment for type 2 diabetes. US FDA. 2022 [cited 2023 Jul 2]. https://www.fda.gov/news-events/press-announcements/fda-approves-novel-dual-targeted-treatment-type-2-diabetes.
42. US FDA. FDA approves new class of medicines to treat pediatric type 2 diabetes. US FDA. 2023 [cited 2023 Jul 2]. https://www.fda.gov/news-events/press-announcements/fda-approves-new-class-medicines-treat-pediatric-type-2-diabetes.
43. US FDA. FDA in brief: FDA issues draft guidance on early clinical studies for certain medical devices to improve glycemic control for type 2 diabetes. US FDA. 2020 [cited 2023 Jul 3]. https://www.fda.gov/news-events/press-announcements/fda-brief-fda-issues-draft-guidance-early-clinical-studies-certain-medical-devices-improve-glycemic.
44. US FDA. Feasibility and early feasibility clinical studies for certain medical devices intended to therapeutically improve glycemic control in patients with type 2 diabetes mellitus guidance for industry and food and drug administration staff. 2022. https://www.fda.gov/media/148765/download.
45. Fractyl health. TYPE 2 DIABETES THERAPIES REINVENTED. [cited 2023 Jul 3]. https://www.fractyl.com/therapies/.
46. USFDA.Heartdisease.USFDA.2023[cited2023Jul3].https://www.fda.gov/consumers/minority-health-and-health-equity-resources/heart-disease.
47. Guo L, Yuan H, Zhang D, Zhang J, Hua Q, Ma X, et al. A multicenter, randomized, double-blind, placebo-parallel controlled trial for the efficacy and safety of shenfuqiangxin pills in the treatment of chronic heart failure (heart-kidney yang deficiency syndrome). Medicine (Baltimore). 2020;99(21):E20271.
48. Shayani Rad M, Moohebati M, Mohammad Ebrahimi S, Motamedshariaty VS, Mohajeri SA. Safety evaluation and biochemical efficacy of celery seed extract (Apium Graveolens) capsules in hypertensive patients: a randomized, triple-blind, placebo-controlled, cross-over, clinical trial. Inflammopharmacology. 2022;30(5):1669–84.
49. Wang K, Cai J-J, Wu Y, Wang Y, Liu L-L, Shi L, et al. Prospective randomized controlled trial study of Luofengning granule in the treatment of unstable angina. Medicine (Baltimore). 2020;99(20):e20025; http://www.ncbi.nlm.nih.gov/pubmed/32443306.
50. US FDA. Petition for a health claim for eicosapentaenoic acid and docosahexaenoic acid and reduction of blood pressure in the general population (Docket No. FDA-2014-Q-1146). US FDA. 2014 [cited 2023 Jul 2]. https://www.fda.gov/media/128043/download.
51. Bucci LR. Selected herbals and human exercise performance. Am J Clin Nutr. 2000;72(2):624S–36S; https://linkinghub.elsevier.com/retrieve/pii/S000291652306745X.
52. Sellami M, Slimeni O, Pokrywka A, Kuvačić G, Hayes DL, Milic M, et al. Herbal medicine for sports: a review. J Int Soc Sports Nutr. 2018;15(1):14. https://doi.org/10.1186/s12970-018-0218-y.

53. Turmeric and exercise-induced muscle damage and oxinflammation. Clinicaltrials.gov.in. 2021 [cited 2023 Jun 27]. https://clinicaltrials.gov/study/NCT04765527.
54. Effects of herbal supplements on endurance exercise performance. Clinicaltrials.gov.in. 2019 [cited 2023 Jun 27]. https://clinicaltrials.gov/study/NCT03929575.
55. Uchimura N. Insomnia and lifestyle-related diseases. Nihon Rinsho. 2012;70(7):1100–6; http://www.ncbi.nlm.nih.gov/pubmed/22844789.
56. Taherzadeh Z, Khaluyan H, Iranshahy M, Rezaeitalab F, Eshaghi Ghalibaf MH, Javadi B. Evaluation of sedative effects of an intranasal dosage form containing saffron, lettuce seeds and sweet violet in primary chronic insomnia: a randomized, double-dummy, double-blind placebo controlled clinical trial. J Ethnopharmacol. 2020;262:113116; https://linkinghub.elsevier.com/retrieve/pii/S0378874120329974.
57. Birling Y, Bensoussan A, Sarris J, Avard N, Zhu X, Ren Z. An Shen capsule for chronic insomnia. Medicine (Baltimore). 2019;98(14):e14853; https://journals.lww.com/00005792-201904050-00004.
58. Marques P, Piqueras L, Sanz M-J. An updated overview of e-cigarette impact on human health. Respir Res. 2021;22(1):151; http://www.ncbi.nlm.nih.gov/pubmed/34006276.
59. Lim Y, Song T-J, Hwang W, Kim JY, Lee D, Kim Y-J, et al. Synergistic effects of Sanghuang-Danshen bioactives on arterial stiffness in a randomized clinical trial of healthy smokers: an integrative approach to in silico network analysis. Nutrients. 2019;11(1):108; http://www.ncbi.nlm.nih.gov/pubmed/30621047.
60. Komiyama M, Ozaki Y, Wada H, Yamakage H, Satoh-Asahara N, Yasoda A, et al. Randomized double-blind placebo-controlled multicenter trial for the effects of a polyherbal remedy, Yokukansan (YiganSan), in smokers with depressive tendencies. BMC Complement Med Ther. 2022;22(1):311; http://www.ncbi.nlm.nih.gov/pubmed/36434692.
61. Lukas SE, Penetar D, Berko J, Vicens L, Palmer C, Mallya G, et al. An extract of the Chinese herbal root kudzu reduces alcohol drinking by heavy drinkers in a naturalistic setting. Alcohol Clin Exp Res. 2005;29(5):756–62. https://doi.org/10.1097/01.ALC.0000163499.64347.92.
62. Lee M-H, Kwak JH, Jeon G, Lee J-W, Seo J-H, Lee H-S, et al. Red ginseng relieves the effects of alcohol consumption and hangover symptoms in healthy men: a randomized crossover study. Food Funct. 2014;5(3):528; http://xlink.rsc.org/?DOI=c3fo60481k.
63. Kim H, Kim YJ, Jeong HY, Kim JY, Choi EK, Chae SW, et al. A standardized extract of the fruit of Hovenia dulcis alleviated alcohol-induced hangover in healthy subjects with heterozygous ALDH2: a randomized, controlled, crossover trial. J Ethnopharmacol. 2017;209:167–74.
64. Fatima N, Siddiqui MA, Riyazuddin M, Shahid A. Efficacy of sharbat-e-Afsanteen (Artemisia absinthium L.) on altered liver enzymes among alcoholics—a randomized single-blind controlled clinical trial. J Herb Med. 2023;1:37.
65. US FDA. JESDUVROQ (daprodustat) tablets, for oral use initial U.S. approval: 2023-label. 2023 [cited 2023 Feb 18]. https://www.accessdata.fda.gov/drugsatfda_docs/label/2023/216951s000lbl.pdf.
66. US FDA. BRENZAVVY (bexagliflozin) tablets, for oral use initial U.S. approval: 2023-label. 2023 [cited 2023 Feb 18]. https://www.accessdata.fda.gov/drugsatfda_docs/label/2023/214373s000lbl.pdf.
67. US FDA. LEQEMBI™ (lecanemab-irmb) injection, for intravenous use initial U.S. approval: 2023-label. 2023 [cited 2023 Feb 18]. https://www.accessdata.fda.gov/drugsatfda_docs/label/2023/761269Orig1s000lbl.pdf.
68. US FDA. ELAHERETM (mirvetuximab soravtansine-gynx) injection, for intravenous use initial U.S. approval: 2022—label. 2022 [cited 2023 Feb 18]. https://www.accessdata.fda.gov/drugsatfda_docs/label/2022/761310s000lbl.pdf.
69. US FDA. TECVAYLI™ (teclistamab-cqyv) injection, for subcutaneous use initial U.S. approval: 2022-label. 2022 [cited 2023 Feb 18]. https://www.accessdata.fda.gov/drugsatfda_docs/label/2022/761291s000lbl.pdf.
70. US FDA. IMJUDO® (Tremelimumab-actl) injection, for intravenous use initial U.S. approval: 2022-label. 2022. https://www.accessdata.fda.gov/drugsatfda_docs/label/2022/761270s000lbl.pdf.
71. US FDA. OMLONTI® (Omidenepag isopropyl ophthalmic solution) 0.002%, for topical ophthalmic use initial U.S. approval: 2022 -label. 2022 [cited 2023 Feb 18]. https://www.accessdata.fda.gov/drugsatfda_docs/label/2022/215092s000lbl.pdf.
72. US FDA. MOUNJAROTM (Tirzepatide) Injection, for subcutaneous use initial U.S. approval: 2022—label. 2022 [cited 2023 Feb 18]. https://www.accessdata.fda.gov/drugsatfda_docs/label/2022/215866s000lbl.pdf.
73. US FDA. QUVIVIQ (Daridorexant) tablets, for oral use, [controlled substance schedule pending] initial U.S. approval: [pending controlled substance scheduling]-label. 2022 [cited 2023 Feb 18]. https://www.accessdata.fda.gov/drugsatfda_docs/label/2022/214985s000lbl.pdf.
74. US FDA. KIMMTRAK® (Tebentafusp-tebn) injection, for intravenous use initial U.S. approval: 2022-label. 2022 [cited 2023 Feb 18]. https://www.accessdata.fda.gov/drugsatfda_docs/label/2022/761228s000lbl.pdf.
75. US FDA. CAMZYOSTM (Mavacamten) capsules for oral use initial U.S. approval: 2022-label. 2022 [cited 2023 Feb 18]. https://www.accessdata.fda.gov/drugsatfda_docs/label/2022/214998s000lbl.pdf.
76. US FDA. ADUHELM™ (Aducanumab-avwa) injection, for intravenous use initial U.S. approval: 2021- label. 2021 [cited 2023 Feb 18]. https://www.accessdata.fda.gov/drugsatfda_docs/label/2021/761178s000lbl.pdf.
77. US FDA. ZEGALOGUE (Dasiglucagon) injection, for subcutaneous use initial U.S. approval: 2021-label. 2021 [cited 2023 Feb 19]. https://www.accessdata.fda.gov/drugsatfda_docs/label/2021/214231s000lbl.pdf.
78. US FDA. VERQUVOTM (Vericiguat) tablets, for oral use initial U.S. approval: 2021-label. 2021 [cited 2023 Feb 20]. https://www.accessdata.fda.gov/drugsatfda_docs/label/2021/214377s000lbl.pdf.
79. US FDA. COSELA (Trilaciclib) for injection, for intravenous use initial U.S. approval: 2021-label. 2021 [cited 2023 Feb 20]. https://www.accessdata.fda.gov/drugsatfda_docs/label/2021/214200s000lbl.pdf.
80. US FDA. BLENREP (Belantamab mafodotin-blmf) for injection, for intravenous use initial U.S. approval: 2020-label. 2020 [cited 2023 Feb 20]. https://www.accessdata.fda.gov/drugsatfda_docs/label/2020/761158s000lbl.pdf.
81. US FDA. ZEPZELCATM (Lurbinectedin) for injection, for intravenous use initial U.S. approval: 2020-label. 2020 [cited 2023 Feb 20]. https://www.accessdata.fda.gov/drugsatfda_docs/label/2020/213702s000lbl.pdf.
82. US FDA. TAUVID™ (Flortaucipir F 18 injection), for intravenous use initial U.S. approval: 2020—label. 2020 [cited 2023 Feb 20]. https://www.accessdata.fda.gov/drugsatfda_docs/label/2020/212123s000lbl.pdf.
83. US FDA. GEMTESA (Vibegron) tablets, for oral use initial U.S. approval: 2020-label. 2020 [cited 2023 Feb 20]. https://www.accessdata.fda.gov/drugsatfda_docs/label/2020/213006s000lbl.pdf.
84. US FDA. UBRELVY (Ubrogepant) tablets, for oral use initial U.S. approval: 2019—label. 2019 [cited 2023 Feb 24]. https://www.

accessdata.fda.gov/drugsatfda_docs/label/2019/211765s000lbl.pdf.
85. US FDA. DAYVIGO® (Lemborexant) tablets, for oral use, CIV initial U.S. approval: 2019—label. 2019 [cited 2023 Feb 24]. https://www.accessdata.fda.gov/drugsatfda_docs/label/2023/212028s007lbl.pdf.
86. US FDA. ACCRUFER (Ferric maltol) capsules, for oral use initial U.S. approval: 2019—label. 2019 [cited 2023 Feb 24]. https://www.accessdata.fda.gov/drugsatfda_docs/label/2019/212320Orig1s000lbl.pdf.
87. US FDA. PIQRAY® (Alpelisib) tablets, for oral use initial U.S. approval: 2019—label. 2019 [cited 2023 Feb 24]. https://www.accessdata.fda.gov/drugsatfda_docs/label/2019/212526s000lbl.pdf.
88. US FDA. NOURIANZ™ (Istradefylline) tablets, for oral use initial U.S. approval: 2019 -label. 2019 [cited 2023 Feb 24]. https://www.accessdata.fda.gov/drugsatfda_docs/label/2019/022075s000lbl.pdf.
89. US FDA. RHOPRESSA® (Netarsudil ophthalmic solution) 0.02%, for topical ophthalmic use initial U.S. approval: 2017—label. 2017 [cited 2023 Feb 24]. https://www.accessdata.fda.gov/drugsatfda_docs/label/2017/208254lbl.pdf.
90. US FDA. GIAPREZA (Angiotensin II) injection for intravenous infusion initial U.S. approval: TBD—label [Internet]. 2017 [cited 2023 Feb 24]. https://www.accessdata.fda.gov/drugsatfda_docs/label/2017/209360s000lbl.pdf.
91. US FDA. STEGLATRO™ (Ertugliflozin) tablets, for oral use initial U.S. approval: 2017—label. 2017 [cited 2023 Feb 24]. https://www.accessdata.fda.gov/drugsatfda_docs/label/2017/209803s000lbl.pdf.
92. US FDA. OZEMPIC (Semaglutide) injection, for subcutaneous use initial U.S. approval: 2017—label. 2017 [cited 2023 Feb 24]. https://www.accessdata.fda.gov/drugsatfda_docs/label/2017/209637lbl.pdf.
93. US FDA. VYZULTA (Latanoprostene bunod ophthalmic solution) 0.024%, for topical ophthalmic use initial U.S. approval: 2017- label. 2017 [cited 2023 Feb 24]. https://www.accessdata.fda.gov/drugsatfda_docs/label/2017/207795Orig1s000lbl.pdf.
94. US FDA. XADAGO (Safinamide) tablets, for oral use initial U.S. approval: 2017—label. 2017 [cited 2023 Feb 24]. https://www.accessdata.fda.gov/drugsatfda_docs/label/2017/207145lbl.pdf.
95. US FDA. AIMOVIGTM (Erenumab-aooe) injection, for subcutaneous use initial U.S. approval: 2018—label. 2018 [cited 2023 Feb 24]. https://www.accessdata.fda.gov/drugsatfda_docs/label/2018/761077s000lbl.pdf.
96. US FDA. YUPELRI (Revefenacin) inhalation solution, for oral inhalation initial U.S. approval: 2018—label. 2018 [cited 2023 Feb 24]. https://www.accessdata.fda.gov/drugsatfda_docs/label/2018/210598s000lbl.pdf.
97. US FDA. AJOVYTM (Fremanezumab-vfrm) injection, for subcutaneous use initial U.S. approval: 2018—label. 2018 [cited 2023 Feb 24]. https://www.accessdata.fda.gov/drugsatfda_docs/label/2018/761089s000lbl.pdf.
98. US FDA. SAXENDA (Liraglutide) injection, for subcutaneous use initial U.S. approval: 2010. US FDA. 2023. https://www.accessdata.fda.gov/drugsatfda_docs/label/2023/206321s016lbl.pdf.
99. Moran-Lev H, Cohen S, Zelber-Sagi S, Mazkeret Mayer E, Anafy A, Yerushalmy-Feler A, et al. Effect of coffee and tea consumption on adolescent weight control: an interventional pilot study. Child Obes. 2023;19(2):121–9.
100. Ye L, Wang X, Konno T, Gong X, Ding H, Yan H, et al. Fat reducing effects of Nelumbo nucifera leaf extract in overweight patients. Nat Prod Res. 2022;36(18):4776–81.
101. Hellmann PH, Bagger JI, Carlander KR, Forman J, Chabanova E, Svenningsen JS, et al. The effect of curcumin on hepatic fat content in individuals with obesity. Diabetes Obes Metab. 2022;24(11):2192–202. https://doi.org/10.1111/dom.14804.
102. Campiche R, Le Riche A, Edelkamp J, Botello AF, Martin E, Gempeler M, et al. An extract of Leontopodium alpinum inhibits catagen development ex vivo and increases hair density in vivo. Int J Cosmet Sci. 2022;44(3):363–76. https://doi.org/10.1111/ics.12783.
103. Dodd F, Kennedy D, Wightman E, Khan J, Patan M, Elcoate R, et al. The chronic effects of a combination of herbal extracts (Euphytose ®) on psychological mood state and response to a laboratory stressor: a randomised, placebo-controlled, double blind study in healthy humans. J Psychopharmacol. 2022;36(11):1243–56. https://doi.org/10.1177/02698811221112933.
104. Cao M-Z, Wei C-H, Wen M-C, Song Y, Srivastava K, Yang N, et al. Clinical efficacy of weight loss herbal intervention therapy and lifestyle modifications on obesity and its association with distinct gut microbiome: a randomized double-blind phase 2 study. Front Endocrinol (Lausanne). 2023;14:1054674. https://doi.org/10.3389/fendo.2023.1054674/full.
105. Sudeep HV, Aman K, Jestin TV, Shyamprasad K. Aframomum melegueta seed extract with standardized content of 6-paradol reduces visceral fat and enhances energy expenditure in overweight adults—a randomized double-blind, placebo-controlled clinical study. Drug Des Devel Ther. 2022;16:3777–91; https://www.dovepress.com/aframomum-melegueta-seed-extract-with-standardized-content-of-6-parado-peer-reviewed-fulltext-article-DDDT.
106. Sirvent P, Chavanelle V, Otero YF, Bargetto M, Le Joubioux F, Boisseau N, et al. TOTUM-63, a plant-based polyphenol-rich extract, improves glycaemic control in subjects with prediabetes or early stage newly-diagnosed type 2 diabetes in a randomized, double-blind, placebo-controlled trial. Diabetes Obes Metab. 2022;24(12):2331–40.
107. Rondanelli M, Gasparri C, Perna S, Petrangolini G, Allegrini P, Fazia T, et al. A 60-day green tea extract supplementation counteracts the dysfunction of adipose tissue in overweight post-menopausal and class I obese women. Nutrients. 2022;14(24):5209; https://www.mdpi.com/2072-6643/14/24/5209.
108. Langade D, Thakare V, Kanchi S, Kelgane S. Clinical evaluation of the pharmacological impact of ashwagandha root extract on sleep in healthy volunteers and insomnia patients: a double-blind, randomized, parallel-group, placebo-controlled study. J Ethnopharmacol. 2021;264:113276.
109. Pachikian BD, Copine S, Suchareau M, Deldicque L. Effects of saffron extract on sleep quality: a randomized double-blind controlled clinical trial. Nutrients. 2021;13(5):1473; https://www.mdpi.com/2072-6643/13/5/1473.
110. Moon S-Y, Jerng UM, Kwon O-J, Jung S-Y, Lee JY, Yoon SW, et al. Comparative effectiveness of Cheonwangbosimdan (tian Wang Bu Xin dan) versus cognitive-behavioral therapy for insomnia in cancer patients: a randomized, controlled, open-label, parallel-group, pilot trial. Integr Cancer Ther. 2020;19:1534735420935643; http://www.ncbi.nlm.nih.gov/pubmed/32627605.
111. Yang X, Xiao H, Zeng Y, Huang L, Ji K, Deng D, et al. Tianwang Buxin granules influence the intestinal flora in perimenopausal insomnia. Biomed Res Int. 2021;2021:9979511; http://www.ncbi.nlm.nih.gov/pubmed/34825005.
112. Pour ZS, Hosseinkhani A, Asadi N, Shahraki HR, Vafaei H, Kasraeian M, et al. Double-blind randomized placebo-controlled trial on efficacy and safety of Lactuca sativa L. seeds on pregnancy-related insomnia. J Ethnopharmacol. 2018;227:176–80; https://linkinghub.elsevier.com/retrieve/pii/S0378874117342046.
113. Ha H, Kim H, Lee K, et al. Efficacy of Polygonatum sibiricum on mild insomnia: a randomized placebo-controlled trial. Nutrients. 2019;11(8):1719; https://www.mdpi.com/2072-6643/11/8/1719.

114. Lopresti AL, Smith SJ, Metse AP, Drummond PD. Effects of saffron on sleep quality in healthy adults with self-reported poor sleep: a randomized, double-blind, placebo-controlled trial. J Clin Sleep Med. 2020;16(6):937–47; http://www.ncbi.nlm.nih.gov/pubmed/32056539.
115. Lee J, Jung H-Y, Lee SI, Choi JH, Kim S-G. Effects of Passiflora incarnata Linnaeus on polysomnographic sleep parameters in subjects with insomnia disorder. Int Clin Psychopharmacol. 2020;35(1):29–35; http://journals.lww.com/00004850-202001000-00004.
116. Shergis JL, Hyde A, Meaklim H, Varma P, Da Costa C, Jackson ML. Medicinal seeds Ziziphus spinosa for insomnia: a randomized, placebo-controlled, cross-over, feasibility clinical trial. Complement Ther Med. 2021;57:102657; http://www.ncbi.nlm.nih.gov/pubmed/33385511.
117. Haybar H, Javid AZ, Haghighizadeh MH, Valizadeh E, Mohaghegh SM, Mohammadzadeh A. The effects of Melissa officinalis supplementation on depression, anxiety, stress, and sleep disorder in patients with chronic stable angina. Clin Nutr ESPEN. 2018;26:47–52; http://www.ncbi.nlm.nih.gov/pubmed/29908682.
118. Wu C, Lee S-L, Taylor C, Li J, Chan Y-M, Agarwal R, et al. Scientific and regulatory approach to botanical drug development: a U.S. FDA perspective. J Nat Prod. 2020;83(2):552–62. https://doi.org/10.1021/acs.jnatprod.9b00949.

Market Overview of Herbal Medicines for Lifestyle Diseases

Bedanta Bhattacharjee, K. Sandhanam, Shatabdi Ghose, Dhunusmita Barman, and Ram Kumar Sahu

Abstract

The global herbal medicine industry is rapidly growing due to increasing consumer demand for natural and alternative healthcare options. Herbal medicines are valued for their antioxidant and anti-inflammatory properties, making them popular choices for addressing health-related issues and promoting overall well-being. The market includes a variety of products, such as pharmaceuticals, functional foods, dietary supplements, and beauty products, all of which are sought after for their nontoxic nature, fewer side effects, and affordability. Despite the promising opportunities, the industry faces challenges related to safety, efficacy, and regulatory considerations. To succeed in this market, herbal market players must focus on educating consumers, investing in research, and providing evidence-based products that meet the needs of health-conscious individuals. Quality control and standardization are crucial for maintaining the consistent quality of herbal products and differentiating them in a competitive market. Collaborating with healthcare professionals can help integrate herbal medicine into mainstream healthcare, gaining acceptance through recommendations and prescriptions. Personalization and customization are important to attract health-conscious consumers seeking targeted solutions for their specific lifestyle-related health issues. Leveraging e-commerce and digital marketing can broaden the reach of herbal market players globally. Sustainability and ethical sourcing resonate with environmentally conscious consumers, enhancing the appeal of herbal products. Exploring international markets, especially in developing countries with a deep-rooted herbal medicine culture, presents significant growth opportunities. Despite challenges and regulations, the herbal medicine market has vast growth potential due to increasing global demand for natural health solutions, the prevalence of lifestyle diseases, growing awareness, research validation, and integration with conventional healthcare. By adhering to regulations, promoting evidence-based practices, and investing in research and quality control, herbal market players can shape the future of healthcare and wellness, contributing to a healthier and more sustainable approach to healthcare.

Keywords

Herbal medicine industry growth · Safety and regulatory challenges · Personalization and customization in herbal products · Global demand for natural health solutions

B. Bhattacharjee
School of Pharmaceutical Sciences, Girijananda Chowdhury University- Tezpur Campus, Tezpur, Assam, India

K. Sandhanam
Department of Pharmacology, SRM College of Pharmacy, SRM Institute of Science and Technology, Chengalpattu, Tamil Nadu, India

S. Ghose
School of Pharmaceutical Sciences, Girijananda Chowdhury University- Guwahati Campus, Guwahati, Assam, India

D. Barman
Department of Pharmacology, NEF College of Pharmaceutical Education and Research, Nagaon, Assam, India

R. K. Sahu (✉)
Department of Pharmaceutical Sciences, Hemvati Nandan Bahuguna Garhwal University (A Central University), Chauras Campus, Tehri Garhwal, Uttarakhand, India

1 Introduction

The herbal medicine industry is experiencing rapid growth worldwide, driven by various factors. Many countries are expanding their inclusion of herbal remedies, responding to purchaser demand for organic goods. Simultaneously, budget cutbacks in the healthcare sector and rising prices of chemically synthesized pharmaceutical medications are two factors in this increase. The significance of medicinal herbs in mainstream healthcare is on the rise, driven by a growing

A. K. Dhara, S. C. Mandal (eds.), *Role of Herbal Medicines*, https://doi.org/10.1007/978-981-99-7703-1_30

number of individuals who are actively seeking safer and more natural remedies and healthcare approaches [1].

According to several market research findings, medicinal herbs, often referred as botanical medicines, play a significant role in addressing health-related issues and promoting overall well-being. They are regarded internationally for their antioxidant, anti-inflammatory, and pharmacological characteristics and are used under a variety of titles, including dietary herbal supplements, herbal drugs, psychotherapy drugs, and complementary treatments. Treatments for chronic diseases including (HIV) Human Immunodeficiency Virus infection / (AIDS) Acquired Immune Deficiency Syndrome, Alzheimer's disease, Tumor, Parasites, and Yellow fever are increasingly using natural therapies. To ensure their safety, quality, and efficacy, herbal medicines are subjected to scientific evaluation [2]. During the preparation process, plant species are converted into standardized active constituents, including phenolics, terpenoids, alkaloids, flavonoids, and lignans, through various extraction methods.

Herbal medicine includes dietary supplements derived from various plant components, such as roots, berries, leaf, wood, nuts, and oil. Its use dates back to ancient times, when it was employed to treat various ailments. Today, thanks to innovative techniques, herbal medicines are available in diverse dosage forms. Herbal drugs, herbal healthy food, herbal dietary supplements, and natural herbal goods are the different segments of the market. These goods are increasingly recognized as being mostly nontoxic, showing fewer side effects, being more compatible with physiological flora, and being readily available at reasonable rates. These factors are principally responsible for the worldwide trend of rise in the usage of these products. These herbal remedies may also include minerals, seashells, live animals, as well as fungi and honeybee products. Phytomedicine and botanical medicine are two more names for herbal medicine [3].

Lifestyle diseases encompass a range of health conditions, including heart disease, obesity, type 2 diabetes, stroke, chronic obstructive pulmonary disease (COPD), asthma, and osteoporosis. Among these, heart disease is a major global health concern, responsible for approximately 31% of all deaths worldwide. The World Health Organization estimates that over 17.9 million people succumb to this disease each year. Early detection and treatment of heart disease significantly improve the chances of avoiding its severe later stages. Obesity, characterized by excess body weight, poses a significant risk for various health disorders, including cardiovascular diseases. An increase in body mass index can elevate the risk of noncommunicable diseases. Type 2 diabetes is a growing problem both in the UK and worldwide, imposing a substantial burden on those affected. Behavioral factors, notably a poor diet and a lack of physical activity, play a crucial role in its development. Evidence suggests that modifying these behaviors can reduce the incidence of type 2 diabetes in individuals at risk. Stroke, a neurological condition, results from diminished blood circulation or blockage inside the head, blood vessels, resulting to cerebral hemorrhage or cerebrovascular tissue damage. It accounts for one in six fatalities from cardiac illnesses and is a major contributor to long-term physical death from cardiovascular disease. Stroke incidence and prevalence have dramatically increased in the last few decades, with fatalities increasing by greater than 40% worldwide [4]. Stroke cases will continue to be impacted by the treatment of stroke-related risk factors, such as high blood pressure, arrhythmia, high cholesterol, and diabetes, as well as the possible interaction risk of acute ischemic stroke and COVID-19 [5].. Taking cardiovascular medications has been associated with reduced risks of coronary heart disease, stroke, and all-cause mortality [6]. However, adherence to medication remains an issue, with approximately 40–50% of patients not following their prescribed cardiovascular treatment plans. Enhancing medication adherence can greatly benefit individuals at risk of stroke and other cardiovascular complications.

COPD ranks as the world's third-leading causes of morbidity and mortality, accounting for 3.23 million deaths in 2019. The majority of COPD-related deaths in individuals under 70 years old occur in nations with low and medium incomes (LMIC). Although COPD is not curable, its condition can be improved by abstaining from smoking, avoiding air pollution, and receiving vaccinations [7, 8].Asthma and allergic diseases can originate in childhood and persist throughout life, but they may also manifest later in life for reasons that are not yet fully understood. These chronic multifactorial respiratory conditions are significant public health concerns for Europe and the United Nations, emphasizing the importance of prevention, early diagnosis, and treatment. As allergy triggers, such as infections, urbanization leading to biodiversity loss, pollution, and climate change, are not anticipated to change in the foreseeable future, there is a critical need to develop, strengthen, and optimize preventive and treatment strategies. Presently, effective treatments for asthma exist, and several risk factors (e.g., allergies, rhinitis, and tobacco smoke) are well-known, along with developed tools to control the disease. Osteoporosis is a bone condition that develops when the structure and strength of bones alter or when bone mineral density and mass decline. This disorder may result in weaker bones, which increases the risk of cracks (broken bones).Osteoporosis is sometimes described as a "silent" illness since it frequently goes unnoticed until a bone fracture occurs and usually has no symptoms. It is a leading cause of fractures in postmenopausal women and older men, with fractures most commonly occurring in the hip, spine vertebrae, and wrist [9].

The global herbal medicine market has experienced significant growth, with a value of USD 151.91 billion in 2021 and projected growth to reach USD 437.59 billion by 2030.

Factors such as increased adoption of products with natural ingredients, rising consumer awareness, and advancements in biotechnology are driving market growth. The market is expected to continue its upward trajectory, reaching USD 347.50 billion by 2029, with a compound annual growth rate (CAGR) of 11.16% during the forecast period. The research found that 62.9% of the individuals surveyed used medicinal herbs based on biological therapy, making them the most common kind of alternative therapy. Commonly used medicinal herbs include bitter gourd, aloe vera, and others. Indian/Ayurvedic medicine was the leading alternative system, with 29.8% of patients opting for it. About one-fifth of the sick people also used religious treatment, psyche techniques, and regional or traditional cures.

The most often utilized alternative medication for diabetic patients included bitter gourd, cardamom, aloe vera, and others based on biological treatments. The use of multiple alternative medications was indicated by around 50% of diabetic patients. Nearly one-fourth of high blood pressure patients said they used herbal remedies, based on biological treatments, and other methods [10].Among patients with hypertension, asafetida, giloy, bitter gourd, ginger, fenugreek, and ginger were often utilized herbal remedies or biologically based treatments. Patients with cardiac asthma and other respiratory illness often utilized medicinal herbs and biologically based treatments, but patients with chronic lung disease were more likely to use homoeopathic and regional treatments. (Table 1).

Table 1 List of herbal drugs with their impact on lifetime diseases

Sl. No.	Botanical Name	Common Name	Medicinal Uses	Medicinal Plant
1.	*Gymnema Sylvestre*	Australian cow plant	Antidiabetic, anti-asthma, and dyslipidemic agent	
2.	*Crataegus oxyacantha*	Hawthorn	Treatment of CHF, antidiabetes, and antianxiety	
3.	*Echinacea purpurea*	Purple coneflower	Treatment of UTI, arthritis, and bronchitis	
4.	*Panax ginseng*	Asian ginseng	Relieves cardiovascular and respiratory disorders, antianxiety, antidepressant, and menopausal hot flashes	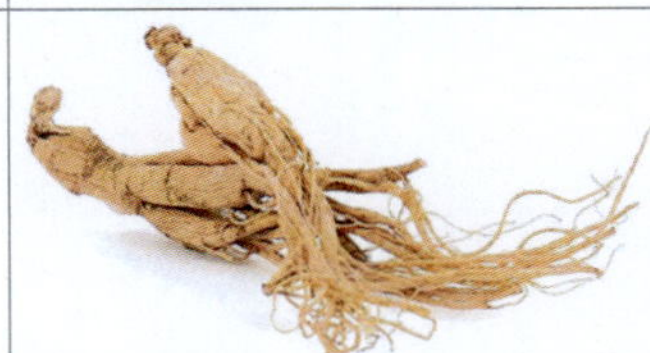
5.	*Taraxacum officinale*	Dandelion root	Treatment for liver and stomach conditions, diuretic, and reduces cancer risk	
6.	*Allium sativum*	Garlic	Atherosclerosis, Hyperlipidemia, and systolic hypertension	

(continued)

Table 1 (continued)

Sl. No.	Botanical Name	Common Name	Medicinal Uses	Medicinal Plant
7.	*Aesculus hippocastanum*	Horsechest nut	Venous insufficiency	
8.	*Salvia miltiorrhiza*	Red sage or Danshen	Angina pectoris	
9.	*Cissus quadrangularis*	Veld grape	Obesity	
10	*Barberry vulgaris*	Berberis	Treatment of arrhythmia	
11.	*Nerium oleander*	Rose Bay Oleander	Treatment of cardiovascular diseases, asthma, and diabetes	
12	*Rosmarinus officinalis*	Rosemary	Treatment of cervical cancer	

The worldwide market for herbal medicines is anticipated to be worth US$135.8 billion in 2022 in the post-COVID-19 corporate environment, and it is expected to hit US$248.6 billion by 2030, rising at a CAGR of 7.9% from 2022 to 2030 [11]. By the conclusion of the study period, the herbal medicinal products market is anticipated to grow at an 8.1% CAGR and reach US$106.6 billion.In addition, understanding the continuing post-pandemic revival, it is expected that the herbal health food market would register a revised 7.8% CAGR over the following 8 years. States' market for ayurvedic medicines is estimated to be worth US$24.6 billion in 2022, while China, as the second-largest economy, is forecast to reach US$50.2 billion by 2030, with a CAGR of 10.3% from 2022 to 2030 [11].Japan and Canada are two more noteworthy geographic markets, with growth rates of 6.8% and 6.9%, respectively, predicted for the years 2022–2030. Germany is projected to expand in Europe at a CAGR of around 6.3%. The Asia-Pacific region, led by countries like Australia, India, and South Korea, is forecast to reach US$40.3 billion by 2030. This chapter provides an in-depth analysis of herbal medicines, encompassing the market overview, its scope, key opportunities for herbal market participants, leading industry players, and the dynamics driving the herbal medicine market.

2 Overview of Complementary and Alternative Medicines (CAM)

Healthcare products and procedures beyond the scope of conventional medical treatment are alternative and supplementary medicine (CAM). Plant-based products and dietary goods, including herbal remedies, nutritional supplements, and vitamins, are widely used in CAM therapies [12].

CAM treatments can be categorized into numerous groups:

- **Traditional alternative medicine:** Acupuncture, homoeopathy, and Chinese treatments are examples of therapies that fall under the heading of this therapies and have been utilized for centuries all throughout the globe [13].
- **Mind-body medicine**: This form of CAM works to improve health by using the mind-body link (meditation, tai chi, and yoga) [13].
- **Manipulative and body-based practices**: Therapies like massage treatment and chiropractic care fall under this type since they physically manipulate the body [14].
- **Energy medicine**: This kind of CAM employs energy fields to advance health (touch therapy and Reiki) [13].

CAM is often utilized by individuals with arthritis, and the reasons for doing so include symptom alleviation, enhancing general well-being, and getting disappointed with traditional therapies, according to a critical evaluation of the topic. According to a comprehensive study of CAM usage across the general public and medical professionals, there has been a steady rise in CAM use throughout time. The general population usually views CAM favorably, although medical experts have differing views on its therapeutic efficacy [15].

2.1 Unani Medicines

Unani medicine is an ancient Greek method of treatment and health management that was subsequently refined and improved by Muslim scholars and practitioners [16]. It is designed to cure the spirit, mind, and body and depends on the principle of the four senses of humor: black bile, yellow bile, phlegm, and blood [17]. One of the most traditional and recognized types of CAM is Unani Tibb. The Unani System of Medicine is a holistic medical system that extensively examines numerous conditions of health and illness [17]. In a study, an Unani herbal decoction's therapeutic and comforting effects on COVID-19 and related respiratory viral diseases were investigated. The phytometabolite in the decoction was proven to have immunomodulatory and antiviral properties [18].

2.2 Ayurvedic Medicine

Ancient Indian medical practice known as Ayurveda emphasizes on a natural, comprehensive strategy for maintaining good bodily and psychological well-being [19]. Ayurvedic medicine includes items (mostly made from plants, but sometimes occasionally from minerals and animals) and promotes physical activity and modifications to lifestyles. Ayurvedic medicine is principally based on the ideas of global interconnection, the body's foundation (Prakriti), and life energies (Doshas). Ayurvedic therapy aims to help the patient by lowering anxiety, relieving symptoms, removing contaminants, relieving symptoms, strengthening resistance to disease, and improving life harmony [19].

2.3 Homeopathy Medicine

Homeopathy is a type of complementary and alternative medicine (CAM) that uses specially formulated, very diluted remedies in an effort to promote the body's natural healing reaction to illness. As in late 1700s, Samuel Hahnemann founded it in Germany [20].Homeopathy should not be used to treat chronic, serious, or potentially serious medical conditions, according to the National Health and Medical Research Council (NHMRC) of Australia [21].

2.4 Siddha Medicine

One of India's oldest healthcare systems is called Siddha, an ancient Indian curative approach that began in South India. The alternative medical systems area includes Siddha medicine as a complementary and alternative medicine (CAM) approach [22]. The Siddha system of medicine is renowned for its effectiveness in treating a number of lifestyle disorders including asthma, obesity, hypertension, and stroke using herbal treatments [23].The ancient philosophical principles of the Siddha system of medicine are still explored and used today by its practitioners.

3 Uses of Herbal Medicine in The Management of Lifestyle Diseases

The list of herbal medicinal plants which are having promising actions against lifestyle diseases is outlined in Table 2.

Table 2 List of herbal medicinal plants for the treatment of lifestyle diseases

Scientific Name	Common Name	Family	Plant Parts Used	Pharmacological Dose	Pharmacological Actions	References
Heart disease						
Digitalis purpurea	Foxglove	Scrophulariaceae	Dried leaves	0.125–0.25 mg	Used as cardiotonic with action by increasing cardiac contractility and decreasing excitability, plasma renin activity, serum aldosterone, and norepinephrine levels	[24]
Drimia maritima	Sea onion	Asparagaceae	Dried slices of bulb	Daily oral doses for adults range from 30–500 mg	Used as cardiotonic via inhibition of Na^+/K^+-ATPase pump and show positive ionotropic activity	[25]
Annona cherimola	Cream Fruit	Annonaceae	Leaves	200 mg/kg and 500 mg/kg body weight (BW)	Increase in HDL-cholesterol and decrease in plasma triglycerides, and LDL-cholesterol levels	[26]
Curcuma longa	Turmeric	Zingiberaceae	Rhizome	100 mg/kg BW	Lowering blood pressure reduces the symptoms of hypertension, and by preventing the production of thromboxane A2, it may enhance myocardial trophic blood flow, blood viscosity, and thrombosis development (TXA2)	[27]
Obesity						
Camellia sinensis	Green tea	Theaceae	Leaves	600-900 mg/day (12 weeks)	It acts on adipose tissue and inhibit ghrelin secretion	[28]
Hibiscus sabdariffa	Roselle	Malvaceae	Flowering plants	150 mg/kg BW	It causes a reduction in body weight and inhibition of lipid accumulation and suppression of adipogenesis	[29]
Hypericum perforatum	St. John's Wort	Hypericaceae	Flowering plants	100 and 200 mg/kg BW	Helps in weight control and inhibiting lipase activity	[30]
Asthma						
Aerva javanica	Bui	Amaranthaceae	White axillary bunches of small woolly flowers	30 and 60 mg/kg BW doses orally	Mast cell degranulation in mice possesses antiasthmatic activity	[31]
Ageratum conyzoides	Goat weed	Asteraceae	Herb	250, 500 and 1000 mg/kg BW	It causes smooth muscle relaxation and prevents mast cell release of histamine	[32]
Asystasia gangetica	Chinese violet	Acanthaceae	Leaves	50 mg/kg BW	Inhibited the contraction evoked by spasmogens	[33]
Type 2 Diabetes						
Morinda citrifolia	Noni	Rubiaceae	Fruit juice	0.2–3.6 mL/kg twice day	It acts through insulin secretagogue activity	[34]
Aloe barbadensis	Aloe	Liliaceae	Leaves	200 and 400 mg/kg BW	Decrease hepatic glucose production	[35]
Vachellia nilotica	Babool	Fabaceae	Bark	100 mg/kg BW	It helps in release of insulin from pancrease	[36]
Stroke						
Tectona grandis	Teck	Verbenaceae	Leaves	50 mg/kg BW	Prevents clot formation	[37]
Ocimum gratissimum	Clove basil and African basil	Lamiaceae	Leaves	–	Prevents blood flow blockages	[38]

Scientific Name	Common Name	Family	Plant Parts Used	Pharmacological Dose	Pharmacological Actions	References
Hypertension						
Veratrum album	White helleborine	Liliaceae	Rhizome	–	It showed antioxidant activity, enhanced nitric oxide production and release, and depleted biogenic amine levels	[39]
Allium cepa	Bulb onion	Liliaceae	Leaves	150 mg/kg BW	It showed reactive oxygen species (ROS) scavenging activity and inhibited angiotensin converting enzyme	[39]
Rauvolfia serpentina	Indian Snakeroot	Apocynaceae	Roots	100 mg/kg BW	It showed vasodilation properties, antioxidant and blocked Ca channel	[40]
Chronic Obstructive Pulmonary Disorders						
Acasia farnesiana	Cassia Flower	Mimoseaceae	Roots and Bark	50 mg/kg BW	Relax airway smooth muscle and cause bronchodilation via increasing intracellular concentration of cyclic adenosine monophosphate	[41]
Chelidonium majus	Greater Celandine	Papaveraceae	Aerial parts and roots	100 and 200 mg/kg BW	Relax airway smooth muscle and cause bronchodilation via increasing intracellular concentration of cyclic adenosine monophosphate	[41]
Malva sylvestris	Mallow	Malvaceae	Flower	–	Relax airway smooth muscle and cause bronchodilation via increasing intracellular concentration of cyclic adenosine monophosphate	[41]
Osteoporosis						
Carthamus tinctorius	Safflower	Asteraceae	Seed	–	Promote bone formation	[42]
Vaccinium angustifolium	Lowbush blueberry	Ericaceae	Seed	–	Inhibit bone resorption	[42]

4 Herbal Medicines Market Dynamics

During the last few decades, there has been a significant global surge in demand for medicinal herbs. In contrast to prior to the pandemic, the COVID-19 pandemic has increased the need for herbal treatment globally beyond what was predicted. The use of herbal treatments has dramatically increased over the years. According to a report observed in the National Centre for Biotechnology Information (NCBI), throughout the past 3 years, the consumption of herbal supplements and therapies has grown significantly, with at least 80% of the population utilizing them globally [43].A study published in PMC found that the prevalence of herbal medicine use in United States adults is higher than the estimates of other national studies, with 35% of respondents reporting current use of at least one herbal medicine [44]. Herbal supplement sales in the United States surpassed $10 billion for the first time in 2020, totaling an estimated $11.261 billion, according to the Nutrition Business Journal (NBJ). Despite the growing popularity of herbal medicines, there are concerns about their safety and efficacy. A review paper on MDPI addresses the approaches and recent trends for determining the possible toxicity of herbal medicines [45]. The nutraceuticals market is vast, encompassing many different products with inconsistent levels of evidence available to support their use. Globally, there has been an increase in the demand for medicinal herbs, with the United States being the biggest market, followed by Europe, Australia, and Canada [46]. The global herbal medicine market is projected to grow from USD 165.66 billion in 2022 to USD 347.50 billion by 2029, exhibiting a CAGR of 11.16% with major market players like Pharmaceutical company Hishimo Pvt. Ltd., Arizona Natural Products, Venus Pharma GmbH, Himalaya Wellness Company, and Bayer AG, Blackmores Ltd. About 960 plant species are employed in India by the herbal business, with 178 of them having large volume uses that surpass 100 metric tons annually [46]..

5 Global Herbal Medicines Market for Management of Lifestyle Diseases

Of the 20,000 herbal (herbal) plants that were found, only 7000–7500 are really used to cure different diseases in the Indian traditional healthcare system. Additionally, Indian traditional methods use around 25,000 plant-based medications. Noncommunicable diseases known as "lifestyle diseases" have been related to unhealthy eating habits, drinking too much alcohol, smoking, sedentary behavior, and other variables. One of the world's epidemics, obesity has become a social problem as a result of the numerous health hazards it poses. One of the main factors contributing to the emergence of metabolic illnesses including diabetes, hypertension, and cardiovascular conditions is obesity. The International Diabetes Federation estimates that 25% of adults worldwide have metabolic syndrome, and that number is expected to rise over the next few decades [47]. Due to their relative lack of adverse effects, traditional medical systems have become increasingly popular, especially in developing nations. These systems are essential to the broader healthcare system [48].

6 Market Analysis and Size

6.1 The Cost of Herbal Medicines on The Market

According to estimates, the Vedic medicine drugs are growing at a 20% yearly rate. China and India are two of the countries that employ medicinal herbs the most. India utilizes over 7000 plant species, whereas Traditional Chinese Medicine uses about 5000. The worldwide market for the trade of medicinal plants is growing at a rate of 7% yearly, according to Export-Import Bank. The use of medicinal plants as beginning points to the manufacture of medications or as models because pharmacologically active chemicals make them crucial for pharmacological research and drug discovery. This is true despite whether plant substances are used directly as medicines.

6.2 Indian Herbal Market

There are said to be 25,000 Indian systems of medicine-licensed pharmacies in the country. Approximately 1000 single medications and 3000 compound formulations have been registered as of this writing. India's herbal business employs 8000 or more medicinal herbs [49]. However, only around 25 of India's approximately 8000 medicine manufacturers may be categorized as large-scale producers. Homoeopathic and Ayurveda medicine generated about $27.7 million USD in yearly revenue from the Indian herbal sector. Again reaching US $ 31.7 million in 1998–1999, the overall turnover of Ayurvedic and herbal goods reached US $ 48.9 million in 1999–2000. India exports herbal medicines for about $80 million [49]. According to their turnover, some of the most popular therapeutic herbs are included in Table 3. The graphical depiction of various popular Indian medicinal herbs vs estimated annual consumption (Tons) is shown in Fig. 1 [50].

Table 3 Indian medicinal plant with reference to sales

Sl. No.	Common name	Part use	Plant name	Consumption (tons)
1.	Aloes	Leaf	*Aloe barbadensis*	200
2.	Kalmegh	Aerial part	*Andrographis paniculata*	250
3.	Satavatri	Root	*Asparagus recemosus*	500
4.	Daru haldi	Root	*Berberis aristata*	500
5.	Guggul Gum	Resin	*Commipphora wrightii*	500
6.	Jatamansi	Root	*Nardostachy gradiflora*	200
7.	Black pepper	Fruit	*Piper nigrum*	150
8.	Chirayita	Whole plant	*Swertia chirayta*	300
9.	Clove	Flower bud	*Syzygium aromaticum*	150
10.	Jaman beej	Seed	*Syzygium cumini*	300
11.	Harar	Fruit	*Terminalia chebula*	500
12.	Guduchi	Stem	*Tinospora cardifolia*	1000
13.	Ginger	Rhizome	*Zingiber officinalis*	500

7 Herbal Medicines Market Regional Analysis/Insights

7.1 Global Herbal Market

Herbs are utilized in a variety of items, including meals, energy drinks, multivitamins, and raw form. Profit is the primary motivator behind the marketing and sales of various herbs in some nations. Herbs, especially those that may be cultivated locally or regionally, can be a significant part in treating specific ailments or diseases more affordably in other nations [51].

7.2 Worldwide Herbal Trade

With an average yearly growth rate of 6.4%, the sale of herbal medications is anticipated to increase. The market for herbal medicines has expanded significantly on a global scale as a result of the contributions of various important elements. Some of these include: customers' desire for natural remedies; worry about unfavorable side effects of contemporary medications; and the notion that ayurvedic

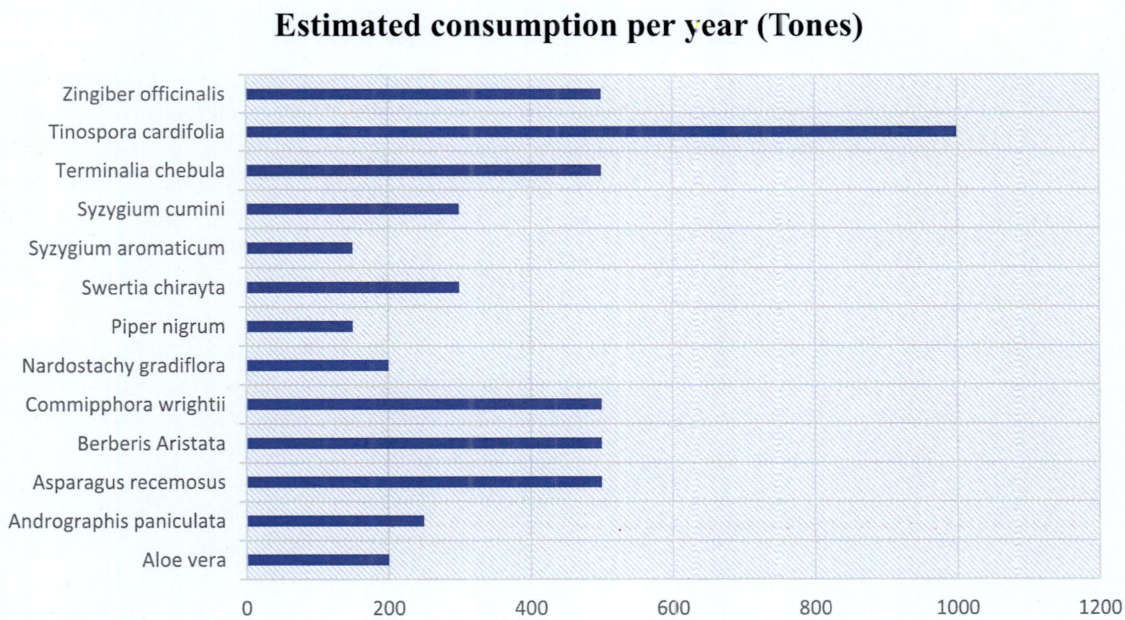

Fig. 1 Indian medicinal plant with reference to sales

pharmaceuticals have no adverse effects because several millions of people worldwide have used them for thousands of years. There is a lot of interest in alternative medicine, a preference for preventive medicine due to the aging of the population, the belief that herbal medicines may be useful in treating some diseases where conventional therapies and medications have been found to be ineffective, a propensity for self-medication, improved quality evidence supporting the safety and effectiveness of herbal medications, and the high cost of artificial drugs [52]. According to estimates from the World Health Organization, the worldwide market for herbal medicines is worth around US$43 billion each year. The herbal medicine industry in Europe was estimated to be worth $6 billion in 1991, with Germany accounting for $3 billion, France for $1.6 billion, and Italy for $0.6 billion, while the market in other nations was at $0.8 billion. The herbal medicine industry was worth around $10 billion in 1996 in European nations, $four million in the USA, $1 billion in India, and $5 billion in other nations. The European market alone was worth roughly $ 7.0 billion in 1997. Nearly half of the $3.5 billion European market is made up of the German market. The next biggest markets are France ($1.8 billion), Italy ($700 million), the United Kingdom ($400 million), Spain ($300 million), and the Netherlands ($100 million) after this one (Table 4) [49].

Table 4 Percent growth of the Pharmaceutical market and ranking of countries

SL. No.	Country	Market in USD		% Growth	Rank
		2005	2015		
1.	Brazil	9	20	122	9
2.	Canada	13	25	92	7
3.	China	13	38	192	2
4.	France	32	46	44	4
5.	Germany	31	38	23	12
6.	India	6	20	233	5
7.	Italy	20	25	25	14
8.	Japan	68	82	21	3
9.	Mexico	10	19	90	10
10.	South Korea	8	15	88	13
11.	Spain	14	25	79	8
12.	Turkey	7	15	114	11
13.	UK	19	32	68	6
14.	USA	248	444	79	1

Source: IMS World Review, Mckinsey Pharma Model (Indian Pharmaceuticals 2015; IMS Health Information; and Consulting Services India Pvt. Ltd)

8 Herbal Medicines Hold Major Market Share

The discovery of novel pharmaceuticals focuses significantly on herbal remedies. WHO reports that around 25% of current pharmaceuticals have been developed from plants used in traditional treatment. Numerous of them are synthetic derivatives based on plant-isolated prototype molecules. Furthermore, the WHO has officially acknowledged herbal medicine as a vital aspect of basic healthcare [53]. Contemporary pharmacotherapy has been completely transformed by plant-based medications. A significant breakthrough in managing hypertension and lowering blood pressure was made in 1953 with a molecule called serpentine discovered from the root plant *Rauwolfia serpentine* [54]. According to reports, between 1950 and 1970, the US pharmaceutical sector introduced around 100 novel plant-based pharmaceuticals, such as vincristine, vinblastine, reserpine, rescinnamine, and deserpidine. Between 1971 and 1990, novel therapeutics were discovered that were derived from plants, including ginkgolides, artemisinin, Z-guggulsterone, plaunotol, nabilone, teniposide, E-guggulsterone, and etoposide. More plant-based pharmaceuticals, such as irinotecan, gomisin, topotecan, and paclitaxel, entered the pharmaceutical market between 1991 and 1995.Colchicine, atropine, ephedrine, aspirin, digoxin, and quinine are just a few of the pharmacological molecules isolated from plant origin that have been identified by multiple investigators in the last few decades [55].The utilization of herbal medicine is presently expanding rapidly throughout industrialized nations as well as remaining prominent in the majority of underdeveloped nations. According to projections, 70% of all doctors in Germany and France routinely prescribe prescriptions for herbal remedies. Furthermore, an exponential increase of patients have switched to herbal treatments [55]. Approximately 30%–50% of all drugs consumed in China are still traditional herbal remedies. According to estimates, more than half of the population in industrialized nations like North America and Europe have taken herbal remedies at least once in their lifetimes. It is reported that 158 million people in the United States utilize herbal medications, and this number is steadily escalating. In summary, the herbal medicine industry worldwide is about US$60 billion yearly and is steadily expanding. Interestingly, older people are more likely to consume herbal and conventional drugs equally than younger demographics. Although chronic illnesses are more prevalent in this demographic, their long-lasting adverse effects frequently prevent long-term usage of advanced standard pharmacological regimens. On the other hand, long-term usage of herbal medications offers treatment without adverse effects. Considering such demands, herbal medicine has attracted attention and is gaining more rapid popularity globally [53].

9 Key Industry Players

9.1 Patanjali Ayurved

Patanjali Ayurved, a global conglomerate holding company, is headquartered in Haridwar, India. Established in 2006 by Balkrishna and Ramdev, the company operates its office in Delhi, while its primary offices and manufacturing facilities are located in the Haridwar industrial area. Patanjali specializes in the production of food, cosmetics, personal care products, and ayurvedic medicines.In 2016, Patanjali experienced one of the highest growth rates among Indian FMCG companies, as reported by CLSA and HSBC. The company's valuation at that time was 3000 crores (equivalent to 43 billion rupees or 540 million US dollars in 2023). For the fiscal year 2016–17, Patanjali anticipated its annual income to reach Rs. 10,216 crore (approximately $1.3 billion).The key manufacturing facility for Patanjali is situated at the Patanjali Food and Herbal Park in Haridwar.

With a 35,000 crore (470 billion or 5.8 billion US dollars) manufacturing capacity, it is increasing to 60,000 crores (860 billion or 11 billion US dollars) in 2023 by building new production facilities in Noida, Nagpur, and Indore [56].

9.2 Herbalife Nutrition Ltd.

An international multilevel marketing (MLM) business that develops and distributes nutritional supplements is known as Herbalife Nutrition Ltd., also known as Herbalife International, Inc. and Herbalife International of America. Mark R. Hughes established the business in 1980 and there are reportedly 9900 workers employed by it globally. Despite having its headquarters in Los Angeles, California, the business was established in the Cayman Islands. Herbalife Nutrition's products include protein bars, teas, aloes, vitamins, and sports hydration, energy, and personal care goods in addition to smoothies with protein for losing weight. The brand's first offering is the soy-based meal replacement smoothie known as Formula 1. The product made its debut in 1980 and was the best-selling item for the company as of 2015, making up around 30% of overall sales [57].

9.3 Nutraceutical Corporation

An integrated manufacturer, marketer, distributor, and retailer of branded nutritional supplements and other natural goods, Nutraceutical, a Better Being company, sells its products largely to and via domestic health and natural food stores. To and via distributors and retailers of natural and health goods, Nutraceutical advertises and distributes branded nutritional supplements across the world. The primary objective of Nutraceutical is to buy, run, and merge natural goods companies that produce, sell, and distribute branded nutritional supplements. Solaray®, KAL®, Nature's Life®, LifeTime®, Natural Balance®, bioAllers®, Herbs for Kids®, NaturalCare®, Health from the Sun®, Life-flo®, Organix South®, Pioneer®, and Monarch NutraceuticalsTM are just a few of the brands that Nutraceutical produces and sells [57].

9.4 Dabur Ltd.

Dabur Ltd., a leading global Indian consumer products corporation, was founded by S. K. Burman and is headquartered in Ghaziabad. The company is renowned as one of India's largest fast-moving consumer goods (FMCG) firms, specializing in the production of natural consumer goods and Ayurvedic medicines. The majority of Dabur's revenue, slightly over 60%, is generated from its consumer products division, while 11% comes from its food division, with the remaining earnings attributed to its foreign business section [58].

10 Key Opportunity for Herbal Market Players

A key opportunity for herbal market players lies in capitalizing on the growing consumer demand for natural and organic products. As people become more health-conscious and seek alternatives to conventional pharmaceuticals, herbal medicine presents a lucrative market for those who can effectively capitalize on this trend. As well as being environmentally aware, there is a significant market potential for herbal products that offer various health benefits and are sourced sustainably. By focusing on research, quality assurance, and innovative product development, herbal market players can position themselves to meet the evolving needs and preferences of consumers, and thereby expand their market share and revenue [59].

Here are some specific areas of opportunity for herbal market players:

10.1 Consumer Awareness and Education

Educating consumers about the benefits, safety, and appropriate usage of herbal remedies is crucial. Market players can invest in educational campaigns, informative websites, and product labeling to build trust and confidence among consumers.

10.2 Research and Product Development

Conducting rigorous scientific research on herbal remedies can help validate their efficacy and safety, leading to evidence-based products. Investing in research and product development will enable market players to offer innovative and reliable herbal solutions for lifestyle diseases.

10.3 Certification and Quality Standards

Obtaining certifications and adhering to quality standards can boost consumer trust and differentiate products in a crowded market. Seeking endorsements from regulatory bodies or third-party organizations can enhance the credibility of herbal products.

10.4 Collaboration with Healthcare Professionals

Partnering with healthcare professionals, such as doctors, naturopaths, and nutritionists, can open up channels for herbal medicine integration into mainstream healthcare. Building relationships with healthcare practitioners can lead to recommendations and prescriptions of herbal remedies.

10.5 Customization and Personalization

Offering personalized herbal products and formulations tailored to individual needs can appeal to health-conscious consumers seeking holistic and targeted solutions for their lifestyle-related health issues.

10.6 E-commerce and Digital Marketing

Embracing e-commerce platforms and digital marketing strategies will enable herbal market players to reach a broader audience and expand their market presence globally. The growth of e-commerce and online retail has opened up new avenues for herbal market players to reach a broader audience. Establishing a strong online presence and leveraging digital marketing strategies can help attract and engage potential customers globally.

10.7 Sustainability and Ethical Sourcing

Promoting sustainable practices and ethical sourcing of herbal ingredients can resonate with environmentally conscious consumers, further enhancing the appeal of herbal products.

10.8 Global Market Expansion

Exploring opportunities to expand into international markets with rising interest in herbal medicine can unlock significant growth potential for market players.

10.9 Collaboration with Traditional Medicine Practitioners

Partnering with traditional medicine practitioners from various cultures can provide valuable insights into herbal remedies and open doors to new markets and customer bases.

10.10 Integration with Modern Healthcare

Collaborating with pharmaceutical companies or healthcare institutions to develop combination therapies that integrate herbal medicine with conventional treatments can offer comprehensive solutions for lifestyle diseases.

10.11 Health Consciousness

Consumers are becoming more health-conscious and seeking products that contribute to their overall well-being. Herbal products are often perceived as healthier alternatives to conventional products due to their natural origins and potential therapeutic benefits. This health-conscious consumer base presents a significant opportunity for herbal market players to cater to a growing segment of the population that prioritizes wellness and preventive healthcare [60].

10.12 Safety and Nontoxicity

Herbal products are often perceived as safer and less likely to cause adverse effects compared to synthetic or chemically derived products. As consumers become more aware of the potential risks associated with certain chemicals and artificial additives, they are turning to herbal alternatives that are seen as gentler and less harmful to the body [45, 61].

10.13 Fewer Side Effects

Traditional medical systems have traditionally used herbal treatments, and many customers think they are less likely to have negative side effects. This perception is attractive to individuals who want to avoid the potential risks and complications associated with some pharmaceutical drugs.

By strategically capitalizing on these opportunities, herbal market players can position themselves as leaders in the rapidly growing herbal medicine industry and contribute to shaping the future of healthcare and wellness. Herbal market players have a unique opportunity to tap into the growing consumer demand for natural, safe, and sustainable products. By understanding consumer preferences, investing in research and quality assurance, and adapting to evolving market trends, herbal companies can position themselves for significant growth and success in the global market.

11 Challenges and Regulations

As the use of herbal medicinal products (HMP) becomes more widespread, there has been an increased awareness of the potential risks associated with these products, despite the public perception of them as relatively safe. These risks stem from various factors, including inherent toxicity of certain herbs, contamination, adulteration, misidentification of plants, and potential combinations with other prescription medicines or natural supplements. The evaluation of safety of HMPs involves examining adverse reaction cases and reviewing published toxicity information. However, this integrated investigation faces several challenges due to the limited volume and high standards of available data. Harmful impacts are often underreported, and the quality of the products themselves may not always meet ideal standards. Additionally, medicinal herbs have intricate chemical makeups, and there is a lack of comprehensive information on the toxicity of various herbs and the ingredients in them. Despite these challenges, there are opportunities to enhance the scientific evidence surrounding HMP safety. Adverse event data from national drug safety databases may be supplemented with information from more recent sources, such as poison control data. Additionally, more sophisticated methods of toxicological evaluation, such as predictive toxicology and genomes, might provide important insights. By integrating all available information, the uncertainty surrounding the evaluation of the safety of herbal medicines can be reduced. A prime example that underscores the challenges and opportunities in this field is the case of aristolochia and aristolochic acids, demonstrating the need for more accurate methods to determine the toxicity of herbal medicines [62].

Challenges and regulations in the herbal medicine market are complex and diverse due to the international nature of the industry, varying national policies, concerns about safety and efficacy, and the need for scientific evidence to support claims. Let's explore these challenges and regulations in detail:

11.1 International Diversity

One of the significant challenges in the herbal medicine market is the international diversity of herbal remedies and traditional practices. Different cultures and regions have their own traditional healing systems, which may involve the use of specific herbs or formulations. As a result, standardizing herbal products and practices across countries becomes difficult, leading to variations in quality, safety, and efficacy. Additionally, international trade and sourcing of herbal ingredients can be complicated by differences in regulatory requirements and quality control standards among countries. Market players may face challenges in navigating import and export regulations and ensuring compliance with different regional guidelines [63].

11.2 National Policies and Regulation

Herbal medicine regulations vary widely from one country to another. Some countries have well-established regulatory frameworks that govern the production, distribution, and marketing of herbal products. In contrast, others may have limited or no specific regulations in place for herbal remedies. The lack of harmonized international regulations can create barriers for market players looking to operate globally. It may require them to comply with different sets of rules and undergo separate registration processes in each market they enter.

11.3 Safety and Efficacy Concerns

Safety and efficacy are paramount concerns in the herbal medicine market. While herbal remedies are generally considered to be natural, they can still interact with conventional medications or cause adverse effects if not used appropriately. The lack of standardized dosages and formulations can also impact their safety.

Efficacy is another challenge, as some herbal products may lack sufficient scientific evidence to support their claimed benefits. The absence of robust clinical trials and scientific studies can lead to skepticism among healthcare professionals and consumers [64].

11.4 Scientific Evidence and Research

The need for scientific evidence to support the efficacy of herbal medicine is crucial for gaining acceptance in mainstream healthcare. However, conducting high-quality research on herbal remedies can be challenging due to various factors, including funding limitations, complex herbal compositions, and the difficulty of conducting placebo-controlled trials.

Obtaining funding for herbal research can be particularly challenging, as pharmaceutical companies may be less inclined to invest in natural remedies that cannot be patented [64].

11.5 Quality Control and Standardization

It ensures that the consistent quality of herbal products is vital for consumer safety and trust. Market players may face challenges in maintaining quality control, especially when sourcing herbs from different suppliers or regions.

Standardization of herbal products is essential to ensure uniformity in potency and composition. However, the inherent variability of natural ingredients can make standardization challenging [65].

The market for herbal medicines is subject to a number of restrictions and obstacles relating to national rules, global variety, safeness, effectiveness, and scientific proof. Market players must navigate these complexities to ensure compliance with regulations, provide safe and effective products, and gain acceptance in the broader healthcare community. Collaborating with regulatory authorities, investing in research, and prioritizing quality control are essential strategies to overcome these challenges and foster the growth of the herbal medicine industry.

12 Herbal Medicine Market Scope

In recent years, the distribution channels of E-commerce have experienced a remarkable surge in the sales of herbal medicinal products. This significant shift can be credited to strategic investments made in enhancing online purchasing experiences through targeted marketing and comprehensive online education. Giants in the online retail space, like Amazon, have taken the initiative to curate specialized portfolios solely dedicated to herbal products, catering to the growing demand. To further elevate customer satisfaction, online distributors have integrated live chat options, enabling them to address individual customer needs and queries effectively. Moreover, both manufacturers and retailers are now offering complimentary access to valuable 'wellness tools.' These tools empower customers to create personalized lists of herbal medicinal products tailored to their specific health conditions and requirements. Overall, these advancements in E-commerce distribution have not only boosted the sales of herbal medicinal products, but have also elevated the overall customer experience, promoting a healthier and more informed approach to wellness [50].

The demand for herbal medicines has increased significantly in recent years, and its scope continues to expand as consumers increasingly seek natural and alternative approaches to health and wellness. Let's delve into the details of the factors contributing to the scope of the herbal medicine market:

12.1 Growing Global Demand

The demand for herbal medicines is rising worldwide due to a shift in consumer preferences toward natural and organic products. Consumers are looking towards herbal medicines due to worries about the negative side effects of artificial pharmaceuticals and a yearning for natural health solutions.

12.2 Prevalence of Lifestyle Diseases

Lifestyle diseases, such as obesity, tachycardia, and hyperglycemia, and stress-related disorders, have become widespread due to sedentary lifestyles and poor dietary choices. Herbal medicine offers potential therapeutic options to manage and prevent these conditions, contributing to the market's scope [66].

The herbal medicine market has a promising scope and continues to grow as consumers seek natural alternatives and complementary treatments for various health conditions. Several factors contribute to the expanding scope of the herbal medicine market.

12.3 Increasing Consumer Demand

As mentioned earlier, there is a rising trend of health-conscious consumers seeking natural and holistic approaches to healthcare. Herbal medicines, with their perceived safety and minimal side effects, are increasingly preferred by individuals looking for alternative remedies [67].

12.4 Growing Awareness and Acceptance

There has been a shift in public perception towards herbal medicines, fueled by the information transmission using the internet and other media channels. As people become more informed about the benefits and historical use of herbal remedies, they are more willing to consider them as viable treatment options.

12.5 Support from Traditional Medicine Systems

Herbal medicine has a rich heritage and is an integral part of traditional medicine systems in many cultures worldwide. The support and endorsement of herbal remedies by traditional medicine practitioners contribute to their credibility and acceptance among the population.

12.6 Research and Scientific Validation

The herbal medicine industry has seen an increase in research and scientific research assessing the effectiveness and safety of various herbal products. As more evidence-based research emerges, it strengthens the legitimacy of herbal medicines and enhances their market scope [68].

12.7 Regulatory Support

In many countries, regulatory bodies have started to recognize the importance of herbal medicines and have established guidelines for their manufacturing, safety, and quality control. This regulatory support provides a framework for the growth and standardization of the herbal medicine market.

12.8 Integration with Conventional Medicine

Herbal medicine is increasingly being integrated into mainstream healthcare systems as complementary or adjunctive therapies. This integration widens the market scope as more healthcare professionals incorporate herbal remedies into their treatment protocols.

12.9 Lifestyle and Wellness Trends

The global wellness movement, coupled with the emphasis on preventive healthcare, has led to an increased demand for herbal products that promote overall well-being, stress relief, and immunity support.

12.10 Expanding Distribution Channels

The availability of herbal medicines has expanded significantly through various distribution channels, including brick-and-mortar stores, online platforms, and specialty health stores. This accessibility enhances the market scope and enables consumers to access herbal products more easily.

12.11 Potential in Developing Countries

In many developing countries, herbal medicine is deeply ingrained in the healthcare systems due to historical and cultural reasons. The market for herbal medicines and conventional treatments is expanding as these areas become more urbanized and enjoy economic expansion.

The scope of the herbal medicine market is continuously expanding due to the growing global demand for natural and holistic health solutions, the prevalence of lifestyle diseases, integration with conventional healthcare, government support, scientific validation, technological advancements, and changing consumer preferences. As the market expands, there will likely be more opportunities for herbal medicine manufacturers, practitioners, and distributors to cater to a diverse and health-conscious consumer base. However, it is essential to strike a balance between promoting herbal remedies responsibly and ensuring evidence-based practices to maintain consumer safety and confidence. As these trends continue to shape the healthcare landscape, the herbal medicine market is expected to witness sustained growth in the foreseeable future.

By strategically capitalizing on these opportunities, herbal market players can position themselves as leaders in the rapidly growing herbal medicine industry and contribute to shaping the future of healthcare and wellness [68].

13 Conclusion

The overview of complementary and alternative medicines, with a specific focus on the use of herbal medicines in the management of lifestyle diseases, highlights the growing significance of these practices in modern healthcare. Lifestyle diseases, which have become increasingly prevalent due to changing habits and urbanization, present a substantial burden on public health systems globally. Conventional treatments for these conditions often involve pharmaceutical interventions with potential side effects. However, there has been a growing interest in exploring alternative and complementary approaches, such as herbal medicine, for managing lifestyle diseases. Herbal medicines have gained popularity as a natural and holistic approach to addressing lifestyle diseases. With growing awareness of the potential side effects of conventional medications, people are turning to herbal remedies for a more sustainable and personalized approach to their health. The market analysis of lifestyle diseases demonstrates a rising trend in the adoption of herbal medicines, with significant potential for continued growth in the foreseeable future. Herbal medicine currently holds a major market share, reflecting the increasing trust in its efficacy and safety. Key industry players have been pivotal in driving research, innovation, and accessibility for herbal products. As the herbal medicine market continues to expand, it presents lucrative opportunities for both established and emerging market players. However, amidst the promising prospects, the herbal medicine market also faces challenges and regulations. Standardization, quality control, and evidence-based practices are essential to guaranteeing the effectiveness and safety of herbal medications.

Regulatory bodies must collaborate with the industry to establish guidelines and standards, fostering a trustworthy environment for consumers. Despite these challenges, the scope of the herbal medicine market remains vast. With ongoing research and technological advancements, herbal medicines are likely to gain even greater recognition as an integral part of the healthcare system. The key opportunity for herbal market players lies in developing innovative, scientifically backed products that meet the evolving needs of consumers. The utilization of herbal medicine for lifestyle diseases is deeply rooted in traditional knowledge and has been practiced across various cultures for centuries. Herbal remedies are derived from plant extracts and are believed to offer a natural and holistic approach to health. These natural remedies have gained popularity due to their perceived safety, efficacy, and minimal adverse effects, in contrast to synthetic drugs. Common lifestyle diseases where herbal medicine is being considered as an adjunct or alternative treatment include diabetes, obesity, hypertension, cardiovascular diseases, and stress-related disorders. The management of lifestyle diseases using herbal medicine involves the identification of specific plants and their active compounds that possess therapeutic properties. Researchers have been exploring the biological mechanisms underlying the effects of herbal remedies and have reported promising findings in preclinical and clinical studies. In conclusion, medicinal herbs have a bright future in the treatment of lifestyle disorders and have the potential to completely alter the medical industry. By addressing challenges, adhering to regulations, and capitalizing on emerging opportunities, the herbal medicine market is poised to make a significant impact on public health and wellness, contributing to a healthier and more sustainable future.

References

1. Abrahams S, Miller HC, Lombard C, van der Westhuizen FH, Bardien S. Curcumin pre-treatment may protect against mitochondrial damage in LRRK2-mutant Parkinson's disease and healthy control fibroblasts. Biochem Biophys Rep. 2021;27:101035.
2. Puri V, Kanojia N, Sharma A, Huanbutta K, Dheer D, Sangnim T. Natural product-based pharmacological studies for neurological disorders. Front Pharmacol. 2022;4675
3. Li Y, Liu X, Guo L, Li J, Zhong D, Zhang Y, et al. Traditional Chinese herbal medicine for treating novel coronavirus (COVID-19) pneumonia: protocol for a systematic review and meta-analysis. Syst Rev. 2020;9:1–6.
4. Harini JA, Luthra A, Madeka S, Shankar P, Mandava P, Pervaje R, et al. Ayurvedic treatment of acute ischemic stroke: a prospective observational study. Global Adv Health Med. 2019;8:2164956119849396.
5. Shaito A, Thuan DTB, Phu HT, Nguyen THD, Hasan H, Halabi S, et al. Herbal medicine for cardiovascular diseases: efficacy, mechanisms, and safety. Front Pharmacol. 2020;11:422.
6. Yusupova Z, Baratjonogli SF, Abduqunduzovna MZ. Medicinal Plants Growing in Our Republic Medicinal Properties. Periodica J Modern Philosophy Soc Sci Humanities. 2023;15:5–7.
7. Agustí A, Celli BR, Criner GJ, Halpin D, Anzueto A, Barnes P, et al. Global initiative for chronic obstructive lung disease 2023 report: GOLD executive summary. Am J Respir Crit Care Med. 2023;207(7):819–37.
8. Soler-Cataluña JJ, Miravitlles M, Fernández-Villar A, Izquierdo JL, García-Rivero JL, Cosio BG, et al. Exacerbations in COPD: a personalised approach to care. Lancet Respir Med. 2023;11(3):224–6.
9. Wang H, Luo Y, Wang H, Li F, Yu F, Ye L. Mechanistic advances in osteoporosis and anti-osteoporosis therapies. MedComm. 2023;4(3):e244.
10. Verma S, Gupta M, Popli H, Aggarwal G. Diabetes mellitus treatment using herbal drugs. Int J Phytomed. 2018;10(1):1–10.
11. Tilburt JC, Kaptchuk TJ. Herbal medicine research and global health: an ethical analysis. Bull World Health Organ. 2008;86:594–9.
12. Agranat S, Baris H, Kedar I. The PDQ Integrative, Alternative, and Complementary Therapies Editorial. 2017.
13. Millstine D. 2022. Types of complementary and alternative medicine. Merck Man Available online: https://www.hopkinsmedicine org/health/wellness-and-prevention/types-of-complementary-and-alternative-medicine. Accessed 24 Mar 2023.
14. Kerezoudis P, Kerezoudi EN, Choudhry A, Himes BT, Parney IF. Complementary and alternative medicine for gliomas: systematic review and critical appraisal of current literature. Neurosurgery. 2023;92(3):464–71.
15. Storr Krogh C, Skovgard L, Lynning M, Steenberg JL. Reasons for Engaging in Complementary and Alternative Medicine Among Highly Educated Women With Multiple Sclerosis. Int J MS Care. 2023;25(3):104–10.
16. Parveen A, Parveen R, Akhatar A, Parveen B, Siddiqui KM, Iqbal M. Concepts and quality considerations in Unani system of medicine. J AOAC Int. 2020;103(3):609–33.
17. Hoosen M, Roman NV, Mthembu TG, Naseer M. Unani Tibb practitioners' perceptions and attitudes towards spirituality and spiritual care in Unani Tibb practice in South Africa. BMC Complementary Med Therapies. 2023;23(1):1–12.
18. Hongal S, Torwane NA, Pankaj G, Chandrashekhar B, Gouraha A. Role of Unani system of medicine in management of orofacial diseases: a review. J Clin Diagn Res. 2014;8(10):ZE12.
19. Weeks LC, Strudsholm T. A scoping review of research on complementary and alternative medicine (CAM) and the mass media: looking back, moving forward. BMC Complement Altern Med. 2008;8(1):1–9.
20. Donelli D, Antonelli M. Homeopathy and Psychological Therapies. Encyclopedia. 2020;1(1):57–64.
21. Yadav R, Jee B, Rao K. How homeopathic medicine works in cancer treatment: deep insight from clinical to experimental studies. J Exp Ther Oncol. 2019;13(1)
22. Tabish SA. Complementary and alternative healthcare: is it evidence-based? Int J Health Sci. 2008;2(1):V.
23. Thas JJ. Siddha medicine—background and principles and the application for skin diseases. Clin Dermatol. 2008;26(1):62–78.
24. Ünlü ES, Kaya Ö, Eker İ, Gürel E. Sequencing, de novo assembly and annotation of Digitalis ferruginea subsp. schischkinii transcriptome. Mol Biol Rep. 2021;48(1):127–37.
25. Badalamenti N, Bruno M, Pavela R, Maggi F, Marinelli O, Zeppa L, et al. Acaricidal Activity of Bufadienolides Isolated from Drimia

pancration against Tetranychus urticae, and Structural Elucidation of Arenobufagin-3-O-α-L-rhamnopyranoside. Plan Theory. 2022;11(13):1629.
26. Dong D-D, Li H, Jiang K, Qu S-J, Tang W, Tan C-H, et al. Diverse lignans with anti-inflammatory activity from Urceola rosea. Fitoterapia. 2019;134:96–100.
27. Fu Y-S, Chen T-H, Weng L, Huang L, Lai D, Weng C-F. Pharmacological properties and underlying mechanisms of curcumin and prospects in medicinal potential. Biomed Pharmacother. 2021;141:111888.
28. Li J, Chen Q, Zhai X, Wang D, Hou Y, Tang M. Green tea aqueous extract (GTAE) prevents high-fat diet-induced obesity by activating fat browning. Food Sci Nutr. 2021;9(12):6548–58.
29. Diab F, Zbeeb H, Baldini F, Portincasa P, Khalil M, Vergani L. The Potential of Lamiaceae Herbs for Mitigation of Overweight, Obesity, and Fatty Liver: Studies and Perspectives. Molecules. 2022;27(15):5043.
30. Chen L, Liu Y, Tang Z, Shi X, Song Z, Cao F, et al. Improvements in estrogen deficiency-induced hypercholesterolemia by Hypericum perforatum L. extract are associated with gut microbiota and related metabolites in ovariectomized (OVX) rats. Biomed Pharmacother. 2021;135:111131.
31. Singh SA, Gowri K, Chitra V. A review on phytochemical constituents and pharmacological activities of the plant: Aerva Lanata. Res J Pharmacy Technol. 2020;13(3):1580–6.
32. Bouabdallah S, Sobhy HA, Adetuyi BO, Eldahshan OA, Egbuna C. Important Antihistaminic Plants and Their Potential Role in Health. Phytochemical Drug Discovery for Central Nervous System Disorders: Biochem Therapeut Effects; 2023. p. 385–96.
33. Barbaza MYU, De Castro-Cruz KA, Hsieh C-L, Tsai P-W. Determination of the chemical constituent contents and antioxidation properties of Asystasia Gangetica. Indian J Pharm Educ Res. 2021;55:863–71.
34. Park SJ, Nam J, Ahn CW, Kim Y. Anti-diabetic properties of different fractions of Korean red ginseng. J Ethnopharmacol. 2019;236:220–30.
35. Khan A, Unnisa A, Sohel M, Date M, Panpaliya N, Saboo SG, et al. Investigation of phytoconstituents of Enicostemma littorale as potential glucokinase activators through molecular docking for the treatment of type 2 diabetes mellitus. In Silico Pharmacol. 2022;10:1–16.
36. Jimenez-Garcia SN, Garcia-Mier L, Vazquez-Cruz MA, Ramirez-Gomez XS, Guevara-Gonzalez RG, Garcia-Trejo JF, et al. Role of Natural Bio-active Compounds as Antidiabetic Agents. Bioact Nat Prod Pharmaceut Appl. 2021:535–61.
37. Balch MH, Nimjee SM, Rink C, Hannawi Y. Beyond the brain: the systemic pathophysiological response to acute ischemic stroke. Journal of Stroke. 2020;22(2):159.
38. Zhang B, Saatman KE, Chen L. Therapeutic potential of natural compounds from Chinese medicine in acute and subacute phases of ischemic stroke. Neural Regen Res. 2020;15(3):416.
39. Verma T, Sinha M, Bansal N, Yadav SR, Shah K, Chauhan NS. Plants used as antihypertensive. Nat Prod Bioprospect. 2021;11:155–84.
40. Dey A, Roy D, Mohture VM, Ghorai M, Rahman MH, Anand U, et al. Biotechnological interventions and indole alkaloid production in Rauvolfia serpentina. Appl Microbiol Biotechnol. 2022;106(13–16):4867–83.
41. Jasemi SV, Khazaei H, Momtaz S, Farzaei MH, Echeverría J. Natural products in the treatment of pulmonary emphysema: Therapeutic effects and mechanisms of action. Phytomedicine. 2022;99:153988.
42. Sharma A, Sharma C, Shah OP, Chigurupati S, Ashokan B, Meerasa SS, et al. Understanding the mechanistic potential of plant based phytochemicals in management of postmenopausal osteoporosis. Biomed Pharmacother. 2023;163:114850.
43. Ekor M. The growing use of herbal medicines: issues relating to adverse reactions and challenges in monitoring safety. Front Pharmacol. 2014;4:177.
44. Rashrash M, Schommer JC, Brown LM. Prevalence and predictors of herbal medicine use among adults in the United States. J Patient Exp. 2017;4(3):108–13.
45. Jităreanu A, Trifan A, Vieriu M, Caba I-C, Mârțu I, Agoroaei L. Current trends in toxicity assessment of herbal medicines: A narrative review. PRO. 2022;11(1):83.
46. Bareetseng S. The Worldwide Herbal Market: Trends and Opportunities. Journal ISSN. 2022;2766:2276.
47. O'Neill S, O'Driscoll L. Metabolic syndrome: a closer look at the growing epidemic and its associated pathologies. Obes Rev. 2015;16(1):1–12.
48. Singh S, Jain V, Jain SK, Chandra K. Medicinal plants and phytochemicals in prevention and management of life style disorders: pharmacological studies and challenges. Asian J Pharm Clin Res. 2021;14(12):1–6.
49. Gunjan M, Naing TW, Saini RS, Ahmad A, Naidu JR, Kumar I. Marketing trends & future prospects of herbal medicine in the treatment of various disease. World J Pharmaceut Res. 2015;4(9):132–55.
50. Sharma A, Shanker C, Tyagi LK, Singh M, Rao CV. Herbal medicine for market potential in India: an overview. Acad J Plant Sci. 2008;1(2):26–36.
51. Inamdar N, Edalat S, Kotwal VB, Pawar S. Herbal drugs in milieu of modern drugs. Int J Green Pharmacy (IJGP). 2008;2(1)
52. Yazarlu O, Iranshahi M, Kashani HRK, Reshadat S, Habtemariam S, Iranshahy M, et al. Perspective on the application of medicinal plants and natural products in wound healing: A mechanistic review. Pharmacol Res. 2021;174:105841.
53. Ahad B, Shahri W, Rasool H, Reshi Z, Rasool S, Hussain T. Medicinal plants and herbal drugs: An overview. In: Medicinal and Aromatic Plants: Healthcare and Industrial Applications; 2021. p. 1–40.
54. Keshvari M, Nedaeinia R, Nedaeinia M, Ferns GA, Nia SN, Asgary S. Assessment of heavy metal contamination in herbal medicinal products consumed in the Iranian market. Environ Sci Pollut Res. 2021;28:33208–13.
55. Chaachouay N, Douira A, Zidane L. COVID-19, prevention and treatment with herbal medicine in the herbal markets of Salé Prefecture, North-Western Morocco. Eur J Integrative Med. 2021;42:101285.
56. Mukherji R, Zarhani SH. The Pandemic and Competitive Authoritarianism in India. In: Democracy, State Capacity and the Governance of COVID-19 in Asia-Oceania. Routledge; 2023. p. 131–52.
57. Zheng Q. Opportunities and Challenges in the Health and Nutrition Industry with the Rise of Younger Consumers. In: Transition and Opportunity: Strategies from Business Leaders on Making the Most of China's Future; 2022. p. 229–37.
58. Sekhon AK, Kathuria LM. Leading by Example: Corporate Social Responsibility Initiatives by Dabur India Limited in North India. Asia-Pacific J Manag Res Innov. 2020;16(1):43–53.
59. Barkat MA, Goyal A, Barkat HA, Salauddin M, Pottoo FH, Anwer ET. Herbal medicine: clinical perspective and regulatory status. Comb Chem High Throughput Screen. 2021;24(10):1573–82.
60. Bodeker G, Pecorelli S, Choy L, Guerra R, Kariippanon K. Well-being and mental wellness. Oxford Research Encyclopedia of Global Public Health; 2020.
61. Liu X, Gong X, Liu Y, Liu J, Zhang H, Qiao S, et al. Application of high-throughput sequencing on the Chinese herbal medicine for the data-mining of the bioactive compounds. Front Plant Sci. 2022;13:900035.

62. Parihar N, Barve K, Bhatt LK. Clinical, ethical, and regulatory issues of botanicals. Phytopharmaceuticals and Herbal Drugs: Elsevier; 2023. p. 265–79.
63. Khakurel D, Uprety Y, Ahn G, Cha J-Y, Kim W-Y, Lee S-H, et al. Diversity, distribution, and sustainability of traditional medicinal plants in Kaski district, western Nepal. Front Pharmacol. 2022;13:1076351.
64. Ali F, Khasimbi S, Sharma K, Trivedi M, Ali A, Ahmad J. Regulatory perspectives of Herbal Medicinal Products. Bioactive Phytochem Drug Discov Prod Develop. 2021;2020:158–75.
65. Balekundri A, Mannur V. Quality control of the traditional herbs and herbal products: A review. Future J Pharmaceut Sci. 2020;6:1–9.
66. Tabish S. Lifestyle diseases: consequences, characteristics, causes and control. J Cardiol Curr Res. 2017;9(3):00326.
67. Abuga I, Sulaiman SF, Wahab RA, Ooi KL, Rasad MSBA. Phytochemical constituents and antibacterial activities of 45 Malay traditional medicinal plants. J Herbal Med. 2022;32:100496.
68. Chaughule RS, Barve RS. Role of herbal medicines in the treatment of infectious diseases. Vegetos. 2023:1–11.

Future Prospective of Herbal Medicines for Lifestyle Diseases

Lolita Rapoliene and Vilma Matuleviciute

Abstract

Millions of people worldwide suffer from chronic lifestyle diseases mostly referred to as non-communicable diseases (NCDs) and conditions. The most frequent of them are cardiovascular disease, stroke, diabetes, obesity, metabolic syndrome, chronic obstructive pulmonary disease, and some types of cancer, followed by depression, dementia, arthritis and osteoporosis, asthma, chronic liver and kidney diseases, and other diseases. NCDs became the leading contributors to morbidity and mortality responsible for 74% of all global deaths. The management of lifestyle diseases consists of risk factors reduction or elimination, early non-pharmacological and pharmacological treatment of disease, preventing its outcomes to preserve workability and quality of life. According to WHO, 85% of people worldwide use medicinal plants for treatment, 80% of people in developing countries use traditional and complementary medicine for their primary health care needs in the prevention, diagnosing, or treatment of physical and mental illness even if not approved by the regulatory agencies. The trends driving integration of HM in contemporary healthcare are: high rates of chronic diseases globally, chronic diseases are not adequately prevented and treated, high rates of adverse side effects, drug interactions because of aging population, multimorbidity and polypharmacy, high cost of healthcare, high utilization of TCAM by general public without informing physicians, inadequate education of health professionals about TCAM methods, the growing interest of physicians in TCAM education for integrative medicine, the increasing amount of research and clinical guidelines from health organizations for the inclusion of TCAM practices.

This chapter will discuss the role of some herbal medicines in the treatment of major NCDs, current challenges, and some future perspectives for herbal medicines.

L. Rapoliene (✉)
Klaipeda University, Klaipeda, Lithuania

Baltic Medics Port Clinic, Klaipeda, Lithuania
e-mail: lolita.rapoliene@inbox.lt

V. Matuleviciute
Klaipeda University, Klaipeda, Lithuania

Irish Register of Herbalists, Dublin, Ireland

The Excelsior School of Herbal Medicine, Belfast, Northern Ireland

Keywords

Health · Herbal medicine · Lifestyle · Lifestyle diseases · Non-communicable diseases · Phytotherapy

1 Introduction

"All that man needs for health and healing has been provided by God in nature, the Challenge of science is to find it." ~ Paracelsus

Lifestyle diseases are caused by prolonged exposure of main similar modifiable behavioral risk factors- smoking, poor diet, and lack of exercise—this leads to the onset of chronic non-infectious origin diseases: diabetes, cardiovascular and pulmonary diseases, obesity, strokes, metabolic syndrome, disease, and some types of cancer [1]. There are many more behavioral, biological, and societal risk factors originating from industrialization, urbanization, intense pace of life, and aging promoting lifestyle diseases like depression, dementia, arthritis and osteoporosis, asthma, chronic liver and kidney diseases, and other diseases (Fig. 1).

Millions of people worldwide suffer from chronic lifestyle diseases mostly referred to as non-communicable diseases (NCDs) and conditions. NCDs became the leading contributors to morbidity and mortality especially, responsible for 74% of all global deaths, and 77% in low and middle income countries, affecting working-age people (45% to

A. K. Dhara, S. C. Mandal (eds.), *Role of Herbal Medicines*, https://doi.org/10.1007/978-981-99-7703-1_31

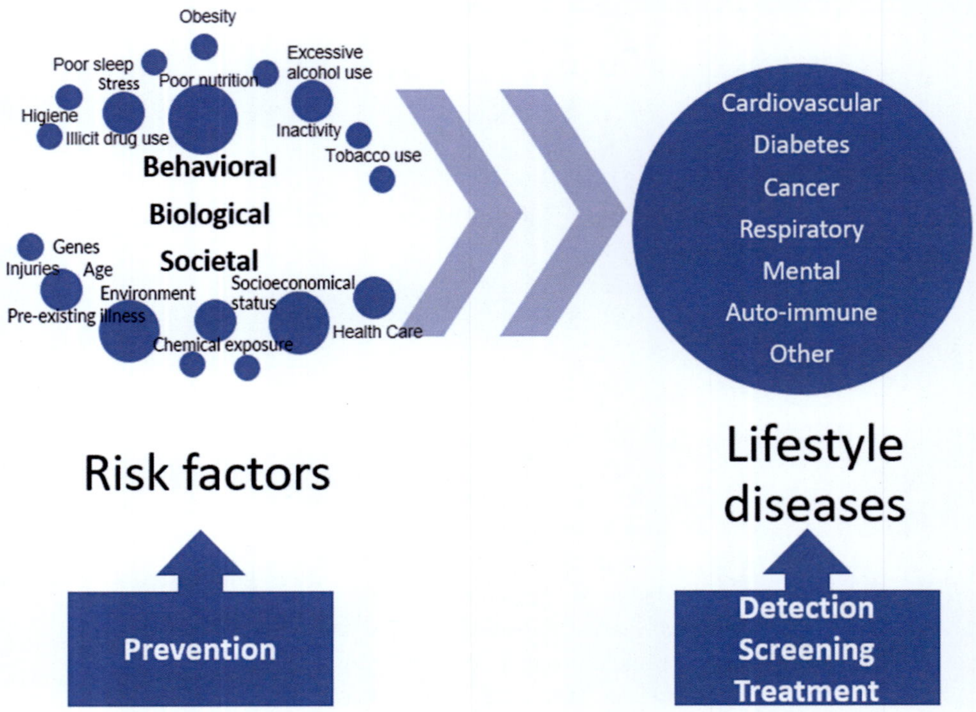

Fig. 1 Risk factors for main lifestyle diseases and key components of the response

people under 60 years). Cardiovascular (17.7 million), cancers (8.8 million), chronic respiratory diseases (3.9 million), and diabetes (1.6 million) account for over 80% of all premature NCD deaths—all around kill 40 million people each year [2]. The vast majority of all death are preventable [3]. NCDs tend to have complex etiology, multiple risk factors, long latency period, non-infectious origin, prolonged course, and functional impairment [4]—if to count—loss in life years due to these diseases range from 7.5 to 23.7 years [5–7].

Main responses to NCDs are prevention, early detection, screening, and treatment. WHO Global action plan to control and prevent non-communicable diseases was extended to 2030 by The World Health Assembly with the aim of making faster progress in preventing and controlling NCDs [8]. As well in the 2030 Agenda for Sustainable Development NCDs are named as a major challenge for sustainable development. To achieve SDG 3.4 goal the greater prioritization of NCDs in health systems and timely and high-quality care are essential. NCD Target 3.4 is linked to eight SDGs by stronger (bold lines) or weakened relationships means that mutually reinforcing progress on others Sustainable Development Goals also will help to achieve SDG Target 3.4 (Fig. 2) [9]. The systematic analysis found 265 different NCDs interventions used and suitable for use in different levels, ranging from screening to prevention to treatment and palliation which differ in regions [10]. The management of lifestyle diseases consists of risk factors reduction or elimination, early non-pharmacological and pharmacological treatment of disease, preventing its outcomes to preserve workability and quality of life. It has been reported that adherence to a healthy lifestyle in the middle of life is associated with a longer life expectancy without the main chronic illness [11].

A very important and still underestimated resource is traditional and complementary medicine (T&CM or TCAM), ethnopharmacology which has many uses, especially in control, prevention, and treatment of lifestyle-related diseases. Traditional medicine (TM) refers to natural indigenous forms of medicines and treatments which thrived in all cultures before modern conventional medicine. In 2022 "complementary, alternative and integrative medicine" (CAIM) included 604 TCAM therapies for the first time in its definition [12]. Due to people beliefs in TCAM therapies as natural, safe, holistic, TCAM is now a more widespread health care choice although some methods or preparations lack scientific evidence because of methodological limitations, clear indications for use and dosage. TCAM includes diet and herbs. Herbalism is a natural medicine modality which utilizes plant medicine and materials, through different products such as tinctures, creams which have plant materials as their active ingredients. Among diverse cultures, some herbal medicines may contain natural animal and mineral materials [13].

According to WHO, 85% of people worldwide use medicinal plants for treatment, 80% of people in developing countries use TCAM for their primary health care needs [14, 15]. TM is used by 60–80% of people in Asia and Africa; India is the country with the highest use of traditional medicine among other Asian countries with 11.7% of people using it as a frequent source of health care; despite the popularity of

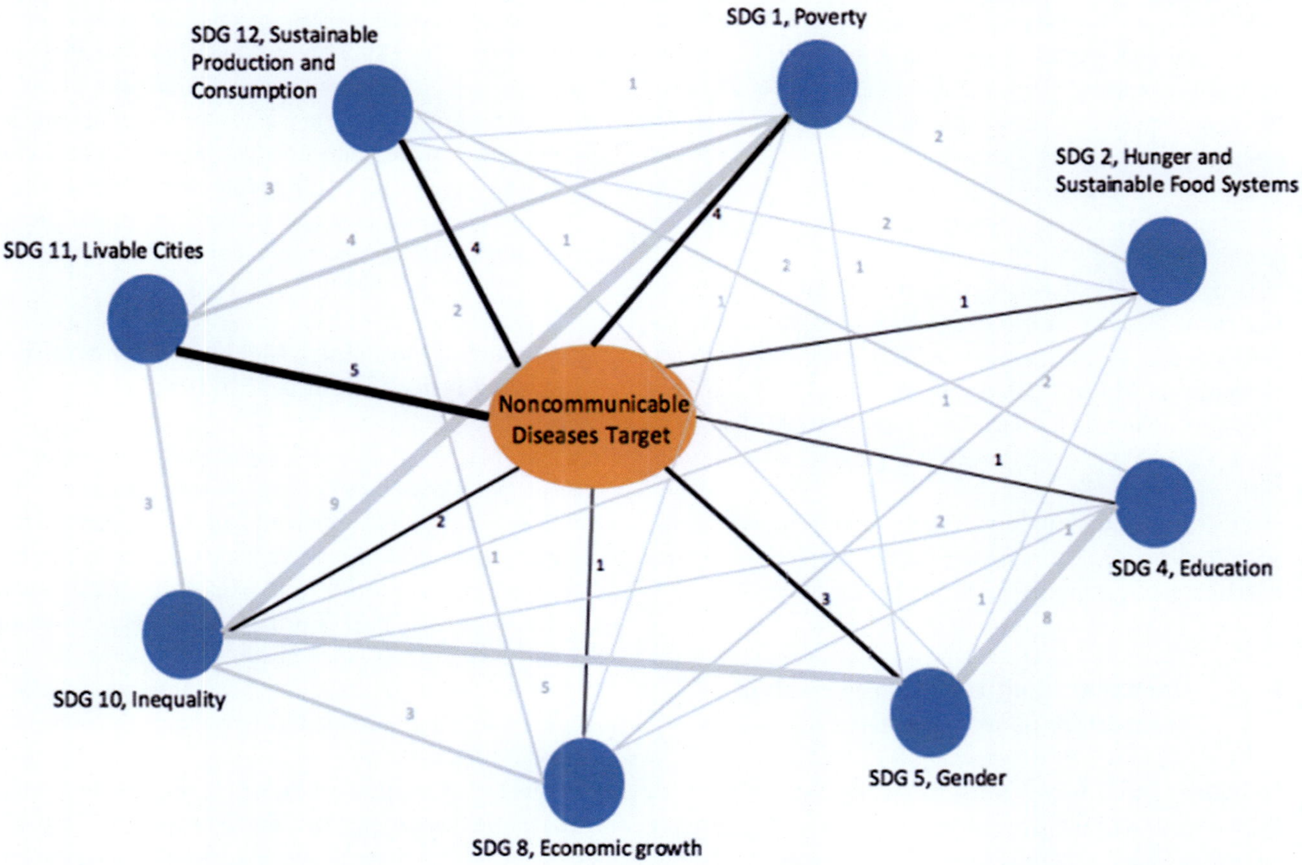

Fig. 2 Links among 10 SDCs and between those SDGs and NCD target 3.4. (Source: Adapted from Leblanc, UN DESA, 2015 [8])

traditional Chinese medicine (TCM) only less than 3% population prefer it; Ghana and South Africa use less than 2% of TM even though it is cheap, affordable, and easily accessible comparing with modern medicine [16]. Many countries have traditional native methods of healing such as Ayurveda, TCM, and Unani Medicine which are also used worldwide [17]. Both China and India are popularly known for their historical use of herbal medicine, all very well recorded.

WHO classifies herbal remedies according to their source of origin, their evolution, and usage as: indigenous herbal medicines, herbal medicines in systems, modified herbal medicines, imported products with a herbal medicine base [18]. Worldwide there are 422,000 plant species recorded, 8% of which are classified as "threatened," and 52,000 (or 12.5%) being utilized for medical purposes [19]. It is thought that almost 61% of all anticancer drugs and as much as 50% of drugs for cardiovascular treatments were developed from herbal medicines [20]. There are many diverse ways of discovering potentials for creating new pharmaceutical medicines from traditional herbal medicines [21].

In response to the World Health Assembly resolution on traditional medicine (WHA62.13) the WHO developed traditional medicine strategy 2014–2023. According to its global report, 88% (170) of the members of the WHA are known to use T&CM. These member states have formal laws and regulations for governing T&CM [22]. *European Union (*EU) legislation on pharmaceutical products for human use also applies to traditional herbal medicines. A simplified registration procedure was introduced by *Directive 2004/24/EC* and is intended for herbal medicinal products with a long tradition of medicinal use (at least 30 years, including 15 in the EU). Also a Committee for Herbal Medicinal Products (HMPC) was established at the European Medicines Agency (EMA). On the basis of the scientific opinion of the HMPC, a list of herbal substances, preparations, and combinations for use in certain traditional herbal medicinal products has been established by *Commission Decision 2008/911/EC* [23]. The common consumers of medicinal plants found in survey (Spain) were young women between 18 and 44 years with higher education; the main reasons for the use were perception that being natural means harmless [24]. According to WHO (2013), there is growing popular use of traditional medicines in Africa, Oceania, the Americas, and Asia. Though the results widely vary on how frequently they are used, many TCIM modalities include herbal medicines. In 2014, a study showed that as many as 18.8% of Europeans have used at least one supplement, which is plant based [25, 26].

TCAM is used in the prevention, diagnosing, or treatment of physical and mental illness even if not approved by the regulatory agencies [27]. In the statistical anlysis of TCM for NCDs was stated that TCM can reduce the symptoms of diseases, the adverse effects caused by conventional treatment and improve the quality of life in NCDs based on NCDs aspect of inflammatory pathways, oxidative stress, and cell apoptosis [28].

Growing sales of herbal supplements reflect increased consumer interest, growing awareness, and demand for botanical and medicinal traditions. The 10 top-selling supplements in 2017 in US Mainstream Multi-Outlet Channel were Horehound, Echinacea, Cranberry, Ivy leaf, Turmeric, Black Cohosh, Garcinia, Green tea, ginger, and fenugreek [29].

This chapter will discuss the role of some herbal medicines in the treatment of major NCDs, current challenges, and some future perspectives for herbal medicines.

2 Herbal Medicine's Role in Lifestyle-Related Diseases: Present and Future

Herbs have been used across the world for millennia to address various health issues. There is an array of intensively studied and chemically diverse secondary plant metabolites which exert numerous physiological and pharmacological effects on tissues and have an important role in maintaining health [30]. The interest in herbs and plant-derived compounds is rapidly growing. Most commonly well researched groups of active metabolites showing strong therapeutic activity are phenols, phytosterols, flavonoids, tannins, saponins, polysaccharides, and terpenoids among others [31]. As oxidative stress underpins pathogenesis of most chronic diseases, addressing inflammation, tissue oxidation and damage by using phytochemicals exhibiting strong antioxidant effects, provides an area of huge potential in prevention and treatment of NCDs [31, 32].

2.1 Herbal Medicines and Cardiovascular Diseases (CVD)

Benefits of purified herbal compounds in prevention and treatment of cardiovascular diseases are extensively recorded, but it is especially important to look at research and clinical trials using whole herbs and their combinations [33, 34]. Numerous studies indicate the efficacy of herbs and benefits in using them to treat hyperlipidemia, atherosclerosis, hypertension, and other aspects of chronic heart diseases [35]. The use of traditional herbal medicines in China is well researched and widely applied in treatment and recovery from ischemic stroke to improve various post-stroke symptoms and neurological deficits [36, 37]. Some extensively researched herbs (*Crataegus* spp., *Allium sativum* L., *Curcuma longa* L., *Salvia miltiorrhiza* Bunge.) at analyzed dosages are safe and efficacious in reducing inflammatory processes and various disease parameters, hence providing strong evidence for their applications in treatment of congestive and chronic heart failure and hypertension [38–45] (Table 1).

2.2 Herbal Medicines, Metabolic Syndrome, and Type-2 Diabetes Mellitus

Research indicates that type-2 diabetes mellitus complications can be addressed and prevented with the use of herbal medicines [59, 60]. Antioxidant plant metabolites, such as polyphenols, flavonoids and terpenoids, can reduce inflammation and protect tissues from oxidative damage caused by glycosylation [61]. Certain herbs (for example, *Cinnamomum* spp., *Allium sativum* L., *Curcuma longa* L., *Silybum marianum,* L. Gaert., etc.) can improve endothelial function and micro-circulation, consequently preventing atherosclerosis.

Clinical trials suggest efficacy of herbal medicines used in metabolic syndrome and diabetes [62–64]. Positive therapeutic interactions between prescribed pharmaceutical treatment (metformin) and some herbs were observed in controlling glycemic index, suggesting synergistic activity between two interventions [65]. Trials using traditional herbal medicine in China also showed tendency to better clinical outcomes, indicating possible synergistic effect, when herbals were combined with conventional medications [66].

More clinical trials are needed, but the use of herbal medicines as adjuncts might facilitate the reduction of prescribed pharmaceuticals for diabetes (Table 2).

2.3 Herbal Medicines and Chronic Pulmonary Diseases

2.3.1 Respiratory

There is notable amount of research on treatments of chronic obstructive pulmonary disease (COPD) with traditional herbal medicine in China and Korea, but studies are lacking elsewhere [107–109]. Ethnobotanical studies give a broad view of traditional herbal uses and provide insights in phytochemical variations of researched plants (Joshi 2014), but more clinical trials are needed to support ethnomedical herbal traditions. Aromatic herbs rich in essential oils contain terpenoids (for example, *Thymus vulgaris* L., *Eucalyptus globulus* Labill.), which show broad spectrum antimicrobial,

Table 1 Examples of herbs used in cardiovascular diseases

Herb/plant name	Mechanism of action	Traditional and clinical use	Studies
Crataegus spp.—Hawthorn	Cardiotonic Anti-inflammatory Inotropic Hypotensive Anti-atherosclerotic Hypolipidemic Antioxidant Diuretic Lipidemic Vasodilator ACE inhibitor	Hypertension Atherosclerosis Chronic heart failure Cardiovascular disease Congestive heart failure	[33–41, 46, 47]
Allium sativum L.—Garlic	Hypotensive Antithrombotic Hypocholesterolaemic Hypolipidemic Antioxidant Antiatherogenic Fibrinolytic Antiplatelet ACE inhibitor	Hypertension Hypercholesterolaemia Hyperlipidemia Atherosclerosis Cardiovascular disease Obesity	[42, 48–50]
Leonurus cardiaca L.—Motherwort	Cardioprotective Cardiotonic Antioxidant Antiarrhythmic Hypotensive	Chronic heart diseases Angina pectoris Hypertension Palpitations Tachycardia	[51–55]
Salvia miltiorrhiza Bunge—Red sage	Antihypertensive Antithrombotic Antiatherogenic Hypolipidemic Antioxidant Vasodilator Anti-inflammatory Cardioprotective	Ischemic stroke Cardiovascular disease Coronary heart disease Congestive heart failure Angina pectoris Arrhythmia Myocardial ischemia Hypertension Atherosclerosis	[45, 56, 57]
Curcuma longa L. and curcumin—turmeric	Antihyperlipidemic Cardioprotective Antioxidant Antihypertensive Anti-inflammatory Anti-obesity	Cardiovascular disease Hypercholesterolemia Atherosclerosis Arrhythmia Hypertension Obesity	[43, 44, 58]

mucolytic and expectorant properties are often used in respiratory diseases. Flavonoids, saponins, and alkaloids are other main groups of the most promising phytochemicals for managing chronic respiratory conditions, relieving exacerbations, reducing inflammatory processes and preventing infections [110].

Clinical trials indicate that including herbal medicines to the prescribed pharmaceuticals can improve the outcomes of treatments for pulmonary diseases [111, 112] (Table 3).

Even though traditional use and ethnomedicine studies identify many plants/herbs with hypoglycemic properties, good quality clinical trials are lacking. Investigations on synergistic herb-drug interactions are especially important, as this would allow to reduce pharmaceutical medications, thus minimizing their side effects.

2.4 Herbal Medicines and Other Lifestyle-Related Diseases

Current research indicates that certain herbal medicines have a beneficial role in alleviating depression and anxiety [137, 138]. There is a volume of studies, including clinical trials comparing herbal medicines with prescribed antidepressants, showing that herbs and pharmaceutical antidepressants have similar efficacy [139]. Traditional Chinese herbal medicine also present some evidence of favorable CHM comparability with antidepressant medications [140, 141].

The effectiveness of herbal medicines in prevention and treatment of dementia, arthritis, liver, and kidney diseases is also supported by clinical studies [62, 142–146].

Table 2 Examples of herbs used in metabolic syndrome and type-2 diabetes mellitus

Herb name	Mechanism of action	Traditional and clinical use	Studies
Trigonella foenum-graecum L.—fenugreek	Antidiabetic Hypoglycemic Antioxidant Anti-inflammatory Hypocholesterolemic	Type-2 diabetes mellitus T2DM Hyperglycemia Hypercholesterolemia Weight loss Obesity	[67–72]
Cinnamomum spp.—cinnamon	Hypolipidemic Hypoglycemic Insulin sensitivity enhancer Antiatherogenic Antioxidant Anti-inflammatory Antihypertensive	Diabetes mellitus Atherosclerosis Hypertension Hyperlipidemia Obesity Metabolic syndrome	[73–77]
Urtica dioica L.—nettle	Hypoglycemic Antidiabetic Anti-inflammatory Antioxidant Hypotensive	Type-2 diabetes mellitus Metabolic syndrome	[78–84]
Allium sativum L.—garlic	Hypoglycemic Antidiabetic Hypolipidemic Anti-inflammatory Antioxidant	Type 2 diabetes mellitus Metabolic syndrome Oxidative stress Dyslipidemia	[49, 85–88]
Galega officinalis L.—Goat's rue	Antidiabetic Hypoglycemic	Type 2 diabetes mellitus	[89–91]
Curcuma longa L.– turmeric and curcumin	Antioxidant Anti-inflammatory Antidiabetic Hypolipidemic	Type 2 diabetes mellitus Metabolic syndrome Inflammation Dyslipidemia NAFLD Oxidative stress Obesity Non-alcoholic steatohepatitis	[85, 92–99]
Silybum marianum L. Gaert.—Milk thistle	Antioxidant Anti-inflammatory Hypolipidemic Antidiabetic Hepatoprotective Anti-atherosclerotic Antihypertensive	Metabolic syndrome NAFLD Type 2 diabetes mellitus Obesity Inflammation Oxidative stress Dyslipidemia	[80, 88, 100, 101]
Zingiber officinale Rosc.—ginger	Cardioprotective Antioxidant Anti-inflammatory Antilipidemic Thermogenic Hypoglycemic	Type 2 diabetes mellitus Cardiovascular disease Obesity	[96, 102–106]

T2DM type-2 diabetes mellitus, *NAFLD* non-alcoholic fatty liver disease

3 Challenges for Herbal Medicine

Recently there is a significant tendency across the world to use complementary and traditional medicine concurrently, and specifically the combined use of the herbal preparations and prescribed medications. The WHO survey (2018) showed that actually 88% of its Member States recognized the use of TCAM, and this increases to more than 90% among the South-East Asian, Western Pacific and Eastern Mediterranean populations. Still, in many countries new herbal products are brought to the market without going through any mandatory toxicology and safety checks, and structures to regulate quality control and manufacturing practices are lacking [147]. There are a number of prospectives and challenges in relation to national health policies, regulation, quality control, safety, and efficacy of TCAM, health insurance and the integration of TCAM in healthcare systems. There is no unified TCAM; patterns of use vary depending on culture, history and regulations and generally are: use in countries where TM is one of the primary sources of health care (Africa), use of TM due to cultural and historical influences (Singapore, Republic of Korea), use of T&CM as complementary therapy (Europe) [148].

Table 3 Examples of herbs used in chronic pulmonary diseases

Herb name	Mechanism of action	Traditional and clinical use	Studies
Thymus vulgaris L.—thyme	Bronchodilator Antimicrobial Antiviral Antispasmodic Antioxidant Expectorant Anti-inflammatory	COPD Asthma Bronchitis Cough Respiratory infections Emphysema	[113–118]
Inula helenium L.—elecampane	Antimicrobial Anti-inflammatory Expectorant Anti-tumor Antibiofilm Antioxidant Antifibrotic	Respiratory infections Bronchitis Asthma Cough Pulmonary fibrosis	[114, 119–121]
Glycyrrhiza glabra L./*uralensis* Fisch.—licorice	Anti-inflammatory Bronchorelaxant	Asthma Bronchitis COPD	[122–124]
Grindelia spp.—gumweed	Anti-inflammatory Expectorant Broncholdilator Antimicrobial Antioxidant Antibacterial Antifungal Immunomodulatory	Asthma COPD Bronchitis Cough	[119, 125–129]
Urtica dioica L.—nettle	Anti-inflammatory Antihistamine Anti-asthmatic Immunomodulatory Antioxidant	Asthma Bronchitis Cough Emphysema	[130–133]
Curcuma longa L.—turmeric	Anti-inflammatory Mast cell stabilizer Anti-allergic Antioxidant Bronchodilator Immunomodulatory	Asthma COPD Emphysema	[131, 134–136]

Effective and safe herbal medicine needs knowledge (therapeutic action, interaction with others, contraindications, dosage), science-based evidence for effectiveness, safety and toxicity, properly prepared resources, good manufacturing practice (GMP), and educated specialists. Incorrect harvesting (plant maturity, inadequate drying, and prolonged storage) is rapidly diminishing the resources. Several factors can affect the product development process from plant sources: correct collection time and place, proper cultivation of medicinal plants, challenging environmental influences, testing and identifying active ingredients, and setting up structures for the production, processing, preparation and storage of plant material [147].

One of the primary goals for WHO is to improve the assessments of safety and efficacy as well as the quality of traditional and complementary medicines, as reported in the WHO Traditional Medicine Strategy: 2014–2023 [148] as well as support countries in overcoming challenges concerning herbal product regulation [147].

The situation on TCAM regulatory issues in WHO Member state countries was established using few surveys. During the period from 1999 to 2018, a steadily increasing number of Member States was noted on setting up the national policy on TCAM—over 50% of Member States (98 countries out of 194) had implemented a specific TCAM policy or incorporated TCAM related policy into existing national policy:

The African and South-East Asia Regions—85%.

The Western Pacific Region—63%.

The Eastern Mediterranean Region—43%.

The Americas—31%.

The European Region—21%.

In 2018, over a half of WHO Member States (56%) set either legal or administrative structure for TCAM; WHO European Region percentage of the region is 40%.

TCAM national educational courses were set only in 40% of Member States:

The South-East Asia Region—91%.

The African Region—72%.

The Western Pacific Region—41%.
The Region of the Americas—37%.
The Eastern Mediterranean Region—19%.
The European Region—13%.

As at 2018, a National Health Board or Committee for TCAM was set in 55% of all Member States:

The South-East Asia Region—91%.
The African Region—72%.
The Eastern Mediterranean Region—52%.
The Western Pacific Region—41%.
The Region of the Americas—34%.
The European Region—28%.

A National Research Institute for TCAM with government or public research funding was established by almost 40% of WHO Member States.

Very few countries (Benin, Brazil, Chile, China, Cuba, Democratic People's Republic of Korea, India, Mali, Oman, Peru, Thailand, and the USA) reported about setting up a national plan for integrating TCAM into their national health service delivery; many other countries reported that they are in the process of developing guidelines to lead the harmonization of TCAM therapies within their health systems.

As of 2018, 64% had laws or regulations on herbal medicines. The regulatory framework for herbal medicines was established in many European countries, followed by the African Region, the Region of the Americas and the Eastern Mediterranean Region, the Western Pacific Region and then the South-East Asia Region. Survey showed that in 2005 the majority of WHO Member States (71%) were regulating herbal products as non-prescription medicines (self-medication and OTC) and this was decreased to 41% by 2012 together with increase in products regulated as "herbal medicines" (from 13 to 40%), which means tighter regulatory criteria and monitoring. By 2012, the number of Member States who sell herbal medicines with claims had increased: medical claims (alleging to cure, treat or prevent certain diseases, or to correct or alter physiological functions; prior to entering the market, these products have to be registered by the medical products agency), health claims (any proposal or hint on the labels or advertising about any specific health benefit of the product); nutrient content claims (suggesting that a specific herbal product has especially high or low contents of nutritional ingredients).

The most frequently encountered claims for herbal products were medical claims, followed by health claims and nutrient claims.

Some form of the national pharmacopeia containing herbal medicines was used by 57% of WHO Member States in 2012; 34% of those stated the pharmacopeia to be legally binding, 48% of the Member States reported using national and other monographs on herbal products. A WHO list of pharmacopeias can be found online. Regarding Good Manufacturing Practice (GMP) only 43% had it in place. Some European countries from the WHO European Region described following European Union (EU) guidelines for GMP, but others reported also using WHO guidelines for GMP.

As of 2018, nearly 65% of Member States had established a national system for registering herbal products:

The South-East Asia Region—91%.
The European Region—85%.
The Eastern Mediterranean Region—81%.
The Region of the Americas—54%.
The African Region—49%.
The Western Pacific Region—41%.

In 2012, 43% of countries had market surveillance system for herbal medicines.

Based on the WHO survey (2012), in Member States TOP- 5 from TCAM acknowledged use were acupuncture (85%), herbal medicines (83%), indigenous traditional medicine (82%), homeopathy (75%), and TCM (75%). The regulation for providers of herbal medicine had only 18% of Member States; 12% of countries had health insurance coverage (partial or full) for it. Regulation for TCAM providers was expanding from 67 in 2012 to 78 Member States in 2018; twice as many were practicing privately than in the public sector, some of them in both sectors; the herbal medicine providers had 46% states in 2012.The TCAM license for the practicing TCAM was needed only in 36% of Member States, some form of TCAM education was available in 47% of states in 2012, that was officially recognized by the government. The country summaries are included in the WHO global report on traditional and complementary medicine (2019) [13].

For the future of herbal medicine and conventional medicine to be successfully integrated, the essence is to overcome difficulties and risks, which have arisen in the WHO Member States. This makes it necessary to achieve a scientific and regulatory basis that can be widely recognized and used worldwide. Difficulties faced by WHO countries regarding regulatory issues could be turned to challenges and are listed in Tables 4 and 5.

Table 4 Challenges of WHO countries regarding regulatory issues and support needed

Challenges[a]	Support from WHO needed
Lack of research data	Expertise on analysis and assessment of TCAM effectiveness, side effects, and value Provision of research databases Evaluation of herb–drug interaction Arrangement of global meetings
Lack of policies controlling and regulating how TCAM is advertised and claims made	Sharing data on regulatory matters Seminars on establishing national procedures and practices for TCAM
Insufficient mechanisms in herbal products regulation	Seminars on developing national policy and programs for TCAM. regulation for HM
Insufficient policies to oversee and control TCAM providers	Capacity building workshops on regulations on TCAM practice Supply guidance for stipulations for basic training of TCAM providers
Insufficient funding for TCAM research	Arrangement of funding projects for the reseaech on TCAM
Insufficient knowledge within the relevant health authorities and regulation bodies	Provide specialist assistance to advance productive and safe usage of native TCAM
Insufficient mechanisms for monitoring safety of TCAM, inc. HM	Capacity building workshops on safety monitoring of HM Seminars on developing national policy and programs for TCAM
Lack of cooperation channels between national health authorities to share information about TCAM	Capacity building workshops on safety monitoring of HM Provision on research databases Capacity building workshops about integrations of TCAM in healthcare centers context
Lack of education and training for TCAM providers	Seminars, workshops on integration of TCAM in health care centers context Provision of research databases Provide guidance for stipulations on basic training of TCAM providers Provide guidelines for self-care, instruction for the general public in healthcare centers

According to importance for WHO countries. (Source: interim data from second WHO TRM global survey, 2012 [13])

Table 5 The risks and harms of herbal medicines and possible solution

Risks	Possible harm	Solution
Poor/no regulatory status of herbal medicines: – no need to match standards set for conventional medicines – fewer requirements to provide information on packaging for patients – no need for research studies – no need for safety monitoring – poor quality control – misleading information – less requirements for TCAM providers	Insufficient details on contraindications, safe usage, and harmful side effects resulting in harm caused by recommended herbs (including self-medication) Common erroneous beliefs that herbal and supplement products are risk-free and well-governed Adverse effects resulting from recommended herbs (including through self-medication)	National/international policy building Scientific research database (efficacy, toxicity studies) Complete information for specialist and customer (composition, dosage, contraindications, interactions), guidance on self-care Safety monitoring, pharmacovigilance implementation TCAM providers education and regulation
Poor/no efficacy and safety studies	Lack of information on safe dosage, toxicity, or contraindications resulting in harm from prescribed substances (including self-prescribing) No guidelines for healthcare providers for the integration of HM Adverse effects resulting from recommended herbs (including through self-medication) Potential negative effect from replacing established effective care	Standardized research protocols required to evaluate the efficiency and safety of herbal products To have scientifical based, conclusive recommendations and guidelines Analysis of single active constituents as well as combinations Scientific research database (efficacy, toxicity studies) Standard system for prediction and evaluation of interactions Safety monitoring

(continued)

Table 5 (continued)

Risks	Possible harm	Solution
Poor/no quality control	Adverse effects resulting from recommended herbs (including through self-medication) Potential negative effect from replacing established effective care Financial harm	Standardized quality specifications for herbal products, identification, standardization, quality control on combinations, good manufacturing practice (GMP), licensing and information on product packaging Development and enforcement of policy and regulations of methodology and criteria for evaluating efficacy Research and development Sharing knowledge of procedures, rules, research, and service profiles Examine the safety of herbal products, and phytochemical and pharmacological research is required for herbal supplements
No safety monitoring	Adverse effects resulting from recommended herbs (including through self-medication) Erroneous assumptions that all herbals are safe and well-regulated Poor pharmacovigilance	Adopt commonly used binomial nomenclatures for medicinal plants Effective collaboration between phytotherapists, botanists, pharmacologists, phytochemists, and other important participants System of recording and collating adverse effects and interactions if self-administered or administered by providers
New herbals/supplement with unknown side effects, interactions	Adverse effects resulting from recommended herbs (including through self-medication) Harm resulting from procedures, interactions	Efficacy, toxicity studies Release to the market only those products which meet National Quality Specification and standard criteria
Limited information for the usage in pregnancy and breast feeding	Adverse effects resulting from recommended herbs (including through self-medication) to mother and child	Research and longitudinal observational studies, collected clinical evidence to support the safety of herbal medicines during pregnancy and breast feeding Greater knowledge of all the chemicals in plant extracts Appropriate labeling of all substances included and information in pregnancy and breast feeding
Using alternative health care to treat medical conditions based on misinformation: – indeceptive advertising/marketing – from unqualified practitioners (misdiagnosis, delayed diagnosis and treatment) – taking herbs for conditions not medically recognized as illness, or misdiagnosed, for example: gut dysbiosis, adrenal fatigue, etc. – using TCAM instead of conventional medicines for curable illnesses	Financial loss by paying for ineffective treatment Harm caused by receiving health advice which was insufficient and lacking knowledge Adverse side effects caused by contraindications Harm resulting from replacing established effective care Adverse effects resulting from recommended herbs (including through self-medication) Adverse side effects from practices Harm resulting from poor diagnosis or delayed treatments	Regulation of information about herbals, medicinal claims, advertising, and marketing Recognizing potential strategies and principles for incorporating TM into national and primary health care Requirements for the education of TCAM providers Continuous education and training of TCAM practitioners Patient insurance due to compensation for damages
Using TCAM treatments alongside conventional medicines without informing the doctor	Adverse effects resulting from recommended herbs (including through self-medication) Herb–drug contraindications	Assignment of responsibilities for self-prescribing Information sharing on regulatory issues Provision of guidance on self-care
Societal economic effect from the financial costs accrued by the use of alternative practitioners or therapists who are not covered by public health care insurance – when covered by third parties	Economic harm: Possible monetary cost through payment for some interventions which were found to be ineffective, or did not work Not getting coverage for the effective conventional treatment or delaying it	Information sharing on regulatory issues Patient insurance due to compensation for damages

Table 5 (continued)

Risks	Possible harm	Solution
Lack of clear and reliable information on dosage recommendations, interactions	Adverse effects resulting from recommended herbs (including through self-medication) Toxic effects when overused or taken in larger doses, or in wrong circumstances	Requirements for the herbal medicine, qualification of practitioners Ability to control and regulate TCAM advertising and claims Obtaining reliable objective information resources for consumers Information of usage for elderly, in pregnancy, possible interactions
Toxicity of certain herbal products/supplements: – St. John's wort, kava, ginger, ginkgo, ginseng, arnica, golden seal, aloe vera, ephedra, black cohosh, feverfew, henna, Licorice root, beta-carotene – Toxicity with specific TCM medicines: Cardiac arrhythmias, gastrointestinal symptoms, respiratory paralysis, dermatological reactions, hematological reactions, nephro/hepatotoxicity	Adverse effects resulting from recommended herbs (including through self-medication)	Scientific research database (efficacy, toxicity studies) Complete information for specialist and customer (composition, dosage, contraindications, interactions),guidance on self-care Requirements for the education of TCAM providers
Toxicity with specific herbal remedies/supplements: – herbal supplements containing unlisted pharmacological ingredients, for example, arsenic, betamethasone, diazepam – adulterated herbal products – herbal supplements containing metals, for example: aluminum, silver, lead, mercury, tin, and zinc – weight loss remedies containing N-nitroso-fenfluramine, chromium picolinate	Adverse effects resulting from recommended herbs (including through self-medication)	National/international policy building Product and quality control, especially assessments of services and safety Standardized processes and examination protocols required to test the quality, efficiency, and safety of herbal medicines
Adverse effects and possible interactions of herbal medicines/products for vulnerable people: Use of herbal medicines/supplements prescribed for pregnant women Prescribed herbal medicines for the elderly	Adverse effects resulting from recommended herbs (including through self-medication)	Product and quality control, especially assessments of services and safety Collection of data, examination, and publishing of herbal–drug interactions Assessing clinical reports and human case studies Consumer information and education
Herbal over exploitation	Harm for nature Effective herbal medicine shortage	Useful herbals cultivation Conserve plant species in the reserve *biosphere* The requirement of plant species to the industries can be fulfilled by medicinal plant farming and in vitro *Micropropagation*

Risk/solution (dosage, side effects, interaction, pregnancy, nanotechnology, etc.) [147, 149–151]

4 Pathway of Ancient Wisdom and Modern Science for the Health of People

Nature is the greatest medicinal apothecary on the planet and the use of plants for the purposes of health predates even human history and has led to the development of much modern medicine [18, 152]. Ancient civilizations created concoctions of seeds, herbs, leaves, fruit, and bark to treat a wide variety of illnesses. Archeological studies have shown that the herbalism dates as far back as 60,000 years ago in Iraq and 8000 years ago in China, there are written records plants dated back 5000 years to the Sumerians, who described uses for laurel, caraway, and thyme [153]. TCM has been healing people for several millennia. As the earliest written evidence of the past from India, Nagpur, is 5000-year-old clay tablets about usage of plants like the poppy and mandrake as drugs; an ancient Egyptian scroll dating back to around 1500 BC, where is written medicines of that time; the Ebers Papyrus where are recommendations about warming herbs for asthma, mint, and sandal wood to improve digestion and juniper for chest pain. It seems the plants are a necessary benefit, and their contribution to human health and vitality has long been known.

To date, 35,000–70,000 plant species have been screened for their medicinal use. 11% of the 252 medicines considered essential by the WHO were exclusively of flowering plant origin. There is increasing evidence that old molecules are finding new applications through better understanding of molecular biology and clinical observations [154]. Herbal medicine is the precursor of modern medicine and drug development [155]. Plants are responsible for as much as 25–50% of all pharmaceutical medications and natural products will always be a dependable source for pharmaceutical drug development. The success rate of the synthetic way of developing new drugs can be 1/10,000; however, the success rate of finding new therapeutic parts based on medicinal plants used in the traditional medical system may be 1/4 or more [156]. The natural products are mostly derived from plants, animal, and microbial resources. The strategies used for the drug development from natural resources are: source identification, collection of material and cultivation, crude extracts preparation and active part choice, bioactivity evaluation, preparation of active fraction and evaluation of bioactivity, lead component identification with mode of action described, preclinical assessment for the efficacy and safety, dose optimization, biochemical evaluation, clinical assessment for mode of administrations, dose optimization, bioavailability and all other request of FDA to approve drug efficacy, safety, treatment planning and monitoring, post marketing surveillance [16].

Special innovative technologies for drug invention from natural resources could be used: High-throughput screening, genomics, proteomics, metabolomics, CRISPR/Cas9 system, computer-aided drug development, artificial intelligence with the aim to address public health issues and achieve health-related sustainable and economic development [16].

The effects of extractions from the whole plant and the purified preparation in most cases the potency of the purified preparation decreases with each fractionation step [157]; since the therapeutic effect may be due to a combination of several compounds present in a medicinal plant, a complex mixture of compounds has a greater effect than isolated compounds. The advantages of a combinatorial approach may include synergy of various components, increased bioavailability, cumulative effects, and effects on the entire pathway network together [158].

The WHO encourages the use of traditional medicine to meet the world's populations healthcare needs because it is economical, safe, and culturally suitable (Organization 2013). Herbal medications appear to have novel clinical effects as a therapy option for NCDs or other disorders. There are many benefits and advantages of herbal medicine, if sustainable and high-quality herbal material is used and it is prescribed by qualified naturopath or herbalist:

1. Concept of holistic or integrative, whole-encompassing healthcare.
2. Taking responsibility for self-care and self-health [159].
3. Herbal medicines are natural, affordable, and accessible.
4. Lower side effects profile.

There is an increasing body of evidence showing that herbal medicines are helpful when used alongside allopathic medications in treatments for NCDs [160]. Evidence for herbal medicine safety is also constantly being discovered. Numerous clinical studies report no significant adverse effects observed [41, 62, 137, 139–141, 143–145, 161, 162]. However, the caution should always be applied to using strong herbal remedies when there is a lack of training and understanding of herbal medicines, and when mixing herbs and pharmaceuticals, especially a short therapeutic window medicines.

The goal of the Herbal medicine is to go toward in maximizing evidences of quality, safety, and effectiveness, to be integrated in national health systems for the promotion of well-being and health of people worldwide. The big efforts from the WHO, Member States, partners, stakeholders must be made.

WHO had created a program for 2002–2005 with four principal goals: policy; safety, efficacy, and quality standards; access and rational use which helped to move forward. Their Traditional Medicine program for 2014–2023 was intended to assist countries in promoting health and protect TCAM consumers by understanding TCAM own national situation, exploring the forms and reasons for TCAM usage, determining present and future needs, developing policies, regulations and guidelines that address TCAM, supervising individual products, practices, and practitioners. WHO invited Member States to direct their projects toward three strategic objectives:

1. to create a knowledge base for active management of TCAM through appropriate national policies;
2. to make clearer quality assurances, safety measurements, and reliable use and effectiveness of TCAM through regulation of products, services and practitioners through better training and education; and,
3. to promote universal health coverage by integrating TCAM services into national health services and self-health care by ensuring users are able to make informed choices.

For the reaching of these goals the inclusion of TCAM into national health systems, preparing of international standards, guidelines and methodologies for research, support of clinical research on effectiveness and safety, formation of knowledge and evidence base, information exchange, reimbursement are needed (Table 6).

Table 6 Strategies and actions for future TCAM

Strategic objective	Actions for countries	Actions for partners and stakeholders
Strengthening of the knowledge	Monitoring of safety Identifying historical, traditional, and scientific sources Determining cost/benefit analysis and cost-effectiveness Promoting study and creativity, developing national research agenda Performing health technology assessments Knowledge generation, translation, dissemination Developing evaluation of safety, efficacy, and quality criteria for products and practitioners Controlling advertising, practice and media Protecting TCAM resources Following global trends and events Publishing TCAM treatment guidelines and list of essential herbal medicines Create mechanisms for the availability of TCAM Create workable standards of care for primary healthcare, NCDs	Supporting countries and WHO in research programs Helping on assessing risks and benefits Fostering communication, documentation, evaluation, innovation Increasing database of knowledge repositories Developing research methodologies with reference on theories and practice Building the capacity, and collaboration for international research
Product regulation	Identifying current models, policies, and rules to guarantee product Efficacy and safety Apply suitable regulatory systems to make safe the use of TCAM Provide for a basic regulatory guideline that ensures quality Safety and efficiency, licensing providers, adverse events reporting Disseminating evidence, frameworks and data which support health claims Setting or adopting standards for T&CM products International regulatory information sharing about TCAM products Defining mechanisms of stakeholders involvement	Encouraging practitioners, TCAM communities and industries, analysts and clients to be included in devising regulations for products Work together to monitor and supervise systems to control risk management
Practice/practitioner regulation	Improving availability of data and expand knowledge of the Quality, safety, and effectiveness of TCAM Create a code of conduct to ensure a high standard of ethics Creation of reference points, principles, and regulations for training, continued education, accreditation, licensing and reimbursement for different forms of TCAM Formalize, license, and accredit health practitioners Setting the range and principles of practice, monitoring Registering TCAM practitioners, including their licensing Designate and build the capabilities of their bodies and schools to apply risk based and context specific regulations Gather a precise inventory of national TCAM resources and involvements that are useful to the requirements of the general public of population	Encourage clients, practitioner organizations, and researchers to support their community by setting standards for and regulating training Developing frameworks and a code of conduct, for the ethical sponsorship of education, training, and practice of TM practitioners Encourage the sharing of information and knowledge and widespread communication and between practitioners Fostering cohesiveness among traditional health practitioners Encouraging and promote accreditation of education and training institutes Promoting respect between TCAM and other health practitioners Encouraging conventional medicine practitioners to become more familiar with the use of TCAM
Universal health coverage promotion	Recognizing TM as useful resource which contributes to the betterment of national health care systems, particularly primary Exploring possible TCAM integration in the national health service Encouraging the development of appropriate health facilities for TCAM Ensuring impartial consideration for safe and accessible TCAM within the current insurance model and in national health financial coverage models Promoting sound analysis of the economic benefits of including TCAM approaches Promoting continuous professional development and education, and also continued assessment, analysis, and proofs into TCAM practices Encouraging insurance companies and health reimbursement agencies, as well as nongovernmental organizations and the for-profit sector, work for greater coverage	

(continued)

Table 6 (continued)

Strategic objective	Actions for countries	Actions for partners and stakeholders
Informed choices for consumers	Developing mechanisms and frameworks for consumer protection, and a growing awareness of the benefits of TCAM Adverse events reporting channels and the rightful utility of TCAM services, supplements and healthcare providers, their promotion and advertising Strengthening education across disciplines and complementary practice between conventional medicine and TCAM Fostering fellowship and opening communication about self-health care among all who are interested Ensure consumers can make informed choices Providing information and advice to consumers on TCAM-potential risk of practices and products Promoting health care practitioner education about TCAM Encouraging patients to declare their use of TCAM for practitioners Advocating responsible and accurate advertising and promotion	Promoting mutual respect, collaboration and sharing between conventional and TCAM practitioners Promote worldwide dialogue among practitioners Regarding integrative models Provide information on TCAM services that have proven efficacy, safety, and quality and where these can be accessed Address medical pluralism

These listed and more steps and actions could be taken for future success in herbal medicine. The starting point could be defining herbal medicine in health and social policies, mobilizing its place in enhancing the health and well-being of the population by ensuring effectiveness, quality, and safety, access and wide coverage for its therapy together with sound analysis and innovations [163].

Research on herbal medicine will translate eastern therapeutic measures into western medical paradigms.

5 Summary

The current dynamics of NSDs deaths worldwide seems decreasing slightly in the world as well as Disability-Adjusted Life Year (DALY) with the big difference in high and low-income countries (Mortality from non-communicable diseases, 2000 to 2019). (Accessed 22 January 2023). [https://ourworldindata.org/grapher/mortality-from-ncds-sdgs?tab=chart] (Figs. 3 and 4).

In 2011 the forecasting model projected that by 240% will be increase in elderly and 24% in obese people, average of lifespan will grow to 78.5 years, deaths from communicable diseases (CD) will decrease by almost 70% by 2060, whereas deaths from both NCDs and injuries will more than double- will be increase in death from diabetes by 205%, cancer- by 108%, mental disease- by 245%. Deaths from NCDs will outnumber deaths from CDs by more than 5 to 1 [164]. DALY (Disability-Adjusted Life Year) represents one lost year of healthy life.

Aboveall, we need to make more effort for both preventive and curative measures. The long history of TCAM, growing scientific evidence is encouraging more attention to be given to HM. As per WHO, phyto medicine or herbal medicine is the sum total of the knowledge, skill, and practices based on the theories and experiences indigenous to different cultures, whether explicable or not, used in the maintenance of health as well as in the prevention, diagnosis, improvement, or treatment of physical and mental illness [165].

Contemporary conventional medicine and HM are inextricably linked because until the Industrial Revolution in eighteenth century and with the advent of synthetic chemistry, all medicines were derived from natural sources, mainly plants. The pursuit of evidence-based medicine in all types of therapeutic approaches means that herbal medicine can be a safe and beneficial part of the integrative medicine of the future. In order to achieve progress for HM, the clear scope of usage, a legal status, constant safety monitoring, and the strengthening of a solid scientific basis by health care and scientific organizations is necessary. The trends driving integration of HM in contemporary healthcare are:

High rates of chronic diseases globally.

Chronic diseases are not adequately prevented and treated.

High rates of adverse side effects, drug interactions because of aging population, multimorbidity, and polypharmacy.

High cost of healthcare.

High utilization of TCAM by general public without informing physicians.

In adequate education of health professionals about TCAM methods.

The growing interest of physicians in TCAM education for integrative medicine.

The increasing amount of research and clinical guidelines from health organizations for the inclusion of TCAM practices [166].

Pharmacological research into innovative drug has been based around herbal medicine. We need to practice strict ecological ethics to conserve biodiversity for research into new drugs. Modern anthropocentrism, which has philosophical implications for sustainable development, should be introduced in herbal medicine processes [21].

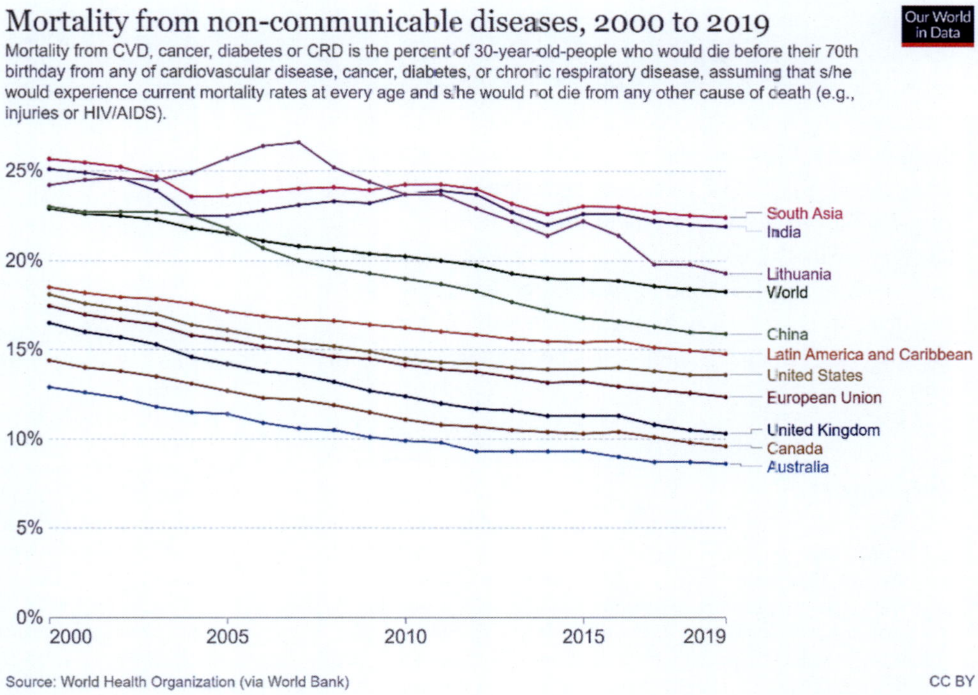

Fig. 3 Mortality from non-communicable diseases, 2000 to 2019 [courtesy to Our World in Data]

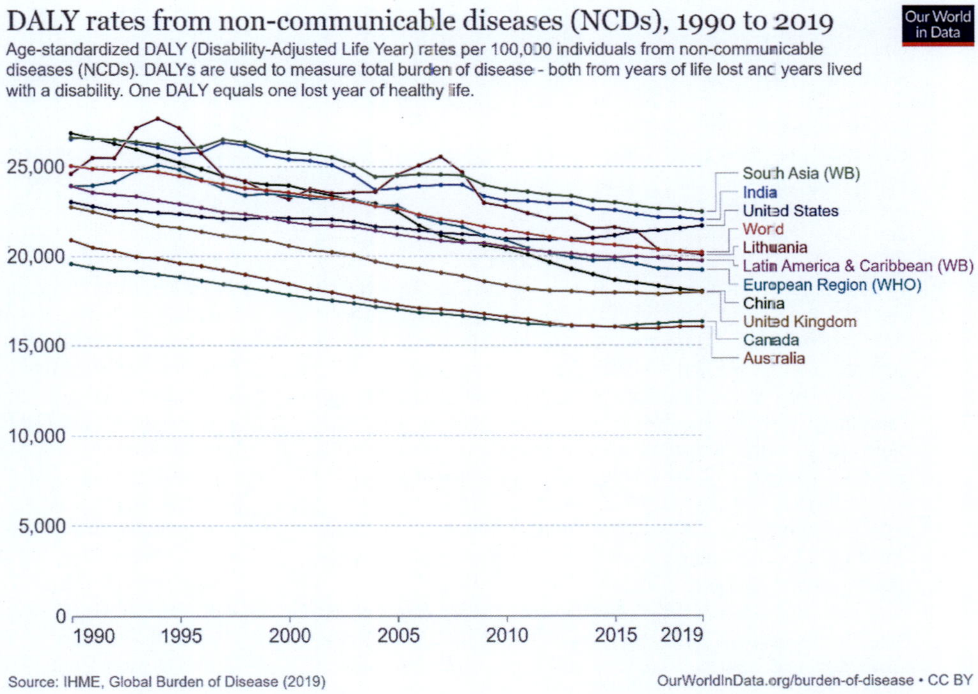

Fig. 4 DALY rates from NCDs. [courtesy to Our World in Data]

TCAM is a crucial and often neglected part of health care and interest in its services is increasing. Good global planning will encourage its appropriate inclusion, control, and supervision to develop a proactive policy toward this part of health care. The pandemic has stimulated a real increase in the use of natural medicines worldwide. Besides supplying demand it is need to put efforts on proactively creating future, exploring ways of herbal medicine to become an effective and radical part future [167].

The worldwide inclusion of traditional medicine into national health systems as imagined by WHO relied upon a broad and wide ranging approach by the following participants: governments, healthcare practitioners from all modalities, manufacturers and the public [168].

The plan has two important aims: to help countries to utilize the contribution of TCAM to health, well-being and health care that is centered on the individual needs: and to promote the safe and effective use of TCAM through ensuring the safety of the products, the practices and the practitioners. These aims can be attained by creating a knowledge base, by advocating for national policies, increased emphasis on safety, quality, and effectiveness through careful control; and promoting universal health coverage (Strategija 2023). To secure the future of herbal medicine for all we need to promote environmentally friendly farming methods, decarbonizing supply chains, choosing green energy and developing new research-based production practices, thus we can help ensure that safe, effective and transparent medicinal plants could be available to anyone who needs it.

References

1. Al-Maskari F. LIFESTYLE DISEASES: an economic burden on the health services. UN Chron. 2010;5:1–2; https://www.un.org/en/chronicle/article/lifestyle-diseases-economic-burden-health-services.
2. World Health Organization. Noncommunicable diseases. 2022; [cited 2023 jan 15]. https://www.who.int/news-room/fact-sheets/detail/noncommunicable-diseases.
3. Kontis V, Mathers CD, Rehm J, Stevens GA, Shield KD, Bonita R, et al. Contribution of six risk factors to achieving the 25x25 non-communicable disease mortality reduction target: a modelling study. Lancet. 2015;384(9941):427–37.
4. Sharma CB. Applied environmental sciences & enginerrings. BFC Publications; 2021. p. 274.
5. Capocaccia R, Gatta G, Dal Maso L. Life expectancy of colon, breast, and testicular cancer patients: an analysis of US-SEER population-based data. Ann Oncol. 2015;26:1263–8. https://doi.org/10.1093/annonc/mdv131; pmid:25735314.
6. Franco OH, Steyerberg EW, Hu FB, Mackenbach J, Nusselder W. Associations of diabetes mellitus with total life expectancy and life expectancy with and without cardiovascular disease. Arch Intern Med. 2007;167:1145–51. https://doi.org/10.1001/archinte.167.11.1145; pmid:17563022.
7. Lee HY, Hwang JS, Jeng JS, Wang JD. Quality-adjusted life expectancy (QALE) and loss of QALE for patients with ischemic stroke and intracerebral hemorrhage: a 13-year follow-up. Stroke. 2010;41:739–44. https://doi.org/10.1161/STROKEAHA.109.573543; pmid:20150543.
8. World Health Organization. Noncommunicable diseases. 2022; [cited 2022 Dec 19]. https://www.who.int/news-room/fact-sheets/detail/noncommunicable-diseases.
9. Nugent R, Bertram MY, Jan S, Niessen LW, Sassi F, Jamison DT, Pier EG, Beaglehole R. Investing in non-communicable disease prevention and management to advance the sustainable development goals. Lancet. 2018;391(10134):2029–35. https://doi.org/10.1016/S0140-6736(18)30667-6; Epub 2018 Apr 5. Erratum in: Lancet 2018 Apr 9; PMID: 29627167.
10. Hategeka C, Adu P, Desloge A, Marten R, Shao R, Tian M, et al. Implementation research on non communicable disease prevention and control interventions in low- and middle-income countries: a systematic review. PLoS Med. 19(7):e1004055. https://doi.org/10.1371/journal.pmed.1004055.
11. Li Y, Schoufour J, Wang DD, Dhana K, et al. Healthy lifestyle and life expectancy free of cancer, cardiovascular disease, and type 2 diabetes: prospective cohort study. BMJ. 2020;368:l6669.
12. Lee EL, Richards N, Harrison J, Barnes J. Prevalence of use of traditional, complementary and alternative medicine by the general population: a systematic review of national studies published from 2010 to 2019. Drug Saf. 2022;45(7):713–35. https://doi.org/10.1007/s40264-022-01189-w; Epub 2022 Jul 5. PMID: 35788539; PMCID: PMC9296440.
13. WHO global report on traditional and complementary medicine 2019. ISBN 978-92-4-151543-6. World Health Organization 2019 [cited 2022 Dec 19].
14. Kennedy DA, Lupattelli A, Koren G, Nordeng H. Safety classification of herbal medicines used in pregnancy in a multinational study. BMC Complement Altern Med. 2016;16(1):102.
15. Traditional, complementary and integrative medicine: International Regulatory Cooperation for Herbal Medicines (IRCH). World Health Organization. Geneva. http://www.who.int/traditional-complementary-integrative-medicine/collaboration/IRCH/en//.
16. Aware CB, Patil DN, Suryawanshi SS, Mali PR, Rane MR, Gurav RG, Jadhav JP. Natural bioactive products as promising therapeutics: a review of natural product-based drug development. S Afr J Bot. 2022;151(part B):512–28.
17. WHO traditional medicine strategy 2014–2023. 2013. https://www.who.int/publications/i/item/9789241506096.
18. Bhardwaj S, Verma R, Gupta J. Challenges and future prospects of herbal medicine. Int Res Med Health Sci. 2018;1(1):12–5.
19. Schippmann U, Leaman DJ, Cunningham ARBA. A comparison of cultivation and wild collection of medicinal and aromatic plants under sustainability aspects. Frontis. 2006;1:75. https://doi.org/10.1007/1-4020-5449-1_6.
20. Newman DJ, Cragg GM. Natural products as sources of new drugs over the nearly four decades from 01/1981 to 09/2019. J Nat Prod. 2020;83(3):770–803.
21. Pan S-Y, Zhou SF, Gao SH, Yu ZL, Zhang SF, Tang MK, et al. New perspectives on how to discover drugs from herbal medicines: CAM's outstanding contribution to modern therapeutics. Evid Based Complement Altern Med. 2013;25:627375. https://doi.org/10.1155/2013/627375.
22. WHO global report on traditional and complementary medicine. 2019. ISBN 978–92–4-151543-6.
23. Herbal medicinal products. European Commission. Public health. https://health.ec.europa.eu/medicinal-products/herbal-medicinal-products_en. Accessed 15 Jan 2023.
24. Sánchez M, González-Burgos E, Iglesias I, Lozano R, Gómez-Serranillos MP. Current uses and knowledge of medicinal plants in the autonomous community of Madrid (Spain): a descriptive cross-sectional study. BMC Complement Med Ther. 2020;20:306. https://doi.org/10.1186/s12906-020-03089-x.
25. Garcia-Alvarez A, Egan B, de Klein S, Dima L, Maggi FM, Isoniemi M, et al. Usage of plant food supplements across six European countries: findings from the PlantLIBRA consumer survey. PLoS One. 2014;9(3):e92265. https://doi.org/10.1371/journal.pone.0092265.
26. World Health Organisation. Key technical issues of herbal medicines with reference to interaction with other medicines. World Health Organization; 2021; ISBN 978–92–4-001914-0 (electronic version).
27. World Health Organization. Traditional, complementary and integrative medicine: International Regulatory Cooperation for Herbal Medicines (IRCH) 2017. http://www.who.int/traditional-complementary-integrative-medicine/collaboration/IRCH/en//.
28. Li Y, Liu Y, Cui J, Zhao H, Liu Y, Huang L. Cohort studies on chronic non-communicable diseases treated with traditional Chinese medicine: a bibliometric analysis. Front Pharmacol. 2021;12:639860. https://doi.org/10.3389/fphar.2021.639860.
29. Smitha T, Kawab K, Ecklb V, Mortonc C, Stredney R. Herbal supplement sales in US increase 8.5% in 2017, topping $8 billion. Market Rep. 2018;119:1; https://static1.squarespace.com/

static/5c468b4bb40b9dbedbc53f12/t/5d141cc6205fd60001 4da424/1561599180850/Herb+MarketReport+2017+Botanicals.pdf.
30. Ganora L. Herbal constituents: foundations in phytochemistry. Colorado: Herbalchem Press; 2009.
31. Zhang YJ, Gan RY, Li S, et al. Antioxidant phytochemicals for the prevention and treatment of chronic diseases. Molecules. 2015;20(12):21138–56.
32. Liguori I, Russo G, Curcio F, Bulli G, Aran L, Della-Morte D, Gargiulo G, Testa G, Cacciatore F, Bonaduce D, Abete P. Oxidative stress, aging, and diseases. Clin Interv Aging. 2018;13:757–72.
33. Shayganni E, Bahmani M, Asgary S, Rafieian-Kopaei M. Inflammaging and cardiovascular disease: management by medicinal plants. Phytomedicine. 2016;23(11):1119–26.
34. Bachheti RK, Worku LA, Gonfa YH, Zebeaman M, Pandey DP, Bachheti A. Prevention and treatment of cardiovascular diseases with plant phytochemicals: a review. Evid Based Complement Alternat Med. 2022;2022:5741198.
35. Shaito A, Thuan DTB, Phu HT, Nguyen THD, Hasan H, Halabi S, Abdelhady S, Nasrallah GK, Eid AH, Pintus G. Herbal medicine for cardiovascular diseases: efficacy, mechanisms, and safety. Front Pharmacol. 2020;11:422.
36. Wei M, Wang D, Kang D, et al. Overview of cochrane reviews on Chinese herbal medicine for stroke. Integr Med Res. 2020;9(1):5–9.
37. Hao D-L, Li J-M, Xie R, Huo H-R, Xiong X-J, Sui F, Wang P-Q. The role of traditional herbal medicine for ischemic stroke: from bench to clinic—a critical review. Phytomedicine. 2023;109:154609.
38. Degenring FH, Suter A, Weber M, Saller R. A randomised double blind placebo controlled clinical trial of a standardised extract of fresh Crataegus berries (Crataegisan®) in the treatment of patients with congestive heart failure NYHA II. Phytomedicine. 2003;10(5):363–9.
39. Guo R, Pittler MH, Ernst E. Hawthorn extract for treating chronic heart failure. Cochrane Database Syst Rev. 2008;1:CD005312.
40. Wang J, Xiong X, Feng B. Effect of crataegus usage in cardiovascular disease prevention: an evidence-based approach. Evid Based Complement Alternat Med. 2013;2013:149363.
41. Cloud A, Vilcins D, McEwen B. The effect of hawthorn (Crataegus spp.) on blood pressure: a systematic review. Adv Integr Med. 2020;7(3):167–75.
42. Imaizumi VM, Laurindo LF, Manzan B, Guiguer EL, Oshiiwa M, Otoboni AM, Araujo AC, Tofano RJ, Barbalho SM. Garlic: a systematic review of the effects on cardiovascular diseases. Crit Rev Food Sci Nutr. 2022;18:1–23.
43. Hadi A, Pourmasoumi M, Ghaedi E, SahebkarA. The effect of curcumin/turmeric on blood pressure modulation: a systematic review and meta-analysis. Pharmacol Res. 2019;150:104505.
44. Latif R, Mumtaz S, Al Sheikh MH, Chathoth S, Nasser Al Naimi S. Effects of turmeric on cardiovascular risk factors, mental health, and serum homocysteine in overweight, obese females. Altern Ther Health Med. 2020;pii:AT6135.
45. Jung I, Kim H, Moon S, Lee H, Kim B. Overview of *Salvia miltiorrhiza* as a potential therapeutic agent for various diseases: AN update on efficacy and mechanisms of action. Antioxidants (Basel). 2020;9(9):857.
46. Lacaille-Dubois MA, Franck U, Wagner H. Search for potential angiotensin converting enzyme (ACE)-inhibitors from plants. Phytomedicine. 2001;8(1):47–52.
47. Dalli E, Colomer E, Tormos MC, Cosín-Sales J, Milara J, Esteban E, Sáez G. Crataegus laevigata decreases neutrophil elastase and has hypolipidemic effect: a randomized, double-blind, placebo-controlled trial. Phytomedicine. 2011;18(8–9):769–75.
48. Kamranirad Z, Javadi B, Hayes A, Karimi G. Potential angiotensin converting enzyme (ACE) inhibitors from Iranian traditional plants described by Avicenna's canon of medicine. Avicenna J Phytomed. 2019;9(4):291–309.
49. Sobenin IA, Nedosugova LV, Filatova LV, Balabolkin MI, Gorchakova TV, Orekhov AN. Metabolic effects of time-released garlic powder tablets in type 2 diabetes mellitus: the results of double-blinded placebo-controlled study. Acta Diabetol. 2007;45:1–6.
50. Batiha GES, Beshbishy AM, Wasef LG, Elewa YHA, Al-Sagan AA, Abd El-Hack ME, Taha AE, Abd-Elhakim YM, Devkota HP. Chemical constituents and pharmacological activities of garlic (Allium sativum L.): a review. Nutrients. 2020;12:872.
51. Yarnell E, Abascal K. Botanicals for regulating heart rhythms. Altern Complement Ther. 2003;9(3):125–9.
52. Savtschenko A, Dhein S, Rauwald HW. The antiarrhythmic effects of lavandulifolioside and ferulic acid from Leonurus cardiaca extracts on cardiac electrophysiology. Zeitschrift für Phytothérapie. 2013;34(S 01):P25.
53. Bernatoniene J, Kopustinskiene DM, Jakstas V, Majiene D, Baniene R, Kuršvietiene L, Masteikova R, Savickas A, Toleikis A, Trumbeckaite S. The effect of Leonurus cardiaca herb extract and some of its flavonoids on mitochondrial oxidative phosphorylation in the heart. Planta Med. 2014;80(07):525–32.
54. Dong Y, Liao J, Yao K, Jiang W, Wang J. Application of traditional Chinese medicine in treatment of atrial fibrillation. Evid Based Complement Alternat Med. 2017;2017:1381732.
55. Shen S, Wu G, Luo W, Li W, Li X, Dai C, Huang W, Liang G. Leonurine attenuates angiotensin II-induced cardiac injury and dysfunction via inhibiting MAPK and NF-κB pathway. Phytomedicine. 2023;108:154519.
56. Liu B, Du Y, Cong L, Jia X, Yang G. Danshen (Salvia miltiorrhiza) compounds improve the biochemical indices of the patients with coronary heart disease. Evid Based Complement Alternat Med. 2016;2016:9781715.
57. Wang L, Ma R, Liu C, et al. Salvia miltiorrhiza: a potential red light to the development of cardiovascular diseases. Curr Pharm Des. 2017;23(7):1077–97.
58. Jain S, Buttar HS, Chintameneni M, Kaur G. Prevention of cardiovascular diseases with anti-inflammatory and anti- oxidant nutraceuticals and herbal products: an overview of pre-clinical and clinical studies. Recent Patents Inflamm Allergy Drug Discov. 2018;12:145–57.
59. Kumar S, Mittal A, Babu D, Mittal A. Herbal medicines for diabetes management and its secondary complications. Curr Diabetes Rev. 2021;17(4):437–56.
60. Park S, Keum D, Kim H. Efficacy and safety of anti-obesity herbal medicine focused on pattern identification: a systematic review and meta-analysis. Medicine (Baltimore). 2022;101(50):e32087.
61. Tian J, Liu Y, Liu Y, Chen K, Lyu S. Cellular and molecular mechanisms of diabetic atherosclerosis: herbal medicines as a potential therapeutic approach. Oxidative Med Cell Longev. 2017;2017:9080869.
62. Liu ZL, Xie LZ, Zhu J, Li GQ, Grant SJ, Liu JP. Herbal medicines for fatty liver diseases. Cochrane Database Syst Rev. 2013;8:CD009059.
63. Jang S, Jang BH, Ko Y, Sasaki Y, Park JS, Hwang EH, Song YK, Shin YC, Ko SG. Herbal medicines for treating metabolic syndrome: a systematic review of randomized controlled trials. Evid Based Complement Alternat Med. 2016;2016:5936402.
64. Payab M, Hasani-Ranjbar S, Shahbal N, Qorbani M, Aletaha A, Haghi-Aminjan H, Soltani A, Khatami F, Nikfar S, Hassani S, Abdollahi M. Effect of the herbal medicines in obesity and metabolic syndrome: a systematic review and meta-analysis of clinical trials. Phytother Res. 2020;34(3):526–45.
65. Prasopthum A, Insawek T, Pouyfung P. Herbal medicine use in Thai patients with type 2 diabetes mellitus and its association

with glycemic control: a cross-sectional evaluation. Heliyon. 2022;8(10):e10790.
66. Yu A, Adelson D, Mills D. Chinese herbal medicine versus other interventions in the treatment of type 2 diabetes: a systematic review of randomized controlled trials. J Evid Based Integr Med. 2018;23:2515690X18781519.
67. Kiss R, Szabó K, Gesztelyi R, et al. Insulin-sensitizer effects of fenugreek seeds in parallel with changes in plasma MCH levels in healthy volunteers. Int J Mol Sci. 2018;19(3):771.
68. Tavakoly R, Maracy MR, Karimifar M, Entezari MH. Does fenugreek (*Trigonella foenum-graecum*) seed improve inflammation, and oxidative stress in patients with type 2 diabetes mellitus? A parallel group randomized clinical trial. European J Integr Med. 2018;18:13–7.
69. Geberemeskel GA, Debebe YG, Nguse NA. Antidiabetic effect of fenugreek seed powder solution (*Trigonella foenum-graecum L.*) on hyperlipidemia in diabetic patients. J Diabetes Res. 2019;2019:8507453.
70. Najdi RA, Hagras MM, Kamel FO, et al. A randomized controlled clinical trial evaluating the effect of Trigonella foenum-graecum (fenugreek) versus glibenclamide in patients with diabetes. Afr Health Sci. 2019;19(1):1594–601.
71. Banerjee S, Bhattacharjee P, Kar A, Mukherjee PK. LC–MS/MS analysis and network pharmacology of Trigonella foenum-graecum–a plant from ayurveda against hyperlipidemia and hyperglycemia with combination synergy. Phytomedicine. 2019;60:152944.
72. Rao AS, Hegde S, Pacioretty LM, DeBenedetto J, Babish JG. *Nigella sativa* and *Trigonella foenum-graecum* supplemented chapatis safely improve HbA1c, body weight, waist circumference, blood lipids, and fatty liver in overweight and diabetic subjects: a twelve-week safety and efficacy study. J Med Food. 2020;23(9):905–19.
73. Deyno S, Eneyew K, Seyfe S, Tuyiringire N, Peter EL, Muluye RA, Tolo CU, Ogwang PE. Efficacy and safety of cinnamon in type 2 diabetes mellitus and pre-diabetes patients: a meta-analysis and meta-regression. Diabetes Res Clin Pract. 2019;156:107815.
74. Zare R, Nadjarzadeh A, Zarshenas MM, Shams M, Heydari M. Efficacy of cinnamon in patients with type II diabetes mellitus: a randomized controlled clinical trial. Clin Nutr. 2019;38(2):549–56.
75. Shang C, Lin H, Fang X, Wang Y, Jiang Z, Qu Y, Xiang M, Shen Z, Xin L, Lu Y, Gao J. Beneficial effects of cinnamon and its extracts in the management of cardiovascular diseases and diabetes. Food Funct. 2021;12(24):12194–220.
76. Shirzad F, Morovatdar N, Rezaee R, Tsarouhas K, Abdollahi MA. Cinnamon effects on blood pressure and metabolic profile: a double-blind, randomized, placebo-controlled trial in patients with stage 1 hypertension. Avicenna J Phytomed. 2021;11(1):91–100.
77. Qin B, Panickar KS, Anderson RA. Cinnamon: potential role in the prevention of insulin resistance, metabolic syndrome, and type 2 diabetes. J Diabetes Sci Technol. 2010;4(3):685–93.
78. Kianbakht S, Khalighi-Sigaroodi F, Dabaghian FH. Improved glycemic control in patients with advanced type 2 diabetes mellitus taking *Urtica dioica* leaf extract: a randomized double-blind placebo-controlled clinical trial. Clin Lab. 2013;59:1071–6.
79. Bahmani M, Zargaran A, Rafieian-Kopaei M, Saki K. Ethnobotanical study of medicinal plants used in the management of diabetes mellitus in the Urmia, Northwest Iran. Asian Pac J Trop Med. 2014;7:S348–54.
80. Khalili N, Fereydoonzadeh R, Mohtashami R, Mehrzadi S, Heydari M, Huseini HF. Silymarin, olibanum, and nettle, a mixed herbal formulation in the treatment of type II diabetes: a randomized, double-blind, placebo-controlled, clinical trial. J Evid Based Complement Altern Med. 2017;22:603.
81. Korani B, Mirzapour A, Moghadamnia AA, Khafri S, Neamati N, Parsian H. The effect of *Urtica dioica* hydro-alcoholic extract on glycemic index and AMP-activated protein kinase levels in diabetic patients: a randomized single-blind clinical trial. Iran Red Crescent Med J. 2017;19(3):e40572.
82. Gohari A, Noorafshan A, Akmali M, Zamani-Garmsiri F, Seghatoleslam A. Urtica Dioica distillate regenerates pancreatic beta cells in Streptozocin-induced diabetic rats. Iran J Med Sci. 2018;43:174.
83. El Haouari M, Rosado JA. Phytochemical, anti-diabetic and cardiovascular properties of Urtica Dioica L. (Urticaceae): a review. Mini Rev Med Chem. 2018;19(1):63–71.
84. Samakar B, Mehri S, Hosseinzadeh H. A review of the effects of *Urtica dioica* (nettle) in metabolic syndrome. Iran J Basic Med Sci. 2022;25(5):543–53.
85. Sukandar EY, Permana H, Adnyana IK, Sigit J, Ilyas RA, Hasimun P, Mardiyah D. Clinical study of turmeric (Curcuma longa L.) and garlic (Allium sativum L.) extracts as antihyperglycemic and anthyperlipidemic agent in type-2 diabetes-dyslipidemia patients. Int J Pharmacol. 2010;6(4):456–63.
86. Kumar R, Chhatwal S, Arora S, et al. Antihyperglycemic, antihyperlipidemic, anti-inflammatory and adenosine deaminase- lowering effects of garlic in patients with type 2 diabetes mellitus with obesity. Diabetes Metab Syndr Obes. 2013;6:49–56.
87. Wang J, Zhang X, Lan H, Wang W. Effect of garlic supplement in the management of type 2 diabetes mellitus (T2DM): a meta-analysis of randomized controlled trials. Food Nutr Res. 2017;61(1):1377571.
88. Bahramsoltani R, Farzaei MH, Ram M, Nikfar S, Rahimi R. Bioactive foods and medicinal plants for cardiovascular complications of type II diabetes: current clinical evidence and future perspectives. Evid Based Complement Alternat Med. 2021;2021:6681540.
89. Lupak M, Khokhla M, Hachkova GY, Kanyuka O, Klymyshyn N, Chajka YP, Skybitska M, Sybirna N. The alkaloid-free fraction from Galega officinalis extract prevents oxidative stress under experimental diabetes mellitus. Ukr Biochem J. 2015;87:78–86.
90. Abtahi-Evari SH, Shokoohi M, Abbasi A, Rajabzade A, Shoorei H, Kalarestaghi H. Protective effect of Galega officinalis extract on streptozotocin-induced kidney damage and biochemical factor in diabetic rats. Crescent J Med Biol Sci. 2017;4:108–14.
91. Shikov AN, Narkevich IA, Akamova AV, Nemyatykh OD, Flisyuk EV, Luzhanin VG, Povydysh MN, Mikhailova IV, Pozharitskaya ON. Medical species used in Russia for the management of diabetes and related disorders. Front Pharmacol. 2021;12:697411.
92. Chuengsamarn S, Rattanamongkolgul S, Luechapudiporn R, Phisalaphong C, Jirawatnotai S. Curcumin extract for prevention of type 2 diabetes. Diabetes Care. 2012;35(11):2121–7.
93. Meng B, Li J, Cao H. Antioxidant and antiinflammatory activities of curcumin on diabetes mellitus and its complications. Curr Pharm Des. 2013;19(11):2101–13.
94. Amin F, Islam N, Anila N, Gilani AH. Clinical efficacy of the co-administration of turmeric and black seeds (Kalongi) in metabolic syndrome–a double blind randomized controlled trial–TAK-MetS trial. Complement Ther Med. 2015;23(2):165–74.
95. Alavinejad P, Farsi F, Husain D, Rezazadeh A. The effect of turmeric on lipid profile, malondialdehyde, liver echogenicity and enzymes among patients with nonalcoholic fatty liver disease: a randomized double blind clinical trial. Diabetol Metab Syndr. 2021;13(1):1–9.

96. Marmitt DJ, Shahrajabian MH, Goettert MI, Rempel C. Clinical trials with plants in diabetes mellitus therapy: a systematic review. Expert Rev Clin Pharmacol. 2021;14(6):735–47.
97. Marton LT, Pescinini-E-Salzedas LM, Camargo MEC, et al. The effects of curcumin on diabetes mellitus: a systematic review. Front Endocrinol (Lausanne). 2021;12:669448.
98. Vafaeipour Z, Razavi BM, Hosseinzadeh H. Effects of turmeric (*Curcuma longa*) and its constituent (curcumin) on the metabolic syndrome: an updated review. J Integr Med. 2022;20(3):193–203.
99. Uchio R, Okuda-Hanafusa C, Saji R, et al. A hot water extract of *Curcuma longa* L. improves fasting serum glucose levels in participants with low-grade inflammation: re-analysis of data from two randomized, double-blind, placebo-controlled trials. Nutrients. 2022;14(18):3763.
100. Tajmohammadi A, Razavi BM, Hosseinzadeh H. Silybum marianum (milk thistle) and its main constituent, silymarin, as a potential therapeutic plant in metabolic syndrome: a review. Phytother Res. 2018;32(10):1933–49.
101. Doostkam A, Fathalipour M, Anbardar MH, Purkhosrow A, Mirkhani H. Therapeutic effects of milk thistle (Silybum marianum L.) and artichoke (Cynara scolymus L.) on non alcoholic fatty liver disease in type 2 diabetic rats. Can J Gastroenterol Hepatol. 2022;2022:2868904.
102. Helal EGE, El-Wahab SMA, Sharaf AMM, Zedan GA. Effect of Zingiber officinale on fatty liver induced by oxytetracycline in albino rats. Egypt J Hosp Med. 2012;46:26–42.
103. Mansourm MS, Ni YM, Roberts AL, Kelleman M, Roychoudhury A, St-Onge MP. Ginger consumption enhances the thermic effect of food and promotes feelings of satiety without affecting metabolic and hormonal parameters in overweight men: a pilot study. Metabolism. 2012;61:1347–52.
104. Mozaffari-Khosravi H, Talaei B, Jalali BA, Najarzadeh A, Mozayan MR. The effect of ginger powder supplementation on insulin resistance and glycemic indices in patients with type 2 diabetes: a randomized, double-blind, placebo-controlled trial. Complement Ther Med. 2014;22:9–16.
105. Athirah ZA, Ibrahim J, Roza D, Fhataheya B. Protective effects of the standardized extract of Zingiber officinale on myocardium against isoproterenol-induced biochemical and histopathological alterations in rats. Pharm Biol. 2015;53:1795–802.
106. Shidfar F, Rajab A, Rahideh T, Khandouzim N, Hosseini S, Shidfar S. The effect of ginger (*Zingiber officinale*) on glycemic markers in patients with type 2 diabetes. J Complement Integr Med. 2015;12:165–70.
107. An X, Zhang AL, May BH, Lin L, Xu Y, Xue CC. Oral Chinese herbal medicine for improvement of quality of life in patients with stable chronic obstructive pulmonary disease: a systematic review. J Altern Complement Med. 2012;18(8):731–43.
108. Li J, Zhang H, Ruan H, et al. Effects of Chinese herbal medicine on acute exacerbations of COPD: a randomized, placebo-controlled study. Int J Chron Obstruct Pulmon Dis. 2020;15:2901–12.
109. Kwon CY, Lee B, Lee BJ, Kim KI, Jung HJ. Herbal medicine compared to placebo for chronic obstructive pulmonary disease: a systematic review and meta-analysis. Front Pharmacol. 2021;12:717570.
110. Timalsina D, Pokhrel KP, Bhusal D. Pharmacologic activities of plant-derived natural products on respiratory diseases and inflammations. Biomed Res Int. 2021;2021:1636816.
111. Moghadam KG, Inançlı HM, Bazazy N, Plinkert PK, Efferth T, Sertel S. Phytomedicine in otorhinolaryngology and pulmonology: clinical trials with herbal remedies. Pharmaceuticals. 2012;5(8):853–74.
112. Li H, Chen H, Morgan L, Li W, Oliver BG. A narrative review of clinical studies of herbal treatment of difficult to manage asthma. Complement Ther Clin Pract. 2021;44:101433.
113. Glastonbury S. Herbal medicine: role in the management of Chronic Obstructive Pulmonary Disease (COPD). Aust J Med Herb. 2009;21(1):3–9.
114. Ram A, Balachandar S, Vijayananth P, Singh VP. Medicinal plants useful for treating chronic obstructive pulmonary disease (COPD): current status and future perspectives. Fitoterapia. 2011;82(2):141–51.
115. Engelbertz J, Lechtenberg M, Studt L, Hensel A, Verspohl EJ. Bioassay-guided fractionation of a thymol-deprived hydrophilic thyme extract and its antispasmodic effect. J Ethnopharmacol. 2012;141(3):848–53.
116. Nabissi M, Marinelli O, Morelli MB, Nicotra G, Iannarelli R, Amantini C, Santoni G, Maggi F. Thyme extract increases mucociliary-beating frequency in primary cell lines from chronic obstructive pulmonary disease patients. Biomed Pharmacother. 2018;105:1248–53.
117. Salehi B, Mishra AP, Shukla I, Sharifi-Rad M, del Mar Contreras M, Segura-Carretero A, Fathi H, Nasrabadi NN, Kobarfard F, Sharifi-Rad J. Thymol, thyme, and other plant sources: health and potential uses. Phytother Res. 2018;32:1688–706.
118. Shakeri F, Ghorani V, Saadat S, Gholamnezhad Z, Boskabady MH. The stimulatory effects of medicinal plants on β2-adrenoceptors of tracheal smooth muscle. Iran J Allergy Asthma Immunol. 2019;18(1):12–26.
119. Gierlikowska B, Gierlikowski W, Bekier K, Skalicka-Woźniak K, Czerwińska ME, Kiss AK. *Inula helenium* and *Grindelia squarrosa* as a source of compounds with anti-inflammatory activity in human neutrophils and cultured human respiratory epithelium. J Ethnopharmacol. 2020;249:112311.
120. Yu B, Jin XQ, Yu WY, Dong YY, Ying HZ, Yu CH. 1β-hydroxyalantolactone from inulae flos alleviated the progression of pulmonary fibrosis via inhibiting JNK/FOXO1/NF-κB pathway. Int Immunopharmacol. 2021;101:108339.
121. Kenny CR, Stojakowska A, Furey A, Lucey B. From monographs to chromatograms: the antimicrobial potential of *Inula helenium* L. (Elecampane) naturalised in Ireland. Molecules. 2022;27(4):1406.
122. Al-Jawad FH, Al-Razzuqi RA, Hashim HM, Al-Bayati NJ. Glycyrrhiza glabra versus Boswellia carterii in chronic bronchial asthma: a comparative study of efficacy. Indian J Allergy Asthma Immunol. 2012;26(1):6.
123. Yue GGL, Chan BCL, Kwok HF, To MH, Hon KL, Fung KP, Lau CBS, Leung PC. Screening for anti-inflammatory and broncho relaxant activities of 12 commonly used Chinese herbal medicines. Phytother Res. 2012;26(6):915–25.
124. Yu D, Liu X, Zhang G, Ming Z, Wang T. Isoliquiritigenin inhibits cigarette smoke-induced COPD by attenuating inflammation and oxidative stress via the regulation of the Nrf2 and NF-kappaB signaling pathways. Front Pharmacol. 2018;9:1001.
125. Krenn L, Wollenweber E, Steyrleuthner K, Görick C, Melzig MF. Contribution of methylated exudate flavonoids to the anti-inflammatory activity of Grindelia robusta. Fitoterapia. 2009;80(5):267–9.
126. La VD, Lazzarin F, Ricci D, Fraternale D, Genovese S, Epifano F, Grenier D. Active principles of Grindelia robusta exert anti-inflammatory properties in a macrophage model. Phytother Res. 2010;24(11):1687–92.
127. Assessment report on Grindelia robusta Nutt., Grindelia squarrosa (Pursh) Dunal, Grindelia humilis hook. Et Arn., Grindelia camporum Greene, herba. European Medicines Agency (EMA) 2012. EMA/HMPC/748218/2011.

128. Nowak S, Lisiecki P, Tomaszczak-Nowak A, Grudzińska E, Olszewska MA, Kicel A. Chemical composition and antimicrobial activity of the essential oils from flowers and leaves of Grindelia integrifolia DC. Nat Prod Res. 2019;33(10):1535–40.
129. Schepetkin IA, Özek G, Özek T, Kirpotina LN, Khlebnikov AI, Quinn MT. Neutrophil immunomodulatory activity of (−)-borneol, a major component of essential oils extracted from Grindelia squarrosa. Molecules. 2022;27(15):4897.
130. Joshi BC, Mukhija M, Kalia AN. Pharmacognostical review of Urtica dioica L. Int J Green Pharm. 2014;8(4):1.
131. Zemmouri H, Sekiou O, Ammar S, El Feki A, Bouaziz M, Messarah M, Boumendjel A. Urtica dioica attenuates ovalbumin-induced inflammation and lipid peroxidation of lung tissues in rat asthma model. Pharm Biol. 2017;55(1):1561–8.
132. Irani M, Choopani R, Esmaeili S, Dargahi T, Athari SM, Athari SS. Effect of nettle seed on immune response in a murine model of allergic asthma. Rev Fr Allergol. 2020;60(5):417–22.
133. Grauso L, de Falco B, Lanzotti V, Motti R. Stinging nettle, Urtica dioica L.: botanical, phytochemical and pharmacological overview. Phytochem Rev. 2020;19:1341–77.
134. Yuan J, Liu R, Ma Y, Zhang Z, Xie Z. Curcumin attenuates airway inflammation and airway remolding by inhibiting NF-κB signaling and COX-2 in cigarette smoke-induced COPD mice. Inflammation. 2018;41:1804–14.
135. Manarin G, Anderson D, Silva JME, Coppede J, Roxo-Junior P, Pereira AM, Carmona F. Curcuma longa L. ameliorates asthma control in children and adolescents: a randomized, double-blind, controlled trial. J Ethnopharmacol. 2019;238:111882.
136. Liu M, Wu K, Lin J, Xie Q, Liu Y, Huang Y, Zeng J, Yang Z, Wang Y, Dong S, Deng W. Emerging biological functions of IL-17A: a new target in chronic obstructive pulmonary disease? Front Pharmacol. 2021;12:695957.
137. Kamalifard M, Khalili AF, Namadian M, Herizchi S, Ranjbar Y. Comparison of the effect of lavender and bitter orange on depression in menopausal women: a triple-blind randomized controlled trial. Int J Womens Health Reprod Sci. 2017;5(3):224–30.
138. Yeung KS, Hernandez M, Mao JJ, Haviland I, Gubili J. Herbal medicine for depression and anxiety: a systematic review with assessment of potential psycho-oncologic relevance. Phytother Res. 2018;32(5):865–91.
139. Araj-Khodaei M, Noorbala AA, Yarani R, Emadi F, Emaratkar E, Faghihzadeh S, Parsian Z, Alijaniha F, Kamalinejad M, Naseri M. A double-blind, randomized pilot study for comparison of Melissa officinalis L. and Lavandula angustifolia mill. With fluoxetine for the treatment of depression. BMC Complement Med Ther. 2020;20(1):1–9.
140. Kou MJ, Chen JX. Integrated traditional and Western medicine for treatment of depression based on syndrome differentiation: a meta-analysis of randomized controlled trials based on the Hamilton depression scale. J Tradit Chin Med. 2012;32(1):1–5.
141. Ren Y, Zhu C, Wu J, Zheng R, Cao H. Comparison between herbal medicine and fluoxetine for depression: a systematic review of randomized controlled trials. Complement Ther Med. 2015;23(5):674–84.
142. Alzobaidi N, Quasimi H, Emad NA, Alhalmi A, Naqvi M. Bioactive compounds and traditional herbal medicine: promising approaches for the treatment of dementia. Degener Neurol Neuromuscul Dis. 2021;12:1–4.
143. Daily JW, Yang M, Park S. Efficacy of turmeric extracts and curcumin for alleviating the symptoms of joint arthritis: a systematic review and meta-analysis of randomized clinical trials. J Med Food. 2016;19(8):717–29.
144. Farpour HR, Rajabi N, Ebrahimi B. The efficacy of harpagophytum procumbens (teltonal) in patients with knee osteoarthritis: a randomized active-controlled clinical trial. Evid Based Complement Alternat Med. 2021;2021:1–8.
145. Rahmani S, Asgary S, Askari G, Keshvari M, Hatamipour M, Feizi A, Sahebkar A. Treatment of non-alcoholic fatty liver disease with curcumin: a randomized placebo-controlled trial. Phytother Res. 2016;30(9):1540–8.
146. Peng A, Gu Y, Lin SY. Herbal treatment for renal diseases. Ann Acad Med Singap. 2005;34(1):44–51.
147. Bhardwaj S. Aware.World Health Organisation. Key technical issues of herbal medicines with reference to interaction with other medicines. 2022. ISBN 978–92–4-001914-0 (electronic version).
148. WHO. World Health Organization traditional medicine strategy: 2014–2023. Hong Kong: SAR, China, World Health Organization; 2013.
149. Bruno LO, Simoes RS, de Jesus M, Simoes M, et al. Pregnancy and herbal medicines: an unnecessary risk for women's health—a narrative review. Phytother Res. 2018;32:796–810; 2014–2023 strategy.
150. Garrett B, Caulfield T, Murdoch B, Brignall M, et al. A taxonomyof risk-associated alternative health practices: a Delphi study. Health Soc Care Community. 2022;30(3):1163–81.
151. Byard RW, Musgrave I, Maker G, Bunce M. What risks do herbal products pose to the Australian community? Med J Aust. 2017;206(2):86–90.
152. Williamson EM, Chan K. Evolution of herbal medicines in Europe and its relationship with modern medicine. In: Phytotherapies: efficacy, safety, and regulation, vol. 24; 2015. p. 183–98.
153. Pan SY, Litscher G, Gao SH, Zhou SF, Yu ZL, Chen HQ, Zhang SF, Tang MK, Sun JN, Ko KM. Historical perspective of traditional indigenous medical practices: the current renaissance and conservation of herbal resources. Evid Based Complement Alternat Med. 2014;2014:27.
154. Veeresham C. Natural products derived from plants as a source of drugs. J Adv Pharm Technol Res. 2012;3(4):200.
155. Da-Yong L, Ting-Ren L. Herbal medicine in new era. Hos Pal Med Int Jnl. 2019;3(4):125–30.
156. Pan SY, Zhou SF, Gao SH, Yu ZL, Zhang SF, Tang MK, Sun JN, Ma DL, Han YF, Fong WF, Ko KM. New perspectives on how to discover drugs from herbal medicines: CAM's outstanding contribution to modern therapeutics. Evid Based Complement Alternat Med. 2013;2013:627375.
157. Rasoanaivo P, Wright CW, Willcox ML, Gilbert B. Whole plant extracts versus single compounds for the treatment of malaria: synergy and positive interactions. Malar J. 2011;10(1):1–2.
158. Wagner H, Ulrich-Merzenich G. Synergyresearch: approaching a new generation of phytopharmaceuticals. Phytomedicine. 2009;16(2–3):97–110.
159. Gale N. The sociology of traditional, complementary and alternative medicine. Sociol Compass. 2014;8(6):805–22.
160. Akrom A, Rizki M. Exercise and consumption of herbal remedies decrease the risk of strokes in hypertension and diabetes mellitus patients undergoing a standard therapy. Pharmaciana. 2019;9(2):237–48.
161. Jang S, Jang BH, Ko Y, Sasaki Y, Park JS, Hwang EH, Song YK, Shin YC, Ko SG. Herbal medicines for treating metabolic syndrome: a systematic review of randomized controlled trials. Evid Based Complement Alternat Med. 2016;2016:1.
162. Li Y, Liu Y, Cui J, Zhao H, Liu Y, Huang L. Cohort studies on chronic non-communicable diseases treated with traditional Chinese medicine: a bibliometric analysis. Front Pharmacol. 2021;19(12):639860.
163. World Health Organisation. Regional framework for harnessing traditional and complementary medicine for achieving health and well-being in the Western Pacific. 2022; ISBN 978 92 9061 988 8.
164. Hughes BB, Kuhn R, Peterson CM, Rothman DS, Solórzano JR, Mathers CD, et al. Projections of global health outcomes from 2005 to 2060 using the international futures integrated forecasting model. Bull World Health Organ. 2011;89(7):478–86. https://doi.

org/10.2471/BLT.10.083766; Epub 2011 Apr 8. PMID: 21734761; PMCID: PMC3127264. https://www.ncbi.nlm.nih.gov/pmc/articles/PMC3127264. Accessed 21 Jan 2023.
165. WHO. World Health OrganizationTraditional medicine strategy: 2014–2023. Hong Kong: SAR, China, World Health Organization; 2013.
166. Seetharaman M, Krishnan G, Schneider RH. The Future of Medicine: frontiers in integrative health and medicine. Medicina. 2021;57:1303. https://doi.org/10.3390/medicina57121303.
167. Karbwang J, Crawley FP, Na-Bangchang K, Maramba-Lazarte C. Herbal medicine development: methodologies, challenges, and issues. Evid Based Complement Alternat Med. 2019;2019:9.
168. Silveira D, Prieto JM, Freitas MM, Mazzari ALDA. Herbal medicine and public healthcare: current and future challenges. In: Cechinel Filho V, editor. Natural products as source of molecules with therapeutic potential. Cham: Springer; 2018. https://doi.org/10.1007/978-3-030-00545-0_13.

SPRINGER NATURE

GPSR Compliance

The European Union's (EU) General Product Safety Regulation (GPSR) is a set of rules that requires consumer products to be safe and our obligations to ensure this.

If you have any concerns about our products, you can contact us on ProductSafety@springernature.com

In case Publisher is established outside the EU, the EU authorized representative is:

Springer Nature Customer Service Center GmbH
Europaplatz 3
69115 Heidelberg, Germany